Teacher Edition

Reveal
ALGEBRA 2®

Volume 1

mheducation.com/prek-12

Cover: (t to b, l to r) Kelley Miller/National Geographic/Getty Images; Westend61/Getty Images; skodonnell/E+/Getty Images; Daniel Viñé Garcia/Moment/Getty Images; Aksonov/E+/Getty Images

Send all inquiries to:
McGraw-Hill Education
8787 Orion Place
Columbus, OH 43240

ISBN: 978-0-07-662600-7 (*Interactive Student Edition*, Volume 1)
MHID: 0-07-662600-8 (*Interactive Student Edition*, Volume 1)
ISBN: 978-0-07-899754-9 (*Interactive Student Edition*, Volume 2)
MHID: 0-07-899754-2 (*Interactive Student Edition*, Volume 2)

ISBN: 978-0-07-899755-6 (*Reveal Algebra 2 Teacher Edition*, Volume 1)
MHID: 0-07-899755-0 (*Reveal Algebra 2 Teacher Edition*, Volume 1)
ISBN: 978-0-07-899756-3 (*Reveal Algebra 2 Teacher Edition*, Volume 2)
MHID: 0-07-899756-9 (*Reveal Algebra 2 Teacher Edition*, Volume 2)

Printed in the United States of America.

5 6 7 8 9 10 WEB 26 25 24 23 22 21

Contents in Brief

Reveal Math Guiding Principles

Academic research and the science of learning provide the foundation for this powerful K–12 math program designed to help reveal the mathematician in every student.

Reveal Math is built on a solid foundation of **RESEARCH** that shaped the **PEDAGOGY** of the program.

Reveal Algebra 1, Reveal Geometry, and *Reveal Algebra 2* (Reveal AGA) used findings from research on teaching and learning mathematics to develop its instructional model. Based on analyses of research findings, these areas form the foundational structure of the program:

- Rigor
- Productive Struggle
- Formative Assessment
- Rich Tasks
- Mathematical Discourse
- Collaborative Learning

Instructional Model

1 Launch

 WARM UP

During the **Warm Up,** students complete exercises to activate prior knowledge and review prerequisite concepts and skills.

 LAUNCH THE LESSON

In **Launch the Lesson**, students view a real-world scenario and image to pique their interest in the lesson content. They are introduced to questions that they will be able to answer at the end of the lesson.

 EXPLORE

During the **Explore** activity, students work in partners or small groups to explore a rich mathematical problem related to the lesson content.

 INDIVIDUAL ACTIVITY

 GROUP ACTIVITY

 CLASS ACTIVITY

Klaus Vedfelt/DigitalVision/Getty Images

2 Explore and Develop

 LEARN

In the **Learn** section, students gain the foundational knowledge needed to actively work through upcoming Examples.

 EXAMPLES & CHECK

Students work through **Examples** related to the key concepts and engage in mathematical discourse.

Students complete a **Check** after several Examples as a quick formative assessment to help teachers adjust instruction as needed.

3 Reflect and Practice

 EXIT TICKET

The **Exit Ticket** gives students an opportunity to convey their understanding of the lesson concepts.

  PRACTICE

Students complete **Practice** exercises individually or collaboratively to solidify their understanding of lesson concepts and build proficiency with lesson skills.

Reveal Math Key Areas of Focus

Reveal Algebra 1, Reveal Geometry, and *Reveal Algebra 2* (Reveal AGA) have a strong focus on rigor—especially the development of conceptual understanding—an emphasis on student mindset, and ongoing formative assessment feedback loops.

Rigor

Reveal AGA has been thoughtfully designed to incorporate a balance of the three elements of rigor: conceptual understanding, procedural skills and fluency, and application.

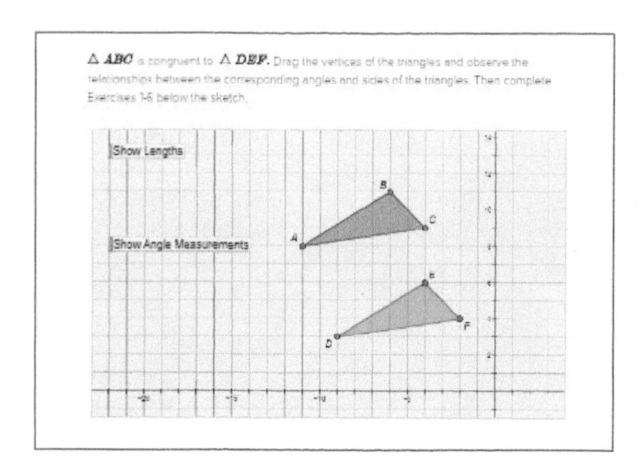

Conceptual Understanding

Explore activities give all students an opportunity to work collaboratively and discuss their thinking as they build conceptual understanding of new concepts. In the **Explore** activity to the left, students use **Web Sketchpad**® to build understanding of the relationships between corresponding sides and angles in congruent triangles.

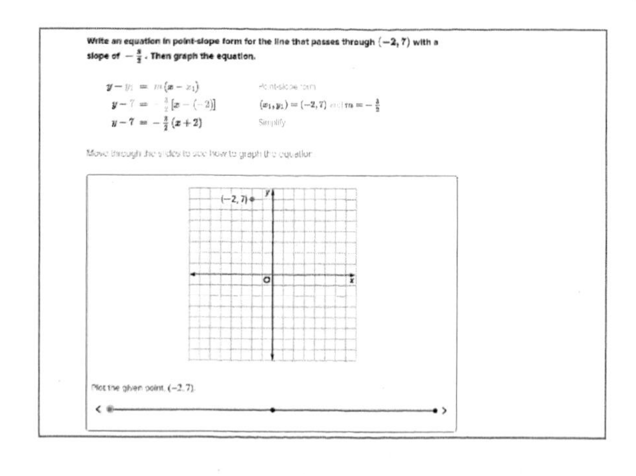

Procedural Skills and Fluency

Students use different strategies and tools to build procedural fluency. In the **Example** shown, students build proficiency with writing equations in point-slope form.

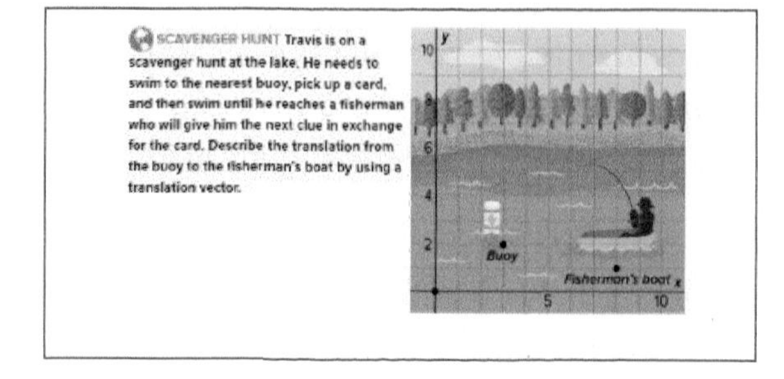

Application

Real-world examples and practice problems are opportunities for students to apply their learning to new situations. In the real-world example shown, students apply their understanding by solving a multi-step problem with translations.

Student Mindset

Mindset Matters tips located in each module provide specific examples of how Reveal AGA content can be used to promote a growth mindset in all students. Another feature focused on promoting a growth mindset is **Ignite! Activities** developed by Dr. Raj Shah to spark student curiosity about why the math works. An **Ignite!** delivers problem sets that are flexible enough so that students with varying background knowledge can engage with the content and motivates them to ask questions, solve complex problems, and develop a can-do attitude toward math.

Mindset Matters

Growth Mindset vs. Fixed Mindset

Everyone has a core belief or mindset about how they learn. People with a *growth mindset* believe that they can grow their intelligence through hard work. Those with a *fixed mindset* believe that while they can learn new things, they cannot increase their intelligence. When a student approaches school, life, and the future workplace with a growth mindset, they are more likely to persevere through challenging problems, learn from their mistakes, and ultimately learn concepts in a deeper, more meaningful way.

How Can I Apply It?

Assign students rich tasks, such as the **Explore** activities, that can help them to develop their intelligence. Encourage them with the thought that each time they learn a new idea, neurons fire electric currents that connect different parts of their brain!

Teacher Edition Mindset Tip

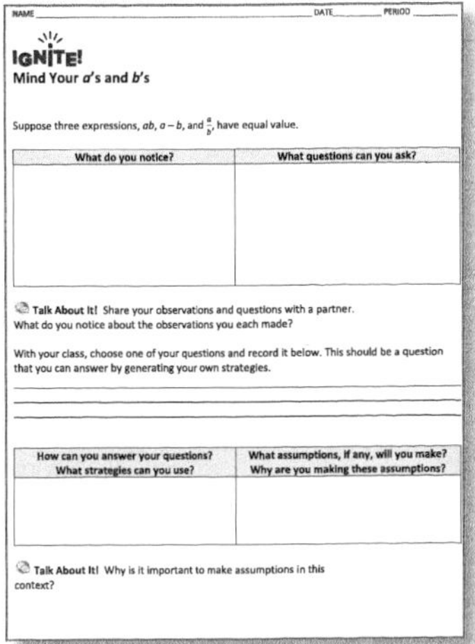

NAME ______ DATE ______ PERIOD ______

IGNITE!

Mind Your *a*'s and *b*'s

Suppose three expressions, ab, $a - b$, and $\frac{a}{b}$, have equal value.

What do you notice?	What questions can you ask?

Talk About It! Share your observations and questions with a partner. What do you notice about the observations you each made?

With your class, choose one of your questions and record it below. This should be a question that you can answer by generating your own strategies.

How can you answer your questions? What strategies can you use?	What assumptions, if any, will you make? Why are you making these assumptions?

Talk About It! Why is it important to make assumptions in this context?

Student Ignite! Activity

Formative Assessment

The key to reaching all learners is to adjust instruction based on each student's understanding. Reveal AGA offers powerful formative assessment tools that help teachers to efficiently and effectively differentiate instruction for all students.

Math Probes

Each module includes a **Cheryl Tobey Formative Assessment Math Probe** that is focused on addressing student misconceptions about key math topics. Students can complete these probes at the beginning, middle, or end of a module. The teacher support includes a list of recommended differentiated resources that teachers assign based on students' responses.

Example Checks

After multiple examples, a formative assessment **Check** that students complete on their own allows teachers to gauge students' understanding of the concept or skill presented. When students complete the Check online, the teacher receives resource recommendations which can be assigned to students.

Learning Target

Formative Assessment

CHERYL TOBEY MATH PROBES

Quantitative Checks and Homework

Observations and Probes

Data

REVEAL the full potential of all students.

A Powerful Blended Learning Experience

The *Reveal Algebra 1, Reveal Geometry,* and *Reveal Algebra 2* (Reveal AGA) blended learning experience was designed to include purposeful print and digital components focused on sparking student curiosity and providing teachers with flexible implementation options.

Reveal AGA has been thoughtfully developed to provide a rich learning experience no matter where a district, school, or classroom falls on the digital spectrum. All of the instructional content can be projected or can be accessed via desktop, laptop, or tablet.

Lesson

1 Launch

WARM UP

The **Warm Up** exercise can be projected on an interactive whiteboard.

LAUNCH THE LESSON

Launch the Lesson can be projected or assigned to students to access on their own devices.

EXPLORE

The **Explore** activity can be projected while students record their observations in the Interactive Student Edition or can be assigned for students to complete on individual devices.

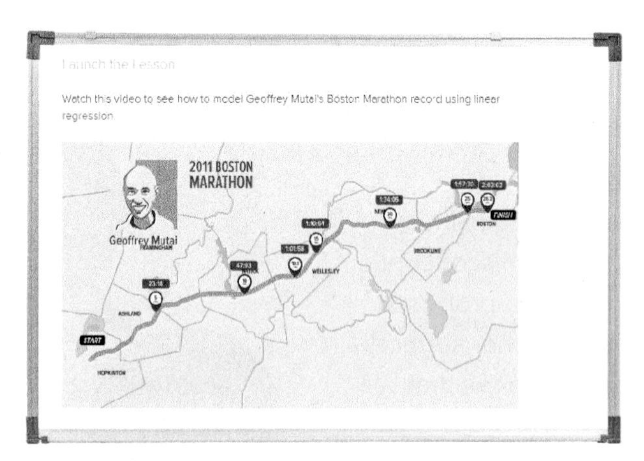

Launch the Lesson

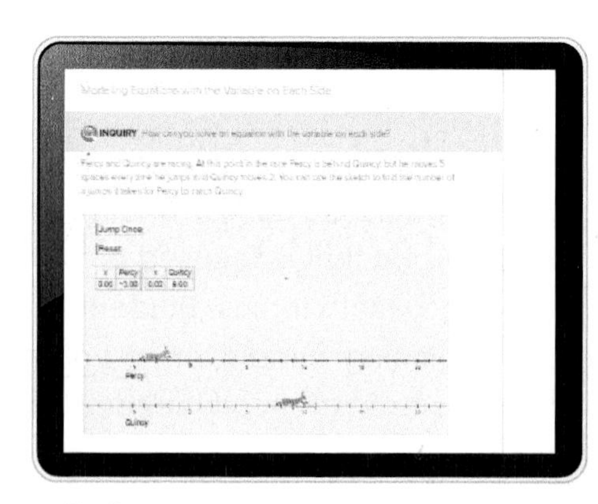

Explore

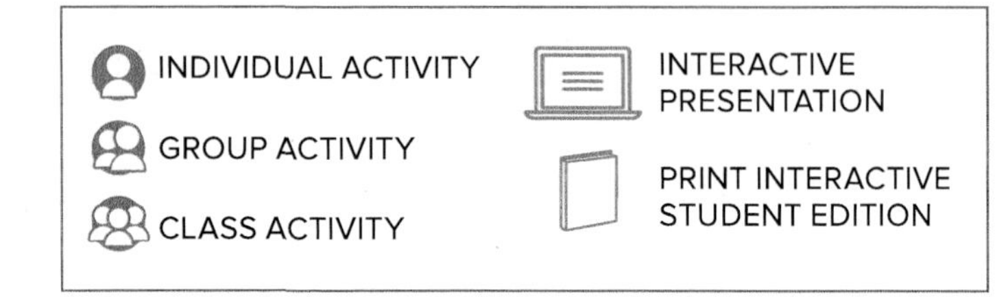

2 Explore and Develop

3 Reflect and Practice

LEARN

As students are introduced to the key lesson concepts, they can progress through the **Learn** by recording guided notes in their Interactive Student Edition or on their own devices.

EXAMPLES & CHECK

In their Interactive Student Edition or on an individual device, students work through one or more **Examples** related to key lesson concepts.

A **Check** follows several Examples in either the Interactive Student Edition or on each student device.

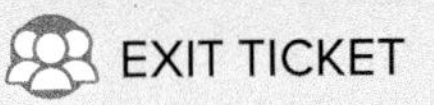

EXIT TICKET

The **Exit Ticket** is projected or accessed via student devices to provide students with lesson closure and an opportunity to revisit the lesson concepts.

PRACTICE

Assign students **Practice** problems from their Interactive Student Edition or create a digital assignment for them to work on their device in class or at home to solidify lesson concepts.

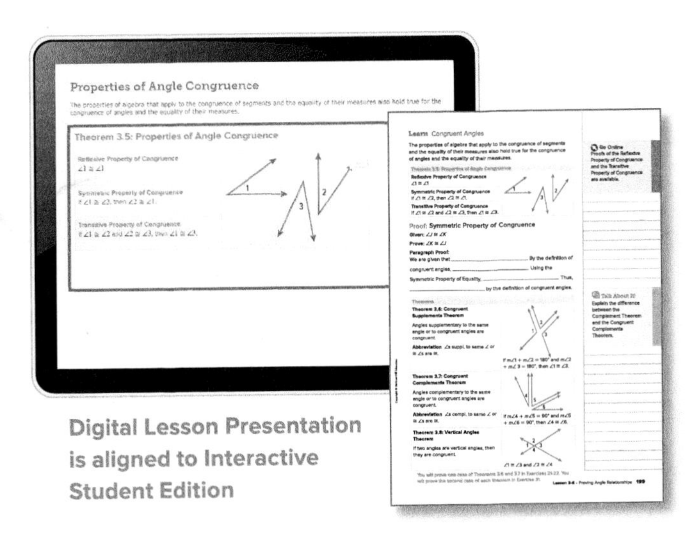

Digital Lesson Presentation is aligned to Interactive Student Edition

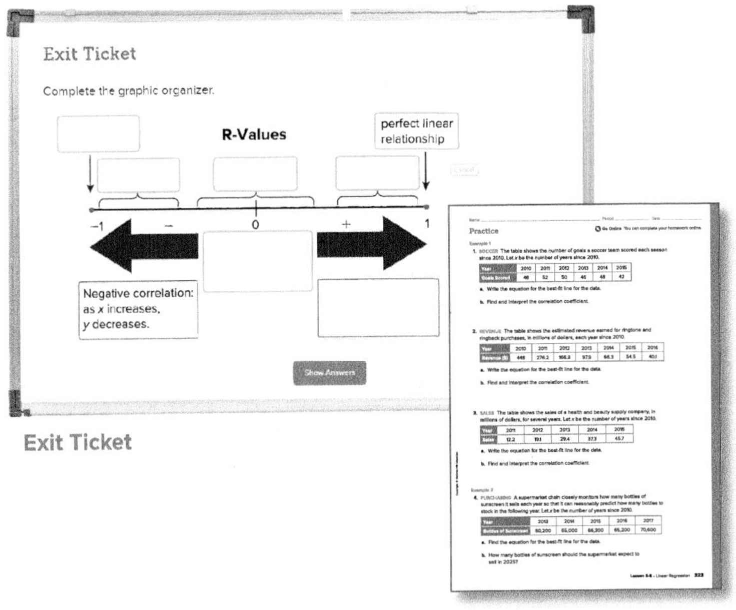

Exit Ticket

Practice

Supporting All Learners

The *Reveal Algebra 1, Reveal Geometry,* and *Reveal Algebra 2* (Reveal AGA) programs were designed so that all students have access to:

- rich tasks that promote productive struggle,
- opportunities to develop proficiency with the habits of mind and thinking strategies of mathematicians, and
- prompts to promote mathematical discourse and build academic language.

Resources for Differentiating Instruction

When needed, resources are available to differentiate math instruction for students who may need to see a concept in a different way, practice prerequisite skills, or are ready to extend their learning.

Approaching Level Resources
- Remediation Activities
- Extra Examples

Beyond Level Resources
- Beyond Level Differentiated Activities
- Extension Activities

Resources for English Language Learners

Reveal AGA also includes student and teacher resources to support students who are simultaneously learning grade-level math and building their English proficiency. Appropriate, research-based language scaffolds are also provided to support students as they engage in rigorous mathematical tasks and discussions.

English Language Learners
- Spanish Interactive Student Edition
- Spanish Personal Tutors
- Math Language-Building Activities
- Language Scaffolds
- *Think About It!* and *Talk About It!* Prompts
- Multilingual eGlossary
- Audio
- Graphic Organizers
- Web Sketchpad, Desmos, and eTools

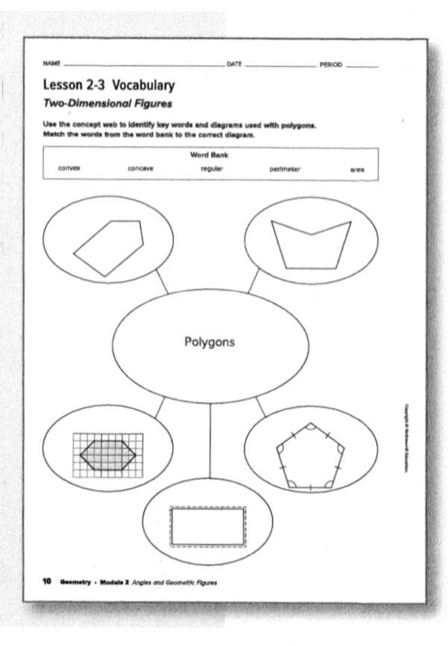
NAME ______ DATE ______ PERIOD ______

Lesson 2-3 Vocabulary
Two-Dimensional Figures

Use the concept web to identify key words and diagrams used with polygons. Match the words from the word bank to the correct diagram.

Word Bank: convex, concave, regular, perimeter, area

Polygons

10 Geometry • Module 2 Angles and Geometric Figures

Developing Mathematical Thinking and Strategic Questioning

Reveal Algebra 1, Reveal Geometry, and *Reveal Algebra 2* (Reveal AGA) are comprised of high-quality math content designed to be accessible and relevant to each student. Throughout the program, students are presented with a variety of thoughtfully designed questioning strategies related to the content. Using these questions provides you with an additional, built-in type of formative assessment that can be used to modify instruction. They also strengthen students' ownership of mathematical content knowledge and daily use of the Standards for Mathematical Practice.

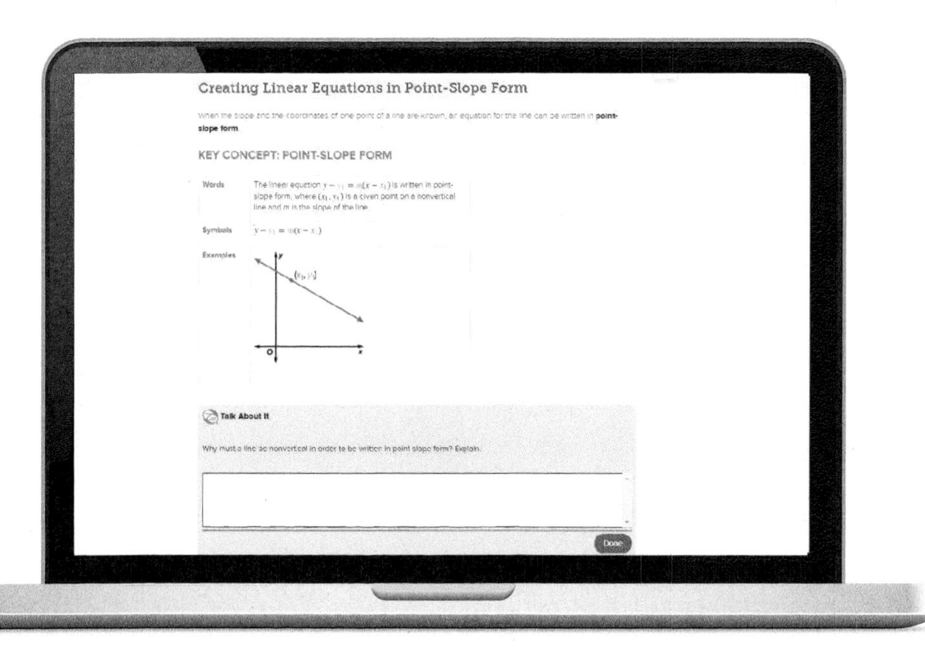

Key Concept Introduction followed by a Talk About It question to discuss with a classmate.

You will find these types of questioning strategies throughout Reveal AGA. The related Standard for Mathematical Practice for each is also indicated.

- **Talk About It** questions encourage students to engage in mathematical discourse with classmates (SMP3)
- **Alternate Method** shows students another way to solve a problem and asks them to compare and contrast the methods and solutions (SMP1)
- **Avoid a Common Error** shows students a problem similar to an example but with a flaw in reasoning, and students have to find and explain the error (SMP3)
- **State Your Assumptions** requires that student state the assumptions they made to solve a problem (SMP4)
- **Use a Source** asks students to find information using an external source, such as the Internet, and use it to pose or solve a problem (SMP5)
- **Think About It** questions help students make sense of mathematical problems (SMP1)
- **Concept Checks** prompt students to analyze how the Key Concepts of the lesson apply to various use cases (SMP3)

Reveal Student Readiness with Individualized Learning Tools

Reveal Algebra 1, Reveal Geometry, and *Reveal Algebra 2* (Reveal AGA) incorporate innovative, technology-based tools that are designed to extend the teacher's reach in the classroom to help address a wide range of knowledge gaps, set and align academic goals, and meet student individualized learning needs.

LEARNSMART®

Topic-Mastery

With embedded **LearnSmart,®** students have a built-in study partner for topic practice and review to prepare for multi-module, mid-year, or end-of-year testing.

LearnSmart's revolutionary adaptive technology measures students' awareness of their own learning, time on topic, answer accuracy, and suggests alternative resources to support student learning, confidence, and topic mastery.

ALEKS®

Individualized Learning Pathways

Learners of all levels benefit from the use of **ALEKS'** adaptive, online math technology designed to pinpoint what each student knows, does not know, and most importantly, what each student is ready to learn.

When paired with Reveal AGA, **ALEKS** is a powerful tool designed to provide integrated instructionally actionable data enabling teachers to utilize Reveal AGA resources for individual students, groups, or the entire classroom.

Activity Report

Scott Dunlap/iStockvectors/Getty Images

Powerful Tools for Modeling Mathematics

Reveal Algebra 1, Reveal Geometry, and *Reveal Algebra 2* (Reveal AGA) have been designed with purposeful, embedded digital tools to increase student engagement and provide unique modeling opportunities.

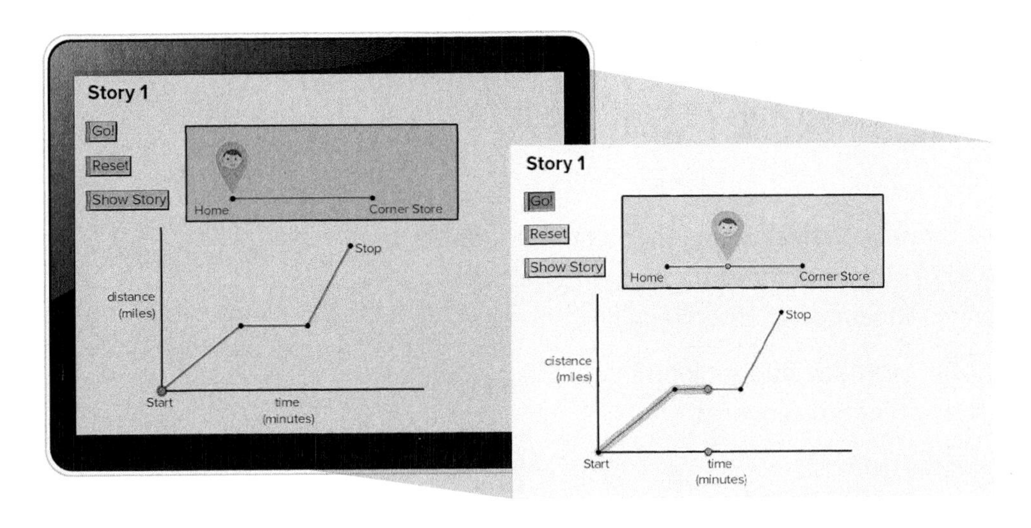

Web Sketchpad® Activities

The leading dynamic mathematics visualization software has now been integrated with **Web Sketchpad Activities** at point of use within Reveal AGA. Student exploration (and practice) using **Web Sketchpad** encourages problem solving and visualization of abstract math concepts.

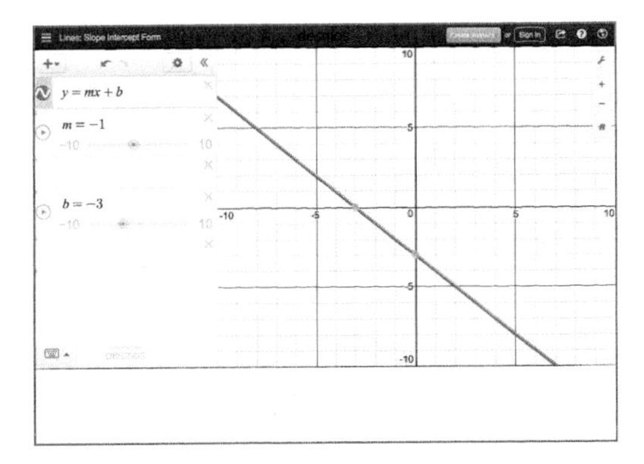

desmos

The powerful **Desmos** graphing calculator is available in Reveal AGA for students to explore, model, and apply math to the real-world.

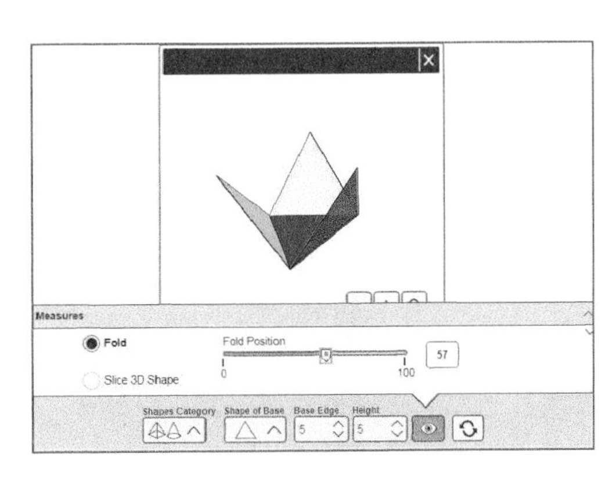

eTools

By using a wide variety of digital **eTools** embedded within Reveal AGA, students gain additional hands-on experience while they learn and teachers have the option to create problem-based learning opportunities.

Technology-Enhanced Items

Embedded within the digital lesson, technology-enhanced items—such as drag-and-drop, flashcard flips, or diagram completion—are strategically placed to give students the practice with common computer functions needed to master computer-based testing.

TYPE

SWIPE

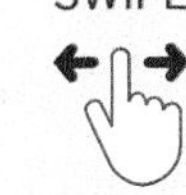

DRAG & DROP

FLASHCARDS

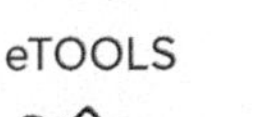

eTOOLS

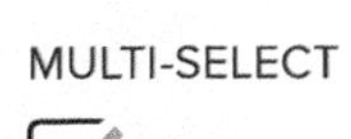

MULTI-SELECT

WATCH

EXPAND

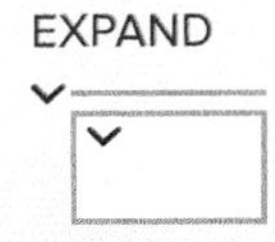

Assessment Tools to Reveal Student Progress and Success

Reveal Algebra 1, Reveal Geometry, and *Reveal Algebra 2* (Reveal AGA) provide a comprehensive array of assessment tools, with both print and digital administration options, to measure student understanding and progress. The digital assessment tools include next-generation assessment items, such as multiple-response, selected-response, and technology-enhanced items.

Assessment Solutions

Reveal AGA provides embedded, regular formative checkpoints to monitor student learning and provide feedback that can be used to modify instruction and help direct student learning using reports and recommendations based on resulting scores.

Summative assessments built in Reveal AGA evaluate student learning at the module conclusion by comparing it against the state standards covered.

Formative Assessment Resources

- Cheryl Tobey Formative Assessment Math Probes
- Checks
- Exit Tickets
- Put It All Together

Summative Assessment Resources

- Module Tests
- Performance Tasks
- End-of-Course Tests
- LearnSmart

Or **Build Your Own** assessments focused on standards or objectives. Access to banks of questions, including those with tech-enhanced capabilities, enable a wide range of options to mirror high-stakes assessment formats.

Reporting

Clear, instructionally actionable data is a click away with the Reveal AGA Reporting Dashboard.

Activity Report Real-time class and student reporting of activities completed by the class. Includes average score, submission rate, and skills covered for the class and each student.

- **Item Analysis Report** A detailed analysis of response rates and patterns, answers, and question types in a class snapshot or by student.

Standards Report Performance data by class or individual student are aggregated by standards, skills, or objectives linked to the related activities completed.

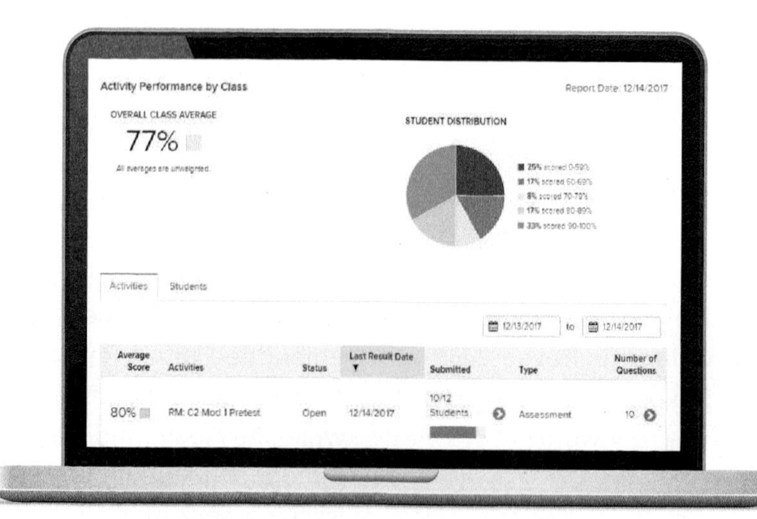

Activity Report

Professional Development Support for Continuous Learning

McGraw-Hill Education supports lifelong learning and demonstrates commitment to teachers with a built-in professional learning environment designed for support during planning or extended learning opportunities.

What You Will Find

- Best-practice resources
- Implementation support
- Teaching Strategies
- Classroom Videos
- Math Misconception Videos
- Content and Pedagogy Videos
- Content Progression Information

Why Professional Development Is so Important

- Research-based understanding of student learning
- Improved student performance
- Evidence-based instructional best practices
- Collaborative content strategy planning
- Extended knowledge of program how-to's

Kevin Dodge/Blend Images/Getty Images

Reveal Math Expert Advisors

Cathy Seeley, Ed.D.

Austin, Texas

Mathematics educator, speaker, and writer, former Senior Fellow at the Charles A. Dana Center at The University of Texas at Austin, past President of NCTM, former Director of K-12 Mathematics for the State of Texas

Areas of expertise:
Mathematics Teaching, Equity, Assessment, STEM Learning, Informal Learning, Upside-Down Teaching, Productive Struggling, Mathematical Practices, Mathematical Habits of Mind, Family and Community Outreach, Mathematics Education Policy, Advocacy

> "We want students to believe deeply that mathematics makes sense–in generating answers to problems, discussing their thinking and other students' thinking, and learning new material."
>
> **—Seeley, 2016, Making Sense of Math**

Cheryl R. Tobey, M.Ed.

Gardiner, Maine

Senior Mathematics Associate at Education Development Center (EDC)

Areas of expertise:
Formative assessment and professional development for mathematics teachers; tools and strategies to uncovering misconceptions

> "Misunderstandings and partial understandings develop as a normal part of learning mathematics. Our job as educators is to minimize the chances of students' harboring misconceptions by knowing the potential difficulties students are likely to encounter, using assessments to elicit misconceptions and implementing instruction designed to build new and accurate mathematical ideas."
>
> **—Tobey, et al 2007, 2009, 2010, 2013, 2104, Uncovering Student Thinking Series**

Nevels Nevels, Ph.D.

Saint Louis, Missouri

PK–12 Mathematics Curriculum Coordinator for Hazelwood School District

Areas of expertise:
Mathematics Teacher Education; Student Agency & Identity; Socio-Cultural Perspective in Mathematics Learning

> "A school building is one setting for learning mathematics. It is understood that all children should be expected to learn meaningful mathematics within its walls. Additionally, teachers should be expected to learn within the walls of this same building. More poignantly, I posit that if teachers are not learning mathematics in their school building, then it is not a school."
>
> **—Nevels, 2018**

Raj Shah, Ph.D.

Columbus, Ohio

Founder of Math Plus Academy, a STEM enrichment program and founding member of The Global Math Project

Areas of expertise:
Sparking student curiosity, promoting productive struggle, and creating math experiences that kids love

> "As teachers, it's imperative that we start every lesson by getting students to ask more questions because curiosity is the fuel that drives engagement, deeper learning and perseverance."
>
> **—Shah, 2017**

(l to r, t to b) Photo by Damon Leo, Shannon Larsen, Matthew McClellan, Raj Shah

Walter Secada, Ph.D.

Coral Gables, Florida

Professor of Teaching and Learning at the University of Miami

Areas of expertise:
Improving education for English language learners, equity in education, mathematics education, bilingual education, school restructuring, professional development of teachers, student engagement, Hispanic dropout and prevention, and reform

"The best lessons take place when teachers have thought about how their individual English language learners will respond not just to the mathematical content of that lesson, but also to its language demands and mathematical practices."

—Secada, 2018

Ryan Baker, Ph.D.

Philadelphia, Pennsylvania

Associate Professor and Director of Penn Center for Learning Analytics at the University of Pennsylvania

Areas of expertise:
Interactions between students and educational software; data mining and learning analytics to understand student learning

"The ultimate goal of the field of Artificial Intelligence in Education is not to promote artificial intelligence, but to promote education... systems that are designed intelligently, and that leverage teachers' intelligence. Modern online learning systems used at scale are leveraging human intelligence to improve their design, and they're bringing human beings into the decision-making loop and trying to inform them."

—Baker, 2016

Chris Dede, Ph.D.

Cambridge, Massachusetts

Timothy E. Wirth Professor in Learning Technologies at Harvard Graduate School of Education

Areas of expertise:
Provides leadership in educational innovation; educational improvements using technology

"People are very diverse in how they prefer to learn. Good instruction is like an ecosystem that has many niches for alternative types of learning: lectures, games, engaging video-based animations, readings, etc. Learners then can navigate to the niche that best fulfills their current needs."

—Dede, 2017

Dinah Zike, M.Ed.

Comfort, Texas

President of Dinah.com in San Antonio, Texas and Dinah Zike Academy

Areas of expertise:
Developing educational materials that include three-dimensional graphic organizers; interactive notebook activities for differentiation; and kinesthetic, cross-curricular manipulatives

"It is education's responsibility to meet the unique needs of students, and not the students' responsibility to meet education's need for uniformity."

—Zike, 2017, InRIGORating Math Notebooks

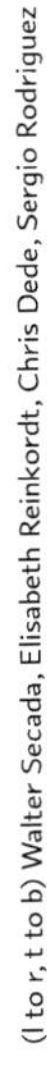

Reveal Everything Needed for Effective Instruction

Reveal Algebra 1, Reveal Geometry, and *Reveal Algebra 2* (Reveal AGA) provide both print and innovative, technology-based tools designed to address a wide range of classrooms. No matter whether you're in a 1:1 district, or have a classroom projector, Reveal AGA provides you with the resources you need for a rich learning experience.

Blended Classrooms

Focused on projection of the **Interactive Presentation**, students follow along, taking notes and working through problems in their Interactive Student Edition during class time. Also included in the Interactive Student Edition is a glossary, selected answers, and a reference sheet.

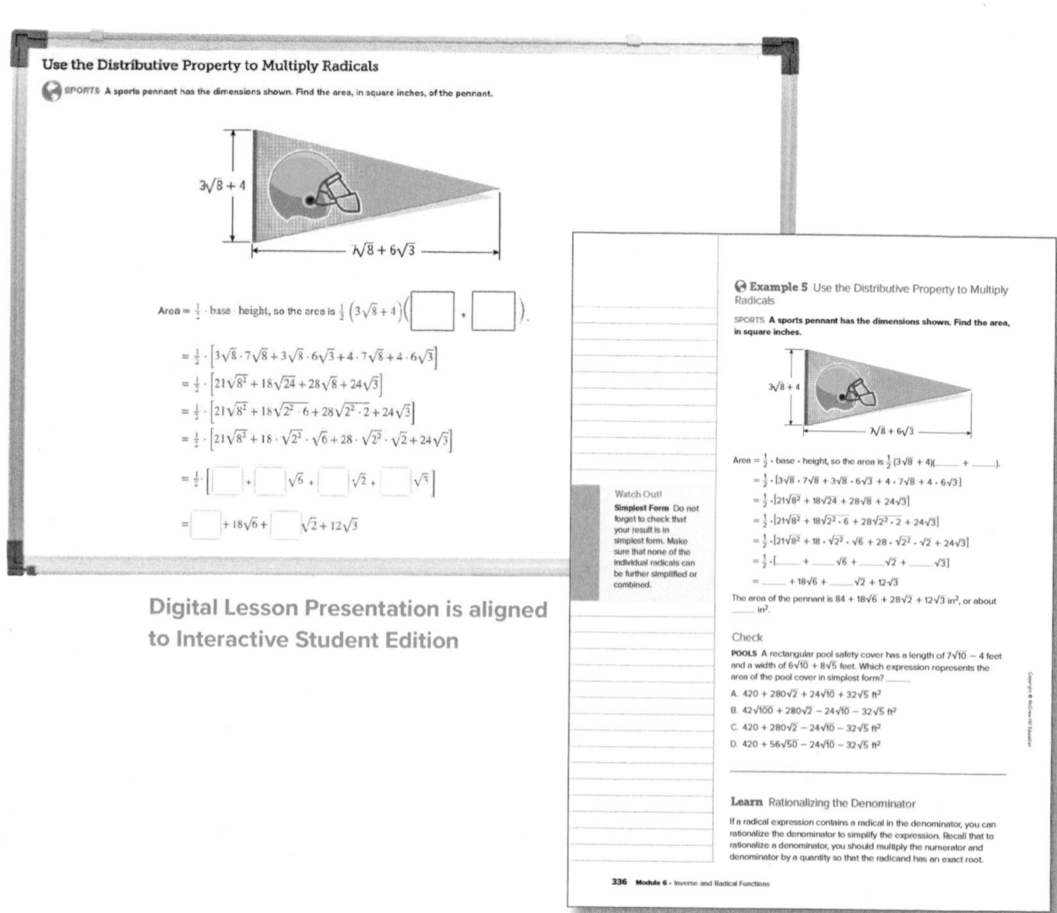

Digital Lesson Presentation is aligned to Interactive Student Edition

Andrey_Kuzmin/iStock/Getty Images

Digital Classrooms

Projection is a focal point for key areas of the course with students interacting with the lesson using their own devices. Each student can access teacher-assigned sections of the lessons for **Explore** activities, **Learn** sections, and **Examples**. Point of use videos, animations, as well as interactive content enable students to experience math in interesting and impactful ways.

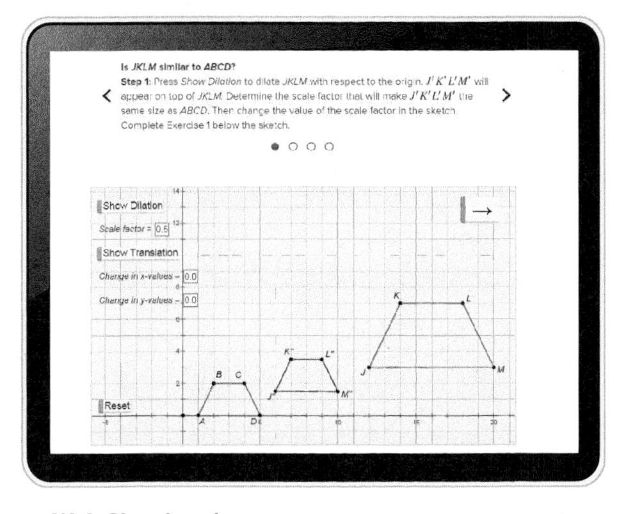

Web Sketchpad

Desmos

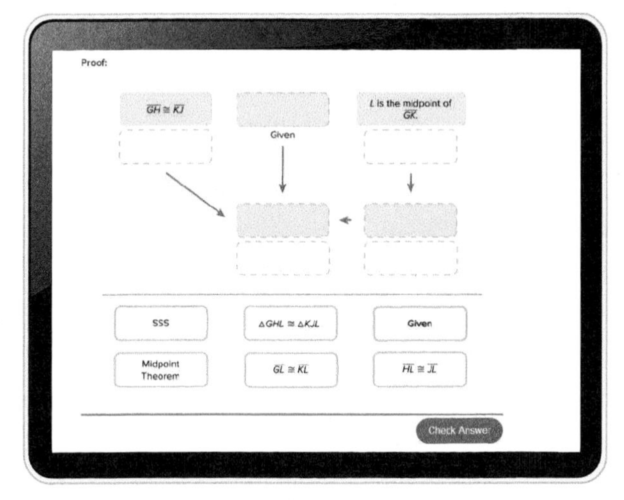

Drag-and-Drop

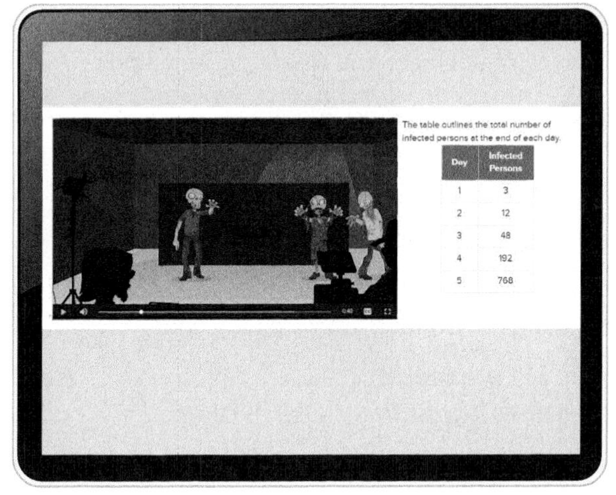

Video

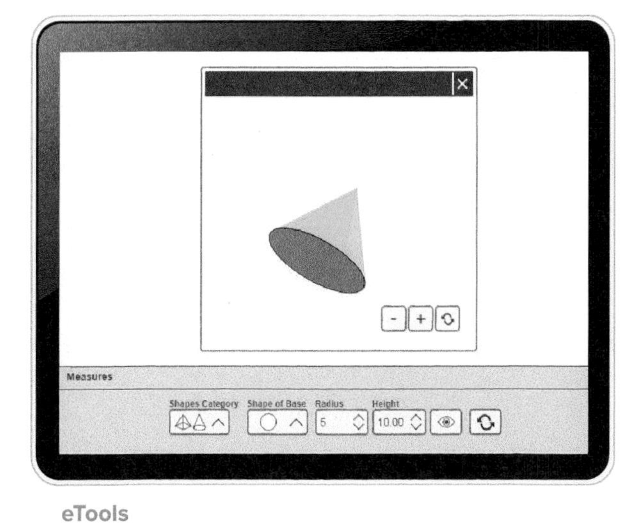

eTools

In Reveal AGA, R is for—

- Research
- Rigor
- Relevant Connections

Are you... READY to start?

Module 1

Relations and Functions

Module 2

Linear Equations, Inequalities, and Systems

Module 3

Quadratic Functions

		Page	CCSS
	What Will You Learn?	**141**	
3-1	**Graphing Quadratic Functions** Explore Transforming Quadratic Functions	**143**	F.IF.4, F.IF.6
3-2	**Solving Quadratic Equations by Graphing** Explore Roots of Quadratic Equations	**153**	A.CED.2, F.IF.4
3-3	**Complex Numbers** Explore Factoring the Sum of Two Squares	**161**	N.CN.1, N.CN.2
3-4	**Solving Quadratic Equations by Factoring** Explore Finding the Solutions of Quadratic Equations by Factoring	**167**	N.CN.7, N.CN.8, F.IF.8a
3-5	**Solving Quadratic Equations by Completing the Square** Explore Using Algebra Tiles to Complete the Square	**173**	N.CN.7, F.IF.8a
3-6	**Using the Quadratic Formula and the Discriminant** Explore The Discriminant	**183**	N.CN.7, N.CN.8, A.SSE.1b
3-7	**Quadratic Inequalities** Explore Graphing Quadratic Inequalities	**191**	A.CED.1, A.CED.3
3-8	**Solving Linear-Nonlinear Systems** Explore Linear-Quadratic Systems	**197**	A.REI.11
	Module 3 Review	**207**	

Module 4

Polynomials and Polynomial Functions

Module 5

Polynomial Equations

Module 6

Inverses and Radical Functions

Module 7

Exponential Functions

Module 8

Logarithmic Functions

Module 9

Rational Functions

Module 10

Inferential Statistics

Module 11

Trigonometric Functions

			CCSS
	What Will You Learn?	**555**	
11-1	**Angles and Angle Measure** Explore Arc Length	**557**	F.TF.1
11-2	**Trigonometric Functions of General Angles**	**565**	F.TF.3
11-3	**Circular and Periodic Functions** Explore Trigonometric Functions of Special Angles	**575**	F.TF.2, F.TF.5
11-4	**Graphing Sine and Cosine Functions**	**585**	F.IF.4, F.IF.7e
11-5	**Graphing Other Trigonometric Functions**	**595**	F.IF.4, F.IF.7e
11-6	**Translations of Trigonometric Graphs** Explore Analyzing Graphs of Trigonometric Functions by Using Technology	**603**	F.IF.7e, F.BF.3
11-7	**Inverse Trigonometric Functions**	**613**	F.TF.7
	Module 11 Review	**617**	

Module 12

Trigonometric Identities and Equations

Standards for Mathematical Content, Algebra II

This correlation shows the alignment of *Reveal Algebra 2* to the Standards for Mathematical Content, Algebra II, from the Common Core State Standards for Mathematics. Lessons in which the standard is the primary focus are indicated in **bold**.

Additional mathematics that students should learn in order to take advanced mathematical courses is indicated by (+).

	Standard	Lesson(s)
Number and Quantity		
The Complex Number System N-CN		
N.CN.1	Perform arithmetic operations with complex numbers. Know there is a complex number i such that $i^2 = -1$, and every complex number has the form $a + bi$ with a and b real.	**3-3**
N.CN.2	Use the relation $i^2 = -1$ and the commutative, associative, and distributive properties to add, subtract, and multiply complex numbers.	**3-3**
N.CN.7	Use complex numbers in polynomial identities and equations. Solve quadratic equations with real coefficients that have complex solutions.	**3-4, 3-5, 3-6**
N.CN.8	(+) Extend polynomial identities to the complex numbers.	3-3, **3-4,** 3-6
N.CN.9	(+) Know the Fundamental Theorem of Algebra; show that it is true for quadratic polynomials.	**5-5**
Algebra		
Seeing Structure in Expressions A-SSE		
A.SSE.1	Interpret the structure of expressions. Interpret expressions that represent a quantity in terms of its context.★ **a.** Interpret parts of an expression, such as terms, factors, and coefficients.	3-1, 4-1, 7-1, 11-4
	b. Interpret complicated expressions by viewing one or more of their parts as a single entity.	1-6, 1-7, 2-2, 2-3, 3-1, **3-6,** 4-2
A.SSE.2	Use the structure of an expression to identify ways to rewrite it.	3-4, 3-5, 6-3, 6-5, 7-2, 8-1, 8-3, 8-4, 8-5
A.SSE.4	Write expressions in equivalent forms to solve problems. Derive the formula for the sum of a finite geometric series (when the common ratio is not 1), and use the formula to solve problems.★	**7-4**
Arithmetic with Polynomials and Rational Expressions A-APR		
A.APR.1	Perform arithmetic operations on polynomials. Understand that polynomials form a system analogous to the integers, namely, they are closed under the operations of addition, subtraction, and multiplication; add, subtract, and multiply polynomials.	**4-3**
A.APR.2	Understand the relationship between zeros and factors of polynomials. Know and apply the Remainder Theorem: For a polynomial $p(x)$ and a number a, the remainder on division by $x - a$ is $p(a)$, so $p(a) = 0$ if and only if $(x - a)$ is a factor of $p(x)$.	**5-4**
A.APR.3	Identify zeros of polynomials when suitable factorizations are available, and use the zeros to construct a rough graph of the function defined by the polynomial.	**5-5**
A.APR.4	Use polynomial identities to solve problems. Prove polynomial identities and use them to describe numerical relationships.	3-4, **5-3**
A.APR.5	(+) Know and apply the Binomial Theorem for the expansion of $(x + y)^n$ in powers of x and y for a positive integer n, where x and y are any numbers, with coefficients determined for example by Pascal's Triangle.	**4-5**

★ Mathematical Modeling Standards

	Standard	Lesson(s)
A.APR.6	Rewrite rational expressions. Rewrite simple rational expressions in different forms; write $\frac{a(x)}{b(x)}$ in the form $q(x) + \frac{r(x)}{b(x)}$, where $a(x)$, $b(x)$, $q(x)$, and $r(x)$ are polynomials with the degree of $r(x)$ less than the degree of $b(x)$, using inspection, long division, or, for the more complicated examples, a computer algebra system.	**4-4**
A.APR.7	(+) Understand that rational expressions form a system analogous to the rational numbers, closed under addition, subtraction, multiplication, and division by a nonzero rational expression; add, subtract, multiply, and divide rational expressions.	**9-1, 9-2**
Creating Equations★ A-CED		
A.CED.1	Create equations that describe numbers or relationships. Create equations and inequalities in one variable and use them to solve problems.	**2-1, 2-2, 2-9,** 3-2, 3-4, 3-5, **3-6, 3-7, 5-1, 5-2,** 5-5, **7-2, 8-2, 8-5, 9-5, 9-6**
A.CED.2	Create equations in two or more variables to represent relationships between quantities; graph equations on coordinate axes with labels and scales.	**1-5, 1-6, 1-7, 2-1, 2-3, 2-4, 2-5, 3-1, 3-2, 4-1, 4-2, 5-1, 6-4, 6-6, 7-1, 7-3, 7-5, 8-1, 8-5, 9-3, 9-4, 9-5, 11-3, 11-4, 11-5, 11-6**
A.CED.3	Represent constraints by equations or inequalities, and by systems of equations and/or inequalities, and interpret solutions as viable or non-viable options in a modeling context.	**1-5,** 1-6, 2-1, 2-2, 2-4, 2-5, 2-6, 2-7, 2-8, 3-7, 8-5, 9-6
A.CED.4	Rearrange formulas to highlight a quantity of interest, using the same reasoning as in solving equations.	**3-6,** 6-2, 8-1, 9-5
Reasoning with Equations and Inequalities A-REI		
A.REI.2	Understand solving equations as a process of reasoning and explain the reasoning. Solve simple rational and radical equations in one variable, and give examples showing how extraneous solutions may arise.	**6-6, 9-6**
A.REI.11	Represent and solve equations and inequalities graphically. Explain why the x-coordinates of the points where the graphs of the equations $y = f(x)$ and $y = g(x)$ intersect are the solutions of the equation $f(x) = g(x)$; find the solutions approximately, e.g., using technology to graph the functions, make tables of values, or find successive approximations. Include cases where $f(x)$ and/or $g(x)$ are linear, polynomial, rational, absolute value, exponential, and logarithmic functions.★	**2-4, 3-8, 5-1, 7-2, 8-3, 9-6**
Functions		
Interpreting Functions F-IF		
F.IF.4	Interpret functions that arise in applications in terms of the context. For a function that models a relationship between two quantities, interpret key features of graphs and tables in terms of the quantities, and sketch graphs showing key features given a verbal description of the relationship. ★	**1-1, 1-2, 1-3, 1-4, 1-5, 1-6, 1-7, 2-1, 3-1, 3-2, 4-1, 4-2, 5-5, 6-2, 6-4, 7-1, 8-1,** 8-4, 8-5, **9-3, 9-4, 11-4, 11-5,** 11-6
F.IF.5	Relate the domain of a function to its graph and, where applicable, to the quantitative relationship it describes.	**1-1, 1-2, 1-6, 3-1, 4-1, 6-2, 6-4, 7-1, 8-1, 9-3, 9-4, 11-4, 11-5**
F.IF.6	Calculate and interpret the average rate of change of a function (presented symbolically or as a table) over a specified interval. Estimate the rate of change from a graph.★	2-3, **3-1,** 4-2, 7-3

★ Mathematical Modeling Standards

	Standard	Lesson(s)
F.IF.7	Analyze functions using different representations. Graph functions expressed symbolically and show key features of the graph, by hand in simple cases and using technology for more complicated cases.★ **b.** Graph square root, cube root, and piecewise-defined functions, including step functions and absolute value functions.	**1-6, 6-4**
	c. Graph polynomial functions, identifying zeros when suitable factorizations are available, and showing end behavior.	1-2, 1-3, **4-1, 4-2, 5-5**
	e. Graph exponential and logarithmic functions, showing intercepts and end behavior, and trigonometric functions, showing period, midline, and amplitude.	**7-1, 8-1, 11-4, 11-5, 11-6**
F.IF.8	Write a function defined by an expression in different but equivalent forms to reveal and explain different properties of the function. **a.** Use the process of factoring and completing the square in a quadratic function to show zeros, extreme values, and symmetry of the graph, and interpret these in terms of a context.	**3-4, 3-5**
	b. Use the properties of exponents to interpret expressions for exponential functions.	7-1, 8-5
F.IF.9	Compare properties of two functions each represented in a different way (algebraically, graphically, numerically in tables, or by verbal descriptions).	**1-4,** 1-7, 3-1, 4-1, 6-4, 7-1, 8-1, 9-4
Building Functions F-BF		
F.BF.1	Build a function that models a relationship between two quantities. **b.** Combine standard function types using arithmetic operations.	**6-1**
F.BF.3	Build new functions from existing functions. Identify the effect on the graph of replacing $f(x)$ by $f(x) + k$, $k\,f(x)$, $f(kx)$, and $f(x + k)$ for specific values of k (both positive and negative); find the value of k given the graphs. Experiment with cases and illustrate an explanation of the effects on the graph using technology.	**1-7, 6-4, 7-1, 8-1, 9-3, 11-6**
F.BF.4	Find inverse functions. **a.** Solve an equation of the form $f(x) = c$ for a simple function f that has an inverse and write an expression for the inverse.	**6-2**
Linear, Quadratic, and Exponential Models F-LE		
F.LE.4	Construct and compare linear and exponential models and solve problems. For exponential models, express as a logarithm the solution to $ab^{ct} = d$ where a, c, and d are numbers and the base b is 2, 10, or e; evaluate the logarithm using technology.	**8-3, 8-4, 8-5**
Trigonometric Functions F-TF		
F.TF.1	Extend the domain of trigonometric functions using the unit circle. Understand radian measure of an angle as the length of the arc on the unit circle subtended by the angle.	**11-1**
F.TF.2	Explain how the unit circle in the coordinate plane enables the extension of trigonometric functions to all real numbers, interpreted as radian measures of angles traversed counterclockwise around the unit circle.	**11-3**
F.TF.5	Model periodic phenomena with trigonometric functions. Choose trigonometric functions to model periodic phenomena with specified amplitude, frequency, and midline.★	**11-3,** 11-4, 11-5, 11-6
F.TF.8	Prove and apply trigonometric identities. Prove the Pythagorean identity $\sin^2(\theta) + \cos^2(\theta) = 1$ and use it to calculate trigonometric ratios.	**12-1, 12-2, 12-5**

★ Mathematical Modeling Standards

Standard		Lesson(s)
Statistics and Probability		
Interpreting Categorical and Quantitative Data S-ID		
S.ID.4	Summarize, represent, and interpret data on a single count or measurement variable. Use the mean and standard deviation of a data set to fit it to a normal distribution and to estimate population percentages. Recognize that there are data sets for which such a procedure is not appropriate. Use calculators, spreadsheets, and tables to estimate areas under the normal curve.	**10-4**
Making Inferences and Justifying Conclusions S-IC		
S.IC.1	Understand and evaluate random processes underlying statistical experiments Understand statistics as a process for making inferences about population parameters based on a random sample from that population.	**10-1**
S.IC.2	Decide if a specified model is consistent with results from a given data-generating process, e.g., using simulation.	**10-2**
S.IC.3	Make inferences and justify conclusions from sample surveys, experiments, and observational studies Recognize the purposes of and differences among sample surveys, experiments, and observational studies; explain how randomization relates to each.	**10-1**
S.IC.4	Use data from a sample survey to estimate a population mean or proportion; develop a margin of error through the use of simulation models for random sampling.	**10-3, 10-5**
S.IC.5	Use data from a randomized experiment to compare two treatments; use simulations to decide if differences between parameters are significant.	**10-2**
S.IC.6	Evaluate reports based on data.	10-1, 10-2, 10-4, 10-5
Using Probability to Make Decisions S-MD		
S.MD.6	(+) Use probabilities to make fair decisions (e.g., drawing by lots, using a random number generator).	10-2
S.MD.7	(+) Analyze decisions and strategies using probability concepts (e.g., product testing, medical testing, pulling a hockey goalie at the end of a game).	10-2

Standards for Mathematical Practice

This correlation shows the alignment of *Reveal Algebra 2* to the Standards for Mathematical Practice, from the Common Core State Standards.

Standard	Lesson(s)
1 Make sense of problems and persevere in solving them. Mathematically proficient students start by explaining to themselves the meaning of a problem and looking for entry points to its solution. They analyze givens, constraints, relationships, and goals. They make conjectures about the form and meaning of the solution and plan a solution pathway rather than simply jumping into a solution attempt. They consider analogous problems, and try special cases and simpler forms of the original problem in order to gain insight into its solution. They monitor and evaluate their progress and change course if necessary. Older students might, depending on the context of the problem, transform algebraic expressions or change the viewing window on their graphing calculator to get the information they need. Mathematically proficient students can explain correspondences between equations, verbal descriptions, tables, and graphs or draw diagrams of important features and relationships, graph data, and search for regularity or trends. Younger students might rely on using concrete objects or pictures to help conceptualize and solve a problem. Mathematically proficient students check their answers to problems using a different method, and they continually ask themselves, "Does this make sense?" They can understand the approaches of others to solving complex problems and identify correspondences between different approaches.	*Reveal Algebra 2* requires students to make sense of problems and persevere in solving them in Examples and Practice throughout the program. Some specific lessons for review are: Lessons 1-1, 1-4, 2-1, 2-3, 2-4, 2-5, 2-8, 2-9, 3-1, 3-3, 3-4, 3-5, 3-6, 3-7, 3-8, 4-1, 5-2, 5-5, 6-1, 6-4, 7-1, 8-1, 9-2, 9-6, 10-1, 11-3, 11-4, 12-2, 12-4
2 Reason abstractly and quantitatively. Mathematically proficient students make sense of quantities and their relationships in problem situations. They bring two complementary abilities to bear on problems involving quantitative relationships: the ability to *decontextualize*—to abstract a given situation and represent it symbolically and manipulate the representing symbols as if they have a life of their own, without necessarily attending to their referents—and the ability to *contextualize*, to pause as needed during the manipulation process in order to probe into the referents for the symbols involved. Quantitative reasoning entails habits of creating a coherent representation of the problem at hand; considering the units involved; attending to the meaning of quantities, not just how to compute them; and knowing and flexibly using different properties of operations and objects.	*Reveal Algebra 2* requires students to reason abstractly and quantitatively in Think About It features and Higher Order Thinking Skills throughout the program. Some specific lessons for review are: Lessons 1-3, 1-7, 2-1, 2-3, 3-1, 4-3, 5-1, 6-2, 6-3, 7-3, 8-3, 8-5, 9-5, 9-6, 10-3, 10-5, 11-1, 11-4, 11-5, 12-1, 12-2
3 Construct viable arguments and critique the reasoning of others. Mathematically proficient students understand and use stated assumptions, definitions, and previously established results in constructing arguments. They make conjectures and build a logical progression of statements to explore the truth of their conjectures. They are able to analyze situations by breaking them into cases, and can recognize and use counterexamples. They justify their conclusions, communicate them to others, and respond to the arguments of others. They reason inductively about data, making plausible arguments that take into account the context from which the data arose. Mathematically proficient students are also able to compare the effectiveness of two plausible arguments, distinguish correct logic or reasoning from that which is flawed, and—if there is a flaw in an argument—explain what it is. Elementary students can construct arguments using concrete referents such as objects, drawings, diagrams, and actions. Such arguments can make sense and be correct, even though they are not generalized or made formal until later grades. Later, students learn to determine domains to which an argument applies. Students at all grades can listen or read the arguments of others, decide whether they make sense, and ask useful questions to clarify or improve the arguments.	*Reveal Algebra 2* requires students to construct viable arguments and critique the reasoning of others in Talk About It features and Practice throughout the program. Some specific lessons for review are: Lessons 1-2, 2-6, 4-4, 5-4, 6-4, 6-6, 7-4, 8-2, 9-1, 9-3, 10-2, 10-5, 11-2, 11-6, 12-3
4 Model with mathematics. Mathematically proficient students can apply the mathematics they know to solve problems arising in everyday life, society, and the workplace. In early grades, this might be as simple as writing an addition equation to describe a situation. In middle grades, a student might apply proportional reasoning to plan a school event or analyze a problem in the community. By high school, a student might use geometry to solve a design problem or use a function to describe how one quantity of interest depends on another. Mathematically proficient students who can apply what they know are comfortable making assumptions and approximations to simplify a complicated situation, realizing that these may need revision later. They are able to identify important quantities in a practical situation and map their relationships using such tools as diagrams, two-way tables, graphs, flowcharts and formulas. They can analyze those relationships mathematically to draw conclusions. They routinely interpret their mathematical results in the context of the situation and reflect on whether the results make sense, possibly improving the model if it has not served its purpose.	*Reveal Algebra 2* requires students to model with mathematics, collaborate, and discuss mathematics in Examples and Practice throughout the program. Some specific lessons for review are: Lessons 1-2, 1-6, 2-2, 2-3, 2-6, 2-7, 3-2, 3-3, 3-6, 3-7, 3-8, 4-2, 4-5, 5-4, 6-1, 6-5, 7-5, 8-4, 9-4, 9-5, 10-3, 11-3, 11-7, 12-5

Standard	Lesson(s)
5 Use appropriate tools strategically. Mathematically proficient students consider the available tools when solving a mathematical problem. These tools might include pencil and paper, concrete models, a ruler, a protractor, a calculator, a spreadsheet, a computer algebra system, a statistical package, or dynamic geometry software. Proficient students are sufficiently familiar with tools appropriate for their grade or course to make sound decisions about when each of these tools might be helpful, recognizing both the insight to be gained and their limitations. For example, mathematically proficient high school students analyze graphs of functions and solutions generated using a graphing calculator. They detect possible errors by strategically using estimation and other mathematical knowledge. When making mathematical models, they know that technology can enable them to visualize the results of varying assumptions, explore consequences, and compare predictions with data. Mathematically proficient students at various grade levels are able to identify relevant external mathematical resources, such as digital content located on a website, and use them to pose or solve problems. They are able to use technological tools to explore and deepen their understanding of concepts.	*Reveal Algebra 2* requires students to use appropriate tools strategically in Explore activities throughout the program. Some specific lessons for review are: Lessons 1-3, 1-5, 2-4, 2-7, 2-9, 3-2, 4-2, 5-1, 6-4, 6-6, 7-1, 7-5, 8-1, 8-3, 8-4, 8-5, 9-4, 9-6, 10-4, 11-1, 11-6, 12-1
6 Attend to precision. Mathematically proficient students try to communicate precisely to others. They try to use clear definitions in discussion with others and in their own reasoning. They state the meaning of the symbols they choose, including using the equal sign consistently and appropriately. They are careful about specifying units of measure, and labeling axes to clarify the correspondence with quantities in a problem. They calculate accurately and efficiently, express numerical answers with a degree of precision appropriate for the problem context. In the elementary grades, students give carefully formulated explanations to each other. By the time they reach high school they have learned to examine claims and make explicit use of definitions.	*Reveal Algebra 2* requires students to attend to precision in Examples and Practice throughout the program. Some specific lessons for review are: Lessons 1-4, 1-5, 1-6, 2-5, 4-1, 5-3, 6-2, 6-3, 7-2, 8-5, 9-3, 9-5, 10-4, 11-2, 11-3, 11-7, 12-4
7 Look for and make use of structure. Mathematically proficient students look closely to discern a pattern or structure. Young students, for example, might notice that three and seven more is the same amount as seven and three more, or they may sort a collection of shapes according to how many sides the shapes have. Later, students will see 7×8 equals the well remembered $7 \times 5 + 7 \times 3$, in preparation for learning about the distributive property. In the expression $x^2 + 9x + 14$, older students can see the 14 as 2×7 and the 9 as $2 + 7$. They recognize the significance of an existing line in a geometric figure and can use the strategy of drawing an auxiliary line for solving problems. They also can step back for an overview and shift perspective. They can see complicated things, such as some algebraic expressions, as single objects or as being composed of several objects. For example, they can see $5 - 3(x - y)^2$ as 5 minus a positive number times a square and use that to realize that its value cannot be more than 5 for any real numbers x and y.	*Reveal Algebra 2* requires students to look for and make use of structure in Explore activities and Higher Order Thinking Skills throughout the program. Some specific lessons for review are: Lessons 1-1, 1-7, 2-2, 3-4, 3-5, 4-5, 5-2, 5-5, 6-1, 6-3, 6-5, 7-2, 7-3, 8-3, 8-4, 9-2, 9-3, 10-1, 11-4, 11-6, 12-3
8 Look for and express regularity in repeated reasoning. Mathematically proficient students notice if calculations are repeated, and look both for general methods and for shortcuts. Upper elementary students might notice when dividing 25 by 11 that they are repeating the same calculations over and over again, and conclude they have a repeating decimal. By paying attention to the calculation of slope as they repeatedly check whether points are on the line through (1, 2) with slope 3, middle school students might abstract the equation $(y - 2)/(x - 1) = 3$. Noticing the regularity in the way terms cancel when expanding $(x - 1)(x + 1)$, $(x - 1)(x^2 + x + 1)$, and $(x - 1)(x^3 + x^2 + x + 1)$ might lead them to the general formula for the sum of a geometric series. As they work to solve a problem, mathematically proficient students maintain oversight of the process, while attending to the details. They continually evaluate the reasonableness of their intermediate results.	*Reveal Algebra 2* requires students to look for and express regularity in repeated reasoning in Concept Check and Think About It features and Higher Order Thinking Skills throughout the program. Some specific lessons for review are: Lessons 2-8, 4-3, 4-4, 5-3, 6-6, 7-4, 8-2, 9-1, 10-2, 11-5, 12-5

Module 1

Relations and Functions

Module Goals

- Students determine if functions are one-to-one and onto.
- Students determine the linearity, intercepts, and symmetry of functions.
- Students analyze and compare graphs.
- Students graph functions and inequalities in two variables.
- Students identify and use transformations of functions.

Focus

Domain: Algebra, Functions

Standards for Mathematical Content:

F.IF.4 For a function that models a relationship between two quantities, interpret key features of graphs and tables in terms of the quantities, and sketch graphs showing key features given in a verbal description of the relationship.

F.IF.7b Graph square root, cube root, and piecewise-defined functions, including step functions and absolute value functions.

Also addresses A.CED.3, F.IF.5, F.IF.7c, F.IF.9, and F.BF.3.

Standards for Mathematical Practice:

All Standards for Mathematical Practice will be addressed in this Module.

Coherence

Vertical Alignment

Previous

Students understood the concept of a function and studied linear and nonlinear functions. **8.F.1, F.IF.1 (Algebra 1)**

Now

Students graph and identify key features of relations and functions. **F.IF.4, F.IF.7**

Next

Students will extend their understanding of functions to study other nonlinear function types, including logarithmic and trigonometric. **F.LE.4, F.TF.1, F.TF.2, F.TF.5**

Rigor

The Three Pillars of Rigor

To help students meet standards, they need to illustrate their ability to use the three pillars of rigor. Students gain conceptual understanding as they move from the Explore to Learn sections within a lesson. Once they understand the concept, they practice procedural skills and fluency and apply their mathematical knowledge as they go through the Examples and Practice.

1 CONCEPTUAL UNDERSTANDING		2 FLUENCY	3 APPLICATION
EXPLORE	LEARN	EXAMPLE & PRACTICE	

Suggested Pacing

Lessons	Standards	45-min classes	90-min classes
Module Pretest and Launch the Module Video		1	0.5
1-1 Functions and Continuity	F.IF.4, F.IF.5	2	1
1-2 Linearity, Intercepts, and Symmetry	F.IF.4, F.IF.5	2	1
1-3 Extrema and End Behavior	F.IF.4, F.IF.7c	1	0.5
Put It All Together: Lessons 1-1 through 1-3		1	0.5
1-4 Sketching Graphs and Comparing Functions	F.IF.4, F.IF.9	1	0.5
1-5 Graphing Linear Functions and Inequalities	A.CED.3, F.IF.4	2	1
1-6 Special Functions	F.IF.4, F.IF.7b	2	1
1-7 Transformations of Functions	F.IF.4, F.BF.3	2	1
Module Review		1	0.5
Module Assessment		1	0.5
	Total Days	**16**	**8**

Formative Assessment Math Probe
Interpreting Functions

Module Resource

NAME ________________ DATE ________ PERIOD ________

Cheryl Tobey Math Probe

Interpreting Functions

The graph of a function is shown.

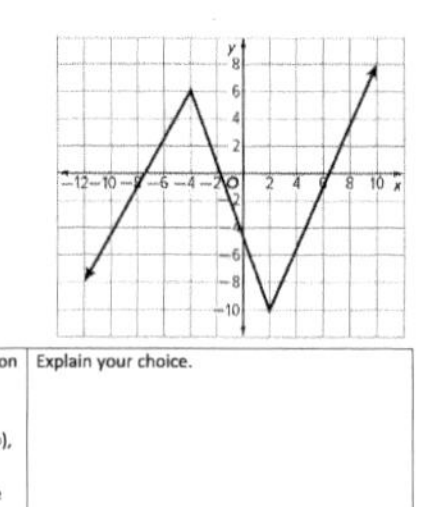

1. Which choice describes where the function is increasing? **A.** As the x values go from negative infinity to positive infinity ($-\infty < x < \infty$), the graph is always increasing. **B.** The graph is increasing from negative infinity to 6 and from −10 to positive infinity ($-\infty < y < 6$ and $-10 < y < \infty$). **C.** The graph is increasing from negative infinity to −4 and from 2 to positive infinity ($-\infty < x < -4$ and $2 < x < \infty$). **D.** The graph is not increasing as part of it goes down. **E.** Cannot tell from the graph.	Explain your choice.
2. Which choice describes where the function is decreasing? **A.** The graph is decreasing from negative infinity to 6 ($-\infty < y < 6$). **B.** The graph is decreasing from −4 to 2 ($-4 < x < 2$). **C.** The graph is decreasing from 6 to −10 ($6 < y < -10$). **D.** The graph is not decreasing as part of it goes up. **E.** Cannot tell from the graph.	Explain your choice.

Cheryl Tobey Math Probe • Interpreting Functions

Correct Answers: 1. C 2. B

Analyze the Probe

Review the probe prior to assigning it to your students.

In this probe, students will determine how to describe the behavior of a graph.

Targeted Concepts Understand how graphical representations show increasing domain intervals.

Targeted Misconceptions

- Students may incorrectly read the graph from top to bottom and use the range for the increasing interval.
- Students may incorrectly use end behaviors to evaluate increasing (and decreasing) intervals.
- Students may not see this function as having increasing/decreasing intervals as it is not continuously increasing/decreasing.

Use the Probe after Lesson 1-4.

Collect and Assess Student Answers

If the student selects these responses...	**Then** the student likely...
1. A **2.** A	interprets the right end behavior (as x approaches infinity) as increasing so considers the whole function is increasing or interprets the left end behavior (as x approaches negative infinity) as decreasing so the left portion is decreasing.
1. B **2.** A, C	is reading the graph from top to bottom and using the y-values for the intervals.
1. D **2.** D	considers only continuously increasing functions as having increasing intervals.
1. E **2.** E	does not know vocabulary and/or cannot make sense of key features of the graph.

Take Action

After the Probe Design a plan to address any possible misconceptions. You may wish to assign the following resources.

- ALEKS® Graphs and Transformations of Functions
- Lesson 1-4, all Learns, all Examples

Revisit the probe at the end of the module to be sure that your students no longer carry these misconceptions.

The Ignite! activities, created by Dr. Raj Shah, cultivate curiosity and engage and challenge students. Use these open-ended, collaborative activities, located online in the module Launch section, to encourage your students to develop a growth mindset towards mathematics and problem solving. Use the teacher notes for implementation suggestions and support for encouraging productive struggle.

Essential Question

At the end of this module, students should be able to answer the Essential Question.

How can analyzing a function help you understand the situation it models? Sample answer: You can analyze a function by observing the domain and range, the intercepts, end behavior, and so on. These observations can help you understand the situation it is modeling by seeing what values make sense in the situation, how the quantities will change as the domain values change, and so on.

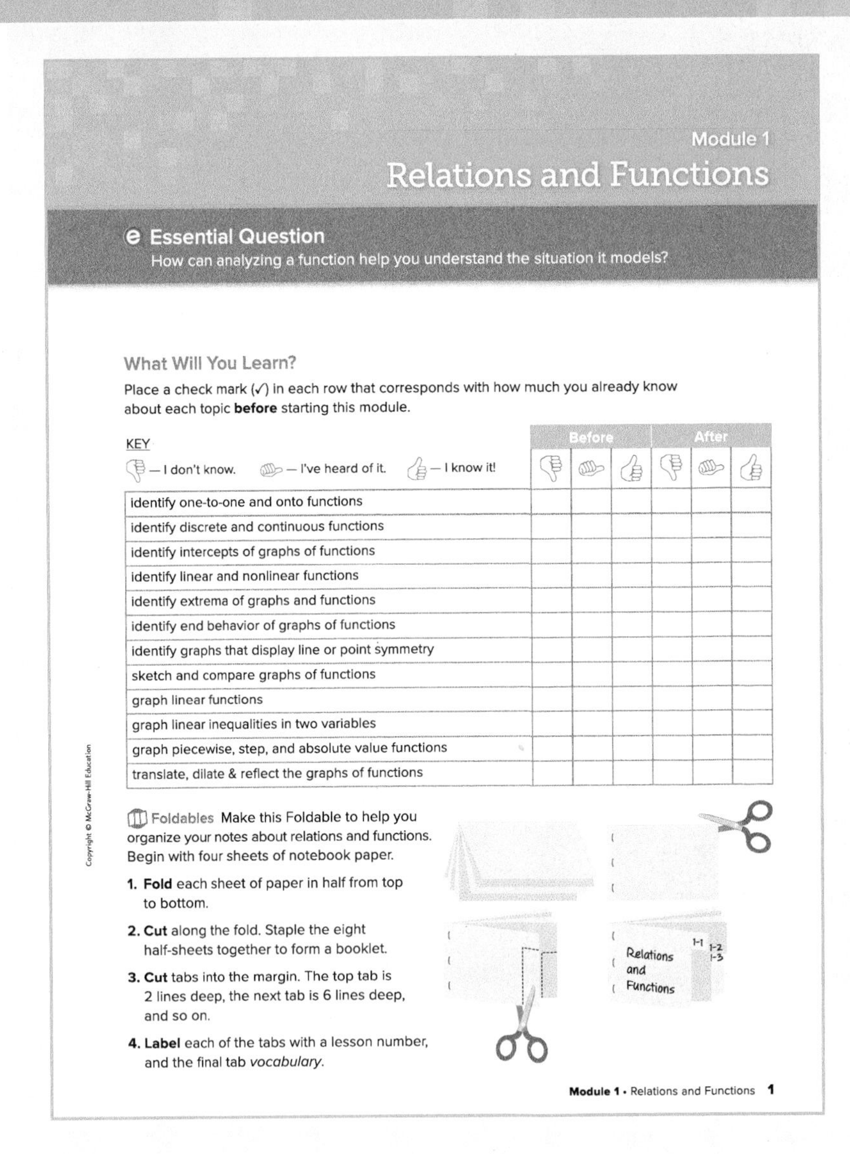

Module 1

Relations and Functions

Essential Question
How can analyzing a function help you understand the situation it models?

What Will You Learn?
Place a check mark (✓) in each row that corresponds with how much you already know about each topic **before** starting this module.

KEY
— I don't know. — I've heard of it. — I know it!

	Before			After		
identify one-to-one and onto functions						
identify discrete and continuous functions						
identify intercepts of graphs of functions						
identify linear and nonlinear functions						
identify extrema of graphs and functions						
identify end behavior of graphs of functions						
identify graphs that display line or point symmetry						
sketch and compare graphs of functions						
graph linear functions						
graph linear inequalities in two variables						
graph piecewise, step, and absolute value functions						
translate, dilate & reflect the graphs of functions						

Foldables Make this Foldable to help you organize your notes about relations and functions. Begin with four sheets of notebook paper.

1. **Fold** each sheet of paper in half from top to bottom.
2. **Cut** along the fold. Staple the eight half-sheets together to form a booklet.
3. **Cut** tabs into the margin. The top tab is 2 lines deep, the next tab is 6 lines deep, and so on.
4. **Label** each of the tabs with a lesson number, and the final tab *vocabulary*.

What Will You Learn?

Prior to beginning this module, have your students rate their knowledge of each item listed. Then, at the end of the module, you will be reminded to have your students return to these pages to rate their knowledge again. They should see that their knowledge and skills have increased.

DINAH ZIKE FOLDABLES

Focus Students read about functions, analyzing and sketching graphs, special functions, and transformations of functions.

Teach Throughout the module, have students take notes under the tabs of their Foldables while working through each lesson. They should include definitions of terms and key concepts. Encourage students to record examples from each lesson in their Foldable.

When to Use It Use the appropriate tabs as students cover each lesson in this module.

Interactive Presentation

Launch the Module

For this module, the Launch the Module video describes everyday activities, such as earning money at a job, using functions. Students learn about how the key features of the graph of a function can be analyzed to make conjectures about future behavior.

What Vocabulary Will You Learn?

Check the box next to each vocabulary term that you may already know.

☐ absolute value function	☐ end behavior	☐ linear function	☐ point of symmetry
☐ algebraic notation	☐ even functions	☐ linear inequality	☐ point symmetry
☐ boundary	☐ extrema	☐ maximum	☐ range
☐ closed half-plane	☐ family of graphs	☐ minimum	☐ reflection
☐ codomain	☐ greatest integer function	☐ nonlinear function	☐ relative maximum
☐ constant function	☐ identity function	☐ odd functions	☐ relative minimum
☐ constraint	☐ intercept	☐ one-to-one function	☐ set-builder notation
☐ continuous function	☐ interval notation	☐ onto function	☐ step function
☐ dilation	☐ line of reflection	☐ open half-plane	☐ symmetry
☐ discontinuous function	☐ line of symmetry	☐ parabola	☐ translation
☐ discrete function	☐ line symmetry	☐ parent function	☐ transformation
☐ domain	☐ linear equation	☐ piecewise-defined function	☐ x-intercept
			☐ y-intercept

Are You Ready?

Complete the Quick Review to see if you are ready to start this module.
Then complete the Quick Check.

Quick Review

Example 1	Example 2
Evaluate $3a^2 - 2ab + b^2$ if $a = 4$ and $b = -3$.	**Solve $3x + 6y = 24$ for y.**
$3a^2 - 2ab + b^2 = 3(4^2) - 2(4)(-3) + (-3)^2$	$3x + 6y = 24$ Original equation
$= 3(16) - 2(4)(-3) + 9$	$3x + 6y - 3x = 24 - 3x$ Subtract $3x$ from each side.
$= 48 - (-24) + 9$	$6y = 24 - 3x$ Simplify.
$= 48 + 24 + 9$	$\frac{6y}{6} = \frac{24}{6} - \frac{3x}{6}$ Divide each side by 6.
$= 81$	$y = 4 - \frac{1}{2}x$ Simplify.

Quick Check

Evaluate each expression if $a = -3$, $b = 4$, and $c = -2$.	Solve each equation for the given variable.
1. $4a - 3$ -15	**5.** $a = 3b + 9$ for b $b = \frac{a}{3} - 3$
2. $2b - 5c$ 18	**6.** $15w - 10 = 5v$ for v $v = 3w - 2$
3. $b^2 - 3b + 6$ 10	**7.** $3x - 4y = 8$ for x $x = \frac{8}{3} + \frac{4}{3}y$
4. $\frac{2a + 4b}{c}$ -5	**8.** $\frac{d}{6} + \frac{f}{3} = 4$ for d $d = -2f + 24$

How Did You Do?

Which exercises did you answer correctly in the Quick Check? Shade those exercise numbers below.

① ② ③ ④ ⑤ ⑥ ⑦ ⑧

What Vocabulary Will You Learn?

ELL As you proceed through the module, introduce the key vocabulary by using the following routine.

Define A graph has line symmetry if each half of the graph maps exactly to the other half.

Example The graph of $y = x^2 - x - 12$.

Ask Does the graph of the function have line symmetry? Explain. Yes; sample answer: The graph of the function is a parabola, and each half of a parabola maps exactly to the other half.

Are You Ready?

Students may need to review the following prerequisite skills to succeed in this module.

- determining functions
- finding domain and range
- using symmetry
- graphing lines and points on lines
- evaluating greatest integer and absolute value expressions
- transforming points

ALEKS®

ALEKS is an adaptive, personalized learning environment that identifies precisely what each student knows and is ready to learn, ensuring student success at all levels.

You may want to use the **Graphs and Functions** section to ensure student success in this module.

Mindset Matters

"Not Yet" Doesn't Mean "Never"

Students with a growth mindset come to understand that just because they haven't yet found a solution, that doesn't mean they can't find one with additional effort and reasoning. It takes time to reason through the different strategies that can be used to solve a problem.

How Can I Apply It?

Assign students the **Math Probes** that are available for each module. Have them complete the probe before starting the module and again at the specified point in the module or at the end of the module so that they can see their progress.

Lesson 1-1

Functions and Continuity

F.IF.4, F.IF.5

LESSON GOAL

Students determine the continuity of functions and whether functions are one-to-one and/or onto.

1 LAUNCH

 Launch the lesson with a **Warm Up** and an introduction.

2 EXPLORE AND DEVELOP

 **Explore:**

- Analyzing Functions Graphically
- Defining and Analyzing Variables

 **Develop:**

Functions

- Domains, Codomains, and Ranges
- Identify One-to-One and Onto Functions from Tables
- Identify One-to-One and Onto Functions from Graphs

Discrete and Continuous Functions

- Determine Continuity from Graphs
- Determine Continuity

Set-Builder and Interval Notation

- Set-Builder and Interval Notation for Continuous Intervals
- Set-Builder and Interval Notation for Discontinuous Intervals

 You may want your students to complete the **Checks** online.

3 REFLECT AND PRACTICE

 Exit Ticket

 Practice

DIFFERENTIATE

 View reports of student progress on the **Checks** after each example.

Resources	AL	OL	BL	ELL
Remediation: Functions	●	●		●
Extension: Correspondence Among Infinite Sets and Subsets		●	●	●

Language Development Handbook

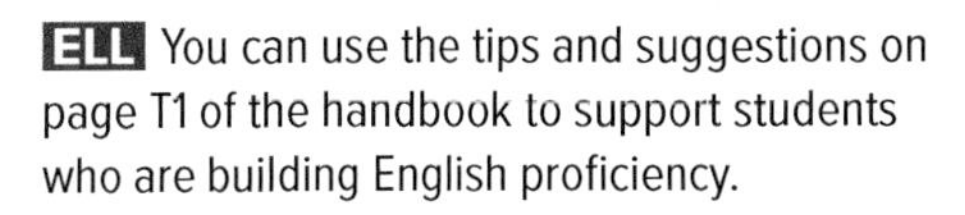

Assign page 1 of the *Language Development Handbook* to help your students build mathematical language related to the continuity of functions, and whether functions are one-to-one and/or onto.

ELL You can use the tips and suggestions on page T1 of the handbook to support students who are building English proficiency.

Suggested Pacing

90 min	1 day
45 min	2 days

Focus

Domain: Functions

Standards for Mathematical Content:

F.IF.4 For a function that models a relationship between two quantities, interpret key features of graphs and tables in terms of the quantities, and sketch graphs showing key features given in a verbal description of the relationship.

F.IF.5 Relate the domain of a function to its graph and, where applicable, to the quantitative relationship it describes.

Standards for Mathematical Practice:

1 Make sense of problems and persevere in solving them.
7 Look for and make use of structure.

Coherence

Vertical Alignment

Previous
Students understood the concept of a function and worked with linear and nonlinear functions. **8.F.1, F.IF.1 (Algebra 1)**

Now
Students determine the continuity of functions and whether functions are one-to-one and/or onto.
F.IF.4, F.IF.5

Next
Students will determine the linearity, intercepts, and symmetry of functions.
F.IF.4, F.IF.5

Rigor

The Three Pillars of Rigor

1 CONCEPTUAL UNDERSTANDING	2 FLUENCY	3 APPLICATION

Conceptual Bridge In this lesson, students build on their understanding of functions to include different types of functions. They develop fluency and apply their understanding by solving real-world problems involving functions.

Mathematical Background

A relation is a mapping of elements of the domain to those of the range. A function is a relation in which each element of the domain is mapped to exactly one element of the range.

Interactive Presentation

Warm Up

Anita is making chicken noodle soup. The recipe calls for 3 times as much broth as meat.

1. Complete the mapping to represent the situation.

Domain	Range
1	
2	
3	
4	

2. Write a function to represent the situation. Let x = the number of cups of meat and y = the number of cups of broth.

Match each function to the situation or mapping it describes.

3. For every dollar a customer spends, Mike earns a $0.15 tip. Let x be the number of dollars a customer spends.

4. Florida has a 6% sales tax. If an item costs $5, you pay $5.30. If an item costs $10, you pay $10.60, and so on. Let x be the price of an item.

A. $f(x) = -x^2 + 4$

B. $f(x) = 0.15x$

C. $f(x) = 1.06x$

Warm Up

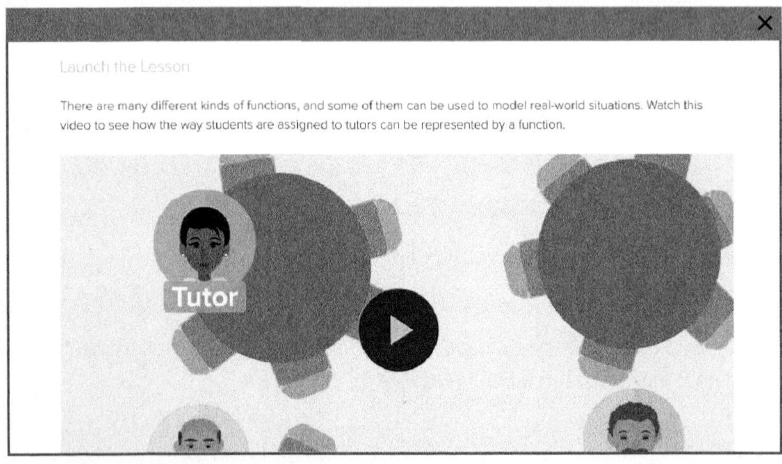

Launch the Lesson

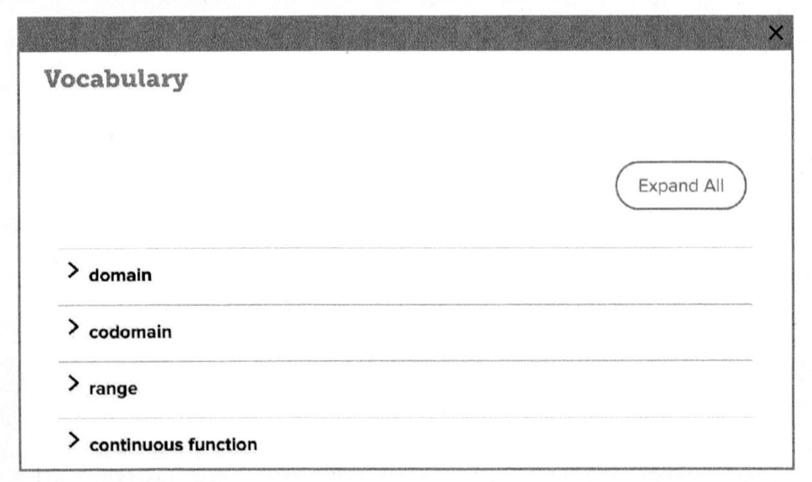

Today's Vocabulary

Warm Up

Prerequisite Skills

The Warm Up exercises address the following prerequisite skill for this lesson:

- determining functions

Answers:

1. 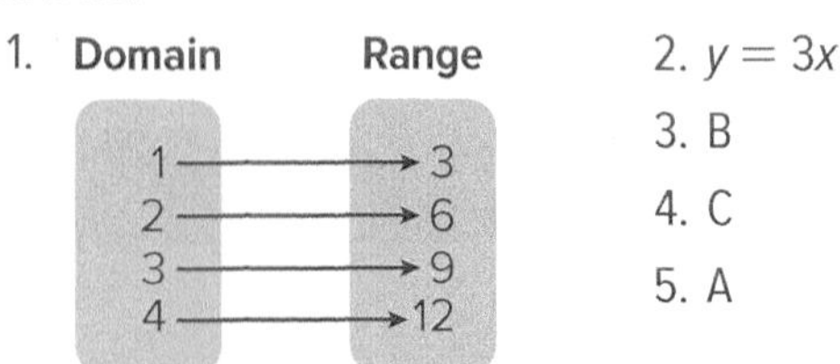

2. $y = 3x$
3. B
4. C
5. A

Launch the Lesson

 **Teaching the Mathematical Practices**

1 Explain Correspondences Encourage students to explain the relationship between the number of students and tutors in the video. Work with students to describe how the situations in the video demonstrate one-to-one and onto functions.

Go Online to find additional teaching notes and questions to promote classroom discourse.

Today's Standards

Tell students that they will be addressing these content and practice standards in this lesson. You may wish to have a student volunteer read aloud *How can I meet these standards?* and *How can I use these practices?*, and connect these to the standards.

See the Interactive Presentation for I Can statements that align with the standards covered in this lesson.

Today's Vocabulary

Tell students that they will be using these vocabulary terms in this lesson. You can expand each row if you wish to share the definitions. Then discuss the questions below with the class.

Explore Analyzing Functions Graphically

Objective

Students use a sketch to explore whether functions are one-to-one and/or onto.

Teaching the Mathematical Practices

5 Use Mathematical Tools Point out that to complete the Explore, students will need to use an interactive sketch to analyze the graphs of several functions.

Ideas for Use

Recommended Use Present the Inquiry Question, or have a student volunteer read it aloud. Have students work in pairs to complete the Explore activity on their devices. Pairs should discuss each of the questions. Monitor student progress during the activity. Upon completion of the Explore activity, have student volunteers share their responses to the Inquiry Question.

What if my students don't have devices? You may choose to project the activity on a whiteboard. A printable worksheet for each Explore is available online. You may choose to print the worksheet so that individuals or pairs of students can use it to record their observations.

Summary of the Activity

Students will complete guiding exercises throughout the Explore activity. Students will use a sketch to apply the vertical line test and horizontal line test to four different functions. Students will analyze the tests to make a conjecture about one-to-one and onto functions. Then, students will answer the Inquiry Question.

Common Error

Students sometimes have trouble remembering which lines are vertical and which are horizontal. Point out that the "V" in Vertical is like an arrow pointing down, so that can help to remember that a vertical line extends straight up and down. On the other hand, the middle bar in "H" in Horizontal is in the same direction as a horizontal line that extends left and right.

(continued on the next page)

Interactive Presentation

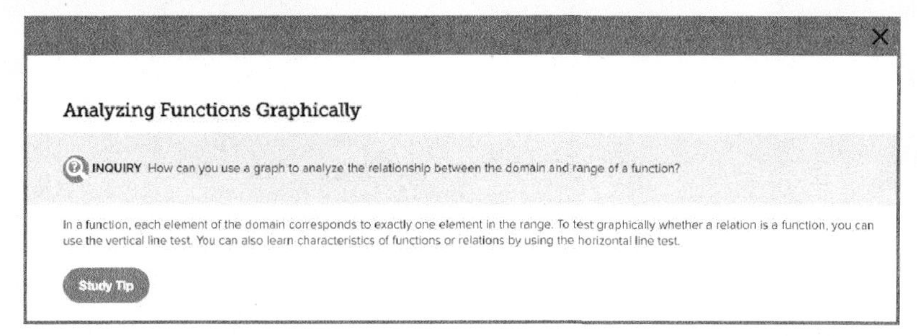

Explore

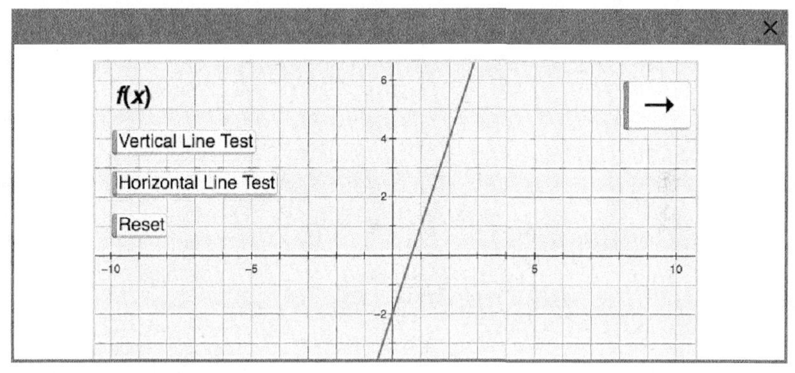

Explore

WEB SKETCHPAD

Students use the sketch to complete an activity in which they examine the intersection of graphed relations with vertical and horizontal lines.

SELECT

Students select answers to complete a table about each function graphed.

TYPE

Students make a conjecture about what it means for a function to be one-to-one and onto.

Interactive Presentation

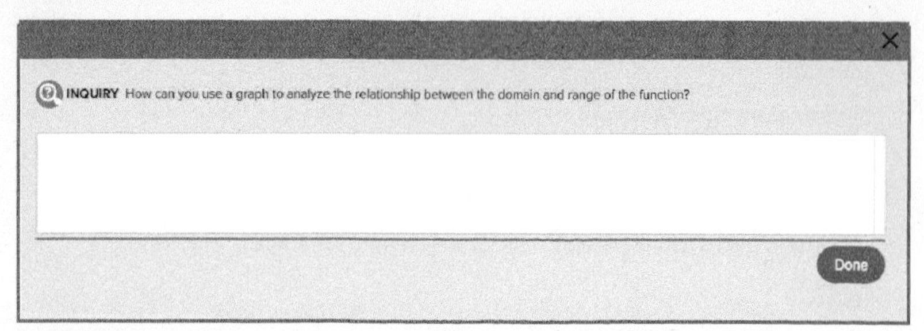

Explore

TYPE

a|

Students respond to the Inquiry Question and can view a sample answer.

Explore Analyzing Functions Graphically (*continued*)

Questions

Have students complete the Explore activity.

Ask:

- How do you tell if a graph is a function? Sample answer: When using the vertical line test, any vertical line only crosses the function at one point.
- What is the maximum number of times the horizontal line will cross function $j(x)$? 3 the minimum number? 1

Inquiry

How can you use a graph to analyze the relationship between the domain and range of a function? Sample answer: You can use the horizontal line test to tell you if elements in a function's range, or y-values, are paired with more than one element from the domain, or x-values.

Go Online to find additional teaching notes and sample answers for the guiding exercises.

Explore Defining and Analyzing Variables

Objective

Students use an expression, equation or function to define variables, making sure to specify units.

Teaching the Mathematical Practices

1 Analyze Givens and Constraints Throughout the Explore encourage students to analyze the givens, constraints, and relationships in each situation to better understand the variables.

Ideas for Use

Recommended Use Present the Inquiry Question, or have a student volunteer read it aloud. Have students work in pairs to complete the Explore activity on their devices. Pairs should discuss each of the questions. Monitor student progress during the activity. Upon completion of the Explore activity, have student volunteers share their responses to the Inquiry Question.

What if my students don't have devices? You may choose to project the activity on a whiteboard. A printable worksheet for each Explore is available online. You may choose to print the worksheet so that individuals or pairs of students can use it to record their observations.

Summary of the Activity

Students will complete guiding exercises throughout the Explore activity. Students will be presented with two situations in which they must define the independent and dependent variables. They move through exercises for each situation considering the different types of variables and which are dependent or independent. Then, students will answer the Inquiry Question.

(continued on the next page)

Interactive Presentation

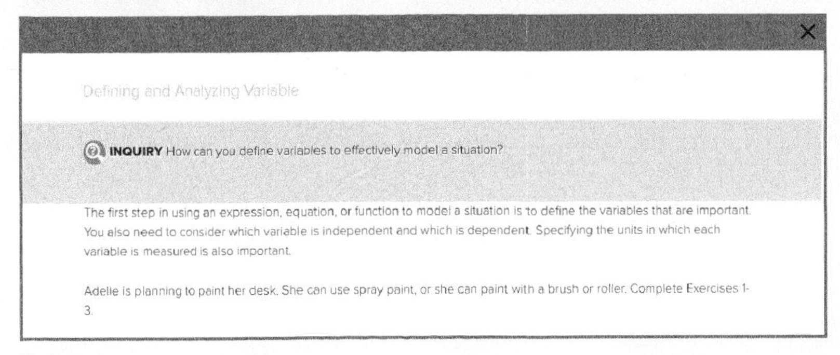

Explore

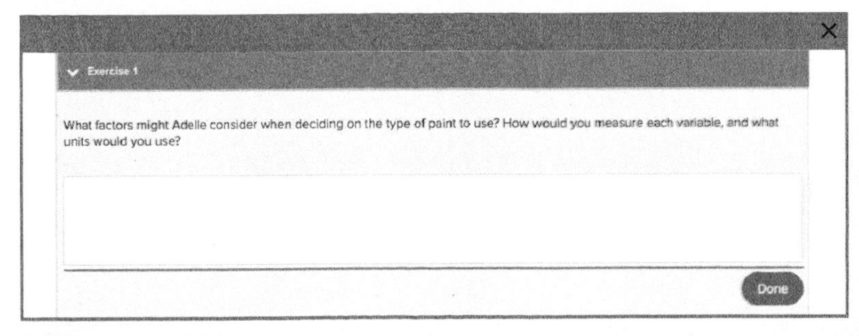

Explore

TYPE

Students answer questions pertaining to defining and analyzing variables.

E TOOLS

Students use the graphing eTool to graph collected data.

Interactive Presentation

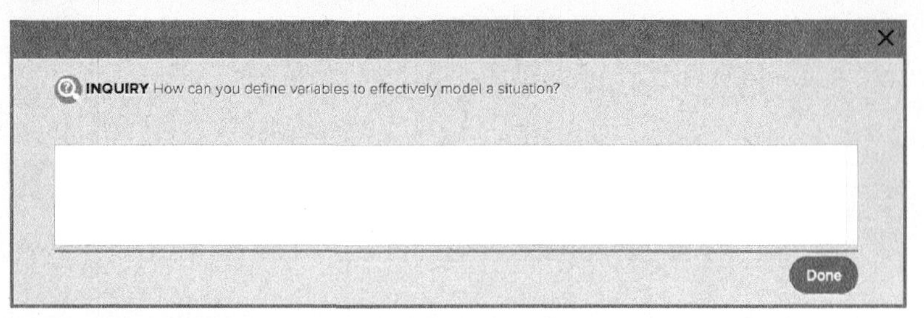

Explore

TYPE

Students respond to the Inquiry Question and can view a sample answer.

Explore Defining and Analyzing Variables (*continued*)

Questions

Have students complete the Explore activity.

Ask:

- How can the dependent variable be identified in a real-world situation? Sample answer: The dependent variable depends on the independent variable, so identify the measurement or output that depends on the quantity being inputted.
- Suppose a new fitness tracker records the time spent exercising and calories burned for 40 volunteers. In this situation, what would be the independent and dependent variables? Time spent exercising is the independent variable, and calories burned is the dependent variable.

Inquiry

How can you define variables to effectively model a situation? Sample answer: List the variables that are important in the situation. Determine the independent and dependent variables. Specify the units in which each variable is measured.

Go Online to find additional teaching notes and sample answers for the guiding exercises.

Learn Functions

Objective

Students determine whether functions are one-to-one and/or onto by examining the graphs of the functions.

Teaching the Mathematical Practices

7 Use Structure Students will use the structure of graphs and tables to analyze the functions they represent.

Common Misconception

Some students believe that if a function is one-to-one it must also be onto, regardless of the graph. Encourage students to write down the definition of one-to-one and onto functions in their notes along with a different example so they can look at those as a reference when working problems.

DIFFERENTIATE

Language Development Activity ELL

Intermediate Instruct a small group of students to draw a graph and write a paragraph to describe how to use the Horizontal Line Test on their graph. Their paragraphs should describe all parts of their graph in their own words. Ask for volunteers to read their paragraphs. Have students ask for clarification as needed.

Example 1 Domains, Codomains, and Ranges

Teaching the Mathematical Practices

7 Use Structure Help students use the structure of the graph of the given function to determine domain, range, and codomain.

Questions for Mathematical Discourse

AL How do you determine the domain of a function? Sample answer: The domain is all the x-values that can be evaluated in the function, so check for any restrictions on x.

OL Describe the difference between the codomain and the range. What might cause a codomain to not be all real numbers? Sample answer: The range is determined by the function, while the codomain can be limited by considerations outside of the function, such as real-world constraints.

BL Why do you think a function with the same codomain and range is called "onto?" Sample answer: When a function is onto, the function maps at least one element of the domain onto each element of the codomain.

Lesson 1-1

Functions and Continuity

Explore Analyzing Functions Graphically

Online Activity Use graphing technology to complete the Explore.

INQUIRY How can you use a graph to analyze the relationship between the domain and range of a function?

Explore Defining and Analyzing Variables

Online Activity Use a real-world situation to complete the Explore.

INQUIRY How can you define variables to effectively model a situation?

Today's Goals
- Determine whether functions are one-to-one and/or onto.
- Determine the continuity, domain, and range of functions.
- Write the domain and range of functions by using set-builder and interval notations.

Today's Vocabulary
domain
codomain
range
one-to-one function
onto function
continuous function
discontinuous function
discrete function
algebraic notation
set-builder notation
interval notation

Learn Functions

A function describes a relationship between input and output values. The **domain** is the set of x-values to be evaluated by a function. The **codomain** is the set of all the y-values that could possibly result from the evaluation of the function. The codomain of a function is assumed to be all real numbers unless otherwise stated. The **range** is the set of y-values that actually result from the evaluation of the function. The range is contained within the codomain.

If each element of a function's range is paired with exactly one element of the domain, then the function is a **one-to-one function**. If a function's codomain is the same as its range, then the function is an **onto function**.

Example 1 Domains, Codomains, and Ranges

Part A Identify the domain, range, and codomain of the graph.

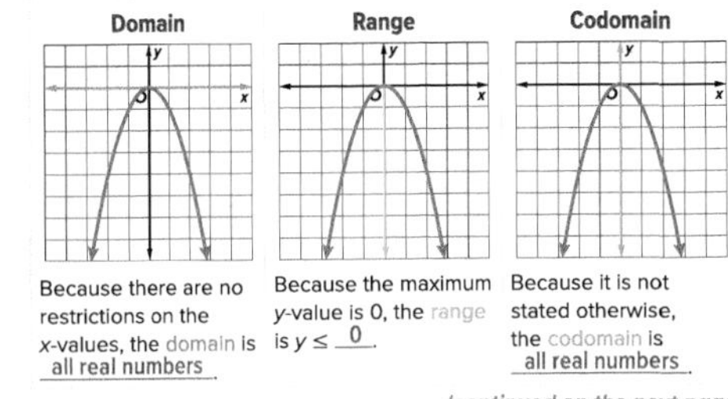

Because there are no restrictions on the x-values, the domain is all real numbers.

Because the maximum y-value is 0, the range is $y \leq$ 0.

Because it is not stated otherwise, the codomain is all real numbers.

(continued on the next page)

Study Tip

Horizontal Line Test Place a pencil at the top of the graph and move it down to represent a horizontal line.
- If there are places where the pencil intersects the graph at more than one point, then more than one element of the range is paired with an element of the domain. The function is not one-to-one.
- If there are places where the pencil does not intersect the graph at all, then there are real numbers that are not paired with an element of the domain. The function is not onto.

Interactive Presentation

Functions

A function describes a relationship between input and output values. For example, suppose you record the number of volunteer hours you still need to complete at the end of each semester before graduation. Your function could be {(1, 40), (2, 36), (3, 24), (4, 18), (5, 16), (6, 10), (7, 3), (8, 0)}.

The sets of x- and y-values in a function have special names.

- The **domain** is the set of x-values to be evaluated by a function. In this function, the domain is the semesters, {1, 2, 3, 4, 5, 6, 7, 8}.
- The **codomain** is the set of all the y-values that could possibly result from the evaluation of the function. The codomain of a function is assumed to be all real numbers unless otherwise stated. In this function, the codomain is nonnegative real numbers because the number of hours may be any number 0 or greater.
- The **range** is the set of y-values that actually result from the evaluation of the function. The range is contained within the codomain. In this function, the range is the remaining volunteer hours, {40, 36, 24, 18, 16, 10, 3, 0}.

Learn

TAP

Students click on each function type to see a definition and example.

Your Notes

Part B Use these values to determine whether the function is onto.

The range is not the same as the codomain because it does not include the positive real numbers. Therefore, the function is not onto.

Check

For what codomain is $f(x)$ an onto function? A

A. $y \leq 3$ **B.** $y \geq 3$

C. all real numbers **D.** $x \leq 3$

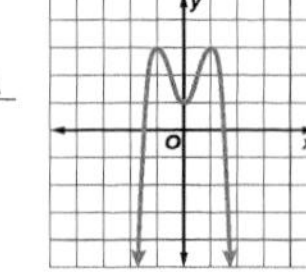

Example 2 Identify One-to-One and Onto Functions from Tables

OLYMPICS **The table shows the number of medals the United States won at five Summer Olympic Games.**

Year	Number of Gold Medals	Number of Silver Medals	Number of Bronze Medals
2016	46	37	38
2012	46	29	29
2008	36	38	36
2004	36	39	26
2000	37	24	32

Analyze the functions that give the number of gold and silver medals won in a particular year. Define the domain and range of each function and state whether it is *one-to-one, onto, both* or *neither.*

Gold Medals

Let $f(x)$ be the function that gives the number of gold medals won in a particular year. The domain is in the column Year, and the range is in the column Number of Gold Medals.

The function is not one-to-one because two values in the domain, 2016 and 2012, share the same value in the range, 46, and two values in the domain, 2008 and 2004, share the same value in the range, 36.

The function is not onto because the range does not include every whole number.

Silver Medals

Let $g(x)$ be the function that gives the number of silver medals won in a particular year. The domain is the column Year, and the range is the column Number of Silver Medals. The function is one-to-one because no two values in the domain share a value in the range.

The function is not onto because the range does not include every whole number.

Go Online You can complete an Extra Example online.

Use a Source

Choose another country and research the number of medals they won in the Summer Olympic Games from 2000-2016. Are the functions that give the number of each type of medal won in a particular year *one-to-one, onto, both,* or *neither*?

Sample answer: The functions that represent the number of each type of medal won by Great Britain given the year are all one-to-one but not onto.

4 Module 1 • Relations and Functions

Interactive Presentation

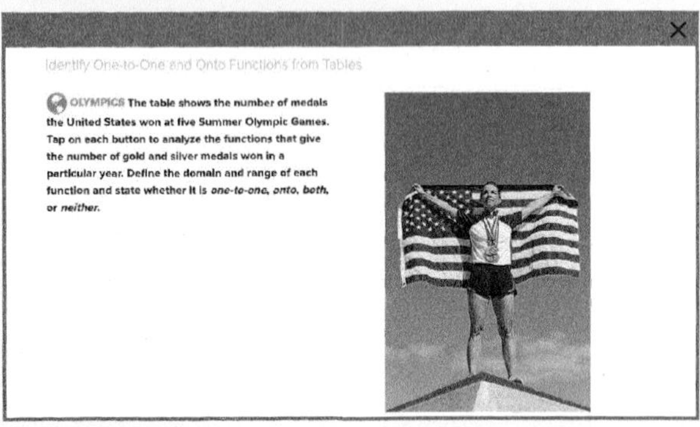

Example 2

TAP

Students tap to analyze whether the function in the table is one-to-one or onto.

TYPE

Students choose a country and research the number of Gold medals won in the Summer Olympics and then decide whether the function is one-to-one, onto, both, or neither.

DIFFERENTIATE

Reteaching Activity AL ELL

IF students are struggling to differentiate between one-to-one and onto functions,

THEN have students place their pencil on the graph horizontally and move the pencil up and down. If the pencil touches the graph more than once at any point, it is not one-to-one. If there are places where the pencil does not touch any graph, then the function is not onto.

Common Error

Because both terms share the word *domain* many students think codomain is part of the domain, and not related to the range. Remind students that the range is a subset of the codomain and codomain is always assumed to be all real numbers unless otherwise stated.

Example 2 Identify One-to-One and Onto Functions from Tables

Teaching the Mathematical Practices

1 Explain Correspondences Encourage students to explain the relationship between the values in the table and the type of function.

Questions for Mathematical Discourse

AL Why is the codomain not all real numbers in this example? You cannot have partial or negative medals.

OL How would the range need to change for the function representing the number of silver medals won in a particular year to not be one-to-one? Sample answer: At least two of the y-values would have to be the same.

BL What additional information would we need in this scenario to describe the function as onto? Sample answer: We would need to know a finite constraint on the possible numbers of a type of medal.

Common Error

Some students think *one-to-one* means each x-value pairs with only one y-value, but this is the definition of a function. Reinforce that one-to-one functions cannot have repeating y-values.

Example 3 Identify One-to-One and Onto Functions from Graphs

Teaching the Mathematical Practices

3 Construct Arguments In Example 3, students will use definitions to classify functions as one-to-one, onto, both, or neither.

Questions for Mathematical Discourse

AL What is the relationship between the codomain and the range of an onto function? The codomain and the range are the same.

OL Why is the horizontal line test useful to determine whether a function is one-to-one when given a graph? Sample answer: A horizontal line has the same y-value at any point on the line, so if the horizontal line touches the function in more than one place, it means that the function has two x-values that result in the same y-value.

BL For $g(x)$, why is the function onto when the horizontal line test fails for $y = 6$? Sample answer: 6 is not an element of the codomain. The horizontal line test should only be applied to elements of the codomain.

Learn Discrete and Continuous Functions

Objective

Students determine the continuity, domain, and range of functions by examining the functions.

Teaching the Mathematical Practices

1 Explain Correspondences Encourage students to explain the relationship between the continuity of a function and the structure of its graph.

Common Misconception

Some students assume that a function is either discrete or continuous. There are numerous functions, especially piece-wise functions, that are discontinuous but not discrete. Encourage students to investigate different types of graphs to identify which ones are continuous, discrete, or neither.

DIFFERENTIATE

Reteaching Activity AL ELL

IF students are having a hard time determining continuity from a graph,
THEN have students place their pencil on the graph and follow the curve. If they can trace the entire graph without picking up their pencil, the graph is continuous.

Example 3 Identify One-to-One and Onto Functions from Graphs

Determine whether each function is *one-to-one, onto, both,* or *neither* for the given codomain.

$f(x)$, where the codomain is all real numbers

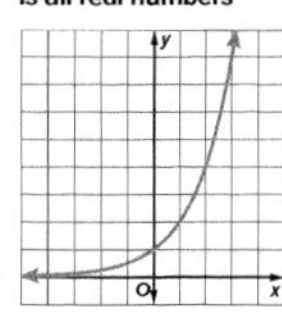

The graph indicates that the domain is all real numbers, and the range is all positive real numbers.

Every x-value is paired with exactly one unique y-value, so the function __is__ one-to-one.

If the codomain is all real numbers, then the range is not equal to the codomain. So, the function __is not__ onto.

$g(x)$, where the codomain is $\{y \mid y \leq 4\}$

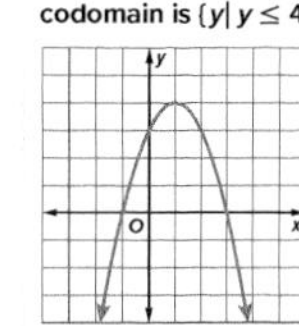

The graph indicates that the domain is all real numbers, and the range is $y \leq 4$.

Each x-value is not paired with a unique y-value; for example, both $x = 0$ and $x = 2$ are paired with $y = 3$. So the function __is not__ one-to one.

The codomain and range are equal, so the function __is__ onto.

$h(x)$, where the codomain is all real numbers

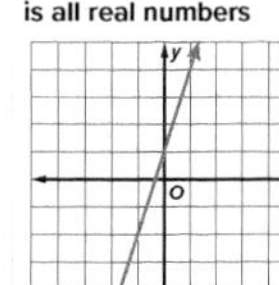

The graph indicates that the domain and range are both all real numbers.

Every x-value is paired with exactly one unique y-value, so the function __is__ one-to-one.

The codomain and range are equal, so the function __is__ onto.

Learn Discrete and Continuous Functions

Functions can be discrete, continuous, or neither. Real-world situations where only some numbers are reasonable are modeled by discrete functions. Situations where all real numbers are reasonable are modeled by continuous functions.

A **continuous function** is graphed with a line or an unbroken curve. A function that is not continuous is a **discontinuous function**. A **discrete function** is a discontinuous function in which the points are not connected. A function that is neither discrete nor continuous may have a graph in which some points are connected, but it is not continuous everywhere.

Study Tip

Intervals An interval is the set of all real numbers between two given numbers. For example, the interval $-2 < x < 5$ includes all values of x greater than -2 but less than 5. Intervals can also continue on infinitely in a direction. For example, the interval $y \geq 1$ includes all values of y greater than or equal to 1. You can use intervals to describe the values of x or y for which a function exists.

Go Online You can complete an Extra Example online.

Lesson 1-1 • Functions and Continuity 5

Interactive Presentation

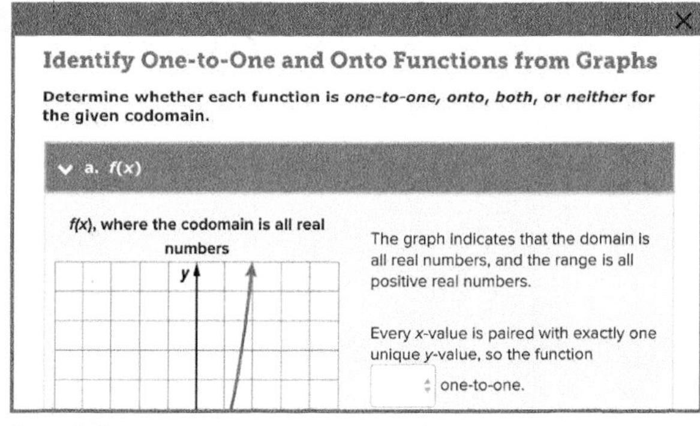

Example 3

SELECT

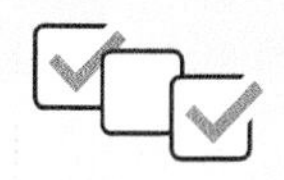

Students select the correct words to complete sentences regarding each function.

CHECK

Students complete the Check online to determine whether they are ready to move on.

Talk About It!
Does the range of the function need to be all real numbers for a function to be continuous? Justify your argument.

No; sample answer: As long as neither the domain nor the range have any discontinuities, the function is continuous.

Problem-Solving Tip
Use a Graph If you are having trouble determining the continuity given the equation of a function, you can graph the function to help visualize the situation.

Study Tip
Accuracy When calculating cost, the result can be any fraction of a dollar or cent, and is therefore continuous. However, because the smallest unit of currency is $0.01, the price you actually pay is rounded to the nearest cent. Therefore, the price you pay is discrete.

Example 4 Determine Continuity from Graphs

Examine the functions. Determine whether each function is *discrete*, *continuous*, or *neither* discrete nor continuous. Then state the domain and range of each function.

a. $f(x)$

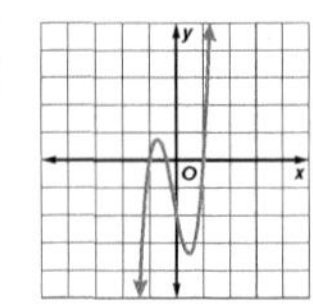

The function is continuous because it is a curve with no breaks or discontinuities.

Because you can assume that the function continues forever, the domain and range are both all real numbers.

b. $g(x)$

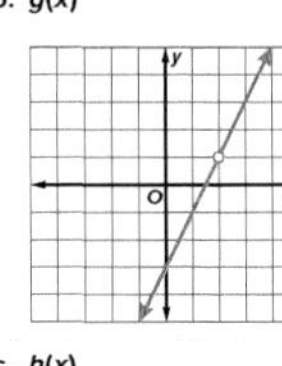

The function is neither because there are continuous sections, but there is a break at (2, 1).

Because the function is not defined for $x = 2$, the domain is all values of x except $x = 2$. The function is not defined for $y = 1$, so the range is all values of y except $y = 1$.

c. $h(x)$

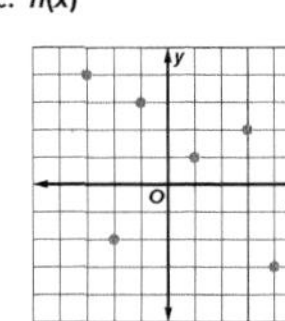

The function is discrete because it is made up of distinct points that are not connected.

The domain is $\{-3, -2, -1, 1, 3, 4\}$ and the range is $\{-3, -2, 1, 2, 3, 4\}$.

Example 5 Determine Continuity

BUSINESS **Determine whether the function that models the cost of coffee beans is *discrete*, *continuous*, or *neither* discrete nor continuous. Then state the domain and range of the function.**

COFFEE

Weight	Price
Up to 2 lbs	$8/lb
2.5 lbs	$20
3 lbs	$22
5 lbs	$35

Because customers can purchase any amount of coffee up to 2 pounds, the function is continuous over the interval $0 \leq x \leq 2$.

Go Online You can complete an Extra Example online.

Interactive Presentation

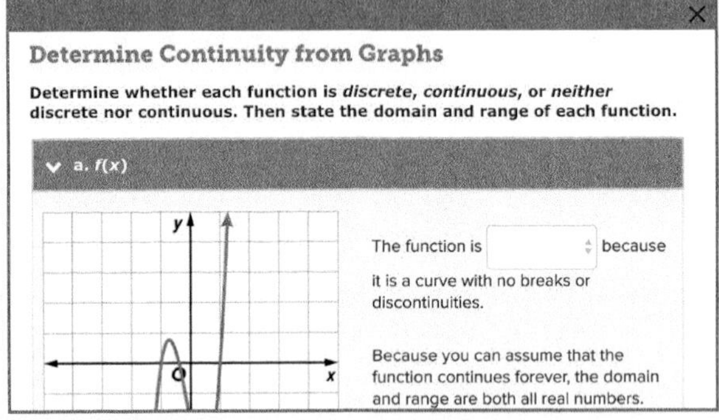

Example 4

SELECT

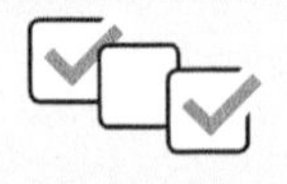

Students select whether each function is continuous.

TYPE

Students consider whether the range of a function must be all real numbers for the function to be continuous.

Example 4 Determine Continuity from Graphs

Teaching the Mathematical Practices

6 Use Definitions Students will use definitions to classify functions as discrete, continuous, or discontinuous.

Questions for Mathematical Discourse

AL What does the graph of a continuous function look like? Sample answer: a line or unbroken curve

OL Could a function that is not an onto function be continuous? Explain. Yes; sample answer: As long as the range does not exclude interior values or contain only a finite set of numbers, it can be continuous. For example, the quadratic function $y = x^2$ is not onto but it is continuous.

BL In part **b**, if the point (2, 2) is added to the graph, would the function be *discrete, continuous*, or *neither*? Explain. Neither; sample answer: There is still a break between the pieces of the graph.

Example 5 Determine Continuity

Teaching the Mathematical Practices

4 Analyze Relationships Mathematically Point out that to classify the function, students will need to analyze the mathematical relationships in the problem to draw a conclusion about the cost of coffee beans.

Questions for Mathematical Discourse

AL What would the graph of the function look like for the interval $0 \leq x \leq 2$? The graph would be a line that is increasing.

OL Would it be possible for a customer to order 3.5 pounds of coffee? Why or why not? No; sample answer: Based on the table, the function is discrete and does not exist for values between 3 and 5 pounds.

BL Suppose the café sold coffee as follows:

up to 2.5 lbs for $8/lb; 2.5 lbs up to 3 lbs for $20; 3 lbs up to 5 lbs for $22; 5 lbs or more for $4/lb

Would the function be continuous? Explain. No; sample answer: the function is still discontinuous because the range values at the ends of each domain interval do not meet.

Learn Set-Builder and Interval Notation

Objective

Students write the domain and range of functions by using set-builder and interval notations.

 Teaching the Mathematical Practices

7 Use Structure Students will use the structure of set-builder and interval notations to write the domain and range of functions.

Important to Know

Sets of numbers can be described using different notations. Set-builder notation is similar to algebraic notation and uses braces to indicate a set. The symbol | means *such that* while the symbol ∈ means *an element of.* Interval notation describes sets using endpoints with parentheses or brackets. Parentheses indicate the endpoint is not included in the interval, while brackets indicate the endpoint is included.

Common Misconception

Many students believe that when writing interval notation, if a number is not included, they should use a number with a lesser value with a bracket. For example, if x can be any real number less than 5, students may write $(-\infty, 4]$ or $(-\infty, 4.9]$. Encourage students to use the correct symbol with the given number, not change the number.

DIFFERENTIATE

Enrichment Activity BL

Ask students how the algebraic notation $x \leq -3$, $0 \leq x < 4$, or $x > 5$ could be written in interval notation.
$(-\infty, -3] \cup [0, 4) \cup (5, \infty)$

For larger quantities, the coffee is sold by distinct amounts. This part of the function is discrete.

Since the domain and range are made up of neither a single interval nor individual points, the function is neither discrete nor continuous.

The domain of the function is $0 \leq x \leq 2$ or $x = 2.5$, 3, 5. This represents the possible weights of coffee beans that customers could purchase. The range of the function is $0 \leq y \leq 16$ or $y = 20$, 22, 35. This represents the possible costs of coffee beans.

Think About It!

Why does the range include values from 0 to 16 instead of 0 to 8?

Sample answer: The price of coffee beans is \$8 per pound for up to 2 pounds, so customers pay up to $2 \cdot \$8 = \$16$.

Learn Set-Builder and Interval Notation

Sets of numbers like the domain and range of a function can be described by using various notations. Set-builder notation, interval notation, and algebraic notation are all concise ways of writing a set of values. Consider the set of values represented by the graph.

−5 −4 −3 −2 −1 0 1 2 3 4 5

- In **algebraic notation**, sets are described using algebraic expressions. Example: $x < 2$
- **Set-builder notation** is similar to algebraic notation. Braces indicate the set. The symbol | is read as *such that.* The symbol ∈ is read *is an element of.* Example: $\{x | x < 2\}$
- In **interval notation** sets are described using endpoints with parentheses or brackets. A parenthesis, (or), indicates that an endpoint *is not* included in the interval. A bracket, [or], indicates that an endpoint is included in the interval. Example: $(-\infty, 2)$

Study Tip

Using Symbols You can use the symbol ℝ to represent all real numbers in set-builder notation. In interval notation, the symbol ∪ indicates the union of two sets. Parentheses are always used with ∞ and −∞ because they do not include endpoints.

Example 6 Set-Builder and Interval Notation for Continuous Intervals

Write the domain and range of the graph in set-builder and interval notation.

Domain

The graph will extend to include all x-values.

The domain is all real numbers.

$\{x | x \in \mathbb{R}\}$

$(-\infty, \infty)$

Range

The least y-value for this function is −6.

The range is all real numbers greater than or equal to −6.

$\{y | y \geq -6\}$

$[-6, \infty)$

Go Online You can complete an Extra Example online.

Example 6 Set-Builder and Interval Notation for Continuous Intervals

 Teaching the Mathematical Practices

1 Explain Correspondences Encourage students to explain the relationship between the graph and the domain and range of the function.

Questions for Mathematical Discourse

AL Could the range also be written as $(\infty, -6]$? Explain. No; The values in the interval must be written in numerical order left to right. Because −6 is less than ∞, it must be written first.

OL Which symbols in set-builder and interval notation are inclusive? Which symbols are non-inclusive? inclusive: ≥ ≤ [ ]; non-inclusive: > < ( )

BL Why do you use parentheses with infinity in interval notation? Sample answer: We are describing values approaching infinity, but infinity is not a specific value to be reached.

Interactive Presentation

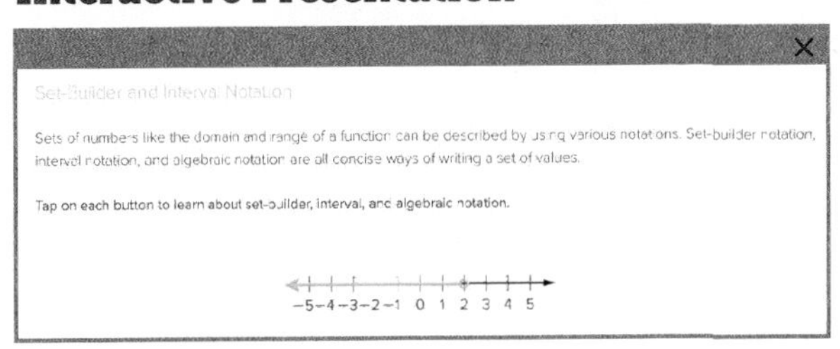

Learn

TAP

Students tap to see how algebraic, set-builder, and interval notation can be used to represent the same number lines

CHECK

Students complete the Check online to determine whether they are ready to move on.

1 CONCEPTUAL UNDERSTANDING **2 FLUENCY** 3 APPLICATION

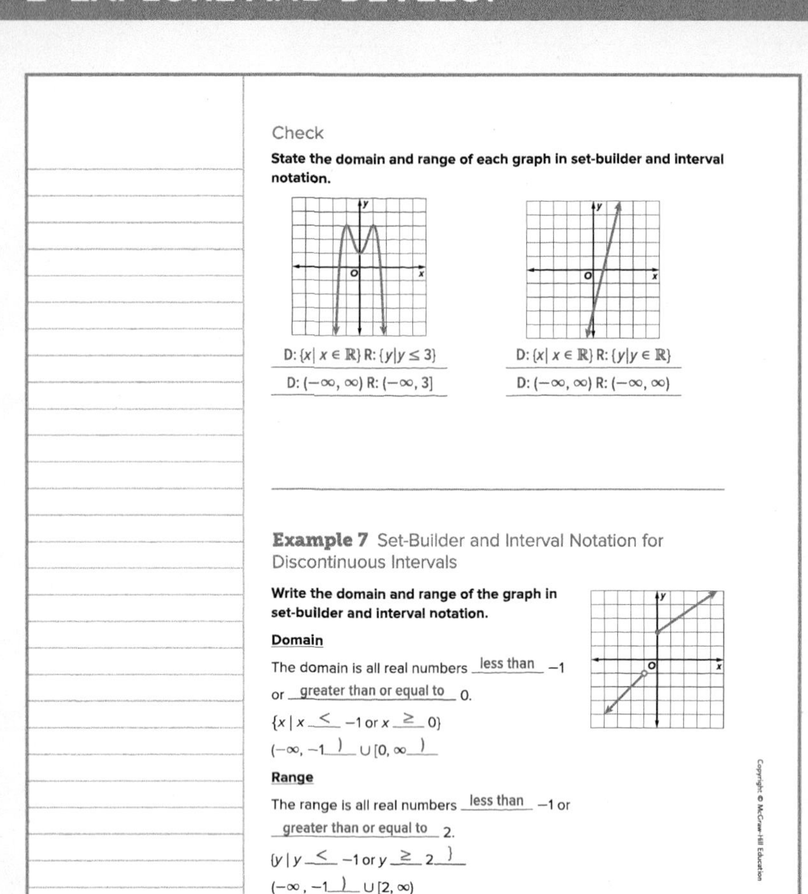

Check

State the domain and range of each graph in set-builder and interval notation.

D: $\{x \mid x \in \mathbb{R}\}$ R: $\{y \mid y \leq 3\}$
D: $(-\infty, \infty)$ R: $(-\infty, 3]$

D: $\{x \mid x \in \mathbb{R}\}$ R: $\{y \mid y \in \mathbb{R}\}$
D: $(-\infty, \infty)$ R: $(-\infty, \infty)$

Example 7 Set-Builder and Interval Notation for Discontinuous Intervals

Write the domain and range of the graph in set-builder and interval notation.

Domain

The domain is all real numbers less than −1 or greater than or equal to 0.

$\{x \mid x < -1 \text{ or } x \geq 0\}$

$(-\infty, -1) \cup [0, \infty)$

Range

The range is all real numbers less than −1 or greater than or equal to 2.

$\{y \mid y < -1 \text{ or } y \geq 2\}$

$(-\infty, -1) \cup [2, \infty)$

Check

State the domain and range of the graph in set-builder and interval notation.

D: $\{x \mid x < -1 \text{ or } x \geq 1\}$ R: $\{y \mid y \in \mathbb{R}\}$

D: $(-\infty, -1) \cup [1, \infty)$ R: $(-\infty, \infty)$

Go Online You can complete an Extra Example online.

Copyright © McGraw-Hill Education

8 Module 1 • Relations and Functions

Interactive Presentation

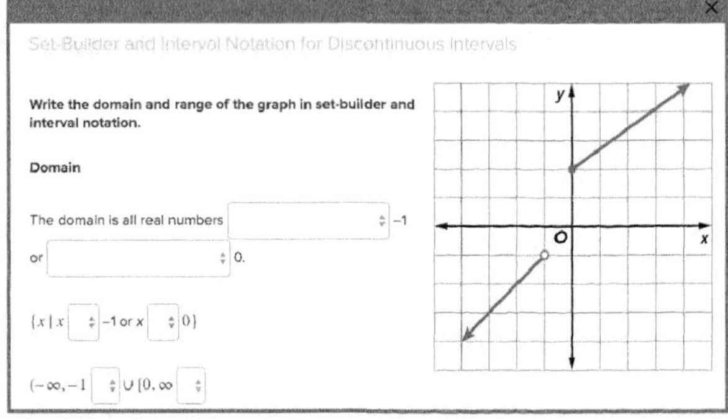

Example 7

SELECT

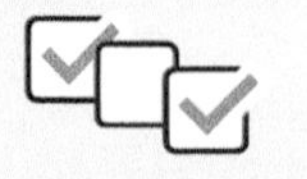

Students select phrases or symbols to write the domain and range over a discontinuous interval.

CHECK

Students complete the Check online to determine whether they are ready to move on.

Example 7 Set-Builder and Interval Notation for Discontinuous Intervals

 **Teaching the Mathematical Practices**

7 Use Structure Help students use the structure of set-builder and interval notation to write the domain and range of the discontinuous function.

Questions for Mathematical Discourse

AL How does the gap in the function affect how you describe the domain and range? Sample answer: There will be two parts for the set of x-values to describe the domain, and two parts for the set of y-values to describe the range.

OL How does the presence of an open circle on the graph affect the domain and range in both set-builder and interval notation? Sample answer: Having an open circle on a graph means that non-inclusive symbols should be used when writing domain and range. For set-builder notation, < or > must be used, while in interval notation (or) must be used.

BL Suppose the first part of the graph ended with an open circle at the point (0, −1) instead of (−1, −1). Would this change the domain or range? Explain. Yes, the domain; sample answer: The domain would be all real numbers because the closed circle fills in the open circle. The range would not change because the y-value is still the same at the new point.

Common Error

Students often interchange the x-values and y-values of a graph when writing domain and range. When the graph starts at the point (0, 2), the domain may be written as $[2, \infty)$ or $x \geq 2$. Remind students that the domain contains only x-values while the range only contains y-values.

DIFFERENTIATE

Language Development Activity ELL

Discuss the relationship between the algebraic notation and the following words: *union, infinity, square bracket,* and *parentheses.*

Exit Ticket

Recommended Use

At the end of class, go online to display the Exit Ticket prompt and ask students to respond using a separate piece of paper. Have students hand you their responses as they leave the room.

Alternate Use

At the end of class, go online to display the Exit Ticket prompt and ask students to respond verbally or by using a mini-whiteboard. Have students hold up their whiteboards so that you can see all student responses. Tap to reveal the answer when most or all students have completed the Exit Ticket.

Practice and Homework

Suggested Assignments

Use the table below to select appropriate exercises.

DOK	Topic	Exercises
1, 2	exercises that mirror the examples	1–21
2	exercises that use a variety of skills from this lesson	22–27
2	exercises that extend concepts learned in this lesson to new contexts	28–34
3	exercises that emphasize higher-order and critical thinking skills	35–39

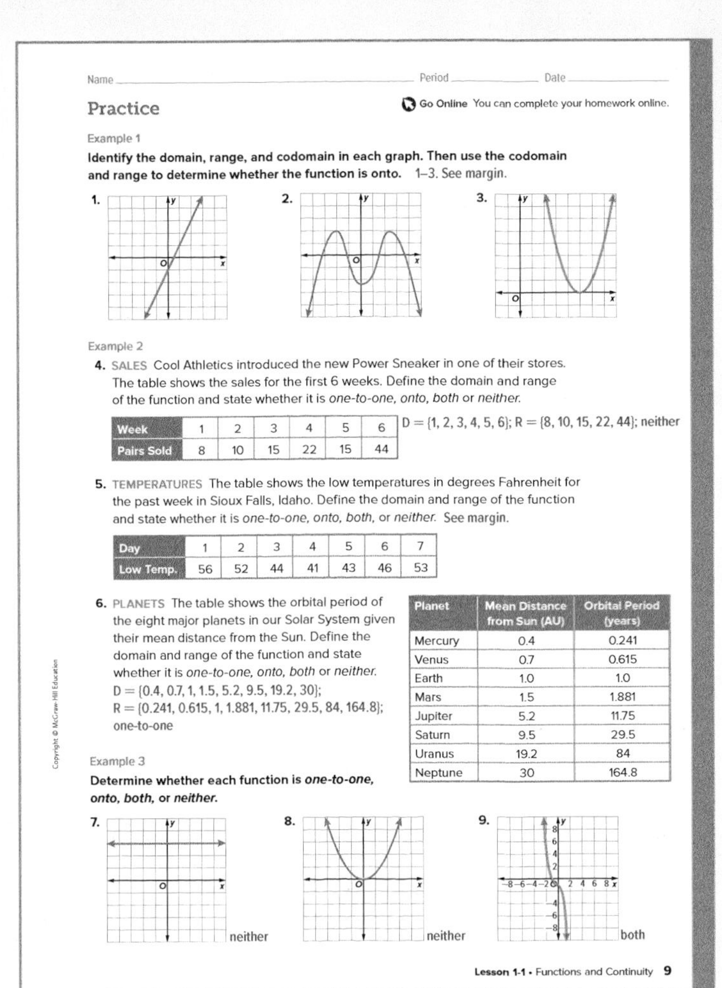

Name ______ Period ______ Date ______

Practice

Go Online You can complete your homework online.

Example 1

Identify the domain, range, and codomain in each graph. Then use the codomain and range to determine whether the function is onto. 1–3. See margin.

1. 2. 3.

Example 2

4. SALES Cool Athletics introduced the new Power Sneaker in one of their stores. The table shows the sales for the first 6 weeks. Define the domain and range of the function and state whether it is *one-to-one, onto, both* or *neither.*

Week	1	2	3	4	5	6
Pairs Sold	8	10	15	22	15	44

D = {1, 2, 3, 4, 5, 6}; R = {8, 10, 15, 22, 44}; neither

5. TEMPERATURES The table shows the low temperatures in degrees Fahrenheit for the past week in Sioux Falls, Idaho. Define the domain and range of the function and state whether it is *one-to-one, onto, both,* or *neither.* See margin.

Day	1	2	3	4	5	6	7
Low Temp.	56	52	44	41	43	46	53

6. PLANETS The table shows the orbital period of the eight major planets in our Solar System given their mean distance from the Sun. Define the domain and range of the function and state whether it is *one-to-one, onto, both* or *neither.*
D = {0.4, 0.7, 1, 1.5, 5.2, 9.5, 19.2, 30};
R = {0.241, 0.615, 1, 1.881, 11.75, 29.5, 84, 164.8};
one-to-one

Planet	Mean Distance from Sun (AU)	Orbital Period (years)
Mercury	0.4	0.241
Venus	0.7	0.615
Earth	1.0	1.0
Mars	1.5	1.881
Jupiter	5.2	11.75
Saturn	9.5	29.5
Uranus	19.2	84
Neptune	30	164.8

Example 3

Determine whether each function is *one-to-one, onto, both,* or *neither.*

7. neither 8. neither 9. both

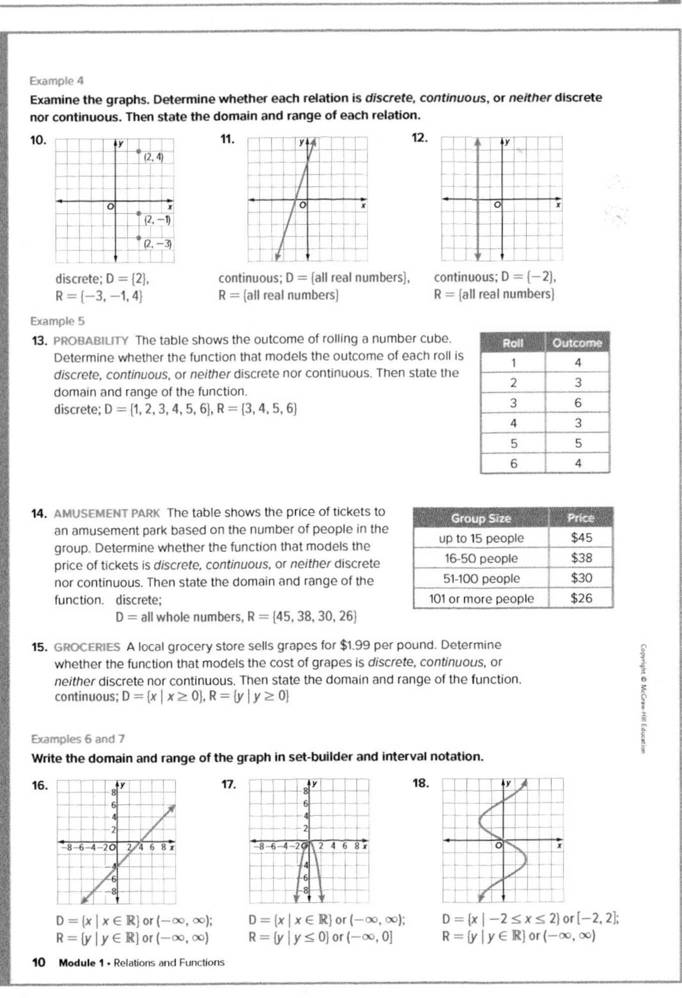

Example 4

Examine the graphs. Determine whether each relation is *discrete, continuous,* or *neither* discrete nor continuous. Then state the domain and range of each relation.

10. discrete; D = {2}, R = {−3, −1, 4}

11. continuous; D = {all real numbers}, R = {all real numbers}

12. continuous; D = {−2}, R = {all real numbers}

Example 5

13. PROBABILITY The table shows the outcome of rolling a number cube. Determine whether the function that models the outcome of each roll is *discrete, continuous,* or *neither* discrete nor continuous. Then state the domain and range of the function.
discrete; D = {1, 2, 3, 4, 5, 6}, R = {3, 4, 5, 6}

Roll	Outcome
1	4
2	3
3	6
4	3
5	5
6	4

14. AMUSEMENT PARK The table shows the price of tickets to an amusement park based on the number of people in the group. Determine whether the function that models the price of tickets is *discrete, continuous,* or *neither* discrete nor continuous. Then state the domain and range of the function. discrete;
D = all whole numbers, R = {45, 38, 30, 26}

Group Size	Price
up to 15 people	\$45
16-50 people	\$38
51-100 people	\$30
101 or more people	\$26

15. GROCERIES A local grocery store sells grapes for \$1.99 per pound. Determine whether the function that models the cost of grapes is *discrete, continuous,* or *neither* discrete nor continuous. Then state the domain and range of the function.
continuous; $D = \{x \mid x \geq 0\}$, $R = \{y \mid y \geq 0\}$

Examples 6 and 7

Write the domain and range of the graph in set-builder and interval notation.

16. $D = \{x \mid x \in \mathbb{R}\}$ or $(-\infty, \infty)$; $R = \{y \mid y \in \mathbb{R}\}$ or $(-\infty, \infty)$

17. $D = \{x \mid x \in \mathbb{R}\}$ or $(-\infty, \infty)$; $R = \{y \mid y \leq 0\}$ or $(-\infty, 0]$

18. $D = \{x \mid -2 \leq x \leq 2\}$ or $[-2, 2]$; $R = \{y \mid y \in \mathbb{R}\}$ or $(-\infty, \infty)$

ASSESS AND DIFFERENTIATE

Use the data from the **Checks** to determine whether to provide resources for extension, remediation, or intervention.

IF students score 90% or more on the Checks, BL
THEN assign:

- Practice, Exercises 1–33 odd, 35–39
- Extension: Correspondence Among Infinite Sets and Subsets
- **ALEKS**® Graphs of Functions; Transformations

IF students score 66%–89% on the Checks, OL
THEN assign:

- Practice, Exercises 1–39 odd
- Remediation, Review Resources: Functions
- Personal Tutors
- Extra Examples 1–7
- **ALEKS**® Sets, Relations, and Functions

IF students score 65% or less on the Checks, AL
THEN assign:

- Practice, Exercises 1–21 odd
- Remediation, Review Resources: Functions
- *Quick Review Math Handbook*: Relations and Functions
- **ALEKS**® Sets, Relations, and Functions

Answers

1. D = {all real numbers}; R = {all real numbers}; Codomain = {all real numbers}; onto
2. D = {all real numbers}; $R = \{y \mid y \leq 2\}$; Codomain = {all real numbers}; not onto
3. D = {all real numbers}, $R = \{y \mid y \geq 0\}$, Codomain = {all real numbers}; not onto
5. D = {1, 2, 3, 4, 5, 6, 7}; R = {56, 52, 44, 41, 43, 46, 53}; one-to-one

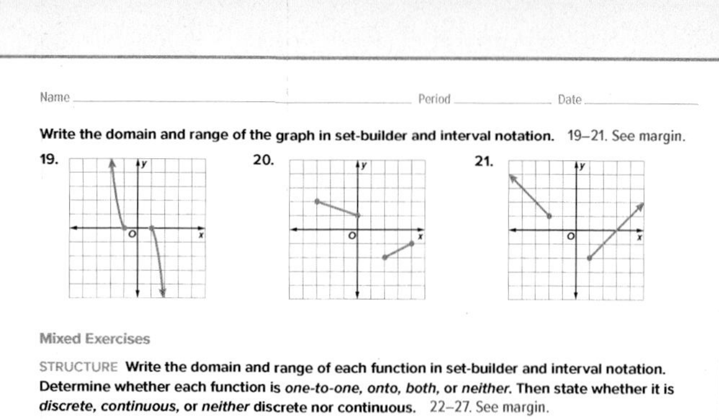

Name ________ Period ________ Date ________

Write the domain and range of the graph in set-builder and interval notation. 19–21. See margin.

19. 20. 21.

Mixed Exercises

STRUCTURE **Write the domain and range of each function in set-builder and interval notation. Determine whether each function is *one-to-one, onto, both,* or *neither.* Then state whether it is *discrete, continuous,* or *neither* discrete nor continuous.** 22–27. See margin.

22. 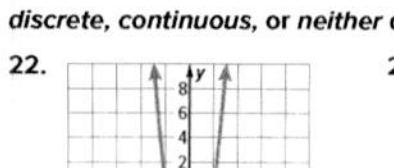

23. 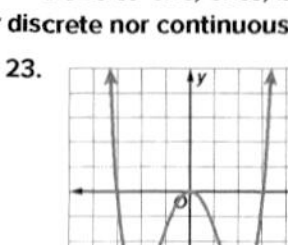

24. 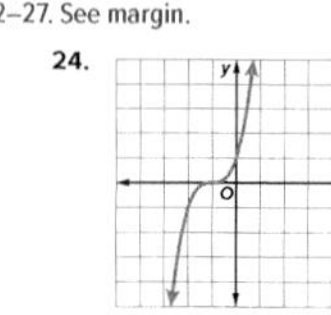

25. 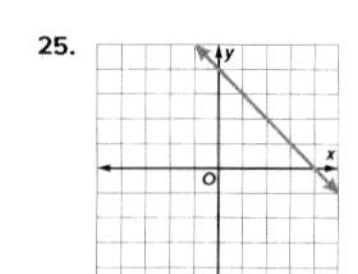

26. 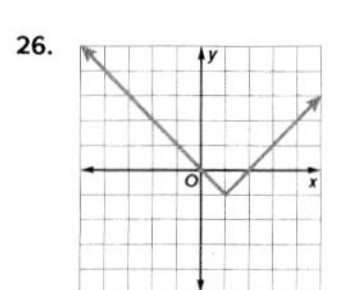

27. 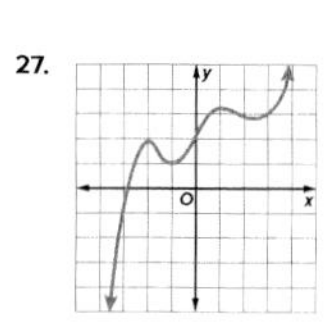

28. USE A SOURCE Research the total number of games won by a professional baseball team each season for five consecutive years. Determine the domain, range, and continuity of the function that models the number of wins. Sample answer: Chicago Cubs: D = {2012, 2013, 2014, 2015, 2016}; R = {61, 66, 73, 97, 103}; discrete

29. SPRINGS When a weight up to 15 pounds is attached to a 4-inch spring, the length L, in inches, that the spring stretches is represented by the function $L(w) = \frac{1}{2}w + 4$, where w is the weight, in pounds, of the object. State the domain and range of the function. Then determine whether it is *one-to-one, onto, both,* or *neither* and whether it is *discrete, continuous,* or *neither* discrete nor continuous. $D = \{w \mid 0 \le w \le 15\}$; $R = \{L \mid 4 \le L \le 11.5\}$; one-to-one; continuous

30. CASHEWS An airport snack stand sells whole cashews for $12.79 per pound. Determine whether the function that models the cost of cashews is *discrete, continuous,* or *neither* discrete nor continuous. Then state the domain and range of the function in set-builder and interval notation. continuous; $D = \{x \mid x \ge 0\}$ or $[0, \infty)$; $R = \{y \mid y \ge 0\}$ or $[0, \infty)$

31. PRICES The Consumer Price Index (CPI) gives the relative price for a fixed set of goods and services. The CPI from September, 2017 to July, 2018 is shown in the graph. Determine whether the function that models the CPI is *one-to-one, onto, both,* or *neither.* Then state whether it is *discrete, continuous,* or *neither* discrete nor continuous. neither one-to-one nor onto; discrete

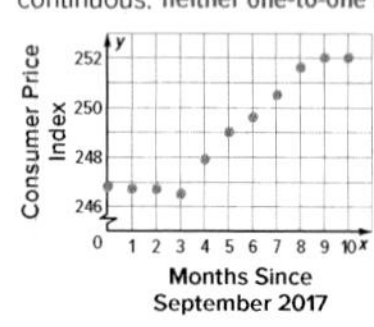

32. LABOR A town's annual jobless rate is shown in the graph. Determine whether the function that models the jobless rate is *one-to-one, onto, both,* or *neither.* Then state whether it is *discrete, continuous,* or *neither* discrete nor continuous. one-to-one; discrete

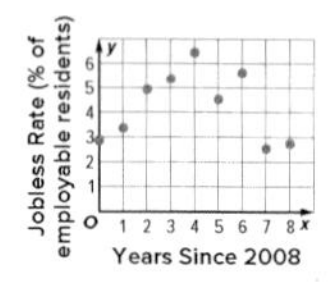

33. COMPUTERS If a computer can do one calculation in 0.0000000015 second, then the function $T(n) = 0.0000000015n$ gives the time required for the computer to do n calculations. State the domain and range of the function. Then determine whether it is *one-to-one, onto, both,* or *neither* and whether it is *discrete, continuous,* or *neither* discrete nor continuous. $D = \{n \mid n \ge 0\}$; $R = \{T(n) \mid T(n) \ge 0\}$; both (within the restrictions of the domain and codomain); continuous

34. SHIPPING The table shows the cost to ship a package based on the weight of the package. Determine whether the function that models the shipping cost is *discrete, continuous,* or *neither* discrete nor continuous. Then state the domain and range of the function in set-builder notation. neither discrete nor continuous; $D = \{x \mid x > 0\}$, $R = \{4, 6\} \cup \{y \mid y > 6.50\}$

Package Weight (lbs)	Cost
up to 5 pounds	$4
5-10 pounds	$6
exceeds 10 pounds	$0.65/lb

Higher-Order Thinking Skills

35. CREATE Sketch the graph of a function that is onto, but not one-to-one, if the codomain is restricted to values greater than or equal to −3. See margin.

36. ANALYZE Determine whether the following statement is *true* or *false.* Explain your reasoning.

If a function is onto, then it must be one-to-one as well. See margin.

37. PERSEVERE Consider $f(x) = \frac{1}{x}$. State the domain and the range of the function. Determine whether the function is *one-to-one, onto, both,* or *neither.* Determine whether the function is *discrete, continuous,* or *neither* discrete nor continuous. $D = \{x \mid x \ne 0\}$ or $(-\infty, 0) \cup (0, \infty)$; $R = \{y \mid y \ne 0\}$ or $(-\infty, 0) \cup (0, \infty)$; one-to-one; neither

38. PERSEVERE Use the domain {−4, −2, 0, 2, 4}, the codomain {−4, −2, 0, 2, 4}, and the range {0, 2, 4} to create a function that is neither one-to-one nor onto. Sample answer: {(−4, 0), (−2, 0), (0, 2), (2, 2), (4, 4)}

39. WRITE Compare and contrast the vertical and horizontal line tests. See margin.

Answers

19. $D = \{x \mid x \le -1 \text{ or } x \ge 1\}$ or $(-\infty, -1] \cup [1, \infty)$; $R = \{y \mid y \in \mathbb{R}\}$ or $(-\infty, \infty)$
20. $D = \{x \mid -3 \le x \le 0 \text{ or } 2 \le x \le 4\}$ or $[-3, 0] \cup [2, 4]$; $R = \{y \mid -2 \le y \le -1 \text{ or } 1 \le y \le 2\}$ or $[-2, -1] \cup [1, 2]$
21. $D = \{x \mid x \le -2 \text{ or } x \ge 1\}$ or $(-\infty, -2] \cup [1, \infty)$; $R = \{y \mid y \ge -2\}$ or $[-2, \infty)$
22. $D = \{x \mid x \in \mathbb{R}\}$ or $(-\infty, \infty)$; $R = \{y \mid y \ge -8\}$ or $[-8, \infty)$; neither one-to-one nor onto; continuous
23. $D = \{x \mid x \in \mathbb{R}\}$ or $(-\infty, \infty)$; $R = \{y \mid y \ge -4\}$ or $[-4, \infty)$; neither one-to-one nor onto; continuous
24. $D = \{x \mid x \in \mathbb{R}\}$ or $(-\infty, \infty)$; $R = \{y \mid y \in \mathbb{R}\}$ or $(-\infty, \infty)$; both; continuous
25. $D = \{x \mid x \in \mathbb{R}\}$ or $(-\infty, \infty)$; $R = \{y \mid y \in \mathbb{R}\}$ or $(-\infty, \infty)$; both; continuous
26. $D = \{x \mid x \in \mathbb{R}\}$ or $(-\infty, \infty)$; $R = \{y \mid y \ge -1\}$ or $[-1, \infty)$; neither one-to-one or onto; continuous
27. $D = \{x \mid x \in \mathbb{R}\}$ or $(-\infty, \infty)$; $R = \{y \mid y \in \mathbb{R}\}$ or $(-\infty, \infty)$; onto; continuous
35. Sample answer:

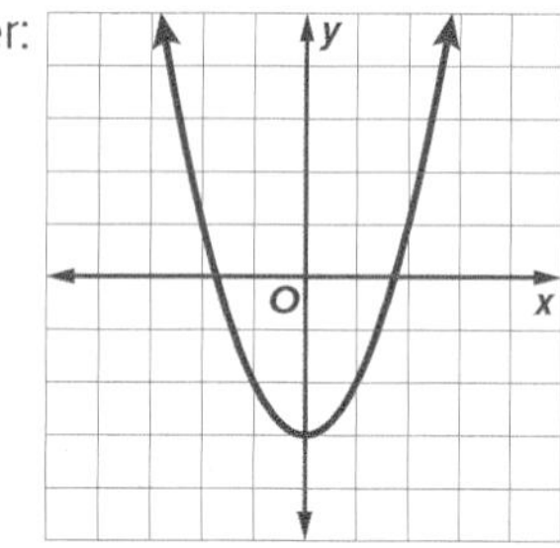

36. False; sample answer: A function is onto and not one-to-one if all of the elements of the domain correspond to an element of the range, but more than one element of the domain corresponds to the same element of the range.
39. Sample answer: The vertical line test is used to determine whether a relation is a function. If no vertical line intersects a graph in more than one point, the graph represents a function. The horizontal line test is used to determine whether a function is one-to-one. If no horizontal line intersects the graph more than once, then the function is one-to one. The horizontal line test can also be used to determine whether a function is onto. If every horizontal line intersects the graph at least once, then the function is onto.

Lesson 1-2

F.IF.4, F.IF.5

Linearity, Intercepts, and Symmetry

LESSON GOAL

Students determine the linearity, intercepts, and symmetry of functions.

1 LAUNCH

 Launch the lesson with a **Warm Up** and an introduction.

2 INQUIRY AND DEVELOP

 **Explore:** Symmetry and Functions

 **Develop:**

Linear and Nonlinear Functions
- Identify Linear Functions from Equations
- Identify Linear Functions from Graphs
- Identify Linear Functions from Tables

Intercepts of Graphs of Functions
- Find Intercepts of a Linear Function
- Find Intercepts of a Nonlinear Function
- Interpret the Meaning of Intercepts

Symmetry of Graphs of Functions
- Identify Types of Symmetry
- Identify Even and Odd Functions

 You may want your students to complete the **Checks** online.

3 REFLECT AND PRACTICE

 Exit Ticket

 Practice

DIFFERENTIATE

 View reports of student progress on the **Checks** after each example.

Resources	AL	OL	BL	ELL
Remediation: Functions	●	●		●
Extension: Even and Odd Functions		●	●	●

Language Development Handbook

Assign page 2 of the *Language Development Handbook* to help your students build mathematical language related to the linearity, intercepts, and symmetry of functions.

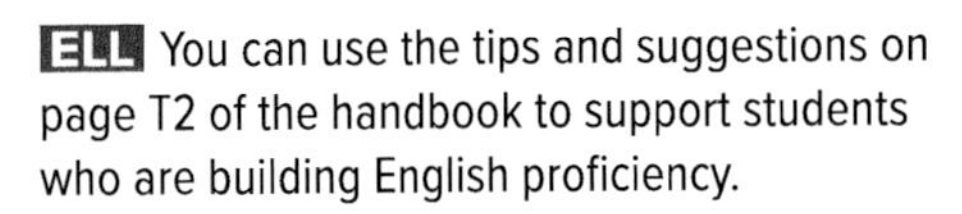

ELL You can use the tips and suggestions on page T2 of the handbook to support students who are building English proficiency.

Suggested Pacing

90 min	1 day
45 min	2 days

Focus

Domain: Functions

Standards for Mathematical Content:

F.IF.4 For a function that models a relationship between two quantities, interpret key features of graphs and tables in terms of the quantities, and sketch graphs showing key features given in a verbal description of the relationship.

F.IF.5 Relate the domain of a function to its graph and, where applicable, to the quantitative relationship it describes.

Standards for Mathematical Practice:

3 Construct viable arguments and critique the reasoning of others.

4 Model with mathematics.

Coherence

Vertical Alignment

Previous
Students determined the continuity of functions and whether functions are one-to-one and/or onto.
F.IF.4, F.IF.5

Now
Students determine the linearity, intercepts, and symmetry of functions.
F.IF.4, F.IF.5

Next
Students will identify extrema and end behavior of functions.
F.IF.4, F.IF.7c

Rigor

The Three Pillars of Rigor

1 CONCEPTUAL UNDERSTANDING	2 FLUENCY	3 APPLICATION

Conceptual Bridge In this lesson, students build on their understanding of functions to include different types of functions, and they build fluency by graphing linear and nonlinear functions. They apply their understanding by solving real-world problems involving functions.

Interactive Presentation

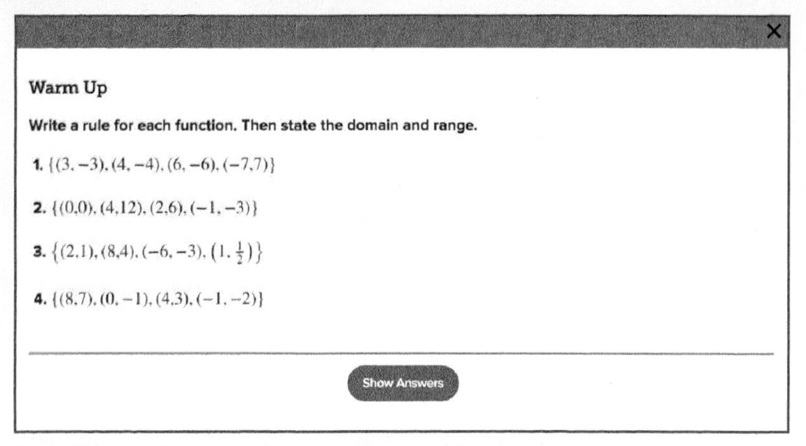

Warm Up

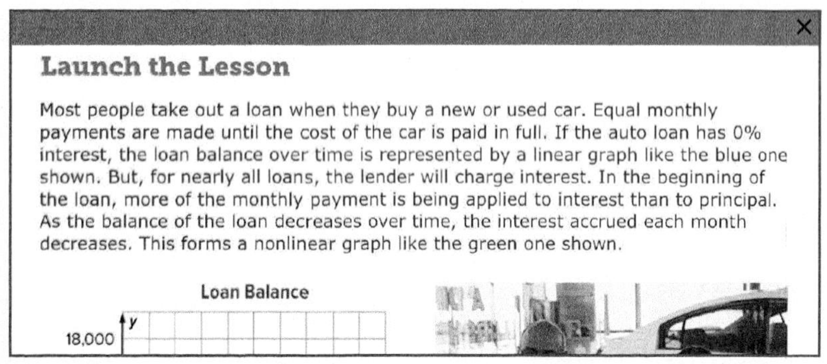

Launch the Lesson

Vocabulary

Expand All Collapse All

> linear function
> linear equation
> intercept
> line symmetry
> point symmetry
> even functions
> odd functions

- How do you determine the most efficient method for graphing a *linear function*?
- By looking at each of these things, how can you tell whether it represents a *linear function*: a) a table, b) an equation, c) a graph?

Today's Vocabulary

Warm Up

Prerequisite Skills

The Warm Up exercises address the following prerequisite skill for this lesson:

- finding domain and range

Answers:

1. $y = -x$; D = $\{-7, 3, 4, 6)$, R = $\{-6, -4, -3, 7\}$
2. $y = 3x$; D = $\{-1, 0, 2, 4\}$, R = $\{-3, 0, 6, 12\}$
3. $y = \frac{1}{2}$; D = $\{-6, 1, 2, 8\}$, R = $\left\{-3, \frac{1}{2}, 1, 4\right\}$
4. $y = x - 1$; D = $\{-1, 0, 4, 8\}$, R = $\{-2, -1, 3, 7\}$

Launch the Lesson

 Teaching the Mathematical Practices

2 Make Sense of Quantities Mathematically proficient students need to be able to make sense of quantities and their relationships. Encourage students to explain why interest and zero-interest loans generate different graphs. Ask students to describe how each type of loan relates to its graph.

Go Online to find additional teaching notes and questions to promote classroom discourse.

Today's Standards

Tell students that they will be addressing these content and practice standards in this lesson. You may wish to have a student volunteer read aloud *How can I meet these standards?* and *How can I use these practices?*, and connect these to the standards.

See the Interactive Presentation for I Can statements that align with the standards covered in this lesson.

Today's Vocabulary

Tell students that they will be using these vocabulary terms in this lesson. You can expand each row if you wish to share the definitions. Then discuss the questions below with the class.

Mathematical Background

When determining whether an equation is linear, examine the structure of the equation. A linear equation has no operations other than addition, subtraction, and multiplication of a variable.

Explore Symmetry and Functions

Objective

Students use a sketch to explore symmetry in graphs of functions.

 **Teaching the Mathematical Practices**

3 Reason Inductively In the Explore, students will use inductive reasoning to make plausible arguments about line and point symmetry.

Ideas for Use

Recommended Use Present the Inquiry Question, or have a student volunteer read it aloud. Have students work in pairs to complete the Explore activity on their devices. Pairs should discuss each of the questions. Monitor student progress during the activity. Upon completion of the Explore activity, have student volunteers share their responses to the Inquiry Question.

What if my students don't have devices? You may choose to project the activity on a whiteboard. A printable worksheet for each Explore is available online. You may choose to print the worksheet so that individuals or pairs of students can use it to record their observations.

Summary of the Activity

Students will complete guiding exercises throughout the Explore activity. Students will use a sketch to investigate symmetry of four different functions. Students will analyze the results to make a conjecture about symmetry of functions. Then, students will answer the Inquiry Question.

(continued on the next page)

Interactive Presentation

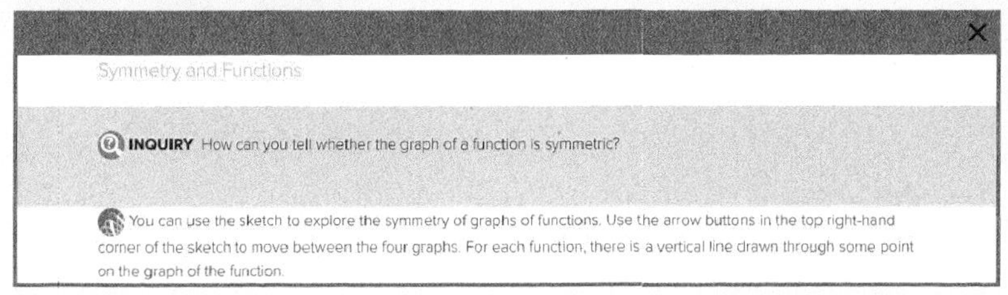

Symmetry and Functions
INQUIRY How can you tell whether the graph of a function is symmetric?
You can use the sketch to explore the symmetry of graphs of functions. Use the arrow buttons in the top right-hand corner of the sketch to move between the four graphs. For each function, there is a vertical line drawn through some point on the graph of the function.

Explore

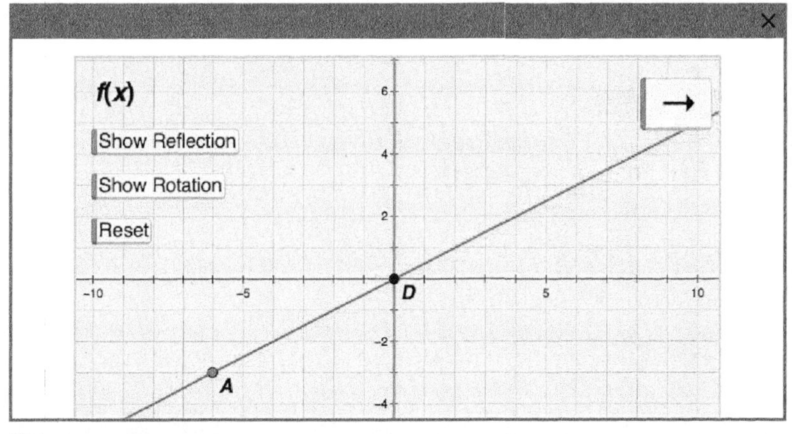

Explore

WEB SKETCHPAD

Students use the sketch to complete an activity in which they explore symmetry in graphs of functions.

SELECT

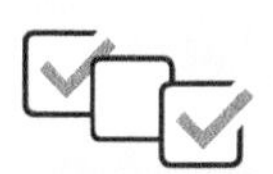

Students select answers to complete a table to analyze the functions.

TYPE

Students record observations and make conjectures about symmetry.

Interactive Presentation

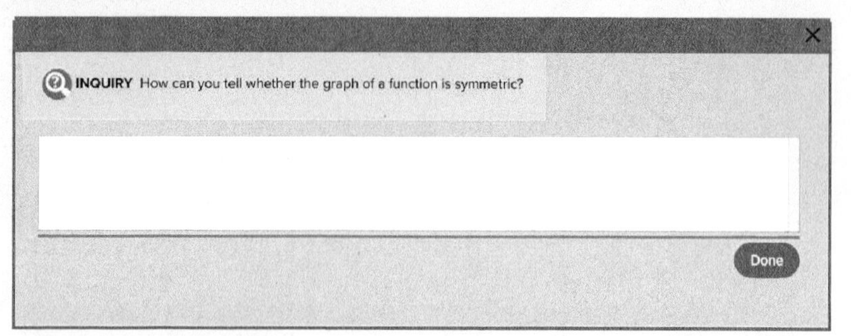

Explore

TYPE

Students respond to the Inquiry Question and can view a sample answer.

Explore Symmetry and Functions (*continued*)

Questions

Have students complete the Explore activity.

Ask:

- For functions that demonstrate line symmetry, do you believe there is only one line of symmetry or could there be many? Explain. Sample answer: There could be many lines of symmetry in which a function is symmetric. For example, a horizontal line has an infinite number of lines of symmetry.
- Is there a vertical line in which the graph of $j(x)$ would be symmetric about? Explain. No; sample answer: No matter where the vertical line is placed, Point A will never reflect over the line to Point B and be on the graph.

Inquiry

How can you tell whether the graph of a function is symmetric? Sample answer: If any point on the graph of a function can be reflected in a vertical line or rotated by 180° about another point on the function and still be on the function, then the graph of the function displays symmetry.

Go Online to find additional teaching notes and sample answers for the guiding exercises.

1 CONCEPTUAL UNDERSTANDING | 2 FLUENCY | 3 APPLICATION

Learn Linear and Nonlinear Functions

Objective

Students identify linear and nonlinear functions by examining equations and graphs.

Teaching the Mathematical Practices

3 Construct Arguments Students will use stated definitions to determine whether functions are linear or nonlinear.

Important to Know

Linear functions can be written in the form $f(x) = mx + b$, where m and b are real numbers. Linear functions can be modeled by linear equations, and the graph is a straight line. If a function is not linear, it is called nonlinear, and graphs of nonlinear functions include a set of points that cannot lie on the same line.

Example 1 Identify Linear Functions from Equations

Teaching the Mathematical Practices

1 Seek Information Mathematically proficient students must be able to transform algebraic expressions to reach solutions. In Example 1, students will rewrite functions in slope-intercept form to determine whether there are linear or nonlinear.

Questions for Mathematical Discourse

AL Is $y = 2$ a linear function? Yes What is the value of m in this function? zero

OL Would the equation $y^2 = 5x + 1$ be linear? Explain. No; sample answer: The variable y is raised to a power of 2, which is greater than one. Thus the equation is nonlinear.

BL Why is a function with x in the denominator not linear? Sample answer: When x is in the denominator of a fraction, it can be rewritten as x^{-1}. Because the power is not one, the function is nonlinear.

DIFFERENTIATE

Reteaching Activity AL ELL

IF students are struggling to differentiate between linear and nonlinear functions,

THEN have students circle all exponents of variables in an equation. If none of the variables are in a denominator or under a radical symbol and the circled exponents are one, the function is most likely linear. If any circled exponent is greater than one, the function is not linear. Graphically, students can use their pencil to determine linearity. If they can place their pencil over the entire graph, the function is linear. If the pencil cannot cover the curve of the whole graph, it is nonlinear.

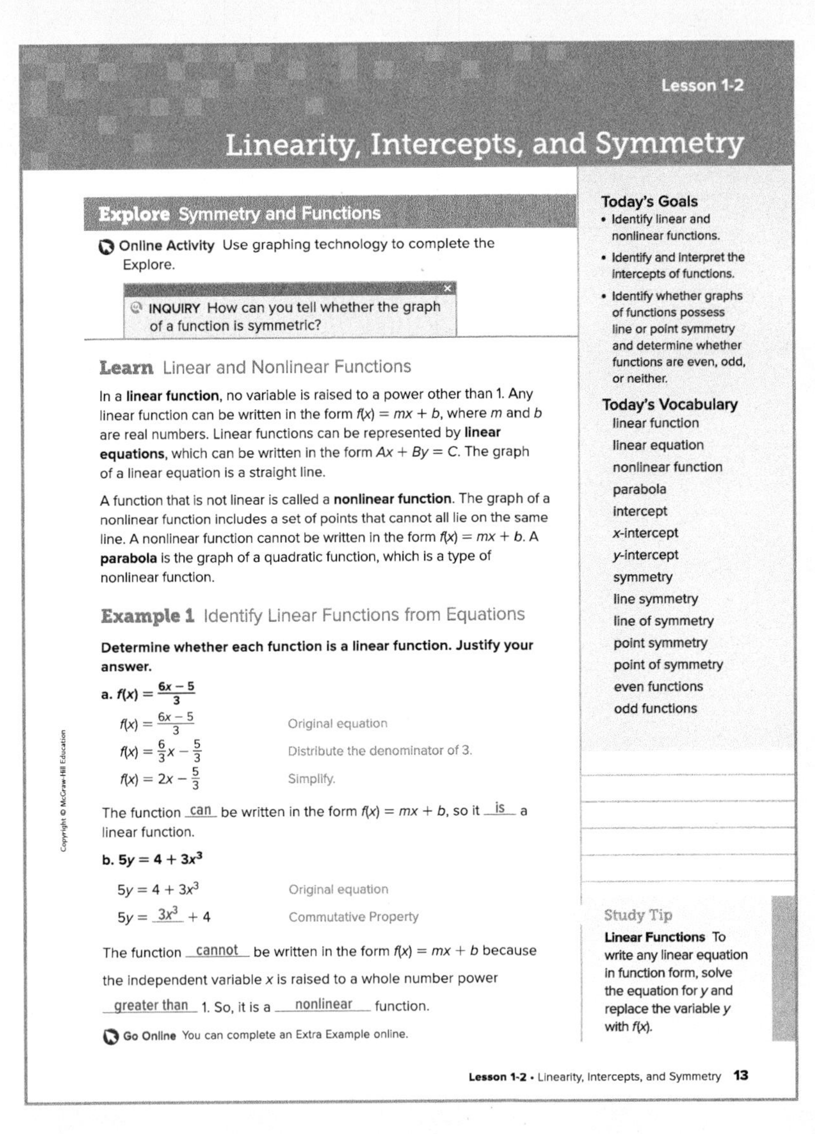

Lesson 1-2

Linearity, Intercepts, and Symmetry

Explore Symmetry and Functions

Online Activity Use graphing technology to complete the Explore.

INQUIRY How can you tell whether the graph of a function is symmetric?

Learn Linear and Nonlinear Functions

In a **linear function**, no variable is raised to a power other than 1. Any linear function can be written in the form $f(x) = mx + b$, where m and b are real numbers. Linear functions can be represented by **linear equations**, which can be written in the form $Ax + By = C$. The graph of a linear equation is a straight line.

A function that is not linear is called a **nonlinear function**. The graph of a nonlinear function includes a set of points that cannot all lie on the same line. A nonlinear function cannot be written in the form $f(x) = mx + b$. A **parabola** is the graph of a quadratic function, which is a type of nonlinear function.

Example 1 Identify Linear Functions from Equations

Determine whether each function is a linear function. Justify your answer.

a. $f(x) = \frac{6x - 5}{3}$

$f(x) = \frac{6x - 5}{3}$ Original equation

$f(x) = \frac{6}{3}x - \frac{5}{3}$ Distribute the denominator of 3.

$f(x) = 2x - \frac{5}{3}$ Simplify.

The function <u>can</u> be written in the form $f(x) = mx + b$, so it <u>is</u> a linear function.

b. $5y = 4 + 3x^3$

$5y = 4 + 3x^3$ Original equation

$5y = \underline{3x^3} + 4$ Commutative Property

The function <u>cannot</u> be written in the form $f(x) = mx + b$ because the independent variable x is raised to a whole number power <u>greater than</u> 1. So, it is a <u>nonlinear</u> function.

Go Online You can complete an Extra Example online.

Today's Goals
- Identify linear and nonlinear functions.
- Identify and interpret the intercepts of functions.
- Identify whether graphs of functions possess line or point symmetry and determine whether functions are even, odd, or neither.

Today's Vocabulary
linear function
linear equation
nonlinear function
parabola
intercept
x-intercept
y-intercept
symmetry
line symmetry
line of symmetry
point symmetry
point of symmetry
even functions
odd functions

Study Tip
Linear Functions To write any linear equation in function form, solve the equation for y and replace the variable y with $f(x)$.

Copyright © McGraw-Hill Education

Lesson 1-2 • Linearity, Intercepts, and Symmetry 13

Interactive Presentation

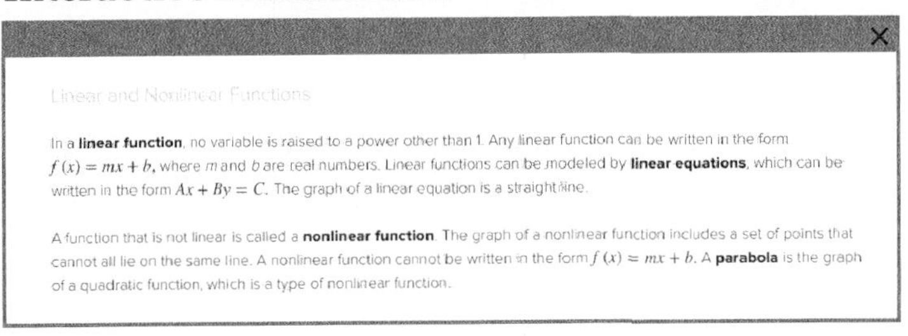

Learn

TAP

Students move through the slides to categorize linear and nonlinear functions.

TYPE

Students explain whether all linear equations represent linear functions.

Your Notes

Think About It!
Why is $f(x) = \sqrt{2x} + 3$ not a linear function?

Sample answer: The variable is under the square root sign, which is the same as having an exponent of $\frac{1}{2}$.

Example 2 Identify Linear Functions from Graphs

Determine whether each graph represents a *linear* or *nonlinear* function.

a.

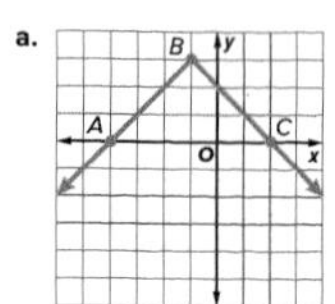

There is no straight line that will contain the chosen points *A*, *B*, and *C*, so this graph represents a nonlinear function.

b.

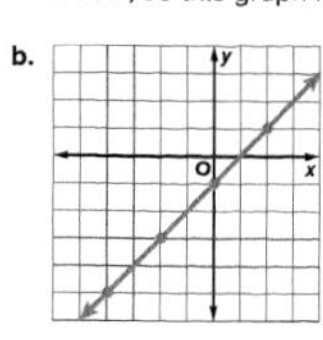

The points on this graph all lie on the same line, so this graph represents a linear function.

Example 3 Identify Linear Functions from Tables

EARNINGS **Makayla has started working part-time at the local hardware store. Her time at work steadily increases for the first five weeks. The table shows her total earnings each of those weeks. Are her weekly earnings modeled by a *linear* or *nonlinear* function?**

Week	1	2	3	4	5
Earnings ($)	85	119	153	187	221

Graph the points that represent the week and total earnings and try to draw a line that contains all the points.

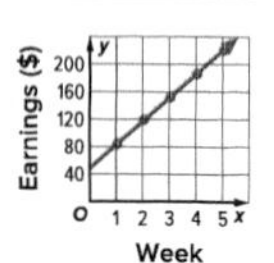

Since there is a line that contains all the points, Makayla's earning can be modeled by a linear function.

Think About It!
Are negative *x*- or *y*-values possible in the context of the situation?

No; sample answer: because Makayla cannot work a negative number of weeks or earn a negative amount of money, the function only exists for nonnegative *x*- and *y*-values.

Go Online You can complete an Extra Example online.

Interactive Presentation

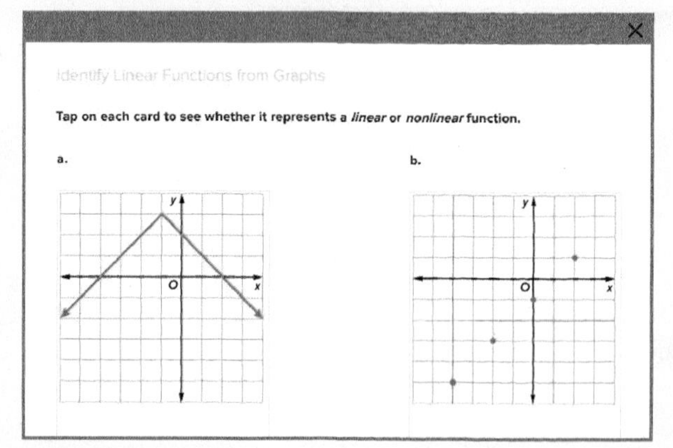

Example 2

TAP

Students tap each card to see whether it represents a linear or nonlinear function.

CHECK

Students complete the Check online to determine whether they are ready to move on.

Example 2 Identify Linear Functions from Graphs

Teaching the Mathematical Practices

3 Construct Arguments Students will use stated assumptions and definitions to classify functions as linear or nonlinear given their graphs.

Questions for Mathematical Discourse

AL How do you know if a graph is a linear function? Sample answer: All points on the graph lie on the same line.

OL If you find three points that lie on the same line on the graph of a function, is the function automatically linear? Explain. No; sample answer: The function may not be linear over its entire domain.

BL While we established that the function in part **a** is nonlinear, how could properties of linear functions be useful in analyzing this function? Sample answer: The function is made up of two linear parts that meet at *B*, so it could be written as two linear functions. In doing so, you could recognize that the slope of one portion is the negative of the other portion.

Example 3 Identify Linear Functions from Tables

Teaching the Mathematical Practices

3 Construct Arguments Students will use stated assumptions and definitions to classify functions as linear or nonlinear given their graphs.

Questions for Mathematical Discourse

AL If a graph contains a curved portion, could it be linear? Explain. No; sample answer: If a graph is curved, all points will not lie on the same line.

OL Without graphing, how can you tell whether a function is linear from a table? Sample answer: If the change in the *y*-values is the same for a set change in *x*-values, the function is linear.

BL Explain how you could determine how much Makayla would need to earn in the 8th week for the function to remain linear. Sample answer: Determine the change in earnings that occurs between each week and add three times that amount to the 5th week earnings.

Go Online

- Find additional teaching notes.
- View performance reports of the Checks.
- Assign or present an Extra Example.

Learn Intercepts of Graphs of Functions

Objective

Students identify and interpret the intercepts of functions by examining graphs or tables.

Teaching the Mathematical Practices

1 Explain Correspondences Students will explain how the intercepts of a function correspond to its graph or table.

Example 4 Find Intercepts of a Linear Function

Teaching the Mathematical Practices

1 Explain Correspondences In Example 4, students will explain the relationship between the graph of a function and its intercepts.

Questions for Mathematical Discourse

AL What number do you always see in the coordinates of intercepts? Why do all intercepts have this coordinate? zero; Sample answer: Intercepts are points where the graph touches an axis. Along the x-axis, y equals zero, and along the y-axis, x equals zero.

OL If a function has a y-intercept at $(0, -3)$, what does that tell you about the graph? Sample answer: The graph crosses the y-axis at -3.

BL Is it possible for a linear function to not have any intercepts? Over a domain of all real numbers, a linear function would at least cross the y-axis. A horizontal line would not have x-intercepts.

Example 5 Find Intercepts of a Nonlinear Function

Teaching the Mathematical Practices

3 Construct Arguments In Example 5, students will use stated definitions to identify the intercepts of a nonlinear function.

Questions for Mathematical Discourse

AL What is different about the intercept $(2, 0)$ compared to the intercepts $(-3, 0)$ and $(-1, 0)$? The graph touches the x-axis but does not cross it.

OL How many x-intercepts can a linear function have? A linear function can have zero or one x-intercept. How many x-intercepts can a nonlinear function have? A nonlinear function can have any number of x-intercepts.

BL Suppose a graph has an x-intercept of 7 and a y-intercept of -1. Can we determine the linearity of the function? Explain. No; sample answer: Because linear and nonlinear functions can each have one x-intercept, we cannot tell the linearity from the information given.

Learn Intercepts of Graphs of Functions

A point at which the graph of a function intersects an axis is called an **intercept**. An ***x*-intercept** is the x-coordinate of a point where the graph crosses the x-axis, and a ***y*-intercept** is the y-coordinate of a point where the graph crosses the y-axis.

A linear function has at most one x-intercept while a nonlinear function may have more than one x-intercept.

Example 4 Find Intercepts of a Linear Function

Use the graph to estimate the *x*- and *y*-intercepts.

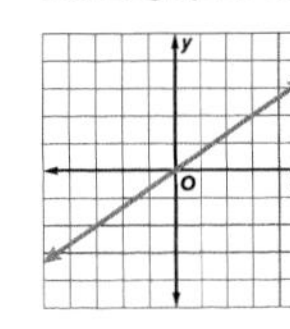

The graph intersects the x-axis at (0, 0), so the x-intercept is 0.

The graph intersects the y-axis at (0, 0), so the y-intercept is 0.

Example 5 Find Intercepts of a Nonlinear Function

Use the graph to estimate the *x*- and *y*-intercepts.

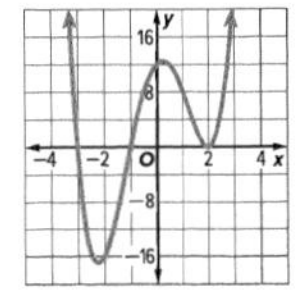

The graph appears to intersect the x-axis at $(-3, 0)$, $(-1, 0)$, and $(2, 0)$, so the function has x-intercepts of -3, -1, and 2.

The graph appears to intersect the y-axis at (0, 12), so the function has a y-intercept of 12.

Check

Estimate the x- and y-intercepts of each graph.

a.

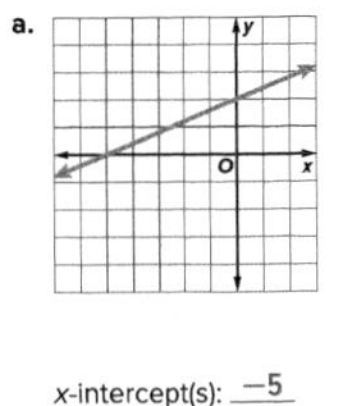

x-intercept(s): -5

y-intercept(s): 2

b.

x-intercept(s): $-4, -2, 0, 3, 5$

y-intercept(s): 0

Go Online You can complete an Extra Example online.

Study Tip

Point or Coordinate *Intercept* may refer to the point or one of its coordinates. The context of the situation will often dictate which form to use.

Think About It!

Describe a line that does not have two distinct intercepts.

Sample answer: A horizontal line has only a y-intercept. A vertical line has only an x-intercept. A line through the origin has the same x- and y-intercepts.

Think About It!

The graph of the nonlinear function has three x-intercepts. Can the graph have more than one y-intercept? Explain your reasoning.

No; sample answer: if there is more than one y-intercept, then the graph would not pass the vertical line test. Thus, it would not be a function.

Interactive Presentation

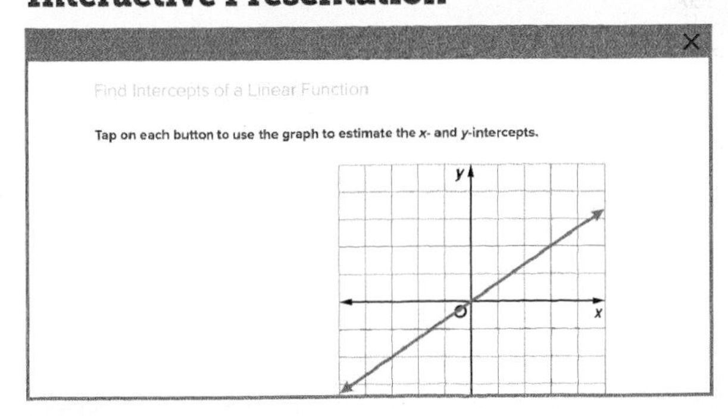

Example 4

TAP

Students tap to see the intercept on the graph.

1 CONCEPTUAL UNDERSTANDING | 2 FLUENCY | 3 APPLICATION

Think About It!
Describe the domain of the function that models the rocket's height over time.

Sample answer: The domain is $0 \leq x \leq 8$ and represents the amount of time that the rocket is in the air.

Watch Out!
Switching Coordinates A common mistake is to switch the coordinates for the intercepts. Remember that for the x-intercept, the y-coordinate is 0, and for the y-intercept, the x-coordinate is 0.

Talk About It
Can the graph of a function be symmetric in a horizontal line? Justify your answer.

Sample answer: Only if the graph of the function is a horizontal line; a horizontal line would be symmetric with respect to itself. However if a graph other than a horizontal line were symmetric about a horizontal line, then the graph would contain two points with the same x-value. The graph would fail the vertical line test. Thus, it could not be a function.

Example 6 Interpret the Meaning of Intercepts

MODEL ROCKETS **Ricardo launches a rocket from a balcony. The table shows the height of the rocket after each second of its flight.**

Time (s)	Height (ft)
0	15
1	60
2	130
3	180
4	210
5	170
6	110
7	55
8	0

Part A Identify the x- and y-intercepts of the function that models the flight of the rocket.

In the table, the x-coordinate when $y = 0$ is 8. Thus, the x-intercept is 8.

In the table, the y-coordinate when $x = 0$ is 15. Thus, the y-intercept is 15.

Part B What is the meaning of the intercepts in the context of the rocket's flight?

The x-intercept is the number of seconds after the rocket is launched that it returns to the ground. The y-intercept is the height of the balcony from which the rocket is launched.

Learn Symmetry of Graphs of Functions

A figure has **symmetry** if there exists a rigid motion—reflection, translation, rotation, or glide reflection—that maps the figure onto itself.

Key Concept • Symmetry

Type of Symmetry	Description	Example
A graph has **line symmetry** if it can be reflected in a vertical line so that each half of the graph maps exactly to the other half.	The line dividing the graph into matching halves is called the **line of symmetry**. Each point on one side is reflected in the line to a point equidistant from the line on the opposite side.	
A graph has **point symmetry** when a figure is rotated 180° about a point and maps onto itself.	The point about which the graph is rotated is called the **point of symmetry**. The image of each point on one side of the point of symmetry can be found on the line through the point and the point of symmetry, equidistant from the point of symmetry.	

Interactive Presentation

Interpret the Meaning of Intercepts

MODEL ROCKETS **Ricardo launches a rocket from a second-floor balcony. The rocket is equipped with an electronic altimeter. The table shows the height given by the altimeter after each second of the rocket's flight.**

Time (s)	Height
0	15
1	60
2	130
3	180
4	210
5	170
6	110
7	55

Example 6

TAP

Students tap to identify the x- and y-intercepts from the table.

TYPE

Students describe the domain of the function modeling the rocket's height over time.

CHECK

Students complete the Check online to determine whether they are ready to move on.

Example 6 Interpret the Meaning of Intercepts

Teaching the Mathematical Practices

4 Interpret Mathematical Results Point out that students must interpret the intercepts of the function in the context of the problem.

Questions for Mathematical Discourse

AL How can you determine whether time or height is represented by the independent variable x? Sample answer: Think about which variable affects the other. The height of the rocket depends on the time since it was launched. So the time is the independent variable, x, and the height is the dependent variable, y.

OL Suppose the first point in the table was 0 seconds, 0 feet. What would this mean in context of the situation? Sample answer: The rocket launched at ground level.

BL Could there be a third x-intercept for the situation? Explain. No; sample answer: It would not make sense for the rocket to hit the ground a second time.

Common Error

Many students struggle to interpret answers in the context of the problem. They may use the wrong unit or description. Encourage students to identify variables before beginning so they know what x and y represent.

Learn Symmetry of Graphs of Functions

Objective

Students identify whether graphs of functions possess line or point symmetry and determine whether functions are even, odd, or neither by analyzing graphs.

Teaching the Mathematical Practices

3 Analyze Cases Students must analyze the different cases of symmetry for the graphs and functions provided.

About the Key Concept

If there exists a rigid motion that maps a figure onto itself, then a figure has symmetry. Line symmetry is when a figure can be reflected in a line so that each half of the graph maps exactly to the other half. A graph has point symmetry when a figure is rotated 180° about a point and maps exactly onto itself. Even functions are symmetric in the y-axis, while odd functions are symmetric about the origin.

Common Misconception

A common misconception students have is that symmetry is only when a graph can be folded about a line. Encourage students to take notes on each type of symmetry with an example to help in future problems.

DIFFERENTIATE

Reteaching Activity AL ELL

IF students are having a hard time identifying the types of symmetry, **THEN** have students find pictures in the real world displaying the two types of symmetry. They can print the pictures and draw the lines or points of symmetry on the images.

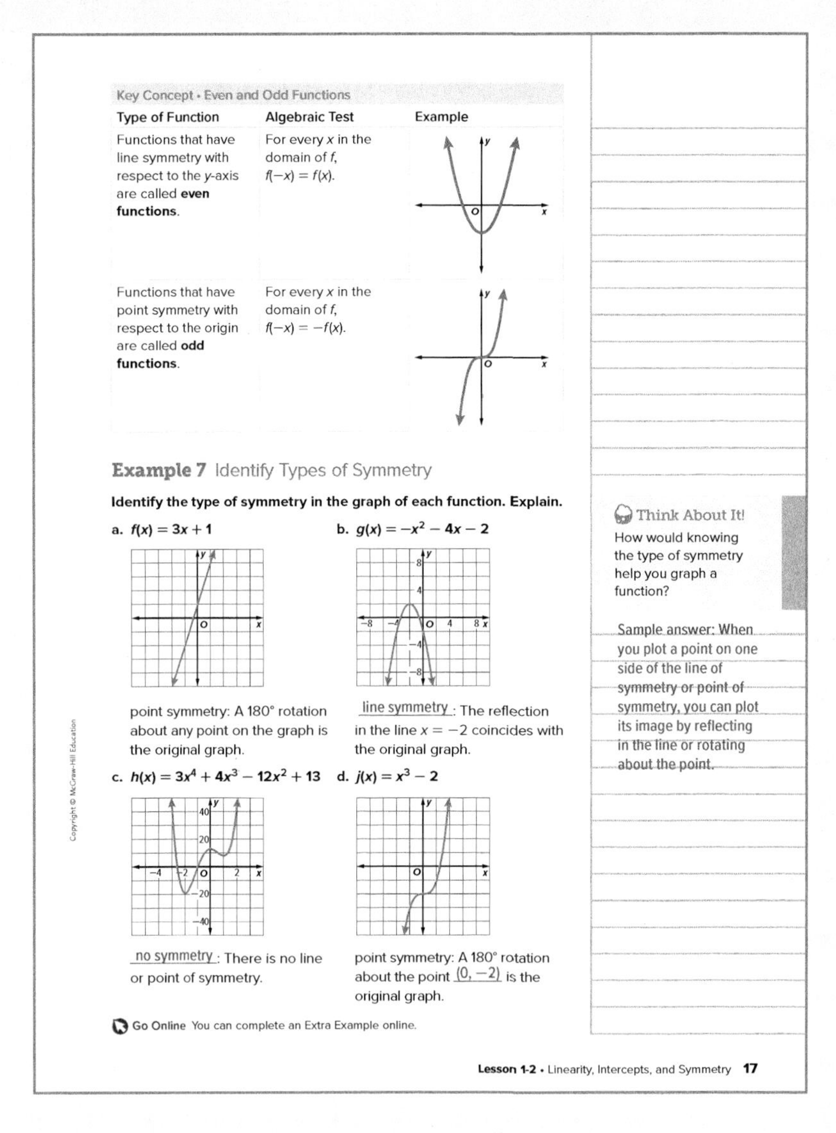

Key Concept • Even and Odd Functions

Type of Function	Algebraic Test	Example
Functions that have line symmetry with respect to the y-axis are called **even functions**.	For every x in the domain of f, $f(-x) = f(x)$.	
Functions that have point symmetry with respect to the origin are called **odd functions**.	For every x in the domain of f, $f(-x) = -f(x)$.	

Example 7 Identify Types of Symmetry

Identify the type of symmetry in the graph of each function. Explain.

a. $f(x) = 3x + 1$

point symmetry: A 180° rotation about any point on the graph is the original graph.

b. $g(x) = -x^2 - 4x - 2$

line symmetry: The reflection in the line $x = -2$ coincides with the original graph.

c. $h(x) = 3x^4 + 4x^3 - 12x^2 + 13$

no symmetry: There is no line or point of symmetry.

d. $j(x) = x^3 - 2$

point symmetry: A 180° rotation about the point $(0, -2)$ is the original graph.

Think About It! How would knowing the type of symmetry help you graph a function?

Sample answer: When you plot a point on one side of the line of symmetry or point of symmetry, you can plot its image by reflecting in the line or rotating about the point.

Go Online You can complete an Extra Example online.

Copyright © McGraw-Hill Education

Lesson 1-2 • Linearity, Intercepts, and Symmetry 17

Example 7 Identify Types of Symmetry

Teaching the Mathematical Practices

3 Make Conjectures In the Think About It! section, students must make a conjecture as to why symmetry is important when graphing functions.

Questions for Mathematical Discourse

AL If you fold a graph in half on a vertical line of symmetry, what happens? Sample answer: The points on one side match up with the points on the other side.

OL What type of symmetry will all quadratic functions display? line symmetry

BL Would all linear functions have point symmetry? Explain. Yes; sample answer: Rotating the line 180° about any point on the line will result in the same line.

Common Error

When identifying line symmetry, students may not stick to vertical lines of symmetry. They may look for any line. Remind students line symmetry is only for vertical lines.

Interactive Presentation

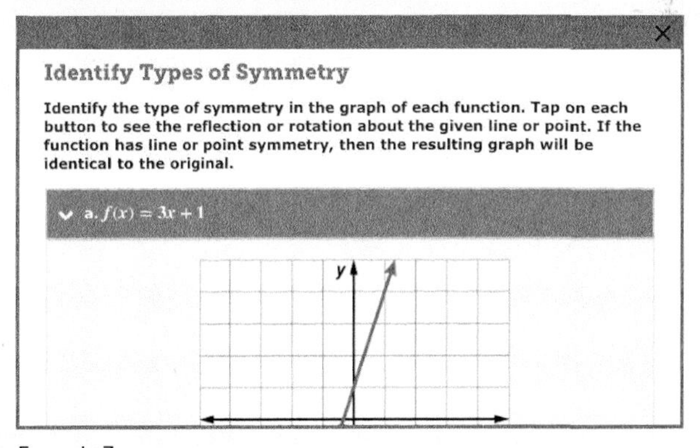

Example 7

TAP

Students move through different line and point symmetries for four functions.

TYPE

Students explain why symmetry helps graph functions.

Example 8 Identify Even and Odd Functions

Determine whether each function is *even*, *odd*, or *neither*. Confirm algebraically. If the function is odd or even, describe the symmetry.

a. $f(x) = x^3 - 4x$

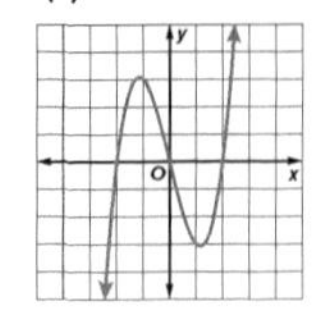

It appears that the graph of $f(x)$ is symmetric about the origin. Substitute $-x$ for x to test this algebraically.

$f(-x) = (\underline{-x})^3 - 4(\underline{-x})$

$= -x^3 + \underline{4x}$ Simplify.

$= -(x^3 - 4x)$ Distribute.

$= \underline{-f(x)}$ $f(x) = x^3 - 4x$

Because $f(-x) = -f(x)$ the function is $\underline{\text{odd}}$ and is symmetric about the $\underline{\text{origin}}$.

b. $g(x) = 2x^4 - 6x^2$

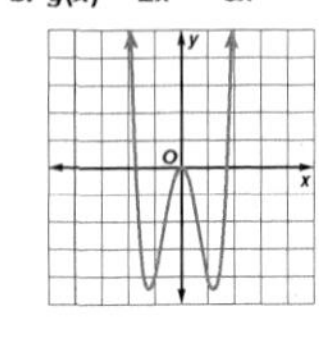

It appears that the graph of $g(x)$ is symmetric about the y-axis. Substitute $-x$ for x to test this algebraically.

$g(-x) = 2(\underline{-x})^4 - 6(\underline{-x})^2$

$= 2x^4 - \underline{6x^2}$ Simplify.

$= g(x)$ $g(x) = 2x^4 - 6x^2$

Because $g(-x) = g(x)$ the function is $\underline{\text{even}}$ and is symmetric in the $\underline{y\text{-axis}}$.

c. $h(x) = x^3 + 0.25x^2 - 3x$

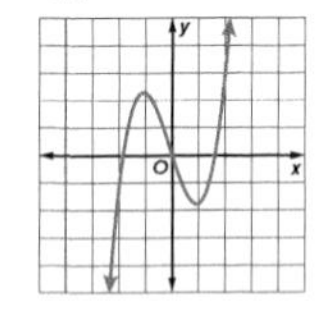

It appears that the graph of $h(x)$ may be symmetric about the origin. Substitute $-x$ for x to test this algebraically.

$h(-x) = (\underline{-x})^3 + 0.25(\underline{-x})^2 - 3(\underline{-x})$

$= \underline{-x^3} - 0.25x^2 + \underline{3x}$

Because $-h(x) = -x^3 - 0.25x^2 + 3x$, the function is $\underline{\text{neither even nor odd}}$ because $h(-x) \neq h(x)$ and $h(-x) \neq -h(x)$.

Watch Out!

Even and Odd Functions Always confirm symmetry algebraically. Graphs that appear to be symmetric may not actually be.

Check

Assume that f is a function that contains the point $(2, -5)$. Which of the given points must be included in the function if f is:

even? $\underline{(-2, -5)}$ odd? $\underline{(-2, 5)}$

$(-2, -5)$ $(-2, 5)$ $(2, 5)$ $(-5, -2)$ $(-5, 2)$

Go Online You can complete an Extra Example online.

Interactive Presentation

Identify Even and Odd Functions

Analyze the graph of each function. Determine whether each function is *even*, *odd*, or *neither*. Confirm algebraically. If the function is odd or even, describe the symmetry.

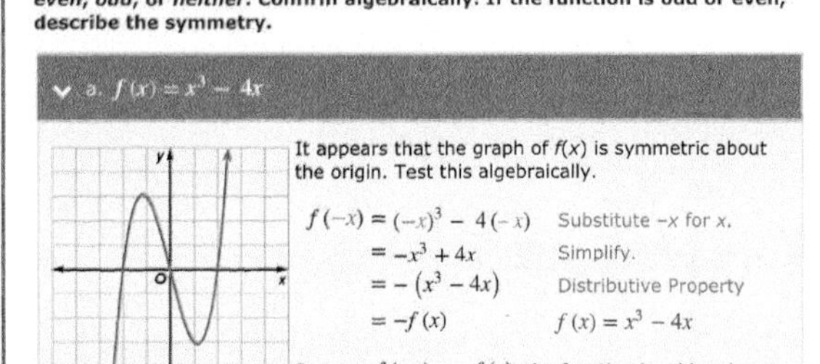
a. $f(x) = x^3 - 4x$
It appears that the graph of $f(x)$ is symmetric about the origin. Test this algebraically.
$f(-x) = (-x)^3 - 4(-x)$ Substitute $-x$ for x.
$= -x^3 + 4x$ Simplify.
$= -(x^3 - 4x)$ Distributive Property
$= -f(x)$ $f(x) = x^3 - 4x$

Example 8

TAP

Students move through the slides to identify the graphs as even, odd, or neither.

CHECK

Students complete the Check online to determine whether they are ready to move on.

Example 8 Identify Even and Odd Functions

Teaching the Mathematical Practices

1 Check Answers Point out that in Example 8, students need to check their answer algebraically to confirm whether each function is even, odd, or neither.

Questions for Mathematical Discourse

AL What is the sign of the result when you take a negative number to an odd power? negative What is the sign of the result when you take a negative number to an even power? positive How do you use this to evaluate $f(-x)$? Sample answer: When you substitute each x with $-x$, look at whether the power of the term is odd or even to determine whether the negative remains when you simplify it.

OL Would the function $y = \frac{1}{x}$ be even, odd, or neither? Justify your response. Odd; sample answer: Because $f(-x) = \frac{1}{-x} = -\frac{1}{x}$, the function is odd.

BL Without graphing or testing algebraically, will the function $f(x) = 2x^4 + 3x^2$ be even, odd, or neither? How did you determine the symmetry? Even; sample answer: because both exponents are even.

Common Error

Students may not want to test functions algebraically to determine whether they are even, odd, or neither, but many graphs can be deceiving. A graph that appears to be even may actually be neither. Reinforce the importance of testing functions algebraically.

DIFFERENTIATE

Differentiate Enrichment Activity BL

Have students investigate patterns in the power functions $f(x) = x^n$ and combinations of power functions like $f(x) = x^6 - x^2$. When are they odd functions or even functions? Does a constant function follow the pattern? Sample answer: Power functions and combinations of power functions where n is even are even functions. Power functions and combinations of power functions where n is odd are odd functions. A constant function is a power function where $n = 0$, so it is an even power and it is an even function. Thus it follows the pattern.

Exit Ticket

Recommended Use

At the end of class, go online to display the Exit Ticket prompt and ask students to respond using a separate piece of paper. Have students hand you their responses as they leave the room.

Alternate Use

At the end of class, go online to display the Exit Ticket prompt and ask students to respond verbally or by using a mini-whiteboard. Have students hold up their whiteboards so that you can see all student responses. Tap to reveal the answer when most or all students have completed the Exit Ticket.

Practice and Homework

Suggested Assignments

Use the table below to select appropriate exercises.

DOK	Topic	Exercises
1, 2	exercises that mirror the examples	1–24
2	exercises that use a variety of skills from this lesson	25–33
2	exercises that extend concepts learned in this lesson to new contexts	34–40
3	exercises that emphasize higher-order and critical thinking skills	41–44

ASSESS AND DIFFERENTIATE

IF students score 90% or more on the Checks, **THEN** assign: BL

- Practice, Exercises 1–39 odd, 41–44
- Extension: Even and Odd Functions
- **ALEKS**® Graphs of Functions; Transformations

IF students score 66%–89% on the Checks, **THEN** assign: OL

- Practice, Exercises 1–43 odd
- Remediation, Review Resources: Functions
- Personal Tutors
- Extra Examples 1–8
- **ALEKS**® Sets, Relations, and Functions

IF students score 65% or less on the Checks, **THEN** assign: AL

- Practice, Exercises 1–23 odd
- Remediation, Review Resources: Functions
- *Quick Review Math Handbook:* Linear Relations and Functions
- **ALEKS**® Sets, Relations, and Functions

Answers

9. The number of inches and corresponding number of feet is a linear function because when graphed, a line contains all of the points shown in the table.

10. *Cassini* 2's model is not a linear function because when graphed, a line does not contains all of the points shown in the table.

17b. The x-intercept represents the number of days until Aksa will run out of money. The y-intercept represents the total amount Aksa had in her lunch account at the beginning of the week.

18b. The x-intercept represents the number of seconds that it takes for the ball to hit the ground. The y-intercept represents the starting height of the ball, in inches.

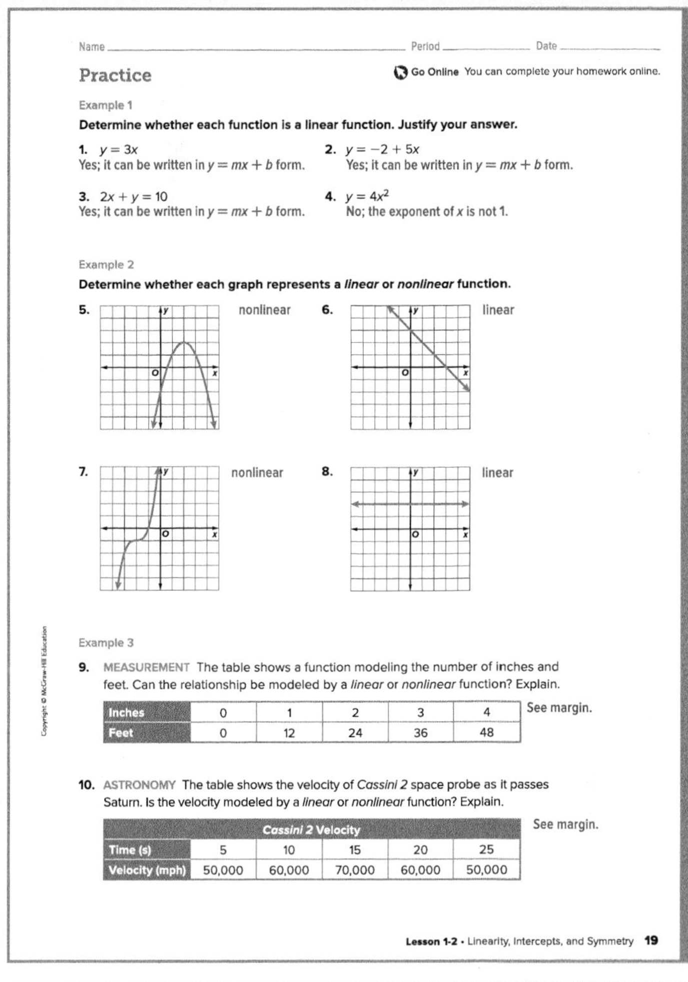

Name ______ Period ______ Date ______

Practice Go Online You can complete your homework online.

Example 1

Determine whether each function is a linear function. Justify your answer.

1. $y = 3x$ Yes; it can be written in $y = mx + b$ form.
2. $y = -2 + 5x$ Yes; it can be written in $y = mx + b$ form.
3. $2x + y = 10$ Yes; it can be written in $y = mx + b$ form.
4. $y = 4x^2$ No; the exponent of x is not 1.

Example 2

Determine whether each graph represents a *linear* or *nonlinear* function.

5. nonlinear
6. linear
7. nonlinear
8. linear

Example 3

9. MEASUREMENT The table shows a function modeling the number of inches and feet. Can the relationship be modeled by a *linear* or *nonlinear* function? Explain. See margin.

Inches	0	1	2	3	4
Feet	0	12	24	36	48

10. ASTRONOMY The table shows the velocity of *Cassini 2* space probe as it passes Saturn. Is the velocity modeled by a *linear* or *nonlinear* function? Explain. See margin.

Cassini 2 Velocity					
Time (s)	5	10	15	20	25
Velocity (mph)	50,000	60,000	70,000	60,000	50,000

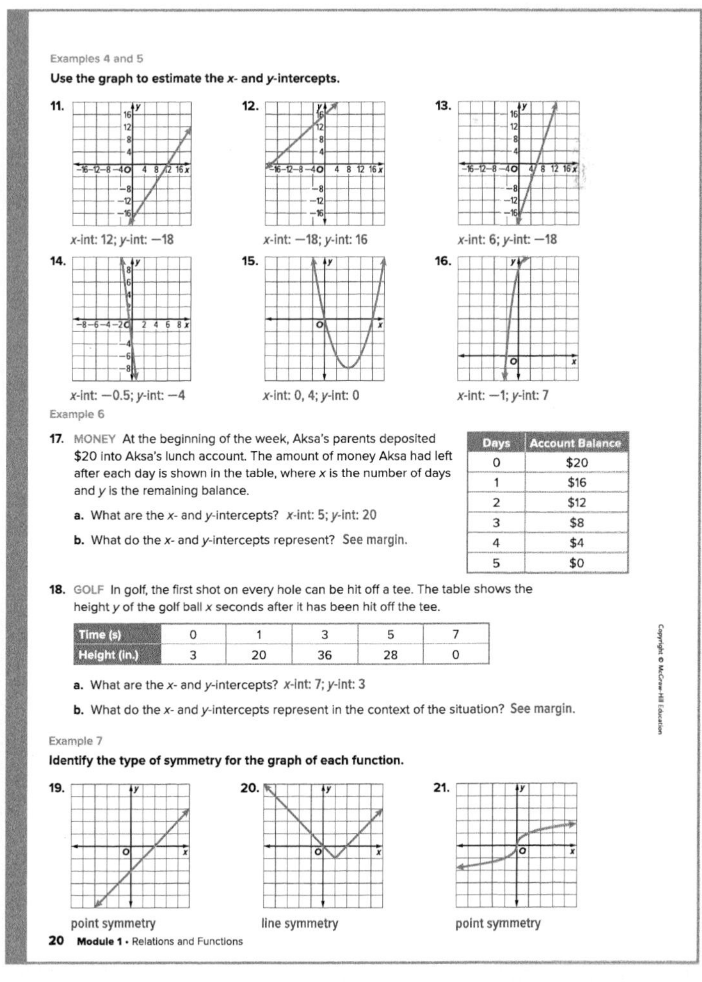

Examples 4 and 5

Use the graph to estimate the x- and y-intercepts.

11. x-int: 12; y-int: −18
12. x-int: −18; y-int: 16
13. x-int: 6; y-int: −18
14. x-int: −0.5; y-int: −4
15. x-int: 0, 4; y-int: 0
16. x-int: −1; y-int: 7

Example 6

17. MONEY At the beginning of the week, Aksa's parents deposited $20 into Aksa's lunch account. The amount of money Aksa had left after each day is shown in the table, where x is the number of days and y is the remaining balance.

a. What are the x- and y-intercepts? x-int: 5; y-int: 20

b. What do the x- and y-intercepts represent? See margin.

Days	Account Balance
0	$20
1	$16
2	$12
3	$8
4	$4
5	$0

18. GOLF In golf, the first shot on every hole can be hit off a tee. The table shows the height y of the golf ball x seconds after it has been hit off the tee.

Time (s)	0	1	3	5	7
Height (in.)	3	20	36	28	0

a. What are the x- and y-intercepts? x-int: 7; y-int: 3

b. What do the x- and y-intercepts represent in the context of the situation? See margin.

Example 7

Identify the type of symmetry for the graph of each function.

19. point symmetry
20. line symmetry
21. point symmetry

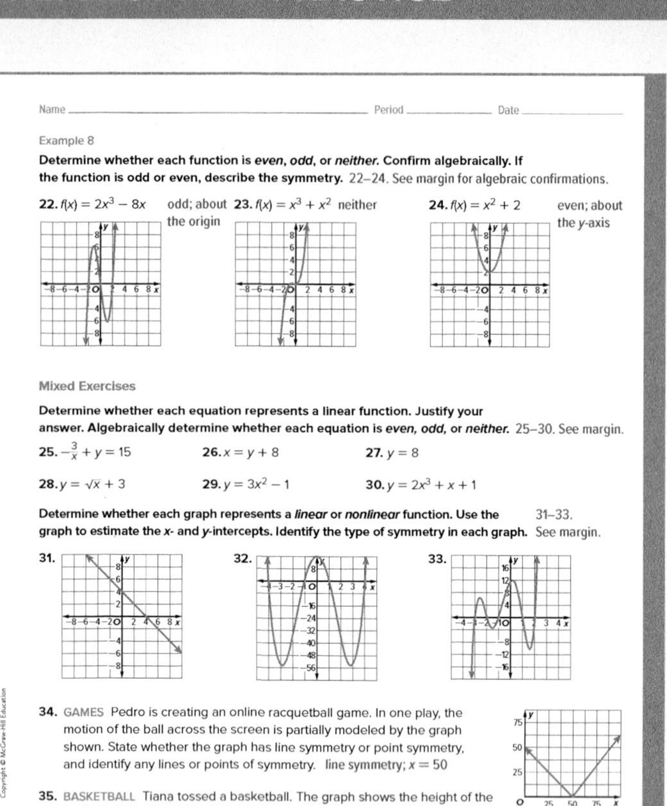

Name ______ Period ______ Date ______

Example 8

Determine whether each function is *even*, *odd*, or *neither*. Confirm algebraically. If the function is odd or even, describe the symmetry. 22–24. See margin for algebraic confirmations.

22. $f(x) = 2x^3 - 8x$ odd; about the origin
23. $f(x) = x^3 + x^2$ neither
24. $f(x) = x^2 + 2$ even; about the y-axis

Mixed Exercises

Determine whether each equation represents a linear function. Justify your answer. Algebraically determine whether each equation is *even*, *odd*, or *neither*. 25–30. See margin.

25. $-\frac{3}{x} + y = 15$
26. $x = y + 8$
27. $y = 8$
28. $y = \sqrt{x} + 3$
29. $y = 3x^2 - 1$
30. $y = 2x^3 + x + 1$

Determine whether each graph represents a *linear* or *nonlinear* function. Use the graph to estimate the x- and y-intercepts. Identify the type of symmetry in each graph. 31–33. See margin.

31.
32.
33.

34. GAMES Pedro is creating an online racquetball game. In one play, the motion of the ball across the screen is partially modeled by the graph shown. State whether the graph has line symmetry or point symmetry, and identify any lines or points of symmetry. line symmetry; $x = 50$

35. BASKETBALL Tiana tossed a basketball. The graph shows the height of the basketball as a function of time. If the graph of the function is extended to include the domain of all real numbers, would it have line or point symmetry? If so, identify the line or point of symmetry. line symmetry; $x = 0.4$

36. PROFIT Stefon charges people $25 to test the air quality in their homes. The device he uses to test air quality cost him $500. The function $y = 25x - 500$ describes Stefon's net profit, y, as a function of the number of clients he gets, x.

a. State whether the function is a linear function. Write *yes* or *no*. Explain. Yes; it is in $y = mx + b$ form.

b. What do the x- and y-intercepts of the function represent in terms of the situation? The x-intercept is the number of clients it will take to break even; the y-intercept is the initial cost of the device.

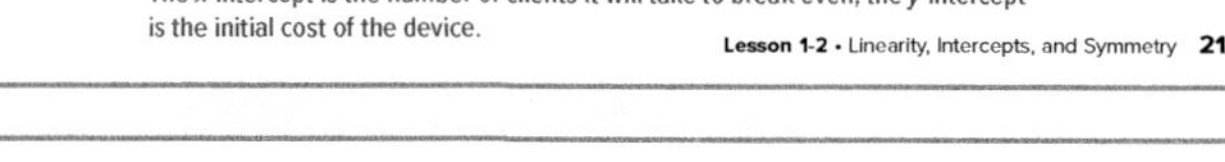

37. PLAYGROUNDS A playground is shaped as shown. The total perimeter is 500 feet.

a. Write an equation that relates x and y. $2x + 2y + 10 = 500$

b. Is the equation that relates x and y linear? Explain. Yes; it can be written in $y = mx + b$ form.

c. Graph the equation. State whether the graph has line symmetry or point symmetry. See margin.

38. POOLS The graph represents a 720-gallon pool being drained.

a. What are the x- and y-intercepts? What do the x- and y-intercepts represent? See margin.

b. Does the graph display line symmetry? Explain why or why not in terms of the situation. No; the water drains at a constant rate, not refilling.

39. VOLUME The function, $f(r) = \frac{4}{3}\pi r^3$ describes the relationship between the volume $f(r)$ and radius r of a sphere. Determine whether the function is *odd*, *even*, or *neither*. Explain your reasoning. Odd; $f(-r) = -f(r)$

40. USE A SOURCE Research online to find an equation that models a car's braking distance in relation to its speed. Then identify and interpret the y-intercept of the equation. See margin.

Higher-Order Thinking Skills

41. FIND THE ERROR Javier claimed that all cubic functions are odd. Is he correct? If not, provide a counterexample. No; sample answer: $f(x) = x^3 + 2x - 5$

42. ANALYZE The table shows a function modeling the number of gifts y Cornell can wrap if he spends x hours wrapping. Can the table be modeled by a *linear* or *nonlinear* function? Explain. See margin.

Hours	0	1	2	3	4
Gifts	0	14	28	42	56

43. PERSEVERE Determine whether an equation of the form $x = a$, where a is a constant, is *sometimes*, *always*, or *never* a linear function. Explain your reasoning. Sample answer: Never; the graph of $x = a$ is a vertical line, so it is not a function.

44. WHICH ONE DOESN'T BELONG? Of the four functions shown, identify the one that does not belong. Explain your reasoning. $y = 2xy$; Sample answer: $y = 2xy$ is not a linear function.

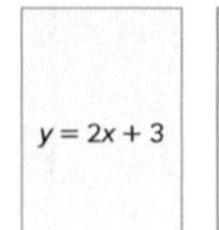

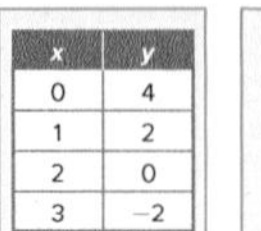

x	y
0	4
1	2
2	0
3	−2

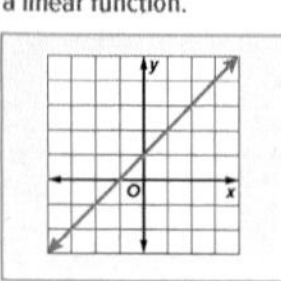

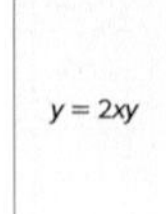

Answers

22. $f(-x) = 2(-x)^3 - 8(-x) = -2x^3 + 8x = -f(x)$
23. $f(-x) = (-x)^3 + (-x)^2 = -x^3 + x^2 \neq f(x)$ and $\neq -f(-x)$
24. $f(-x) = (-x)^2 + 2 = x^2 + 2 = f(x)$
25. No; x is in a denominator. The equation is neither even nor odd.
26. Yes; it can be written in $y = mx + b$ form. The equation is neither even nor odd.
27. Yes; it can be written in $y = mx + b$ form. The equation is even.
28. No; x is inside a square root. The equation is neither even nor odd.
29. No; there is an x^2-term. The equation is even.
30. No; there is an x^3-term. The equation is neither even nor odd.
31. linear; x-int: 4; y-int: 4; point symmetry
32. nonlinear; x-int: −4, −1, 1, 4; y-int: 16; line symmetry
33. nonlinear; x-int: −3, −2, −1, 1, 2; y-int: 12; neither point nor line symmetry

37c. point symmetry;

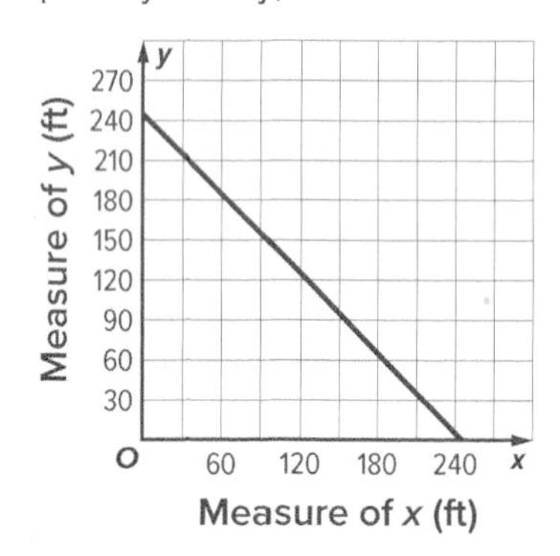

38a. x-int: 90; y-int: 720; The x-intercept represents the number of minutes it takes for there to be 0 gallons of water in the pool. The y-intercept represents the initial amount of water in the pool.

40. The formula for the stopping distance is $d = \frac{v^2}{2\mu g}$, where d is the stopping distance in meters, v is the velocity of the car in meters per second, μ is the coefficient of friction (unitless), and g is the acceleration due to gravity (9.80 m/s²). The y-intercept is 0, meaning that a car that is not moving takes no distance to stop.

42. Linear; sample answer: The number of gifts wrapped over time is a linear function because when graphed, a line contains all of the points shown in the table.

Extrema and End Behavior

F.IF.4, F.IF.7c

LESSON GOAL

Students identify extrema and end behavior of functions.

1 LAUNCH

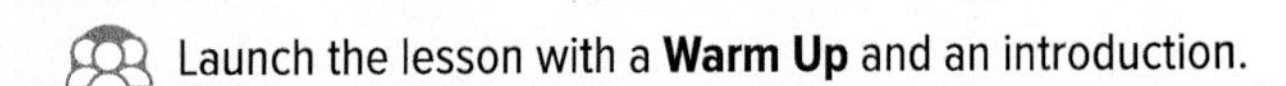
Launch the lesson with a **Warm Up** and an introduction.

2 EXPLORE AND DEVELOP

Develop:

Extrema of Functions
- Find Extrema from Graphs
- Find and Interpret Extrema

Explore: End Behavior of Linear and Quadratic Functions

Develop:

End Behavior of Graphs of Functions
- End Behavior of Linear Functions
- End Behavior of Nonlinear Functions
- Determine and Interpret End Behavior

You may want your students to complete the **Checks** online.

3 REFLECT AND PRACTICE

Exit Ticket

Practice

DIFFERENTIATE

View reports of student progress on the **Checks** after each example.

Resources	AL	OL	BL	ELL
Remediation: Symmetry of Graphs of Functions	●	●		●
Extension: End Behavior of Rational Functions		●	●	●

Language Development Handbook

Assign page 3 of the *Language Development Handbook* to help your students build mathematical language related to the extrema and end behavior of functions.

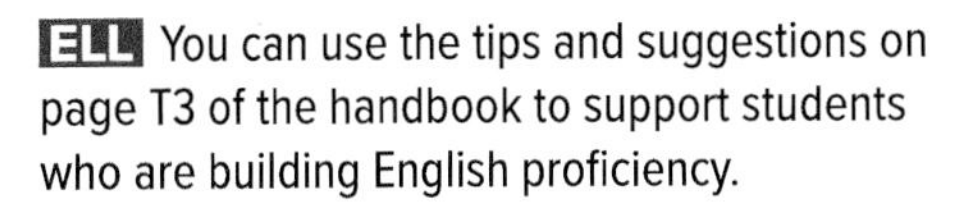
ELL You can use the tips and suggestions on page T3 of the handbook to support students who are building English proficiency.

Suggested Pacing

90 min	0.5 day
45 min	1 day

Focus

Domain: Functions

Standards for Mathematical Content:

F.IF.4 For a function that models a relationship between two quantities, interpret key features of graphs and tables in terms of the quantities, and sketch graphs showing key features given in a verbal description of the relationship.

F.IF.7c Graph functions expressed symbolically and show key features of the graph, by hand in simple cases and using technology for more complicated cases. Graph polynomial functions, identifying zeros when suitable factorizations are available, and showing end behavior.

Standards for Mathematical Practice:

2 Reason abstractly and quantitatively.

5 Use mathematical tools strategically.

Coherence

Vertical Alignment

Previous
Students determined the linearity, intercepts, and symmetry of functions.
F.IF.4, F.IF.5

Now
Students identify extrema and end behavior of functions.
F.IF.4, F.IF.7c

Next
Students will sketch graphs of functions and compare two functions in different ways. **F.IF.4, F.IF.9**

Rigor

The Three Pillars of Rigor

1 CONCEPTUAL UNDERSTANDING	2 FLUENCY	3 APPLICATION

Conceptual Bridge In this lesson, students build on their understanding of functions to include the behavior of graphs of functions. They build fluency by graphing different types of functions, and they apply their understanding by solving real-world problems involving functions.

Interactive Presentation

Warm Up

Assume that p is a relation that contains the point $(3, -4)$. What other point must be included in the relation if p is

1. even?
2. odd?
3. symmetric about the x-axis?
4. symmetric about the y-axis?

Warm Up

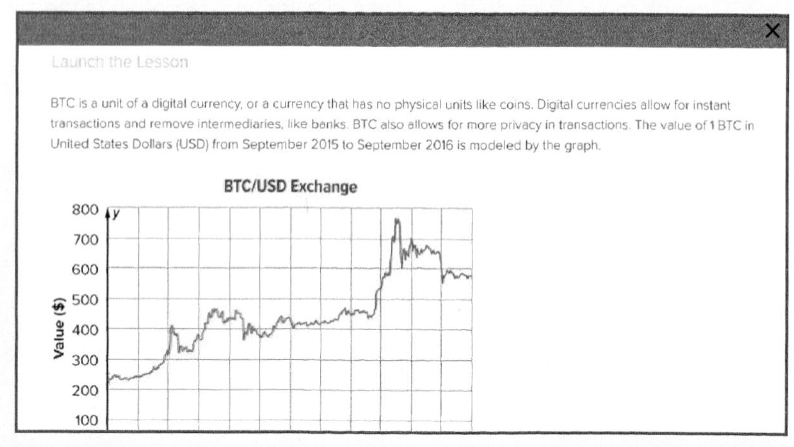

Launch the lesson

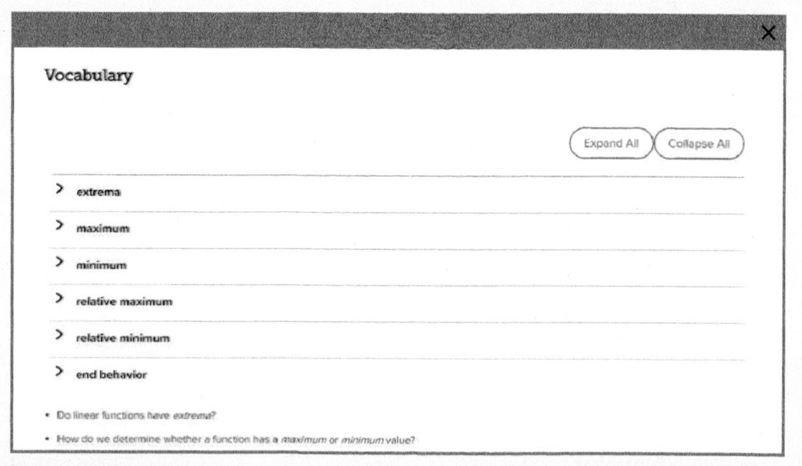

Today's Vocabulary

Warm Up

Prerequisite Skills

The Warm Up exercises address the following prerequisite skill for this lesson:

- using symmetry

Answers:

1. $(-3, -4)$
2. $(-3, 4)$
3. $(3, 4)$
4. $(-3, -4)$

Launch the Lesson

 **Teaching the Mathematical Practices**

4 Interpret Mathematical Results Encourage students to identify other relative high and low points on the graph and interpret them in the context of the situation.

Go Online to find additional teaching notes and questions to promote classroom discourse.

Today's Standards

Tell students that they will be addressing these content and practice standards in this lesson. You may wish to have a student volunteer read aloud *How can I meet these standards?* and *How can I use these practices?*, and connect these to the standards.

See the Interactive Presentation for I Can statements that align with the standards covered in this lesson.

Today's Vocabulary

Tell students that they will be using these vocabulary terms in this lesson. You can expand each row if you wish to share the definitions. Then discuss the questions below with the class.

Mathematical Background

Tables of values can be used to explore two types of changes in the values of a polynomial function. A change of signs in the value of $f(x)$ from one value of x to the next indicates that the graph of the function crosses the x-axis between the two x-values. A change between increasing values and decreasing values indicates that the graph is turning for that interval. A turning point on a graph is a relative maximum or minimum.

Explore End Behavior of Linear and Quadratic Functions

Objective
Students use a sketch to explore the end behavior of linear and quadratic functions.

 **Teaching the Mathematical Practices**

3 Construct Arguments In the Explore, students will use previously established results to make a conjecture about how to determine the end behavior of linear and nonlinear functions.

Ideas for Use
Recommended Use Present the Inquiry Question, or have a student volunteer read it aloud. Have students work in pairs to complete the Explore activity on their devices. Pairs should discuss each of the questions. Monitor student progress during the activity. Upon completion of the Explore activity, have student volunteers share their responses to the Inquiry Question.

What if my students don't have devices? You may choose to project the activity on a whiteboard. A printable worksheet for each Explore is available online. You may choose to print the worksheet so that individuals or pairs of students can use it to record their observations.

Summary of the Activity
Students will complete guiding exercises throughout the Explore activity. Students will use a sketch to investigate the end behavior of linear and quadratic functions. Then students will analyze the graphs to answer questions regarding the value of the function as x goes to positive or negative infinity. Then, students will answer the Inquiry Question.

(continued on the next page)

Interactive Presentation

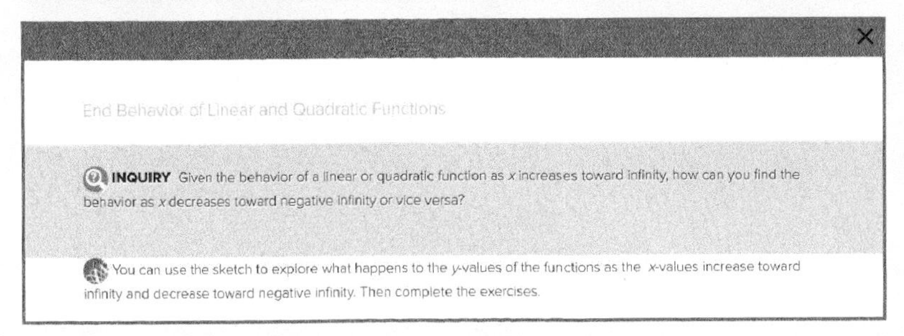

Explore

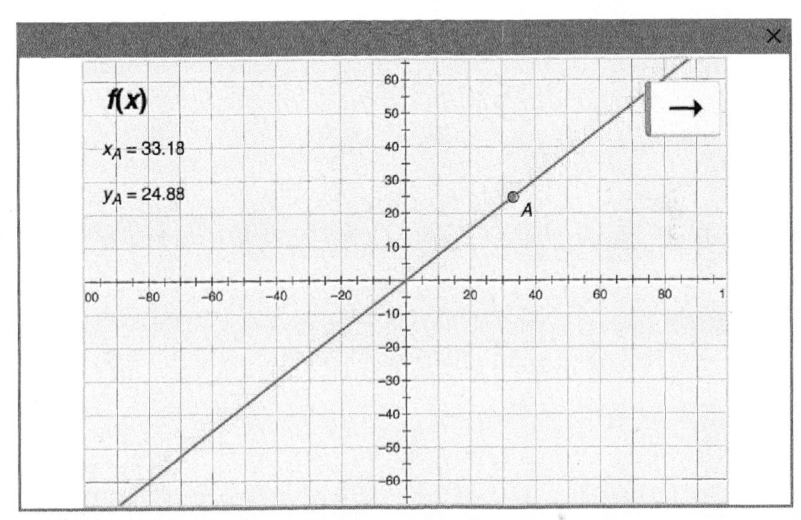

Explore

WEB SKETCHPAD

Students use the sketch to complete an activity in which they explore the end behavior of linear and quadratic functions.

TYPE

Students answer questions regarding end behavior for each function.

Interactive Presentation

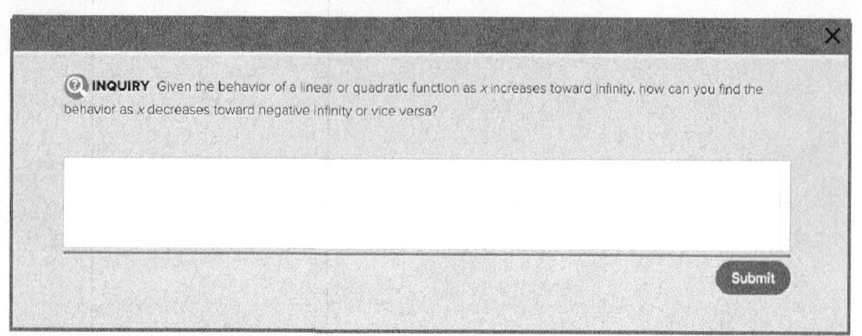

Explore

TYPE

Students respond to the Inquiry Question and can view a sample answer.

Explore End Behavior of Linear and Quadratic Functions (*continued*)

Questions

Have students complete the Explore activity.

Ask:

- Consider a linear function with a negative slope. What will happen to the value of the function as x approaches negative infinity? The value of the function will approach infinity.
- What do you think end behavior of a function represents? Sample answer: It represents the value of the function as x approaches positive or negative infinity.

Inquiry

Given the behavior of a linear or quadratic function as x increases toward infinity, how can you find the behavior as x decreases toward negative infinity or vice versa? Sample answer: For a linear function that is not constant, the behavior as x decreases toward negative infinity will be the opposite of the behavior as x increases toward infinity. For a quadratic function, the behavior as x decreases toward negative infinity will be the same as the behavior as x increases toward infinity.

Go Online to find additional teaching notes and sample answers for the guiding exercises.

1 CONCEPTUAL UNDERSTANDING | 2 FLUENCY | 3 APPLICATION

Learn Extrema of Functions

Objective

Students identify extrema of functions by examining their graphs.

Teaching the Mathematical Practices

1 Explain Correspondences Encourage students to explain the relationship between graphs of functions and the extrema of the functions.

Essential Question Follow-Up

Students have begun identifying extrema of functions.

Ask:

Why would the maximum or minimum value of a function be important? Sample answer: The maximum or minimum value may represent a desired amount in the real world. For example, if a business graphed its profit equation, they would want to know their maximum profit, or if it was a cost equation, the minimum cost.

Example 1 Find Extrema From Graphs

Teaching of Mathematical Practices

8 Attend to Details In Example 1, encourage students to evaluate the reasonableness of their answers. Students should check that the behavior of the function surrounding the extrema corresponds to the type of extrema they identify.

Questions for Mathematical Discourse

AL If a function decreases towards a point and then increases away from the point, would the point be a maximum or minimum? minimum

OL How do you tell if a point is a maximum or a relative maximum? Sample answer: For both, the graph will be increasing towards the point and then decreasing away from the point. But for a maximum, the graph will be higher at that point than anywhere else with consideration of the end behavior.

BL Could a graph have a maximum and a different point that is a relative maximum? Explain. Yes; sample answer: It is possible for a graph to have an absolute highest point as well as a relative maximum that is not as high. A quartic function could be an example.

Common Error

When a graph has two relative maxima, students may believe one is a maximum while one is a relative maximum. Remind students that maxima occur only when there is nothing higher on the graph, so they must check the entire graph before deciding.

Lesson 1-3

Extrema and End Behavior

Learn Extrema of Functions

Graphs of functions can have high and low points where they reach a maximum or minimum value. The maximum and minimum values of a function are called **extrema.**

The **maximum** is at the highest point on the graph of a function. The **minimum** is at the lowest point on the graph of a function.

A **relative maximum** is located at a point on the graph of a function where no other nearby points have a greater y-coordinate. A **relative minimum** is located at a point on the graph of a function where no other nearby points have a lesser y-coordinate.

Example 1 Find Extrema from Graphs

Identify and estimate the x- and y-values of the extrema. Round to the nearest tenth if necessary.

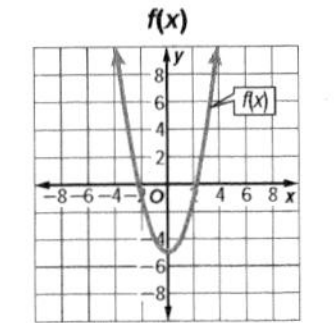

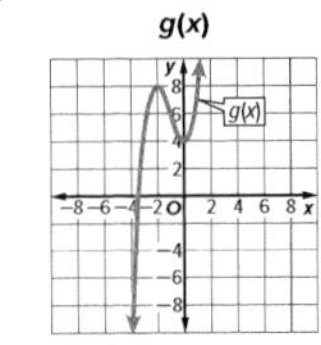

$f(x)$: The function $f(x)$ is decreasing as it approaches $x = 0$ from the left and increasing as it moves away from $x = 0$. Further, $(0, -5)$ is the lowest point on the graph, so $(0, -5)$ is a minimum.

$g(x)$: The function $g(x)$ is increasing as it approaches $x = -2$ from the left and decreasing as it moves away from $x = -2$. Further, there are no greater y-coordinates surrounding $(-2, 8)$. However, $(-2, 8)$ is not the highest point on the graph, so $(-2, 8)$ is a relative maximum.

The function $g(x)$ is decreasing as it approaches $x = 0$ from the left and increasing as it moves away from $x = 0$. Further, there are no lesser y-coordinates surrounding $(0, 4)$. However, $(0, 4)$ is not the lowest point on the graph, so $(0, 4)$ is a relative minimum.

Go Online You can complete an Extra Example online.

Today's Goals
- Identify extrema of functions.
- Identify end behavior of graphs.

Today's Vocabulary
extrema
maximum
minimum
relative maximum
relative minimum
end behavior

Watch Out!
No Extrema Some functions, like $f(x) = x^3$, have no extrema.

Study Tip
Reading in Math In this context, *extrema* is the plural form of *extreme point*. The plural of *maximum* and *minimum* are *maxima* and *minima*, respectively.

Think About It!
Why are the extrema identified on the graph of $g(x)$ relative maxima and minima instead of maxima and minima?

Sample answer: The extrema are not the absolute highest or lowest points on the graphs, but they are the highest or lowest in relation to nearby points.

Lesson 1-3 • Extrema and End Behavior 23

Interactive Presentation

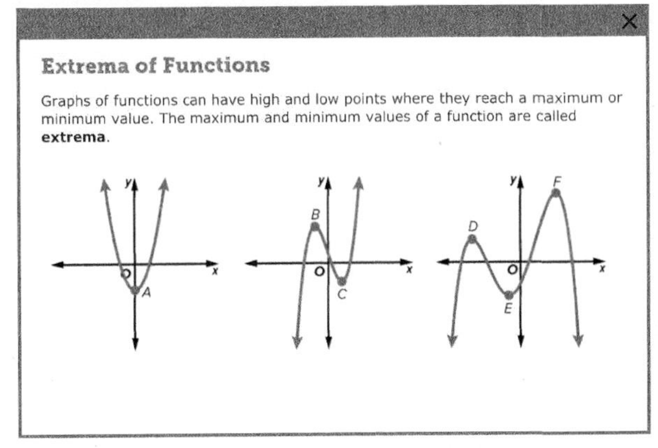

Learn

TAP

Students move through each type of extrema to see it interpreted in context.

Your Notes

Example 2 Find and Interpret Extrema

SOCIAL MEDIA **Use the table and graph to estimate the extrema of the function that relates the number hours since midnight x to the number of posts being uploaded y. Describe the meaning of the extrema in the context of the situation.**

x	y
0	2.8
4	1.8
8	3.1
12	11.5
14	9.1
16	10.2
20	5.8
24	2.8

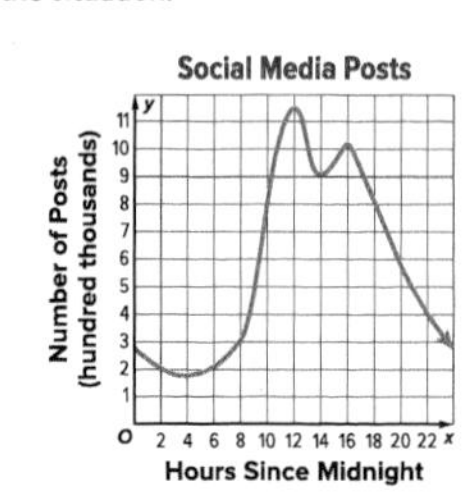

maxima The number of posts sent 12 hours after midnight is greater than the number of posts made at any other time during the day. The highest point on the graph occurs when $x = 12$. Therefore, the maximum number of posts sent is about 1,150,000 at noon.

minima The number of posts sent 4 hours after midnight is less than the number of posts made at any other time during the day. The lowest point on the graph occurs when $x = 4$. Therefore, the minimum number of posts sent is about 180,000 at 4:00 A.M.

relative maxima The number of posts sent 16 hours after midnight is greater than the number of posts during surrounding times, but is not the greatest number sent during the day. The graph has a relative peak when $x = 16$. Therefore, there is a relative peak in the number of posts sent, or relative maximum, at 4:00 P.M. of about 1,020,000 posts.

relative minima The number of posts sent 14 hours after midnight is less than the number of posts during surrounding times, but is not the least number sent during the day. The graph dips when $x = 14$. Therefore, there is a relative low in the number of posts sent, or relative minimum, at 2:00 P.M. of about 910,000 posts.

Explore End Behavior of Linear and Quadratic Functions

Online Activity Use graphing technology to complete the Explore.

INQUIRY Given the behavior of a linear or quadratic function as x increases towards infinity, how can you find the behavior as x decreases toward negative infinity or vice versa?

Go Online
You can complete an Extra Example online.

Interactive Presentation

Find and Interpret Extrema

SOCIAL MEDIA The number of posts sent on a popular social media site varies throughout the day. Move through the categories to see how to use the table and graph to estimate the extrema of the function that models the number of posts in hundred thousands x given the number hours since midnight y. Describe the meaning of the extrema in the context of the situation.

Example 2

TAP

Students move through each type of extrema to see it interpreted in context.

CHECK

Students complete the Check online to determine whether they are ready to move on.

DIFFERENTIATE

Reteaching Activity AL ELL

IF students are unable to differentiate between absolute maximum/minimum values and relative maximum/minimum values,
THEN have students highlight any "peaks" or "valleys" of a graph. Guide them to determine whether the graph goes higher or lower elsewhere. If not, the values are absolute maximum/minimum values.

Example 2 Find and Interpret Extrema

 Teaching the Mathematical Practices

2 Attend to Quantities Point out that it is important to note the meaning of the quantities used in Example 2 when describing the extrema of the function.

Questions for Mathematical Discourse

AL Why do you think the minimum value of the function occurs at $x = 4$? Sample answer: Many people are sleeping at 4:00 A.M so there would not be as many people making posts at that time.

OL Can you find the extrema from the table alone? Sample answer: If you only had the table and not the graph, you would not be able to know how the function behaves between the points given in the table.

BL Does it make sense for a y-value to be a decimal in this context? Sample answer: In this case, the decimal y-values given in the table represent whole numbers because the y-axis is in hundreds of thousands. However, the graph does indicate that the function models with a continuous function that would also include all real numbers greater than 0.

DIFFERENTIATE

Language Development Activity ELL

Ask students the difference between *minimum/maximum/extreme* and *minima/maxima/extrema*. Instruct them to find other examples of this in English and mathematics. Discuss the definition and examples of *relative* in the following contexts: general, academic, and technical (mathematical).

Learn End Behavior of Graphs of Functions

Objective

Students identify end behavior of functions by examining their graphs.

 Teaching the Mathematical Practices

6 Communicate Precisely Encourage students to routinely write or explain their solutions methods. Use the Think About It! question to encourage students to use clear definitions when answering the question.

Important to Know

The end behavior of a graph is what the graph is doing as x approaches positive or negative infinity. When moving right along a graph, the values of x are increasing toward infinity, represented by $x \to \infty$. When moving left along a graph, the values of x are decreasing toward negative infinity, represented by $x \to -\infty$.

Common Misconception

Some students do not separate the horizontal and vertical movement of a graph. They believe that if x is approaching positive infinity, y must also be doing the same. Encourage students to look at where the graph is headed, up or down, to determine what the function is approaching as x approaches positive or negative infinity.

Example 3 End Behavior of Linear Functions

Teaching the Mathematical Practices

3 Analyze Cases Help students to analyze the case when $m < 0$ and when $m > 0$ to answer the question in the Talk About It! feature.

Questions for Mathematical Discourse

AL What is different about the end behavior of $f(x)$ compared to $g(x)$? Sample answer: For $f(x)$, end behavior approaches positive and negative infinity. For $g(x)$, the end behavior is a constant value.

OL Suppose the graph of $f(x)$ had a positive slope. How would the end behavior change? As x decreases, $f(x)$ decreases, and as x increases, $f(x)$ increases.

BL Why can we not find the end behavior for a vertical line? Sample answer: A vertical line is not a function so we cannot find the end behavior. Also, x is constant and will not approach positive or negative infinity.

Example 4 End Behavior of Nonlinear Functions

Teaching the Mathematical Practices

6 State Meaning of Symbols Help students to use the correct symbols when describing the end behavior of the functions.

Questions for Mathematical Discourse

AL What would be the end behavior if you reflected $f(x)$ so that it opens down? Both sides of the graph would approach negative infinity.

OL What can you conclude about the end behavior of odd functions? Sample answer: The end behavior of odd functions will be opposite, meaning if one side of the graph approaches negative infinity, the other will approach infinity.

BL What is another type of end behavior that a function could have? Sample answer: A function could approach a specific value but never reach the value. For example, if $f(x) = \frac{1}{x}$, increasing x values would make $f(x)$ approach but never reach zero.

Learn End Behavior of Graphs of Functions

End behavior is the behavior of a graph as x approaches positive or negative infinity. As you move right along the graph, the values of x are increasing toward infinity. This is denoted as $x \to \infty$. At the left end, the values of x are decreasing toward negative infinity, denoted as $x \to -\infty$. When a function $f(x)$ increases without bound, it is denoted as $f(x) \to \infty$. When a function $f(x)$ decreases without bound, it is denoted as $f(x) \to -\infty$.

Example 3 End Behavior of Linear Functions

Describe the end behavior of each linear function.

a. $f(x)$

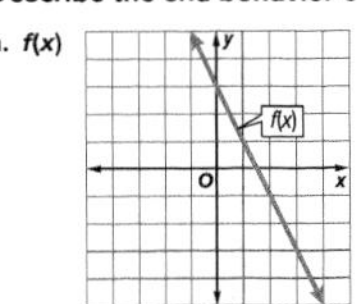

As x decreases, $f(x)$ increases, and as x increases $f(x)$ decreases. Thus, as $x \to -\infty$, $f(x) \to \infty$ and as $x \to \infty$, $f(x) \to -\infty$.

b. $g(x)$

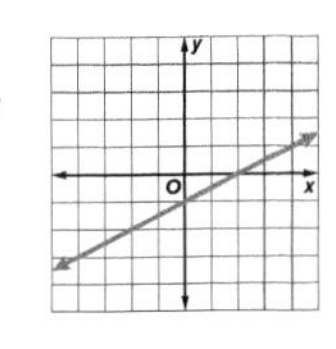

As x decreases or increases, $g(x) = 2$. Thus, as $x \to -\infty$, $g(x) = 2$, and as $x \to \infty$, $g(x) = 2$.

Check

Use the graph to describe the end behavior of the function.

As $x \to -\infty$, $f(x) \to -\infty$, and as $x \to \infty$ $f(x) \to \infty$.

Example 4 End Behavior of Nonlinear Functions

Describe the end behavior of each nonlinear function.

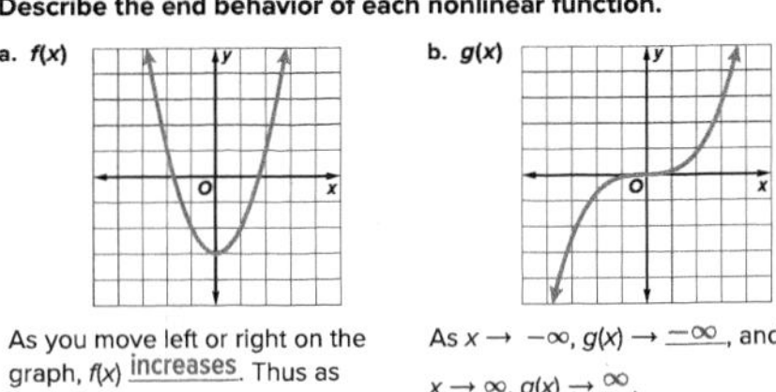

a. $f(x)$

As you move left or right on the graph, $f(x)$ increases. Thus as $x \to -\infty$, $f(x) \to \infty$, and as $x \to \infty$, $f(x) \to \infty$.

b. $g(x)$

As $x \to -\infty$, $g(x) \to -\infty$, and as $x \to \infty$, $g(x) \to \infty$.

Go Online You can complete an Extra Example online.

Think About It! For $f(x) = a$, where a is a real number, describe the end behavior of $f(x)$ as $x \to \infty$ and as $x \to -\infty$.

Sample answer: If $f(x) = a$, where a is a real number, then $f(x) = a$ as $x \to \infty$ and as $x \to -\infty$, because the function is equal to a for all values of x.

Talk About It! In part **a**, the function's end behavior as $x \to -\infty$ is the opposite of the end behavior as $x \to \infty$. Do you think this is true for all linear functions where $m \neq 0$? Explain your reasoning.

Yes; sample answer: If $m > 0$, then $f(x)$ increases as x increases and $f(x)$ decreases as x decreases. If $m < 0$, then $f(x)$ decreases as x increases and $f(x)$ increases as x decreases.

Math History Minute

Júlio César de Mello e Souza (1895–1974) was a Brazilian mathematician who is known for his books on recreational mathematics. His most famous book, *The Man Who Counted*, includes problems, puzzles, and curiosities about math. The State Legislature of Rio de Janeiro declared that his birthday, May 6, be Mathematician's Day.

Interactive Presentation

End Behavior of Graphs of Functions

End behavior is the behavior of a graph as x approaches positive or negative infinity. As you move right along the graph, the values of x are increasing toward infinity. This is denoted as $x \to \infty$. At the left end, the values of x are decreasing toward negative infinity, denoted as $x \to -\infty$.

Learn

TAP

Students tap on each button to learn about end behavior for the given functions.

TYPE

Students describe the end behavior for a constant function.

Think About It!
If the graph of a function is symmetric about a vertical line, what do you think is true about the end behavior of $f(x)$ as $x \to -\infty$ and as $x \to \infty$?

Sample answer: If a function has vertical symmetry, then the end behavior of $f(x)$ as $x \to -\infty$ and $x \to \infty$ must be the same, because both sides of the graph will approach the same value if the function is symmetric.

Study Tip

Assumptions Assuming that the drone can continue to fly for an infinite amount of time and to an infinite altitude lets us analyze the end behavior as $x \to \infty$. While there are maximum legal altitudes that a drone can fly as well as limited battery life, assuming that the time and altitude will continue to increase allows us to describe the end behavior.

Go Online to practice what you've learned about analyzing graphs in the Put It All Together over Lessons 1-1 through 1-3.

Check

Use the graph to describe the end behavior of the function.

As $x \to -\infty$, $f(x) \to -\infty$, and as $x \to \infty$, $f(x) \to -\infty$.

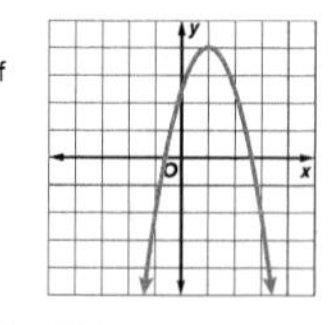

Example 5 Determine and Interpret End Behavior

DRONES **The graph shows the altitude of a drone above the ground $f(x)$ after x minutes. Describe the end behavior of $f(x)$ and interpret it in the context of the situation.**

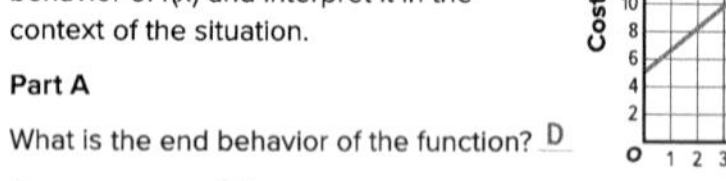

Since the drone cannot travel for a negative amount of time, the function is not defined for $x < 0$. So, there is no end behavior as $x \to -\infty$.

As $x \to \infty$, $f(x) \to \infty$. The drone is expected to continue to fly higher.

Flight of a Drone (Altitude (ft) vs. Time (min))

Check

RIDESHARING Mika and her friends are using a ride-sharing service to take them to a concert. The function models the cost of the ride $f(x)$ after x miles. Describe the end behavior of $f(x)$ and interpret it in the context of the situation.

Ridesharing (Cost ($) vs. Miles)

Part A

What is the end behavior of the function? D

A. as $x \to -\infty$, $f(x) \to -\infty$; as $x \to \infty$, $f(x) \to -\infty$

B. as $x \to -\infty$, $f(x) \to \infty$; as $x \to \infty$, $f(x) \to \infty$

C. as $x \to \infty$, $f(x) \to -\infty$; $f(x)$ is not defined for $x < 0$

D. as $x \to \infty$, $f(x) \to \infty$; $f(x)$ is not defined for $x < 0$

Part B

What does the end behavior represents in the context of the situation?

The farther Mika and her friends travel, the more the ride costs.

Go Online You can complete an Extra Example online.

26 Module 1 • Relations and Functions

Interactive Presentation

Example 5

CHECK

Students complete the Check online to determine whether they are ready to move on.

Example 5 Determine and Interpret End Behavior

Teaching the Mathematical Practices

4 Apply Mathematics In Example 5, students apply what they have learned about end behavior to a real-world situation.

Questions for Mathematical Discourse

AL What are the real-world limitations on time and altitude in this example? Time and altitude cannot be negative.

OL Will the graph head toward infinity forever? Explain. No; sample answer: Eventually the drone will drain its battery and fall back to the ground, or the person flying the drone will land it safely at some point.

BL What would the graph look like if the drone could only rise at a constant rate? The sections of the graph that represent the drone rising would be lines.

Common Error

In a real-world context, it usually does not make sense for x or $f(x)$ to approach negative infinity. Remind students to consider the context of real-world problems before writing end behavior.

Go Online

- Find additional teaching notes.
- View performance reports of the Checks.
- Assign or present an Extra Example.

Exit Ticket

Recommended Use

At the end of class, go online to display the Exit Ticket prompt and ask students to respond using a separate piece of paper. Have students hand you their responses as they leave the room.

Alternate Use

At the end of class, go online to display the Exit Ticket prompt and ask students to respond verbally or by using a mini-whiteboard. Have students hold up their whiteboards so that you can see all student responses. Tap to reveal the answer when most or all students have completed the Exit Ticket.

Practice and Homework

Suggested Assignments

Use the table below to select appropriate exercises.

DOK	Topic	Exercises
1, 2	exercises that mirror the examples	1–10
2	exercises that use a variety of skills from this lesson	11–21
2	exercises that extend concepts learned in this lesson to new contexts	22–30
3	exercises that emphasize higher-order and critical thinking skills	31–35

ASSESS AND DIFFERENTIATE

Use the data from the **Checks** to determine whether to provide resources for extension, remediation, or intervention.

IF students score 90% or more on the Checks,
THEN assign:

BL

- Practice, Exercises 1–29 odd, 31–35
- Extension: End Behavior of Rational Functions
- ALEKS® Graphs of Functions; Transformations

IF students score 66%–89% on the Checks,
THEN assign:

- Practice, Exercises 1–35 odd
- Remediation, Review Resources: Linearity and Continuity of Graphs
- Personal Tutors
- Extra Examples 1–5
- ALEKS® Sets, Relations, and Functions; Equations of Lines

IF students score 65% or less on the Checks,
THEN assign:

- Practice, Exercises 1–10
- Remediation, Review Resources: Linearity and Continuity of Graphs
- ALEKS® Sets, Relations, and Functions; Equations of Lines

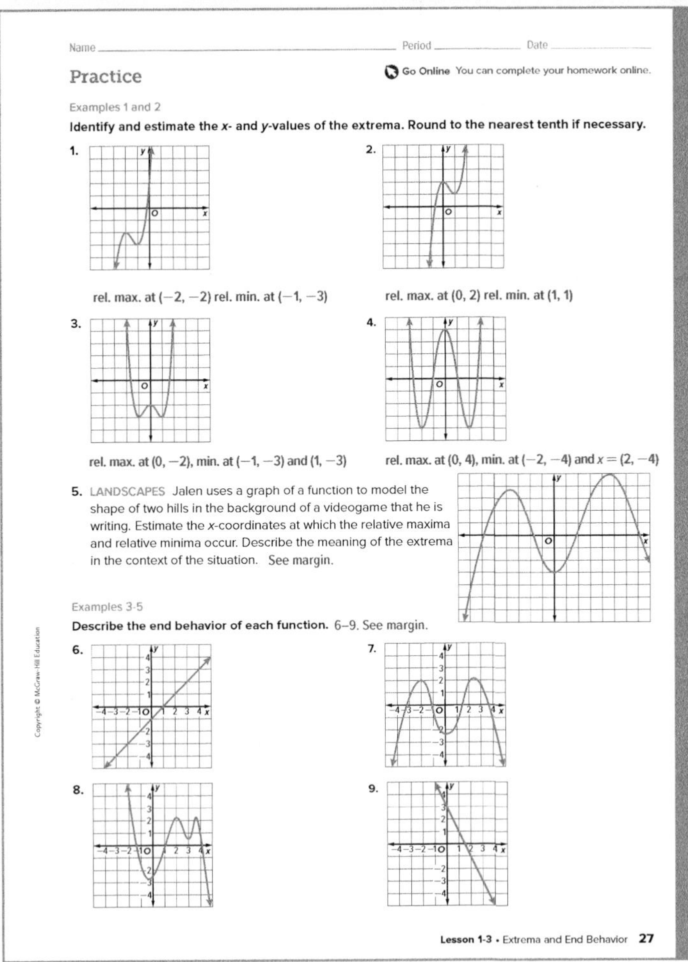

Name ______ Period ______ Date ______

Practice

Go Online You can complete your homework online.

Examples 1 and 2

Identify and estimate the x- and y-values of the extrema. Round to the nearest tenth if necessary.

1. rel. max. at (−2, −2) rel. min. at (−1, −3)

2. rel. max. at (0, 2) rel. min. at (1, 1)

3. rel. max. at (0, −2), min. at (−1, −3) and (1, −3)

4. rel. max. at (0, 4), min. at (−2, −4) and $x = (2, -4)$

5. LANDSCAPES Jalen uses a graph of a function to model the shape of two hills in the background of a videogame that he is writing. Estimate the x-coordinates at which the relative maxima and relative minima occur. Describe the meaning of the extrema in the context of the situation. See margin.

Examples 3-5

Describe the end behavior of each function. 6–9. See margin.

6. 7. 8. 9.

Copyright © McGraw-Hill Education

Lesson 1-3 • Extrema and End Behavior 27

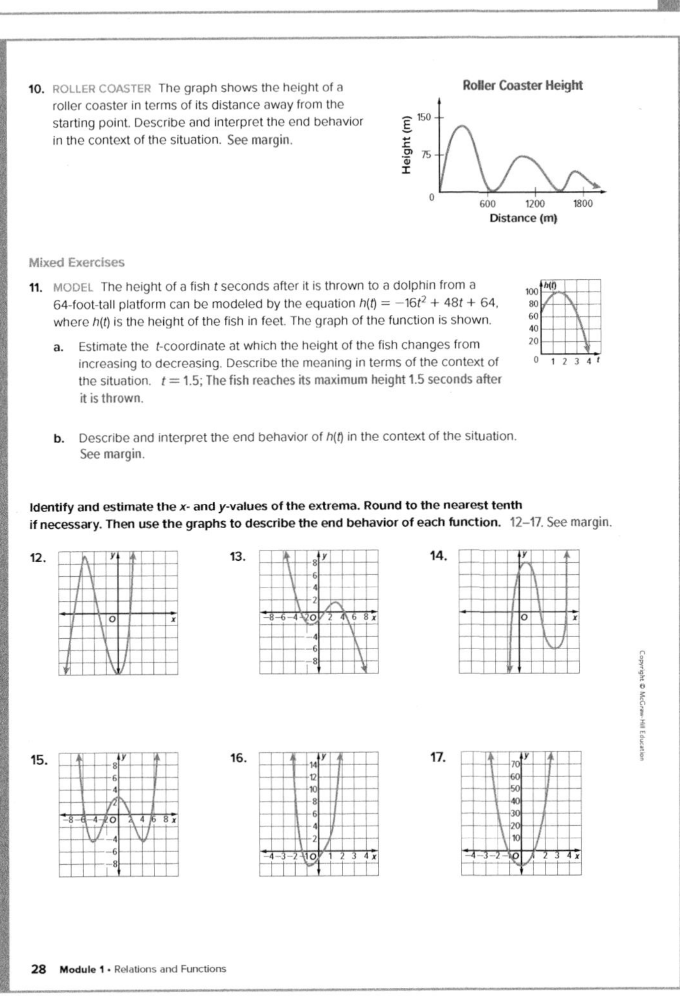

10. ROLLER COASTER The graph shows the height of a roller coaster in terms of its distance away from the starting point. Describe and interpret the end behavior in the context of the situation. See margin.

Mixed Exercises

11. MODEL The height of a fish t seconds after it is thrown to a dolphin from a 64-foot-tall platform can be modeled by the equation $h(t) = -16t^2 + 48t + 64$, where $h(t)$ is the height of the fish in feet. The graph of the function is shown.
    a. Estimate the t-coordinate at which the height of the fish changes from increasing to decreasing. Describe the meaning in terms of the context of the situation. $t = 1.5$; The fish reaches its maximum height 1.5 seconds after it is thrown.
    b. Describe and interpret the end behavior of $h(t)$ in the context of the situation. See margin.

Identify and estimate the x- and y-values of the extrema. Round to the nearest tenth if necessary. Then use the graphs to describe the end behavior of each function. 12–17. See margin.

12. 13. 14. 15. 16. 17.

Copyright © McGraw-Hill Education

28 Module 1 • Relations and Functions

Answers

5. The relative maxima occur at $x = -3.7$ and $x = 4.5$, and the relative minimum occurs at $x = 0$. The relative maxima at $x = -3.7$ and $x = 4.5$ represents the top of two hills. The relative minimum at $x = 0$ represents a valley between the hills.
6. As $x \to -\infty, y \to -\infty$ and as $x \to \infty, y \to \infty$.
7. As $x \to -\infty, y \to -\infty$ and as $x \to \infty, y \to -\infty$.
8. As $x \to -\infty, y \to \infty$ and as $x \to \infty, y \to -\infty$.
9. As $x \to -\infty, y \to \infty$ and as $x \to \infty, y \to -\infty$.

F.IF.4, F.IF.7c

1 CONCEPTUAL UNDERSTANDING | 2 FLUENCY | 3 APPLICATION

Name ______ Period ______ Date ______

18. BUBBLES The volume of a soap bubble can be estimated by the formula $V = 4\pi r^2$, where r is its radius. The graph shows the function of the bubble's volume. Describe the end behavior of the graph.
As $r \to \infty$, $V \to \infty$.

19. SCIENCE The table shows the density of water at its saturation pressure for various temperatures. Interpret the end behavior of the graph of the function as temperature increases.
As temperature increases, density decreases.

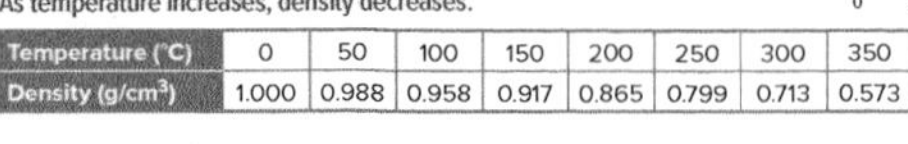

Temperature (°C)	0	50	100	150	200	250	300	350
Density (g/cm³)	1.000	0.988	0.958	0.917	0.865	0.799	0.713	0.573

Identify and estimate the x- and y-values of the extrema. Round to the nearest tenth if necessary. Then describe the end behavior of each function. 20–21. See margin.

20.

21.

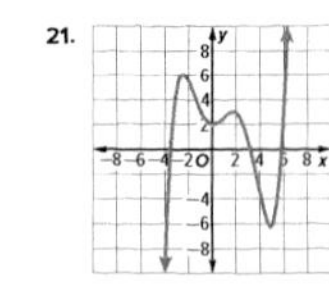

USE ESTIMATION **Use a graphing calculator to estimate the x-coordinates at which any extrema occur for each function. Round to the nearest hundredth.**

22. $f(x) = x^3 + 3x^2 - 6x - 6$
rel. max: $x = -2.73$; rel. min: $x = 0.73$

23. $f(x) = -2x^3 + 8$
no relative max or min

24. $f(x) = -2x^4 + 5x^3 - 4x^2 + 3x - 7$
rel. max: $x = 1.34$; no rel. min

25. $f(x) = x^5 - 4x^3 + 3x^2 - 8x - 6$
rel. max: $x = -1.87$; rel. min: $x = 1.52$

26. CONSTRUCT ARGUMENTS Sheena says that in the graph of $f(x)$ shown below, the graph has relative maxima at B and G, and a relative minimum at A. Is she correct? Explain.

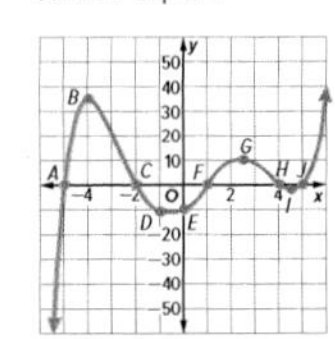

Sheena is partially correct; while B and G are relative maxima, A is not a minimum, whereas I is.

27. CHEMISTRY Dynamic pressure is generated by the velocity of a moving fluid and is given by $q(v) = \frac{1}{2}pv^2$, where p is the density of the fluid and v is its velocity. Water has a density of 1 g/cm³. What happens to the dynamic pressure of water when the velocity continuously increases?

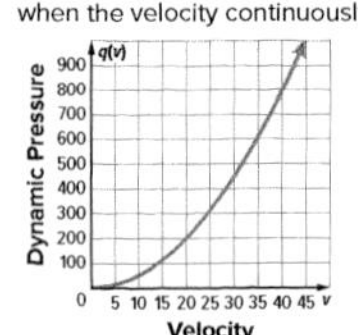

The dynamic pressure would approach ∞.

28. ENGINEERING Several engineering students built a catapult for a class project. They tested the catapult by launching a watermelon and modeled the height h of the watermelon in feet over time t in seconds.
   a. Considering the context of the problem, what is an appropriate domain for $h(t)$? Explain your reasoning. See margin.
   b. Use the graph of $h(t)$ to find the maximum height of the watermelon. When does the watermelon reach the maximum height? Explain your reasoning. See margin.

29. DRILLING The volume of a drill bit can be estimated by the formula for a cone, $V = \frac{1}{3}\pi hr^2$, where h is the height of the bit and r is its radius. Substituting $\frac{\sqrt{3}}{3}r$ for h, the volume of the drill bit is estimated as $\frac{\sqrt{3}}{9}\pi r^3$. The graph shows the function of drill bit volume. Describe the end behavior. as $r \to \infty$, $V \to \infty$

30. The table shows the values of a function. Use the table to describe the end behavior of the function.
as $x \to -\infty$, $y \to -\infty$ and as $x \to \infty$, $y \to \infty$

x	y
−1000	−1,001,000,000
−100	−1,010,000
−10	−1100
−1	−2
1	0
10	900
100	990,000
1000	999,000,000

Higher-Order Thinking Skills

31. WRITE Describe what the end behavior of a graph is and how it is determined. See margin.

32. CREATE Sketch a graph of a linear function and a nonlinear function with the following end behavior: as $x \to -\infty$, $f(x) \to \infty$ and as $x \to \infty$, $f(x) \to -\infty$. See margin.

33. ANALYZE A catalyst is used to increase the rate of a chemical reaction. The reaction rate, or the speed at which the reaction is occurring, is given by $R(x) = \frac{0.5x}{x + 10}$, where x is the concentration of the catalyst solution in milligrams of solute per liter. What does the end behavior of the graph mean in the context of this experiment?
As the concentration of the catalyst is increased, the reaction rate approaches 0.5.

34. PERSEVERE Sketch a graph with the following characteristics: See margin.
   - 2 relative maxima
   - 2 relative minima
   - end behavior: $x \to \infty$, $f(x) \to \infty$ and as $x \to -\infty$, $f(x) \to -\infty$

35. FIND THE ERROR Joshua states that the end behavior of the graph is: as $x \to -\infty$, $f(x) \to -\infty$ and as $x \to \infty$, $f(x) \to \infty$. What error did he make? Joshua switched the $f(x)$ values. He read the graph from right to left instead of left to right.

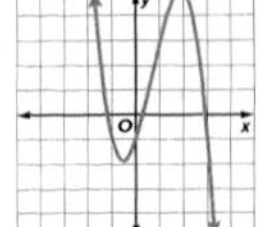

Answers

10. Because the roller coaster cannot travel a negative distance from the starting point, the function is not defined for $x < 0$. So, there is no end behavior as $x \to -\infty$. As $x \to \infty$, $f(x) \to 0$.

11b. As $t \to -\infty$, $h(t) \to -\infty$ and as $t \to \infty$, $h(t) \to -\infty$; The height cannot be negative because we are considering the path of the fish above the surface of the water, $h = 0$. Time cannot be negative because we're measuring from the initial time the fish was thrown at $t = 0$.

12. rel. max. at $x = -2.8$, $y = 4.6$, rel. min. at $x = 0$, $y = -5$; As $x \to -\infty$, $y \to -\infty$ and as $x \to \infty$, $y \to \infty$.

13. rel. max. at $x = 2.7$, $y = 1.7$, rel. min. at $x = -1.2$, $y = -2$; As $x \to -\infty$, $y \to \infty$ and as $x \to \infty$, $y \to -\infty$.

14. rel. max. at $x = \frac{1}{2}$, $y = 4$, rel. min. at $x = 3$, $y = -3$; As $x \to -\infty$, $y \to -\infty$ and as $x \to \infty$, $y \to \infty$.

15. rel. max. at $x = 0$, $y = 2.6$, rel. min. at $x = -4.4$, $y = -4.2$ and $x = 4.2$, $y = -4.2$; As $x \to -\infty$, $y \to \infty$ and as $x \to \infty$, $y \to \infty$.

16. no rel. max, min: $x = -0.9$, $y = -2.1$; As $x \to -\infty$, $y \to \infty$ and as $x \to \infty$, $y \to \infty$.

17. no rel. max, min: $x = 0$, $y = -11$; As $x \to -\infty$, $y \to \infty$ and as $x \to \infty$, $y \to \infty$.

20. rel. max: (0, −3); rel. min: (−2, −5), (2, −5); As $x \to -\infty$, $y \to \infty$ and as $x \to \infty$, $y \to \infty$.

21. rel. max: (−2.8, 6), (1.8, 3); rel. min: (0, 2), (5, −6); As $x \to -\infty$, $y \to -\infty$ and as $x \to \infty$, $y \to \infty$.

28a. $0 \le t \le 2.5$; an appropriate domain is all times from the launch to when the watermelon hits the ground or 0 seconds to 2.5 seconds.

28b. 25 feet at 1.25 seconds; the graph of $h(t)$ has a maximum at (1.25, 25), which represents the time at which the watermelon reaches it maximum height.

31. Sample answer: The end behavior of a graph describes the output values as the input values approach negative and positive infinity. It can be determined by examining the graph.

32. Sample answer:

linear:

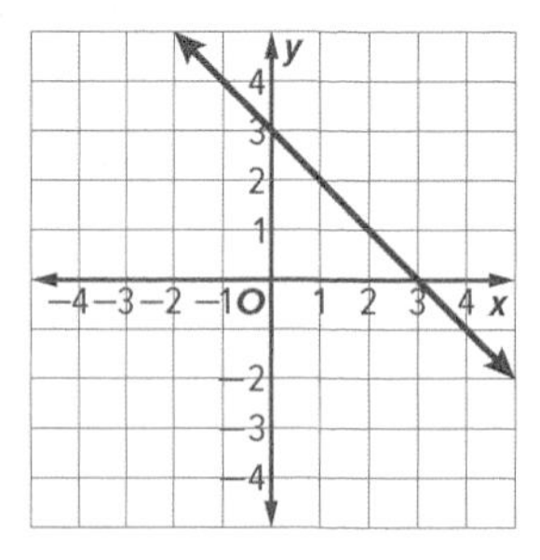

nonlinear:

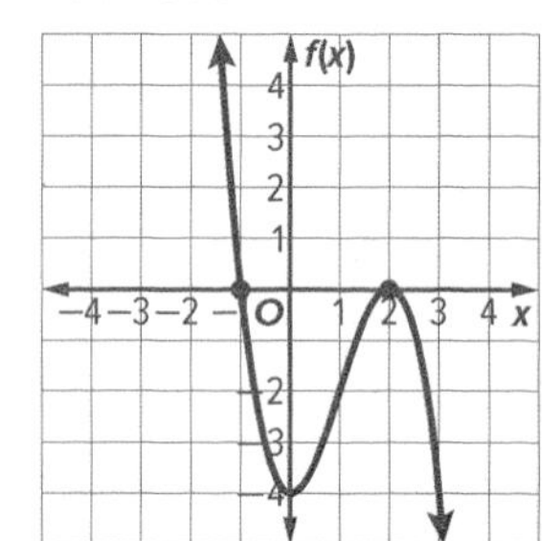

34. Sample answer:

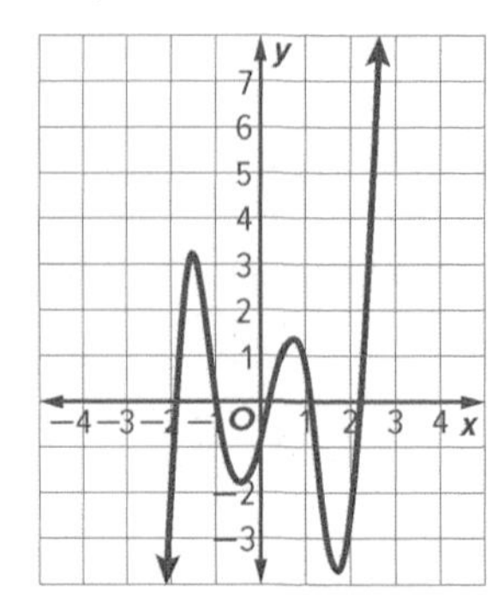

Sketching Graphs and Comparing Functions

LESSON GOAL

Students sketch graphs of functions and compare two functions represented in different ways.

1 LAUNCH

Launch the lesson with a **Warm Up** and an introduction.

2 EXPLORE AND DEVELOP

Explore: Using Technology to Examine Key Features of Graphs

Develop:

Sketching Graphs of Functions
- Sketch a Linear Function
- Sketch a Nonlinear Function
- Sketch a Real-World Function

Comparing Functions
- Compare Properties of Linear Functions
- Compare Properties of Nonlinear Functions

You may want your students to complete the **Checks** online.

3 REFLECT AND PRACTICE

Exit Ticket

Practice

Formative Assessment Math Probe

DIFFERENTIATE

View reports of student progress on the **Checks** after each example.

Resources	AL	OL	BL	ELL
Remediation: End Behavior of Graphs of Functions	●	●		●
Extension: Turning Points		●	●	●

Language Development Handbook

Assign page 4 of the *Language Development Handbook* to help your students build mathematical language related to sketching the graphs of functions and comparing two functions represented in different ways.

ELL You can use the tips and suggestions on page T4 of the handbook to support students who are building English proficiency.

Suggested Pacing

90 min	0.5 day
45 min	1 day

Focus

Domain: Functions

Standards for Mathematical Content:

F.IF.4 For a function that models a relationship between two quantities, interpret key features of graphs and tables in terms of the quantities, and sketch graphs showing key features given in a verbal description of the relationship.

F.IF.9 Compare properties of two functions each represented in a different way (algebraically, graphically, numerically in tables, or by verbal descriptions).

Standards for Mathematical Practice:

1 Make sense of problems and persevere in solving them.

6 Attend to precision.

Coherence

Vertical Alignment

Previous
Students determined the linearity, intercepts, and symmetry of functions.
F.IF.4, F.IF.7c

Now
Students sketch graphs of functions and compare two functions represented in different ways.
F.IF.4, F.IF.9

Next
Students will graph linear functions and inequalities in two variables.
A.CED.3, F.IF.4

Rigor

The Three Pillars of Rigor

1 CONCEPTUAL UNDERSTANDING	2 FLUENCY	3 APPLICATION

Conceptual Bridge In this lesson, students expand on their understanding of graphs of functions and build fluency by using key features to sketch graphs. They apply their understanding by solving real-world problems that require them to compare and interpret the key features of graphs.

Interactive Presentation

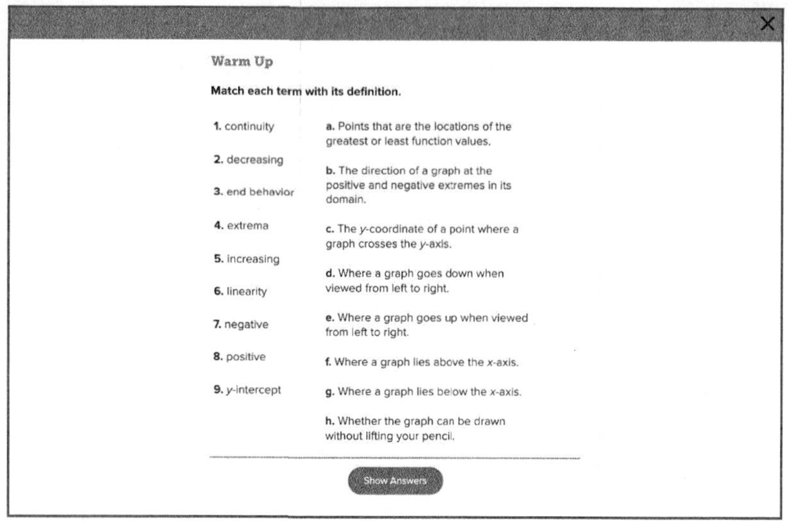

Warm Up

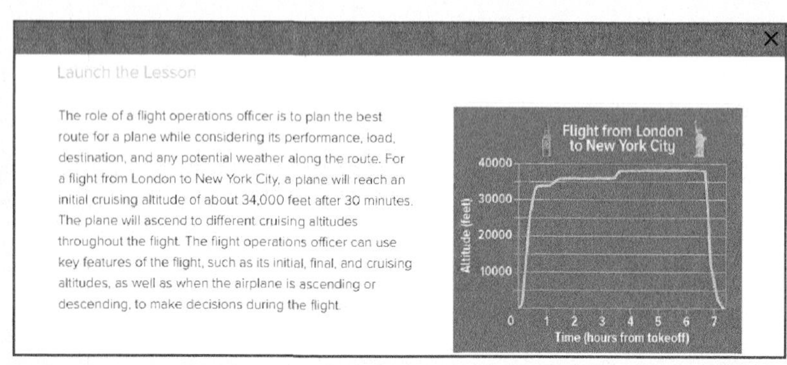

Launch the Lesson

Warm Up

Prerequisite Skills

The Warm Up exercises address prerequisite skill for this lesson.

- key features of a function

Answers:

1. h
2. d
3. b
4. a
5. e
6. i
7. g
8. f
9. c

Launch the Lesson

 Teaching the Mathematical Practices

1 Explain Correspondences Encourage students to explain the relationship between the graph and flight plan in the given situation.

Go Online to find additional teaching notes and questions to promote classroom discourse.

Today's Standards

Tell students that they will be addressing these content and practice standards in this lesson. You may wish to have a student volunteer read aloud *How can I meet these standards?* and *How can I use these practices?*, and connect these to the standards.

See the Interactive Presentation for I Can statements that align with the standards covered in this lesson.

Mathematical Background

Graphs can be sketched by creating a function table. Choose values for x and find the corresponding y-values. Consider creating an input and output table to create points to sketch graphs.

Explore Using Technology to Examine Key Features of Graphs

Objective

Students use a graphing calculator to explore the key features of functions.

 Teaching the Mathematical Practices

5 Analyze Graphs Help students analyze the key features of the graph they have generated using graphing calculators. Point out that to see the entire graph, students may need to adjust the viewing window.

Ideas for Use

Recommended Use Present the Inquiry Question, or have a student volunteer read it aloud. Have students work in pairs to complete the Explore activity on their devices. Pairs should discuss each of the questions. Monitor student progress during the activity. Upon completion of the Explore activity, have student volunteers share their responses to the Inquiry Question.

What if my students don't have devices? You may choose to project the activity on a whiteboard. A printable worksheet for each Explore is available online. You may choose to print the worksheet so that individuals or pairs of students can use it to record their observations.

Summary of the Activity

Students will complete guiding exercises throughout the Explore activity. Students will use a graphing calculator to investigate the key features of the graph of a function. Then students will answer questions about the key features. Then, students will answer the Inquiry Question.

Go Online to find additional teaching notes and sample answers for the guiding exercises.

(continued on the next page)

Interactive Presentation

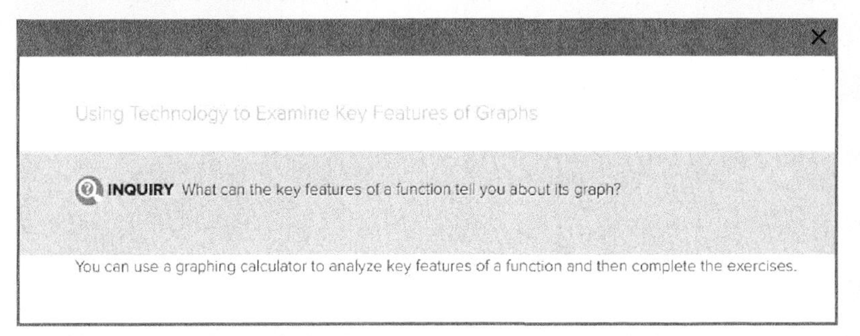

Explore

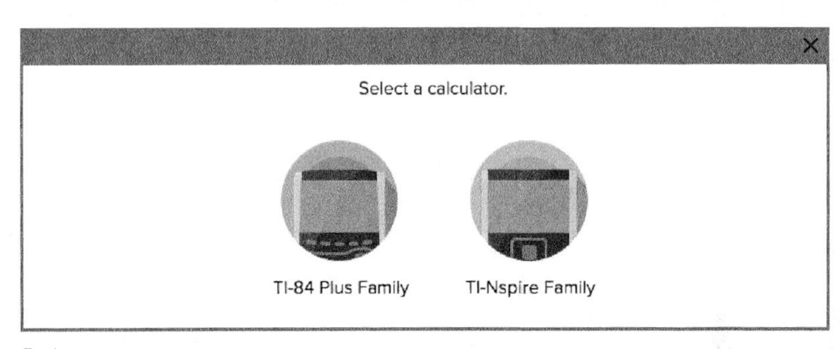

Explore

TAP

Students move through the slides to see how to find the key features of a graph using a graphing calculator.

TYPE

Students answer questions regarding the key features of the graph.

Interactive Presentation

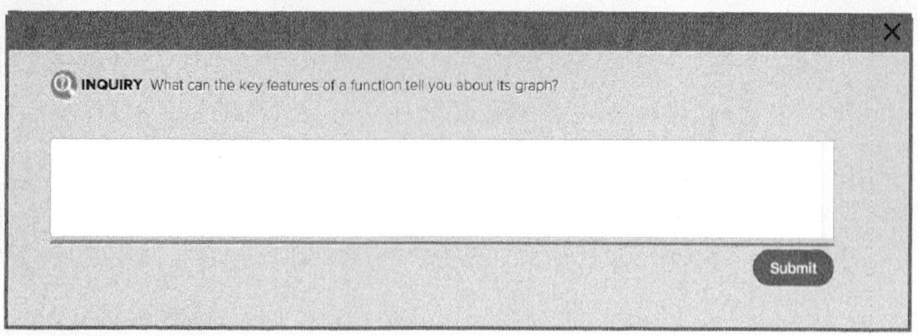

Explore

TYPE

a|

Students respond to the Inquiry Question and can view a sample answer.

Explore Using Technology to Examine Key Features of Graphs (*continued*)

Questions

Have students complete the Explore activity.

Ask:

- Why is a graphing calculator a useful tool when finding key features of functions? Sample answer: A graphing calculator can quickly graph more complex functions and find better estimations of key features.
- Besides extrema, end behavior, domain and range, and intercepts, what is another key feature for a function? Sample answer: intervals where the function is increasing, decreasing, positive, negative

Inquiry

What can the key features of a function tell you about its graph? Sample answer: Key features can tell you where the graph will have high points and low points, where and how many times it will intersect the axes, and for what values of x and y the graph exists. They can also indicate the behavior of the function on the extreme left and right sides as x increases or decreases, as well as how it behaves between values of x.

Go Online to find additional teaching notes and sample answers for the guiding exercises.

Learn Sketching Graphs of Functions

Objective

Students sketch graphs of functions by using key features.

Teaching the Mathematical Practices

1 Explain Correspondences Encourage students to explain the relationships between the key features and how they are represented on graphs.

Common Misconception

Some students may draw an extremely rough sketch of the graph without any degree of precision. Remind students that *sketch* does not mean a graph can lack identifiable information. Instead *sketch* means to graph a function clearly indicating the key features.

Essential Question Follow-Up

Students have begun learning about the key features of functions.

Ask:

Why are the key features of a function important when graphing? Sample answer: A graph gives information about a function, so clearly showing the key features makes the information more accurate and applicable.

Example 1 Sketch a Linear Function

Teaching the Mathematical Practices

6 Communicate Precisely Encourage students to routinely write or explain their solution methods. Point out that they should use clear definitions when they answer the question in the Talk About It! feature.

Questions for Mathematical Discourse

AL How do you find the x-intercept for this function? Sample answer: An x-intercept is a point on the function where $y = 0$. Because you know the function is a line and is positive for $x < -30$, this tells you that the function crosses the x-axis at -30.

OL For a linear function that is decreasing for all values of x, what do you know about the function? The slope of the line is negative.

BL Explain how to write an equation for this function given the key features. Sample answer: the y-intercept is $(0, -70)$, and the x-intercept would be $(-30, 0)$. Because the function is linear, I can use these two points to write the equation in slope intercept form. The equation is $y = -\frac{7}{3}x - 70$.

Common Error

When drawing a graph, many students begin with a standard 10 by 10 grid. In this case, the scale will not accommodate the key features of the graph. Encourage students to choose a scale that will incorporate the given key features before drawing a grid.

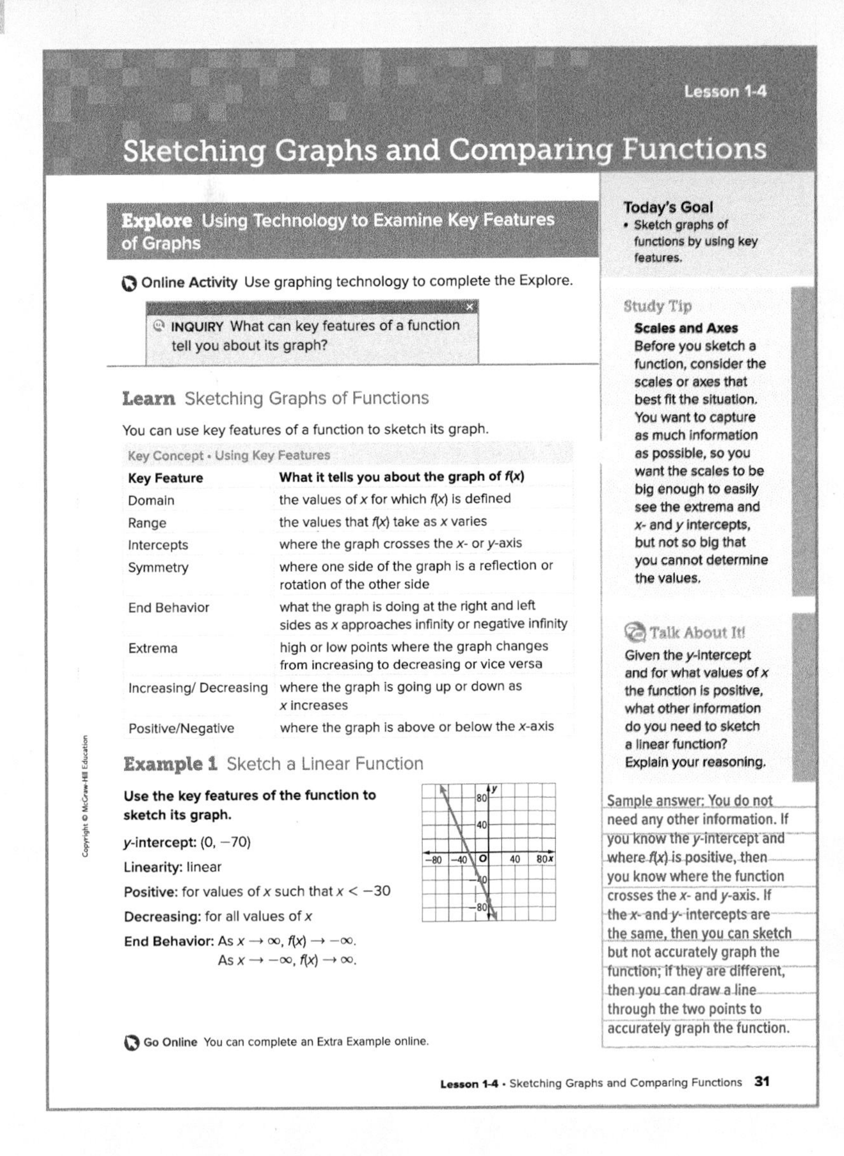

Lesson 1-4

Sketching Graphs and Comparing Functions

Explore Using Technology to Examine Key Features of Graphs

Online Activity Use graphing technology to complete the Explore.

INQUIRY What can key features of a function tell you about its graph?

Learn Sketching Graphs of Functions

You can use key features of a function to sketch its graph.

Key Concept • Using Key Features

Key Feature	What it tells you about the graph of $f(x)$
Domain	the values of x for which $f(x)$ is defined
Range	the values that $f(x)$ take as x varies
Intercepts	where the graph crosses the x- or y-axis
Symmetry	where one side of the graph is a reflection or rotation of the other side
End Behavior	what the graph is doing at the right and left sides as x approaches infinity or negative infinity
Extrema	high or low points where the graph changes from increasing to decreasing or vice versa
Increasing/ Decreasing	where the graph is going up or down as x increases
Positive/Negative	where the graph is above or below the x-axis

Example 1 Sketch a Linear Function

Use the key features of the function to sketch its graph.

y-intercept: $(0, -70)$

Linearity: linear

Positive: for values of x such that $x < -30$

Decreasing: for all values of x

End Behavior: As $x \to \infty$, $f(x) \to -\infty$.
As $x \to -\infty$, $f(x) \to \infty$.

Go Online You can complete an Extra Example online.

Today's Goal
- Sketch graphs of functions by using key features.

Study Tip

Scales and Axes Before you sketch a function, consider the scales or axes that best fit the situation. You want to capture as much information as possible, so you want the scales to be big enough to easily see the extrema and x- and y intercepts, but not so big that you cannot determine the values.

Talk About It!

Given the y-intercept and for what values of x the function is positive, what other information do you need to sketch a linear function? Explain your reasoning.

Sample answer: You do not need any other information. If you know the y-intercept and where $f(x)$ is positive, then you know where the function crosses the x- and y-axis. If the x- and y-intercepts are the same, then you can sketch but not accurately graph the function; if they are different, then you can draw a line through the two points to accurately graph the function.

Lesson 1-4 • Sketching Graphs and Comparing Functions 31

Interactive Presentation

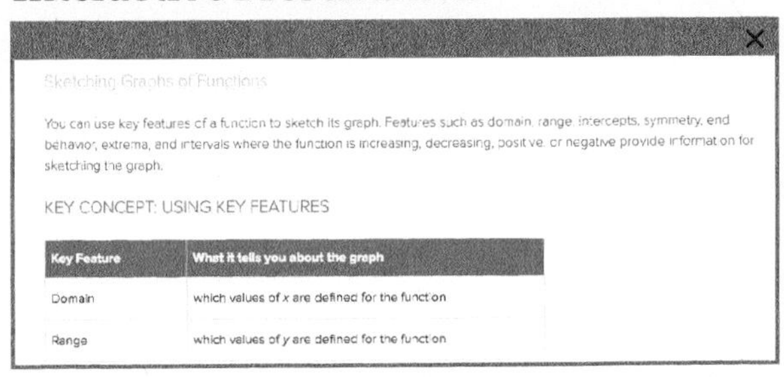

Sketching Graphs of Functions
KEY CONCEPT: USING KEY FEATURES
Key Feature
What it tells you about the graph
Domain
which values of x are defined for the function
Range
which values of y are defined for the function

Learn

TAP

Students move through the slides to sketch a graph given the key features.

Your Notes

Study Tip

Assumptions When sketching the function using the given key features, assumptions must be made. As in this example, the same key features could describe many different graphs. The key features could also be represented by a parabola, a curve that is narrower or wider, or an absolute value function.

Example 2 Sketch a Nonlinear Function

Use the key features of the function to sketch its graph.

y-intercept: (0, 3)

Linearity: nonlinear

Continuity: continuous

Positive: for all values of x

Decreasing: for all values of x such that $x < 0$

Extrema: minimum at (0, 3)

End Behavior: As $x \to \infty$, $f(x) \to \infty$. As $x \to -\infty$, $f(x) \to \infty$.

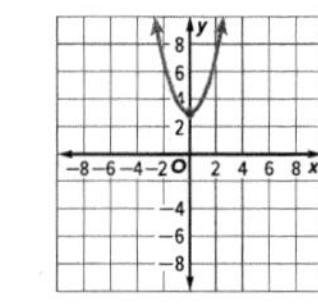

Think About It!

Explain why the end behavior is not defined in the context of this situation.

Sample answer: Because Hae cannot have a negative speed or time, the graph cannot extend into other quadrants.

Think About It!

Based on the graph, the speed of the car at 10 seconds is 40 miles per hour. Is it appropriate to assume that the car is traveling that exact speed at a specific time? Explain.

No; sample answer: Because the graph only represents an approximation of the speed of the car over time, it is not possible to determine a specific speed at a specific time.

Example 3 Sketch a Real-World Function

TEST DRIVE **Hae is test driving a car she is thinking of buying. She decides to accelerate to 60 miles per hour and then decelerate to a stop to test its acceleration and brakes. It takes her 15 seconds to reach her maximum speed and 15 additional seconds to come to a stop. Use the key features to sketch a graph that shows the speed y as a function of time x.**

y-intercept: Hae starts her test drive at a speed of 0 miles per hour.

Linear or Nonlinear: The function that models the situation is nonlinear.

Extrema: Hae's maximum speed is 60 miles per hour, which she reaches 15 seconds into her test drive.

Increasing: Hae increases the speed at a uniform rate for the first 15 seconds.

Decreasing: Hae decreases the speed at a uniform rate for the next 15 seconds until she reaches a stop.

End Behavior: Because Hae starts at 0 miles per hour and ends at 0 miles per hour, there is no end behavior.

Before sketching, consider the constraints of the situation. Hae cannot drive a negative speed or for a negative amount of time. Therefore, the graph only exists for positive x- and y-values.

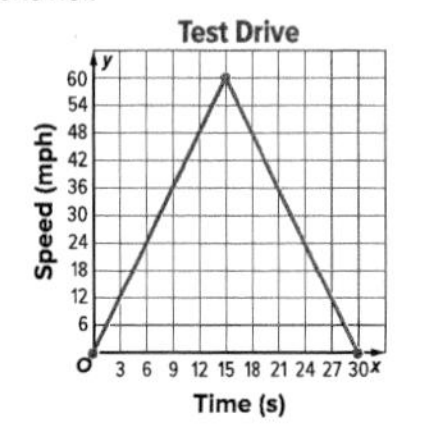

 Go Online You can complete an Extra Example online.

32 Module 1 • Relations and Functions

DIFFERENTIATE

Reteaching Activity AL ELL

IF students are struggling to identify the key features of a function, **THEN** have them sketch the graph of the function $y = x^2 - 4$. To represent domain, have them highlight the x-axis in one color and then write the domain in the same color. For range, have students highlight the y-axis in a different color and then write the range in the same color. Students can place a star on the x- and y-intercept, and then circle the minimum point, or they can draw arrows pointing downward along the decreasing intervals and arrows pointing upward on the increasing intervals.

Interactive Presentation

Sketch a Nonlinear Function

Use the key features of the function to sketch its graph.

Hide Key Features

y-Intercept: (0, 3)
Linearity: nonlinear
Continuity: continuous
Positive: for all values of x
Decreasing: for values of x such that $x < 0$
Extrema: relative minimum at (0, 3)
End Behavior: As $x \to \infty, f(x) \to \infty$. As $x \to -\infty, f(x) \to \infty$.

Example 2

TAP

Students move through the slides to sketch a graph using the key features.

CHECK

Students complete the Check online to determine whether they are ready to move on.

Example 2 Sketch a Nonlinear Function

Teaching the Mathematical Practices

1 Understand Different Approaches Use the Study Tip feature to encourage students to consider other graphs that may be described by the given key features.

Questions for Mathematical Discourse

AL What makes a function nonlinear? All the points on the graph do not lie in a line; the graph can be curved.

OL Could the graph have an x-intercept? Explain your reasoning. No; sample answer: The graph has a minimum at (0, 3). So, the y-value is never 0.

BL It is given that the function is decreasing for all values of x such that $x < 0$. Does this mean that the function must be increasing for $x \geq 0$? No; sample answer: y-value could be constant over a finite range of x-values and still meet the conditions of the key features given.

Example 3 Sketch a Real-World Function

Teaching the Mathematical Practices

4 Use Tools In Example 3, students will need to identify important quantities in the problem and use a graph to map the relationship between speed and time.

Questions for Mathematical Discourse

AL What is the minimum value of the function representing the speed of the car at a particular time? 0

OL What does "uniform rate" mean? The speed is changing at a constant rate, which indicates a linear relationship.

BL Suppose during the test drive, Hae drove a constant speed for 5 seconds. How would this be represented on the graph? This would be represented by a horizontal line over a 5-second interval of x-values.

1 CONCEPTUAL UNDERSTANDING | **2 FLUENCY** | 3 APPLICATION

Example 4 Compare Properties of Linear Functions

 **Teaching the Mathematical Practices**

6 Communicate Precisely Encourage students to use clear definitions and mathematical terms when they answer the question in the Think About It! feature.

Questions for Mathematical Discourse

AL What are two key features that can be used to compare two functions? Sample answer: domain, range, intercepts, extrema, increasing/decreasing intervals, positive/negative intervals, symmetry, end behavior

OL How does the end behavior of the two functions compare? Sample answer: The end behavior is the same for each function. As $x \rightarrow \infty$, $f(x) \rightarrow \infty$, and as $x \rightarrow -\infty$, $f(x) \rightarrow -\infty$.

BL Suppose that $h(x)$ increases at the same rate as $g(x)$ and has a y-intercept of 2 and an x-intercept of −1. How can you describe the relationship between $g(x)$ and $h(x)$? They are parallel.

Go Online

- Find additional teaching notes.
- View performance reports of the Checks.
- Assign or present an Extra Example.

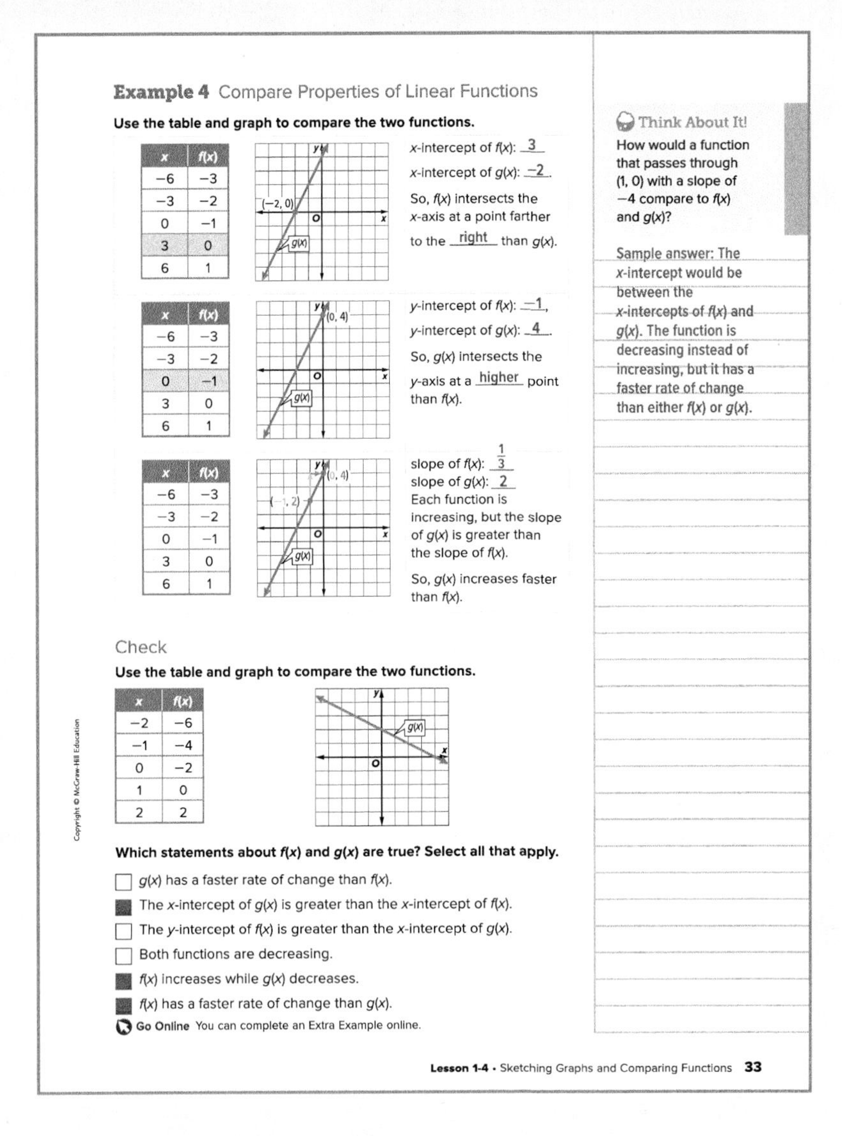

Example 4 Compare Properties of Linear Functions

Use the table and graph to compare the two functions.

x	f(x)
−6	−3
−3	−2
0	−1
3	0
6	1

x-intercept of $f(x)$: 3
x-intercept of $g(x)$: −2
So, $f(x)$ intersects the x-axis at a point farther to the right than $g(x)$.

x	f(x)
−6	−3
−3	−2
0	−1
3	0
6	1

y-intercept of $f(x)$: −1
y-intercept of $g(x)$: 4
So, $g(x)$ intersects the y-axis at a higher point than $f(x)$.

x	f(x)
−6	−3
−3	−2
0	−1
3	0
6	1

slope of $f(x)$: $\frac{1}{3}$
slope of $g(x)$: 2
Each function is increasing, but the slope of $g(x)$ is greater than the slope of $f(x)$.
So, $g(x)$ increases faster than $f(x)$.

Think About It!
How would a function that passes through (1, 0) with a slope of −4 compare to $f(x)$ and $g(x)$?

Sample answer: The x-intercept would be between the x-intercepts of $f(x)$ and $g(x)$. The function is decreasing instead of increasing, but it has a faster rate of change than either $f(x)$ or $g(x)$.

Check

Use the table and graph to compare the two functions.

x	f(x)
−2	−6
−1	−4
0	−2
1	0
2	2

Which statements about $f(x)$ and $g(x)$ are true? Select all that apply.

- ☐ $g(x)$ has a faster rate of change than $f(x)$.
- ■ The x-intercept of $g(x)$ is greater than the x-intercept of $f(x)$.
- ☐ The y-intercept of $f(x)$ is greater than the x-intercept of $g(x)$.
- ☐ Both functions are decreasing.
- ■ $f(x)$ increases while $g(x)$ decreases.
- ■ $f(x)$ has a faster rate of change than $g(x)$.

Go Online You can complete an Extra Example online.

Lesson 1-4 • Sketching Graphs and Comparing Functions 33

Interactive Presentation

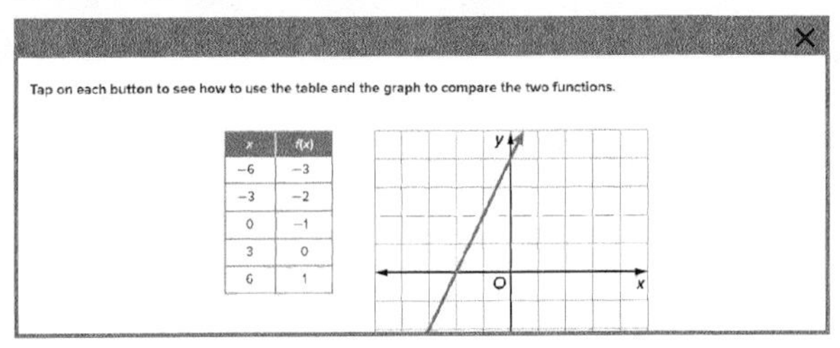

x	f(x)
−6	−3
−3	−2
0	−1
3	0
6	1

Example 4

TAP

Students tap each button to compare key features of the table and graph.

TYPE

Students describe how a function passing through a given point with a given slope compares to the two functions in the problem.

Example 5 Compare Properties of Nonlinear Functions

Examine the categories to see how to use the description and the graph to identify key features of each function. Then complete the statements to compare the two functions.

$f(x)$	$g(x)$
x-intercept: $(-3.4, 0)$ y-intercept: $(0, 1.5)$ relative maximum: $(-2.3, 4.7)$ relative minimum: $(-0.4, 1.1)$ end behavior: as $x \to -\infty$, $f(x) \to -\infty$ and as $x \to \infty$, $f(x) \to \infty$	[graph of $g(x)$]

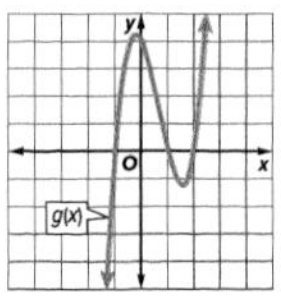

x-intercept(s)	
$f(x)$ intersects the x-axis once at $(-3.4, 0)$.	$g(x)$ intersects the x-axis three times at $(-1, 0)$, $(1, 0)$, and $(2, 0)$.
y-intercept	
$f(x)$ intersects the y-axis at $(0, 1.5)$.	$g(x)$ intersects the y-axis at $(0, 4)$.
Extrema	
$f(x)$ has a relative maximum of 4.7 and a relative minimum of 1.1.	$g(x)$ has a relative maximum of about 4.2 and a relative minimum of about -1.2.
End Behavior	
As $x \to -\infty$, $f(x) \to -\infty$, and as $x \to \infty$, $f(x) \to \infty$.	As $x \to -\infty$, $g(x) \to -\infty$, and as $x \to \infty$, $g(x) \to \infty$.

- The x-intercept of $f(x)$ is <u>less than</u> any of the x-intercepts of $g(x)$.
- The graph of $g(x)$ intersects the x-axis <u>more</u> times than $f(x)$.
- The y-intercept of $f(x)$ is <u>less than</u> the y-intercept of $g(x)$.
- So, $g(x)$ intersects the y-axis at a <u>higher</u> point than $f(x)$.
- The relative maximum of $f(x)$ is <u>greater than</u> the relative maximum of $g(x)$. The relative minimum of $f(x)$ is <u>greater than</u> the relative minimum of $g(x)$.
- The two functions have <u>the same</u> end behavior.

Go Online You can complete an Extra Example online.

Interactive Presentation

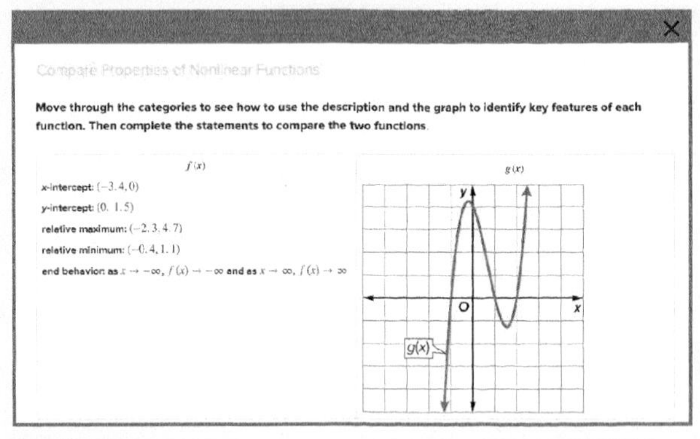

Example 5

TAP

Students move through the categories to see how to compare two functions.

SELECT

Students select the correct word or phrase to compare key features of the two functions.

CHECK

Students complete the Check online to determine whether they are ready to move on.

Example 5 Compare Properties of Nonlinear Functions

Teaching the Mathematical Practices

1 Explain Correspondences Guide students to use the relationships between the verbal description and graph used in this example to compare the nonlinear functions.

Questions for Mathematical Discourse

AL How can you determine where $f(x)$ is increasing and where it is decreasing? The relative maximum and relative minimum indicate points where the graph changes direction. The end behavior indicates that the function is increasing from $-\infty$ until it reaches the relative maximum at $(-2.3, 4.7)$. From that point it decreases to the relative minimum of $(-0.4, 1.1)$, and then it increases again.

OL What additional information is needed to determine the domain and range of $f(x)$? We need to know whether $f(x)$ is continuous for all values of x. If it is, then the functions would have the same domain and range, all real numbers, which is indicated by the end behavior of both functions.

BL If $h(x) = f(x) - 1.1$, how many x-intercepts would $h(x)$ have? $h(x)$ would have 2 x-intercepts.

If $j(x) = g(x) + 1.2$, how many x-intercepts would $j(x)$ have? We cannot determine the number of x-intercepts for $j(x)$ exactly because the relative minimum for $g(x)$ is estimated from the graph.

Common Error

When presented with a verbal description and a graph, students may find it difficult to compare the two nonlinear functions. Encourage students to either write a similar verbal description for the graph, or graph the verbal description given. Then both functions will be expressed in the same format, and comparison may be easier.

Exit Ticket

Recommended Use

At the end of class, go online to display the Exit Ticket prompt and ask students to respond using a separate piece of paper. Have students hand you their responses as they leave the room.

Alternate Use

At the end of class, go online to display the Exit Ticket prompt and ask students to respond verbally or by using a mini-whiteboard. Have students hold up their whiteboards so that you can see all student responses. Tap to reveal the answer when most or all students have completed the Exit Ticket.

Practice and Homework

Suggested Assignments

Use the table below to select appropriate exercises.

DOK	Topic	Exercises
1, 2	exercises that mirror the examples	1–16
2	exercises that use a variety of skills from this lesson	17–22
3	exercises that emphasize higher-order and critical thinking skills	23–27

ASSESS AND DIFFERENTIATE

Use the data from the **Checks** to determine whether to provide resources for extension, remediation, or intervention.

IF students score 90% or more on the Checks, **THEN** assign: BL

- Practice, Exercises 1–21 odd, 23–27
- Extension: Turning Points
- **ALEKS**® Graphs of Functions; Transformations

IF students score 66%–89% on the Checks, **THEN** assign: OL

- Practice, Exercises 1–27 odd
- Remediation, Review Resources: Shapes of Graphs
- Personal Tutors
- Extra Examples 1–5
- **ALEKS**® Interpreting Graphs: Extrema and End Behavior

IF students score 65% or less on the Checks, **THEN** assign: AL

- Practice, Exercises 1–15 odd
- Remediation, Review Resources: Shapes of Graphs
- **ALEKS**® Interpreting Graphs: Extrema and End Behavior

Answers

1.

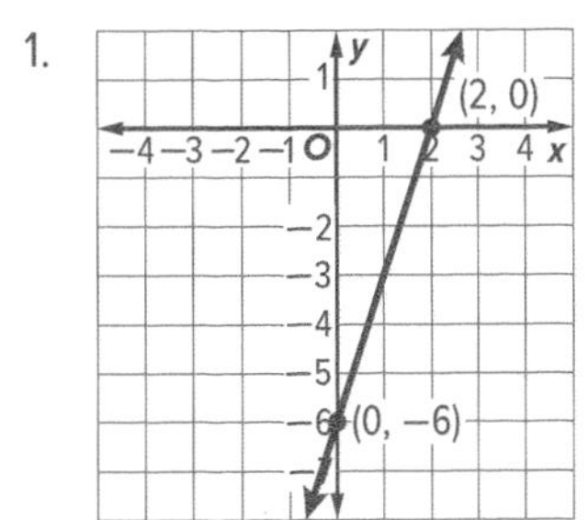

2.

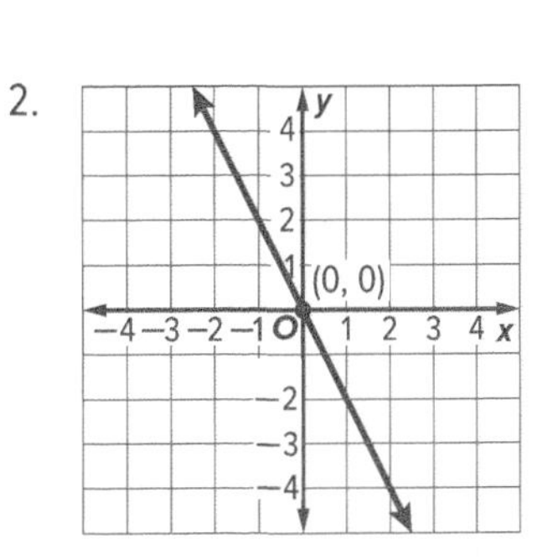

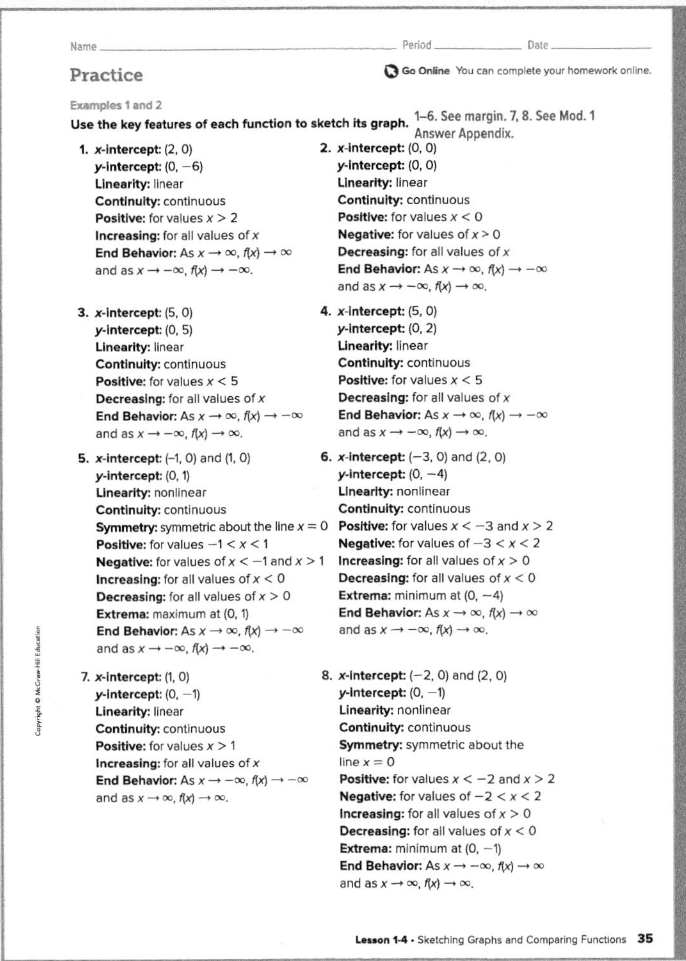

Name ______ Period ______ Date ______

Practice

Go Online You can complete your homework online.

Examples 1 and 2

Use the key features of each function to sketch its graph. 1–6. See margin. 7, 8. See Mod. 1 Answer Appendix.

1. **x-intercept:** (2, 0)
   y-intercept: (0, −6)
   Linearity: linear
   Continuity: continuous
   Positive: for values $x > 2$
   Increasing: for all values of x
   End Behavior: As $x \to \infty$, $f(x) \to \infty$ and as $x \to -\infty$, $f(x) \to -\infty$.

2. **x-intercept:** (0, 0)
   y-intercept: (0, 0)
   Linearity: linear
   Continuity: continuous
   Positive: for values $x < 0$
   Negative: for values of $x > 0$
   Decreasing: for all values of x
   End Behavior: As $x \to \infty$, $f(x) \to -\infty$ and as $x \to -\infty$, $f(x) \to \infty$.

3. **x-intercept:** (5, 0)
   y-intercept: (0, 5)
   Linearity: linear
   Continuity: continuous
   Positive: for values $x < 5$
   Decreasing: for all values of x
   End Behavior: As $x \to \infty$, $f(x) \to -\infty$ and as $x \to -\infty$, $f(x) \to \infty$.

4. **x-intercept:** (5, 0)
   y-intercept: (0, 2)
   Linearity: linear
   Continuity: continuous
   Positive: for values $x < 5$
   Decreasing: for all values of x
   End Behavior: As $x \to \infty$, $f(x) \to -\infty$ and as $x \to -\infty$, $f(x) \to \infty$.

5. **x-intercept:** (−1, 0) and (1, 0)
   y-intercept: (0, 1)
   Linearity: nonlinear
   Continuity: continuous
   Symmetry: symmetric about the line $x = 0$
   Positive: for values $-1 < x < 1$
   Negative: for values of $x < -1$ and $x > 1$
   Increasing: for all values of $x < 0$
   Decreasing: for all values of $x > 0$
   Extrema: maximum at (0, 1)
   End Behavior: As $x \to \infty$, $f(x) \to -\infty$ and as $x \to -\infty$, $f(x) \to -\infty$.

6. **x-intercept:** (−3, 0) and (2, 0)
   y-intercept: (0, −4)
   Linearity: nonlinear
   Continuity: continuous
   Positive: for values $x < -3$ and $x > 2$
   Negative: for values of $-3 < x < 2$
   Increasing: for all values of $x > 0$
   Decreasing: for all values of $x < 0$
   Extrema: minimum at (0, −4)
   End Behavior: As $x \to \infty$, $f(x) \to \infty$ and as $x \to -\infty$, $f(x) \to \infty$.

7. **x-intercept:** (1, 0)
   y-intercept: (0, −1)
   Linearity: linear
   Continuity: continuous
   Positive: for values $x > 1$
   Increasing: for all values of x
   End Behavior: As $x \to -\infty$, $f(x) \to -\infty$ and as $x \to \infty$, $f(x) \to \infty$.

8. **x-intercept:** (−2, 0) and (2, 0)
   y-intercept: (0, −1)
   Linearity: nonlinear
   Continuity: continuous
   Symmetry: symmetric about the line $x = 0$
   Positive: for values $x < -2$ and $x > 2$
   Negative: for values of $-2 < x < 2$
   Increasing: for all values of $x > 0$
   Decreasing: for all values of $x < 0$
   Extrema: minimum at (0, −1)
   End Behavior: As $x \to -\infty$, $f(x) \to \infty$ and as $x \to \infty$, $f(x) \to \infty$.

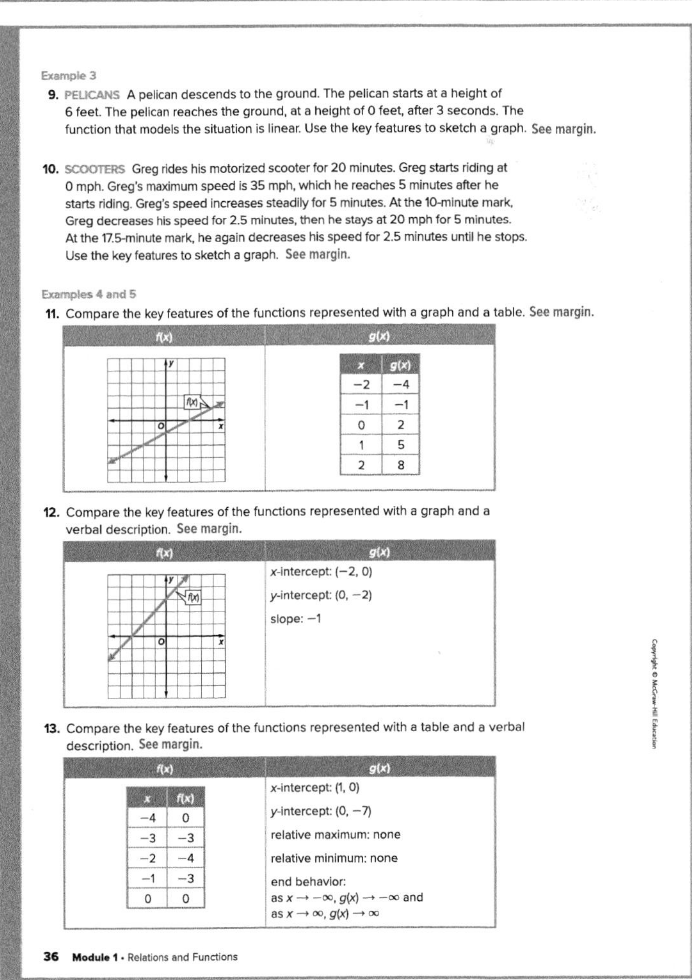

Example 3

9. PELICANS A pelican descends to the ground. The pelican starts at a height of 6 feet. The pelican reaches the ground, at a height of 0 feet, after 3 seconds. The function that models the situation is linear. Use the key features to sketch a graph. See margin.

10. SCOOTERS Greg rides his motorized scooter for 20 minutes. Greg starts riding at 0 mph. Greg's maximum speed is 35 mph, which he reaches 5 minutes after he starts riding. Greg's speed increases steadily for 5 minutes. At the 10-minute mark, Greg decreases his speed for 2.5 minutes, then he stays at 20 mph for 5 minutes. At the 17.5-minute mark, he again decreases his speed for 2.5 minutes until he stops. Use the key features to sketch a graph. See margin.

Examples 4 and 5

11. Compare the key features of the functions represented with a graph and a table. See margin.

$f(x)$: [graph: y, x, O, f(x)]

$g(x)$:

x	$g(x)$
−2	−4
−1	−1
0	2
1	5
2	8

12. Compare the key features of the functions represented with a graph and a verbal description. See margin.

$f(x)$	$g(x)$
[graph: y, x, O, f(x)]	x-intercept: (−2, 0) y-intercept: (0, −2) slope: −1

13. Compare the key features of the functions represented with a table and a verbal description. See margin.

$f(x)$:

x	$f(x)$
−4	0
−3	−3
−2	−4
−1	−3
0	0

$g(x)$:
x-intercept: (1, 0)
y-intercept: (0, −7)
relative maximum: none
relative minimum: none
end behavior: as $x \to -\infty$, $g(x) \to -\infty$ and as $x \to \infty$, $g(x) \to \infty$

Name ______ Period ______ Date ______

14. Compare the key features of the functions represented with a graph and a verbal description. **See margin.**

f(x)

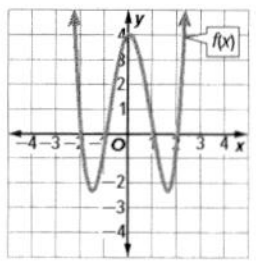

g(x)

x-intercept: (−1, 0), (1, 0), (2, 0)

y-intercept: (0, −4)

relative maximum: (1.37, 0.35)

relative minimum: (−0.37, −4.85), (2, 0)

end behavior:
as $x \to -\infty$, $g(x) \to \infty$ and
as $x \to \infty$, $g(x) \to \infty$

15. Compare the key features of the functions represented with a table and a verbal description. **See Mod. 1 Answer Appendix.**

f(x)

x	$f(x)$
$-\frac{2}{3}$	1
$-\frac{1}{3}$	$\frac{3}{4}$
0	$\frac{1}{2}$
$\frac{1}{3}$	$\frac{1}{4}$
$\frac{2}{3}$	0

g(x)

x-intercept: $\left(\frac{3}{8}, 0\right)$

y-intercept: $\left(0, \frac{1}{2}\right)$

slope: $-\frac{4}{3}$

16. Compare the key features of the functions represented with a graph and a table. **See Mod. 1 Answer Appendix.**

f(x)

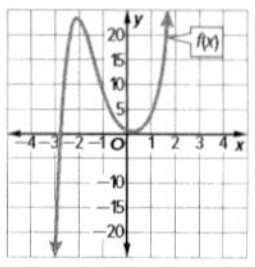

g(x)

x	$g(x)$
−1	7
−0.56	0
0	−3
1.89	0
2	1

Mixed Exercises

17. USE A MODEL Sketch the graph of a linear function with the following key features. The x-intercept is 2. The y-intercept is 2. The function is decreasing for all values of x. The function is positive for $x < 2$. As $x \to -\infty$, $f(x) \to \infty$ and as $x \to \infty$, $f(x) \to -\infty$. **See Mod. 1 Answer Appendix.**

18. WATER Sia filled a pitcher with water. The pitcher started with 0 ounces of water. After 8 seconds the pitcher contains 64 ounces of water. The function that models the situation is linear.

a. Use the key features to sketch a graph. **See Mod. 1 Answer Appendix.**

b. What is the end behavior of the graph? Explain. **There is no end behavior because the pitcher starts with 0 ounces and ends with 64 ounces.**

19. USE TOOLS Monica walks for 60 minutes. She starts walking from her house. The maximum distance Monica is from her house is 2 miles, which she reaches 30 minutes after she starts walking. At the 30-minute mark, Monica starts walking back to her house for 30 minutes until she reaches her house. Use the key features to sketch a graph. **See Mod. 1 Answer Appendix.**

20. USE A SOURCE Research the value of a new car after it is purchased. Use the information you collect to describe key features of a graph that represents the value of a new car x years after it is purchased. Then use the key features to sketch a graph. **See students' work.**

21. SKI LIFTS A ski lift descends at a steady pace down a mountainside from a height of 1800 feet to ground level. If it makes no stops along the way to load or unload passengers, then the time it takes to complete its descension is 4 minutes. **a-b. See Mod. 1 Answer Appendix.**

a. Is the graph that relates the lift's height as a function of time linear or nonlinear? Explain.

b. Use the key features to sketch a graph.

22. CONSTRUCT ARGUMENTS Keisha babysits for her aunt for an hourly rate of \$9. The graph shows Keisha's earnings y as a function of hours spent babysitting x. Explain why the graph only exists for positive x- and y-values. **See Mod. 1 Answer Appendix.**

Keisha's Earnings

Higher-Order Thinking Skills

23. CREATE Choose a function and create a list of key features to describe the function. Then sketch the function. **See Mod. 1 Answer Appendix.**

24. WRITE Describe the relationship between the slope of a linear function and when the function is increasing or decreasing. **See Mod. 1 Answer Appendix.**

25. ANALYZE Determine whether the statement is *always*, *sometimes*, or *never* true. **See Mod. 1 Answer Appendix.**

A graph that has more than one x-intercept is represented by a nonlinear function.

26. PERSEVERE Deborah filled an empty tub with water for 30 minutes. The maximum amount of water in the tub is 50 gallons, which is reached 10 minutes after Deborah starts filling the tub. The amount of water in the tub increases steadily for 10 minutes. At the 10-minute mark, the amount of water in the tub starts decreasing for 20 minutes until there is no water left in the tub.

a. Use the key features to sketch a graph. **See Mod. 1 Answer Appendix.**

b. Describe an event that could have occurred at the 10-minute mark if Deborah continues filling the tub at the same rate from the 10-minute mark to the 30-minute mark as the rate from the 0-minute mark to the 10-minute mark. **See Mod. 1 Answer Appendix.**

27. FIND THE ERROR Linda and Rubio sketched a graph with the following key features. The x-intercept is 2. The y-intercept is −9. The function is positive for $x > 2$. As $x \to -\infty$, $f(x) \to -\infty$ and as $x \to \infty$, $f(x) \to \infty$. Is either graph correct based on the key features? Explain your reasoning. **See Mod. 1 Answer Appendix.**

Linda's Graph

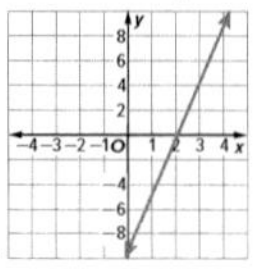

Rubio's Graph

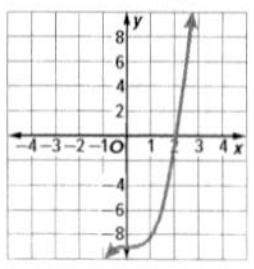

Answers

3.

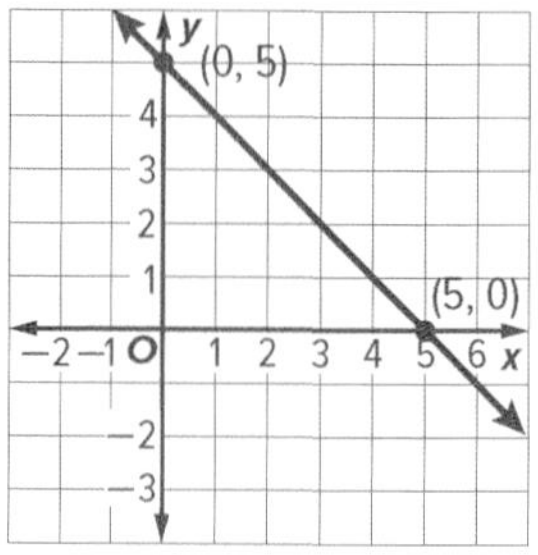

4.

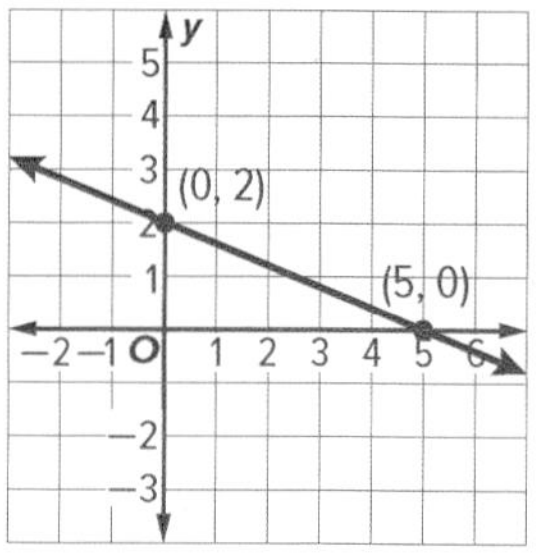

5.

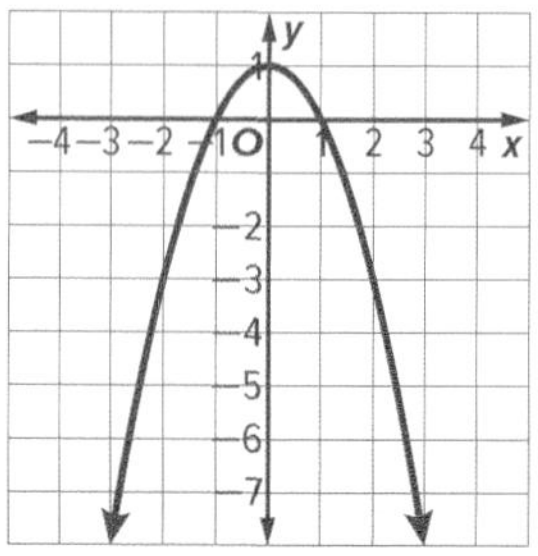

6.

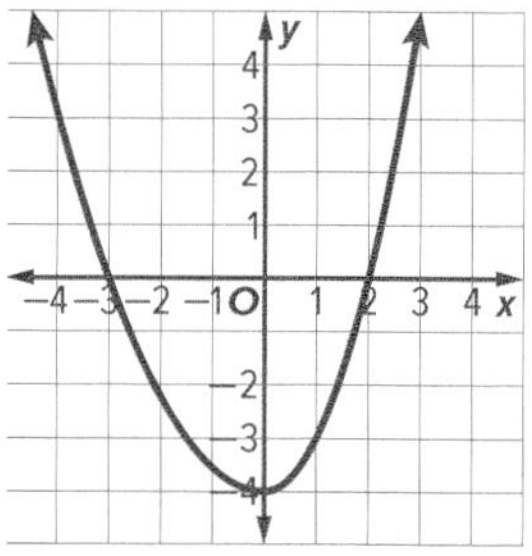

9. **Pelican's Height**

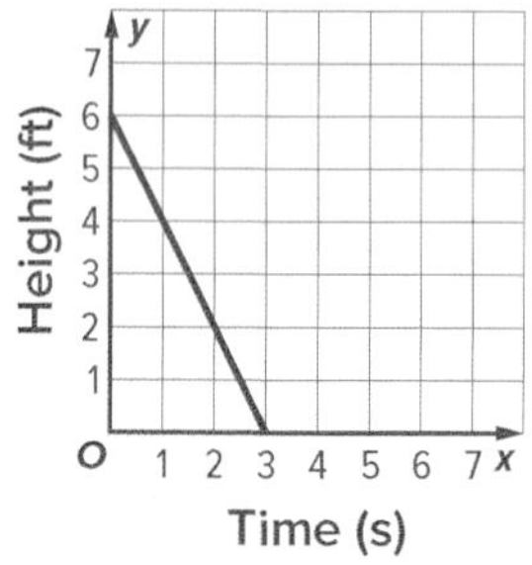

10. **Greg's Ride**

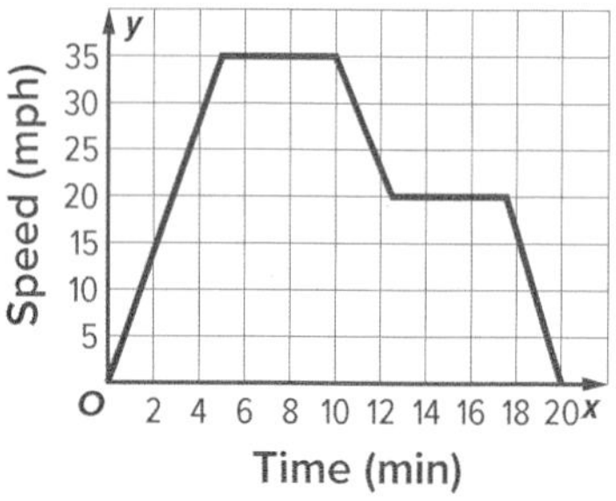

11. The x-intercept of $f(x)$ is 2, and the x-intercept of $g(x)$ is $-\frac{2}{3}$. The x-intercept of $f(x)$ is greater than the x-intercept of $g(x)$. So, $f(x)$ intersects the x-axis at a point farther to the right than $g(x)$. The y-intercept of $f(x)$ is −1, and the y-intercept of $g(x)$ is 2. The y-intercept of $g(x)$ is greater than the y-intercept of $f(x)$. So, $g(x)$ intersects the y-axis at a higher point than $f(x)$. The slope of $f(x)$ is $\frac{1}{2}$ and the slope of $g(x)$ is 3. Each function is increasing, but the slope of $g(x)$ is greater than the slope of $f(x)$. So, $g(x)$ increases faster than $f(x)$.

12. The x-intercept of $f(x)$ is −3, and the x-intercept of $g(x)$ is −2. The x-intercept of $g(x)$ is greater than the x-intercept of $f(x)$. So, $g(x)$ intersects the x-axis at a point farther to the right than $f(x)$. The y-intercept of $f(x)$ is 3, and the y-intercept of $g(x)$ is −2. The y-intercept of $f(x)$ is greater than the y-intercept of $g(x)$. So, $f(x)$ intersects the y-axis at a higher point than $g(x)$. The slope of $f(x)$ is 1 and the slope of $g(x)$ is −1. $f(x)$ is increasing and $g(x)$ is decreasing. $f(x)$ increases at the same rate that $g(x)$ decreases.

13. Both x-intercepts of $f(x)$ are less than the x-intercept of $g(x)$. The graph of $f(x)$ intersects the x-axis more times than $g(x)$. The y-intercept of $g(x)$ is less than the y-intercept of $f(x)$. So, $f(x)$ intersects the y-axis at a higher point than $g(x)$. Neither function has a relative maximum. $f(x)$ has a minimum at (−2, −4). The two functions have the opposite end behaviors as $x \to -\infty$. The two functions have the same end behavior as $x \to \infty$.

14. Both $f(x)$ and $g(x)$ have x-intercepts of −1, 1, and 2. The graph of $f(x)$ intersects the x-axis more times than $g(x)$. The y-intercept of $g(x)$ is less than the y-intercept of $f(x)$. So, $f(x)$ intersects the y-axis at a higher point than $g(x)$. The relative maximum of $f(x)$ is greater than the relative maximum of $g(x)$. One relative minimum of $f(x)$ is between the two relative minima of $g(x)$. The functions have the same end behavior.

Lesson 1-5

A.CED.3, F.IF.4

Graphing Linear Functions and Inequalities

LESSON GOAL

Students graph linear functions and inequalities in two variables.

1 LAUNCH

Launch the lesson with a **Warm Up** and an introduction.

2 EXPLORE AND DEVELOP

Develop:

Graphing Linear Functions
- Graph by Using a Table
- Graph by Using Intercepts
- Graph by Using the Slope and *y*-Intercept

Explore: Shading Graphs of Linear Inequalities

Develop:

Graphing Linear Inequalities in Two Variables
- Graph an Inequality with an Open Half-Plane
- Graph an Inequality with a Closed Half-Plane
- Linear Inequalities

You may want your students to complete the **Checks** online.

3 REFLECT AND PRACTICE

Exit Ticket

Practice

DIFFERENTIATE

View reports of student progress on the **Checks** after each example.

Resources	AL	OL	BL	ELL
Remediation: Graphing Linear Functions	●	●		●
Extension: Graphing Special Inequalities		●	●	●

Language Development Handbook

Assign page 5 of the *Language Development Handbook* to help your students build mathematical language related to graphing linear functions and inequalities in two variables.

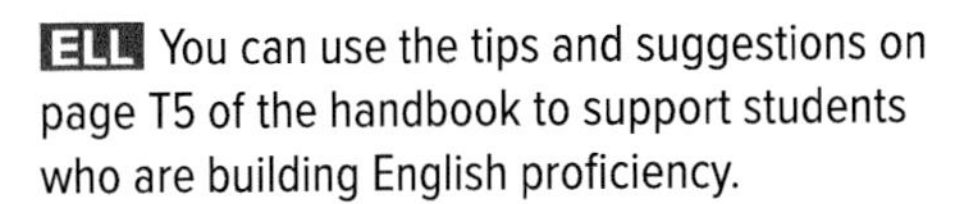

ELL You can use the tips and suggestions on page T5 of the handbook to support students who are building English proficiency.

Suggested Pacing

90 min	1 day
45 min	2 days

Focus

Domain: Algebra, Functions

Standards for Mathematical Content:

A.CED.3 Represent constraints by equations or inequalities, and by systems of equations and/or inequalities, and interpret solutions as viable or non-viable options in a modeling context.

F.IF.4 For a function that models a relationship between two quantities, interpret key features of graphs and tables in terms of the quantities, and sketch graphs showing key features given in a verbal description of the relationship.

Standards for Mathematical Practice:

5 Use appropriate tools strategically.

6 Attend to precision.

Coherence

Vertical Alignment

Previous
Students sketched graphs of functions and compared two functions represented in different ways.
F.IF.4, F.IF.9

Now
Students graph linear functions and inequalities in two variables.
A.CED.3, F.IF.4

Next
Students will write and graph piecewise-defined, step, and absolute value functions.
F.IF.4, F.IF.7b

Rigor

The Three Pillars of Rigor

1 CONCEPTUAL UNDERSTANDING	2 FLUENCY	3 APPLICATION

Conceptual Bridge In this lesson, students expand on their understanding of and fluency with linear functions to graphing linear functions and inequalities. They apply their understanding of linear functions and inequalities by solving real-world problems.

Interactive Presentation

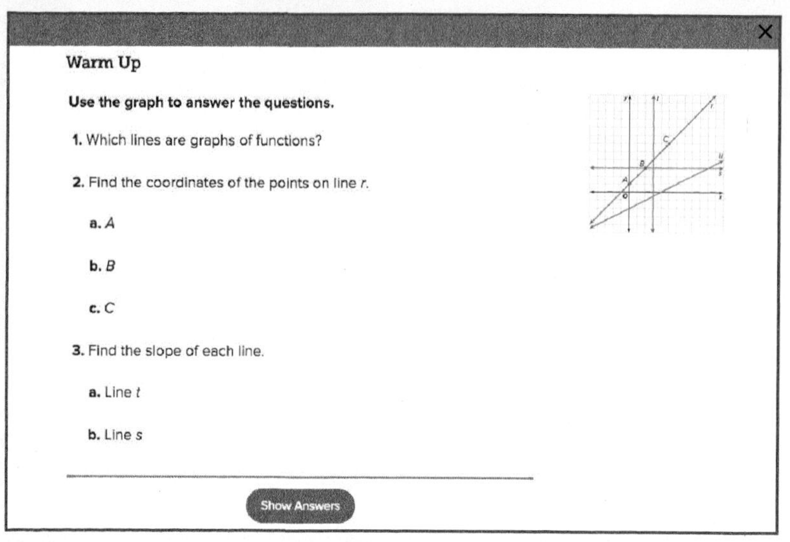

Warm Up

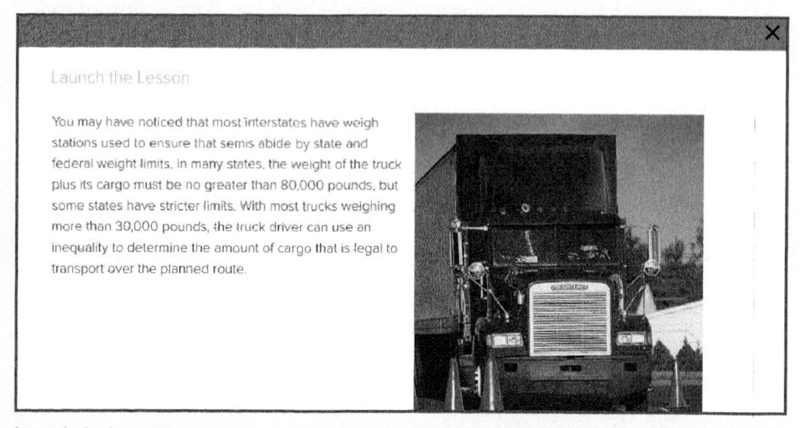

Launch the Lesson

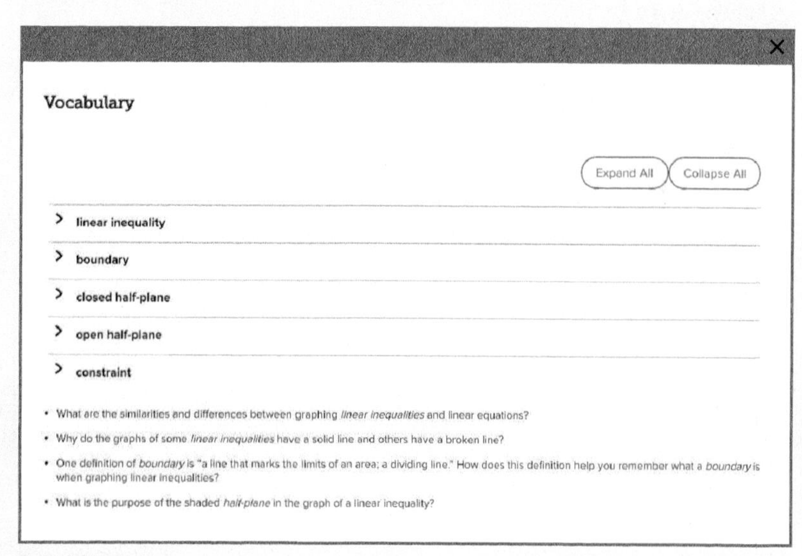

Today's Vocabulary

Warm Up

Prerequisite Skills

The Warm Up exercises address the following prerequisite skill for this lesson:

- graphing lines and points on lines

Answers:

1. r, s, and u

2a. (0, 1)

2b. (2, 3)

2c. (5, 6)

3a. undefined

3b. 0

Launch the Lesson

Teaching the Mathematical Practices

2 Create Representations Guide students to write an inequality that models the situation in the Launch the Lesson. Ask them to use the inequality to determine the maximum cargo load of a truck that weighs a given amount.

Go Online to find additional teaching notes and questions to promote classroom discourse.

Today's Standards

Tell students that they will be addressing these content and practice standards in this lesson. You may wish to have a student volunteer read aloud *How can I meet these standards?* and *How can I use these practices?*, and connect these to the standards.

See the Interactive Presentation for I Can statements that align with the standards covered in this lesson.

Today's Vocabulary

Tell students that they will be using these vocabulary terms in this lesson. You can expand each row if you wish to share the definitions. Then discuss the questions below with the class.

Mathematical Background

The solution set of a linear inequality is the set of all ordered pairs that satisfy the inequality.

Explore Shading Graphs of Linear Inequalities

Objective

Students use a sketch to explore inequalities by testing points.

Teaching the Mathematical Practices

5 Use Mathematical Tools Point out that to complete the Explore, students will need to use an interactive sketch. Work with students to explore and deepen their understanding of the solutions and graphs of inequalities.

Ideas for Use

Recommended Use Present the Inquiry Question, or have a student volunteer read it aloud. Have students work in pairs to complete the Explore activity on their devices. Pairs should discuss each of the questions. Monitor student progress during the activity. Upon completion of the Explore activity, have student volunteers share their responses to the Inquiry Question.

What if my students don't have devices? You may choose to project the activity on a whiteboard. A printable worksheet for each Explore is available online. You may choose to print the worksheet so that individuals or pairs of students can use it to record their observations.

Summary of the Activity

Students will complete guiding exercises throughout the Explore activity. Students will use a sketch to investigate graphing linear inequalities. Then students will analyze the graphs to answer questions regarding the solutions. Then, students will answer the Inquiry Question.

(continued on the next page)

Interactive Presentation

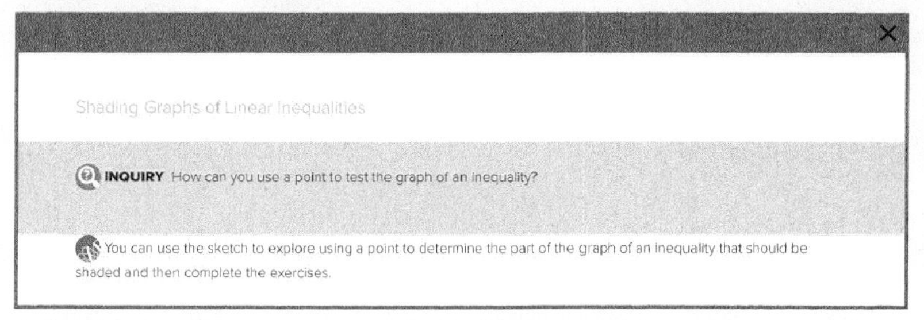

Explore

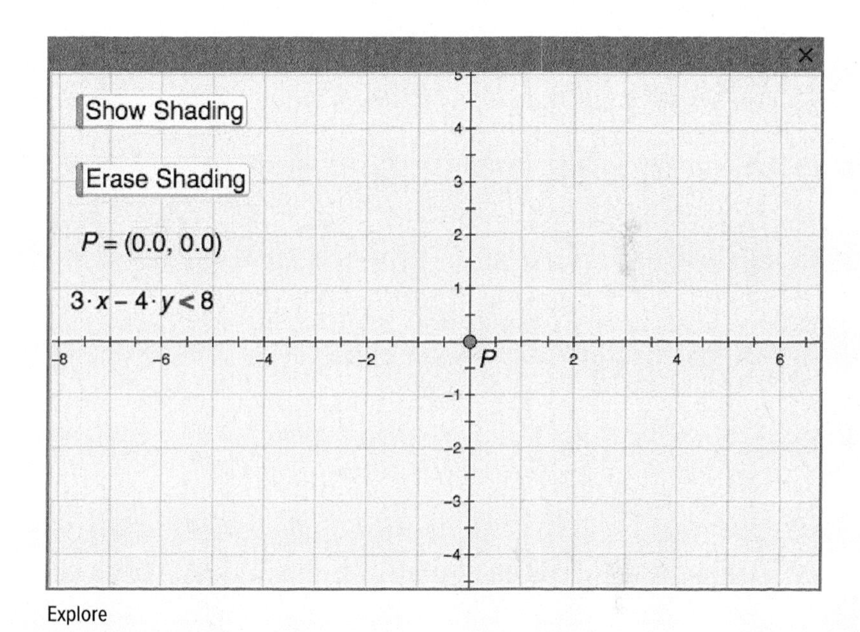

Explore

WEB SKETCHPAD

Students use the sketch to complete an activity in which they explore shading linear inequalities.

TYPE

Students answer questions regarding the test points and the inequality.

DRAG & DROP

Students drag and drop points which make an equation or inequality true.

Interactive Presentation

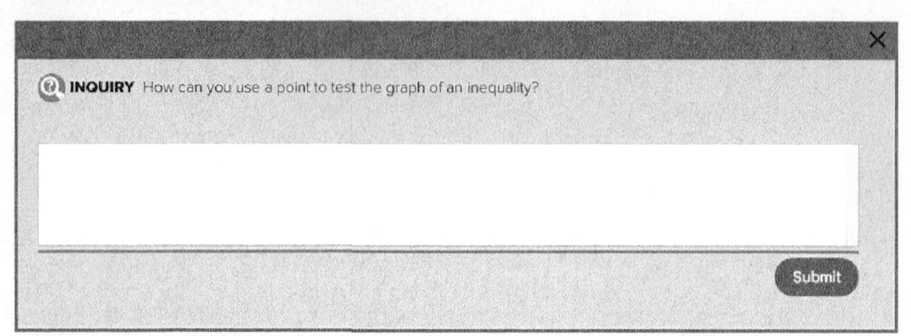

Explore

TYPE

Students respond to the Inquiry Question and can view a sample answer.

Explore Shading Graphs of Linear Inequalities (*continued*)

Questions

Have students complete the Explore activity.

Ask:

- Which inequality symbol should be used in $3x - 4y$ __?__ 8 if $(-1, 1)$ is a solution? $<$
- Write the inequality $3x - 4y < 8$ in slope-intercept form. $y > \frac{3}{4}x - 2$

Inquiry

How can you use a point to test the graph of an inequality? Sample answer: I can use the coordinates of a point to see whether an inequality is true or false for the point.

Go Online to find additional teaching notes and sample answers for the guiding exercises.

Learn Graphing Linear Functions

Objective

Students graph linear functions by using a table, intercepts, or the slope and intercept.

Teaching the Mathematical Practices

1 Understand Different Approaches Explain the different approaches to graphing linear functions: using a table, using intercepts, and using the slope and y-intercept. Work with students to compare the approaches.

Example 1 Graph by Using a Table

Teaching the Mathematical Practices

6 Use Precision In Example 1, encourage students to select values of x that allow them to calculate accurately and efficiently.

Questions for Mathematical Discourse

AL What does a table of values represent? Sample answer: coordinates on the graph of a function

OL What is the minimum number of points needed to graph a linear function? two

BL If you solved the equation for x and substituted y-values to find corresponding x-values, would the resulting graph be different? No; the graph would be the same.

Why do we typically solve for y when graphing a function? Sample answer: We solve for y because y typically represents the dependent variable, and thus each x-value should only correspond to one y-value. The reverse is not true, as there are functions where one y-value can correspond to more than one x-value.

Example 2 Graph by Using Intercepts

Teaching the Mathematical Practices

1 Check Answers Use the Think About It! feature to encourage students to check their graph using a different method.

Questions for Mathematical Discourse

AL Can you graph any function by only finding the intercepts? Explain your reasoning. No; sample answer: This works for a linear function because only two points are needed to draw a line. Other types of functions will require more points to graph accurately.

OL What is an advantage of finding intercepts to graph a linear function compared to using other points on the line? Sample answer: Substituting a zero removes a term and the resulting equation is simpler to solve. This method would be less beneficial when the intercept values are not integers.

BL Given any linear equation in the form $Ax + By = C$, how can you identify whether graphing by using intercepts would be a good method to use? Sample answer: Determine whether A and B are factors of C.

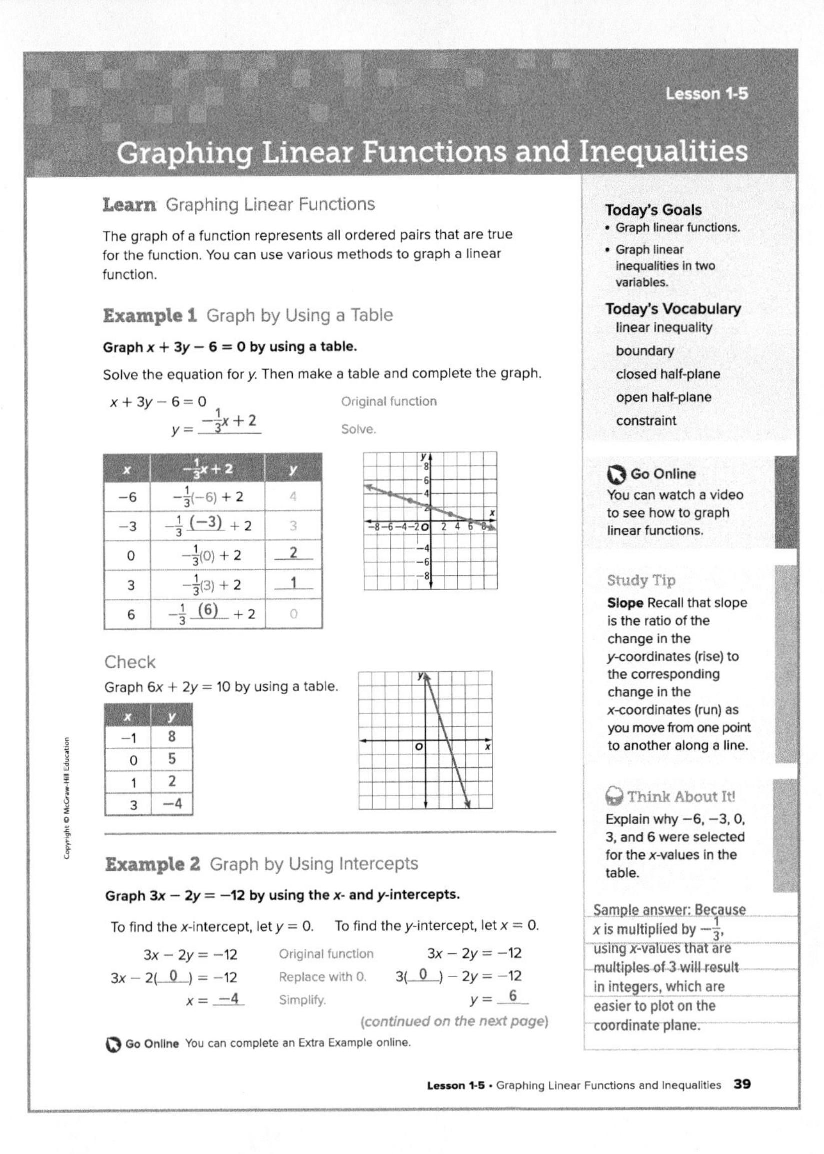

Lesson 1-5

Graphing Linear Functions and Inequalities

Learn Graphing Linear Functions

The graph of a function represents all ordered pairs that are true for the function. You can use various methods to graph a linear function.

Example 1 Graph by Using a Table

Graph $x + 3y - 6 = 0$ by using a table.

Solve the equation for y. Then make a table and complete the graph.

$x + 3y - 6 = 0$ Original function

$y = -\frac{1}{3}x + 2$ Solve.

x	$-\frac{1}{3}x + 2$	y
−6	$-\frac{1}{3}(-6) + 2$	4
−3	$-\frac{1}{3}(-3) + 2$	3
0	$-\frac{1}{3}(0) + 2$	2
3	$-\frac{1}{3}(3) + 2$	1
6	$-\frac{1}{3}(6) + 2$	0

Check

Graph $6x + 2y = 10$ by using a table.

x	y
−1	8
0	5
1	2
3	−4

Example 2 Graph by Using Intercepts

Graph $3x - 2y = -12$ by using the x- and y-intercepts.

To find the x-intercept, let $y = 0$. To find the y-intercept, let $x = 0$.

$3x - 2y = -12$ Original function $3x - 2y = -12$

$3x - 2(0) = -12$ Replace with 0. $3(0) - 2y = -12$

$x = -4$ Simplify. $y = 6$

(continued on the next page)

Go Online You can complete an Extra Example online.

Today's Goals
- Graph linear functions.
- Graph linear inequalities in two variables.

Today's Vocabulary
linear inequality
boundary
closed half-plane
open half-plane
constraint

Go Online You can watch a video to see how to graph linear functions.

Study Tip

Slope Recall that slope is the ratio of the change in the y-coordinates (rise) to the corresponding change in the x-coordinates (run) as you move from one point to another along a line.

Think About It!

Explain why −6, −3, 0, 3, and 6 were selected for the x-values in the table.

Sample answer: Because x is multiplied by $-\frac{1}{3}$, using x-values that are multiples of 3 will result in integers, which are easier to plot on the coordinate plane.

Lesson 1-5 • Graphing Linear Functions and Inequalities 39

Interactive Presentation

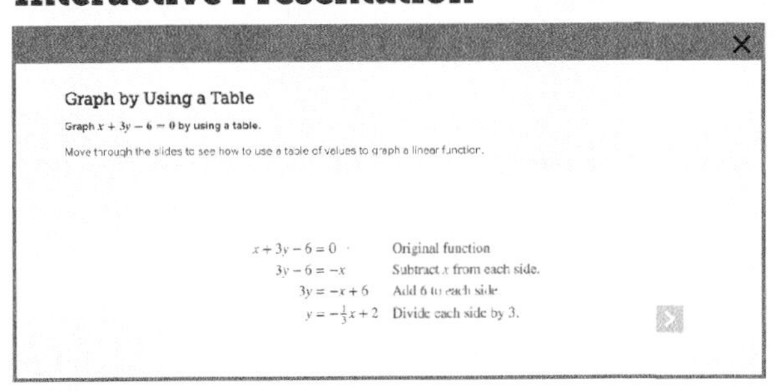

Example 1

TAP

Students move through the slides to see how to use a table of values to graph a linear function.

TYPE

Students explain why specific x-values were selected for the table.

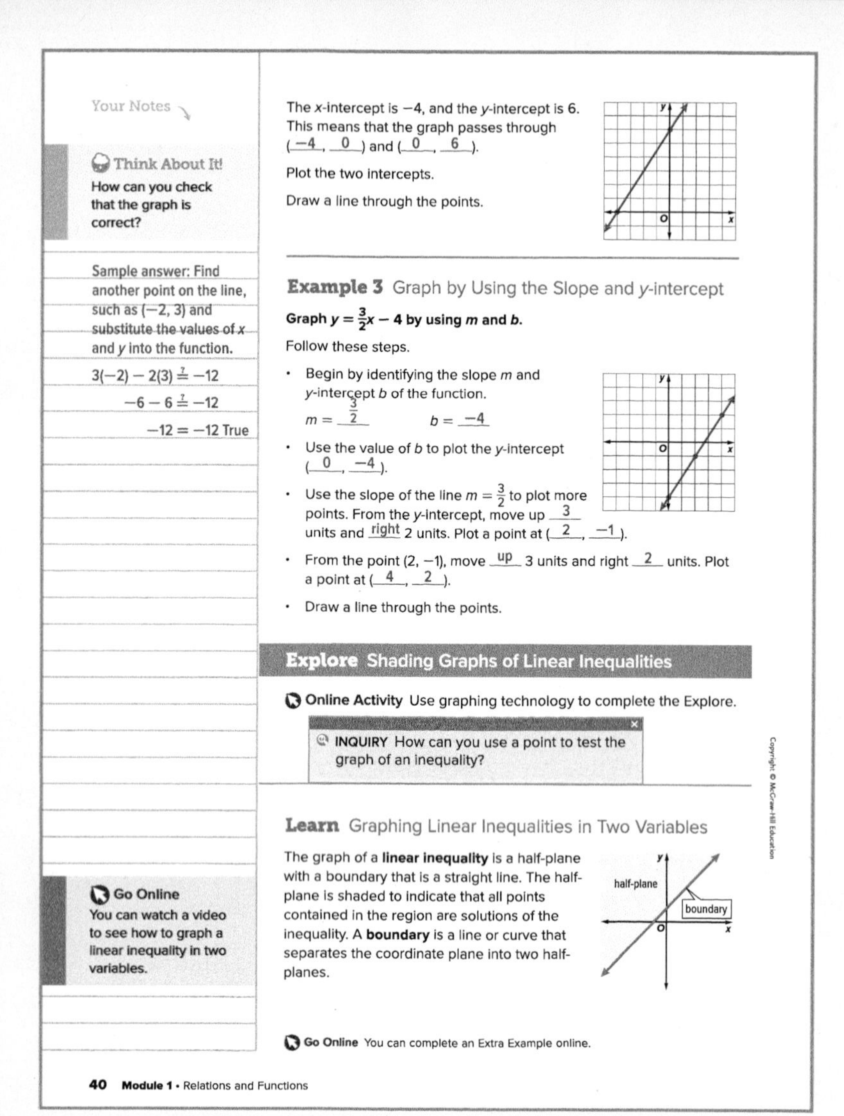

Your Notes

Think About It!
How can you check that the graph is correct?

Sample answer: Find another point on the line, such as (−2, 3) and substitute the values of x and y into the function.

$3(-2) - 2(3) \stackrel{?}{=} -12$

$-6 - 6 \stackrel{?}{=} -12$

$-12 = -12$ True

The x-intercept is −4, and the y-intercept is 6. This means that the graph passes through (−4, 0) and (0, 6).

Plot the two intercepts.

Draw a line through the points.

Example 3 Graph by Using the Slope and y-intercept

Graph $y = \frac{3}{2}x - 4$ by using m and b.

Follow these steps.

- Begin by identifying the slope m and y-intercept b of the function.
  $m = \frac{3}{2}$ $b = -4$
- Use the value of b to plot the y-intercept (0, −4).
- Use the slope of the line $m = \frac{3}{2}$ to plot more points. From the y-intercept, move up 3 units and right 2 units. Plot a point at (2, −1).
- From the point (2, −1), move up 3 units and right 2 units. Plot a point at (4, 2).
- Draw a line through the points.

Explore Shading Graphs of Linear Inequalities

Online Activity Use graphing technology to complete the Explore.

INQUIRY How can you use a point to test the graph of an inequality?

Learn Graphing Linear Inequalities in Two Variables

The graph of a **linear inequality** is a half-plane with a boundary that is a straight line. The half-plane is shaded to indicate that all points contained in the region are solutions of the inequality. A **boundary** is a line or curve that separates the coordinate plane into two half-planes.

half-plane

boundary

Go Online You can watch a video to see how to graph a linear inequality in two variables.

Go Online You can complete an Extra Example online.

Copyright © McGraw-Hill Education

40 Module 1 • Relations and Functions

Interactive Presentation

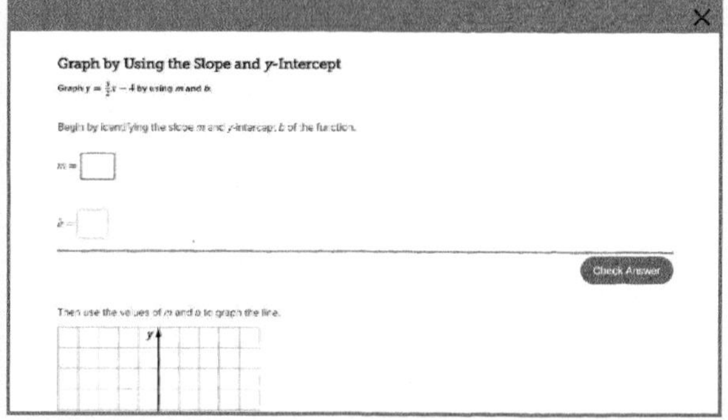

Example 3

TYPE

Students identify the slope and y-intercept of the function.

TAP

Students move through the steps to see how to graph a function.

CHECK

Students complete the Check online to determine whether they are ready to move on.

Common Error

Students may try and plot the x-intercept on the y-axis and vice versa. Remind students the x-intercept goes on the x-axis because the coordinate represents where the graph crosses the x-axis.

DIFFERENTIATE

Language Development Activity ELL

Discuss with students the difference between the words *intersect* and *intercept*.

Example 3 Graph by Using the Slope and y-Intercept

Teaching the Mathematical Practices

2 Make Sense of Quantities Mathematically proficient students need to be able to make sense of quantities and their relationships. In Example 3, notice the relationship between the slope and y-intercept and the graph of the function.

Questions for Mathematical Discourse

AL If you have a linear function with a slope of $m = 4$ and a given point, how would you plot the next point? Sample answer: Move up 4 units and over 1 unit, and plot a point.

OL Why does the constant b represent the y-intercept when a linear equation is written in slope-intercept form? The y-intercept occurs where $x = 0$. Substituting zero into $y = mx + b$ results in $y = b$.

BL How would the graph change if $m = -\frac{3}{2}$? Sample answer: The new graph would be decreasing as x increases.

Learn Graphing Linear Inequalities in Two Variables

Objective

Students graph linear inequalities in two variables by using the related equations.

Teaching the Mathematical Practices

6 Communicate Precisely Encourage students to routinely write or explain their solution methods. Point out that they should use clear definitions when they discuss their solutions to the Talk About It! feature.

Go Online

- Find additional teaching notes.
- View performance reports of the Checks.
- Assign or present an Extra Example.

Example 4 Graph an Inequality with an Open Half-Plane

Teaching the Mathematical Practices

5 Use Mathematical Tools Point out that to graph the inequality in Example 4, students will need to use pencil and paper.

Questions for Mathematical Discourse

AL Suppose you tested the point (0, 4), which is above the boundary. Would the inequality result in a true or false statement? How does this help you determine which region to shade? Substituting (0, 4) results in $16 < 12$, which is a false statement. This means that (0, 4) is not a solution to the inequality, so you would not shade the region that contains this point.

OL Why is (0, 0) a useful point to use as a test point? When should it not be used as a test point? Sample answer: Using (0, 0) as a test point makes the calculations simpler. It would not be a valid test point when it lies on the boundary of the inequality.

BL If you could not use (0, 0) as a test point, what point would you consider using next? Explain your reasoning. Sample answer: (0, 1) or (1, 0) will still usually result in simple calculations. You could strategically choose which of these two points to use by determining which term of the inequality would be better to result in zero.

Example 5 Graph an Inequality with a Closed Half-Plane

Teaching the Mathematical Practices

1 Seek Information Mathematically proficient students must be able to transform algebraic expressions to reach solutions. Point out that to graph inequalities in two variables, it is often easier to rewrite the inequality by solving for y.

Questions for Mathematical Discourse

AL What happens if you use the test point (3, 5)? Sample answer: This results in $24 \leq 24$, which means this point is on the boundary. This does not help you determine which region to shade.

OL Does it make sense to say the slope of the inequality is $\frac{8}{3}$ and the y-intercept of the inequality is -3? Explain. No; sample answer: the slope and y-intercept are relevant to the related function that defines the boundary.

BL Suppose you graphed two linear inequalities and there was a region where the shading of the inequalities overlapped. What would this mean? The points in the region are valid solutions to both inequalities.

The boundary is solid when the inequality contains $\leq$ or $\geq$ to indicate that the points on the boundary are included in the solution, creating a **closed half-plane**. The boundary is dashed when the inequality contains $<$ or $>$ because the points on the boundary do not satisfy the inequality. This results in an **open half-plane**.

A **constraint** is a condition that a solution must satisfy. Each solution of the inequality represents a viable, or possible, option that satisfies the constraint.

Example 4 Graph an Inequality with an Open Half-Plane

Graph $12 - 4y > x$.

Step 1 Graph the boundary.

$12 - 4y > x$ — Original inequality

$\underline{-4y} > x - 12$ — Subtract 12 from each side.

$y < \underline{-\frac{1}{4}}\,x + \underline{3}$ — Divide each side by -4, and reverse the inequality symbol.

The boundary of the graph is $y \underline{\ =\ } -\frac{1}{4}x + 3$. Because the inequality symbol is $>$, the boundary is <u>dashed</u>.

Step 2 Use a test point and shade.

Test (0, 0).

$12 - 4y > x$ — Original inequality

$12 - 4(\underline{0}) \overset{?}{>} \underline{0}$ — Substitute.

$12 > 0$ — True

Because (0, 0) is a solution of the inequality, shade the half-plane that contains the test point.

Check: Check by selecting another point in the shaded region to test.

Example 5 Graph an Inequality with a Closed Half-Plane

Graph $9 + 3y \leq 8x$.

Step 1 Graph the boundary.

Solve for y in terms of x and graph the related function.

$9 + 3y \leq 8x$ — Original inequality

$3y \leq 8x \underline{-9}$ — Subtract 9 from each side.

$y \leq \underline{\frac{8}{3}}\,x - \underline{3}$ — Divide each side by 3.

The related equation of $y \leq \frac{8}{3}x - 3$ is $y = \frac{8}{3}x - 3$, and the boundary is solid.

(continued on the next page)

Study Tip

Above or Below Usually the shaded half-plane of a linear inequality is said to be *above* or *below* the line of the related equation. However, if the equation of the boundary is $x = c$ for some constant c, then the function is a vertical line. In this case, the shading is considered to be *to the left* or *to the right* of the boundary.

Talk About It!

Can a linear inequality ever be a function? Explain your reasoning.

No; sample answer: For any value of x, there are infinitely many values of y in the solution set of a linear inequality. Therefore, a linear inequality cannot be a function.

Think About It!

Why should you not test a point that is on the boundary?

Sample answer: Testing a point on the boundary does not indicate whether you have properly shaded the solution set. It will only indicate whether the boundary should be included.

Lesson 1-5 • Graphing Linear Functions and Inequalities 41

Interactive Presentation

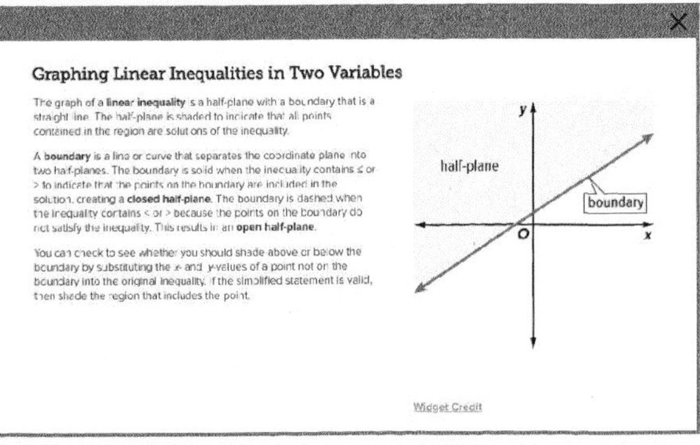

Learn

WATCH

Students watch the video to see how to graph a linear inequality in two variables.

TYPE

Students consider whether a linear inequality can be a function.

Think About It!
Is (3, 5) a solution of the inequality? Explain.

Yes; sample answer: (3, 5) is on the solid boundary, so it is included in the solution.

Step 2 Use a test point and shade.

Select a test point, such as (0, 0).

$9 + 3y \le 8x$ — Original inequality

$9 + 3(0) \stackrel{?}{\le} 8(0)$ — $(x, y) = (0, 0)$.

$9 \not\le 0$ — False

Shade the side of the graph that does not contain the test point.

Go Online
You can watch a video to see how to graph an inequality using a graphing calculator.

Apply Example 6 Linear Inequalities

GRADES **Malik's algebra teacher determines semester grades by finding the sum of 70% of a student's test grade average and 30% of a student's homework grade average. If Malik wants a semester grade of 90% or better, write and graph the inequality that represents the constraints for Malik's test grade x and homework grade y.**

1 What is the task?
Describe the task in your own words. Then list any questions that you may have. How can you find answers to your questions?
Sample answer: Use the description to write the inequality. Find points on the boundary and use a test point to create the graph.

2 How will you approach the task? What have you learned that you can use to help you complete the task?
Sample answer: Write and graph an inequality to represent the constraints on Malik's grades. How do the test and homework grades relate to the semester grade?

3 What is your solution?
Use your strategy to solve the problem. What inequality represents the constraints for Malik's test and homework grades? Use the grid to graph the inequality.
$0.7x + 0.3y \ge 0.9$

Which of these are viable solutions for Malik's test and homework grades?

- ■ 88% test, 100% homework
- ■ 90% test, 90% homework
- □ 90% test, 80% homework
- □ 95% test, 70% homework
- ■ 95% test, 80% homework
- ■ 100% test, 70% homework

Go Online
You can complete an Extra Example online.

4 How can you know that your solution is reasonable?

Write About It! Write an argument that can be used to defend your solution. Sample answer: I can select a point in the shaded region, such as (0.95, 0.8) and test it in the inequality.

42 Module 1 • Relations and Functions

Interactive Presentation

Linear Inequalities

GRADES Malik's algebra teacher determines semester grades by finding the sum of 70% of a student's test grade average and 30% of a student's homework grade average. If Malik wants a semester grade of 90% or better, write and graph the inequality that represents the constraints for Malik's test grade x and homework grade y.

Apply Example 6

SELECT

Students select viable solutions to the inequality.

CHECK

Students complete the Check online to determine whether they are ready to move on.

DIFFERENTIATE

Reteaching Activity AL ELL

IF students cannot correctly shade the graph of an inequality, **THEN** have students write the inequality in the form $y \square mx + b$ with the appropriate inequality symbol in the box. After graphing the inequality, tell students to imagine themselves standing on a point on the boundary. If the symbol is $>$ or $\ge$, they will pretend to climb up from the point, which means they will shade above the boundry. If the symbol is $<$ or $\le$, they will pretend to climb down from the point, which means they will shade below the boundry.

Apply Example 6 Linear Inequalities

Teaching the Mathematical Practices

1 Make Sense of Problems and Persevere in Solving Them, 4 Model with Mathematics Students will be presented with a task. They will first seek to understand the task, and then determine possible entry points to solving it. As students come up with their own strategies, they may propose mathematical models to aid them. As they work to solve the problem, encourage them to evaluate their model and/or progress, and change direction, if necessary.

Recommended Use

Have students work in pairs or small groups. You may wish to present the task, or have a volunteer read it aloud. Then allow students the time to make sure they understand the task, think of possible strategies, and work to solve the problem.

Encourage Productive Struggle

As students work, monitor their progress. Instead of instructing them on a particular strategy, encourage them to use their own strategies to solve the problem and to evaluate their progress along the way. They may or may not find that they need to change direction or try out several strategies.

Signs of Non-Productive Struggle

If students show signs of non-productive struggle, such as feeling overwhelmed, frustrated, or disengaged, intervene to encourage them to think of alternate approaches to the problem. Some sample questions are shown.

- How do you know which inequality symbol to use?
- If Malik's homework grade average is 95% and his test grade average is 85%, how much will this score contribute to his overall grade? Is this enough to earn a 90% or higher overall?

Write About It!

Have students share their responses with another pair/group of students or the entire class. Have them clearly state or describe the mathematical reasoning they can use to defend their solution.

Exit Ticket

Recommended Use

At the end of class, go online to display the Exit Ticket prompt and ask students to respond using a separate piece of paper. Have students hand you their responses as they leave the room.

Alternate Use

At the end of class, go online to display the Exit Ticket prompt and ask students to respond verbally or by using a mini-whiteboard. Have students hold up their whiteboards so that you can see all student responses. Tap to reveal the answer when most or all students have completed the Exit Ticket.

Practice and Homework

Suggested Assignments

Use the table below to select appropriate exercises.

DOK	Topic	Exercises
1, 2	exercises that mirror the examples	1–34
2	exercises that use a variety of skills from this lesson	35–56
2	exercises that extend concepts learned in this lesson to new contexts	57–60
3	exercises that emphasize higher-order and critical thinking skills	61–65

ASSESS AND DIFFERENTIATE

Use the data from the **Checks** to determine whether to provide resources for extension, remediation, or intervention.

IF students score 90% or more on the Checks, **BL**
THEN assign:

- Practice Exercises 1–59 odd, 61–65
- Extension: Graphing Special Inequalities
- **ALEKS**® Graphs of Lines; Linear Inequalities with Two Variables

IF students score 66%–89% on the Checks, **OL**
THEN assign:

- Practice Exercises 1–65 odd
- Remediation, Review Resources: Graphing Linear Functions
- Personal Tutors
- Extra Examples 1–6
- **ALEKS**® Ordered Pairs, Graphing Lines

IF students score 65% or less on the Checks, **AL**
THEN assign:

- Practice, Exercises 1–33 odd
- Remediation, Review Resources: Graphing Linear Functions
- *Quick Review Math Handbook*: Linear Relations and Functions
- **ALEKS**® Ordered Pairs, Graphing Lines

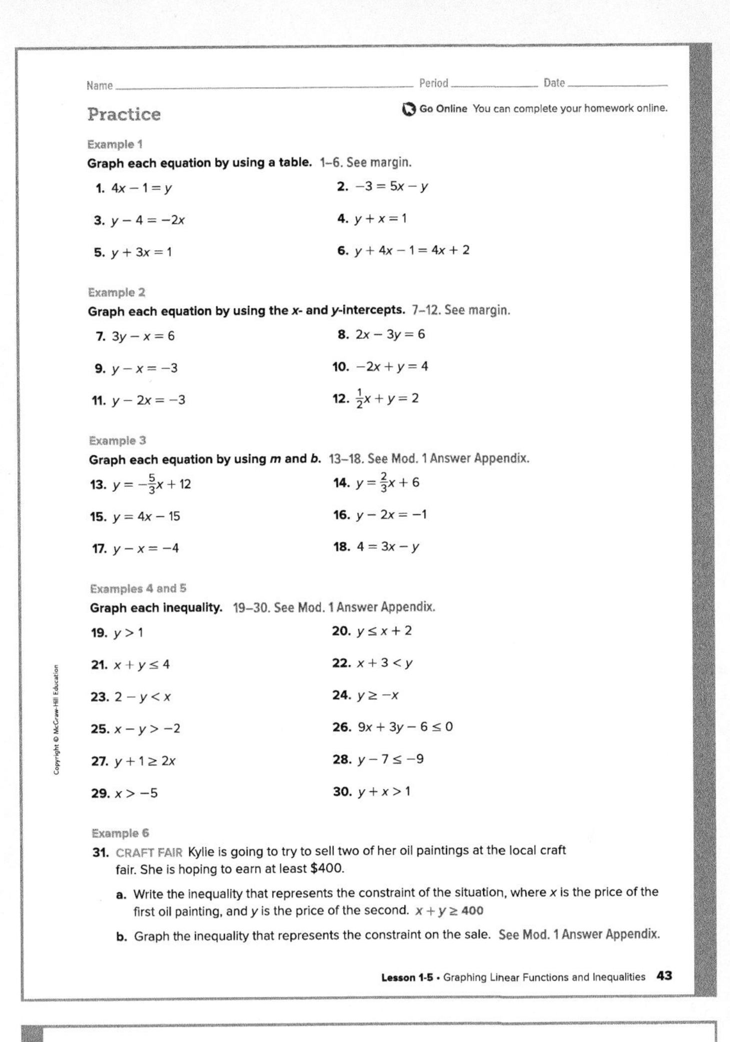

Name ______ Period ______ Date ______

Practice

Go Online You can complete your homework online.

Example 1

Graph each equation by using a table. 1–6. See margin.

1. $4x - 1 = y$
2. $-3 = 5x - y$
3. $y - 4 = -2x$
4. $y + x = 1$
5. $y + 3x = 1$
6. $y + 4x - 1 = 4x + 2$

Example 2

Graph each equation by using the *x*- and *y*-intercepts. 7–12. See margin.

7. $3y - x = 6$
8. $2x - 3y = 6$
9. $y - x = -3$
10. $-2x + y = 4$
11. $y - 2x = -3$
12. $\frac{1}{2}x + y = 2$

Example 3

Graph each equation by using *m* and *b*. 13–18. See Mod. 1 Answer Appendix.

13. $y = -\frac{5}{3}x + 12$
14. $y = \frac{2}{3}x + 6$
15. $y = 4x - 15$
16. $y - 2x = -1$
17. $y - x = -4$
18. $4 = 3x - y$

Examples 4 and 5

Graph each inequality. 19–30. See Mod. 1 Answer Appendix.

19. $y > 1$
20. $y \leq x + 2$
21. $x + y \leq 4$
22. $x + 3 < y$
23. $2 - y < x$
24. $y \geq -x$
25. $x - y > -2$
26. $9x + 3y - 6 \leq 0$
27. $y + 1 \geq 2x$
28. $y - 7 \leq -9$
29. $x > -5$
30. $y + x > 1$

Example 6

31. CRAFT FAIR Kylie is going to try to sell two of her oil paintings at the local craft fair. She is hoping to earn at least $400.
    a. Write the inequality that represents the constraint of the situation, where *x* is the price of the first oil painting, and *y* is the price of the second. $x + y \geq 400$
    b. Graph the inequality that represents the constraint on the sale. See Mod. 1 Answer Appendix.

32. BUILDING CODES A city has a building code that limits the height of buildings around the central park. The code says that the height of a building must be less than $0.1x$, where *x* is the distance in hundreds of feet of the building from the center of the park. Assume that the park center is located at $x = 0$. Graph the inequality that represents the building code. See Mod. 1 Answer Appendix.

33. WEIGHT A delivery crew is going to load a truck with tables and chairs. The trucks weight limitations are represented by the inequality $200t + 60c < 1200$, where *t* is the number of tables and *c* is the number of chairs. Graph this inequality. See Mod. 1 Answer Appendix.

34. ART An artist can sell each drawing for $100 and each watercolor for $400. He hopes to make at least $2000 every month.
    a. Write an inequality that expresses how many drawings and/or watercolors the artist needs to sell each month to reach his goal. $100d + 400w \geq 2000$
    b. Graph the inequality. See Mod. 1 Answer Appendix.
    c. If the artist sells three watercolors one month, how many drawings would he have to sell in the same month to reach $2000? 8

Mixed Exercises

Graph each equation or inequality. 35–52. See Mod. 1 Answer Appendix.

35. $x + y = 1$
36. $y \geq -3x - 2$
37. $x + 2y > 6$
38. $y + 2 = 3x + 3$
39. $y + 3 = 0$
40. $4x - 3y > 12$
41. $x + y = 3$
42. $2y - x = 2$
43. $4x + 3y = 12$
44. $\frac{1}{2}x + \frac{1}{4}y = 8$
45. $-\frac{1}{2}x + y = -2$
46. $y \geq \frac{3}{4}x + 6$
47. $y = -2x + 3$
48. $2x - y = 1$
49. $2y + 3 \leq 11$
50. $y + 2 = -x + 1$
51. $6x + 4y \leq -24$
52. $-2x + 5y = 2$

53. REASONING Name the *x*- and *y*-intercept for the linear equation given by $6x - 2y = 12$. Use the intercepts to graph the equation and describe the graph as *increasing*, *decreasing*, or *constant*. *x*-int $= 2$, *y*-int $= -6$; graph is increasing; See Mod. 1 Answer Appendix for graph.

54. ANIMALS During the winter, a horse requires about 36 liters of water per day and a sheep requires about 3.6 liters per day. A farmer is able to supply his horses and sheep with a total of 300 liters of water each day.
    a. Write an equation or inequality that represents the possible number of each type of animal that the farmer can keep. Let $x =$ horses; let $y =$ sheep; $36x + 3.6y \leq 300$.
    b. Graph the equation or inequality. See Mod. 1 Answer Appendix.

1 CONCEPTUAL UNDERSTANDING | 2 FLUENCY | 3 APPLICATION

Name ______ Period ______ Date ______

55. COMPUTERS A school system is buying new computers. They will buy desktop computers costing $1000 per unit, and notebook computers costing $1200 per unit. The total cost of the computers cannot exceed $80,000.

a. Write an inequality that describes this situation.
Let x = number of desktops; let y = number of notebooks; $1000x + 1200y \leq 80{,}000$.

b. Graph the inequality. See Mod. 1 Answer Appendix.

c. If the school wants to buy 50 desktop computers and 25 notebook computers, will they have enough money? Explain. Yes; sample answer: the point (50, 25) is on the line, which is a part of the viable region.

56. BAKED GOODS Mary sells giant chocolate chip and peanut butter cookies for $1.25 and $1.00, respectively, at a local bake shop. She wants to make at least $25 a day.

a. Write and graph an inequality that represents the number of cookies Mary needs to sell each day. Let x = number of chocolate chip cookies; let y = number of peanut butter cookies; $\$1.25x + \$1.00y \geq \$25$. See Mod. 1 Answer Appendix for graph.

b. If Mary decides to charge $1.50 for chocolate chip cookies rather than $1.25, what impact will this have on the graph of the solution set? Give an (x, y) pair that is not in the original solution set, but is in the solution set of the new revised scenario. See Mod. 1 Answer Appendix.

c. How does the graph of the inequality change if Mary wants to make at least $50 a day? How does the graph of the inequality change if Mary wants to make no more than $25 a day? See Mod. 1 Answer Appendix.

57. FUNDRAISING The school drama club is putting on a play to raise money. Suppose it will cost $400 to put on the play and that 300 students and 150 adults will attend.

a. Write an equation to represent revenue from ticket sales if the club wants to raise $1400 after expenses.
Let x = cost of a student ticket; let y = cost of an adult ticket; $300x + 150y = 1800$.

b. Graph your equation. Then determine four possible prices that could be charged for student and adult tickets to earn $1400 in profit. See Mod. 1 Answer Appendix.

58. CONSTRUCTION You want to make a rectangular sandbox area in your backyard. You plan to use no more than 20 linear feet of lumber to make the sides of the sandbox.

a. Write and graph a linear inequality to describe this situation.
$2\ell + 2w \leq 20$; See Mod. 1 Answer Appendix for graph.

b. What are two possible sizes for the sandbox? Sample answer: 5 ft by 4 ft; 4 ft by 6 ft

c. Can you make a sandbox that is 7 feet by 6 feet? Justify your answer. No; the point (7, 6) is not in the shaded region of the graph and therefore is not a solution of the inequality.

d. What can you conclude about the intercepts of your graph? They cannot represent the side lengths because if one pair of sides = 20 feet, then the other pair of sides = 0 feet, which is not possible.

59. SPIRITWEAR A company makes long-sleeved and short-sleeved shirts. The profit on a long-sleeved shirt is $7 and the profit on a short-sleeved shirt is $4. How many shirts must the company sell to make a profit of at least $280?

a. Write and graph a linear inequality to describe this situation.
Let x = long-sleeved shirts; let y = short-sleeved shirts; $7x + 4y \geq 280$; See Mod. 1 Answer Appendix for graph.

b. Write two possible solutions to the problem.
See Mod. 1 Answer Appendix.

c. Which values are reasonable for the domain and for the range? Explain.
See Mod. 1 Answer Appendix.

d. The point (−10, 90) is in the shaded region. Is it a solution of the problem? Explain your reasoning. No, you cannot buy −10 long-sleeved shirts.

60. MONEY Gemma buys candles and soaps online. The scented candles cost $9, and the hand soaps cost $4. To qualify for free shipping, Gemma needs to spend at least $50.

a. Write an inequality that represents the constraints on the number of candles x and the number of soaps y that Gemma must buy in order to qualify for free shipping. $9x + 4y \geq 50$

b. Graph the inequality. b–e. See Mod. 1 Answer Appendix.

c. Suppose Gemma decides not to buy any soaps. Determine the number of candles she needs to buy in order to qualify for free shipping. Explain.

d. If Gemma decides not to buy any candles how many soaps will she need to buy in order to qualify for free shipping? Explain.

e. Will Gemma qualify for free shipping if she buys 2 candles and 8 soaps? Explain how you can be sure.

Higher-Order Thinking Skills

61. FIND THE ERROR Paulo and Janette are graphing $x - y \geq 2$. Is either of them correct? Explain your reasoning. Paulo; Janette shaded the incorrect region.

Paulo

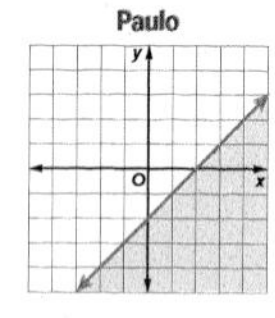

Janette

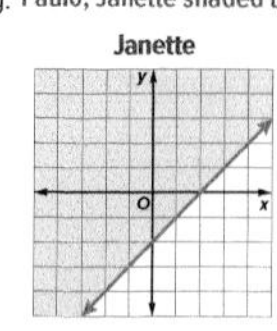

62. CREATE Write an inequality that has a graph with a dashed boundary line. Then graph the inequality. See Mod. 1 Answer Appendix.

63. WRITE You can graph a line by making a table, using the x- and y-intercepts, or by using m and b. Which method do you prefer? Explain your reasoning. See Mod. 1 Answer Appendix.

64. ANALYZE Write a counterexample to show that the following statement is false. *Every point in the first quadrant is a solution for $3y > -x + 6$.* Sample answer: (2, 1)

65. PERSEVERE Write an equation of the line that has the same slope as $2x - 8y = 7$ and the same y-intercept as $4x + 3y = 15$. $y = \frac{1}{4}x + 5$

Answers

1.
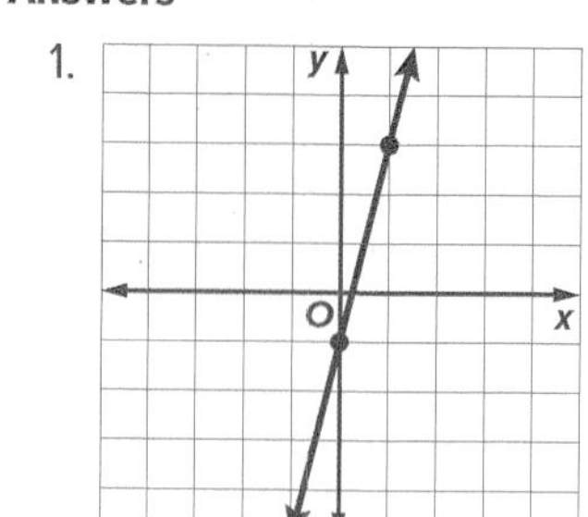

2.
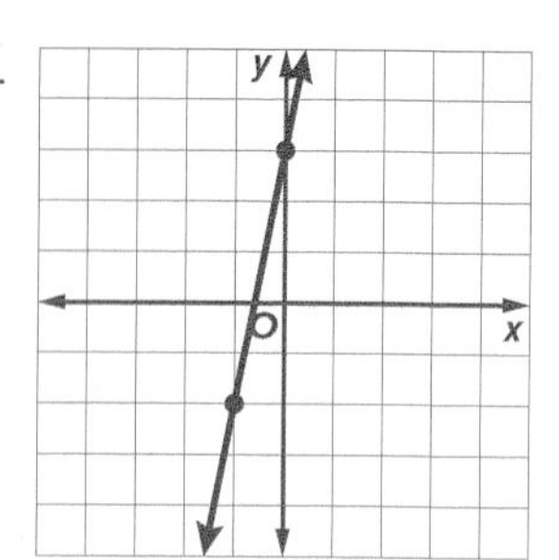

3.
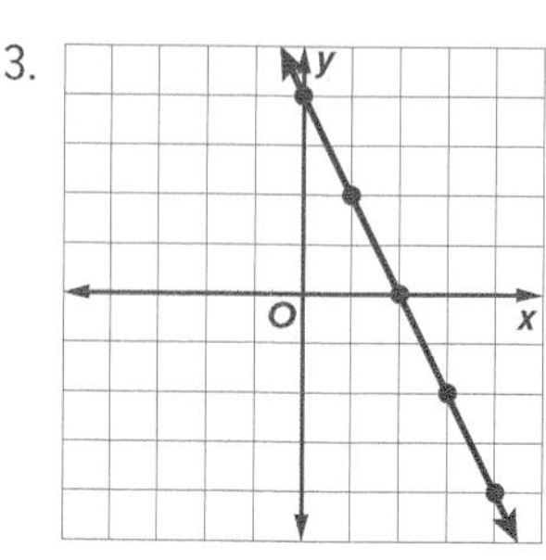

4.
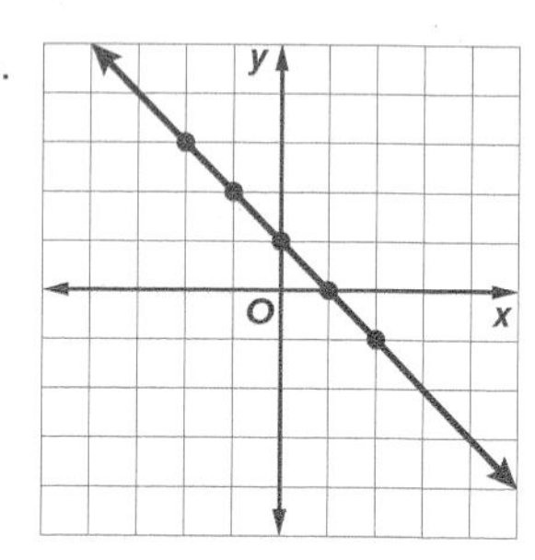

5.
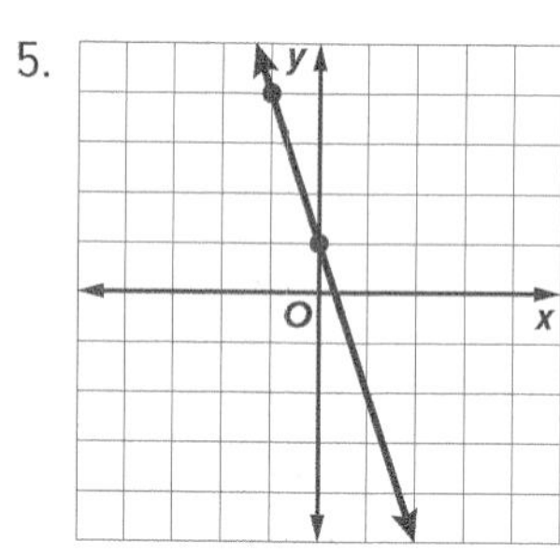

6.
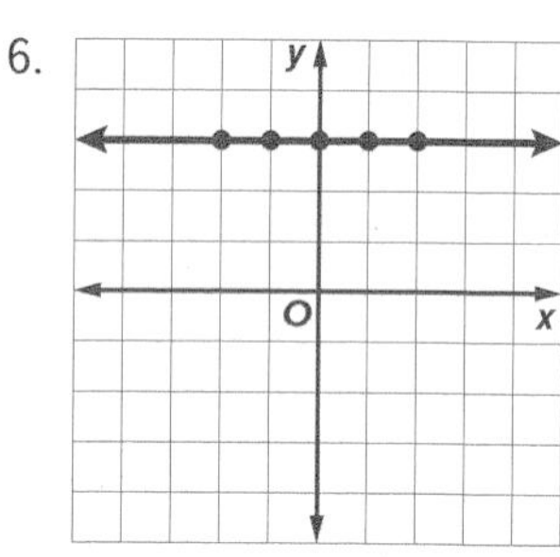

7.
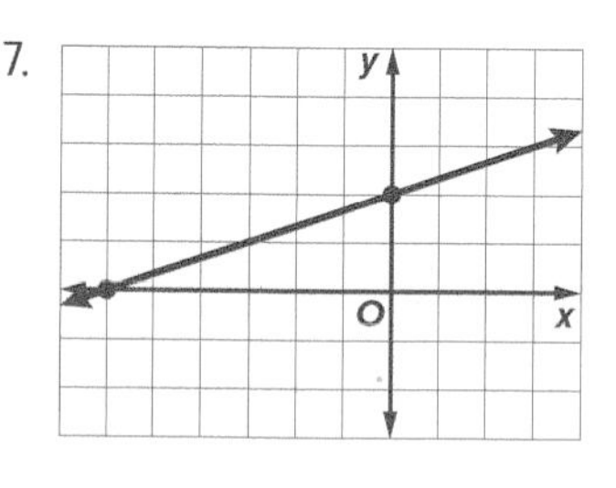

8.
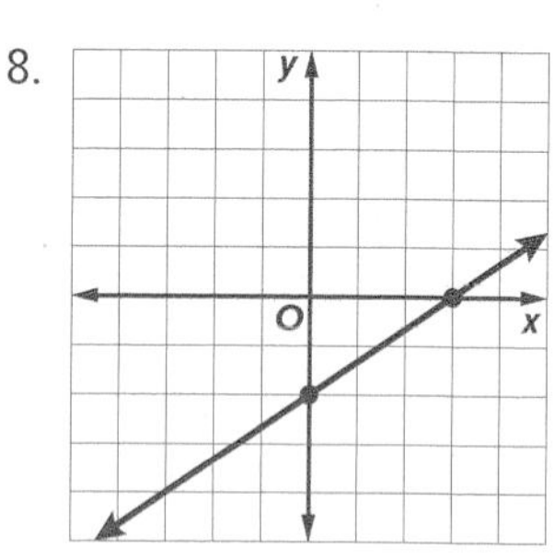

9.
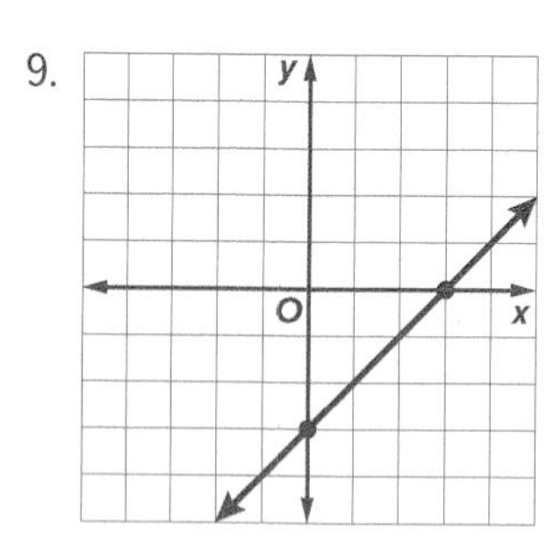

10.
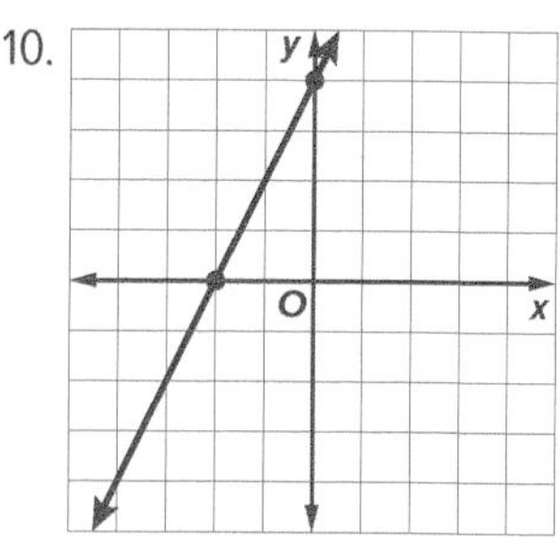

11.
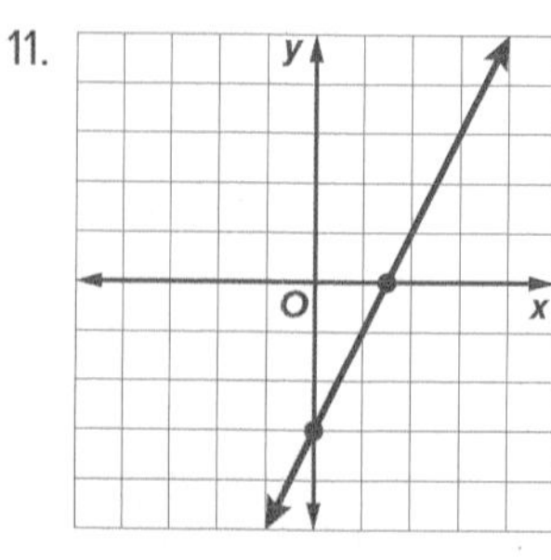

12.
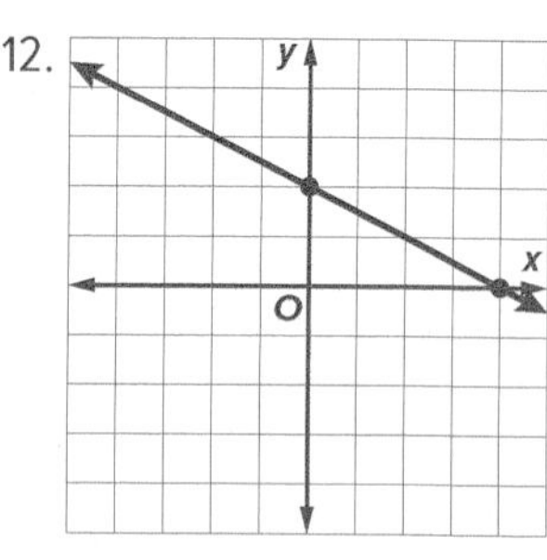

Lesson 1-6

Special Functions

F.IF.4, F.IF.7b

LESSON GOAL

Students write and graph piecewise-defined, step, and absolute value functions.

1 LAUNCH

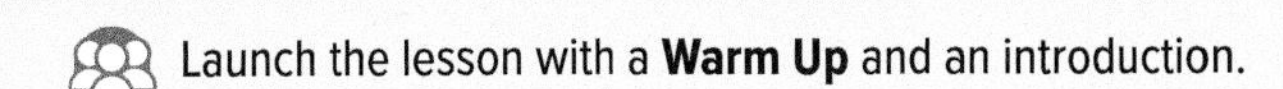
Launch the lesson with a **Warm Up** and an introduction.

2 EXPLORE AND DEVELOP

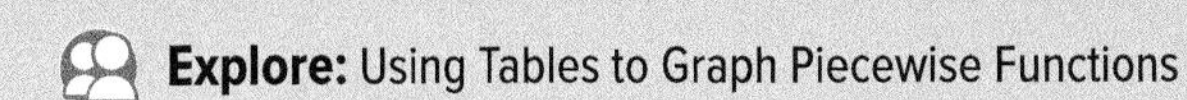
Explore: Using Tables to Graph Piecewise Functions

Develop

Graphing Piecewise-Defined Functions
- Graph a Piecewise-Defined Function
- Model by Using a Piecewise-Defined Function

Graphing Step Functions
- Graph a Step Function
- Graph a Greatest Integer Function

Graphing Absolute Value Functions
- Graph an Absolute Value Function, Positive Coefficient
- Graph an Absolute Value Function, Negative Coefficient

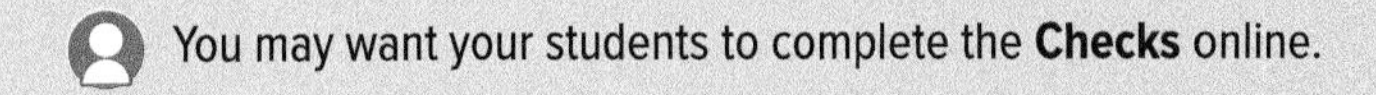
You may want your students to complete the **Checks** online.

3 REFLECT AND PRACTICE

Exit Ticket

Practice

DIFFERENTIATE

View reports of student progress on the **Checks** after each example.

Resources	AL	OL	BL	ELL
Remediation: Expressions Involving Absolute Value	●	●		●
Extension: Graphing Greatest Integer Functions		●	●	●

Language Development Handbook

Assign page 6 of the *Language Development Handbook* to help your students build mathematical language related to writing and graphing piecewise-defined, step, and absolute value functions.

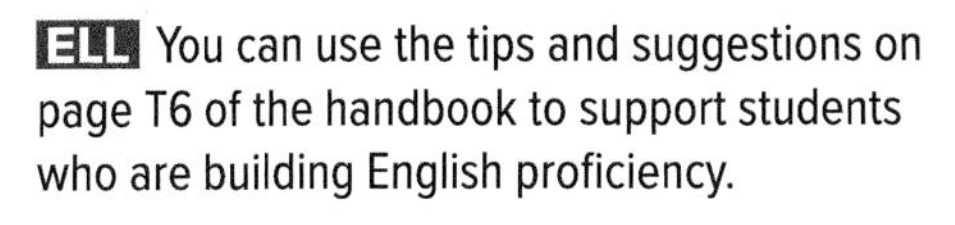
ELL You can use the tips and suggestions on page T6 of the handbook to support students who are building English proficiency.

Suggested Pacing

90 min	1 day
45 min	2 days

Focus

Domain: Functions

Standards for Mathematical Content:

F.IF.4 For a function that models a relationship between two quantities, interpret key features of graphs and tables in terms of the quantities, and sketch graphs showing key features given in a verbal description of the relationship.

F.IF.7b Graph functions expressed symbolically and show key features of the graph, by hand in simple cases and using technology for more complicated cases. Graph square root, cube root, and piecewise-defined functions, including step functions and absolute value functions.

Standards for Mathematical Practice:

4 Model with mathematics.

6 Attend to precision.

Coherence

Vertical Alignment

Previous
Students graphed linear functions and inequalities in two variables.
A.CED.3, F.IF.4

Now
Students write and graph piecewise-defined, step, and absolute value functions.
F.IF.4, F.IF.7b

Next
Students will identify and use transformations of functions.
F.IF.4, F.BF.3

Rigor

The Three Pillars of Rigor

1 CONCEPTUAL UNDERSTANDING	2 FLUENCY	3 APPLICATION

Conceptual Bridge In this lesson, students extend their understanding of linear functions to piecewise, step, and absolute value functions. They build fluency by graphing these types of functions, and they apply their understanding by solving real-world problems related to these functions.

Interactive Presentation

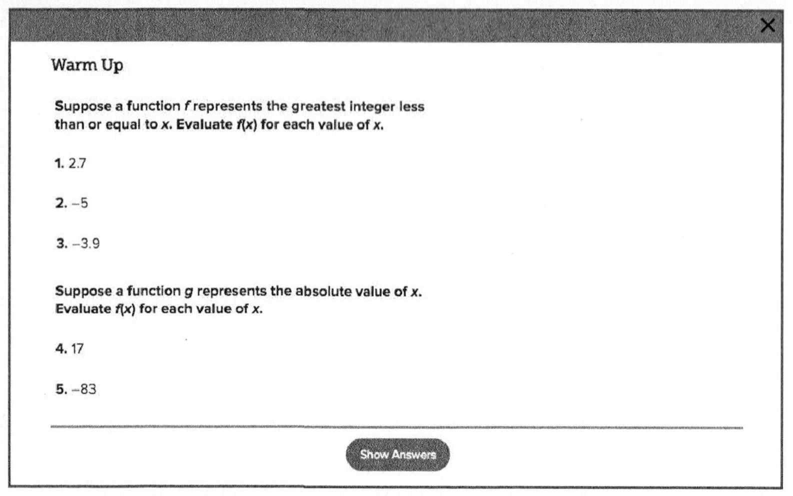

Warm Up

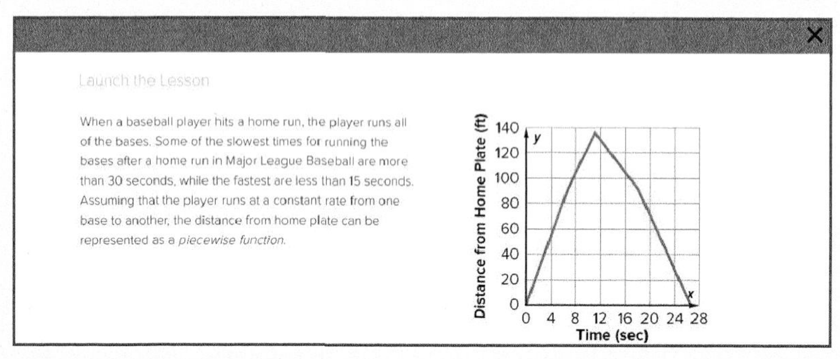

Launch the Lesson

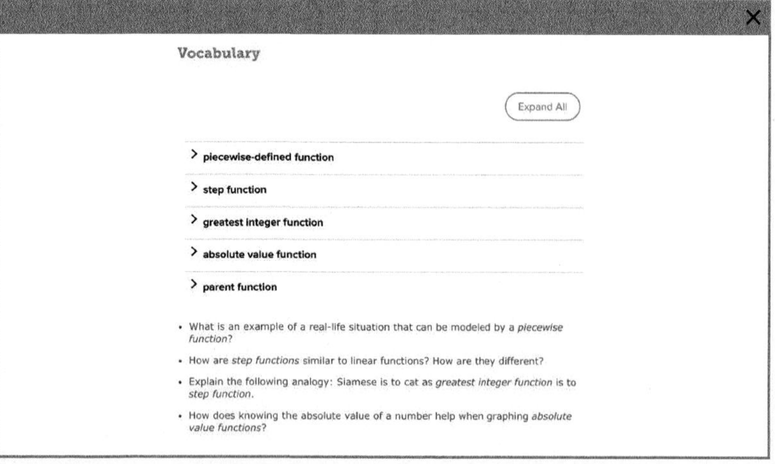

Today's Vocabulary

Warm Up

Prerequisite Skills

The Warm Up exercises address the following prerequisite skill for this lesson:

- evaluating greatest integer and absolute value expressions

Answers:

1. 2
2. −5
3. −4
4. 17
5. 83

Launch the Lesson

Teaching the Mathematical Practices

2 Create Representations In the Launch the Lesson, guide students to write equations that model each part of the graph.

Go Online to find additional teaching notes and questions to promote classroom discourse.

Today's Standards

Tell students that they will be addressing these content and practice standards in this lesson. You may wish to have a student volunteer read aloud *How can I meet these standards?* and *How can I use these practices?*, and connect these to the standards.

See the Interactive Presentation for I Can statements that align with the standards covered in this lesson.

Today's Vocabulary

Tell students that they will be using these vocabulary terms in this lesson. You can expand each row if you wish to share the definitions. Then discuss the questions below with the class.

Mathematical Background

Piecewise-defined functions are also called multi-part functions or split-domain functions.

Explore Using Tables to Graph Piecewise Functions

Objective

Students write and graph a piecewise function given a table.

 Teaching the Mathematical Practices

5 Use Mathematical Tools Point out that to complete the Explore, students will need to use an interactive sketch. Work with students to explore and deepen their understanding of piecewise-defined functions.

Ideas for Use

Recommended Use Present the Inquiry Question, or have a student volunteer read it aloud. Have students work in pairs to complete the Explore activity on their devices. Pairs should discuss each of the questions. Monitor student progress during the activity. Upon completion of the Explore activity, have student volunteers share their responses to the Inquiry Question.

What if my students don't have devices? You may choose to project the activity on a whiteboard. A printable worksheet for each Explore is available online. You may choose to print the worksheet so that individuals or pairs of students can use it to record their observations.

Summary of the Activity

Students will complete guiding exercises throughout the Explore activity. Students will analyze the relationships between the x-values and the function values for three different subsets of the domain. Then students will use the sketch to graph the piecewise defined function and then write the function. Then, students will answer the Inquiry Question.

DIFFERENTIATE

Language Development Activity ELL
Have students define *piecewise* within their native language, including visual graphic representation. Share with the class.

(continued on the next page)

Interactive Presentation

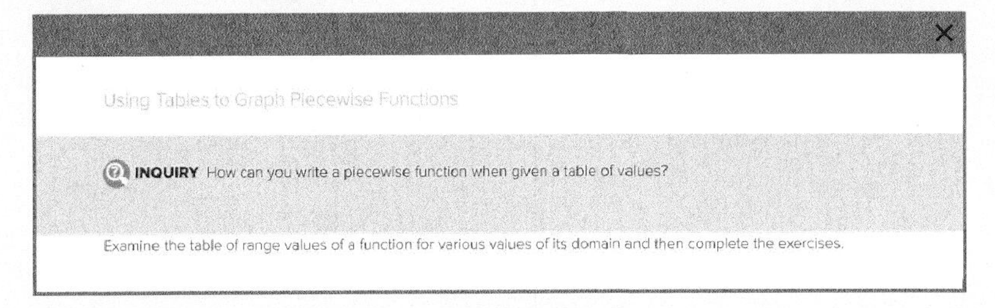

Explore

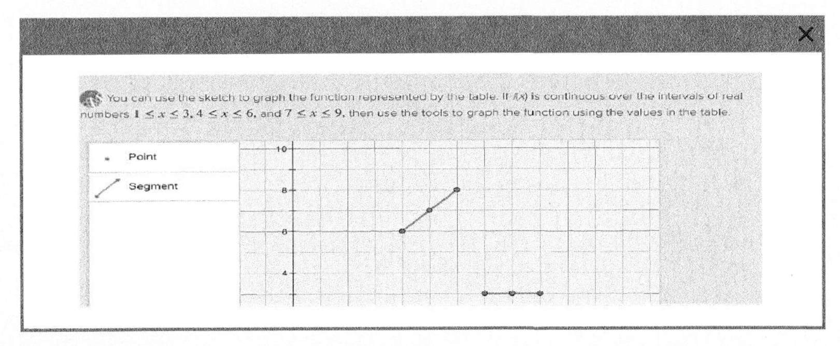

Explore

TYPE

Students enter the relationship between the x-values and $f(x)$ for the subset of x-values.

WEB SKETCHPAD

Students use the sketch to graph a piecewise function.

TYPE

Students enter the expressions that make up the piecewise function.

Interactive Presentation

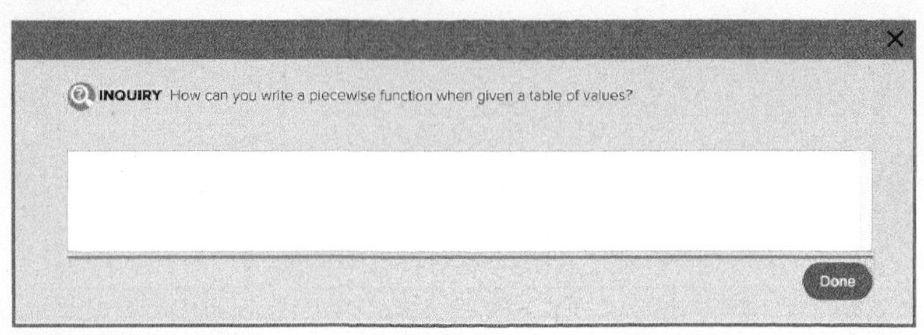

Explore

TYPE

Students respond to the Inquiry Question and can view a sample answer.

Explore Using Tables to Graph Piecewise Functions (*continued*)

Questions

Have students complete the Explore activity.

Ask:

- Why were the intervals of the domain $1 \leq x \leq 3$, $4 \leq x \leq 6$, and $7 \leq x \leq 9$? Sample answer: Because in those intervals the x-values had a specific relationship with the function values.
- What do you think creates a piecewise function? Sample answer: A function that is made up of different parts over intervals of the domain.

Inquiry

How can you write a piecewise function when given a table of values? Sample answer: You can write a piecewise function by determining the function that represents each interval and then writing the functions and intervals as a piecewise function.

Go Online to find additional teaching notes and sample answers for the guiding exercises.

Learn Graphing Piecewise-Defined Functions

Objective

Students write and graph piecewise-defined functions by analyzing intervals of the domain.

Teaching the Mathematical Practices

1 Explain Correspondences Encourage students to explain the relationships between dots and circles and the solutions of piecewise-defined functions.

Common Misconception

Some students believe piecewise graphs are not functions. They may link the concept of continuous with the idea of functions, and most piecewise functions are not continuous; thus, students may believe they are not functions. Encourage students to use the vertical line test on graphs to determine function status, not the concept of continuity.

Essential Question Follow-Up

Students have begun graphing piecewise-defined functions.

Ask:

Why is it important to define some functions over a specific interval? Sample answer: Not all functions are valid over all real numbers, so specific domain intervals are required. For example, shipping rates may depend on the weight of the package, so each weight class has a different price.

Example 1 Graph a Piecewise-Defined Function

Teaching the Mathematical Practices

3 Construct Arguments In the Talk About It! feature, students will use stated assumptions, definitions, and previously established results to construct an argument about the form of the piecewise-defined function.

Questions for Mathematical Discourse

AL Why is the domain of the function all real numbers if the function is not continuous? The domain is all real numbers because x can be any real number, including $x = 1$, because one of the expressions includes 1 in its domain.

OL Suppose the domain of the first expression was $x < 1$. How would the domain of the function change? The domain would be all real numbers except 1, because neither domain interval would contain 1.

BL Can you describe a piecewise-defined function that contains only linear expressions and has both a domain and range of all real numbers? Yes; sample answer: If the second expression was $2x - 4$ for $x > 1$, the open circle endpoint would occur at $(1, -2)$, which is an included point for the other expression.

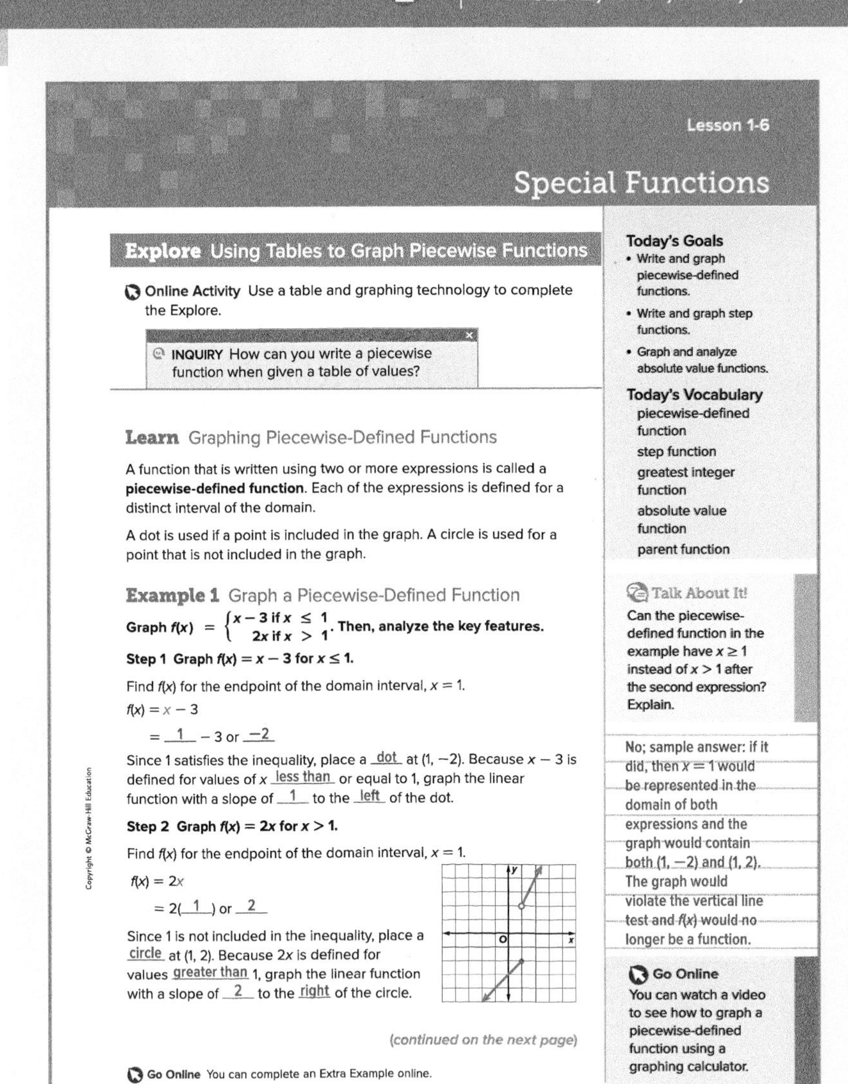

Lesson 1-6

Special Functions

Explore Using Tables to Graph Piecewise Functions

Online Activity Use a table and graphing technology to complete the Explore.

INQUIRY How can you write a piecewise function when given a table of values?

Learn Graphing Piecewise-Defined Functions

A function that is written using two or more expressions is called a **piecewise-defined function**. Each of the expressions is defined for a distinct interval of the domain.

A dot is used if a point is included in the graph. A circle is used for a point that is not included in the graph.

Example 1 Graph a Piecewise-Defined Function

Graph $f(x) = \begin{cases} x - 3 \text{ if } x \leq 1 \\ 2x \text{ if } x > 1 \end{cases}$. **Then, analyze the key features.**

Step 1 Graph $f(x) = x - 3$ for $x \leq 1$.

Find $f(x)$ for the endpoint of the domain interval, $x = 1$.

$f(x) = x - 3$

$= \underline{1} - 3$ or $\underline{-2}$

Since 1 satisfies the inequality, place a __dot__ at $(1, -2)$. Because $x - 3$ is defined for values of x __less than__ or equal to 1, graph the linear function with a slope of __1__ to the __left__ of the dot.

Step 2 Graph $f(x) = 2x$ for $x > 1$.

Find $f(x)$ for the endpoint of the domain interval, $x = 1$.

$f(x) = 2x$

$= 2(\underline{1})$ or $\underline{2}$

Since 1 is not included in the inequality, place a __circle__ at $(1, 2)$. Because $2x$ is defined for values __greater than__ 1, graph the linear function with a slope of __2__ to the __right__ of the circle.

(continued on the next page)

Go Online You can complete an Extra Example online.

Today's Goals
- Write and graph piecewise-defined functions.
- Write and graph step functions.
- Graph and analyze absolute value functions.

Today's Vocabulary
piecewise-defined function
step function
greatest integer function
absolute value function
parent function

Talk About It!
Can the piecewise-defined function in the example have $x \geq 1$ instead of $x > 1$ after the second expression? Explain.

No; sample answer: If it did, then $x = 1$ would be represented in the domain of both expressions and the graph would contain both $(1, -2)$ and $(1, 2)$. The graph would violate the vertical line test and $f(x)$ would no longer be a function.

Go Online
You can watch a video to see how to graph a piecewise-defined function using a graphing calculator.

Copyright © McGraw-Hill Education

Lesson 1-6 • Special Functions 47

Interactive Presentation

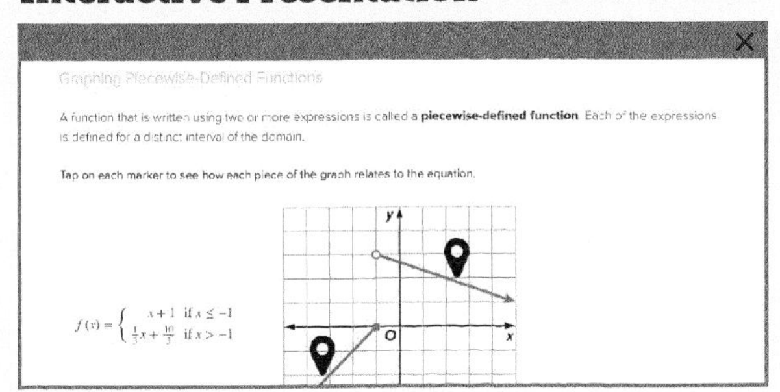

Learn

TAP

Students tap to learn more about each segment of the piecewise graph.

TYPE

Students state the domain of the function and if the function is continuous.

Your Notes

Step 3 Analyze key features.

The function is defined for all values of x, so the domain is all real numbers.

The range is all real numbers less than or equal to -2 and all real numbers greater than 2. This can be represented symbolically as $\{f(x) \mid f(x) \leq -2 \text{ or } f(x) > 2\}$.

The y-intercept is -3, and there is no x-intercept.

The function is increasing for all values of x.

Check

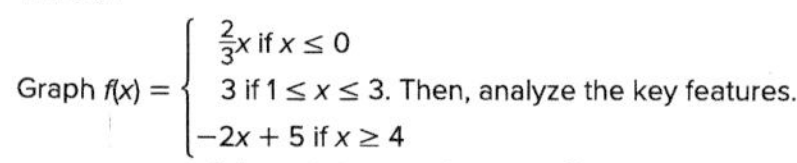

Graph $f(x) = \begin{cases} \frac{2}{3}x \text{ if } x \leq 0 \\ 3 \text{ if } 1 \leq x \leq 3 \\ -2x + 5 \text{ if } x \geq 4 \end{cases}$. Then, analyze the key features.

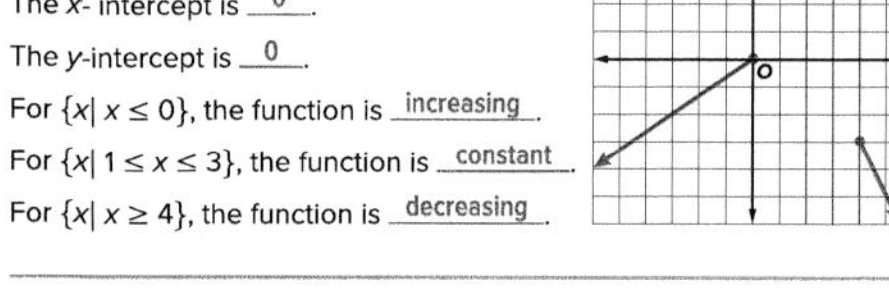

The domain is $\{x \mid x \leq 0, 1 \leq x \leq 3, \text{ or } x \geq 4\}$.

The range is $\{f(x) \mid f(x) \leq 0 \text{ or } f(x) = 3\}$.

The x- intercept is 0.

The y-intercept is 0.

For $\{x \mid x \leq 0\}$, the function is increasing.

For $\{x \mid 1 \leq x \leq 3\}$, the function is constant.

For $\{x \mid x \geq 4\}$, the function is decreasing.

Think About It! What do the domain and range represent in the context of this situation?

Sample answer: The domain is the number of jerseys that the coach can purchase. The range represents the cost of the jerseys.

Example 2 Model by Using a Piecewise-Defined Function

UNIFORMS **The football coach is ordering new jerseys for the new season. The manufacturer charges \$88 for each jersey when five or fewer are ordered, \$75 each for an order of six to 11 jerseys, \$65 each for an order of 12 to 29 jerseys, and \$56 each when thirty or more jerseys are ordered.**

Part A Write a piecewise-defined function describing the cost of the jerseys.

$$f(x) = \begin{cases} 88x & \text{if } 0 < x \leq 5 \\ 75x & \text{if } 5 < x \leq 11 \\ 65x & \text{if } 11 < x \leq 29 \\ 56x & \text{if } x > 29 \end{cases}$$

Part B Evaluate the function.

What would it cost to purchase 11 jerseys? \$825

What would it cost to purchase 25 jerseys? \$1625

Evaluate $f(29)$. $f(29) =$ \$1885

Watch Out! **Evaluating Endpoints of Intervals** When evaluating a piecewise-defined function for a value of x that is an endpoint for two consecutive intervals, be careful to evaluate the function that contains that point.

Go Online You can complete an Extra Example online.

48 **Module 1** • Relations and Functions

Interactive Presentation

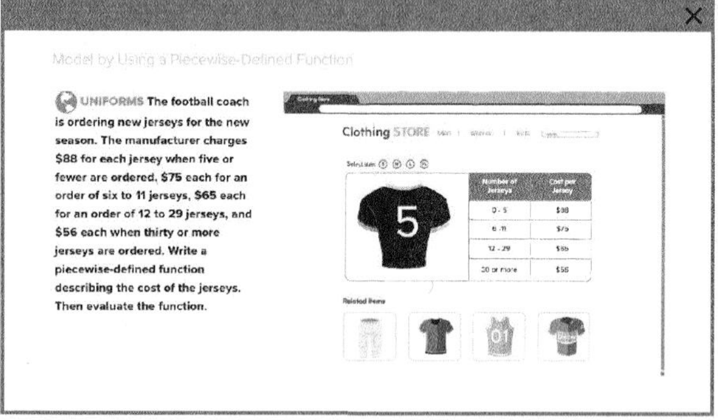

Example 2

DRAG & DROP

Students drag and drop each domain interval to the correct expression to form the piecewise-defined function.

TYPE

Students enter the function values for different x-values.

CHECK

Students complete the Check online to determine whether they are ready to move on.

Common Error

When graphing piecewise-defined functions many students place arrows on both ends of each graphed expression rather than where appropriate. Encourage students to check the domain interval to see if an arrow is appropriate and if so, where to place the arrow.

DIFFERENTIATE

Reteaching Activity AL ELL

IF students are struggling to graph piecewise-defined functions, **THEN** have students generate a table of values for each expression of the function. Start by listing out all the values of the specified domain intervals including a few values before and after. Students highlight each expression and the corresponding domain in the table using different colors, and then complete the table. Students then graph the points represented by the table.

Example 2 Model by Using a Piecewise-Defined Function

Teaching the Mathematical Practices

4 Interpret Mathematical Results In Example 2, point out that to evaluate the piecewise-defined function, students should interpret their results in the context of the problem.

Questions for Mathematical Discourse

AL Does it make sense for the domain to be all real numbers greater than 0? Explain. No; sample answer: It is not possible to order a decimal of a jersey, like 14.32 jerseys. The domain is limited to whole numbers.

OL What would the graph representing $f(x)$ look like? The graph would have four sections that are increasing over each interval of x. Each section increases at a smaller rate than the previous segment.

BL Could the interval for $f(x) = 75x$ also be written as $6 \leq x \leq 11$? Yes. Why can you do this in this case but not for all piecewise-defined functions? This would not work for a continuous function. In this example, $f(x)$ has a domain limited to whole numbers.

Common Error

When evaluating a piecewise-defined function, students may use the wrong expression when asked to evaluate an endpoint. Remind students that in a function, only one expression will include the x-value, so use the expression with the correct inequality symbol.

Go Online

- Find additional teaching notes.
- View performance reports of the Checks.
- Assign or present an Extra Example.

Learn Graphing Step Functions

Objective

Students write and graph step functions by analyzing intervals of the domain.

Teaching the Mathematical Practices

6 Communicate Precisely Encourage students to routinely write or explain their solution methods. Point out that they should use clear definitions when they answer the Think About It! feature.

Important To Know

One common type of piecewise-linear function is called a step function, for which the graph is a series of horizontal line segments. The function is defined by a set of constant functions, and the domain is a set of real numbers while the range is a discrete set of real numbers. The greatest integer function is one kind of step function where the function rounds an x-value to the greatest integer less than or equal to x.

Example 3 Graph a Step Function

Teaching the Mathematical Practices

6 State Meaning of Symbols Guide students to define variables in order to set up and graph the step function in Example 3. Help students identify the independent and dependent variables.

Questions for Mathematical Discourse

AL Why does each interval on the graph start with an open circle and end with a closed circle? Sample answer: An open circle means that the point is not included, and a closed circle means that the point is included. When mailing a letter, the weights are up to a given amount; therefore, the end is included in the interval. The start of the next interval does not include the endpoint of the previous interval.

OL This problem uses the phrase "not over" to indicate whether the endpoint is included. What are more examples of ways to express whether an endpoint is included? Sample answer: "or more" would include the given number as an endpoint for the range of numbers above it. "Smaller than" would not include the given number as an endpoint for the range of numbers below it.

BL Consider a letter within the weight range of $C(x)$. Is it possible to have a letter with a weight that would make it cheaper to split and mail as two separate letters? No. There is no combination of two weights that would result in a cost cheaper than the cost for the total weight.

Learn Graphing Step Functions

A common type of piecewise function is a step function. A **step function** has a graph that is a series of horizontal line segments that may resemble a staircase.

A step function is defined by a set of constant functions. The domain of a step function is an interval of real numbers. The range of a step function is a discrete set of real numbers. The graph of a step function is discontinuous because it cannot be drawn without lifting your pencil.

The **greatest integer function**, written $f(x) = [\![x]\!]$, is one kind of step function in which $f(x)$ is the greatest integer less than or equal to x. For example, $[\![10.7]\!] = 10$, $[\![-6.35]\!] = -7$, and $[\![5]\!] = 5$.

Think About It! Why is the range of the function not expressed as $\{y \mid -3 \le y \le 3\}$?

Sample answer: Not all of the real numbers from -3 to 3 are used as range values. Only the integers -3, -1, and 3 are used.

Think About It! Explain why the value of $[\![4.3]\!]$ is 4, but the value of $[\![-4.3]\!]$ is -5.

Sample answer: The value of $[\![-4.3]\!]$ is -5 instead of -4 because -4 is greater than -4.3.

Example 3 Graph a Step Function

POSTAL RATES **The cost of mailing a first-class letter is determined by rates adopted by the U.S. Postal Service. The rates adopted in 2016 charge \$0.47 for letters not over 1 ounce, \$0.68 if not over 2 ounces, \$0.89 if not over 3 ounces, and \$1.10 if not over 3.54 ounces. Complete the table and draw a graph that represents the charges. State the domain and range.**

Step 1 Make a table.

Let x be the weight of a first-class letter and $C(x)$ represent the cost for mailing it. Use the given rates to make a table.

x	$C(x)$
$0 < x \le 1$	\$0.47
$1 < x \le 2$	\$0.68
$2 < x \le 3$	\$0.89
$3 < x \le 3.54$	\$1.10

Step 2 Make a graph.

Graph the first step of the function. Place a circle at (0, 0.47) since there is no charge for not mailing a letter. Place a dot at (1, 0.47) since a letter weighing one ounce will cost \$0.47 to mail. Draw a segment that connects the points.

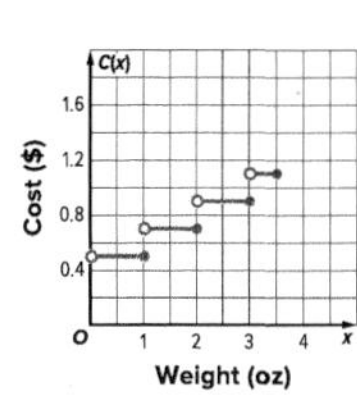

Graph the remaining steps.

Place a circle on the left end of each segment as that domain value is included with the segment below it.

(continued on the next page)

Interactive Presentation

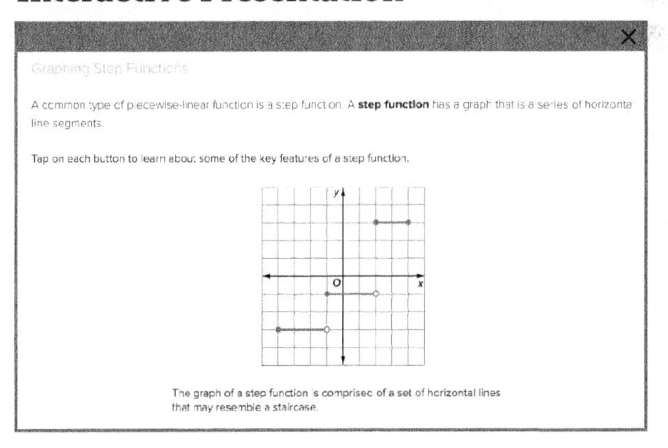

Learn

TAP

Students tap each button to learn some key features about a step function.

TYPE

Students explain the notation of the range of the function.

Step 3 State the domain and range.

The constraints for the weight of a first-class letter are more than 0 ounces up to and including 3.54 ounces. Therefore, the domain is $\{x \mid 0 < x \leq 3.54\}$.

Because the only viable solutions for the cost of mailing a first-class letter are \$0.47, \$0.68, \$0.89, and \$1.10, the range is $\{C(x) = 0.47, 0.68, 0.89, 1.10\}$.

Think About It!
Would $C(x)$ still be a function if the open points at (0, 0.47), (1, 0.68), (2, 0.89), and (3, 1.10) were closed points? Justify your argument.

No; sample answer: If the points were closed, then there would be two values for $C(x)$ for $x = 1$, 2, and 3, and $C(x)$ would not be a function.

Check

FIGURINES Chris and Joaquin design figurines for board game and toy companies. The rate they charge $R(x)$ depends on the number of hours x they spend creating the figurines. Draw a graph that represents the charges. State the domain and range.

x	$R(x)$
$0 < x \leq 5$	500
$5 < x \leq 15$	1400
$15 < x \leq 30$	2500
$30 < x \leq 50$	4000

$D = \{x \mid 0 < x \leq 50\}$;

$R = \{y \mid y = 500, 1400, 2500, 4000\}$

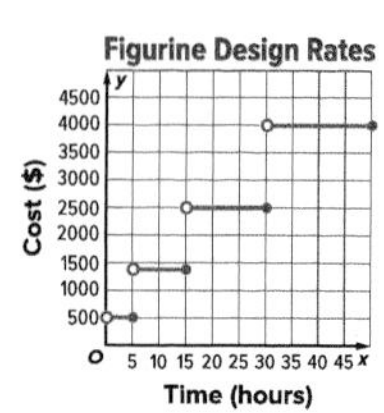

Example 4 Graph a Greatest Integer Function

Complete the table and graph $f(x) = [\![2x - 1]\!]$. State the domain and range.

Step 1 Make a table.

Make a table of the intervals of x and associated values of $f(x)$.

x	$f(x)$
$-1 \leq x < -0.5$	-3
$-0.5 \leq x < 0$	-2
$0 \leq x < 0.5$	-1
$0.5 \leq x < 1$	0
$1 \leq x < 1.5$	1
$1.5 \leq x < 2$	2

Step 2 Make a graph.

Graph the first step. Place a dot at $(-1, -3)$, because $f(-1) = [\![2(-1) - 1]\!] = [\![-3]\!] = -3$. Place a circle at $(-0.5, -3)$, since every decimal value greater than -1 and up to but not including -0.5 produces an $f(x)$ value of -3.

Graph the remaining steps. Place a dot on the left end of each segment as that point is included with the segment, and place a circle on the right end because that domain value is included with the segment above it.

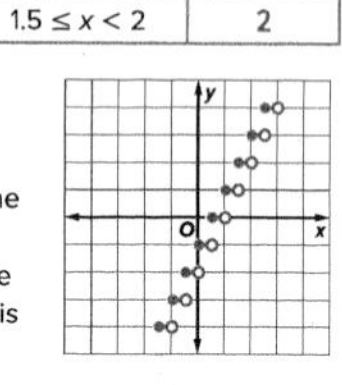

Step 3 State the domain and range.

The domain of $f(x) = [\![2x - 1]\!]$ is all real numbers. The range is all integers.

Think About It!
Will the range of a greatest integer function always be all integers? If not, provide a counterexample.

No; sample answer: If $f(x) = [\![x]\!] + 0.1$, then $f(x) = [\![0]\!] + 0.1 = 0 + 0.1 = 0.1$. Therefore, there is an element in the range that is not an integer.

Go Online You can watch a video to see how to graph step functions.

Go Online You can complete an Extra Example online.

50 **Module 1** • Relations and Functions

Interactive Presentation

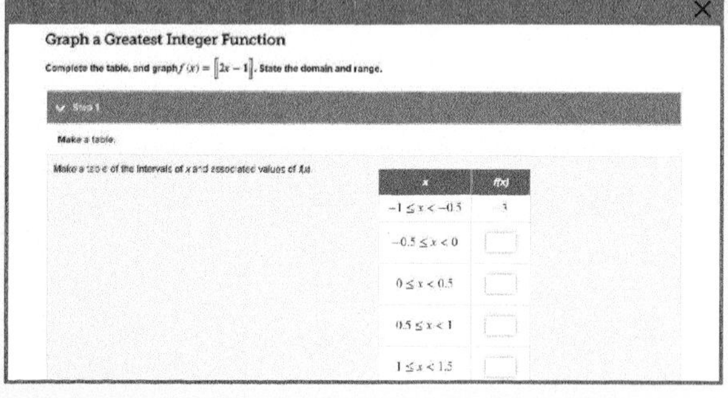

Example 4

TAP

Students move through the steps of graphing a greatest integer function.

TYPE

Students discuss the range of a greatest integer function.

CHECK

Students complete the Check online to determine whether they are ready to move on.

DIFFERENTIATE

Reteaching Activity AL ELL

IF students struggle to graph step functions,

THEN have students draw a coordinate plane for each expression in the step function. They should graph one constant function on each of the grids. Students then cut out the section of each constant function for the domain interval and tape them together to form the graph. Students can then add open and closed circles as dictated by the inequality symbols.

Example 4 Graph a Greatest Integer Function

Teaching the Mathematical Practices

6 Use Precision To graph the greatest integer function, students must calculate accurately and efficiently and express numerical answers with an appropriate degree of precision.

Questions for Mathematical Discourse

AL Why do the domain intervals include the left endpoint but not the right endpoint? The greatest integer function rounds the function value at any x down to the greatest integer less than or equal to that value. Each interval of x will round down to the left endpoint, so it will be included.

OL Why do the domain intervals increase by half a unit rather than a whole unit? Sample answer: Due to x being multiplied by 2, a domain interval of a whole unit results in function values that round to two different integers.

BL Would the graph change if the function was $[\![2x - 1]\!]$? No. Would this always be the case if you moved any constant out of the greatest integer part of the function? No, the graphs would only remain the same if the constant is an integer.

Common Error

Students may believe the domain is limited to the intervals set in the example. Remind students that the table is a helpful tool to graph the greatest integer function but cannot possibly incorporate every possible value of x.

Learn Graphing Absolute Value Functions

Objective

Students graph and analyze absolute value functions.

Teaching the Mathematical Practices

6 Communicate Precisely Encourage students to routinely write or explain their solution methods. Point out that they should use clear definitions when they answer the question in the Think About It! feature.

About the Key Concept

A function that contains an algebraic expression within absolute value symbols is called an absolute value function. These functions can be defined and graphed as a piecewise function. The parent function is represented as $f(x) = |x|$ and can be written as $f(x) = \begin{cases} x \text{ if } x \geq 0 \\ -x \text{ if } x < 0 \end{cases}$. The domain is all real numbers and the range is all nonnegative real numbers. The x- and y-intercept is (0, 0).

DIFFERENTIATE

Enrichment Activity BL

What would be the range of $f(x) = |ax + b| + c$? $\{f(x) \mid f(x) \geq c\}$

What would be the range of $f(x) = -|ax + b| + c$? $\{f(x) \mid f(x) \leq c\}$

Example 5 Graph an Absolute Value Function, Positive Coefficient

Teaching the Mathematical Practices

1 Explain Correspondences Encourage students to explain the relationships between the equation, table, and graph used in this example.

Questions for Mathematical Discourse

AL What is the slope of the portion of the function where $x < 0$? How does it compare to the slope of the other portion of the function? The slope of the function where $x < 0$ is $-\frac{3}{4}$, which is the opposite of the slope of the other portion.

OL Consider a piecewise representation for $f(x)$. Why do you replace the absolute value symbol with a negative when $x < 0$? In this part of the domain, x is negative, so the term inside the absolute value symbols will be negative. The absolute value operation will then make this term positive. Multiplying a negative by a negative yields a positive, so the absolute value symbols can be replaced by a negative for this portion of the function.

BL Given $x + 1 > x$ is always true, does this mean $|x + 1| > |x|$ is always true? Explain. No; sample answer: When $x = -0.5$, the two sides are equal. When $x < -0.5$, $|x + 1| < |x|$ because $-x + 1 < -x$.

Learn Graphing Absolute Value Functions

An **absolute value function** is a function that contains an algebraic expression within absolute value symbols. It can be defined and graphed as a piecewise function.

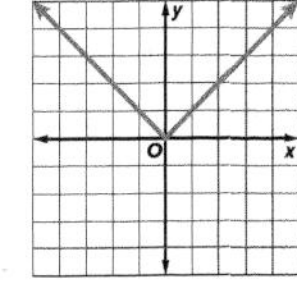

For an absolute value function, $f(x) = |x|$ is the **parent function**, which is the simplest of the functions in a family.

Key Concept • Parent Function of Absolute Value Functions	
parent function	$f(x) = |x|$ or $f(x) = \begin{cases} x \text{ if } x \geq 0 \\ -x \text{ if } x < 0 \end{cases}$
domain	all real numbers
range	all nonnegative real numbers
intercepts	x-intercept: $x = 0$, y-intercept: $y = 0$

Think About It! Describe the line of symmetry of any absolute value function and compare it with the line of symmetry of the parent function.

Sample answer: The line of symmetry passes through the vertex of the absolute value graph. For the parent function, the line of symmetry is $x = 0$, or the y-axis.

Example 5 Graph an Absolute Value Function, Positive Coefficient

Graph $f(x) = \left|\frac{3}{4}x\right| + 3$. State the domain and range.

Create a table of values. Plot the points and connect them with two rays.

x	$f(x) = \left\lvert\frac{3}{4}x\right\rvert + 3$
−4	$\left\lvert\frac{3}{4}(-4)\right\rvert + 3 = 6$
−2	$\left\lvert\frac{3}{4}(-2)\right\rvert + 3 = 4\frac{1}{2}$
0	$\left\lvert\frac{3}{4}(0)\right\rvert + 3 = 3$
2	$\left\lvert\frac{3}{4}(2)\right\rvert + 3 = 4\frac{1}{2}$
4	$\left\lvert\frac{3}{4}(4)\right\rvert + 3 = 6$

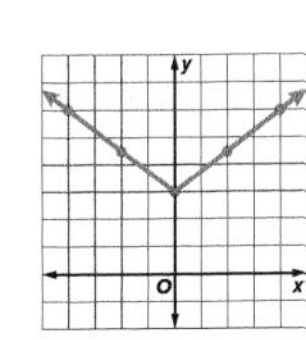

The function is defined for all values of x, so the domain is all real numbers. The function is defined only for values of $f(x)$ such that $f(x) \geq 3$, so the range is $\{f(x) \mid f(x) \geq 3\}$.

Check

Graph $f(x) = |x - 1| + 3$. State the domain and range.

D = all real numbers; R = $\{f(x) \mid f(x) \geq 3\}$

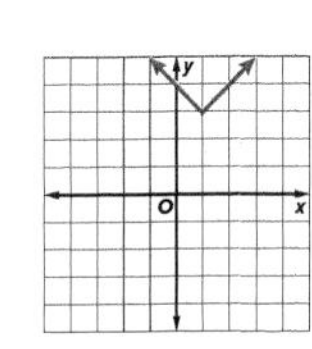

Go Online An alternate method is available for this example.

Go Online You can complete an Extra Example online.

Lesson 1-6 • Special Functions 51

Interactive Presentation

Graphing Absolute Value Functions

An **absolute value function** is a function that contains an algebraic expression within absolute value symbols. It can be defined and graphed as a piecewise function.

For an absolute value function $f(x) = |x|$ is the **parent function**, which is the simplest of the functions in a family.

KEY CONCEPT: PARENT FUNCTION OF ABSOLUTE VALUE FUNCTIONS

Tap on each button to learn more about $f(x) = |x|$.

Learn

TAP

Students move through the key concepts of the parent function of absolute value functions.

TYPE

a

Students describe the line of symmetry of any absolute value function and compare it to the parent function.

Example 6 Graph an Absolute Value Function, Negative Coefficient

Graph $f(x) = -2|x + 1|$. State the domain and range.

Determine the two related linear equations using the two possible cases for the expression inside of the absolute value.

Case 1: $(x + 1)$ is positive.

$f(x) = -2(x + 1)$ — $x + 1$ is positive, so $|x + 1| = x + 1$.

$= -2x - 2$ — Simplify.

Case 2: $(x + 1)$ is negative.

$f(x) = -2[-(x + 1)]$ — $x + 1$ is negative, so $|x + 1| = -(x + 1)$.

$= -2(-x - 1)$ — Distributive Property

$= 2x + 2$ — Simplify.

The x-coordinate of the vertex is the value of x where the two cases of the absolute value are equal.

$-2x - 2 = 2x + 2$ — Set Case 1 equal to Case 2.

$-2x = 2x + 4$ — Add 2 to each side.

$-4x = 4$ — Subtract $2x$ from each side.

$x = -1$ — Divide each side by -4.

The x-coordinate of the vertex represents the constraint of the piecewise-defined function. Write the piecewise-defined function that describes the function and use it to graph the absolute value function.

$$f(x) = \begin{cases} 2x + 2 & \text{if } x < -1 \\ -2x - 2 & \text{if } x \geq -1 \end{cases}$$

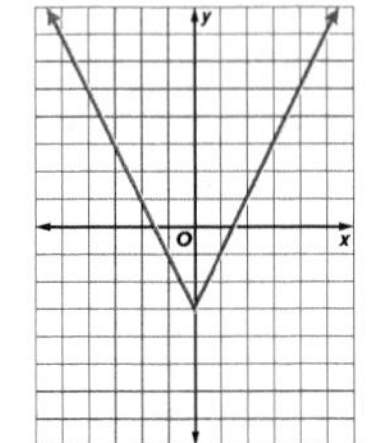

The function is defined for all values of x, so the domain is all real numbers. The function is defined only for values of $f(x)$ such that $f(x) \leq 0$, so the range is $\{f(x) \mid f(x) \leq 0\}$.

Think About It! How does multiplying the absolute value by a negative number affect the shape of the graph? the range?

Sample answer: The graph is reflected in the x-axis. The range is all numbers less than the y-value of the vertex.

Check

Graph $f(x) = 0.25|8x| - 3$. State the domain and range.

D = all real numbers

R = $\{f(x) \mid f(x) \geq -3\}$

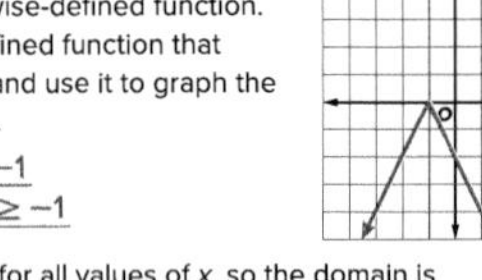

Go Online You can complete an Extra Example online.

52 Module 1 • Relations and Functions

Interactive Presentation

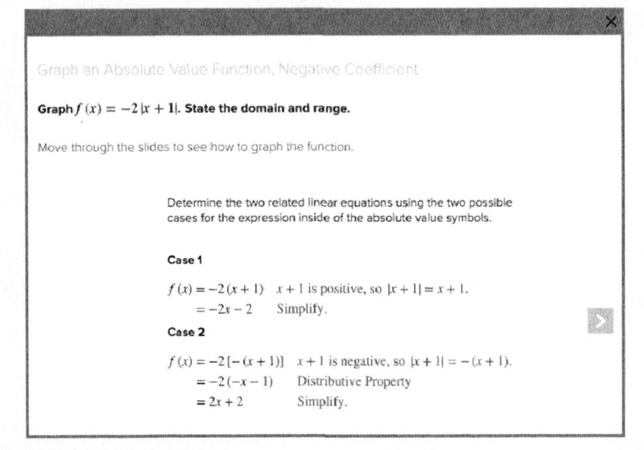

Example 6

TAP

Students move through the steps of writing an absolute value function as a piecewise-defined function and graphing.

TYPE

Students explain how multiplying the absolute value by a negative number affects the shape and range of the graph.

CHECK

Students complete the Check online to determine whether they are ready to move on.

Example 6 Graph an Absolute Value Function, Negative Coefficient

Teaching the Mathematical Practices

3 Analyze Cases Example 6 guides students to examine the two cases for the expression inside the absolute value. Encourage students to familiarize themselves with these cases when graphing absolute value functions.

Questions for Mathematical Discourse

AL How can you tell the range of $f(x)$ without looking at the graph of the function? The entire expression is multiplied by a negative that is outside of the absolute value symbols, so only negative numbers or zero could result from any x input.

OL Does it matter which expression of the piecewise function includes the x-value of the vertex? Explain. No. Sample answer: Since the vertex is part of both linear expressions, it does not matter which expression includes the value, but only one can.

BL Suppose $g(x)$ is similar to $f(x)$ but with the -2 inside the absolute value, $g(x) = |-2(x + 1)|$. How would the graph of $g(x)$ compare to the graph of $f(x)$? $g(x)$ would be a reflection of $f(x)$ across the x-axis. Now consider $h(x)$, which is similar to the graph of $g(x)$ but with the negative sign removed. How would the graph of $h(x)$ compare to the graph of $g(x)$? Explain. The functions would look the same; sample answer: The negative affects both terms in the absolute value, and the absolute value would always make the negative result positive. If the negative only affected one of the terms in the absolute value, this would not be the case.

Common Error

Students may believe the absolute value function will turn a negative coefficient into a positive value. Reinforce that only values inside the absolute value expressions will be positive.

Exit Ticket

Recommended Use

At the end of class, go online to display the Exit Ticket prompt and ask students to respond using a separate piece of paper. Have students hand you their responses as they leave the room.

Alternate Use

At the end of class, go online to display the Exit Ticket prompt and ask students to respond verbally or by using a mini-whiteboard. Have students hold up their whiteboards so that you can see all student responses. Tap to reveal the answer when most or all students have completed the Exit Ticket.

Practice and Homework

Suggested Assignments

Use the table below to select appropriate exercises.

DOK	Topic	Exercises
1, 2	exercises that mirror the examples	1–20
2	exercises that use a variety of skills from this lesson	21–45
2	exercises that extend concepts learned in this lesson to new contexts	46–48
3	exercises that emphasize higher-order and critical thinking skills	49–54

ASSESS AND DIFFERENTIATE

Use the data from the **Checks** to determine whether to provide resources for extension, remediation, or intervention.

IF students score 90% or more on the Checks, **THEN** assign: BL

- Practice Exercises 1–47 odd, 49–54
- Extension: Graphing Greatest Integer Functions
- **ALEKS**® Graphs of Functions; Transformations

IF students score 66%–89% on the Checks, **THEN** assign: OL

- Practice Exercises 1–53 odd
- Remediation, Review Resources: Expressions Involving Absolute Value
- Personal Tutors
- Extra Examples 1–6
- **ALEKS**® Operations with Signed Numbers

IF students score 65% or less on the Checks, **THEN** assign: AL

- Practice Exercises 1–19 odd
- Remediation, Review Resources: Expressions Involving Absolute Value
- *Quick Review Math Handbook*: Special Functions
- **ALEKS**® Operations with Signed Numbers

Name ______ Period ______ Date ______

Practice

Go Online You can complete your homework online.

Examples 1 and 2

Graph each function. Then, analyze the key features. 1–4. See margin.

1. $f(x) = \begin{cases} -1 & \text{if } x \leq 0 \\ 2x & \text{if } 0 < x \leq 3 \\ 6 & \text{if } x > 3 \end{cases}$

2. $f(x) = \begin{cases} -x & \text{if } x < -1 \\ 0 & \text{if } -1 \leq x \leq 1 \\ x & \text{if } x > 1 \end{cases}$

3. $f(x) = \begin{cases} x & \text{if } x < 0 \\ 2 & \text{if } x \geq 0 \end{cases}$

4. $h(x) = \begin{cases} 3 & \text{if } x < -1 \\ x + 1 & \text{if } x > 1 \end{cases}$

5. TILE Mark is purchasing new tile for his bathrooms. The home improvement store charges $48 for each box of tiles when three or fewer boxes are purchased, $45 for each box when 4 to 8 boxes are purchased, $42 for each box when 9 to 19 boxes are purchased, and $38 for each box when more than nineteen boxes are purchased.

a. Write a piecewise-defined function describing the cost of the boxes of tile. See margin.

b. What is the cost of purchasing 5 boxes of tile? What is the cost of purchasing 19 boxes of tile? $225; $798

6. BOOKLETS A digital media company is ordering booklets to promote their business. The manufacturer charges $0.50 for each booklet when 50 or fewer are ordered, $0.45 for each booklet when 51 to 100 booklets are ordered, $0.40 for each booklet when 101 to 250 booklets are ordered, and $0.35 for each booklet when 251 or more booklets are ordered. Each order consists of a $10 shipping charge, no matter the size of the order.

a. Write a piecewise-defined function describing the cost of ordering booklets. See margin.

b. What is the cost of purchasing 132 booklets? What is the cost of purchasing 518 booklets? $62.80; $191.30

Examples 3 and 4

Graph each function. State the domain and range. 7. See margin. 8–10. See Mod. 1 Answer Appendix.

7. $f(x) = [\![x]\!] - 6$

8. $h(x) = [\![3x]\!] - 8$

9. $f(x) = [\![x + 1]\!]$

10. $f(x) = [\![x - 3]\!]$

11. PARKING The rates at a short-term parking garage are $5.00 for 2 hours or less, $10.00 for 4 hours or less, $15.00 for 6 hours or less, and $20.00 for 8 hours or less. Draw a graph that represents the charges. State the domain and range. See Mod. 1 Answer Appendix.

12. BOWLING The bowling alley offers special team rates. They charge $30 for one hour or less of team bowling, $45 for 2 hours or less, and $60 for unlimited bowling after 2 hours of play. Draw a graph that represents the charges. State the domain and range. See Mod. 1 Answer Appendix.

Examples 5 and 6

Graph each function. State the domain and range. 13–20. See Mod. 1 Answer Appendix.

13. $f(x) = |x - 5|$

14. $g(x) = |x + 2|$

15. $h(x) = |2x| - 8$

16. $k(x) = |-3x| + 3$

17. $f(x) = 2|x - 4| + 6$

18. $h(x) = -3\,|0.5x + 1| - 2$

19. $g(x) = 2|x|$

20. $f(x) = |x| + 1$

Mixed Exercises

Graph each function. State the domain and range. 21–32. See Mod. 1 Answer Appendix.

21. $f(x) = \begin{cases} -3x & \text{if } x \leq -4 \\ x & \text{if } 0 < x \leq 3 \\ 8 & \text{if } x > 3 \end{cases}$

22. $f(x) = \begin{cases} 2x & \text{if } x \leq -6 \\ 5 & \text{if } -6 < x \leq 2 \\ -2x + 1 & \text{if } x > 4 \end{cases}$

23. $g(x) = \begin{cases} 2x + 2 & \text{if } x < -6 \\ x & \text{if } -6 \leq x \leq 2 \\ -3 & \text{if } x > 2 \end{cases}$

24. $g(x) = \begin{cases} -2 & \text{if } x < -4 \\ x - 3 & \text{if } -1 \leq x \leq 5 \\ 2x - 15 & \text{if } x > 7 \end{cases}$

25. $f(x) = \begin{cases} -0.5x + 1.5 & \text{if } x \leq 1 \\ x - 4 & \text{if } x > 1 \end{cases}$

26. $f(x) = |x - 2|$

27. $f(x) = [\![x + 2]\!]$

28. $g(x) = 2[\![0.5x + 4]\!]$

29. $f(x) = [\![|0.5x|]\!]$

30. $g(x) = |[\![2x]\!]|$

31. $g(x) = \begin{cases} [\![x]\!] & \text{if } x < -4 \\ x + 1 & \text{if } -4 \leq x \leq 3 \\ -x & \text{if } x > 3 \end{cases}$

32. $h(x) = \begin{cases} -|x| & \text{if } x < -6 \\ |x| & \text{if } -6 \leq x \leq 2 \\ |-x| & \text{if } x > 2 \end{cases}$

33. Identify the domain and range of $h(x) = |x + 4| + 2$. D = {all real numbers}; R = $\{h(x) \mid h(x) \geq 2\}$

34. FINANCE For every transaction, a certain financial advisor gets a 5% commission, regardless of whether the transaction is a deposit or withdrawal. Write a formula using the absolute value function for the advisor's commission C. Let D represent the value of one transaction. $C = 0.05\,|D|$ or $C = |0.05D|$

35. GAMING The graph shows the monthly fee that an an online gaming site charges based on the average number of hours spent online per day. Write the function represented by the graph. See Mod. 1 Answer Appendix.

Monthly Fee ($)

Gaming Time (h)

36. ROUNDING A science teacher instructs students to round their measurements as follows: If the decimal portion of a measurement is less than 0.5 mm, round down to the nearest whole millimeter. If the decimal portion of a measurement is exactly 0.5 or greater, round up to the next whole millimeter. Write a formula to represent the rounded measurements. $l = [\![x + 0.5]\!]$ millimeters

37. REUNIONS The cost to reserve a banquet hall is $500. The catering cost per guest is $17.50 for the first 40 guests and $14.75 for each additional guest.

a. Write a piecewise-defined function describing the cost C of the reunion. $C(x) = \begin{cases} 500 + 17.50x & \text{if } 0 \leq x \leq 40 \\ 1200 + 14.75(x - 40) & \text{if } x \geq 41 \end{cases}$

b. Use a graphing calculator to graph the function. See Mod. 1 Answer Appendix.

c. If the Cramers can spend up to $900 on an event, what is the greatest number of guests that can attend? Explain. See Mod. 1 Answer Appendix.

Name ______ Period ______ Date ______

38. SAVINGS Nathan puts \$200 into a checking account when he gets his paycheck each month. The value of his checking account is modeled by $v = 200[\![m]\!]$, where m is the number of months that Nathan has been working. After 105 days, how much money is in the account? \$600

39. POLITICS The approval rating $R(t)$, measured as a percent, of a class officer during her 9-month term starting in September is described by the graph, where t is her time in office.

   a. Formulate a piecewise-defined function $R(t)$ describing the approval rating of this class officer. Then, identify the range. See Mod. 1 Answer Appendix.

   b. During which months is the approval rating increasing? The graph is increasing from $t = 0$ to $t = 3$. This corresponds to the months of September, October, and November.

40. STRUCTURE Consider the functions $f(x) = 3[\![x]\!]$ and $g(x) = [\![3x]\!]$ for $0 \leq x \leq 2$.

   a. Graph each function. a–e. See Mod. 1 Answer Appendix.

   b. What effect does this 3 appear to have on the graphs?

   c. Consider the functions $f(x) = 4[\![x]\!]$ and $g(x) = [\![4x]\!]$ for $0 \leq x \leq 2$. Graph each function.

   d. What effect does this 4 appear to have on the graphs?

   e. Generalize your findings from **parts a through d** to explain the differences between $f(x) = n[\![x]\!]$ and $g(x) = [\![nx]\!]$ for $0 \leq x \leq 2$, where n is any positive integer greater than or equal to 2.

41. USE TOOLS Use a graphing calculator to graph the absolute value of the greatest integer of x, or $f(x) = |[\![x]\!]|$. Is the graph what you expected? Explain. See Mod. 1 Answer Appendix.

Write a piecewise-defined function for each graph.

42.

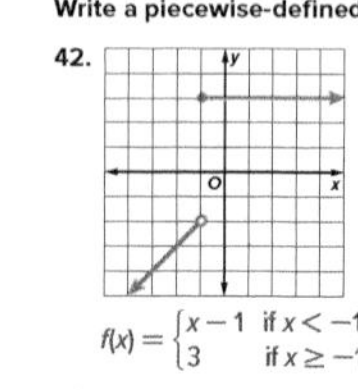

$$f(x) = \begin{cases} x - 1 & \text{if } x < -1 \\ 3 & \text{if } x \geq -1 \end{cases}$$

43.

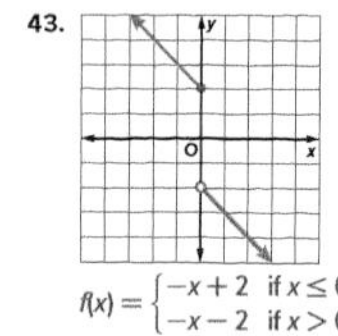

$$f(x) = \begin{cases} -x + 2 & \text{if } x \leq 0 \\ -x - 2 & \text{if } x > 0 \end{cases}$$

44.

$$f(x) = \begin{cases} 2x & \text{if } x \leq 1 \\ -3 & \text{if } x \geq 2 \end{cases}$$

45.

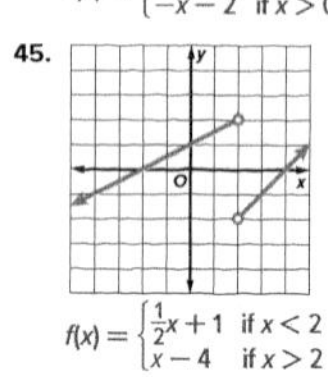

$$f(x) = \begin{cases} \frac{1}{2}x + 1 & \text{if } x < 2 \\ x - 4 & \text{if } x > 2 \end{cases}$$

46. SKYSCRAPERS To clean windows of skyscrapers, some companies use a carriage. A carriage is mounted on a railing on the roof of a skyscraper and moves up and down using cables. A crew plans to start at the 12th floor and move the carriage down as they clean the windows on the west side of a building. If the crew members clean windows at a constant rate of 0.75 floor per hour, the absolute value function $f(t) = |0.75t - 12|$ represents the the number of floors above ground level that the carriage is after t hours. Graph the function. How far above ground level is the carriage after 4 hours? 9 floors; See Mod. 1 Answer Appendix for graph.

47. TAXIS The table shows the cost C of a taxi ride of m miles. Graph the function. State the domain and range. See Mod. 1 Answer Appendix.

m	C
$0 < m \leq 1$	\$2.00
$1 < m \leq 2$	\$4.00
$2 < m \leq 3$	\$6.00
$3 < m \leq 4$	\$8.00
$4 < m \leq 5$	\$10.00
$5 < m \leq 6$	\$12.00

48. WALKING Jackson left his house and walked at a constant rate. After 20 minutes, he was 2 miles from his house. Jackson then walked back towards his house at a constant rate. After another 30 minutes he arrived at his house.

   a. Jackson wants to write a function to model the distance from his house d as after t minutes. Should Jackson write an absolute value function or a piecewise-defined function? Explain your reasoning. a–d. See Mod. 1 Answer Appendix.

   b. Write an appropriate function to model the distance from his house d as after t minutes.

   c. Graph the function.

   d. State the domain and range.

Higher-Order Thinking Skills

49. CREATE Write an absolute value relation in which the domain is all nonnegative numbers and the range is all real numbers. Sample answer: $|y| = x$

50. PERSEVERE Graph $|y| = 2|x + 3| - 5$. See Mod. 1 Answer Appendix.

51. ANALYZE Find a counterexample to the statement and explain your reasoning.

   In order to find the greatest integer function of x when x is not an integer, round x to the nearest integer. See margin.

52. CREATE Write an absolute value function in which $f(5) = -3$. Sample answer: $f(x) = -|x - 2|$

53. WRITE Explain how piecewise functions can be used to accurately represent real-world problems. See margin.

54. WRITE Explain the difference between a piecewise function and step function. See margin.

Answers

1. The function is defined for all values of x, so the domain is all real numbers. The range is -1 and all real numbers greater than 0 and less than or equal to 6, which is also represented as $\{f(x) \mid f(x) = -1 \text{ or } 0 < f(x) \leq 6\}$. The y-intercept is -1, and there is no x-intercept. The function is increasing when $0 < x \leq 3$.

2. The function is defined for all values of x, so the domain is all real numbers. The range is 0 and all real numbers greater than 1, which is also represented as $\{f(x) \mid f(x) = 0 \text{ or } f(x) > 1\}$. The y-intercept is 0, and the function intersects the x-axis over $-1 \leq x \leq 1$. The function is increasing when $x < -1$ and $x > 1$.

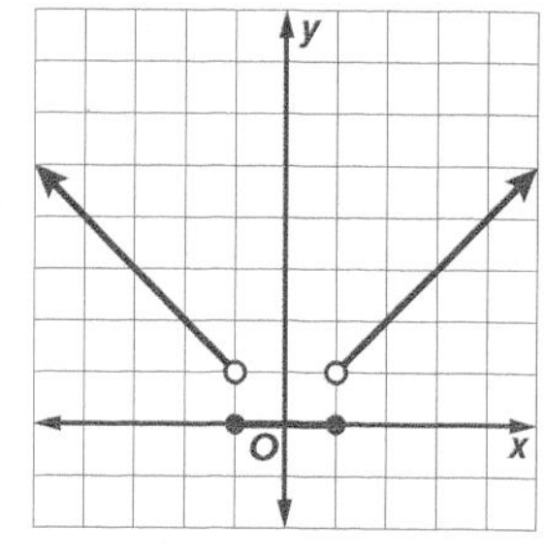

3. The function is defined for all values of x, so the domain is all real numbers. The range is 2 and all real numbers less than 0, which is also represented as $\{f(x) \mid f(x) = 0 \text{ or } f(x) < 0\}$. The y-intercept is 2, and there is no x-intercept. The function is decreasing when $x < 0$.

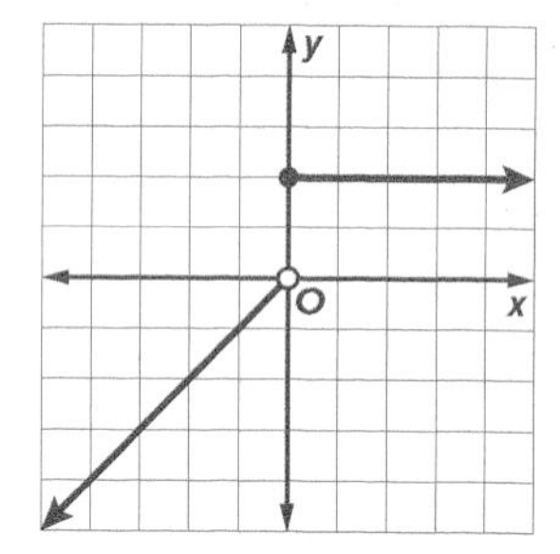

4. The domain is all real numbers less than -1 or all real numbers greater than 1, or $\{x \mid x < -1 \text{ or } x > 1\}$. The range is all real numbers greater than 2, which is also represented as $\{h(x) \mid h(x) > 2\}$. There is no y-intercept, and there is no x-intercept. The function is increasing when $x > 1$.

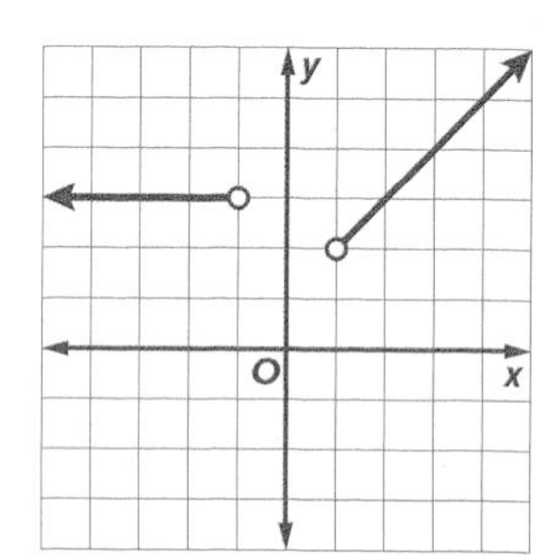

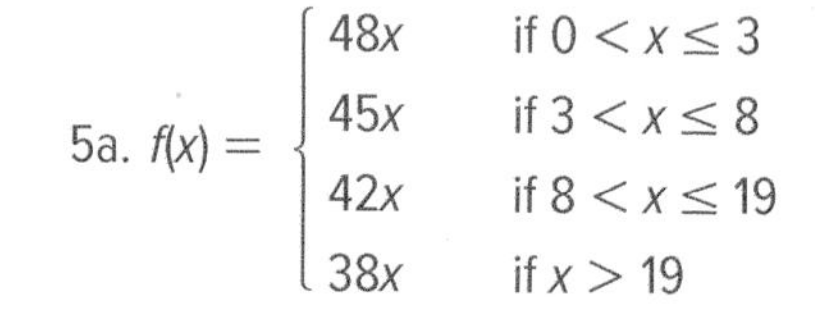

5a. $$f(x) = \begin{cases} 48x & \text{if } 0 < x \leq 3 \\ 45x & \text{if } 3 < x \leq 8 \\ 42x & \text{if } 8 < x \leq 19 \\ 38x & \text{if } x > 19 \end{cases}$$

6a. $$f(x) = \begin{cases} 0.50x + 10 & \text{if } 0 < x \leq 50 \\ 0.45x + 10 & \text{if } 51 \leq x \leq 100 \\ 0.40x + 10 & \text{if } 101 \leq x \leq 250 \\ 0.35x + 10 & \text{if } x > 250 \end{cases}$$

51. 8.6; Sample answer: The greatest integer function asks for the greatest integer less than or equal to the given value; thus 8 is the greatest integer. If we were to round this value to the nearest integer, we would round up to 9.

53. Sample answer: Piecewise functions can be used to represent the cost of items when purchased in quantities, such as a dozen eggs.

54. Sample answer: A piecewise function represents any function over different parts of a domain, where a step function represents a constant over different parts of a domain.

Lesson 1-7

Transformations of Functions

F.IF.4, F.BF.3

LESSON GOAL

Students identify and use transformations of functions.

1 LAUNCH

 Launch the lesson with a **Warm Up** and an introduction.

2 EXPLORE AND DEVELOP

 **Develop:**

Translations of Functions
- Translations
- Identify Translated Functions from Graphs

Dilations and Reflections of Functions
- Vertical Dilations
- Horizontal Dilations

 **Explore:** Using Technology to Transform Functions

 **Develop:**

Transformations of Functions
- Multiple Transformations of Functions
- Apply Transformations of Functions
- Identify an Equation from a Graph

 You may want your students to complete the **Checks** online.

3 REFLECT AND PRACTICE

 Exit Ticket

 Practice

DIFFERENTIATE

 View reports of student progress on the **Checks** after each example.

Resources	AL	OL	BL	ELL
Remediation: Transformations of Linear Functions	●	●		●
Extension: Transformations of Points on the Coordinate Plane		●	●	●

Language Development Handbook

Assign page 7 of the *Language Development Handbook* to help your students build mathematical language related to the transformations of functions.

ELL You can use the tips and suggestions on page T7 of the handbook to support students who are building English proficiency.

Suggested Pacing

90 min	1 day
45 min	2 days

Focus

Domain: Functions

Standards for Mathematical Content:

F.IF.4 For a function that models a relationship between two quantities, interpret key features of graphs and tables in terms of the quantities, and sketch graphs showing key features given in a verbal description of the relationship.

F.BF.3 Identify the effect on the graph of replacing $f(x)$ by $f(x) + k$, $kf(x)$, $f(kx)$, and $f(x + k)$ for specific values of k (both positive and negative); find the value of k given the graphs. Experiment with cases and illustrate an explanation of the effects on the graph using technology.

Standards for Mathematical Practice:

2 Reason abstractly and quantitatively.
7 Look for and make use of structure.

Coherence

Vertical Alignment

Previous
Students graphed linear functions and inequalities in two variables.
A.CED.3, F.IF.4

Now
Students identify and use transformations of functions.
F.IF.4, F.BF.3

Next
Students will identify the effect of transformations of the graphs of nonlinear functions, including logarithmic and trigonometric functions. **F.BF.3**

Rigor

The Three Pillars of Rigor

1 CONCEPTUAL UNDERSTANDING	2 FLUENCY	3 APPLICATION

Conceptual Bridge In this lesson, students develop an understanding of transformations of functions. They build fluency by describing transformations and identifying transformed functions. They apply their understanding by solving real-world problems involving transformations.

Interactive Presentation

Warm Up

Find the coordinates of the point $(-1, 4)$ after each transformation.

1. horizontal translation of 2 units right
2. horizontal translation of 6 units left
3. vertical translation of −5 units
4. reflection across x-axis
5. reflection across y-axis

Show Answers

Warm Up

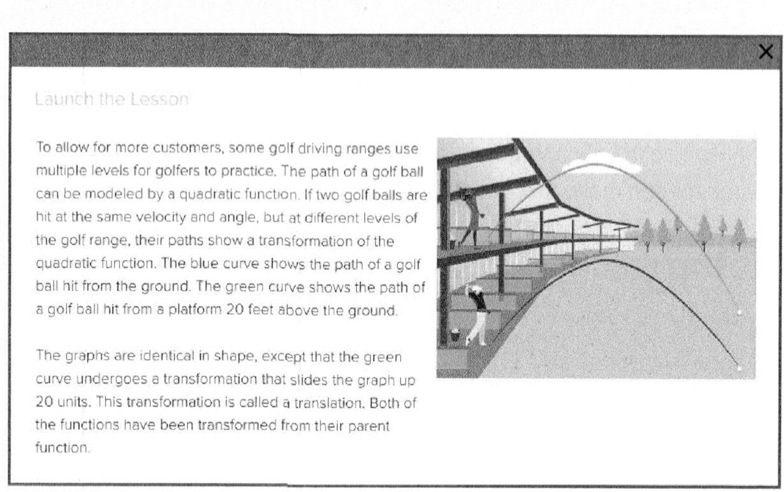

Launch the Lesson

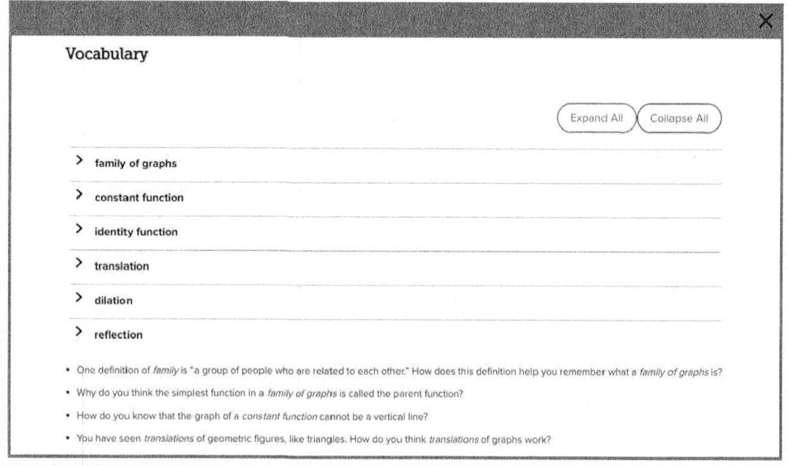

Today's Vocabulary

Warm Up

Prerequisite Skills

The Warm Up exercises address the following prerequisite skill for this lesson:

- transforming points

Answers:

1. (1, 4)
2. (−7, 4)
3. (−1, −1)
4. (−1, −4)
5. (1, −4)

Launch the Lesson

 Teaching the Mathematical Practices

1 Explain Correspondences Help students to see how the paths can be represented by a quadratic function. Encourage students to explain the relationship between paths of the golf balls.

Go Online to find additional teaching notes and questions to promote classroom discourse.

Today's Standards

Tell students that they will be addressing these content and practice standards in this lesson. You may wish to have a student volunteer read aloud *How can I meet these standards?* and *How can I use these practices?*, and connect these to the standards.

See the Interactive Presentation for I Can statements that align with the standards covered in this lesson.

Today's Vocabulary

Tell students that they will be using these vocabulary terms in this lesson. You can expand each row if you wish to share the definitions. Then discuss the questions below with the class.

Mathematical Background

A family of graphs is a group of graphs that display one or more similar characteristics. The parent graph, which is the graph of the parent function, is the simplest of the graphs in a family.

Explore Using Technology to Transform Functions

Objective

Students use a sketch to explore how changing the parameters changes the graphs of functions.

Teaching the Mathematical Practices

5 Decide When to Use Tools Mathematically proficient students can make sound decisions about when to use mathematical tools such as interactive sketches. Help them see why using these tools will help to explore transformations of functions and what the limitations are of using the tools.

Ideas for Use

Recommended Use Present the Inquiry Question, or have a student volunteer read it aloud. Have students work in pairs to complete the Explore activity on their devices. Pairs should discuss each of the questions. Monitor student progress during the activity. Upon completion of the Explore activity, have student volunteers share their responses to the Inquiry Question.

What if my students don't have devices? You may choose to project the activity on a whiteboard. A printable worksheet for each Explore is available online. You may choose to print the worksheet so that individuals or pairs of students can use it to record their observations.

Summary of the Activity

Students will complete guiding exercises throughout the Explore activity. Students will use the sketch to investigate how each parameter of a function transformation affects the graph of a parent function. Students will analyze each transformation. Then, students will answer the Inquiry Question.

DIFFERENTIATE

Language Development Activity ELL
Discuss the choice to use (h, k), where h is an x-value (which means it can only change or move along the **h**orizontal axes left and right) while k is a y-value (which means it can only change or move along the "verti**k**al" axes up and down).

(continued on the next page)

Interactive Presentation

Using Technology to Transform Functions

INQUIRY How does performing an operation on a function change its graph?

If $f(x)$ is the parent function, changing the parameters a, h, and k in a function $g(x) = a \cdot f(x - h) + k$ changes the graph from the graph of the parent function. You can use the sketch to explore the effect that changing the parameters has on the shape, vertex, and axis of symmetry of linear, quadratic, and absolute value functions. Then complete the exercises.

Explore

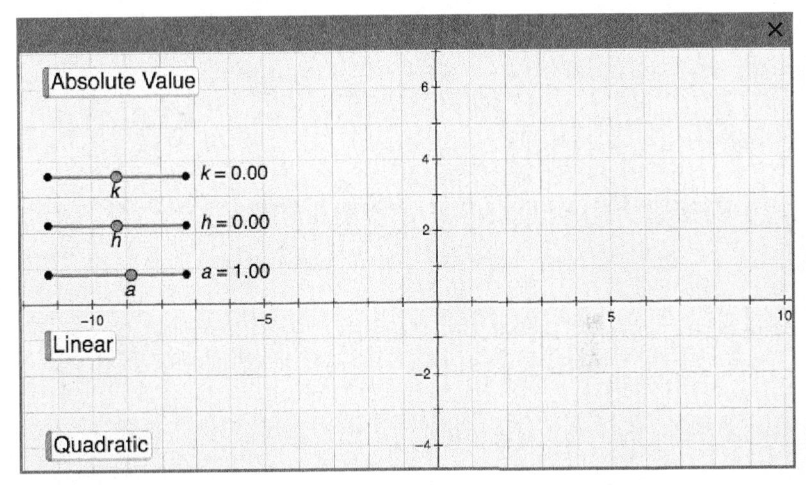

Explore

WEB SKETCHPAD

Students use the sketch to explore function transformations.

TYPE

Students describe similarities and differences among each transformation and the parent function.

Interactive Presentation

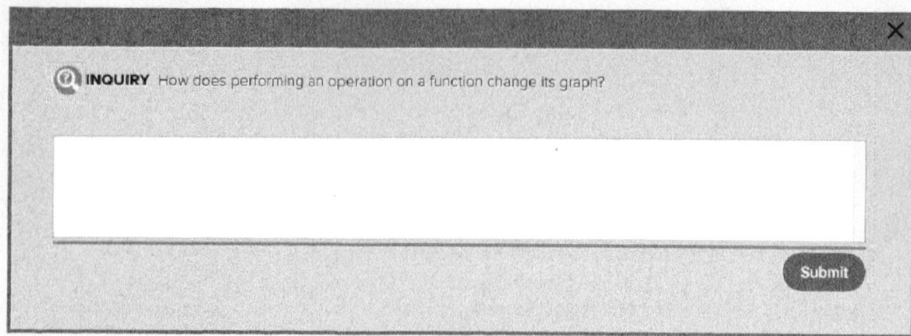

Explore

TYPE

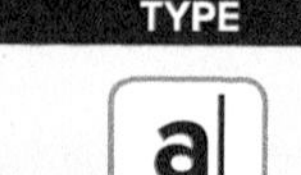

Students respond to the Inquiry Question and can view a sample answer.

Explore Using Technology to Transform Functions (*continued*)

Questions

Have students complete the Explore activity.

Ask:

- Consider the graph of the function $f(x) = x^2$. How will the graph of $g(x) = (x - 2)^2$ be different? Sample answer: The graph of $g(x)$ will be the same as the graph of $f(x)$ but shifted right 2 units.
- If the graph of $p(x)$ was shifted up 5 units to create $h(x)$, what will be the equation of $h(x)$? $h(x) = p(x) + 5$

Inquiry

How does performing an operation on a function change its graph? Sample answer: Adding a value to the function moves the graph up or down. Subtracting a value from x moves the graph left or right. Multiplying the function by a value makes the graph wider or narrower or flips it over the x-axis.

Go Online to find additional teaching notes and sample answers for the guiding exercises.

Learn Translations of Functions

Objective

Students identify the effect on the graphs of functions by replacing $f(x)$ with $f(x) + k$ and $f(x - h)$ for positive and negative values.

Teaching the Mathematical Practices

7 Use Structure Help students to explore how the structure of functions can be used to identify the translation in the function.

About the Key Concept

Graphs and equations of graphs that have at least one characteristic in common are a family of graphs. When the parent graph is transformed, it creates other members in the family. The constant function is $f(x) = a$, where a is any real number. The identity function, $f(x) = x$, is the parent function to most linear functions. Functions in the form $f(x) = |x|$ are absolute value functions, while $f(x) = x^2$ is a quadratic function. When parent graphs are translated, the graphs are slid from one position to another without being turned. When a constant k is added to a function, the graph moves up or down k units. When a constant h is subtracted from x before evaluation, the function is moved right or left h units.

Common Misconception

A common misconception students have about translated functions is when a constant is added to or subtracted from x, the translation is in the same direction as the sign. For example, students think $f(x + 1)$ would be a shift of one unit to the right, when the shift is to the left. Reinforce the general form $f(x - h)$ since students may need to rewrite the given transformation as $f(x - [-1])$ to identify the transformation.

Go Online

- Find additional teaching notes.
- View performance reports of the Checks.
- Assign or present an Extra Example.

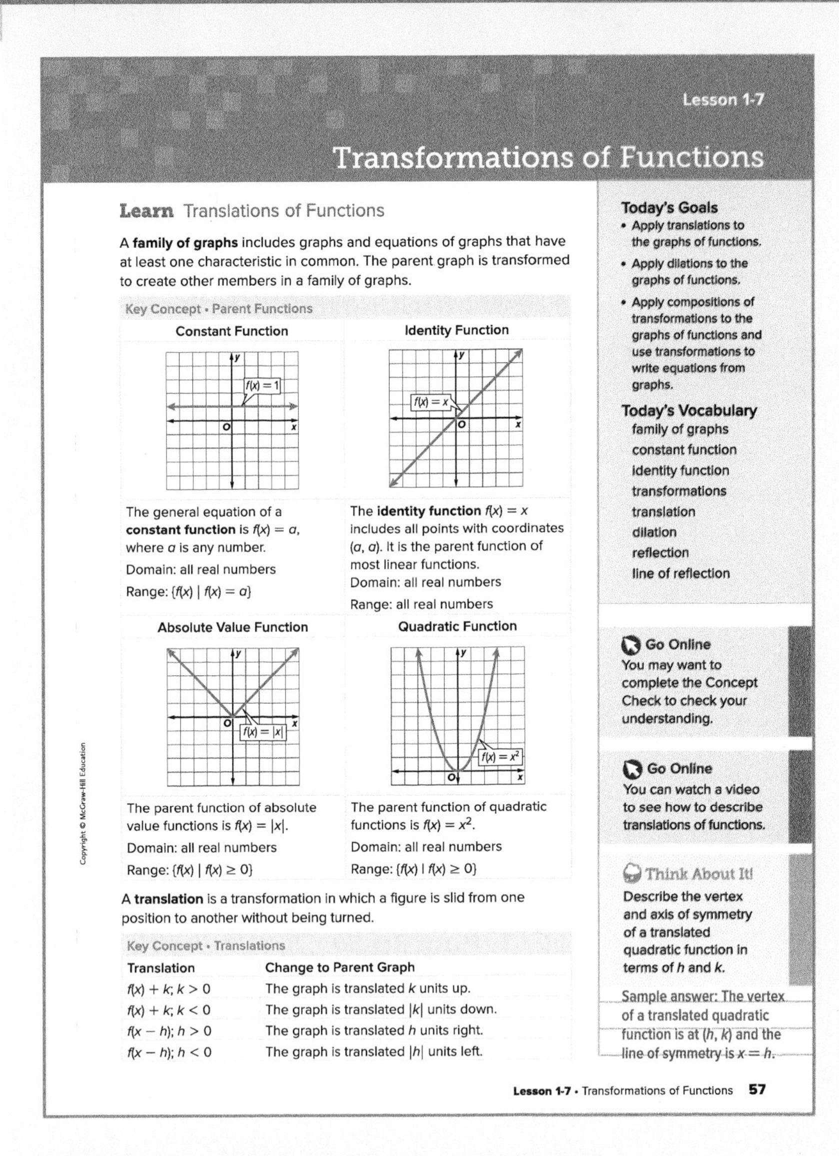

Lesson 1-7

Transformations of Functions

Learn Translations of Functions

A **family of graphs** includes graphs and equations of graphs that have at least one characteristic in common. The parent graph is transformed to create other members in a family of graphs.

Key Concept • Parent Functions

The general equation of a **constant function** is $f(x) = a$, where a is any number.
Domain: all real numbers
Range: $\{f(x) \mid f(x) = a\}$

The **identity function** $f(x) = x$ includes all points with coordinates (a, a). It is the parent function of most linear functions.
Domain: all real numbers
Range: all real numbers

The parent function of absolute value functions is $f(x) = |x|$.
Domain: all real numbers
Range: $\{f(x) \mid f(x) \geq 0\}$

The parent function of quadratic functions is $f(x) = x^2$.
Domain: all real numbers
Range: $\{f(x) \mid f(x) \geq 0\}$

A **translation** is a transformation in which a figure is slid from one position to another without being turned.

Key Concept • Translations

Translation	Change to Parent Graph
$f(x) + k; k > 0$	The graph is translated k units up.
$f(x) + k; k < 0$	The graph is translated $\|k\|$ units down.
$f(x - h); h > 0$	The graph is translated h units right.
$f(x - h); h < 0$	The graph is translated $\|h\|$ units left.

Today's Goals
- Apply translations to the graphs of functions.
- Apply dilations to the graphs of functions.
- Apply compositions of transformations to the graphs of functions and use transformations to write equations from graphs.

Today's Vocabulary
family of graphs
constant function
identity function
transformations
translation
dilation
reflection
line of reflection

Go Online You may want to complete the Concept Check to check your understanding.

Go Online You can watch a video to see how to describe translations of functions.

Think About It! Describe the vertex and axis of symmetry of a translated quadratic function in terms of h and k.

Sample answer: The vertex of a translated quadratic function is at (h, k) and the line of symmetry is $x = h$.

Lesson 1-7 • Transformations of Functions 57

Interactive Presentation

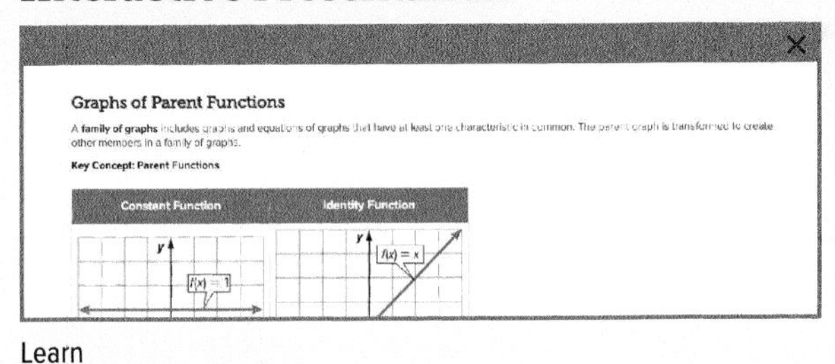

Learn

TAP

Students move through the slides to see how different translations affect linear, absolute, and quadratic functions.

SELECT

Students select the direction of different translations on a parent function.

TYPE

Students compare the graph of a translated function with the parent graph.

Study Tip

Signs Although the translated function uses subtraction, the translation is in the form $f(x) + k$ because a constant is being added to the parent function. The value of k is negative because adding a negative number is the same as subtracting the positive opposite of that number. The translation $p(x) = (x - 3)^2$ is in the form $f(x - h)$, because the constant is being subtracted from the variable before evaluating.

Example 1 Translations

Describe the translation in $g(x) = (x + 2)^2 - 4$ as it relates to the graph of the parent function.

Since $g(x)$ is quadratic, the parent function is $f(x) = \underline{x^2}$.

Since $f(x) = x^2$, $g(x) = f(x - h) + k$, where $h = \underline{-2}$ and $k = \underline{-4}$.

The constant k is added to the function after it has been evaluated, so k affects the output, or y-values. The value of k is less than 0, so the graph of $f(x) = x^2$ is translated <u>down</u> 4 units.

The value of h is subtracted from x before it is evaluated and is less than 0, so the graph of $f(x) = x^2$ is also translated 2 units <u>left</u>.

The graph of $g(x) = (x + 2)^2 - 4$ is the translation of the graph of the parent function <u>2</u> units left and 4 units <u>down</u>.

Problem-Solving Tip

Use a Graph When writing the equation of a graph, use the key features of the graph to determine transformations. Notice how the maximum, minimum, intercepts, and axis of symmetry have changed from the parent function in order to determine which transformations have been applied.

Example 2 Identify Translated Functions from Graphs

Use the graph of the function to write its equation.

The graph is an absolute value function with a parent function of $\underline{f(x) = |x|}$. Notice that the vertex of the function has been shifted both vertically and horizontally from the parent function.

To write the equation of the graph, determine the values of h and k in $g(x) = |x - h| + k$.

The translated graph has been shifted 2 units up and 1 unit right. So, $h = \underline{1}$ and $k = \underline{2}$. Thus, $g(x) = \underline{|x - 1| + 2}$.

Learn Dilations and Reflections of Functions

A **dilation** is a transformation that stretches or compresses the graph of a function. Multiplying a function by a constant dilates the graph with respect to the x- or y-axis.

Key Concept • Dilations

Dilation	Change to Parent Graph
$af(x)$, $\lvert a\rvert > 1$	The graph is stretched vertically.
$af(x)$, $0 < \lvert a\rvert < 1$	The graph is compressed vertically.
$f(ax)$, $\lvert a\rvert > 1$	The graph is compressed horizontally.
$f(ax)$, $0 < \lvert a\rvert < 1$	The graph is stretched horizontally.

Go Online You may want to complete the Concept Check to check your understanding.

Go Online You can complete an Extra Example online.

Interactive Presentation

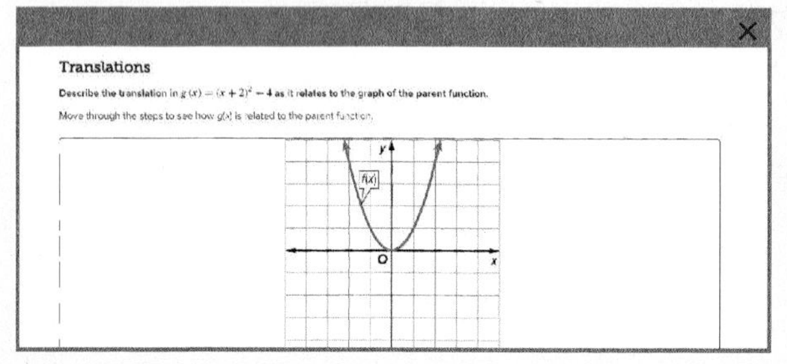

Example 1

TAP

Students move through the steps to graph a translation of a parent function.

TYPE

Students describe the vertex and axis of symmetry of a translated quadratic function in terms of h and k.

CHECK

Students complete the Check online to determine whether they are ready to move on.

Example 1 Translations

Teaching the Mathematical Practices

7 Interpret Complicated Expressions Mathematically proficient students can see complicated expressions as being composed of several objects. Guide students to see what information they can gather about the function just from looking at it.

Questions for Mathematical Discourse

AL How can you verify that you performed the translation correctly? Test points in the function and see if they match the translated graph.

OL Without using the given form to identify the effects of a translation, how can you tell that subtracting 4 will shift the graph down? Sample answer: The vertex of the parent function occurs at (0, 0). Consider $(x + 2)^2 = q^2$. Then $g(x) = q^2 - 4$. When $q = 0$, $g(x) = -4$.

BL Without using the given form to identify the effects of a translation, how can you tell that adding 2 to x results in a shift to the left? Sample answer: The vertex of the parent function occurs at (0, 0). Consider $(x + 2)^2 = q^2$. For $f(q)$ we know the vertex occurs at $q = 0$. When $q = 0$, $x = -2$, thus the x-coordinate of the vertex is -2.

DIFFERENTIATE

Reteaching Activity AL ELL

IF students are having difficulty identifying translations,
THEN have students create tables and graph $f(x) = x^2$ in black, $g(x) = x^2 + 2$ in red, and $h(x) = (x - 3)^2$ in blue on the same coordinate plane. They can then identify the shift in $f(x)$ from each translation.

Example 2 Identify Translated Functions from Graphs

Teaching the Mathematical Practices

1 Explain Correspondences Guide students as they use the graph in Example 2 to write a function that represents it.

Questions for Mathematical Discourse

AL How did the translation affect the domain and range of the function compared to the parent function? The domain is the same, all real numbers. The range changed from $[0, \infty)$ to $[2, \infty)$.

OL If a quadratic parent function were shifted using the same translations, what would be the equation of the transformed quadratic function? $g(x) = (x - 1)^2 + 2$

BL For what values of h and k would a translated absolute value function have a section that overlaps the parent function? (Assume no reflections or dilations). The functions have a section that overlaps when $|h| = |k|$.

Common Error

Students may switch the value of h and k when writing the transformed equation. Remind students the horizontal movement, h, goes inside the function with x while the vertical movement goes outside the function.

Learn Dilations and Reflections of Functions

Objective

Students identify the effect on the graphs of functions by replacing $f(x)$ with $af(x)$ and $f(ax)$ for positive and negative values.

Teaching the Mathematical Practices

1 Explain Correspondences Encourage students to explain the relationships between the equations and graphs of functions that have been dilated or reflected.

About the Key Concept

A dilation is a transformation that stretches or compresses the graph of a function. When a function is multiplied by a constant, the graph stretches or compresses with respect to the x- or y-axis. When $|a| > 1$, then $af(x)$ will vertically stretch the graph, but if $0 < |a| < 1$, then the graph will be vertically compressed. When $|a| > 1$, then $f(ax)$ will horizontally compress the graph, but if $0 < |a| < 1$, then the graph will be horizontally stretched. $-f(x)$ will reflect the graph in the x-axis, but $f(-x)$ will be reflected in the y-axis.

Common Misconception

When working with dilations, some students think that both the vertical and horizontal component will be stretched if $|a|>1$. Remind students that the horizontal component has the inverse effect as the vertical component when dealing with dilations. For example, $2f(x)$ will stretch the function vertically, but $f(2x)$ will compress the function horizontally.

Example 3 Vertical Dilations

Teaching the Mathematical Practices

3 Construct Arguments In Example 3, students will use definitions and previously established results to describe the dilation and reflection in the function.

Questions for Mathematical Discourse

AL What is the slope of the parent function? 1 What is the slope of the transformed function? $-\frac{2}{5}$

OL Will dilations and reflections affect the domain or range of a linear function? Explain. Only if the graph of the function is a horizontal line; sample answer: A reflection in the x-axis will change the range of a constant function. But for a linear function that is not a constant function, the domain and range will always be all real numbers regardless of stretches or compressions or reflections.

BL For a linear function, how do you tell the difference between a vertical dilation and a horizontal compression? Explain. There is no difference. Sample answer: For a linear function, a horizontal compression by a is the same as a vertical dilation by a.

A **reflection** is a transformation where a figure, line, or curve, is flipped in a **line of reflection**. Often the reflection is in the x- or y-axis.

When a parent function $f(x)$ is multiplied by -1, the result $-f(x)$ is a reflection of the graph in the x-axis. When only the variable is multiplied by -1, the result $f(-x)$ is a reflection of the graph in the y-axis.

Go Online You can watch videos to see how to describe dilations or reflections of functions.

Key Concept • Reflections

Reflection	Change to Parent Graph
$-f(x)$	reflection in the x-axis
$f(-x)$	reflection in the y-axis

Example 3 Vertical Dilations

Describe the dilation and reflection in $g(x) = -\frac{2}{5}x$ as it relates to the parent function.

Since $g(x)$ is a linear function, the parent function is $f(x) =$ __x__.

Since $f(x) = x$, $g(x) = -1 \cdot a \cdot f(x)$ where $a =$ __$\frac{2}{5}$__.

The function is multiplied by -1 and the constant a after it has been evaluated. $0 < |a| < 1$, so the graph is compressed vertically and reflected in the x-axis.

The graph of $g(x) = -\frac{2}{5}x$ is the graph of the parent function __compressed__ vertically and reflected in the __x-axis__.

Think About It! Describe the effect of multiplying the same value of a, $-\frac{2}{5}$, by a different parent function such as $f(x) = |x|$.

Sample answer: a would have a similar effect, a vertical compression and reflection in the x-axis.

Example 4 Horizontal Dilations

Describe the dilation and reflection in $g(x) = (-2.5x)^2$ as it relates to the parent function.

Since $g(x)$ is a quadratic function, the parent function is $f(x) =$ __x^2__. Since $f(x) = x^2$, $g(x) = f(-1 \cdot a \cdot x)$, where $a =$ __2.5__.

x is multiplied by __-1__ and the constant a before the function is performed and $|a|$ is greater than 1, so the graph of $f(x) = x^2$ is compressed __horizontally__ and reflected in the __y-axis__.

Think About It! Why does the graph of $g(x) = (-2.5x)^2$ appear the same as $j(x) = (2.5x)^2$?

Sample answer: Because both positive and negative a are positive after being squared, so both graphs have the same y-values.

Go Online You can complete an Extra Example online.

Lesson 1-7 • Transformations of Functions 59

Interactive Presentation

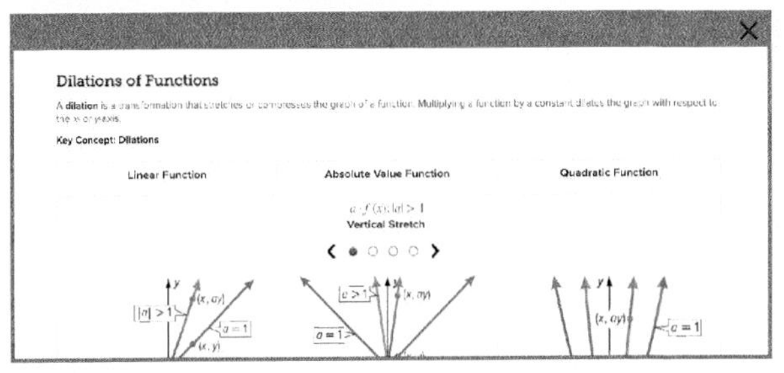

Learn

TAP

Students move through the slides to see how dilations or reflections affect linear, absolute value, and quadratic functions.

SELECT

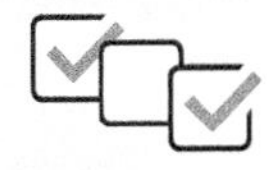

Students select the change and effect of dilations and reflections to a parent function.

WATCH

Students watch a video to see how to describe dilations or reflections of functions.

1 CONCEPTUAL UNDERSTANDING | 2 FLUENCY | 3 APPLICATION

Explore Using Technology to Transform Functions

Online Activity Use graphing technology to complete the Explore.

INQUIRY How does performing an operation on a function change its graph?

Think About It!
Do the values of a, h, and k affect various parent functions in different ways? Explain.

No; sample answer: a always stretches or compresses the parent function, h always shifts the parent function left or right, and k always shifts the function up or down.

Go Online
You can watch a video to see how to graph transformations of functions using a graphing calculator.

Learn Transformations of Functions

The general form of a function is $g(x) = a \cdot f(x - h) + k$, where $f(x)$ is the parent function. Each constant in the equation affects the parent graph.

- The value of $|a|$ stretches or compresses (dilates) the parent graph.
- When the value of a is negative, the graph is reflected across the x-axis.
- The value of h shifts (translates) the parent graph left or right.
- The value of k shifts (translates) the parent graph up or down.

In $g(x) = a \cdot f(x - h) + k$, each constant affects the graph of $f(x) = x^2$.

Key Concept • Transformations of Functions

Dilation, a
If $|a| > 1$, the graph of $f(x)$ is stretched vertically. If $0 < |a| < 1$, the graph of $f(x)$ is compressed vertically.

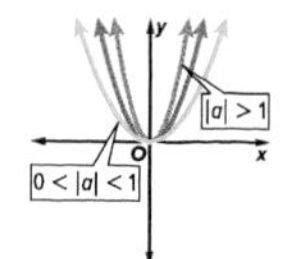

Reflection, a
If $a > 0$, the graph of $f(x)$ opens up. If $a < 0$, the graph of $f(x)$ opens down.

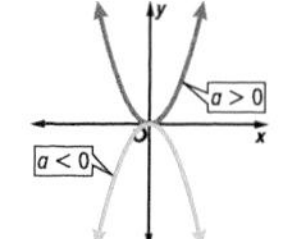

Horizontal Translation, h
If $h > 0$, the graph of $f(x)$ is translated h units right. If $h < 0$, the graph of $f(x)$ is translated $|h|$ units left.

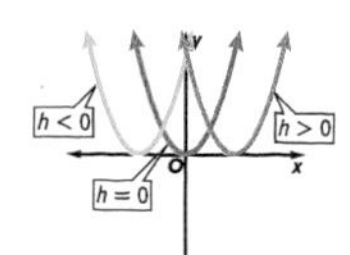

Vertical Translation, k
If $k > 0$, the graph of $f(x)$ is translated k units up. If $k < 0$, the graph of $f(x)$ is translated $|k|$ units down.

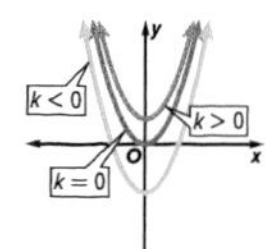

60 Module 1 • Relations and Functions

Interactive Presentation

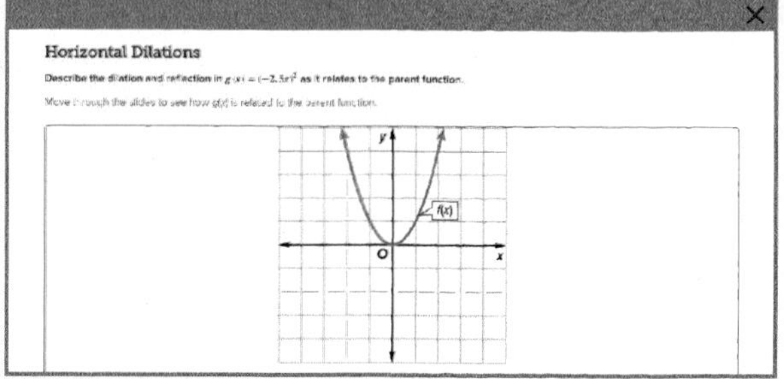

Example 4

TAP

Students move through the steps to graph a dilation and reflection of a parent function.

TYPE

Students explain why two different transformations produce the same graph.

CHECK

Students complete the Check online to determine whether they are ready to move on.

Example 4 Horizontal Dilations

Teaching the Mathematical Practices

7 Use Structure Help students to use the structure of the function to describe the transformations in the function.

Questions for Mathematical Discourse

AL How do we know $g(x)$ will be reflected in the y-axis? Sample answer: The x-part of the function is being multiplied by a negative number. For $f(-x)$, the function will be reflected in the y-axis.

OL Consider the functions $g(x) = 4x^2$ and $h(x) = (2x)^2$. How do their graphs compare? Sample answer: Their graphs will be the same because $g(x)$ and $h(x)$ are the same function since $(2x)^2 = 4x^2$.

BL For a linear function, a vertical dilation and a horizontal compression by the same value of a result in the same graph. Is this true for a quadratic function? Explain. No; sample answer: For a quadratic function, a horizontal compression by a would be equivalent to a vertical dilation of a^2.

DIFFERENTIATE

Reteaching Activity 

IF students are having difficulty identifying horizontal dilations, **THEN** have students make tables and graph $f(x) = |x|$ in black, $g(x) = |3x|$ in red, and $h(x) = \left|\frac{1}{4}x\right|$ in blue on the same coordinate plane. They can then see the stretch or compression of $f(x)$ from each dilation.

Learn Transformations of Functions

Objective

Students identify the effect on the graphs of functions by replacing $f(x)$ with $af(x - h) + k$ and will use transformations to write equations from graphs.

Teaching the Mathematical Practices

3 Analyze Cases Use the Key Concept to encourage students to familiarize themselves with the possible cases for each type of transformation.

About the Key Concept

The general form of the equation of a function is $g(x) = af(x - h) + k$, where $f(x)$ is the parent function. The value of $|a|$ vertically stretches or compresses the parent graph. If the value of a is negative, the graph is reflected over the x-axis. The value of h shifts the parent graph left or right, while the value of k shifts the graph up or down.

Example 5 Multiple Transformations of Functions

Teaching the Mathematical Practices

7 Use Structure Guide students to use the structure of the function in Example 5 to describe the transformations performed on the absolute value parent function.

Questions for Mathematical Discourse

AL What value of x makes $|x + 3| = 0$? -3 When $x = -3$, what is $g(x)$? 1 What is the point $(-3, 1)$ in this example? It is the vertex of $g(x)$.

OL How do you use the values of a, h, and k to draw the graph of $g(x)$? Sample answer: Start with the parent function vertex of (0, 0). Move h units horizontally and k units vertically to find the new vertex. The negative sign outside the absolute value reflects the graph in the x-axis, so the function will open down. Find additional points on each portion of the function by moving down 2 on the y-axis and over 3 on the x-axis in each direction, then draw rays connecting each point to the vertex.

BL If the same transformations were applied to a quadratic function, would you be able to graph the function using the same process as the absolute value function? Explain. Sample answer: You would be able to use the same process to find the new vertex, determine the direction in which the graph opens, and determine in what way the graph is dilated. From this you could draw an approximation, but you would need to plot points from the function to increase the accuracy of the graph.

Example 6 Apply Transformations of Functions

Teaching the Mathematical Practices

2 Attend to Quantities Point out that it is important to note the meaning of the quantities used in this problem in order to interpret the parameters of the function.

Questions for Mathematical Discourse

AL In the context of the problem, why must $a < 0$? Sample answer: When a dolphin jumps out of the water, the shape of its path increases and then decreases, which is a parabola opening downward.

OL Will the domain of the transformed function be the same as the domain of the parent function? Explain. No; sample answer: The transformed function represents the path of a jumping dolphin, and the distance the dolphin can travel during a jump is constrained. The domain is a set interval of positive real numbers.

BL What do you need to know to confirm that the vertex (12, 18) indicates that the maximum height of the jump occurs 12 feet from the starting point of the jump? You need to know whether (0, 0) is a point on the function. In this case $g(0)$ does equal zero, so the interpretation is accurate.

Example 5 Multiple Transformations of Functions

Describe how the graph of $g(x) = -\frac{2}{3}|x + 3| + 1$ is related to the graph of the parent function.

The parent function is $f(x) = |x|$.

Since $f(x) = |x|$, $g(x) = af(x - h) + k$ where $a = -\frac{2}{3}$, $h = -3$ and $k = 1$.

The graph of $g(x) = -\frac{2}{3}|x + 3| + 1$ is the graph of the parent function compressed vertically, reflected in the x-axis, and translated 3 units left and 1 unit up.

Check

Describe how $g(x) = -(0.4x + 2)$ is related to the graph of the parent function. graph of $f(x) = x$ stretched horizontally, reflected in the x-axis, and translated 2 units left

Example 6 Apply Transformations of Functions

DOLPHINS **Suppose the path of a dolphin during a jump is modeled by $g(x) = -0.125(x - 12)^2 + 18$, where x is the horizontal distance traveled by the dolphin and $g(x)$ is its height above the surface of the water. Describe how $g(x)$ is related to its parent function and interpret the function in the context of the situation.**

Because $f(x) = x^2$ is the parent function, $g(x) = af(x - h) + k$, where $a = -0.125$, $h = 12$, and $k = 18$.

Translations

$12 > 0$, so the graph of $f(x) = x^2$ is translated 12 units right.

$18 > 0$, so the graph of $f(x) = x^2$ is translated 18 units up.

Dilation and Reflection

$0 < |-0.125| < 1$, so the graph of $f(x) = x^2$ is compressed vertically.

$a < 0$, so the graph of $f(x) = x^2$ is a reflection in the x-axis.

Interpret the Function

Because a is negative, the path of the dolphin is modeled by a parabola that opens down. This means that the vertex of the parabola (h, k) represents the maximum height of the dolphin, 18 feet, at 12 feet from the starting point of the jump.

Go Online You can complete an Extra Example online.

Go Online You can watch a video to see how to use transformations to graph an absolute value function.

Think About It! Write an equation for a quadratic function that opens down, has been stretched vertically by a factor of 4, and is translated 2 units right and 5 units down.

Sample answer: $g(x) = -4(x - 2)^2 - 5$

Study Tip **Interpretations** When interpreting transformations, analyze how each value influences the function and alters the graph. Then determine what you think each value might mean in the context of the situation.

Interactive Presentation

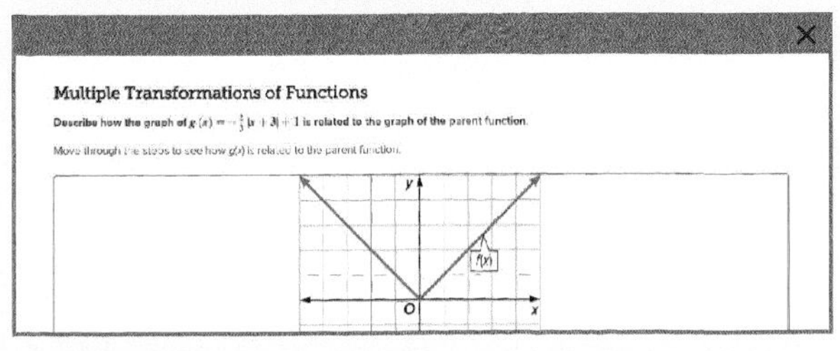

Example 5

TAP

Students move through the steps to graph transformations of a parent function.

WATCH

Students watch a video to see how transformations are used to graph an absolute value function.

TYPE

Students write an equation for a quadratic function with given transformations.

Example 7 Identify an Equation from a Graph

Write an equation for the function.

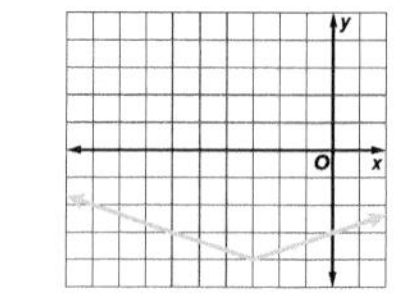

Step 1 Analyze the graph.

The graph is an absolute value function with a parent function of $f(x) = |x|$. Analyze the graph to make a prediction about the values of a, h, and k in the equation $y = a|x - h| + k$.

The graph appears to be wider than the parent function, implying a vertical compression, and is not reflected. So, a is positive and $0 < |a| < 1$.

The graph has also been shifted left and down from the parent graph. So, h < 0 and k < 0.

Step 2 Identify the translation(s).

Identify the horizontal and vertical translations to find the values of h and k.

The vertex is shifted 3 units left, so $h = -3$.

It is also shifted 4 units down, so $k = -4$.

$y = a|x - h| + k$ — General form of the equation

$y = a|x - (-3)| + (-4)$ — $h = -3$ and $k = -4$

$y = a|x + 3| - 4$ — Simplify.

Step 3 Identify the dilation and/or reflection.

Use the equation from Step 2 and a point on the graph to find the value of a.

The point (0, −3) lies on the graph. Substitute the coordinates in for x and y to solve for a.

$y = a|x + 3| - 4$ — Original equation

$-3 = a|0 + 3| - 4$ — $(0, -3) = (x, y)$

$-3 = a|3| - 4$ — Add.

$-3 = 3a - 4$ — Evaluate the absolute value.

$\frac{1}{3} = a$ — Solve.

Step 4 Write an equation for the function.

Since $a = \frac{1}{3}$, $h = -3$ and $k = -4$, the equation is $g(x) = \frac{1}{3}|x + 3| - 4$.

Check

Use the graph to write an equation for $g(x)$.

$g(x) = 5|x - 3| + 2$

Go Online You can complete an Extra Example online.

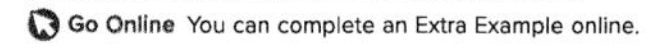

Watch Out!

Choosing a Point When substituting for x and y in the equation, use a point other than the vertex.

Think About It!

How does the equation you found compare to the prediction you made in step 1?

Sample answer: The values of a, h, and k match my prediction: $a = \frac{1}{3}$ and $0 < \frac{1}{3} < 1$; $h = -3$ and $-3 < 0$; $k = -4$ and $-4 < 0$.

62 Module 1 • Relations and Functions

Interactive Presentation

Identify an Equation from a Graph

Write an equation for the function.

Tap on the graph to see the parent function.

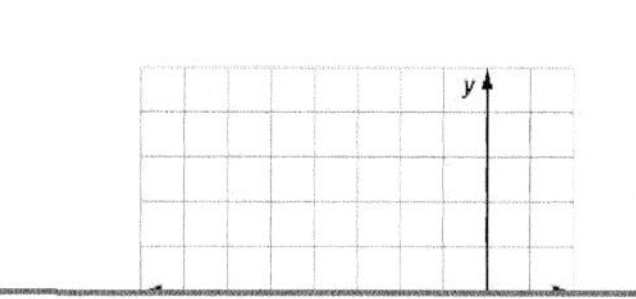

Example 7

TAP

Students tap on the graph of the transformed function to see the parent function, and then move through the steps to writing the equation of the transformation.

TYPE

Students compare the equation to their prediction.

CHECK

Students complete the Check online to determine whether they are ready to move on.

Example 7 Identify an Equation from a Graph

Teaching the Mathematical Practices

2 Create Representations Guide students to write an equation that models the graph in Example 7.

Questions for Mathematical Discourse

AL Rather than substituting the coordinates of a point into the function to find a, could you find a by looking at the graph? Yes, you could count how many units up and over the next integer point is.

OL Why can you not substitute the coordinates of the vertex into the equation when solving for a? Sample answer: If I used the coordinates of the vertex, then everything would cancel out and I would not be able to find a.

BL Why is making a prediction important when writing the equation of a transformed function? Sample answer: Making a prediction helps me to check my answer for reasonableness. Knowing that h and k must be less than zero lets me check my final answer to see if the equation matches that characteristic.

Common Error

Students may use 1 or −1 for the value of a depending on whether the graph opens up or down. Remind students that a has two possible effects, dilation and reflection. In order to identify the dilation aspect, the equation must be set up and solved for a.

Exit Ticket

Recommended Use

At the end of class, go online to display the Exit Ticket prompt and ask students to respond using a separate piece of paper. Have students hand you their responses as they leave the room.

Alternate Use

At the end of class, go online to display the Exit Ticket prompt and ask students to respond verbally or by using a mini-whiteboard. Have students hold up their whiteboards so that you can see all student responses. Tap to reveal the answer when most or all students have completed the Exit Ticket.

Practice and Homework

Suggested Assignments

Use the table below to select appropriate exercises.

DOK	Topic	Exercises
1, 2	exercises that mirror the examples	1–36
2	exercises that use a variety of skills from this lesson	37–48
2	exercises that extend concepts learned in this lesson to new contexts	49–51
3	exercises that emphasize higher-order and critical thinking skills	52–59

ASSESS AND DIFFERENTIATE

Use the data from the **Checks** to determine whether to provide resources for extension, remediation, or intervention.

BL

IF students score 90% or more on the Checks,
THEN assign:

- Practice Exercises 1–51 odd, 52–59
- Extension: Transformations of Points on the Coordinate Plane
- **ALEKS**® Graphs and Transformations of Functions

OL

IF students score 66%–89% on the Checks,
THEN assign:

- Practice Exercises 1–59 odd
- Remediation, Review Resources: Transformations of Linear Functions
- Personal Tutors
- Extra Examples 1–7
- **ALEKS**® Transforming Points

AL

IF students score 65% or less on the Checks,
THEN assign:

- Practice Exercises 1–35 odd
- Remediation, Review Resources: Transformations of Linear Functions
- *Quick Review Math Handbook:* Parent Functions and Transformations
- **ALEKS**® Transforming Points

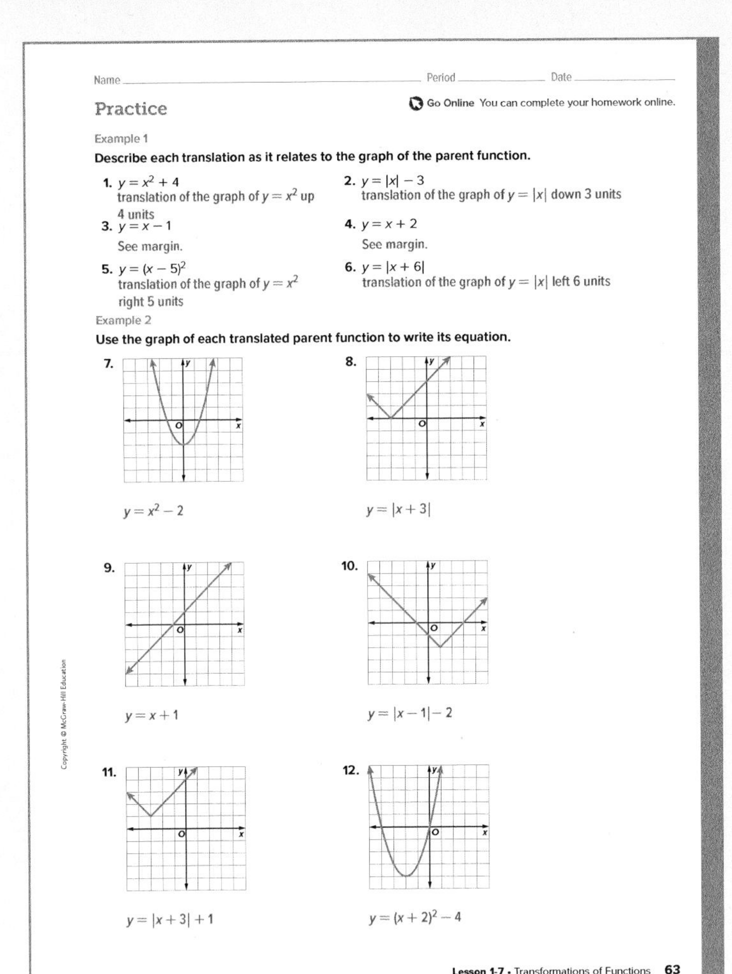

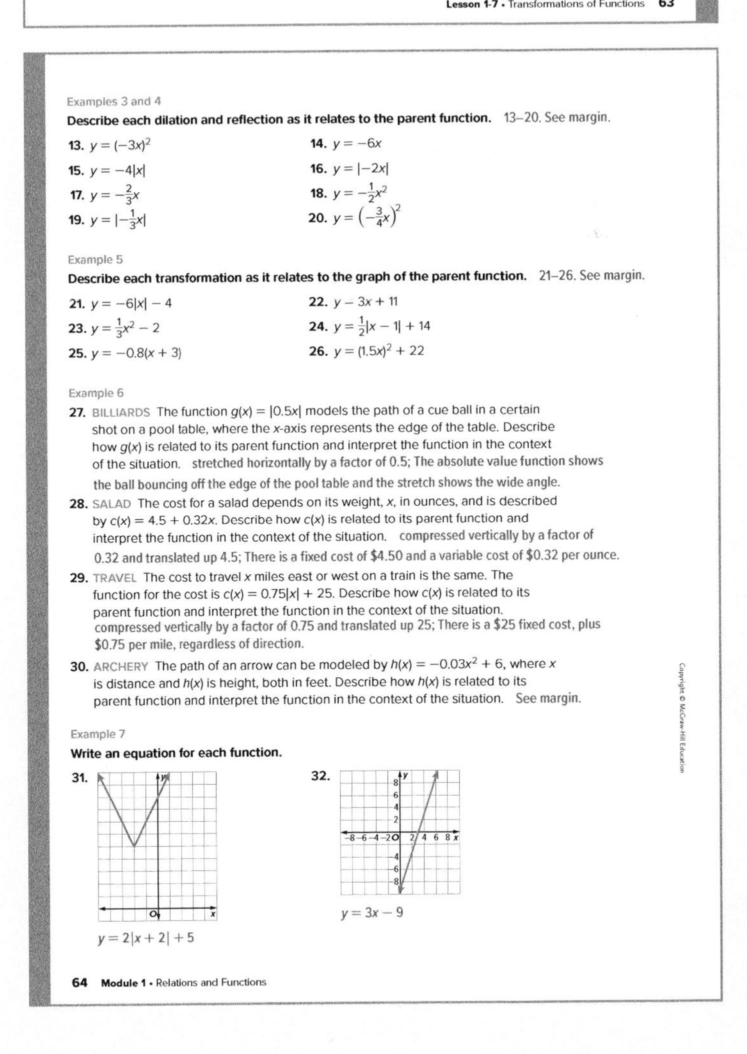

Answers

3. translation of the graph of $y = x$ down 1 unit
4. translation of the graph of $y = x$ up 2 units
13. compressed horizontally and reflected in the y-axis
14. stretched vertically and reflected in the x-axis
15. stretched vertically and reflected in the x-axis
16. compressed horizontally and reflected in the y-axis
17. compressed vertically and reflected in the x-axis
18. compressed vertically and reflected in the x-axis
19. stretched horizontally and reflected in the y-axis
20. stretched horizontally and reflected in the y-axis

Name ______ Period ______ Date ______

Write an equation for each function.

33. $y = -|x| - 3$

34. $y = 0.5x^2$

35. $y = -(x - 4)^2$

36. $y = -x + 7$

Mixed Exercises

Describe each transformation as it relates to the graph of the parent function. Then graph the function. 37–42. See margin.

37. $y = |x| - 2$ **38.** $y = (x + 1)^2$ **39.** $y = -x$

40. $y = -|x|$ **41.** $y = 3x$ **42.** $y = 2x^2$

43. Describe the translation in $y = x^2 - 4$ as it relates to the parent function. translation of $y = x^2$ down 4 units

44. Describe the reflection in $y = -x^3$ as it relates to the parent function. reflection of $y = x^3$ in the x-axis

45. Describe the type of transformation in the function $f(x) = (5x)^2$. horizontal compression of the graph of $y = x^2$

46. ARCHITECTURE The cross-section of a roof is shown in the figure. Write an absolute value function that models the shape of the roof. $y = 6 - |x - 3|$

47. SPEED The speedometer in Henry's car is broken. The function $y = |x - 8|$ represents the difference y between the car's actual speed x and the displayed speed.

a. Describe the translation. Then graph the function. See margin.

b. Interpret the function and the translation in terms of the context of the situation. Sample answer: The speedometer is stuck at 8 mph.

48. GRAPHIC DESIGN Kassie sketches the function $f(x) = -1.25(x - 1)^2 + 18.75$ as part of a new logo design. Describe the transformations she applied to the parent function in creating her function. See margin.

49. GEOMETRY Chen made a graph to show how the perimeter of a square changes as the length of sides increase. How is this graph related to the parent function $y = x$? Sample answer: stretched vertically by a factor of 4

50. REASONING Compare the graph of the parent function $f(x) = |x|$ with the graphs of $g(x) = |x + 2|$ and $h(x) = |x - 3|$. How are the graphs similar? How are they different? See Mod. 1 Answer Appendix.

51. BUSINESS Maria earns $10 an hour working as a lifeguard. She drew the graph to show the relation of her income as a function of the hours she works. How did she modify the function $y = x$ to create her graph? Maria stretched the function vertically by a factor of 10.

Higher-Order Thinking Skills

52. ANALYZE What determines whether a transformation will affect the graph vertically or horizontally? Use the family of quadratic functions as an example. See Mod. 1 Answer Appendix.

53. PERSEVERE Laura sketches the path of a model rocket that she launches.

a. What type of function does the graph show? quadratic

b. In which axis has the parent function been reflected? x-axis

c. How has the graph been translated? Assume that the function has not been dilated. right 25 units and up 81 units

d. What is the equation for the curve shown on the graph? $y = -(x - 25)^2 + 81$

54. ANALYZE Graph $g(x) = -3|x + 5| - 1$. Describe the transformations of the parent function $f(x) = |x|$ that produce the graph of $g(x)$. What are the domain and range? See Mod. 1 Answer Appendix.

55. ANALYZE Consider $f(x) = |2x|$, $g(x) = x + 2$, $h(x) = 2x^2$, and $k(x) = 2x^3$. See Mod. 1 Answer Appendix.

a. Graph each function and its reflection in the y-axis.

b. Analyze the functions and the graphs. Determine whether each function is *odd*, *even*, or *neither*.

c. Recall that if for all values of x, $f(-x) = f(x)$ the function $f(x)$ is an even function. If for all values of x, $f(-x) = -f(x)$ the function $f(x)$ is an odd function. Explain why this is true.

56. PERSEVERE Explain why performing a horizontal translation followed by a vertical translation has the same results as performing a vertical translation followed by a horizontal translation. See Mod. 1 Answer Appendix.

57. CREATE Draw a graph in Quadrant II. Use any of the transformations you learned in this lesson to move your figure to Quadrant IV. Describe your transformation. See Mod. 1 Answer Appendix.

58. ANALYZE Study the parent graphs at the beginning of this lesson. Select a parent graph with positive y-values when $x \to -\infty$ and positive y-values when $x \to \infty$. See Mod. 1 Answer Appendix.

59. WRITE Explain why the graph of $g(x) = (-x)^2$ appears the same as the graph of $f(x) = x^2$. Is this true for all reflections of quadratic functions? If not, describe a case when it is false. See Mod. 1 Answer Appendix.

Answers

21. reflected in the x-axis, stretched vertically, and translated down 4 units
22. stretched vertically and translated up 11 units
23. compressed vertically and translated down 2 units
24. compressed vertically, translated right 1 unit, and translated up 14 units
25. reflected in the x-axis, compressed vertically, and translated left 3 units
26. compressed horizontally and translated up 22 units
30. reflected in the x-axis, compressed vertically by a factor of 0.03, and translated up 6 units; the parabola opens down, so the maximum height of the arrow is the vertex. The translation up 6 units indicates the initial height of the arrow when $x = 0$.

37. translation of $y = |x|$ down 2 units

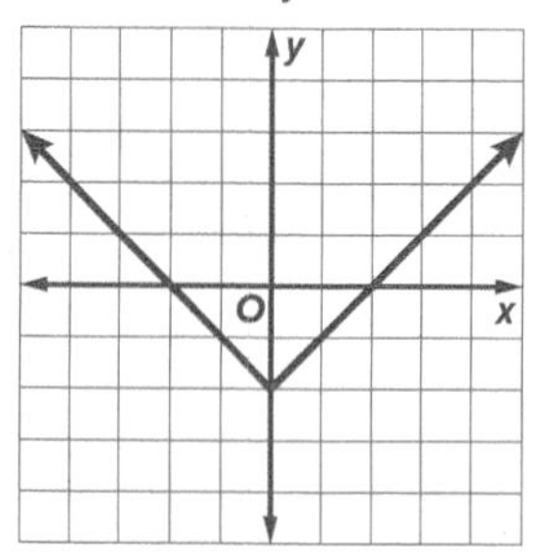

38. translation of $y = x^2$ left 1 unit

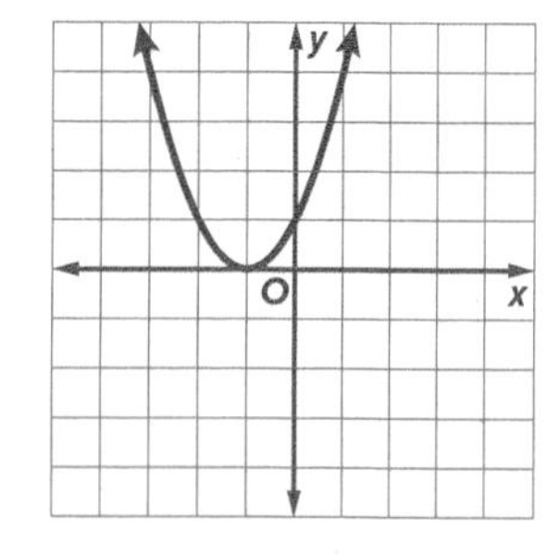

39. reflection of $y = x$ in the x-axis

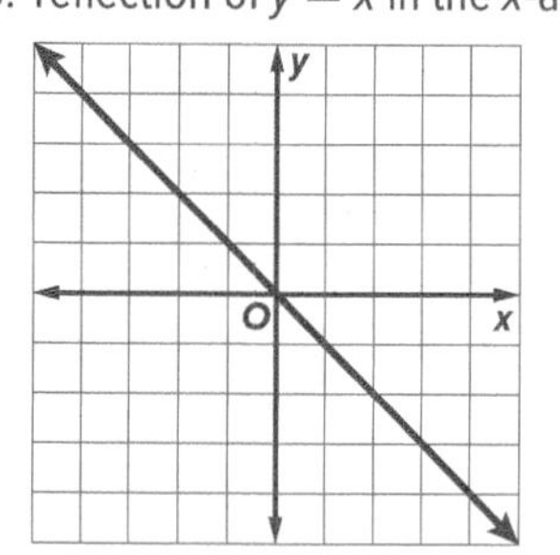

40. reflection of $y = |x|$ in the x-axis

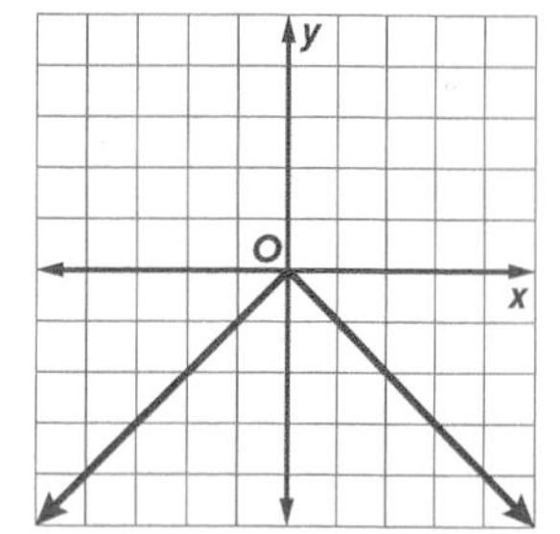

41. vertical stretch of $y = x$

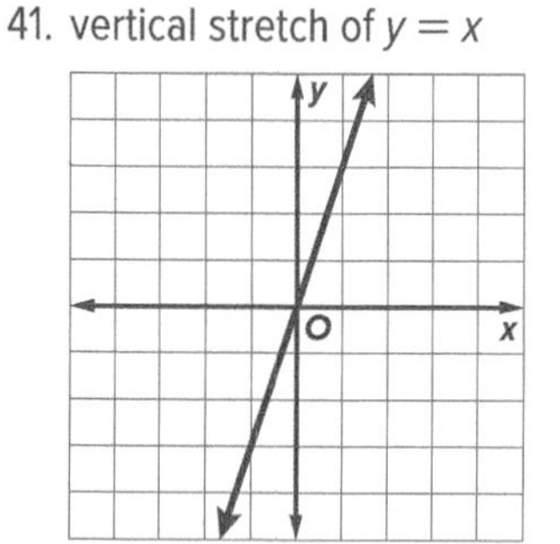

42. vertical stretch of $y = x^2$

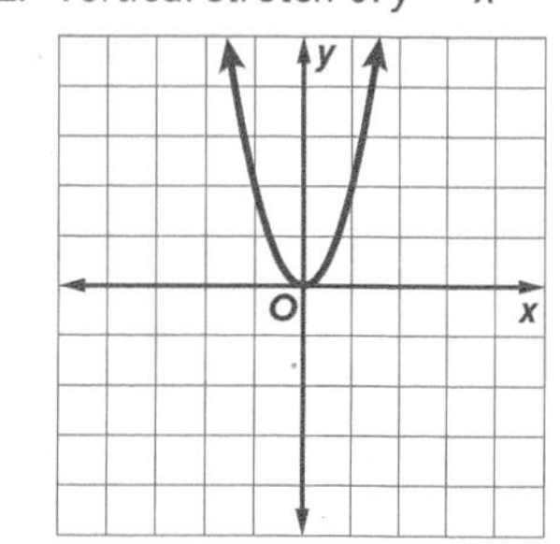

47a. translation of $y = |x|$ right 8 units

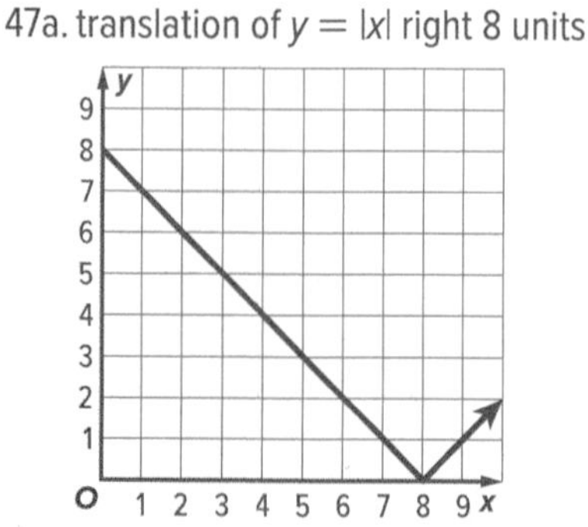

48. The transformations used to create her function include a reflection in the x-axis, a vertical stretch by a scale factor of 1.25, a horizontal shift to the right 1 unit, and a vertical shift up 18.75 units.

Review

Rate Yourself! 

Have students return to the Module Opener to rate their understanding of the concepts presented in this module. They should see that their knowledge and skills have increased. After completing the chart, have them respond to the prompts in their *Interactive Student Edition* and share their responses with a partner.

Answering the Essential Question

Before answering the Essential Question, have students review their answers to the Essential Question Follow-Up questions found throughout the module.

- Why would the maximum or minimum value of a function be important?
- Why are the key features of a function important when graphing?
- Why is it important to define some functions over a specific interval?

Then have them write their answer to the Essential Question in the space provided.

DINAH ZIKE FOLDABLES

ELL A completed Foldable for this module should include the key concepts related to relations and functions.

LS **LearnSmart** Use LearnSmart as part of your test preparation plan to measure student topic retention. You can create a student assignment in LearnSmart for additional practice on these topics for **Modeling with Functions**.

- Comparing Function Models
- Piecewise, Absolute Value, and Step Functions

Module 1 • Relations and Functions

Review

Essential Question

How can analyzing a function help you understand the situation it models?

Sample answer: You can analyze a function by observing the domain and range, the intercepts, end behavior, and so on. These observations can help you understand the situation it is modeling by seeing what values make sense in the situation, how the quantities will change as the domain values change, and so on.

Module Summary

Lesson 1-1 through 1-3

Function Behavior

- The graph of a continuous function is a line or curve. The domain of a continuous function is a single interval of all real numbers.
- A linear function is a function in which no independent variable is raised to a whole number power greater than 1.
- If a vertical line intersects the graph of a relation more than once, then the relation is not a function.
- An x-intercept occurs when the graph intersects the x-axis, and a y-intercept occurs when the graph intersects the y-axis.
- A graph has line symmetry if each half of the graph on either side of a line matches the other side exactly.
- A graph has point symmetry when a figure is rotated 180° about a point and maps exactly onto the other part.
- A point is a relative maximum if there are no other nearby points with a greater y-coordinate. A point is a relative minimum if there are no other nearby points with a lesser y-coordinate.
- End behavior is the behavior of the graph at its ends. At the right end, the values of x are increasing toward infinity. This is denoted as $x \to \infty$. At the left end, the values of x are decreasing toward negative infinity, denoted as $x \to -\infty$.

Lessons 1-4 through 1-7

Graphs of Functions

- You can use key features of a function to sketch its graph. Features such as intercepts, symmetry, end behavior, extrema, and intervals where the function is increasing, decreasing, positive, or negative provide information for sketching the graph.
- A function that is written using two or more expressions is a piecewise-defined function.
- A step function has a graph that is a series of horizontal line segments.
- An absolute value function is a function that contains an algebraic expression within absolute value symbols.
- A translation moves a figure up, down, left, or right.
- A dilation shrinks or enlarges a figure proportionally. Multiplying a function by a constant dilates the graph with respect to the x- or y-axis.
- A reflection is a transformation that flips a figure in a line of reflection.

Study Organizer

Foldables

Use your Foldable to review this module. Working with a partner can be helpful. Ask for clarification of concepts as needed.

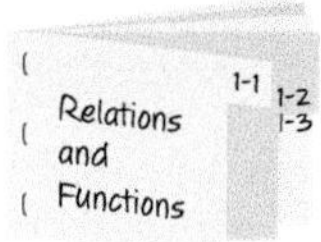

Module 1 Review • Relations and Functions 67

Name _______________ Period _______ Date _______

Test Practice

1. **MULTIPLE CHOICE** What is the domain of the function shown? (Lesson 1-1)

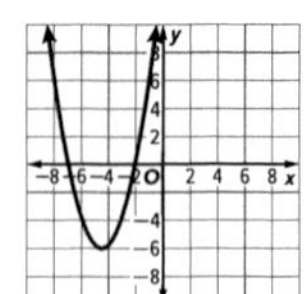

Ⓐ $(-8, 0)$

Ⓑ $[-6, \infty)$

● $(-\infty, \infty)$

Ⓓ $(-\infty, 0)$

2. **MULTIPLE CHOICE** Salvatore is a plumber. He charges \$100 for all work that is completed in less than 2 hours. He charges \$250 for work that requires 2 to 5 hours, and he charges \$400 for work that takes between 5 and 8 hours.

Which best describes the domain of the function that models Salvatore's price scale? (Lesson 1-1)

Ⓐ The domain is $\{100, 150, 400\}$.

● The domain is $\{x \mid 0 < x \leq 8\}$.

Ⓒ The domain is $\{x \mid 100 \leq x \leq 400\}$.

Ⓓ The domain is $\{x \mid x = 2, 5, 8\}$.

3. **OPEN RESPONSE** The table shows the amount of money Tia owed her friend over time after borrowing the money to go to a theme park. (Lesson 1-2)

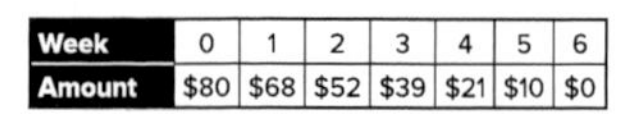

Week	0	1	2	3	4	5	6
Amount	\$80	\$68	\$52	\$39	\$21	\$10	\$0

What are the coordinates of the *x*-intercept and the coordinates of the *y*-intercept? What is the meaning of the intercepts in context of the situation?

Sample answer: The *x*-intercept is (6, 0). This means that after 6 weeks, Tia owes her friend \$0. The *y*-intercept is (0, 80). This means that Tia initially owed her friend \$80, or that Tia borrowed \$80 from her friend to go to a theme park.

4. **MULTIPLE CHOICE** What type of symmetry is shown? (Lesson 1-2)

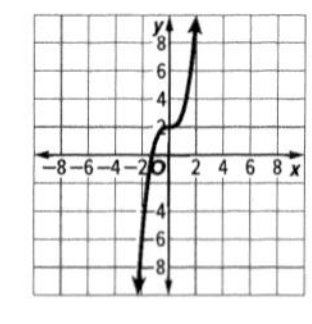

Ⓐ line symmetry

● point symmetry

Ⓒ both line and point symmetry

Ⓓ no symmetry

Review and Assessment Options

The following online review and assessment resources are available for you to assign to your students. These resources include technology-enhanced questions that are auto-scored, as well as essay questions.

Review Resources

Put It All Together: Lessons 1-1 through 1-3

Vocabulary Activity

Module Review

Assessment Resources

Vocabulary Test

AL Module Test Form B

OL Module Test Form A

BL Module Test Form C

Performance Task*

*The module-level performance task is available online as a printable document. A scoring rubric is included.

Test Practice

You can use these pages to help your students review module content and prepare for online assessments. Exercises 1–16 mirror the types of questions your students will see on online assessments.

Question Type	Description	Exercise(s)
Multiple Choice	Students select one correct answer.	1, 2, 4, 8, 9, 15
Table Item	Students complete a table by entering in the correct values.	5, 12
Graph	Students graph on a coordinate plane.	11
Open Response	Students construct their own response in the area provided.	3, 6, 7, 10, 13, 14

To ensure that students understand the standards, check students' success on individual exercises.

Standard(s)	Lesson(s)	Exercise(s)
F.IF.4	1-2, 1-3, 1-4	3–6, 8
F.IF.5	1-1	1, 2
F.IF.7a	1-5	9
F.IF.7b	1-6	11
F.IF.9	1-4	7
F.BF.3	1-7	12–15
A.CED.3	1-5	10

Name ____________ Period ________ Date ________

5. TABLE ITEM Indicate whether each of the following x-values is a *relative maximum*, *relative minimum* or *neither*. (Lesson 1-3)

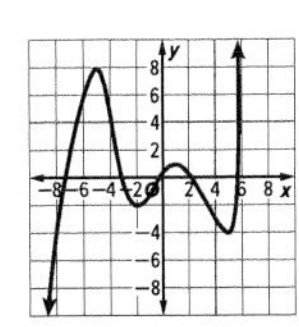

x	Relative Maximum	Relative Minimum	Neither
−5	X		
−4			X
−2		X	
0			X
1	X		
5		X	

6. OPEN RESPONSE The graph shows the height of a ball after being thrown from a height of 26 feet.

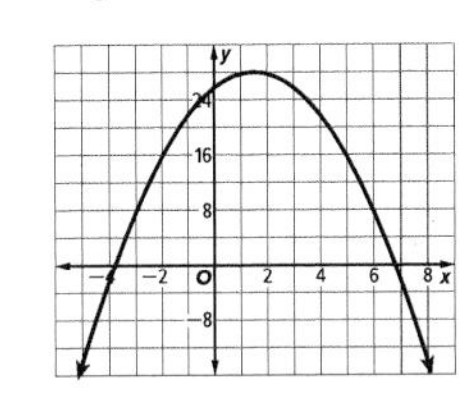

Explain why the end behavior does or does not make sense in this context. (Lesson 1-3)

Sample answer: The end behavior does not make sense because time and height cannot be negative.

7. OPEN RESPONSE Compare the key features of $f(x)$ and $g(x)$. (Lesson 1-4)

$f(x)$	$g(x)$
$f(x) = -2x + 4$	x-intercept: (1, 0) y-intercept: (0, −5) slope: 5

The x-intercept of $f(x)$ is 2, and the x-intercept of $g(x)$ is 1. The x-intercept of $f(x)$ is greater than the x-intercept of $g(x)$. So, $f(x)$ intersects the x-axis at a point farther to the right than $g(x)$. The y-intercept of $f(x)$ is 4, and the y-intercept of $g(x)$ is −5. The y-intercept of $f(x)$ is greater than the y-intercept of $g(x)$. So, $f(x)$ intersects the y-axis at a higher point than $g(x)$. The slope of $f(x)$ is −2 and the slope of $g(x)$ is 5. $f(x)$ is decreasing and $g(x)$ is increasing. The slope of $g(x)$ is greater than the slope of $f(x)$.

8. MULTIPLE CHOICE Sofia is sketching the graph of a function. She knows that as $x \to \infty, y \to -\infty$ and that the function has a y-intercept at (0, 8). Which other feature fits the sketch of the graph? (Lesson 1-4)

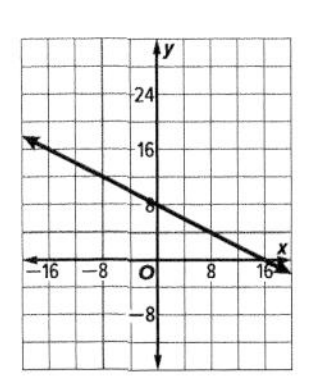

Ⓐ as $x \to -\infty, y \to -\infty$

Ⓑ x-intercept at (−14, 0)

Ⓒ increases for y in the interval $8 < x < 16$

Ⓓ decreases for y in the interval $-16 < x < 16$

Name ______________ Period ________ Date ________

9. MULTIPLE CHOICE A guidance counselor tells Rebekah that she needs a combined score of at least 1210 on the two portions of her college entrance exam to be eligible for the college of her choice. Suppose x represents Rebekah's score on the verbal portion and y represents her score on the math portion. Choose the equation or inequality that represents the scores Rebekah needs for eligibility. (Lesson 1-5)

Ⓐ $x + y \geq 1210$

Ⓑ $x + y = 1210$

Ⓒ $x + y > 1210$

Ⓓ $x + y \leq 1210$

10. OPEN RESPONSE The value of the Kim's portfolio is found by finding the sum of 60% of Stock A's value and 40% of Stock B's value. Kim wants her portfolio to be more than $500,000. Write the inequality that represents the constraints for Stock A's value x and Stock B's value y. (Lesson 1-5)

$0.6x + 0.4y > 500{,}000$

11. GRAPH Graph $f(x) = \left[\!\left[\frac{1}{3}x\right]\!\right] + 2$. (Lesson 1-6)

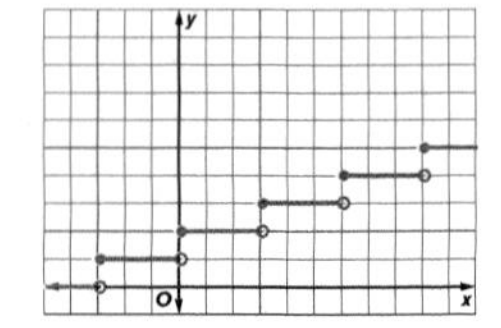

12. TABLE ITEM Match each function with the transformation that occurred from the parent function $f(x) = x^2$. (Lesson 1-7)

$g(x) = -x^2$

$h(x) = (3x)^2$

$j(x) = (x - 4)^2$

Function	Translation	Reflection	Dilation
$g(x)$		X	
$h(x)$			X
$j(x)$	X		

13. OPEN RESPONSE Describe the transformation(s) from the parent function to $f(x) = 3(x - 2)^2 + 9$. (Lesson 1-7)

Sample answer: The graph of the parent function $f(x) = x^2$ has been stretched vertically, translated 2 units to the right, and translated up 9 units.

14. OPEN RESPONSE If $g(x) = f(0.75x)$ then how is the graph of $g(x)$ related to the graph of $f(x)$? (Lesson 1-7)

Sample answer: $g(x)$ is a horizontal stretch of $f(x)$.

15. MULTIPLE CHOICE Which function represents a vertical compression, reflection in the x-axis, and a translation down 3 units in relation to the parent function? (Lesson 1-7)

Ⓐ $y = -\frac{2}{3}(x - 3)^2$

Ⓑ $y = -\left(\frac{2}{3}x\right)^2 - 3$

Ⓒ $y = -\frac{2}{3}x^2 - 3$

Ⓓ $y = \frac{2}{3}(-x + 3)^2$

Lesson 1-4

7.

8.

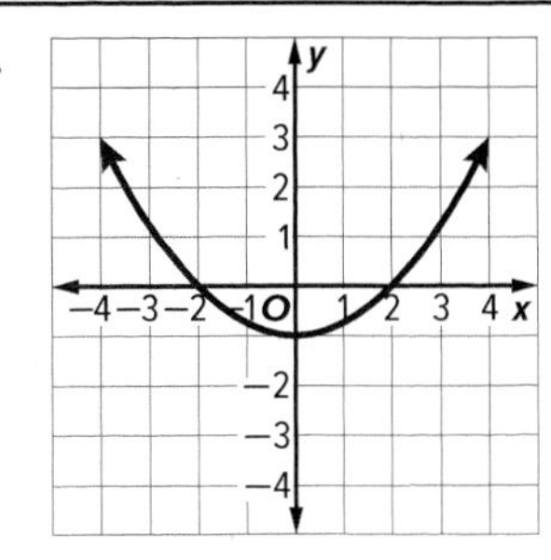

15. The x-intercept of $f(x)$ is $\frac{2}{3}$, and the x-intercept of $g(x)$ is $\frac{3}{8}$. The x-intercept of $f(x)$ is greater than the x-intercept of $g(x)$. So, $f(x)$ intersects the x-axis at a point farther to the right than $g(x)$. The y-intercept of $f(x)$ and $g(x)$ is $\frac{1}{2}$. So, $f(x)$ and $g(x)$ intersect the y-axis at the same point. The slope of $f(x)$ is $-\frac{3}{4}$ and the slope of $g(x)$ is $-\frac{4}{3}$. Each function is decreasing, but the slope of $g(x)$ is less than the slope of $f(x)$. So, $g(x)$ decreases faster than $f(x)$.

16. The x-intercept of $f(x)$ is less than both x-intercepts of $g(x)$. The graph of $g(x)$ intersects the x-axis more times than $f(x)$. The y-intercept of $g(x)$ is less than the y-intercept of $f(x)$. So, $f(x)$ intersects the y-axis at a higher point than $g(x)$. The relative minimum of $f(x)$ is greater than the minimum of $g(x)$. $f(x)$ has a relative maximum, but $g(x)$ does not have a relative maximum. The two functions have the opposite end behaviors as $x \to -\infty$. The two functions have the same end behavior as $x \to \infty$.

17.

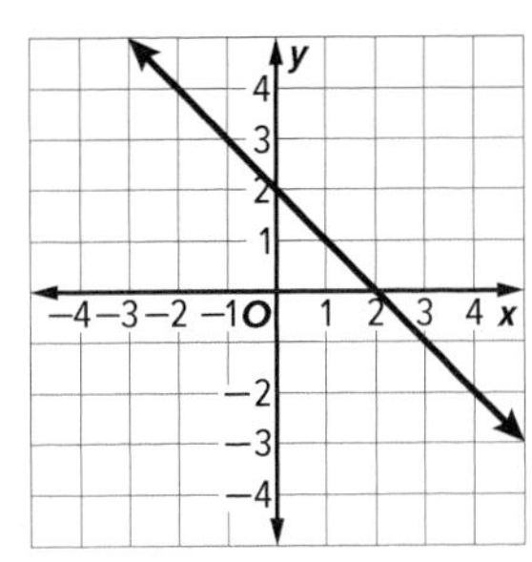

18a.

19.

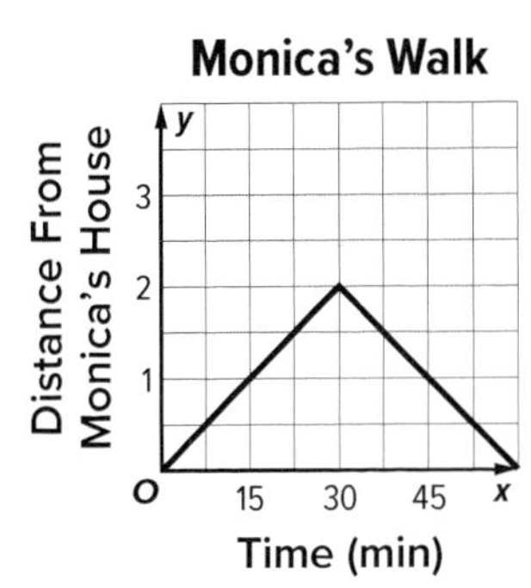

21a. Linear; sample answer: It is linear because it makes no stops along the way, and it descends at a steady pace, which indicates a constant rate of change, or slope.

21b.

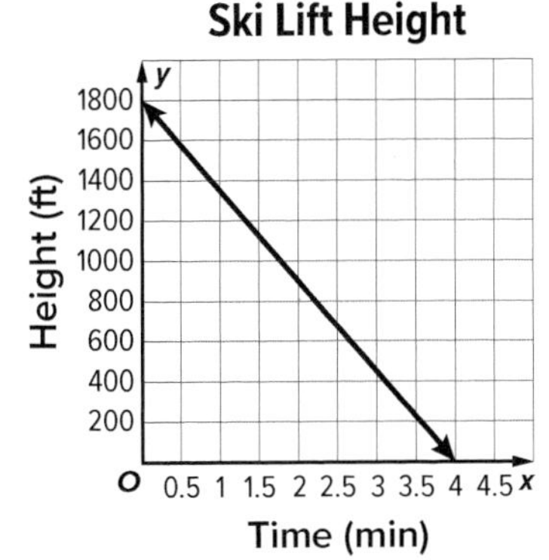

22. Sample answer: The graph only exists for positive x-values because Keisha cannot babysit for a negative number of hours. The graph only exists for positive y-values because Keisha cannot earn a negative amount of money.

23. Sample answer: The function is continuous. The function has three x-intercepts, at $x = -4$, $x = -1.7$, and $x = 3.2$, and a y-intercept at -3. The function has a relative maximum at $(-3, 1)$. The function has a relative minimum at $(1.4, -4)$. As $x \to -\infty$, $f(x) \to -\infty$ and as $x \to \infty$, $f(x) \to \infty$.

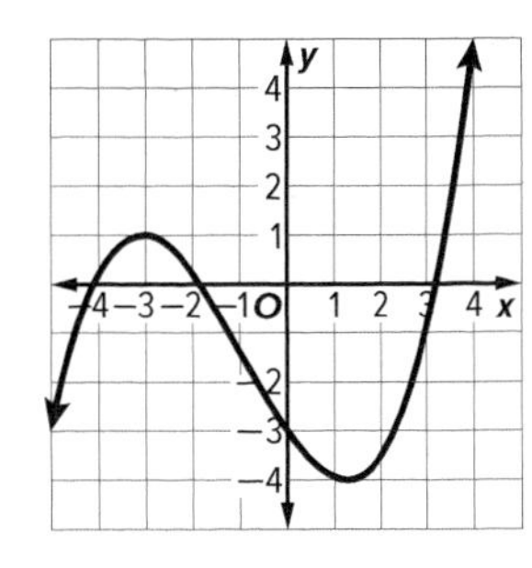

24. Sample answer: If the slope of a linear function is positive, the function is increasing. If the slope of a linear function is negative, the function is decreasing.

25. Always; sample answer: A linear function cannot cross the x-axis more than once. So, if a function has more than one x-intercept, it is a nonlinear function.

26a.

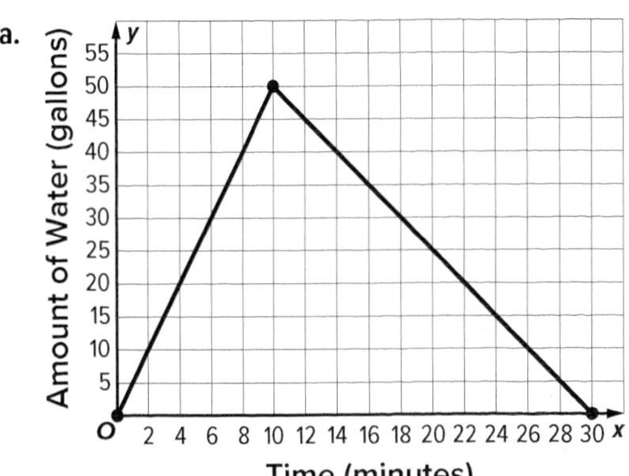

26b. Sample answer: A drain in the tub could have been opened at the 10-minute mark. If the rate at which the water was draining is greater than the rate of the water filling the tub, the amount of water in the tub would be decreasing.

27. Both Linda and Rubio sketched correct graphs. Both graphs have an x-intercept at 2, a y-intercept at -9, are positive for $x > 2$, and have end behavior described by as $x \to -\infty$, $f(x) \to -\infty$ and as $x \to \infty$, $f(x) \to \infty$.

Lesson 1-5

13.

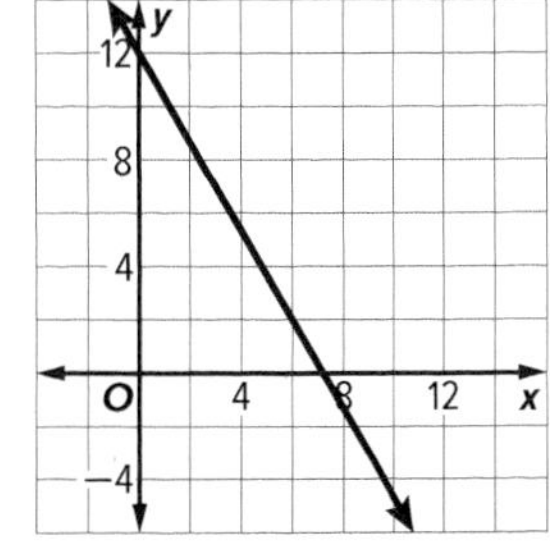

14.

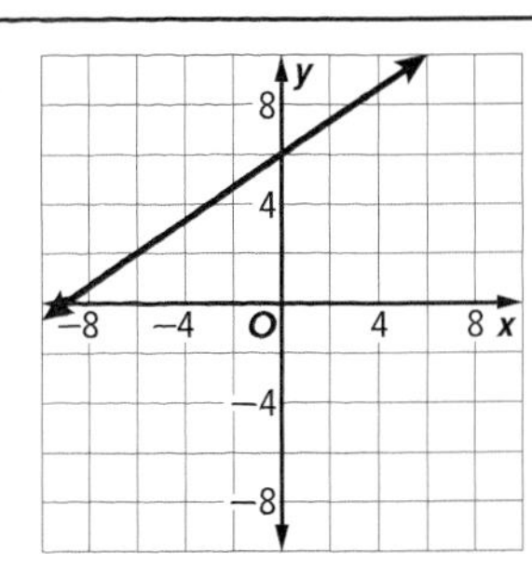

15.

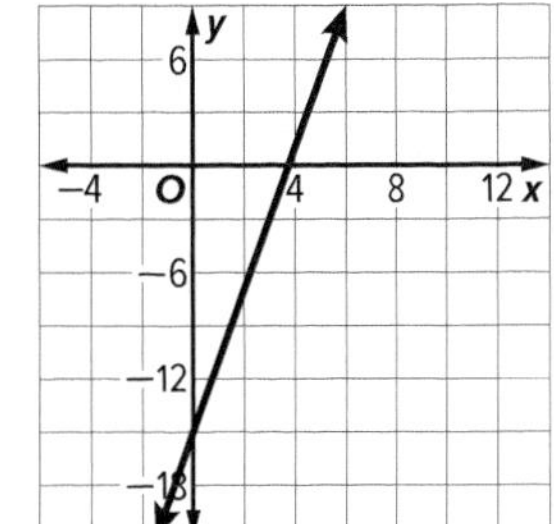

16.

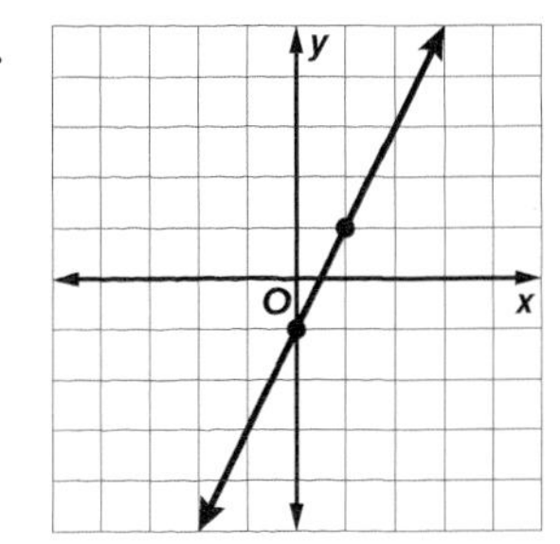

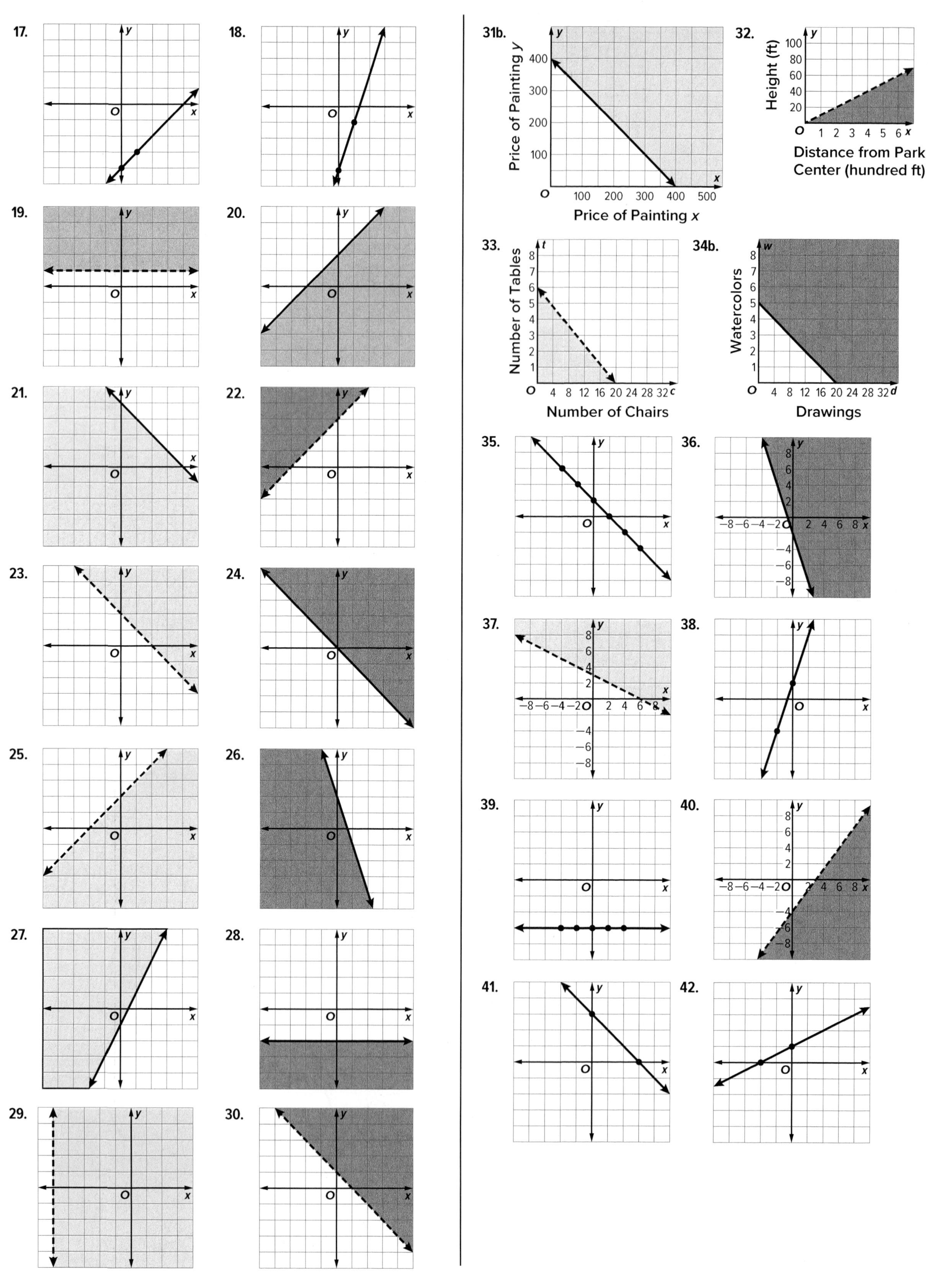

17.
18.
19.
20.
21.
22.
23.
24.
25.
26.
27.
28.
29.
30.
31b.
Price of Painting y
Price of Painting x
32.
Height (ft)
Distance from Park Center (hundred ft)
33.
Number of Tables
Number of Chairs
34b.
Watercolors
Drawings
35.
36.
37.
38.
39.
40.
41.
42.

43.

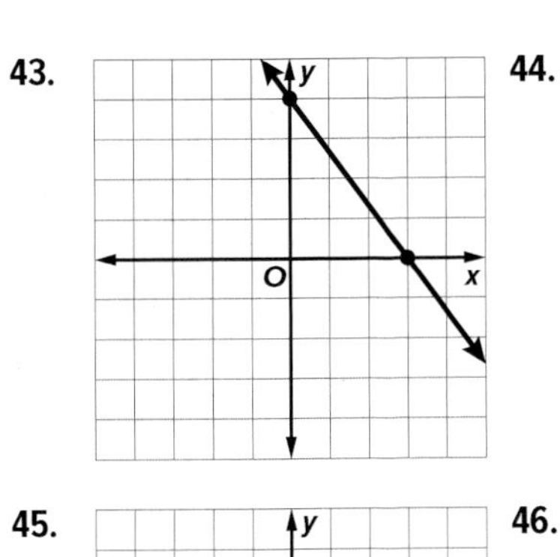

44.

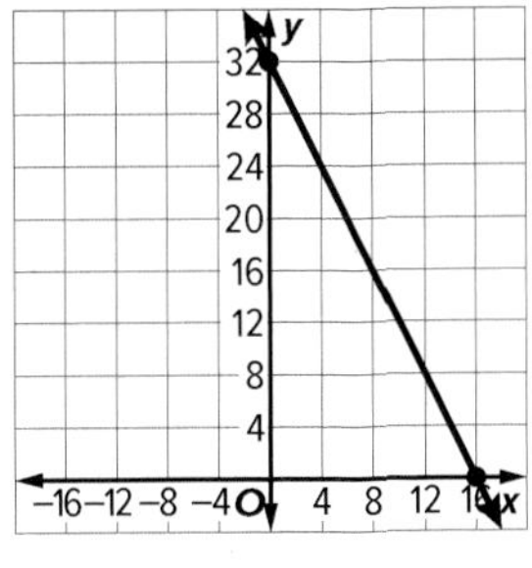

45.

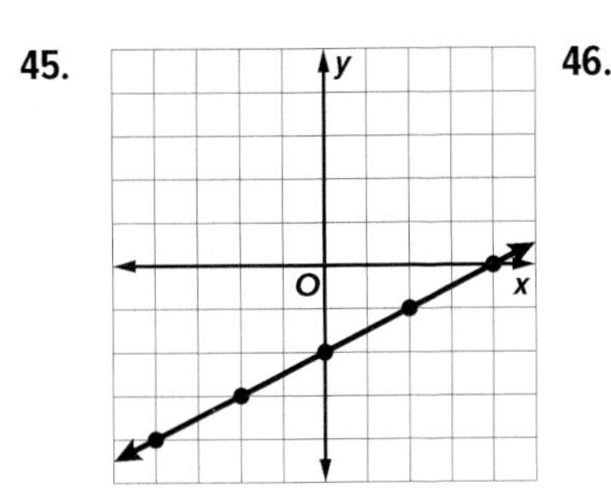

46.

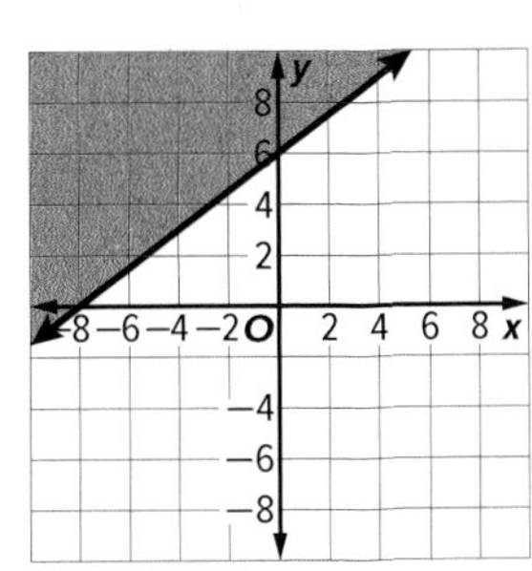

47.

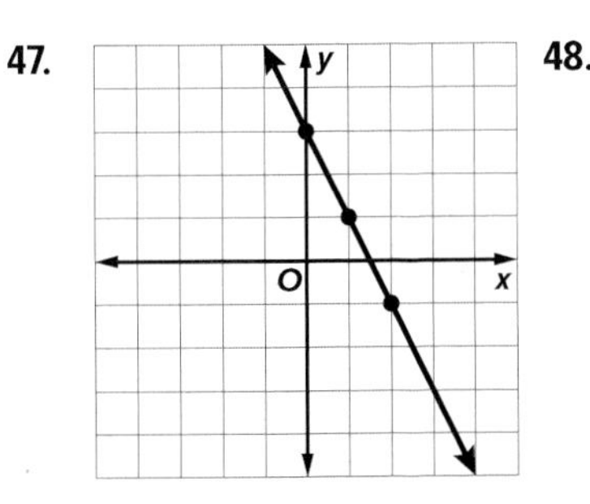

48.

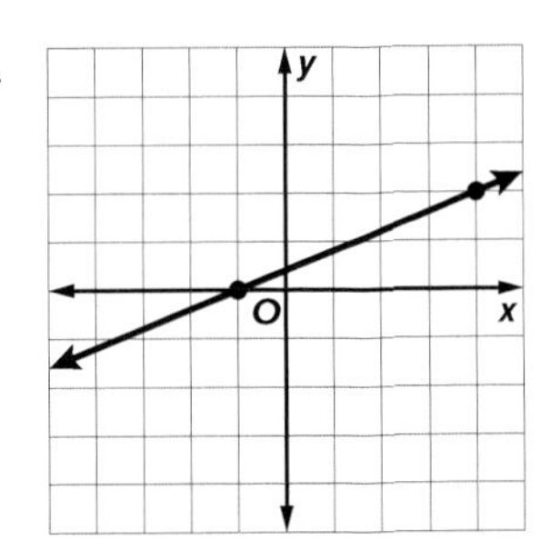

49.

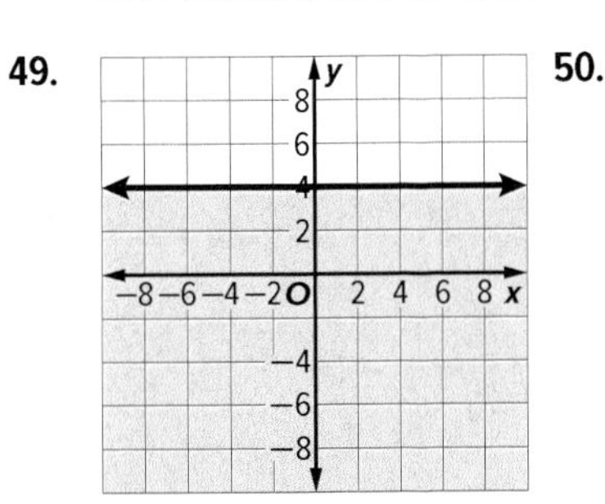

50.

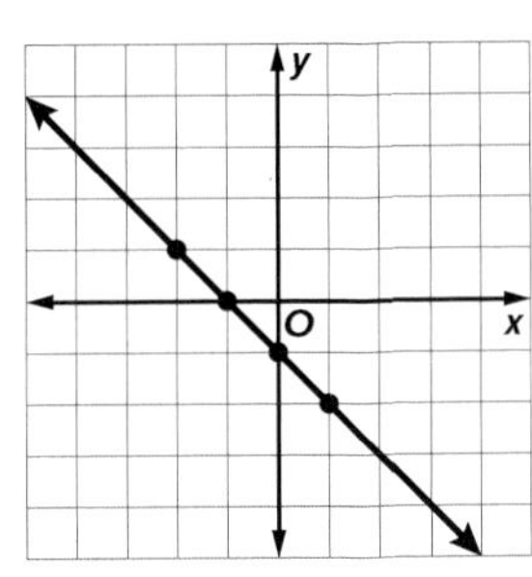

51.

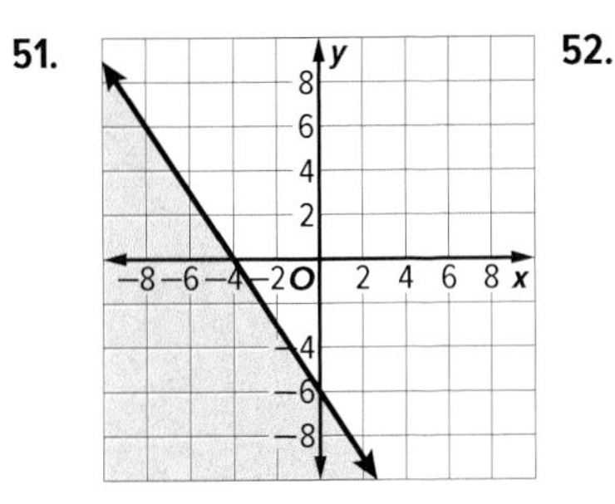

52.

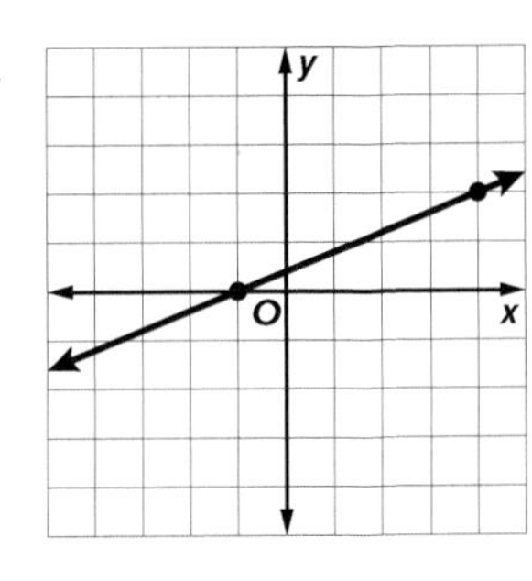

53.

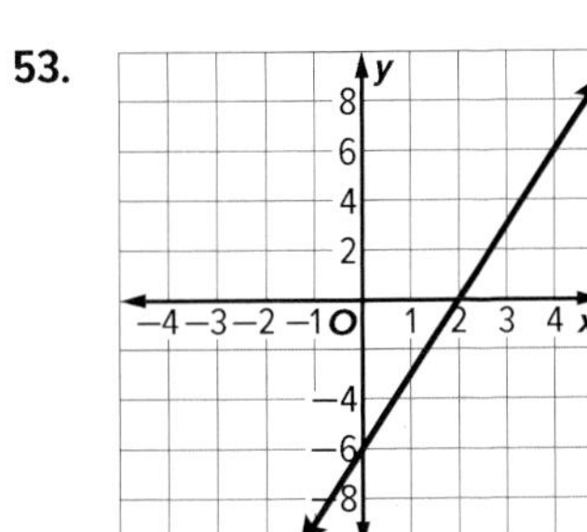

54b.

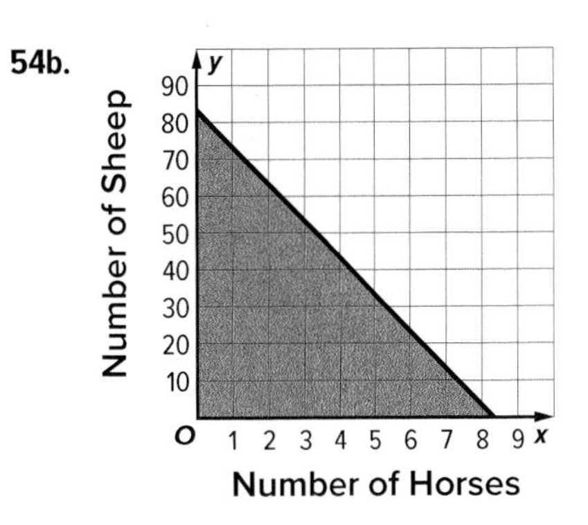

55b.

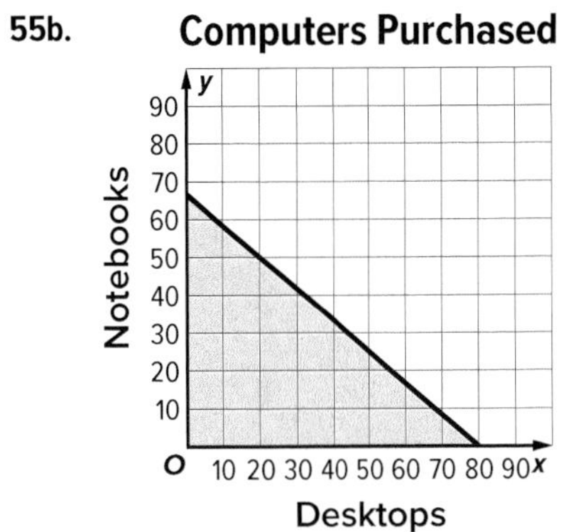

56a.

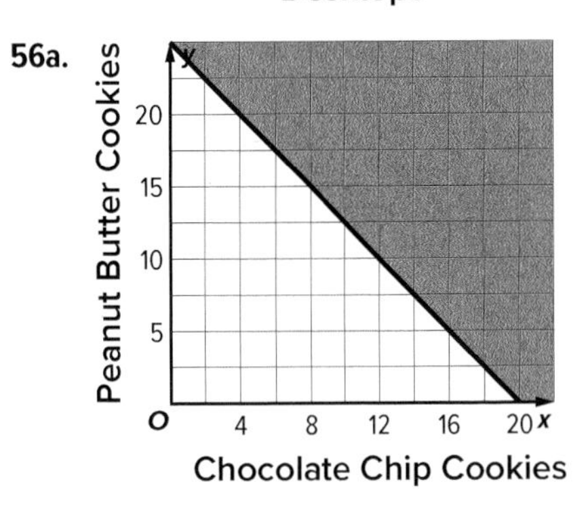

56b. Sample answer: The new model is $\$1.50x + \$1.00y \geq \$25$. This changes the slope of the boundary line from -1.25 to -1.5. It also changes the x-intercept from 20 to 16.67. The y-intercept remains the same. The point (16, 4) is not in the original set, because $(1.25)(16) + (1.00)(4) = 24$, which is less than 25. However, (16, 4) is in the solution set of the revised scenario, because $(1.50)(16) + (1.00)(4) = 28$, which is greater than 25.

56c. Sample answer: If Mary wants to make at least \$50 each day, then the boundary line is translated up by 25. The region of viable solutions is still the area above the boundary line. If Mary wants to make no more than \$25 a day, the boundary line is still solid because it still represents viable solutions, but now the graph is shaded below the boundary line. Since Mary cannot sell a negative number of cookies, the graph is restricted to the first quadrant.

57b.

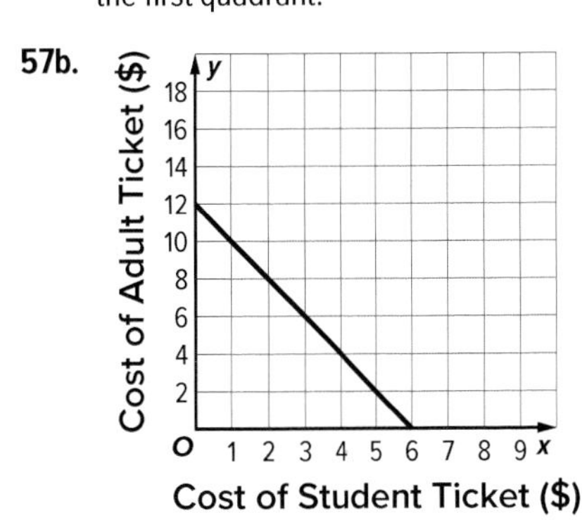

Sample answer: $x = \$2.00$, $y = \$8.00$; $x = \$3.00$, $y = \$6.00$; $x = \$4.60$, $y = \$2.80$; $x = \$6.00$, $y = \$0.00$

58a.

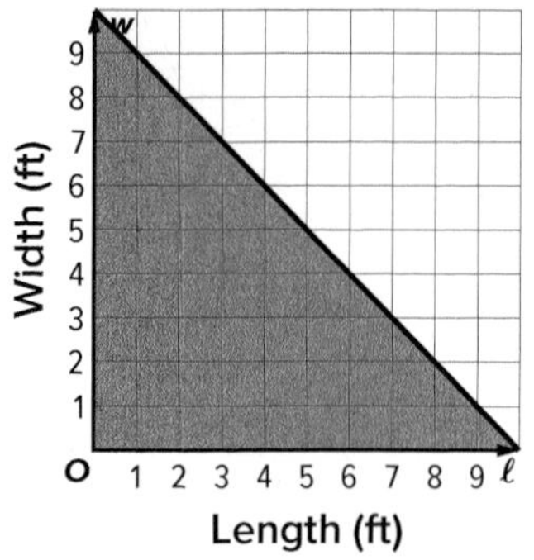

Width (ft)

Length (ft)

59a.

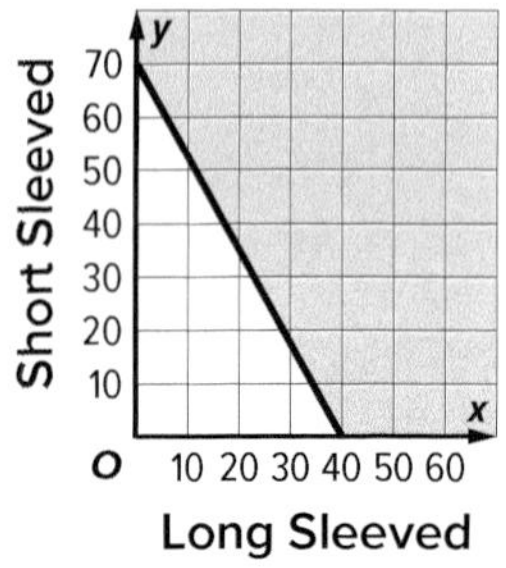

Short Sleeved

Long Sleeved

59b. Sample answer: 30 long-sleeved and 50 short-sleeved shirts; 60 long-sleeved and 40 short-sleeved shirts.

59c. Domain and range values must be positive integers because you cannot buy a negative number of shirts or a portion of a shirt.

60b.

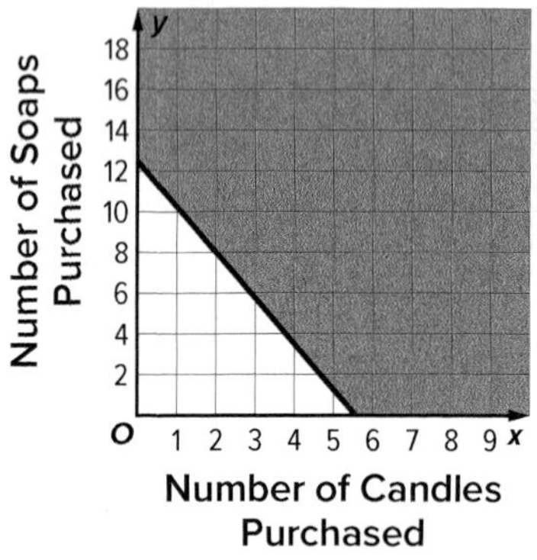

Number of Soaps Purchased

Number of Candles Purchased

60c. 6; Sample answer: This can be determined from looking at the x-intercept, which is between 5 and 6. Because partial candles cannot be purchased, it must be 6.

60d. 13; Sample answer: This can be determined from looking at the y-intercept, which appears to be between 12 and 13. Again, the number of hand soaps purchased must be a whole number, therefore it is 13.

60e. Yes; sample answer: It appears the line passes through (2, 8). Because the line is solid, then this combination of items would qualify for free shipping. To be certain, substitute 2 for x and 8 for y into the inequality from part a as follows: $9x + 4y \geq 50$; $9(2) + 4(8) \geq 50$?; $18 + 32 \geq 50$; $50 \geq 50$ ✓

62. Sample answer: $6 - 2y < 3x$

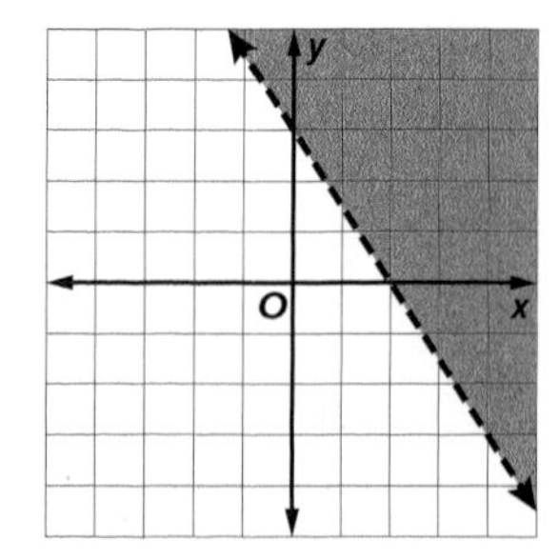

63. Sample answer: If given the x- and y-intercepts of a linear function, I already know two points on the graph. To graph the equation, I only need to graph those two points and connect them with a straight line.

Lesson 1-6

7. D = {all real numbers}; R = {all integers}

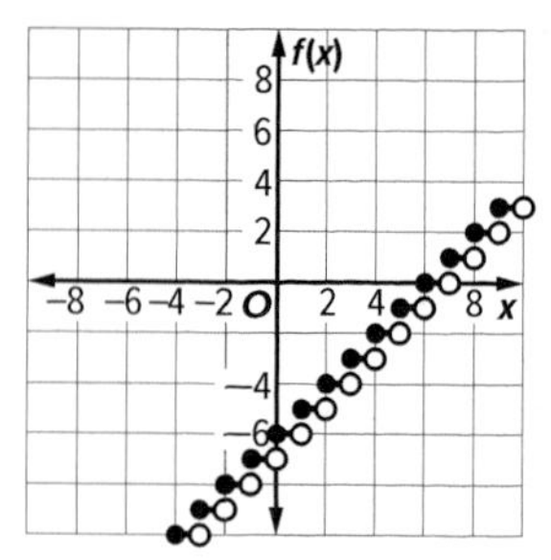

8. D = {all real numbers}; R = {all integers}

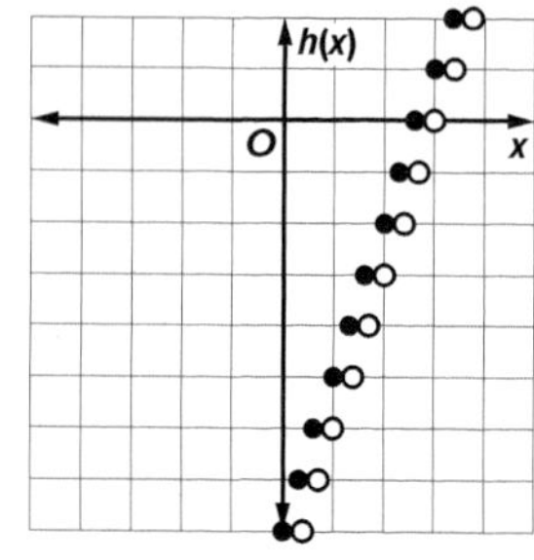

9. D = {all real numbers}; R = {all integers}

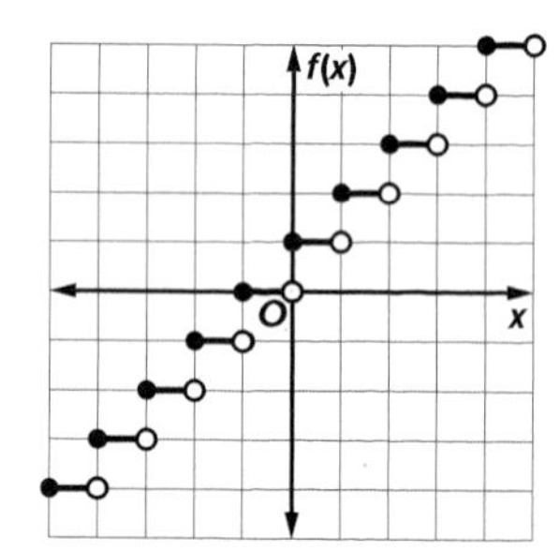

10. D = {all real numbers}; R = {all integers}

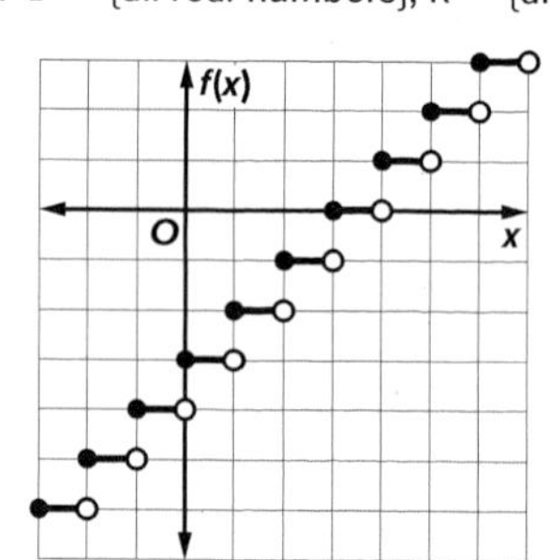

11. D = $\{x \mid 0 < x \leq 8\}$; R $\{y \mid y = 5.00, 10.00, 15.00, 20.00\}$

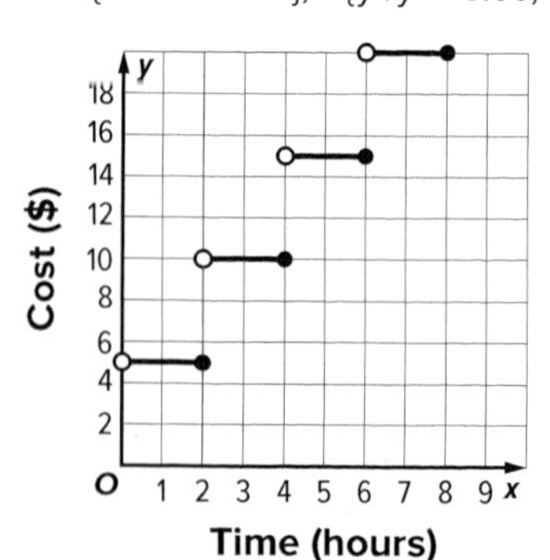

12. D = $\{x \mid x > 0\}$; R $\{y \mid y = 30, 45.00, 60.00\}$

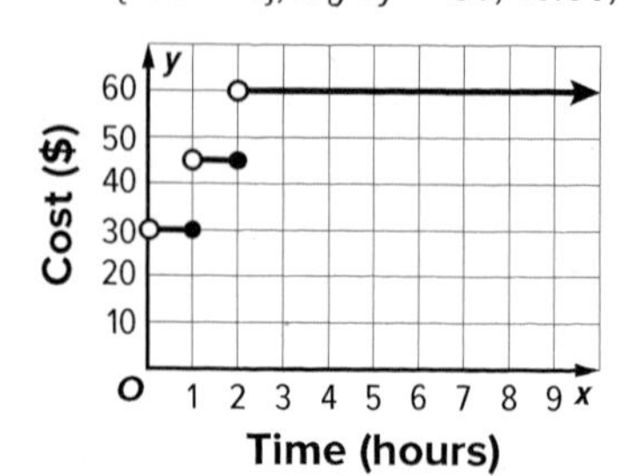

13. D = {all real numbers}; R = $\{f(x) \mid f(x) \geq 0\}$

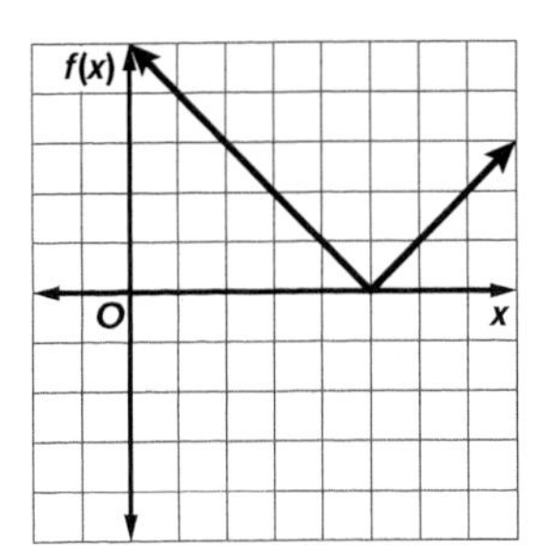

14. D = {all real numbers}; R = $\{g(x) \mid g(x) \geq 0\}$

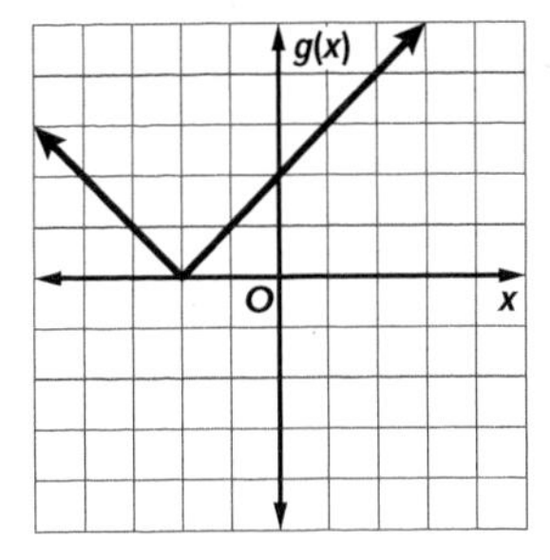

15. D = {all real numbers}; R = $\{h(x) \mid h(x) \geq -8\}$

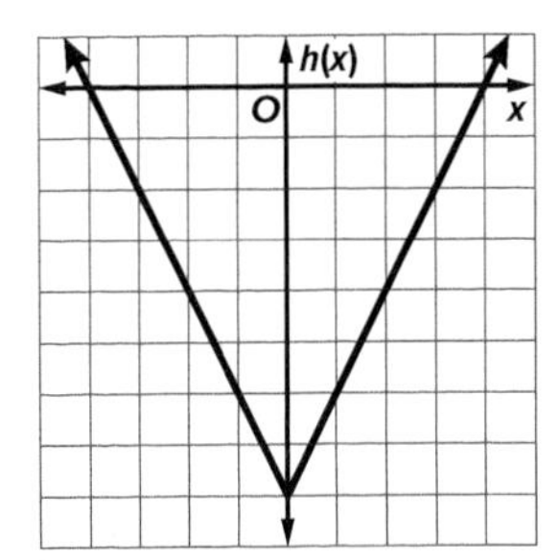

16. D = {all real numbers}; R = $\{k(x) \mid k(x) \geq 3\}$

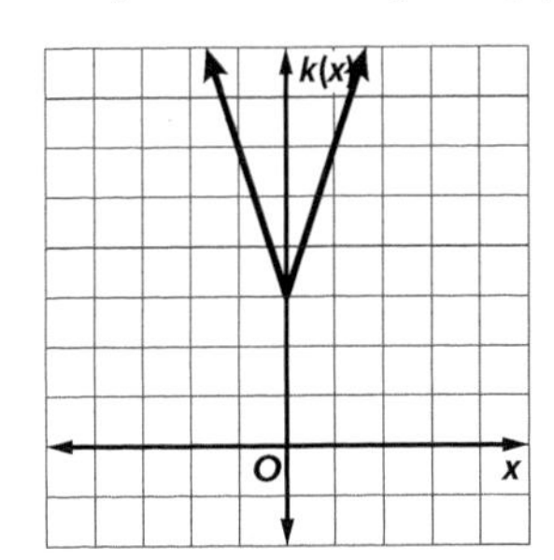

17. D = {all real numbers}; R = $\{f(x) \mid f(x) \geq 6\}$

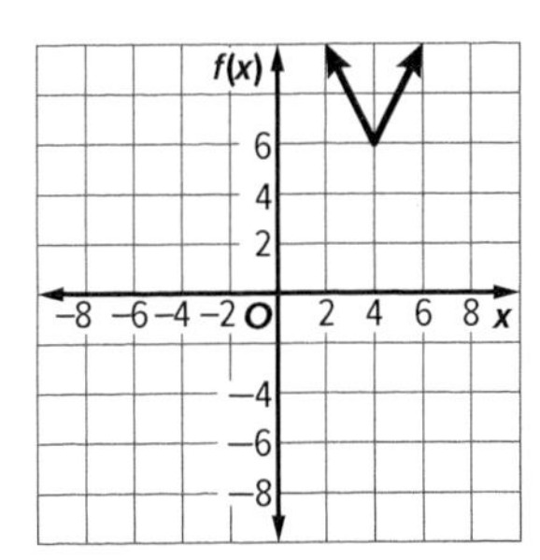

18. D = {all real numbers}; R = $\{h(x) \mid h(x) \leq -2\}$

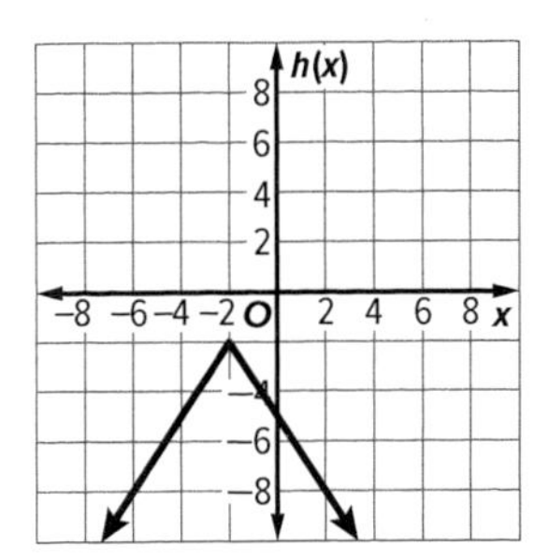

19. D = {all real numbers}; R = $\{g(x) \mid g(x) \geq 0\}$

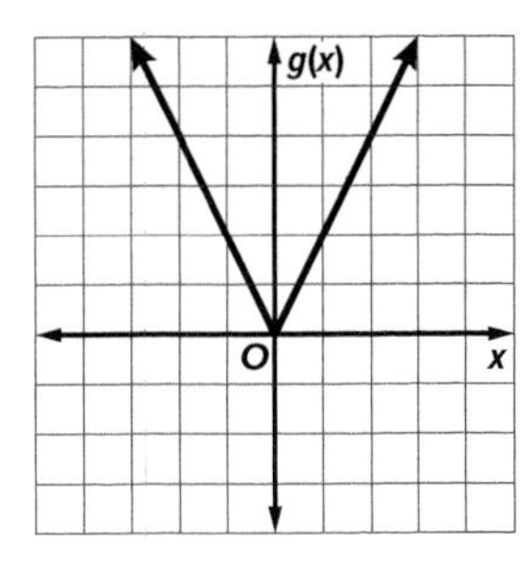

20. D = {all real numbers}; R = $\{f(x) \mid f(x) \geq 1\}$

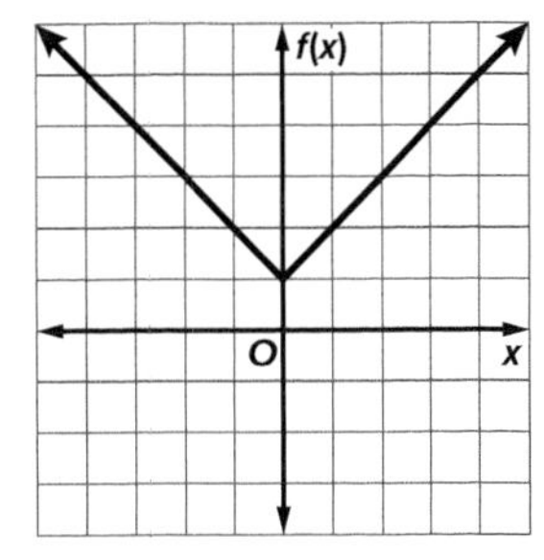

21. D = $\{x \mid x \leq -4 \text{ or } 0 < x\}$; R = $\{f(x) \mid f(x) \geq 12, f(x) = 8, \text{ or } 0 < f(x) \leq 3\}$

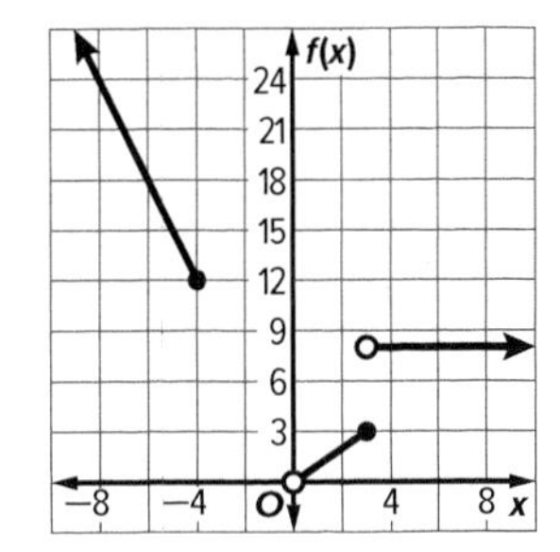

22. D = $\{x \mid x \leq 2 \text{ or } x > 4\}$; R = $\{f(x) \mid f(x) < -7, \text{ or } f(x) = 5\}$

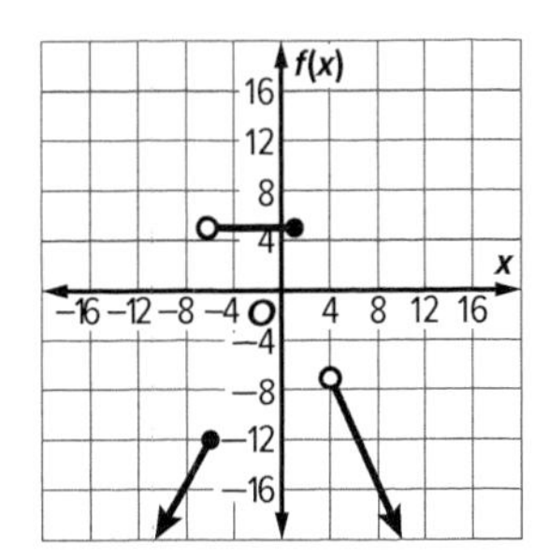

23. D = {all real numbers}; R = $\{g(x) \mid g(x) < -10 \text{ or } -6 \leq g(x) \leq 2\}$

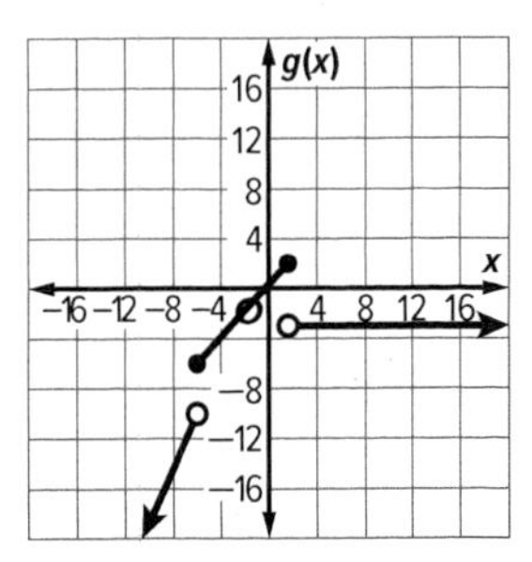

24. D = $\{x \mid x < -4, -1 \leq x \leq 5, \text{ or } x > 7\}$; R = $\{g(x) \mid g(x) \geq -4\}$

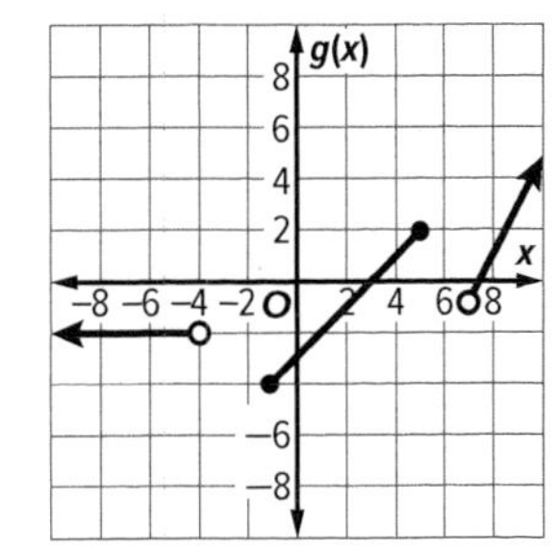

25. D = {all real numbers}; R = $\{f(x) \mid f(x) > -3\}$

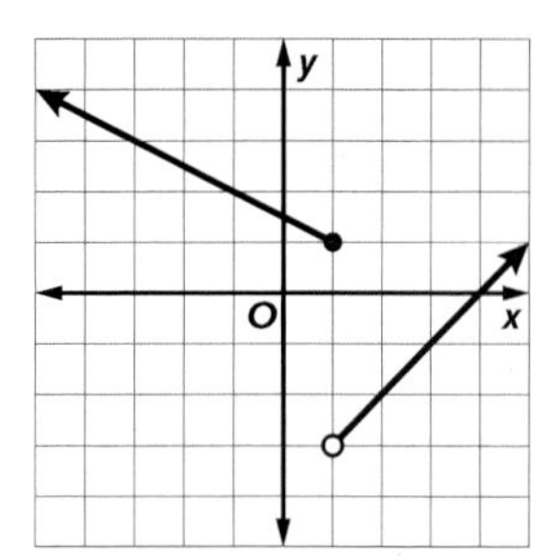

26. D = {all real numbers}; R = $\{y \mid y \geq 0\}$

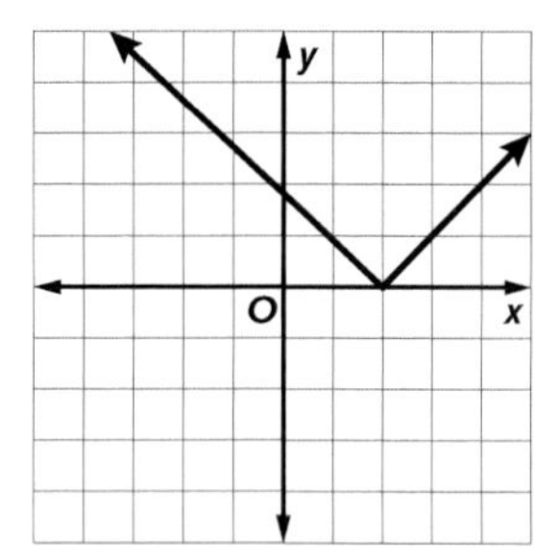

27. D = {all real numbers}; R = {all integers}

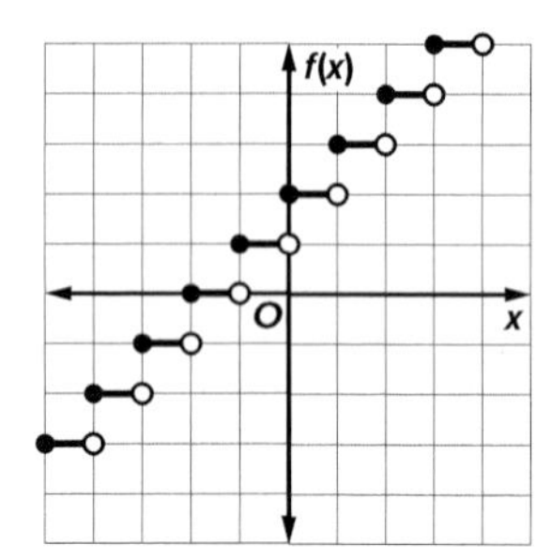

28. D = {all real numbers}; R = {all even integers}

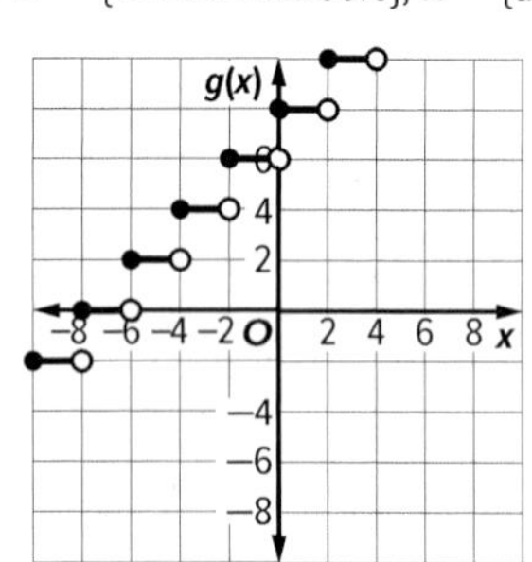

29. D = {all real numbers}; R = {all whole numbers}

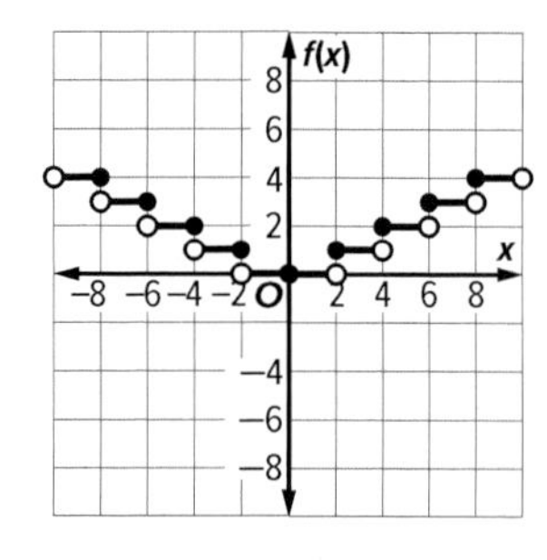

30. D = {all real numbers}; R = {non-negative integers}

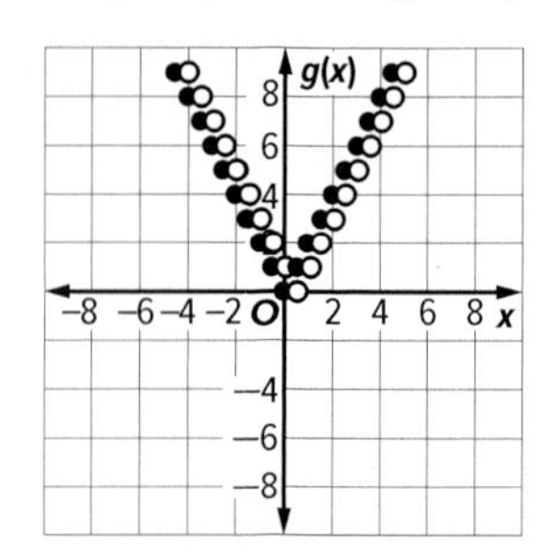

31. D = {all real numbers}; R = $\{g(x) \mid g(x) \leq 4\}$

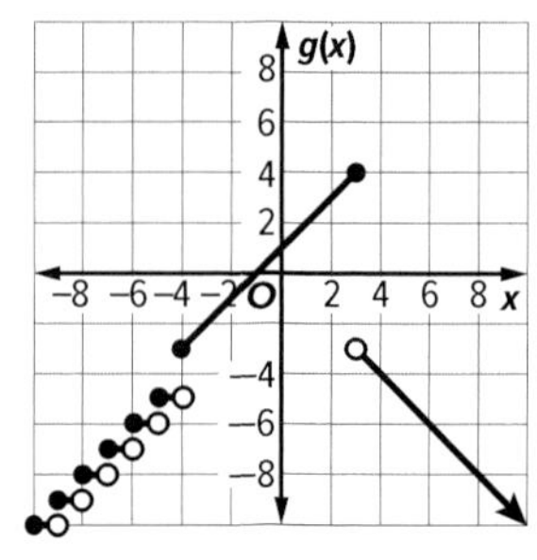

32. D = {all real numbers}; R = $\{h(x) \mid h(x) \leq -6 \text{ or } 0 \leq h(x)\}$

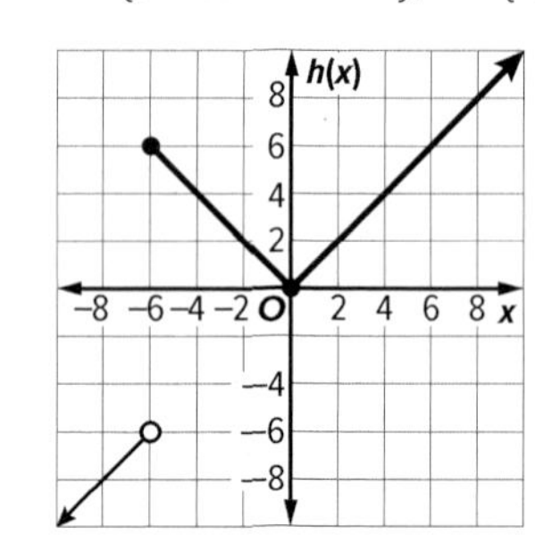

35. $C(x) = \begin{cases} 5 & \text{if } 0 \leq x \leq 1 \\ 7.5 & \text{if } 1 < x \leq 2 \\ 10 & \text{if } 2 < x \leq 3 \\ 12.5 & \text{if } 3 < x \leq 4 \\ 15 & \text{if } 4 < x \leq 24 \end{cases}$

37b.

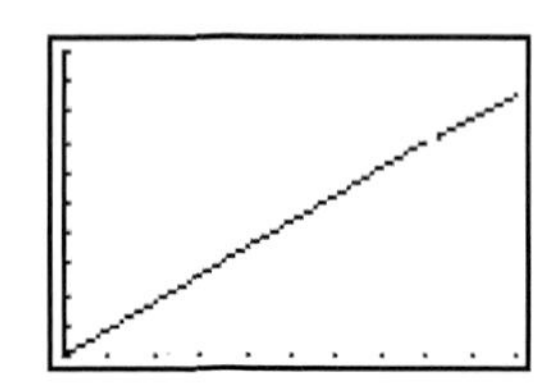

[0, 50] scl: 5 by [500, 1500] scl: 100

37c. Because it costs $1200 for 40 guests to attend, use the first expression in the function $C(x)$. Solve the equation $500 + 17.50x = 900$ to obtain about 22.9. Because there cannot be a fraction of a guest, at most 22 guests can be invited to the reunion.

39a. $R(t) = \begin{cases} \frac{20}{3}t + 30 & \text{if } 0 \leq t \leq 3 \\ 60 & \text{if } 3 < t < 4 \\ 80 & \text{if } 4 \leq t \leq 5 \\ 60 & \text{if } 5 < t < 6 \\ -\frac{50}{3}t + 160 & \text{if } 6 \leq t \leq 9 \end{cases}$

Range = $[10, 60] \cup \{80\}$

40a.

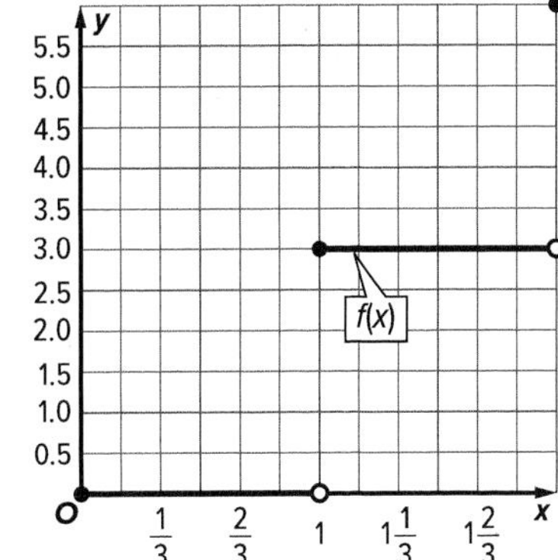

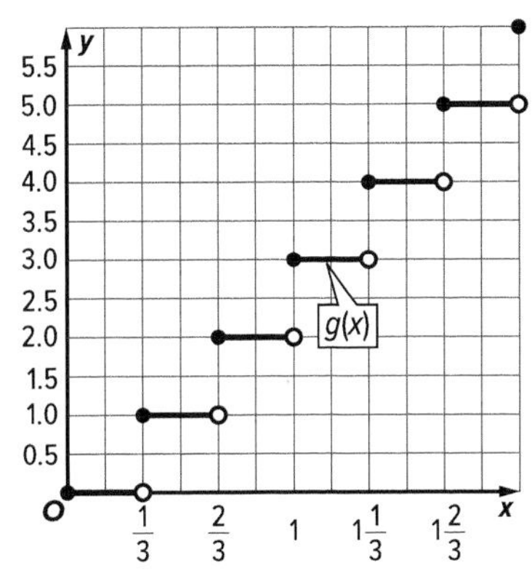

40b. For $f(x) = 3[\![x]\!]$, the jumps between steps increased from 1 unit (which occurs in the graph of the parent function $[\![x]\!]$) to 3 units, but the lengths of the segments are all 1 unit. For $g(x) = [\![3x]\!]$, the length of each segment decreased by a factor of 3, but each of the jumps remained 1 unit, the same as the parent function.

40c.

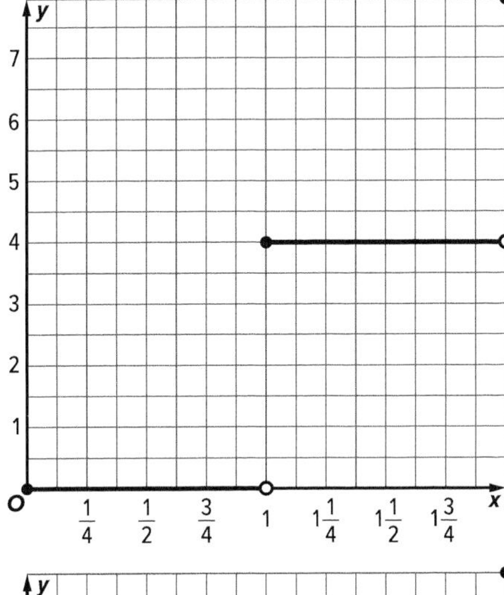

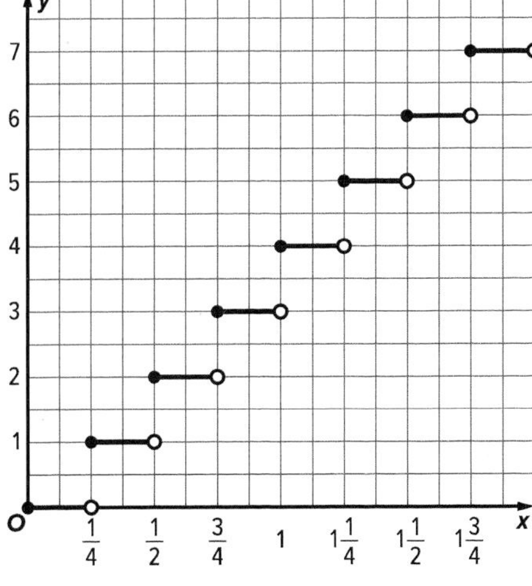

40d. The same as part (b), except now the jump is 4 units instead of 3 for $f(x)$, and the segment lengths are decreased by a factor of 4 instead of 3 for $g(x)$.

40e. For $f(x) = n[\![x]\!]$, the jumps between steps increased from 1 unit (which occurs in the graph of the parent function $[\![x]\!]$) to n units, but the lengths of the segments are all 1 unit. For $g(x) = [\![nx]\!]$, the length of each segment decreased by a factor of n, but each of the jumps remained as 1 unit.

41. Because the absolute value takes negative $f(x)$-values and makes them positive, the graph retains the step-like nature of the greatest integer function, but it also has the "v" shape of the absolute value.

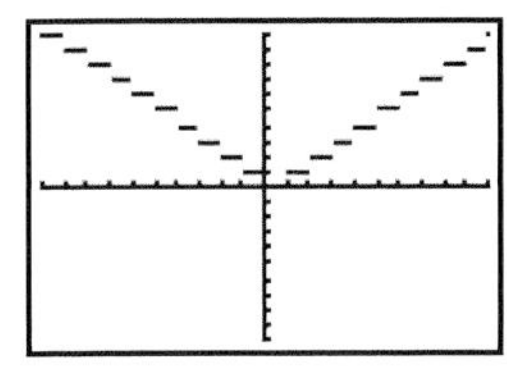

$[-10, 10]$ scl: 1 by $[-10, 10]$ scl: 1

46.

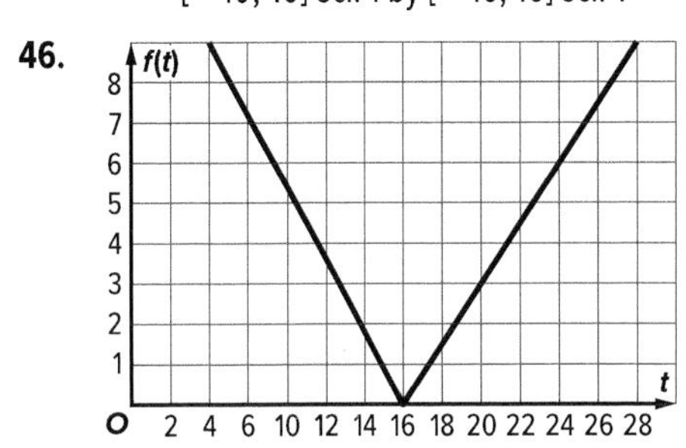

47. D = $\{m \mid 0 < t \leq 6\}$; R = $\{C \mid C = 2.00, 4.00, 6.00, 8.00, 10.00, 12.00\}$

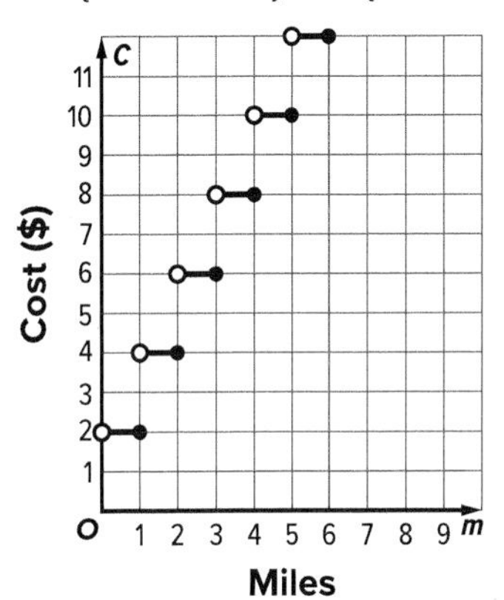

48a. Sample answer: The rate at which Jackson walks away from his house is greater than the rate at which Jackson walks towards his house. Therefore, the graph will not be symmetrical. So, Jackson should write a piecewise-defined function.

48b. $d(t) = \begin{cases} \frac{1}{10}t & \text{if } 0 \leq t < 20 \\ -\frac{1}{15}t + \frac{10}{3} & \text{if } 20 \leq t \leq 50 \end{cases}$

48c.

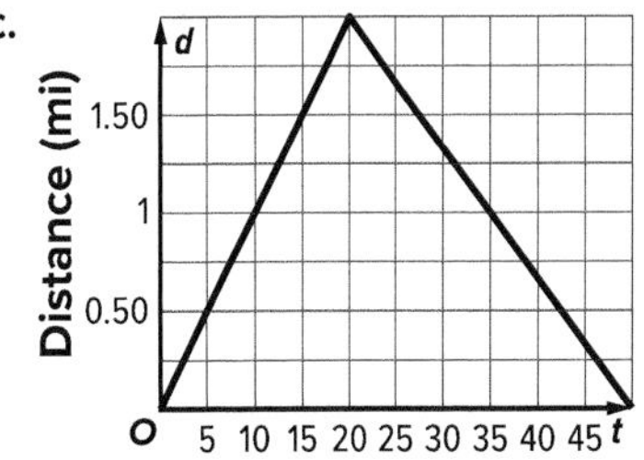

48d. D = $\{t \mid 0 \leq t \leq 50\}$; R = $\{d \mid 0 \leq d \leq 2\}$

50.

Lesson 1-7

50. All of the graphs have a similar shape. The graph of $g(x)$ is the graph of the parent function $f(x)$ translated left 2 units, and that the graph of $h(x)$ is the graph of $f(x)$ translated right 3 units.

52. A transformation will affect the graph vertically or horizontally based on whether it is applied to the variable x before or after the variable expression has been evaluated. For example, the transformation $(x + 1)^2$ adds one to the variable before it is evaluated and affects the graph horizontally. The transformation $x^2 + 1$ adds one to the function after the variable expression has been evaluated and has a vertical effect.

54. The graph is stretched vertically, translated to the left 5 units and down 1 unit, and reflected in the x-axis. The domain is all real numbers and the range is real numbers less than or equal to –1.

55a.

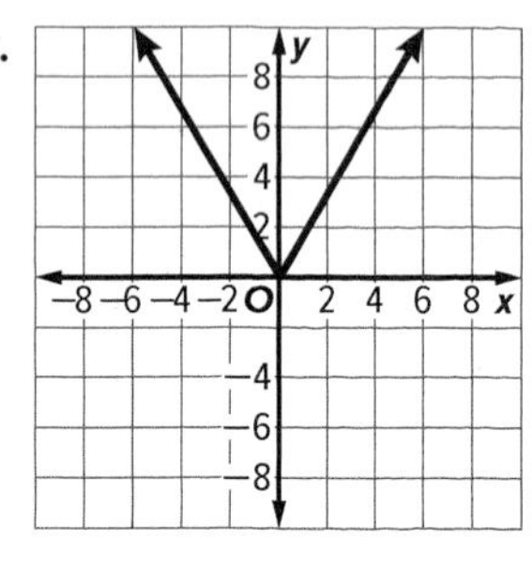

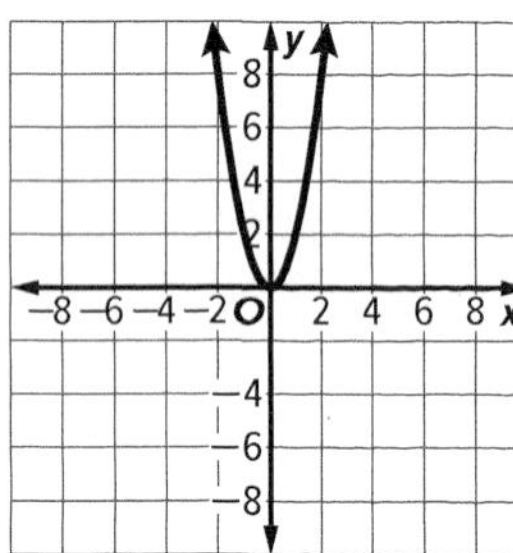

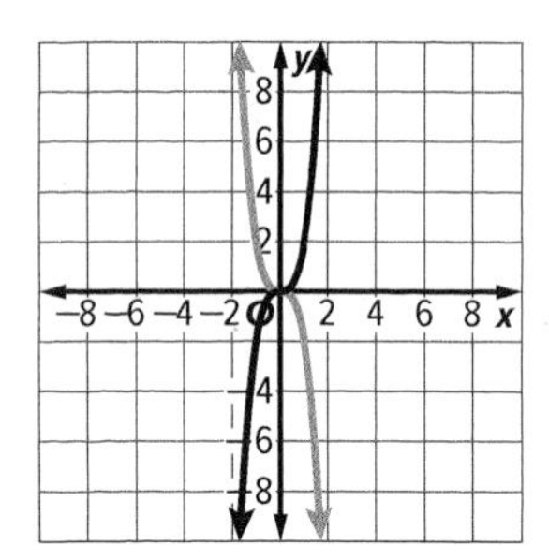

55b. $f(x)$ and $h(x)$ are even, $g(x)$ is neither, and $k(x)$ is odd.

55c. Even functions are symmetric in the y-axis. If $f(-x) = f(x)$, then the graphs of $f(-x)$ and $f(x)$ coincide. If the graph of a function coincides with its own reflection in the y-axis, then the graph is symmetric in the y-axis. Odd functions are symmetric in the origin, which means that the graph of an odd function coincides with its rotation of 180° about the origin. A rotation of 180° is equivalent to reflection in two perpendicular lines. $f(-x)$ is a reflection in the y-axis and $-f(x)$ is a reflection in the x-axis. Thus if the graphs of $f(-x)$ and $-f(x)$ coincide, $f(x)$ is symmetric about the origin.

56. Sample answer: Because a vertical translation affects the y-values and a horizontal translation affects x-values, order is irrelevant.

57. Sample answer: The graph in Quadrant II has been reflected in the x-axis and moved right 10 units.

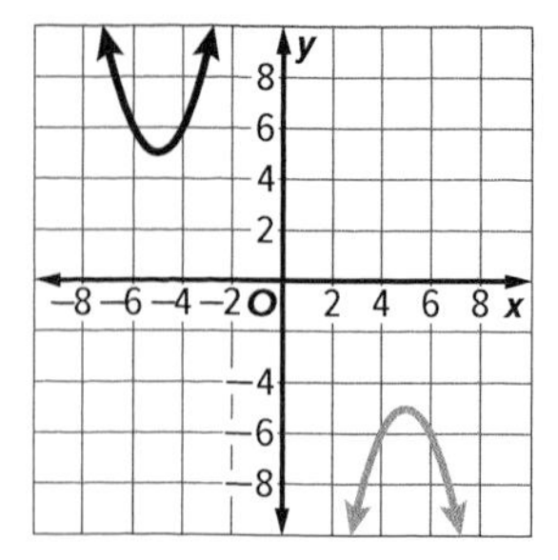

58. Sample answer: The graph of $y = x^2$ is positive when $x \rightarrow -\infty$ and when $x \rightarrow \infty$.

59. Sample answer: Because the graph of $g(x)$ is symmetric about the y-axis, reflecting in the y-axis results in a graph that appears the same. It is not true for all quadratic functions. When the axis of symmetry of the parabola is not along the y-axis, the graph and the graph reflected in the y-axis will be different.

Module 2

Linear Equations, Inequalities, and Systems

Module Goals

- Students solve linear equations and inequalities in one variable.
- Students solve systems of equations by graphing, substitution or elimination.
- Students solve equations and inequalities involving absolute value.

Focus

Domain: Algebra, Functions

Standards for Mathematical Content:

A.CED.3 Represent constraints by equations or inequalities, and by systems of equations and/or inequalities, and interpret solutions as viable or non-viable options in a modeling context.

F.IF.6 Calculate and interpret the average rate of change of a function (presented symbolically or as a table) over a specified interval. Estimate the rate of change from a graph.

Also addresses A.CED.1, A.CED.2, A.REI.1, and A.REI.11.

Standards for Mathematical Practice:

All Standards for Mathematical Practice will be addressed in this Module.

Coherence

Vertical Alignment

Previous

Students analyzed and solved systems of linear equations and inequalities.

8.EE.8, A.REI.5, A.REI.6, A.REI.7, A.REI.12 (Algebra 1)

Now

Students solve linear equations and inequalities in one, two, or three variables algebraically and by graphing.

A.CED.2, A.REI.6, A.REI.11

Next

Students will solve quadratic equations and inequalities and systems with linear and nonlinear equations.

A.REI.4, A.REI.7

Rigor

The Three Pillars of Rigor

To help students meet standards, they need to illustrate their ability to use the three pillars of rigor. Students gain conceptual understanding as they move from the Explore to Learn sections within a lesson. Once they understand the concept, they practice procedural skills and fluency and apply their mathematical knowledge as they go through the Examples and Independent Practice.

1 CONCEPTUAL UNDERSTANDING		2 FLUENCY	3 APPLICATION
EXPLORE	LEARN	EXAMPLE & PRACTICE	

Suggested Pacing

Lessons	Standards	45-min classes	90-min classes
Module Pretest and Launch the Module Video		1	0.5
2-1 Solving Linear Equations and Inequalities	A.CED.1, A.CED.2	1	0.5
2-2 Solving Absolute Value Equations and Inequalities	A.CED.1, A.CED.3	2	1
2-3 Equations of Linear Functions	A.CED.2, F.IF.6	2	1
2-4 Solving Systems of Equations Graphically	A.CED.3, A.REI.11	2	1
2-5 Solving Systems of Equations Algebraically	A.CED.3	2	1
Put It All Together: Lessons 2-1 through 2-5		1	0.5
2-6 Solving Systems of Inequalities	A.CED.3	1	0.5
2-7 Optimization with Linear Programming	A.CED.3	1	0.5
2-8 Systems of Equations in Three Variables	A.CED3	1	0.5
2-9 Solving Absolute Value Equations and Inequalities by Graphing	A.CED.1	1	0.5
Module Review		1	0.5
Module Assessment		1	0.5
	Total Days	**17**	**8.5**

Formative Assessment Math Probe
Absolute Value Inequalities

Module Resource

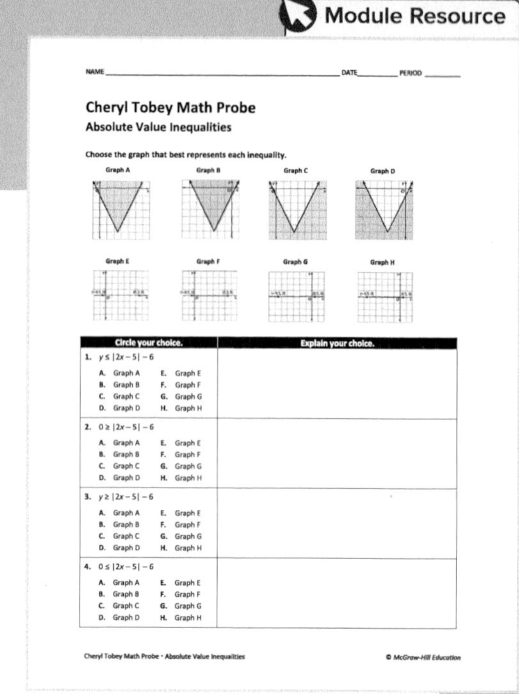

NAME ____________ DATE ________ PERIOD ________

Cheryl Tobey Math Probe
Absolute Value Inequalities

Choose the graph that best represents each inequality.

Graph A Graph B Graph C Graph D

Graph E Graph F Graph G Graph H

Circle your choice.	Explain your choice.
1. $y \leq \|2x-5\|-6$ A. Graph A E. Graph E B. Graph B F. Graph F C. Graph C G. Graph G D. Graph D H. Graph H	
2. $0 \geq \|2x-5\|-6$ A. Graph A E. Graph E B. Graph B F. Graph F C. Graph C G. Graph G D. Graph D H. Graph H	
3. $y \geq \|2x-5\|-6$ A. Graph A E. Graph E B. Graph B F. Graph F C. Graph C G. Graph G D. Graph D H. Graph H	
4. $0 \leq \|2x-5\|-6$ A. Graph A E. Graph E B. Graph B F. Graph F C. Graph C G. Graph G D. Graph D H. Graph H	

Cheryl Tobey Math Probe • Absolute Value Inequalities © McGraw-Hill Education

Correct Answers: 1. C 2. F 3. A 4. E

Analyze the Probe

Review the probe prior to assigning it to your students.

In this probe, students will choose graphs that best represent a set of inequalities.

Targeted Concepts Understand the connection between symbolic and graphic representations of absolute value inequalities.

Targeted Misconceptions

- Students shade the incorrect region when they lack understanding of the relationship between the graphical representation of an inequality and the solutions of the algebraic representation.
- Students do not distinguish an absolute value inequality in one variable from an absolute value inequality in two variables.
- Students incorrectly transform the absolute value from the parent function $y = |x|$.
- Students incorrectly solve the absolute value inequality, leading to choosing the incorrect representation.

Use the Probe after Lesson 2-9.

Collect and Assess Student Answers

If the student selects these responses...	**Then** the student likely...
1. A **3.** C	is having difficulty choosing the graphical region that represents correct solutions of the absolute value inequality.
2. A **4.** C	is confusing a two-variable absolute value inequality with a one-variable absolute value inequality.
1. D **2.** H **3.** B **4.** G	fails to use the opposite value of the number associated with x to determine the direction of the shift from the parent function $y = |x|$.

Take Action

After the Probe Design a plan to address any possible misconceptions. You may wish to assign the following resources.

- ALEKS® Absolute Value Equations and Inequalities
- Lesson 2-9, Learn, Solving Absolute Value Inequalities by Graphing, Examples 4-5

Revisit the probe at the end of the module to be sure that your students no longer carry these misconceptions.

The Ignite! activities, created by Dr. Raj Shah, cultivate curiosity and engage and challenge students. Use these open-ended, collaborative activities, located online in the module Launch section, to encourage your students to develop a growth mindset towards mathematics and problem solving. Use the teacher notes for implementation suggestions and support for encouraging productive struggle.

Essential Question

At the end of this module, students should be able to answer the Essential Question.

How are equations, inequalities, and systems of equations or inequalities best used to model real-world situations? Equations allow you to find quantities that fit constraints, such as costs for given numbers of items. Inequalities represent situations involving intervals of numbers, such as quantities for which cost is below a given value. Systems allow you to have multiple constraints for the same situation.

What Will You Learn?

Prior to beginning this module, have your students rate their knowledge of each item listed. Then, at the end of the module, you will be reminded to have your students return to these pages to rate their knowledge again. They should see that their knowledge and skills have increased.

DINAH ZIKE FOLDABLES

Focus Students read about equations and inequalities, including linear and absolute value functions, solving systems of equations and inequalities, linear programming, and systems in three variables.

Teach Throughout the module, have students take notes under the tabs of their Foldables while working through each lesson. They should include definitions of terms and key concepts. Encourage students to record examples from each lesson in their Foldable.

When to Use It Use the appropriate tabs as students cover each lesson in this module.

Launch the Module

For this module, the Launch the Module video shows how, through linear programming, a system of inequalities can be used to model certain recommendations for daily diet, exercise, and sleep. Students learn about how algebra is applied in daily life.

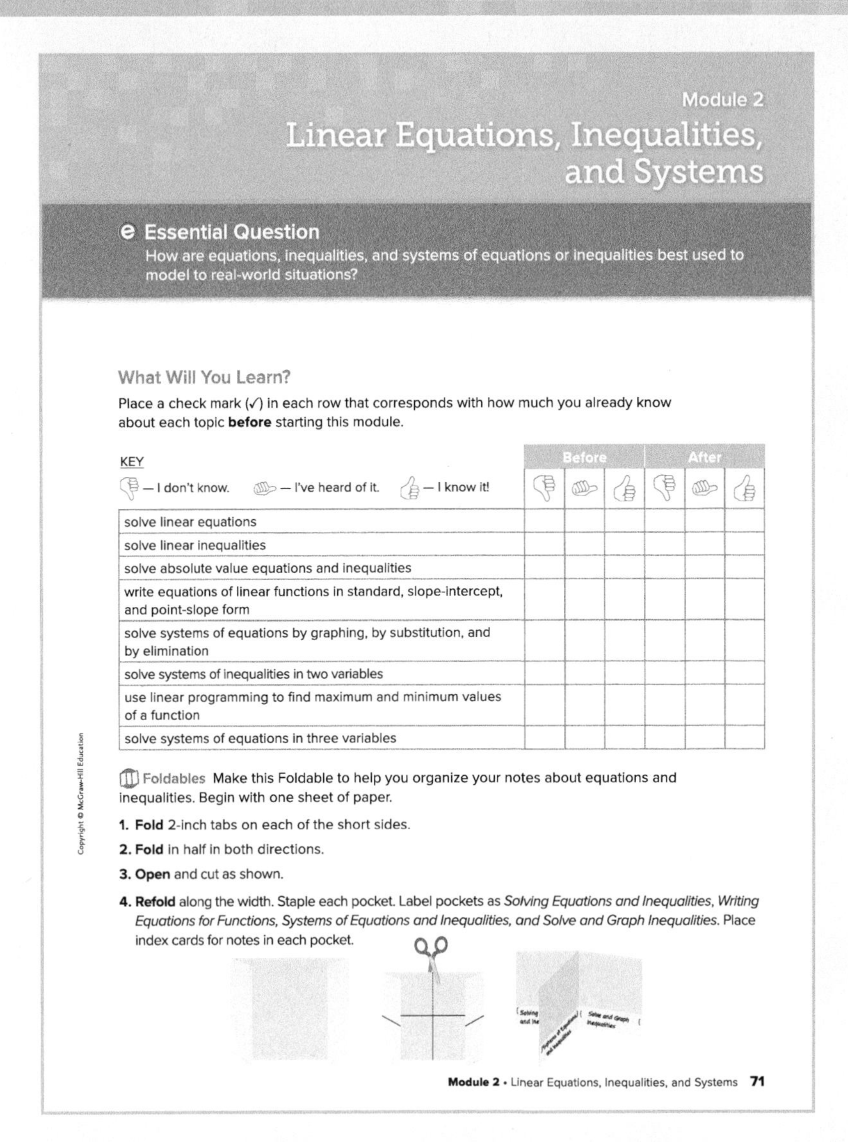

Module 2

Linear Equations, Inequalities, and Systems

Essential Question
How are equations, inequalities, and systems of equations or inequalities best used to model to real-world situations?

What Will You Learn?

Place a check mark (✓) in each row that corresponds with how much you already know about each topic **before** starting this module.

KEY: — I don't know. — I've heard of it. — I know it!

	Before			After		
	I don't know	I've heard of it	I know it!	I don't know	I've heard of it	I know it!
solve linear equations						
solve linear inequalities						
solve absolute value equations and inequalities						
write equations of linear functions in standard, slope-intercept, and point-slope form						
solve systems of equations by graphing, by substitution, and by elimination						
solve systems of inequalities in two variables						
use linear programming to find maximum and minimum values of a function						
solve systems of equations in three variables						

Foldables Make this Foldable to help you organize your notes about equations and inequalities. Begin with one sheet of paper.

1. **Fold** 2-inch tabs on each of the short sides.
2. **Fold** in half in both directions.
3. **Open** and cut as shown.
4. **Refold** along the width. Staple each pocket. Label pockets as *Solving Equations and Inequalities, Writing Equations for Functions, Systems of Equations and Inequalities, and Solve and Graph Inequalities.* Place index cards for notes in each pocket.

Interactive Presentation

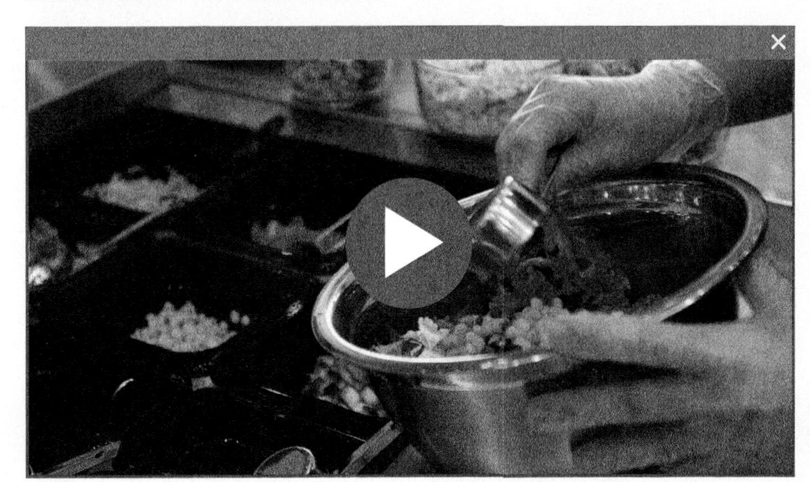

What Vocabulary Will You Learn?

Check the box next to each vocabulary term that you may already know.

□ absolute value	□ feasible region	□ root
□ bounded	□ inconsistent	□ solution
□ consistent	□ independent	□ substitution
□ dependent	□ inequality	□ system of equations
□ elimination	□ linear programming	□ system of inequalities
□ empty set	□ optimization	□ unbounded
□ equation	□ ordered triple	□ zero
□ extraneous solution		

Are You Ready?

Complete the Quick Review to see if you are ready to start this module. Then complete the Quick Check.

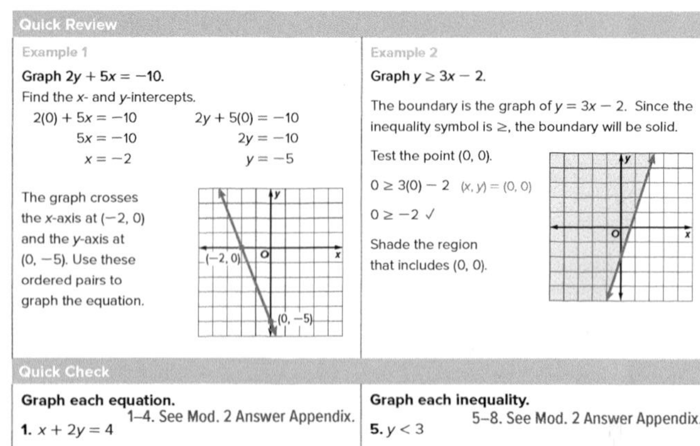

Quick Review

Example 1

Graph $2y + 5x = -10$.

Find the x- and y-intercepts.

$2(0) + 5x = -10$ $2y + 5(0) = -10$

$5x = -10$ $2y = -10$

$x = -2$ $y = -5$

The graph crosses the x-axis at $(-2, 0)$ and the y-axis at $(0, -5)$. Use these ordered pairs to graph the equation.

Example 2

Graph $y \geq 3x - 2$.

The boundary is the graph of $y = 3x - 2$. Since the inequality symbol is $\geq$, the boundary will be solid.

Test the point (0, 0).

$0 \geq 3(0) - 2$ $(x, y) = (0, 0)$

$0 \geq -2$ ✓

Shade the region that includes (0, 0).

Quick Check

Graph each equation. 1–4. See Mod. 2 Answer Appendix.	Graph each inequality. 5–8. See Mod. 2 Answer Appendix.
1. $x + 2y = 4$	5. $y < 3$
2. $y = -x + 6$	6. $x + y \geq 1$
3. $3x + 5y = 15$	7. $3x - y > 6$
4. $3y - 2x = -12$	8. $x + 2y \leq 5$

How Did You Do?

Which exercises did you answer correctly in the Quick Check? Shade those exercise numbers below.

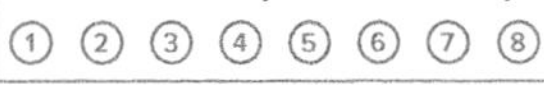

What Vocabulary Will You Learn?

ELL As you proceed through the module, introduce the key vocabulary by using the following routine.

Define Any linear function can be written in standard form, $y = mx + b$, where m is the slope b is the y-intercept.

Example $y = -\frac{4}{3}x + 2$

Ask Is the function written in the form $y = mx + b$ where m is the slope and b is the y-intercept? yes, $m = -\frac{4}{3}$ and $b = 2$

Are You Ready?

Students may need to review the following prerequisite skills to succeed in this module.

- solving open sentences
- finding absolute value and distance
- evaluating slopes of lines
- evaluating equations of lines
- adding polynomials
- writing equations and inequalities
- writing in set-builder and interval notation

ALEKS®

ALEKS is an adaptive, personalized learning environment that identifies precisely what each student knows and is ready to learn, ensuring student success at all levels.

You may want to use the **Real Numbers and Linear Equations, Lines and Functions, Systems of Linear Equations and Matrices** sections to ensure student success in this module.

Mindset Matters

Reward Effort, Not Talent

When adults praise students for their hard work toward a solution, rather than praising them for being smart or talented, it supports students' development of a growth mindset. Reward *actions* like hard work, determination, and perseverance instead of *traits* like inherent skill or talent.

How Can I Apply It?

Have students complete the **Performance Task** for the module. Allow students a forum to discuss their process or strategy that they used and give them positive feedback on their diligence in completing the task.

Lesson 2-1

A.CED.1, A.CED.2

Solving Linear Equations and Inequalities

LESSON GOAL

Students solve linear equations and inequalities in one variable.

1 LAUNCH

Launch the lesson with a **Warm Up** and an introduction.

2 EXPLORE AND DEVELOP

Develop:

Solving Linear Equations
- Solve a Linear Equation
- Write and Solve an Equation
- Solve for a Variable

Solving Linear Equations by Graphing
- Solve a Linear Equation by Graphing
- Estimate Solutions by Graphing

Explore: Comparing Linear Equations and Inequalities

Develop:

Solving Linear Inequalities
- Solve a Linear Inequality
- Write and Solve an Inequality

You may want your students to complete the **Checks** online.

3 REFLECT AND PRACTICE

Exit Ticket

Practice

DIFFERENTIATE

View reports of student progress on the **Checks** after each example.

Resources	AL	OL	BL	ELL
Remediation: Writing and Interpreting Equations	●	●		●
Extension: United States' Gross National Product		●	●	●

Language Development Handbook

Assign page 8 of the *Language Development Handbook* to help your students build mathematical language related to linear equations and inequalities in one variable.

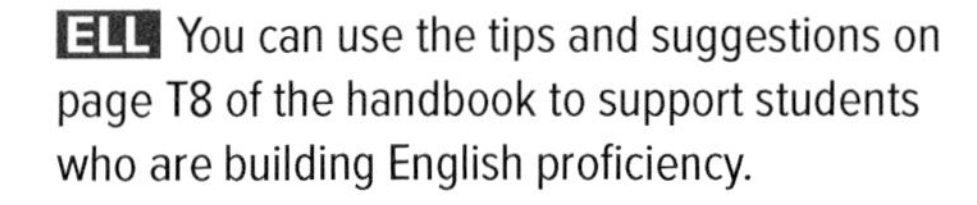

ELL You can use the tips and suggestions on page T8 of the handbook to support students who are building English proficiency.

Suggested Pacing

90 min	0.5 day
45 min	1 day

Focus

Domain: Algebra

Standards for Mathematical Content:

A.CED.1 Create equations and inequalities in one variable and use them to solve problems.

A.CED.2 Create equations in two or more variables to represent relationships between quantities; graph equations on coordinate axes with labels and scales.

Standards for Mathematical Practice:

1 Make sense of problems and persevere in solving them.

2 Reason abstractly and quantitatively.

Coherence

Vertical Alignment

Previous
Students graphed linear functions and inequalities in two variables.
F.IF.4, F.BF.3

Now
Students solve linear equations and inequalities in one variable.
A.CED.1, A.CED.2

Next
Students will solve equations and inequalities involving absolute value algebraically.
A.CED.1, A.CED.3

Rigor

The Three Pillars of Rigor

1 CONCEPTUAL UNDERSTANDING	2 FLUENCY	3 APPLICATION

Conceptual Bridge In this lesson, students review their understanding of and build fluency by solving equations and inequalities. They apply their understanding by solving real-world problems using equations and inequalities.

Mathematical Background

Adding the same number to, or subtracting the same number from each side of an inequality does not affect the truth of the inequality. Multiplying or dividing each side of an inequality by a positive number also does not affect the truth of the inequality. However, multiplying or dividing each side of an inequality by negative values does affect the truth of an inequality.

Interactive Presentation

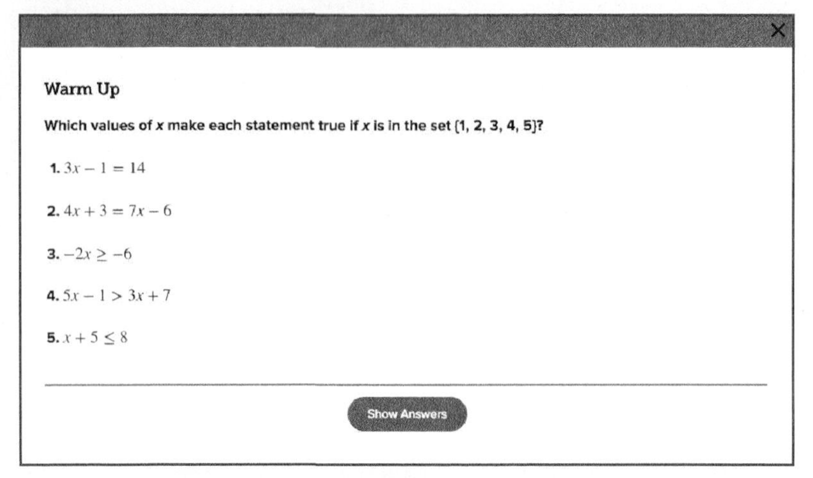

Warm Up

Launch the Lesson

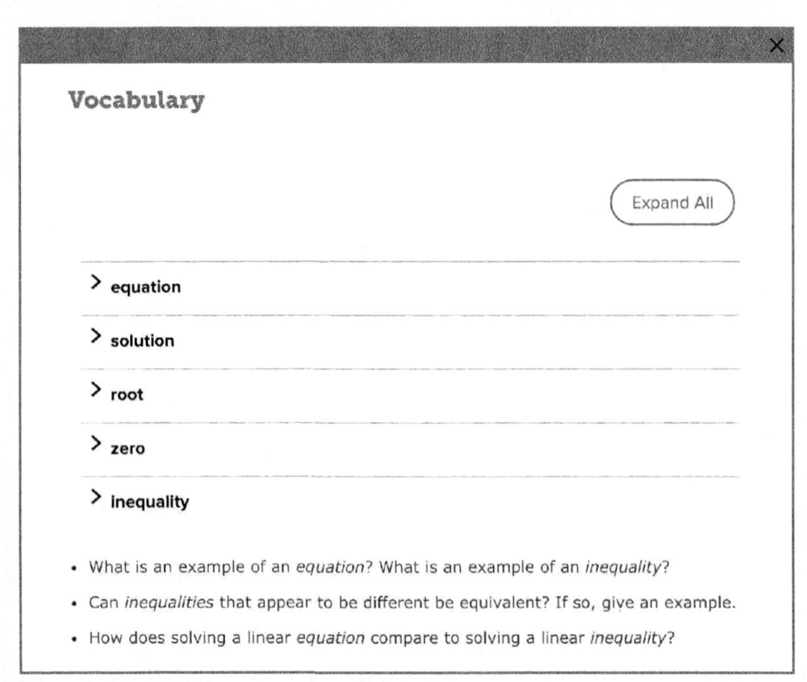

Today's Vocabulary

Warm Up

Prerequisite Skills

The Warm Up exercises address the following prerequisite skill for this lesson:

- solving open sentences

Answers:

1. 5
2. 3
3. 1, 2, 3
4. 5
5. 1, 2, 3

Launch the Lesson

Teaching the Mathematical Practices

2 Make Sense of Quantities Mathematically proficient students need to be able to make sense of quantities and their relationships. In the Launch the Lesson, notice the relationships between U.S. dollars and other currencies. Help students to use these relationships to write an equation that can be used to answer the question in the infographic.

Go Online to find additional teaching notes and questions to promote classroom discourse.

Today's Standards

Tell students that they will be addressing these content and practice standards in this lesson. You may wish to have a student volunteer read aloud *How can I meet these standards?* and *How can I use these practices?*, and connect these to the standards.

See the Interactive Presentation for I Can statements that align with the standards covered in this lesson.

Today's Vocabulary

Tell students that they will be using these vocabulary terms in this lesson. You can expand each row if you wish to share the definitions. Then discuss the questions below with the class.

Explore Comparing Linear Equations and Inequalities

Objective

Students compare solving linear equations and inequalities in one variable.

 Teaching the Mathematical Practices

6 Communicate Precisely Encourage students to routinely write and explain their reasoning. Point out that they should use clear mathematical language and definitions when completing the exercises in the Explore activity.

Ideas for Use

Recommended Use Present the Inquiry Question, or have a student volunteer read it aloud. Have students work in pairs to complete the Explore activity on their devices. Pairs should discuss each of the questions. Monitor student progress during the activity. Upon completion of the Explore activity, have student volunteers share their responses to the Inquiry Question.

What if my students don't have devices? You may choose to project the activity on a whiteboard. A printable worksheet for each Explore is available online. You may choose to print the worksheet so that individuals or pairs of students can use it to record their observations.

Summary of the Activity

Students will complete guiding exercises throughout the Explore activity. Students will use algebra to identify the solution of a linear equation and inequality in one variable. Then students will compare the solutions to the equation and inequality. Then, students will answer the Inquiry Question.

(continued on the next page)

Interactive Presentation

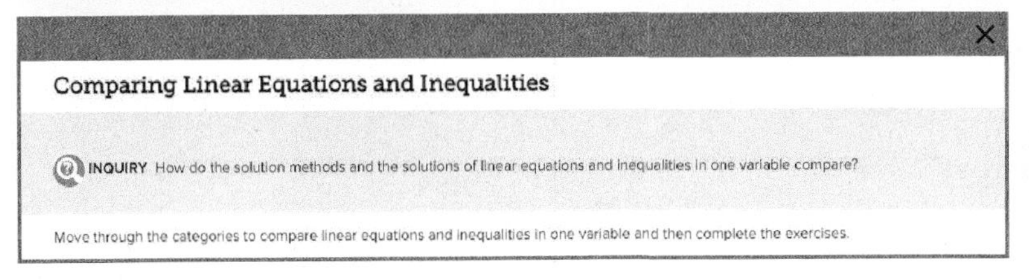

Explore

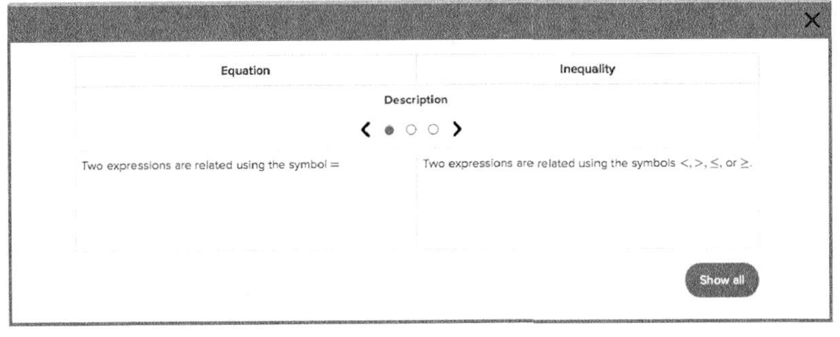

Explore

EXPAND

Students tap to compare and contrast equations and inequalities

Interactive Presentation

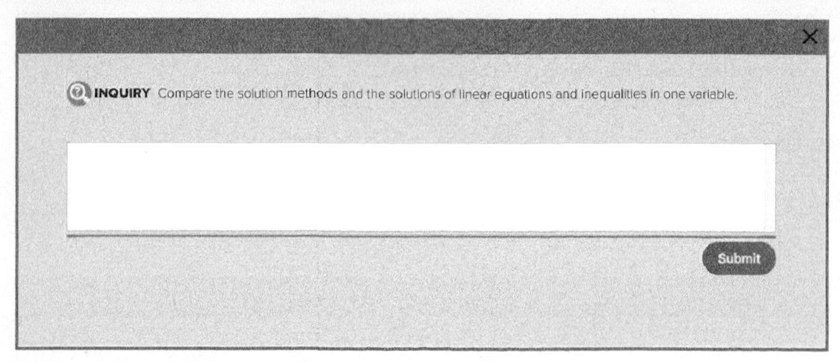

Explore

TYPE

Students respond to the Inquiry Question and can view a sample answer.

Explore Comparing Linear Equations and Inequalities (*continued*)

Questions

Have students complete the Explore activity.

Ask:

- How do the solutions of $4x - 1 = 7$ and $4x - 1 \leq 7$ compare? Sample answer: The solutions both contain the number 2, but the inequality also contains all real numbers less than two.
- Joseph wants to determine how many pencils he could buy if he has \$12 and each pencil is \$0.60. Should he write an equation or inequality to solve? Explain. Sample answer: He should write an inequality since he could buy 20 pencils or less. He does not necessarily have to buy just 20.

Inquiry

How do the solution methods and the solutions of linear equations and inequalities in one variable compare? Sample answer: The steps used to solve are the same, but if you multiply or divide by a negative value when solving an inequality, the symbol needs to be reversed. A linear equation has only one solution, while the solution of a linear inequality is a set of values.

Go Online to find additional teaching notes and sample answers for the guiding exercises.

1 CONCEPTUAL UNDERSTANDING | 2 FLUENCY | 3 APPLICATION

Learn Solving Linear Equations

Objective

Students solve linear equations by applying the properties of equality.

Teaching the Mathematical Practices

2 Different Properties Mathematically proficient students are familiar with and can use different properties of operations. Encourage students to explain how the properties of equality can be used to solve linear equations.

DIFFERENTIATE

Reteaching Activity AL ELL

IF students are struggling to solve equations in one variable,
THEN have them use algebra tiles to model and solve the equation. Remind students that they must add the same number of positive or negative tiles to each side.

Example 1 Solve a Linear Equation

Teaching the Mathematical Practices

1 Seek Information Mathematically proficient students must be able to transform algebraic expressions to reach solutions. In Example 1, students must transform each side of the equation to isolate the variable.

Questions for Mathematical Discourse

AL It is often possible to solve an equation via multiple approaches. What must you keep in mind when deciding what step to take next when solving an equation? Each step must be consistent with the order of operations.

OL How could the fractions be eliminated first, and what would be the resulting equation? The entire equation could be multiplied by 3, resulting in $2x - 57 + 6 - x = -12$.

BL What is another approach you could take to solving this equation? What do you look for to determine the best approach to solving an equation? Sample answer: Both terms in the parentheses are multiplied by the same number, so you could combine the inside terms and get $\frac{1}{3}(x - 51) = -4$. It is advantageous to pick the approach that results in the simplest or least amount of calculations.

Go Online

- Find additional teaching notes.
- View performance reports of the Checks.
- Assign or present an Extra Example.

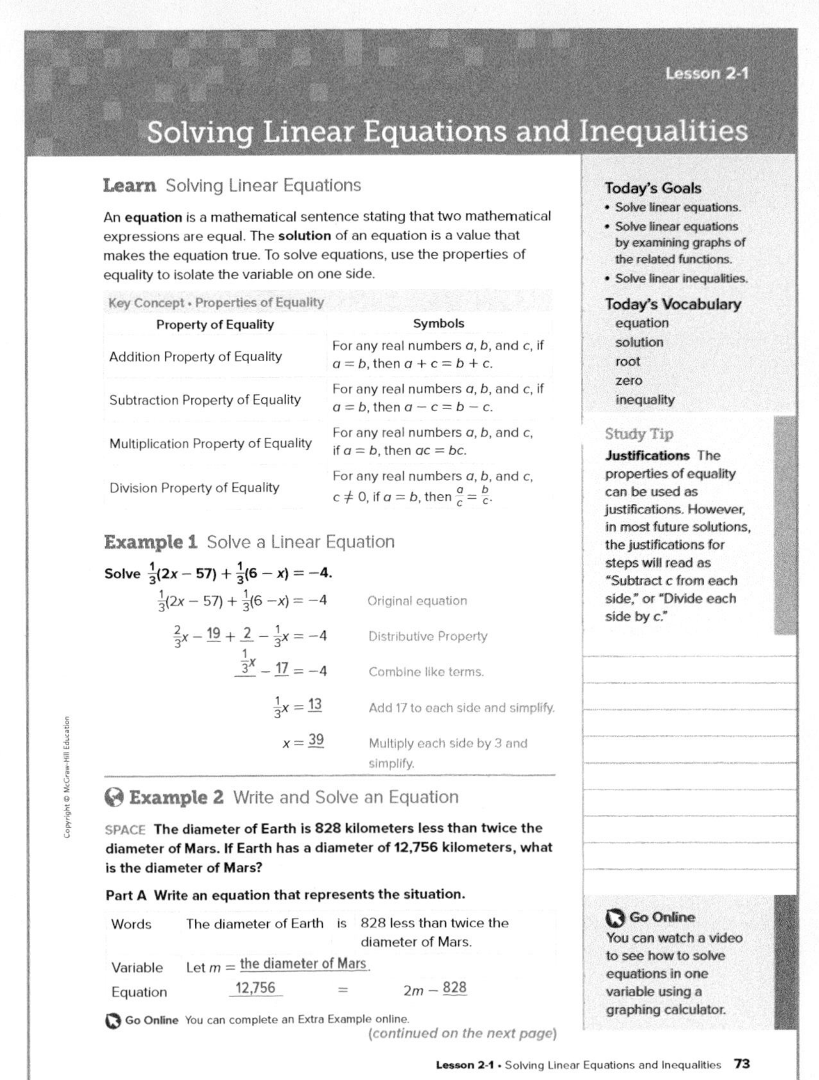

Lesson 2-1

Solving Linear Equations and Inequalities

Learn Solving Linear Equations

An **equation** is a mathematical sentence stating that two mathematical expressions are equal. The **solution** of an equation is a value that makes the equation true. To solve equations, use the properties of equality to isolate the variable on one side.

Key Concept • Properties of Equality

Property of Equality	Symbols
Addition Property of Equality	For any real numbers a, b, and c, if $a = b$, then $a + c = b + c$.
Subtraction Property of Equality	For any real numbers a, b, and c, if $a = b$, then $a - c = b - c$.
Multiplication Property of Equality	For any real numbers a, b, and c, if $a = b$, then $ac = bc$.
Division Property of Equality	For any real numbers a, b, and c, $c \neq 0$, if $a = b$, then $\frac{a}{c} = \frac{b}{c}$.

Example 1 Solve a Linear Equation

Solve $\frac{1}{3}(2x - 57) + \frac{1}{3}(6 - x) = -4$.

$\frac{1}{3}(2x - 57) + \frac{1}{3}(6 - x) = -4$	Original equation
$\frac{2}{3}x - \underline{19} + \underline{2} - \frac{1}{3}x = -4$	Distributive Property
$\underline{\frac{1}{3}x} - \underline{17} = -4$	Combine like terms.
$\frac{1}{3}x = \underline{13}$	Add 17 to each side and simplify.
$x = \underline{39}$	Multiply each side by 3 and simplify.

Example 2 Write and Solve an Equation

SPACE **The diameter of Earth is 828 kilometers less than twice the diameter of Mars. If Earth has a diameter of 12,756 kilometers, what is the diameter of Mars?**

Part A Write an equation that represents the situation.

Words	The diameter of Earth	is	828 less than twice the diameter of Mars.
Variable	Let m = the diameter of Mars.		
Equation	12,756	=	$2m - 828$

Go Online You can complete an Extra Example online.

(continued on the next page)

Today's Goals
- Solve linear equations.
- Solve linear equations by examining graphs of the related functions.
- Solve linear inequalities.

Today's Vocabulary
equation
solution
root
zero
inequality

Study Tip
Justifications The properties of equality can be used as justifications. However, in most future solutions, the justifications for steps will read as "Subtract c from each side," or "Divide each side by c."

Go Online You can watch a video to see how to solve equations in one variable using a graphing calculator.

Lesson 2-1 • Solving Linear Equations and Inequalities 73

Interactive Presentation

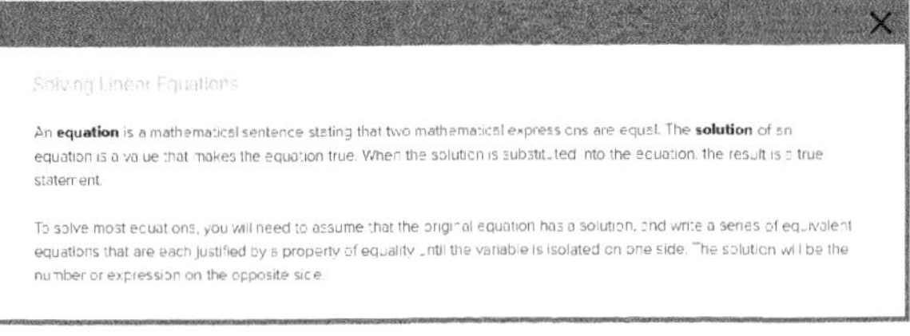

Learn

DRAG & DROP

Students drag the Properties of Equality to the definitions.

WATCH

Students can watch a video that explains how to use algebra tiles to solve equations in one variable.

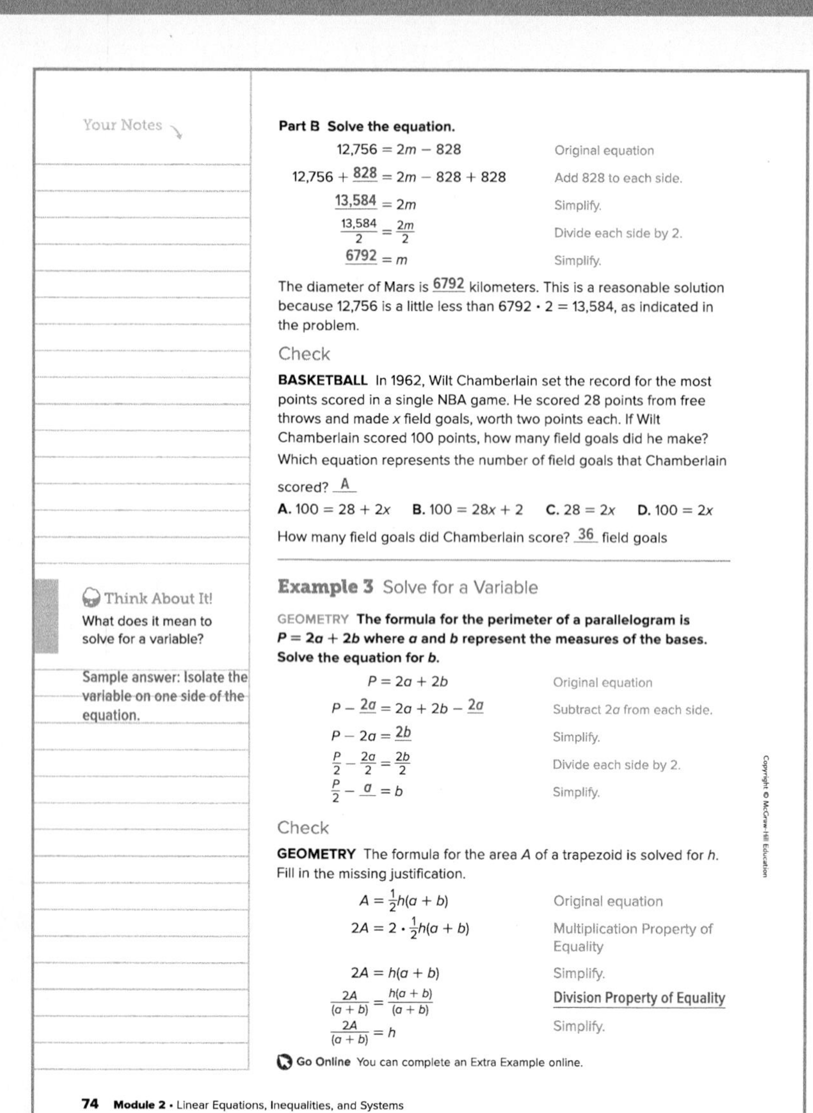

Your Notes

Part B **Solve the equation.**

$12{,}756 = 2m - 828$ Original equation

$12{,}756 + 828 = 2m - 828 + 828$ Add 828 to each side.

$13{,}584 = 2m$ Simplify.

$\frac{13{,}584}{2} = \frac{2m}{2}$ Divide each side by 2.

$6792 = m$ Simplify.

The diameter of Mars is 6792 kilometers. This is a reasonable solution because 12,756 is a little less than $6792 \cdot 2 = 13{,}584$, as indicated in the problem.

Check

BASKETBALL In 1962, Wilt Chamberlain set the record for the most points scored in a single NBA game. He scored 28 points from free throws and made x field goals, worth two points each. If Wilt Chamberlain scored 100 points, how many field goals did he make? Which equation represents the number of field goals that Chamberlain scored? A

A. $100 = 28 + 2x$ **B.** $100 = 28x + 2$ **C.** $28 = 2x$ **D.** $100 = 2x$

How many field goals did Chamberlain score? 36 field goals

Think About It! What does it mean to solve for a variable?

Sample answer: Isolate the variable on one side of the equation.

Example 3 Solve for a Variable

GEOMETRY **The formula for the perimeter of a parallelogram is $P = 2a + 2b$ where a and b represent the measures of the bases. Solve the equation for b.**

$P = 2a + 2b$ Original equation

$P - 2a = 2a + 2b - 2a$ Subtract $2a$ from each side.

$P - 2a = 2b$ Simplify.

$\frac{P}{2} - \frac{2a}{2} = \frac{2b}{2}$ Divide each side by 2.

$\frac{P}{2} - a = b$ Simplify.

Check

GEOMETRY The formula for the area A of a trapezoid is solved for h. Fill in the missing justification.

$A = \frac{1}{2}h(a + b)$ Original equation

$2A = 2 \cdot \frac{1}{2}h(a + b)$ Multiplication Property of Equality

$2A = h(a + b)$ Simplify.

$\frac{2A}{(a + b)} = \frac{h(a + b)}{(a + b)}$ Division Property of Equality

$\frac{2A}{(a + b)} = h$ Simplify.

Go Online You can complete an Extra Example online.

Copyright © McGraw-Hill Education

74 **Module 2** • Linear Equations, Inequalities, and Systems

Interactive Presentation

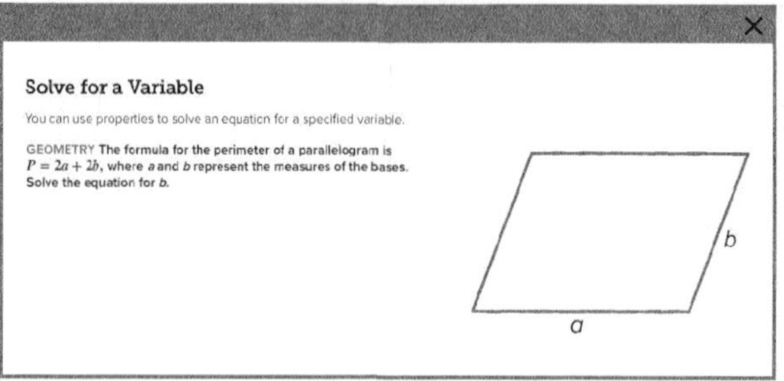

Example 3

TYPE

Students answer a question to determine whether they understand how to solve for a variable.

CHECK

Students complete the Check online to determine whether they are ready to move on.

Example 2 Write and Solve an Equation

Teaching the Mathematical Practices

2 Create Representations Guide students to write an equation that models the situation in Example 2. Then use the equation to solve the problem.

Questions for Mathematical Discourse

AL In **Part A**, how do you use the words to write the equation? Sample answer: The first part of the description, "The diameter of Earth is," means that the value given for the diameter of Earth goes on one side of the equation. We do not know the diameter of Mars, so it will be a variable, m, and it will go on the other side of the equation. Twice the diameter of Mars is represented as $2m$, and 828 kilometers less than that is $2m - 828$.

OL When writing an equation for a real-world example, what should you check before you put numerical values into the equation? Check to see if the units are given in the same system of measurement and determine whether you need to make any conversions.

BL How would you write the equation for the diameter of Mars as it relates to the diameter of Earth if you were not given the value for the diameter of Earth? Sample answer: Instead of 12,756 in the original equation, use variable e. Next, solve the equation for m.

Example 3 Solve for a Variable

Teaching the Mathematical Practices

2 Attend to Quantities Point out that it is important to note the meaning of the variables used in Example 3. Ask students to explain what the resulting equation represents.

Questions for Mathematical Discourse

AL In the third step, why are all the terms divided by 2? Sample answer: We need to isolate b, which has a coefficient of 2. If we divide that term by 2, the result is b. When we divide one term of the equation, we must perform the same operation on the rest of the terms to maintain equality.

OL How would the process and solution change if you had to solve for a? Sample answer: The coefficients of a and b are the same in the original equation, so the process would be the same and the solution would be in the same form, but with a and b switched.

BL How would the approach change if the initial equation contained more than one term with b? Sample answer: The terms with b would need to be consolidated before b could be isolated.

Common Error

Students may try to isolate the specified variable by first eliminating the coefficient of the variable. If students choose this path, they may not divide all terms by the coefficient, but rather one on each side. Encourage students to use the reverse order of operations rather than starting with the variable itself.

1 CONCEPTUAL UNDERSTANDING | 2 FLUENCY | 3 APPLICATION

Learn Solving Linear Equations by Graphing

Objective

Students solve linear equations by examining graphs of the related functions.

Teaching the Mathematical Practices

1 Explain Correspondences Encourage students to explain the relationships between the equation, related function, and graph as well as the relationship between the zero of the function and the solution of the equation.

What Students Are Learning

A root is a solution to an equation. Graphing a related function is one method for finding the root. An equation is solved for zero to find the related function, and then the zero is replaced with $f(x)$. Any values of x for which $f(x) = 0$ are called zeros, which are also the x-intercepts of the graph.

Example 4 Solve a Linear Equation by Graphing

Teaching the Mathematical Practices

3 Construct Arguments In the Think About It! feature students will use definitions and previously established results to construct an argument about the zero of the function.

Questions for Mathematical Discourse

AL What is the difference between the equation and its related function? Sample answer: The equation contains one variable that has a dependent solution. The related function allows different values of the variable to be examined by graphing to see what value makes the original equation true.

OL Why does the x-intercept represent the solution to the equation? The equation was set equal to zero to find the related function, thus the x-value that makes the function zero is the valid solution to the original equation.

BL If the related function $f(x) = -\frac{1}{2}x + 3$ were used instead, how would the graph of this function compare to the other related function. Sample answer: Both functions would cross the x-axis at 6, but the y-intercept is different and the slope is negative.

Common Error

Students may include y-intercepts as zeros of a function. Remind students that zeros are the x-intercepts of a graph.

Learn Solving Linear Equations by Graphing

The solution of an equation is called a **root**. You can find the root of an equation by examining the graph of its related function $f(x)$. A related function is found by solving the equation for 0 and then replacing 0 with $f(x)$. A related function for $2x + 13 = 9$ is $f(x) = 2x + 4$ or $y = 2x + 4$.

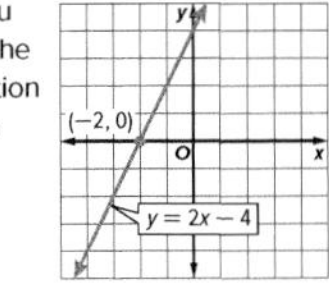

Values of x for which $f(x) = 0$ are called **zeros** of the function f. The zero of a function is the x-intercept of its graph. The solution and root of a linear equation are the same as the zero and x-intercept of its related function.

Talk About It!
Because there is typically more than one way to solve an equation for 0, there may be more than one related function for an equation. What is another possible related function of $2x + 13 = 9$? How does the zero of this function compare to the zero of $f(x) = 2x + 4$?

Sample answer: $f(x) = -2x - 4$; It is the same.

Example 4 Solve a Linear Equation by Graphing

Solve $\frac{1}{2}x - 11 = -8$ by graphing.

Step 1 Find a related function.

Rewrite the equation with 0 on the right side.

$\frac{1}{2}x - 11 = -8$ Original equation.

$\frac{1}{2}x - 11 + 8 = -8 + 8$ Add 8 to each side.

$\frac{1}{2}x - 3 = 0$ Simplify.

Replacing 0 with $f(x)$ gives the related function, $f(x) = \frac{1}{2}x - 3$.

Step 2 Graph the related function.

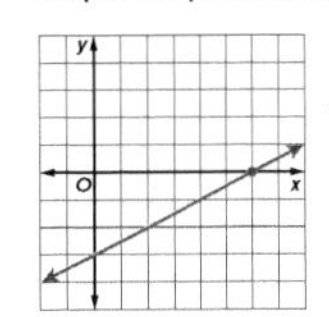

Since the graph of $f(x) = \frac{1}{2}x - 3$ intersects the x-axis at 6, the solution of the equation is 6.

Think About It!
Explain why −3 is *not* a zero of the function.

Sample answer: The x-intercept, not the y-intercept, represents the zero of a function. −3 is the y-intercept.

Check

Graph the related function of $2x - 5 = -11$. Use the graph to solve the equation.

The related function is $y = 2x + 6$.

The solution is −3.

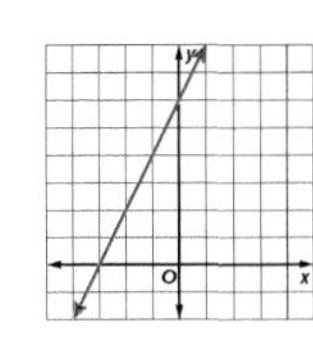

Watch Out!
Intercepts Be careful not to mistake y-intercepts for zeros of functions. The y-intercept on a graph occurs when $x = 0$. The x-intercepts are the zeros of a function because they are where $f(x) = 0$.

Go Online You can complete an Extra Example online.

Interactive Presentation

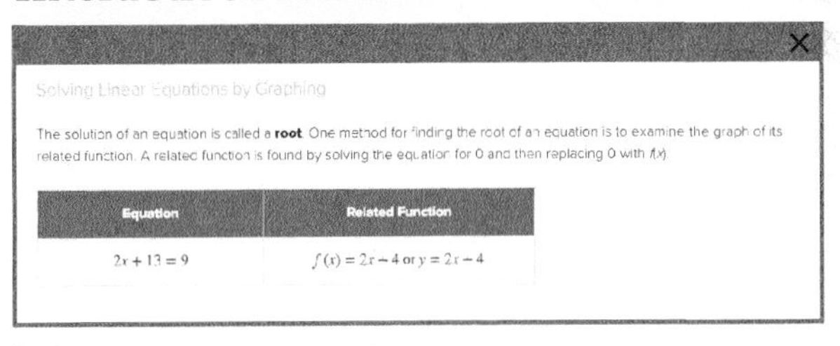

Learn

TYPE

Students answer a question about the possibility of multiple related functions.

Use a Source
Use available resources to find Paul Crake's actual time. How does this compare to your solution?

Sample answer: Paul Crake finished the race in 9 minutes and 33 seconds. This is close to my estimate and the same as the solution I found algebraically.

Study Tip
Assumptions Assuming that the rate at which Paul Crake climbed the stairs was constant allows us to represent the situation with a linear equation. While the rate at which he climbed likely varied throughout the race, using constant rates allows for reasonable graphs and solutions.

Example 5 Estimate Solutions by Graphing

TOWER RACE **The Empire State Building Run-Up is a race in which athletes run up the building's 1576 stairs. In 2003, Paul Crake set the record for the fastest time, running up an average of about 165 stairs per minute. The function $c = 1576 - 165m$ represents the number of steps Crake had left to climb c after m minutes. Find the zero of the function and interpret its meaning in the context of the situation.**

Step 1 Graph the function.

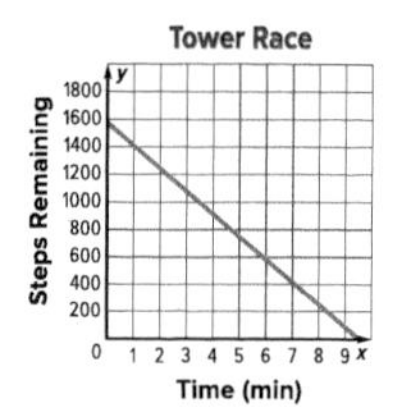

Step 2 Estimate the zero.

The graph appears to intersect the x-axis at about 9.5. This means that Paul Crake finished the race in about 9.5 minutes, or 9 minutes and 30 seconds.

Step 3 Solve algebraically.

Use the equation. Substitute 0 for c, and solve algebraically to check your solution.

$c = 1576 - 165m$ Original equation

$0 = 1576 - 165m$ Replace c with 0.

$165m = 1576$ Add $165m$ to each side.

$m \approx 9.55$ Divide each side by 165.

The solution is about 9.55. So, Paul Crake completed the Empire State Building Run-Up in about 9.55 minutes, or 9 minutes and 33 seconds. This is close to the estimated time of 9.5 minutes.

Check

DOG WALKING Bethany spends \$480 on supplies to start a dog walking service for which she plans to charge \$23 per hour. The function $y = 23x - 480$ represents Bethany's profit after x hours of dog walking.

Part A The graph appears to intersect the x-axis at about 20.9.

Part B Solve algebraically to verify your estimate. Round to the nearest hundredth. 20.87

Go Online You can complete an Extra Example online.

76 Module 2 • Linear Equations, Inequalities, and Systems

Interactive Presentation

Example 5

TYPE

Students use a source to compare their solution with the actual time.

DIFFERENTIATE

Reteaching Activity AL ELL

IF students are struggling to set up related functions,
THEN have students draw a straight line down through the equal sign of the equation. Have them decide the side on which they want the variable to be, the left or the right. Having the visual cue of the line may help students move one side of the equation to the other, setting it equal to zero.

Example 5 Estimate Solutions by Graphing

Teaching the Mathematical Practices

2 Represent a Situation Symbolically Guide students to see the connection between the situation and the equation in Example 5. Help students to identify the independent and dependent variables. Then work with them to find the other relationships in the problem.

Questions for Mathematical Discourse

AL What does the negative slope of the graph represent? The negative slope represents that Paul is decreasing the number of steps left in his run.

OL What is an advantage of solving by graphing? What is a disadvantage? Sample answer: Graphing can allow you to quickly approximate a solution, but it does not provide the exact solution.

BL How would the graph change if instead of counting the number of steps remaining, the graph showed the number of steps climbed? The slope would switch signs and the x-intercept would be $x = 0$.

Common Error

Students may feel uncomfortable estimating solutions from a graph and try and be too precise about the x-intercept. Remind students the graph is used to approximate the solution but algebraic methods can always be used to check the answer.

1 CONCEPTUAL UNDERSTANDING | 2 FLUENCY | 3 APPLICATION

Learn Solving Linear Inequalities

Objective

Students solve linear inequalities by applying the properties of inequality.

Teaching the Mathematical Practices

1 Explain Correspondences Encourage students to explain the relationship between the solution set of an inequality expressed in set-builder notation and graphed on a number line.

About the Key Concept

For any two real numbers a and b, either $a < b$, $a > b$ or $a = b$. An inequality is a mathematical sentence that contains the symbol $<$, $\leq$, $>$, or $\geq$. If an inequality is multiplied or divided by a negative number, the inequality sign must be reversed to keep the resulting inequality true. The solution set of an inequality can be expressed using set-builder notation and can be represented as a graph on a number line. A circle means the point is not included in the solution set, while a dot indicates the point is included.

Common Misconception

A common misconception students often have is that a linear inequality is just like a linear equation, and the solutions are the same. Even though students can solve a linear inequality in the same way as a linear equation, dividing or multiplying by a negative value has an important effect on an inequality, but not an equation. Also, the solution of an inequality is often a set of real numbers, while the solution of an equation is one real number.

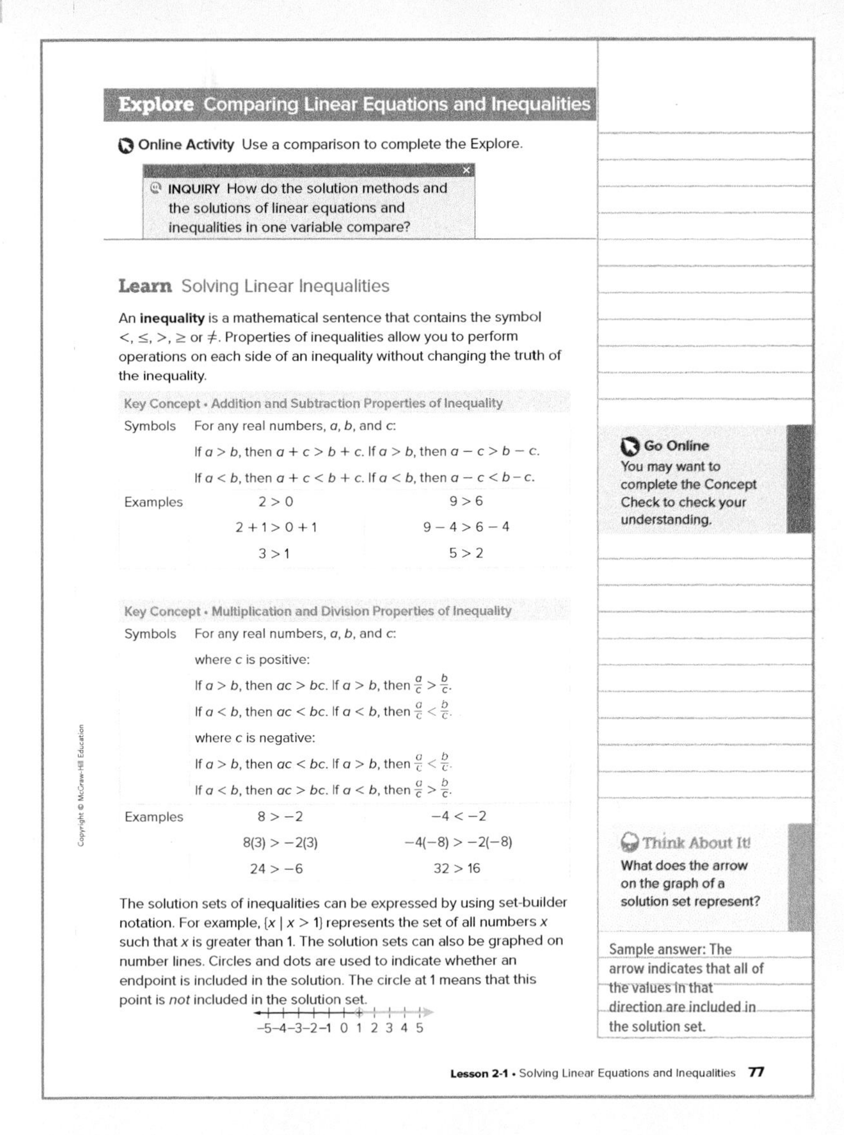

Explore Comparing Linear Equations and Inequalities

Online Activity Use a comparison to complete the Explore.

INQUIRY How do the solution methods and the solutions of linear equations and inequalities in one variable compare?

Learn Solving Linear Inequalities

An **inequality** is a mathematical sentence that contains the symbol $<$, $\leq$, $>$, $\geq$ or $\neq$. Properties of inequalities allow you to perform operations on each side of an inequality without changing the truth of the inequality.

Key Concept • Addition and Subtraction Properties of Inequality

Symbols For any real numbers, a, b, and c:

If $a > b$, then $a + c > b + c$. If $a > b$, then $a - c > b - c$.

If $a < b$, then $a + c < b + c$. If $a < b$, then $a - c < b - c$.

Examples	
$2 > 0$	$9 > 6$
$2 + 1 > 0 + 1$	$9 - 4 > 6 - 4$
$3 > 1$	$5 > 2$

Key Concept • Multiplication and Division Properties of Inequality

Symbols For any real numbers, a, b, and c:

where c is positive:

If $a > b$, then $ac > bc$. If $a > b$, then $\frac{a}{c} > \frac{b}{c}$.

If $a < b$, then $ac < bc$. If $a < b$, then $\frac{a}{c} < \frac{b}{c}$.

where c is negative:

If $a > b$, then $ac < bc$. If $a > b$, then $\frac{a}{c} < \frac{b}{c}$.

If $a < b$, then $ac > bc$. If $a < b$, then $\frac{a}{c} > \frac{b}{c}$.

Examples	
$8 > -2$	$-4 < -2$
$8(3) > -2(3)$	$-4(-8) > -2(-8)$
$24 > -6$	$32 > 16$

The solution sets of inequalities can be expressed by using set-builder notation. For example, $\{x \mid x > 1\}$ represents the set of all numbers x such that x is greater than 1. The solution sets can also be graphed on number lines. Circles and dots are used to indicate whether an endpoint is included in the solution. The circle at 1 means that this point is *not* included in the solution set.

−5 −4 −3 −2 −1 0 1 2 3 4 5

Go Online You may want to complete the Concept Check to check your understanding.

Think About It! What does the arrow on the graph of a solution set represent?

Sample answer: The arrow indicates that all of the values in that direction are included in the solution set.

Lesson 2-1 • Solving Linear Equations and Inequalities 77

Interactive Presentation

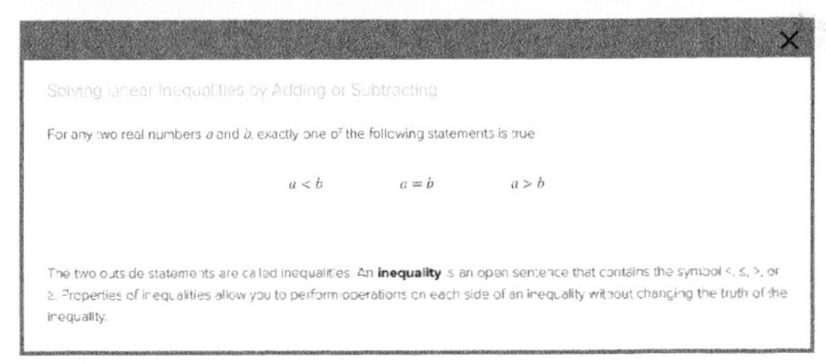

Learn

TYPE

Students examine cases to determine whether they represent counterexamples.

Think About It!

What does a dot on the graph of a solution set indicate?

Sample answer: The endpoint is included in the solution set.

Study Tip

Reversing the Inequality Symbol Adding the same number to, or subtracting the same number from, each side of an inequality does not change the truth of the inequality. Multiplying or dividing each side of an inequality by a positive number does not change the truth of the inequality. However, multiplying or dividing each side of an inequality by a negative number requires that the order of the inequality be reversed. In Example 6, $\geq$ was replaced with $\leq$.

Watch Out!

Reading Math Be sure to always read problems carefully. The term *at least* is used here, and can be confusing since it actually means *greater than or equal to*, and is represented by $\geq$. In this instance, Jake should intake at least 1300 mg, which means he must intake an amount greater than or equal to 1300 mg.

Example 6 Solve a Linear Inequality

Solve $-5.6n + 12.9 \geq -71.1$. Graph the solution set on a number line.

$-5.6n + 12.9 \geq -71.1$	Original inequality.
$-5.6n + 12.9 - 12.9 \geq -71.1 - 12.9$	Subtract 12.9 from each side.
$-5.6n \geq -84$	Simplify.
$\frac{-5.6n}{-5.6} \leq \frac{-84}{-5.6}$	Divide each side by -5.6, reversing the inequality symbol.
$n \leq 15$	Simplify.

The solution set is $\{n \mid n \leq 15\}$. Graph the solution set.

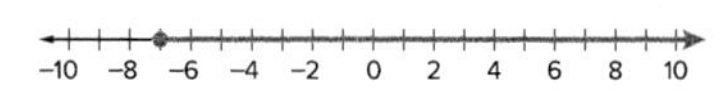

Check

What is the solution of $-p - 3 \geq -4(p + 6)$? $p \geq -7$

Graph the solution set.

Example 7 Write and Solve an Inequality

NUTRITION **The recommended daily intake of calcium for teens is 1300 mg. Jake gets 237 mg of calcium from a multivitamin he takes each morning and 302 mg from each glass of skim milk that he drinks. How many glasses of milk would Jake need to drink to meet the recommendation?**

Step 1 Write an inequality to represent the situation.

Let g = the number of glasses of milk Jake needs.

$237 + 302g \geq 1300$

Step 2 Solve the inequality.

$237 + 302g \geq 1300$	Original inequality
$302g \geq 1063$	Subtract 237 from each side.
$g \geq 3.52$	Divide each side by 302.

Step 3 Interpret the solution in the context of the situation.

Jake will need to drink slightly more than 3.5 glasses of milk to intake at least the recommended daily amount of calcium. This is a viable solution because Jake can pour part of a full glass of milk.

Go Online You can complete an Extra Example online.

Interactive Presentation

Solve a Linear Inequality

Solve $-5.6n + 12.9 \geq -71.1$. Graph the solution set on a number line.

$-5.6n + 12.9 \geq -71.1$	Original inequality
$-5.6n + 12.9 - 12.9 \geq -71.1 - 12.9$	Subtract 12.9 from each side.
$-5.6n \geq -84$	Simplify.
$\frac{-5.6n}{-5.6} \leq \frac{-84}{-5.6}$	Divide each side by -5.6, reversing the inequality symbol.
$n \leq 15$	Simplify.

The solution set is $\{n \mid n \leq 15\}$. The graph of the solution set is shown.

Example 6

TYPE

Students answer a question to determine whether they understand the graph of an inequality.

CHECK

Students complete the Check online to determine whether they are ready to move on.

Example 6 Solve a Linear Inequality

Teaching the Mathematical Practices

2 Different Properties Guide students to use the properties of inequality to correctly solve the inequality.

Questions for Mathematical Discourse

AL Why is the point on the number line at $x = 15$ closed? Sample answer: Because the inequality uses $\leq$, this indicates that 15 is included in the solution.

OL How would the solution set have changed if the coefficient of n were 5.6 rather than -5.6? Sample answer: The solution set would be $n \geq -15$ since the inequality symbol would not be reversed.

BL How many values satisfy this inequality? How many do not? An infinite number of values satisfy the inequality, and an infinite number of values do not satisfy the inequality.

Example 7 Write and Solve an Inequality

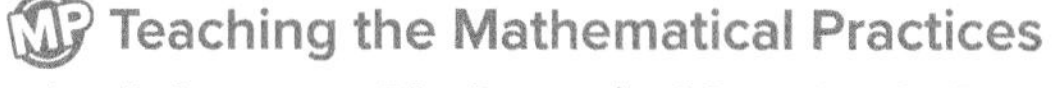

Teaching the Mathematical Practices

4 Interpret Mathematical Results In Example 7, point out that to solve the problem, student should interpret their mathematical results in the context of the problem.

Questions for Mathematical Discourse

AL Why is g multiplied by 302 rather than 237? Each glass of skim milk has 302 mg of calcium, and g represents the number of glasses of skim milk Jake drinks.

OL Why is the inclusive inequality used rather than the non-inclusive? The question asks for how many glasses of milk Jake needs to drink to meet the recommendation.

BL How would you word the problem if you wanted to use a non-inclusive inequality? You could ask to find how many glasses of milk Jake needs to drink to exceed the recommendation.

Exit Ticket

Recommended Use

At the end of class, go online to display the Exit Ticket prompt and ask students to respond using a separate piece of paper. Have students hand you their responses as they leave the room.

Alternate Use

At the end of class, go online to display the Exit Ticket prompt and ask students to respond verbally or by using a mini-whiteboard. Have students hold up their whiteboards so that you can see all student responses. Tap to reveal the answer when most or all students have completed the Exit Ticket.

Practice and Homework

Suggested Assignments

Use the table below to select appropriate exercises.

DOK	Topic	Exercises
1, 2	exercises that mirror the examples	1–32
2	exercises that use a variety of skills from this lesson	33–38
2	exercises that extend concepts learned in this lesson to new contexts	39, 40
3	exercises that emphasize higher-order and critical thinking skills	41–46

ASSESS AND DIFFERENTIATE

Use the data from the **Checks** to determine whether to provide resources for extension, remediation, or intervention.

BL

IF students score 90% or more on the Checks,
THEN assign:

- Practice Exercises 1–39 odd, 41–46
- Extension: United States' Gross National Product
- **ALEKS**® Linear Equations, Linear Inequalities

OL

IF students score 66%–89% on the Checks,
THEN assign:

- Practice Exercises 1–45 odd
- Remediation, Review Resources: Writing and Interpreting Equations
- Personal Tutors
- Extra Examples 1–7
- **ALEKS**® Writing Expressions and Equations

AL

IF students score 65% or less on the Checks,
THEN assign:

- Practice Exercises 1–31 odd
- Remediation, Review Resources: Writing and Interpreting Equations
- *Quick Review Math Handbook*: Equations and Inequalities in One Variable
- **ALEKS**® Writing Expressions and Equations

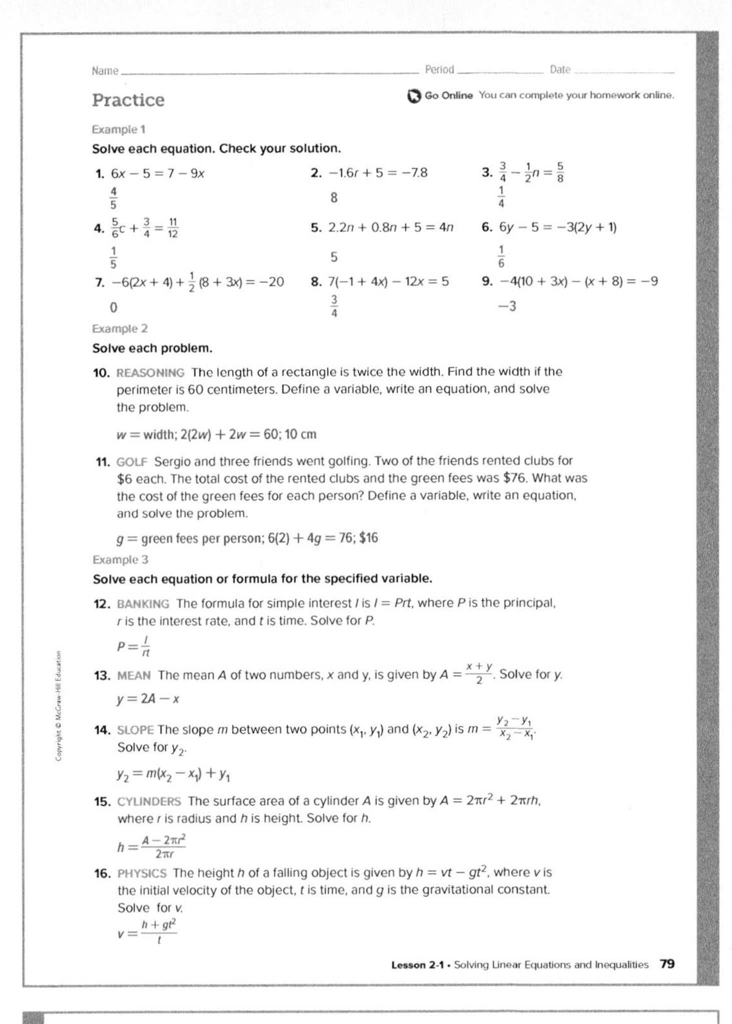

Name ______ Period ______ Date ______

Practice

Go Online You can complete your homework online.

Example 1

Solve each equation. Check your solution.

1. $6x - 5 = 7 - 9x$
   $\frac{4}{5}$
2. $-1.6r + 5 = -7.8$
   8
3. $\frac{3}{4} - \frac{1}{2}n = \frac{5}{8}$
   $\frac{1}{4}$
4. $\frac{5}{6}c + \frac{3}{4} = \frac{11}{12}$
   $\frac{1}{5}$
5. $2.2n + 0.8n + 5 = 4n$
   5
6. $6y - 5 = -3(2y + 1)$
   $\frac{1}{6}$
7. $-6(2x + 4) + \frac{1}{2}(8 + 3x) = -20$
   0
8. $7(-1 + 4x) - 12x = 5$
   $\frac{3}{4}$
9. $-4(10 + 3x) - (x + 8) = -9$
   -3

Example 2

Solve each problem.

10. REASONING The length of a rectangle is twice the width. Find the width if the perimeter is 60 centimeters. Define a variable, write an equation, and solve the problem.
    w = width; $2(2w) + 2w = 60$; 10 cm
11. GOLF Sergio and three friends went golfing. Two of the friends rented clubs for \$6 each. The total cost of the rented clubs and the green fees was \$76. What was the cost of the green fees for each person? Define a variable, write an equation, and solve the problem.
    g = green fees per person; $6(2) + 4g = 76$; \$16

Example 3

Solve each equation or formula for the specified variable.

12. BANKING The formula for simple interest I is $I = Prt$, where P is the principal, r is the interest rate, and t is time. Solve for P.
    $P = \frac{I}{rt}$
13. MEAN The mean A of two numbers, x and y, is given by $A = \frac{x + y}{2}$. Solve for y.
    $y = 2A - x$
14. SLOPE The slope m between two points (x_1, y_1) and (x_2, y_2) is $m = \frac{y_2 - y_1}{x_2 - x_1}$. Solve for y_2.
    $y_2 = m(x_2 - x_1) + y_1$
15. CYLINDERS The surface area of a cylinder A is given by $A = 2\pi r^2 + 2\pi rh$, where r is radius and h is height. Solve for h.
    $h = \frac{A - 2\pi r^2}{2\pi r}$
16. PHYSICS The height h of a falling object is given by $h = vt - gt^2$, where v is the initial velocity of the object, t is time, and g is the gravitational constant. Solve for v.
    $v = \frac{h + gt^2}{t}$

Lesson 2-1 • Solving Linear Equations and Inequalities 79

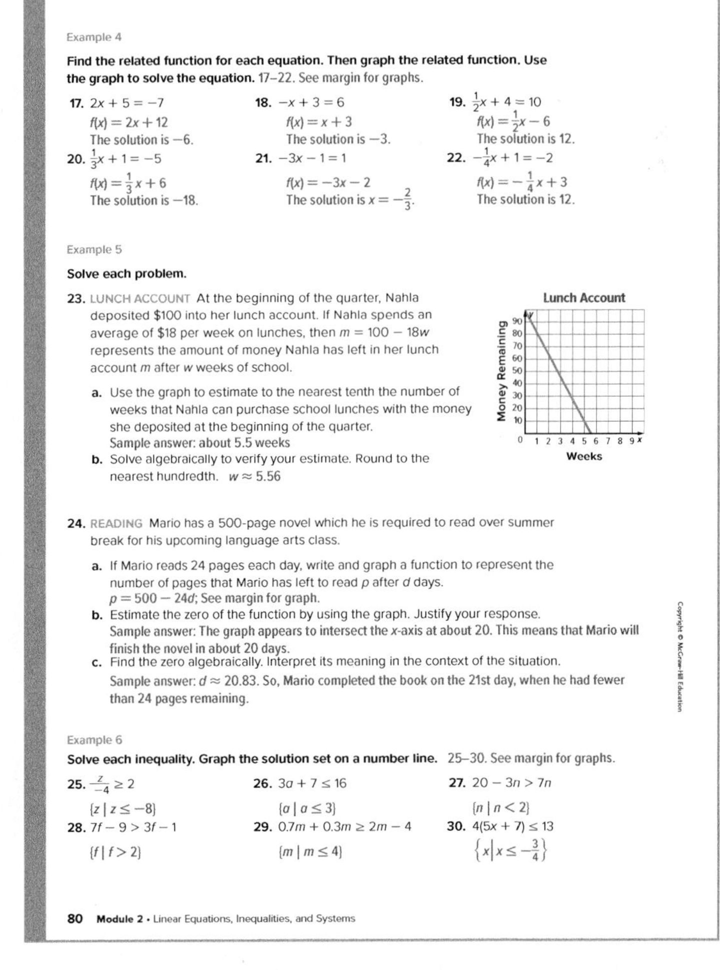

Example 4

Find the related function for each equation. Then graph the related function. Use the graph to solve the equation. 17–22. See margin for graphs.

17. $2x + 5 = -7$
    $f(x) = 2x + 12$
    The solution is -6.
18. $-x + 3 = 6$
    $f(x) = x + 3$
    The solution is -3.
19. $\frac{1}{2}x + 4 = 10$
    $f(x) = \frac{1}{2}x - 6$
    The solution is 12.
20. $\frac{1}{3}x + 1 = -5$
    $f(x) = \frac{1}{3}x + 6$
    The solution is -18.
21. $-3x - 1 = 1$
    $f(x) = -3x - 2$
    The solution is $x = -\frac{2}{3}$.
22. $-\frac{1}{4}x + 1 = -2$
    $f(x) = -\frac{1}{4}x + 3$
    The solution is 12.

Example 5

Solve each problem.

23. LUNCH ACCOUNT At the beginning of the quarter, Nahla deposited \$100 into her lunch account. If Nahla spends an average of \$18 per week on lunches, then $m = 100 - 18w$ represents the amount of money Nahla has left in her lunch account m after w weeks of school.

    Lunch Account

    a. Use the graph to estimate to the nearest tenth the number of weeks that Nahla can purchase school lunches with the money she deposited at the beginning of the quarter.
       Sample answer: about 5.5 weeks
    b. Solve algebraically to verify your estimate. Round to the nearest hundredth. $w \approx 5.56$
24. READING Mario has a 500-page novel which he is required to read over summer break for his upcoming language arts class.
    a. If Mario reads 24 pages each day, write and graph a function to represent the number of pages that Mario has left to read p after d days.
       $p = 500 - 24d$; See margin for graph.
    b. Estimate the zero of the function by using the graph. Justify your response.
       Sample answer: The graph appears to intersect the x-axis at about 20. This means that Mario will finish the novel in about 20 days.
    c. Find the zero algebraically. Interpret its meaning in the context of the situation.
       Sample answer: $d \approx 20.83$. So, Mario completed the book on the 21st day, when he had fewer than 24 pages remaining.

Example 6

Solve each inequality. Graph the solution set on a number line. 25–30. See margin for graphs.

25. $\frac{z}{-4} \geq 2$
    $\{z \mid z \leq -8\}$
26. $3a + 7 \leq 16$
    $\{a \mid a \leq 3\}$
27. $20 - 3n > 7n$
    $\{n \mid n < 2\}$
28. $7f - 9 > 3f - 1$
    $\{f \mid f > 2\}$
29. $0.7m + 0.3m \geq 2m - 4$
    $\{m \mid m \leq 4\}$
30. $4(5x + 7) \leq 13$
    $\left\{x \middle| x \leq -\frac{3}{4}\right\}$

80 **Module 2** • Linear Equations, Inequalities, and Systems

Name ______________ Period ______ Date ______

Example 7

Solve each problem.

31. INCOME Manuel takes a job translating English instruction manuals to Spanish. He will receive \$15 per page plus \$100 per month. Manuel plans to work for 3 months during the summer and wants to make at least \$1500. Write and solve an inequality to find the minimum number of pages Manuel must translate in order to reach his goal. Then, interpret the solution in the context of the situation.
$15P + 300 \geq 1500$; $P \geq 80$; Manuel must translate at least 80 pages.

32. STRUCTURE On a conveyor belt, there can only be two boxes moving at a time. The total weight of the boxes cannot be more than 300 pounds. Let x and y represent the weights of two boxes on the conveyor belt.

a. Write an inequality that describes the weight limitation in terms of x and y. $x + y \leq 300$

b. Write an inequality that describes the limit on the average weight a of the two boxes. $a \leq 150$

c. Two boxes are to be placed on the conveyor belt. The first box weighs 175 pounds. What is the maximum weight of the second box? 125 pounds

Mixed Exercises 33–35. See students' graphs.

Solve each equation. Check your solution by graphing the related function.

33. $-3b + 7 = -15 + 2b$ $\quad \frac{22}{5}$

34. $a - \frac{2a}{5} = 3$ $\quad 5$

35. $2.2n + 0.8n + 5 = 4n$ $\quad 5$

36–38. See Mod. 2 Answer Appendix for graphs.

Solve each inequality. Graph the solution set on a number line.

36. $\frac{4x - 3}{2} \geq -3.5$ $\quad \{x \mid x \geq -1\}$

37. $1 + 5(x - 8) \leq 2 - (x + 5)$ $\quad \{x \mid x \leq 6\}$

38. $-36 - 2(w + 77) > -4(2w + 52)$ $\quad \{w \mid w > -3\}$

39. REASONING An ice rink offers open skating several times a week. An annual membership to the skating rink costs \$60. The table shows the cost of one session for members and non-members.

Open Ice Skating Sessions	
members	\$6
non-members	\$10

Kaliska plans to spend no more than \$90 on skating this year. Define a variable then write and solve inequalities to find the number of sessions she can attend with and without buying a membership. Should Kaliska buy a membership?

Sample answer: Let x represent the number of skating sessions. With a membership: $6x + 60 \leq 90$; $x \leq 5$. Without a membership: $10x \leq 90$; $x \leq 9$. She should not buy a membership.

40. PRECISION The formula to convert temperature in degrees Fahrenheit to degrees Celsius is $\frac{5}{9}(F - 32) = C$.

a. Solve the equation for F. $F = \frac{9}{5}C + 32$

b. Use your result from part a to determine the temperature in degrees Fahrenheit when the Celsius temperature is 30. 86°F

c. At a certain temperature, a Fahrenheit thermometer and a Celsius thermometer will read the same temperature. Write and solve an equation to find the temperature. $\frac{5}{9}(F - 32) = F$; -40°F

Higher-Order Thinking Skills

41. FIND THE ERROR Steven and Jade are solving $A = \frac{1}{2}h(b_1 + b_2)$ for b_2. Is either of them correct? Explain your reasoning. Jade; sample answer: In the last step, when Steven subtracted b_1 from each side, he mistakenly put b_1 in the numerator instead of subtracting it from the fraction.

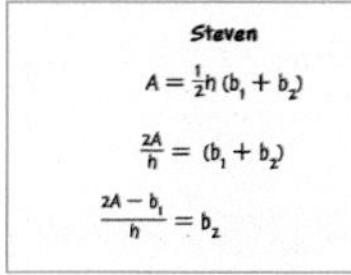

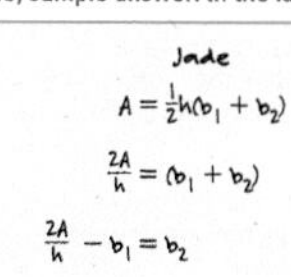

42. CREATE Write an equation involving the Distributive Property that has no solution and another example that has infinitely many solutions.
Sample answer: $3(x - 4) = 3x + 5$; $2(3x - 1) = 6x - 2$

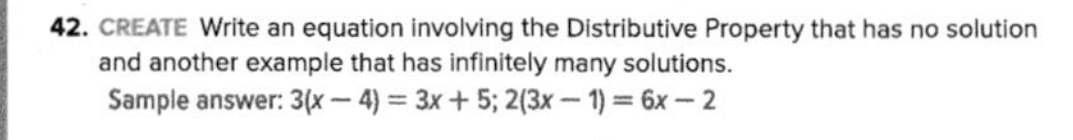

43. PERSEVERE Solve $d = \sqrt{(x_2 - x_1)^2 + (y_2 - y_1)^2}$ for y_1. $y_1 = y_2 - \sqrt{d^2 - (x_2 - x_1)^2}$

44. ANALYZE Vivek's teacher made the statement, "Four times a number is less than three times a number." Vivek quickly responded that the answer is *no solution*. Do you agree with Vivek? Write and solve an inequality to justify your argument.
No; sample answer: Let n represent the number. The teacher's statement can be written as an inequality as $4n < 3n$; subtract $3n$ from each side to get $n < 0$. So, the statement is true for any negative real number n.

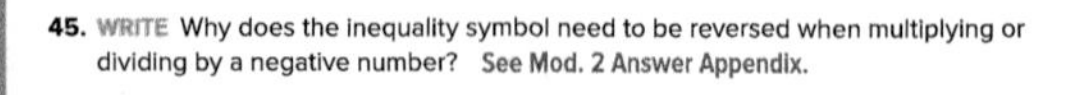

45. WRITE Why does the inequality symbol need to be reversed when multiplying or dividing by a negative number? See Mod. 2 Answer Appendix.

46. PERSEVERE Given $\triangle ABC$ with sides $AB = 3x + 4$, $BC = 2x + 5$, and $AC = 4x$, determine the values of x such that $\triangle ABC$ exists. See Mod. 2 Answer Appendix.

Answers

17.

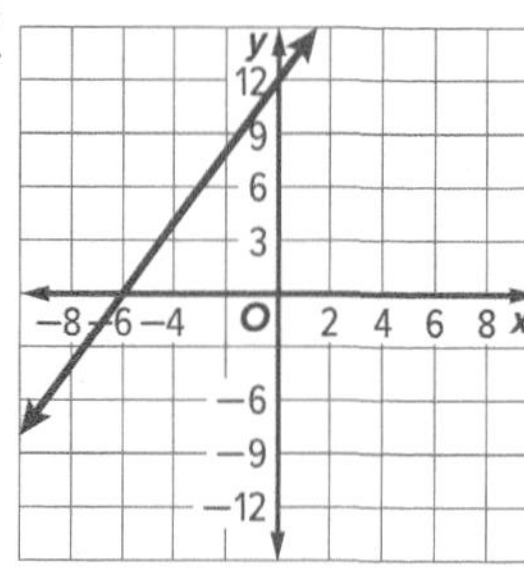

18.
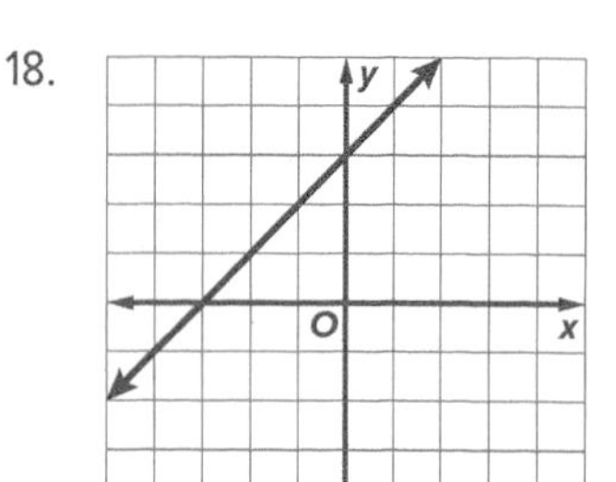

19.
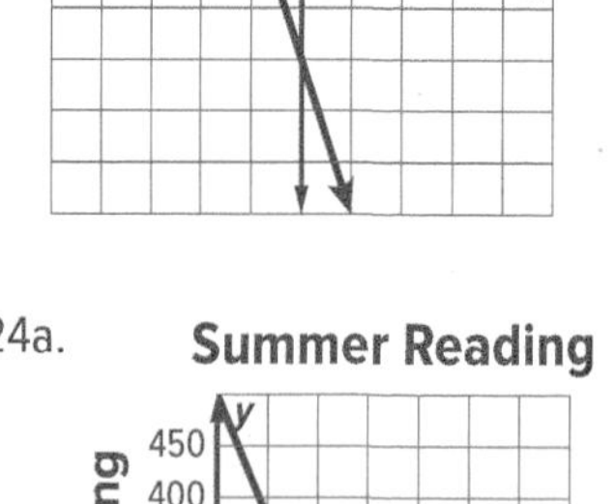

20.
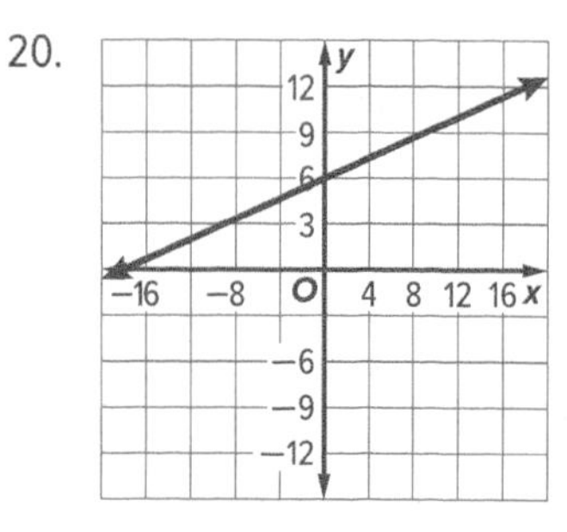

21.
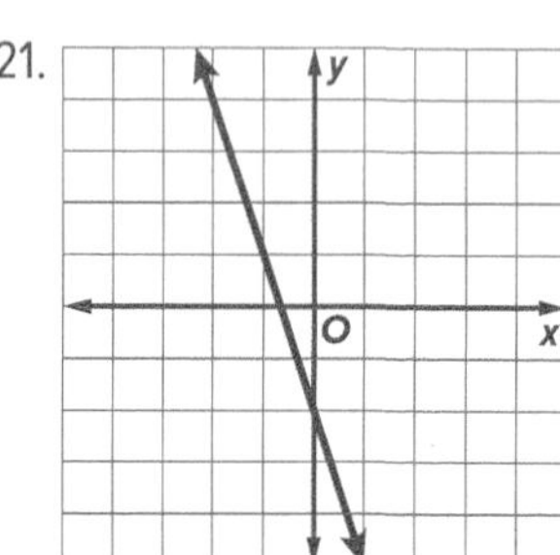

22.
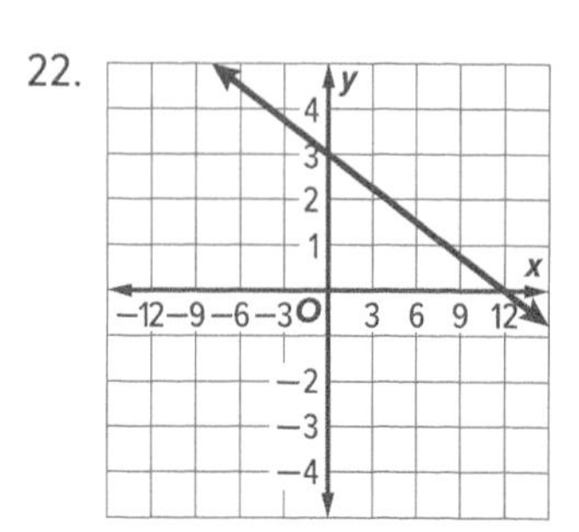

24a.
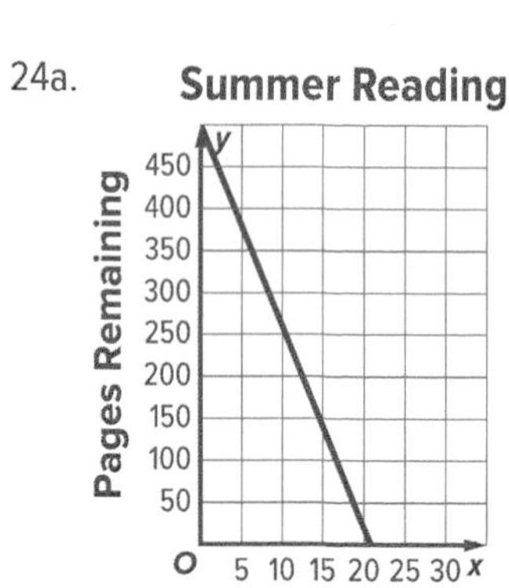

25.
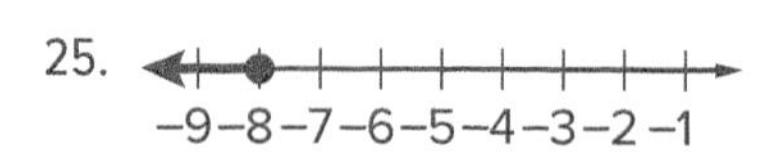

26.

27.

28.
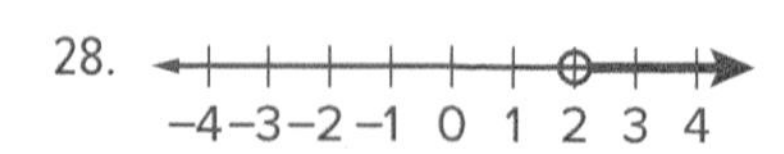

29.

30.
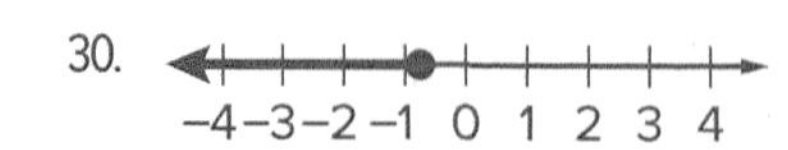

Lesson 2-2 A.CED.1, A.CED.3

Solving Absolute Value Equations and Inequalities

LESSON GOAL

Students solve equations and inequalities involving absolute value algebraically.

1 LAUNCH

 Launch the lesson with a **Warm Up** and an introduction.

2 EXPLORE AND DEVELOP

 **Develop:**

Solving Absolute Value Equations Algebraically
- Solve an Absolute Value Equation
- Extraneous Solution
- The Empty Set
- Write and Solve an Absolute Value Equation

Solving Absolute Value Inequalities Algebraically
- Solve an Absolute Value Inequality ($<$ or $\leq$)
- Solve an Absolute Value Inequality ($>$ or $\geq$)
- Write and Solve an Absolute Value Inequality

 You may want your students to complete the **Checks** online.

3 REFLECT AND PRACTICE

 Exit Ticket

 Practice

DIFFERENTIATE

 View reports of student progress on the **Checks** after each example.

Resources	AL	OL	BL	ELL
Remediation: Solving Equations Involving Absolute Value	●	●		●
Extension: Diophantine Equations		●	●	●
ELL Support				

Language Development Handbook

Assign page 9 of the *Language Development Handbook* to help your students build mathematical language related to solving equations and inequalities involving absolute value algebraically.

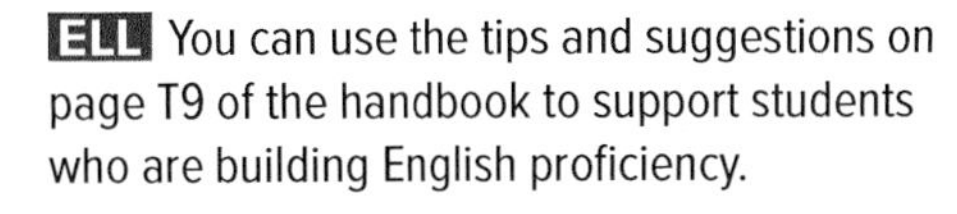
ELL You can use the tips and suggestions on page T9 of the handbook to support students who are building English proficiency.

Suggested Pacing

90 min	1 day
45 min	2 days

Focus

Domain: Algebra

Standards for Mathematical Content:

A.CED.1 Create equations and inequalities in one variable and use them to solve problems.

A.CED.3 Represent constraints by equations or inequalities, and by a system of equations and/or inequalities, and interpret solutions as viable or non-viable options in a modeling context.

Standards for Mathematical Practice:

4 Model with mathematics.

7 Look for and make use of structure.

Coherence

Vertical Alignment

Previous
Students solved linear equations and inequalities in one variable.
A.CED.1, A.CED.2

Now
Students solve equations and inequalities involving absolute value algebraically.
A.CED.1, A.CED.3

Next
Students will write equations in standard, slope-intercept, and point-slope forms.
A.CED.2, F.IF.6

Rigor

The Three Pillars of Rigor

1 CONCEPTUAL UNDERSTANDING	2 FLUENCY	3 APPLICATION

Conceptual Bridge In this lesson, students expand their understanding of absolute value expressions to build fluency with solving equations and inequalities that involve absolute value. They apply their understanding of solving absolute value equations and inequalities by solving real-world problems.

Mathematical Background

The absolute value of a number represents the distance of that number from 0 on a number line. So, the statement "the absolute value of x is always x" is not true. If $x = -3$, then the absolute value of x is 3.

Interactive Presentation

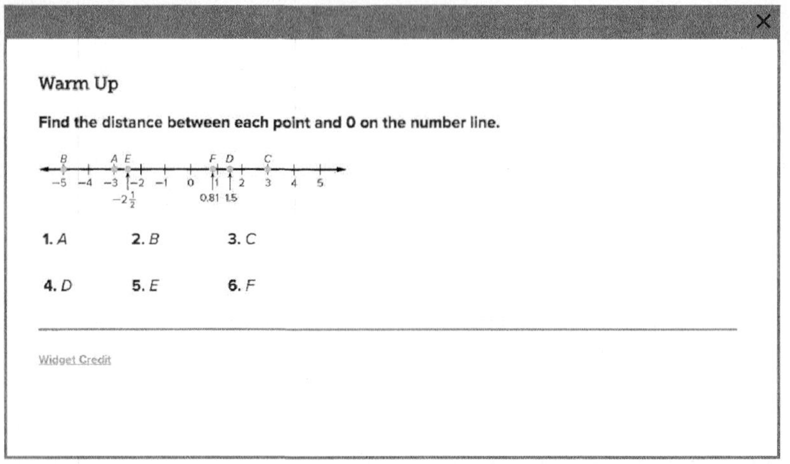

Warm Up

Launch the Lesson

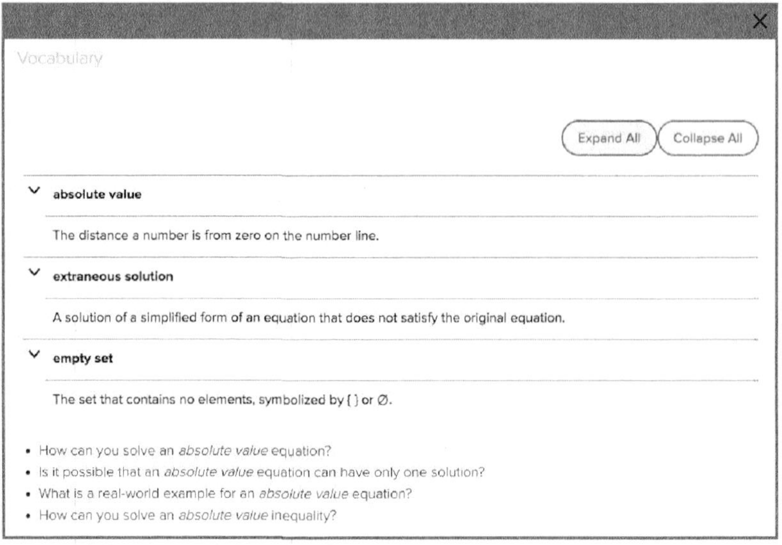

Today's Vocabulary

Warm Up

Prerequisite Skills

The Warm Up exercises address the following prerequisite skill for this lesson:

- determining absolute value and distance

Answers:

1. 3
2. 5
3. 3
4. 1.5
5. $2\frac{1}{2}$
6. 0.81

Launch the Lesson

 **Teaching the Mathematical Practices**

4 Model with Mathematics Encourage students to apply the concept of absolute value to the rising and setting of the sun. Help students to write an equation to represent the earliest and latest times that the Sun will rise or set in your town today.

Go Online to find additional teaching notes and questions to promote classroom discourse.

Today's Standards

Tell students that they will be addressing these content and practice standards in this lesson. You may wish to have a student volunteer read aloud *How can I meet these standards?* and *How can I use these practices?,* and connect these to the standards.

See the Interactive Presentation for I Can statements that align with the standards covered in this lesson.

Today's Vocabulary

Tell students that they will be using these vocabulary terms in this lesson. You can expand each row if you wish to share the definitions. Then discuss the questions below with the class.

1 CONCEPTUAL UNDERSTANDING | 2 FLUENCY | 3 APPLICATION

Learn Solving Absolute Value Equations Algebraically

Objective

Students write and solve absolute value equations by constructing two cases for the equation, and then graph the solutions on a number line.

Teaching the Mathematical Practices

1 Special Cases Work with students to look at the two cases that must be considered when solving an absolute value equation. Reiterate to students that their solutions will not be complete if both cases are not considered.

Common Misconception

A common misconception students often have is that since the absolute value of a number must be nonnegative, then solutions to absolute value equations cannot be negative. They will either deem negative answers as extraneous solutions or change them to a positive. Remind students that the solution can be negative because when they check their answers, the negative number can be substituted inside of an absolute value.

DIFFERENTIATE

Reteaching Activity AL ELL

IF students are struggling to set up the two equations for absolute value, **THEN** have students write the two cases with the positive and negative signs placed on the inside of the absolute value, rather than the outside. For example $|x| = 3$ would be written as $x = 3$ and $-x = 3$, which yields $x = -3$.

Example 1 Solve an Absolute Value Equation

Teaching the Mathematical Practices

7 Interpret Complicated Expressions Mathematically proficient students can see complicated expressions as single objects or as being composed of several objects. In Example 1, guide students to see the absolute value equation as two separate equations.

Questions for Mathematical Discourse

AL Why do you need to write and solve two equations to solve an absolute value equation? Sample answer: For any linear expression in x within an absolute value, it is possible to have a value and its negative as the results from two different x-values. The absolute value then makes the value and its negative into the same value.

OL How is the graph of the solution of an absolute value equation different from the graph of the solution of a linear equation? The solution graph of an absolute value equation could have two, one, or no points, while a linear equation could have one or no points.

BL Why should we isolate the absolute value expression before writing the two cases? Sample answer: Isolating the absolute value reduces the chance of making a sign error when finding the solutions. The expression inside the absolute value can either be the positive or negative version of itself before the absolute value is taken, so having the other terms on the opposite side of the equation helps you only apply the negative to the correct terms.

Lesson 2-2

Solving Absolute Value Equations and Inequalities

Today's Goals

- Write and solve absolute value equations, and graph the solutions on a number line.
- Write and solve absolute value inequalities, and graph the solutions on a number line.

Today's Vocabulary

absolute value
extraneous solution
empty set

Learn Solving Absolute Value Equations Algebraically

The **absolute value** of a number is its distance from zero on the number line. The definition of absolute value can be used to solve equations that contain absolute value expressions by constructing two cases. For any real numbers a and b, if $|a| = b$ and $b \geq 0$, then $a = b$ or $a = -b$.

Step 1 Isolate the absolute value expression on one side of the equation.

Step 2 Write the two cases.

Step 3 Use the properties of equality to solve each case.

Step 4 Check your solutions.

Absolute value equations may have one, two, or no solutions.

- An absolute value equation has one solution if one of the answers does not meet the constraints of the problem. Such an answer is called an **extraneous solution**.
- An absolute value equation has no solution if there is no answer that meets the constraints of the problem. The solution set of this type of equation is called the **empty set**, symbolized by {} or Ø.

Example 1 Solve an Absolute Value Equation

Solve $2|5x + 1| - 9 = 4x + 17$. Check your solutions. Then graph the solution set.

$2|5x + 1| - 9 = 4x + 17$ Original equation

$2|5x + 1| = 4x + 26$ Add 9 to each side.

$|5x + 1| = 2x + 13$ Divide each side by 2.

Case 1	Case 2
$5x + 1 = 2x + 13$	$5x + 1 = -(2x + 13)$
$3x + 1 = 13$	$7x + 1 = -13$
$x = 4$	$x = -2$

CHECK Substitute each value in the original equation.

$2|5(4) + 1| - 9 \stackrel{?}{=} 4(4) + 17$ → $33 = 33$ True

$2|5(-2) + 1| - 9 \stackrel{?}{=} 4(-2) + 17$ → $9 = 9$ True

Both solutions make the equation true. Thus, the solution set is {4, −2}.

The solution set can be graphed by graphing each solution on a number line.

−5 −4 −3 −2 −1 0 1 2 3 4 5

Watch Out!

Distribute the Negative For Case 2, remember to use the Distributive Property to multiply the entire expression on the right side of the equation by −1.

Go Online You can complete an Extra Example online.

Interactive Presentation

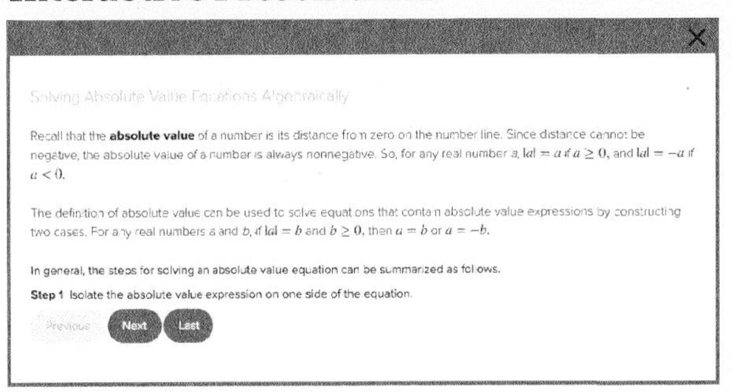

Learn

TAP

Students move through the steps to solve an absolute value equation.

Your Notes

Check

Graph the solution set of $|5x - 3| - 6 = -2x + 12$.

−10 −9 −8 −7 −6 −5 −4 −3 −2 −1 0 1 2 3 4 5 6 7 8 9 10

Think About It!

What would the graph of the solution set of an absolute value equation with only one solution look like?

a single point graphed on a number line

Example 2 Extraneous Solution

Solve $2|x + 1| - x = 3x - 4$. Check your solutions.

$2|x + 1| - x = 3x - 4$ Original equation

$2|x + 1| = 4x - 4$ Add x to each side.

$|x + 1| = 2x - 2$ Divide each side by 2.

Case 1

$x + 1 = 2x - 2$

$1 = x - 2$

$3 = x$

Case 2

$x + 1 = -(2x - 2)$

$x + 1 = -2x + 2$

$3x + 1 = 2$

$x = \frac{1}{3}$

There appear to be two solutions, 3 and $\frac{1}{3}$.

CHECK Substitute each value in the original equation.

$2|3 + 1| - 3 \stackrel{?}{=} 3(3) - 4$

$5 = 5$ True

$2\left|\frac{1}{3} + 1\right| - \left(\frac{1}{3}\right) \stackrel{?}{=} 3\left(\frac{1}{3}\right) - 4$

$\frac{7}{3} \neq -3$ False

Because $\frac{7}{3} \neq -3$, the only solution is 3. Thus, the solution set is $\{3\}$.

Talk About It!

Is the following statement *always*, *sometimes*, or *never* true? Justify your argument. For real numbers a, b, and c, $|ax + b| = -c$ has no solution.

Sometimes; sample answer: If $c > 0$, then the equation has no solution, but if $c \leq 0$, the equation may have a solution.

Example 3 The Empty Set

Solve $|4x - 7| + 10 = 2$.

$|4x - 7| + 10 = 2$ Original equation

$|4x - 7| = -8$ Subtract 10 from each side.

Because the absolute value of a number is always positive or zero, this sentence is never true. The solution is $\varnothing$.

Check

Solve each absolute value equation.

a. $|x + 10| = 4x - 8$ $\{6\}$

b. $3|4x - 11| + 1 = 9x + 13$ $\{15, 1\}$

c. $|2x + 5| - 18 = -3$ $\{5, -10\}$

d. $-5|7x - 2| + 3x = 3x + 10$ $\varnothing$

Go Online You can complete an Extra Example online.

84 Module 2 • Linear Equations, Inequalities, and Systems

Interactive Presentation

Extraneous Solution

Solve $2|x + 1| - x = 3x - 4$. Check your solutions.

Step 1 Isolate the absolute value expression.

$2|x + 1| - x = 3x - 4$ Original equation

$2|x + 1| = 4x - 4$ Add x to each side.

$|x + 1| = 2x - 2$ Divide each side by 2.

Example 2

TYPE

Students answer a question to determine whether they understand what the graph of the solution set of an absolute value equation with only one solution will look like.

Example 2 Extraneous Solution

Teaching the Mathematical Practices

1 Check Answers Mathematically proficient students continually ask themselves, "Does this make sense?" Point out that in Example 2, they should check their answers to identify any extraneous solutions.

Questions for Mathematical Discourse

AL What is an extraneous solution? Sample answer: a solution that does not make the equation true when substituted back in

OL Does solving $2|x + 1| = 4x - 4$ and $2|x + 1| = -(4x - 4)$ give the same solution? yes

BL Is it always true that $2|x + 1| = |2x + 2|$? If so, explain why. If not, give a counterexample. Is the same true for $-2|x + 1| = |-2x - 2|$? Yes, $2|x + 1| = |2x + 2|$ because $|2| = 2$. No, $-2|x + 1| \neq |-2x - 2|$ because $-2 \neq |-2|$.

Common Error

Students often skip checking solutions after solving an equation, but when dealing with absolute value equations, this can result in a solution that does not work. Reinforce the importance of checking solutions especially with absolute value equations.

Example 3 The Empty Set

Teaching the Mathematical Practices

3 Make Conjectures Use the Talk About It! feature to have students make a conjecture about the given statement. Then guide students to build a logical progression of statements to validate their conjecture.

Questions for Mathematical Discourse

AL Why can't an absolute value be a negative number? Absolute value is the distance a number is from zero, and distance cannot be negative.

OL Consider if the equation were $|4x - 7| + 10 = 10$. Would there be a solution? Explain. Yes; sample answer: Absolute value can equal zero, just not a negative value. There will be one solution $x = \frac{7}{4}$.

BL Can an absolute value equation ever be set equal to a negative value and have a solution? Yes; sample answer: There would be a solution if the other terms on the absolute value side of the equation would result in the isolated absolute value term being equal to a positive number or zero.

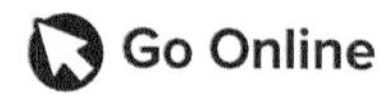

- Find additional teaching notes.
- View performance reports of the Checks.
- Assign or present an Extra Example.

Apply Example 4 Write and Solve an Absolute Value Equation

Teaching the Mathematical Practices

1 Make Sense of Problems and Persevere in Solving Them, 4 Model with Mathematics Students will be presented with a task. They will first seek to understand the task, and then determine possible entry points to solving it. As students come up with their own strategies, they may propose mathematical models to aid them. As they work to solve the problem, encourage them to evaluate their model and/or progress, and change direction, if necessary.

Recommended Use

Have students work in pairs or small groups. You may wish to present the task, or have a volunteer read it aloud. Then allow students the time to make sure they understand the task, think of possible strategies, and work to solve the problem.

Encourage Productive Struggle

As students work, monitor their progress. Instead of instructing them on a particular strategy, encourage them to use their own strategies to solve the problem and to evaluate their progress along the way. They may or may not find that they need to change direction or try out several strategies.

Signs of Non-Productive Struggle

If students show signs of non-productive struggle, such as feeling overwhelmed, frustrated, or disengaged, intervene to encourage them to think of alternate approaches to the problem. Some sample questions are shown.

- If the NFL allowed the air pressure to be 13 PSI plus or minus 1 PSI, would the range of acceptable weights for footballs be greater or lower?
- How many solutions do you need to answer the question?

Write About It!

Have students share their responses with another pair/group of students or the entire class. Have them clearly state or describe the mathematical reasoning they can use to defend their solution.

Common Error

Students often struggle with where to place numbers in an absolute value equation. They may flip flop the range with the central value in the equation. Reinforce the general form explained in the example.

Apply Example 4 Write and Solve an Absolute Value Equation

FOOTBALL **The NFL regulates the inflation, or air pressure, of footballs used during games. It requires that footballs have an air pressure of 13 pounds per square inch (PSI), plus or minus 0.5 PSI. What is the greatest and least acceptable air pressure of a regulation NFL football?**

1 What is the task?

Describe the task in your own words. Then list any questions that you may have. How can you find answers to your questions?

Sample answer: I need to find the greatest and least acceptable air pressure for an NFL football. How can I write an absolute value equation to find the solution? Will there be one solution or two that make sense in this problem? I can find the answers to my questions by referencing other examples in the lesson and by checking my solutions.

2 How will you approach the task? What have you learned that you can use to help you complete the task?

Sample answer: I will write an equation to represent the situation. I have learned how to write and solve an equation involving absolute value.

3 What is your solution?

Use your strategy to solve the problem.

What absolute value equation represents the greatest and least acceptable air pressure?

$|x - 13| = 0.5$

How are the solutions of the equation represented on a graph?

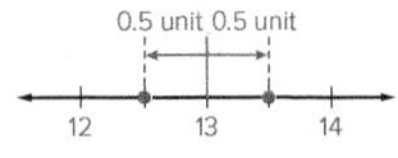

Interpret your solution. What are the greatest and least acceptable air pressures for an NFL football?

The greatest air pressure an NFL football can have is 13.5 PSI and the least is 12.5 PSI.

4 How can you know that your solution is reasonable?

Write About It! Write an argument that can be used to defend your solution. Sample answer: Because the distance between 13 and each solution is 0.5, both solutions satisfy the constraints of the equation. I can substitute each solution back into the original equation and check that the value makes the equation true. The pressures of 12.5 PSI and 13.5 PSI are within 0.5 PSI of 13 PSI.

Go Online You can complete an Extra Example online.

Lesson 2-2 • Solving Absolute Value Equations and Inequalities **85**

Interactive Presentation

Write and Solve an Absolute Value Equation

FOOTBALL **The NFL regulates the inflation, or air pressure, of footballs used during games. It requires that footballs have an air pressure of 13 pounds per square inch (PSI), plus or minus 0.5 PSI. What is the greatest and least acceptable air pressure of a regulation NFL football? Write and solve an equation, and graph the solution set.**

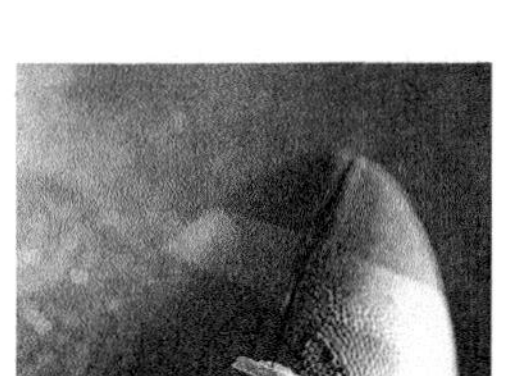

Apply Example 4

TYPE

Students answer questions about their strategies to complete the task.

CHECK

Students complete the check online to determine if they are ready to move on.

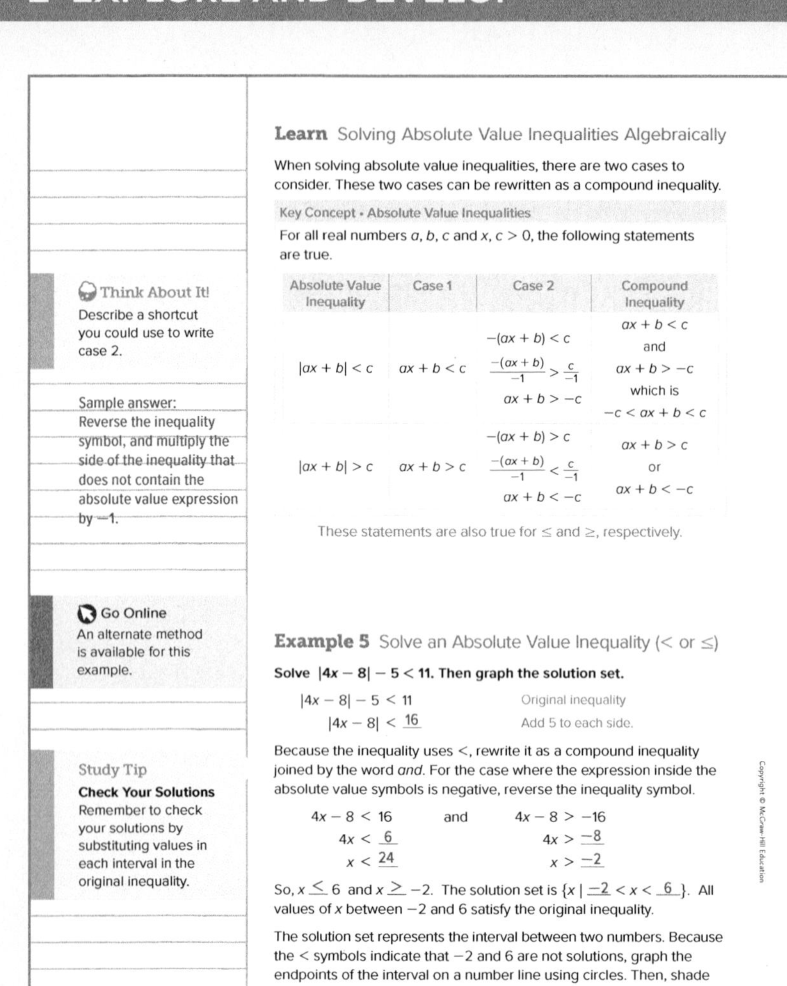

Think About It!
Describe a shortcut you could use to write case 2.

Sample answer: Reverse the inequality symbol, and multiply the side of the inequality that does not contain the absolute value expression by −1.

Go Online
An alternate method is available for this example.

Study Tip
Check Your Solutions Remember to check your solutions by substituting values in each interval in the original inequality.

Learn Solving Absolute Value Inequalities Algebraically

When solving absolute value inequalities, there are two cases to consider. These two cases can be rewritten as a compound inequality.

Key Concept • Absolute Value Inequalities

For all real numbers a, b, c and x, $c > 0$, the following statements are true.

Absolute Value Inequality	Case 1	Case 2	Compound Inequality
$\lvert ax + b\rvert < c$	$ax + b < c$	$-(ax + b) < c$ $\frac{-(ax+b)}{-1} > \frac{c}{-1}$ $ax + b > -c$	$ax + b < c$ and $ax + b > -c$ which is $-c < ax + b < c$
$\lvert ax + b\rvert > c$	$ax + b > c$	$-(ax + b) > c$ $\frac{-(ax+b)}{-1} < \frac{c}{-1}$ $ax + b < -c$	$ax + b > c$ or $ax + b < -c$

These statements are also true for ≤ and ≥, respectively.

Example 5 Solve an Absolute Value Inequality (< or ≤)

Solve $\lvert 4x - 8\rvert - 5 < 11$. Then graph the solution set.

$\lvert 4x - 8\rvert - 5 < 11$ — Original inequality
$\lvert 4x - 8\rvert < \underline{16}$ — Add 5 to each side.

Because the inequality uses <, rewrite it as a compound inequality joined by the word *and*. For the case where the expression inside the absolute value symbols is negative, reverse the inequality symbol.

$4x - 8 < 16$	and	$4x - 8 > -16$
$4x < \underline{6}$		$4x > \underline{-8}$
$x < \underline{24}$		$x > \underline{-2}$

So, $x \underline{\leq} 6$ and $x \underline{\geq} -2$. The solution set is $\{x \mid \underline{-2} < x < \underline{6}\}$. All values of x between −2 and 6 satisfy the original inequality.

The solution set represents the interval between two numbers. Because the < symbols indicate that −2 and 6 are not solutions, graph the endpoints of the interval on a number line using circles. Then, shade the interval from −2 to 6.

−10 −8 −6 −4 −2 0 2 4 6 8 10

Go Online You can complete an Extra Example online.

Copyright © McGraw-Hill Education

86 Module 2 • Linear Equations, Inequalities, and Systems

Learn Solving Absolute Value Inequalities Algebraically

Objective

Students write and solve absolute value inequalities by constructing compound inequalities, and then graph the solutions on a number line.

Teaching the Mathematical Practices

7 Use Structure In the Think About It! feature, help students to use the structure of the inequality in case 2 to determine a shortcut for writing the case.

Common Misconception

A common misconception students often have is that the inequality symbol of the absolute value inequality remains the same for both cases. This may be due to how absolute value equations are handled. Remind students that when an inequality is multiplied or divided by a negative value, the inequality sign must reverse, so the second inequality must have a different inequality sign.

Example 5 Solve an Absolute Value Inequality (< or ≤)

Teaching the Mathematical Practices

1 Special Cases Work with students to evaluate the two cases shown. Encourage them to familiarize themselves with when to write the cases as a compound inequality with the word *and* or with the word *or*.

Questions for Mathematical Discourse

AL Why is the first step to isolate the absolute value? Sample answer: This helps you write the positive and negative cases for the expression inside the absolute value.

OL How would the graph of the solution change if the < sign was ≤? The shading would be the same, but the open circles would now be closed.

BL Why do you reverse the sign for the negative case? Sample answer: We are considering when the expression inside the absolute value has a negative result, so applying the absolute value to that would be the same as applying a negative to the negative to yield a positive. We could first say $-(4x - 8) < 16$ and see that dividing both sides of the negative then reverses the inequality.

Common Error

When presented with an equation, many students will begin solving the equation without pausing to consider if there is a solution. Encourage students to isolate the absolute value expression and then consider if the resulting equation is possible, not to immediately set up the equations for the two cases.

Interactive Presentation

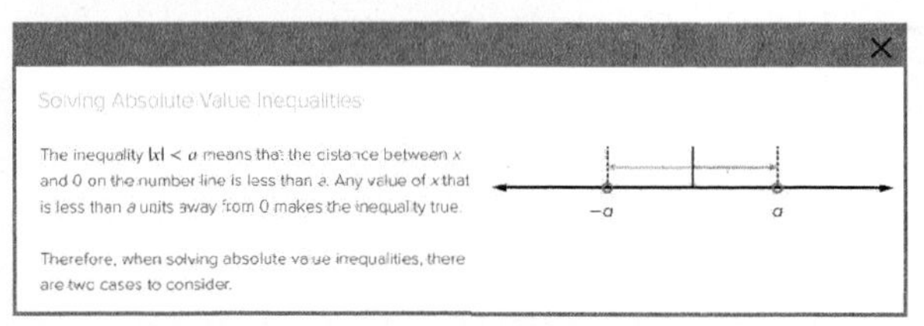

Learn

WATCH

Students can watch a video that explains how to solve compound inequalities.

Example 6 Solve an Absolute Value Inequality (> or ≥)

Teaching the Mathematical Practices

7 Interpret Complicated Expressions Mathematically proficient students can see complicated expressions as single objects or as being composed of several objects. Students must see the absolute value inequality as two separate inequalities.

Questions for Mathematical Discourse

AL Why is zero not a solution to the inequality? When we substitute zero for x in the original inequality, we get $5 \geq 14$, which is a false statement.

OL Why are absolute value inequalities containing < or ≤ compound inequalities joined by the word *and* while > or ≥ are compound inequalities joined by the word *or*? Sample Answer: When the absolute value inequality contains < or ≤, the solutions are valid between the two end points on the number line, thus they satisfy both inequalities. When the absolute value inequality contains > or ≥, the solutions are valid in two different parts of the number line, so each solution can only satisfy one of the inequalities.

BL What would the solution set look like if the 5 was changed to 14 in the original equation? There would only be one range of valid x-values, $x \geq -\frac{1}{2}$.

Common Error

Students may try and shade between the two values of the inequalities, instead of shading to the outside. Have students write the solution set and then consider if the values between the two points or outside the points make the inequalities true or false. This will help identify the solution shading.

DIFFERENTIATE

Reteaching Activity AL ELL

IF students cannot determine if the absolute value inequality is an "and" or "or" compound inequality,
THEN tell students the trick:
Less ThAND < ≤
GreatOR > ≥

Example 6 Solve an Absolute Value Inequality (> or ≥)

Solve $\frac{|6x+3|}{2} + 5 \geq 14$. Then graph the solution set.

$\frac{|6x+3|}{2} + 5 \geq 14$	Original inequality
$\frac{|6x+3|}{2} \geq \underline{9}$	Subtract 5 from each side.
$|6x+3| \geq \underline{18}$	Multiply each side by 2.

Since the inequality uses ≥, rewrite it as a compound inequality joined by the word *or*. For the case where the expression inside the absolute value symbols is negative, reverse the inequality symbol.

$6x + 3 \geq \underline{18}$ or $6x + 3 \leq \underline{-18}$
$6x \geq \underline{15}$ $\quad 6x \leq \underline{-21}$
$x \geq \underline{\frac{5}{2}}$ $\quad x \leq \underline{-\frac{7}{2}}$

So, $x \geq \underline{\frac{5}{2}}$ or $x \leq \underline{-\frac{7}{2}}$. The solution set is $\left\{x \mid x \leq -\frac{7}{2} \underline{\text{ or }} x \geq \frac{5}{2}\right\}$.

All values of x less than or equal to $\underline{-\frac{7}{2}}$ as well as values of x greater than $\underline{\frac{5}{2}}$ satisfy the constraints of the original inequality.

The solution set represents the union of two intervals. Since the ≤ and ≥ symbols indicate that $-\frac{7}{2}$ and $\frac{5}{2}$ are solutions, graph the endpoints of the interval on a number line using dots. Then, shade all points less than $-\frac{7}{2}$ and all points greater than $\frac{5}{2}$.

−5 −4 −3 −2 −1 0 1 2 3 4 5

Watch Out!
Isolate the Expression Remember to isolate the absolute value expression on one side of the inequality symbol before determining whether to rewrite an absolute value inequality using *and* or *or*. When transforming the inequality, you might divide or multiply by a negative number, causing the inequality symbol to be reversed.

Check

Match each solution set with the appropriate absolute value inequality.

$-8|x + 14| + 7 \geq -17$ _F_

$\frac{|2x-8|}{3} - 10 < 6$ _E_

$5|2x + 28| + 6 \geq -24$ _A_

$\frac{|3x-12|}{4} - 13 > 5$ _C_

A. $\{x \mid x \text{ is a real number.}\}$ B. $\{x \mid 28 < x < -20\}$
C. $\{x \mid x < -20 \text{ or } x > 28\}$ D. $\{x \mid x \leq -11 \text{ or } x \geq -17\}$
E. $\{x \mid -20 < x < 28\}$ F. $\{x \mid -17 \leq x \leq -11\}$

Go Online You can complete an Extra Example online.

Lesson 2-2 • Solving Absolute Value Equations and Inequalities 87

Interactive Presentation

Solve an Absolute Value Inequality (> or ≥)

Solve $\frac{|6x+3|}{2} + 5 \geq 14$. Then graph the solution set.

Part A
Solve the inequality.

$\frac{|6x+3|}{2} + 5 \geq 14$	Original equation
$\frac{|6x+3|}{2} \geq 9$	Subtract 5 from each side.
$|6x+3| \geq 18$	Multiply each side by 2.

Example 6

EXPAND

Students can tap to see how to isolate an expression in an absolute value inequality.

Example 7 Write and Solve an Absolute Value Inequality

SLEEP **You can find how much sleep you need by going to sleep without turning on an alarm. Once your sleep pattern has stabilized, record the amount of time you spend sleeping each night. The amount of time you sleep plus or minus 15 minutes is your sleep need. Suppose you sleep 8.5 hours per night. Write and solve an inequality to represent your sleep need, and graph the solution on a number line.**

Part A Write an absolute value inequality to represent the situation.

The difference between your actual sleep need and the amount of time you sleep is less than or equal to 15 minutes. So, 8.5 hours is the central value and 15 minutes, or 0.25 hour, is the acceptable range.

The difference between your actual sleep need and 8.5 hours is 0.25 hour. Let n = your actual sleep need.

$|n - 8.5| \leq 0.25$

Part B Solve the inequality and graph the solution set.

Rewrite $|n - 8.5| \leq 0.25$ as a compound inequality.

$n - 8.5 \leq 0.25$ and $n - 8.5 \geq -0.25$

$n \leq 8.75$ $\qquad n \geq 8.25$

The solution set represents the interval between two numbers. Since the $\leq$ and $\geq$ symbols indicate that 8.25 and 8.75 are solutions, graph the endpoints of the interval on a number line using dots. Then, shade the interval from 8.25 to 8.75.

7 7.25 7.5 7.75 8 8.25 8.5 8.75 9

This means that you need between 8.25 and 8.75 hours of sleep per night, inclusive.

Check

FOOD A survey found that 58% of American adults eat at a restaurant at least once a week. The margin of error was within 3 percentage points.

Part A Write an absolute value inequality to represent the range of the percent of American adults who eat at a restaurant once a week, where x is the actual percent. $|x - 58| \leq 3$

Part B Use your inequality from Part **A** to find the range of the percent of American adults who eat at a restaurant once a week.

The actual percent of American adults who eat out at least once a week is $\{x \mid 55 \leq x \leq 61\}$.

Go Online You can complete an Extra Example online.

Interactive Presentation

Write and Solve an Absolute Value Inequality

An inequality can be viewed as a constraint in a problem situation. Each solution of the inequality represents a value that meets the constraint. In real-world situations, solutions are often restricted to nonnegative or whole numbers.

SLEEP **You can find how much sleep you need by going to sleep without turning on an alarm. Once your sleep pattern has stabilized, record the amount of time you spend sleeping each night. The amount of time you sleep plus or minus 15 minutes is your sleep need. Suppose you sleep 8.5 hours per night. Write and solve an inequality to represent your sleep need, and graph the solution on a number line.**

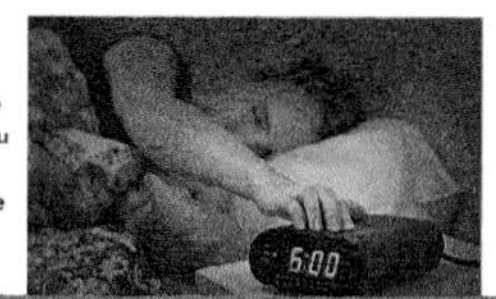

Example 7

TYPE

Students complete the absolute value equation.

CHECK

Students complete the Check online to determine whether they are ready to move on.

Example 7 Write and Solve an Absolute Value Inequality

Teaching the Mathematical Practices

4 Interpret Mathematical Results In Example 7, point out that to solve the problem, students should interpret their mathematical results in the context of the problem.

Questions for Mathematical Discourse

AL Why must we convert 15 minutes to 0.25 hour? The central value is given in hours and both numbers must be expressed in the same units.

OL How do the constraints of the problem help you know which direction the inequality goes? We know the sleep need must satisfy both inequalities to be a finite range, so the absolute value expression must be $\leq$ the acceptable range.

BL Karl sleeps an average of x hours per night, and Gina sleeps and average of y hours per night. If they want to go to bed and wake at the same time, how would you find the acceptable lengths of sleep that work for both? Sample answer: Graph the sleep need for each person and find where the shaded regions overlap

Common Error

Students may write the given values in the wrong location in the absolute value equation. They may subtract the acceptable range from the variable instead of the central value. Encourage students to write the general form $|x - \text{central value}|$ $(\leq, \geq, <, >)$ acceptable range in their notes.

Exit Ticket

Recommended Use

At the end of class, go online to display the Exit Ticket prompt and ask students to respond using a separate piece of paper. Have students hand you their responses as they leave the room.

Alternate Use

At the end of class, go online to display the Exit Ticket prompt and ask students to respond verbally or by using a mini-whiteboard. Have students hold up their whiteboards so that you can see all student responses. Tap to reveal the answer when most or all students have completed the Exit Ticket.

Practice and Homework

Suggested Assignments

Use the table below to select appropriate exercises.

DOK	Topic	Exercises
1, 2	exercises that mirror the examples	1–20
2	exercises that use a variety of skills from this lesson	21–28
2	exercises that extend concepts learned in this lesson to new contexts	29–39
3	exercises that emphasize higher-order and critical thinking skills	40–47

ASSESS AND DIFFERENTIATE

Use the data from the **Checks** to determine whether to provide resources for extension, remediation, or intervention.

IF students score 90% or more on the Checks, 
THEN assign:

- Practice, Exercises 1–39 odd, 40–47
- Extension: Diophantine Equations
- **ALEKS**® Absolute Value Equations; Absolute Value Inequalities

IF students score 66%–89% on the Checks, 
THEN assign:

- Practice, Exercises 1–47, odd
- Remediation, Review Resources: Solving Equations Involving Absolute Value
- Personal Tutors
- Extra Examples 1–7
- **ALEKS**® Absolute Value Equations

IF students score 65% or less on the Checks, AL
THEN assign:

- Practice, Exercises 1–19, odd
- Remediation, Review Resources: Solving Equations Involving Absolute Value
- **ALEKS**® Absolute Value Equations

Name ______ Period ______ Date ______

Practice

Examples 1–3

Solve each equation. Check your solutions.

1. $|8 + p| = 2p - 3$ $\{11\}$
2. $|4w - 1| = 5w + 37$ $\{-4\}$
3. $4|2y - 7| + 13 = 9$ $\varnothing$
4. $-2|7 - 3y| - 6 = -14$ $\left\{1, 3\frac{2}{3}\right\}$
5. $2|4 - n| = -3n$ $\{-8\}$
6. $5 - 3|2 + 2w| = -7$ $\{-3, 1\}$
7. $5|2r + 3| - 5 = 0$ $\{-2, -1\}$
8. $3 - 5|2d - 3| = 4$ $\varnothing$

Example 4

Solve each problem.

9. WEATHER The packaging of a thermometer claims that the thermometer is accurate within 1.5 degrees of the actual temperature in degrees Fahrenheit. Write and solve an absolute value equation to find the least and greatest possible temperature if the thermometer reads 87.4° F. $|x - 87.4| = 1.5$; 85.9° F; 88.9° F

10. OPINION POLLS Public opinion polls reported in newspapers are usually given with a margin of error. A poll for a local election determined that Candidate Morrison will receive 51% of the votes. The stated margin of error is ±3%. Write and solve an absolute value equation to find the minimum and maximum percent of the vote that Candidate Morrison can expect to receive. $|x - 51| = 3$; 48%; 54%

Examples 5 and 6

Solve each inequality. Graph the solution set on a number line. 11–18. See margin for graphs.

11. $|2x + 2| - 7 \leq -5$ $\{x \mid -2 \leq x \leq 0\}$
12. $\left|\frac{x}{2} - 5\right| + 2 > 10$ $\{x \mid x < -6 \text{ or } x > 26\}$
13. $|3b + 5| \leq -2$ $\varnothing$
14. $|x| > x - 1$ all real numbers
15. $|4 - 5x| < 13$ $\{x \mid -1.8 < x < 3.4\}$
16. $|3n - 2| - 2 < 1$ $\left\{n \mid -\frac{1}{3} < n < \frac{5}{3}\right\}$
17. $|3x + 1| > 2$ $\left\{x \mid x < -1 \text{ or } x > \frac{1}{3}\right\}$
18. $|2x - 1| < 5 + 0.5x$ $\{x \mid -1.6 < x < 4\}$

Example 7

Solve each problem.

19. RAINFALL For 90% of the last 30 years, the rainfall at Shell Beach has varied no more than 6.5 inches from its mean value of 24 inches. Write and solve an absolute value inequality to describe the rainfall in the other 10% of the last 30 years, and graph the solution on a number line. $|r - 24| > 6.5$; $\{r \mid r < 17.5 \text{ or } r > 30.5\}$; See margin for graph.

20. MANUFACTURING A food manufacturer's guidelines state that each can of soup produced cannot vary from its stated volume of 14.5 fluid ounces by more than 0.08 fluid ounce. Write and solve an absolute value inequality to describe acceptable volumes, and graph the solution on a number line. $|v - 14.5| \leq 0.08$; $\{v \mid 14.42 \leq v \leq 14.58\}$; See margin for graph.

Mixed Exercises

Solve. Check your solutions.

21. $8x = 2|6x - 2|$ $\left\{1, \frac{1}{5}\right\}$
22. $-6y + 4 = |4y + 12|$ $\left\{-\frac{4}{5}\right\}$
23. $8z + 20 > -|2z + 4|$ $\left\{z \mid z > -\frac{8}{3}\right\}$
24. $-3y - 2 \leq |6y + 25|$ $\left\{y \mid y \leq -7\frac{2}{3} \text{ or } y \geq -3\right\}$

REASONING **Write an absolute value equation to represent each situation. Then solve the equation and discuss the reasonableness of your solution given the constraints of the absolute value equation.** 25, 26. See margin.

25. The absolute value of the sum of 4 times a number and 7 is the sum of 2 times a number and 3.

26. The sum of 7 and the absolute value of the difference of a number and 8 is −2 times a number plus 4.

27. MODELING A carpenter cuts lumber to the length of 36 inches. For her project, the lumber must be accurate within 0.125 inch.
    a. Write an inequality to represent the acceptable length of the lumber. Explain your reasoning. $|x - 36| \leq 0.125$; Sample answer: The inequality shows that the length of the lumber x could be as much as 0.125 inch greater than 36 inches or 0.125 inch less than 36 inches.
    b. Solve the inequality. Then state the maximum and minimum length for the lumber. $\{x \mid 35.875 \leq x \leq 36.125\}$; The length of the lumber can range from 35.875 inches to 36.125 inches.

28. SAND A home improvement store sells bags of sand, which are labeled as weighing 35 pounds. The equipment used to package the sand produces bags with a weight that is within 8 ounces of the labeled weight.
    a. Write an absolute value equation to represent the maximum and minimum weight for the bags of sand. $|x - 35| = 0.5$
    b. Solve the equation and interpret the result. $x = 35.5$; $x = 34.5$; The bags of sand weigh no less than 34.5 lbs and no more than 35.5 lbs.

Name ______ Period ______ Date ______

29. CONSTRUCT ARGUMENTS Megan and Yuki are solving the equation $|x - 9| = |5x + 6|$. Megan says that there are 4 cases to consider because there are two possible values for each absolute value expression. Yuki says only 2 cases need to be considered. With which person, do you agree? Will they both get the same solution(s)? Yuki; Sample answer: Yuki is correct because if $|a| = |b|$, then either $a = b$ or $a = -b$. They will get the same answers because $a = -b$ and $b = -a$ and $a = b$ and $-a = -b$ are equivalent equations.

Solve each inequality. Graph the solution set on a number line. 30–35. See margin for graphs.

30. $3|2z - 4| - 6 > 12$ $\{z \mid z < -1 \text{ or } z > 5\}$

31. $6|4p + 2| - 8 < 34$ $\left\{p \mid -\frac{9}{4} < p < \frac{5}{4}\right\}$

32. $\frac{|5f - 2|}{6} > 4$ $\left\{f \mid f < -\frac{22}{5} \text{ or } f > \frac{26}{5}\right\}$

33. $\frac{|2w + 8|}{5} \geq 3$ $\left\{w \mid w \leq -\frac{23}{2} \text{ or } w \geq \frac{7}{2}\right\}$

34. $-\frac{3x|6x + 1|}{5} < 12x$
all real numbers

35. $-\frac{7}{8}|2x + 5| > 14$
$\varnothing$

36. TIRES The recommended inflation of a car tire is no more than 35 pounds per square inch. Depending on weather conditions, the actual reading of the tire pressure could fluctuate up to 3.4 psi. Write and solve an absolute value equation to find the maximum and minimum tire pressure. $|x - 35| = 3.4$; 38.4 psi; 31.6 psi

37. PROJECTILE An object is launched into the air and then falls to the ground. Its velocity is modeled by the equation $v = 200 - 32t$, where the velocity v is measured in feet per second and time t is measured in seconds. The object's speed is the absolute value of its velocity. Write and solve a compound inequality to determine the time intervals in which the speed of the object will be between 40 and 88 feet per second. Interpret your solution in the context of the situation. $40 < |200 - 32t| < 88$; $3.5 < t < 5$ or $7.5 < t < 9$; The speed is between 40 and 88 feet per second in the intervals from 3.5 to 5 seconds going up and from 7.5 to 9 seconds coming down.

38. USE A SOURCE Research to find a poll with a margin of error. Describe the poll then write an absolute value inequality to represent the actual results. Answers will vary. Sample answer: According to a 2018 poll conducted by CBS News, the favorite genre of music in America is country with 21% of participants preferring country. There is a 4% margin of error; $|x - 21| \leq 4$.

39. CONSTRUCT ARGUMENTS Roberto claims that the solution to $|3c - 4| > -4.5$ is the same as the solution to $|3c - 4| \geq 0$, because an absolute value is always greater than or equal to zero. Is he correct? Explain your reasoning. Roberto is correct. Sample answer: The solution set for each inequality is all real numbers. For any value of c (positive, negative, or zero), each inequality will be true.

Higher-Order Thinking Skills

40. WRITE Summarize the difference between *and* compound inequalities and *or* compound inequalities. See Mod. 2 Answer Appendix.

41. WHICH ONE DOESN'T BELONG? Identify the compound inequality that does not share the same characteristics as the other three. Justify your conclusion. $x > 5$ and $x < 1$; Sample answer: Each of these has a non-empty solution set except for $x > 5$ and $x < 1$. There are no values of x that are simultaneously greater than 5 and less than 1.

$-3 < x < 5$	$x > 2$ and $x < 3$	$x > 5$ and $x < 1$	$x > -4$ and $x > -2$

42. FIND THE ERROR Ana and Ling are solving $|3x + 14| = -6x$. Is either of them correct? Explain your reasoning. Ling; Sample answer: Ana included an extraneous solution. She would have noticed this error if she had checked to see if her answers were correct by substituting the values into the original equation.

Ana

$|3x + 14| = -6x$

$3x + 14 = -6x$ or $3x + 14 = 6x$

$9x = -14$ $\quad$ $14 = 3x$

$x = -\frac{14}{9}$ ✓ $\quad$ $x = \frac{14}{3}$ ✓

Ling

$|3x + 14| = -6x$

$3x + 14 = -6x$ or $3x + 14 = 6x$

$9x = -14$ $\quad$ $14 = 3x$

$x = -\frac{14}{9}$ ✓ $\quad$ $x = \frac{14}{3}$ X

43. PERSEVERE Solve $|2x - 1| + 3 = |5 - x|$. List all cases and resulting equations. See Mod. 2 Answer Appendix.

ANALYZE **If *a*, *x*, and *y* are real numbers, determine whether each statement is *sometimes*, *always*, or *never* true. Justify your argument.** 44–46. See Mod. 2 Answer Appendix.

44. If $|a| > 7$, then $|a + 3| > 10$.

45. If $|x| < 3$, then $x + 3 > 0$.

46. If y is between 1 and 5, then $|y - 3| \leq 2$.

47. CREATE Write an absolute value inequality with a solution of $a \leq x \leq b$.
Sample answer: $\left|x - \frac{a + b}{2}\right| \leq b - \frac{a + b}{2}$

Answers

11. −4 −3 −2 −1 0 1 2 3 4

12. −10 −5 0 5 10 15 20 25 30

13. −4 −3 −2 −1 0 1 2 3 4

14. −4 −3 −2 −1 0 1 2 3 4

15. −3 −2 −1 0 1 2 3 4 5 6

16. −4 −3 −2 −1 0 1 2 3 4

17. -1 $-\frac{2}{3}$ $-\frac{1}{3}$ 0 $\frac{1}{3}$ $\frac{2}{3}$ 1

18. −3 −2 −1 0 1 2 3 4 5 6

19. 15 20 25 30 35

20. 14.4 14.5 14.6

25. $|4x + 7| = 2x + 3$; $x = -2$, $x = -\frac{5}{3}$; The absolute value equation is valid when $2x + 3 \geq 0$, so the equation is valid when $x \geq -\frac{3}{2}$. Since neither value of x is greater than or equal to $-\frac{3}{2}$, both solutions are extraneous.

26. $7 + |x - 8| = -2x + 4$; $x = \frac{5}{3}$, $x = -11$; The absolute value equation is valid when $-2x + 4 \geq 0$, so $x \leq 2$. So, $x = \frac{5}{3}$; is an extraneous solution.

30. −5 −4 −3 −2 −1 0 1 2 3 4 5

31. −3 −2 −1 0 1 2 3

32. −10 −8 −6 −4 −2 0 2 4 6 8 10

33. −16 −14 −12 −10 −8 −6 −4 −2 0 2 4

34. −5 −4 −3 −2 −1 0 1 2 3 4 5

35. −5 −4 −3 −2 −1 0 1 2 3 4 5

Lesson 2-3

Equations of Linear Functions

A.CED.2, F.IF.6

LESSON GOAL

Students write linear equations in standard, slope-intercept, and point-slope forms.

1 LAUNCH

Launch the lesson with a **Warm Up** and an introduction.

2 EXPLORE AND DEVELOP

Explore: Arithmetic Sequences

Develop:

Linear Equations in Standard Form

- Write Linear Equations in Standard Form

Linear Equations in Slope-Intercept Form

- Write Linear Equations in Slope-Intercept Form
- Interpret an Equation in Slope-Intercept Form
- Use a Linear Equation in Slope-Intercept Form

Linear Equations in Point-Slope Form

- Point-Slope Form Given Slope and One Point
- Point-Slope Form Given Two Points
- Write and Interpret a Linear Equation in Point-Slope Form

You may want your students to complete the **Checks** online.

3 REFLECT AND PRACTICE

Exit Ticket

Practice

DIFFERENTIATE

View reports of student progress on the **Checks** after each example.

Resources	AL	OL	BL	ELL
Remediation: Writing Equations in Slope-Intercept Form	●	●		●
Extension: Two-Intercept Form of a Linear Equation		●	●	●
ELL Support				

Language Development Handbook

Assign page 10 of the *Language Development Handbook* to help your students build mathematical language related to writing linear equations in standard, slope-intercept, and point-slope form.

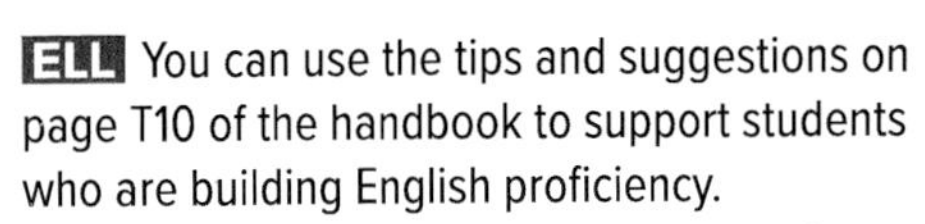

ELL You can use the tips and suggestions on page T10 of the handbook to support students who are building English proficiency.

Suggested Pacing

90 min	1 day
45 min	2 days

Focus

Domain: Algebra, Functions

Standards for Mathematical Content:

A.CED.2 Create equations in two or more variables to represent relationships between quantities; graph equations on coordinate axes with labels and scales.

F.IF.6 Calculate and interpret the average rate of change of a function over a specified interval.

Standards for Mathematical Practice:

1 Make sense of problems and persevere in solving them.

2 Reason abstractly and quantitatively.

4 Model with mathematics.

Coherence

Vertical Alignment

Previous

Students solved linear equations and inequalities in one variable.

A.CED.1, A.CED.2

Now

Students write equations in standard, slope-intercept, and point-slope forms.

A.CED.2, F.IF.6

Next

Students will solve systems of equations by graphing.

A.CED.3, A.REI.11

Rigor

The Three Pillars of Rigor

1 CONCEPTUAL UNDERSTANDING	2 FLUENCY	3 APPLICATION

Conceptual Bridge In this lesson, students build on their understanding of linear equations. They build fluency by writing equations in several forms and apply their understanding by using linear equations to solve real-world problems.

Mathematical Background

The point-slope form of a linear equation is related to the slope formula. As an alternative to memorizing point-slope form, replace (x_2, y_2) with (x, y) in the slope formula, then multiply each side by $(x - x_1)$.

Interactive Presentation

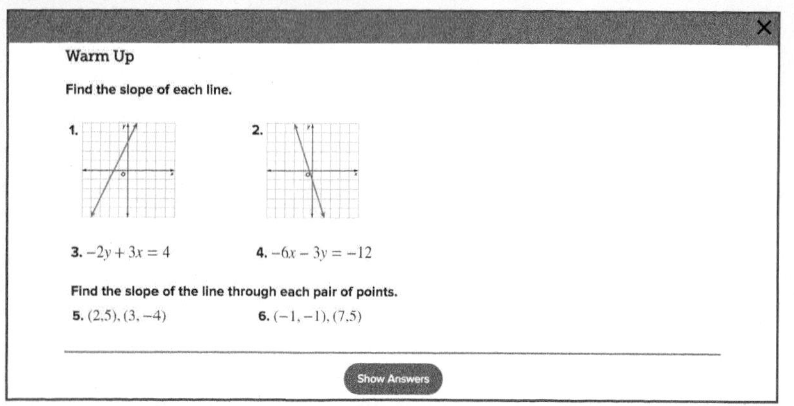

Warm Up

Warm Up

Prerequisite Skills

The Warm Up exercises address the following prerequisite skill for this lesson:

- determining slopes of lines

Answers:

1. 2
2. -3
3. $\frac{3}{2}$
4. -2
5. -9
6. $\frac{3}{4}$

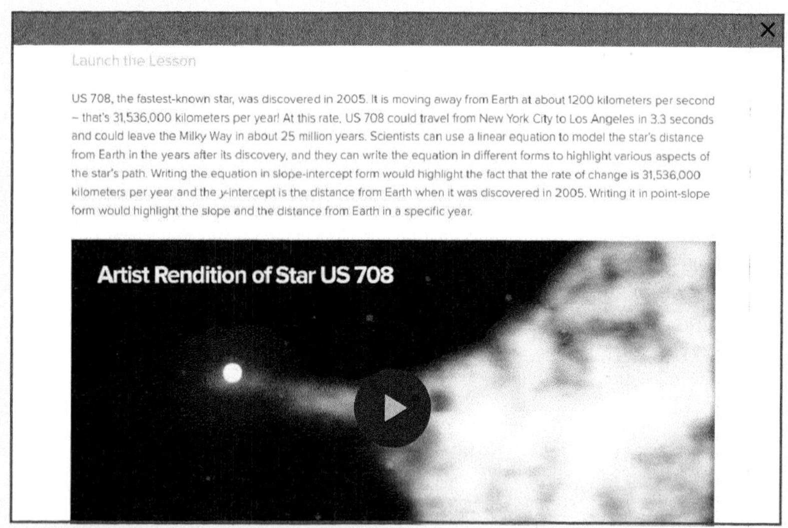

Launch the Lesson

Launch the Lesson

 **Teaching the Mathematical Practices**

2 Attend to Quantities Encourage students to consider the two forms for the linear equation representing the star's distance from Earth and consider the meaning of the quantities in each equation.

Go Online to find additional teaching notes and questions to promote classroom discourse.

Today's Standards

Tell students that they will be addressing these content and practice standards in this lesson. You may wish to have a student volunteer read aloud *How can I meet these standards?* and *How can I use these practices?*, and connect these to the standards.

See the Interactive Presentation for I Can statements that align with the standards covered in this lesson.

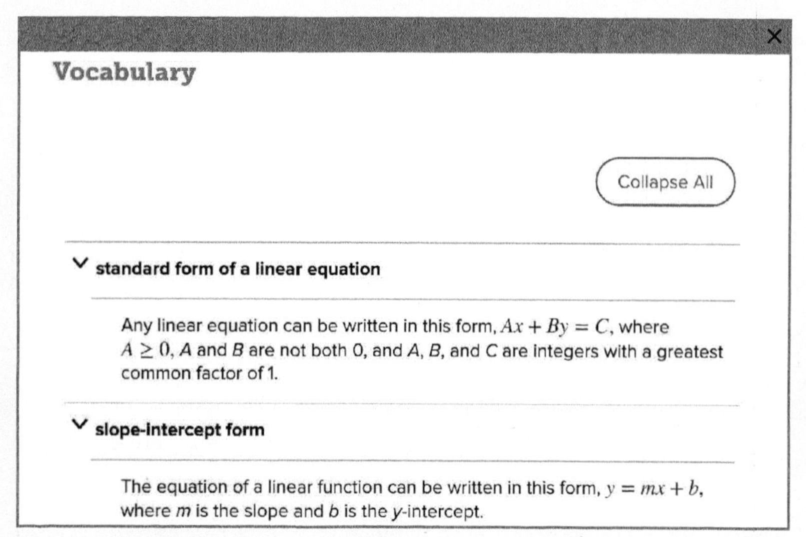

Today's Vocabulary

Today's Vocabulary

Tell students that they will be using these vocabulary terms in this lesson. You can expand each row if you wish to share the definitions. Then discuss the questions below with the class.

DIFFERENTIATE

Language Development Activity ELL

Standard form is a more general form where A, B, and C represent constants. Slope-intercept form and point-slope form are more specific forms where m is a constant that also represents a specific slope; b is a constant that represents a specific y-intercept; x_1 and y_1 are also constants that, together (x_1, y_1), represent a specific point.

Explore Arithmetic Sequences

Objective

Students use explicit and recursive formulas to explore arithmetic sequences.

Teaching the Mathematical Practices

8 Look for a Pattern Encourage students to look for the pattern in the remaining balance to write general formulas for the situation.

Recommended Use

Present the Inquiry Question to set up the Explore activity, or have a student read it aloud.

Have students work in pairs to complete the Explore activity on their devices, or project the activity on a whiteboard. You may choose to have each student record his or her observations using their device, or have each pair record using one device. Monitor student progress, making sure they answer all of the questions.

Upon completion of the Explore activity, have student volunteers share their responses to the Inquiry Question.

Summary of the Activity

Students will complete guiding exercises throughout the Explore activity. Students will analyze a real-world situation to determine the difference in the remaining balance owed as monthly payments are made. They will complete guiding exercises to represent the pattern using recursive and explicit formulas, and use the formulas to find the remaining balance. Then, students will answer the Inquiry Question.

(continued on the next page)

Interactive Presentation

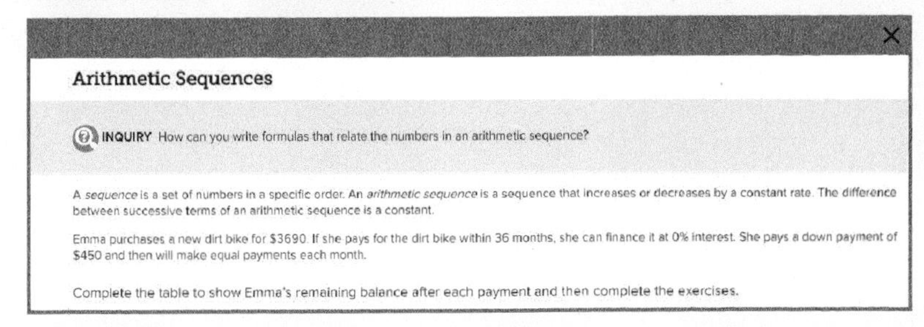

Explore

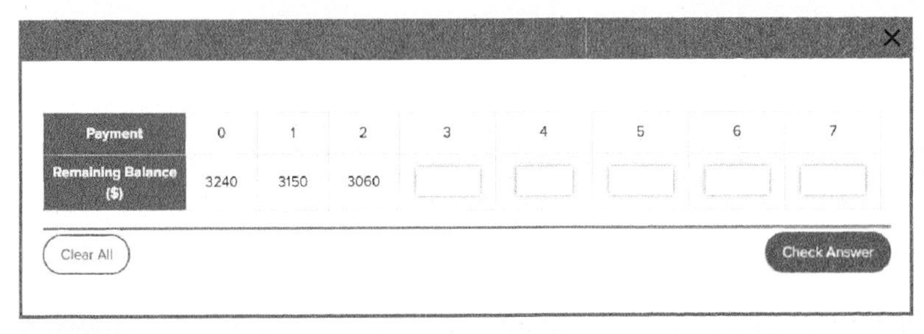

Explore

TYPE

Students complete the table to determine the remaining balance after each payment.

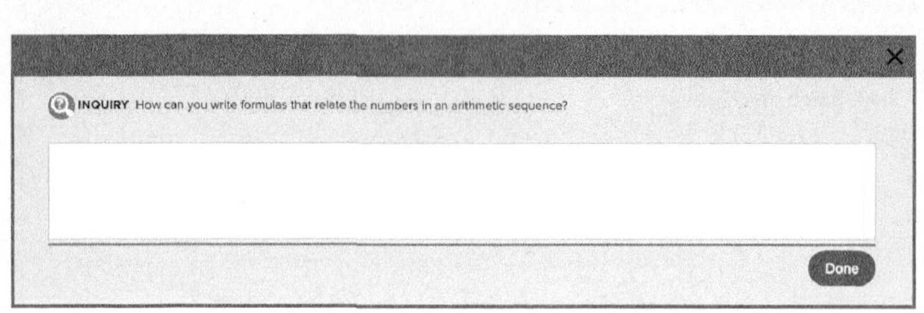

Explore

TYPE

Students respond to the Inquiry Question and can view a sample answer.

Explore Arithmetic Sequences (*continued*)

Questions

Have students complete the Explore activity.

Ask:

- What are the benefits of using an explicit formula to represent a situation that models an arithmetic sequence? Sample answer: An explicit formula allows you to quickly determine the value of any term by substituting the term number for n.
- How could you use the explicit formula to determine after how many payments the dirt bike will be paid off? Substitute 0 for a_n and solve for n.

Inquiry

How can you write formulas that relate the numbers in an arithmetic sequence? Sample answer: I can use the common difference between the terms and the first term of the sequence to write a recursive or explicit formula.

Go Online to find additional teaching notes and sample answers for the guiding exercises.

1 CONCEPTUAL UNDERSTANDING | 2 FLUENCY | 3 APPLICATION

Learn Linear Equations in Standard Form

Objective

Students write linear equations in standard form and identify values of A, B, and C by using the properties of equality.

Teaching the Mathematical Practices

6 Communicate Precisely Students should use clear definitions and mathematical language to answer the question in the Think About It! feature.

Example 1 Write Linear Equations in Standard Form

Teaching the Mathematical Practices

1 Seek Information In Example 1, students must transform the equation to rewrite it in standard form and get the information they need.

Questions for Mathematical Discourse

AL If you subtract a term from one side of an equation, what must you do to the other side to maintain equality? Subtract the same term from the other side.

OL Why are you able to multiply the equation by -5 to make the coefficient of x an integer ≥ 0? Sample answer: You can do this because of the multiplication property of equality. If $a = b$, $ac = bc$.

BL If the coefficient of x is π, what would need to be true to be able to write the equation in standard form? The other terms would need to have π as a factor, otherwise there would be irrational coefficients that cannot be multiplied to become integers.

Learn Linear Equations in Slope-Intercept Form

Objective

Students create linear equations in slope-intercept form by rewriting given equations and by using the coordinates of two points.

Teaching the Mathematical Practices

1 Explain Correspondences Encourage students to explain the relationships between the slope and intercept of a function and its graph.

Go Online

- Find additional teaching notes.
- View performance reports of the Checks.
- Assign or present an Extra Example.

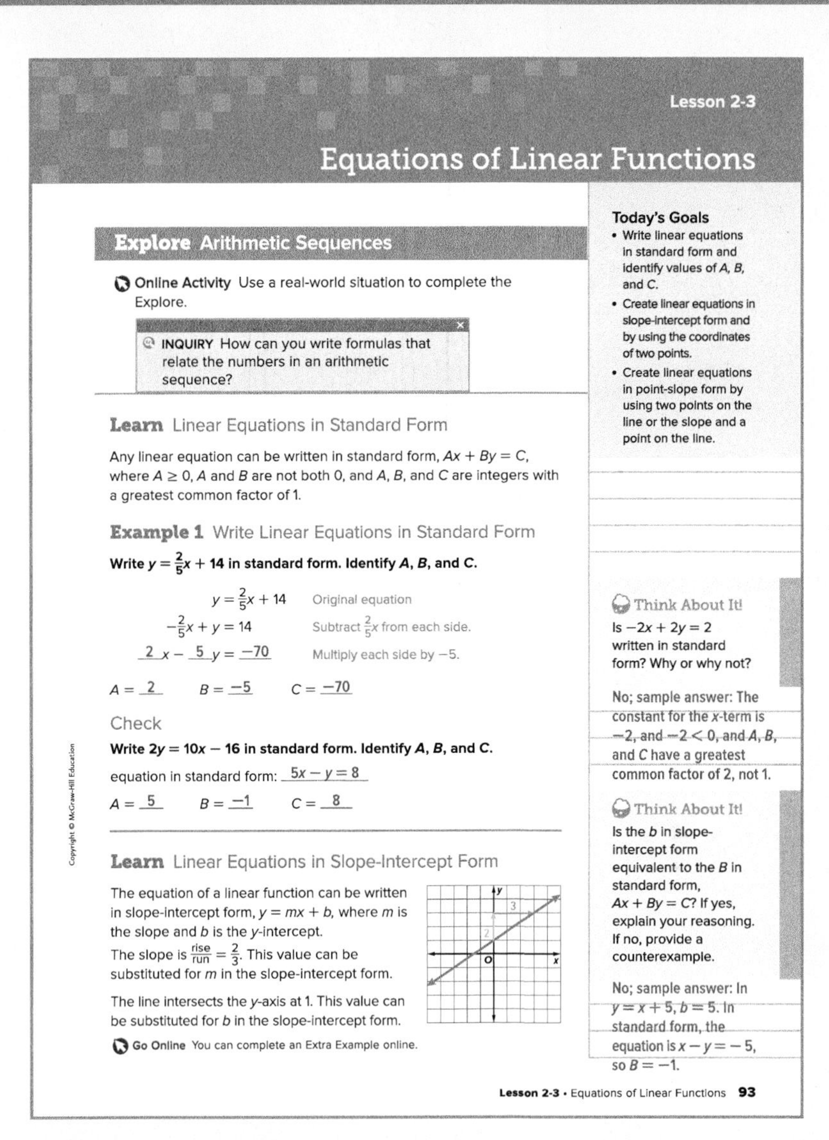

Lesson 2-3

Equations of Linear Functions

Explore Arithmetic Sequences

Online Activity Use a real-world situation to complete the Explore.

INQUIRY How can you write formulas that relate the numbers in an arithmetic sequence?

Learn Linear Equations in Standard Form

Any linear equation can be written in standard form, $Ax + By = C$, where $A \geq 0$, A and B are not both 0, and A, B, and C are integers with a greatest common factor of 1.

Example 1 Write Linear Equations in Standard Form

Write $y = \frac{2}{5}x + 14$ in standard form. Identify A, B, and C.

$y = \frac{2}{5}x + 14$ Original equation

$-\frac{2}{5}x + y = 14$ Subtract $\frac{2}{5}x$ from each side.

$\underline{2}\,x - \underline{5}\,y = \underline{-70}$ Multiply each side by -5.

$A = \underline{2}$ $B = \underline{-5}$ $C = \underline{-70}$

Check

Write $2y = 10x - 16$ in standard form. Identify A, B, and C.

equation in standard form: $\underline{5x - y = 8}$

$A = \underline{5}$ $B = \underline{-1}$ $C = \underline{8}$

Learn Linear Equations in Slope-Intercept Form

The equation of a linear function can be written in slope-intercept form, $y = mx + b$, where m is the slope and b is the y-intercept.

The slope is $\frac{\text{rise}}{\text{run}} = \frac{2}{3}$. This value can be substituted for m in the slope-intercept form.

The line intersects the y-axis at 1. This value can be substituted for b in the slope-intercept form.

Go Online You can complete an Extra Example online.

Today's Goals
- Write linear equations in standard form and identify values of A, B, and C.
- Create linear equations in slope-intercept form and by using the coordinates of two points.
- Create linear equations in point-slope form by using two points on the line or the slope and a point on the line.

Think About It!
Is $-2x + 2y = 2$ written in standard form? Why or why not?

No; sample answer: The constant for the x-term is -2, and $-2 < 0$, and A, B, and C have a greatest common factor of 2, not 1.

Think About It!
Is the b in slope-intercept form equivalent to the B in standard form, $Ax + By = C$? If yes, explain your reasoning. If no, provide a counterexample.

No; sample answer: In $y = x + 5$, $b = 5$. In standard form, the equation is $x - y = -5$, so $B = -1$.

Lesson 2-3 • Equations of Linear Functions 93

Interactive Presentation

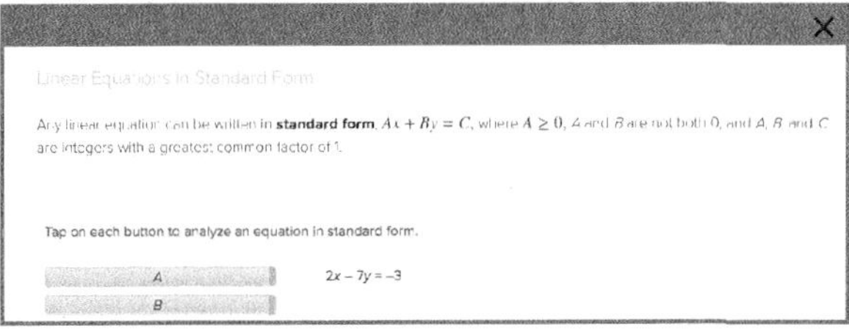

Learn

TAP

Students tap to analyze an equation in standard form.

TYPE

Students answer a question to determine whether they understand why a given equation is or is not in standard form.

CHECK

Students complete the Check online to determine whether they are ready to move on.

Your Notes

Example 2 Write Linear Equations in Slope-Intercept Form

Write $12x - 4y = 24$ in slope-intercept form. Identify the slope m and y-intercept b.

$12x - 4y = 24$	Original equation
$-4y = -12x + 24$	Subtract 12x from each side.
$y = \underline{3}x - \underline{6}$	Divide each side by −4.

$m = \underline{3}$ $b = \underline{-6}$

Check

Write $4x = -2y + 22$ in slope-intercept form. $y = -\underline{2}x + \underline{11}$

Example 3 Interpret an Equation in Slope-Intercept Form

SHOES **The equation $3246x - 2y = -152{,}722$ can be used to estimate shoe sales in Europe from 2010 to 2015, where x is the number of years after 2010 and y is the revenue in millions of dollars.**

Part A Write the equation in slope-intercept form.

$3246x - 2y = -152{,}722$	Original equation
$-2y = -3246x - 152{,}722$	Subtract 3246x from each side.
$y = \underline{1623}x + \underline{76{,}361}$	Divide each side by −2.

Part B Interpret the parameters in the context of the situation.

1623 represents that sales increased by $\underline{\$1623 \text{ million}}$ each year.

76,361 represents that in year 0, or in $\underline{2010}$, sales were $\underline{\$76{,}361 \text{ million}}$

Study Tip

Assumptions Assuming that the rate at which the number of smartphone users increases is constant allows us to represent the situation using a linear equation. While the rate at which the number of smartphone users increases may vary each year, using a constant rate allows for a reasonable equation that can be used to estimate future data.

Example 4 Use a Linear Equation in Slope-Intercept Form

SMARTPHONES **In 2013, there were 1.31 billion smartphone users worldwide. By 2017, there were 2.38 billion smartphone users. Write and use an equation to estimate the number of users in 2025.**

Step 1 Define the variables. Because you want to estimate the number of users in 2025, write an equation that represents the number of smartphone users y after x years. Let x be the number of years after 2013 and let y be the number of billions of smartphone users.

Step 2 Find the slope. Since x is the years after 2013, (0, 1.31) and (4, 2.38) represent the number of smartphone users in 2013 and 2017, respectively. Round to the nearest hundredth.

$$m = \frac{2.38 - 1.31}{4 - 0} = \underline{0.27}$$

So, the number of users is increasing at a rate of $\underline{0.27}$ billion per year.

Go Online You can complete an Extra Example online.

Think About It!

When using the equation to estimate the number of smartphone users in the future, what constraint does the world's population place on the possible number of users?

Sample answer: The number of estimated users cannot be greater than the world's population.

Interactive Presentation

Use a Linear Equation in Slope-Intercept Form

SMARTPHONES In 2013, there were 1.31 billion smartphone users worldwide. By 2017, there were 2.38 billion smartphone users. Write and use an equation to estimate the number of users in 2025.

Example 4

TAP

Students tap to move through the steps to represent the number of smartphone users as a linear equation in slope-intercept form.

TYPE

Students answer a question to describe a constraint the world's population places on the possible number of smartphone users.

Example 2 Write Linear Equations in Slope-Intercept Form

Questions for Mathematical Discourse

AL Can m or b be a fraction? Explain. Yes; sample answer: The slope represents the change in y over the change in x, and the graph can cross the y-axis anywhere, so both could be a fraction.

OL Why is b the y-intercept? When $x = 0$, $y = b$.

BL Suppose the coefficient of y was a fraction. What could be done to the equation to remove the coefficient? Sample answer: Multiply both sides of the equation by the reciprocal of the fraction.

Example 3 Interpret an Equation in Slope-Intercept Form

Teaching the Mathematical Practices

2 Attend to Quantities Point out that it is important to note the meaning of the quantities used in this problem. Students should interpret the meaning of the slope and intercept in the context of the situation.

Questions for Mathematical Discourse

AL What does $x = 3$ represent? the year 2013

OL How do you know that revenue is increasing? Sample answer: The slope, which represents the rate of change in sales, is positive.

BL What scale would you use for the y-axis if you graphed the equation? Sample answer: 100 millions

Example 4 Use a Linear Equation in Slope-Intercept Form

Teaching the Mathematical Practices

4 Analyze Mathematical Relationships Point out that to solve the problem in Example 4, student still need to analyze the mathematical relationships in the problem to draw conclusions about the number of smartphone users.

Questions for Mathematical Discourse

AL Why does $x = 0$ represent the year 2013? Sample answer: $x = 0$ represents 2013 because we are interested in the years after 2013, so we defined the variable this way.

OL Why is the scale for a variable important to consider when modeling? Sample answer: Optimizing the chosen scale allows you to construct graphs that are useful in how they visually represent the data.

BL If the total number of smartphone users is limited by the population of the world, what might the graph look like as x continues to increase? Sample answer: The graph might continue to increase, but at a slower rate than is related to the population increase of the world.

1 CONCEPTUAL UNDERSTANDING | 2 FLUENCY | 3 APPLICATION

Learn Linear Equations in Point-Slope Form

Objective

Students create linear equations in point-slope form by using two points on the line or the slope and a point on the line.

Teaching the Mathematical Practices

2 Attend to Quantities Point out that it is important to note the meaning of x_1 and y_1 used in the point-slope form of a line.

Example 5 Point-Slope Form Given Slope and One Point

Teaching the Mathematical Practices

6 Use Definitions In this example, students will use the definition of point-slope form to write an equation given the slope and a point on the line.

Questions for Mathematical Discourse

AL When the y-coordinate is substituted into the point-slope formula, why does the expression become $y + 5$ instead of $y - 5$? Sample answer: The formula subtracts the y-coordinate from y, and the y-coordinate is negative. Subtracting a negative simplifies to a positive.

OL Joe says the y-intercept for this equation is -8. Is he correct? Explain. Joe is not correct because he did not distribute the 11 to the -3 term in the parenthesis.

BL Could point-slope form have been defined as $y_1 - y = m(x_1 - x)$? Explain. Yes; sample answer: This version of the equation is equivalent to the original because multiplying both sides of the new equation by -1 yields the original equation.

Example 6 Point-Slope Form Given Two Points

Teaching the Mathematical Practices

3 Construct Arguments Students will use definitions and previously established results to construct an argument and answer the question in the Talk About It! feature.

Questions for Mathematical Discourse

AL Why are the y-values in the numerator and the x-values in the denominator? The slope, m, represents the change in y-values per change in x-values, or the rise over run.

OL How would the slope change if all the coordinates are the negative of the original coordinates? The slope would not change.

BL How can you confirm that using either point in the point-slope form of the equation produces equivalent equations? Sample answer: If I rewrite each equation in slope-intercept or standard form, they should be the same.

Step 3 Find the y-intercept. The y-intercept represents the number of smartphone users when $x = 0$, or in 2013. So, $b = $ 1.31

Step 4 Write an equation. Use $m = 0.27$ and $b = 1.31$ to write the equation.

$y = 0.27x + 1.31$ — $m = 0.27, b = 1.31$

Step 5 Estimate. Since 2025 is 12 years after 2013, substitute 12 for x.

$y = 0.27(12) + 1.31; y = 4.55$

If the trend continues, there will be about 4.55 billion users in 2025.

Think About It! Suppose the data spanned 2 years instead of 4 years. That is, there were 1.31 billion smartphone users in 2013 and 2.38 billions users in 2015. How would this affect the rate of change abnd your estimate in Step 5?

Sample answer: It would take half as long for the number of users to increase from 1.31 billion to 2.38 billion. So, the rate at which the number of users increases would double to about 0.54 billion users per year. This would cause the estimate in Step 5 to increase from 4.55 billion users in 2025 to 7.79 billion users.

Learn Linear Equations in Point-Slope Form

The equation of a linear function can be written in point-slope form, $y - y_1 = m(x - x_1)$, where m is the slope and (x_1, y_1) are the coordinates of a point on the line.

Example 5 Point-Slope Form Given Slope and One Point

Write the equation of a line that passes through (3, −5) and has a slope of 11 in point-slope form.

$y - y_1 = m(x - x_1)$	Point-slope form
$y - (-5) = 11(x - 3)$	$m = 11$; $(x_1, y_1) = (3, -5)$
$y + 5 = 11(x - 3)$	Simplify.

Check

Write the equation of a line that passes through (13, −5) and has a slope of 4.5 in point-slope form.

$y + 5 = 4.5(x - 13)$

Example 6 Point-Slope Form Given Two Points

Write an equation of a line that passes through (1, 1) and (7, 13) in point-slope form.

Step 1 Find the slope.

$m = \frac{y_2 - y_1}{x_2 - x_1}$	Slope formula
$= \frac{13 - 1}{7 - 1}$	$(x_1, y_1) = (1, 1)$; $(x_2, y_2) = (7, 13)$
$= \frac{12}{6}$	Simplify.
$= 2$	Simplify.

(continued on the next page)

Go Online You can complete an Extra Example online.

Interactive Presentation

Point-Slope Form Given Slope and One Point

Write the equation of a line that passes through (3, −5) and has a slope of 11 in point-slope form.

$y - y_1 = m(x - x_1)$	Point-slope form
$y - (-5) = 11(x - 3)$	$m = 11; (x_1, y_1) = (3, -5)$
$y + 5 = 11(x - 3)$	Simplify.

Example 5

Students answer a question to determine whether they understand how to write an equation in slope-intercept and standard form.

CHECK

Students complete the Check online to determine whether they are ready to move on.

Talk About It!
What other values would you need to write the equation of this line in slope-intercept form? Could you determine those values from the given information?

Sample answer: To write the equation in slope-intercept form, I do not need to know any other values. I can use the Distributive Property and solve for y.

Step 2 Write an equation.

Substitute the slope for m and the coordinates of either of the given points for (x_1, y_1) in the point-slope form.

$y - y_1 = m(x - x_1)$ Point-slope form

$y - 1 = 2(x - 1)$ $m = 2$; $(x_1, y_1) = (1, 1)$

Check B, H

Select all the equations for the line that passes through (−1, 1) and (−2, 13).

A. $x - 1 = -12(y + 1)$ B. $y - 1 = -12(x + 1)$ C. $x + 1 = -12(y - 1)$
D. $y + 1 = -12(x - 1)$ E. $y - 2 = -12(x + 13)$ F. $x - 2 = -12(y + 13)$
G. $x + 2 = -12(y - 13)$ H. $y - 13 = -12(x + 2)$

Example 7 Write and Interpret a Linear Equation in Point-Slope Form

ARCHITECTURE **The Tower of Pisa began tilting during its construction in 1178 and continued to move until a restoration effort reduced the lean and stabilized the structure. The Tower of Pisa leaned 5.4 meters in 1993 compared to a lean of just 1.4 meters in 1350. Write an equation in point-slope form that represents the lean y in meters of the Tower of Pisa x years after its construction in 1178.**

Step 1 Find the slope. Round to the nearest hundredth.

The tower was leaning 1.4 meters in 1350, 172 years after 1178.

The tower was leaning 5.4 meters in 1993, 815 years after 1178.

$m = \frac{5.4 - 1.4}{815 - 172} = 0.006$ The lean of the Tower of Pisa increased at a rate of 0.006 meter per year.

Step 2 Write an equation.

Substitute the slope for m and the coordinates of either of the given points for (x_1, y_1) in the point-slope form.

$y - y_1 = m(x - x_1)$ Point-slope form

$y - 1.4 = 0.006(x - 172)$ $m = 0.006$; $(x_1, y_1) = (172, 1.4)$

Think About It!
Could this equation be used to estimate the lean of the Tower of Pisa for any year? Explain your reasoning.

No; sample answer: Because the Tower of Pisa was eventually stabilized and its lean decreased, this equation could only be used to estimate the lean of the Tower of Pisa from the year it began leaning until the year when restoration began.

Check

SOCIAL MEDIA In 2011, the Miami Marlins had about 11,000 followers on a social media site. In 2016, they had about 240,000 followers. Which equation represents the number of followers y the Miami Marlin's had x years after they joined the site in 2009? A

A. $y - 11{,}000 = 45{,}800(x - 2)$ **B.** $y - 45{,}800 = 11{,}000(x - 2)$
C. $y - 11{,}000 = 45{,}800(x - 2011)$ **D.** $y - 2 = 45{,}800(x - 11{,}000)$

Go Online You can complete an Extra Example online.

96 Module 2 • Linear Equations, Inequalities, and Systems

Interactive Presentation

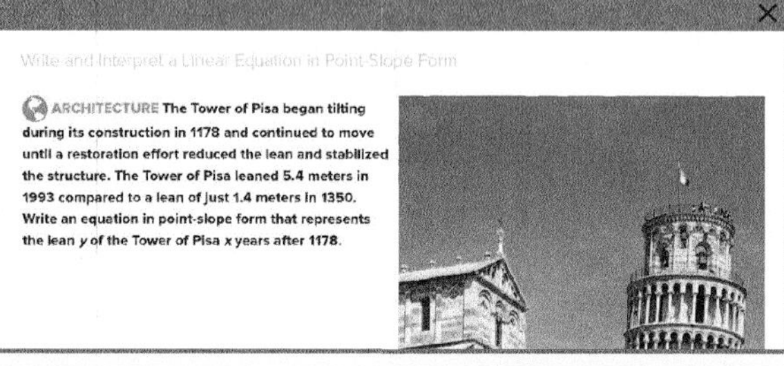

Example 7

TAP

Students tap to write and interpret a linear equation in point-slope form.

TYPE

Students complete the calculations to find the slope of the line representing the lean of the Tower of Pisa.

CHECK

Students complete the Check online to determine whether they are ready to move on.

DIFFERENTIATE

Language Development Activity ELL

Beginning Reinforce the use of visual context to derive meaning through examples of environmental print such as store or restaurant signs. Can a linear function represent the products and prices shown? Pantomime or elicit one-word responses to the meaning derived from such images.

Intermediate Provide real-world signs to illustrate problems that students can work in pairs to solve. Move around the room to monitor progress.

Advanced High Provide a sign and have students write an opinion or explanation of the pricing strategy. Have volunteers share their observations with their group.

Example 7 Write and Interpret a Linear Equation in Point-Slope Form

Teaching the Mathematical Practices

2 Create Representations In Example 7, guide students to write an equation that models the lean of the Tower of Pisa over time.

Questions for Mathematical Discourse

AL Why is 1178 subtracted from the years 1350 and 1993? Sample answer: The linear equation represents the lean of the tower after 1178. Subtracting the starting date makes $x = 0$ represent the year 1178.

OL Why does a positive slope make sense for this example? Sample answer: The dependent variable is the lean, which is measured as a distance, and the lean increases over time.

BL Is the y-intercept positive, negative, or zero? What does this mean? The y-intercept is positive, which means the tower was leaning in the year it was constructed.

Common Error

When a word problem contains years, many students do not see that is may be easier to define the variable as the number of years since a given year. By defining the variable in this way, calculations will be simpler.

Exit Ticket

Recommended Use

At the end of class, go online to display the Exit Ticket prompt and ask students to respond using a separate piece of paper. Have students hand you their responses as they leave the room.

Alternate Use

At the end of class, go online to display the Exit Ticket prompt and ask students to respond verbally or by using a mini-whiteboard. Have students hold up their whiteboards so that you can see all student responses. Tap to reveal the answer when most or all students have completed the Exit Ticket.

Practice and Homework

Suggested Assignments

Use the table below to select appropriate exercises.

DOK	Topic	Exercises
1, 2	exercises that mirror the examples	1–26
2	exercises that use a variety of skills from this lesson	27–31
2	exercises that extend concepts learned in this lesson to new contexts	32–36
3	exercises that emphasize higher-order and critical thinking skills	37–43

ASSESS AND DIFFERENTIATE

Use the data from the **Checks** to determine whether to provide resources for extension, remediation, or intervention.

BL

IF students score 90% or more on the Checks,
THEN assign:

- Practice Exercises 1–35 odd, 37–43
- Extension: Two-Intercept Form of a Linear Equation
- **ALEKS**® Applications of Linear Equations with Two Variables

OL

IF students score 66%–89% on the Checks,
THEN assign:

- Practice Exercises 1–43 odd
- Remediation, Review Resources: Writing Equations in Slope-Intercept Form
- Personal Tutors
- Extra Examples 1–7
- **ALEKS**® Equations of Lines

AL

IF students score 65% or less on the Checks,
THEN assign:

- Practice Exercises 1–25 odd
- Remediation, Review Resources: Writing Equations in Slope-Intercept Form
- *Quick Review Math Handbook*: Writing Linear Equations
- **ALEKS**® Equations of Lines

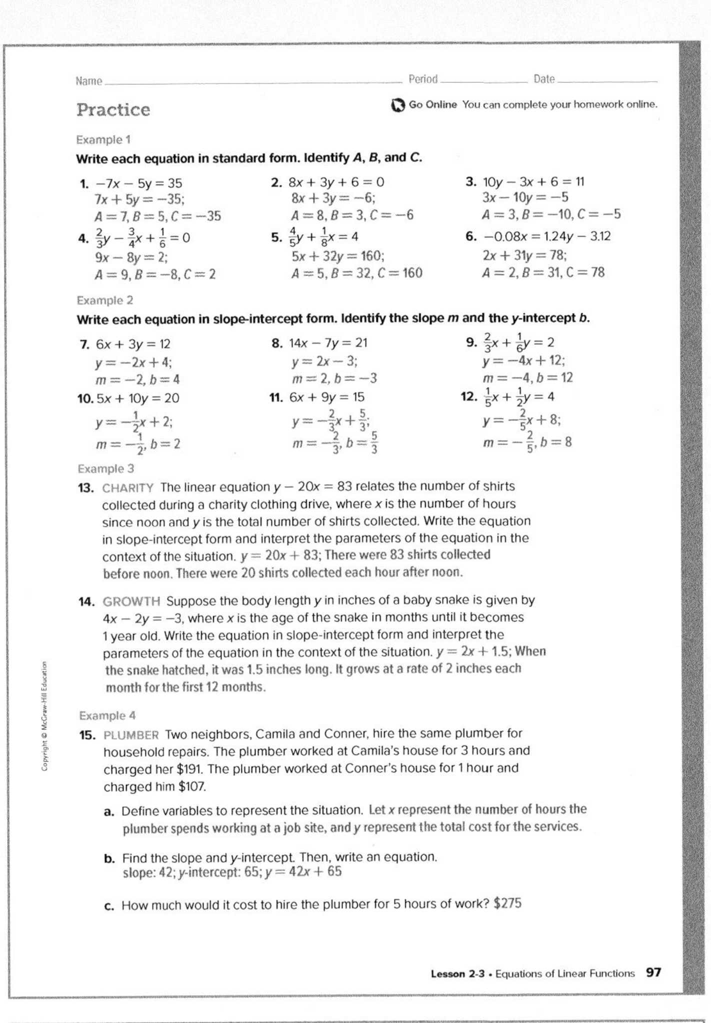

Name ________ Period ________ Date ________

Practice

Go Online You can complete your homework online.

Example 1

Write each equation in standard form. Identify A, B, and C.

1. $-7x - 5y = 35$
   $7x + 5y = -35$;
   $A = 7, B = 5, C = -35$
2. $8x + 3y + 6 = 0$
   $8x + 3y = -6$;
   $A = 8, B = 3, C = -6$
3. $10y - 3x + 6 = 11$
   $3x - 10y = -5$
   $A = 3, B = -10, C = -5$
4. $\frac{2}{3}y - \frac{3}{4}x + \frac{1}{6} = 0$
   $9x - 8y = 2$;
   $A = 9, B = -8, C = 2$
5. $\frac{4}{5}y + \frac{1}{8}x = 4$
   $5x + 32y = 160$;
   $A = 5, B = 32, C = 160$
6. $-0.08x = 1.24y - 3.12$
   $2x + 31y = 78$;
   $A = 2, B = 31, C = 78$

Example 2

Write each equation in slope-intercept form. Identify the slope m and the y-intercept b.

7. $6x + 3y = 12$
   $y = -2x + 4$;
   $m = -2, b = 4$
8. $14x - 7y = 21$
   $y = 2x - 3$;
   $m = 2, b = -3$
9. $\frac{2}{3}x + \frac{1}{6}y = 2$
   $y = -4x + 12$;
   $m = -4, b = 12$
10. $5x + 10y = 20$
    $y = -\frac{1}{2}x + 2$;
    $m = -\frac{1}{2}, b = 2$
11. $6x + 9y = 15$
    $y = -\frac{2}{3}x + \frac{5}{3}$;
    $m = -\frac{2}{3}, b = \frac{5}{3}$
12. $\frac{1}{5}x + \frac{1}{2}y = 4$
    $y = -\frac{2}{5}x + 8$;
    $m = -\frac{2}{5}, b = 8$

Example 3

13. CHARITY The linear equation $y - 20x = 83$ relates the number of shirts collected during a charity clothing drive, where x is the number of hours since noon and y is the total number of shirts collected. Write the equation in slope-intercept form and interpret the parameters of the equation in the context of the situation. $y = 20x + 83$; There were 83 shirts collected before noon. There were 20 shirts collected each hour after noon.

14. GROWTH Suppose the body length y in inches of a baby snake is given by $4x - 2y = -3$, where x is the age of the snake in months until it becomes 1 year old. Write the equation in slope-intercept form and interpret the parameters of the equation in the context of the situation. $y = 2x + 1.5$; When the snake hatched, it was 1.5 inches long. It grows at a rate of 2 inches each month for the first 12 months.

Example 4

15. PLUMBER Two neighbors, Camila and Conner, hire the same plumber for household repairs. The plumber worked at Camila's house for 3 hours and charged her \$191. The plumber worked at Conner's house for 1 hour and charged him \$107.
    a. Define variables to represent the situation. Let x represent the number of hours the plumber spends working at a job site, and y represent the total cost for the services.
    b. Find the slope and y-intercept. Then, write an equation. slope: 42; y-intercept: 65; $y = 42x + 65$
    c. How much would it cost to hire the plumber for 5 hours of work? \$275

16. HIKING Tim began a mountain hike near Big Bear Lake, California at 9:00 A.M. By 10:30 A.M., his elevation is 7200 feet above sea level. At 11:15 A.M., he is at an elevation of 7425 feet above sea level.
    a. Define variables to represent the situation. Let y be his altitude above sea level in feet and let x be the number of hours Tim has been hiking.
    b. Write an equation in slope-intercept form that represent Tim's elevation since he began hiking. $y = 300x + 6750$
    c. If Tim's altitude continues to increase at the same rate, estimate his altitude at 12:30 P.M. 7800 feet above sea level

Example 5

Write an equation in point-slope form for the line that satisfies each set of conditions.

17. slope of -5, passes through $(-3, -8)$
    $y + 8 = -5(x + 3)$
18. slope of $\frac{4}{5}$, passes through $(10, -3)$
    $y + 3 = \frac{4}{5}(x - 10)$
19. slope of $-\frac{2}{3}$, passes through $(6, -8)$
    $y + 8 = -\frac{2}{3}(x - 6)$
20. slope of 0, passes through $(0, -10)$
    $y + 10 = 0(x - 0)$

Example 6

Write an equation in point-slope form for a line that passes through each set of points.

21. $(2, -3)$ and $(1, 5)$
    $y + 3 = -8(x - 2)$
    or $y - 5 = -8(x - 1)$
22. $(3, 5)$ and $(-6, -4)$
    $y - 5 = 1(x - 3)$
    or $y + 4 = 1(x + 6)$
23. $(-1, -2)$ and $(-3, 1)$
    $y + 2 = -\frac{3}{2}(x + 1)$
    or $y - 1 = -\frac{3}{2}(x + 3)$
24. $(-2, -4)$ and $(1, 8)$
    $y + 4 = 4(x + 2)$
    or $y - 8 = 4(x - 1)$

Example 7

Solve each problem.

25. SALES Light truck is a vehicle classification for trucks weighing up to 8500 pounds. In 2011, 5.919 million light trucks were sold in the U.S. In 2017, 11.055 million light trucks were sold. Write an equation in point-slope form that represents the number of light trucks y sold x years after 2010.
    $y - 5.919 = 0.856(x - 1)$ or $y - 11.055 = 0.856(x - 7)$

26. RESTAURANTS In 2012, a popular pizza franchise had 2483 restaurants. In 2017, there were 2606 franchised restaurants. Write an equation in point-slope form that represents the number of restaurants y that are franchised x years after 2010.
    $y - 2483 = 24.6(x - 2)$ or $y - 2606 = 24.6(x - 7)$

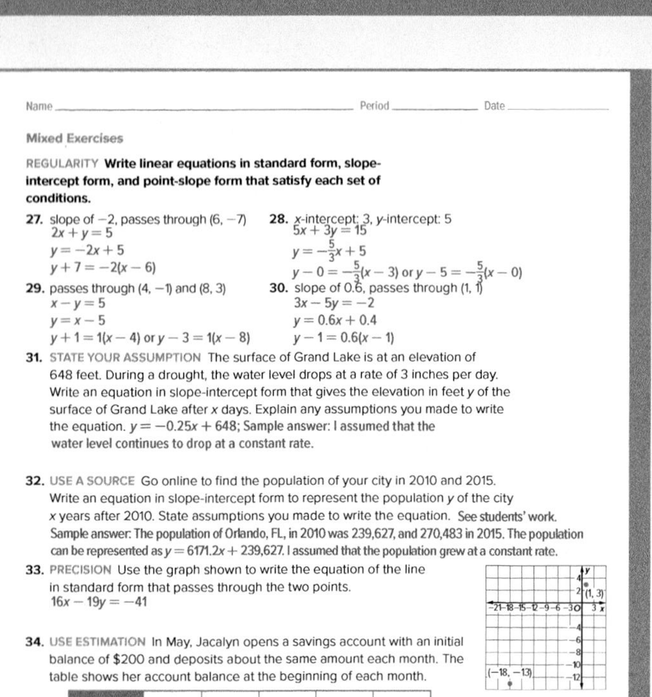

Name ______ Period ______ Date ______

Mixed Exercises

REGULARITY **Write linear equations in standard form, slope-intercept form, and point-slope form that satisfy each set of conditions.**

27. slope of -2, passes through $(6, -7)$
$2x + y = 5$
$y = -2x + 5$
$y + 7 = -2(x - 6)$

28. x-intercept: 3, y-intercept: 5
$5x + 3y = 15$
$y = -\frac{5}{3}x + 5$
$y - 0 = -\frac{5}{3}(x - 3)$ or $y - 5 = -\frac{5}{3}(x - 0)$

29. passes through $(4, -1)$ and $(8, 3)$
$x - y = 5$
$y = x - 5$
$y + 1 = 1(x - 4)$ or $y - 3 = 1(x - 8)$

30. slope of 0.6, passes through (1, 1)
$3x - 5y = -2$
$y = 0.6x + 0.4$
$y - 1 = 0.6(x - 1)$

31. STATE YOUR ASSUMPTION The surface of Grand Lake is at an elevation of 648 feet. During a drought, the water level drops at a rate of 3 inches per day. Write an equation in slope-intercept form that gives the elevation in feet y of the surface of Grand Lake after x days. Explain any assumptions you made to write the equation. $y = -0.25x + 648$; Sample answer: I assumed that the water level continues to drop at a constant rate.

32. USE A SOURCE Go online to find the population of your city in 2010 and 2015. Write an equation in slope-intercept form to represent the population y of the city x years after 2010. State assumptions you made to write the equation. See students' work. Sample answer: The population of Orlando, FL, in 2010 was 239,627, and 270,483 in 2015. The population can be represented as $y = 6171.2x + 239{,}627$. I assumed that the population grew at a constant rate.

33. PRECISION Use the graph shown to write the equation of the line in standard form that passes through the two points.
$16x - 19y = -41$

34. USE ESTIMATION In May, Jacalyn opens a savings account with an initial balance of \$200 and deposits about the same amount each month. The table shows her account balance at the beginning of each month.

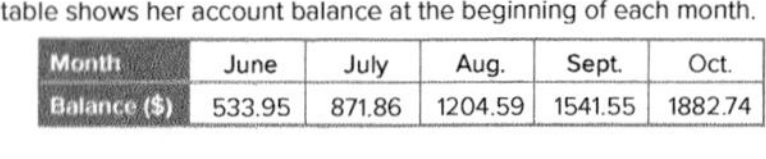

Month	June	July	Aug.	Sept.	Oct.
Balance (\$)	533.95	871.86	1204.59	1541.55	1882.74

a. Write a linear equation in slope-intercept form that relates the balance of her savings account y in months x since she opened her account. Sample answer: $y = 337.20x + 200$

b. Estimate the balance of Jacalyn's account at the beginning of December. Sample answer: \$2560.40

35. CONSTRUCT ARGUMENTS Consider the line that passes through (3, 1) and (0, 7).

a. Given the two points, find a third point on the line. Justify your argument. Sample answer: $(8, -9)$; I used the given two points to write an equation. Then, I substituted 8 for x and solved for y.

b. Explain why both $y - 1 = -2(x - 3)$ and $y - 7 = -2(x - 0)$ can be used to represent the line. Sample answer: The equations are equivalent when simplified.

Lesson 2-3 • Equations of Linear Functions 99

36. USE A MODEL Joe and Alisha are reading novels for book reports. Joe records the number of pages he has remaining to read after each day in the table below. Alisha records the number of pages she has remaining each day on the graph at the right.

Joe

Days	0	1	2	3	4	5
Pages Remaining	585	520	455	390	325	260

a. Describe the function that models the number of pages remaining for each student.
Sample answer: Both functions are linear and decreasing.

b. What is the y-intercept for each function? Interpret its meaning in the context of the problem. b–e. See margin.

c. Write a linear equation in slope-intercept form for the function that can be used to model the pages remaining for each student. Then write each equation in standard form.

d. After how many days will each finish with their reading? What feature of the function represents this event? Explain your answer.

e. Who is the faster reader and by how many pages per day? Support your answer.

Higher-Order Thinking Skills

37. CONSTRUCT ARGUMENTS Determine whether an equation in the form $x = a$, where a is a constant is *sometimes*, *always*, or *never* a function. Explain your reasoning. Never; sample answer: The graph of $x = a$ is a vertical line.

38. PERSEVERE Write an equation in point-slope form of a line that passes through $(a, 0)$ and $(0, b)$. $y - b = -\frac{b}{a}(x - 0)$ or $y - 0 = -\frac{b}{a}(x - a)$

39. CREATE Write an equation in point-slope form of a line with an x-intercept of 3. Sample answer: $y - 0 = 2(x - 3)$

40. WRITE Consider the relationship between hours worked and earnings. When would this situation represent a linear relationship? When would this situation represent a nonlinear relationship? Explain your reasoning.

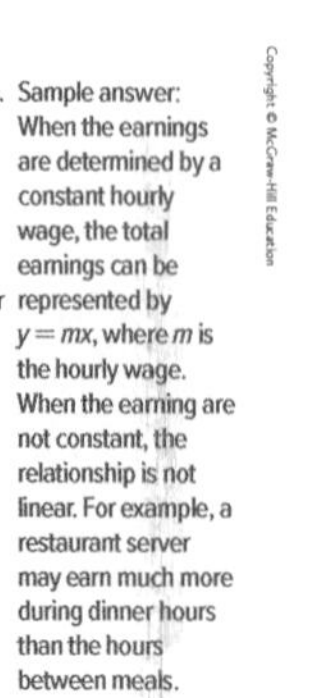

40. Sample answer: When the earnings are determined by a constant hourly wage, the total earnings can be represented by $y = mx$, where m is the hourly wage. When the earning are not constant, the relationship is not linear. For example, a restaurant server may earn much more during dinner hours than the hours between meals.

41. FIND THE ERROR Dan claims that since $y = x + 1$ and $y = 3x + 2$ are both linear functions, the function $y = (x + 1)(3x + 2)$ must also be linear. Is he correct? Explain your reasoning. See margin.

42. PERSEVERE Write $y = ax + b$ in point-slope form.
$y - 0 = a\left(x + \frac{b}{a}\right)$ or $y - b = a(x - 0)$

43. WRITE Why is it important to be able to represent linear equations in more than one form? See margin.

100 Module 2 • Linear Equations, Inequalities, and Systems

Answers

36b. Joe: y-int $= 585$, Alisha: y-int $= 450$. Sample answer: The book Joe is reading has 585 pages and the book Alisha is reading has 450 pages.

36c. Joe: $y = -65x + 585 \rightarrow 65x + y = 585$, Alisha: $y = -70x + 450 \rightarrow 70x + y = 450$

36d. Joe will finish in 9 days. Alisha will finish in 7 days. Sample answer: The books will be finished when the pages remaining is 0, which is the x-intercept for each function.

36e. Alisha, by 5 pages; Sample answer: The equation that models Joe's reading has a slope of -65 and the equation that models Alisha's reading has a slope of -70.

41. No; sample answer: You can choose points on the graph and show on a coordinate plane that they do not fall on a single line. For instance, the points (0, 2), (1, 10), (2, 24), and (3, 44) do not lie on a straight line.

43. Sample answer: Depending on what information is given and what the problem is, it might be easier to represent a linear equation in one form over another. For example, if you are given the slope and the y-intercept, you could represent the equation in slope-intercept form. If you are given a point and the slope, you could represent the equation in point-slope form. If you are trying to graph an equation using the x- and y-intercepts, you could represent the equation in standard form.

Lesson 2-4

A.CED.3, A.REI.11

Solving Systems of Equations Graphically

LESSON GOAL

Students solve systems of equations by graphing.

1 LAUNCH

Launch the lesson with a **Warm Up** and an introduction.

2 EXPLORE AND DEVELOP

Explore: Solutions of Systems of Equations

Develop:

Solving Systems of Equations in Two Variables by Graphing

- Classify Systems of Equations
- Solve a System of Equations by Graphing
- Solve a System of Equations
- Write and Solve a System of Equations by Graphing

Solving Systems of Equations in Two Variables by Using Technology

- Solve a System by Using Technology
- Solve a Linear Equation by Using a System

You may want your students to complete the **Checks** online.

3 REFLECT AND PRACTICE

Exit Ticket

Practice

DIFFERENTIATE

View reports of student progress on the **Checks** after each example.

Resources	AL	OL	BL	ELL
Remediation: Write Equations in Standard and Point-Slope Forms	●	●		●
Extension: Systems of Equations and Absolute Value		●	●	●

Language Development Handbook

Assign page 11 of the *Language Development Handbook* to help your students build mathematical language related to solving systems of equations by graphing.

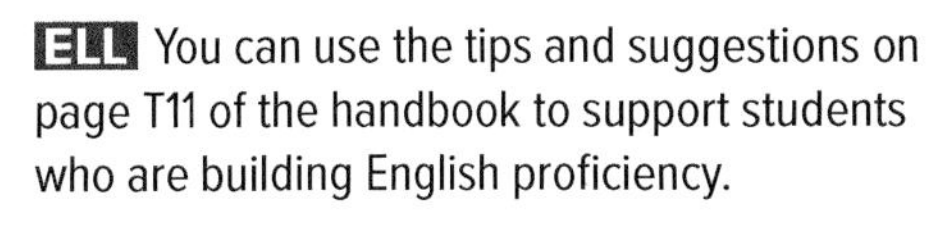

ELL You can use the tips and suggestions on page T11 of the handbook to support students who are building English proficiency.

Suggested Pacing

90 min	1 day
45 min	2 days

Focus

Domain: Algebra

Standards for Mathematical Content:

A.CED.3 Represent constraints by equations or inequalities, and by systems of equations and/or inequalities, and interpret solutions as viable or nonviable options in a modeling context..

A.REI.11 Explain why the x-coordinates of the points where the graphs of the equations $y = f(x)$ and $y = g(x)$ intersect are the solutions of the equation $f(x) = g(x)$; find the solutions approximately, e.g. using technology to graph the functions, make tables of values, or find successive approximations. Include cases where $f(x)$ and $g(x)$ are linear, polynomial, rational, absolute value, exponential, and logarithmic functions.

Standards for Mathematical Practice:

1 Make sense of problems and persevere in solving them.

5 Use appropriate tools strategically.

Coherence

Vertical Alignment

Previous

Students were introduced to solving systems of linear equations by graphing.
8.EE.8, A.REI.5, A.REI.6, A.REI.7, A.REI.12 (Algebra 1)

Now

Students solve systems of equations by graphing.
A.CED.3, A.REI.11

Next

Students will solve systems of equations algebraically.
A.CED.3

Rigor

The Three Pillars of Rigor

1 CONCEPTUAL UNDERSTANDING	2 FLUENCY	3 APPLICATION

Conceptual Bridge In this lesson, students expand their understanding of graphing linear equations to build fluency with graphing systems of linear equations. They apply their understanding of solving systems of linear equations by solving real-world problems.

Mathematical Background

Systems of linear equations can be solved by graphing. Graph each line and look for the point of intersection of the two lines. The point of intersection will be the solution to the system.

Interactive Presentation

Warm Up

Write the equation of a line in point-slope, slope-intercept, and standard forms that passes through the given point and has the given slope.

1. $(0, 3), m = 6$
2. $(3, -3), m = 8$
3. $\left(1, \frac{1}{2}\right), m = -2$

Show Answers

Warm Up

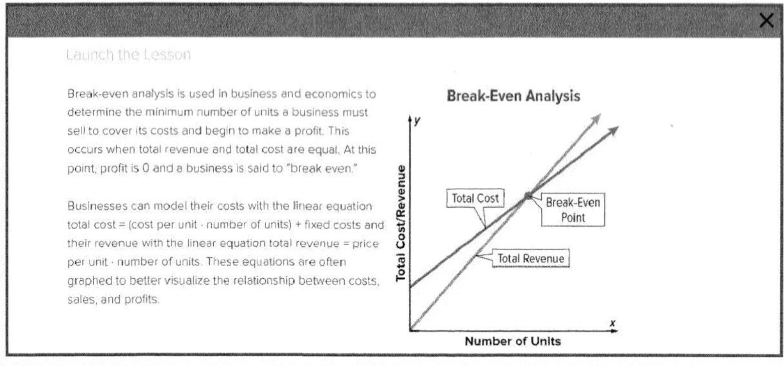

Launch the Lesson

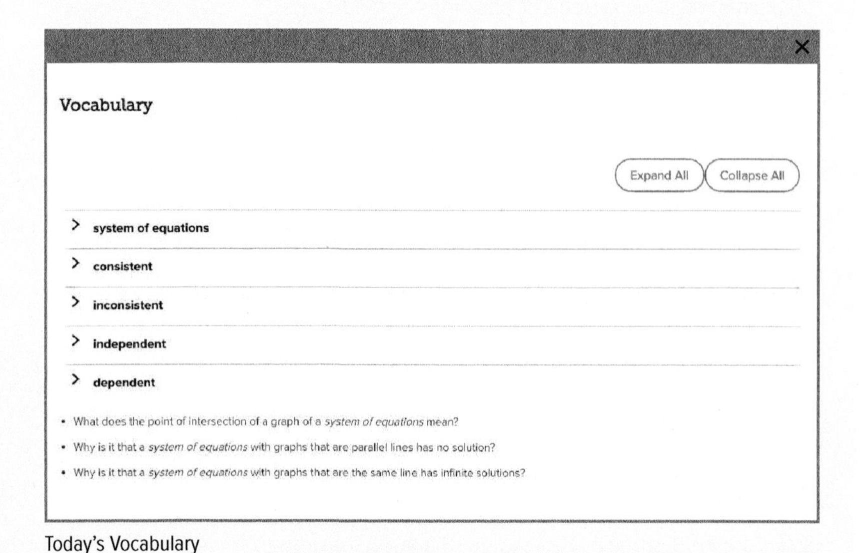

Today's Vocabulary

Warm Up

Prerequisite Skills

The Warm Up exercises address the following prerequisite skill for this lesson:

- writing equations of lines

Answers:

1. $y - 3 = 6x$, $y = 6x + 3$, $6x - y = -3$
2. $y + 3 = 8(x - 3)$, $y = 8x - 27$, $8x - y = 27$
3. $y - \frac{1}{2} = -2(x - 1)$, $y = -2x + 2\frac{1}{2}$, $4x + 2y = 5$

Launch the Lesson

Teaching the Mathematical Practices

4 Apply Mathematics In the Launch the Lesson, students will learn how graphing systems of equations can be used to solve real-world problems in business and economics.

Go Online to find additional teaching notes and questions to promote classroom discourse.

Today's Standards

Tell students that they will be addressing these content and practice standards in this lesson. You may wish to have a student volunteer read aloud *How can I meet these standards?* and *How can I use these practices?*, and connect these to the standards.

See the Interactive Presentation for I Can statements that align with the standards covered in this lesson.

Today's Vocabulary

Tell students that they will be using these vocabulary terms in this lesson. You can expand each row if you wish to share the definitions. Then discuss the questions below with the class.

Explore Solutions of Systems of Equations

Objective

Students use a sketch to explore how the solution of a system of equations is represented on a graph.

Teaching the Mathematical Practices

5 Use Mathematical Tools Point out that to complete the exercises in the Explore, students will need to use a sketch. Work with students to explore how systems of equations can be solved by graphing.

Ideas for Use

Recommended Use Present the Inquiry Question, or have a student volunteer read it aloud. Have students work in pairs to complete the Explore activity on their devices. Pairs should discuss each of the questions. Monitor student progress during the activity. Upon completion of the Explore activity, have student volunteers share their responses to the Inquiry Question.

What if my students don't have devices? You may choose to project the activity on a whiteboard. A printable worksheet for each Explore is available online. You may choose to print the worksheet so that individuals or pairs of students can use it to record their observations.

Summary of the Activity

Students will complete guiding exercises throughout the Explore activity. They will use a sketch to graph a system of equations, and then complete a table of values for each linear function. They will answer a series of questions leading to the solution of the system of equations. Then, students will answer the Inquiry Question.

(continued on the next page)

Interactive Presentation

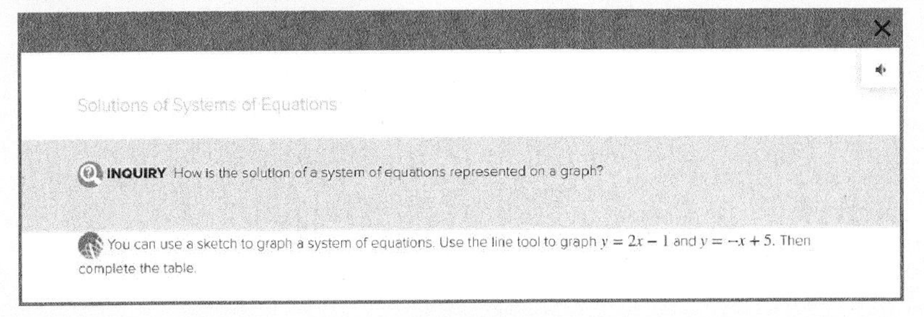

Explore

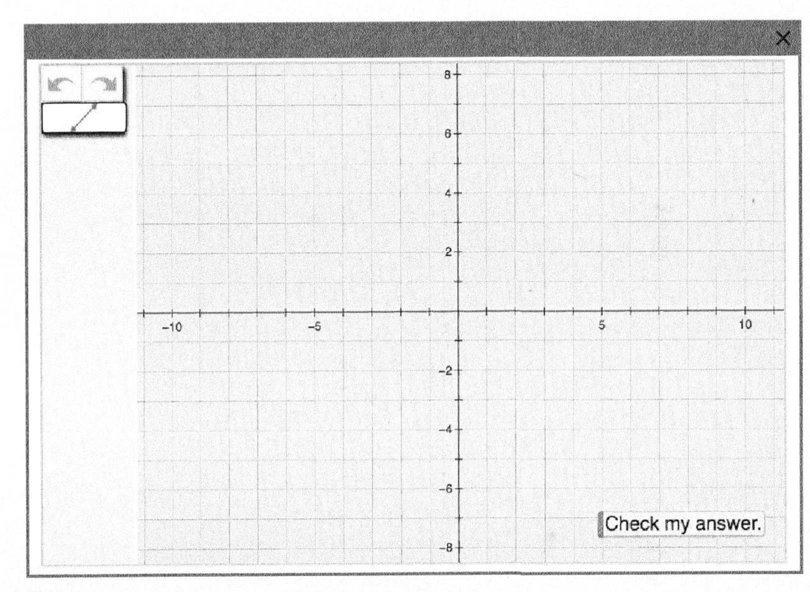

Explore

WEB SKETCHPAD

Students will use a sketch to graph a system of equations.

TYPE

Students complete a table of values for two linear functions.

TYPE

Students answer a series of questions leading to the solution of the system.

Interactive Presentation

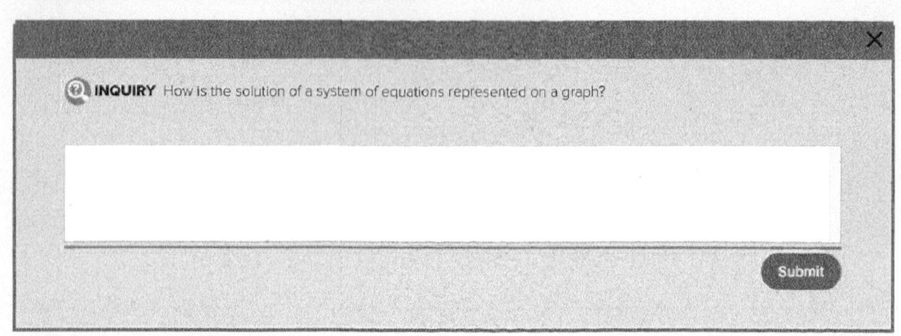

Explore

TYPE

Students respond to the Inquiry Question and can view a sample answer.

Explore Solutions of Systems of Equations (*continued*)

Questions

Have students complete the Explore activity.

Ask:

- If the slope of each linear equation was 2, would the graph of the two lines intersect? Explain your reasoning. Sample answer: The lines would not intersect because they would be parallel. Parallel lines do not intersect.
- If a system of two linear equations intersects, is it possible for the system to have more than one solution? Explain. No; sample answer: Lines can only cross each other once if they intersect.

Inquiry

How is the solution of a system of equations represented on a graph? Sample answer: The solution is the point where the graphs of the two lines intersect.

Go Online to find additional teaching notes and sample answers for the guiding exercises.

1 CONCEPTUAL UNDERSTANDING | 2 FLUENCY | 3 APPLICATION

Learn Solving Systems of Equations in Two Variables by Graphing

Objective

Students solve systems of linear equations by identifying the intersection of their graphs.

Teaching the Mathematical Practices

1 Explain Correspondences Encourage students to explain the relationship between the graph and solution(s) of a system of equations.

Essential Question Follow-Up

Students have begun learning about solving a system of equations by graphing.

Ask:

Why are systems of equations useful when solving real-world situations? Sample answer: Systems of equations are two or more equations with the same variable, so a business may need to write more than one equation to describe their cost, profit, or revenue. Systems of equations allow multiple constraints to be considered at one time.

DIFFERENTIATE

Reteaching Activity AL ELL

IF students are having difficulties classifying systems of equations, **THEN** have them write the following chart in their notes to help them identify the type of system.

The slopes are:	AND the y-intercepts are:	THEN the system is:
different	the same or different	consistent and independent
the same	different	inconsistent
the same	the same	consistent and dependent

Go Online

- Find additional teaching notes.
- View performance reports of the Checks.
- Assign or present an Extra Example.

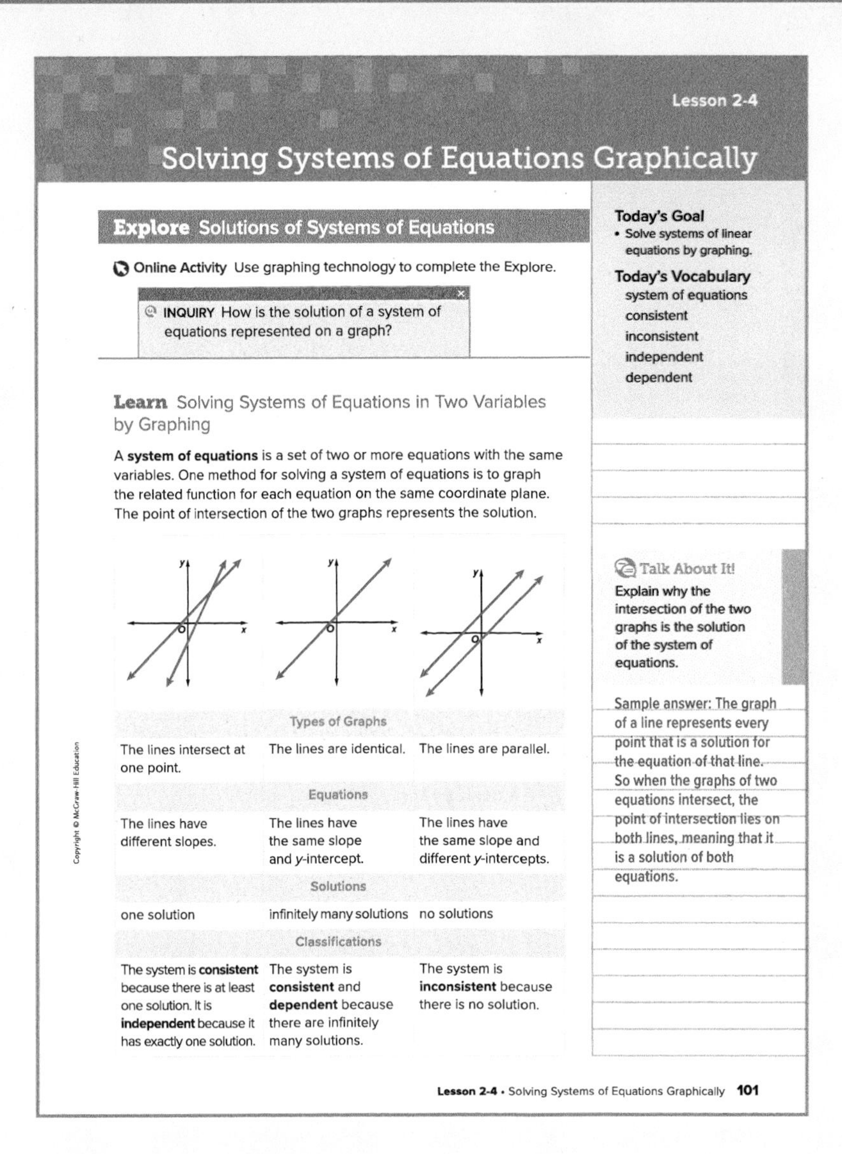

Lesson 2-4

Solving Systems of Equations Graphically

Explore Solutions of Systems of Equations

Online Activity Use graphing technology to complete the Explore.

INQUIRY How is the solution of a system of equations represented on a graph?

Learn Solving Systems of Equations in Two Variables by Graphing

A **system of equations** is a set of two or more equations with the same variables. One method for solving a system of equations is to graph the related function for each equation on the same coordinate plane. The point of intersection of the two graphs represents the solution.

Types of Graphs		
The lines intersect at one point.	The lines are identical.	The lines are parallel.
Equations		
The lines have different slopes.	The lines have the same slope and y-intercept.	The lines have the same slope and different y-intercepts.
Solutions		
one solution	infinitely many solutions	no solutions
Classifications		
The system is **consistent** because there is at least one solution. It is **independent** because it has exactly one solution.	The system is **consistent** and **dependent** because there are infinitely many solutions.	The system is **inconsistent** because there is no solution.

Today's Goal
- Solve systems of linear equations by graphing.

Today's Vocabulary
system of equations
consistent
inconsistent
independent
dependent

Talk About It!
Explain why the intersection of the two graphs is the solution of the system of equations.

Sample answer: The graph of a line represents every point that is a solution for the equation of that line. So when the graphs of two equations intersect, the point of intersection lies on both lines, meaning that it is a solution of both equations.

Copyright © McGraw-Hill Education

Lesson 2-4 • Solving Systems of Equations Graphically 101

Interactive Presentation

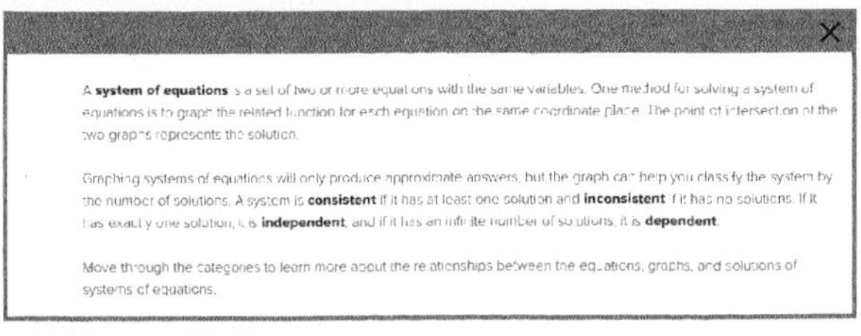

Learn

TAP

Students tap to see the type of graph, solution, and classification for different examples.

TYPE

Students answer a question to determine whether they understand that the intersection of a graphed system of equations is the solution.

Your Notes

Study Tip

Number of Solutions By first determining the number of solutions a system has, you can make decisions about whether further steps need to be taken to solve the system. If a system has one solution, you can graph to find it. If a system has infinitely many solutions or no solution, no further steps are necessary. However, you can graph the system to confirm.

Example 1 Classify Systems of Equations

Determine the number of solutions the system has. Then state whether the system of equations is *consistent* or *inconsistent* and whether it is *independent* or *dependent*.

$2y = 6x - 14$
$3x - y = 7$

Solve each equation for y.

$2y = 6x - 14 \rightarrow y = 3x - 7$

$3x - y = 7 \rightarrow y =$ <u>$3x - 7$</u>

The equations have the same slope and y-intercept. Thus, both equations represent the same line and the system has <u>infinitely many solutions</u>. The system is <u>consistent</u> and <u>dependent</u>.

Check

Determine the number of solutions and classify the system of equations. <u>one solution; consistent and independent</u>

$3x - 2y = -7$

$4y = 9 - 6x$

Example 2 Solve a System of Equations by Graphing

Solve the system of equations.

$5x - y = 3$
$-x + y = 5$

Solve each equation for y. They have different slopes, so there is one solution. Graph the system.

$5x - y = 3 \rightarrow$ <u>$y = 5x - 3$</u>

$-x + y = 5 \rightarrow$ <u>$y = x + 5$</u>

The lines appear to intersect at one point, (<u>2</u>, <u>7</u>).

CHECK Substitute the coordinates into each original equation.

$5x - y = 3$	Original equation	$-x + y = 5$
$5(2) - 7 \stackrel{?}{=} 3$	$x = 2$ and $y = 7$	$-(2) + 7 \stackrel{?}{=} 5$
$3 = 3$	True	$5 = 5$

The solution is (2, 7).

Check

Solve the system of equations by graphing.

$2y + 14x = -6$
$8x - 4y = -24$

The solution is (<u>−1, 4</u>).

102 **Module 2** • Linear Equations, Inequalities, and Systems

Interactive Presentation

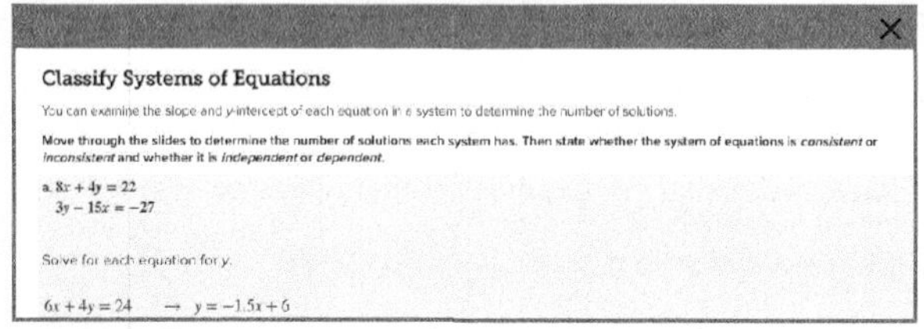

Example 1

SELECT

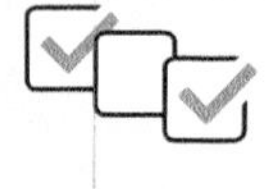

Students select the correct classification for each system of equations.

Example 1 Classify Systems of Equations

Teaching the Mathematical Practices

6 Communicate Precisely Encourage students to routinely write and explain their solution methods. Point out that they should use clear definitions when classifying the system of equation in Example 1.

Questions for Mathematical Discourse

AL If a system of two linear equations is classified as consistent and dependent, what does the graph of the lines look like? They would be the same line.

OL If a system of two linear equations is classified as inconsistent, what does the graph of the lines look like? The graph would be two parallel lines with different y-intercepts.

BL Can you decide what type of system two linear equations are by comparing the y-intercepts? Why or why not? No; sample answer: two lines with the same y-intercept could be the same line or intersecting lines. Two lines with different y-intercepts could be parallel lines or intersecting lines.

Common Error

When two lines have the same slope, students may automatically believe the lines are parallel, without comparing the y-intercepts. Reinforce to students that they must check the slope and y-intercept of both lines before deciding what type of lines are given.

Example 2 Solve a System of Equations by Graphing

Teaching the Mathematical Practices

1 Check Answers Mathematically proficient students continually ask themselves, "Does this make sense?" Point out that in Example 2, students need to check their answer using a different method.

Questions for Mathematical Discourse

AL What will be the solution of a system of equations when the lines intersect on a graph? the coordinates of the intersection

OL If a system of two linear equations has different slopes but the same y-intercept, what would be the appropriate classification? consistent and independent

BL How could you change the second equation to make the system of equations inconsistent? Make the slope 5.

Common Error

When identifying the coordinates of the solution, students may switch the x- and y-values in the ordered pair. Remind students that the x-value always goes first while the y-value always goes second in an ordered pair.

Example 3 Solve a System of Equations

Teaching the Mathematical Practices

1 Analyze Givens and Constraints In Example 3, guide students to analyze the given system of equations to make a conjecture about the solution. Then work with them to confirm the solution by graphing.

Questions for Mathematical Discourse

AL How many solutions does a system of parallel lines have? There are no solutions.

OL Why do parallel lines never intersect? Sample answer: Parallel lines have the same slope, but they start at different points. If both lines rise and run at the same rate, they will never intersect.

BL Is there another way you could tell these equations are parallel without solving for y? Sample answer: The coefficients of x and y in the second equation are -3 times the coefficients of the first equation, but this is not true for the constant term.

Example 4 Write and Solve a System of Equations by Graphing

Teaching the Mathematical Practices

4 Use Tools Point out that to solve the problem in Example 4, students will need to identify the important quantities in the situation and use an equation and graph.

Questions for Mathematical Discourse

AL How can you check your solution estimated from the graph? Plug the x-value into the equations and see if you get approximately the same y-value for both.

OL If the cost of purchasing the gasoline car was $30,000, would there ever be a point where the total cost would be the same? No. How is this possible when the slopes are different? The intersection of the lines would be outside of the valid domain and range for the scenario.

BL The average life expectancy of a car is 150,000 miles. How could you determine a purchase price for an electric car that would make the cost equivalency point happen at 150,000 miles? Sample answer: Plug 150,000 into the gasoline equation to find the y-value on the line. We want to find a new equation for the electric car that has the same slope but contains this point. Write the new electric car equation in point-slope form, then plug zero in for x to find the new y-intercept.

Example 3 Solve a System of Equations

Solve the system of equations.

$7x + 2y = 16$

$-21x - 6y = 24$

Solve each equation for y to determine the number of solutions the system has.

$7x + 2y = 16 \rightarrow y = \underline{-3.5}x + \underline{8}$

$-21x - 6y = 24 \rightarrow y = \underline{-3.5}x + \underline{-4}$

The equations have the ___same___ slope and ___different___ y-intercepts. So, these equations represent ___parallel___ lines, and there is ___no solution___.

You can graph each equation on the same grid to confirm that they do not intersect.

Study Tip

Parallel Lines Graphs of lines with the same slope and different intercepts are, by definition, parallel.

Think About It!

What would the graph of a system of linear equations with infinitely many solutions look like? Explain your reasoning.

Sample answer: For a system of linear equations to have infinitely many solutions, both equations must represent the same line. Therefore, the graph of the system would be a single line.

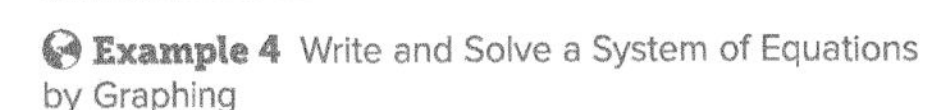

Example 4 Write and Solve a System of Equations by Graphing

CARS **Suppose an electric car costs $29,000 to purchase and $0.036 per mile to drive, and a gasoline-powered car costs $19,000 to purchase and $0.08 per mile to drive. Estimate after how many miles of driving the total cost of each car will be the same.**

Part A Write equations for the total cost of owning each type of car.

Let y = the total cost of owning the car and x = the number of miles driven.

So, the equation is $y = \underline{0.036}x + \underline{29{,}000}$ for the electric car and $y = \underline{0.08}x + \underline{19{,}000}$ for the gasoline car.

Part B Examine the graph to estimate the number of miles you would have to drive before the cost of owning each type of car would be same.

The graphs appear to intersect at approximately (225,000, 37,500). This means that after driving about 225,000 miles, the cost of owning each car will be the same.

Gasoline Car vs. Electric Car

Think About It!

Explain what the two equations represent in the context of the situation.

Sample answer: Each equation represents the cost of owning a car after driving x miles. The equation represents the sum of the initial cost of the car and the cost of driving each mile based on the price of its energy source.

(continued on the next page)

Interactive Presentation

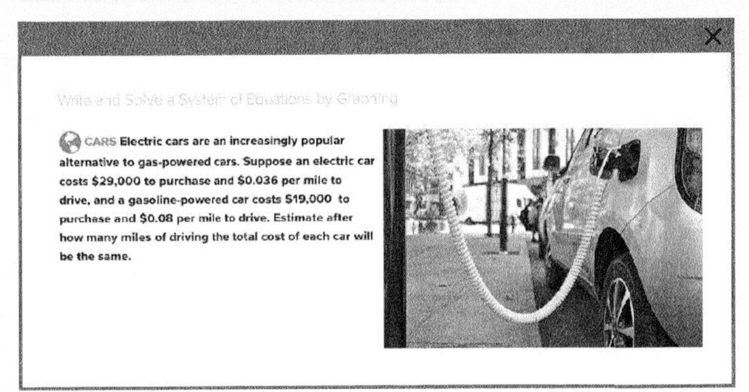

Example 4

TYPE

Students answer a question to determine whether they understand how to write and solve a system of equations by graphing.

CHECK

Students complete the Check online to determine whether they are ready to move on.

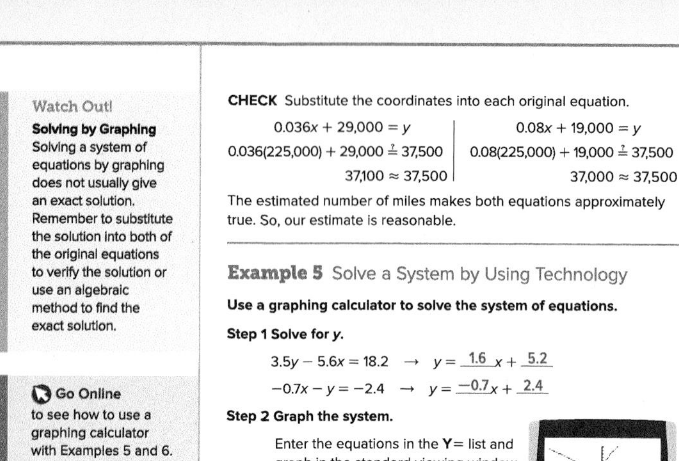

Watch Out!

Solving by Graphing Solving a system of equations by graphing does not usually give an exact solution. Remember to substitute the solution into both of the original equations to verify the solution or use an algebraic method to find the exact solution.

Go Online to see how to use a graphing calculator with Examples 5 and 6.

CHECK Substitute the coordinates into each original equation.

$0.036x + 29{,}000 = y$	$0.08x + 19{,}000 = y$
$0.036(225{,}000) + 29{,}000 \stackrel{?}{=} 37{,}500$	$0.08(225{,}000) + 19{,}000 \stackrel{?}{=} 37{,}500$
$37{,}100 \approx 37{,}500$	$37{,}000 \approx 37{,}500$

The estimated number of miles makes both equations approximately true. So, our estimate is reasonable.

Example 5 Solve a System by Using Technology

Use a graphing calculator to solve the system of equations.

Step 1 Solve for *y*.

$3.5y - 5.6x = 18.2 \rightarrow y = \underline{1.6}\,x + \underline{5.2}$

$-0.7x - y = -2.4 \rightarrow y = \underline{-0.7}\,x + \underline{2.4}$

Step 2 Graph the system.

Enter the equations in the **Y=** list and graph in the standard viewing window.

Step 3 Find the intersection.

Use the **intersect** feature from the **CALC** menu to find the coordinates of the point of intersection. When prompted, select each line. Press enter to see the intersection.

[−10, 10] scl: 1 by [−10, 10] scl: 1

The solution is approximately $(\underline{-1.22, 3.25})$.

Study Tip

Window Dimensions If the point of intersection is not visible in the standard viewing window, zoom out or adjust the window settings manually until it is visible. If the lines appear to be parallel, zoom out to verify that they do not intersect.

Example 6 Solve a Linear Equation by Using a System

Use a graphing calculator to solve $4.5x - 3.9 = 6.5 - 2x$ by using a system of equations.

Step 1 Write a system.

Set each side of $4.5x - 3.9 = 6.5 - 2x$ equal to y to create a system of equations.

$y = \underline{4.5x - 3.9}$

$y = \underline{6.5 - 2x}$

Step 2 Graph the system.

Enter the equations in the **Y=** list and graph in the standard viewing window.

Step 3 Find the intersection.

The solution is the x-coordinate of the intersection, which is $\underline{1.6}$.

[−10, 10] scl: 1 by [−10, 10] scl: 1

Go Online You can complete an Extra Example online.

Interactive Presentation

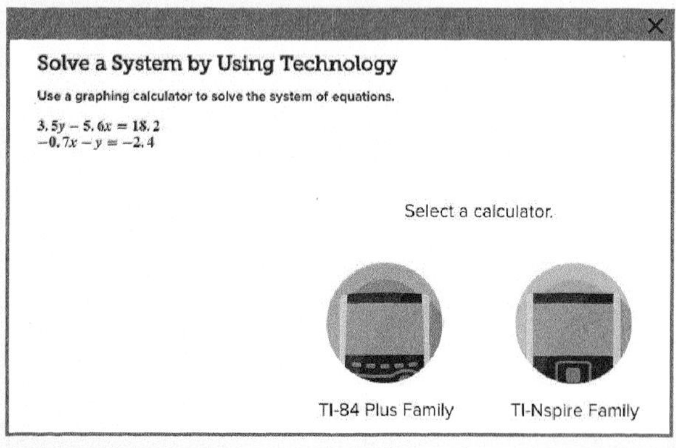

Example 5

SELECT

Students select the calculator they will use to solve the system of equations by graphing and then move through the steps.

CHECK

Students complete the Check online to determine whether they are ready to move on.

Example 5 Solve a System by Using Technology

Teaching the Mathematical Practices

5 Decide When to Use Tools Help students see why using a graphing calculator will help to solve the system of equations. Use the Watch Out! feature to point out the limitations of using the tool.

Questions for Mathematical Discourse

AL Why do the lines have to be written in slope-intercept form in order to graph using the calculator? Sample answer: because the calculator has $y =$ and this indicates slope-intercept form

OL If we check our solution, what can we expect? Sample answer: We can expect the two sides of each equation to be close to the same number, but not exactly the same.

BL How do you think the calculator could increase accuracy when finding an intersection point? Sample answer: The calculator calculates many points in an interval. If you increase the number of points calculated in an interval, the accuracy increases.

Example 6 Solve a Linear Equation by Using a System

Questions for Mathematical Discourse

AL Why is the intersection of the two lines the solution to the original equation? When you set each side of the equation equal to y and graph functions, the intersection occurs at the x-value where the y-value is the same for each function. Thus, the x-value is where both sides of the original equation are equal.

OL If you added 3.9 to each side of the original equation, would the graph change? Would the solution change? Explain. The graph would change. The intersection would occur at a different y-value, but the same x-value. The solution to the original equation is the x-value of the intersection, so the solution is the same.

BL Why might you want to solve an equation by graphing a system rather than solving algebraically? Sample answer: This method may be faster when the calculations are more complicated.

Exit Ticket

Recommended Use

At the end of class, go online to display the Exit Ticket prompt and ask students to respond using a separate piece of paper. Have students hand you their responses as they leave the room.

Alternate Use

At the end of class, go online to display the Exit Ticket prompt and ask students to respond verbally or by using a mini-whiteboard. Have students hold up their whiteboards so that you can see all student responses. Tap to reveal the answer when most or all students have completed the Exit Ticket.

Practice and Homework

Suggested Assignments

Use the table below to select appropriate exercises.

DOK	Topic	Exercises
1, 2	exercises that mirror the examples	1–21
2	exercises that use a variety of skills from this lesson	22–24
2	exercises that extend concepts learned in this lesson to new contexts	25, 26
3	exercises that emphasize higher-order and critical thinking skills	27–30

ASSESS AND DIFFERENTIATE

Use the data from the **Checks** to determine whether to provide resources for extension, remediation, or intervention.

IF students score 90% or more on the Checks, **THEN** assign: BL

- Practice Exercises 1–25 odd, 27–30
- Extension: Systems of Equations and Absolute Value
- **ALEKS**® Systems of Linear Equations in Two Variables

IF students score 66%–89% on the Checks, **THEN** assign: OL

- Practice Exercises 1–29 odd
- Remediation, Review Resources: Writing Equations in Standard and Point-Slope Forms
- Personal Tutors
- Extra Examples 1–6
- **ALEKS**® Equations of Lines

IF students score 65% or less on the Checks, **THEN** assign: AL

- Practice Exercises 1–21 odd
- Remediation, Review Resources: Writing Equations in Standard and Point-Slope Forms
- *Quick Review Math Handbook:* Solving Systems of Equations Graphically
- **ALEKS**® Equations of Lines

Name ______ Period ______ Date ______

Practice

Go Online You can complete your homework online.

Example 1

Determine the number of solutions for each system. Then state whether the system of equations is *consistent* or *inconsistent* and whether it is *independent* or *dependent*.

1. $y = 3x$
   $y = -3x + 2$
   1; consistent and independent
2. $y = x - 5$
   $-2x + 2y = -10$
   infinitely many consistent and dependent
3. $2x - 5y = 10$
   $3x + y = 15$
   1; consistent and independent
4. $3x + y = -2$
   $6x + 2y = 10$
   0; inconsistent
5. $x + 2y = 5$
   $3x - 15 = -6y$
   infinitely many consistent and dependent
6. $3x - y = 2$
   $x + y = 6$
   1; consistent and independent

Examples 2 and 3

Solve the system of equations by graphing. 7–12. See margin for graphs.

7. $x - 2y = 0$
   $y = 2x - 3$ (2, 1)
8. $-4x + 6y = -2$
   $2x - 3y = 1$ infinitely many solutions
9. $2x + y = 3$
   $y = \frac{1}{2}x - \frac{9}{2}$ (3, −3)
10. $y - x = 3$
    $y = 1$ (−2, 1)
11. $2x - 3y = 0$
    $4x - 6y = 3$ no solution
12. $5x - y = 4$
    $-2x + 6y = 4$ (1, 1)

Example 4

Solve each problem.

13. USE ESTIMATION Mr. Lycan is considering buying clay from two art supply companies. Company A sells 50-pound containers of clay for $24, plus $42 to ship the total order. Company B sells the same clay for $28, plus $25 to ship the total order.
    a. Write equations for the total cost of ordering clay from each company. Company A: $y = 24x + 42$; Company B: $y = 28x + 25$
    b. Graph the equations on the same coordinate plane. Examine the graph to estimate how much Mr. Lycan would have to order for the cost of ordering clay from each company to be the same. about 4 containers; see margin for graph.
    c. Check your estimate by substituting into each original equation. How reasonable is your estimation? Justify your reasoning.

13c. Sample answer: I estimated that the cost would be the same when ordering 4 containers. By substituting $x = 4$ in the equations, the cost at Company A is $138 and the cost at Company B is $137. These values are approximately equal, so the estimate is reasonable.

14. USE ESTIMATION Two moving truck companies offer the same vehicle at different rates. At Haul-n-Save, the truck can be rented for $30, plus $0.79 per mile. At Rent It Trucks, the truck can be rented for $75, plus $0.55 per mile.
    a. Write equations for the total cost of renting a truck from each company. Haul-n-Save: $y = 0.79x + 30$; Rent It Trucking: $y = 0.55x + 75$
    b. Graph the equations on the same coordinate plane. Examine the graph to estimate after how many miles of driving the total rental cost will be the same from each company. about 190 miles; see margin for graph.
    c. Check your estimate. How reasonable is your estimation? Justify your reasoning. Sample answer: I estimated the total cost would be the same at 190 miles. By substituting $x = 190$ in the equations, the cost of renting from Haul-n-Save is $180.10 and the cost of renting from Rent It Trucking is $179.50. These values are approximately equal, so the estimate is reasonable.

Example 5

USE TOOLS **Use a graphing calculator to solve each system of equations. Round the coordinates to the nearest hundredth, if necessary.**

15. $12y = 5x - 15$
$4.2y + 6.1x = 11$
$(2.07, -0.39)$

16. $-3.8x + 2.9y = 19$
$6.6x - 5.4y = -23$
$(-26.01, -27.54)$

17. $5.8x - 6.3y = 18$
$-4.3x + 8.8y = 32$
$(15.03, 10.98)$

Example 6

USE TOOLS **Use a graphing calculator to solve each equation by using a system of equations. Round to the nearest hundredth, if necessary.**

18. $-4.7x + 16 = 16.79x - 80.2$
4.48

19. $0.0019x + 3.55 = 0.27x + 2.81$
2.76

20. $471 - 63x = -50.5x + 509$
-3.04

21. $-47.83x - 9 = 33x + 71.019$
-0.99

Mixed Exercises

Solve each system of equations by graphing. 22-24. See Mod. 2 Answer Appendix for graphs.

22. $x - 3y = 6$
$2x - y = -3$ $(-3, -3)$

23. $2x - y = 3$
$x + 2y = 4$ $(2, 1)$

24. $4x + y = -2$
$2x + \frac{y}{2} = -1$ infinitely many

25. LASERS A machinist programs a laser cutting machine to focus two laser beams at the same point. One beam is programmed to follow the path $y = 0.5x - 3.15$ and the other is programmed to follow $10x + 5y = 63$. Graph both equations and find the point at which the lasers are focused.
(6.3, 0); See Mod. 2 Answer Appendix for graph.

26. REASONING A high school band was selling ride tickets for the school fair. On the first day, 250 children's tickets and 150 adult tickets were sold for a total of \$550. On the second day, 180 children's tickets and 120 adult tickets were sold for a total of \$420. What is the price for each child ticket and each adult ticket?

a. Write a system of equations to represent this situation. Let x represent the cost of children's tickets and y represent the cost of adult tickets. $250x + 150y = 550$; $180x + 120y = 420$

b. Graph the system of equations. See Mod. 2 Answer Appendix for graph.

c. Find the intersection of the graphs. What does the point of intersection represent? (1, 2); The x-value represents the cost of children's tickets, \$1. The y-value represents the cost of an adult ticket, \$2.

Higher-Order Thinking Skills

27. ANALYZE For linear functions a, b, and c, if a is consistent and dependent with b, b is inconsistent with c, and c is consistent and independent with d, then a will *sometimes*, *always*, or *never* be consistent and independent with d. Explain your reasoning. See Mod. 2 Answer Appendix.

28. WRITE Explain how to find the solution to a system of linear equations by graphing. See Mod. 2 Answer Appendix.

29. ANALYZE Determine if the following statement is *sometimes*, *always*, or *never* true. Explain your reasoning. See Mod. 2 Answer Appendix.

A system of linear equations in two variables can have exactly two solutions.

30. CREATE Write a system of equations that has no solution.
Sample answer: $y = 2x + 1$ and $y = 2x - 3$

106 Module 2 • Linear Equations, Inequalities, and Systems

Answers

7.
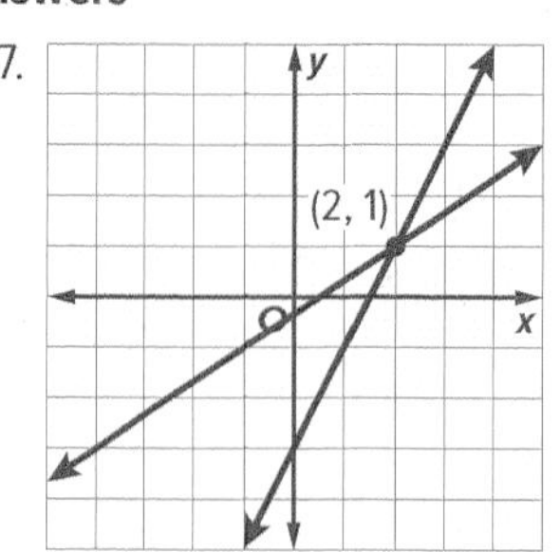

8.
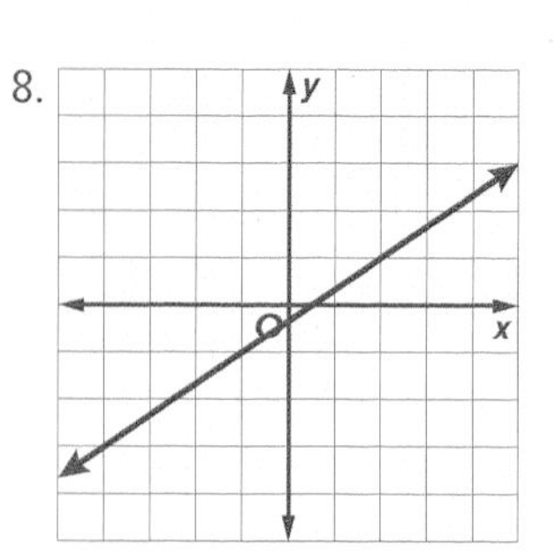

9.
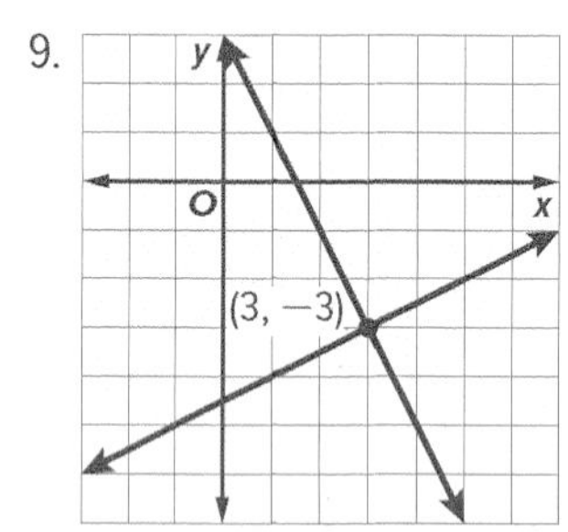

10.
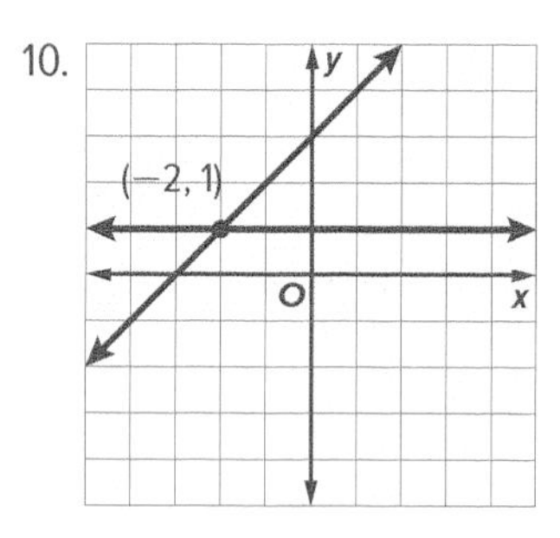

11.
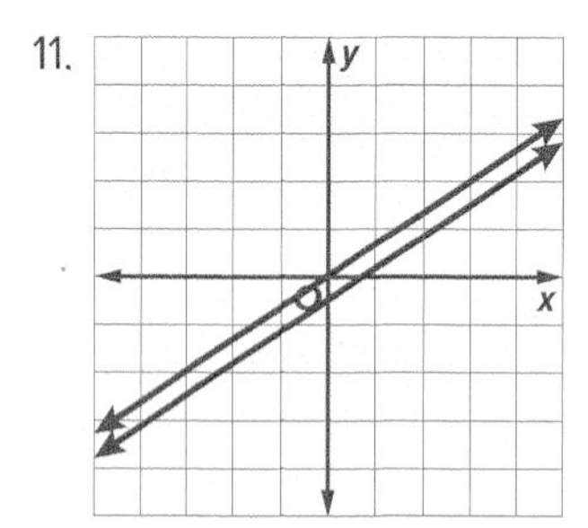

12.
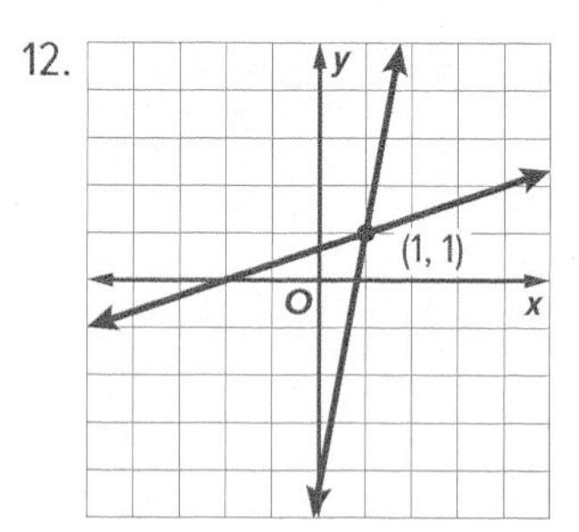

13b.
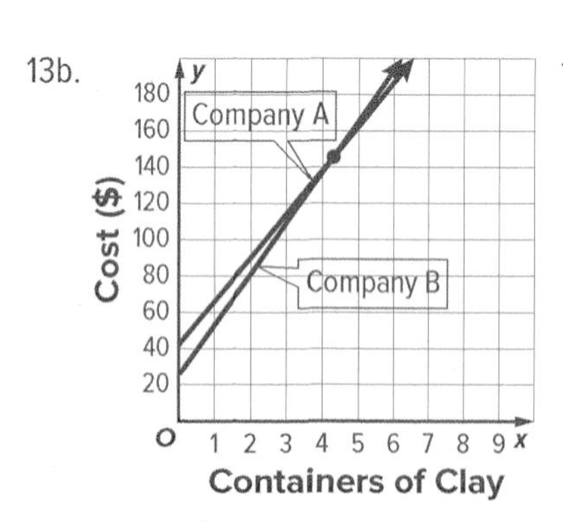

14b.
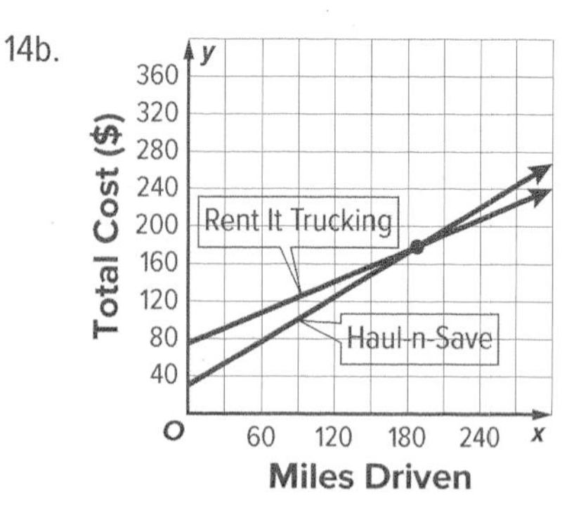

Lesson 2-5

Solving Systems of Equations Algebraically

A.CED.3

LESSON GOAL

Students solve systems of equations by using substitution or elimination.

1 LAUNCH

 Launch the lesson with a **Warm Up** and an introduction.

2 EXPLORE AND DEVELOP

 **Develop:**

Solving Systems of Equations in Two Variables by Substitution

- Substitution When There Is One Solution
- Substitution When There Is Not Exactly One Solution
- Apply the Substitution Method

Solving Systems of Equations in Two Variables by Elimination

- Elimination When There Is One Solution
- Multiply Both Equations Before Using Elimination
- Elimination When There Is Not Exactly One Solution

 You may want your students to complete the **Checks** online.

3 REFLECT AND PRACTICE

 Exit Ticket

 Practice

DIFFERENTIATE

 View reports of student progress on the **Checks** after each example.

Resources	AL	OL	BL	ELL
Remediation: Substitution	●	●		●
Extension: Creative Designs		●	●	●
ELL Support				

Language Development Handbook

Assign page 12 of the *Language Development Handbook* to help your students build mathematical language related to solving systems of equations by using substitution or elimination.

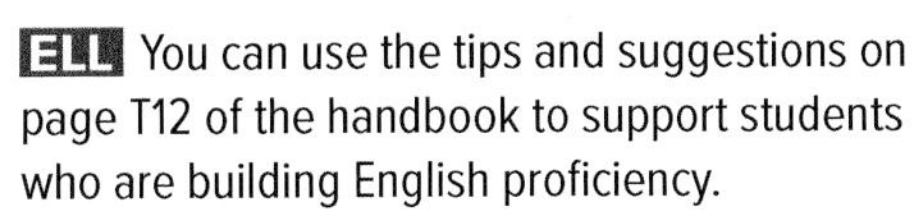
ELL You can use the tips and suggestions on page T12 of the handbook to support students who are building English proficiency.

Suggested Pacing

90 min	1 day
45 min	2 days

Focus

Domain: Algebra

Standards for Mathematical Content:

A.CED.3 Represent constraints by equations or inequalities, and by a system of equations and/or inequalities, and interpret solutions as viable or non-viable options in a modeling context.

Standards for Mathematical Practice:

1 Make sense of problems and persevere in solving them.

6 Attend to precision.

Coherence

Vertical Alignment

Previous

Students were introduced to solving systems of linear equations algebraically. **8.EE.8, A.REI.5, A.REI.6, A.REI.7, A.REI.12 (Algebra 1)**

Now

Students solve systems of equations by using substitution or elimination.
A.CED.3

Next

Students will solve systems of inequalities in two variables.
A.CED.3

Rigor

The Three Pillars of Rigor

1 CONCEPTUAL UNDERSTANDING	2 FLUENCY	3 APPLICATION

Conceptual Bridge In this lesson, students build on their understanding of using algebraic methods to solve systems of linear equations. They build fluency by using substitution and elimination to solve systems of equations, and they apply their understanding by solving real-world problems.

Mathematical Background

If the coefficients for one of the variables in the equations are already the same or opposite, that variable can be easily eliminated by subtracting or adding the given equations. If this is not the case, it may be a better choice to use the method of substitution. Substitution and elimination will both solve a system of equations.

Interactive Presentation

Warm Up

Add or subtract.

1. $\begin{array}{r} 4x + 3y - 5 \\ (+)\ 2x - 5y + 8 \\ \hline \end{array}$

2. $\begin{array}{r} 7x - 4y - 12 \\ (-)\ 5x - 6y + 8 \\ \hline \end{array}$

3. $\begin{array}{r} 9x + 2y - 8 \\ (+)\ 6x - 2y - 7 \\ \hline \end{array}$

4. $\begin{array}{r} 5x + 4y - 16 \\ (+)\ x + 4y + 21 \\ \hline \end{array}$

Warm Up

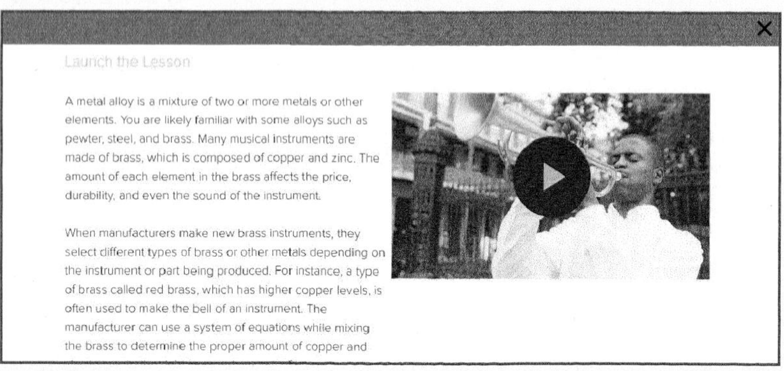

Launch the Lesson

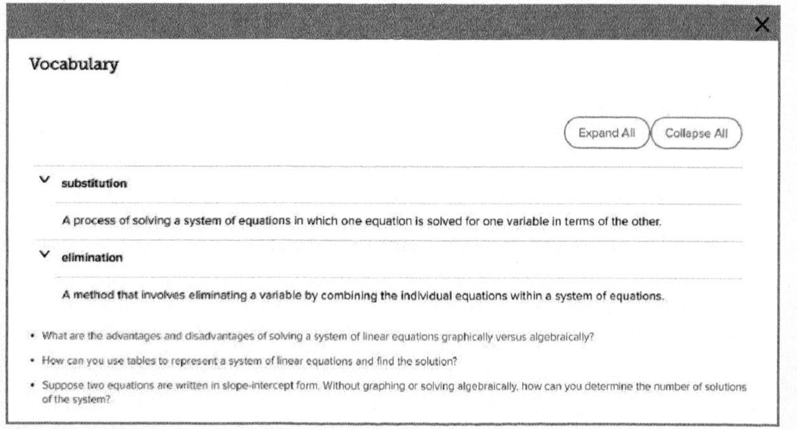

Today's Vocabulary

Warm Up

Prerequisite Skills

The Warm Up exercises address the following prerequisite skill for this lesson:

- adding polynomials

Answers:

1. $6x - 2y + 3$
2. $2x + 2y - 20$
3. $15x - 15$
4. $6x + 8y + 5$

Launch the Lesson

 **Teaching the Mathematical Practices**

4 Apply Mathematics In the Launch the Lesson, students will learn about how a system of equations in two variables can be used to determine the amount of copper and zinc needed to make brass instruments.

Go Online to find additional teaching notes and questions to promote classroom discourse.

Today's Standards

Tell students that they will be addressing these content and practice standards in this lesson. You may wish to have a student volunteer read aloud *How can I meet these standards?* and *How can I use these practices?*, and connect these to the standards.

See the Interactive Presentation for I Can statements that align with the standards covered in this lesson.

Today's Vocabulary

Tell students that they will be using these vocabulary terms in this lesson. You can expand each row if you wish to share the definitions. Then discuss the questions below with the class.

1 CONCEPTUAL UNDERSTANDING | 2 FLUENCY | 3 APPLICATION

Learn Solving Systems of Equations in Two Variables by Substitution

Objective

Students solve systems of equations by using the substitution method to find the value of each variable.

Teaching the Mathematical Practices

2 Different Properties Mathematically proficient students look for different ways to solve problems. To answer the question in the Talk About It! feature, students must consider two methods for solving a system of equations and choose the method that works best.

DIFFERENTIATE

Reteaching Activity AL ELL

IF students cannot distinguish between no solution and infinitely many solutions,

THEN tell students that if the variables cancel out and a false statement remains, like $6 = 0$ or $5 = -2$, then there is no solution because the statement will never be true. If the variables cancel out and a true statement remains, like $6 = 6$ or $-2 = -2$, then there are infinitely many solutions because the statement will always be true.

Example 1 Substitution When There Is One Solution

Teaching the Mathematical Practices

1 Seek Information In Example 1, students must transform the system of equations to apply the substitution method and reach a solution.

Questions for Mathematical Discourse

AL What does it mean to substitute? Sample answer: to plug in an equivalent value or quantity for another value

OL If the system of equations were graphed, where would they intersect? at the point $(-2, -5)$

BL What is the classification for the given system of equations? consistent and independent

Go Online

- Find additional teaching notes.
- View performance reports of the Checks.
- Assign or present an Extra Example.

Lesson 2-5

Solving Systems of Equations Algebraically

Learn Solving Systems of Equations in Two Variables by Substitution

One algebraic method to solve a system of equations is a process called **substitution,** in which one equation is solved for one variable in terms of the other.

Key Concept • Substitution Method

Step 1 When necessary, solve at least one equation for one of the variables.

Step 2 Substitute the resulting expression from Step 1 into the other equation to replace the variable. Then solve the equation.

Step 3 Substitute the value from Step 2 into either equation, and solve for the other variable. Write the solution as an ordered pair.

Example 1 Substitution When There Is One Solution

Use substitution to solve the system of equations.

$8x - 3y = -1$	Equation 1
$x + 2y = -12$	Equation 2

Step 1 Solve one equation for one of the variables.

Because the coefficient of x in Equation 2 is 1, solve for x in that equation.

$x + 2y = -12$	Equation 2
$x = \underline{-2y - 12}$	Subtract $2y$ from each side.

Step 2 Substitute the expression. Substitute for x. Then solve for y.

$8x - 3y = -1$	Equation 1
$8(\underline{-2y - 12}) - 3y = -1$	$x = -2y - 12$
$\underline{-16}y - \underline{96} - 3y = -1$	Distributive Property
$\underline{-19}y - 96 = -1$	Simplify.
$-19y = \underline{95}$	Add 96 to each side.
$y = \underline{-5}$	Divide each side by -19.

Step 3 Substitute to solve. Use one of the original equations to solve for x.

$x + 2y = -12$	Equation 2
$x + 2(\underline{-5}) = -12$	$y = -5$
$x = \underline{-2}$	Simplify.

The solution is $(-2, -5)$. Substitute into the original equations to check.

Go Online You can complete an Extra Example online.

Today's Goals
- Solve systems of equations by using the substitution method.
- Solve systems of equations by using the elimination method.

Today's Vocabulary
substitution
elimination

Go Online You can watch a video to see how to use algebra tiles to solve a system of equations by using substitution.

Talk About It! Describe the benefit of solving a system of equations by substitution instead of graphing when the coefficients are not integers.

Sample answer: If the coefficients are not integers, it may be difficult to find the exact solution using a graph. Since an exact solution can be calculated using substitution, it would be a better method.

Interactive Presentation

Solving Systems of Equations in Two Variables by Substitution

Some systems of equations have solutions that are difficult to determine from a graph or a table. In these cases, algebraic methods are used to solve for the variables.

One algebraic method is a process called **substitution**, in which one equation is solved for one variable in terms of the other.

Learn

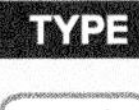

TYPE

Students answer a question to determine whether they understand when to solve by graphing or algebraically.

WATCH

Students can watch a video that explains how algebra tiles can be used to solve a system of equations by substitution.

Your Notes

Check

Use substitution to solve the system of equations.

$-5x + y = -3$

$3x - 8y = 24$ $\underline{(0, -3)}$

Example 2 Substitution When There Is Not Exactly One Solution

Use substitution to solve the system of equations.

$-5x + 2.5y = -15$ Equation 1
$y = 2x - 11$ Equation 2

Equation 2 is already solved for y, so substitute $2x - 11$ for y in Equation 1.

$-5x + 2.5y = -15$ Equation 1
$-5x + 2.5(\underline{2x - 11}) = -15$ $y = 2x - 11$
$-5x + \underline{5}x - \underline{27.5} = -15$ Distributive Property
$\underline{-27.5} = -15$ False

This system has $\underline{\text{no solution}}$ because $-27.5 = -15$ is not true.

Think About It! What can you conclude about the slopes and y-intercepts of the equations when a system of equations has no solution? when a system of equations has infinitely many solutions?

Sample answer: When a system of equations has no solution, the slopes of the equations are the same, but the y-intercepts are different. When a system of equations has infinitely many solutions, the slopes and y-intercepts of the equations are the same.

Example 3 Apply the Substitution Method

CHEMISTRY **Ms. Washington is preparing a hydrochloric acid (HCl) solution. She will need 300 milliliters of a 5% HCl solution for her class to use during a lab. If she has a 3.5% HCl solution and a 7% HCl solution, how much of each solution should she use in order to make the solution needed?**

Step 1 Write two equations in two variables.

Let x be the amount of 3.5% solution and y be the amount of 7% solution.

$x + y = 300$ Equation 1
$0.035x + 0.07y = 0.05(300)$ Equation 2

Step 2 Solve one equation for one of the variables.

$x + y = 300$ Equation 1
$x = \underline{-y} + 300$ Subtract y from each side.

Step 3 Substitute the resulting expression and solve.

$0.035x + 0.07y = 15$ Equation 2
$0.035(\underline{-y + 300}) + 0.07y = 15$ $x = -y + 300$
$\underline{-0.035}y + \underline{10.5} + 0.07y = 15$ Distributive Property
$0.035y = 4.5$ Simplify.
$y \approx \underline{128.57}$ Divide each side by 0.035.

Go Online You can complete an Extra Example online.

Think About It! Explain what approximations were made while solving this problem and how they affect the solution.

Sample answer: The value of y is rounded in Step 3 since it would not be practical to measure to the exact value of y. Because x is calculated based on the value of y, it is also an approximation. When two rounded values are used to check the solution, the equations are only approximately equal.

(continued on the next page)

Interactive Presentation

Apply the Substitution Method

CHEMISTRY For an upcoming chemistry lab, Ms. Washington is preparing a hydrochloric acid (HCl) solution. She will need 300 milliliters of a 5% HCl solution for her class to use during the lab. If she has a 3.5% HCl solution and a 7% HCl solution, how much of each solution should she use in order to make the solution needed?

Example 3

SELECT

Students select the values to complete the statement about an ordered pair in its real-world context.

TYPE

Students answer a question to determine whether they understand using the substitution method.

CHECK

Students complete the Check online to determine whether they are ready to move on.

Example 2 Substitution When There Is Not Exactly One Solution

Teaching the Mathematical Practices

7 Use Structure Work with students to look at the Think About It! feature. Guide students to use the structure of the equations in a system to draw conclusions about the number of solutions.

Questions for Mathematical Discourse

AL What is always the first step when solving a system using substitution? Solve one equation for one variable.

OL Suppose your first step was solving the first equation for y. Would you have to substitute to be able to tell there are no solutions? Sample answer: No, you could see that the lines have the same slope and different y-intercepts, and thus are parallel.

BL How could you change Equation 2 so that there are infinitely many solutions? Sample answer: You could make the -11 a -6 instead. Then $2.5(-6) = -15$, and the two sides will be equal.

Example 3 Apply the Substitution Method

Teaching the Mathematical Practices

4 Interpret Mathematical Results To solve the problem, students should interpret their mathematical results in the context of the problem.

6 State Meaning of Symbols Guide students to define variables to solve the problem in Example 3. Work with them to find the relationships in the problem and represent them using variables and equations.

Questions for Mathematical Discourse

AL What does the 300 mL represent in the context of the problem? the total volume of the solution needed

OL Why is 300 mL multiplied by 0.05 in the second equation? Sample answer: Because the second equation represents the amount of acid in the solution, we have to multiply the amount of liquid by the percent of acid.

BL Suppose Ms. Washington only had a 3.5% HCl solution and a 4.5% HCl solution. Would it be possible to make the needed 5% HCl solution? Explain. No; sample answer: Because both percentages of the HCl solutions are less than 5%, there is no way to mix those two acid solutions to get a percentage greater than 4%.

1 CONCEPTUAL UNDERSTANDING | 2 FLUENCY | 3 APPLICATION

Learn Solving Systems of Equations in Two Variables by Elimination

Objective

Students solve systems of equations by using the elimination method to find the value of each variable.

Teaching the Mathematical Practices

3 Analyze Cases The Think About It! feature asks students to examine the cases when elimination by addition should be used to solve a system and when elimination by subtraction should be used. Encourage students to familiarize themselves with both of the cases.

About the Key Concept

Elimination is the algebraic process used to solve a system of equations by adding or subtracting to eliminate one variable. To solve using elimination, each equation must be multiplied by a number that results in two equations containing opposite or equal terms. The resulting equations are then added or subtracted which will eliminate one variable. After the resulting equation is solved, the value found is substituted back into either equation and then solved for the remaining variable. The solution is written as an ordered pair.

Example 4 Elimination When There Is One Solution

Teaching the Mathematical Practices

6 Communicate Precisely Encourage students to routinely write or explain their solution methods. Point out that they should use clear mathematical language when answering the question in the Think About It! feature.

Questions for Mathematical Discourse

AL Did you have to multiply the second equation by -1 to use elimination? No, you could also subtract one equation from the other because subtracting a value is the same as adding the negative of the value.

OL If the system of equations were graphed, where would the lines intersect? $(-1, 3)$ If the system of equations were solved using substitution, what would be the solution? $(-1, 3)$

BL Can you find an exact solution to a system of equations with 3 variables by using elimination? Yes, you can if you have at least 3 equations and an exact solution exists.

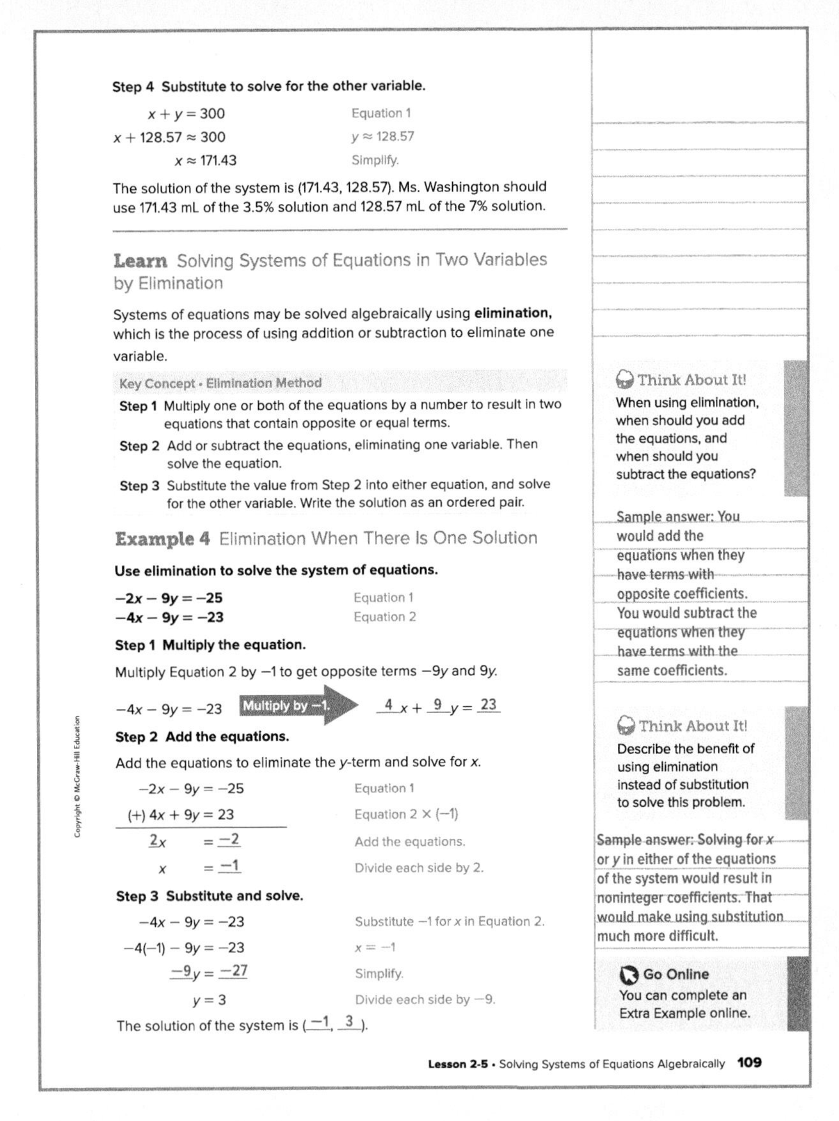

Step 4 Substitute to solve for the other variable.

$x + y = 300$ — Equation 1

$x + 128.57 \approx 300$ — $y \approx 128.57$

$x \approx 171.43$ — Simplify.

The solution of the system is (171.43, 128.57). Ms. Washington should use 171.43 mL of the 3.5% solution and 128.57 mL of the 7% solution.

Learn Solving Systems of Equations in Two Variables by Elimination

Systems of equations may be solved algebraically using **elimination**, which is the process of using addition or subtraction to eliminate one variable.

Key Concept • Elimination Method

Step 1 Multiply one or both of the equations by a number to result in two equations that contain opposite or equal terms.

Step 2 Add or subtract the equations, eliminating one variable. Then solve the equation.

Step 3 Substitute the value from Step 2 into either equation, and solve for the other variable. Write the solution as an ordered pair.

Example 4 Elimination When There Is One Solution

Use elimination to solve the system of equations.

$-2x - 9y = -25$ — Equation 1

$-4x - 9y = -23$ — Equation 2

Step 1 Multiply the equation.

Multiply Equation 2 by -1 to get opposite terms $-9y$ and $9y$.

$-4x - 9y = -23$ → Multiply by -1. → $\underline{4}x + \underline{9}y = \underline{23}$

Step 2 Add the equations.

Add the equations to eliminate the y-term and solve for x.

$-2x - 9y = -25$ — Equation 1

$(+)\ 4x + 9y = 23$ — Equation 2 × (−1)

$\underline{2x} = \underline{-2}$ — Add the equations.

$x = \underline{-1}$ — Divide each side by 2.

Step 3 Substitute and solve.

$-4x - 9y = -23$ — Substitute −1 for x in Equation 2.

$-4(-1) - 9y = -23$ — $x = -1$

$\underline{-9}y = \underline{-27}$ — Simplify.

$y = 3$ — Divide each side by −9.

The solution of the system is $(\underline{-1}, \underline{3})$.

Think About It!
When using elimination, when should you add the equations, and when should you subtract the equations?

Sample answer: You would add the equations when they have terms with opposite coefficients. You would subtract the equations when they have terms with the same coefficients.

Think About It!
Describe the benefit of using elimination instead of substitution to solve this problem.

Sample answer: Solving for x or y in either of the equations of the system would result in noninteger coefficients. That would make using substitution much more difficult.

Go Online
You can complete an Extra Example online.

Copyright © McGraw-Hill Education

Lesson 2-5 • Solving Systems of Equations Algebraically 109

Interactive Presentation

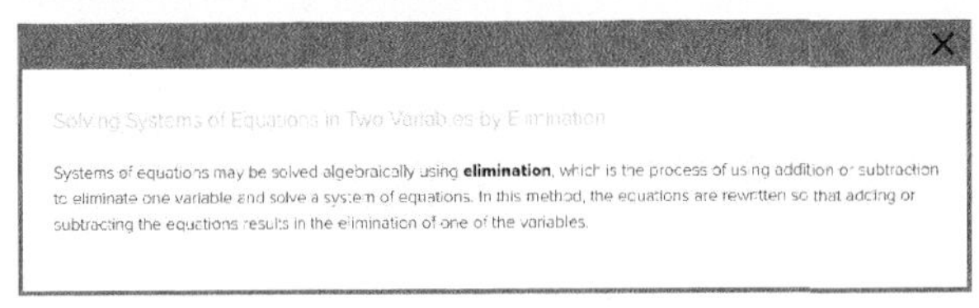

Learn

TYPE

Students answer a question to show they understand solving by using elimination.

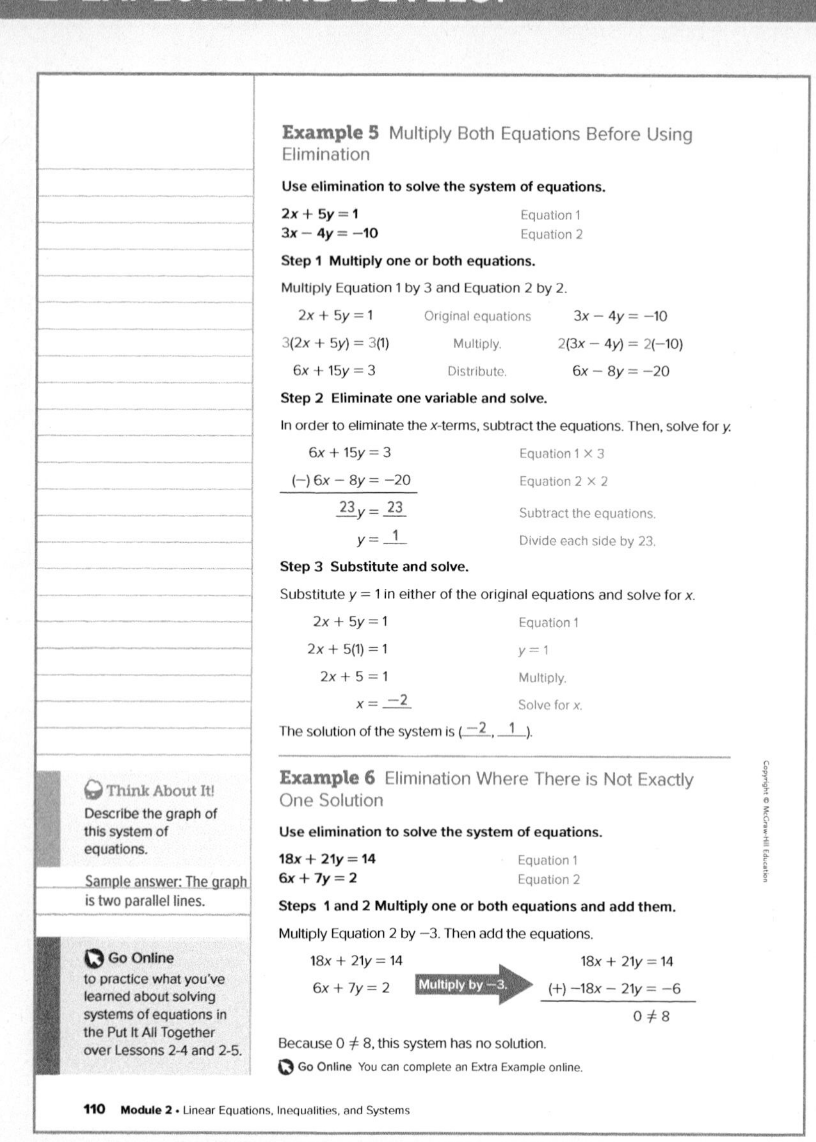

Example 5 Multiply Both Equations Before Using Elimination

Use elimination to solve the system of equations.

$2x + 5y = 1$ Equation 1
$3x - 4y = -10$ Equation 2

Step 1 Multiply one or both equations.

Multiply Equation 1 by 3 and Equation 2 by 2.

$2x + 5y = 1$	Original equations	$3x - 4y = -10$
$3(2x + 5y) = 3(1)$	Multiply.	$2(3x - 4y) = 2(-10)$
$6x + 15y = 3$	Distribute.	$6x - 8y = -20$

Step 2 Eliminate one variable and solve.

In order to eliminate the x-terms, subtract the equations. Then, solve for y.

$6x + 15y = 3$ Equation 1 × 3
$(-)\ 6x - 8y = -20$ Equation 2 × 2
$23y = 23$ Subtract the equations.
$y = 1$ Divide each side by 23.

Step 3 Substitute and solve.

Substitute $y = 1$ in either of the original equations and solve for x.

$2x + 5y = 1$ Equation 1
$2x + 5(1) = 1$ $y = 1$
$2x + 5 = 1$ Multiply.
$x = -2$ Solve for x.

The solution of the system is $(-2, 1)$.

Example 6 Elimination Where There is Not Exactly One Solution

Use elimination to solve the system of equations.

$18x + 21y = 14$ Equation 1
$6x + 7y = 2$ Equation 2

Steps 1 and 2 Multiply one or both equations and add them.

Multiply Equation 2 by −3. Then add the equations.

$18x + 21y = 14$ → $18x + 21y = 14$
$6x + 7y = 2$ —Multiply by −3.→ $(+) -18x - 21y = -6$
$0 \neq 8$

Because $0 \neq 8$, this system has no solution.

Go Online You can complete an Extra Example online.

Think About It! Describe the graph of this system of equations.

Sample answer: The graph is two parallel lines.

Go Online to practice what you've learned about solving systems of equations in the Put It All Together over Lessons 2-4 and 2-5.

Copyright © McGraw-Hill Education

110 Module 2 • Linear Equations, Inequalities, and Systems

Interactive Presentation

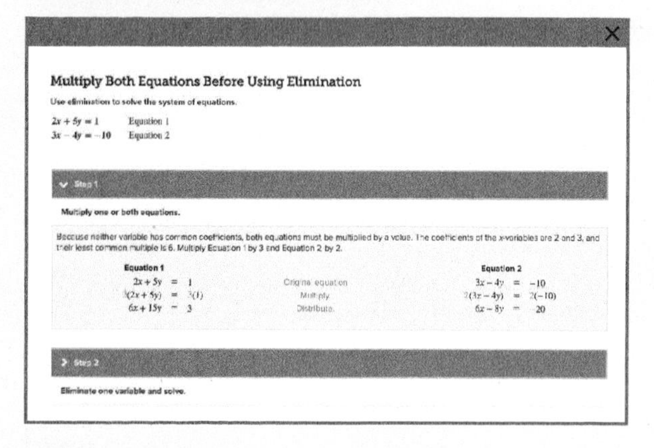

Example 5

TAP

Students move through the steps to solving a system of equations using elimination.

CHECK

Students complete the Check online to determine whether they are ready to move on.

Example 5 Multiply Both Equations Before Using Elimination

Teaching the Mathematical Practices

1 Monitor and Evaluate When using elimination, students should stop and evaluate their progress and change course if necessary. In Example 5, students should check their intermittent solutions.

Questions for Mathematical Discourse

AL Why do you have to multiply both equations to use elimination in this example? None of the coefficients are factors of the corresponding coefficients.

OL If the y-variables are eliminated, by what value would the equations be multiplied? Equation 1 would be multiplied by 4 while Equation 2 would be multiplied by 5.

BL Explain why eliminating the y-terms may have been an easier path. Sample answer: Because the y-terms have opposite signs, the equations could be added. Adding the equations is easier because a negative does not need to be distributed.

Example 6 Elimination When There Is Not Exactly One Solution

Teaching the Mathematical Practices

1 Explain Correspondences Use the Think About It! feature to encourage students to explain the relationships between the system of equations, the solution, and the graph of the system.

Questions for Mathematical Discourse

AL What does it mean when you get a false statement? It means there is no solution that satisfies both equations.

OL Once Equation 2 is multiplied by −3, how could you identify that there is no solution? Sample answer: The left side of one equation is the negative of the other, while the right side is not. This means the variables will all eliminate leaving a false statement.

BL Describe how to identify that the system of equations will have no solution without graphing, using substitution, or elimination. Sample answer: Equation 1 is three times Equation 2 on the left side, but the right side does not follow that rule. Therefore, there will be no solution.

Exit Ticket

Recommended Use

At the end of class, go online to display the Exit Ticket prompt and ask students to respond using a separate piece of paper. Have students hand you their responses as they leave the room.

Alternate Use

At the end of class, go online to display the Exit Ticket prompt and ask students to respond verbally or by using a mini-whiteboard. Have students hold up their whiteboards so that you can see all student responses. Tap to reveal the answer when most or all students have completed the Exit Ticket.

Practice and Homework

Suggested Assignments

Use the table below to select appropriate exercises.

DOK	Topic	Exercises
1, 2	exercises that mirror the examples	1–14
2	exercises that use a variety of skills from this lesson	15–20
2	exercises that extend concepts learned in this lesson to new contexts	21–24
3	exercises that emphasize higher-order and critical thinking skills	25–27

ASSESS AND DIFFERENTIATE

Use the data from the **Checks** to determine whether to provide resources for extension, remediation, or intervention.

IF students score 90% or more on the Checks,
THEN assign:

- Practice, Exercises 25–27
- Extension: Cramer's Rule
- ALEKS® Systems of Linear Equations in Two Variables

IF students score 66%–89% on the Checks,
THEN assign:

- Practice, Exercises 1–23, odd
- Remediation, Review Resources: Substitution
- Personal Tutors
- Extra Examples 1–5
- ALEKS® Systems of Linear Equations

AL

IF students score 65% or less on the Checks,
THEN assign:

- Practice, Exercises 1–13, odd
- Remediation, Review Resources: Substitution
- *Quick Review Math Handbook:* Solving Systems of Equations Algebraically
- ALEKS® Systems of Linear Equations

Name ______ Period ______ Date ______

Practice

Go Online You can complete your homework online.

Examples 1 and 2

Use substitution to solve each system of equations.

1. $2x - y = 9$
   $x + 3y = -6$
   $(3, -3)$
2. $2x - y = 7$
   $6x - 3y = 14$
   no solution
3. $2x + y = 5$
   $3x - 3y = 3$
   $(2, 1)$
4. $3x + y = 7$
   $4x + 2y = 16$
   $(-1, 10)$
5. $4x - y = 6$
   $2x - \frac{y}{2} = 4$
   no solution
6. $2x + y = 8$
   $3x + \frac{3}{2}y = 12$
   infinitely many

Example 3

Solve each problem.

7. BAKE SALE Cassandra and Alberto are selling pies for a fundraiser. Cassandra sold 3 small pies and 14 large pies for a total of \$203. Alberto sold 11 small pies and 11 large pies for a total of \$220. Determine the cost of each pie.
   a. Write a system of equations and solve by using substitution.
   Cassandra: $3x + 14y = 203$; Alberto: $11x + 11y = 220$; $x = 7$, $y = 13$
   b. What does the solution represent in terms of this situation?
   The cost of each small pie is \$7. The cost of each large pie is \$13.
   c. How can you verify that the solution is correct? Sample answer: By substituting the solution into each equation in the system, you can verify that it is correct. $3(7) + 14(13) = 203$, and $11(7) + 11(13) = 220$.

8. STOCKS Ms. Patel invested a total of \$825 in two stocks. At the time of her investment, one share of Stock A was valued at \$12.41 and a share of Stock B was valued at \$8.62. She purchased a total of 79 shares.
   a. Write a system of equations and solve by substitution.
   $a + b = 79$; $12.41a + 8.62b = 825$; $a = 38$, $b = 41$
   b. How many shares of each stock did Ms. Patel buy? How much did she invest in each of the two stocks?
   Ms. Patel bought 38 shares of Stock A for a total of \$471.58 and 41 shares of Stock B for a total of \$353.42.

Examples 4-6

Use elimination to solve each system of equations.

9. $3x - 2y = 4$
   $5x + 3y = -25$
   $(-2, -5)$
10. $5x + 2y = 12$
    $-6x - 2y = -14$
    $(2, 1)$
11. $7x + 2y = -1$
    $21x + 6y = -9$
    no solution
12. $3x - 5y = -9$
    $-7x + 3y = 8$
    $\left(-\frac{1}{2}, \frac{3}{2}\right)$
13. $x - 3y = -12$
    $2x + y = 11$
    $(3, 5)$
14. $6w - 8z = 16$
    $3w - 4z = 8$
    infinitely many

Mixed Exercises

Use substitution or elimination to solve each system of equations.

15. $0.5x + 2y = 5$
$x - 2y = -8$
$(-2, 3)$

16. $h - z = 3$
$-3h + 3z = 6$
no solution

17. $-r + t = 5$
$-2r + t = 4$
$(1, 6)$

18. $3r - 2t = 1$
$2r - 3t = 9$
$(-3, -5)$

19. $5g + 4k = 10$
$-3g - 5k = 7$
$(6, -5)$

20. $4m - 2p = 0$
$-3m + 9p = 5$
$\left(\frac{1}{3}, \frac{2}{3}\right)$

21. The sum of two numbers is 12. The difference of the same two numbers is −4. Find the two numbers. 4, 8

22. Twice a number minus a second number is −1. Twice the second number added to three times the first number is 9. Find the two numbers. 1, 3

23. REASONING Mr. Janson paid for admission to the high school football game for his family. He purchased 3 adult tickets and 2 student tickets for a total of $22. Ms. Pham purchased 5 adult tickets and 3 student tickets for a total of $35.75. What is the cost of each adult ticket and each student ticket?
adult ticket $5.50; student ticket $2.75

24. USE A MODEL The Newton City Park has 11 basketball courts, which are all in use. There are 54 people playing basketball. Some are playing one-on-one, and some are playing in teams. A one-on-one game requires 2 players, and a team game requires 10 players.

a. Write a system of equations that represents the number of one-on-one and team games being played.
Let x be one-on-one games and y be team games. $x + y = 11$; $2x + 10y = 54$

b. Solve the system of equations and interpret your results.
$x = 7$, $y = 4$; There are 7 one-on-one games and 4 team games being played.

Higher-Order Thinking Skills

25. FIND THE ERROR Gloria and Syreeta are solving the system $6x - 4y = 26$ and $-3x + 4y = -17$. Is either of them correct? Explain your reasoning. See margin.

Gloria

$6x - 4y = 26$
$-3x + 4y = -17$
$3x = 9$
$x = 3$

$-3(3) + 4y = -17$
$-9 + 4y = -17$
$4y = -8$
$y = -2$

The solution is $(3, -2)$.

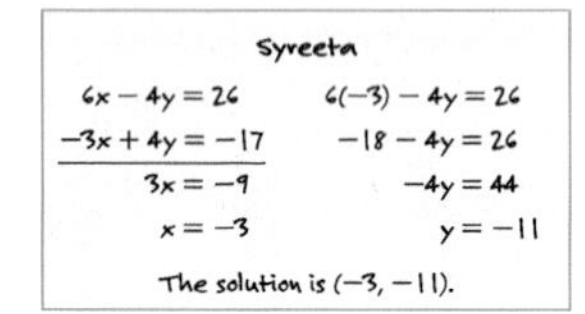

26. CREATE Write a system of equations in which one equation should be multiplied by 3 and the other should be multiplied by 4 in order to solve the system with elimination. Then solve your system. See margin.

27. WRITE Why is substitution sometimes more helpful than elimination? See margin.

Answers

25. Gloria is correct; sample answer: Syreeta subtracted 26 from 17 instead of 17 from 26 and got $3x = -9$ instead of $3x = 9$. She proceeded to get a value of −11 for y. She would have found her error if she had substituted the solution into the original equations.

26. Sample answer:

$$\begin{aligned} 4x + 5y &= 21 \\ 3x - 2y &= 10 \end{aligned} \rightarrow \begin{aligned} 3(4x + 5y &= 21) \\ 4(3x - 2y &= 10) \end{aligned} \rightarrow$$

$$\begin{aligned} 12x + 15y &= 63 \\ (-)12x - 8y &= 40 \\ \hline 23y &= 23 \\ y &= 1 \end{aligned} \qquad \begin{aligned} 4x + 1(5) &= 21 \\ 4x + 5 &= 21 \\ 4x &= 16 \\ x &= 4 \end{aligned}$$

27. Sample answer: It is more helpful to use substitution when one of the variables has a coefficient of 1 or if a coefficient can easily be reduced to 1.

Solving Systems of Inequalities

A.CED.3

LESSON GOAL

Students solve systems of inequalities in two variables.

1 LAUNCH

 Launch the lesson with a **Warm Up** and an introduction.

2 EXPLORE AND DEVELOP

 **Explore:** Solutions of Systems of Inequalities

 **Develop:**

Solving Systems of Inequalities in Two Variables
- Unbounded Region
- Bounded Region
- Use Systems of Inequalities

 You may want your students to complete the **Checks** online.

3 REFLECT AND PRACTICE

 Exit Ticket

 Practice

DIFFERENTIATE

 View reports of student progress on the **Checks** after each example.

Resources	AL	OL	BL	ELL
Remediation: Elimination Using Addition and Subtraction	●	●		●
Extension: Creative Designs		●	●	●
ELL Support				

Language Development Handbook

Assign page 13 of the *Language Development Handbook* to help your students build mathematical language related to solving systems of inequalities in two variables.

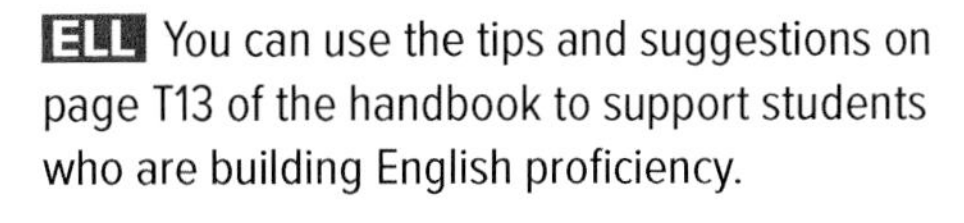

ELL You can use the tips and suggestions on page T13 of the handbook to support students who are building English proficiency.

Suggested Pacing

90 min	0.5 day
45 min	1 day

Focus

Domain: Algebra

Standards fowr Mathematical Content:

A.CED.3 Represent constraints by equations or inequalities, and by a system of equations and/or inequalities, and interpret solutions as viable or non-viable options in a modeling context.

Standards for Mathematical Practice:

3 Construct viable arguments and critique the reasoning of others.

4 Model with mathematics.

Coherence

Vertical Alignment

Previous

Students were introduced to solving systems of linear inequalities.

A.REI.12 (Algebra 1)

Now

Students solve systems of inequalities in two variables.

A.CED.3

Next

Students will use linear programming to find maximum or minimum values of a function.

A.CED.3

Rigor

The Three Pillars of Rigor

1 CONCEPTUAL UNDERSTANDING	2 FLUENCY	3 APPLICATION

Conceptual Bridge In this lesson, students expand their understanding of graphing linear inequalities to build fluency with graphing systems of linear inequalities. They apply their understanding of graphing systems of linear inequalities by solving real-world problems.

Mathematical Background

The inequalities in the system are graphed on the same coordinate plane, and the ordered pairs that satisfy all of the inequalities in the system are in the region that is common to the inequalities. If the regions do not intersect, the solution is the empty set, and no solution exists.

Interactive Presentation

Warm Up

Students are selling caps c and bumper stickers *b* for a park project. Write an equation or inequality to express each constraint.

1. They cannot buy more than 144 caps.
2. They must buy at least 200 bumper stickers.
3. They want to sell a combined total of at least 400 items.
4. The total cost is *C*, when caps cost \$2 each and bumper stickers cost \$0.75 each.
5. They earn \$5 in revenue *R* for each cap and \$2 for each bumper sticker.

Show Answers

Warm Up

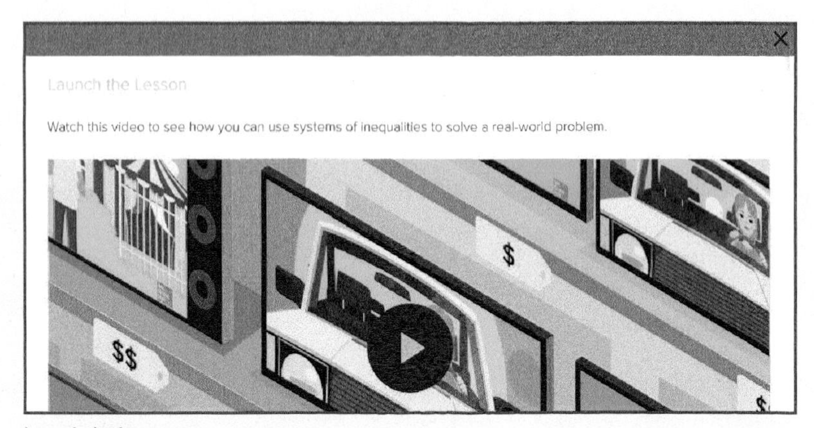

Launch the Lesson

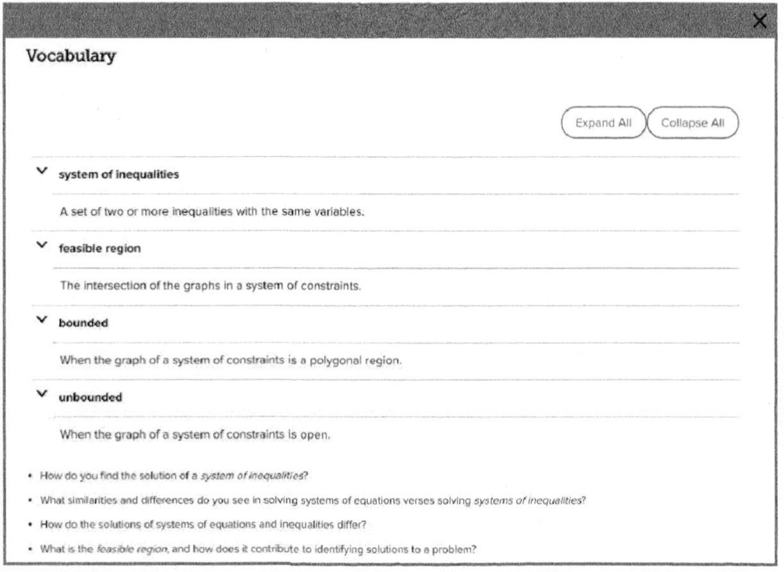

Today's Vocabulary

Warm Up

Prerequisite Skills

The Warm Up exercises address the following prerequisite skill for this lesson:

- writing equations and inequalities

Answers:

1. $c \leq 144$
2. $b \geq 200$
3. $c + b \geq 400$
4. $C = 2c + 0.75b$
5. $R = 5c + 2b$

Launch the Lesson

 Teaching the Mathematical Practices

1 Analyze Givens and Constraints In the Launch the Lesson, students will learn how constraints and relationships must be analyzed when using a system of inequalities to solve a real-world problem.

Go Online to find additional teaching notes and questions to promote classroom discourse.

Today's Standards

Tell students that they will be addressing these content and practice standards in this lesson. You may wish to have a student volunteer read aloud *How can I meet this standard?* and *How can I use these practices?*, and connect these to the standards.

See the Interactive Presentation for I Can statements that align with the standards covered in this lesson.

Today's Vocabulary

Tell students that they will be using these vocabulary terms in this lesson. You can expand each row if you wish to share the definitions. Then discuss the questions below with the class.

Explore Solutions of Systems of Inequalities

Objective

Students use a graph to explore the solutions of a system of inequalities.

 Teaching the Mathematical Practices

3 Construct Arguments Throughout the Explore, students will use previously established results to construct arguments about the solutions of systems of inequalities.

Ideas for Use

Recommended Use Present the Inquiry Question, or have a student volunteer read it aloud. Have students work in pairs to complete the Explore activity on their devices. Pairs should discuss each of the questions. Monitor student progress during the activity. Upon completion of the Explore activity, have student volunteers share their responses to the Inquiry Question.

What if my students don't have devices? You may choose to project the activity on a whiteboard. A printable worksheet for each Explore is available online. You may choose to print the worksheet so that individuals or pairs of students can use it to record their observations.

Summary of the Activity

Students will complete guiding exercises throughout the Explore activity. They will learn about solutions to a system of inequalities using a graph. They will answer a series of questions about the points contained in solution regions. Then, students will answer the Inquiry Question.

(continued on the next page)

Interactive Presentation

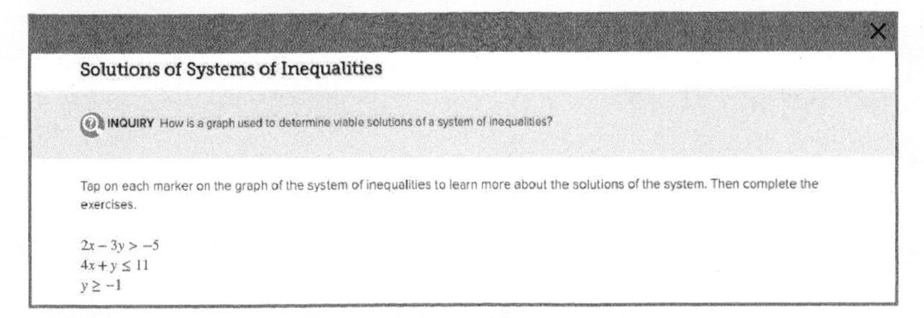

Explore

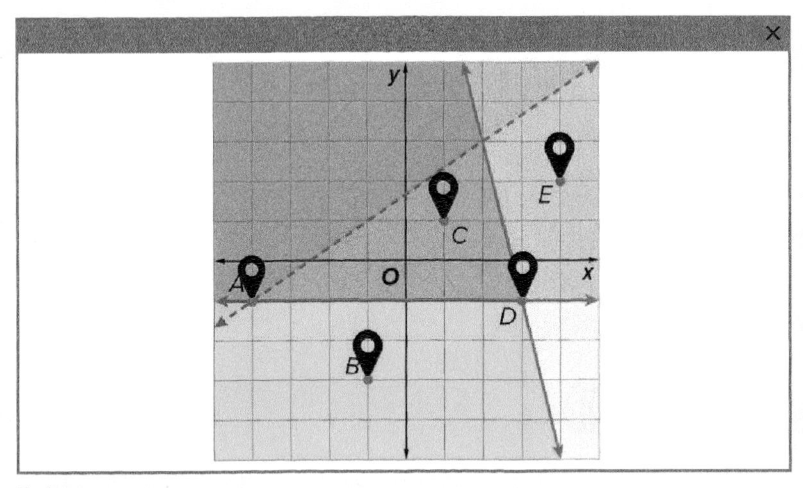

Explore

TAP Students tap on each marker on the graph to learn more about solutions of the system.

TYPE Students answer questions to show they understand solutions of systems of inequalities.

Interactive Presentation

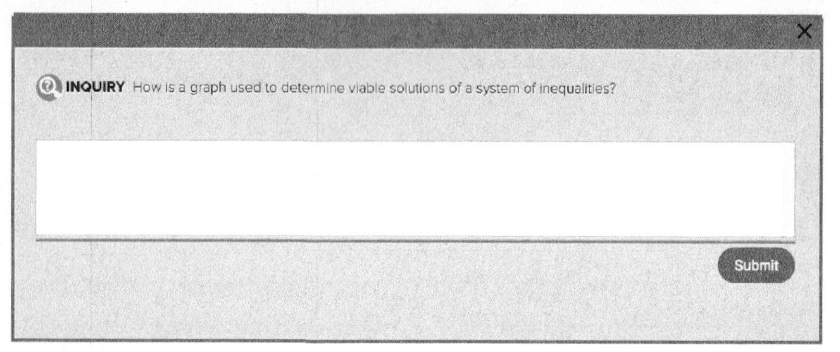

Explore

TYPE

Students respond to the Inquiry Question and can view a sample answer.

Explore Solutions of Systems of Inequalities *(continued)*

Teaching the Mathematical Practices

3 Construct Viable Arguments and Critique the Reasoning of Others Throughout the Explore, encourage students use the graph to make reasonable conclusions about the solutions of systems of inequalities.

Questions

Have students complete the Explore activity.

Ask:

- What type of boundary line is used for the symbols $>$ or $<$? a dashed line
- Why is point A not included in the solution region? Sample answer: Point A is on a boundary line that is dashed, which means the line does not include that point. Point A does not satisfy all three inequalities.

Inquiry

How is a graph used to determine viable solutions of a system of inequalities? Sample answer: The solution of the system is the region where the individual inequalities intersect.

Go Online to find additional teaching notes and sample answers for the guiding exercises.

DIFFERENTIATE

Language Development Activity ELL AL

Discuss *bounded* and *unbounded* regions relative to native language. *Boundaries* or *borders* may be general or academic synonyms, but not technical, mathematical synonyms.

1 CONCEPTUAL UNDERSTANDING | 2 FLUENCY | 3 APPLICATION

Learn Solving Systems of Inequalities in Two Variables

Objective

Students solve systems of linear inequalities in two variables by using the related equations.

 ### Teaching the Mathematical Practices

1 Explain Correspondences Encourage students to explain the relationships between the inequalities in a system, their related equations, and the graph of the system.

About the Key Concept

A system of inequalities has a solution set contained in a feasible region, which is the intersection of the graphs in the system of constraints. Ordered pairs in the feasible region satisfy all constraints in the system and are viable solutions. To solve a system of inequalities, each inequality must be graphed and the correct region shaded. The feasible region that is shaded for all of the inequalities represents the solution set.

DIFFERENTIATE

Reteaching Activity AL ELL

IF students are struggling to identify the feasible region,
THEN have students graph each inequality using a different colored pencil. They should lightly shade the region for each inequality, and then find the area where all colors are represented. This is the feasible region.

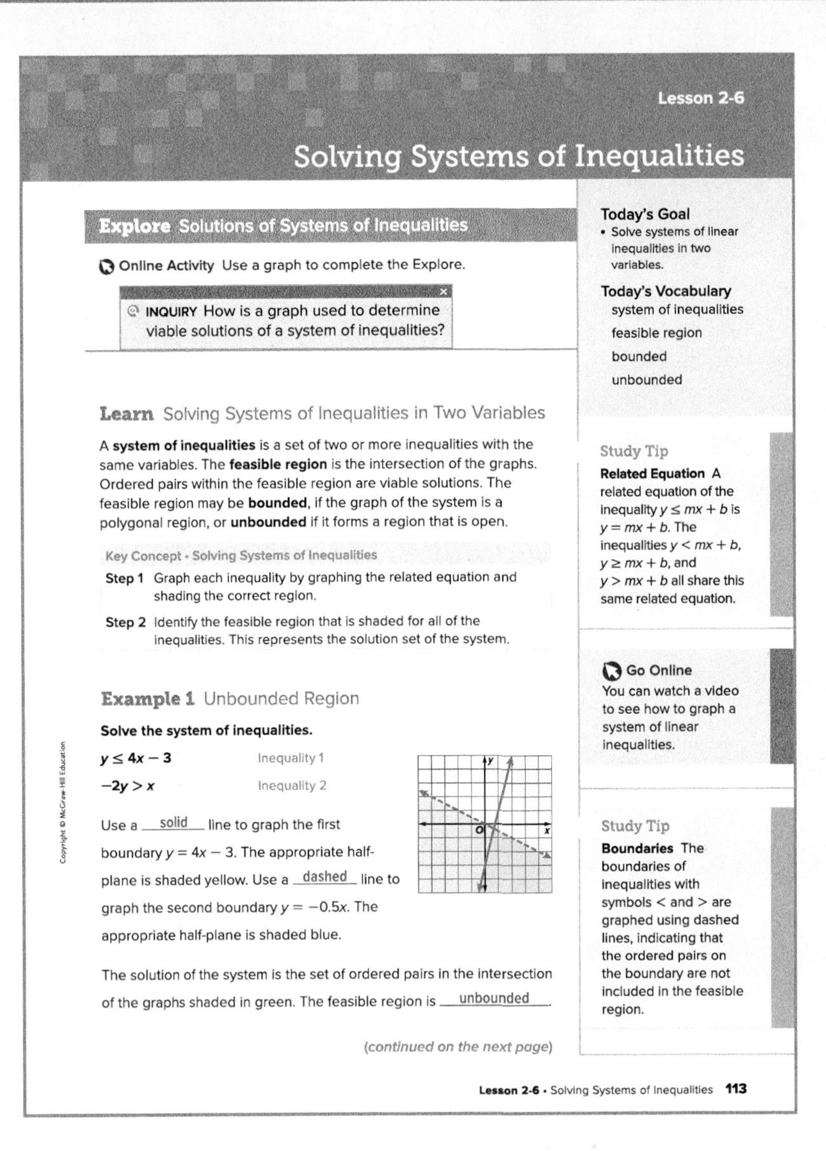

Lesson 2-6

Solving Systems of Inequalities

Explore Solutions of Systems of Inequalities

Online Activity Use a graph to complete the Explore.

INQUIRY How is a graph used to determine viable solutions of a system of inequalities?

Learn Solving Systems of Inequalities in Two Variables

A **system of inequalities** is a set of two or more inequalities with the same variables. The **feasible region** is the intersection of the graphs. Ordered pairs within the feasible region are viable solutions. The feasible region may be **bounded**, if the graph of the system is a polygonal region, or **unbounded** if it forms a region that is open.

Key Concept • Solving Systems of Inequalities

Step 1 Graph each inequality by graphing the related equation and shading the correct region.

Step 2 Identify the feasible region that is shaded for all of the inequalities. This represents the solution set of the system.

Example 1 Unbounded Region

Solve the system of inequalities.

$y \le 4x - 3$ Inequality 1

$-2y > x$ Inequality 2

Use a solid line to graph the first boundary $y = 4x - 3$. The appropriate half-plane is shaded yellow. Use a dashed line to graph the second boundary $y = -0.5x$. The appropriate half-plane is shaded blue.

The solution of the system is the set of ordered pairs in the intersection of the graphs shaded in green. The feasible region is unbounded.

(continued on the next page)

Today's Goal
- Solve systems of linear inequalities in two variables.

Today's Vocabulary
system of inequalities
feasible region
bounded
unbounded

Study Tip
Related Equation A related equation of the inequality $y \le mx + b$ is $y = mx + b$. The inequalities $y < mx + b$, $y \ge mx + b$, and $y > mx + b$ all share this same related equation.

Go Online
You can watch a video to see how to graph a system of linear inequalities.

Study Tip
Boundaries The boundaries of inequalities with symbols < and > are graphed using dashed lines, indicating that the ordered pairs on the boundary are not included in the feasible region.

Example 1 Unbounded Region

 ### Teaching the Mathematical Practices

1 Check Answers Mathematically proficient students continually ask themselves, "Does this make sense?" Point out that in Example 1, students need to check their answer. They should ask themselves whether their answer makes sense and whether they have answered the problem question.

Questions for Mathematical Discourse

AL What is a feasible region? Sample answer: the area where the shading of all inequalities in a system intersect

OL Is $\left(\frac{2}{3}, -\frac{1}{3}\right)$ included in the solution set?
No, it is a solution of the first inequality, but not the second.

BL Could a system of two linear inequalities be classified as consistent and independent? Explain. No; sample answer: Independent means a system has exactly one solution and the solution to a system of inequalities either contains infinite points or none at all.

Go Online

- Find additional teaching notes.
- View performance reports of the Checks.
- Assign or present an Extra Example.

Interactive Presentation

Solving Systems of Inequalities in Two Variables

A **system of inequalities** is a set of two or more inequalities with the same variables. Solving a system of inequalities involves finding the **feasible region**, which is the intersection of the graphs in a system of constraints. Ordered pairs within the feasible region satisfy all of the inequalities in the system and are viable solutions.

The feasible region may be **bounded**, in which the graph of a system of constraints is a polygonal region, or **unbounded** if it forms a region that is open.

Learn

TAP

Students move through the steps to solve a system of inequalities by graphing.

TYPE

Students answer a question to determine whether they understand the feasible region for a given graph of inequalities.

WATCH

Students can watch a video that explains how to solve a system of linear inequalities.

1 CONCEPTUAL UNDERSTANDING | **2 FLUENCY** | 3 APPLICATION

CHECK

Test the solution by substituting the coordinate of a point in the unbounded region, such as (2, −3), into the system of inequalities. If the point is viable for both inequalities, it is a solution of the system.

$y \le 4x - 3$	Original inequality	$-2(y) > x$
$-3 \overset{?}{\le} 4(2) - 3$	$x = 2$ and $y = -3$	$-2(-3) \overset{?}{>} 2$
$-3 \le 5$	True	$6 > 2$

Check

Graph the solution of the system of inequalities.

$y \le \frac{1}{3}x + 2$

$y > x$

$y \le 1$

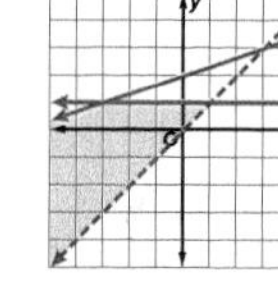

Example 2 Bounded Region

Solve the system of inequalities.

$y < -\frac{4}{3}x + 5$ Inequality 1

$y \ge x - 2$ Inequality 2

$x > 1$ Inequality 3

Use a dashed line to graph the first boundary $y = -\frac{4}{3}x + 5$.

The appropriate shaded area contains regions 1, 2, 3, and 7.

Use a solid line to graph the second boundary $y = x - 2$. The appropriate shaded area contains regions 1, 5, 6, and 7.

Use a dashed line to graph the third boundary $x = 1$. The appropriate shaded area contains regions 3, 4, 5 and 7.

The solution of the system is the set of ordered pairs in the intersection of the graphs, represented by region 7. The feasible region is bounded.

Go Online You can complete an Extra Example online.

Think About It!

How can you find the coordinates of the vertices of a polygon formed by the system of inequalities?

Sample answer: I can make systems of equations using the related equations of the intersecting boundaries and solve for x and y.

114 Module 2 • Linear Equations, Inequalities, and Systems

Interactive Presentation

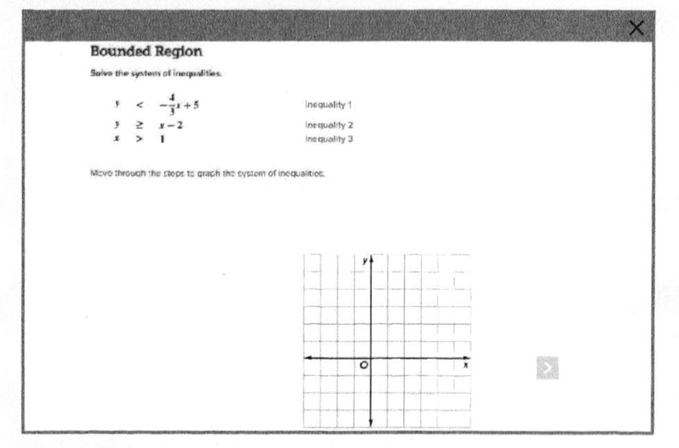

Example 2

SWIPE

Students move through the steps to see how to graph a system of inequalities.

Example 2 Bounded Region

Teaching the Mathematical Practices

3 Construct Arguments In the Think About It! feature, students will use stated assumptions, definitions, and previously established results to make a conjecture about identifying the vertices of a bounded region.

Leveled Discussion Questions

AL What shape is the bounded feasible region? a triangle

OL Will any of the vertices of the polygon region be included in the solution set? Explain. No; sample answer: Because two of the three boundary lines are dashed, none of the vertices are included.

BL Is the solution set of the bounded region finite? No, there are an infinite number of points within the bounded region. Is it possible to have a finite set of solutions within a bounded region of inequalities? Yes, if the domain and range are restricted by the constraints of the problem, such as being limited to whole numbers.

Common Error

Students are often in the habit of drawing a solid line, but with linear inequalities, not all boundary lines are solid.

Encourage students to check the inequality symbol to determine if the boundary line is solid or dashed before connecting the plotted points.

DIFFERENTIATE

Reteaching Activity AL

When there are multiple inequalities, it may be difficult for students to find the region shaded by all of them. This is especially challenging for students when the feasible region is empty. A different approach is to have students leave the solutions unshaded and shade the region on the other side of the line. Then the unshaded region is the solution set for the system.

Example 3 Use Systems of Inequalities

Teaching the Mathematical Practices

4 Analyze Relationships Mathematically Point out that to solve the problem in Example 3, students will need to analyze the mathematical relationships between the number of tickets, capacity of the boat, price of tickets, and operating cost to draw conclusions.

Leveled Discussion Questions

AL Why is the graph only in the first quadrant? A negative number of tickets does not make sense.

OL Suppose a local school rented a boat for a field trip and bought 500 children's tickets. How many chaperones could purchase a ticket? Sample answer: Any number from 0 to 92 adults could purchase a ticket.

BL Where does the greatest profit occur? Explain. The greatest profit occurs at (592, 0) because adult tickets cost more than child tickets.

Common Error

In real-world situations, many times the graph is limited to the first quadrant. Students may try and graph on a regular coordinate plane and be confused when the feasible region is unbounded. Remind students to consider domain and range restrictions before graphing.

Common Error

Students may forget that when they are putting an inequality in slope-intercept form they must watch out when they divide or multiply both sides by a negative number. They must reverse the inequality symbol if both sides are multiplied or divided by a negative.

Check

Graph the solution of the system of inequalities.

$y \geq \frac{4}{5}x - 3$

$y < -\frac{2}{3}x + 2$

$x \geq 0$

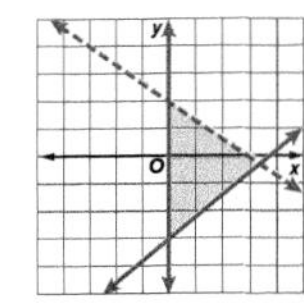

Your Notes

Example 3 Use Systems of Inequalities

TOURS **A Niagara Falls boat tour company charges $19.50 for adult tickets and $11 for children's tickets. Each boat has a capacity of 600 passengers, including 8 crew members. Suppose the company's operating cost for one boat tour is $2750. Write and graph a system of inequalities to represent the situation so the company will make a profit on each tour. Then, identify some viable solutions.**

Part A Write the system of inequalities.

Let a represent the number of adult tickets and c represent the number of children's tickets.

Inequality 1: $a + c + 8 \leq 600$ Inequality 2: $19.5a + 11c > 2750$

Inequality 3: $a \geq 0$ Inequality 4: $c \geq 0$

Part B Graph the system of inequalities.

Graph the inequalities.

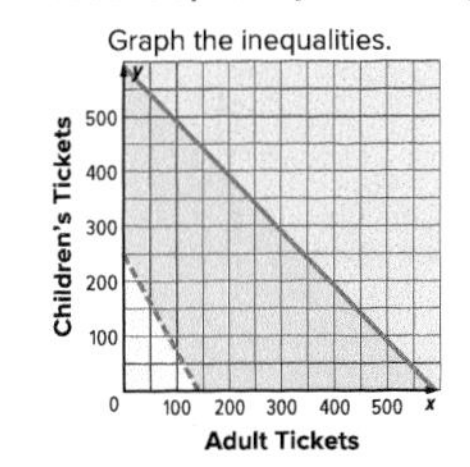

Identify feasible region.

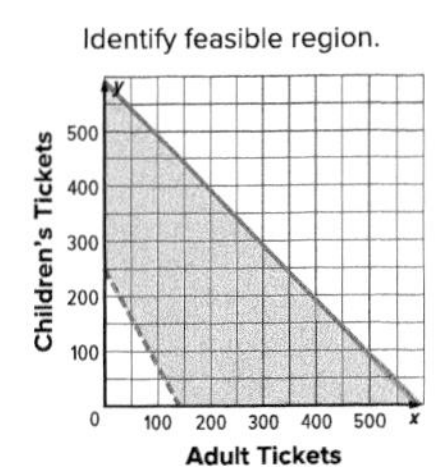

Part C Identify some viable solutions.

Passengers	Viable	Nonviable
60 adults, 100 children	☐	☒
210 adults, 350 children	☒	☐
415 adults, 200 children	☐	☒
390 adults, 240 children	☐	☒
550 adults, 0 children	☒	☐

Go Online You can complete an Extra Example online.

Talk About It! Why is it important to label the axes given the context of this problem? Explain.

Sample answer: Because either adult tickets or children's tickets could represent the independent variable, the graph could be made with either variable on the x-axis. It is important to label the graph so that viable solutions can be determined correctly.

Study Tip

Consider the Context While the feasible region represents the viable solutions, the solution may be limited to only integers or only positive numbers. In this case, the touring company cannot sell a fraction of a ticket. So the solution must be given as whole numbers.

Interactive Presentation

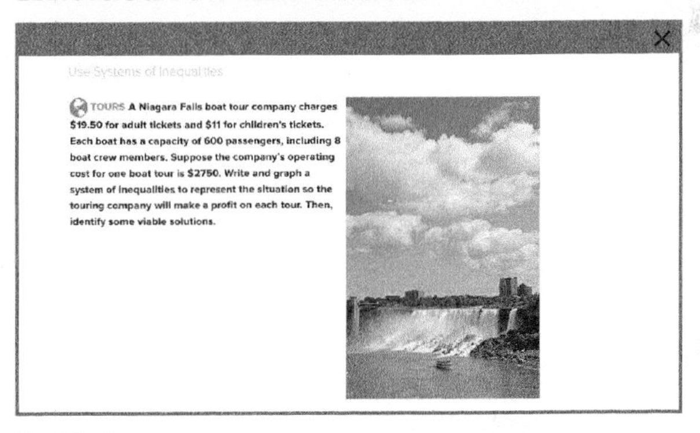

Example 3

TAP

Students move through the steps to write and solve a system of inequalities in a real-world situation.

SELECT

Students select whether each passenger group is viable or nonviable.

TYPE

Students answer a question to determine whether they understand the importance of labeling the axes given the context.

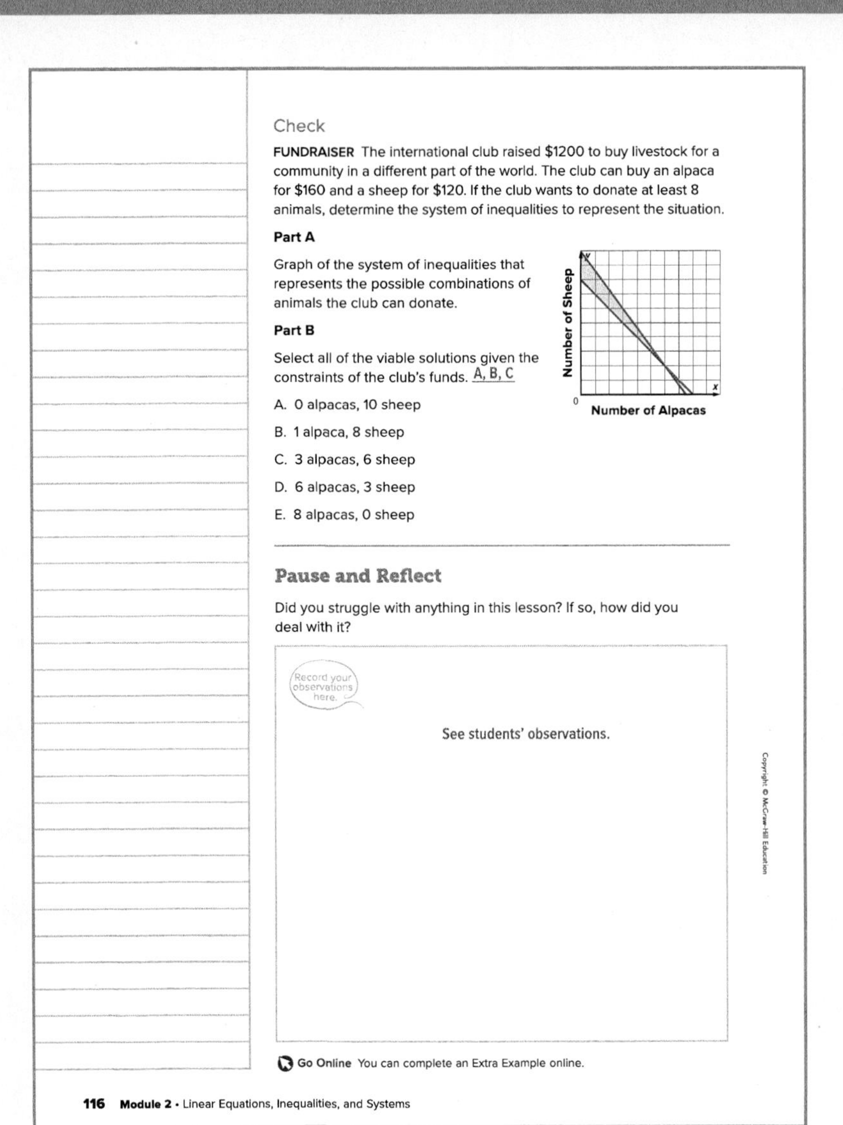

Check
FUNDRAISER The international club raised $1200 to buy livestock for a community in a different part of the world. The club can buy an alpaca for $160 and a sheep for $120. If the club wants to donate at least 8 animals, determine the system of inequalities to represent the situation.
Part A
Graph of the system of inequalities that represents the possible combinations of animals the club can donate.
Part B
Select all of the viable solutions given the constraints of the club's funds. A, B, C
A. 0 alpacas, 10 sheep
B. 1 alpaca, 8 sheep
C. 3 alpacas, 6 sheep
D. 6 alpacas, 3 sheep
E. 8 alpacas, 0 sheep
Number of Sheep
Number of Alpacas
Pause and Reflect
Did you struggle with anything in this lesson? If so, how did you deal with it?
Record your observations here.
See students' observations.
Go Online You can complete an Extra Example online.
116 Module 2 • Linear Equations, Inequalities, and Systems

Exit Ticket

Recommended Use

At the end of class, have students respond to the Exit Ticket prompt using a separate piece of paper. Have students hand you their responses as they leave the room.

Alternate Use

At the end of class, have students respond to the Exit Ticket prompt verbally or by using a mini-whiteboard. Have students hold up their whiteboards so that you can see all student responses. Tap to reveal the answer when most or all students have completed the Exit Ticket.

CHECK

Students complete the Check online to determine whether they are ready to move on.

Practice and Homework

Suggested Assignments

Use the table below to select appropriate exercises.

DOK	Topic	Exercises
1, 2	exercises that mirror the examples	1–14
2	exercises that use a variety of skills from this lesson	15–18
2	exercises that extend concepts learned in this lesson to new contexts	19–20
3	exercises that emphasize higher-order and critical thinking skills	21–24

ASSESS AND DIFFERENTIATE

Use the data from the **Checks** to determine whether to provide resources for extension, remediation, or intervention.

IF students score 90% or more on the Checks, **THEN** assign: BL

- Practice Exercises 21–24
- Extension: Creative Designs
- **ALEKS**® Systems of Inequalities and Linear Programming

IF students score 66%–89% on the Checks, **THEN** assign: OL

- Practice Exercises 1–17 odd
- Remediation, Review Resources: Elimination Using Addition and Subtraction
- Personal Tutors
- Extra Examples 1–3
- **ALEKS**® Systems of Linear Equations

IF students score 65% or less on the Checks, **THEN** assign: AL

- Practice Exercises 1–13 odd
- Remediation, Review Resources: Elimination Using Addition and Subtraction
- *Quick Review Math Handbook*: Solving Systems of Inequalities by Graphing
- **ALEKS**® Systems of Linear Equations

Name ______ Period ______ Date ______

Practice

Go Online You can complete your homework online.

Example 1

Solve each system of inequalities. 1–6. See margin.

1. $x - y \leq 2$, $x + 2y \geq 1$
2. $3x - 2y \leq -1$, $x + 4y \geq -12$
3. $y \geq \frac{x}{2} - 3$, $y < 2x$
4. $y < \frac{x}{3} + 2$, $y < -2x + 1$
5. $x + y \geq 4$, $2x - y > 2$
6. $x + 3y < 3$, $x - 2y \geq 4$

Example 2

Solve each system of inequalities. 7–12. See margin.

7. $y \geq -3x + 7$, $y > \frac{1}{2}x$, $y < 2$
8. $x > -3$, $y < -\frac{1}{3}x + 3$, $y > x - 1$
9. $y < -\frac{1}{2}x + 3$, $y > \frac{1}{2}x + 1$, $y < 3x + 10$
10. $y \leq 0$, $x \leq 0$, $y \geq -x - 1$
11. $y \leq 3 - x$, $y \geq 3$, $x \geq -5$
12. $x \geq -2$, $y \geq x - 2$, $x + y \leq 2$

Example 3

13. TICKETS The high school auditorium has 800 seats. Suppose that the drama club has a goal of making at least \$3400 each night of their spring play. Student tickets are \$4 and adult tickets are \$7.
    a. Write a system of inequalities to represent the situation. Let x be student tickets and y be adult tickets. $x + y \leq 800$; $4x + 7y \geq 3400$
    b. Graph the system of inequalities. In which quadrant(s) is the solution? See margin.
    c. Could the club meet its goal by selling 200 adult and 475 student tickets? Explain. No; they would only make \$3300.

14. CONSTRUCT ARGUMENTS Anthony charges \$15 an hour for tutoring and \$10 an hour for babysitting. He can work no more than 14 hours a week. How many hours should Anthony spend on each job if he wants to earn at least \$125 each week? b–c. See margin.
    a. Write a system of inequalities to represent this situation. Let x represent hours of tutoring and y represent hours of babysitting. $x + y \leq 14$; $15x + 10y \geq 125$
    b. Graph the system of inequalities and highlight the solution.
    c. Determine whether (4, 5), (7, 6), and (5, 10) are viable solutions given the constraints of the situation. Explain.

Lesson 2-6 • Solving Systems of Inequalities 117

Answers

1.

2.

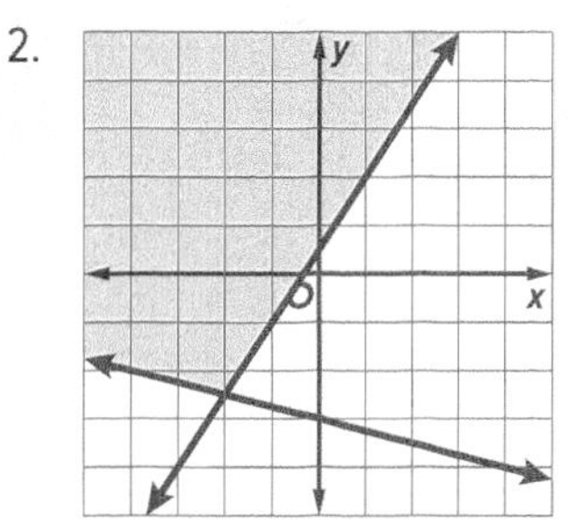

3.

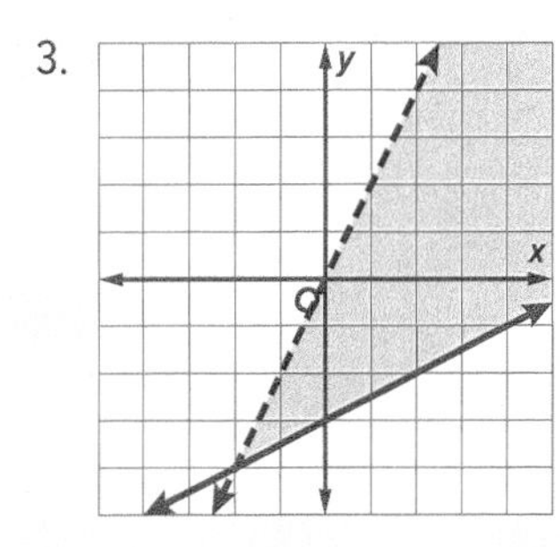

4.

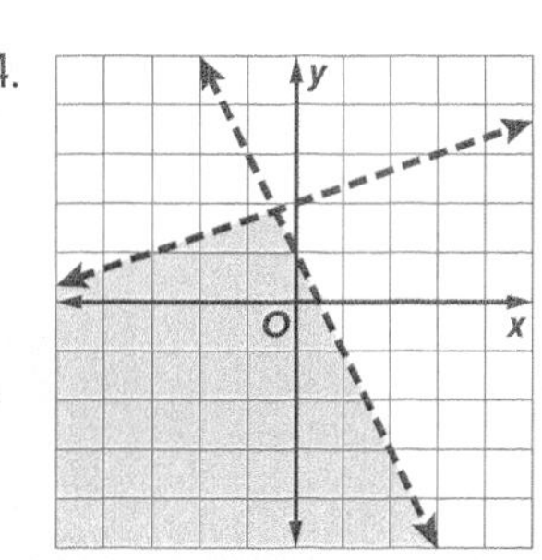

5.

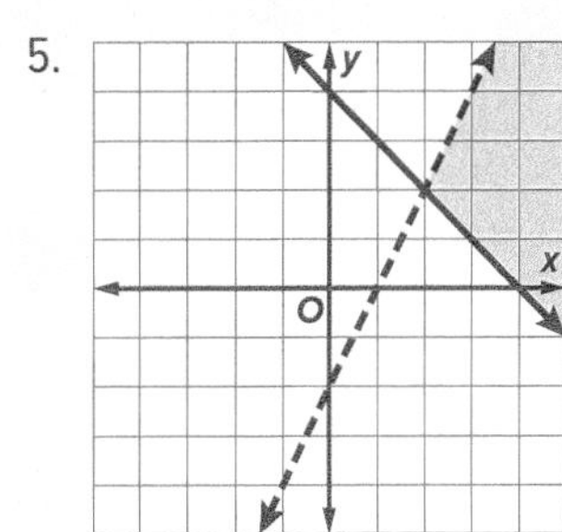

6.

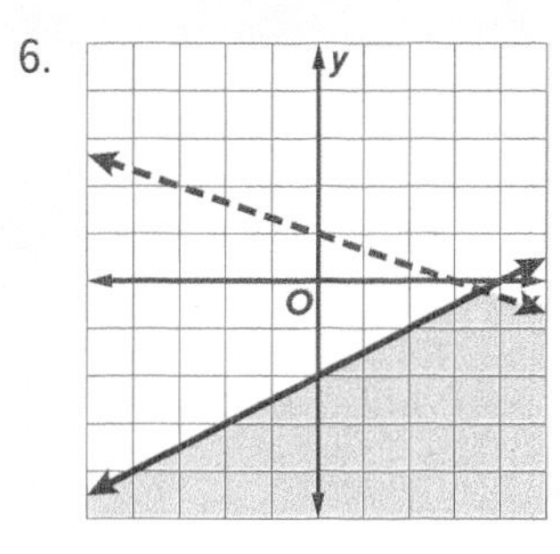

Mixed Exercises

Solve each system of inequalities. 15–18. See Mod. 2 Answer Appendix.

15. $y \geq |2x + 4| - 2$
$3y + x \leq 15$

16. $y \geq |6 - x|$
$y \leq 4$

17. $y > -3x + 1$
$4y \leq x - 8$
$3x - 5y < 20$

18. $|x| > y$
$y \leq 6$
$y \geq -2$

19. FINANCE Sheila plans to invest \$2000 or less in two different accounts. The low risk account pays 3% annual simple interest, and the high risk account pays 12% annual simple interest. Sheila wants to make at least \$150 in interest this year.

a. Define the variables, then write and graph a system of inequalities to show how Sheila can split her investment between the accounts.
Let x represent the low risk investment and y represent the high risk investment. $x + y \leq 2000$; $0.03x + 0.12y \geq 150$; See Mod. 2 Answer Appendix for graph.

b. Explain why your graph for this situation is restricted to Quadrant I.
Sample answer: Because Sheila cannot invest a negative amount of money, the graph is limited to positive values of x and y.

c. Give three viable solutions to meet the constraints of Sheila's investments.
Sample answer: \$200 in low risk and \$1700 in high risk; \$400 in low risk and \$1600 in high risk; \$1000 in low risk and \$1000 in high risk

20. STRUCTURE Write a system of inequalities for the graph shown.
$x + y \leq 3$; $x \geq -3$; $y \geq -2$

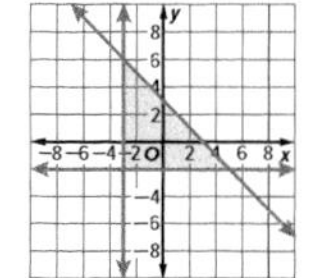

Higher Order Thinking Skills

21. PERSEVERE Find the area of the region defined by the following inequalities. 75 units^2

$y \geq -4x - 16$
$4y \leq 26 - x$
$3y + 6x \leq 30$
$4y - 2x \geq -10$

22. CREATE Write systems of two inequalities in which the solution:

a. lies only in the third quadrant. Sample answer: $y < -2, x < -1$

b. does not exist. Sample answer: $y > 2, y < -2$

c. lies only on a line. Sample answer: $y \geq x, y \leq x$

23. ANALYZE Determine whether the statement is *true* or *false*. Justify your argument.

A system of two linear inequalities has either no points or infinitely many points in its solution. True; sample answer: The feasible region is the intersection of the graphs of the inequalities. If the graphs intersect, there are infinitely many points in the feasible region. If the graphs do not intersect, they contain no common points and there is no solution.

24. WRITE Explain how you would determine whether (−4, 6) is a solution of a system of inequalities. See margin.

118 Module 2 • Linear Equations, Inequalities, and Systems

Answers

7.

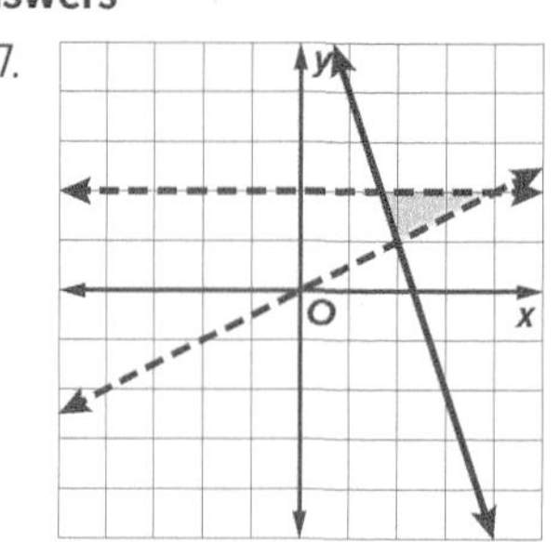

8.

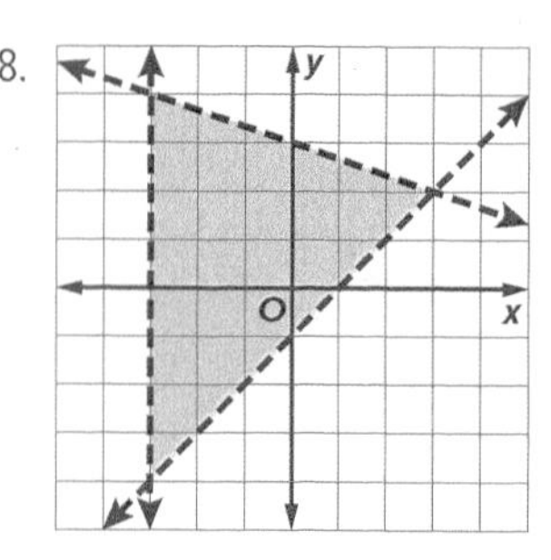

9.

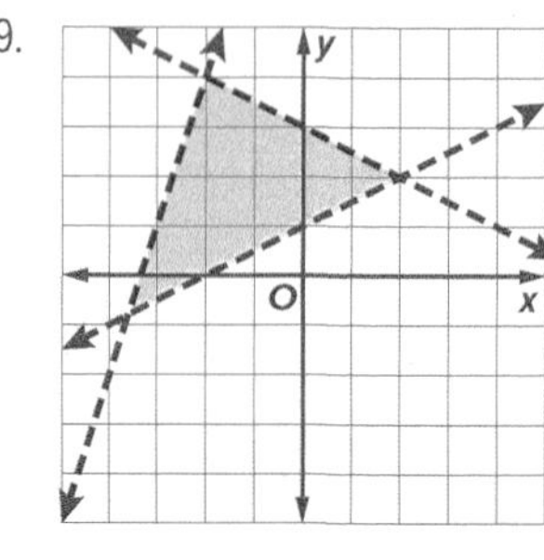

10.

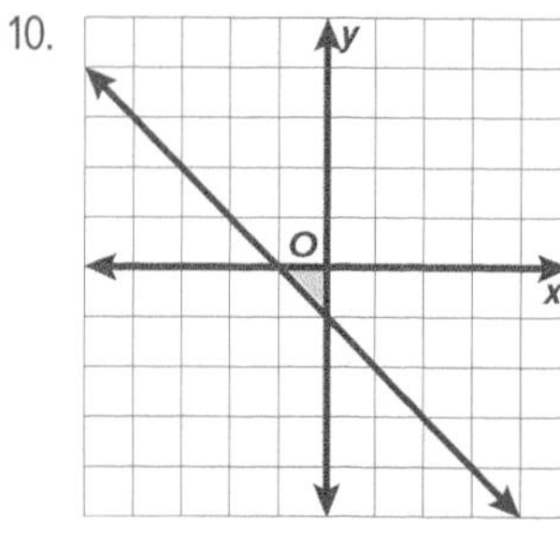

11.

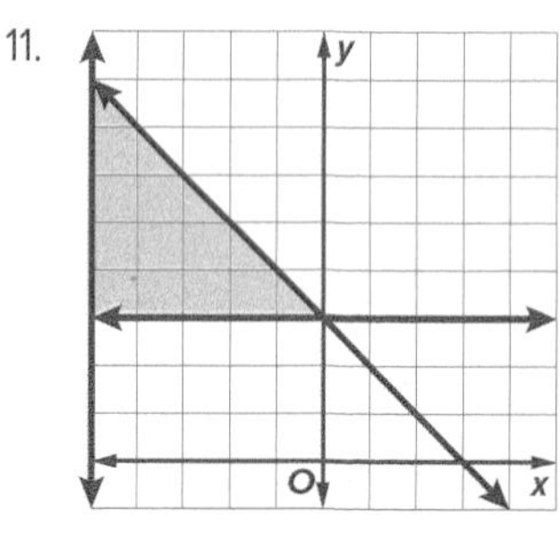

12.

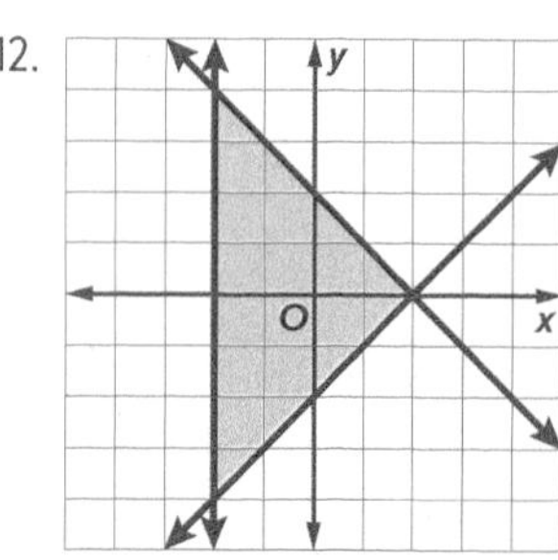

13b.

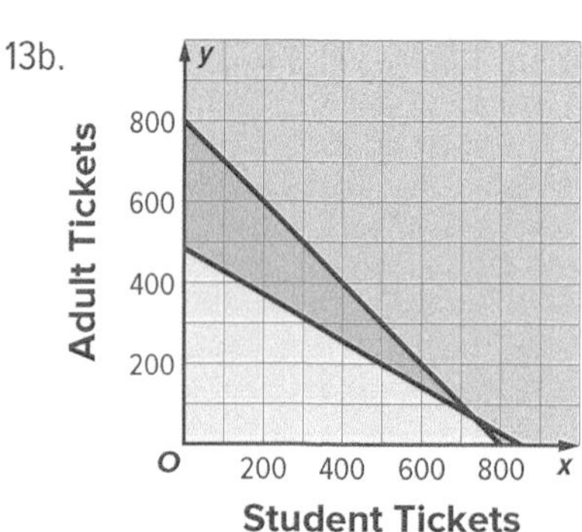

14b.

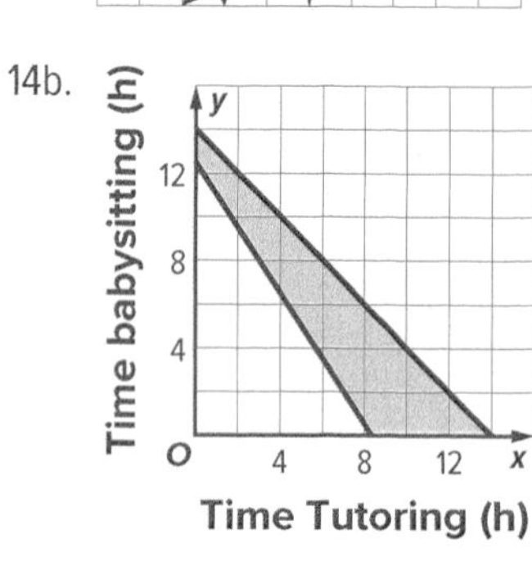

14c. (4, 5): no, not in feasible region and is only \$110; (7, 6): yes, in the feasible region and is 13 hours, \$165; (5, 10): no, not in the feasible region and 15 hours is too many hours

24. Sample answer: Determine whether the point falls in the shaded area of the graphs and/or determine whether the values satisfy each inequality.

Lesson 2-7

A.CED.3

Optimization with Linear Programming

LESSON GOAL

Students use linear programming to find maximum and minimum values of a function.

1 LAUNCH

 Launch the lesson with a **Warm Up** and an introduction.

2 EXPLORE AND DEVELOP

 **Develop:**

Finding Maximum and Minimum Values
- Maximum and Minimum Values for a Bounded Region
- Maximum and Minimum Values for an Unbounded Region

 **Explore:** Using Technology with Linear Programming

 **Develop:**

Linear Programming
- Optimizing with Linear Programming

You may want your students to complete the **Checks** online.

3 REFLECT AND PRACTICE

 Exit Ticket

 Practice

DIFFERENTIATE

 View reports of student progress on the **Checks** after each example.

Resources	AL	OL	BL	ELL
Remediation: Systems of Inequalities	●	●		●
Extension: Sensitivity Analysis		●	●	●
ELL Support				

Language Development Handbook

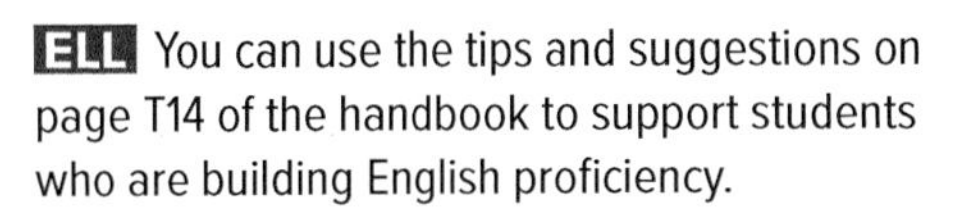

Assign page 14 of the *Language Development Handbook* to help your students build mathematical language related to using linear programming to find maximum and minimum values of a function.

ELL You can use the tips and suggestions on page T14 of the handbook to support students who are building English proficiency.

Suggested Pacing

90 min	0.5 day
45 min	1 day

Focus

Domain: Algebra

Standards for Mathematical Content:

A.CED.3 Represent constraints by equations or inequalities, and by a system of equations and/or inequalities, and interpret solutions as viable or non-viable options in a modeling context.

Standards for Mathematical Practice:

4 Model with mathematics.

5 Use appropriate tools strategically.

Coherence

Vertical Alignment

Previous
Students solved linear equations in two variables.
A.REI.5, A.REI.6, A.REI.7 (Algebra 1, Algebra 2)

Now
Students use linear programming to find maximum or minimum values of a function.
A.CED.3

Next
Students will solve systems of equations in three variables.
A.CED.3

Rigor

The Three Pillars of Rigor

1 CONCEPTUAL UNDERSTANDING	2 FLUENCY	3 APPLICATION

Conceptual Bridge In this lesson, students build on their understanding of solving systems of linear inequalities to optimizing with linear programming. They build fluency and apply their understanding by using linear programming to solve real-world problems.

Mathematical Background

The process of finding maximum or minimum values of a function for a region defined by linear inequalities is called linear programming.

Interactive Presentation

Warm Up

Write a system of inequalities to model each situation.

1. **NUMBERS** The sum of two positive numbers is less than 15, and the difference between the numbers is greater than 6.

2. **JEWELRY** A jeweler wants to use gold beads and pearls to make some necklaces. The gold beads cost $10 each, and the pearls cost $15 each. He wants to spend no more than $300 per necklace for beads, and each necklace must use at least 15 beads. Let x represent the number of gold beads, and let y represent the number of pearls.

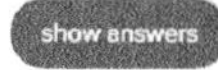

Warm Up

Warm Up

Prerequisite Skills

The Warm Up exercises address the following prerequisite skill for this lesson:

- writing systems of inequalities

Answers:

1. $x + y < 15, x - y > 6, x > 0, y > 0$
2. $10x + 15y \leq 300, x + y \geq 15, x \geq 0, y \geq 0$

Launch the Lesson

For many hospitals, one of the biggest challenges is staffing the optimal number of nurses for each shift. There need to be enough nurses to provide care to patients without overstaffing, which costs the hospital money. Hospitals can use past records to determine the average number of patients by day and shift as well as the cost associated with the nursing staff. Then, using this information, they can determine the appropriate medical staff to maximize patient care and safety while minimizing cost.

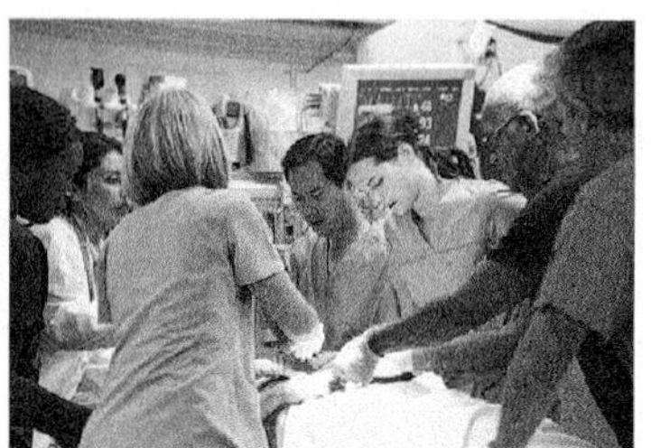

Launch the Lesson

Launch the Lesson

Teaching the Mathematical Practices

4 Apply Mathematics In the Launch the Lesson, students will learn how linear programming and optimization can be used by hospitals to determine the optimal number of staff members.

Go Online to find additional teaching notes and questions to promote classroom discourse.

Today's Standards

Tell students that they will be addressing these content and practice standards in this lesson. You may wish to have a student volunteer read aloud *How can I meet these standards?* and *How can I use these practices?*, and connect these to the standards.

See the Interactive Presentation for I Can statements that align with the standards covered in this lesson.

Vocabulary

Collapse All

linear programming

The process of finding the maximum or minimum values of a function for a region defined by a system of linear inequalities.

optimization

Today's Vocabulary

Today's Vocabulary

Tell students that they will be using these vocabulary terms in this lesson. You can expand each row if you wish to share the definitions. Then discuss the questions below with the class.

Explore Using Technology with Linear Programming

Objective

Students use a graphing calculator to explore maximum and minimum values over a region.

Teaching the Mathematical Practices

5 Analyze Graphs In the Explore, students will analyze the graph they have generated using graphing calculators. They will locate the vertices of the bounded region and evaluate the given function at the vertices.

Ideas for Use

Recommended Use Present the Inquiry Question, or have a student volunteer read it aloud. Have students work in pairs to complete the Explore activity on their devices. Pairs should discuss each of the questions. Monitor student progress during the activity. Upon completion of the Explore activity, have student volunteers share their responses to the Inquiry Question.

What if my students don't have devices? You may choose to project the activity on a whiteboard. A printable worksheet for each Explore is available online. You may choose to print the worksheet so that individuals or pairs of students can use it to record their observations.

Summary of the Activity

Students will complete guiding exercises throughout the Explore activity. They will learn how to use the graphing calculator to solve linear programming questions. They will use the graphing calculator to identify the feasible region and the vertices of the region. They will evaluate the vertices in the given function to identify the maximum and minimum value. Then, students will answer the Inquiry Question.

(continued on the next page)

Interactive Presentation

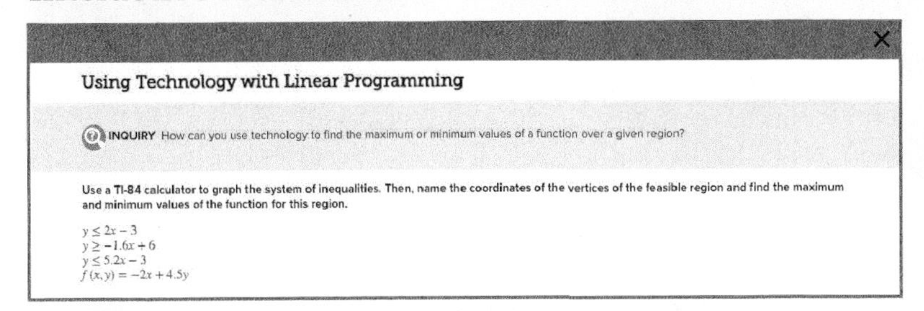

Explore

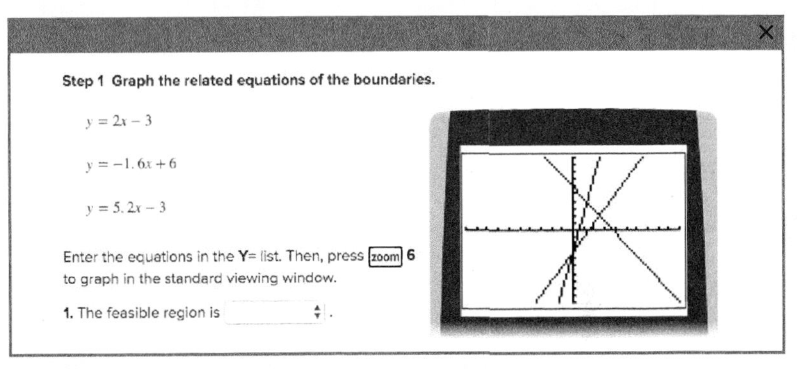

Explore

SELECT

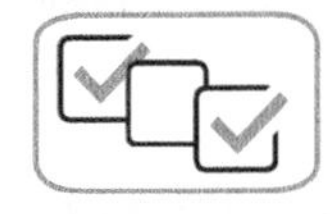

Students select the correct type of feasible region, the maximum of the feasible region, and the minimum.

TYPE

Students complete the calculations for the coordinates of the vertices of the feasible region and the function value at each vertex.

Interactive Presentation

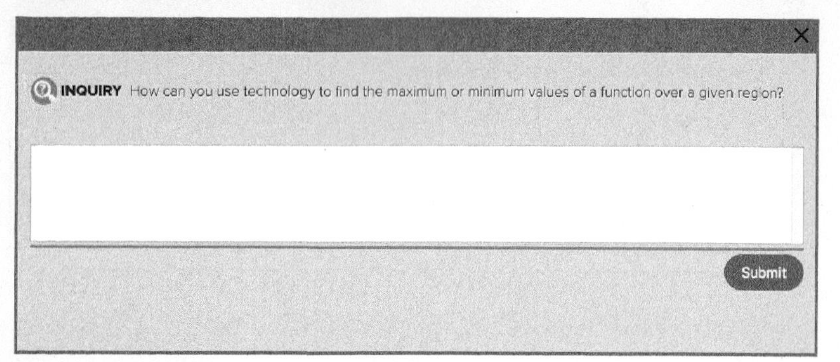

Explore

TYPE

a|

Students respond to the Inquiry Question and can view a sample answer.

Explore Using Technology with Linear Programming (*continued*)

Teaching the Mathematical Practices

5 Use Appropriate Tools Strategically Throughout the Explore, encourage students to use the graphing calculator to explore linear programming, and analyze the graph and its feasible region.

Questions

Have students complete the Explore activity.

Ask:

- What is the shape of the feasible region? a triangle
- In a real-world setting, what other inequalities are usually included in the system of inequalities and why? $x \geq 0$ and $y \geq 0$; because we only graph in the first quadrant since negative values do not make sense

Inquiry

How can you use technology to find the maximum or minimum values of a function over a given region? Sample answer: I can use a graphing calculator to graph the related equations and find the coordinates of the vertices of the feasible region. Then, I can use the calculator to evaluate each vertex to determine the maximum and minimum.

Go Online to find additional teaching notes and sample answers for the guiding exercises.

Learn Finding Maximum and Minimum Values

Objective

Students find maximum and minimum values of a function over a region by using its vertices.

Teaching the Mathematical Practices

1 Special Cases Work with students to look at the two cases in Step 4 of the Key Concept. Encourage students to familiarize themselves with both cases, and know when to use each one.

About the Key Concept

Linear programming is the process of finding the maximum or minimum values of a function for a region defined by a system of inequalities. Each inequality is a constraint of the system, and the vertices of the feasible region and the function are used to find the maximum or minimum value. First the inequalities must be graphed and the coordinates of the vertices found. Next, the function is evaluated at each vertex. If the region is bounded, the maximum or minimum can be determined by the function values, but if the region is unbounded, other points within the feasible region must be tested to determine which vertex represents the maximum or minimum value.

Example 1 Maximum and Minimum Values for a Bounded Region

Teaching the Mathematical Practices

1 Explain Correspondences Encourage students to explain the relationships between the inequalities, function, graph, and table used in Example 1.

Questions for Mathematical Discourse

AL Why is $f(x, y)$ not graphed with the inequalities? The function cannot be graphed because it cannot be solved for y. How is $f(x, y)$ a function if it cannot be solved for y? Any point on the xy-plane results in only one value of $f(x, y)$.

OL Why do the constraints need inclusive inequality symbols in linear programming problems? Sample answer: Because the vertices of the feasible region lie on the boundary lines, the inequalities must include the points on the line. Otherwise, the vertices would not be included in the feasible region.

BL How could you graphically represent $f(x, y)$, and what the function look like? You would need a 3-dimensional coordinate system. $f(x, y)$ would be a surface with each (x, y) only resulting in one value for $f(x, y)$.

Common Error

One of the biggest mistakes students make when graphing a large number of inequalities, is going the wrong direction with slopes. For a negative slope, students may go in the positive direction. Encourage students to pay attention to details like slope so the inequalities will intersect at the correct point.

Lesson 2-7

Optimization with Linear Programming

Learn Finding Maximum and Minimum Values

Linear programming is the process of finding the maximum or minimum values of a function for a region defined by a system of linear inequalities.

Key Concept • Linear Programming

Step 1 Graph the inequalities.

Step 2 Determine the coordinates of the vertices.

Step 3 Evaluate the function at each vertex.

Step 4 For a bounded region, determine the maximum and minimum. For an unbounded region, test other points within the feasible region to determine which vertex represents the maximum or minimum.

Today's Goals
- Find maximum and minimum values of a function over a region.
- Solve real-world optimization problems by graphing systems of inequalities maximizing or minimizing constraints.

Today's Vocabulary
linear programming
optimization

Example 1 Maximum and Minimum Values for a Bounded Region

Graph the system of inequalities. Name the coordinates of the vertices of the feasible region. Find the maximum and minimum values of the function for this region.

$-2 \leq x \leq 4$

$y \leq x + 2$

$y \geq -0.5x - 3$

$f(x, y) = -2x + 6y$

Steps 1 and 2 Graph the inequalities and determine the vertices.

The vertices of the feasible region are (−2, −2), (−2, 0), (4, −5) and (4, 6).

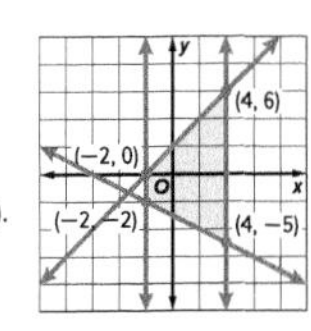

Step 3 Evaluate the function at each vertex.

(x, y)	$-2x + 6y$	$f(x, y)$
(−2, −2)	−2(−2) + 6(−2)	−8
(−2, 0)	−2(−2) + 6(0)	4
(4, −5)	−2(4) + 6(−5)	−38
(4, 6)	−2(4) + 6(6)	28

Step 4 Determine the maximum and minimum.

The maximum value is 28 at (4, 6). The minimum value is −38 at (4, −5).

Study Tip

Unbounded Regions An unbounded feasible region does not necessarily contain a maximum or minimum.

Study Tip

Feasible Region To determine the feasible region, you can shade the solution set of each inequality individually, and then find where they all overlap. Shading each inequality using a different color or shading style can help you easily determine the feasible region.

Interactive Presentation

Finding Maximum and Minimum Values

Linear programming is the process of finding the maximum or minimum values of a function for a region defined by a system of inequalities. Each inequality represents a constraint of the system. Once the inequalities are graphed, the vertices of the feasible region and the function can be used to find the maximum or minimum value.

Learn

TAP

Students tap to move through the steps to linear programming.

Your Notes

Study Tip

Feasible Region To determine whether an unbounded region has a maximum or minimum for the function $f(x, y)$, you need to test several points in the feasible region to see if any values of $f(x, y)$ are greater than or less than the values of $f(x, y)$ for the vertices.

Talk About It!

The function in the Example has a minimum but no maximum on the unbounded region. Is there a function that has a maximum, but no minimum on the region? Does the function $f(x) = y$ have a maximum and/or a minimum on the region? Justify your reasoning.

Sample answer: $f(x) = 2x + 3y$ has a maximum of 3 at $(-3, 3)$ but no minimum.; $f(x) = y$ has a minimum of 1 at $(-4, 1)$ and other points on the boundary and a maximum of 3 at $(-3, 3)$ and other points on the boundary.

Example 2 Maximum and Minimum Values for an Unbounded Region

Graph the system of inequalities. Name the coordinates of the vertices of the feasible region. Find the maximum and minimum values of the function for this region.

$1 \le y \le 3$

$y \le -x$

$y \ge 0.5x + 3$

$f(x, y) = -x + y$

Steps 1 and 2 Graph the inequalities and determine the vertices.

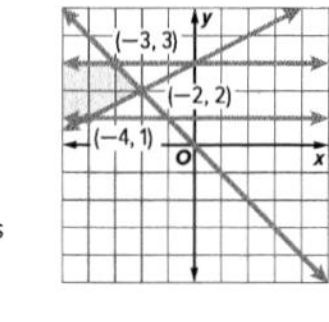

The vertices of the feasible region are $(-3, 3)$, $(-2, 2)$, and $(-4, 1)$.

Notice that the region is unbounded. This may indicate that there is no minimum or maximum value.

Step 3 Evaluate the function.

Evaluate at each vertex and a point in the feasible region.

(x, y)	$-x + y$	$f(x, y)$
$(-3, 3)$	$-(-3) + 3$	6
$(-2, 2)$	$-(-2) + 2$	4
$(-4, 1)$	$-(-4) + 1$	5
$(-10, 2)$	$-(-10) + 2$	12

Step 4 The minimum value is 4 at $(-2, 2)$. As shown by the test point $(-10, 2)$, there is no maximum value.

Explore Using Technology with Linear Programming

Online Activity Use graphing technology to complete the Explore.

INQUIRY How can you use technology to find the maximum or minimum values of a function over a given region?

Go Online You can complete an Extra Example online.

Interactive Presentation

Maximum and Minimum Values for an Unbounded Region

Graph the system of inequalities. Name the coordinates of the vertices of the feasible region. Find the maximum and minimum values of the function for this region.

$1 \le y \le 3$
$y \le -x$
$y \ge 0.5x + 3$
$f(x, y) = -x + y$

Example 2

TAP

Students tap to graph the inequalities and determine the vertices.

TYPE

Students answer a question to determine whether they understand the difference between a bounded and unbounded region.

CHECK

Students complete the Check online to determine whether they are ready to move on.

Example 2 Maximum and Minimum Values for an Unbounded Region

Teaching the Mathematical Practices

6 Communicate Precisely Encourage students to routinely write or explain their reasoning. Point out that they should use clear definitions when they discuss their answer to the question in the Talk About It! feature.

Questions for Mathematical Discourse

AL How do you find the vertices of the feasible region? Sample answer: Graph the related equation of each inequality. Shade the inequalities using test points and find where the shaded regions overlap for all the inequalities. The vertices are the points where the boundaries of the inequalities intersect. You can identify them from the graph or solve for them using the related equations.

OL Suppose the point $(-5, 2)$ were used as the test point. Would the result have been different? Explain. No; sample answer: Even though $(-5, 2)$ is closer to the other vertices, it still has a function value larger than the other coordinates. Its value is 7, so there is still no maximum.

BL Suppose the third inequality were $y \le 0.5x + 3$. How would this change the results? Sample answer: The feasible region would now be bounded, so there would be a maximum and a minimum value.

Common Error

When the feasible region is unbounded, students may forget to test a point in the region or they may not understand what the function value of the test point reveals about the maximum or minimum. Remind students the test point helps you determine if there is a maximum or minimum since an unbounded region often does not have both. If the function value for the test point is either greater or less than all of the other values, it indicates that the function has no maximum or minimum, respectively.

Go Online

- Find additional teaching notes.
- View performance reports of the Checks.
- Assign or present an Extra Example.

Learn Linear Programming

Objective

Students solve real-world optimization problems by graphing systems of inequalities and using the vertices of the feasible region to maximize or minimize constraints.

 Teaching the Mathematical Practices

4 Apply Mathematics Students will learn how to apply linear programming to real-world problems to optimize the situation.

What Students Are Learning

Linear programming can be applied to real-world situations where decisions to improve systems are needed, such as cost, production, or investments. Optimization is the process of seeking the optimal price or amount desired to minimize costs or maximize profits.

Essential Question Follow-Up

Students have begun learning about linear programming and optimization.

Ask:

Why is linear programming important in the real world?

Sample answer: Linear programming can help businesses identify the right combination of variables to maximize their profit or minimize their cost.

Example 3 Optimizing with Linear Programming

 Teaching the Mathematical Practices

4 Interpret Mathematical Results In Example 3, point out that to solve the problem, students should interpret their mathematical results in the context of the problem. The Think About It! feature asks students to consider whether the results makes sense.

Questions for Mathematical Discourse

AL What is the slope-intercept form for the inequality $c + t \geq 16$? $t \geq -c + 16$

OL What are the function values for the vertices? 304, 338, 336

BL Suppose the function instead represented the profit Avoree could make on each cucumber and lettuce plant. What value would she be interested in now, and why? the maximum; sample answer: Avoree would be interested in making the most money from her garden, so she would care about the greatest function value.

Common Error

When students are faced with the long word problems optimization problems present, they may feel overwhelmed and unsure of how to begin. Encourage students to always read the problem twice and define any variables before trying to write inequalities.

Learn Linear Programming

Optimization is the process of seeking the optimal value of a function subject to given constraints.

Key Concept • Optimization with Linear Programming

Step 1 Define the variables.

Step 2 Write a system of inequalities.

Step 3 Graph the system of inequalities.

Step 4 Find the coordinates of the vertices of the feasible region.

Step 5 Write a linear function to be maximized or minimized.

Step 6 Evaluate the function at each vertex by substituting the coordinates into the function.

Step 7 Interpret the results.

Math History Minute

In the 1960s, **Christine Darden (1942–)** became one of the "human computers" who crunched numbers for engineers at NASA's Langley Research Center. After earning a doctorate degree in mechanical engineering, Darden became one of few female aerospace engineers at NASA Langley. For most of her career, her focus was sonic boom minimization.

Example 3 Optimizing with Linear Programming

GARDENING **Avoree has a 30-square-foot plot in the school greenhouse and wants to plant lettuce and cucumbers while minimizing the amount of water she uses for them. Each cucumber requires 2.25 square feet of space and uses 25 gallons of water over the lifetime of the plant. Each lettuce plant requires 1.5 square feet of space and uses 17 gallons of water. She wants to grow at least 4 of each type of plant and at least 16 plants in total. Determine how many of each plant Avoree should plot in order to minimize her water usage.**

Step 1 Define the variables.

Because the number of plants of different types determine the water usage, the independent variables should be the numbers of plants. The dependent variable in the function to be minimized should be total water used. Let c represent the number of cucumber plants and t represent the number of lettuce plants. Let $f(c, t)$ represent the water used for c cucumber plants and t lettuce plants.

Step 2 Write a system of inequalities.

Avoree wants to have at least 4 of each type of plant, so 4 must be included as minimums for both c and t in the inequalities. The total number of plants must be at least 16. Each cucumber requires 2.25 square feet of space and each lettuce plant requires 1.5 square feet of space. The total planting area of the plants must be less than or equal to 30 ft^2.

$c \geq 4$

$t \geq \underline{4}$

$c + t \geq \underline{16}$

$2.25c + 1.5t \leq 30$

(continued on the next page)

Go Online You can complete an Extra Example online.

Interactive Presentation

Linear Programming

The principles of linear programming are applied to schedule airline crews, improve production and distribution for manufacturers, and to make sound financial investments. In each case, the decision maker is trying to determine the best ways to optimize the business or situation. **Optimization** is the process of seeking the optimal price or amount that is desired to minimize costs or maximize profits. The decision maker must also consider constraints and limitations of resources in order to find viable solutions.

Learn

EXPAND

Students can tap to see about a real-world mathematician.

Step 3 Graph the system of inequalities.

Step 4 Find the coordinates of the vertices of the feasible region.

The vertices of the feasible region are (4, 12), (4, 14), and (8, 8).

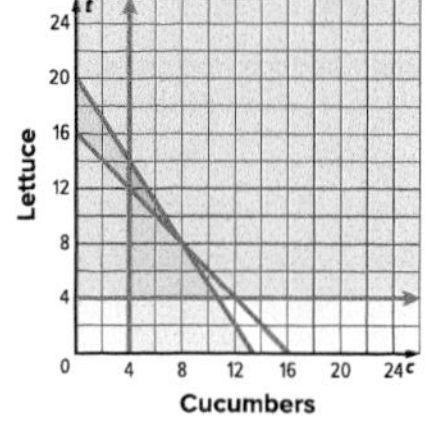

Step 5 Write a linear function to be minimized.

Because Avoree wants to minimize her water usage, the linear function will be the sum of the water usage for each plant.

$f(c, t) = \underline{25}\,c + \underline{17}\,t$

Step 6 Evaluate the function at each vertex.

(c, t)	25c + 17t	f(c, t)
(4, 12)	25(4) + 17(12)	304
(4, 14)	25(4) + 17(14)	338
(8, 8)	25(8) + 17(8)	336

Step 7 Interpret the results.

Avoree should plant __4__ cucumber plants and __12__ lettuce plants to minimize her water usage.

Think About It! Does this solution seem reasonable? Explain.

Yes; sample answer: Because cucumber plants use much more water and space than lettuce plants, it makes sense that only the minimum number of cucumber plants should be used.

Check

SOCCER A new soccer team is being created for a professional soccer league, and they need to hire at least ten new players. They need five to eight defenders and seven to ten forwards, and they want to minimize the amount they spend on these players' salaries so they have enough money remaining to hire goalkeepers and midfielders. Determine the number of forwards f and defenders d they should hire to minimize the cost.

Position	Minimum	Maximum	Salary per Player ($)
forward f	5	8	120,000
defender d	7	10	100,000

The least amount of money that the team can spend is $ __1,300,000__ by hiring __5__ forwards and __7__ defenders.

Go Online You can complete an Extra Example online.

122 Module 2 • Linear Equations, Inequalities, and Systems

Interactive Presentation

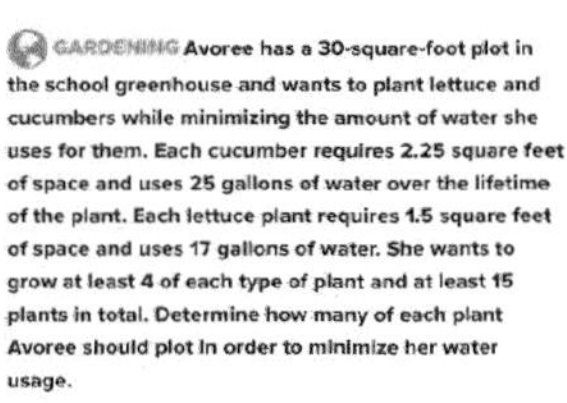

Example 3

TYPE

Students complete the calculations to solve the problem.

TAP

Students tap to move through the steps to solve the optimization problem.

CHECK

Students complete the Check online to determine whether they are ready to move on.

Common Misconception

Some students may skip graphing the system of inequalities and simply find the intersection of all the constraints. Remind students that the feasible region does not always occur between the intersection of all the inequalities, and graphing the system is a requirement.

DIFFERENTIATE

Reteaching Activity AL ELL

IF students are having problems writing the system of linear inequalities,

THEN have students define the variables and select a color to represent each variable. They will then underline or highlight any information in the problem related to each variable in the appropriate color. This helps students process the information and categorize the values to help write the inequalities.

Common Error

Students often believe the maximum or minimum value occurs where the two-variable inequalities intersect. Remind students that all vertices are possible solutions, and they must evaluate each vertex before determining the maximum or minimum value.

Exit Ticket

Recommended Use

At the end of class, go online to display the Exit Ticket prompt and ask students to respond using a separate piece of paper. Have students hand you their responses as they leave the room.

Alternate Use

At the end of class, go online to display the Exit Ticket prompt and ask students to respond verbally or by using a mini-whiteboard. Have students hold up their whiteboards so that you can see all student responses. Tap to reveal the answer when most or all students have completed the Exit Ticket.

Practice and Homework

Suggested Assignments

Use the table below to select appropriate exercises.

DOK	Topic	Exercises
1, 2	exercises that mirror the examples	1–11
2	exercises that use a variety of skills from this lesson	12
2	exercises that extend concepts learned in this lesson to new contexts	13, 14
3	exercises that emphasize higher-order and critical thinking skills	15–18

ASSESS AND DIFFERENTIATE

Use the data from the **Checks** to determine whether to provide resources for extension, remediation, or intervention.

IF students score 90% or more on the Checks, **THEN** assign: BL

- Practice Exercises 1–13 odd, 15–18
- Extension: Sensitivity Analysis
- **ALEKS**® Systems of Linear Inequalities and Linear Programming

IF students score 66%–89% on the Checks, **THEN** assign: OL

- Practice Exercises 1–17 odd
- Remediation, Review Resources: Systems of Inequalities
- Personal Tutors
- Extra Examples 1–3
- **ALEKS**® Systems of Linear Inequalities

IF students score 65% or less on the Checks, **THEN** assign: AL

- Practice Exercises 1–11 odd
- Remediation, Review Resources: Systems of Inequalities
- *Quick Review Math Handbook*: Optimization with Linear Programming
- **ALEKS**® Systems of Linear Inequalities

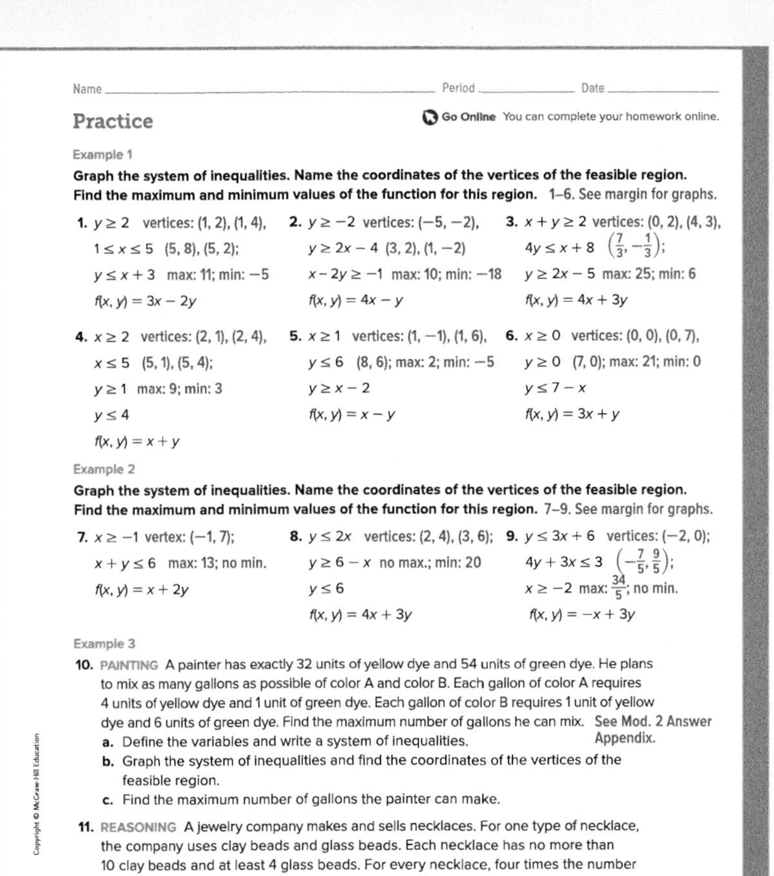

Name ______ Period ______ Date ______

Practice

Go Online You can complete your homework online.

Example 1

Graph the system of inequalities. Name the coordinates of the vertices of the feasible region. Find the maximum and minimum values of the function for this region. 1–6. See margin for graphs.

1. $y \geq 2$, $1 \leq x \leq 5$, $y \leq x + 3$, $f(x, y) = 3x - 2y$ vertices: (1, 2), (1, 4), (5, 8), (5, 2); max: 11; min: −5
2. $y \geq -2$, $y \geq 2x - 4$, $x - 2y \geq -1$, $f(x, y) = 4x - y$ vertices: (−5, −2), (3, 2), (1, −2) max: 10; min: −18
3. $x + y \geq 2$, $4y \leq x + 8$, $y \geq 2x - 5$, $f(x, y) = 4x + 3y$ vertices: (0, 2), (4, 3), $\left(\frac{7}{3}, -\frac{1}{3}\right)$; max: 25; min: 6
4. $x \geq 2$, $x \leq 5$, $y \geq 1$, $y \leq 4$, $f(x, y) = x + y$ vertices: (2, 1), (2, 4), (5, 1), (5, 4); max: 9; min: 3
5. $x \geq 1$, $y \leq 6$, $y \geq x - 2$, $f(x, y) = x - y$ vertices: (1, −1), (1, 6), (8, 6); max: 2; min: −5
6. $x \geq 0$, $y \geq 0$, $y \leq 7 - x$, $f(x, y) = 3x + y$ vertices: (0, 0), (0, 7), (7, 0); max: 21; min: 0

Example 2

Graph the system of inequalities. Name the coordinates of the vertices of the feasible region. Find the maximum and minimum values of the function for this region. 7–9. See margin for graphs.

7. $x \geq -1$, $x + y \leq 6$, $f(x, y) = x + 2y$ vertex: (−1, 7); max: 13; no min.
8. $y \leq 2x$, $y \geq 6 - x$, $y \leq 6$, $f(x, y) = 4x + 3y$ vertices: (2, 4), (3, 6); no max.; min: 20
9. $y \leq 3x + 6$, $4y + 3x \leq 3$, $x \geq -2$, $f(x, y) = -x + 3y$ vertices: (−2, 0); $\left(-\frac{7}{5}, \frac{9}{5}\right)$; max: $\frac{34}{5}$; no min.

Example 3

10. PAINTING A painter has exactly 32 units of yellow dye and 54 units of green dye. He plans to mix as many gallons as possible of color A and color B. Each gallon of color A requires 4 units of yellow dye and 1 unit of green dye. Each gallon of color B requires 1 unit of yellow dye and 6 units of green dye. Find the maximum number of gallons he can mix. See Mod. 2 Answer Appendix.
    a. Define the variables and write a system of inequalities.
    b. Graph the system of inequalities and find the coordinates of the vertices of the feasible region.
    c. Find the maximum number of gallons the painter can make.

11. REASONING A jewelry company makes and sells necklaces. For one type of necklace, the company uses clay beads and glass beads. Each necklace has no more than 10 clay beads and at least 4 glass beads. For every necklace, four times the number of glass beads is less than or equal to 8 more than twice the number of clay beads. Each clay bead costs \$0.20 and each glass bead costs \$0.40. The company wants to find the minimum cost to make a necklace with clay and glass beads and find the combination of clay and glass beads in a necklace that costs the least to make. See Mod. 2 Answer Appendix.
    a. Define the variables and write a system of inequalities. Then write an equation for the cost C.
    b. Graph the system of inequalities and find the coordinates of the vertices of the feasible region.
    c. Find the number of clay beads and glass beads in a necklace that costs the least to make.

Mixed Exercises

12. REASONING Juan has 8 days to make pots and plates to sell at a local craft fair. Each pot weighs 2 pounds and each plate weighs 1 pound. Juan cannot carry more than 50 pounds to the fair. Each day, he can make at most 5 plates and 3 pots. He will make \$12 profit for every plate and \$25 profit for every pot that he sells. See Mod. 2 Answer Appendix.

a. Write linear inequalities to represent the number of pots p and plates a Juan can bring to the fair.

b. List the coordinates of the vertices of the feasible region.

c. How many pots and plates should Juan make to maximize his potential profit?

13. USE A MODEL A trapezoidal park is built on a slight incline. The ground elevation above sea level is given by $f(x, y) = x - 3y + 20$ feet. What are the coordinates of the highest point in the park and what is the elevation at that point? (5, 2); 19 feet

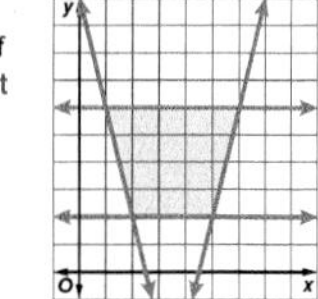

14. FOOD A zoo is mixing two types of food for the animals. Each serving is required to have at least 60 grams of protein and 30 grams of fat. Custom Foods has 15 grams of protein and 10 grams of fat and costs 80 cents per unit. Zookeeper's Best contains 20 grams of protein and 5 grams of fat and costs 50 cents per unit. See Mod. 2 Answer Appendix.

a. The zoo wants to minimize their costs. Define the variables and write the inequalities that represent the constraints of the situation.

b. Graph the inequalities. What does the unbound region represent? Determine how much of each type of food should be used to minimize costs.

Higher-Order Thinking Skills

15. ANALYZE Determine whether the following statement is *sometimes*, *always*, or *never* true. Explain your reasoning. See Mod. 2 Answer Appendix.

An unbounded region will not have both a maximum and minimum value.

16. WHICH ONE DOESN'T BELONG? Identify the system of inequalities that is not the same as the other three. Explain your reasoning.
b; The feasible region of Graph b is unbounded while the other three are bounded.

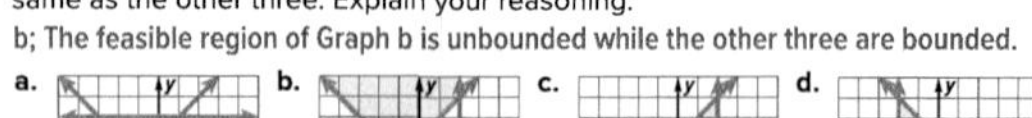

a. 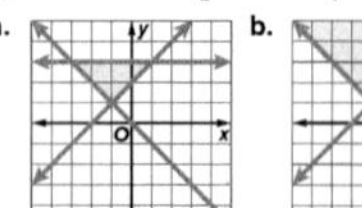 b. 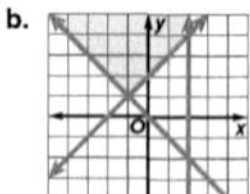 c. 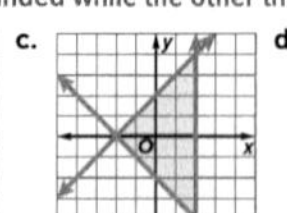 d. 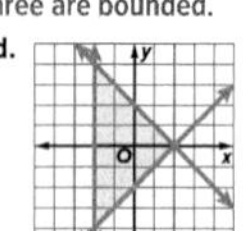

17. WRITE Upon determining a bounded feasible region, Kelvin noticed that vertices $A(-3, 4)$ and $B(5, 2)$ yielded the same maximum value for $f(x, y) = 16y + 4x$. Kelvin confirmed that his constraints were graphed correctly and his vertices were correct. Then he said that those two points were not the only maximum values in the feasible region. Explain how this could have happened. See Mod. 2 Answer Appendix.

18. CREATE Create a set of inequalities that forms a bounded region with an area of 20 units^2 and lies only in the fourth quadrant. Sample answer: $-2 \geq y \geq -6, 4 \leq x \leq 9$

Answers

1.
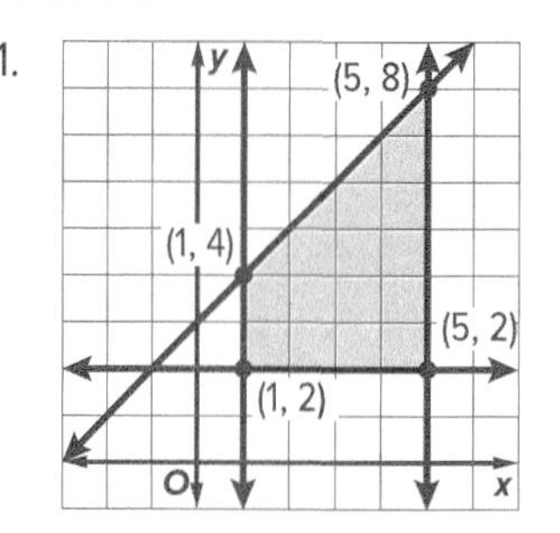

2.
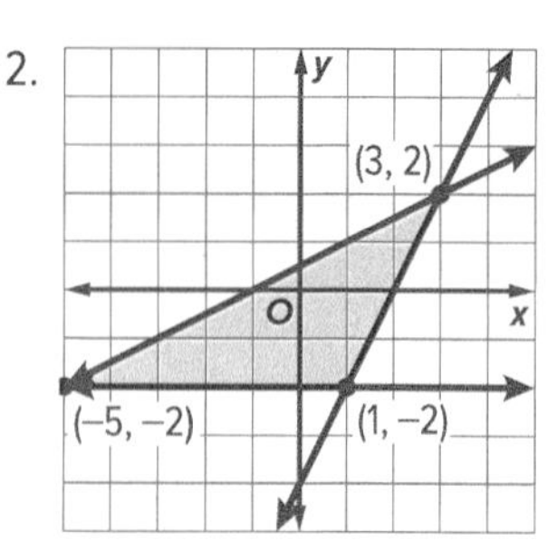

3.
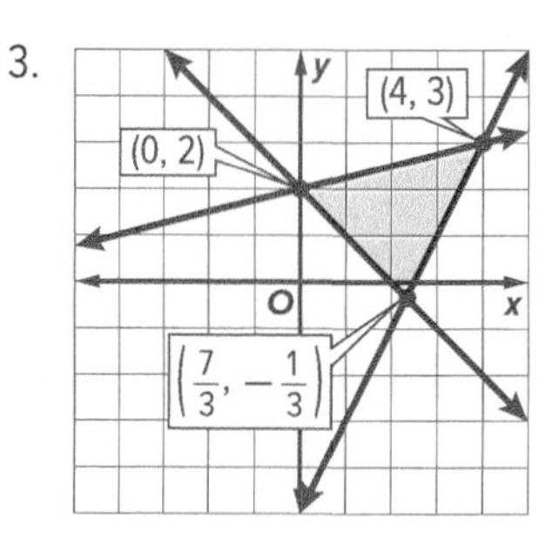

4.
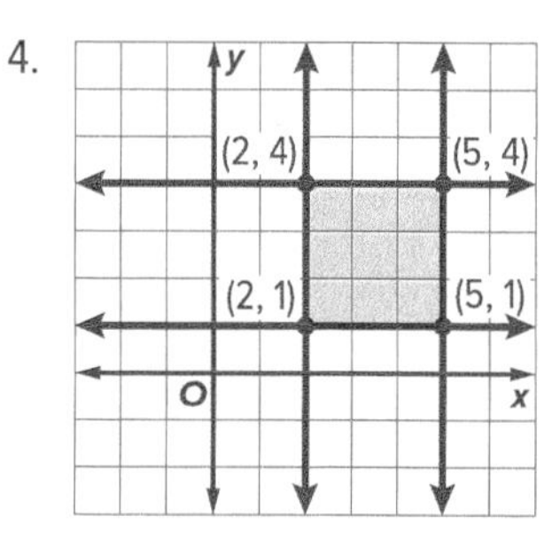

5.
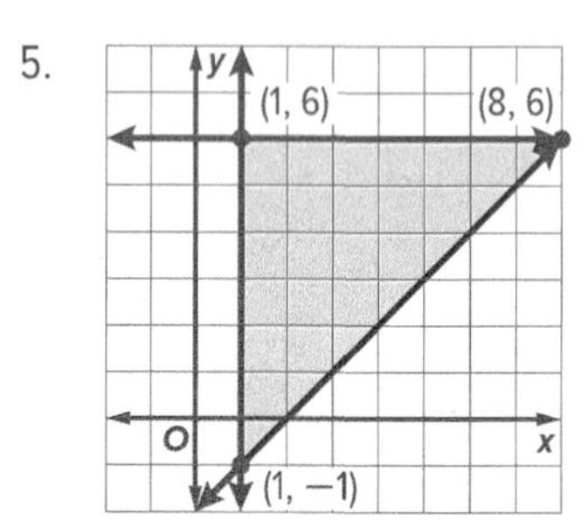

6.
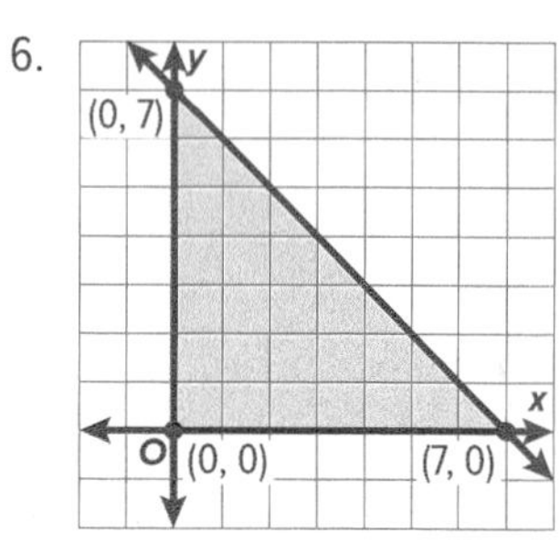

7.
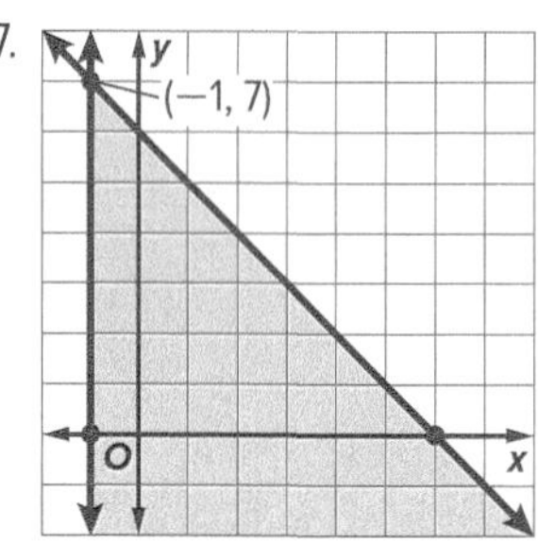

8.
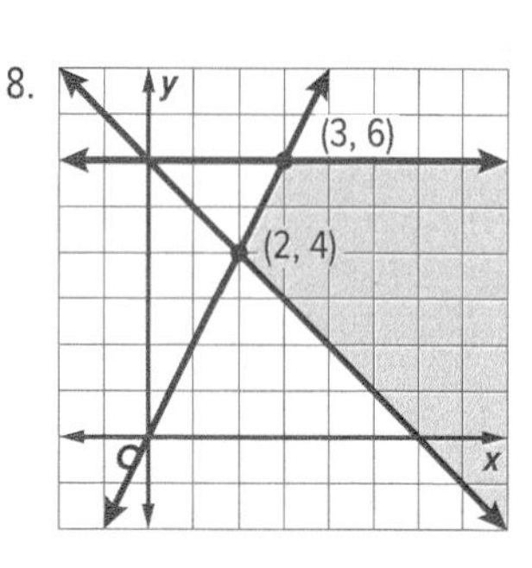

9.
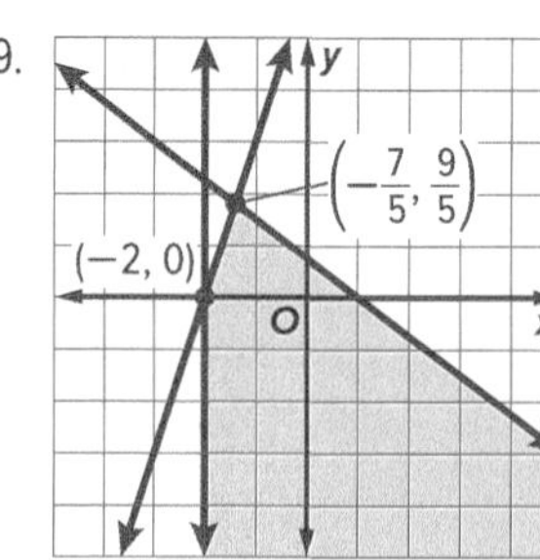

Lesson 2-8

A.CED.3

Systems of Equations in Three Variables

LESSON GOAL

Students solve systems of equations in three variables.

1 LAUNCH

 Launch the lesson with a **Warm Up** and an introduction.

2 EXPLORE AND DEVELOP

 **Explore:** Systems of Equations Represented as Lines and Planes

 **Develop:**

Solving Systems of Equations in Three Variables

- Solve a System with One Solution
- Solve a System with Infinitely Many Solutions
- Solve a System with No Solution
- Write and Solve a System of Equations

 You may want your students to complete the **Checks** online.

3 REFLECT AND PRACTICE

 Exit Ticket

 Practice

DIFFERENTIATE

 View reports of student progress on the **Checks** after each example.

Resources	AL	OL	BL	ELL
Remediation: Elimination Using Multiplication	●	●		●
Extension: Homogeneous Systems		●	●	●
ELL Support				

Language Development Handbook

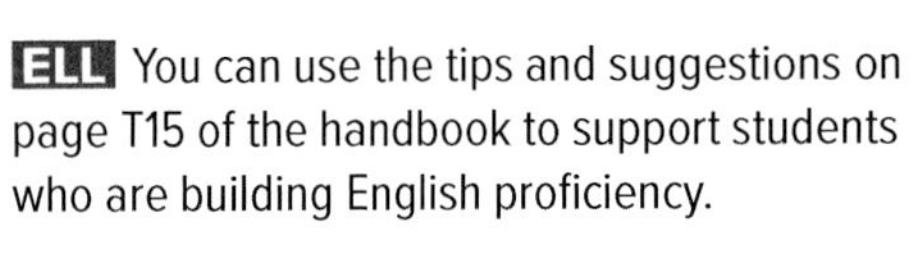

Assign page 15 of the *Language Development Handbook* to help your students build mathematical language related to solving systems of equations in three variables.

ELL You can use the tips and suggestions on page T15 of the handbook to support students who are building English proficiency.

Suggested Pacing

90 min	0.5 day
45 min	1 day

Focus

Domain: Algebra

Standards for Mathematical Content:

A.CED.3 Represent constraints by equations or inequalities, and by a system of equations and/or inequalities, and interpret solutions as viable or non-viable options in a modeling context.

Standards for Mathematical Practice:

1 Make sense of problems and persevere in solving them.

8 Look for and express regularity in repeated reasoning.

Coherence

Vertical Alignment

Previous

Students solved linear equations in two variables.

A.REI.5, A.REI.6, A.REI.7 (Algebra 1, Algebra 2)

Now

Students solve systems of equations in three variables.

A.CED.3

Next

Students will solve equations and inequalities involving absolute value by graphing.

A.CED.1

Rigor

The Three Pillars of Rigor

1 CONCEPTUAL UNDERSTANDING	2 FLUENCY	3 APPLICATION

Conceptual Bridge In this lesson, students extend their understanding of solving systems of equations in two variables to three equations in three variables. They build fluency and apply their understanding by solving real-world problems that involve three equations in three variables.

Mathematical Background

To graph equations in three variables, it is necessary to add a third dimension to the coordinate system. Ordered triples such as (x, y, z) represent a point in space just as ordered pairs such as (x, y) represent a point in a plane. The graph of an equation of the form $Ax + By + Cz = D$, where A, B, C, and D are not all equal to zero, is a plane.

Interactive Presentation

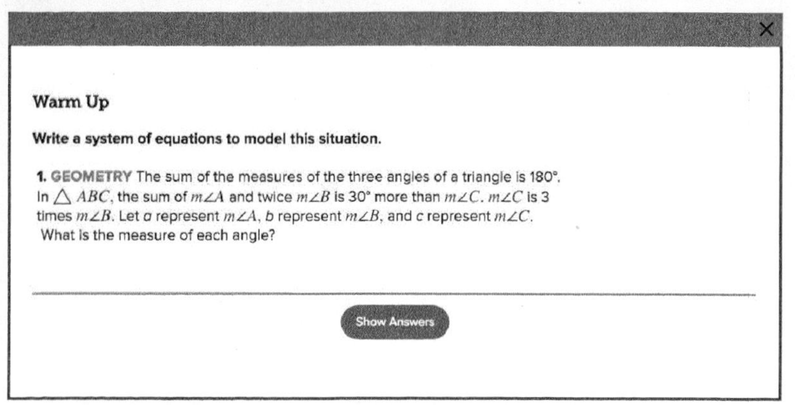

Warm Up

Warm Up

Prerequisite Skills

The Warm Up exercises address the following prerequisite skill for this lesson:

- writing systems of three equations

Answer:

$a + b + c = 180$, $a + 2b = c + 30$, $c = 3b$

Launch the Lesson

Launch the Lesson

 Teaching the Mathematical Practices

2 Represent a Situation Symbolically In the Launch the Lesson, guide students to identify the three variables in the situation. Then ask them how they could use those variables to write a system of equations.

Go Online to find additional teaching notes and questions to promote classroom discourse.

Today's Standards

Tell students that they will be addressing these content and practice standards in this lesson. You may wish to have a student volunteer read aloud *How can I meet these standards?* and *How can I use these practices?*, and connect these to the standards.

See the Interactive Presentation for I Can statements that align with the standards covered in this lesson.

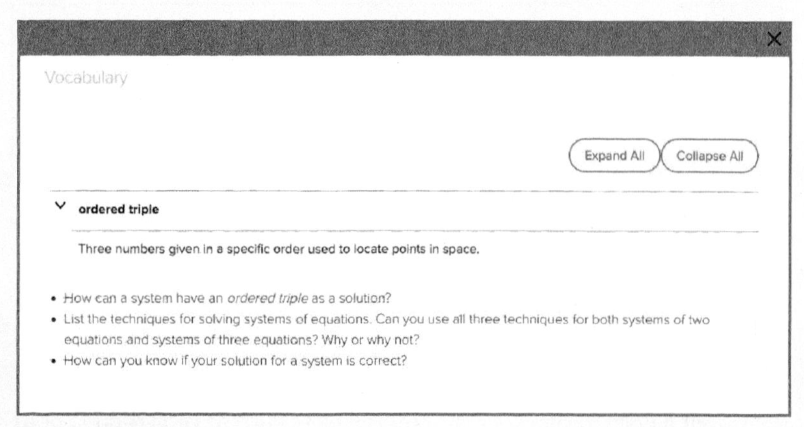

Today's Vocabulary

Today's Vocabulary

Tell students that they will be using this vocabulary term in this lesson. You can expand the row if you wish to share the definition. Then discuss the questions below with the class.

Explore Systems of Equations Represented as Lines and Planes

Objective

Students explore the solutions of systems of equations in two and three variables by analyzing lines and planes.

Teaching the Mathematical Practices

1 Explain Correspondences Throughout the Explore, students must explain the relationships between the graphs of systems of equations in two and three variables and the solutions of the system.

Ideas for Use

Recommended Use Present the Inquiry Question, or have a student volunteer read it aloud. Have students work in pairs to complete the Explore activity on their devices. Pairs should discuss each of the questions. Monitor student progress during the activity. Upon completion of the Explore activity, have student volunteers share their responses to the Inquiry Question.

What if my students don't have devices? You may choose to project the activity on a whiteboard. A printable worksheet for each Explore is available online. You may choose to print the worksheet so that individuals or pairs of students can use it to record their observations.

Summary of the Activity

Students will complete guiding exercises throughout the Explore activity. Students will move through multiple graphs of systems of two and three equations. Students will decide the correct number of solutions for each system. Then, students will answer the Inquiry Question.

(continued on the next page)

Interactive Presentation

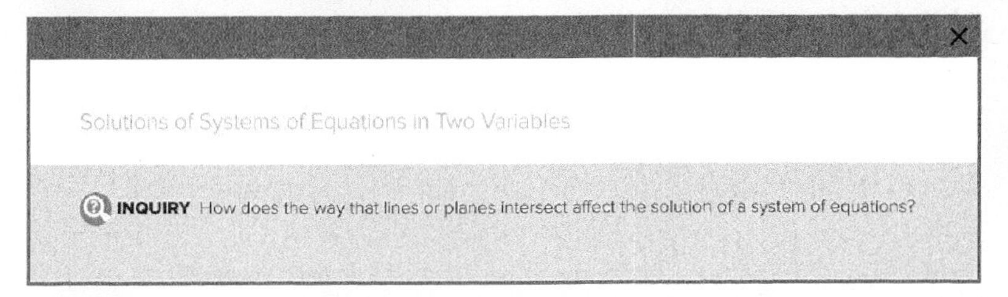

Explore

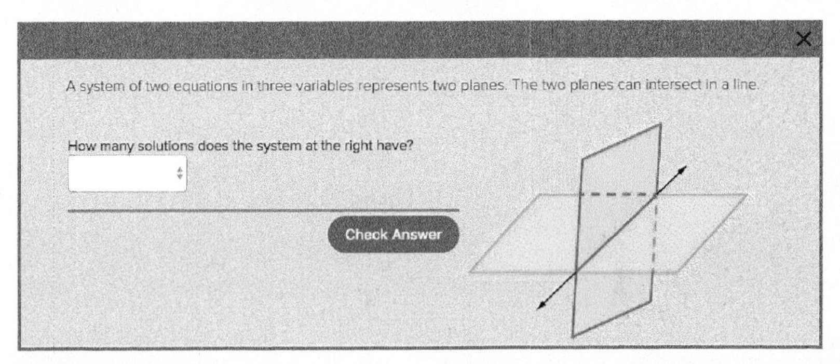

Explore

SELECT

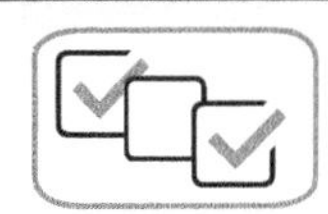

Students select the correct number of solutions for systems of two and three equations.

Interactive Presentation

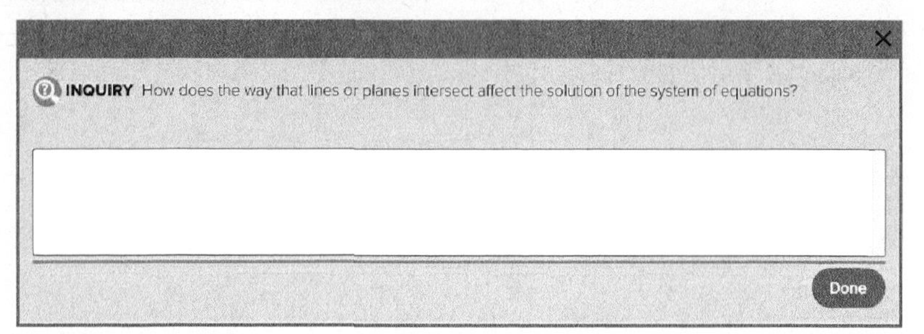

Explore

TYPE

Students respond to the Inquiry Question and can view a sample answer.

Explore Systems of Equations Represented as Lines and Planes (*continued*)

Teaching the Mathematical Practices

3 Construct Arguments Students must use previously established results to construct an argument about how solutions of systems of equations relate to the intersections of lines and planes.

Question

Have students complete the Explore activity.

Ask:

- If a system of two equations has infinitely many solutions, what does that reveal about the intersection of the graph of the system? The intersection is the line itself.
- If a system of three equations has infinitely many solutions, what does that reveal about the intersection of the graph of the system? The system could intersect at a line or a plane.

Inquiry

How does the way that lines or planes intersect affect the solution of a system of equations? Sample answer: When lines or planes intersect at one point, the system has one solution. When lines or planes coincide or intersect to form a line, there are infinitely many solutions. When all of the lines or planes of the system do not intersect, there is no solution.

Go Online to find additional teaching notes and sample answers for the guiding exercises.

1 CONCEPTUAL UNDERSTANDING | 2 FLUENCY | 3 APPLICATION

Learn Solving Systems of Equations in Three Variables

Objective

Students solve systems of linear equations in three variables by using algebraic methods.

Teaching the Mathematical Practices

3 Justify Conclusions The Talk About It! feature asks students to justify their conclusion to answer the question.

Important to Know

The graph of a linear equation in three variables is a plane. When a system of equations in three variables is graphed, the intersection could be a point, line, plane, or none at all. When the three planes intersect at one point, the solution is written as an ordered triple. If the solution is a line or plane, there are infinite solutions. If the planes do not all intersect, there is no solution. If a system of equations in three variables intersects at a point, this can be found algebraically using elimination and substitution.

Essential Question Follow-Up

Students have begun solving systems of equations in three variables.

Ask:

Why is solving a system of equations in three variables similar to solving a system in two variables? Sample answer: To solve a system of equation in three variables, you have to eliminate a variable to create a system of equations in two variables.

Example 1 Solve a System with One Solution

Teaching the Mathematical Practices

8 Attend to Details Use the Problem-Solving Tip to point out that when solving a system of equation in three variables, students should evaluate the reasonableness of their answer after each step.

(continued on the next page)

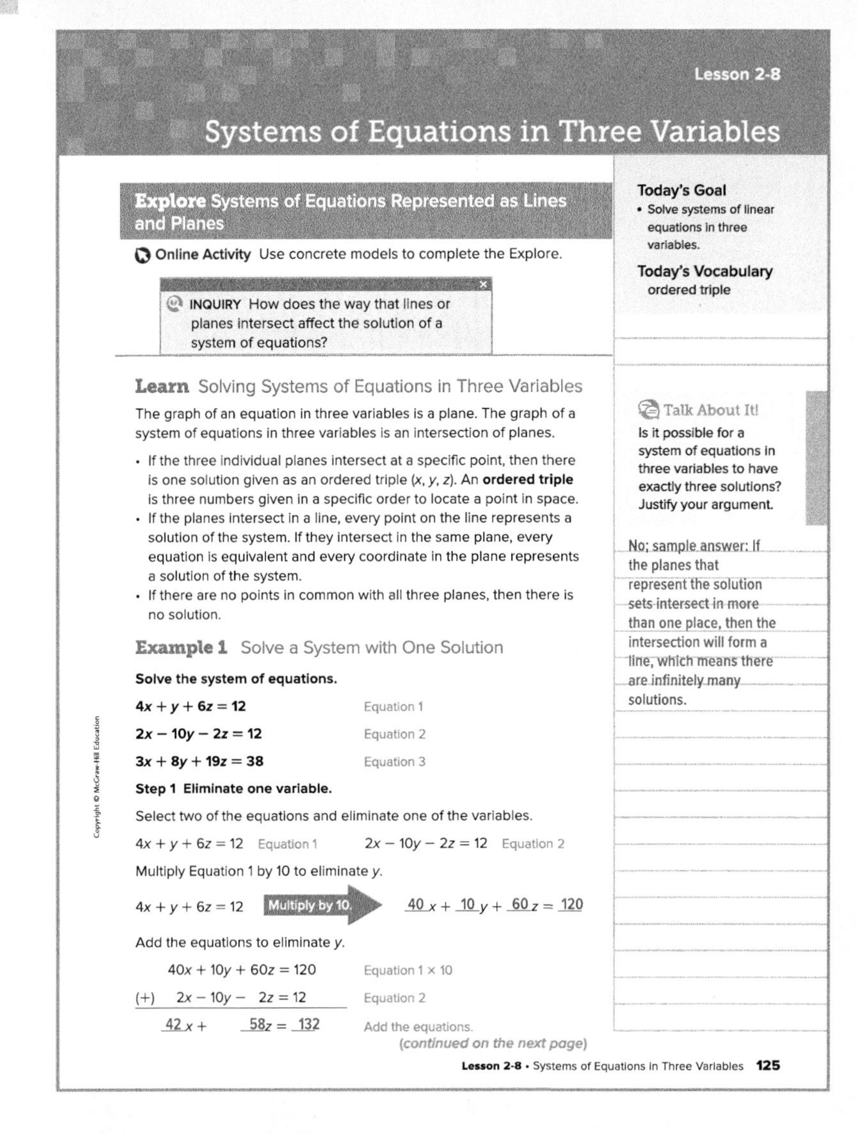

Lesson 2-8

Systems of Equations in Three Variables

Today's Goal
- Solve systems of linear equations in three variables.

Today's Vocabulary
ordered triple

Explore Systems of Equations Represented as Lines and Planes

Online Activity Use concrete models to complete the Explore.

INQUIRY How does the way that lines or planes intersect affect the solution of a system of equations?

Learn Solving Systems of Equations in Three Variables

The graph of an equation in three variables is a plane. The graph of a system of equations in three variables is an intersection of planes.

- If the three individual planes intersect at a specific point, then there is one solution given as an ordered triple (x, y, z). An **ordered triple** is three numbers given in a specific order to locate a point in space.
- If the planes intersect in a line, every point on the line represents a solution of the system. If they intersect in the same plane, every equation is equivalent and every coordinate in the plane represents a solution of the system.
- If there are no points in common with all three planes, then there is no solution.

Talk About It!
Is it possible for a system of equations in three variables to have exactly three solutions? Justify your argument.

No; sample answer: If the planes that represent the solution sets intersect in more than one place, then the intersection will form a line, which means there are infinitely many solutions.

Example 1 Solve a System with One Solution

Solve the system of equations.

$4x + y + 6z = 12$ Equation 1

$2x - 10y - 2z = 12$ Equation 2

$3x + 8y + 19z = 38$ Equation 3

Step 1 Eliminate one variable.

Select two of the equations and eliminate one of the variables.

$4x + y + 6z = 12$ Equation 1 $2x - 10y - 2z = 12$ Equation 2

Multiply Equation 1 by 10 to eliminate y.

$4x + y + 6z = 12$ Multiply by 10. $40x + 10y + 60z = 120$

Add the equations to eliminate y.

$40x + 10y + 60z = 120$ Equation 1 × 10

$(+)\ 2x - 10y - 2z = 12$ Equation 2

$42x + 58z = 132$ Add the equations.

(continued on the next page)

Copyright © McGraw-Hill Education

Lesson 2-8 • Systems of Equations in Three Variables 125

Interactive Presentation

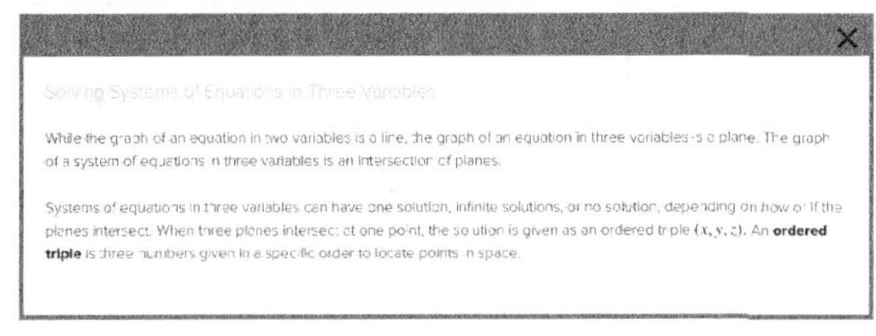

Learn

TYPE

Students answer a question to determine whether they understand if it is possible for a system of equations in three variables to have exactly three solutions.

Your Notes

Problem-Solving Tip

Identify Subgoals Before you begin solving a system of equations in 3 variables, try to identify which variable would be easiest to eliminate in Step 1. Be sure that it makes sense to select that variable to eliminate from the set of equations. As you work through the problem, check your work after each step. Check for reasonableness before moving on to the next subgoal.

Think About It!

Suppose you initially eliminated x in Step 1. How would that affect the solution? Explain your reasoning.

Sample answer: The solution would not be affected. Because there is only one solution to the system of equations, solving for the variables in any order will result in the same ordered triple.

Use a different combination of the original equations to create another equation in two variables. Eliminate y again.

$4x + y + 6z = 12$ Equation 1

$3x + 8y + 19z = 38$ Equation 3

Multiply Equation 1 by -8 and add the equations to eliminate y.

$4x + y + 6z = 12$ → Multiply by −8. → $-32x - \underline{8}y - 48z = \underline{-96}$

Add the equations to eliminate y.

	$-32x - 8y - 48z = -96$	Equation 1 × (−8).
(+)	$3x + 8y + 19z = 38$	Equation 3
	$\underline{-29}x + \underline{-29}z = \underline{-58}$	Add the equations.

Step 2 Solve the system of two equations.

Multiply the second equation by 2 and add the equations to eliminate z.

$42x + 58z = 132$ → $42x + 58z = 132$

$-29x - 29z = -58$ → Multiply by 2. → (+) $-58x - 58z = -116$

$-16x = 16$

$x = -1$

Use substitution to solve for z.

$42x + 58z = 132$	Equation in two variables
$42(\underline{-1}) + 58z = 132$	$x = -1$
$-42 + 58z = 132$	Multiply.
$58z = \underline{174}$	Add 42 to each side.
$z = \underline{3}$	Divide each side by 58.

The result is $x = \underline{-1}$ and $z = \underline{3}$.

Step 3 Solve for y.

Substitute the two values into one of the original equations to find y.

$4x + y + 6z = 12$	Equation 1
$4(\underline{-1}) + y + 6(\underline{3}) = 12$	$x = -1, z = 3$
$\underline{-4} + y + \underline{18} = 12$	Multiply.
$y = \underline{-2}$	Subtract 14 from each side.

The ordered triple is $(\underline{-1}, \underline{-2}, \underline{3})$.

Go Online You can complete an Extra Example online.

Interactive Presentation

Solve a System with One Solution

Solve the system of equations.

$4x + y + 6z = 12$ Equation 1
$2x - 10y - 2z = 12$ Equation 2
$3x + 8y + 19z = 38$ Equation 3

Example 1

TAP

Students move through the steps to solve a system of equations in three variables with one solution.

TYPE

Students answer a question to determine whether they understand how to solve a system of equations with one solution.

Questions for Mathematical Discourse

AL To eliminate a variable in two different equations, what steps must be taken? Sample answer: First each equation must be multiplied by the necessary number so the variables have the same coefficient, but opposite signs. Then the two equations are added together.

OL When checking your answer, do you have to substitute the ordered triple into all three equations, or can you just check one? Explain. All three; sample answer: The ordered triple could possibly work in one equation but not the other two, so it should always be checked in each equation.

BL How could substitution be used as the first step instead of elimination? Sample answer: The first equation could be solved for y and then substituted into Equations 2 and 3. Then the process would continue as normal.

Common Misconception

A common misconception students often have is that to solve a system of equations in three variables, they will either use elimination or substitution, as they did with systems in two variables. Remind students that systems with three variables require more algebraic work and therefore elimination and substitution are often both used when solving the system. Students must remain open to using multiple solution methods in one problem.

DIFFERENTIATE

Reteaching Activity AL

IF students are struggling to solve a system of equations in three variables,

THEN have students practice solving systems in two variables to gain the fluency necessary to solve systems in three variables.

Common Error

Students may understand how to solve systems of equations in three variables, but make careless or simple mistakes along the way. There are lots of places where students could make a sign error, add a number incorrectly, or forget to use the Distributive Property. Encourage students to take their time and check over their work along the way.

1 CONCEPTUAL UNDERSTANDING | **2 FLUENCY** | 3 APPLICATION

Example 2 Solve a System with Infinitely Many Solutions

Teaching the Mathematical Practices

1 Monitor and Evaluate As students solve the system of equations in three variables, they should stop and evaluate their progress to ensure they stay on the correct solution path.

Questions for Mathematical Discourse

AL When solving a system of equations in two variables, if all of the variables cancel and a true statement remains, what will be the solution? There will be infinitely many solutions because all points on the line are solutions.

OL Suppose a student chose to use Equations 1 and 3 first to eliminate *x*. Would this have changed the solution? Explain. No; sample answer: No matter which equations a student begins with, the solution will always be the same.

BL Which has more points, the intersection line of two planes, or two planes that are the same? Both have an infinite number of points.

Common Error

When solving systems of equations in three variables, there are many different choices to start the process. Students could select different variables to eliminate or different equations to use. This can make it difficult to check in on student progress or diagnose student errors. Suggest to students before they begin that they select the easiest variable to eliminate so that most students will be on a similar solution pathway.

 **Go Online**

- Find additional teaching notes.
- View performance reports of the Checks.
- Assign or present an Extra Example.

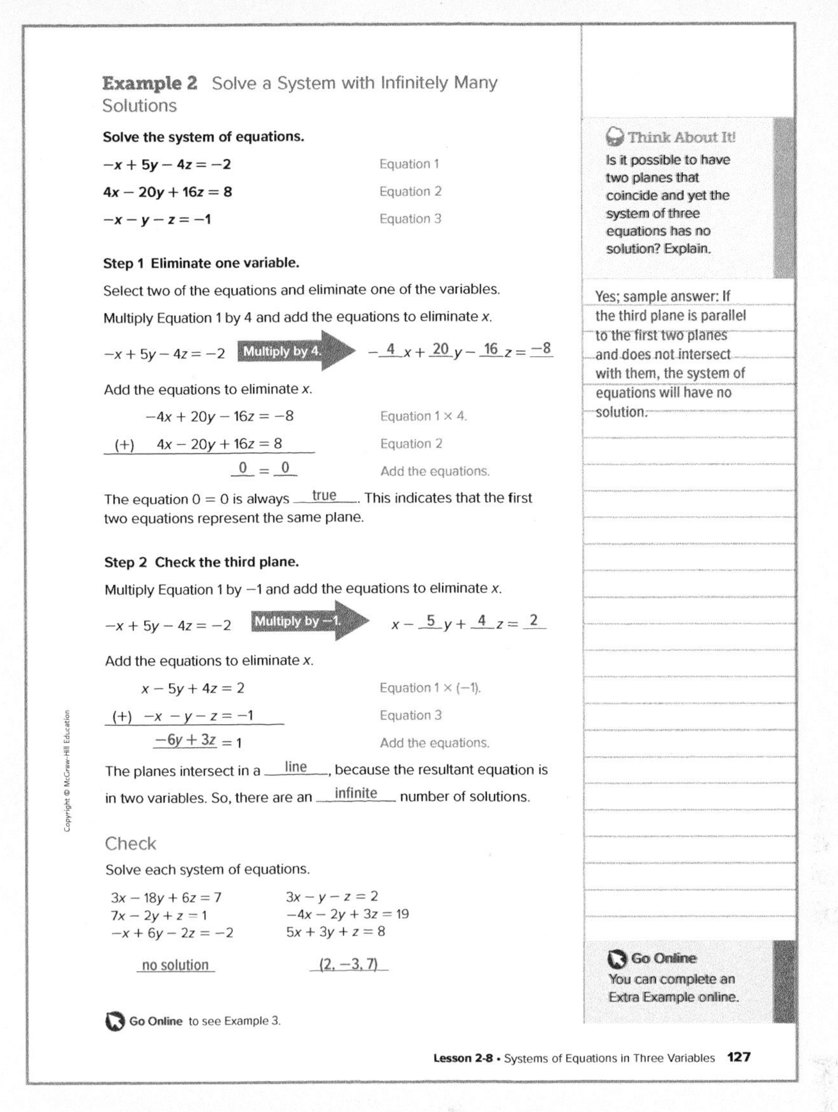

Example 2 Solve a System with Infinitely Many Solutions

Solve the system of equations.

$-x + 5y - 4z = -2$ Equation 1
$4x - 20y + 16z = 8$ Equation 2
$-x - y - z = -1$ Equation 3

Step 1 Eliminate one variable.

Select two of the equations and eliminate one of the variables.

Multiply Equation 1 by 4 and add the equations to eliminate *x*.

$-x + 5y - 4z = -2$ → Multiply by 4. → $-4x + 20y - 16z = -8$

Add the equations to eliminate *x*.

$-4x + 20y - 16z = -8$ Equation 1 × 4.
$(+)\ 4x - 20y + 16z = 8$ Equation 2
$0 = 0$ Add the equations.

The equation $0 = 0$ is always true. This indicates that the first two equations represent the same plane.

Step 2 Check the third plane.

Multiply Equation 1 by −1 and add the equations to eliminate *x*.

$-x + 5y - 4z = -2$ → Multiply by −1. → $x - 5y + 4z = 2$

Add the equations to eliminate *x*.

$x - 5y + 4z = 2$ Equation 1 × (−1).
$(+)\ -x - y - z = -1$ Equation 3
$-6y + 3z = 1$ Add the equations.

The planes intersect in a line, because the resultant equation is in two variables. So, there are an infinite number of solutions.

Check

Solve each system of equations.

$3x - 18y + 6z = 7$
$7x - 2y + z = 1$
$-x + 6y - 2z = -2$
no solution

$3x - y - z = 2$
$-4x - 2y + 3z = 19$
$5x + 3y + z = 8$
(2, −3, 7)

Go Online to see Example 3.

Think About It! Is it possible to have two planes that coincide and yet the system of three equations has no solution? Explain.

Yes; sample answer: If the third plane is parallel to the first two planes and does not intersect with them, the system of equations will have no solution.

Go Online You can complete an Extra Example online.

Copyright © McGraw-Hill Education

Lesson 2-8 • Systems of Equations in Three Variables 127

Interactive Presentation

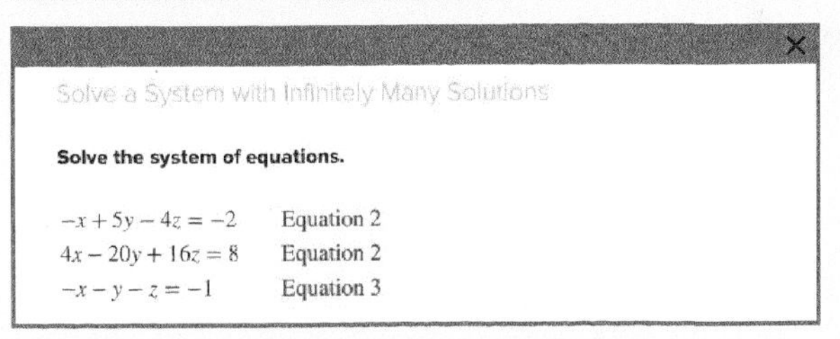

Example 2

TAP

Students move through the steps to solve a system of equations in three variables with infinite solutions.

TYPE

Students answer a question to determine whether they understand solving a system with infinitely many solutions.

Example 4 Write and Solve a System of Equations

MUSEUM MEMBERSHIPS In 2016, Dali Museum in St. Petersburg, Florida offered individual, dual, and family memberships, which cost \$60, \$80, and \$100, respectively. Suppose in one month the museum sells a total of 81 new memberships, for a total of \$6420. The number of dual memberships purchased is twice that of individual memberships. Write and solve a system of equations to determine the number of new individual memberships x, dual memberships y, and family memberships z.

Step 1 Write the system of equations.

a total of 81 new memberships: $x + y + z = 81$

The number of dual memberships purchased is twice that of individual memberships: $y = 2x$

Fees for individual, dual, and family, memberships which cost \$60, \$80, and \$100, respectively, for a total of \$6420: $60x + 80y + 100z = 6420$

Step 2 Eliminate one variable.

Substitute $y = 2x$ into Equation 1 and Equation 3 to eliminate y.

$x + y + z = 81$ Equation 1
$x + \underline{2x} + z = 81$ $y = 2x$
$\underline{3}x + z = 81$ Add.

$60x + 80y + 100z = 6420$ Equation 3
$60x + 80(\underline{2x}) + 100z = 6420$ $y = 2x$
$\underline{220}x + \underline{100}z = 6420$ Simplify.

Step 3 Solve the resulting system of two equations.

$-300x - 100z = -8100$ Multiply new Equation 1 by -100.
$(+)\ 220x + 100z = 6420$
$-80x = -1680$ Add to eliminate z.
$x = 21$ Solve for x.

Step 4 Substitute to find z.

$3x + z = 81$ Remaining equation in two variables
$3(\underline{21}) + z = 81$ $x = 21$
$z = 18$ Simplify.

Step 5 Substitute to find y.

$y = 2x$ Equation 2
$y = 2(\underline{21})$ $x = 21$
$y = \underline{42}$ Multiply.

The solution is (21, 42, 18). So, the museum sold 21 individual memberships, 42 dual memberships, and 18 family memberships.

Go Online You can complete an Extra Example online.

Think About It!
Is the solution reasonable? Explain.

Yes; sample answer: The solution is whole, positive numbers, which makes sense in the context of the problem. I can substitute the ordered triple into one of the original equations to check.

128 Module 2 • Linear Equations, Inequalities, and Systems

Interactive Presentation

Write and Solve a System of Equations

MUSEUM MEMBERSHIPS The Dali Museum in St. Petersburg, Florida, features the work of Salvador Dali as well as other artists. In 2016, the museum offered three types of basic membership, individual, dual, and family, which cost \$60, \$80, and \$100, respectively. Suppose in one month the museum sells a total of 81 new memberships. The number of dual memberships purchased is twice that of individual memberships. The total amount that the museum earned from the new memberships is \$6420. Write and solve a system of equations to determine the number of new individual members x, dual members y, and family members z.

Example 4

TAP

Students move through the steps to write and solve a system of equations in three variables.

TYPE

Students answer a question to determine whether that their solution is reasonable.

CHECK

Students complete the Check online to determine whether they are ready to move on.

Example 4 Write and Solve a System of Equations

Teaching the Mathematical Practices

4 Analyze Relationships Mathematically Point out that to solve the problem in Example 4, students will need to analyze the mathematical relationships in the problem to draw a conclusion about the number of each type of membership sold.

Questions for Mathematical Discourse

AL Why is substitution a better first choice when solving the system than elimination? Sample answer: Substitution is better because one equation is already solved for a variable.

OL How many equations do you need to get one solution to a system of equations in three variables? 3

BL Could you find one solution for four equations in three variables? yes Describe what one such system might look like when graphed. Sample answer: You could have three vertical planes that intersect on the same line and one horizontal plane.

Common Error

Translating verbal descriptions to equations can be a daunting task for some students. They struggle to identify which pieces of information form an equation. Remind students that variables should always be defined first. Then encourage students to read the problem and consider how one sentence may form one equation. If there are three variables, there will usually be three equations.

Exit Ticket

Recommended Use

At the end of class, go online to display the Exit Ticket prompt and ask students to respond using a separate piece of paper. Have students hand you their responses as they leave the room.

Alternate Use

At the end of class, go online to display the Exit Ticket prompt and ask students to respond verbally or by using a mini-whiteboard. Have students hold up their whiteboards so that you can see all student responses. Tap to reveal the answer when most or all students have completed the Exit Ticket.

Practice and Homework

Suggested Assignments

Use the table below to select appropriate exercises.

DOK	Topic	Exercises
1, 2	exercises that mirror the examples	1–15
2	exercises that use a variety of skills from this lesson	16–18
2	exercises that extend concepts learned in this lesson to new contexts	20–22
3	exercises that emphasize higher-order and critical thinking skills	23–25

ASSESS AND DIFFERENTIATE

Use the data from the **Checks** to determine whether to provide resources for extension, remediation, or intervention.

BL

IF students score 90% or more on the Checks,
THEN assign:

- Practice Exercises 1–22 odd, 23–25
- Extension: Homogeneous Systems
- **ALEKS**® Other Topics Available: Linear Systems

OL

IF students score 66%–89% on the Checks,
THEN assign:

- Practice Exercises 1–25 odd
- Remediation, Review Resources: Elimination Using Multiplication
- Personal Tutors
- Extra Examples 1–4
- **ALEKS**® Systems of Linear Equations

AL

IF students score 65% or less on the Checks,
THEN assign:

- Practice Exercises 1–15 odd
- Remediation, Review Resources: Elimination Using Multiplication
- *Quick Review Math Handbook*: Systems of Equations in Three Variables
- **ALEKS**® Systems of Linear Equations

Name ________________ Period ________ Date ________

Practice

Go Online You can complete your homework online.

Examples 1–3

Solve each system of equations.

1. $2x + 3y - z = 0$
$x - 2y - 4z = 14$
$3x + y - 8z = 17$ $(4, -3, -1)$

2. $2p - q + 4r = 11$
$p + 2q - 6r = -11$
$3p - 2q - 10r = 11$ $(2, -5, \frac{1}{2})$

3. $a - 2b + c = 8$
$2a + b - c = 0$
$3a - 6b + 3c = 24$ infinitely many solutions

4. $3s - t - u = 5$
$3s + 2t - u = 11$
$6s - 3t + 2u = -12$ $\left(\frac{2}{3}, 2, -5\right)$

5. $2x - 4y - z = 10$
$4x - 8y - 2z = 16$
$3x + y + z = 12$ no solution

6. $p - 6q + 4r = 2$
$2p + 4q - 8r = 16$
$p - 2q = 5$ infinitely many solutions

7. $2a + c = -10$
$b - c = 15$
$a - 2b + c = -5$ $(5, -5, -20)$

8. $x + y + z = 3$
$13x + 2z = 2$
$-x - 5z = -5$ $(0, 2, 1)$

9. $2m + 5n + 2p = 6$
$5m - 7n = -29$
$p = 1$ $(-3, 2, 1)$

10. $f + 4g - h = 1$
$3f - g + 8h = 0$
$f + 4g - h = 10$ no solution

11. $-2c = -6$
$2a + 3b - c = -2$
$a + 2b + 3c = 9$ $(2, -1, 3)$

12. $3x - 2y + 2z = -2$
$x + 6y - 2z = -2$
$x + 2y = 0$ $(-2, 1, 3)$

Example 4

13. ANIMAL NUTRITION A veterinarian wants to make a food mix for guinea pigs that contains 23 grams of protein, 6.2 grams of fat, and 16 grams of moisture. The composition of three available mixtures are shown in the table. How many grams of each mix should be used to make the desired new mix?
60 grams of Mix A, 50 grams of Mix B, and 40 grams of Mix C

	Protein (g)	Fat (g)	Moisture (g)
Mix A	0.2	0.02	0.15
Mix B	0.1	0.06	0.10
Mix C	0.15	0.05	0.05

14. ENTERTAINMENT At the arcade, Marcos, Sara, and Darius played video racing games, pinball, and air hockey. Marcos spent $6 for 6 racing games, 2 pinball games, and 1 game of air hockey. Sara spent $12 for 3 racing games, 4 pinball games, and 5 games of air hockey. Darius spent $12.25 for 2 racing games, 7 pinball games, and 4 games of air hockey. How much did each of the games cost?
racing game: $0.50; pinball: $0.75; air hockey: $1.50

15. FOOD A natural food store makes its own brand of trail mix from dried apples, raisins, and peanuts. A one-pound bag of the trail mix costs $3.18. It contains twice as much peanuts by weight as apples. If a pound of dried apples costs $4.48, a pound of raisins is $2.40, and a pound of peanuts is $3.44, how many ounces of each ingredient are contained in 1 pound of the trail mix?
3 oz of apples, 7 oz of raisins, 6 oz of peanuts

Mixed Exercises

Solve each system of equations.

16. $-x - 5z = -5$
$y - 3x = 0$
$13x + 2z = 2$ (0, 0, 1)

17. $-3x + 2z = 1$
$4x + y - 2z = -6$
$x + y + 4z = 3$ (1, −6, 2)

18. $x - y + 3z = 3$ no solution
$-2x + 2y - 6z = 6$
$y - 5z = -3$

19. REASONING A newspaper company has three printing presses that together can produce 3500 newspapers each hour. The fastest printer can print 100 more than twice the number of papers as the slowest press. The two slower presses combined produce 100 more papers than the fastest press. How many newspapers can each printing press produce in 1 hour? fastest press: 1700 papers; slower press: 1000 papers; slowest press: 800 papers

20. USE A SOURCE A shop is having a sale on pool accessories. The table shows the orders of three customers and their total price before tax. Research the sales tax in your area to determine whether a customer who has $200 could buy 1 chlorine filter, 1 raft, and 1 large lounge chair after sales tax is applied. Justify your response. See margin.

Combo	Price Before Tax
1 Raft and 2 Chlorine Filters	$220
1 Chlorine Filter and 2 Large Lounge Chairs	$245
1 Raft and 4 Large Lounge Chairs	$315

21. TICKETS Three kinds of tickets are available for a concert: orchestra seating, mezzanine seating, and balcony seating. The orchestra tickets cost $2 more than the mezzanine tickets, while the mezzanine tickets cost $1 more than the balcony tickets. Twice the cost of an orchestra ticket is $1 less than 3 times the cost of a balcony ticket. Determine the price of each kind of ticket. orchestra ticket: $10; mezzanine ticket: $8; balcony ticket: $7

22. CONSTRUCT ARGUMENTS Consider the following system. Prove that if $b = c = -a$, then $ty = a$. See margin.

$$rx + ty + vz = a$$
$$rx - ty + vz = b$$
$$rx + ty - vz = c$$

Higher-Order Thinking Skills

23. PERSEVERE The general form of an equation for a parabola is $y = ax^2 + bx + c$, where (x, y) is a point on the parabola. If three points on a parabola are (2, −10), (−5, −101), and (6, −90), determine the values of a, b, and c and write the equation of the parabola in general form. $a = -3, b = 4, c = -6; y = -3x^2 + 4x - 6$

24. WRITE Use your knowledge of solving a system of three linear equations with three variables to explain how to solve a system of four equations with four variables. See margin.

25. CREATE Write a system of three linear equations that has a solution of (−5, −2, 6). Show that the ordered triple satisfies all three equations. See margin.

130 Module 2 • Linear Equations, Inequalities, and Systems

Answers

20. No; sample answer: Let x = cost of 1 raft, y = cost of 1 chlorine filter, and z = cost of 1 large lounge chair. $x + 2y = 220$, $y + 2z = 245$, $x + 4z = 315$; Solve the first equation for y and the third for z to get $y = \frac{1}{2}(220 - x) = 110 - \frac{1}{2}x$, $z = \frac{1}{4}(315 - x) = \frac{315}{4} - \frac{1}{4}x$. Substitute to find $x = 22.50$, $y = 98.75$ and $z = 73.125$. The cost of buying one each, before sales tax, is $194.38. If the sales tax is 6%, the total is $206.04, which exceeds $200.

22. Sample answer:

$a + b = (rx + ty + vz) + (rx - ty + vz)$

Replace a with $rx + ty + vz$, and b with $rx - ty + vz$.

$a + b = 2rx + 2vz$	Simplify.
$a + (-a) = 2rx + 2vz$	Replace b with $-a$.
$0 = 2rx + 2vz$	Simplify.
$0 = rx + vz$	Divide each side by 2.
$rx + ty + vz = a$	Given
$ty + (rx + vz) = a$	Commutative and Associative Properties of Addition
$ty + 0 = a$	Substitution
$ty = a$	Simplify.

24. Sample answer: First, combine two of the original equations using elimination to form a new equation with three variables. Next, combine a different pair of the original equations using elimination to eliminate the same variable and form a second equation with three variables. Repeat the process with a third pair of the original equations. You now have a system of three equations with three variables. Once you solve for the three variables substitute them into one of the original equations to find the fourth variable.

25. Sample answer:

$$3x + 4y + z = -17$$
$$2x - 5y - 3z = -18$$
$$-x + 3y + 8z = 47$$
$$3x + 4y + z = -17$$
$$3(-5) + 4(-2) + 6 = -17$$
$$-15 + (-8) + 6 = -17$$
$$-17 = -17$$
$$2x - 5y - 3z = -18$$
$$2(-5) - 5(-2) - 3(6) = -18$$
$$-10 + 10 - 18 = -18$$
$$-18 = -18$$
$$-x + 3y + 8z = 47$$
$$-(-5) + 3(-2) + 8(6) = 47$$
$$5 - 6 + 48 = 47$$
$$47 = 47$$

Lesson 2-9

A.CED.1

Solving Absolute Value Equations and Inequalities by Graphing

LESSON GOAL

Students solve equations and inequalities involving absolute value by graphing.

1 LAUNCH

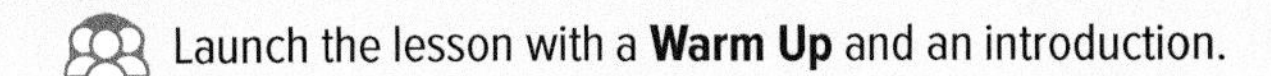
Launch the lesson with a **Warm Up** and an introduction.

2 EXPLORE AND DEVELOP

Develop:

Solving Absolute Value Equations by Graphing
- Solve an Absolute Value Equation by Graphing
- Solve an Absolute Value Equation by Using Technology
- Confirm Solutions by Using Technology

Solving Absolute Value Inequalities by Graphing
- Solve an Absolute Value Inequality by Graphing
- Solve an Absolute Value Inequality by Using Technology

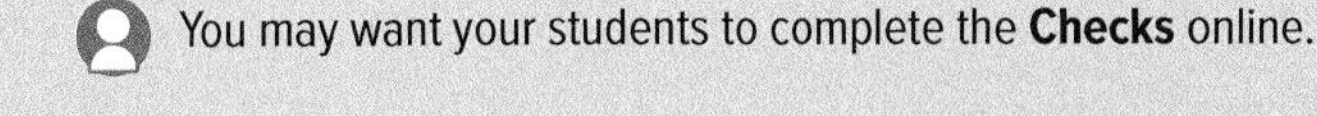
You may want your students to complete the **Checks** online.

3 REFLECT AND PRACTICE

Exit Ticket

Practice

Formative Assessment Math Probe

DIFFERENTIATE

View reports of student progress on the **Checks** after each example.

Resources	AL	OL	BL	ELL
Remediation: Functions and Continuity	●	●		●
Extension: The Sign Function		●	●	●
ELL Support				

Language Development Handbook

Assign page 16 of the *Language Development Handbook* to help your students build mathematical language related to solving equations and inequalities involving absolute value by graphing.

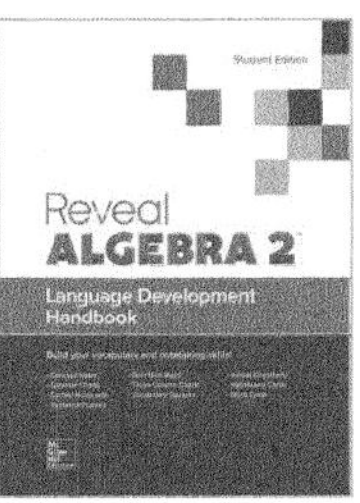

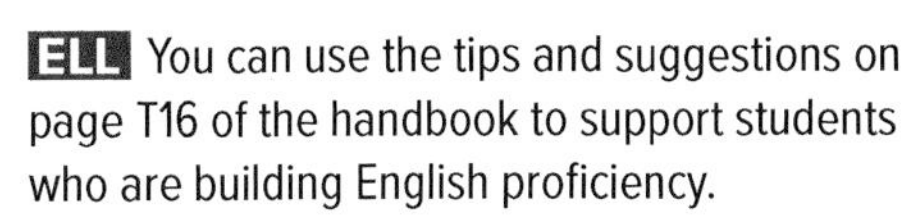
ELL You can use the tips and suggestions on page T16 of the handbook to support students who are building English proficiency.

Suggested Pacing

90 min	0.5 day
45 min	1 day

Focus

Domain: Algebra

Standards for Mathematical Content:

A.CED.1 Create equations and inequalities in one variables and use them to solve problems.

Standards for Mathematical Practice:

1 Make sense of problems and persevere in solving them.
5 Use appropriate tools strategically.

Previous
Students solved linear equations in two variables.
A.REI.5, A.REI.6, A.REI.7 (Algebra 1, Algebra 2)

Now
Students solve equations and inequalities involving absolute value by graphing.
A.CED.1

Next
Students will graph and solve quadratic equations and inequalities.
A.CED1. A.CED.3

Rigor

The Three Pillars of Rigor

1 CONCEPTUAL UNDERSTANDING	2 FLUENCY	3 APPLICATION

Conceptual Bridge In this lesson, students extend their understanding of solving systems of equations to solving equations by graphing each side and finding the intersection. They build fluencyby doing this with absolute value equations, and they apply their understanding by solving real-world problems that involve absolute value.

Mathematical Background

Absolute value equations and inequalities in one variable can be solved by graphing the related functions on the coordinate grid. The solution is given on the x-axis.

Interactive Presentation

Warm Up

Write each inequality in set-builder and interval notation.

1.

2.

3.

4.

5.

Show Answers

Warm Up

Launch the Lesson

Many local television news stations throughout the country boast the accuracy of their weather forecasts by offering a "three-degree guarantee." The meteorologists forecast the high temperature each day with the guarantee that the actual temperature will be within three degrees of the predicted temperature. Some stations even offer community prizes based the meteorologists' accuracy. Absolute value equations and inequalities can be used to find the range within which the actual temperate could be while still upholding the guarantee.

Launch the Lesson

Warm Up

Prerequisite Skills

The Warm Up exercises address the following prerequisite skill for this lesson:

- writing in set-builder and interval notation

Answers:

1. $\{x \mid x > -3\}$; $(-3, \infty)$
2. $\{x \mid x \leq 2\}$; $(-\infty, 2]$
3. $\{x \mid -1 < x \leq 4\}$; $(-1, 4]$
4. $\{x \mid x \leq -5 \text{ or } x > 1\}$; $(-\infty, -5] \cup (1, \infty)$
5. $\{x \mid x \neq -4\}$; $(-\infty, -4) \cup (-4, \infty)$

Launch the Lesson

Teaching the Mathematical Practices

4 Apply Mathematics In the Launch the Lesson, students will learn about how the accuracy of weather forecasts can be modeled using absolute value equations and inequalities.

Go Online to find additional teaching notes and questions to promote classroom discourse.

Today's Standards

Tell students that they will be addressing these content and practice standards in this lesson. You may wish to have a student volunteer read aloud *How can I meet these standards?* and *How can I use these practices?*, and connect these to the standards.

See the Interactive Presentation for I Can statements that align with the standards covered in this lesson.

1 CONCEPTUAL UNDERSTANDING | 2 FLUENCY | 3 APPLICATION

Learn Solving Absolute Value Equations by Graphing

Objective

Students solve absolute value equations by examining graphs of related functions.

Teaching the Mathematical Practices

1 Explain Correspondences Encourage students to explain the relationships between the solutions of absolute value equations and the graphs and x-intercepts of the related functions.

What Students are Learning

To solve an equation involving absolute value, the graph of a related function can be used. The graph of an absolute value function may intersect the x-axis up to two times, or it may not intersect at all. The number of x-intercepts is the same as the number of solutions to the equation.

Example 1 Solve an Absolute Value Equation by Graphing

Teaching the Mathematical Practices

2 Different Properties Mathematically proficient students look for different ways to solve problems. Use the Think About It! feature to encourage students to consider another way to solve the problem.

Questions for Mathematical Discourse

AL How can the solution be checked for accuracy? Substitute the value back into the original equation and see if the statement is true.

OL Why can the 5 in $5 + |2x + 6|$ not be combined with the 6 inside the absolute value? Sample answer: The absolute value affects the terms on the inside but not on the outside, so combining them could change the result for some values of x. For $x = -4$, the left side of the equation equals 7, but with the 5 pulled into the absolute value it equals 3.

BL Could you solve an absolute value equation by graphing without setting the equation equal to zero? Explain. Yes; sample answer: You would need to find the intersection of both sides of the equation. You could graph $y = 5 + |2x + 6|$ and $y = 5$, then find the intersection. Using zeros is a convention that often makes calculations and vizualization simpler.

Go Online

- Find additional teaching notes.
- View performance reports of the Checks.
- Assign or present an Extra Example.

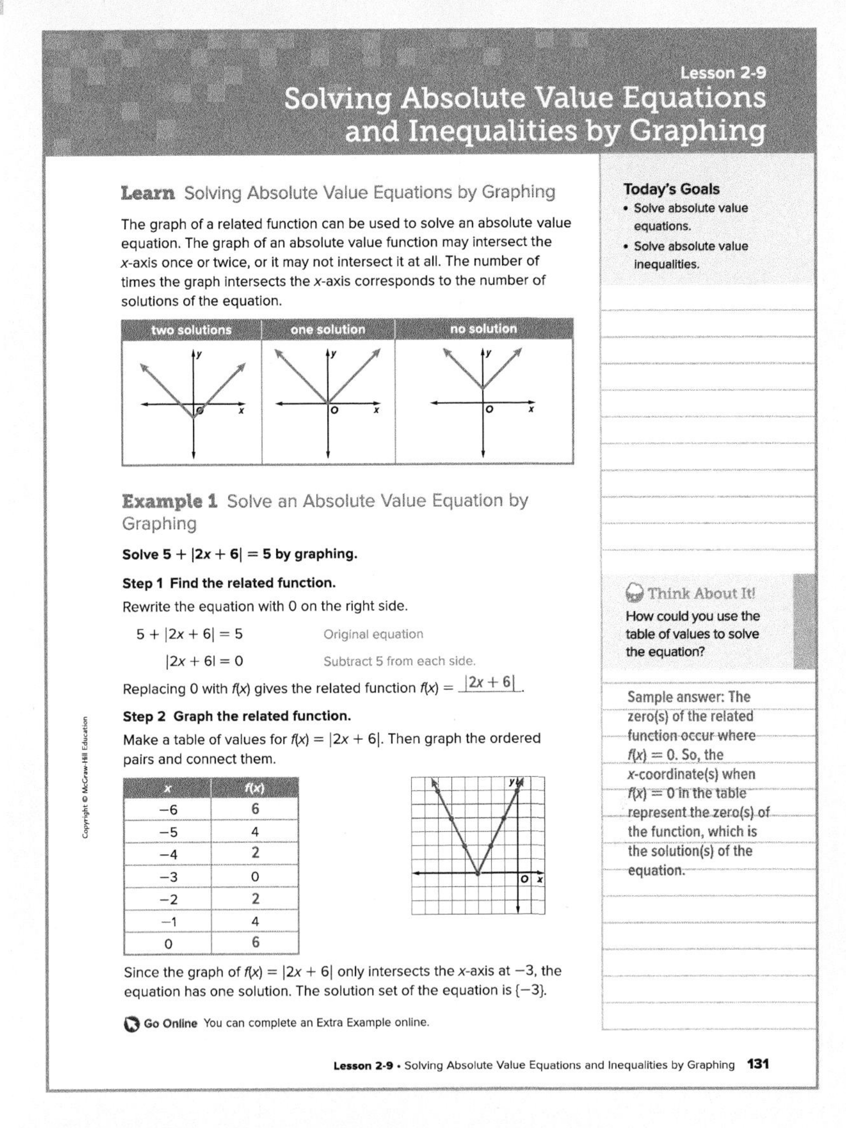

Lesson 2-9

Solving Absolute Value Equations and Inequalities by Graphing

Learn Solving Absolute Value Equations by Graphing

The graph of a related function can be used to solve an absolute value equation. The graph of an absolute value function may intersect the x-axis once or twice, or it may not intersect it at all. The number of times the graph intersects the x-axis corresponds to the number of solutions of the equation.

Example 1 Solve an Absolute Value Equation by Graphing

Solve $5 + |2x + 6| = 5$ by graphing.

Step 1 Find the related function.

Rewrite the equation with 0 on the right side.

$5 + |2x + 6| = 5$ Original equation

$|2x + 6| = 0$ Subtract 5 from each side.

Replacing 0 with $f(x)$ gives the related function $f(x) = |2x + 6|$.

Step 2 Graph the related function.

Make a table of values for $f(x) = |2x + 6|$. Then graph the ordered pairs and connect them.

x	$f(x)$
−6	6
−5	4
−4	2
−3	0
−2	2
−1	4
0	6

Since the graph of $f(x) = |2x + 6|$ only intersects the x-axis at −3, the equation has one solution. The solution set of the equation is {−3}.

Go Online You can complete an Extra Example online.

Today's Goals
- Solve absolute value equations.
- Solve absolute value inequalities.

Think About It! How could you use the table of values to solve the equation?

Sample answer: The zero(s) of the related function occur where $f(x) = 0$. So, the x-coordinate(s) when $f(x) = 0$ in the table represent the zero(s) of the function, which is the solution(s) of the equation.

Interactive Presentation

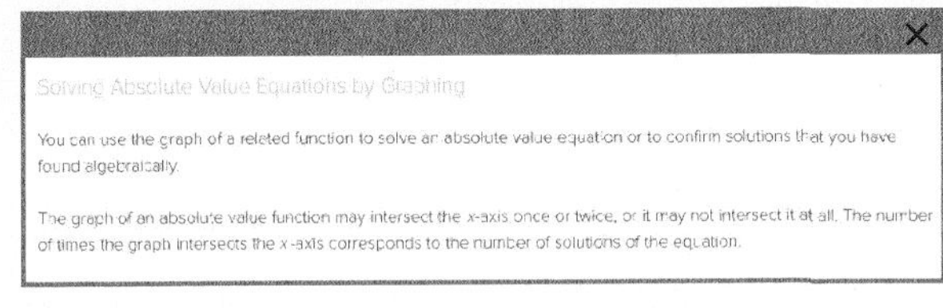

Learn

Your Notes

Check

Solve $|x + 1| + 9 = 13$ by graphing.

Part A Graph the related function.

Part B What is the solution set of the equation? $\{-5, 3\}$

Go Online to see how to use a graphing calculator with Examples 2 and 3.

Example 2 Solve an Absolute Value Equation by Using Technology

Use a graphing calculator to solve $\frac{4}{5}|x - 1| + 8 = 11$.

Rewrite the equation as the system $f(x) = \frac{4}{5}|x - 1| + 8$ and $g(x) = 11$.

Enter the functions in the **Y** = list and graph. To enter the absolute value symbols, press [math] and select **abs(** from the **NUM** menu.

Use the **intersect** feature from the **CALC** menu to find the x-coordinates of the points of intersection. When prompted, use the arrow keys to move the cursor close to each point of intersection, and press [enter] three times.

[−10, 10] scl: 1 by [−5, 15] scl: 1

Go Online An alternate method is available for this example.

The graphs intersect where $x =$ −2.75 and $x =$ 4.75. So, the solution set of the equation is {−2.75, 4.75}.

Check

Use a graphing calculator to solve $-\left|\frac{2}{3}x + 5\right| - 16 = -10$. Ø

Example 3 Confirm Solutions by Using Technology

Solve $-3|x + 7| + 9 = 14$. Check your solutions graphically.

$-3|x + 7| + 9 = 14$ Original equation

$-3|x + 7| = 5$ Subtract 9 from each side.

$|x + 7| = -\frac{5}{3}$ Divide each side by −3.

Because the absolute value of a number is always positive or zero, this sentence is *never* true. The solution is Ø.

(continued on the next page)

Go Online You can complete an Extra Example online.

Interactive Presentation

Solve an Absolute Value Equation by Using Technology

Use a graphing calculator to solve $\frac{4}{5}|x - 1| + 8 = 11$.

Select a calculator.

Example 2

TAP

Students tap to select a calculator and follow the steps to solving an absolute value equation by using technology.

CHECK

Students complete the Check online to determine whether they are ready to move on.

DIFFERENTIATE

Reteaching Activity AL ELL

IF students are struggling to identify absolute value solutions from a graph,

THEN have students circle any point where the graph of the related function crosses the x-axis. The x-values of these coordinates are the solutions to the equation.

Example 2 Solve an Absolute Value Equation by Using Technology

Teaching the Mathematical Practices

5 Analyze Graphs Help students analyze the graphs they have generated using graphing calculators to identify the solutions of the absolute value equation in Example 2.

Questions for Mathematical Discourse

AL What does the caption under the graph image mean? ([−10, 10] scl: 1 by [−5, 15] scl: 1) Each tick mark on the graph represents one unit for both axes, and the interval on the x-axis is −10 to 10 while the interval on the y-axis is −5 to 15.

OL Why are the solutions of the original equation the x-coordinates of the points of intersection of the system? Sample answer: These are the values that have the same function values for $f(x)$ and $g(x)$. So these are the values that make the expressions used to create the functions equal, and therefore make the equation true.

BL Why do you need to make a guess or set bounds for the graphing calculator to find the zeros? Sample answer: The calculator finds the zeros by doing many calculations over the function. Making a guess or setting bounds limits the number of calculations that need to be performed to reach the answer, which optimizes the process.

Example 3 Confirm Solutions by Using Technology

Teaching the Mathematical Practices

5 Decide When to Use Tools Mathematically proficient students can make sound decisions about when to use mathematical tools such as graphing calculators. Help them to see why using graphing calculators will help them solve or confirm the solutions of absolute value equations.

Questions for Mathematical Discourse

AL What does the graph of the related function look like for this example? Sample answer: The graph opens downward and does not cross the x-axis.

OL What transformations could you perform on the graph to make it intercept the x-axis? Sample answer: you could translate the graph up by at least 5, or reflect it across $y = -5$.

BL What could you change in the original equation so that the absolute value equation has at least one solution? Sample answer: Either make the coefficient of the absolute value expression a positive three or change the 9 to a 14 or larger.

Learn Solving Absolute Value Inequalities by Graphing

Objective

Students solve absolute value inequalities by examining graphs of related functions.

 Teaching the Mathematical Practices

3 Analyze Cases The Learn guides students to look at the cases of absolute value inequalities. Encourage students to familiarize themselves with all of the cases and know how to solve each type of inequality.

Example 4 Solve an Absolute Value Inequality by Graphing

 Teaching the Mathematical Practices

1 Explain Correspondences Encourage students to explain the relationships between the inequality, table, and graph used in Example 4.

Questions for Mathematical Discourse

AL Will the x-intercepts be included in a solution if the inequality symbol is $<$ or $>$? Explain. No; sample answer: The symbol $<$ or $>$ means the x-intercepts are not solutions to the inequality

OL What does the symbol $\geq$ indicate about the solution set of an absolute value inequality in terms of the graph of the related function? The solution set will be where the graph is above the x-axis, including the x-intercepts.

BL Could an absolute value inequality have no solution? Explain. Yes; sample answer: If the entire graph of the related function was above the x-axis but the inequality was $\leq$ or $<$, there would be no solution since the graph is never below the x-axis.

Use a graphing calculator to confirm that there is no solution.

Step 1 Find and graph the related function.

Rewriting the equation results in the related function $f(x) = -3|x + 7| - 5$. Enter the related function in the **Y=** list and graph.

Step 2 Find the zeros.

The graph does not appear to intersect the x-axis. Use the **ZOOM** feature or adjust the window manually to see this more clearly. Since the related function never intersects the x-axis, there are no real zeros. This confirms that the equation has no solution.

[−40, 40] scl: 1 by [−40, 40] scl: 1

Think About It! How could you use a calculator to confirm the solutions of an equation with one or two real solutions?

Sample answer: Use the calculator to locate the zeros and make sure that they are the same as the algebraic solutions.

Learn Solving Absolute Value Inequalities by Graphing

The related functions of absolute value inequalities are found by solving the inequality for 0, replacing the inequality symbol with an equals sign, and replacing 0 with $f(x)$.

For $<$ and $\leq$, identify the x-values for which the graph of the related function lies below the x-axis. For $\leq$, include the x-intercepts in the solution. For $>$ and $\geq$, identify the x-values for which the graph of the related function lies *above* the x-axis. For $\geq$, include the x-intercepts in the solution.

Example 4 Solve an Absolute Value Inequality by Graphing

Solve $|3x - 9| - 6 \leq 0$ by graphing.

The solution set consists of x-values for which the graph of the related function lies *below* the x-axis, including the x-intercepts. The related function is $f(x) = |3x - 9| - 6$. Graph $f(x)$ by making a table.

x	$f(x)$
0	3
1	0
2	−3
3	−6
4	−3
5	0
6	3

The graph lies below the x-axis between $x =$ 1 and $x =$ 5. Thus, the solution set is $\{x \mid 1 \leq x \leq 5\}$ or $[1, 5]$. All values of x between 1 and 5 satisfy the constraints of the original inequality.

Talk About It! Would the solution change if the inequality symbol was changed from $\leq$ to $\geq$? Explain your reasoning.

Yes; sample answer: The solution set would include all x-values for which the graph of the related function lies *above* the x-axis instead of *below*. The x-intercepts would still be included in the solution set.

Go Online You can complete an Extra Example online.

Interactive Presentation

Solving Absolute Value Inequalities by Graphing

Absolute value inequalities can be solved by using the graphs of related absolute value functions. The related functions are found by solving the inequality for 0, replacing the inequality symbol with an equals sign, and replacing 0 with $f(x)$.

The inequality symbol indicates what to look for on the graph of the related function. Move through the categories to learn more about how to solve absolute value inequalities by graphing.

Learn

TAP

Students tap to compare and contrast the graphs of greater than and less than inequalities.

Check

Solve $|x - 2| - 3 \leq 0$ by graphing.

Part A Graph a related function.

Part B What is the solution set of $|x - 2| - 3 \leq 0$?

$x \geq -1$ and $x \leq 5$

Think About It!

The inequality in the example is solved for 0. What additional step(s) would you need to take if the given inequality had a nonzero term on the right side of the inequality symbol?

Sample answer: I would have to rewrite the inequality so that one side is 0 before I could determine the direction of the inequality and find the related function.

Go Online to see how to use a graphing calculator with this example.

Example 5 Solve an Absolute Value Equation by Using Technology

Use a graphing calculator to solve $\left|\frac{5}{7}x + 2\right| - 3 > 0$.

Step 1 Graph the related function.

Rewriting the inequality results in the related function $f(x) = \left|\frac{5}{7}x + 2\right| - 3$.

Step 2 Find the zeros.

The $>$ symbol indicates that the solution set consists of x-values for which the graph of the related function lies *above* the x-axis, not including the x-intercepts.

Use the **zero** feature from the **CALC** menu to find the zeros, or x-intercepts.

[−10, 10] scl: 1 by [−10, 10] scl: 1

The zeros are located at $x =$ __−7__ and $x =$ __1.4__. The graph lies above the x-axis when $x <$ __−7__ and $x >$ __1.4__.

So the solution set is $\{x \mid x <$ __−7__ __or__ $x >$ __1.4__$\}$.

Check

Use a graphing calculator to solve $\frac{1}{2}|4x + 1| - 5 > 0$.

$x < -2.75$ or $x > 2.25$

Go Online You can complete an Extra Example online.

Example 5 Solve an Absolute Value Inequality by Using Technology

Teaching the Mathematical Practices

5 Analyze Graphs Help students analyze the graphs they have generated using graphing calculators to identify the solutions of the absolute value inequality in Example 5.

Questions for Mathematical Discourse

AL Will the solution set contain the x-intercepts? no

OL How could the absolute value inequality be solved algebraically? Sample answer: First isolate the absolute value expression, and then make two equations representing the two cases. Because the original inequality was $>$, the solution will be two separate inequalities joined by "or."

BL What would be the solution if the -3 was $+3$? all real numbers

DIFFERENTIATE

Reteaching Activity AL ELL

IF students are having difficulties graphing absolute value inequalities, **THEN** have students select a test point inside and outside the shape of the graphed function to see where to shade for the solution set. The shading illustrates the domain interval of the solution set.

Interactive Presentation

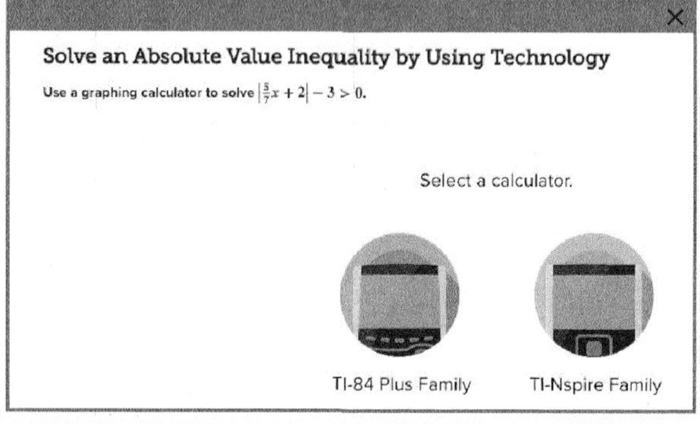

Example 5

SELECT

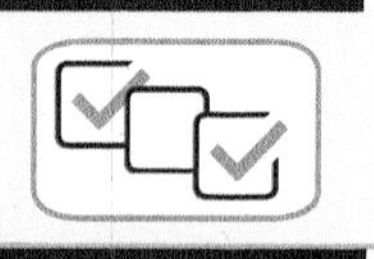

Students select the calculator they will use to solve the absolute value inequality.

CHECK

Students complete the Check online to determine whether they are ready to move on.

Exit Ticket

Recommended Use

At the end of class, go online to display the Exit Ticket prompt and ask students to respond using a separate piece of paper. Have students hand you their responses as they leave the room.

Alternate Use

At the end of class, go online to display the Exit Ticket prompt and ask students to respond verbally or by using a mini-whiteboard. Have students hold up their whiteboards so that you can see all student responses. Tap to reveal the answer when most or all students have completed the Exit Ticket.

Practice and Homework

Suggested Assignments

Use the table below to select appropriate exercises.

DOK	Topic	Exercises
1, 2	exercises that mirror the examples	1–30
2	exercises that use a variety of skills from this lesson	31–39
2	exercises that extend concepts learned in this lesson to new contexts	40–42
3	exercises that emphasize higher-order and critical thinking skills	43–45

ASSESS AND DIFFERENTIATE

Use the data from the **Checks** to determine whether to provide resources for extension, remediation, or intervention.

IF students score 90% or more on the Checks, **THEN** assign: BL

- Practice Exercises 1–41 odd, 42–45
- Extension: The Sign Function
- **ALEKS®** Absolute Value Equations; Absolute Value Inequalities

IF students score 66%–89% on the Checks, **THEN** assign: 

- Practice Exercises 1–45 odd
- Remediation, Review Resources: Functions and Continuity
- Personal Tutors
- Extra Examples 1–4
- **ALEKS®** Graphs of Functions

IF students score 65% or less on the Checks, **THEN** assign: 

- Practice Exercises 1–29 odd
- Remediation, Review Resources: Functions and Continuity
- **ALEKS®** Graphs of Functions

Name ______ Period ______ Date ______

Practice

Go Online You can complete your homework online.

Example 1

Solve each equation by graphing. See margin for graphs.

1. $|x-4|=5$ $\{-1, 9\}$
2. $|2x-3|=17$ $\{10, -7\}$
3. $3+|2x+1|=3$ $\left\{-\frac{1}{2}\right\}$
4. $|x-1|+6=4$ $\varnothing$
5. $7+|3x-1|=7$ $\left\{\frac{1}{3}\right\}$
6. $|x+2|+5=13$ $\{-10, 6\}$

Example 2 See Mod. 2 Answer Appendix for graphs.

USE TOOLS **Use a graphing calculator to solve each equation.**

7. $\frac{1}{2}|x-1|+5=9$ $\{-7, 9\}$
8. $\frac{3}{4}|x+1|+1=7$ $\{-9, 7\}$
9. $\frac{2}{3}|x-2|-4=4$ $\{-10, 14\}$
10. $2|x+2|=10$ $\{-7, 3\}$
11. $\frac{1}{5}|x+6|-1=9$ $\{-56, 44\}$
12. $3|x+5|-1=11$ $\{-9, -1\}$

Example 3

USE TOOLS **Solve each equation algebraically. Use a graphing calculator to check your solutions.**

13. $|3x-6|=42$ $\{-12, 16\}$
14. $7|x+3|=42$ $\{-9, 3\}$
15. $-3|4x-9|=24$ $\varnothing$
16. $-6|5-2x|=-9$ $\{1.75, 3.25\}$
17. $5|2x+3|-5=0$ $\{-2, -1\}$
18. $|15-2x|=45$ $\{-15, 30\}$

Example 4

Solve each inequality by graphing. See Mod. 2 Answer Appendix for graphs.

19. $|2x-6|-4\le 0$ $\{x \mid 1\le x\le 5\}$
20. $|x-1|-3\le 0$ $\{x \mid -2\le x\le 4\}$
21. $|2x-1|\ge 4$ $\left\{x \mid x\le -\frac{3}{2} \text{ or } x\ge \frac{5}{2}\right\}$
22. $|3x+2|\ge 6$ $\left\{x \mid x\le -\frac{8}{3} \text{ or } x\ge \frac{4}{3}\right\}$
23. $2|x+2|<8$ $\{x \mid -6<x<2\}$
24. $3|x-1|<12$ $\{x \mid -3<x<5\}$

Example 5

USE TOOLS **Use a graphing calculator to solve each inequality.** See Mod. 2 Answer Appendix for graphs.

25. $\left|\frac{1}{4}x+4\right|-1>0$ $\{x \mid x<-20 \text{ or } x>-12\}$
26. $\frac{2}{5}|x-5|+1>0$ all real numbers
27. $|3x-1|<2$ $\left\{x \mid -\frac{1}{3}<x<1\right\}$
28. $|4x+1|\le 1$ $\left\{x \mid -\frac{1}{2}\le x\le 0\right\}$
29. $\frac{1}{6}|x-1|+1\le 0$ $\varnothing$
30. $\frac{1}{4}|x+5|-1\le 1$ $\{x \mid -13\le x\le 3\}$

Mixed Exercises

Solve by graphing. 31–36. See Mod. 2 Answer Appendix for graphs.

31. $0.4|x-1| = 0.2$ $\{0.5, 1.5\}$ **32.** $0.16|x+1| = 4.8$ $\{-31, 29\}$ **33.** $0.78|2x+0.1| + 2.3 = 0$ $\varnothing$

34. $\left|\frac{1}{3}x+3\right| + 1 = 0$ $\varnothing$

35. $\frac{1}{2}|6-2x| \leq 1$ $\{x \mid 2 \leq x \leq 4\}$

36. $|3x-2| < \frac{1}{2}$ $\left\{x \middle| \frac{1}{2} < x < \frac{5}{6}\right\}$

USE TOOLS **Solve each equation or inequality algebraically. Use a graphing calculator to check your solutions.**

37. $\left|\frac{5}{9}x+1\right| - 5 > 0$ $\left\{x \middle| x < -10\frac{4}{5} \text{ or } x > 7\frac{1}{5}\right\}$

38. $\frac{2}{7}\left|\frac{1}{2}x-1\right| < 1$ $\{x \mid -5 < x < 9\}$

39. $0.28|0.4x-2| = 10.08$ $\{-85, 95\}$

40. REASONING A pet store sells bags of dog food that are labeled as weighing 50 pounds. The equipment used to package the dog food produces bags with a weight that is within 0.75 lbs of the advertised weight. Write an absolute value equation to determine the acceptable maximum and minimum weight for the bags of dog food. Then, use a graph to find the minimum and maximum weights.
$|x-50| = 0.75$; $x = 50.75$, $x = 49.25$: The minimum weight is 49.25 lbs and the maximum is 50.75 lbs.

41. SPACE The mean distance from Mars to the Sun is 1.524 astronomical units (au). The distance varies during the orbit of Mars by 0.147 au. Write an absolute value inequality to represent the distance of Mars from the Sun as it completes an orbit. Then, use a graph to solve the inequality. $|x-1.524| \leq 0.147$; $\{x \mid 1.377 \leq x \leq 1.671\}$; See Mod. 2 Answer Appendix for graph.

42. CONSTRUCT ARGUMENTS Ms. Uba asked her students to write an absolute value equation with the solutions and related function shown in the graph. Sawyer wrote $|x+2| = 6$ and Kaleigh wrote $2|x+2| = 12$. Is either student correct? Justify your argument.
Sawyer is correct. Sample answer: Both equations have the solution set $\{-8, 4\}$ but only Sawyer's equation has the same related function shown in the graph.

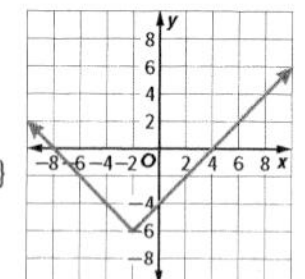

Higher-Order Thinking Skills

43. WRITE Compare and contrast the solution sets of absolute value equations and absolute value inequalities. See Mod. 2 Answer Appendix.

44. ANALYZE How can you tell that an absolute value equation has no solutions without graphing or completely solving it algebraically? Sample answer: After isolating the absolute value, if the other side of the equation is a negative value, then there are no solutions.

45. CREATE Create an absolute value equation for which the solution set is $\{9, 11\}$.
Sample answer: $|x-10| = 1$

136 Module 2 • Linear Equations, Inequalities, and Systems

Answers

1.

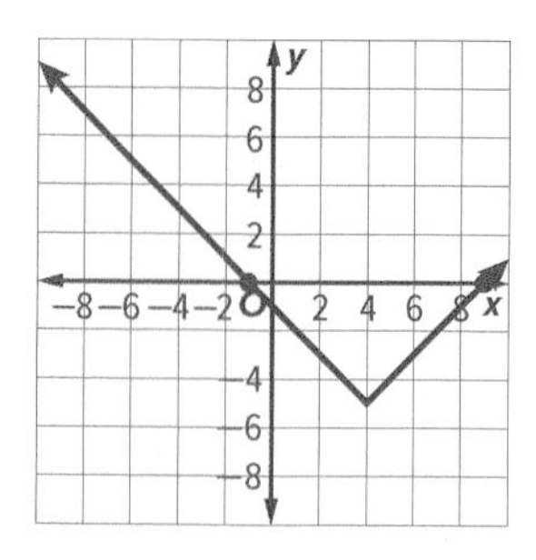

2.

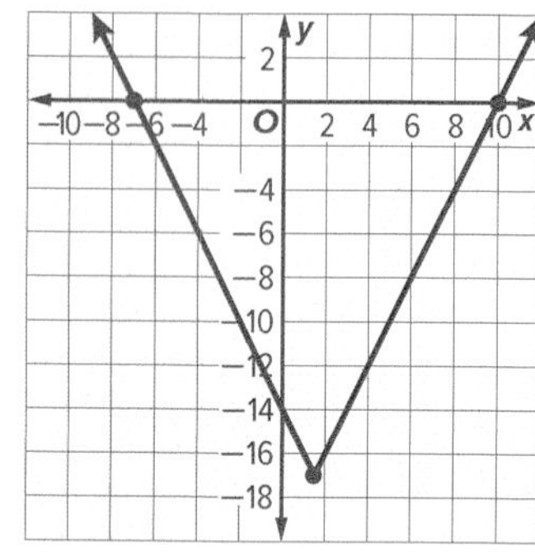

3.

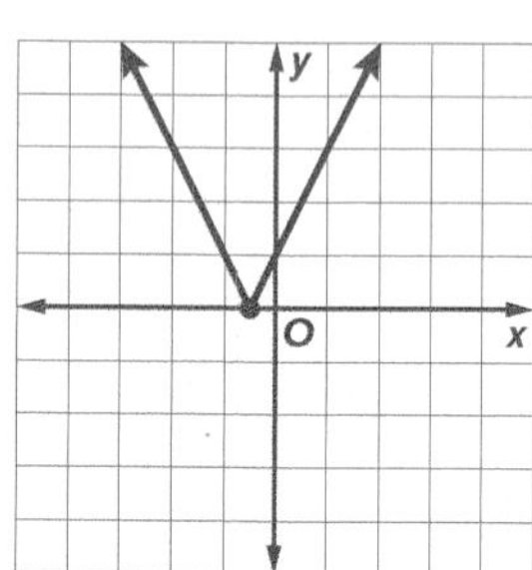

4.

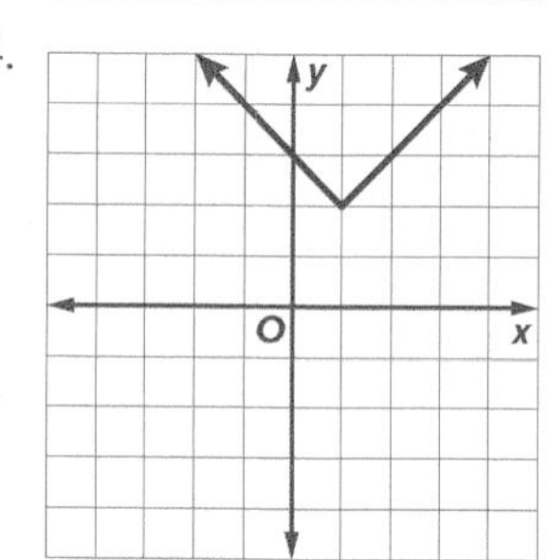

5.

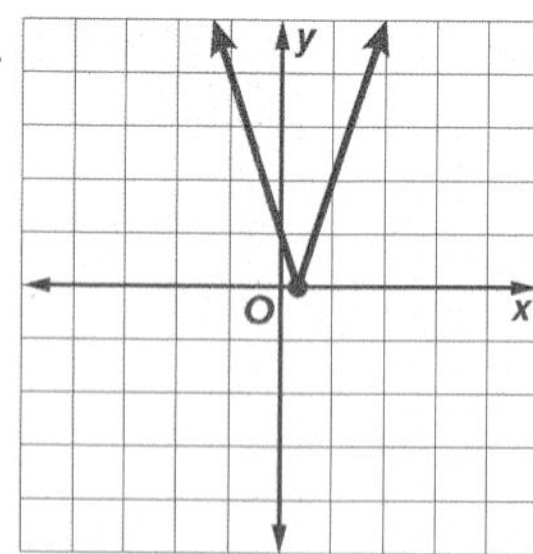

6.

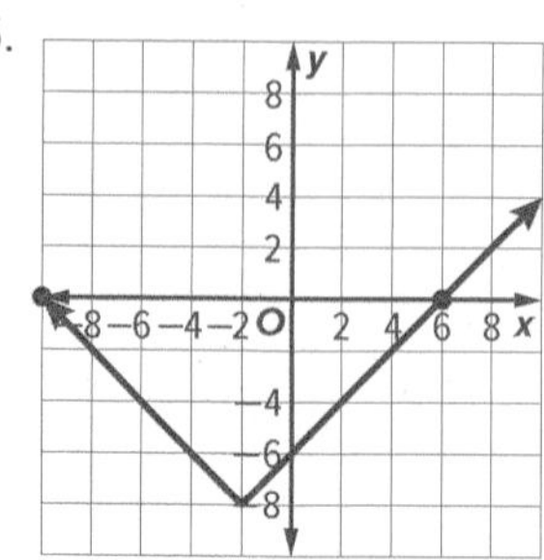

Review

Rate Yourself!

Have students return to the Module Opener to rate their understanding of the concepts presented in this module. They should see that their knowledge and skills have increased. After completing the chart, have them respond to the prompts in their *Interactive Student Edition* and share their responses with a partner.

Answering the Essential Question

Before answering the Essential Question, have students review their answers to the Essential Question Follow-Up questions found throughout the module.

- Why are systems of equations useful when solving real-world situations?
- Why is linear programming important in the real world?
- Why is solving a system of equations in three variables similar to solving a system in two variables?

Then have them write their answer to the Essential Question in the space provided.

DINAH ZIKE FOLDABLES

ELL A completed Foldable for this module should include the key concepts related to equations and inequalities in two and three variables.

LS **LearnSmart** Use LearnSmart as part of your test preparation plan to measure student topic retention. You can create a student assignment in LearnSmart for additional practice on these topics for **Modeling with Functions**.

- Building Functions

Review

Essential Question

How are equations, inequalities, and systems of equations or inequalities best used to model to real-world situations?

Sample answer: Equations allow you to find quantities that fit constraints, such as costs for given numbers of items. Inequalities represent situations involving intervals of numbers, such as quantities for which cost is below a given value. Systems allow you to have multiple constraints for the same situation.

Module Summary

Lessons 2-1 and 2-2

Linear and Absolute Value Equations and Inequalities

- The solution of an equation or an inequality is any value that, when substituted into the equation, results in a true statement.
- An absolute value equation or inequality is solved by writing it as two cases. An absolute value equation may have 0, 1, or 2 solutions.

Lesson 2-3

Equations of Linear Functions

- Standard form: $Ax + By = C$, where A, B, and C are integers with a greatest common factor of 1, $A \geq 0$, and A and B are not both 0
- Slope-intercept form: $y = mx + b$, where m is the slope and b is the y-intercept
- Point-slope form: $y - y_1 = m(x - x_1)$, where m is the slope and (x_1, y_1) are the coordinates of a point on the line

Lessons 2-4 through 2-8

Systems of Equations and Inequalities

- The point of intersection of the two graphs a system of equations represents the solution.
- In the substitution method, one equation is solved for a variable and substituted to find the value of another variable.
- In the elimination method, one variable is eliminated by adding or subtracting the equations.
- To solve a system of inequalities, graph each inequality. Viable solutions to the system of inequalities are in the intersection of all the graphs.
- Linear programming is a method for finding maximum or minimum values of a function over a given system of inequalities with each inequality representing a constraint.
- Systems of equations in three variables can have infinitely many solutions, no solution, or one solution which is written as an ordered triple (x, y, z).
- Systems of equations in three variables can be solved by using elimination and substitution.

Lesson 2-9

Solving Absolute Value Equations and Inequalities by Graphing

- The graph of the related absolute value function can be used to solve an equation or inequality. The x-intercept(s) of the function give the solution(s) of the equation.

Study Organizer

Foldables

Use your Foldable to review this module. Working with a partner can be helpful. Ask for clarification of concepts as needed.

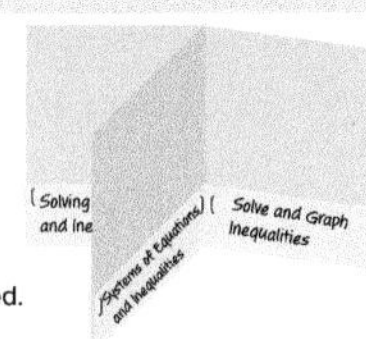

Name _______________ Period _______ Date _______

Test Practice

1. **OPEN RESPONSE** Explain each step in solving the equation $3(2x + 9) + \frac{1}{2}(4x - 8) = 55$. (Lesson 2-1)

   Sample answer: First, use the Distributive Property: $6x + 27 + 2x - 4 = 55$. Combine like terms: $8x + 23 = 55$. Then use the Subtraction Property of Equality: $8x = 32$. Finally, use the Division Property of Equality: $x = 4$.

2. **MULTIPLE CHOICE** In a single football game, a team scored 14 points from touchdowns and made x field goals, which are worth three points each. The team scored 23 points. How many field goals did the team make? (Lesson 2-1)

   Ⓐ 2
   Ⓑ 3 (answer)
   Ⓒ 4
   Ⓓ 5

3. **OPEN RESPONSE** A temperature in degrees Celsius, C, is equal to five-ninths times the difference of the temperature in degrees Fahrenheit, F, and 32. Write an equation to relate the temperature in degrees Celsius to the temperature in degrees Fahrenheit. Then determine the Fahrenheit temperature that is equivalent to 25°C. (Lesson 2-1)

   $C = \frac{5}{9}(F - 32)$; 77°F

4. **OPEN RESPONSE** The temperature of an oven varies by as much as 7.5 degrees from the temperature shown on its display. If the oven is set to 425°, write an equation to find the minimum and maximum actual temperature of the oven. (Lesson 2-2)

   $|x - 425| \leq 7.5$ or $417.5 \leq x \leq 432.5$

5. **OPEN RESPONSE** The equation $2x + y = 10$ can be used to find the number of miles Allie has left to jog this week to reach her goal, where x is the number of days and y is the number of miles left to reach the goal. Write the equation in slope-intercept form. Then interpret the parameters in the context of the situation. (Lesson 2-3)

   $y = -2x + 10$; -2 represents the miles Allie is running each day. 10 represents that Allie's goal at the beginning of the week is to jog 10 miles.

6. **MULTIPLE CHOICE** Thomas is driving his truck at a constant speed. The table below gives the distance remaining to his destination y in miles x minutes after he starts driving.

x	15	30	45
y	186.5	173	159.5

   Which equation models the distance remaining after any number of minutes? (Lesson 2-3)

   Ⓐ $y = \frac{9}{10}x - 200$
   Ⓑ $y = -\frac{10}{9}x + 200$
   Ⓒ $y = \frac{10}{9}x - 200$
   Ⓓ $y = -\frac{9}{10}x + 200$ (answer)

Review and Assessment Options

The following online review and assessment resources are available for you to assign to your students. These resources include technology-enhanced questions that are auto-scored, as well as essay questions.

Review Resources

Put It All Together: Lessons 2-1 through 2-5
Vocabulary Activity
Module Review

Assessment Resources

Vocabulary Test
AL Module Test Form B
OL Module Test Form A
BL Module Test Form C
Performance Task*

*The module-level performance task is available online as a printable document. A scoring rubric is included.

Practice

You can use these pages to help your students review module content and prepare for online assessments. Exercises 1-16 mirror the types of questions your students will see on online assessments.

Question Type	Description	Exercise(s)
Multiple Choice	Students select one correct answer.	2, 6, 7, 8, 10, 12
Multi-Select	Multiple answers may be correct. Students must select all correct answers.	13, 14
Open Response	Students construct their own response in the area provided.	1, 3, 4, 5, 9, 11, 15, 16

To ensure that students understand the standards, check students' success on individual exercises.

Standard(s)	Lesson(s)	Exercise(s)
A.CED.1	2-1, 2-2	2, 4
A.CED.2	2-1, 2-3	3, 6
A.CED.3	2-6, 2-7, 2-8	10, 11, 12, 13, 14, 15
F.LE.5	2-3	5
A.REI.1	2-1	1
A.REI.6	2-4, 2-5	7, 8, 9
A.REI.11	2-9	16

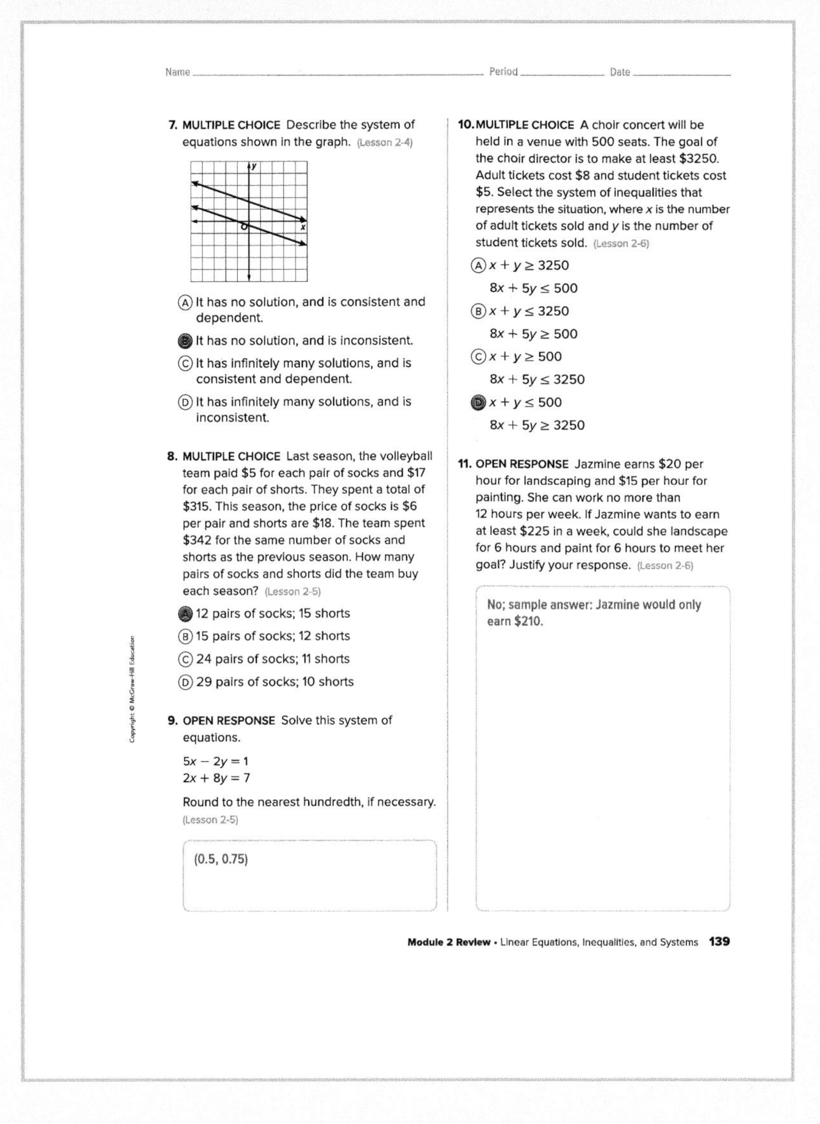

Name ______ Period ______ Date ______

7. MULTIPLE CHOICE Describe the system of equations shown in the graph. (Lesson 2-4)

Ⓐ It has no solution, and is consistent and dependent.
Ⓑ It has no solution, and is inconsistent.
Ⓒ It has infinitely many solutions, and is consistent and dependent.
Ⓓ It has infinitely many solutions, and is inconsistent.

8. MULTIPLE CHOICE Last season, the volleyball team paid \$5 for each pair of socks and \$17 for each pair of shorts. They spent a total of \$315. This season, the price of socks is \$6 per pair and shorts are \$18. The team spent \$342 for the same number of socks and shorts as the previous season. How many pairs of socks and shorts did the team buy each season? (Lesson 2-5)

Ⓐ 12 pairs of socks; 15 shorts
Ⓑ 15 pairs of socks; 12 shorts
Ⓒ 24 pairs of socks; 11 shorts
Ⓓ 29 pairs of socks; 10 shorts

9. OPEN RESPONSE Solve this system of equations.

$5x - 2y = 1$
$2x + 8y = 7$

Round to the nearest hundredth, if necessary. (Lesson 2-5)

(0.5, 0.75)

10. MULTIPLE CHOICE A choir concert will be held in a venue with 500 seats. The goal of the choir director is to make at least \$3250. Adult tickets cost \$8 and student tickets cost \$5. Select the system of inequalities that represents the situation, where x is the number of adult tickets sold and y is the number of student tickets sold. (Lesson 2-6)

Ⓐ $x + y \geq 3250$, $8x + 5y \leq 500$
Ⓑ $x + y \leq 3250$, $8x + 5y \geq 500$
Ⓒ $x + y \geq 500$, $8x + 5y \leq 3250$
Ⓓ $x + y \leq 500$, $8x + 5y \geq 3250$

11. OPEN RESPONSE Jazmine earns \$20 per hour for landscaping and \$15 per hour for painting. She can work no more than 12 hours per week. If Jazmine wants to earn at least \$225 in a week, could she landscape for 6 hours and paint for 6 hours to meet her goal? Justify your response. (Lesson 2-6)

No; sample answer: Jazmine would only earn \$210.

Name ______________ Period ________ Date ________

12. MULTIPLE CHOICE Which system of inequalities represents the graph shown? (Lesson 2-6)

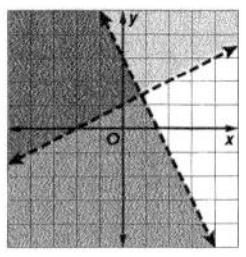

- (A) $x - 2y < -2$, $2x + y < 3$
- (B) $x - 2y < -2$, $2x + y > 3$
- (C) $x - 2y > -2$, $2x + y < 3$
- (D) $x - 2y > -2$, $2x + y > 3$

13. MULTI-SELECT The shaded region represents the feasible region for a linear programming problem.

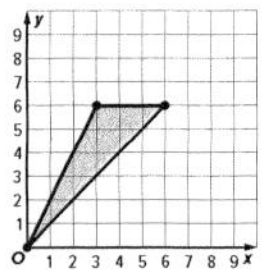

Select all the points at which a minimum or maximum value may occur. (Lesson 2-7)

- (A) (0, 0)
- (B) (1, 5)
- (C) (2, 6)
- (D) (3, 6)
- (E) (4, 5)
- (F) (5, 3)
- (G) (6, 6)

14. MULTI-SELECT A shoe manufacturer makes two types of athletic shoes—a cross-training shoe and a running shoe. Each pair of shoes is assembled by machine and then finished by hand. For the cross-training shoes, it takes 15 minutes for machine assembly and 6 minutes by hand. For the running shoes, it takes 9 minutes on the machine and 12 minutes by hand. The company can allocate no more than 900 machine hours and 500 hand hours each day. The profit is $10 for each pair of cross-training shoes and $15 for each pair of running shoes. Let x represent the number of cross-training shoes and y represent the number of running shoes manufactured each day.

Select the inequalities that form the boundary of the feasible region. (Lesson 2-7)

- (A) $x \geq 0$
- (B) $y \geq 0$
- (C) $0.1x + 0.2y \leq 500$
- (D) $0.25x + 0.15y \leq 900$
- (E) $6x + 12y \leq 500$
- (F) $15x + 9y \leq 900$

15. OPEN RESPONSE Last month, Jeremy took 2 piano lessons, 3 guitar lessons, and 1 drum lesson and spent a total of $285. D'Asia took 1 piano lesson, 5 guitar lessons, and 2 drum lessons and spent a total of $400. Raj took 3 piano lessons, 2 guitar lessons, and 4 drum lessons and spent a total of $440. Write a system of equations that could be used to find the prices of each piano lesson, x, guitar lesson, y, and drum lesson, z. (Lesson 2-8)

$2x + 3y + z = 285$
$x + 5y + 2z = 400$
$3x + 2y + 4z = 440$

16. OPEN RESPONSE Use a graphing calculator to solve the equation $2|x - 3| - 1 = 5$. (Lesson 2-9)

$\{0, 6\}$

Quick Check

1.

2. 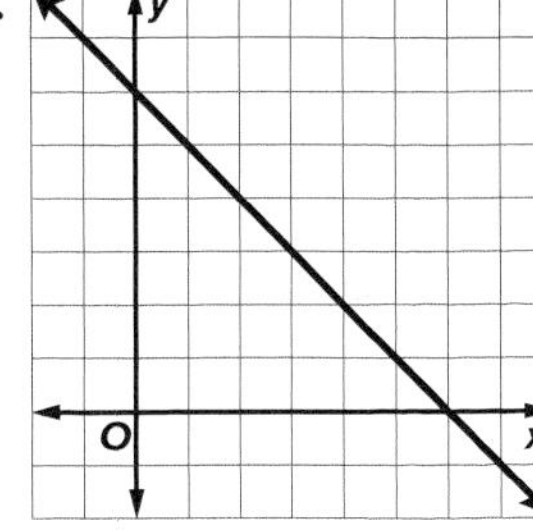

3. 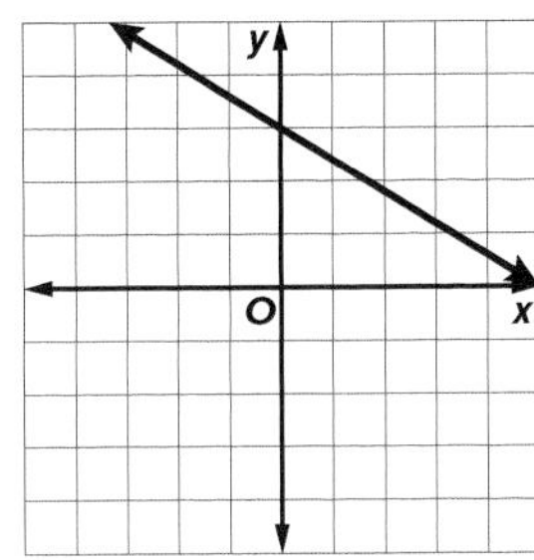

4. 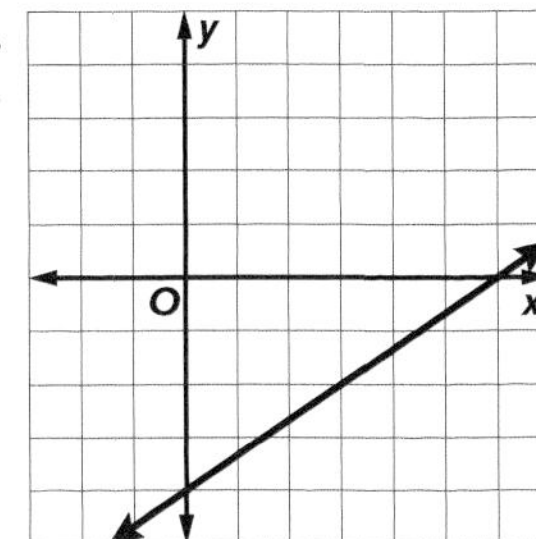

5. 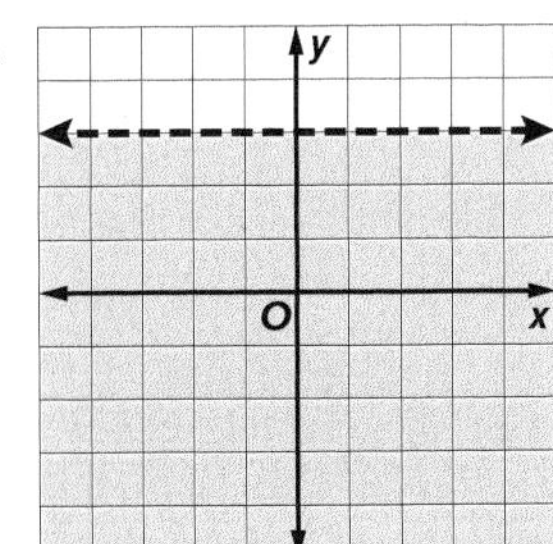

6. 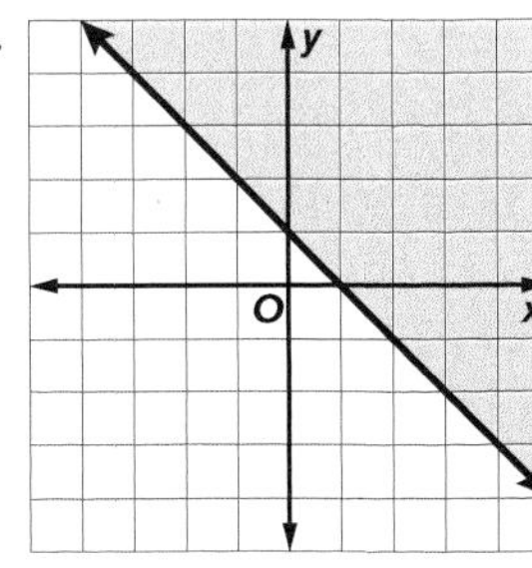

7. 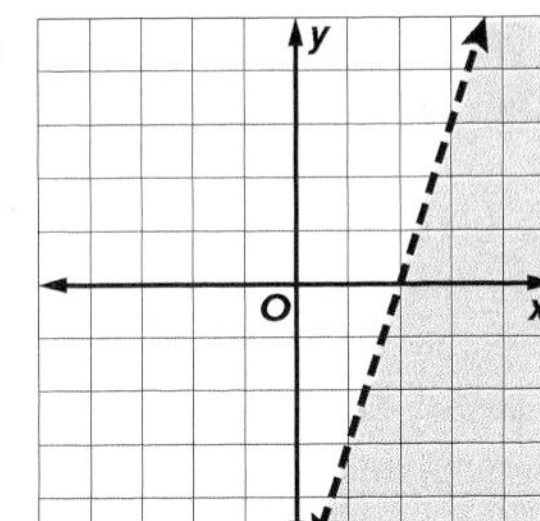

8. 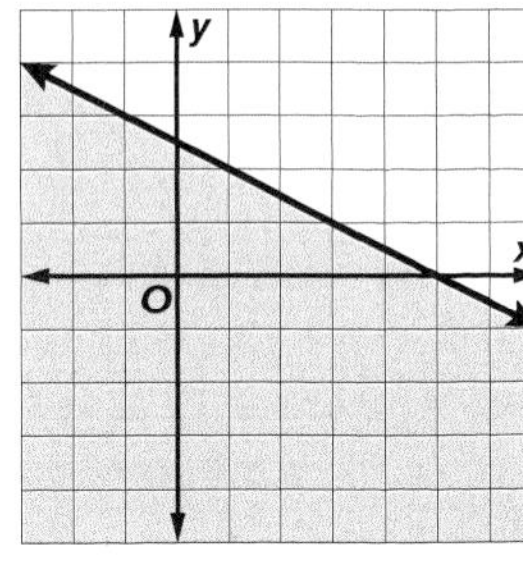

Lesson 2-1

36. −4 −3 −2 −1 0 1 2 3 4

37. 0 1 2 3 4 5 6 7 8

38. 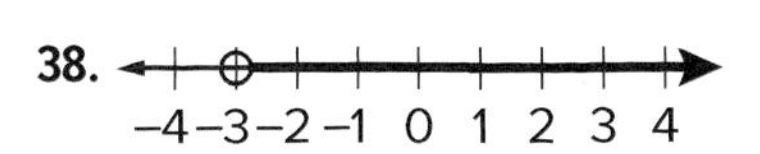

45. Sample answer: When one number is greater than another number, it is either more positive or less negative than that number. When these numbers are multiplied by a negative value, their roles are reversed. That is, the number that was more positive is now more negative than the other number. Thus, it is now *less than* that number and the inequality symbol needs to be reversed.

46. Using the Triangle Inequality Theorem, we know that the sum of the lengths of any 2 sides of a triangle must be greater than the length of the remaining side. This generates 3 inequalities to examine.

$3x + 4 + 2x + 5 > 4x$ $\quad x > -9$

$3x + 4 + 4x > 2x + 5$ $\quad x > 0.2$

$2x + 5 + 4x > 3x + 4$ $\quad x > -\frac{1}{3}$

In order for all 3 conditions to be true, x must be greater than 0.2.

Lesson 2-2

40. Sample answer: A compound inequality that contains *and* is true if and only if both individual inequalities are true, while an inequality containing *or* only needs one of the individual inequalities to be true.

43. The 4 potential solutions are:

1. $(2x - 1) \geq 0$ and $(5 - x) \geq 0$
2. $(2x - 1) \geq 0$ and $(5 - x) < 0$
3. $(2x - 1) < 0$ and $(5 - x) \geq 0$
4. $(2x - 1) < 0$ and $(5 - x) < 0$

The resulting equations corresponding to these cases are:

1. $2x - 1 + 3 = 5 - x : x = 1$
2. $2x - 1 + 3 = x - 5 : x = -7$
3. $1 - 2x + 3 = 5 - x : x = -1$
4. $1 - 2x + 3 = x - 5 : x = 3$

The solutions from case 1 and case 3 work. The others are extraneous. The solution set is $\{-1, 1\}$.

44. Sometimes; this is only true for certain values of a. For example, it is true for $a = 8$; if $8 > 7$, then $11 > 10$. However it is not true for $a = -8$; if $8 > 7$, then $5 \not> 10$.

45. Always; if $|x| < 3$, then x is between -3 and 3. Adding 3 to the absolute value of any of the numbers in this set will produce a positive number.

46. Always; starting with numbers between 1 and 5 and subtracting 3 will produce numbers between -2 and 2. These all have an absolute value less than or equal to 2.

Lesson 2-4

22. 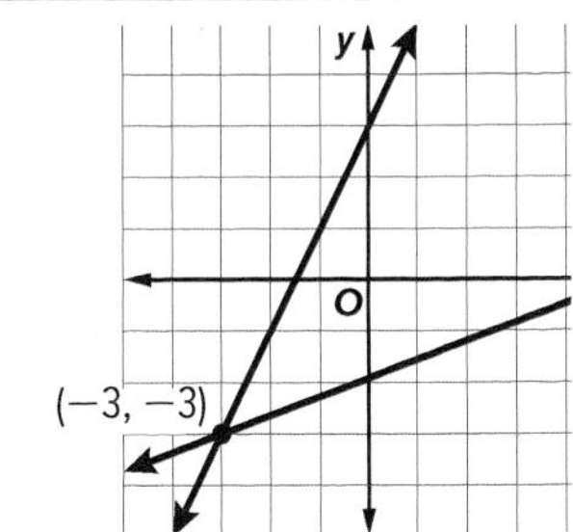

23. 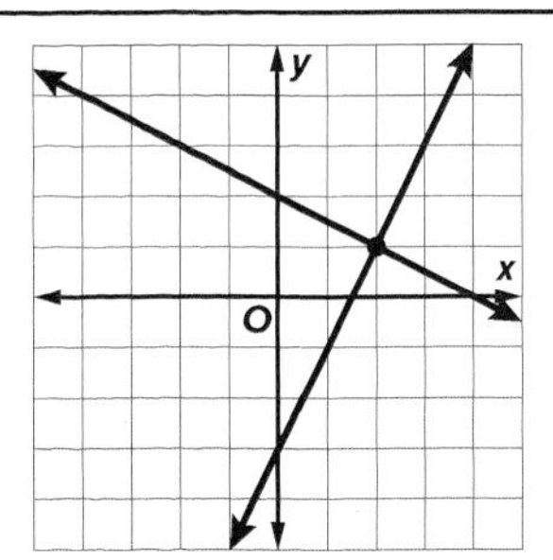

24. 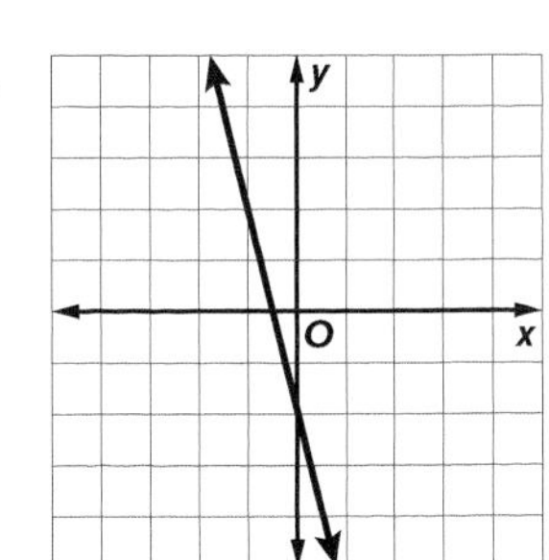

25. 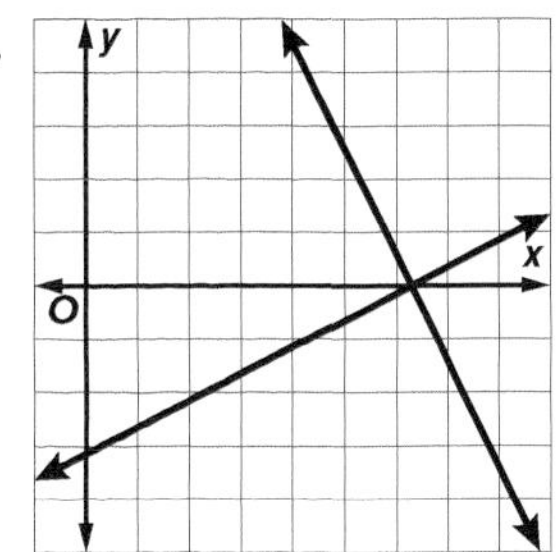

26b. 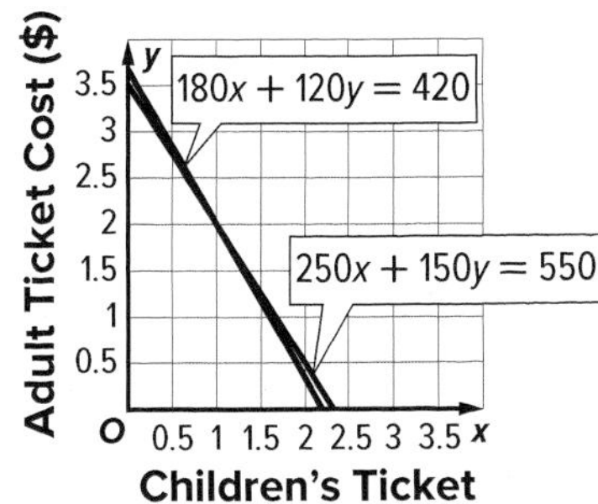

MODULE 2 ANSWER APPENDIX

27. Always; sample answer: a and b are the same line. b is parallel to c, so a is also parallel to c. Since c and d are consistent and independent, then c is not parallel to d and, thus, intersects d. Since a and c are parallel, then a cannot be parallel to d, so, a must intersect d and must be consistent and independent with d.

28. Sample answer: Write each equation of the system in slope-intercept form. Graph each equation. The point where the lines intersect is the solution of the system of equations. If the lines coincide, the system of equations has an infinite number of solutions. If the lines never intersect, the system of equations has no solution.

29. Never; sample answer: Lines cannot intersect at exactly two distinct points. Lines intersect once (one solution), coincide (infinite solutions), or never intersect (no solution).

Lesson 2-6

15.

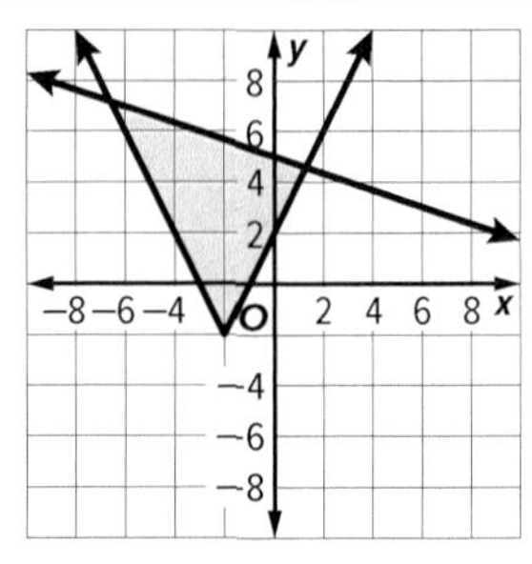

16.

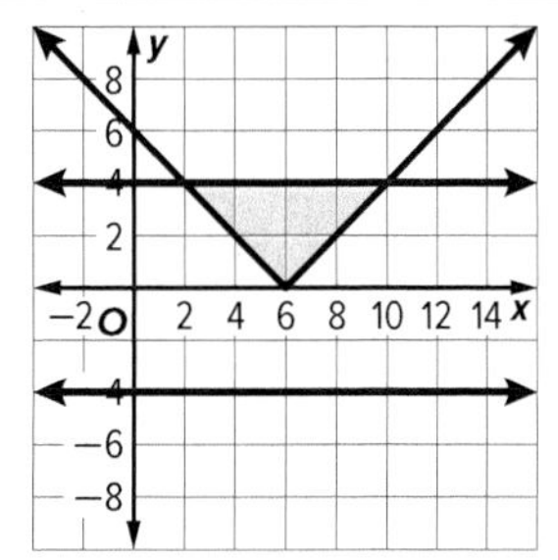

17.

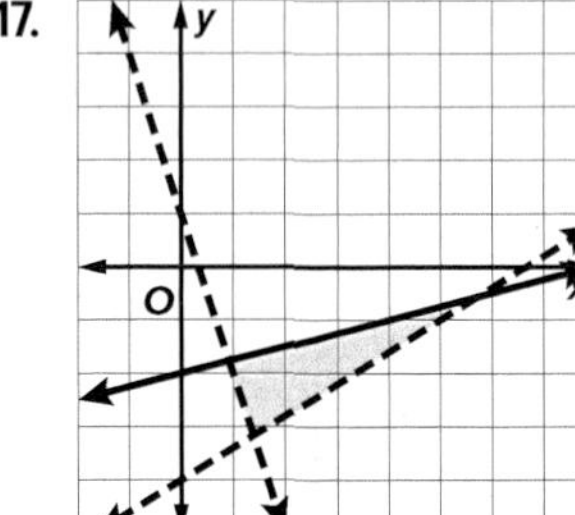

18.

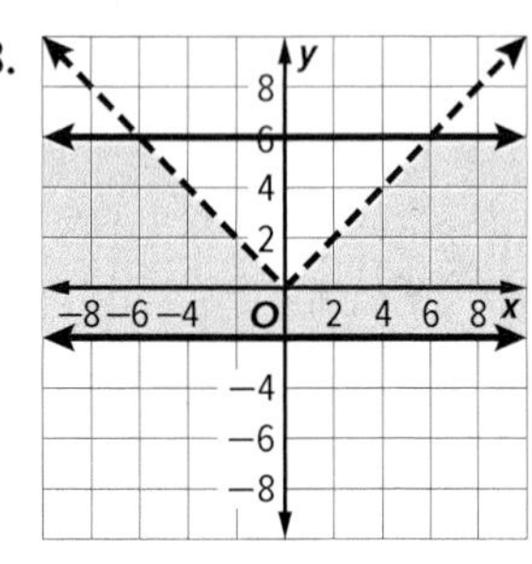

19a.

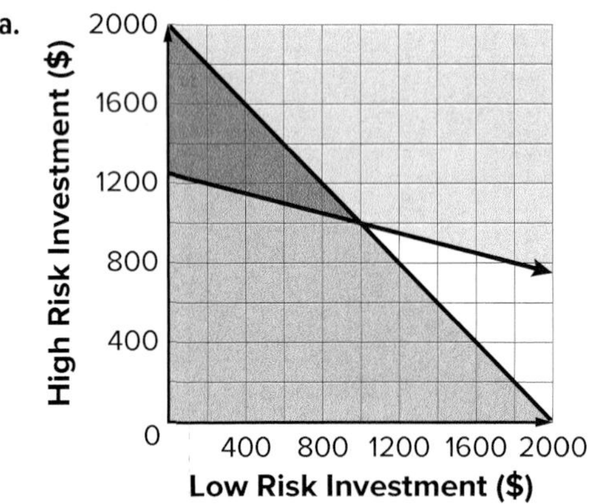

Lesson 2-7

10a. a = the number of gallons of color A made; b = the number of gallons of color B made; $4a + b \leq 32$; $a + 6b \leq 54$

10b. The vertices of the feasible region are (0, 0), (0, 9), (6, 8), and (8, 0).

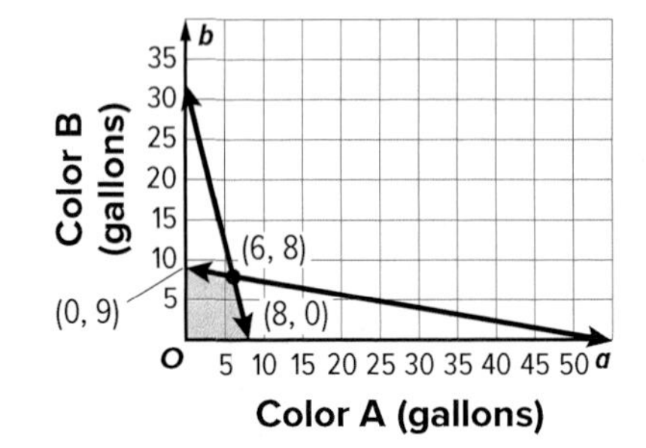

10c. 14 gallons total; 6 of color A, 8 of color B

11a. Let x represent clay beads and y represent glass beads; $0 \leq x \leq 10$; $y \geq 4$; $4y \leq 2x + 8$
$C = 0.20x + 0.40y$; The total cost equals 0.20 times the number of clay beads plus 0.40 times the number of glass beads.

11b. (4, 4), (10, 4), and (10, 7)

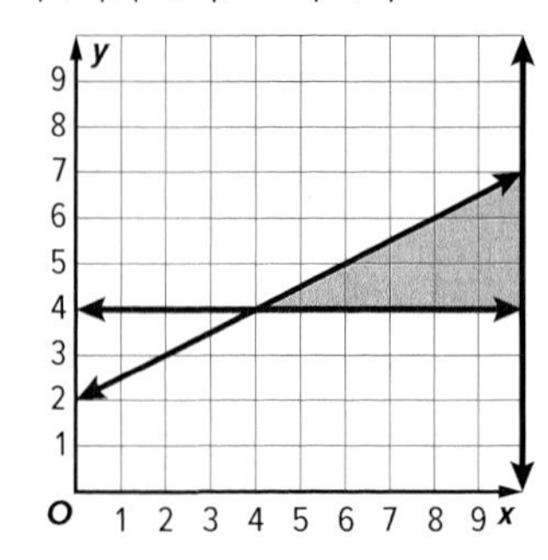

11c. Substitute into $C = 0.20x + 0.40y$: (4, 4) yields \$2.40, (10, 4) yields \$3.60, and (10, 7) yields \$4.80.
The minimum cost would be \$2.40 at (4, 4), which represents 4 clay beads and 4 glass beads.

12a. $a \geq 0, p \geq 0, 2p + a \leq 50, a \leq 40$, and $p \leq 24$

12b. (0, 0), (0, 24), (40, 0), (2, 24), (40, 5) where the horizontal axis represents a

12c. 2 plates and 24 pots

14a. Let x represent the number of units for Custom Foods, and y represent the number of units for Zookeeper's Best. The constraints are represented by the inequalities, $x \geq 0$, $y \geq 0$ because the number of units purchased must be greater than or equal to zero, $15x + 20y \geq 60$ because the food must have at least 60 grams of protein, and $10x + 5y \geq 30$ because the food must have at least 30 grams of fat. The cost of the food is given by $C = 0.80x + 0.50y$.

14b. $C = 0.80x + 0.50y$, $0.80(0) + 0.5(6) = \$3$, $0.80(2.4) + 0.5(1.2) = \$2.52$, $0.80(4) + 0.5(0) = \$3.20$; To minimize costs, the zoo should use 2.4 parts Custom Foods and 1.2 parts Zookeeper's Best.
The unbound region represents the infinite amount of combinations of the two food types.

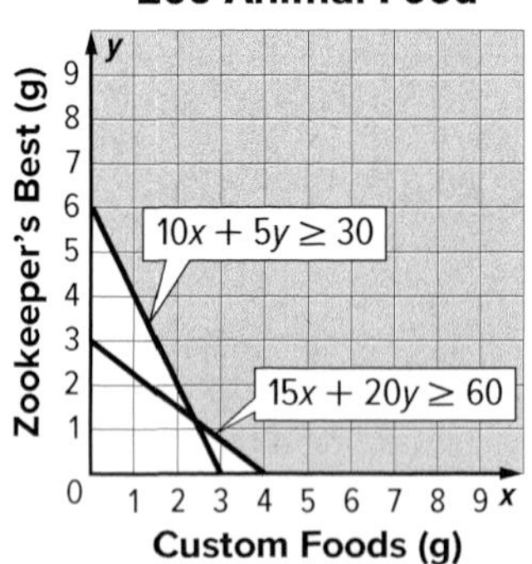

15. Always; sample answer: If a point on the unbounded region forms a minimum, then a maximum cannot also be formed because of the unbounded region. There will always be a value in the solution that will produce a higher value than any projected maximum.

17. Sample answer: Even though the region is bounded, multiple maximums occur at A and B and all of the points on the boundary of the feasible region containing both A and B. This happened because that boundary of the region has the same slope as the function.

Lesson 2-9

7.

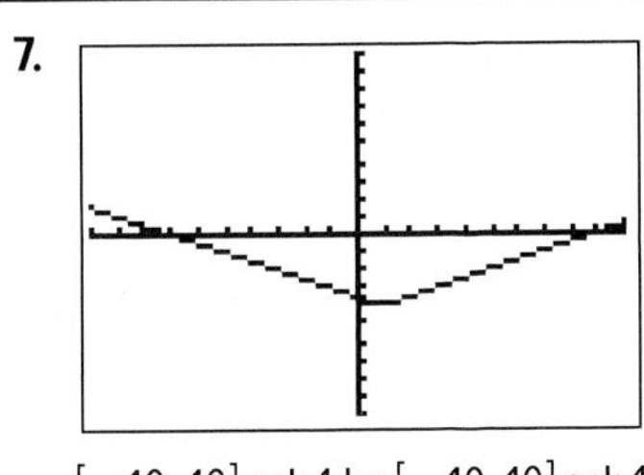

[−10, 10] scl: 1 by [−10, 10] scl: 1

8.

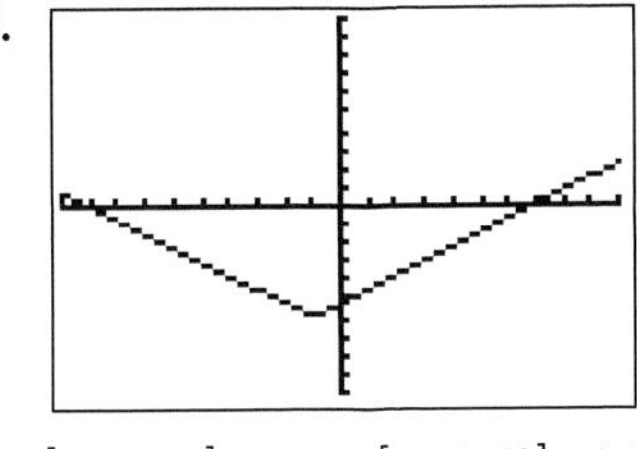

[−10, 10] scl: 1 by [−10, 10] scl: 1

9.

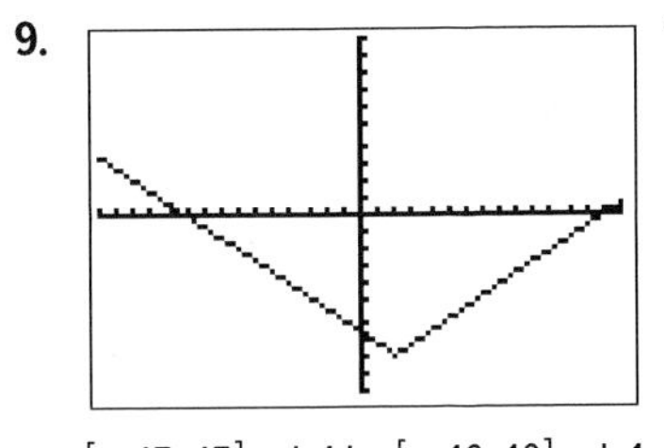

[−15, 15] scl: 1 by [−10, 10] scl: 1

10.

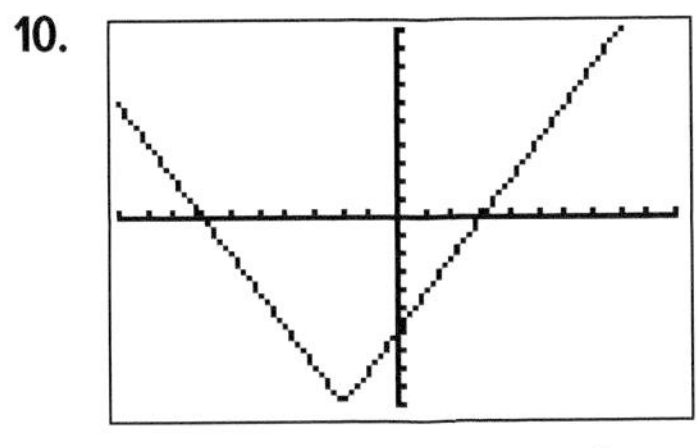

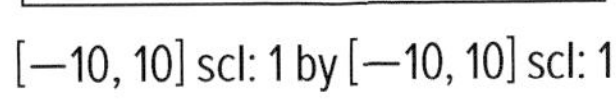

[−10, 10] scl: 1 by [−10, 10] scl: 1

11.

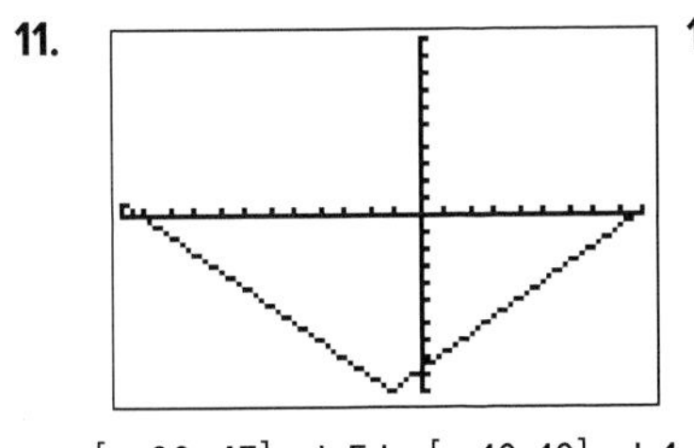

[−60, 45] scl: 5 by [−10, 10] scl: 1

12.

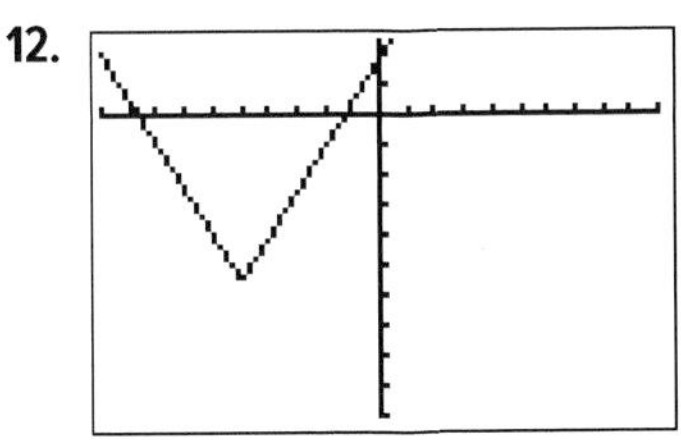

[−10, 10] scl: 1 by [−20, 4] scl: 2

19.

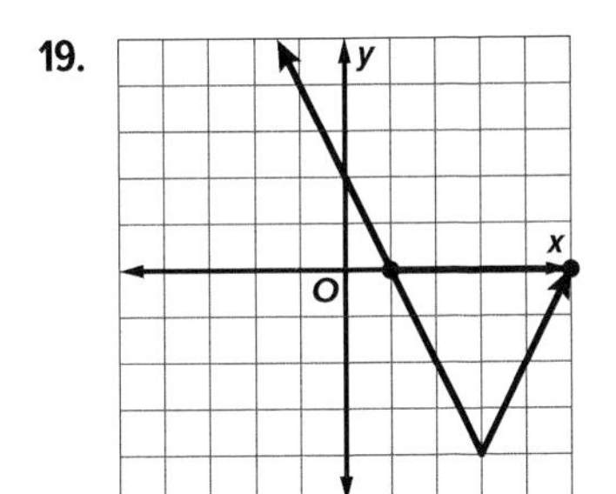

20.

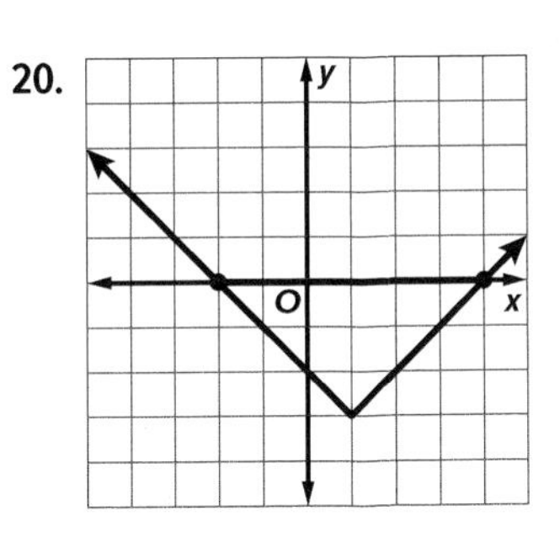

21.

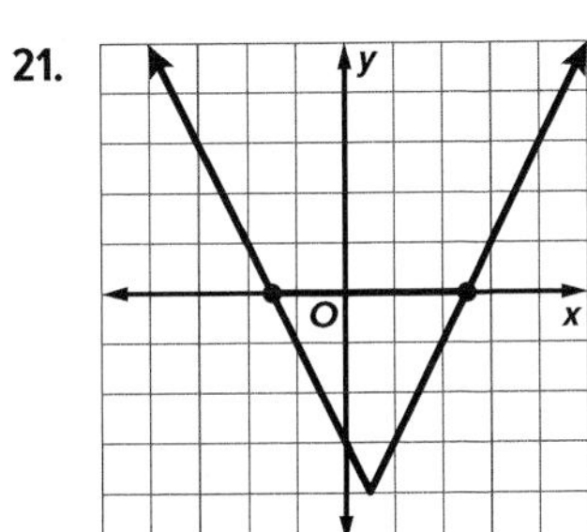

22.

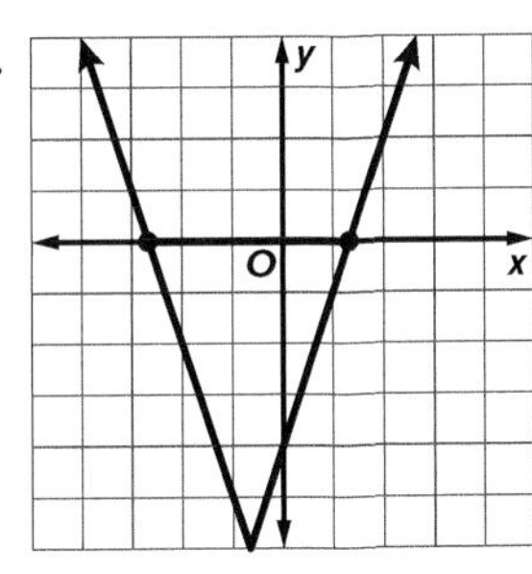

23.

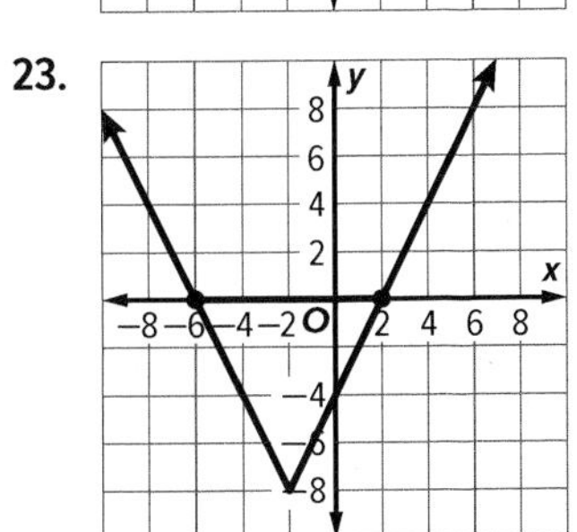

24.

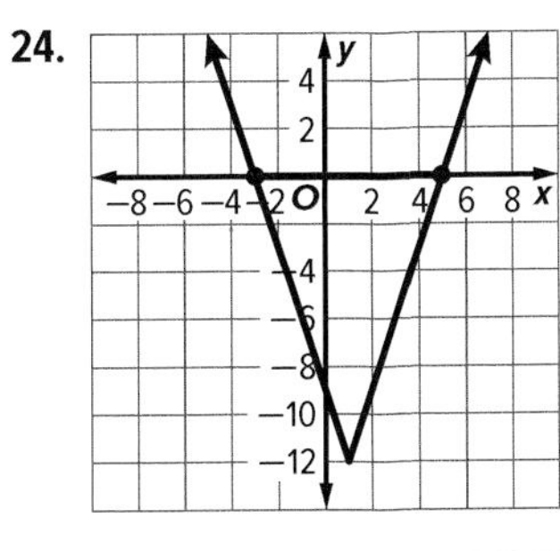

25.

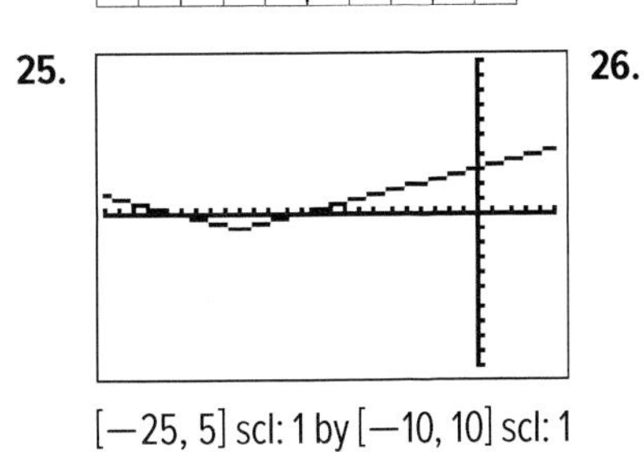

[−25, 5] scl: 1 by [−10, 10] scl: 1

26.

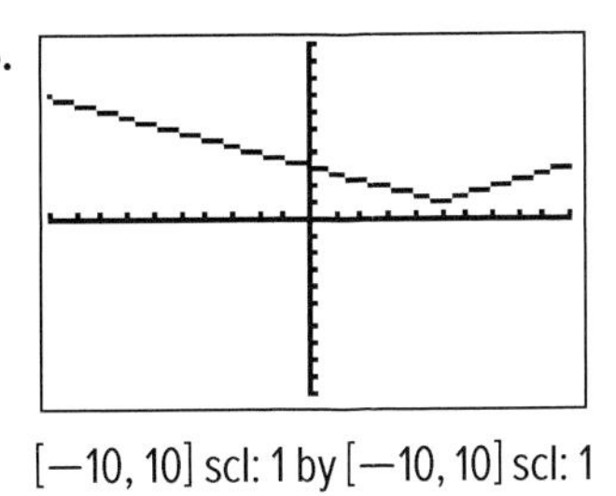

[−10, 10] scl: 1 by [−10, 10] scl: 1

27.

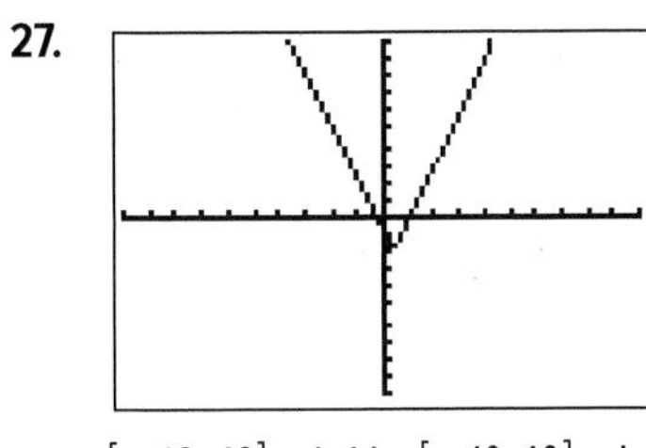

[−10, 10] scl: 1 by [−10, 10] scl: 1

28.

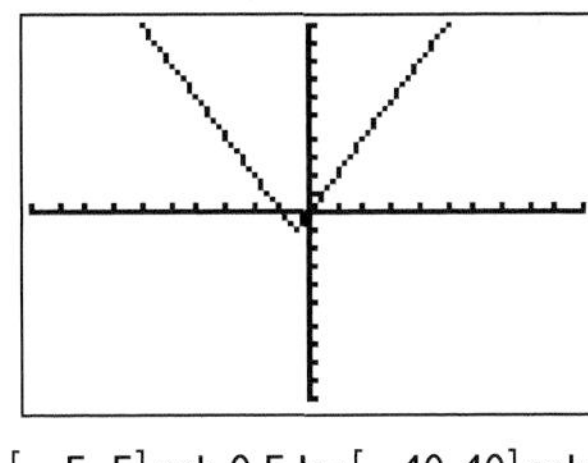

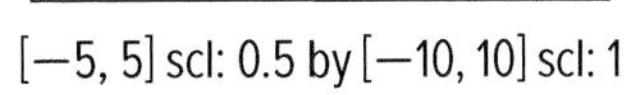

[−5, 5] scl: 0.5 by [−10, 10] scl: 1

29.

[−10, 10] scl: 1 by [−10, 10] scl: 1

30.

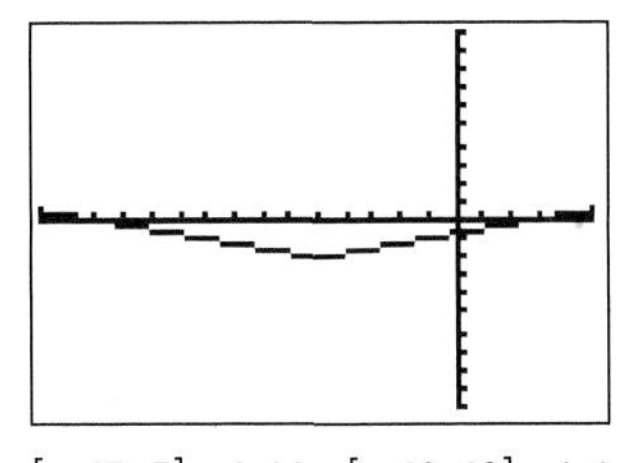

[−15, 5] scl: 1 by [−10, 10] scl: 1

31.

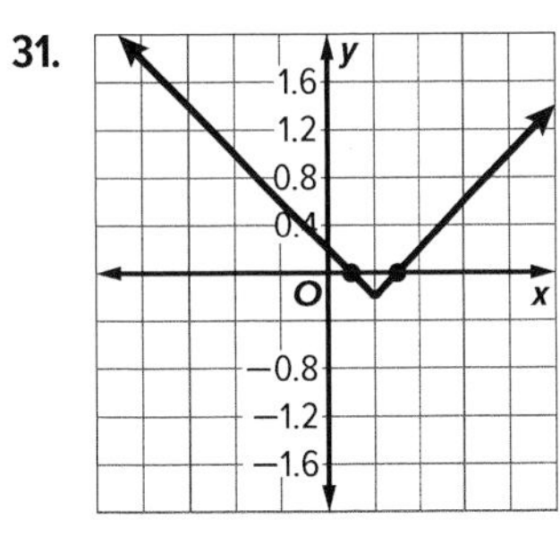

32.

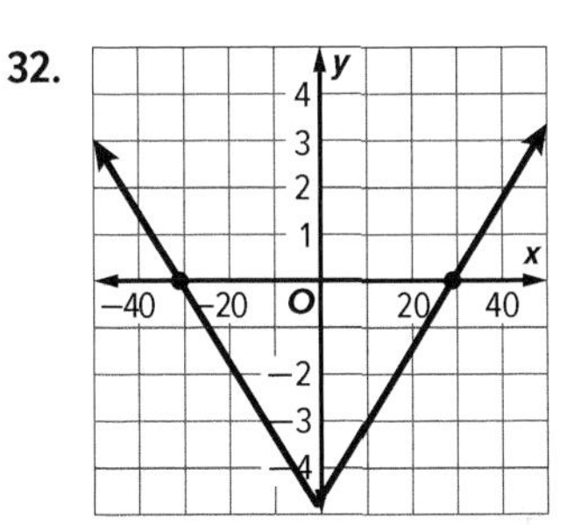

33.

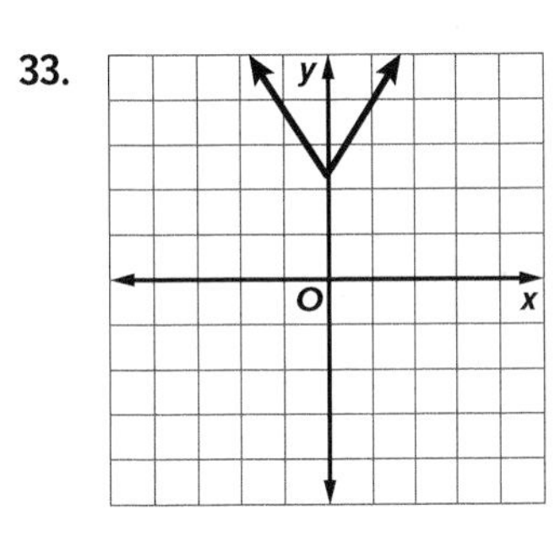

34.

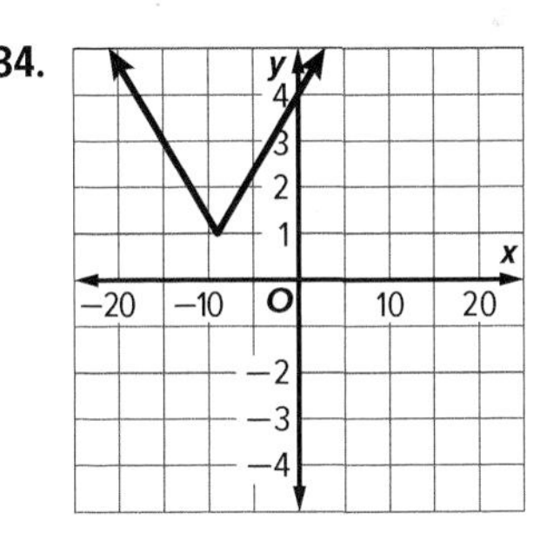

35.

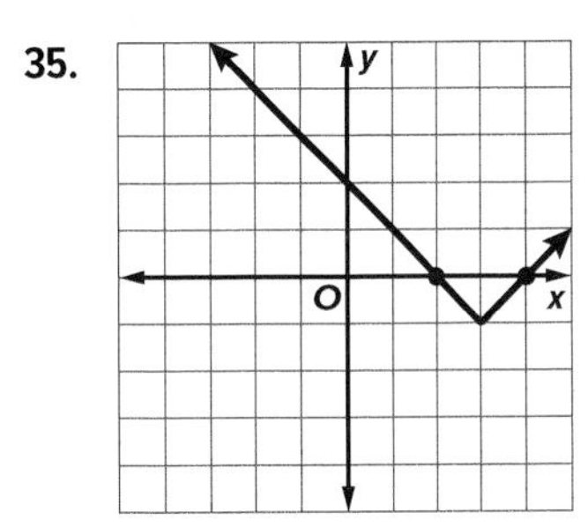

36.

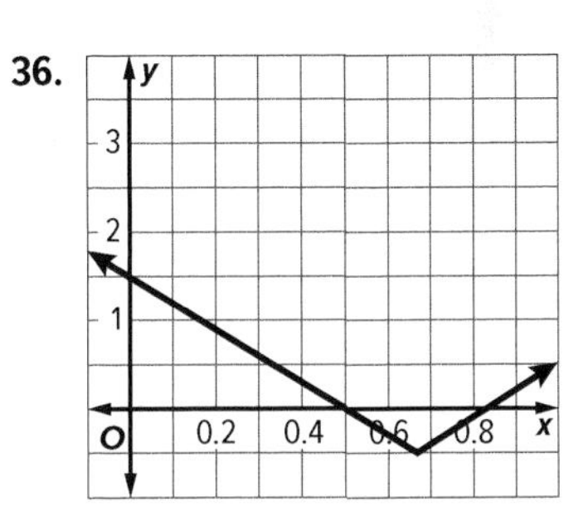

41.

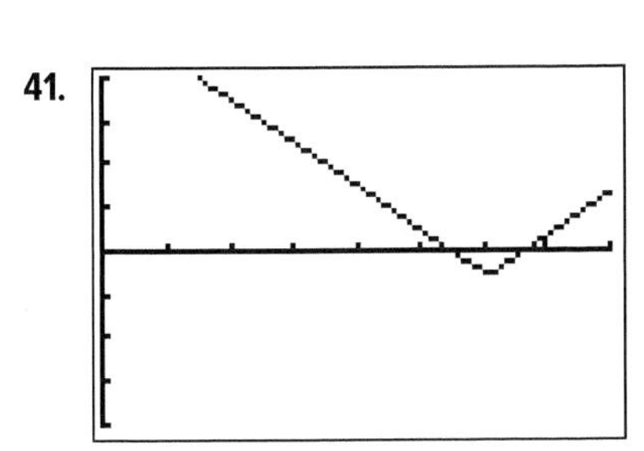

[0, 2] scl: 0.25 by [−1, 1] scl: 0.25

43. Sample answer: The process by which the equation or inequality is set to zero to represent $f(x)$ and making a table of values is the same. However, the graph of an absolute value equation is restricted to having either 0, 1, or 2 solutions; whereas the absolute value inequalities can have infinitely many solutions.

Module 3

Quadratic Functions

Module Goals

- Students graph quadratic functions and inequalities.
- Students solve quadratic equations and inequalities using a variety of methods.
- Students use quadratic functions to model real-world situations and solve problems.

Focus

Domain: Algebra, Functions, Number and Quantity

Standards for Mathematical Content:

A.CED.1 Create equations and inequalities in one variable and use them to solve problems.

F.IF.8a Use the process of factoring and completing the square in a quadratic function to show zeros, extreme values, and symmetry of the graph, and interpret these in terms of a context.

Also addresses N.CN.1, N.CN.2, N.CN.7, N.CN.8, A.SSE.1b, A.CED.2, A.CED.3, A.REI.11, F.IF.4, and F.IF.6.

Standards for Mathematical Practice:

All Standards for Mathematical Practice will be addressed in this module.

Coherence

Vertical Alignment

Previous
Students studied quadratic equations and functions, limited to real number solutions. **A.REI.4, F.IF.7a (Algebra 1)**

Now
Students solve quadratic equations and inequalities algebraically and by graphing, including those with complex solutions.
A.CED.3, F.IF.8a, N.CN.7

Next
Students will graph and identify key features of polynomial functions.
F.IF.4, F.IF.7c

Rigor

The Three Pillars of Rigor

To help students meet standards, they need to illustrate their ability to use the three pillars of rigor. Students gain conceptual understanding as they move from the Explore to Learn sections within a lesson. Once they understand the concept, they practice procedural skills and fluency and apply their mathematical knowledge as they go through the Examples and Practice.

1 CONCEPTUAL UNDERSTANDING | 2 FLUENCY | 3 APPLICATION

EXPLORE → LEARN → EXAMPLE & PRACTICE

Suggested Pacing

Lessons	Standards	45-min classes	90-min classes
Module Pretest and Launch the Module Video		1	0.5
3-1 Graphing Quadratic Functions	F.IF.4, F.IF.6	2	1
3-2 Solving Quadratic Equations by Graphing	A.CED.2, F.IF.4	1	0.5
3-3 Complex Numbers	N.CN.1, N.CN.2	2	1
3-4 Solving Quadratic Equations by Factoring	N.CN.7, N.CN.8, F.IF.8a	2	1
3-5 Solving Quadratic Equations by Completing the Square	N.CN.7, F.IF.8a	3	1.5
3-6 Using the Quadratic Formula and the Discriminant	N.CN.7, N.CN.8, A.SSE.1b	1	0.5
Put It All Together: Lessons 3-1 through 3-6		1	0.5
3-7 Quadratic Inequalities	A.CED.1, A.CED.3	1	0.5
3-8 Solving Linear-Nonlinear Systems	A.REI.11	2	1
Module Review		1	0.5
Module Assessment		1	0.5
	Total Days	18	9

Formative Assessment Math Probe
Factoring

Analyze the Probe

Review the probe prior to assigning it to your students.

In this probe, students will determine the correct ways to factor polynomial expressions.

Targeted Concepts Understand how polynomial expressions, such as the difference of squares, are factored.

Targeted Misconceptions

- Students may confuse the variable's exponent with a common factor.
- Students do not understand that the expanded form of a squared binomial is a trinomial.
- Students only take the square root of the variable and leave numbers as squares.
- Students do not understand that one cannot distribute an exponent over addition and subtraction.
- Students may not recognize the difference of squares.
- Students do not factor completely.

Use the Probe after Lesson 3-4.

Module Resource

NAME ____________ DATE ________ PERIOD ________

Cheryl Tobey Math Probe
Factoring

Circle your choice.	Explain your choice.
1. Which is the completely factored form of $x^2 - 16$? **A.** $2(x - 8)$ **B.** $(x - 4)^2$ **C.** $(x - 4)(x - 4)$ **D.** $(x + 4)(x - 4)$ **E.** Cannot be factored	
2. Which is the completely factored form of $4x^2 - 9$? **A.** $(2x - 3)^2$ **B.** $(2x + 3)(2x - 3)$ **C.** $(2x - 3)(2x - 3)$ **D.** $(4x - 9)^2$ **E.** Cannot be factored	
3. Which is the completely factored form of $x^4 - 81$? **F.** $(x^2 - 9)^2$ **G.** $(x^2 + 9)(x^2 - 9)$ **H.** $(x^2 - 9)(x^2 - 9)$ **I.** $(x + 9)(x + 3)(x - 3)$ **J.** Cannot be factored	

Cheryl Tobey Math Probe • Factoring

Correct Answers:
1. D 2. B 3. D

Collect and Assess Student Answers

If the student selects these responses...	**Then** the student likely...
1. A	is using the square as a common factor with 16.
1. B **2.** A **3.** A	is incorrectly viewing factoring as taking the square root of each term and then squaring the expression. This error is related to a common error in the binomial expansion $(x - 4)^2 \neq x^2 - 16$.
1. C **2.** C **3.** C	is not considering how the middle term is eliminated by multiplying conjugates.
2. D	is only taking the square root of the variable and leaving the numbers as perfect squares. They are also not eliminating the linear term by considering a squared binomial.
3. B	does not recognize that factoring the original expression produces a difference of squares binomial $(x^2 - 9)$ that can also be factored.

Take Action

After the Probe Design a plan to address any possible misconceptions. You may wish to assign the following resources.

- ALEKS® Quadratic Equations
- Lesson 3-4, all Learns, all Examples

Revisit the probe at the end of the module to be sure that your students no longer carry these misconceptions.

The Ignite! activities, created by Dr. Raj Shah, cultivate curiosity and engage and challenge students. Use these open-ended, collaborative activities, located online in the module Launch section, to encourage your students to develop a growth mindset towards mathematics and problem solving. Use the teacher notes for implementation suggestions and support for encouraging productive struggle.

Essential Question

At the end of this module, students should be able to answer the Essential Question.

What are important characteristics of a quadratic function, and what real-world situations can be modeled by quadratic functions and equations? Sample answer: Quadratic functions have an independent variable that is squared as the highest degree. These quadratic functions are a parabola shape when graphed, opening either up or down. Many physical situations follow a quadratic pattern, including the motion of a thrown object falling due to gravity.

What Will You Learn?

Prior to beginning this module, have your students rate their knowledge of each item listed. Then, at the end of the module, you will be reminded to have your students return to these pages to rate their knowledge again. They should see that their knowledge and skills have increased.

DINAH ZIKE FOLDABLES

Focus Students write about the different ways quadratic functions and relations can be solved as these methods are presented in the lessons of this module.

Teach Have students make and label their Foldables as illustrated. Students should fill in the appropriate sections with their notes, diagrams, and examples as they cover each lesson in this module. Have students use the appropriate tabs as they cover each lesson in this module.

When to Use It Encourage students to add to their Foldables as they work through the module and to use them to review for the module test.

Launch the Module

For this module, the Launch the Module video uses a trip to Mars to explore quadratic functions and quadratic models. Students learn how to connect quadratic graphs to real-world situations.

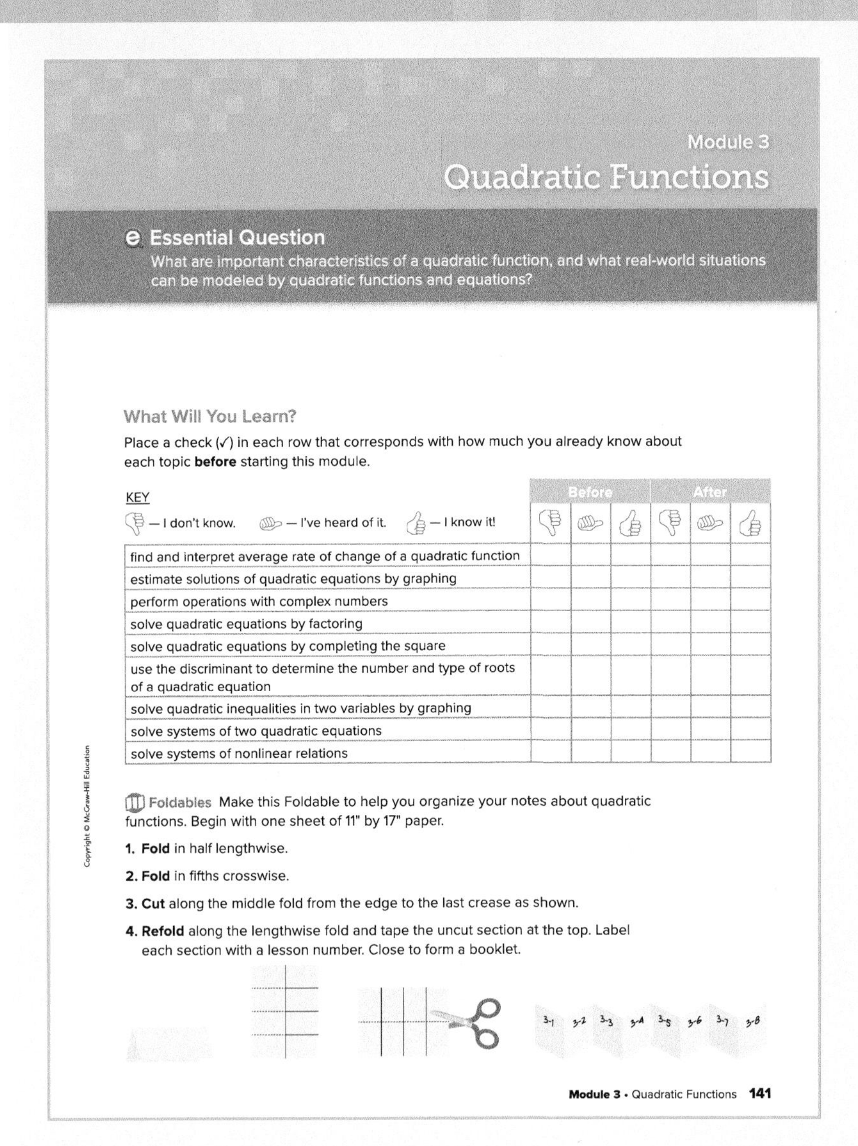

Module 3

Quadratic Functions

Essential Question
What are important characteristics of a quadratic function, and what real-world situations can be modeled by quadratic functions and equations?

What Will You Learn?

Place a check (✓) in each row that corresponds with how much you already know about each topic **before** starting this module.

KEY
— I don't know. — I've heard of it. — I know it!

	Before			After		
find and interpret average rate of change of a quadratic function						
estimate solutions of quadratic equations by graphing						
perform operations with complex numbers						
solve quadratic equations by factoring						
solve quadratic equations by completing the square						
use the discriminant to determine the number and type of roots of a quadratic equation						
solve quadratic inequalities in two variables by graphing						
solve systems of two quadratic equations						
solve systems of nonlinear relations						

Foldables Make this Foldable to help you organize your notes about quadratic functions. Begin with one sheet of 11" by 17" paper.

1. **Fold** in half lengthwise.
2. **Fold** in fifths crosswise.
3. **Cut** along the middle fold from the edge to the last crease as shown.
4. **Refold** along the lengthwise fold and tape the uncut section at the top. Label each section with a lesson number. Close to form a booklet.

Module 3 • Quadratic Functions **141**

Interactive Student Presentation

What Vocabulary Will You Learn?

Check the box next to each vocabulary term that you may already know.

- ☐ average rate of change
- ☐ axis of symmetry
- ☐ completing the square
- ☐ complex conjugates
- ☐ complex number
- ☐ difference of squares
- ☐ discriminant
- ☐ factored form
- ☐ imaginary unit *i*
- ☐ maximum
- ☐ minimum
- ☐ perfect square trinomials
- ☐ projectile motion problems
- ☐ pure imaginary number
- ☐ quadratic equation
- ☐ quadratic function
- ☐ quadratic inequality
- ☐ quadratic relations
- ☐ rate of change
- ☐ rationalizing the denominator
- ☐ standard form of a quadratic equation
- ☐ vertex
- ☐ vertex form

Are You Ready?

Complete the Quick Review to see if you are ready to start this module.
Then complete the Quick Check.

Quick Review

Example 1

Given $f(x) = -2x^2 + 3x - 1$ and $g(x) = 3x^2 - 5$, find each value.

a. $f(2)$

$f(x) = -2x^2 + 3x - 1$ Original function

$f(2) = -2(2)^2 + 3(2) - 1$ Substitute 2 for x.

$= -8 + 6 - 1$ or -3 Simplify.

b. $g(-2)$

$g(x) = 3x^2 - 5$ Original function

$g(-2) = 3(-2)^2 - 5$ Substitute −2 for x.

$= 12 - 5$ or 7 Simplify.

Example 2

Factor $2x^2 - x - 3$ completely. If the polynomial is not factorable, write *prime*.

To find the coefficients of the x-terms, you must find two numbers whose product is $2(-3)$ or -6, and whose sum is -1. The two coefficients must be 2 and -3 since $2(-3) = -6$ and $2 + (-3) = -1$. Rewrite the expression and factor by grouping.

$2x^2 - x - 3$

$= 2x^2 + 2x - 3x - 3$ Substitute $2x - 3x$ for $-x$.

$= (2x^2 + 2x) + (-3x - 3)$ Associative Property

$= 2x(x + 1) + -3(x + 1)$ Factor out the GCF.

$= (2x - 3)(x + 1)$ Distributive Property

Quick Check

Given $f(x) = 2x^2 + 4$ and $g(x) = -x^2 - 2x + 3$, find each value.

1. $f(3)$ 22
2. $f(0)$ 4
3. $g(4)$ -21
4. $g(-3)$ 0

Factor completely. If the polynomial is not factorable, write *prime*.

5. $x^2 - 10x + 21$ $(x - 3)(x - 7)$
6. $2x^2 + 7x - 4$ $(2x - 1)(x + 4)$
7. $2x^2 - 7x - 15$ $(2x + 3)(x - 5)$
8. $x^2 - 11x + 15$ prime

How Did You Do?

Which exercises did you answer correctly in the Quick Check? Shade those exercise numbers below.

① ② ③ ④ ⑤ ⑥ ⑦ ⑧

What Vocabulary Will You Learn?

ELL As you proceed through the module, introduce the key vocabulary by using the following routine.

Define A quadratic inequality is an inequality that includes a quadratic expression.

Example Consider $y < x^2 - 5x + 4$.

Ask Based on your knowledge of graphing inequalities and quadratic functions, what will the boundary of the graph of the quadratic inequality look like? a dashed parabola

Are You Ready?

Students may need to review the following prerequisite skills to succeed in this module.

- finding function values
- finding x-intercepts
- simplifying expressions
- factoring
- factoring perfect squares
- evaluating methods for solving quadratic equations
- determining whether an ordered pair is a solution
- solving linear systems of equations
- solving linear-quadratic systems of equations

ALEKS®

ALEKS is an adaptive, personalized learning environment that identifies precisely what each student knows and is ready to learn, ensuring student success at all levels.

You may want to use the **Quadratic and Polynomial Functions** section to ensure student success in this module.

Mindset Matters

Mistakes = Learning

When a student makes a mistake and goes on to learn from it, he or she can actually build new connections in their brain while figuring out a new path or process.

How Can I Apply It?

ALEKS is a great tool to not only individualize learning for each student, but to also help students understand that making mistakes and trying new problems will help them to learn and grow over the long run. Have students keep track of their ALEKS Pie Chart to see their progress.

Graphing Quadratic Functions

LESSON GOAL

Students graph quadratic functions.

1 LAUNCH

 Launch the lesson with a **Warm Up** and an introduction.

2 EXPLORE AND DEVELOP

 **Explore:** Transforming Quadratic Functions

 **Develop:**

Graphing Quadratic Functions
- Graph a Quadratic Function by Using a Table
- Compare Quadratic Functions
- Use Quadratic Functions

Finding and Interpreting Average Rate of Change
- Find Average Rate of Change from an Equation
- Find Average Rate of Change from a Table
- Find Average Rate of Change from Graph

 You may want your students to complete the **Checks** online.

3 REFLECT AND PRACTICE

 Exit Ticket

 Practice

DIFFERENTIATE

 View reports of student progress on the **Checks** after each example.

Resources	AL	OL	BL	ELL
Remediation: Graphing Quadratic Functions	●	●		●
Extension: Parametric Equations		●	●	●

Language Development Handbook

Assign page 17 of the Language *Development Handbook* to help your students build mathematical language related to graphing quadratic functions.

ELL You can use the tips and suggestions on page T17 of the handbook to support students who are building English proficiency.

Suggested Pacing

90 min	1 day
45 min	2 days

Focus

Domain: Functions

Standards for Mathematical Content:

F.IF.4 For a function that models a relationship between two quantities, interpret key features of graphs and tables in terms of the quantities, and sketch graphs showing key features given a verbal description of the relationship.

F.IF.6 Calculate and interpret the average rate of change of a function (presented symbolically or as a table) over a specified interval. Estimate the rate of change from a graph.

Standards for Mathematical Practice:

1 Make sense of problems and persevere in solving them.

2 Reason abstractly and quantitatively.

Coherence

Vertical Alignment

Previous
Students graphed linear and quadratic functions.
F.IF.7a (Algebra 1)

Now
Students graph quadratic functions and identify key features of the graphs.
F.IF.4, F.IF.6

Next
Students will solve quadratic functions by finding zeros using graphs.
A.CED.2, F.IF.4

Rigor

The Three Pillars of Rigor

1 CONCEPTUAL UNDERSTANDING	2 FLUENCY	3 APPLICATION

Conceptual Bridge In this lesson, students develop an understanding of quadratic functions and use it to build fluency by graphing quadratic functions. They apply their understanding of quadratic functions by solving real-world problems.

Mathematical Background

Graphs of quadratic functions of the form $f(x) = ax^2 + bx + c$, where $a \neq 0$, are called parabolas. All parabolas have an axis of symmetry, a vertex, and a y-intercept. Some parabolas open up, and others open down.

Interactive Presentation

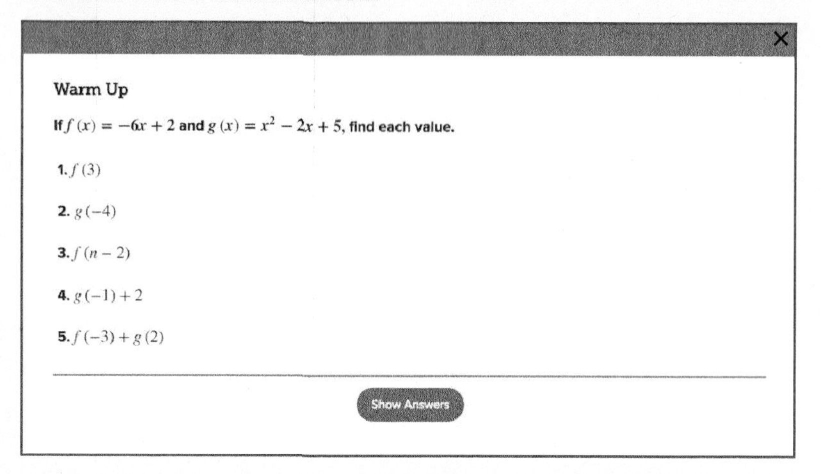

Warm Up

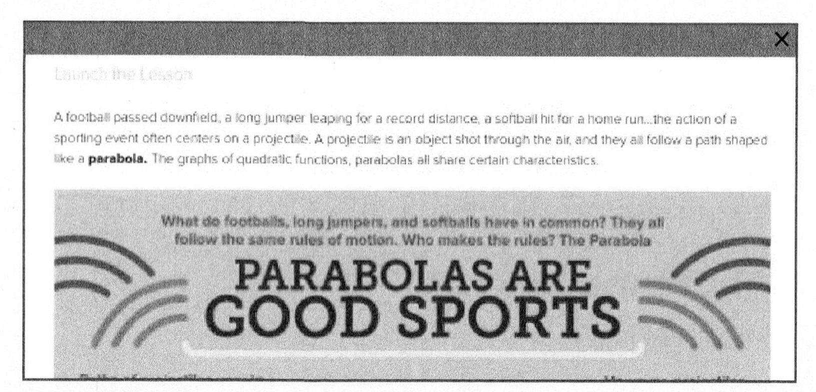

Launch the Lesson

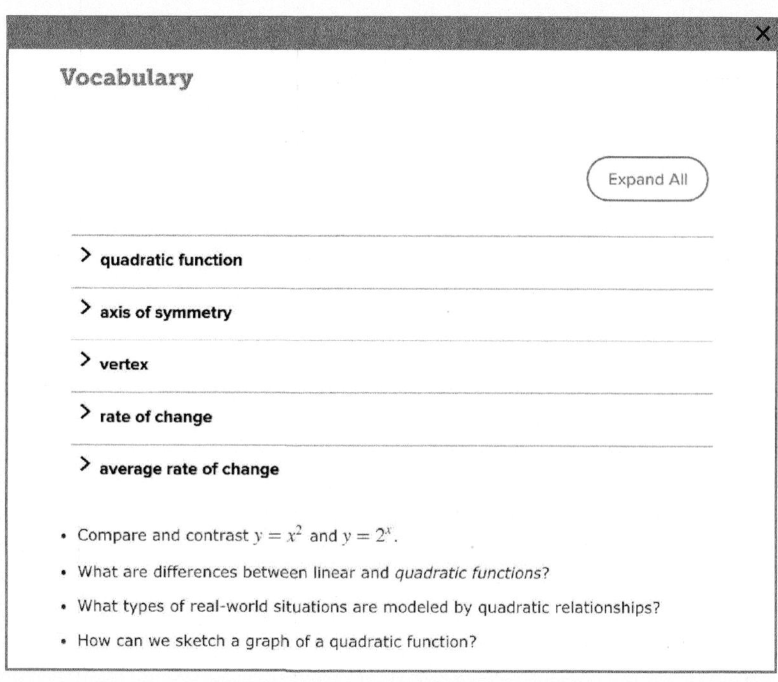

Today's Vocabulary

Warm Up

Prerequisite Skills

The Warm Up exercises address the following prerequisite skill for this lesson:

- finding function values

Answers:

1. -16
2. 29
3. $-6n + 14$
4. 10
5. 25

Launch the Lesson

Teaching the Mathematical Practices

4 Analyze Relationships Mathematically In the Launch the Lesson, encourage students to explain the relationships between the data about the athletes and the graph of a parabola. Guide students to see how the length of the long jump, height of a basket toss, and hang time of a slam dunk relate to the key features of a parabola described in the infographic.

Go Online to find additional teaching notes and questions to promote classroom discourse.

Today's Standards

Tell students that they will be addressing these content and practice standards in this lesson. You may wish to have a student volunteer read aloud *How can I meet these standards?* and *How can I use these practices?*, and connect these to the standards.

See the Interactive Presentation for I Can statements that align with the standards covered in this lesson.

Today's Vocabulary

Tell students that they will be using these vocabulary terms in this lesson. You can expand each row if you wish to share the definitions. Then discuss the questions below with the class.

Explore Transforming Quadratic Functions

Objective

Students use a sketch to explore graphing quadratic functions.

Teaching the Mathematical Practices

5 Use Mathematical Tools Point out that to complete the Explore, students will need to use the sketch to draw conclusions. Work with students to explore and deepen their understanding of how the parameters of a quadratic function change the graph of the function.

Ideas for Use

Recommended Use Present the Inquiry Question, or have a student volunteer read it aloud. Have students work in pairs to complete the Explore activity on their devices. Pairs should discuss each of the questions. Monitor student progress during the activity. Upon completion of the Explore activity, have student volunteers share their responses to the Inquiry Question.

What if my students don't have devices? You may choose to project the activity on a whiteboard. A printable worksheet for each Explore is available online. You may choose to print the worksheet so that individuals or pairs of students can use it to record their observations.

Summary of the Activity

Students will complete guiding exercises throughout the Explore activity. Students interact with a sketch of a quadratic function by adjusting the constants and coefficients *a*, *b*, and *c*. They make observations about how changes to these values affect the graph of the quadratic function. Then, students will answer the Inquiry Question.

(continued on the next page)

Interactive Presentation

Transforming Quadratic Functions

INQUIRY How can you use the values of *a*, *b*, and *c* in the equation of a quadratic function $y = ax^2 + bx + c$ to visualize its graph?

Explore

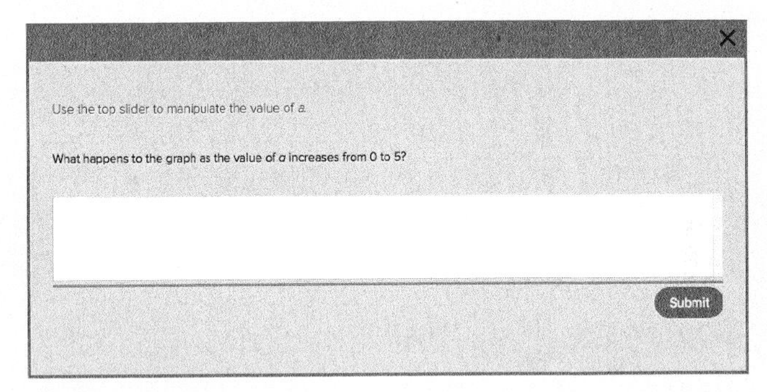

Use the top slider to manipulate the value of a.
What happens to the graph as the value of a increases from 0 to 5?
Submit

Explore

WEB SKETCHPAD

Students will use a sketch to explore the features of a quadratic graph.

Interactive Presentation

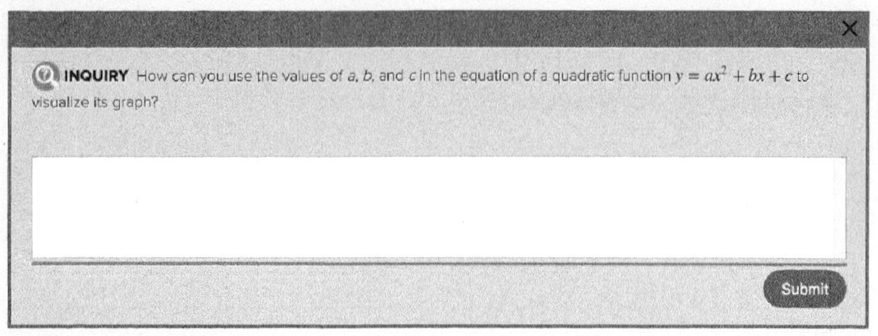

Explore

TYPE

Students respond to the Inquiry Question and can view a sample answer.

Explore Transforming Quadratic Functions (*continued*)

Questions

Have students complete the Explore activity.

Ask:

- What happens to the standard form of the quadratic function when? $x = 0$? How does this help you to describe c? Sample answer: When $x = 0$, the equation becomes $y = 0$. This reminds me that c is the y-intercept for the function.
- How does knowing the value of a help you graph a quadratic function? Sample answer: If a is positive, I know that the graph will open up and the vertex will be the minimum value. If a is negative, I know that the graph will open down and the vertex will be the maximum value.

Inquiry

How can you use the values of a, b, and c in the equation of a quadratic function $y = ax^2 + bx + c$ to visualize its graph? Sample answer: The value of a tells you how wide the graph will be and whether it opens up or down. The value of b tells you whether the graph is shifted horizontally. The value of c tells you the y-intercept.

Go Online to find additional teaching notes and sample answers for the guiding exercises.

Learn Graphing Quadratic Functions

Objective

Students graph quadratic functions by making a table of values.

Teaching the Mathematical Practices

1 Explain Correspondences Encourage students to explain the relationships between the parameters, key features, and graph of a quadratic function.

Common Misconception

Students may think all nonlinear functions are quadratic. Consider discussing different types of nonlinear functions that are not represented by parabolas to help to prepare for future work with polynomial, radical, and exponential functions.

Common Error

Make sure students realize that $f(x)$ and y can be used interchangeably. They should also realize that the maximum or minimum value of the quadratic function is given by the y-coordinate of the vertex of the parabola.

Go Online

- Find additional teaching notes.
- View performance reports of the Checks.
- Assign or present an Extra Example.

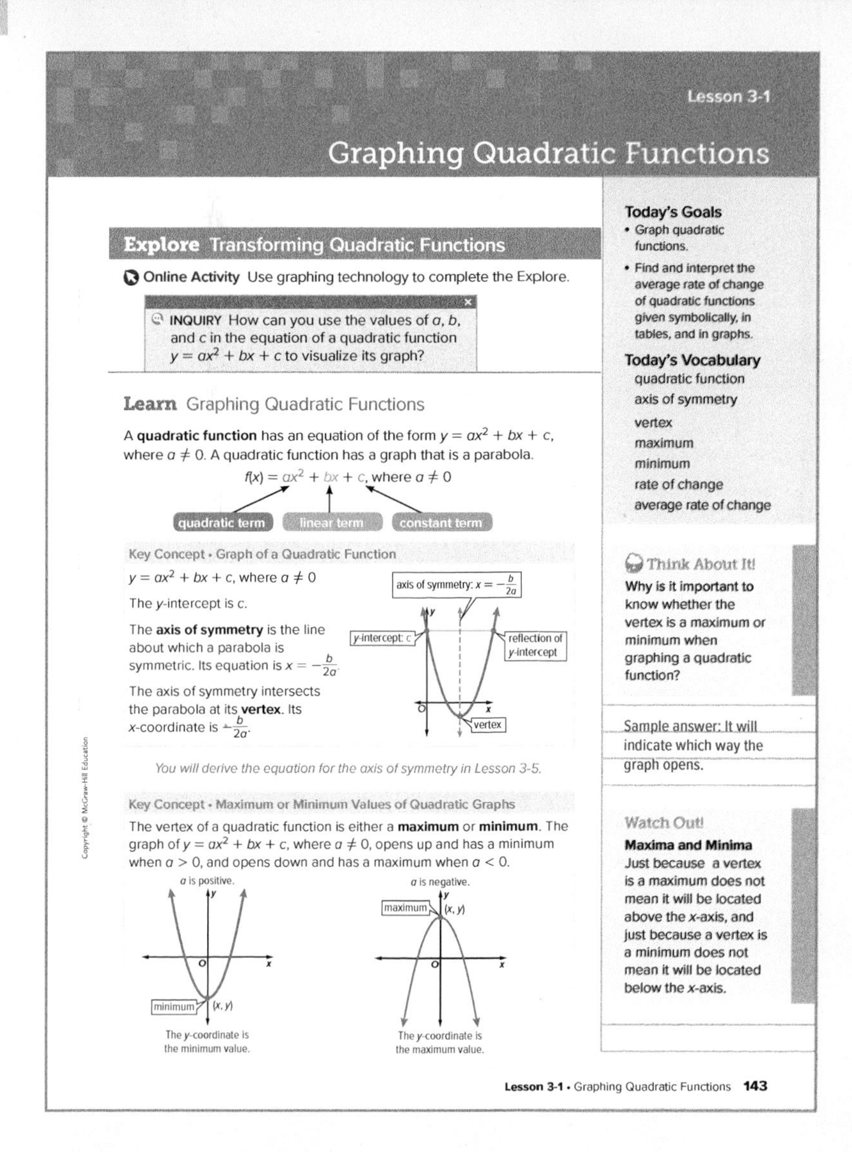

Interactive Presentation

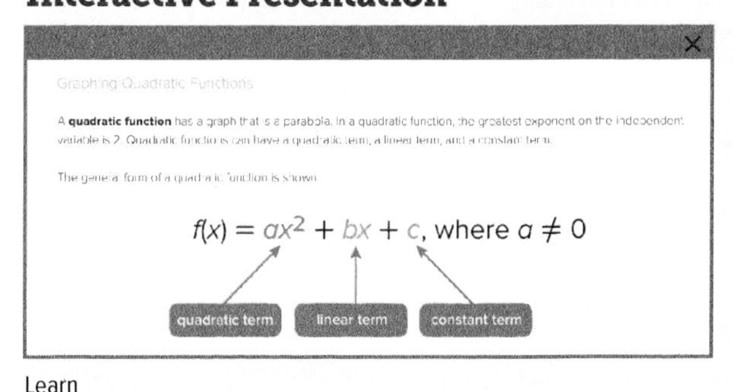

Learn

TYPE

Students answer a question about the importance of the vertex when graphing a quadratic function.

Your Notes

Go Online
You can watch videos to see how to graph quadratic functions by using a table or its key features.

Think About It!
If you know that $f(-4) = 5$, find $f(2)$ without substituting 2 for x in the function. Justify your argument.

5; Sample answer: Because $f(x)$ is symmetric about $x = -1$, $f(-4) = f(2)$.

Think About It!
Compare the end behavior of $f(x)$ and $g(x)$.

Sample answer: Both functions have the same end behavior, as $x \to -\infty$, $f(x) \to -\infty$ and as $x \to \infty$, $f(x) \to -\infty$.

Example 1 Graph a Quadratic Function by Using a Table

Graph $f(x) = x^2 + 2x - 3$. State the domain and range.

Step 1 Analyze the function.

For $f(x) = x^2 + 2x - 3$, $a =$ 1, $b =$ 2, and $c =$ −3.
c is the y-intercept, so the y-intercept is −3.
Find the axis of symmetry.

$x = -\frac{b}{2a}$ — Equation of the axis of symmetry

$= -\frac{2}{2(1)}$ — $a = 1, b = 2$

$= -1$ — Simplify.

The equation of the axis of symmetry is $x = -1$, so the x-coordinate of the vertex is −1. Because $a > 0$, the vertex is a minimum.

Step 2 Graph the function.

x	$x^2 + 2x - 3$	$(x, f(x))$
−3	$(-3)^2 + 2(-3) - 3$	(−3, 0)
−2	$(-2)^2 + 2(-2) - 3$	(−2, −3)
−1	$(-1)^2 + 2(-1) - 3$	(−1, −4)
0	$(0)^2 + 2(0) - 3$	(0, −3)
1	$(1)^2 + 2(1) - 3$	(−3, 0)

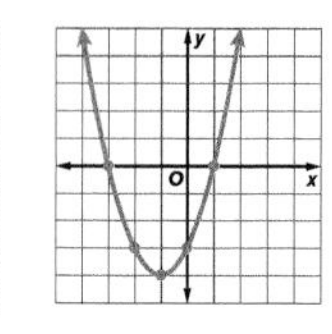

Step 3 Analyze the graph.

The parabola extends to positive and negative infinity, so the domain is all real numbers. The range is $\{y \mid y \geq -4\}$.

Example 2 Compare Quadratic Functions

Compare the graph of $f(x)$ to a quadratic function $g(x)$ with a y-intercept of −1 and a vertex at (1, 2). Which function has a greater maximum?

From the graph, $f(x)$ appears to have a maximum of 5. Graph $g(x)$ using the given information.

The vertex is at (1, 2), so the axis of symmetry is $x =$ 1.

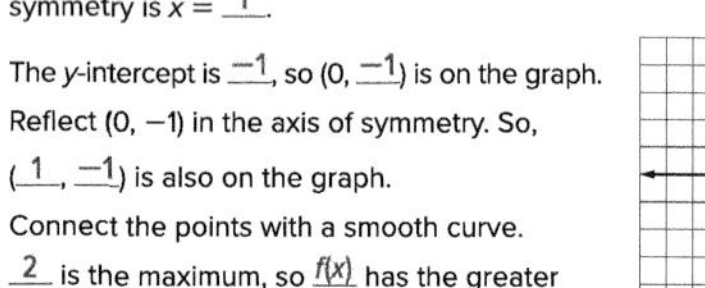

The y-intercept is −1, so (0, −1) is on the graph. Reflect (0, −1) in the axis of symmetry. So, (1, −1) is also on the graph.
Connect the points with a smooth curve.
2 is the maximum, so $f(x)$ has the greater maximum.

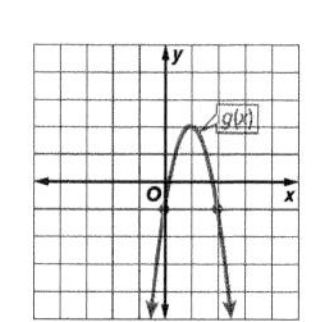

Example 1 Graph a Quadratic Function by Using a Table

Teaching the Mathematical Practices

7 Interpret Complicated Expressions Mathematically proficient students can see complicated expressions as single objects or as being composed of several objects. In Example 1, guide students to see what information they can gather about the graph just by looking at the quadratic function.

Questions for Mathematical Discourse

AL How is the axis of symmetry related to the vertex of a quadratic function? The x-coordinate for all points on the axis of symmetry and the x-coordinate for the vertex are the same.

OL How would the graph change if a were −1? The axis of symmetry shifts to 1, the vertex becomes a maximum, and the y-intercept remains the same.

BL Can a parabola ever exist in only one quadrant? Explain. No; sample answer: All parabolas extend to infinity and negative infinity, so it will always exist in at least two quadrants, unless additional constraints are present.

Example 2 Compare Quadratic Functions

Teaching the Mathematical Practices

1 Explain Correspondences Encourage students to explain the relationships between the key features of the verbal description and graph used in Example 2.

Questions for Mathematical Discourse

AL Is the vertex of $g(x)$ is a maximum or a minimum? Explain your reasoning. Maximum; sample answer: The y-coordinate of the vertex is greater than the y-intercept, which means the parabola opens down and has a maximum.

OL If the y-intercept of a quadratic function is the same as the y-coordinate of the vertex, what is the x-coordinate of the vertex? 0

BL Could you determine which function has the greater maximum without graphing $g(x)$? Explain. Yes; sample answer: Because the y-coordinate of the vertex gives the value of the maximum or minimum, once I confirmed that the function had a maximum, I could just compare the y-coordinate of the vertex of $g(x)$ to the maximum of $f(x)$.

Interactive Presentation

Graph a Quadratic Function by Using a Table

Graph $f(x) = x^2 + 2x - 3$. State the domain and range.

Step 1 — Analyze the function.

Step 2 — Graph the function.

Step 3 — Analyze the graph.

Example 1

EXPAND

Students step through the process of graphing a quadratic function using a table.

TYPE

Students answer a question about the graph of a quadratic function.

Common Error

Make sure students understand that a parabola that opens up is the graph of a function with a minimum value, and that a parabola that opens down is the graph of a function with a maximum value. Compare these parabolas to valleys (where the altitude of the valley floor is a minimum) and hills (where the peak of the hill is the maximum altitude).

Example 3 Use Quadratic Functions

Teaching the Mathematical Practices

2 Represent Situations Symbolically Guide students to define variables to solve the problem in Example 3. Help students to identify the independent and dependent variables. Then work with them to find other relationships in the problem and represent them with an equation.

Questions for Mathematical Discourse

AL What about the equation indicates there will be a maximum amount of money generated? The a value is negative.

OL What does the y-intercept represent in this situation? The money generated last year. This is their expected profit if the price stays the same.

BL Assuming this model works for all values of x, how would you determine how many price increases would lead to a total of 0 tickets sold? Replace $P(x)$ with 0 and solve for x or use the calculator to find the x-intercept.

Example 3 Use Quadratic Functions

SKIING **A ski resort has extended hours on one holiday weekend per year. Last year, the resort sold 680 ski passes at $120 per holiday weekend pass. This year, the resort is considering a price increase. They estimate that for each $5 increase, they will sell 20 fewer holiday weekend passes.**

Part A How much should they charge in order to maximize profit?

Step 1 Define the variables.

Let x represent the number of $5 price increases, and let $P(x)$ represent the total amount of money generated. $P(x)$ is equal to the price of each pass (120 + __5x__) times the total number of passes sold (__680__ − 20x).

Step 2 Write an equation.

$P(x) = (120 + 5x)(680 - 20x)$ — Original equation

$= 81{,}600 - 2400x + 3400x - 100x^2$ — Multiply.

$= -100x^2 + 1000x + 81{,}600$ — Simplify.

Step 3 Find the axis of symmetry.

Because a is negative, the vertex is a maximum.

$x = -\frac{b}{2a}$ — Formula for the axis of symmetry

$= -\frac{1000}{2(-100)}$ — $a = -100, b = 1000$

$= 5$ — Simplify.

Step 4 Interpret the results.

The ski resort will make the most money with __5__ price increases, so they should charge 120 + 5(__5__) or $__145__ for each holiday weekend pass.

Part B Find the domain and range in the context of the situation.

The domain is $\{x \mid 0 \le x \le 34\}$ because the number of price increases, and the number of passes sold cannot be negative. The range is $\{y \mid 0 \le y \le 84{,}100\}$ because the amount of money generated cannot be negative, and the maximum amount of money generated is $P(5) = -100(5)^2 + 1000(5) + 81{,}600$ or __84,100__.

Study Tip

Assumptions You assumed that the ski resort has the ability to increase the price indefinitely and that every price increase will be $5 and will cause the resort to lose sales from exactly 20 holiday weekend passes.

Think About It!

Why is the maximum amount of money generated from holiday weekend passes $P(5)$?

Sample answer: Because the x-coordinate of the vertex is 5, the y-coordinate can be found by substituting 5 for x in the original equation.

Check

CONCERTS Last year, a ticket provider sold 1350 lawn seats for a concert at $70 per ticket. This year, the provider is considering increasing the price. They estimate that for each $2 increase, they will sell 30 fewer tickets.

Part A How much should the ticket provider charge in order to maximize profit? $__80__

Part B Find the domain and range in the context of the situation.

D = $\{x \mid x \ge 0\}$; R = $\{y \mid 0 \le y \le 96{,}000\}$

Go Online You can complete an Extra Example online.

Interactive Presentation

Use Quadratic Functions

SKIING A ski resort has extended hours on one holiday weekend per year. Last year, the resort sold 680 ski passes at $120 per holiday weekend pass. This year, the resort is considering a price increase. They estimate that for each $5 increase, they will sell 20 fewer holiday weekend passes.

Example 3

EXPAND

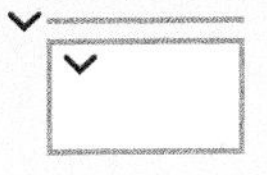

Students step through the process of analyzing a quadratic function and its graph in a real-world context.

TYPE

Students answer a question about the maximum of a real-world quadratic model.

CHECK

Students complete the Check online to determine whether they are ready to move on.

Learn Finding and Interpreting Average Rate of Change

A function's **rate of change** is how a quantity is changing with respect to a change in another quantity. For nonlinear functions, the rate of change is not the same over the entire function. You can calculate the **average rate of change** of a nonlinear function over an interval.

Key Concept • Average Rate of Change

The average rate of change of a function $f(x)$ is equal to the change in the value of the dependent variable $f(b) - f(a)$ divided by the change in the value of the independent variable $b - a$ over the interval $[a, b]$.

$$\frac{f(b) - f(a)}{b - a}$$

Think About It!
Find the average rate of change of the function over the interval $[-3, 1]$. Compare it to your results from the interval $[-4, 4]$.

Sample answer: $\frac{f(1) - f(-3)}{1 - (-3)} = \frac{0 - (-16)}{1 - (-3)}$ or 4. The rate of change is greater over the interval $[-3, 1]$ than over the interval $[-4, 4]$.

Example 4 Find Average Rate of Change from an Equation

Determine the average rate of change of $f(x) = -x^2 + 2x - 1$ over the interval $[-4, 4]$.

The average rate of change is equal to $\frac{f(4) - f(-4)}{4 - (-4)}$.

First find $f(-4)$ and $f(4)$.

$f(4) = -(4)^2 + 2(4) - 1$ or $\underline{-9}$ $\quad$ $f(-4) = -(-4)^2 + 2(-4) - 1$ or $\underline{-25}$

Then use $f(-4)$ and $f(4)$ to find the average rate of change.

$\frac{f(4) - f(-4)}{4 - (-4)} = \frac{-9 - (-25)}{4 - (-4)}$ or $\underline{2}$

The average rate of change of the function over the interval $[-4, 4]$ is $\underline{2}$.

Check

Find the average rate of change of $f(x) = -2x^2 - 5x + 7$ over the interval $[-5, 5]$.

average rate of change = $\underline{-5}$

Talk About It!
Without graphing, how can you tell that this function is nonlinear? Justify your argument.

Sample answer: The average rate of change is not the same over all intervals of the function. For example, the average rate of change over the interval $[-2, 1]$ is -15, and the average rate of change over the interval $[-3, 3]$ is -12.

Example 5 Find Average Rate of Change from a Table

Determine the average rate of change of $f(x)$ over the interval $[-3, 3]$.

x	$f(x)$
−3	48
−2	21
−1	0
0	−15
1	−24
2	−27
3	−24

The average rate of change is equal to $\frac{f(3) - f(-3)}{3 - (-3)}$.

First find $f(3)$ and $f(-3)$ from the table.

$f(3) = \underline{-24}$ $\quad$ $f(-3) = \underline{48}$

$\frac{f(3) - f(-3)}{3 - (-3)} = \frac{-24 - 48}{3 - (-3)}$ or $\underline{-12}$

The average rate of change of the function over the interval $[-3, 3]$ is $\underline{-12}$.

Go Online You can complete an Extra Example online.

Interactive Presentation

Finding and Interpreting Average Rate of Change

A function's **rate of change** is how a quantity is changing with respect to a change in another quantity. Recall that for linear functions, the rate of change is the same over the entire function. Therefore, to find the average rate of change for a linear function, you can choose any two points on the function and calculate their slope.

For nonlinear functions, the rate of change is not the same over the entire function. However, you can calculate the **average rate of change** of a nonlinear function over an interval. A function's average rate of change over an interval is equal to the change in the value of the dependent variable divided by the change in the value of the independent variable.

Learn

TYPE

Students answer a question to determine if they understand average rate of change of a linear function.

Learn Finding and Interpreting Average Rate of Change

Teaching the Mathematical Practices

2 Make Sense of Quantities Mathematically proficient students need to make sense of quantities and their relationships. Guide students to see how rate of change represents the relationship between problem variables.

Example 4 Find Average Rate of Change from an Equation

Teaching the Mathematical Practices

7 Interpret Complicated Expressions Mathematically proficient students can see complicated expressions as single objects or as being composed of several objects. In Example 4, guide students to first calculate the value of the function for −4 and for 4 before calculating the average rate of change.

Questions for Mathematical Discourse

AL Does the shape of the function over an interval matter when determining the average rate of change for that interval? Explain. No; sample answer: Only the function values at the interval endpoints are used to determine the average rate of change.

OL If the average rate of change over an interval is positive, what do you know about the end of the interval compared to the beginning of the interval? The function will have a greater output at the end of the interval.

BL If a function has a positive average rate of change over the interval $[a, b]$, but a negative average rate of change over the interval $[a, c]$, where $c > b$, what can you conclude about the extrema of the function? The function has a maximum in the interval $[a, c]$.

Example 5 Find Average Rate of Change from a Table

Teaching the Mathematical Practices

3 Construct Arguments Point out that to answer the question in the Talk About It! feature, students will need to use definitions and previously established results to construct an argument.

Questions for Mathematical Discourse

AL Over which interval is the average rate of change positive? $[2, 3]$

OL What does it mean if the average rate of change over an interval is 0? Sample answer: The function has the same value at the beginning and end of the interval.

BL How could you use the average rate of change to estimate where an extrema occurs? Find where the average rate of change switches from positive to negative (maximum) or from negative to positive (minimum).

1 CONCEPTUAL UNDERSTANDING | 2 FLUENCY | 3 APPLICATION

Example 6 Find Average Rate of Change from a Graph

Teaching the Mathematical Practices

5 Use a Source Guide students to find external information to answer the question posed in the Use a Source feature.

Questions for Mathematical Discourse

AL If the value of a function is greater at the end of an interval than it is at the beginning of the interval, what can you tell about the average rate of change over that interval? The average rate of change will be positive over that interval.

OL Is the amount spent on sports footwear from 2005 to 2015 linear or nonlinear? Explain. Nonlinear; the average rate of change is not constant over each interval.

BL How would the average rate of change from 2005 to 2015 differ if the spending each year between 2005 and 2015 increased? The average rate of change would remain the same.

Common Error

Students may subtract in the wrong direction, ending up with a rate of change that is the negative of what they expect. Encourage students to look at the interval and see if they expect a positive or negative average rate of change over the interval, and then use that to check their solution.

DIFFERENTIATE

Language Development ELL

Discuss how the average rate of change and slope are similar concepts. Average rate of change is the change in values of the dependent variable divided by the change in values of the independent variable. Remind students that slope is found by taking the difference of the y-coordinates divided by the difference of the x-coordinates. Students should understand that change in the values of the dependent variable is another, equivalent way of saying the difference of the y-coordinates, and that the change in the values of the independent variable is another, equivalent way of saying the difference of the x-coordinates.

Check

TESTING The table shows the number of students who took the ACT between 2011 and 2015.

Year	Number of Students
2011	1,623,112
2012	1,666,017
2013	1,799,243
2014	1,845,787
2015	1,924,436

Part A

Find the average rate of change in the number of students taking the ACT from 2011 to 2015.

average rate of change = 75,331

Part B

Interpret your results in the context of the situation.

From 2011 to 2015, the number of students taking the ACT increased by an average of 75,331 students per year.

Example 6 Find Average Rate of Change from a Graph

FOOTWEAR **The graph shows the amount of money the United States has spent on sports footwear since 2005.**

Part A

Use the graph to estimate the average rate of change of spending on sports footwear from 2005 to 2015. Then check your results algebraically.

Spending on Sports Footwear
Spending (billions of dollars)
(10, 20.99)
(0, 15.72)
Years Since 2005

Estimate

From the graph, the change in the y-values is approximately 5.5, and the change in the x-values is 10.

So, the rate of change is approximately $\frac{5.5}{10}$ or 0.55.

(continued on the next page)

Use a Source

Research the sales of another industry over a ten-year period. Then find the average rate of change during that time.

See students' work.

Interactive Presentation

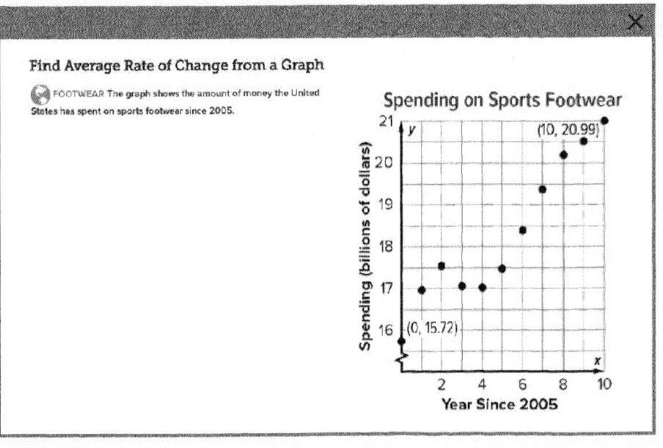

Example 6

TYPE

Students answer a question to determine whether they understand average rate of change.

Algebraically

The average rate of change is equal to $\frac{f(10) - f(0)}{10 - 0}$.

$\frac{f(10) - f(0)}{10 - 0} = \frac{20.99 - 15.72}{10 - 0}$ or $\underline{0.527}$

Part B

Interpret your results in the context of the situation.

From <u>2005</u> to 2015, the amount of money spent on sports footwear in the United States <u>increased</u> by an average of <u>$527 million</u> per year.

Check

SUPER BOWL The graph shows the number of television viewers of the Super Bowl since 2006.

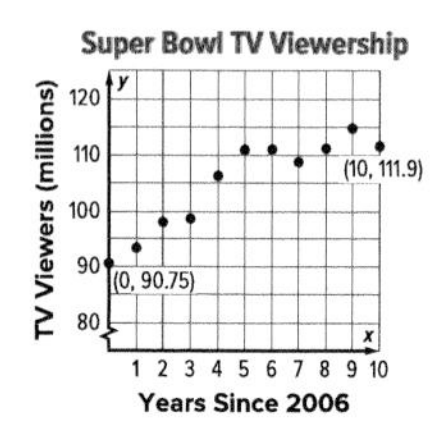

Part A

Use the graph to estimate the average rate of change in Super Bowl viewers from 2006 to 2016 to the nearest hundredth of a million. Then check your results algebraically.

estimate = <u>2.05</u> million

average rate of change = <u>2.115</u> million per year

Part B

Interpret your results in the context of the situation.

From <u>2006</u> to <u>2016</u>, the number of Super Bowl viewers <u>increased</u> by an average of <u>2.115</u> million viewers per year.

Go Online You can complete an Extra Example online.

CHECK

Students complete the Check online to determine whether they are ready to move on.

Exit Ticket

Recommended Use

At the end of class, go online to display the Exit Ticket prompt and ask students to respond using a separate piece of paper. Have students hand you their responses as they leave the room.

Alternate Use

At the end of class, go online to display the Exit Ticket prompt and ask students to respond verbally or by using a mini-whiteboard. Have students hold up their whiteboards so that you can see all student responses. Tap to reveal the answer when most or all students have completed the Exit Ticket.

Practice and Homework

Suggested Assignments

Use the table below to select appropriate exercises.

DOK	Topic	Exercises
1, 2	exercises that mirror the examples	1–26
2	exercises that use a variety of skills from this lesson	27–38
2	exercises that extend concepts learned in this lesson to new contexts	39–44
3	exercises that emphasize higher-order and critical thinking skills	45–48

ASSESS AND DIFFERENTIATE

Use the data from the **Checks** to determine whether to provide resources for extension, remediation, or intervention.

BL

IF students score 90% or more on the Checks,
THEN assign:

- Practice, Exercises 1–43 odd, 45–48
- Extension: Parametric Equations
- **ALEKS®** Quadratic Functions

OL

IF students score 66%–89% on the Checks,
THEN assign:

- Practice, Exercises 1–47 odd
- Remediation, Review Resources: Graphing Quadratic Functions
- Personal Tutors
- Extra Examples 1–5
- **ALEKS®** Quadratic Functions

AL

IF students score 65% or less on the Checks,
THEN assign:

- Practice, Exercises 1–25 odd
- Remediation, Review Resources: Graphing Quadratic Functions
- *Quick Review Math Handbook:* Graphing Quadratic Functions
- **ALEKS®** Quadratic Functions

Name ______ Period ______ Date ______

Practice

Go Online You can complete your homework online.

Example 1

Graph each function. Then state the domain and range. 1–6. See margin for graphs.

1. $f(x) = x^2 + 6x + 8$ D = {all real numbers}, R = $\{y \mid y \geq -1\}$
2. $f(x) = -x^2 - 2x + 2$ D = {all real numbers}, R = $\{y \mid y \leq 3\}$
3. $f(x) = 2x^2 - 4x + 3$ D = {all real numbers}, R = $\{y \mid y \geq 1\}$
4. $f(x) = -2x^2$ D = {all real numbers}, R = $\{y \mid y \leq 0\}$
5. $f(x) = x^2 - 4x + 4$ D = {all real numbers}, R = $\{y \mid y \geq 0\}$
6. $f(x) = x^2 - 6x + 8$ D = {all real numbers}, R = $\{y \mid y \geq -1\}$

Example 2

7. Compare the graph of $f(x)$ to a quadratic function $g(x)$ with a y-intercept of 1 and a vertex at (1, 3). Which function has a greater maximum? Explain.

$g(x)$; Sample answer: Its vertex is a maximum point at (1, 3), which is 4 units above the vertex of $f(x)$ which is (1, −1).

8. Compare the graph of $f(x)$ to a quadratic function $g(x)$ with a y-intercept of 0.5 and a vertex at (−1, −5). Which function has a lesser minimum? Explain.

$g(x)$; Sample answer: Its vertex is a minimum point at (−1, −5), which is 4 units below the vertex of $f(x)$ which is (−1, −1).

9. Compare $f(x) = x^2 - 10x + 5$ to the quadratic function $g(x)$ shown in the table. Which function has the lesser minimum? Explain.

x	$g(x)$
−10	170
−5	70
0	10
5	−10
10	10

$f(x)$; Sample answer: Its vertex is a minimum point at (5, −20), which is 10 units below the vertex of $g(x)$ which is (5, −10).

10. Compare $f(x) = -x^2 + 6x - 15$ to the quadratic function $g(x)$ shown in the table. Which function has the greater maximum? Explain.

x	$g(x)$
−6	−26
−3	−11
0	−2
3	1
6	−2

$g(x)$; Sample answer: Its vertex is a maximum point at (3, 1), which is 7 units above the vertex of $f(x)$ which is (3, −6).

Example 3

11. FISHING A county park sells annual permits to its fishing lake. Last year, the county sold 480 fishing permits for \$80 each. This year, the park is considering a price increase. They estimate that for each \$4 increase, they will sell 16 fewer annual fishing permits.
    a. Define variables and write a function to represent the situation. For x = the number of price increases and y = revenue, $y = (80 + 4x)(480 - 16x)$.
    b. How much would the park have to charge for each permit in order to maximize its revenue from fishing permits? \$100 per permit
    c. Find the domain and range in the context of this situation. $D = \{x \mid 0 \leq x \leq 30\}$, $R = \{y \mid 0 \leq y \leq 40{,}000\}$

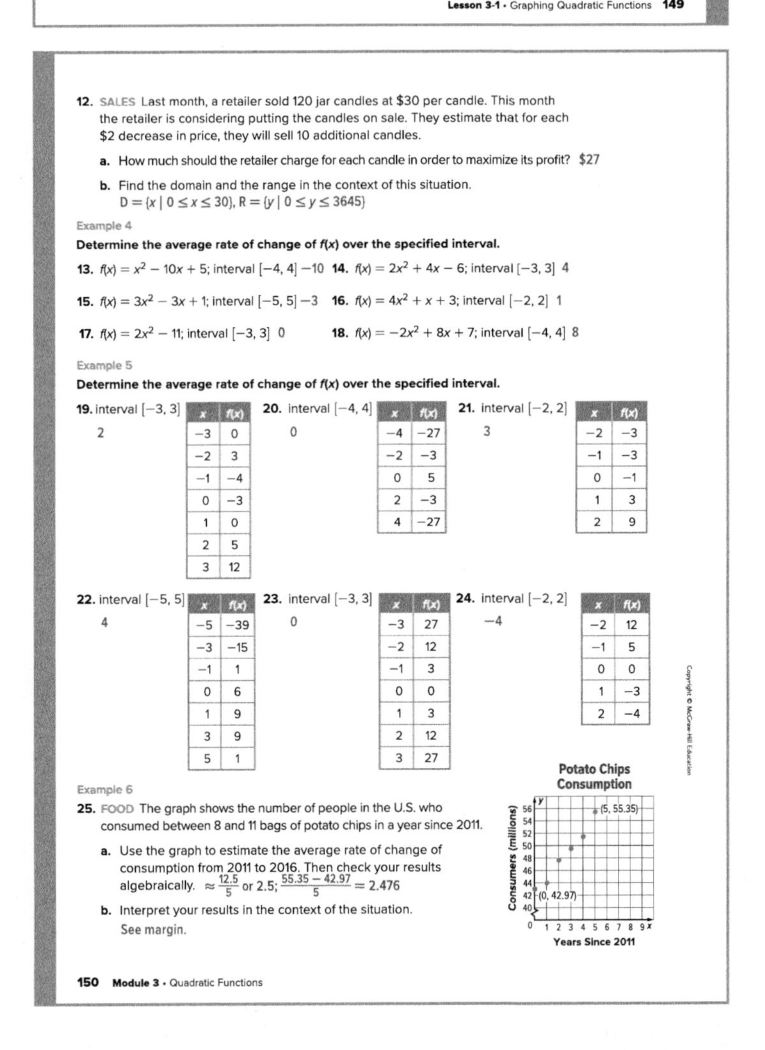

12. SALES Last month, a retailer sold 120 jar candles at \$30 per candle. This month the retailer is considering putting the candles on sale. They estimate that for each \$2 decrease in price, they will sell 10 additional candles.
    a. How much should the retailer charge for each candle in order to maximize its profit? \$27
    b. Find the domain and the range in the context of this situation. $D = \{x \mid 0 \leq x \leq 30\}$, $R = \{y \mid 0 \leq y \leq 3645\}$

Example 4

Determine the average rate of change of $f(x)$ over the specified interval.

13. $f(x) = x^2 - 10x + 5$; interval $[-4, 4]$ −10
14. $f(x) = 2x^2 + 4x - 6$; interval $[-3, 3]$ 4
15. $f(x) = 3x^2 - 3x + 1$; interval $[-5, 5]$ −3
16. $f(x) = 4x^2 + x + 3$; interval $[-2, 2]$ 1
17. $f(x) = 2x^2 - 11$; interval $[-3, 3]$ 0
18. $f(x) = -2x^2 + 8x + 7$; interval $[-4, 4]$ 8

Example 5

Determine the average rate of change of $f(x)$ over the specified interval.

19. interval $[-3, 3]$ 2

x	$f(x)$
−3	0
−2	3
−1	−4
0	−3
1	0
2	5
3	12

20. interval $[-4, 4]$ 0

x	$f(x)$
−4	−27
−2	−3
0	5
2	−3
4	−27

21. interval $[-2, 2]$ 3

x	$f(x)$
−2	−3
−1	−3
0	−1
1	3
2	9

22. interval $[-5, 5]$ 4

x	$f(x)$
−5	−39
−3	−15
−1	1
0	6
1	9
3	9
5	1

23. interval $[-3, 3]$ 0

x	$f(x)$
−3	27
−2	12
−1	3
0	0
1	3
2	12
3	27

24. interval $[-2, 2]$ −4

x	$f(x)$
−2	12
−1	5
0	0
1	−3
2	−4

Example 6

25. FOOD The graph shows the number of people in the U.S. who consumed between 8 and 11 bags of potato chips in a year since 2011.
    a. Use the graph to estimate the average rate of change of consumption from 2011 to 2016. Then check your results algebraically. $\approx \frac{12.5}{5}$ or 2.5; $\frac{55.35 - 42.97}{5} = 2.476$
    b. Interpret your results in the context of the situation. See margin.

Name ______ Period ______ Date ______

26. EARNINGS The graph shows the amount of money Sheila earned each year since 2008. a, b. See margin.

a. Use the graph to estimate the average rate of change of Sheila's earnings from 2008 to 2018. Then check your results algebraically.

b. Interpret your results in the context of the situation.

Sheila's Earnings

Mixed Exercises

Complete parts a-c for each quadratic function. 27–32. See Mod. 3 Answer Appendix.

a. Find the y-intercept, the equation of the axis of symmetry, and the x-coordinate of the vertex.

b. Make a table of values that includes the vertex.

c. Use this information to graph the function.

27. $f(x) = 2x^2 - 6x - 9$
28. $f(x) = -3x^2 - 9x + 2$
29. $f(x) = -4x^2 + 5x$
30. $f(x) = 2x^2 + 11x$
31. $f(x) = 0.25x^2 + 3x + 4$
32. $f(x) = -0.75x^2 + 4x + 6$

Determine whether each function has a maximum or a minimum value. Then find and use the x-coordinate of the vertex to determine the maximum or minimum.

33. $f(x) = -9x^2 - 12x + 19$ max; $\frac{-2}{3}$; 23
34. $f(x) = 7x^2 - 21x + 8$ min; 1.5; −7.75
35. $f(x) = -5x^2 + 14x - 6$ max; 1.4; 3.8
36. $f(x) = 2x^2 - 13x - 9$ min; 3.25; −30.125
37. $f(x) = 9x - 1 - 18x^2$ max; 0.25; 0.125
38. $f(x) = -16 - 18x - 12x^2$ max; −0.75; −9.25

39. HEALTH CLUBS Last year, the Sports Time Athletic Club charged $20 per month to participate in an aerobics class. Seventy people attended the classes. The club wants to increase the class price. They expect to lose one customer for each $1 increase in the price.

a. Define variables and write an equation for a function that represents the situation. Make a table and graph the function. See Mod. 3 Answer Appendix.

b. Find the vertex of the function and interpret it in the context of the situation. Does it seem reasonable? Explain. See Mod. 3 Answer Appendix.

40. TICKETS The manager of a community symphony estimates that the symphony will earn $-40P^2 + 1100P$ dollars per concert if they charge P dollars for tickets. What ticket price should the symphony charge in order to maximize its profits? $13.75

41. REASONING On Friday nights, the local cinema typically sells 200 tickets at $6.00 each. The manager estimates that for each $0.50 increase in the ticket price, 10 fewer people will come to the cinema. a–c. See Mod. 3 Answer Appendix.

a. Define the variables x and y. Then write and graph a function to represent the expected revenue from ticket sales, and determine the domain of the function for the situation.

b. What price should the manager set for a ticket in order to maximize the revenue? Justify your reasoning.

c. Explain why the graph decreases from $x = 4$ to $x = 20$, and interpret the meaning of the x-intercept of the graph.

42. USE A MODEL From 4 feet above a swimming pool, Tomas throws a ball upward with a velocity of 32 feet per second. The height $h(t)$ of the ball t seconds after Tomas throws it is given by $h(t) = -16t^2 + 32t + 4$. For $t \geq 0$, find the maximum height reached by the ball and the time that this height is reached. 20 ft; 1 second

43. TRAJECTORIES At a special ceremonial reenactment, a cannonball is launched from a cannon on the wall of Fort Chambly, Quebec. If the path of the cannonball is traced on a graph so that the cannon is situated on the y-axis, the equation that describes the path is $y = -\frac{1}{1600}x^2 + \frac{1}{2}x + 20$, where x is the horizontal distance from the cliff and y is the vertical distance above the ground in feet. How high above the ground is the cannon? 20 ft

44. CONSTRUCT ARGUMENTS Which function has a greater maximum: $f(x) = -2x^2 + 6x - 7$ or the function shown in the graph at the right? Explain your reasoning using a graph. Graph $f(x)$ on the same coordinate plane and compare; $g(x)$ has the greater maximum value.

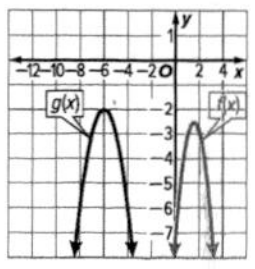

Higher-Order Thinking Skills

45. FIND THE ERROR Lucas thinks that the functions $f(x)$ and $g(x)$ have the same maximum. Madison thinks that $g(x)$ has a greater maximum. Is either of them correct? Explain your reasoning.

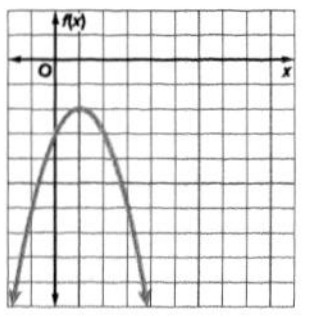

$g(x)$ is a quadratic function with x-intercepts of 4 and 2 and a y-intercept of −8.

Madison; sample answer: $f(x)$ has a maximum of −2. $g(x)$ has a maximum of 1.

46. PERSEVERE The table at the right represents some points on the graph of a quadratic function.

a. Find the values of a, b, c, and d. $a = 22$; $b = 26$; $c = -6$; $d = 2$

b. What is the x-coordinate of the vertex of the function? 0

c. Does the function have a maximum or a minimum? maximum

x	y
−20	−377
c	−13
−5	−2
−1	22
d − 1	a
5	a − 24
7	−b
15	−202
14 − c	−377

47. WRITE Describe how you determine whether a function is quadratic and if it has a maximum or minimum value. See Mod. 3 Answer Appendix

48. CREATE Give an example of a quadratic function with each characteristic.

a. maximum of 8 Sample answer: $f(x) = -x^2 + 8$

b. minimum of −4 Sample answer: $f(x) = x^2 - 4$

c. vertex of (−2, 6) Sample answer: $f(x) = x^2 + 4x + 10$

Answers

1.

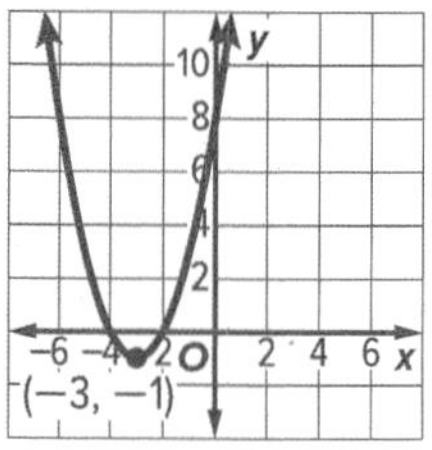

2.

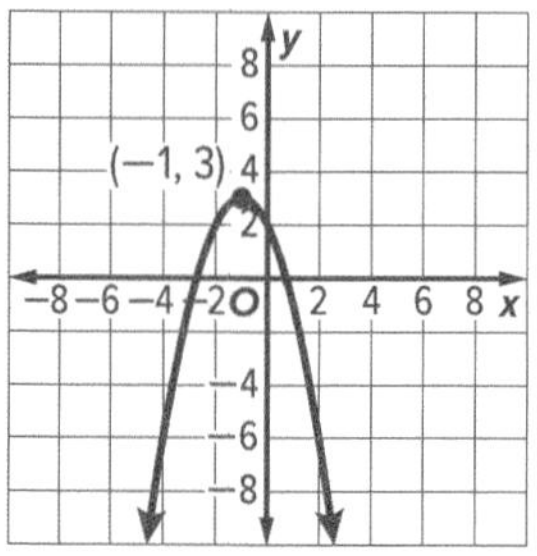

3.

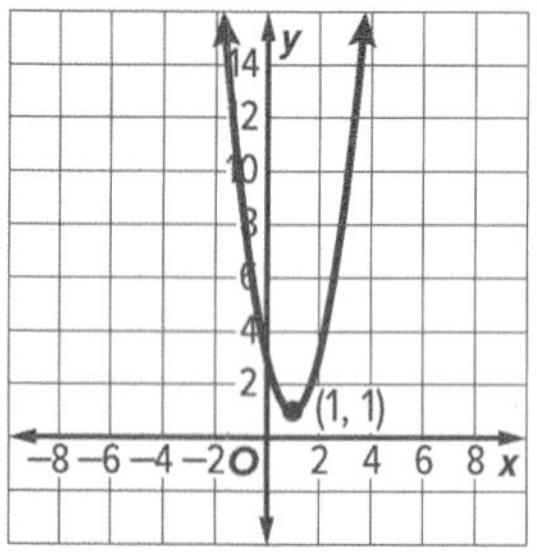

4.

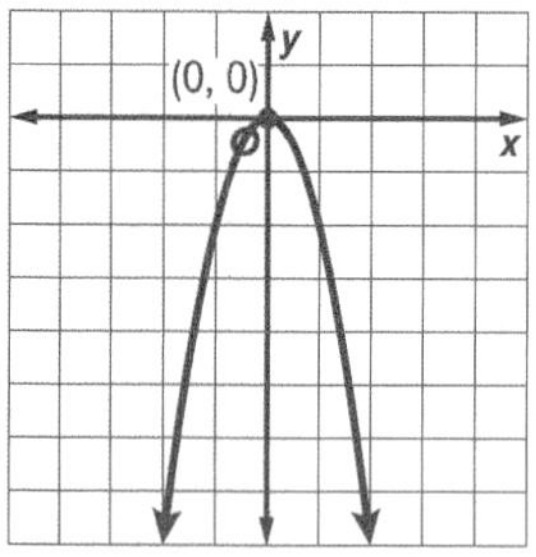

5.

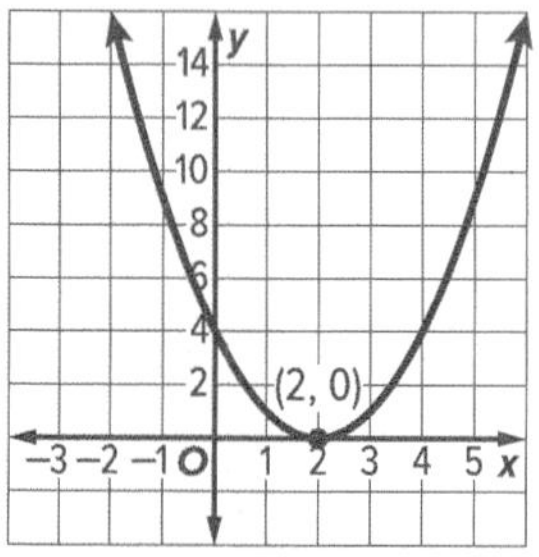

6.

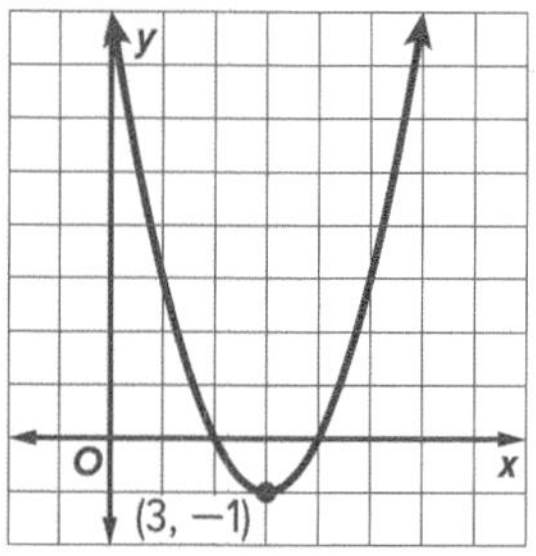

25b. Sample answer: From 2011–2016, the number of people in the U.S. who consumed between 8 and 11 bags increased by an average of 2.476 million people per year.

26a. $\approx \frac{5}{10}, \approx \frac{1}{2} \approx \$5000; \frac{7.2 - 2.1}{10} = \frac{5.1}{10} = \$5100$

26b. Sample answer: From 2008–2018, the amount of money Sheila earned increased by an average of $5100 per year.

Lesson 3-2

A.CED.2, F.IF.4

Solving Quadratic Equations by Graphing

LESSON GOAL

Students solve quadratic equations by graphing.

1 LAUNCH

 Launch the lesson with a **Warm Up** and an introduction.

2 EXPLORE AND DEVELOP

 **Explore:** Roots of Quadratic Equations

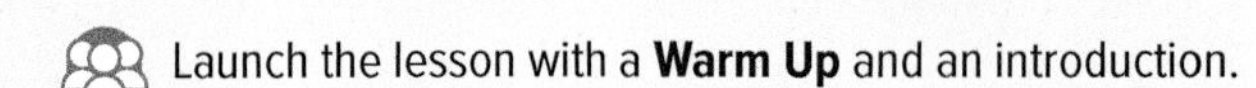

 **Develop:**

Solving Quadratic Equations by Graphing

- One Real Solution
- Two Real Solutions
- Estimate Roots
- Solve by Using a Table
- Solve by Using a Calculator

 You may want your students to complete the **Checks** online.

3 REFLECT AND PRACTICE

 Exit Ticket

 Practice

DIFFERENTIATE

View reports of student progress on the **Checks** after each example.

Resources	AL	OL	BL	ELL
Remediation: Solving Quadratic Equations by Graphing	●	●		●
Extension: Solving Absolute Value Equations by Graphing		●	●	●

Language Development Handbook

Assign page 18 of the *Language Development Handbook* to help your students build mathematical language related to solving quadratic equations by factoring.

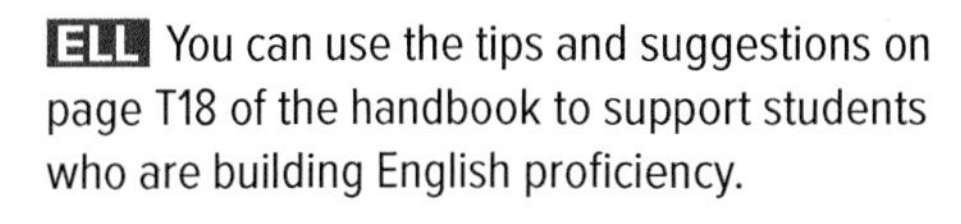

ELL You can use the tips and suggestions on page T18 of the handbook to support students who are building English proficiency.

Suggested Pacing

90 min	0.5 day
45 min	1 day

Focus

Domain: Algebra, Functions

Standards for Mathematical Content:

A.CED.2 Create equations in two or more variables to represent relationships between quantities; graph equations on coordinate axes with labels and scales.

F.IF.4 For a function that models a relationship between two quantities, interpret key features of graphs and tables in terms of the quantities, and sketch graphs showing key features given a verbal description of the relationship.

Standards for Mathematical Practice:

4 Model with mathematics.

5 Use appropriate tools strategically.

Coherence

Vertical Alignment

Previous
Students graphed linear and quadratic functions.
F.IF.7a (Algebra 1)

Now
Students solve quadratic equations by graphing.
A.CED.2, F.IF.4

Next
Students will extend their understanding of solutions of quadratic equations to complex numbers. **N.CN.7**

Rigor

The Three Pillars of Rigor

1 CONCEPTUAL UNDERSTANDING	2 FLUENCY	3 APPLICATION

Conceptual Bridge In this lesson, students expand on their understanding of graphing quadratic functions by using the graphs to solve related quadratic equations. They build fluency by finding solutions by graphing, and they apply their understanding by using graphs to solve real-world problems.

Interactive Presentation

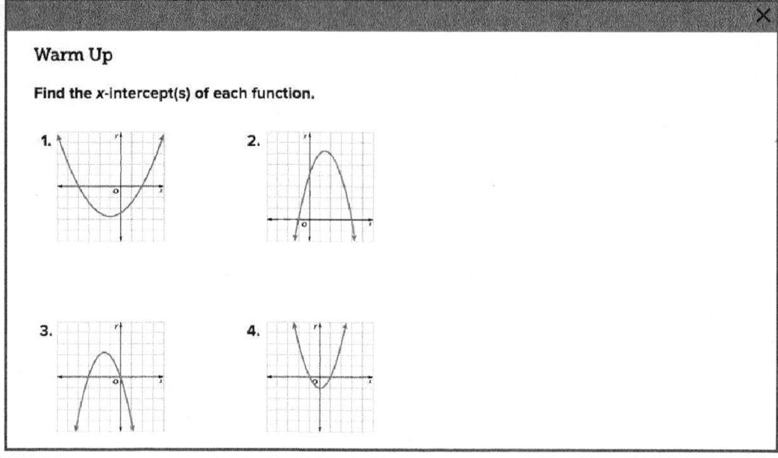

Warm Up

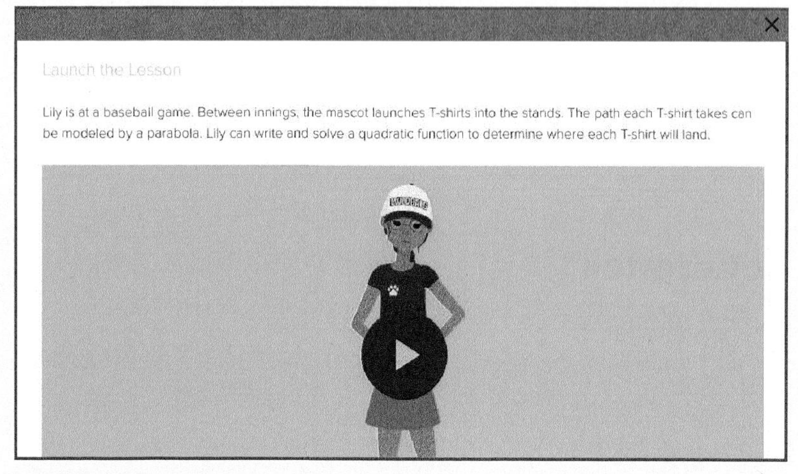

Launch the Lesson

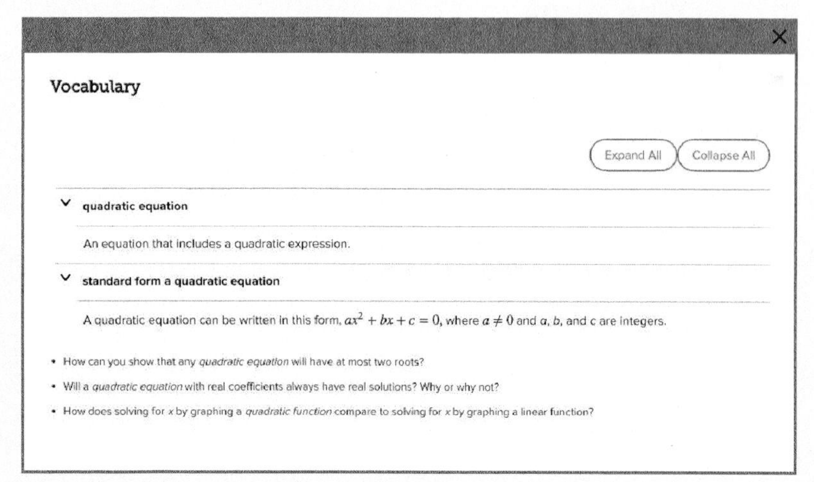

Today's Vocabulary

Warm Up

Prerequisite Skills

The Warm Up exercises address the following prerequisite skill for this lesson:

- finding x-intercepts

Answers:

1. $-4, 2$
2. $-1, 4$
3. $-3, 0$
4. $-1, 1$

Launch the Lesson

Teaching the Mathematical Practices

4 Apply Mathematics In the Launch the Lesson, students will watch an animation about a real-world situation that an be modeled and solved by using a quadratic function.

Go Online to find additional teaching notes and questions to promote classroom discourse.

Today's Standards

Tell students that they will be addressing these content and practice standards in this lesson. You may wish to have a student volunteer read aloud *How can I meet these standards*? and *How can I use these practices*?, and connect these to the standards.

See the Interactive Presentation for I Can statements that align with the standards covered in this lesson.

Today's Vocabulary

Tell students that they will be using these vocabulary terms in this lesson. You can expand each row if you wish to share the definitions. Then discuss the questions below with the class.

Mathematical Background

If the graph of the related quadratic function has one x-intercept, then there is one real solution (double root). If the graph of the related quadratic function has two x-intercepts, then there are two real solutions. If the graph of the related quadratic function does not intersect the x-axis, then there are no real solutions, but two imaginary solutions.

Explore Roots of Quadratic Equations

Objective

Students use a sketch to explore solving quadratic equations by graphing.

Teaching the Mathematical Practices

1 Explain Correspondences Throughout the Explore, students will explain the relationships between the graphs and roots of quadratic functions and the solutions of the related quadratic equations.

Ideas for Use

Recommended Use Present the Inquiry Question, or have a student volunteer read it aloud. Have students work in pairs to complete the Explore activity on their devices. Pairs should discuss each of the questions. Monitor student progress during the activity. Upon completion of the Explore activity, have student volunteers share their responses to the Inquiry Question.

What if my students don't have devices? You may choose to project the activity on a whiteboard. A printable worksheet for each Explore is available online. You may choose to print the worksheet so that individuals or pairs of students can use it to record their observations.

Summary of the Activity

Students will complete guiding exercises throughout the Explore activity. The students will use the sketch to graph various quadratic functions and identify their roots. They will observe the relationship between the zeros of the functions and the solutions of the related equations. Then, students will answer the Inquiry Question.

(continued on the next page)

Interactive Presentation

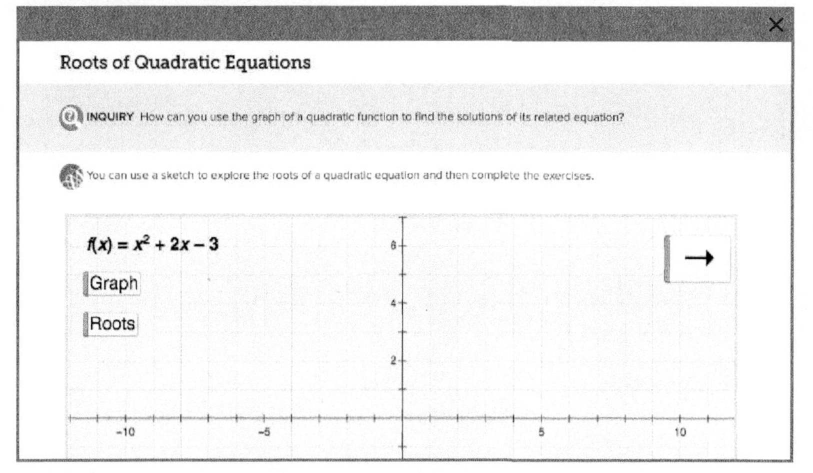

Explore

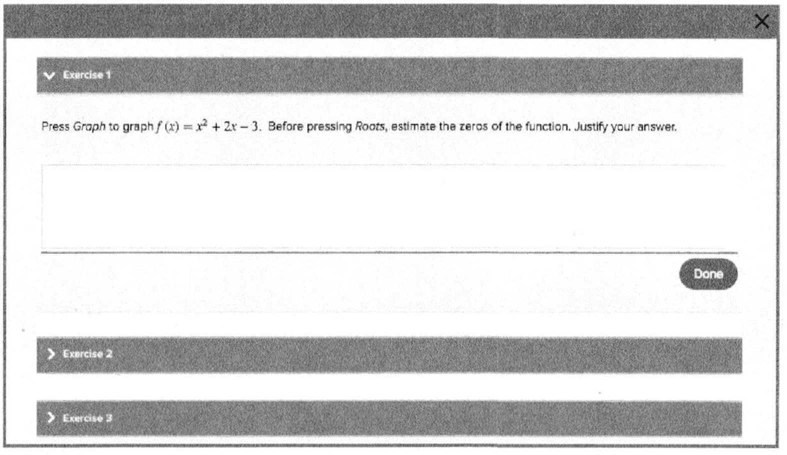

Explore

WEB SKETCHPAD

Students use a sketch to explore the roots of quadratic equations.

EXPAND

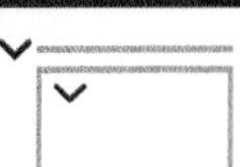

Students tap to see guiding exercises about finding the roots of quadratic equations.

Interactive Presentation

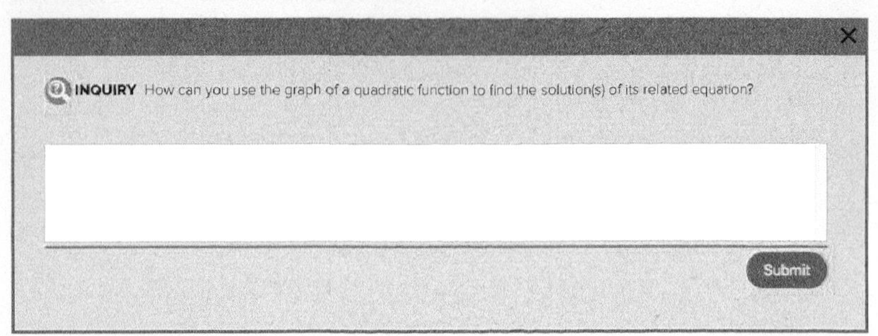

Explore

TYPE

Students respond to the Inquiry Question and can view a sample answer.

Explore Roots of Quadratic Equations (*continued*)

Questions

Have students complete the Explore activity.

Ask:

- What does it mean to be the solution to a quadratic equation? Sample answer: The solution to any equation is a value or values of x that make the given equation true.
- How does changing the equation $6 = x^2 - x$ into $0 = x^2 - x - 6$ make it easier to find the solutions? Sample answer: By setting the equation equal to zero, you can graph $y = x^2 - x - 6$ and find where the graph is equal to zero, or where it crosses the x-axis.

Inquiry

How can you use the graph of a quadratic function to find the solution(s) of its related equation? Sample answer: I can find the x-intercepts of the graph of a quadratic function. Then I can substitute the coordinates of the x-intercepts into the related equation to see if they can make the equation true. If so, they are solutions of the related equation.

Go Online to find additional teaching notes and sample answers for the guiding exercises.

1 CONCEPTUAL UNDERSTANDING | 2 FLUENCY | 3 APPLICATION

Learn Solving Quadratic Equations by Graphing

Objective

Students solve quadratic equations by graphing.

Teaching the Mathematical Practices

3 Analyze Cases The Study Tip features introduces students to the cases where a quadratic equation has zero, one, or two solutions. Ask students to describe what the graph of the related quadratic function would look like for each case.

Essential Question Follow-Up

Students have explored solving quadratic equations by graphing.

Ask:

How can the graph of a quadratic function help you solve the corresponding quadratic equation? Sample answer: From the placement of the graph, you can see whether there are 0, 1, or 2 real solutions. The symmetry of the graph helps you identify a second solution when one solution is found.

Example 1 One Real Solution

Teaching the Mathematical Practices

1 Explain Correspondences Encourage students to explain the relationships between the equation, table, and graph used in Example 1.

Questions for Mathematical Discourse

AL If there is only one real solution, how is the vertex related to that solution? The x-coordinate of the vertex is the solution.

OL For a quadratic equation that has only one real solution, does the graph of the related function always have a minimum? Explain. No; sample answer: If you rearrange the terms in the equation such that the x^2-term is negative, the parabola would open down and have a maximum.

BL Suppose you rearranged the terms so that the other side of the equation was 0. How would this affect your solution? Sample answer: Rearranging the terms so that the other side of the equation is 0 gives $-x^2 - 4x - 4 = 0$. The graph of this related function opens down, but has the same axis of symmetry and zero. Therefore, the solution is the same.

Go Online

- Find additional teaching notes.
- View performance reports of the Checks.
- Assign or present an Extra Example.

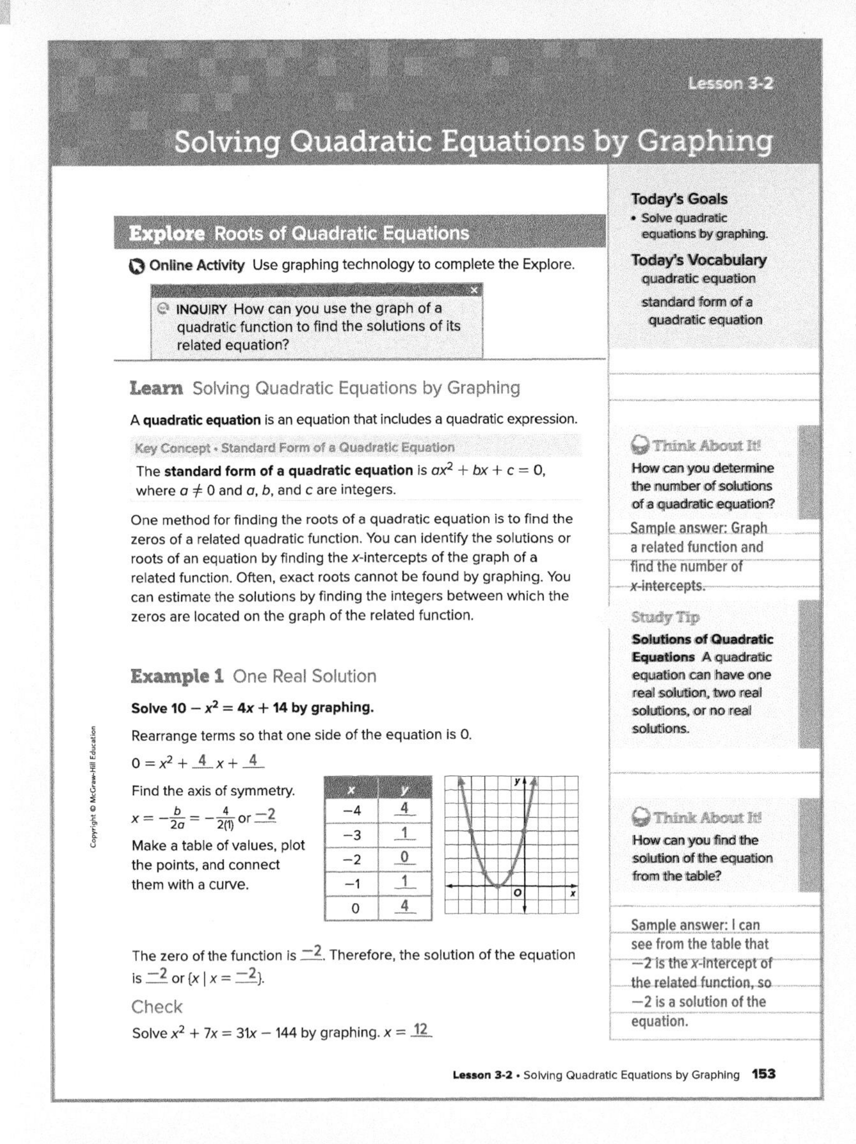

Lesson 3-2

Solving Quadratic Equations by Graphing

Today's Goals
- Solve quadratic equations by graphing.

Today's Vocabulary
quadratic equation
standard form of a quadratic equation

Explore Roots of Quadratic Equations

Online Activity Use graphing technology to complete the Explore.

INQUIRY How can you use the graph of a quadratic function to find the solutions of its related equation?

Learn Solving Quadratic Equations by Graphing

A **quadratic equation** is an equation that includes a quadratic expression.

Key Concept • Standard Form of a Quadratic Equation

The **standard form of a quadratic equation** is $ax^2 + bx + c = 0$, where $a \neq 0$ and a, b, and c are integers.

One method for finding the roots of a quadratic equation is to find the zeros of a related quadratic function. You can identify the solutions or roots of an equation by finding the x-intercepts of the graph of a related function. Often, exact roots cannot be found by graphing. You can estimate the solutions by finding the integers between which the zeros are located on the graph of the related function.

Think About It! How can you determine the number of solutions of a quadratic equation? Sample answer: Graph a related function and find the number of x-intercepts.

Study Tip **Solutions of Quadratic Equations** A quadratic equation can have one real solution, two real solutions, or no real solutions.

Example 1 One Real Solution

Solve $10 - x^2 = 4x + 14$ by graphing.

Rearrange terms so that one side of the equation is 0.

$0 = x^2 + \underline{4}x + \underline{4}$

Find the axis of symmetry.

$x = -\frac{b}{2a} = -\frac{4}{2(1)}$ or $\underline{-2}$

Make a table of values, plot the points, and connect them with a curve.

x	y
−4	4
−3	1
−2	0
−1	1
0	4

Think About It! How can you find the solution of the equation from the table? Sample answer: I can see from the table that −2 is the x-intercept of the related function, so −2 is a solution of the equation.

The zero of the function is $\underline{-2}$. Therefore, the solution of the equation is $\underline{-2}$ or $\{x \mid x = \underline{-2}\}$.

Check

Solve $x^2 + 7x = 31x - 144$ by graphing. $x = \underline{12}$

Copyright © McGraw-Hill Education

Lesson 3-2 • Solving Quadratic Equations by Graphing 153

Interactive Presentation

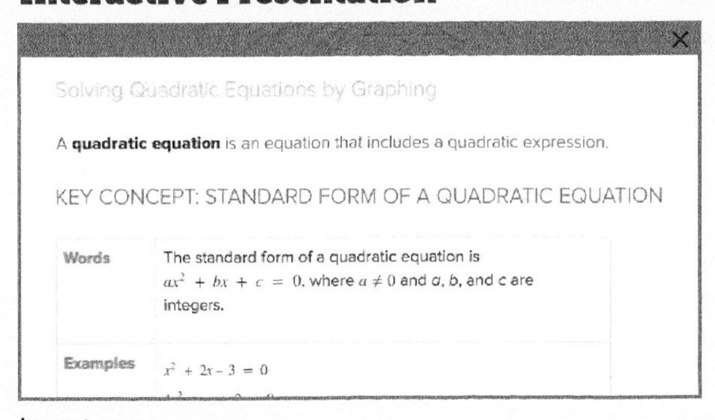

Learn

TYPE

Students will answer a question about determining the number of solutions of a quadratic equation.

Your Notes

Example 2 Two Real Solutions

Use a quadratic equation to find two real numbers with a sum of 24 and a product of 143.

Let x represent one of the numbers. Then $\underline{24} - x$ will represent the other number. So $x(24 - x) = \underline{143}$.

What do you need to find?
x and $24 - x$

Step 1 Solve the equation for 0.

$x(24 - x) = 143$	Original equation
$24x - x^2 = 143$	Distributive Property
$0 = x^2 - 24x + 143$	Subtract $24x - x^2$ from each side.

Go Online You can watch a video to see how to solve quadratic equations by graphing on a graphing calculator.

Step 2 Find the axis of symmetry.

$x = -\frac{b}{2a}$	Equation of the axis of symmetry
$x = -\frac{-24}{2(1)}$	$a = 1, b = -24$
$x = \underline{12}$	Simplify.

Step 3 Make a table of values and graph the function.

x	y
14	3
13	0
12	−1
11	0
10	3

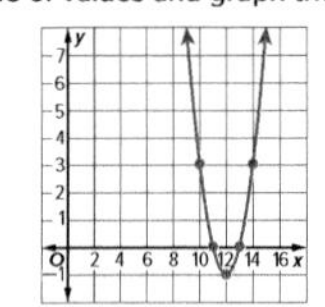

Think About It!
Explain why 9 and 15 cannot be solutions, even though their sum is 24.

Sample answer: The product of 9 and 15 is 135, not 143. Using them would result in a different equation with different solutions.

Steps 4 and 5 Find the zero(s) and determine the solution.

The zeros of the function are $\underline{11}$ and $\underline{13}$.

$x = 11$ or $x = 13$, so $24 - x = 13$ or $24 - x = 11$. Thus, the two numbers with a sum of 24 and a product of 143 are $\underline{11}$ and $\underline{13}$.

Check

Use a quadratic equation to find two real numbers with a sum of −43 and a product of 306. $\underline{-9}$ and $\underline{-34}$

Go Online You can complete an Extra Example online.

Example 2 Two Real Solutions

Teaching the Mathematical Practices

1 Analyze Givens and Constraints In Example 2, guide students through the use of the four-step plan to identify the meaning of the problem and look for entry points to its solution.

Questions for Mathematical Discourse

AL How can you tell from the graph that there will be two solutions? The graph crosses the x-axis in two places.

OL When the first number is defined as x, why is the other number $24 - x$? $x + (24 - x) = 24$

BL Given one root and the x-coordinate of the vertex, how can you find the other root? Sample answer: Because the line of symmetry passed through the vertex, I can determine how far the root is from the x-coordinate of the vertex and then go that far from the vertex in the other direction.

Common Error

Students may identify the y-intercept of the quadratic function as the a solution. Be sure to emphasize that it is only the x-intercepts that relate to the solutions. Emphasize the relationship of $f(x) = 0$ with the graph to reinforce the reason why these intercepts are the solutions.

Interactive Presentation

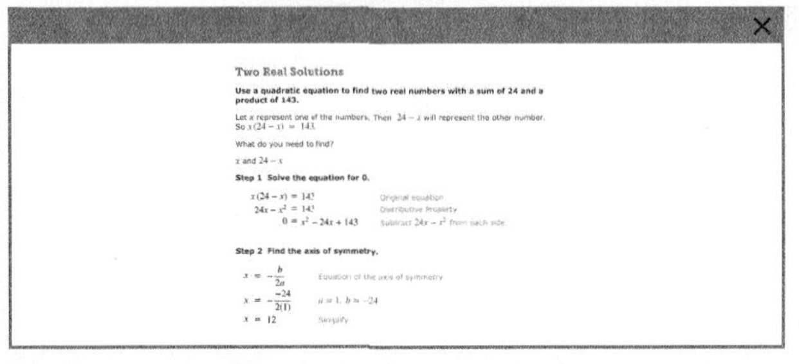

Example 2

DRAG & DROP

Students drag to sort the steps needed to solve the problem.

TYPE

Students explain why other possible pairs do not meet the criteria.

WATCH

Students watch a video to learn how to solve quadratic equations by graphing them on a calculator.

Example 3 Estimate Roots

Teaching the Mathematical Practices

2 Different Properties Mathematically proficient students look for different ways to solve problems. Use the Talk About It! feature to encourage students to consider how the table can be used to estimate the solutions.

Questions for Mathematical Discourse

AL Would this method work if there was only one real solution? Explain. Yes; sample answer: You would just look for the two consecutive integers the one solution is between.

OL How can you tell from the table that a root is between $x = 4$ and $x = 5$? When $x = 4$, the value of y is positive and when $x = 5$, the value of y is negative, which means the graph crosses the x-axis between these two values.

BL When making a table of values and graphing, why is it important to calculate and plot several points? Sample answer: You need to plot enough points over a large enough range to capture the zeros.

Example 4 Solve by Using a Table

Teaching the Mathematical Practices

5 Use Estimation In Example 4, students estimate the solutions of the equation by using a table. Guide students to try smaller intervals to get more accurate estimates.

Questions for Mathematical Discourse

AL Why is $x = 0.209$ selected as the approximation of the zero rather than $x = 0.208$? The y-value when $x = 0.209$ is closer to 0 than the y-value when $x = 0.208$.

OL Between which consecutive integers is the second zero of the function located? 4 and 5

BL How does each change in **ΔTbl** make the approximation more accurate? With each change in **ΔTbl**, the approximation is accurate to another decimal place.

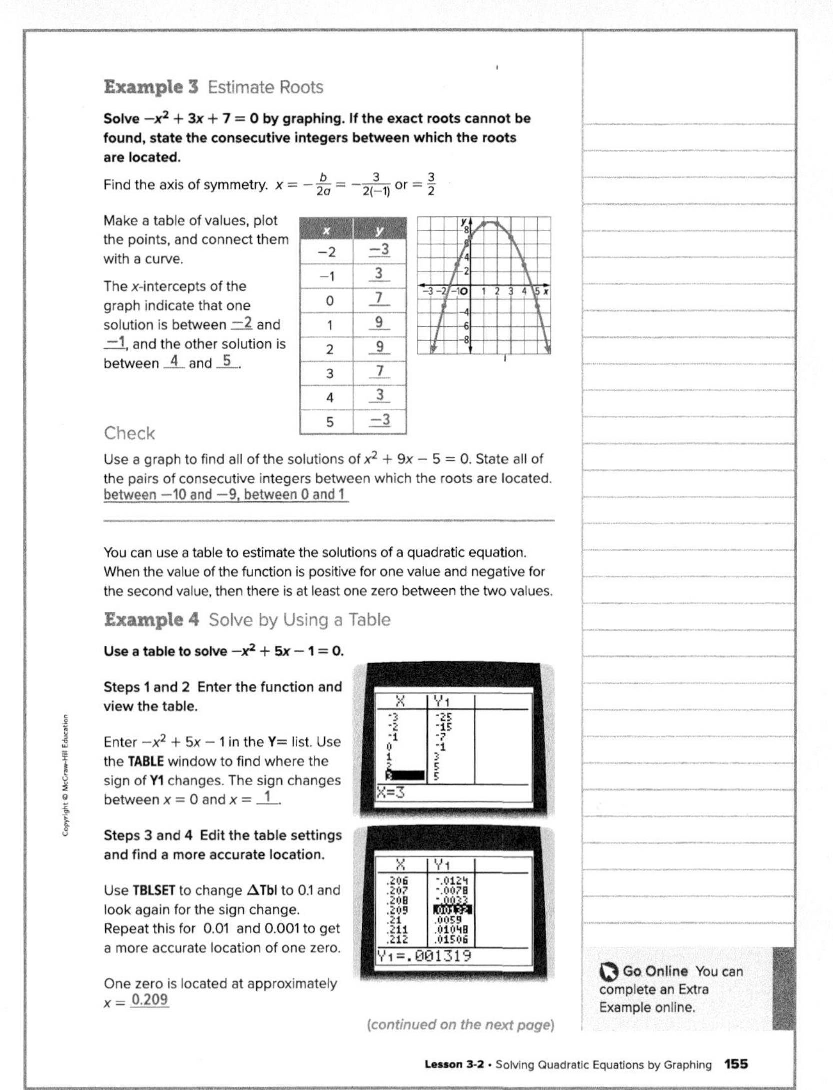

Example 3 Estimate Roots

Solve $-x^2 + 3x + 7 = 0$ by graphing. If the exact roots cannot be found, state the consecutive integers between which the roots are located.

Find the axis of symmetry. $x = -\frac{b}{2a} = -\frac{3}{2(-1)}$ or $= \frac{3}{2}$

Make a table of values, plot the points, and connect them with a curve.

x	y
−2	−3
−1	3
0	7
1	9
2	9
3	7
4	3
5	−3

The x-intercepts of the graph indicate that one solution is between −2 and −1, and the other solution is between 4 and 5.

Check

Use a graph to find all of the solutions of $x^2 + 9x - 5 = 0$. State all of the pairs of consecutive integers between which the roots are located. between −10 and −9, between 0 and 1

You can use a table to estimate the solutions of a quadratic equation. When the value of the function is positive for one value and negative for the second value, then there is at least one zero between the two values.

Example 4 Solve by Using a Table

Use a table to solve $-x^2 + 5x - 1 = 0$.

Steps 1 and 2 Enter the function and view the table.

Enter $-x^2 + 5x - 1$ in the **Y=** list. Use the **TABLE** window to find where the sign of **Y1** changes. The sign changes between $x = 0$ and $x = 1$.

Steps 3 and 4 Edit the table settings and find a more accurate location.

Use **TBLSET** to change **ΔTbl** to 0.1 and look again for the sign change. Repeat this for 0.01 and 0.001 to get a more accurate location of one zero.

One zero is located at approximately $x = 0.209$

(continued on the next page)

Go Online You can complete an Extra Example online.

Lesson 3-2 • Solving Quadratic Equations by Graphing 155

Interactive Presentation

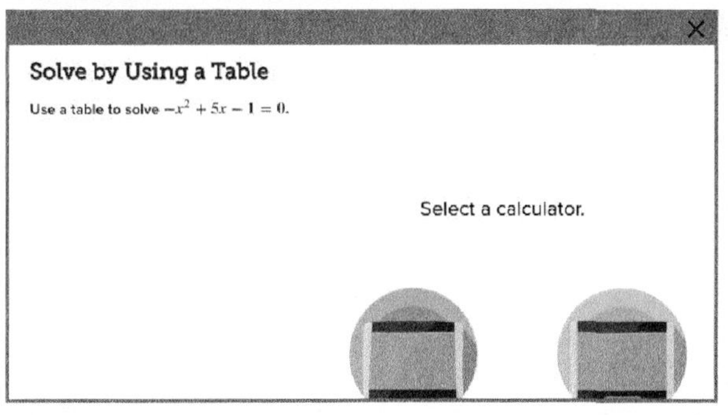

Example 4

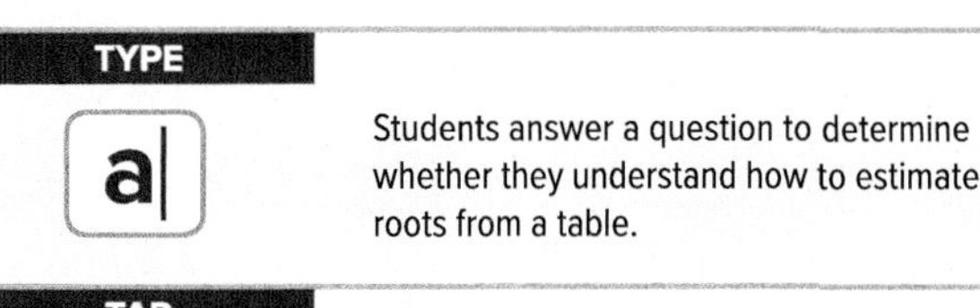

TYPE

Students answer a question to determine whether they understand how to estimate roots from a table.

TAP

Students select a calculator to solve a quadratic equation by using a table.

Think About It!
How can you check your solutions?

Sample answer: I can substitute the values for x in $-x^2 + 5x - 1 = 0$ and see if it is a true statement.

Steps 5 and 6 **Find the other zero and determine the solutions of the equation.**

Repeat the process to find the second zero of the function.

The zeros of the function are at approximately 0.209 and 4.791, so the solutions to the equation are approximately 0.209 and 4.791.

X	Y1
4.788	.01506
4.789	.01048
4.79	.0059
4.791	.00132
4.792	-.0033
4.793	-.0078
4.794	-.0124

Y1=.001319

Check

Use a table to find all of the solutions of $-x^2 - 3x + 8 = 0$.
−4.702 and 1.702

Watch Out!
Graphing Calculator If you cannot see the graph of the function on your graphing calculator, you may need to adjust the viewing window. Having the proper viewing window will also make it easier to see the zeros.

Example 5 Solve by Using a Calculator

FOOTBALL **A kicker punts a football. The height of the ball after t seconds is given by $h(t) = -16t^2 + 50t + h_0$, where h_0 is the initial height. If the ball is 1.5 feet above the ground when the punter's foot meets the ball, how long will it take the ball to hit the ground?**

We know that h_0 is the initial height, so $h_0 = 1.5$. We need to find t when $h(t)$ is 0. Use a graphing calculator to graph the related function $h(t) = -16t^2 + 50t + 1.5$.

Step 1 Enter the function in the **Y=** list, and press graph.

Step 2 Use the zero feature from the **CALC** menu to find the positive zero.

Step 3 Find the left bound by placing the cursor to the left of the intercept.

Step 4 Find the right bound.

Step 5 Find and interpret the solution.

Y1=-16X²+50X+1.5
Left Bound?
X=3.0851064 Y=3.4692168

The zero is approximately 3.15. Thus, the ball hit the ground approximately 3.15 seconds after it was punted.

Think About It!
Why did you only find the positive zero?

Sample answer: The negative zero is not a solution in the context of the situation because you cannot have a negative number of seconds.

Check

SOCCER A goalie punts a soccer ball. If the ball is 1 foot above the ground when her foots meets the ball, find how long it will take, to the nearest hundredth of a second, for the ball to hit the ground. Use the formula $h(t) = -16t^2 + 35t + h_0$, where t is the time in seconds and h_0 is the initial height.
2.22 seconds

Go Online to see how to use a TI-Nspire graphing calculator with this example.

Go Online You can complete an Extra Example online.

Interactive Presentation

Solve by Using a Calculator

FOOTBALL A kicker punts a football. If the ball is 1.5 feet above the ground when his foot meets the ball, how long will it take the ball to hit the ground? Use the formula $h(t) = -16t^2 + 50t + h_0$, where t is the time in seconds and h_0 is the initial height.

Example 5

TYPE

Students answer a question about the viability of the zero of a quadratic equation in a real-world situation.

CHECK

Students complete the Check online to determine whether they are ready to move on.

Example 5 Solve by Using a Calculator

Teaching the Mathematical Practices

4 Interpret Mathematical Results In Example 5, point out that to solve the problem, students should interpret their mathematical results in the context of the problem.

Questions for Mathematical Discourse

AL What does $h(0)$ represent? the height of the ball when the punter's foot meets the ball

OL What does it mean when $h(t) = 0$? The ball has a height of 0 feet and is on the ground.

BL Describe two ways to check your solution. Sample answer: substitute 3.15 into the original function to check that $h(3.15) \approx 0$ or use the table function on my calculator with an interval of 0.01 or smaller to verify that $y = 0$ when $x = 3.15$ in the table.

Common Error

Remind students that when solving real-world problems, there may be solutions that do not make sense within the context.

DIFFERENTIATE

Language Development Activity ELL
Tell students the meaning of *quadratic* comes from French and Latin words for *square*. A quadratic equation must have an x^2-term. Help them recall that a square with side length x has an area of x^2.

Exit Ticket

Recommended Use

At the end of class, go online to display the Exit Ticket prompt and ask students to respond using a separate piece of paper. Have students hand you their responses as they leave the room.

Alternate Use

At the end of class, go online to display the Exit Ticket prompt and ask students to respond verbally or by using a mini-whiteboard. Have students hold up their whiteboards so that you can see all student responses. Tap to reveal the answer when most or all students have completed the Exit Ticket.

Practice and Homework

Suggested Assignments

Use the table below to select appropriate exercises.

DOK	Topic	Exercises
1, 2	exercises that mirror the examples	1–34
2	exercises that use a variety of skills from this lesson	35–49
2	exercises that extend concepts learned in this lesson to new contexts	50–58
3	exercises that emphasize higher-order and critical thinking skills	59–63

ASSESS AND DIFFERENTIATE

Use the data from the **Checks** to determine whether to provide resources for extension, remediation, or intervention.

IF students score 90% or more on the Checks, 
THEN assign:

- Practice Exercises 1–57 odd, 59–63
- Extension: Solving Absolute Value Equations by Graphing
- ALEKS® Quadratic Equations

IF students score 66%–89% on the Checks, OL
THEN assign:

- Practice Exercises 1–63 odd
- Remediation, Review Resources: Solving Quadratic Equations by Graphing
- Personal Tutors
- Extra Examples 1–3
- ALEKS® Quadratic Equations

IF students score 65% or less on the Checks, AL
THEN assign:

- Practice Exercises 1–33 odd
- Remediation, Review Resources: Solving Quadratic Equations by Graphing
- *Quick Review Math Handbook*: Solving Quadratic Equations by Graphing
- ALEKS® Quadratic Equations

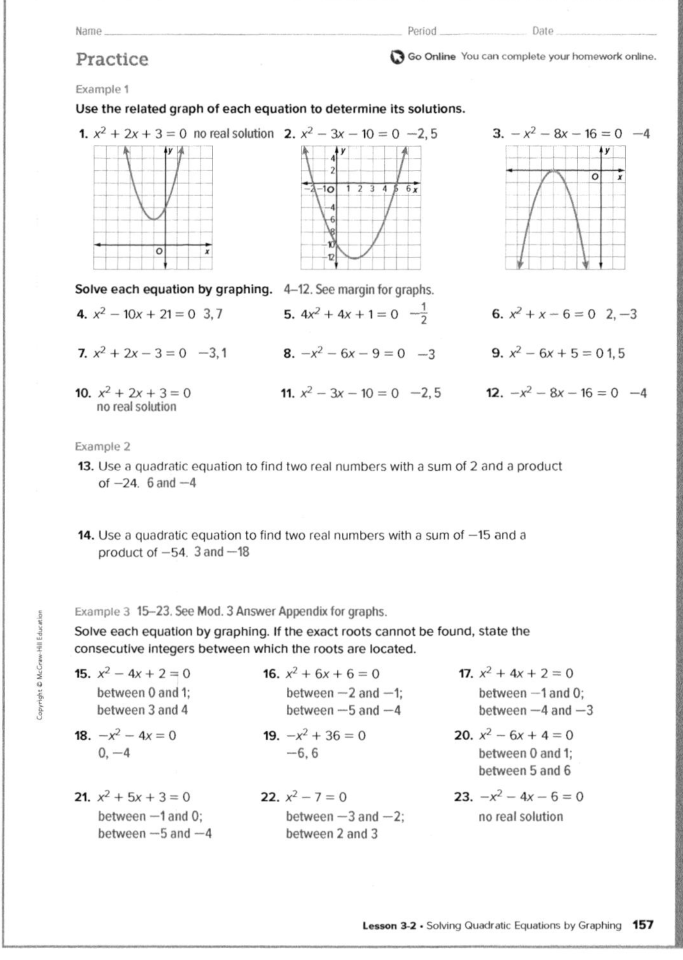

Name ______ Period ______ Date ______

Practice

Go Online You can complete your homework online.

Example 1

Use the related graph of each equation to determine its solutions.

1. $x^2 + 2x + 3 = 0$ no real solution
2. $x^2 - 3x - 10 = 0$ −2, 5
3. $-x^2 - 8x - 16 = 0$ −4

Solve each equation by graphing. 4–12. See margin for graphs.

4. $x^2 - 10x + 21 = 0$ 3, 7
5. $4x^2 + 4x + 1 = 0$ $-\frac{1}{2}$
6. $x^2 + x - 6 = 0$ 2, −3
7. $x^2 + 2x - 3 = 0$ −3, 1
8. $-x^2 - 6x - 9 = 0$ −3
9. $x^2 - 6x + 5 = 0$ 1, 5
10. $x^2 + 2x + 3 = 0$ no real solution
11. $x^2 - 3x - 10 = 0$ −2, 5
12. $-x^2 - 8x - 16 = 0$ −4

Example 2

13. Use a quadratic equation to find two real numbers with a sum of 2 and a product of −24. 6 and −4

14. Use a quadratic equation to find two real numbers with a sum of −15 and a product of −54. 3 and −18

Example 3 15–23. See Mod. 3 Answer Appendix for graphs.

Solve each equation by graphing. If the exact roots cannot be found, state the consecutive integers between which the roots are located.

15. $x^2 - 4x + 2 = 0$ between 0 and 1; between 3 and 4
16. $x^2 + 6x + 6 = 0$ between −2 and −1; between −5 and −4
17. $x^2 + 4x + 2 = 0$ between −1 and 0; between −4 and −3
18. $-x^2 - 4x = 0$ 0, −4
19. $-x^2 + 36 = 0$ −6, 6
20. $x^2 - 6x + 4 = 0$ between 0 and 1; between 5 and 6
21. $x^2 + 5x + 3 = 0$ between −1 and 0; between −5 and −4
22. $x^2 - 7 = 0$ between −3 and −2; between 2 and 3
23. $-x^2 - 4x - 6 = 0$ no real solution

Lesson 3-2 • Solving Quadratic Equations by Graphing 157

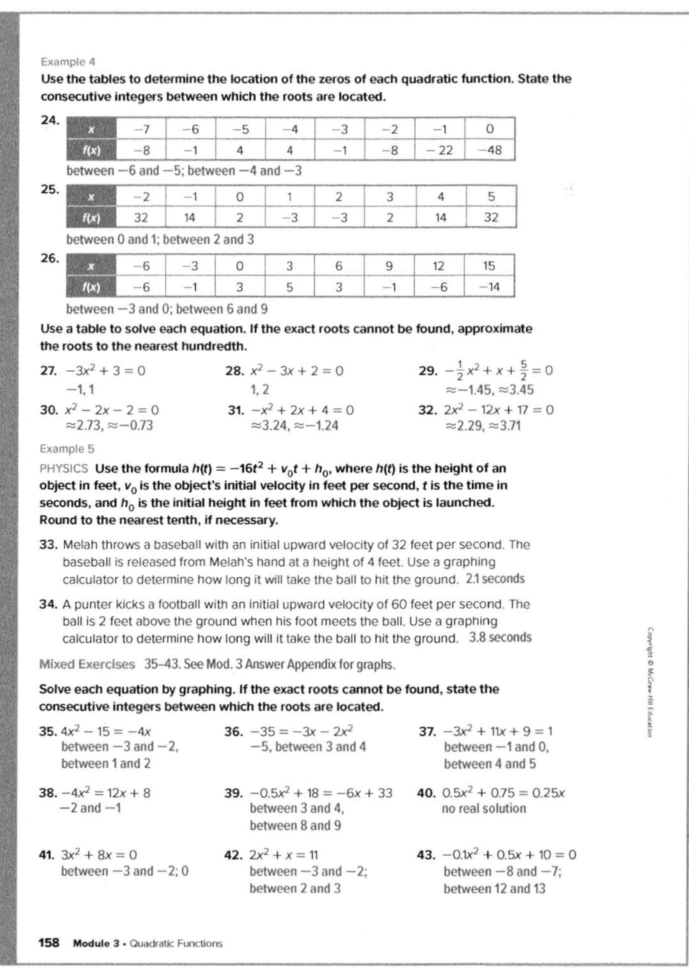

Example 4

Use the tables to determine the location of the zeros of each quadratic function. State the consecutive integers between which the roots are located.

24.

x	−7	−6	−5	−4	−3	−2	−1	0
f(x)	−8	−1	4	4	−1	−8	− 22	−48

between −6 and −5; between −4 and −3

25.

x	−2	−1	0	1	2	3	4	5
f(x)	32	14	2	−3	−3	2	14	32

between 0 and 1; between 2 and 3

26.

x	−6	−3	0	3	6	9	12	15
f(x)	−6	−1	3	5	3	−1	−6	−14

between −3 and 0; between 6 and 9

Use a table to solve each equation. If the exact roots cannot be found, approximate the roots to the nearest hundredth.

27. $-3x^2 + 3 = 0$ −1, 1
28. $x^2 - 3x + 2 = 0$ 1, 2
29. $-\frac{1}{2}x^2 + x + \frac{5}{2} = 0$ ≈−1.45, ≈3.45
30. $x^2 - 2x - 2 = 0$ ≈2.73, ≈−0.73
31. $-x^2 + 2x + 4 = 0$ ≈3.24, ≈−1.24
32. $2x^2 - 12x + 17 = 0$ ≈2.29, ≈3.71

Example 5

PHYSICS **Use the formula $h(t) = -16t^2 + v_0t + h_0$, where $h(t)$ is the height of an object in feet, v_0 is the object's initial velocity in feet per second, t is the time in seconds, and h_0 is the initial height in feet from which the object is launched. Round to the nearest tenth, if necessary.**

33. Melah throws a baseball with an initial upward velocity of 32 feet per second. The baseball is released from Melah's hand at a height of 4 feet. Use a graphing calculator to determine how long it will take the ball to hit the ground. 2.1 seconds

34. A punter kicks a football with an initial upward velocity of 60 feet per second. The ball is 2 feet above the ground when his foot meets the ball. Use a graphing calculator to determine how long will it take the ball to hit the ground. 3.8 seconds

Mixed Exercises 35–43. See Mod. 3 Answer Appendix for graphs.

Solve each equation by graphing. If the exact roots cannot be found, state the consecutive integers between which the roots are located.

35. $4x^2 - 15 = -4x$ between −3 and −2, between 1 and 2
36. $-35 = -3x - 2x^2$ −5, between 3 and 4
37. $-3x^2 + 11x + 9 = 1$ between −1 and 0, between 4 and 5
38. $-4x^2 = 12x + 8$ −2 and −1
39. $-0.5x^2 + 18 = -6x + 33$ between 3 and 4, between 8 and 9
40. $0.5x^2 + 0.75 = 0.25x$ no real solution
41. $3x^2 + 8x = 0$ between −3 and −2; 0
42. $2x^2 + x = 11$ between −3 and −2; between 2 and 3
43. $-0.1x^2 + 0.5x + 10 = 0$ between −8 and −7; between 12 and 13

158 Module 3 • Quadratic Functions

Name ______ Period ______ Date ______

Use a graph or table to solve each equation. If exact roots cannot be found, state the consecutive integers between which the roots are located.

44. $x^2 + 4x = 0$ $-4, 0$

45. $-2x^2 - 4x - 5 = 0$ no real solution

46. $0.5x^2 - 2x + 2 = 0$ 2

47. $-0.25x^2 - x - 1 = 0$ -2

48. $x^2 - 6x + 11 = 0$ no real solution

49. $-0.5x^2 + x + 6 = 0$ between -3 and -2, between 4 and 5

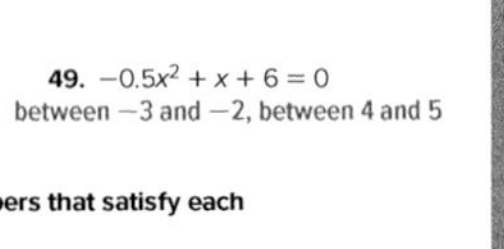

REGULARITY **Use a quadratic equation to find two real numbers that satisfy each situation, or show that no such numbers exist.**

50. Their sum is 4, and their product is -117. 13 and -9

51. Their sum is 12, and their product is -85. -5 and 17

52. Their sum is -13, and their product is 42. -6 and -7

53. Their sum is -8, and their product is -209. 11 and -19

54. BRIDGES In 1895, a brick arch railway bridge was built on North Avenue in Baltimore, Maryland. The arch is described by the equation $h = 9 - \frac{1}{50}x^2$, where h is the height in yards and x is the distance in yards from the center of the bridge. Graph this equation and describe, to the nearest yard, where the bridge touches the ground. See Mod. 3 Answer Appendix for graph. The bridge touches the ground approximately 21 yards from the center of the bridge on either side at $(-21, 0)$ and $(21, 0)$.

55. RADIO TELESCOPES The cross section of a large radio telescope is a parabola. The equation that describes the cross section is $y = \frac{2}{75}x^2 - \frac{4}{3}x - \frac{32}{3}$, where y is the depth of the dish in meters at a point x meters from the center of the dish. If $y = 0$ represents the top of the dish, what is the width of the dish? Solve by graphing. 64 m; See Mod. 3 Answer Appendix for graph.

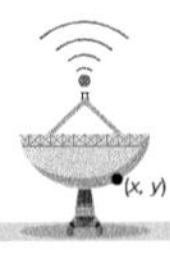

56. VOLCANOES A volcanic eruption blasts a boulder upward with an initial velocity of 240 feet per second. The height $h(t)$ of the boulder in feet, t seconds after the eruption can be modeled by the function $h(t) = -16t^2 + v_0t$. How long will it take the boulder to hit the ground if it lands at the same elevation from which it was ejected? 15 seconds

57. TRAJECTORIES Daniela hit a golf ball from ground level. The function $h = 80t - 16t^2$ represents the height of the ball in feet, where t is the time in seconds after Daniela hit it. Use the graph of the function to determine how long it took for the ball to reach the ground. 5 seconds

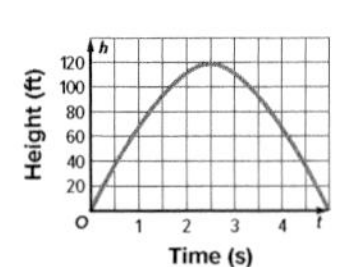

58. HIKING Antonia is hiking and reaches a steep part of the trail that runs along the edge of a cliff. In order to descend more safely, she drops her heavy backpack over the edge of the cliff so that it will land on a lower part of the trail, 38.75 feet below. The height $h(t)$ of an object t seconds after it is dropped straight down can also be modeled by the function $h(t) = -16t^2 + v_0t + h_0$, where v_0 is the initial velocity of the object, and h_0 is the initial height.

a. Write a quadratic function that can be used to determine the amount of time t that it takes for the backpack to land on the trail below the cliff after Antonia drops it. $h(t) = -16t^2 + 38.75$

b. Use a graphing calculator to determine how long until the backpack hits the ground. Round to the nearest tenth. 1.6 seconds

Higher-Order Thinking Skills

59. FIND THE ERROR Hakeem and Nandi were asked to find the location of the roots of the quadratic function represented by the table. Is either of them correct? Explain.

x	-4	-2	0	2	4	6	8	10
$f(x)$	52	26	8	-2	-4	2	16	38

Hakeem	Nandi
The roots are between 4 and 6 because f(x) stops decreasing and begins to increase between x = 4 and x = 6.	The roots are between -2 and 0 because x changes signs at that location.

No; sample answer: Hakeem is right about the location of one of the roots, but his reason is not accurate. The roots are located where $f(x)$ changes signs.

60. PERSEVERE Find the value of a positive integer k such that $f(x) = x^2 - 2kx + 55$ has roots at $k + 3$ and $k - 3$. $k = 8$

61. ANALYZE If a quadratic function has a minimum at $(-6, -14)$ and a root at $x = -17$, what is the other root? Explain your reasoning. 5; Sample answer: The intercepts are equidistant from the axis of symmetry.

62. CREATE Write a quadratic function with a maximum at (3, 125) and roots at -2 and 8. $f(x) = -5x^2 + 30x + 80$

63. WRITE Explain how to solve a quadratic equation by graphing its related quadratic function. Sample answer: Graph the function using the axis of symmetry. Determine where the graph intersects the x-axis. The x-coordinates of those points are solutions of the quadratic equation.

Answers

4.

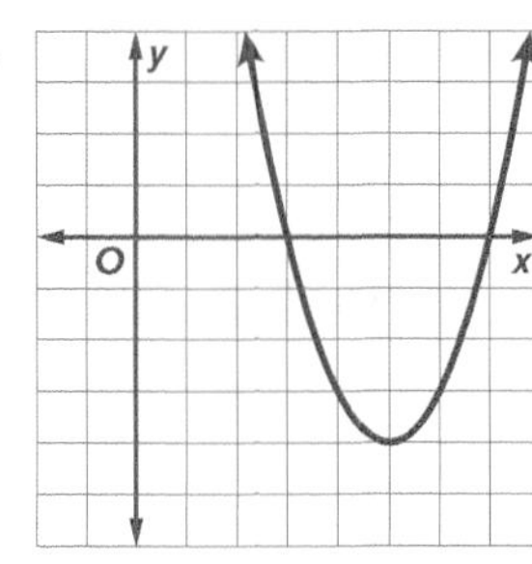

5.

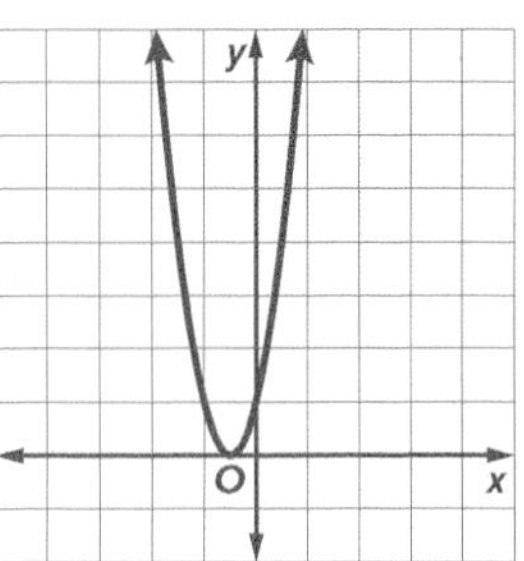

6.

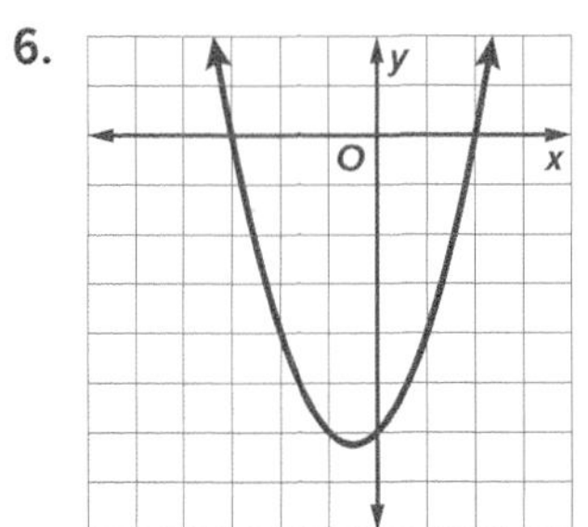

7.

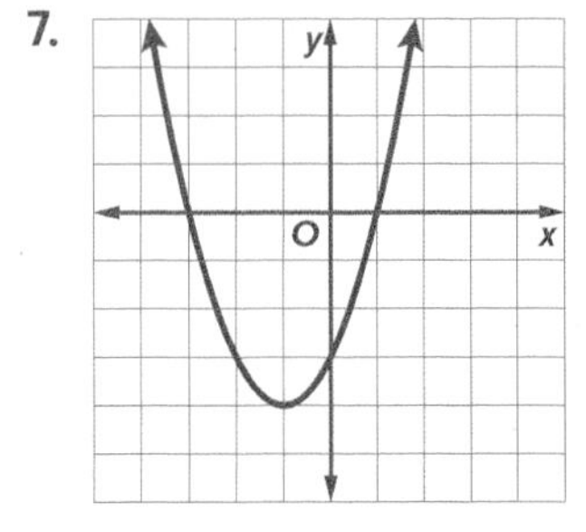

8.

9.

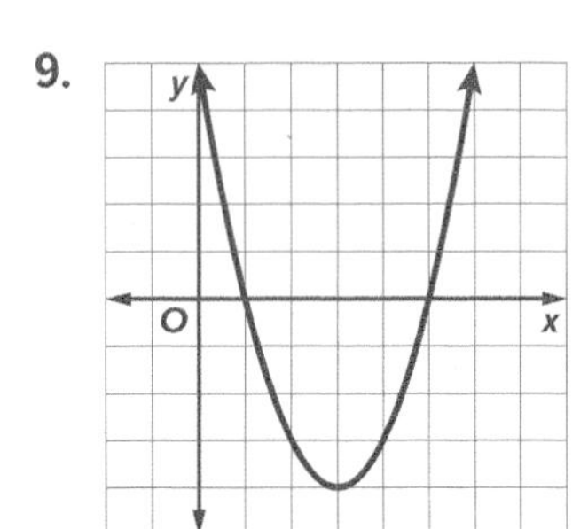

10.

11.

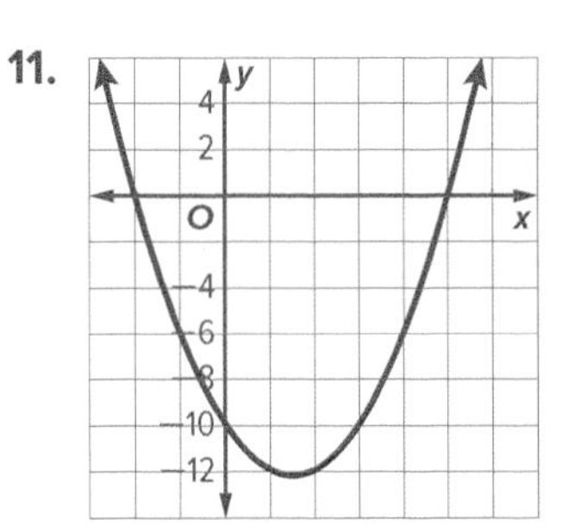

12.

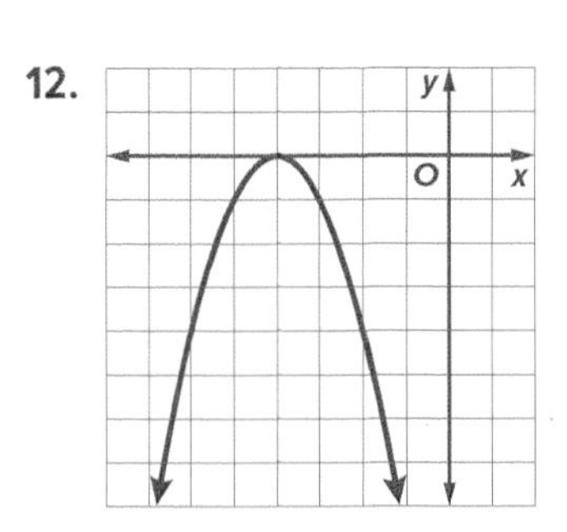

Lesson 3-3

Complex Numbers

N.CN.1, N.CN.2

LESSON GOAL

Students perform operations with pure imaginary and complex numbers.

1 LAUNCH

 Launch the lesson with a **Warm Up** and an introduction.

2 EXPLORE AND DEVELOP

 **Develop:**

Pure Imaginary Numbers
- Square Roots of Negative Numbers
- Products of Pure Imaginary Numbers
- Equation with Pure Imaginary Solutions

 **Explore:** Factoring the Sum of Two Squares

 **Develop:**

Complex Numbers
- Equate Complex Numbers
- Add or Subtract Complex Numbers
- Multiply Complex Numbers
- Divide Complex Numbers

 You may want your students to complete the **Checks** online.

3 REFLECT AND PRACTICE

 Exit Ticket

 Practice

DIFFERENTIATE

 View reports of student progress on the **Checks** after each example.

Resources	AL	OL	BL	ELL
Remediation: Properties of Real Numbers	●	●		●
Extension: Conjugates and Absolute Value		●	●	●

Language Development Handbook

Assign page 19 of the *Language Development Handbook* to help your students build mathematical language related to performing operations with imaginary and complex numbers.

ELL You can use the tips and suggestions on page T19 of the handbook to support students who are building English proficiency.

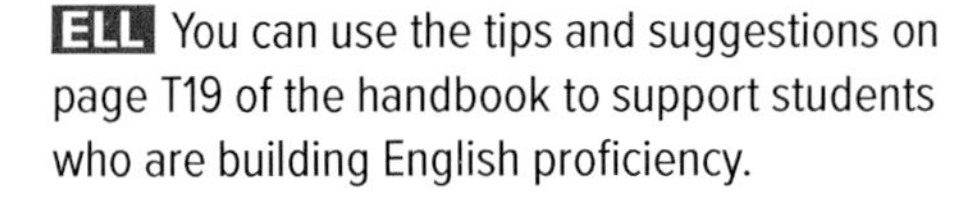

Suggested Pacing

90 min	1 day
45 min	2 days

Focus

Domain: Number and Quantity

Standards for Mathematical Content:

N.CN.1 Know there is a complex number i such that $i^2 = -1$, and every complex number has the form $a + bi$ with a and b real.

N.CN.2 Use the relation $i^2 = -1$ and the commutative, associative, and distributive properties to add, subtract, and multiply complex numbers.

Standards for Mathematical Practice:

3 Construct viable arguments and critique the reasoning of others.
7 Look for and make use of structure.
8 Look for and express regularity in repeated reasoning.

Coherence

Vertical Alignment

Previous
Students studied quadratic equations and functions, limited to real number solutions and roots. **A.REI.4, F.IF.7a (Algebra 1)**

Now
Students perform operations with pure imaginary and complex numbers. **N.CN.1, N.CN.2**

Next
Students will solve quadratic equations with real coefficients that have complex solutions. **N.CN.7**

Rigor

The Three Pillars of Rigor

1 CONCEPTUAL UNDERSTANDING	2 FLUENCY	3 APPLICATION

Conceptual Bridge In this lesson, students develop an understanding of imaginary and complex numbers and build fluency by performing operations with complex numbers and solving equations with complex solutions. They apply their understanding by solving real-world problems.

Mathematical Background

A complex number is any number that can be written in the form $a + bi$, where a and b are real numbers and i is the imaginary unit. If $b = 0$, the complex number is a real number. If $b \neq 0$, the complex number is imaginary. If $a = 0$, the complex number is a pure imaginary number. Pure imaginary and real numbers are both subsets of the set of complex numbers.

Interactive Presentation

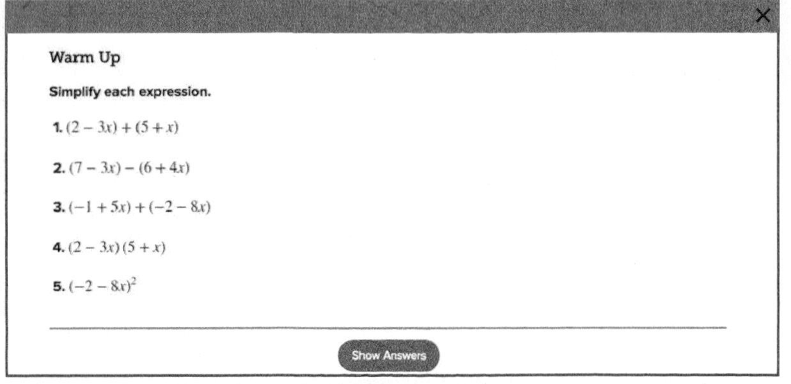

Warm Up

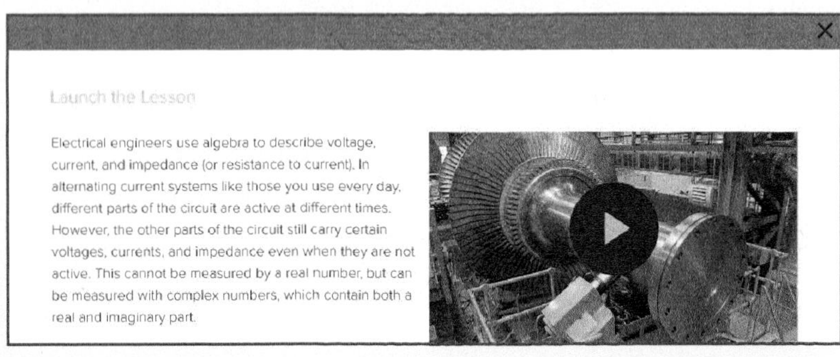

Launch the Lesson

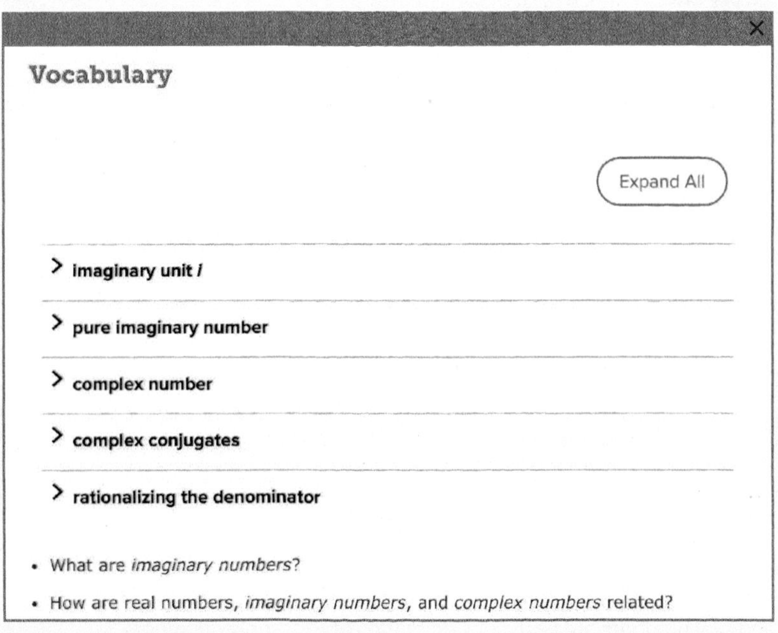

Today's Vocabulary

Warm Up

Prerequisite Skills

The Warm Up exercises address the following prerequisite skill for this lesson:

- simplifying expressions

Answers:

1. $7 - 2x$
2. $1 - 7x$
3. $-3 - 3x$
4. $10 - 13x - 3x^2$
5. $4 + 32x + 64x^2$

Launch the Lesson

Teaching the Mathematical Practices

4 Apply Mathematics In the Launch the Lesson, students will learn how complex numbers are used by engineers to measure circuits.

 **Go Online** to find additional teaching notes and questions to promote classroom discourse.

Today's Standards

Tell students that they will be addressing these content and practice standards in this lesson. You may wish to have a student volunteer read aloud *How can I meet these standards?* and *How can I use these practices?*, and connect these to the standards.

See the Interactive Presentation for I Can statements that align with the standards covered in this lesson.

Today's Vocabulary

Tell students that they will be using these vocabulary terms in this lesson. You can expand each row if you wish to share the definitions. Then discuss the questions below with the class.

Explore Factoring the Sum of Two Squares

Objective

Students explore imaginary numbers by factoring the sum of two squares.

Teaching the Mathematical Practices

7 Use Structure Guide students to use the structure of polynomials to rewrite them in a form that can be factored using the imaginary number *i*.

Ideas for Use

Recommended Use Present the Inquiry Question, or have a student volunteer read it aloud. Have students work in pairs to complete the Explore activity on their devices. Pairs should discuss each of the questions. Monitor student progress during the activity. Upon completion of the Explore activity, have student volunteers share their responses to the Inquiry Question.

What if my students don't have devices? You may choose to project the activity on a whiteboard. A printable worksheet for each Explore is available online. You may choose to print the worksheet so that individuals or pairs of students can use it to record their observations.

Summary of the Activity

Students will complete guiding exercises throughout the Explore activity. Students will explore how defining the imaginary number $i = \sqrt{-1}$ allows you to factor the sum of two squares. Then, students will answer the Inquiry Question.

(continued on the next page)

Interactive Presentation

Factoring the Sum of Two Squares

INQUIRY Can you factor a polynomial of the form $a^2 + b^2$?

Explore

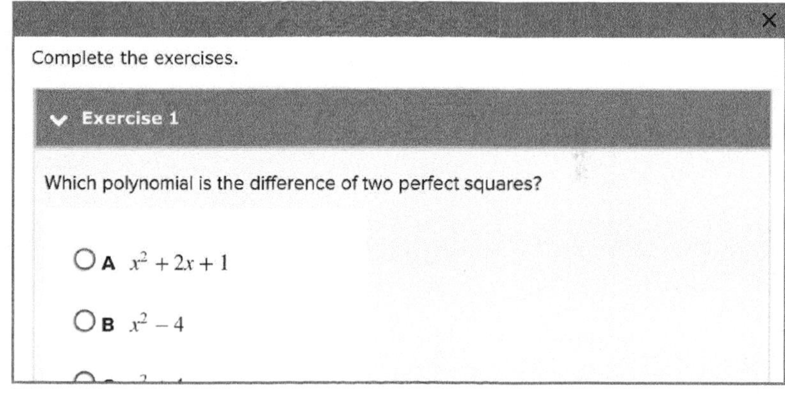

Explore

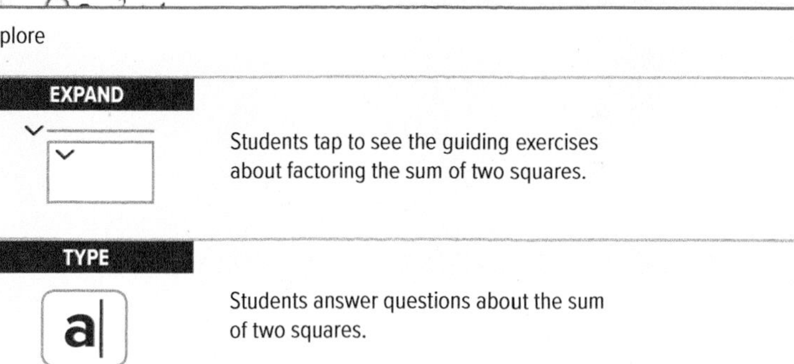

Students tap to see the guiding exercises about factoring the sum of two squares.

Students answer questions about the sum of two squares.

Interactive Presentation

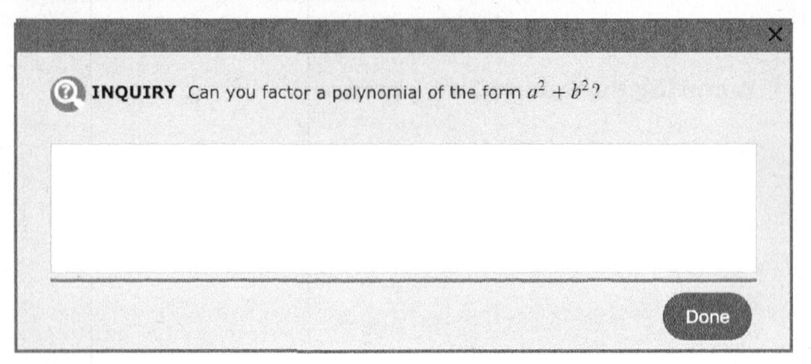

Explore

TYPE

Students respond to the Inquiry Question and can view a sample answer.

Explore Factoring the Sum of Two Squares (*continued*)

Questions

Have students complete the Explore activity.

Ask:

- Why is it helpful to factor a polynomial? Sample answer: Factoring a polynomial can help find the solution to an equation or the zeros of a function.
- How does factoring a prime polynomial like the sum of two squares relate to the phrase "no real solution"? Sample answer: In order to factor a prime polynomial, you have to use imaginary numbers. So, there are no "real" solutions because they are imaginary.

Inquiry

Can you factor a polynomial of the form $a^2 + b^2$? Sample answer: Yes, if you use an imaginary number i defined such that $i = \sqrt{-1}$.

Go Online to find additional teaching notes and sample answers to the guiding exercises.

Learn Pure Imaginary Numbers

Objective

Students perform operations with pure imaginary numbers.

 **Teaching the Mathematical Practices**

8 Look for a Pattern Help students to see the pattern in the powers of *i*. Ask them to look for a general rule for determining the values of powers of *i*.

About the Key Concept

Draw a distinction between the square root of a negative and a negative square root. Explore with students what happens when you square examples of both types, like $-\sqrt{5}$ and $\sqrt{-5}$, to help reinforce this difference.

DIFFERENTIATE

Language Development Activity AL BL ELL

IF students need help remembering the mathematical characteristics of *i*, **THEN** have students write poems about the imaginary number *i* and the repeating values of its powers, perhaps including wordplay with the terms *real* and *imaginary*. The content of the poems should be helpful for remembering the mathematical characteristics of *i*.

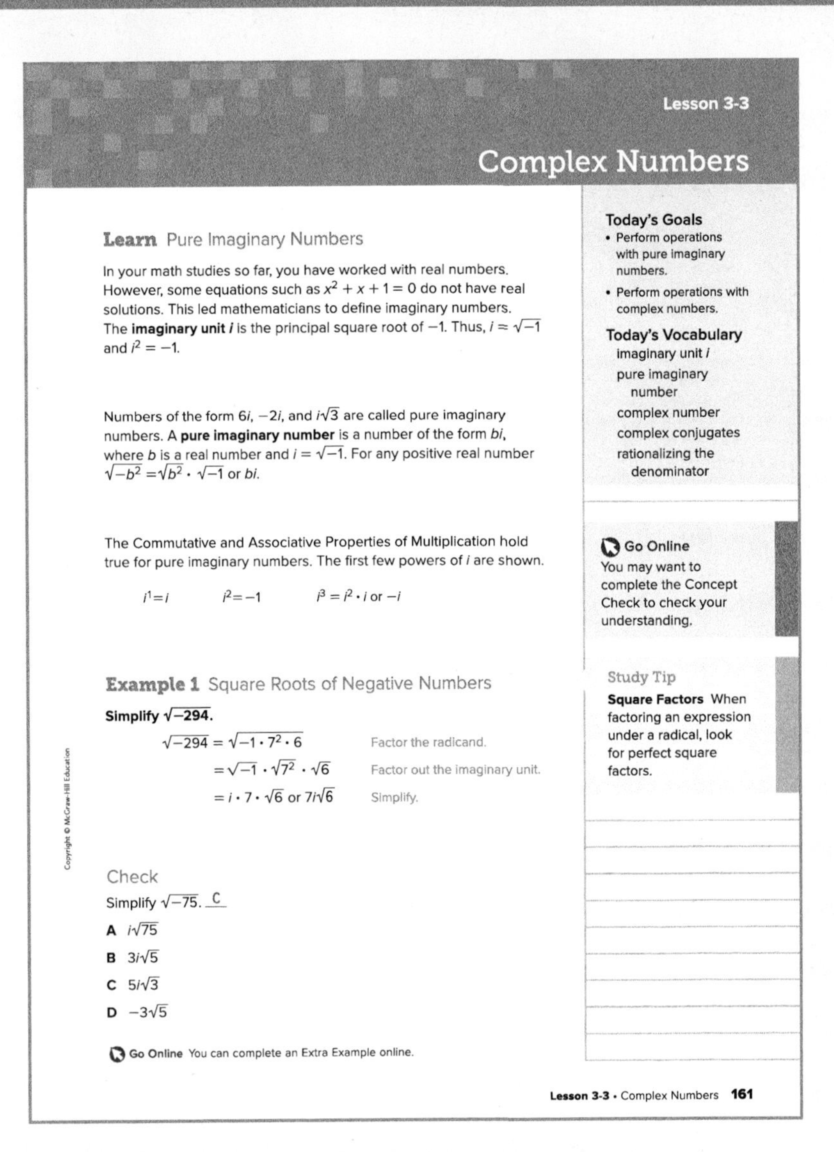

Lesson 3-3

Complex Numbers

Learn Pure Imaginary Numbers

In your math studies so far, you have worked with real numbers. However, some equations such as $x^2 + x + 1 = 0$ do not have real solutions. This led mathematicians to define imaginary numbers. The **imaginary unit** i is the principal square root of -1. Thus, $i = \sqrt{-1}$ and $i^2 = -1$.

Numbers of the form $6i$, $-2i$, and $i\sqrt{3}$ are called pure imaginary numbers. A **pure imaginary number** is a number of the form bi, where b is a real number and $i = \sqrt{-1}$. For any positive real number $\sqrt{-b^2} = \sqrt{b^2} \cdot \sqrt{-1}$ or bi.

The Commutative and Associative Properties of Multiplication hold true for pure imaginary numbers. The first few powers of i are shown.

$i^1 = i$ $\quad i^2 = -1$ $\quad i^3 = i^2 \cdot i$ or $-i$

Example 1 Square Roots of Negative Numbers

Simplify $\sqrt{-294}$.

$\sqrt{-294} = \sqrt{-1 \cdot 7^2 \cdot 6}$	Factor the radicand.
$= \sqrt{-1} \cdot \sqrt{7^2} \cdot \sqrt{6}$	Factor out the imaginary unit.
$= i \cdot 7 \cdot \sqrt{6}$ or $7i\sqrt{6}$	Simplify.

Check

Simplify $\sqrt{-75}$. C

A $i\sqrt{75}$

B $3i\sqrt{5}$

C $5i\sqrt{3}$

D $-3\sqrt{5}$

Go Online You can complete an Extra Example online.

Today's Goals
- Perform operations with pure imaginary numbers.
- Perform operations with complex numbers.

Today's Vocabulary
imaginary unit i
pure imaginary number
complex number
complex conjugates
rationalizing the denominator

Go Online You may want to complete the Concept Check to check your understanding.

Study Tip

Square Factors When factoring an expression under a radical, look for perfect square factors.

Lesson 3-3 • Complex Numbers 161

Interactive Presentation

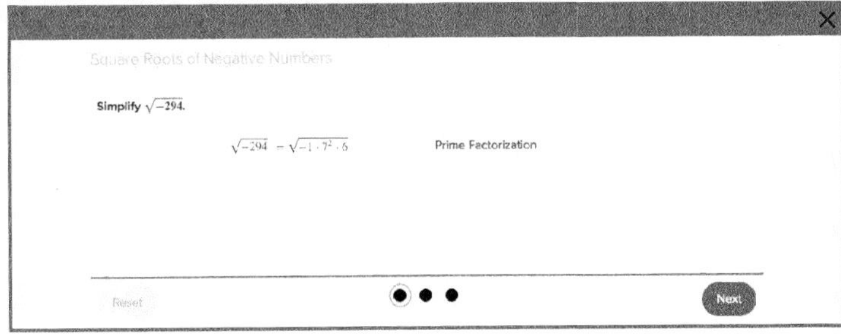

Example 1

Example 1 Square Roots of Negative Numbers

 **Teaching the Mathematical Practices**

7 Use Structure Help students to use the structure of the radicand to rewrite it so that it can be simplified using imaginary numbers.

Questions for Mathematical Discourse

AL Why is there not a negative sign under the radical in the answer? The $\sqrt{-1}$ has been replaced with *i*.

OL How can you tell when there will be an *i* in your answer just by looking at a square root? when the radicand is negative

BL Does $\sqrt{-1} \cdot \sqrt{-1} = \sqrt{((-1)(-1))} = \sqrt{(-1)^2}$? Explain. No; sample answer: $\sqrt{-1} \cdot \sqrt{-1} = (\sqrt{-1})^2 = -1$, and $\sqrt{(-1)(-1)} = \sqrt{(-1)^2} = \sqrt{1} = 1$.

Go Online

- Find additional teaching notes.
- View performance reports of the Checks.
- Assign or present an Extra Example.

Your Notes

Example 2 Products of Pure Imaginary Numbers

Simplify $\sqrt{-10} \cdot \sqrt{-15}$.

$\sqrt{-10} \cdot \sqrt{-15} = i\sqrt{10} \cdot i\sqrt{15}$ — $i = \sqrt{-1}$

$= i^2 \cdot \sqrt{150}$ — Multiply.

$= -1 \cdot \sqrt{25} \cdot \sqrt{6}$ — Simplify.

$= -5\sqrt{6}$ — Multiply.

Check

Simplify $\sqrt{-16} \cdot \sqrt{-25}$.

-20

Talk About It
How can an expression with two imaginary expressions, $\sqrt{-10}$ and $\sqrt{-15}$, have a product that is real?

Sample answer: Because $\sqrt{-10}$ and $\sqrt{-15}$ are both imaginary, you can factor out the imaginary unit i from each. The product of $i \cdot i = i^2 = (\sqrt{-1})^2 = -1$. So, the product of two imaginary numbers is a negative real number.

Example 3 Equation with Pure Imaginary Solutions

Solve $x^2 + 81 = 0$.

$x^2 + 81 = 0$ — Original equation

$x^2 = -81$ — Subtract 81 from each side.

$x = \pm\sqrt{-81}$ — Take the square root of each side.

$x = \pm 9i$ — Simplify.

Check

Solve $3x^2 + 27 = 0$.

$x = 3i$, $x = -3i$

Explore Factoring the Sum of Two Squares

Online Activity Use guiding exercises to complete the Explore.

INQUIRY Can you factor a polynomial of the form $a^2 + b^2$?

Learn Complex Numbers

Key Concept • Complex Numbers

A **complex number** is any number that can be written in the form $a + bi$, where a and b are real numbers and i is the imaginary unit. a is called the real part, and bi is called the imaginary part.

Go Online You can complete an Extra Example online.

162 Module 3 • Quadratic Functions

Interactive Presentation

Products of Pure Imaginary Numbers

Simplify $\sqrt{-10} \cdot \sqrt{-15}$.

$\sqrt{-10} \cdot \sqrt{-15} = i\sqrt{10} \cdot i\sqrt{15}$ — $i = \sqrt{-1}$

Example 2

SWIPE

Students move through the steps to see how to multiply two imaginary numbers.

TYPE

Students answer a question about the product of two imaginary numbers.

CHECK

Students complete the Check online to determine whether they are ready to move on.

Example 2 Products of Pure Imaginary Numbers

Teaching the Mathematical Practices

3 Construct Arguments In the Talk About It! feature, students will use definitions and previously established results to construct an argument.

Questions for Mathematical Discourse

AL What is the prime factorization of 150? $5 \cdot 5 \cdot 2 \cdot 3$

OL List the possible values for all powers of i when simplified. $i, -1, -i, 1$

BL Determine a rule you can use to simplify all powers of i? Sample answer: After you divide the exponent by 4, if the remainder is 0 it is equal to 1, if the remainder is 1 it is equal to i, if the remainder is 2 it is equal to -1, and if the remainder is 3 it is equal to $-i$.

Example 3 Equation with Pure Imaginary Solutions

Teaching the Mathematical Practices

2 Attend to Quantities Remind students a quadratic equation has two solutions, which is indicated by the $\pm$ symbol.

Questions for Mathematical Discourse

AL How can you check your solution? Sample answer: Substitute $-9i$ and then $9i$ into the original equation and check that the equation is true.

OL How are the roots of $x^2 - 81$ and $x^2 + 81$ different? $x^2 - 81$ has two real roots and $x^2 + 81$ has two imaginary roots.

BL Would the graph of $f(x) = x^2 + 81$ ever cross the x-axis? How do you know? No; sample answer: Because the equation $x^2 + 81 = 0$ does not have any real solutions, the graph of the related function does not have any x-intercepts.

Common Error

Make sure students understand that when they take the square root of each side of an equation, they must use the $\pm$ symbol in front of the radical sign.

1 CONCEPTUAL UNDERSTANDING | 2 FLUENCY | 3 APPLICATION

Learn Complex Numbers

Objective

Students perform operations with complex numbers.

Teaching the Mathematical Practices

3 Construct Arguments In the Think About It! feature, students will use the Venn diagram to construct arguments about the complex number system.

About the Key Concept

Be sure that students recognize that complex conjugates are not inverses of each other, but must have the same real components. Confirm that the product of complex conjugates is always a real number by working through examples. Keep in mind that these problems parallel the difference of squares that will be introduced in Lesson 3-4.

Essential Question Follow-Up

Students have explored complex numbers.

Ask:

How do complex numbers relate to solving quadratic equations? Sample answer: When the graph of the related quadratic function does not cross the x-axis, the solutions to a quadratic equation are complex numbers.

Example 4 Equate Complex Numbers

Teaching the Mathematical Practices

7 Interpret Complicated Expressions Guide students to see the equation as being composed of real and imaginary parts.

Questions for Mathematical Discourse

AL Why is $5x - 7$ set equal to 13? They are real parts of each side of the equation.

OL Why do you relate the real and imaginary parts separately? Sample answer: Because two complex numbers are equal if and only if their real parts are equal and their imaginary parts are equal, you need to find the value that makes the real parts of each side equal and the value that makes the imaginary parts of each side equal.

BL Could you have an x-term in both the real and imaginary parts of the equation? Yes. How would this affect the solution? The x-value of the solution would have to satisfy both parts to be valid.

Common Error

Emphasize that two complex numbers are equal if and only if their real parts are equal and their imaginary parts are equal.

The Venn diagram shows the set of complex numbers. Notice that all of the real numbers are part of the set of complex numbers.

Complex Numbers ($a + bi$)

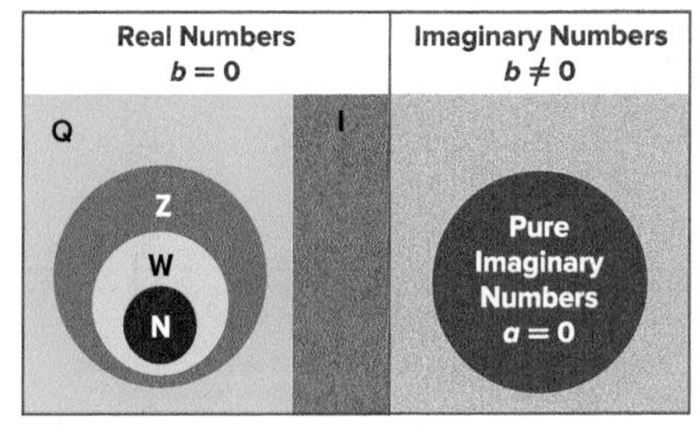

Study Tip

These abbreviations represent the sets of real numbers.

Letter	Set
Q	rationals
I	irrationals
Z	integers
W	wholes
N	naturals

Think About It!

Compare and contrast the subsets of the complex number system using the Venn diagram.

Sample answer: For $a + bi$, if $b = 0$ then the number is real and can be irrational, rational, an integer, whole, and/or a natural number. If $b \neq 0$ then the number is imaginary. It is pure imaginary if $a = 0$.

Two complex numbers are equal if and only if their real parts are equal and their imaginary parts are equal. The Commutative and Associative Properties of Multiplication and Addition and the Distributive Property hold true for complex numbers. To add or subtract complex numbers, combine like terms. That is, combine the real parts, and combine the imaginary parts.

Two complex numbers of the form $a + bi$ and $a - bi$ are called **complex conjugates**. The product of complex conjugates is always a real number.

A radical expression is in simplest form if no radicands contain fractions and no radicals appear in the denominator of a fraction. Similarly, a complex number is in simplest form if no imaginary numbers appear in the denominator of a fraction. You can use complex conjugates to simplify a fraction with a complex number in the denominator. This process is called **rationalizing the denominator**.

Example 4 Equate Complex Numbers

Find the values of x and y that make $5x - 7 + (y + 4)i = 13 + 11i$ true.

Use equations relating the real and imaginary parts to solve for x and y.

$5x - 7 = 13$ Real parts

$5x = 20$ Add 7 to each side.

$x = 4$ Divide each side by 5.

$y + 4 = 11$ Imaginary parts

$y = 7$ Subtract 4 from each side.

Check

Find the values of x and y that make $5x + 13 + (2y - 7)i = -2 + i$ true.

$x = -3$ $y = 4$

Go Online You can complete an Extra Example online.

Interactive Presentation

Complex Numbers

KEY CONCEPT: COMPLEX NUMBERS

Words — A complex number is any number that an be written in the form $a + bi$, where a and b are real numbers and i is the imaginary unit. a is called the real part, and b is called the imaginary part.

Learn

TYPE

Students answer a question about complex numbers.

Go Online You can watch a video to see how to add or subtract complex numbers.

Study Tip
Imaginary Unit Complex numbers are often used with electricity. In these problems, j is usually used in place of i.

Example 5 Add or Subtract Complex Numbers

Simplify $(8 + 3i) - (4 - 10i)$.

$(8 + 3i) - (4 - 10i) = (8 - \underline{4}) + [3 - (\underline{-10})]i$ Commutative and Associative Properties

$= \underline{4} + 13i$ Simplify.

Check

Simplify $(-5 + 5i) - (-3 + 8i)$.

$\underline{-2} + \underline{-3}i$

Example 6 Multiply Complex Numbers

ELECTRICITY **The voltage V of an AC circuit can be found using the formula $V = CI$, where C is current and I is impedance. If $C = 3 + 2j$ amps and $I = 7 - 5j$ ohms, determine the voltage.**

$V = CI$ Voltage Formula

$= (3 + 2j)(7 - 5j)$ $C = 3 + 2j$ and $I = 7 - 5j$

$= 3(\underline{7}) + 3(\underline{-5j}) + 2j(\underline{7}) + 2j(\underline{-5j})$ FOIL Method

$= \underline{21} - 15j + \underline{14j} - 10j^2$ Multiply.

$= 21 - \underline{j} - 10(\underline{-1})$ $j^2 = -1$

$= \underline{31} - j$ Add.

The voltage is $\underline{31 - j}$ volts.

Example 7 Divide Complex Numbers

Simplify $\frac{5i}{3 + 2i}$.

Rationalize the denominator to simplify the fraction.

$\frac{5i}{3+2i} = \frac{5i}{3+2i} \cdot \frac{3-2i}{3-2i}$ $3 + 2i$ and $3 - 2i$ are complex conjugates.

$= \frac{15i - 10i^2}{9 - 4i^2}$ Multiply the numerator and denominator.

$= \frac{15i - 10(-1)}{9 - 4(-1)}$ $i^2 = -1$

$= \frac{15i + 10}{13}$ Simplify.

$= \frac{10}{13} + \frac{15}{13}i$ $a + bi$ form

Check

Simplify $\frac{2i}{-4+3i}$. $\underline{\frac{6-8i}{25} \text{ or } \frac{6}{25} - \frac{8}{25}i}$

Go Online You can complete an Extra Example online.

164 Module 3 • Quadratic Functions

Interactive Presentation

Add or Subtract Complex Numbers

Simplify $(8 + 3i) - (4 - 10i)$.

$(8 + 3i) - (4 - 10i) = (8 - 4) + [3 - (-10)]i$ Commutative and Associative Properties

Example 5

SWIPE

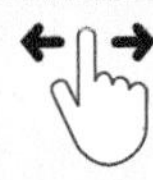

Students move through the steps to see how to subtract two complex numbers.

WATCH

Students watch a video to learn how to add and subtract complex numbers.

CHECK

Students complete the Check online to determine whether they are ready to move on.

Example 5 Add or Subtract Complex Numbers

Teaching the Mathematical Practices

1 Seek Information Mathematically proficient students must be able to transform algebraic expression to reach solutions. In Example 5, students must rearrange and combine terms to simplify the expression.

Questions for Mathematical Discourse

AL When adding or subtracting two complex numbers, how many operations will you need to perform? At least 2.

OL How would removing the parentheses change the expression? Sample answer: Removing the first set does not change the expression. Removing the second set would make the imaginary term $-7i$.

BL Can adding or subtracting complex numbers result in a real number? Explain. Yes, when the imaginary parts cancel.

Example 6 Multiply Complex Numbers

Teaching the Mathematical Practices

4 Apply Mathematics In Example 6, students apply what they have learned about complex numbers to solving a real-world problem.

Questions for Mathematical Discourse

AL Why do you use the FOIL method to multiply C and I? Each expression has two terms.

OL What is the product of a non-zero real number and an imaginary number? An imaginary number

BL Why can you combine real and imaginary parts with multiplication but not with addition? Sample answer: A real term plus an imaginary term is already in its simplest form, but the like factors of real and imaginary terms can be simplified when multiplying. (Similarly, $7 + 3x$ is in its simplest form, but $7(3x)$ can be simplified to $21x$.)

Example 7 Divide Complex Numbers

Teaching the Mathematical Practices

7 Use Structure Help students to use the structure of the expression to rationalize the denominator and simplify the expression.

Questions for Mathematical Discourse

AL What is the complex conjugate of $a + bi$? $a - bi$

OL How do you rewrite $\frac{a+bi}{c}$ in the form $a + bi$? $\frac{a}{c} + \frac{b}{c}i$

BL Why is the product of two complex conjugates a real number? Sample answer: When you distribute the multiplication by using the FOIL method, the outer and inner terms with i cancel one another.

Exit Ticket

Recommended Use

At the end of class, go online to display the Exit Ticket prompt and ask students to respond using a separate piece of paper. Have students hand you their responses as they leave the room.

Alternate Use

At the end of class, go online to display the Exit Ticket prompt and ask students to respond verbally or by using a mini-whiteboard. Have students hold up their whiteboards so that you can see all student responses. Tap to reveal the answer when most or all students have completed the Exit Ticket.

Practice and Homework

Suggested Assignments

Use the table below to select appropriate exercises.

DOK	Topic	Exercises
1, 2	exercises that mirror the examples	1–37
2	exercises that use a variety of skills from this lesson	38–42
2	exercises that extend concepts learned in this lesson to new contexts	43–46
3	exercises that emphasize higher-order and critical thinking skills	47–51

ASSESS AND DIFFERENTIATE

Use the data from the **Checks** to determine whether to provide resources for extension, remediation, or intervention.

IF students score 90% or more on the Checks, **THEN** assign: BL

- Practice Exercises 1–45 odd, 47–51
- Extension: Conjugates and Absolute Value
- **ALEKS**® Complex Numbers and Complex Zeros of Polynomial Functions

IF students score 66%–89% on the Checks, **THEN** assign: OL

- Practice Exercises 1–51 odd
- Remediation, Review Resources: Properties of Real Numbers
- Personal Tutors
- Extra Examples 1–7
- **ALEKS**® Properties of Real Numbers

IF students score 65% or less on the Checks, **THEN** assign: AL

- Practice Exercises 1–37 odd
- Remediation, Review Resources: Properties of Real Numbers
- *Quick Review Math Handbook*: Complex Numbers
- **ALEKS**® Properties of Real Numbers

Name ______ Period ______ Date ______

Practice

Go Online You can complete your homework online.

Examples 1 and 2

Simplify.

1. $\sqrt{-48}$ $4i\sqrt{3}$
2. $\sqrt{-63}$ $3i\sqrt{7}$
3. $\sqrt{-72}$ $6i\sqrt{2}$
4. $\sqrt{-24}$ $2i\sqrt{6}$
5. $\sqrt{-84}$ $2i\sqrt{21}$
6. $\sqrt{-99}$ $3i\sqrt{11}$
7. $\sqrt{-23} \cdot \sqrt{-46}$ $-23\sqrt{2}$
8. $\sqrt{-6} \cdot \sqrt{-3}$ $-3\sqrt{2}$
9. $\sqrt{-5} \cdot \sqrt{-10}$ $-5\sqrt{2}$
10. $(3i)(-2i)(5i)$ $30i$
11. i^{11} $-i$
12. $4i(-6i)^2$ $-144i$

Example 3

Solve each equation.

13. $5x^2 + 45 = 0$ $\pm 3i$
14. $4x^2 + 24 = 0$ $\pm i\sqrt{6}$
15. $-9x^2 = 9$ $\pm i$
16. $7x^2 + 84 = 0$ $\pm 2i\sqrt{3}$
17. $5x^2 + 125 = 0$ $\pm 5i$
18. $8x^2 + 96 = 0$ $\pm 2i\sqrt{3}$

Example 4

Find the values of *x* and *y* that make each equation true.

19. $9 + 12i = 3x + 4yi$ 3, 3
20. $x + 1 + 2yi = 3 - 6i$ 2, −3
21. $2x + 7 + (3 - y)i = -4 + 6i$ $-\frac{11}{2}, -3$
22. $5 + y + (3x - 7)i = 9 - 3i$ $\frac{4}{3}, 4$
23. $20 - 12i = 5x + (4y)i$ 4, −3
24. $x - 16i = 3 - (2y)i$ 3, 8

Examples 5 and 6

Simplify.

25. $(6 + i) + (4 - 5i)$ $10 - 4i$
26. $(8 + 3i) - (6 - 2i)$ $2 + 5i$
27. $(5 - i) - (3 - 2i)$ $2 + i$
28. $(-4 + 2i) + (6 - 3i)$ $2 - i$
29. $(6 - 3i) + (4 - 2i)$ $10 - 5i$
30. $(-11 + 4i) - (1 - 5i)$ $-12 + 9i$
31. $(2 + i)(3 - i)$ $7 + i$
32. $(5 - 2i)(4 - i)$ $18 - 13i$
33. $(4 - 2i)(1 - 2i)$ $-10i$
34. ELECTRICITY Using the formula $V = CI$, find the voltage V in a circuit when the current $C = 3 - j$ amps and the impedance $I = 3 + 2j$ ohms. $11 + 3j$ volts

Example 7

Simplify.

35. $\frac{5}{3+i}$ $\frac{3}{2} - \frac{1}{2}i$
36. $\frac{7-13i}{2i}$ $-\frac{13}{2} - \frac{7}{2}i$
37. $\frac{6-5i}{3i}$ $-\frac{5}{3} - 2i$

Mixed Exercises

STRUCTURE **Simplify.**

38. $(1 + i)(2 + 3i)(4 - 3i)$ $11 + 23i$

39. $\frac{4 - i\sqrt{2}}{4 + i\sqrt{2}}$ $\frac{7}{9} - \frac{4i\sqrt{2}}{9}$

40. $\frac{2 - i\sqrt{3}}{2 + i\sqrt{3}}$ $\frac{1}{7} - \frac{4\sqrt{3}}{7}i$

41. Find the sum of $ix^2 - (4 + 5i)x + 7$ and $3x^2 + (2 + 6i)x - 8i$. $(3 + i)x^2 + (-2 + i)x - 8i + 7$

42. Simplify $[(2 + i)x^2 - ix + 5 + i] - [(-3 + 4i)x^2 + (5 - 5i)x - 6]$. $(5 - 3i)x^2 + (-5 + 4i)x + i + 11$

ELECTRICITY **Use the formula $V = CI$, where V is the voltage, C is the current, and I is the impedance.**

43. The current in a circuit is $2 + 4j$ amps, and the impedance is $3 - j$ ohms. What is the voltage? $10 + 10j$ volts

44. The voltage in a circuit is $24 - 8j$ volts, and the impedance is $4 - 2j$ ohms. What is the current? $\frac{28}{5} + \frac{4j}{5}$

45. CIRCUITS The impedance in one part of a series circuit is $1 + 3j$ ohms and the impedance in another part of the circuit is $7 - 5j$ ohms. Add these complex numbers to find the total impedance in the circuit. $8 - 2j$ ohms

46. ELECTRICAL ENGINEERING The standard electrical voltage in Europe is 220 volts.

a. Find the impedance in a standard European circuit if the current is $22 - 11i$ amps. $8 + 4i$ ohms

b. Find the current in a standard European circuit if the impedance is $10 - 5i$ ohms. $17.6 + 8.8i$ amps

c. Find the impedance in a standard European circuit if the current is $20i$ amps. $-11i$ ohms

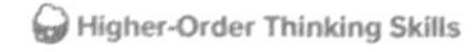

Higher-Order Thinking Skills

47. FIND THE ERROR Jose and Zoe are simplifying $(2i)(3i)(4i)$. Is either of them correct? Explain your reasoning. Zoe; $i^3 = -i$, not -1.

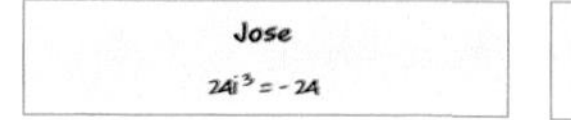

Zoe

$24i^3 = -24i$

48. PERSEVERE Simplify $(1 + 2i)^3$. $-11 - 2i$

49. ANALYZE Determine whether the following statement is *always*, *sometimes*, or *never* true. Explain your reasoning. See margin.

Every complex number has both a real part and an imaginary part.

50. CREATE Write two complex numbers with a product of 20. Sample answer: $(4 + 2i)(4 - 2i)$

51. WRITE Explain how complex numbers are related to quadratic equations. See margin.

Answers

49. Always; sample answer: The value of 5 can be represented by $5 + 0i$, and the value of $3i$ can be represented by $0 + 3i$.

51. Some quadratic equations have complex solutions and cannot be solved using only the real numbers.

N.CN.7, N.CN.8, F.IF.8a

Solving Quadratic Equations by Factoring

LESSON GOAL

Students solve quadratic equations by factoring.

1 LAUNCH

Launch the lesson with a **Warm Up** and an introduction.

2 EXPLORE AND DEVELOP

Explore: Finding the Solutions of Quadratic Equations by Factoring

Develop:

Solving Quadratic Equations by Factoring
- Factor by Using the Distributive Property
- Factor a Trinomial
- Solve an Equation by Factoring
- Factor a Trinomial Where *a* is Not 1

Solving Quadratic Equations by Factoring Special Products
- Factor a Difference of Squares
- Factor a Perfect Square Trinomial
- Complex Solutions

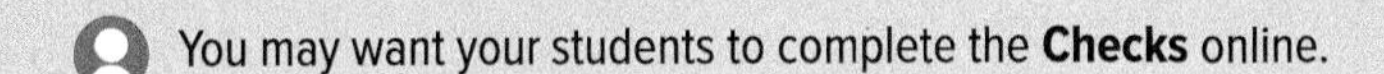

You may want your students to complete the **Checks** online.

3 REFLECT AND PRACTICE

Exit Ticket

Practice

Formative Assessment Math Probe

DIFFERENTIATE

View reports of student progress on the **Checks** after each example.

Resources	AL	OL	BL	ELL
Remediation: Using the Distributive Property	●	●		●
Extension: Using Identities to Factor		●	●	●

Language Development Handbook

Assign page 20 of the *Language Development Handbook* to help your students build mathematical language related to solving quadratic equations by factoring.

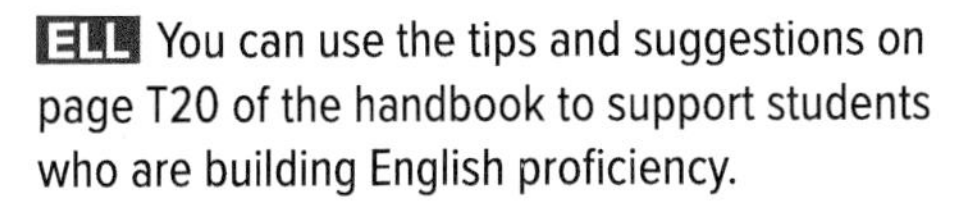

ELL You can use the tips and suggestions on page T20 of the handbook to support students who are building English proficiency.

Suggested Pacing

90 min	1 day
45 min	2 days

Focus

Domain: Number and Quantity, Functions

Standards for Mathematical Content:

N.CN.7 Solve quadratic equations with real coefficients that have complex solutions.

N.CN.8 Extend polynomial identities to the complex numbers.

F.IF.8a Use the properties of exponents to interpret expressions for exponential functions.

Standards for Mathematical Practice:

1 Make sense of problems and persevere in solving them.

7 Look for and make use of structure.

Coherence

Vertical Alignment

Previous
Students performed operations with pure imaginary and complex numbers.
N.CN.1, N.CN.2

Now
Students solve quadratic equations by factoring.
N.CN.7, N.CN.8, F.IF.8a

Next
Students will simplify quadratic equations by using the Square Root Property and completing the square.
N.CN.7, F.IF.8a

Rigor

The Three Pillars of Rigor

1 CONCEPTUAL UNDERSTANDING	2 FLUENCY	3 APPLICATION

Conceptual Bridge In this lesson, students draw on their understanding of factoring to solve quadratic equations. They build fluency by factoring to find solutions, and they apply their understanding of factoring by solving real-world problems.

Interactive Presentation

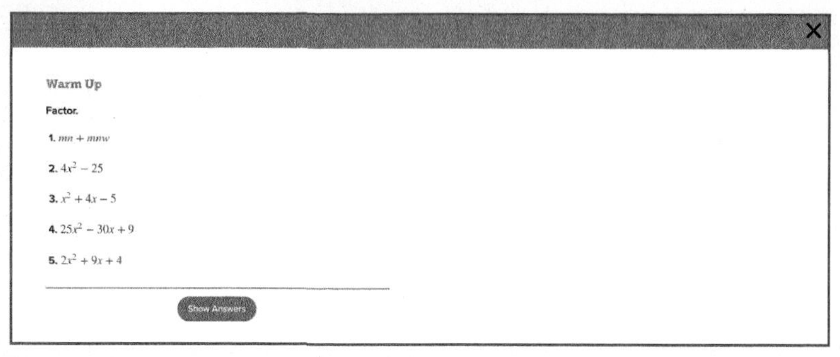

Warm Up

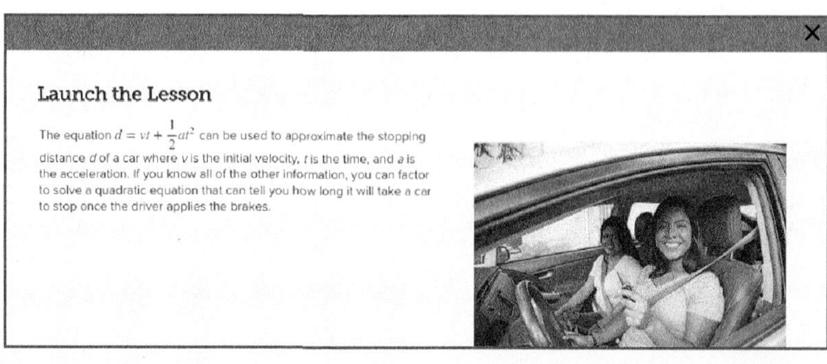

Launch the Lesson

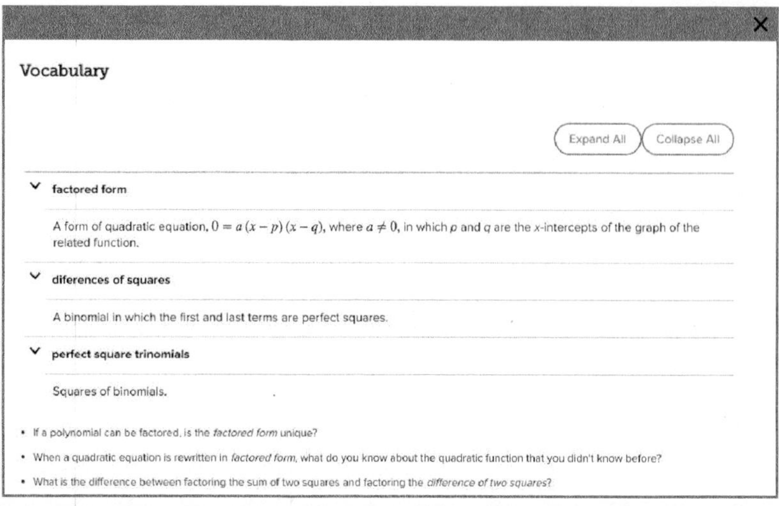

Today's Vocabulary

Warm Up

Prerequisite Skills

The Warm Up exercises address the following prerequisite skill for this lesson:

- factoring

Answers:

1. $mn(1 + w)$
2. $(2x + 5)(2x - 5)$
3. $(x + 5)(x - 1)$
4. $(5x - 3)^2$
5. $(2x + 1)(x + 4)$

Launch the Lesson

Teaching the Mathematical Practices

2 Attend to Quantities Point out that it is important to note the meaning of the variables used in the equation for the stopping distance of a car.

Go Online to find additional teaching notes and questions to promote classroom discourse.

Today's Standards

Tell students that they will be addressing these content and practice standards in this lesson. You may wish to have a student volunteer read aloud *How can I meet these standards?* and *How can I use these practices?,* and connect these to the standards.

See the Interactive Presentation for I Can statements that align with the standards covered in this lesson.

Today's Vocabulary

Tell students that they will be using these vocabulary terms in this lesson. You can expand each row if you wish to share the definitions. Then discuss the questions below with the class.

Mathematical Background

Quadratic equations can be solved using several different methods. Factoring can be a quick method. Once a polynomial has been factored, the Zero Product Property may be used to find the roots of the equation. If the polynomial is difficult to factor or is not factorable, then other methods must be used.

Explore Finding the Solutions of Quadratic Equations by Factoring

Objective

Students use a sketch to explore solving quadratic equations by factoring.

 Teaching the Mathematical Practices

1 Explain Correspondences Encourage students to explain the relationships between the factors and solutions of quadratic equations and the graphs of the related quadratic functions.

Ideas for Use

Recommended Use Present the Inquiry Question, or have a student volunteer read it aloud. Have students work in pairs to complete the Explore activity on their devices. Pairs should discuss each of the questions. Monitor student progress during the activity. Upon completion of the Explore activity, have student volunteers share their responses to the Inquiry Question.

What if my students don't have devices? You may choose to project the activity on a whiteboard. A printable worksheet for each Explore is available online. You may choose to print the worksheet so that individuals or pairs of students can use it to record their observations.

Summary of the Activity

Students will complete guiding exercises throughout the Explore activity. Students review a variety of equations and graphs, connecting the solutions to the factors of the quadratic equation. Then, students will answer the Inquiry Question.

(continued on the next page)

Interactive Presentation

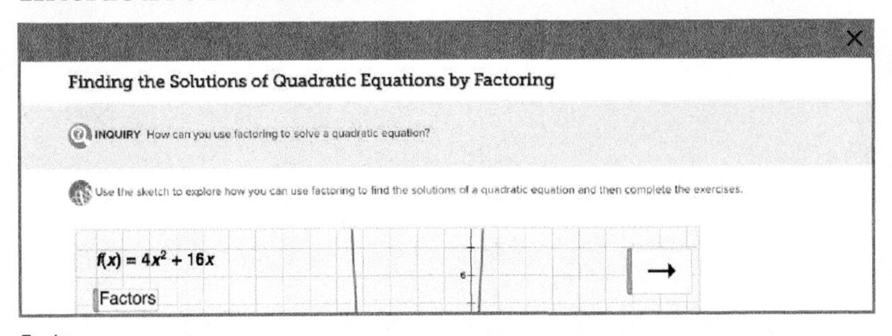

Explore

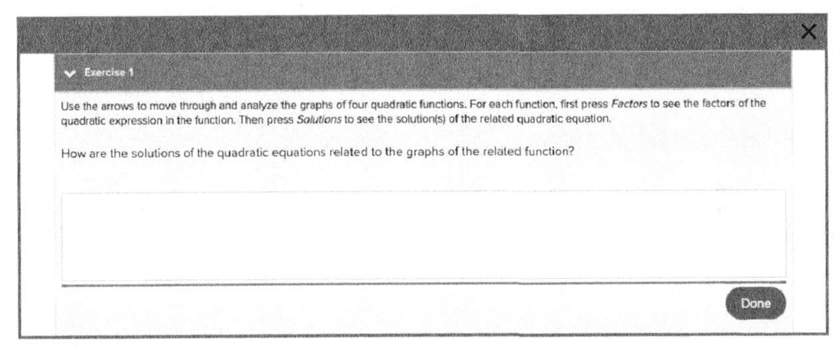

Explore

WEB SKETCHPAD

Students will use a sketch to explore how to solve quadratic equations by factoring.

TYPE

Students will answer questions about the solutions and factors of quadratic equations.

Interactive Presentation

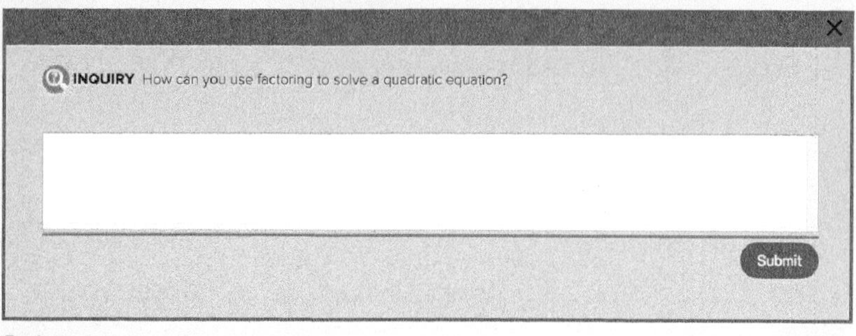

Explore

TYPE

Students respond to the Inquiry Question and can view a sample answer.

Explore Finding the Solutions of Quadratic Equations by Factoring (*continued*)

Questions

Have students complete the Explore activity.

Ask:

- Which function is the difference of two squares? Sample answer: $h(x)$ is the difference of two squares, 49 and x^2.
- When might you want to solve by factoring instead of by graphing? Sample answer: When I am looking for an exact answer rather than an estimate.

Inquiry

How can you use factoring to solve a quadratic equation? Sample answer: Rewrite the equation so that one side is 0, factor the expression in the equation, set each factor equal to 0, and solve.

Go Online to find additional teaching notes and sample answers for the guiding exercises.

Learn Solving Quadratic Equations by Factoring

Objective

Students solve quadratic equations by factoring.

Teaching the Mathematical Practices

7 Use the Distributive Property Point out that the Distributive Property is one of the most-used properties in algebra. Students should know that whenever they see a number outside of a sum or difference within parentheses, they should apply the Distributive Property.

Example 1 Factor by Using the Distributive Property

Teaching the Mathematical Practices

1 Seek Information Mathematically proficient students must be able to transform algebraic expressions to reach solutions. In Example 1, students must transform the quadratic equation to reach a solution.

Questions for Mathematical Discourse

AL What is a greatest common factor? the greatest single factor that can be factored out of multiple terms

OL How is the Zero Product Property used to identify the solutions to this equation? Sample answer: Because the equation can be rewritten as $3x(4x - 1) = 0$, the Zero Product Property means the solutions come from $3x = 0$ and $4x - 1 = 0$.

BL Can you tell one of the roots just by looking at the original equation? Explain. Yes; sample answer: Zero is a root because all the terms have a factor of x.

DIFFERENTIATE

Language Development Activity ELL

Beginning Ask questions about the lesson content to elicit yes/no answers: "Consider the trinomial $x^2 - 2x - 3$ and its factored form $(x - 3)(x + 1)$ given in the Learn section. Is x^2 the product of the first terms of the two binomials?" yes

Intermediate/Advanced Ask questions about the lesson content to elicit short answers: "Consider the trinomial $x^2 - 2x - 3$ and its factored form $(x - 3)(x + 1)$ given in the Learn section. What is the product of the inner terms of the two binomials?" $(-3)(x)$ "What two terms can be combined in the product?" x and $-3x$

Advanced High Ask questions about lesson content to elicit complete sentences: "What is the Zero Product Property?" If the product of two numbers is 0, one or both of the numbers is 0. "How does the Zero Product Property help you solve equations?" If you know that the product of two expressions is 0, one or both expressions equals 0. So finding the values where either expression is 0 solves the equation.

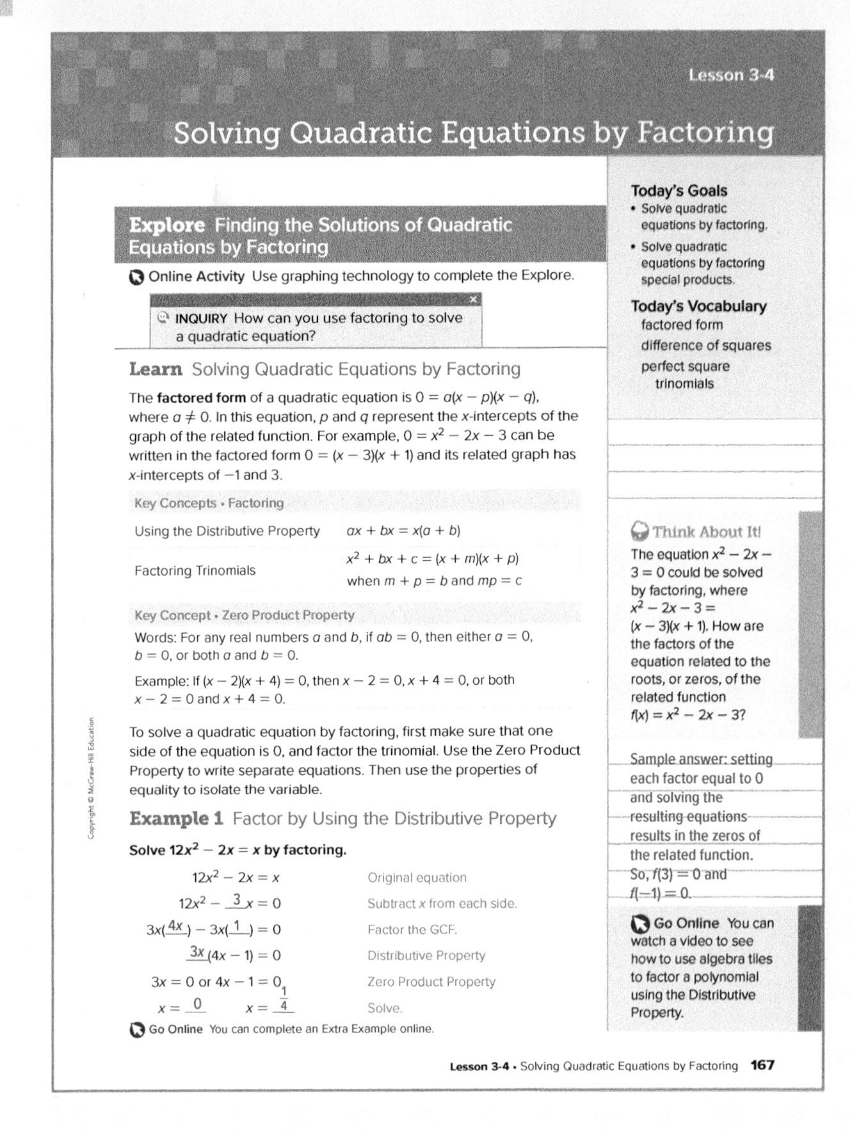

Lesson 3-4

Solving Quadratic Equations by Factoring

Today's Goals
- Solve quadratic equations by factoring.
- Solve quadratic equations by factoring special products.

Today's Vocabulary
factored form
difference of squares
perfect square trinomials

Explore Finding the Solutions of Quadratic Equations by Factoring

Online Activity Use graphing technology to complete the Explore.

INQUIRY How can you use factoring to solve a quadratic equation?

Learn Solving Quadratic Equations by Factoring

The **factored form** of a quadratic equation is $0 = a(x - p)(x - q)$, where $a \neq 0$. In this equation, p and q represent the x-intercepts of the graph of the related function. For example, $0 = x^2 - 2x - 3$ can be written in the factored form $0 = (x - 3)(x + 1)$ and its related graph has x-intercepts of -1 and 3.

Key Concepts • Factoring

Using the Distributive Property	$ax + bx = x(a + b)$
Factoring Trinomials	$x^2 + bx + c = (x + m)(x + p)$ when $m + p = b$ and $mp = c$

Key Concept • Zero Product Property

Words: For any real numbers a and b, if $ab = 0$, then either $a = 0$, $b = 0$, or both a and $b = 0$.

Example: If $(x - 2)(x + 4) = 0$, then $x - 2 = 0$, $x + 4 = 0$, or both $x - 2 = 0$ and $x + 4 = 0$.

Think About It! The equation $x^2 - 2x - 3 = 0$ could be solved by factoring, where $x^2 - 2x - 3 = (x - 3)(x + 1)$. How are the factors of the equation related to the roots, or zeros, of the related function $f(x) = x^2 - 2x - 3$?

Sample answer: setting each factor equal to 0 and solving the resulting equations results in the zeros of the related function. So, $f(3) = 0$ and $f(-1) = 0$.

To solve a quadratic equation by factoring, first make sure that one side of the equation is 0, and factor the trinomial. Use the Zero Product Property to write separate equations. Then use the properties of equality to isolate the variable.

Example 1 Factor by Using the Distributive Property

Solve $12x^2 - 2x = x$ by factoring.

$12x^2 - 2x = x$	Original equation
$12x^2 - \underline{3}x = 0$	Subtract x from each side.
$3x(\underline{4x}) - 3x(\underline{1}) = 0$	Factor the GCF.
$\underline{3x}(4x - 1) = 0$	Distributive Property
$3x = 0$ or $4x - 1 = 0$	Zero Product Property
$x = \underline{0}$ $\quad x = \underline{\frac{1}{4}}$	Solve.

Go Online You can complete an Extra Example online.

Go Online You can watch a video to see how to use algebra tiles to factor a polynomial using the Distributive Property.

Copyright © McGraw-Hill Education

Lesson 3-4 • Solving Quadratic Equations by Factoring 167

Interactive Presentation

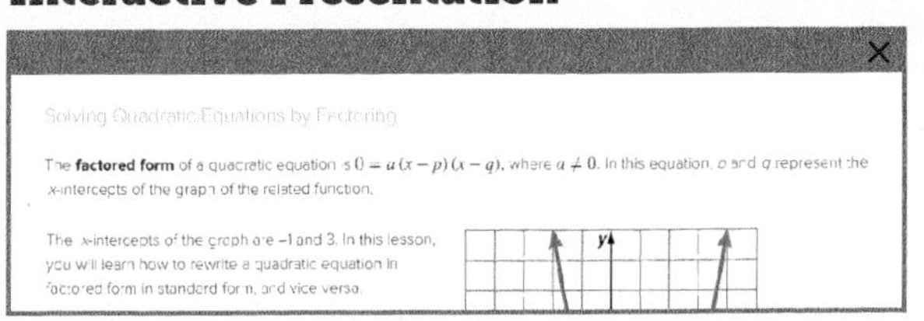

Learn

TAP

Students tap to see the Study Tip.

Your Notes

Think About It!
Choose two integers and write an equation in standard form with these roots. How would the equation change if the signs of the two roots were switched?

Sample answer: 2 and 5; $x^2 - 7x + 10 = 0$; -2 and -5; $x^2 + 7x + 10 = 0$; the linear term is the only term that changes signs.

Example 2 Factor a Trinomial

Solve $x^2 - 6x - 9 = 18$ by factoring. Check your solution.

$x^2 - 6x - 9 = 18$	Original equation
$x^2 - 6x - \underline{27} = 0$	Subtract 18 from each side.
$(x + \underline{3})(x - \underline{9}) = 0$	Factor the trinomial.
$x + 3 = 0$ or $x - 9 = 0$	Zero Product Property
$x = \underline{-3}$ $x = \underline{9}$	Solve.

Apply Example 3 Solve an Equation by Factoring

ACCELERATION **The equation $d = vt + \frac{1}{2}at^2$ represents the displacement d of a car traveling at an initial velocity v where the acceleration a is constant over a given time t. Find how long it takes a car to accelerate from 30 mph to 45 mph if the car moved 605 feet and accelerated slowly at a rate of 2 feet per second squared.**

1 What is the task?

Describe the task in your own words. Then list any questions that you may have. How can you find answers to your questions?
Sample answer: Solve the equation to find the time for the car to accelerate. The acceleration is given in feet per second squared and the velocity is given in miles per hour. How do I address the difference in units?

2 How will you approach the task? What have you learned that you can use to help you complete the task?
Sample answer: Convert the velocity to feet per second. Then substitute the distance, velocity, and acceleration into the formula and solve for time.

3 What is your solution?

Use your strategy to solve the problem.

What is the velocity in feet per second? 44 fps

How long does it take the car to accelerate from 30 mph to 45 mph? 11 s

4 How can you know that your solution is reasonable?

Write About It! Write an argument that can be used to defend your solution.
Sample answer: The solutions of the equation are -55 and 11. Because time cannot be negative, $t = 11$ is the only viable solution in the context of the situation.

Think About It!
Why did you not use 45 mph to solve this problem?

Sample answer: The equation uses the initial velocity and 45 mph is the final velocity.

Go Online You can complete an Extra Example online.

168 Module 3 • Quadratic Functions

Interactive Presentation

Solve Equations by Factoring

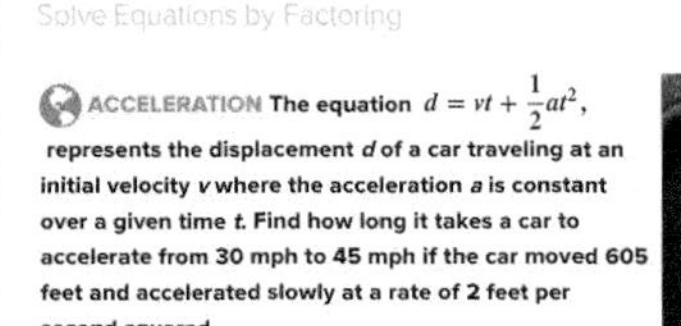

Apply Example 3

DRAG & DROP

Students drag to sort the steps needed to solve the problem.

Example 2 Factor a Trinomial

Teaching the Mathematical Practices

7 Use the Distributive Property In Example 2, students must use the Distributive Property in reverse to factor the trinomial and solve the equation.

Questions for Mathematical Discourse

AL What is the first step before factoring any quadratic equation? Solve the equation so that one side is 0.

OL How do you find the factors of the trinomial? Sample answer: Look for two numbers that have a sum of -6 and a product of -27.

BL What would you change in the original equation so that the solution were 3 and -9? Change $-6x$ to $6x$.

Apply Example 3 Solve an Equation by Factoring

Teaching the Mathematical Practices

1 Make Sense of Problems and Persevere in Solving Them, 4 Model with Mathematics Students will be presented with a task. They will first seek to understand the task, and then determine possible entry points to solving it. As students come up with their own strategies, they may propose mathematical models to aid them. As they work to solve the problem, encourage them to evaluate their model and/or progress, and change direction, if necessary.

Recommended Use

Have students work in pairs or small groups. You may wish to present the task, or have a volunteer read it aloud. Then allow students the time to make sure they understand the task, think of possible strategies, and work to solve the problem.

Encourage Productive Struggle

As students work, monitor their progress. Instead of instructing them on a particular strategy, encourage them to use their own strategies to solve the problem and to evaluate their progress along the way. They may or may not find that they need to change direction or try out several strategies.

Signs of Non-Productive Struggle

If students show signs of non-productive struggle, such as feeling overwhelmed, frustrated, or disengaged, intervene to encourage them to think of alternate approaches to the problem. Some sample questions are shown.

- The car's speed is measured in mph and the distance is measured in feet. What first step must you take to solve the exercise?
- What type of equation is obtained and how can this type of equation be solved?

Write About It!

Have students share their responses with another pair/group of students or the entire class. Have them clearly state or describe the mathematical reasoning they can use to defend their solution.

Example 4 Factor a Trinomial Where a Is Not 1

Teaching the Mathematical Practices

6 Communicate Precisely Encourage students to routinely write or explain their solution methods. The Think About It! feature asks students to explain how they determined the values of m and p when factoring the equation.

Questions for Mathematical Discourse

AL How can you check that your factored expression is correct? Sample answer: Use the FOIL method or the Distributive Property to check that the product of the factors is equal to the original quadratic expression.

OL What must be true about both b and c for $\frac{b}{a}$ and $\frac{c}{a}$ to both be integers? Both b and c would have to be multiples of a.

BL If you first divided the equation by 3, would you find the same solutions? Explain. Yes; sample answer: The solutions would be the same, but the factoring process may be more complicated.

Learn Factoring Special Products

Objective

Students solve quadratic equations by factoring special products.

Teaching the Mathematical Practices

7 Look for a Pattern Help students to see how to apply the patterns for factoring special products.

Go Online

- Find additional teaching notes.
- View performance reports of the Checks.
- Assign or present an Extra Example.

Check

SALES A clothing store is analyzing their market to determine the profitability of their new dress design. If $P(x) = -16x^2 + 1712x - 44{,}640$ represents the store's profit when x is the price of each dress, find the prices at which the store makes no profit on the design. B

A. \$11.25 and \$15.50

B. \$45 and \$62

C. \$50 and \$54

D. \$180 and \$248

Example 4 Factor a Trinomial Where a is Not 1

Solve $3x^2 + 5x + 15 = 17$ by factoring. Check your solution.

$3x^2 + 5x + 15 = 17$	Original equation
$3x^2 + 5x - 2 = 0$	Subtract 17 from each side.
$(3x - 1)(x + 2) = 0$	Factor the trinomial.
$3x - 1 = 0$ or $x + 2 = 0$	Zero Product Property
$x = \frac{1}{3}$ $\quad x = -2$	Solve.

Check

Solve $4x^2 + 12x - 27 = 13$ by factoring. Check your solution.

$x =$ -5, 2

Learn Solving Quadratic Equations by Factoring Special Products

Key Concept • Factoring Differences of Squares

Words: To factor $a^2 - b^2$, find the square roots of a^2 and b^2. Then apply the pattern.

Symbols: $a^2 - b^2 = (a + b)(a - b)$

Key Concept • Factoring Perfect Square Trinomials

Words: To factor $a^2 + 2ab + b^2$, find the square roots of a^2 and b^2. Then apply the pattern.

Symbols: $a^2 + 2ab + b^2 = (a + b)^2$

Not all quadratic equations have solutions that are real numbers. In some cases, the solutions are complex numbers of the form $a + bi$, where $b \neq 0$. For example, you know that the solution of $x^2 = -4$ must be complex because there is no real number for which its square is -4. If you take the square root of each side, $x = 2i$ or $-2i$.

Talk About It!

Explain how to determine which values should be chosen for m and p when factoring a polynomial of the form $ax^2 + bx + c$.

Sample answer: Find two numbers, m and p, with a product of ac and a sum of b.

Math History Minute

English mathematician and astronomer **Thomas Harriot (1560–1621)** was one of the first, if not the first, to consider the imaginary roots of equations. Harriot advanced the notation system for algebra and studied negative and imaginary numbers.

Interactive Presentation

Factor a Trinomial Where a Is Not 1

Solve $3x^2 + 5x + 15 = 17$ by factoring. Check your solution.

$3x^2 + 5x + 15 = 17$ Original equation

Example 4

SWIPE

Students move through the steps to see how to factor a trinomial.

CHECK

Students complete the Check online to determine whether they are ready to move on.

Example 5 Factor a Difference of Squares

Solve $81 = x^2$ by factoring. Check your solution.

$81 = x^2$	Original equation
$81 - x^2 = 0$	Subtract x^2 from each side.
$9^2 - x^2 = 0$	Write in the form $a^2 - b^2$.
$(9 + x)(9 - x) = 0$	Factor the difference of squares.
$9 + x = 0$ or $9 - x = 0$	Zero Product Property
$x = -9$ $\quad x = 9$	Solve.

Go Online You can watch a video to see how to use algebra tiles to factor a difference of squares.

Check

Solve $x^2 = 529$ by factoring. Check your solution.

$x = -23, 23$

Example 6 Factor a Perfect Square Trinomial

Solve $16y^2 - 22y + 23 = 26y - 13$ by factoring. Check your solution.

$16y^2 - 22y + 23 = 26y - 13$	Original equation
$16y^2 - 48y + 23 = -13$	Subtract $26y$ from each side.
$16y^2 - 48y + 36 = 0$	Add 13 to each side.
$(4y)^2 + 2(4y)(-6) + (-6)^2 = 0$	Factor the perfect square trinomial.
$(4y - 6)^2 = 0$	Simplify.
$y = \frac{3}{2}$	Take the square root of each side and solve.

Think About It!
Why does this equation have one solution instead of two?

Sample answer: The factor $4y - 6$ is repeated, so the equation only has one solution.

Check

Solve $16x^2 - 22x + 15 = 10x - 1$ by factoring. Check your solution.

$x = 1$

Example 7 Complex Solutions

Solve $x^2 = -144$ by factoring. Check your solution.

$x^2 = -144$	Original equation
$x^2 + 144 = 0$	Add 144 to each side.
$x^2 - (-144) = 0$	Write as a difference of squares.
$x^2 - (12i)^2 = 0$	$\sqrt{-144} = 12i$
$(x + 12i)(x - 12i) = 0$	Factor the difference of squares.
$x + 12i = 0$ or $x - 12i = 0$	Zero Product Property
$x = -12i$ $\quad x = 12i$	Solve.

Think About It!
Explain why both $(-12i)^2$ and $(12i)^2$ equal -144.

Sample answer: $(-12i)(-12i) = 144i^2$ or -144, and $(12i)(12i) = 144i^2$ or -144

Watch Out!
Complex Numbers Remember i^2 equals -1, not 1.

Go Online You can complete an Extra Example online.

170 **Module 3** • Quadratic Functions

Interactive Presentation

Factor a Perfect Square Trinomial

Solve $16y^2 - 22y + 23 = 26y - 13$ by factoring. Check your solution.

$16y^2 - 22y + 23 = 26y - 13$	Original equation
$16y^2 - 48y + 23 = -13$	Subtract $26y$ from each side.
$16y^2 - 48y + 36 = 0$	Add 13 to each side.
$(4y)^2 - 2(4y)(6) + 6^2 = 0$	Factor the perfect square trinomial.
$(4y - 6)^2 = 0$	Simplify.
$\sqrt{(4y - 6)^2} = \sqrt{0}$	Take the square root of each side.
$y = \frac{6}{4}$ or $\frac{3}{2}$	Simplify.

Example 6

TYPE

Students answer a question to determine whether they understand how to factor perfect square trinomials.

CHECK

Students complete the Check online to determine whether they are ready to move on.

Example 5 Factor a Difference of Squares

Teaching the Mathematical Practices

7 Use Structure Help students to use the structure of the equation to apply the pattern for factoring a difference of squares.

Questions for Mathematical Discourse

AL What does it mean for a number to be a perfect square? The square root is an integer.

OL When multiplying $(x - a)(x + a)$, what always happens to the two resulting x-terms? They add to 0 and cancel out.

BL How would you use this method to solve $(x + 2)^2 = 81$? Sample answer: Let $u = (x + 2)$, solve for u, then substitute $x + 2$ back in for u.

Example 6 Factor a Perfect Square Trinomial

Teaching the Mathematical Practices

1 Seek Information Mathematically proficient students must be able to transform algebraic expressions to reach solutions. In Example 6, students must transform the quadratic equation to apply a pattern for factoring and reach a solution.

Questions for Mathematical Discourse

AL How do you check for a perfect square trinomial? Sample answer: Once all the terms are simplified on one side of the equation, check to see if the constant and the coefficient of the x^2-term are perfect squares. If they are, check that the coefficient of the x-term is -2 times the square roots of the perfect squares.

OL What is the coefficient of the x-term when you simplify $(x - a)^2$? $-2a$

BL Why would the root of a perfect square trinomial be called a double root? Sample answer: The trinomial has two roots that are the same value.

Example 7 Complex Solutions

Teaching the Mathematical Practices

6 Use Definitions In the Think About It! feature, students will use the definition of i to justify their solutions in Example 7.

Questions for Mathematical Discourse

AL If $x^2 = m$ has a pure imaginary number as a solution, what do you know about m? m is a negative number.

OL What is the relationship between $(x + 12i)$ and $(x - 12i)$? They are complex conjugates.

BL Why can you write -144 as $(12i)^2$? $12^2 = 144$ and $i^2 = -1$, so $(12i)^2 = 144(-1) = -144$.

Exit Ticket

Recommended Use

At the end of class, go online to display the Exit Ticket prompt and ask students to respond using a separate piece of paper. Have students hand you their responses as they leave the room.

Alternate Use

At the end of class, go online to display the Exit Ticket prompt and ask students to respond verbally or by using a mini-whiteboard. Have students hold up their whiteboards so that you can see all student responses. Tap to reveal the answer when most or all students have completed the Exit Ticket.

Practice and Homework

Suggested Assignments

Use the table below to select appropriate exercises.

DOK	Topic	Exercises
1, 2	exercises that mirror the examples	1–32
2	exercises that use a variety of skills from this lesson	33–40
2	exercises that extend concepts learned in this lesson to new contexts	41, 42
3	exercises that emphasize higher-order and critical thinking skills	43–47

ASSESS AND DIFFERENTIATE

Use the data from the **Checks** to determine whether to provide resources for extension, remediation, or intervention.

IF students score 90% or more on the Checks, **THEN** assign: BL

- Practice Exercises 1–41 odd, 43–47
- Extension: Using Identities to Factor
- **ALEKS**® Solving Quadratic Equations by Factoring

IF students score 66%–89% on the Checks, **THEN** assign: OL

- Practice Exercises 1–47 odd
- Remediation, Review Resources: Using the Distributive Property
- Personal Tutors
- Extra Examples 1–7
- **ALEKS**® Factoring Using the GCF; Factoring by Grouping

IF students score 65% or less on the Checks, **THEN** assign: AL

- Practice Exercises 1–31 odd
- Remediation, Review Resources: Using the Distributive Property
- *Quick Review Math Handbook:* Solving Quadratic Equations by Factoring
- **ALEKS**® Factoring Using the GCF; Factoring by Grouping

Name ______ Period ______ Date ______

Practice

Go Online You can complete your homework online.

Examples 1 and 2

Solve each equation by factoring. Check your solution.

1. $6x^2 - 2x = 0$ $0, \frac{1}{3}$
2. $x^2 = 7x$ $0, 7$
3. $20x^2 = -25x$ $0, -\frac{5}{4}$
4. $x^2 + x - 30 = 0$ $5, -6$
5. $x^2 + 14x + 33 = 0$ $-11, -3$
6. $x^2 - 3x = 10$ $5, -2$

Example 3

7. GEOMETRY The length of a rectangle is 2 feet more than its width. Find the dimensions of the rectangle if its area is 63 square feet. 7 ft by 9 ft
8. PHOTOGRAPHY The length and width of a 6-inch by 8-inch photograph are reduced by the same amount to make a new photograph with area that is half that of the original. By how many inches will the dimensions of the photograph have to be reduced? 2 in.

Example 4

Solve each equation by factoring. Check your solution.

9. $2x^2 - x - 3 = 0$ $\frac{3}{2}, -1$
10. $6x^2 - 5x - 4 = 0$ $-\frac{1}{2}, \frac{4}{3}$
11. $5x^2 + 28x - 12 = 0$ $\frac{2}{5}, -6$
12. $12x^2 - 8x + 1 = 0$ $\frac{1}{6}, \frac{1}{2}$
13. $2x^2 - 11x - 40 = 0$ $8, -\frac{5}{2}$
14. $3x^2 + 2x = 21$ $\frac{7}{3}, -3$

Examples 5-7

Solve each equation by factoring. Check your solution.

15. $x^2 = 64$ $-8, 8$
16. $x^2 - 100 = 0$ $-10, 10$
17. $289 = x^2$ $-17, 17$
18. $x^2 + 14 = 50$ $-6, 6$
19. $x^2 - 169 = 0$ $-13, 13$
20. $124 = x^2 + 3$ $-11, 11$
21. $4x^2 - 28x + 49 = 0$ $\frac{7}{2}$
22. $9x^2 + 6x = -1$ $-\frac{1}{3}$
23. $16x^2 - 24x + 13 = 4$ $\frac{3}{4}$
24. $81x^2 + 36x = -4$ $-\frac{2}{9}$
25. $25x^2 + 80x + 64 = 0$ $-\frac{8}{5}$
26. $9x^2 + 60x + 95 = -5$ $-\frac{10}{3}$
27. $x^2 + 12 = -13$ $5i, -5i$
28. $x^2 + 100 = 0$ $10i, -10i$
29. $x^2 = -225$ $15i, -15i$
30. $x^2 + 4 = 0$ $2i, -2i$
31. $36x^2 = -25$ $\frac{5}{6}i, -\frac{5}{6}i$
32. $64x^2 = -49$ $\frac{7}{8}i, -\frac{7}{8}i$

Lesson 3-4 • Solving Quadratic Equations by Factoring 171

Mixed Exercises

STRUCTURE **Solve each equation by factoring. Check your solution.**

33. $10x^2 + 25x = 15$ $-3, \frac{1}{2}$
34. $27x^2 + 5 = 48x$ $\frac{5}{3}, \frac{1}{9}$
35. $x^2 + 81 = 0$ $9i, -9i$
36. $45x^2 - 3x = 2x$ $0, \frac{1}{9}$
37. $80x^2 = -16$ $\frac{\sqrt{5}}{5}i, -\frac{\sqrt{5}}{5}i$
38. $16x^2 + 8x = -1$ $-\frac{1}{4}$

39. USE A MODEL The drawing *Maisons prés de la mer* by Claude Monet is approximately 10 inches by 13 inches. Jennifer wants to make an art piece inspired by the drawing, and with an area 60% greater. If she chooses the size of her artwork by increasing both the width and height of Monet's work by same amount, what will be the dimensions of Jennifer's artwork? 13 inches by 16 inches

40. ANIMATION A computer graphics animator would like to make a realistic simulation of a tossed ball. The animator wants the ball to follow the parabolic trajectory represented by $f(x) = -0.2(x + 5)(x - 5)$.
    a. What are the solutions of $f(x) = 0$? $-5, 5$
    b. If the animator changes the equation to $f(x) = -0.2x^2 + 20$, what are the solutions of $f(x) = 0$? $-10, 10$

41. Find two consecutive even positive integers that have a product of 624. 24, 26

42. Find two consecutive odd positive integers that have a product of 323. 17, 19

Higher-Order Thinking Skills

43. FIND THE ERROR Jade and Mateo are solving $-12x^2 + 5x + 2 = 0$. Is either of them correct? Explain your reasoning. See margin.

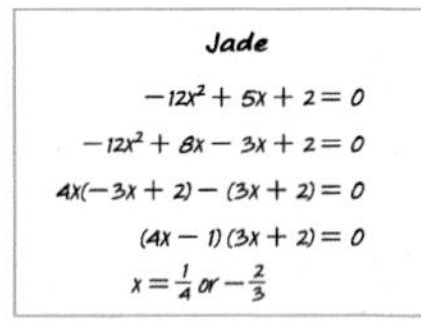
Jade
$-12x^2 + 5x + 2 = 0$
$-12x^2 + 8x - 3x + 2 = 0$
$4x(-3x + 2) - (3x + 2) = 0$
$(4x - 1)(3x + 2) = 0$
$x = \frac{1}{4}$ or $-\frac{2}{3}$

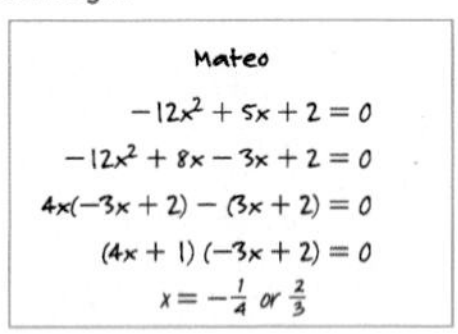
Mateo
$-12x^2 + 5x + 2 = 0$
$-12x^2 + 8x - 3x + 2 = 0$
$4x(-3x + 2) - (3x + 2) = 0$
$(4x + 1)(-3x + 2) = 0$
$x = -\frac{1}{4}$ or $\frac{2}{3}$

44. PERSEVERE The rule for factoring a difference of cubes is shown below. Use this rule to factor $40x^5 - 135x^2y^3$. $5x^2(2x - 3y)(4x^2 + 6xy + 9y^2)$

$$a^3 - b^3 = (a - b)(a^2 + ab + b^2)$$

45. CREATE Choose two integers. Then write an equation in standard form with those roots. How would the equation change if the signs of the two roots were switched? See margin.

46. ANALYZE Determine whether the following statement is *sometimes*, *always*, or *never* true. Explain your reasoning. See margin.

In a quadratic equation in standard form where a, b, and c are integers, if b is odd, then the quadratic cannot be a perfect square trinomial.

47. WRITE Explain how to factor a trinomial in standard form with $a > 1$. See margin.

172 Module 3 • Quadratic Functions

Answers

43. Neither; both students made mistakes in Step 3.
45. Sample answer: 3 and 6 $\rightarrow x^2 - 9x + 18 = 0$. -3 and $-6 \rightarrow x^2 + 9x + 18 = 0$. The linear term changes sign.
46. Always; Sample answer: To factor using perfect square trinomials, the coefficient of the linear term, bx must be a multiple of 2, or even.
47. Sample answer: Standard form is $ax^2 + bx + c$. Multiply a and c. Then find a pair of integers, g and h, that multiply to equal ac and add to equal b. Then write the quadratic expression, substituting the middle term, bx, with $gx + hx$. The expression is now $ax^2 + gx + hx + c$. Then factor the GCF from the first two terms and factor the GCF from the second two terms. So, the expression becomes $GCF(x - q) + GCF_2(x - q)$. Simplify to get $(GCF + GCF_2)(x - q)$ or $(x - p)(x - q)$.

N.CN.7, F.IF.8a

Solving Quadratic Equations by Completing the Square

LESSON GOAL

Students simplify quadratic expressions by using the Square Root Property and completing the square.

1 LAUNCH

Launch the lesson with a **Warm Up** and an introduction.

2 EXPLORE AND DEVELOP

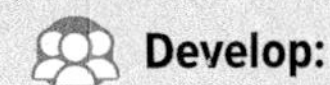

Develop:

Solving Quadratic Equations by Using the Square Root Property

- Solve a Quadratic Equation with Rational Roots
- Solve a Quadratic Equation with Irrational Roots
- Solve a Quadratic Equation with Complex Solutions

Explore: Using Algebra Tiles to Complete the Square

Develop:

Solving Quadratic Equations by Completing the Square

- Complete the Square
- Solve by Completing the Square
- Solve a Problem by Completing the Square (online)
- Solve When *a* Is Not 1
- Solve Equations with Imaginary Solutions

Quadratic Functions in Vertex Form

- Write Functions in Vertex Form
- Determine the Vertex and Axis of Symmetry
- Model with a Quadratic Function

You may want your students to complete the **Checks** online.

3 REFLECT AND PRACTICE

Exit Ticket

Practice

DIFFERENTIATE

View reports of student progress on the **Checks** after each example.

Resources	AL	OL	BL	ELL
Remediation: Factoring Special Products	●	●		●
Extension: The Golden Quadratic Equations		●	●	●

Suggested Pacing

90 min	1.5 days
45 min	3 days

Focus

Domain: Number and Quantity, Functions

Standards for Mathematical Content:

N.CN.7 Solve quadratic equations with real coefficients that have complex solutions.

F.IF.8a Use the properties of exponents to interpret expressions for exponential functions.

Standards for Mathematical Practice:

1 Make sense of problems and persevere in solving them.

7 Look for and make use of structure.

Coherence

Vertical Alignment

Previous
Students solved quadratic equations by factoring.
N.CN.7, N.CN.8, F.IF.8a

Now
Students use completing the square to solve quadratic equations.
N.CN.7, F.IF.8a

Next
Students will solve quadratic equations with the quadratic formula and use the discriminant to determine the number of real roots.
N.CN.7, A.SSE.1b

Rigor

The Three Pillars of Rigor

1 CONCEPTUAL UNDERSTANDING	2 FLUENCY	3 APPLICATION

Conceptual Bridge In this lesson, students develop an understanding of the process of completing the square. They build fluency by using this process to solve equations, and they apply their understanding by solving real-world problems.

Language Development Handbook

Assign page 21 of the *Language Development Handbook* to help your students build mathematical language related to simplifying quadratic formula and discriminant to solve quadratic equations.

ELL You can use the tips and suggestions on page T21 of the handbook to support students who are building English proficiency.

Interactive Presentation

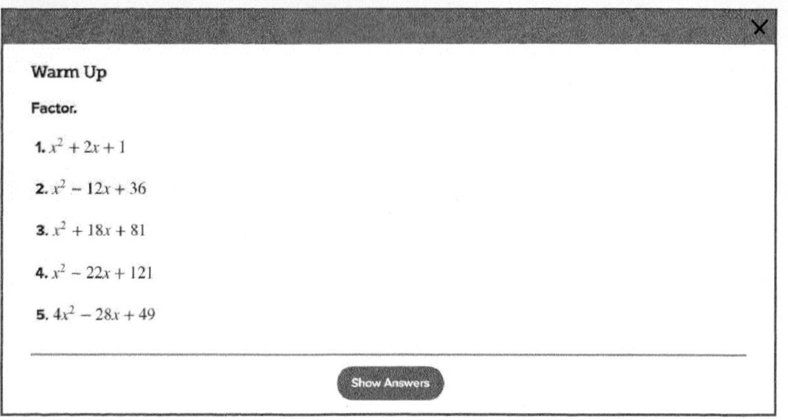

Warm Up

Warm Up

Prerequisite Skills

The Warm Up exercises address the following prerequisite skill for this lesson:

- factoring perfect squares

Answers:

1. $(x + 1)^2$
2. $(x - 6)^2$
3. $(x + 9)^2$
4. $(x - 11)^2$
5. $(2x - 7)^2$

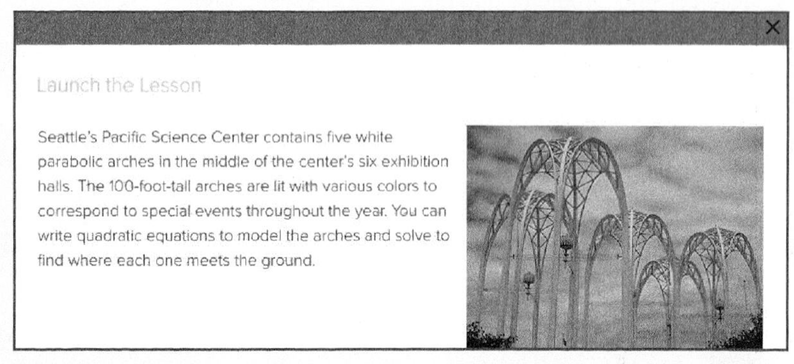

Launch the Lesson

Launch the Lesson

 Teaching the Mathematical Practices

4 Apply Mathematics In the Launch the Lesson, students will learn how quadratic equations can be used to model parabolic arches in a real-world situation.

Go Online to find additional teaching notes and questions to promote classroom discourse.

Today's Standards

Tell students that they will be addressing these content and practice standards in this lesson. You may wish to have a student volunteer read aloud *How can I meet these standards?* and *How can I use these practices?*, and connect these to the standards.

See the Interactive Presentation for I Can statements that align with the standards covered in this lesson.

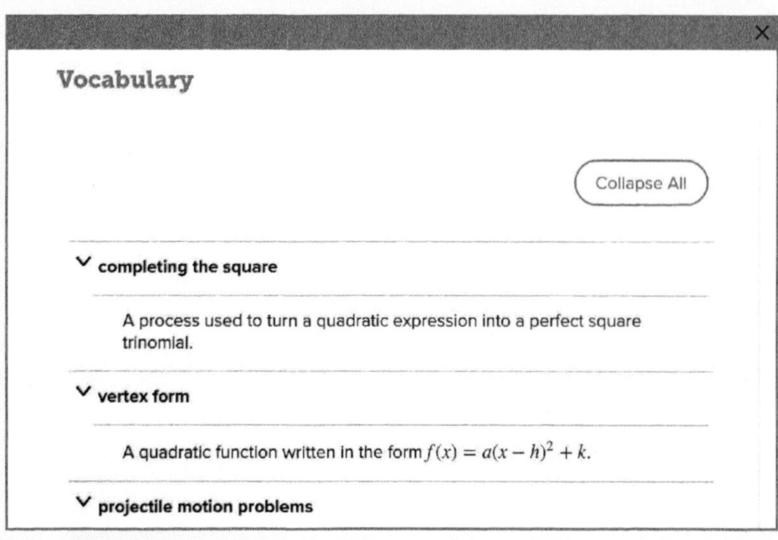

Today's Vocabulary

Today's Vocabulary

Tell students that they will be using these vocabulary terms in this lesson. You can expand each row if you wish to share the definitions. Then discuss the questions below with the class.

Mathematical Background

To complete the square for an expression of the form $x^2 + bx$, add the square of one half of the coefficient b to $x^2 + bx$. Written in symbols:

$$x^2 + bx + \left(\frac{b}{2}\right)^2 = \left(x + \frac{b}{2}\right)^2$$

Explore Using Algebra Tiles to Complete the Square

Objective

Students use algebra tiles to explore simplifying quadratic expressions by completing the square.

 Teaching the Mathematical Practices

5 Use Mathematical Tools To complete the exercises in the Explore, students will need to use algebra tiles. Work with students to explore and deepen their understanding of completing the square.

Ideas for Use

Recommended Use Present the Inquiry Question, or have a student volunteer read it aloud. Have students work in pairs to complete the Explore activity on their devices. Pairs should discuss each of the questions. Monitor student progress during the activity. Upon completion of the Explore activity, have student volunteers share their responses to the Inquiry Question.

What if my students don't have devices? You may choose to project the activity on a whiteboard. A printable worksheet for each Explore is available online. You may choose to print the worksheet so that individuals or pairs of students can use it to record their observations.

Summary of the Activity

Students will complete guiding exercises throughout the Explore activity. The students will watch a video describing the method of using algebra tiles to complete the square. Students will use algebra tiles and the method shown in the video to complete the square for quadratic expressions. Then, students will answer the Inquiry Question.

(*continued on the next page*)

Interactive Presentation

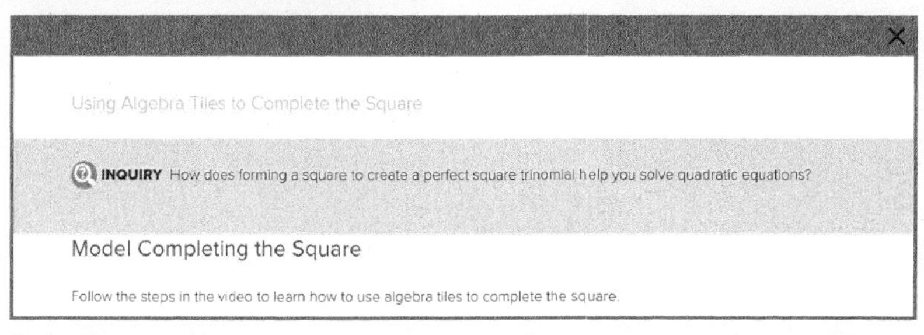

Explore

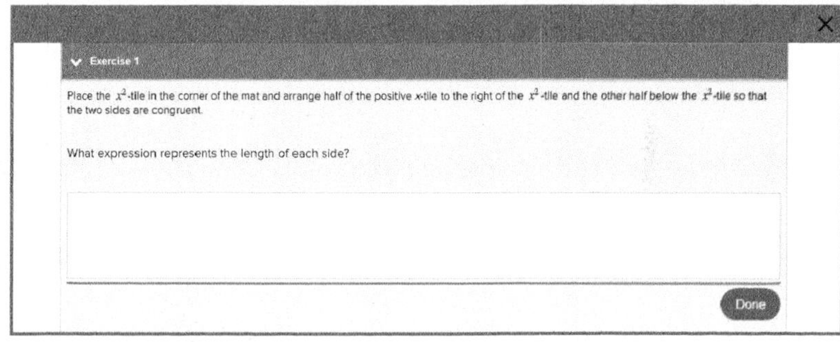

Explore

Students can watch a video that explains how to complete the square to make a perfect square trinomial.

Interactive Presentation

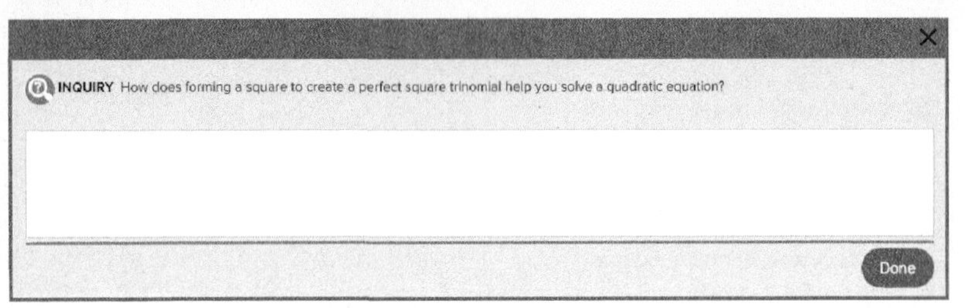

Explore

TYPE

Students respond to the Inquiry Question and can view a sample answer.

Explore Using Algebra Tiles to Complete the Square (*continued*)

Questions

Have students complete the Explore activity.

Ask:

- How does the shape of the algebra tiles help to complete the square? Sample answer: It is easy to evenly distribute the x-tiles around the x^2-tile so that the sides of the product-rectangle are equal, making it a square. The rest can be filled in with 1-tiles to complete the fourth corner of the square.
- Can you model completing the square with algebra tiles for $x^2 + 5x$? Why or why not? No; sample answer: There is an odd number of x-tiles. The fifth x-tile cannot be broken in half, so this would need to be done algebraically.

Inquiry

How does forming a square to create a perfect square trinomial help you solve quadratic equations? Sample answer: You can use the side length to solve for the variable.

Go Online to find additional teaching notes and sample answers for the guiding exercises.

1 CONCEPTUAL UNDERSTANDING | 2 FLUENCY | 3 APPLICATION

Learn Solving Quadratic Equations by Using the Square Root Property

Objective

Students solve quadratic equations by using the Square Root Property.

 **Teaching the Mathematical Practices**

7 Use Structure Help students to explore the structure of equations that can be solved by using the Square Root Property.

Common Misconception

Students often forget that taking the square root of a number results in both positive and negative values. When applying the Square Root Property, you will want to reinforce this. If confusion remains, verify this by showing that both the positive and negative square roots yield the same answer when they are each squared.

Example 1 Solve a Quadratic Equation with Rational Roots

 **Teaching the Mathematical Practices**

1 Understand the Approaches of Others In Example 1, encourage students to consider another way to solve the equation. Work with students to look at the Alternate Method. Ask students to compare and contrast the original method and the alternate method.

Questions for Mathematical Discourse

AL What type of trinomial is the left side of the equation? A perfect square trinomial

OL Why is the $\pm$ sign placed before the radical? Sample answer: The radical refers to the principal square root of 25. Without including $\pm$ before the radical, you are only finding the positive root.

BL If you set each side of the equation equal to y and graphed, what would be the coordinates of the vertex of the quadratic function and the points of intersection of the graphs? vertex: (2, 0); points of intersection: (−3, 25) and (7, 25)

Important to Know

Point out that both constants in the Example 1 equation, 4 and 25, are perfect squares. Establishing this will help solidify the difference between rational and irrational roots when it is addressed in Example 2.

Go Online

- Find additional teaching notes.
- View performance reports of the Checks.
- Assign or present an Extra Example.

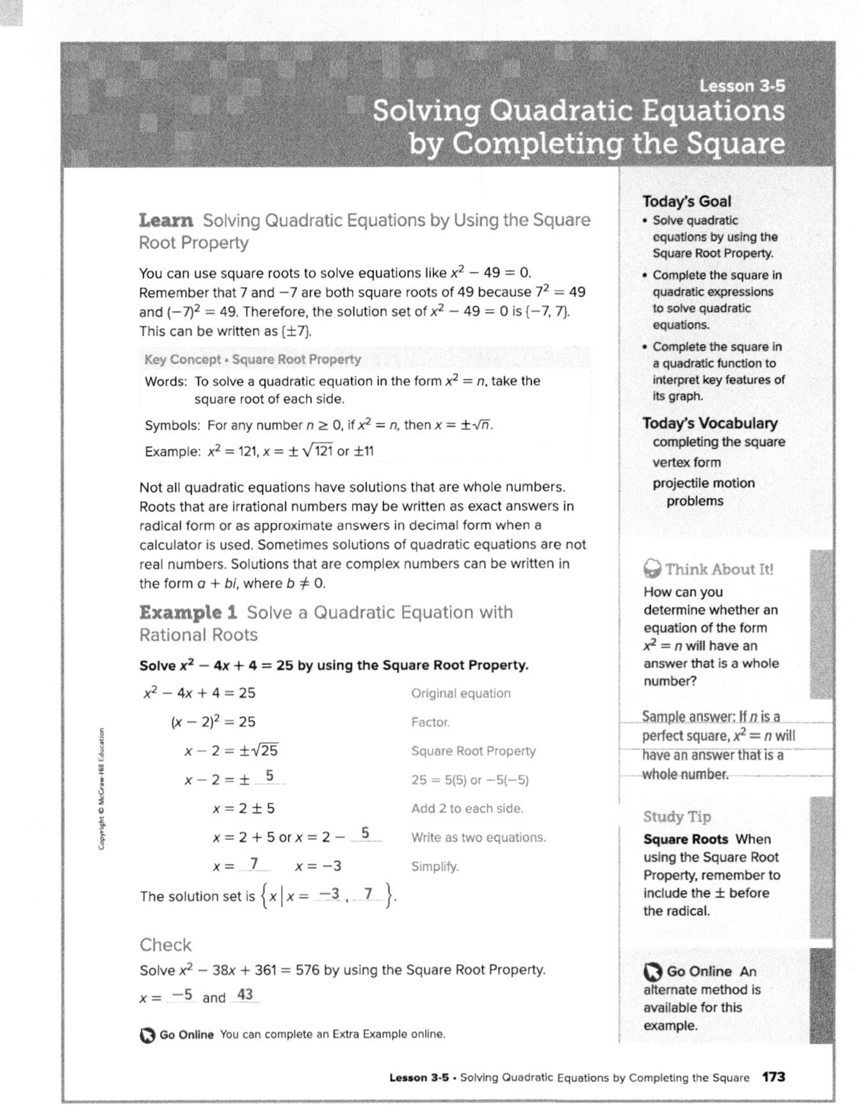

Lesson 3-5
Solving Quadratic Equations by Completing the Square

Learn Solving Quadratic Equations by Using the Square Root Property

You can use square roots to solve equations like $x^2 - 49 = 0$. Remember that 7 and −7 are both square roots of 49 because $7^2 = 49$ and $(-7)^2 = 49$. Therefore, the solution set of $x^2 - 49 = 0$ is {−7, 7}. This can be written as {±7}.

Key Concept • Square Root Property

Words: To solve a quadratic equation in the form $x^2 = n$, take the square root of each side.

Symbols: For any number $n \geq 0$, if $x^2 = n$, then $x = \pm\sqrt{n}$.

Example: $x^2 = 121$, $x = \pm\sqrt{121}$ or ± 11

Not all quadratic equations have solutions that are whole numbers. Roots that are irrational numbers may be written as exact answers in radical form or as approximate answers in decimal form when a calculator is used. Sometimes solutions of quadratic equations are not real numbers. Solutions that are complex numbers can be written in the form $a + bi$, where $b \neq 0$.

Example 1 Solve a Quadratic Equation with Rational Roots

Solve $x^2 - 4x + 4 = 25$ by using the Square Root Property.

$x^2 - 4x + 4 = 25$	Original equation
$(x - 2)^2 = 25$	Factor.
$x - 2 = \pm\sqrt{25}$	Square Root Property
$x - 2 = \pm$ 5	25 = 5(5) or −5(−5)
$x = 2 \pm 5$	Add 2 to each side.
$x = 2 + 5$ or $x = 2 -$ 5	Write as two equations.
$x =$ 7 $\quad x = -3$	Simplify.

The solution set is $\{x \mid x = -3, 7\}$.

Check

Solve $x^2 - 38x + 361 = 576$ by using the Square Root Property.

$x =$ −5 and 43

Go Online You can complete an Extra Example online.

Today's Goal
- Solve quadratic equations by using the Square Root Property.
- Complete the square in quadratic expressions to solve quadratic equations.
- Complete the square in a quadratic function to interpret key features of its graph.

Today's Vocabulary
completing the square
vertex form
projectile motion problems

Think About It!
How can you determine whether an equation of the form $x^2 = n$ will have an answer that is a whole number?

Sample answer: If n is a perfect square, $x^2 = n$ will have an answer that is a whole number.

Study Tip
Square Roots When using the Square Root Property, remember to include the ± before the radical.

Go Online An alternate method is available for this example.

Copyright © McGraw-Hill Education

Lesson 3-5 • Solving Quadratic Equations by Completing the Square 173

Interactive Presentation

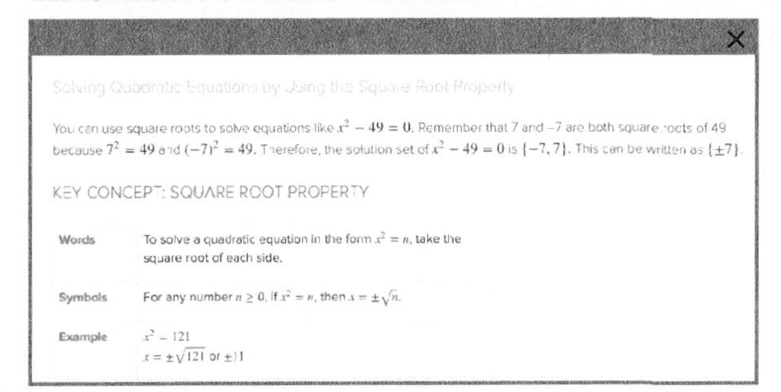

Learn

TYPE

Students answer a question to determine whether they understand solving quadratic equations by using the Square Root Property.

Your Notes

Watch Out!
Perfect Squares The constant, 192, on the right side of the equation is not a perfect square. This means that the roots will be irrational numbers.

Example 2 Solve a Quadratic Equation with Irrational Roots

Solve $x^2 + 24x + 144 = 192$ by using the Square Root Property.

$x^2 + 24x + 144 = 192$	Original equation
$(x + 12)^2 = 192$	Factor.
$x + 12 = \pm\sqrt{192}$	Square Root Property
$x + 12 = \pm 8\sqrt{3}$	$\sqrt{192} = 8\sqrt{3}$
$x = -12 \pm 8\sqrt{3}$	Subtract 12 from each side.
$x = -12 + 8\sqrt{3}$ or $-12 - 8\sqrt{3}$	Write as two equations.
$x \approx 1.86, -25.86$	Use a calculator.

The exact solutions are $-12 - 8\sqrt{3}$ and $-12 + 8\sqrt{3}$. The approximate solutions are -25.86 and 1.86.

Example 3 Solve a Quadratic Equation with Complex Solutions

Solve $2x^2 - 92x + 1058 = -72$ by using the Square Root Property.

$2x^2 - 92x + 1058 = -72$	Original equation
$x^2 - 46x + 529 = -36$	Divide each side by 2.
$(x - 23)^2 = -36$	Factor.
$x - 23 = \pm\sqrt{-36}$	Square Root Property
$x - 23 = \pm 6i$	$\sqrt{-36} = 6i$
$x = 23 \pm 6i$	Add 23 to each side.
$x = 23 + 6i$ or $23 - 6i$	Write as two equations.

The solutions are $23 + 6i$ and $23 - 6i$.

Think About It!
Can you solve a quadratic equation by completing the square if the coefficient of the x^2-term is not 1? Justify your argument.

Yes; sample answer: First divide the equation by the coefficient.

Explore Using Algebra Tiles to Complete the Square

Online Activity Use algebra tiles to complete the Explore.

INQUIRY How does forming a square to create a perfect square trinomial help you solve quadratic equations?

Go Online You can complete an Extra Example online.

174 Module 3 • Quadratic Functions

Interactive Presentation

Solve a Quadratic Equation with Complex Solutions

Solve $2x^2 - 92x + 1058 = -72$ by using the Square Root Property.

$2x^2 - 92x + 1058 = -72$ Original equation

Example 3

SWIPE

Students move through the steps to solve a quadratic equation with complex solutions.

TYPE

Students answer a question to show they understand complex solutions of quadratic equations.

CHECK

Students complete the Check online to determine whether they are ready to move on.

Example 2 Solve a Quadratic Equation with Irrational Roots

Teaching the Mathematical Practices

6 Use Precision In Example 2, students calculate the solutions of the quadratic equation accurately and efficiently. Point out that the exact answers are in radical form, but a calculator may be used to find the approximate solutions.

Questions for Mathematical Discourse

AL Is there an exact decimal answer to this example? No; sample answer Irrational numbers cannot be written as a terminating or repeating decimal.

OL Are irrational numbers real numbers? Yes.

BL When would you want to use the exact solutions versus using the approximate solutions? Sample answer: If you will be doing further calculations with the solutions, you should use the exact solutions. Approximate solutions are useful for the final step but can introduce error in intermediate steps.

Example 3 Solve a Quadratic Equation with Complex Solutions

Teaching the Mathematical Practices

1 Seek Information Students must transform the quadratic equation to reach solutions in Example 3.

Questions for Mathematical Discourse

AL Why is there an imaginary term in each solution? Sample answer: There is a negative under the square root.

OL How can you tell the solutions will be complex rather than purely imaginary? Sample answer: The quadratic expression factors to a square term with a constant subtracted from x. This constant will have to be added to the imaginary part.

BL When substituting the solutions for x in the original equation, how is it possible that the left side simplifies to a real number? Sample answer: The squared x-term will have an imaginary component that will cancel the imaginary component of the linear x-term.

Learn Solving Quadratic Equations by Completing the Square

Objective

Students complete the square in quadratic expressions to solve quadratic equations.

Teaching the Mathematical Practices

7 Use Structure Work with students to examine the structure of the completing the square method.

Important to Know

When discussing the steps for completing the square, emphasize that the coefficient of the quadratic term must be 1.

Example 4 Complete the Square

 Teaching the Mathematical Practices

1 Explain Correspondences Encourage students to explain the relationships between the equations used in this example.

Questions for Mathematical Discourse

AL What formula is used to find c? $\left(\frac{b}{2}\right)^2$

OL How does $\left(\frac{b}{2}\right)^2$ change if b is negative? It does not change.

BL What would be the denominator of c if b is odd? 4

Common Error

If $\frac{b}{2}$ does not simplify to a whole number, remind students that to square a fraction, you must square the numerator and denominator.

Example 5 Solve by Completing the Square

 Teaching the Mathematical Practices

7 Use Structure Help students to use the structure of the equation in Example 5 to identify the ways to rewrite the equation to solve by completing the square.

Questions for Mathematical Discourse

AL Why do you want to find $\left(\frac{b}{2}\right)^2$? to make the quadratic a perfect square trinomial

OL Why is $\left(\frac{b}{2}\right)^2$ added to each side of the equation? to keep both sides equal to each other

BL How could you change the original equation so that the solution set contains only integers? Sample answer: Change the value of c, −4, so that when you add 81 to each side, the result is a perfect square. For example, change c to −19.

DIFFERENTIATE

Language Development Activity AL ELL

Have students solve the equation $x^2 + 6x - 40 = 0$ by completing the square. Then have them discuss with a partner as many ways as they can to check their solutions.

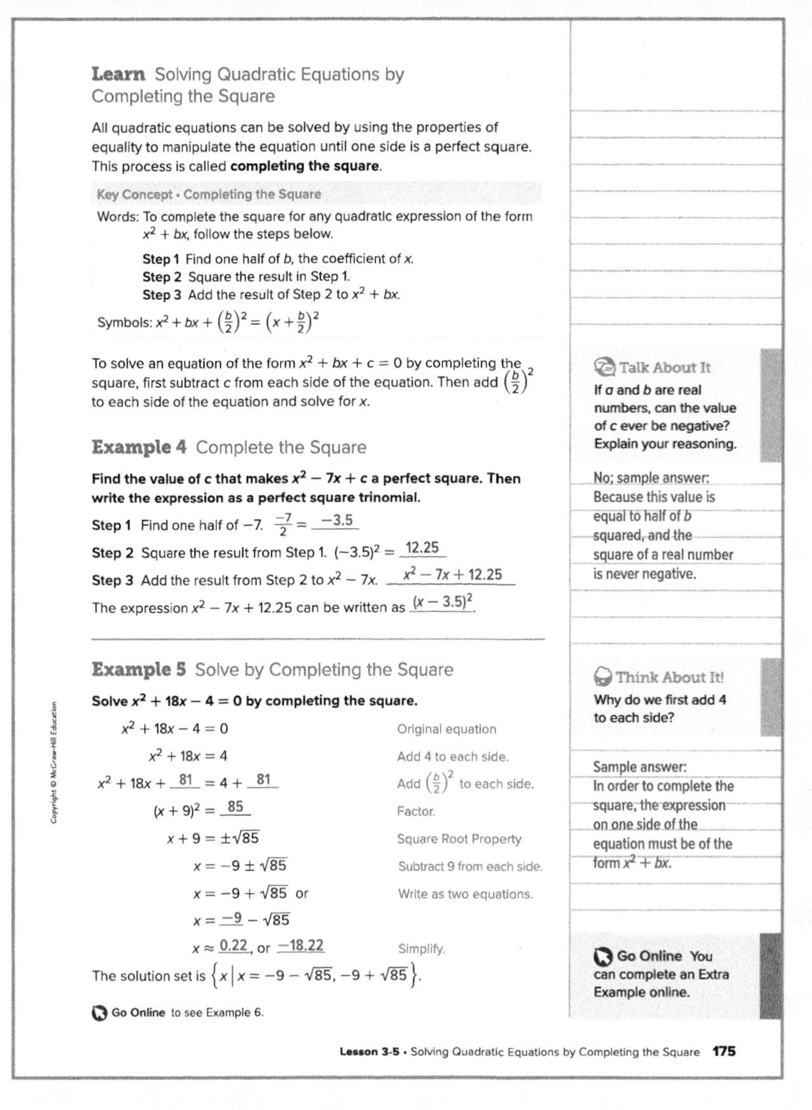

Learn Solving Quadratic Equations by Completing the Square

All quadratic equations can be solved by using the properties of equality to manipulate the equation until one side is a perfect square. This process is called **completing the square**.

Key Concept • Completing the Square

Words: To complete the square for any quadratic expression of the form $x^2 + bx$, follow the steps below.

Step 1 Find one half of b, the coefficient of x.
Step 2 Square the result in Step 1.
Step 3 Add the result of Step 2 to $x^2 + bx$.

Symbols: $x^2 + bx + \left(\frac{b}{2}\right)^2 = \left(x + \frac{b}{2}\right)^2$

To solve an equation of the form $x^2 + bx + c = 0$ by completing the square, first subtract c from each side of the equation. Then add $\left(\frac{b}{2}\right)^2$ to each side of the equation and solve for x.

Example 4 Complete the Square

Find the value of c that makes $x^2 - 7x + c$ a perfect square. Then write the expression as a perfect square trinomial.

Step 1 Find one half of −7. $\frac{-7}{2}$ = −3.5

Step 2 Square the result from Step 1. $(-3.5)^2$ = 12.25

Step 3 Add the result from Step 2 to $x^2 - 7x$. $x^2 - 7x + 12.25$

The expression $x^2 - 7x + 12.25$ can be written as $(x - 3.5)^2$.

Example 5 Solve by Completing the Square

Solve $x^2 + 18x - 4 = 0$ by completing the square.

$x^2 + 18x - 4 = 0$	Original equation
$x^2 + 18x = 4$	Add 4 to each side.
$x^2 + 18x + 81 = 4 + 81$	Add $\left(\frac{b}{2}\right)^2$ to each side.
$(x + 9)^2 = 85$	Factor.
$x + 9 = \pm\sqrt{85}$	Square Root Property
$x = -9 \pm \sqrt{85}$	Subtract 9 from each side.
$x = -9 + \sqrt{85}$ or $x = -9 - \sqrt{85}$	Write as two equations.
$x \approx 0.22$, or -18.22	Simplify.

The solution set is $\{x \mid x = -9 - \sqrt{85}, -9 + \sqrt{85}\}$.

Go Online to see Example 6.

Talk About It
If a and b are real numbers, can the value of c ever be negative? Explain your reasoning.
No; sample answer: Because this value is equal to half of b squared, and the square of a real number is never negative.

Think About It!
Why do we first add 4 to each side?
Sample answer: In order to complete the square, the expression on one side of the equation must be of the form $x^2 + bx$.

Go Online You can complete an Extra Example online.

Interactive Presentation

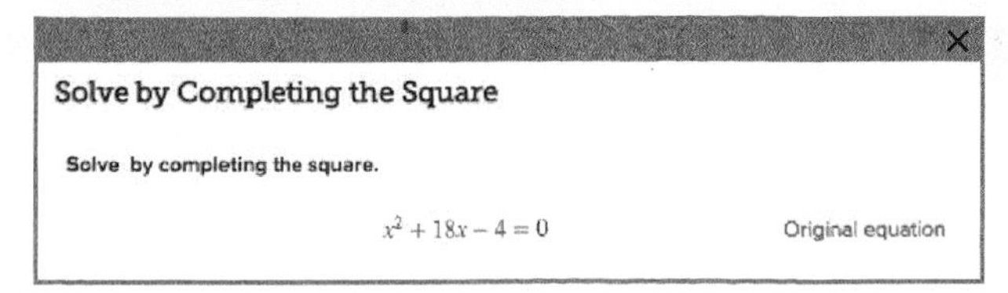

Example 5

SWIPE

Students move through the steps to solve a quadratic equation by completing the square.

Go Online for Example 6.

Example 7 Solve When a Is Not 1

Solve $4x^2 - 12x - 27 = 0$ by completing the square.

$4x^2 - 12x - 27 = 0$	Original equation
$x^2 - \underline{3}\,x - \frac{27}{4} = 0$	Divide each side by 4.
$x^2 - 3x = \underline{\frac{27}{4}}$	Add $\frac{27}{4}$ to each side.
$x^2 - 3x + \underline{\frac{9}{4}} = \frac{27}{4} + \underline{\frac{9}{4}}$	Add $\left(\frac{b}{2}\right)^2$ or $\frac{9}{4}$ to each side.
$\left(x - \frac{3}{2}\right)^2 = \underline{9}$	Factor.
$x - \frac{3}{2} = \pm\,\underline{3}$	Square Root Property
$x = \underline{\frac{3}{2}} \pm 3$	Add $\frac{3}{2}$ to each side.
$x = \frac{3}{2} + 3$ or $x = \frac{3}{2} - \underline{3}$	Write as two equations.
$x = \underline{\frac{9}{2}}$ $\quad x = \underline{-\frac{3}{2}}$	Simplify.

The solution set is $\left\{x \mid x = -\frac{3}{2}, \frac{9}{2}\right\}$.

Check

Solve $6x^2 - 21x + 9 = 0$ by completing the square.

$x = \underline{3}, \underline{0.5}$

Think About It!

Compare and contrast the solutions of this equation and the ones in the previous example. Explain.

Sample answer: Both equations have two solutions. This equation has imaginary solutions because I took the square root of a negative number, and the previous equation had real solutions because I took the square root of a positive number.

Example 8 Solve Equations with Imaginary Solutions

Solve $3x^2 - 72x + 465 = 0$ by completing the square.

$3x^2 - 72x + 465 = 0$	Original equation
$x^2 - \underline{24}\,x + \underline{155} = 0$	Divide each side by 3.
$x^2 - 24x = \underline{-155}$	Subtract 155 from each side.
$x^2 - 24x + \underline{144} = -155 + \underline{144}$	Add $\left(\frac{b}{2}\right)^2$ to each side.
$(x - \underline{12})^2 = \underline{-11}$	Factor.
$x - 12 = \pm\sqrt{-11}$	Square Root Property
$x - 12 = \pm i\sqrt{11}$	$\sqrt{-1} = i$
$x = \underline{12} \pm i\sqrt{11}$	Add 12 to each side.
$x = 12 + i\sqrt{11}$ or $12 - i\sqrt{11}$	Write as two equations.

The solution set is $\left\{x \mid x = 12 + i\sqrt{11}, 12 - i\sqrt{11}\right\}$.

Go Online You can complete an Extra Example online.

176 **Module 3** • Quadratic Functions

Interactive Presentation

Solve When a Is Not 1

Solve $4x^2 - 12x - 27 = 0$ by completing the square

Example 7

SWIPE

Students move through the steps to solve a problem by completing the square.

CHECK

Students complete the Check online to determine whether they are ready to move on.

Example 7 Solve When a Is Not 1

Teaching the Mathematical Practices

1 Seek Information Mathematically proficient students must be able to transform algebraic expression to reach solutions. In Example 7, students must transform the quadratic equation to apply the method of completing the square.

Questions for Mathematical Discourse

AL Why doesn't the right side of the equation change when dividing each side by 4? $\frac{0}{4} = 0$

OL How do you take the square root of a fraction? Take the square root of the numerator and divide by the square root of the denominator.

BL What formula could you use to find c just by looking at the original equation? $\left(\frac{b}{2a}\right)^2$

Example 8 Solve Equations with Imaginary Solutions

Teaching the Mathematical Practices

3 Construct Arguments In the Think About! It feature, students will use previously established results to construct arguments about the solutions in Examples 7 and 8.

Questions for Mathematical Discourse

AL Can a be negative? Yes. How would the first step change if $a = -3$? The result would be $x^2 + 24x - 155 = 0$.

OL Does the positive constant in the original equation guarantee a complex solution? Explain. No; sample answer: It becomes a negative number when moved to the other side of the equation, it is also affected by the term added when completing the square. The solution will only be complex if the right side is negative after $\left(\frac{b}{2}\right)^2$ is added to each side.

BL What is an alternate method that can be used in step 2 to get a constant term of 144 on the left side of the equation? Subtract 11 from each side.

DIFFERENTIATE

Language Development **ELL**

Point out that when a quadratic equation has imaginary roots, the graph of its related function does not appear to cross the x-axis, but the word *imaginary* implies the possibility that the graph crosses the x-axis in an imaginary location that can't be seen.

1 CONCEPTUAL UNDERSTANDING | 2 FLUENCY | 3 APPLICATION

Learn Quadratic Functions in Vertex Form

Objective

Students complete the square in a quadratic function to interpret key features of its graph.

Teaching the Mathematical Practices

7 Interpret Complicated Expressions Mathematically proficient students can see complicated expressions as single objects or as being composed of several objects. Guide students to see what information they can gather about quadratic functions just by looking at the vertex form of the function.

Common Misconception

The squared expression in the vertex form has the structure $(x - h)$, so it is easy for students to get confused about the sign on the value of h. Go over a series of examples of vertex forms where values of h are positive and negative, so students become familiar with the idea of transforming an addition expression into subtracting a negative h.

Example 9 Write Functions in Vertex Form

Teaching the Mathematical Practices

1 Seek Information Mathematically proficient students must be able to transform algebraic expressions to reach solutions. In Example 9, students must transform the function to rewrite it in vertex form.

Questions for Mathematical Discourse

AL Why does –1 need to be factored out in the third step? Vertex form requires the coefficient of x^2 to be 1.

OL Why is it useful to put the function in vertex form? Sample answer: It makes it easier to graph because you can quickly see the vertex and the axis of symmetry of the graph.

BL When completing the square, why do you add and subtract $\left(\frac{b}{2}\right)^2$?

Sample answer: We cannot change the value of y, so we must add and subtract $\left(\frac{b}{2}\right)^2$ so that we are really adding 0.

Important to Know

As $|a|$ increases, the graph gets narrower. Because the quantity $(x - h)^2$ is multiplied by a, as $|a|$ increases, the corresponding y-value has a greater positive value or lesser negative value. This results in a steeper (and thus narrower) graph.

Learn Quadratic Functions in Vertex Form

When a function is given in standard form, $y = ax^2 + bx + c$, you can complete the square to write it in vertex form.

Key Concept • Vertex Form of a Quadratic Function	
Words: The vertex form of a quadratic function is $y = a(x - h)^2 + k$.	
Symbols: Standard Form $y = ax^2 + bx + c$	Vertex Form $y = a(x - h)^2 + k$ The vertex is (h, k).
Example: Standard Form $y = 2x^2 + 12x + 16$	Vertex Form $y = 2(x + 3)^2 - 2$ The vertex is $(-3, -2)$.

After completing the square and writing a quadratic function in vertex form, you can analyze key features of the function. The vertex is (h, k) and $x = h$ is the equation of the axis of symmetry. The shape of the parabola and the direction that it opens are determined by a. The value of k is a minimum value if $a > 0$ or a maximum value if $a < 0$.

The path that an object travels when influenced by gravity is called a *trajectory*, and trajectories can be modeled by quadratic functions. The formula below relates the height of the object $h(t)$ and time t, where g is acceleration due to gravity, v is the initial velocity of the object, and h_0 is the initial height of the object.

$$h(t) = -\frac{1}{2}gt^2 + vt + h_0$$

The acceleration due to gravity g is 9.8 meters per second squared or 32 feet per second squared. Problems that involve objects being thrown or dropped are called **projectile motion problems**.

Think About It!
What is the minimum value of $y = 2(x - 3)^2 - 1$? How do you know that this value is a minimum?

–1; Sample answer: Because a is positive, the graph has a minimum, and the vertex is $(3, -1)$.

Example 9 Write Functions in Vertex Form

Write $y = -x^2 - 12x - 9$ in vertex form.

$y = -x^2 - 12x - 9$ Original function

$y = (-x^2 - 12x) - 9$ Group $ax^2 + bx$.

$y = -(x^2 + 12x) - 9$ Factor out –1.

$y = -(x^2 + 12x + \underline{36}) - 9 - (\underline{-1})(\underline{36})$ Complete the square.

$y = -(x + \underline{6})^2 + \underline{27}$ Simplify.

Watch Out!
The coefficient of the x^2-term must be 1 before you can complete the square.

Check

Write each function in vertex form.

a. $y = x^2 + 8x - 3$
$y = (x \underline{+4})^2 \underline{-19}$

b. $y = -3x^2 - 6x - 5$
$y = \underline{-3}(x \underline{+1})^2 \underline{-2}$

Go Online You can complete an Extra Example online.

Interactive Presentation

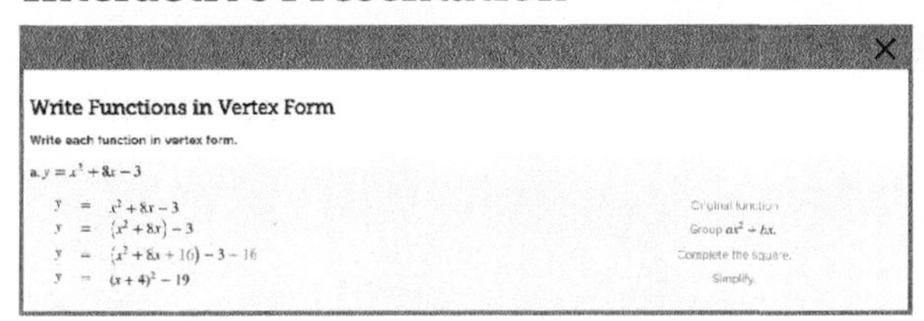

Example 9

TYPE

Students answer a question to determine whether they understand how to write functions in vertex form.

Think About It!

How would your equation for the axis of symmetry change if the vertex form of the equation was $y = 3(x + 2)^2 - 7$? Justify your argument.

Sample answer: The axis of symmetry would be $x = -2$ because the vertex form of the function can be rewritten as $y = [x - (-2)]^2 - 7$.

Example 10 Determine the Vertex and Axis of Symmetry

Consider $y = 3x^2 - 12x + 5$.

Part A Write the function in vertex form.

$y = 3x^2 - 12x + 5$ — Original equation

$y = (3x^2 - 12x) + 5$ — Group $ax^2 + bx$.

$y = 3(x^2 - 4x) + 5$ — Factor.

$y = 3(x^2 - 4x + 4) + 5 - 3(4)$ — Complete the square.

$y = 3(x - 2)^2 - 7$ — Simplify.

Part B Find the axis of symmetry.

The axis of symmetry is $x = h$ or $x =$ <u>2</u>.

Part C Find the vertex, and determine if it is a maximum or minimum.

The vertex is (h, k) or (<u>2</u>, <u>−7</u>). Because $a > 0$, this is a <u>minimum</u>.

Think About It!

If the firework reaches a height of 241 feet after 3 seconds, what is the height of the firework after 5 seconds? Justify your answer.

241 feet; Sample answer: Because the axis of symmetry is $t = 4$, the height of the firework will be the same after 3 seconds and after 5 seconds.

Example 11 Model with a Quadratic Function

FIREWORKS **If a firework is launched 1 foot off the ground at a velocity of 128 feet per second, write a function for the situation. Then find and interpret the axis of symmetry and vertex.**

Step 1 Write the function.

$h(t) = -\frac{1}{2}gt^2 + vt + h_0$ — Function for projectile motion

$h(t) = -\frac{1}{2}(32)t^2 + 128t + 1$ — $g = 32\,\frac{ft}{s^2}$, $v = 128\,\frac{ft}{s}$, $h_0 = 1$ ft

$h(t) = \underline{-16}\,t^2 + \underline{128}\,t + \underline{1}$ — Simplify.

Step 2 Rewrite the function in vertex form.

$h(t) = (-16t^2 + 128t) + 1$ — Group $ax^2 + bx$.

$h(t) = -16(t^2 - 8t) + 1$ — Factor.

$h(t) = -16(t^2 - 8t + 16) + 1 - 16(-16)$ — Complete the square.

$h(t) = \underline{-16}(t - \underline{4})^2 + \underline{257}$ — Simplify.

Step 3 Find and interpret the axis of symmetry.

Because the axis of symmetry divides the function into two equal halves, the firework will be at the same height after 2 seconds as it is after <u>6</u> seconds.

Step 4 Find and interpret the vertex.

The vertex is the maximum of the function because $a < 1$. So the firework reached a maximum height of <u>257</u> feet after <u>4</u> seconds.

Study Tip

Vertex When you interpret the vertex of a function, it is important to also consider the value of a when the function is in vertex or standard form. The value of a will tell you whether the vertex is a maximum or minimum.

Go Online You can complete an Extra Example online.

178 Module 3 • Quadratic Functions

Interactive Presentation

Model with a Quadratic Function

FIREWORKS **The height of a firework at an Independence Day celebration can be modeled by a quadratic function. If the firework is launched 1 foot off the ground at a velocity of 128 feet per second, write a function to represent the situation. Then find the axis of symmetry and the vertex and interpret their meaning in the context of the situation.**

Example 11

TAP

Students tap to see the steps to modeling with a quadratic function.

TYPE

Students answer a question to determine whether they understand modeling with a quadratic function.

CHECK

Students complete the Check online to determine whether they are ready to move on.

Example 10 Determine the Vertex and Axis of Symmetry

Teaching the Mathematical Practices

7 Use Structure Help students to use the structure of the function in vertex form to identify key features.

Questions for Mathematical Discourse

AL What are two features of the graph that can be found from the vertex form of the equation? Sample answer: the vertex and axis of symmetry

OL What does a negative value of a indicate about the graph? The parabola opens down.

BL As the absolute value of a increases, what happens to the parabola? It becomes narrower.

Common Error

When a quadratic expression is multiplied by a common factor, remind students to take that factor into account when adding a constant term to complete the square.

Example 11 Model with a Quadratic Function

Teaching the Mathematical Practices

2 Create Representations Guide students to write a function that models the projectile motion in Example 11. Then use the function to analyze the situation.

Questions for Mathematical Discourse

AL What does the vertex of the parabola represent in this situation? The maximum height of the firework.

OL In **Step 2**, why is the coefficient of the t^2-term negative and the coefficient of the t-term positive? Sample answer: The coefficient of the t^2-term represents the acceleration due to gravity, which pulls in the negative direction (down). The coefficient of the t-term represents the change in height due to the initial velocity, which is in the positive direction (up).

BL Why is transforming the standard form of the quadratic equation into the vertex form helpful? Sample answer: The vertex form can be used to identify the vertex of the function, which gives the time and height of the highest point in the path of the firework.

Exit Ticket

Recommended Use

At the end of class, go online to display the Exit Ticket prompt and ask students to respond using a separate piece of paper. Have students hand you their responses as they leave the room.

Alternate Use

At the end of class, go online to display the Exit Ticket prompt and ask students to respond verbally or by using a mini-whiteboard. Have students hold up their whiteboards so that you can see all student responses. Tap to reveal the answer when most or all students have completed the Exit Ticket.

Practice and Homework

Suggested Assignments

Use the table below to select appropriate exercises.

DOK	Topic	Exercises
1, 2	exercises that mirror the examples	1–51
2	exercises that use a variety of skills from this lesson	52–66
2	exercises that extend concepts learned in this lesson to new contexts	67–71
3	exercises that emphasize higher-order and critical thinking skills	72–77

ASSESS AND DIFFERENTIATE

Use the data from the **Checks** to determine whether to provide resources for extension, remediation, or intervention.

IF students score 90% or more on the Checks,
THEN assign: 

- Practice Exercises 1–71 odd, 72–77
- Extension: The Golden Quadratic Equations
- ALEKS® Completing the Square and the Quadratic Formula

IF students score 66%–89% on the Checks,
THEN assign: 

- Practice Exercises 1–75 odd
- Remediation, Review Resources: Factoring Special Products
- Personal Tutors
- Extra Examples 1–10
- ALEKS® Factoring Special Products

IF students score 65% or less on the Checks,
THEN assign: 

- Practice, Exercises 1–51 odd
- Remediation, Review Resources: Factoring Special Products
- *Quick Review Math Handbook:* Completing the Square
- ALEKS® Factoring Special Products

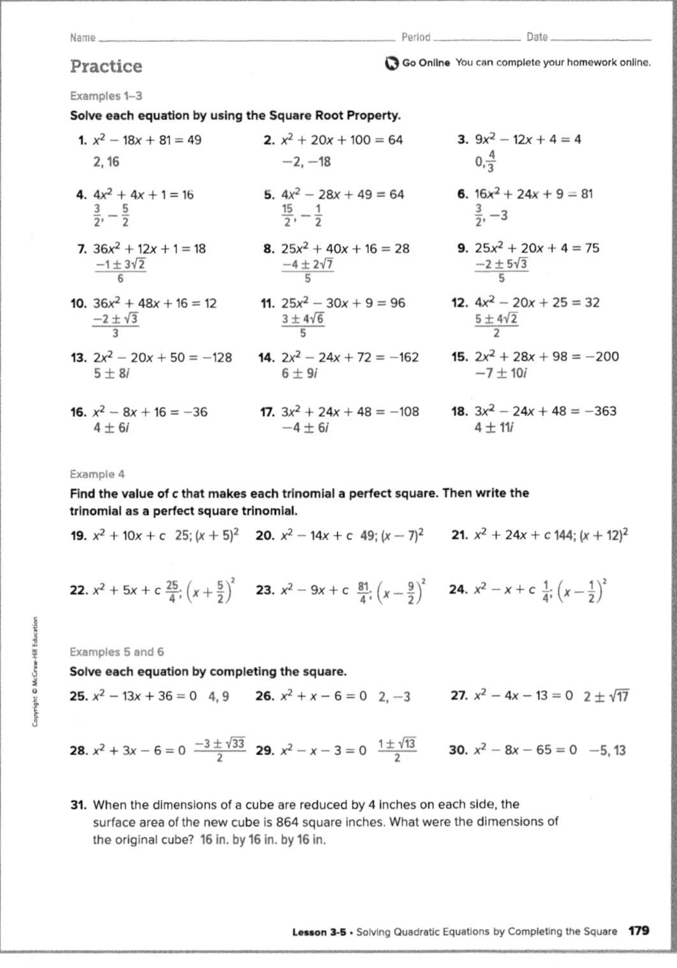

Name ______ Period ______ Date ______

Practice Go Online You can complete your homework online.

Examples 1–3

Solve each equation by using the Square Root Property.

1. $x^2 - 18x + 81 = 49$ 2, 16
2. $x^2 + 20x + 100 = 64$ −2, −18
3. $9x^2 - 12x + 4 = 4$ $0, \frac{4}{3}$
4. $4x^2 + 4x + 1 = 16$ $\frac{3}{2}, -\frac{5}{2}$
5. $4x^2 - 28x + 49 = 64$ $\frac{15}{2}, -\frac{1}{2}$
6. $16x^2 + 24x + 9 = 81$ $\frac{3}{2}, -3$
7. $36x^2 + 12x + 1 = 18$ $\frac{-1 \pm 3\sqrt{2}}{6}$
8. $25x^2 + 40x + 16 = 28$ $\frac{-4 \pm 2\sqrt{7}}{5}$
9. $25x^2 + 20x + 4 = 75$ $\frac{-2 \pm 5\sqrt{3}}{5}$
10. $36x^2 + 48x + 16 = 12$ $\frac{-2 \pm \sqrt{3}}{3}$
11. $25x^2 - 30x + 9 = 96$ $\frac{3 \pm 4\sqrt{6}}{5}$
12. $4x^2 - 20x + 25 = 32$ $\frac{5 \pm 4\sqrt{2}}{2}$
13. $2x^2 - 20x + 50 = -128$ $5 \pm 8i$
14. $2x^2 - 24x + 72 = -162$ $6 \pm 9i$
15. $2x^2 + 28x + 98 = -200$ $-7 \pm 10i$
16. $x^2 - 8x + 16 = -36$ $4 \pm 6i$
17. $3x^2 + 24x + 48 = -108$ $-4 \pm 6i$
18. $3x^2 - 24x + 48 = -363$ $4 \pm 11i$

Example 4

Find the value of c that makes each trinomial a perfect square. Then write the trinomial as a perfect square trinomial.

19. $x^2 + 10x + c$ 25; $(x + 5)^2$
20. $x^2 - 14x + c$ 49; $(x - 7)^2$
21. $x^2 + 24x + c$ 144; $(x + 12)^2$
22. $x^2 + 5x + c$ $\frac{25}{4}; \left(x + \frac{5}{2}\right)^2$
23. $x^2 - 9x + c$ $\frac{81}{4}; \left(x - \frac{9}{2}\right)^2$
24. $x^2 - x + c$ $\frac{1}{4}; \left(x - \frac{1}{2}\right)^2$

Examples 5 and 6

Solve each equation by completing the square.

25. $x^2 - 13x + 36 = 0$ 4, 9
26. $x^2 + x - 6 = 0$ 2, −3
27. $x^2 - 4x - 13 = 0$ $2 \pm \sqrt{17}$
28. $x^2 + 3x - 6 = 0$ $\frac{-3 \pm \sqrt{33}}{2}$
29. $x^2 - x - 3 = 0$ $\frac{1 \pm \sqrt{13}}{2}$
30. $x^2 - 8x - 65 = 0$ −5, 13
31. When the dimensions of a cube are reduced by 4 inches on each side, the surface area of the new cube is 864 square inches. What were the dimensions of the original cube? 16 in. by 16 in. by 16 in.

Examples 7 and 8

Solve each equation by completing the square.

32. $2x^2 - 8x - 24 = 0$ 6, −2
33. $2x^2 - 3x + 1 = 0$ $1, \frac{1}{2}$
34. $2x^2 - 13x - 7 = 0$ $-\frac{1}{2}, 7$
35. $25x^2 + 40x - 9 = 0$ $\frac{1}{5}, -\frac{9}{5}$
36. $2x^2 + 7x - 4 = 0$ $-4, \frac{1}{2}$
37. $3x^2 + 2x - 1 = 0$ $\frac{1}{3}, -1$
38. $x^2 - 4x + 12 = 0$ $2 \pm 2i\sqrt{2}$
39. $2x^2 - 3x + 5 = 0$ $\frac{3 \pm i\sqrt{31}}{4}$
40. $2x^2 + 5x + 7 = 0$ $\frac{-5 \pm i\sqrt{31}}{4}$
41. $x^2 - 2x + 3 = 0$ $1 \pm i\sqrt{2}$
42. $x^2 = -24$ $\pm 2i\sqrt{6}$
43. $x^2 - 2x + 4 = 0$ $1 \pm i\sqrt{3}$

Examples 9 and 10

Write each function in vertex form. Find the axis of symmetry. Then find the vertex, and determine if it is a *maximum* or *minimum*.

44. $y = x^2 + 2x - 5$ $y = (x + 1)^2 - 6; x = -1;$ $(h, k) = (-1, -6)$; minimum
45. $y = x^2 + 6x + 1$ $y = (x + 3)^2 - 8; x = -3;$ $(h, k) = (-3, -8)$; minimum
46. $y = -x^2 + 4x + 2$ $y = -(x - 2)^2 + 6; x = 2$ $(h, k) = (2, 6)$; maximum
47. $y = -x^2 - 8x - 5$ $y = -(x + 4)^2 + 11; x = -4;$ $(h, k) = (-4, 11)$; maximum
48. $y = 2x^2 + 4x + 3$ $y = 2(x + 1)^2 + 1; x = -1;$ $(h, k) = (-1, 1)$; minimum
49. $y = 3x^2 + 6x - 1$ $y = 3(x + 1)^2 - 4; x = -1$ $(h, k) = (-1, -4)$; minimum

Example 11

50. FIREWORKS The height of a firework at an amusement park celebration can be modeled by a quadratic function. Suppose the firework is launched from a platform 2 feet off the ground at a velocity of 96 feet per second. Use $h(t) = -\frac{1}{2}gt^2 + vt + h_0$, where $g = 32\ \frac{ft}{s^2}$.
    a. Write a function to represent this situation. $h(t) = -16t^2 + 96t + 2$
    b. Rewrite the function in vertex form. $h(t) = -16(t - 3)^2 + 146$
    c. Find the axis of symmetry and the vertex and interpret their meaning in the context of the situation. See margin.

51. DIVING Malik is participating in a diving championship. For each of his dives, his height above the water can be modeled by a quadratic function. The diving board is 7.5 meters above the water and Malik jumps with a velocity of 4.18 meters per second. Use $h(t) = -\frac{1}{2}gt^2 + vt + h_0$, where $g = 9.8\ \frac{m}{s^2}$. a, b. See margin.
    a. Write a function in vertex form to represent this situation.
    b. Find the axis of symmetry and the vertex and interpret their meaning in the context of the situation.

Answers

50c. axis of symmetry is $t = 3$; because the axis of symmetry divides the function into two equal halves, the firework will be at the same height after 1 second as it is after 5 seconds. The vertex is (3, 146). This is the maximum value of the function because $a < 0$. So the firework reached a maximum height of 146 feet after 3 seconds.

51a. $h(t) = -4.9(t - 0.427)^2 + 8.391$

51b. $t = 0.427$; points equidistant from the axis of symmetry represent the times when the diver will be at the same height during his dive. Vertex = (0.427, 8.391); Malik reaches a maximum height of about 8.391 meters approximately 0.427 second after he begins his dive.

Name ______________ Period ________ Date ________

Mixed Exercises

PRECISION **Solve each equation. Round to the nearest hundredth, if necessary.**

52. $4x^2 - 28x + 49 = 5$
$2.38, 4.62$

53. $9x^2 + 30x + 25 = 11$
$-2.77, -0.56$

54. $x^2 + x + \frac{1}{3} = \frac{2}{3}$
$-1.26, 0.26$

55. $x^2 + 1.2x + 0.56 = 0.91$
$-1.44, 0.24$

56. $x^2 + 0.7x + 4.1225 = 0$
$-0.35 \pm 2i$

57. $x^2 - 3.2x = -3.46$
$1.6 \pm 0.9i$

58. $x^2 - 1.8x + 11.24 = 2.43$
$0.9 \pm 2.83i$

59. $-0.3x^2 - 0.78x - 5.514 = 0$
$-1.3 \pm 4.09i$

60. $1.1x^2 - 8.8x + 22 = 2.2$
$4 \pm 1.41i$

61. FREE FALL A rock falls from the top of a cliff that is 25.8 meters high. Use the formula $h(t) = -\frac{1}{2}gt^2 + vt + h_0$, where $g = 9.8\,\frac{\text{m}}{\text{s}^2}$, to write a quadratic function that models the situation. Determine to the nearest tenth of a second the amount of time it takes the rock to strike the ground. Explain your reasoning. See margin.

62. REACTION TIME Tela was eating lunch when she saw her friend Jori approach. The room was crowded and Jori had to lift his tray to avoid obstacles. Suddenly, a glass on Jori's lunch tray tipped and fell off the tray. Tela lunged forward and managed to catch the glass just before it hit the ground. The height h, in feet, of the glass t seconds after it was dropped is given by $h = -16t^2 + 4.5$. If Tela caught the glass when it was six inches off the ground, how long was the glass in the air before she caught it? 0.5 second

63. INVESTMENTS The amount of money A in an account in which P dollars are invested for 2 years is given by the formula $A = P(1 + r)^2$, where r is the interest rate compounded annually. If an investment of \$800 in the account grows to \$882 in two years, at what interest rate was it invested, to the nearest percent? 5%

64. INVESTMENTS Niyati invested \$1000 in a savings account with interest compounded annually. After two years the balance in the account is \$1210. Use the compound interest formula $A = P(1 + r)^t$ to find the annual interest rate, to the nearest percent. 10%

Write each function in vertex form. Then find the vertex.

65. $y = x^2 - 10x + 28$
$y = (x - 5)^2 + 3$; $(5, 3)$

66. $y = x^2 + 16x + 65$
$y = (x + 8)^2 + 1$; $(-8, 1)$

67. $y = x^2 - 20x + 104$
$y = (x - 10)^2 + 4$; $(10, 4)$

68. $y = x^2 - 8x + 17$
$y = (x - 4)^2 + 1$; $(4, 1)$

69. AUDITORIUM SEATING The seats in an auditorium are arranged in a square grid pattern. There are 45 rows and 45 columns of chairs. For a special concert, organizers decide to increase seating by adding n rows and n columns to make a square pattern of seating $45 + n$ seats on a side.

a. How many seats are added in the expansion? $n^2 + 90n$

b. What is n if organizers wish to add 1000 seats? 10

70. DECK DESIGN The Rayburns current deck is 12 feet by 12 feet. They decide they would like to expand their deck and maintain its square shape. How much larger will each side need to be for the deck to have an area of 200 square feet? about 2.14 feet

71. VOLUME A piece of sheet metal has a length that is three times its width. It is used to make a box with an open top by cutting out 2-inch by 2-inch squares from each corner, then folding up the sides.

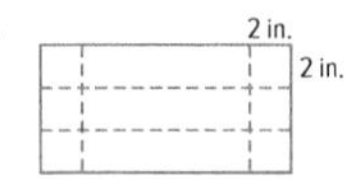

a. Define variables and write a quadratic function that represents the volume of the box in cubic inches.
w = width, $V(w)$ = the volume; $V(w) = 2(w - 4)(3w - 4) = 6w^2 - 32w + 32$

b. What are the dimensions of the metal sheet that results in a box with a volume of 1.125 cubic feet? 20.7 by 62.1 in.

Higher-Order Thinking Skills

72. CONSTRUCT ARGUMENTS Explain why the equation for the axis of symmetry for a quadratic function $y = ax^2 + bx + c$ is $x = -\frac{b}{2a}$. See margin.

73. FIND THE ERROR Alonso and Aika are solving $x^2 + 8x - 20 = 0$ by completing the square. Is either of them correct? Explain your reasoning. Alonso; Aika did not add 16 to each side; she added it only to the left side.

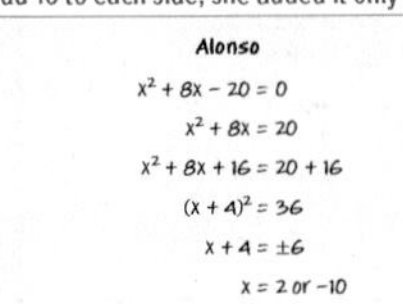

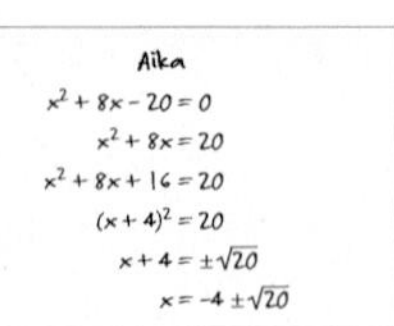

74. PERSEVERE Solve $x^2 + bx + c = 0$ by completing the square. Your answer will be an expression for x in terms of b and c. $x = \frac{-b}{2} \pm \sqrt{\frac{b^2}{4} - c}$

75. ANALYZE Without solving, determine how many unique solutions there are for each equation. Are they *rational*, *real*, or *complex*? Justify your reasoning. See margin.

a. $(x + 2)^2 = 16$
b. $(x - 2)^2 = 16$
c. $-(x - 2)^2 = 16$
d. $36 - (x - 2)^2 = 16$
e. $16(x + 2)^2 = 0$
f. $(x + 4)^2 = (x + 6)^2$

76. CREATE Write a perfect square trinomial equation in which the linear coefficient is negative and the constant term is a fraction. Then solve the equation.
Sample answer: $x^2 - \frac{2}{3}x + \frac{1}{9} = \frac{1}{4}$; $\left\{\frac{5}{6}, -\frac{1}{6}\right\}$

77. WRITE Explain what is means to complete the square. Describe each step. Sample answer: Completing the square is rewriting one side of a quadratic equation in the form of a perfect square. Once in this form, the equation can be solved by using the Square Root Property.

Answers

61. $h(t) = -4.9t^2 + 25.8$; when the rock hits the ground, its height above the ground will be zero, so the time can be determined by setting $-4.9t^2 + 25.8$ equal to zero and solving for t; $-4.9t^2 + 25.8 = 0 \rightarrow 25.8 = 4.9t^2 \rightarrow t^2 = \frac{25.8}{4.9} \rightarrow t = 2.3$ seconds.

72. Sample answer: You can complete the square to write $y = ax^2 + bx + c$ in vertex form, $y = a\left(x + \frac{b}{2a}\right)^2 + \frac{4ac - b^2}{4a}$. So, $h = -\frac{b}{2a}$ and the equation of the axis of symmetry is $x = -\frac{b}{2a}$.

75a. 2; rational; 16 is a perfect square, so $x + 2$ and x are rational.

75b. 2; rational; 16 is a perfect square, so $x - 2$ and x are rational.

75c. 2; complex; if the opposite of a square is positive, the square is negative. The square root of a negative number is complex.

75d. 2; real; the square must equal 20. Since that is positive but not a perfect square, the solutions will be real but not rational.

75e. 1; rational; the expression must be equal to 0 and only -2 makes the expression equal to 0.

75f. 1; rational; the expressions $(x + 4)$ and $(x + 6)$ must either be equal or opposites. No value makes them equal, -5 makes them opposites. The only solution is -5.

Using the Quadratic Formula and the Discriminant

LESSON GOAL

Students use the Quadratic Formula and discriminant to solve quadratic equations and determine the number of real roots.

1 LAUNCH

Launch the lesson with a **Warm Up** and an introduction.

2 EXPLORE AND DEVELOP

Develop:

Using the Quadratic Formula
- Real Roots, c Is Positive
- Real Roots, c Is Negative
- Complex Roots

Explore: The Discriminant

Develop:

Using the Discriminant
- The Discriminant, Real Roots
- The Discriminant, Complex Roots

You may want your students to complete the **Checks** online.

3 REFLECT AND PRACTICE

Exit Ticket

Practice

DIFFERENTIATE

View reports of student progress on the **Checks** after each example.

Resources	AL	OL	BL	ELL
Remediation: Solving Quadratic Equations by Using the Quadratic Formula	●	●		●
Extension: Sum and Product of Roots		●	●	●

Language Development Handbook

Assign page 22 of the *Language Development Handbook* to help your students build mathematical language related to using the quadratic formula and discriminant to solve quadratic equations.

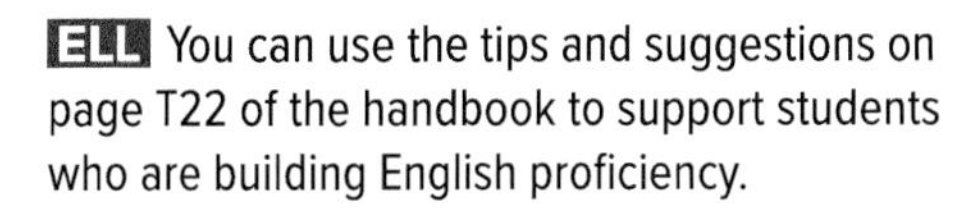

ELL You can use the tips and suggestions on page T22 of the handbook to support students who are building English proficiency.

Suggested Pacing

90 min	0.5 day
45 min	1 day

Focus

Domain: Algebra

Standards for Mathematical Content:

N.CN.7 Solve quadratic equations with real coefficients that have complex solutions.

N.CN.8 Extend polynomial identities to the complex numbers.

A.SSE.1b Interpret complicated expressions by viewing one or more of their parts as a single entity.

Standards for Mathematical Practice:

1 Make sense of problems and persevere in solving them.

4 Model with mathematics.

Coherence

Vertical Alignment

Previous
Students solved quadratic equations by completing the square.
N.CN.7, F.IF.8a

Now
Students solve quadratic equations with the Quadratic Formula and use the discriminant to determine the number of real roots.
N.CN.7, A.SSE.1b

Next
Students will graph and solve quadratic inequalities.
A.CED.1, A.CED.3

Rigor

The Three Pillars of Rigor

1 CONCEPTUAL UNDERSTANDING	2 FLUENCY	3 APPLICATION

Conceptual Bridge In this lesson, students develop an understanding of the Quadratic Formula. They build fluency by using this formula to solve equations, and they apply their understanding by using the formula by solving real-world problems.

Mathematical Background

Any quadratic equation written in the form $ax^2 + bx + c = 0$, where $a \neq 0$, can be solved using the Quadratic Formula: $x = \frac{-b \pm \sqrt{b^2 - 4ac}}{2a}$ where $a \neq 0$.

Interactive Presentation

Warm Up

Match each quadratic equation with the best method for solving it. Be prepared to justify your answers.

Equation:	Method:
1. $x^2 - 6x + 9 = 16$	A. Completing the square
2. $x^2 + 2x = 7$	B. factoring
3. $x^2 - 7x + 12 = 0$	C. graphing
4. $x^2 - 1 = 0$	D. Square Root Property

Show Answers

Warm Up

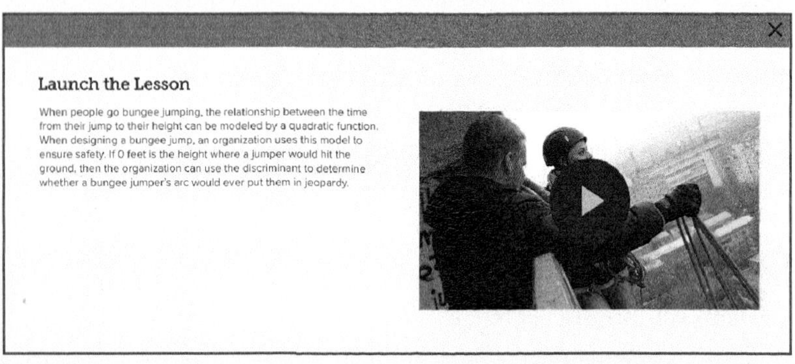

Launch the Lesson

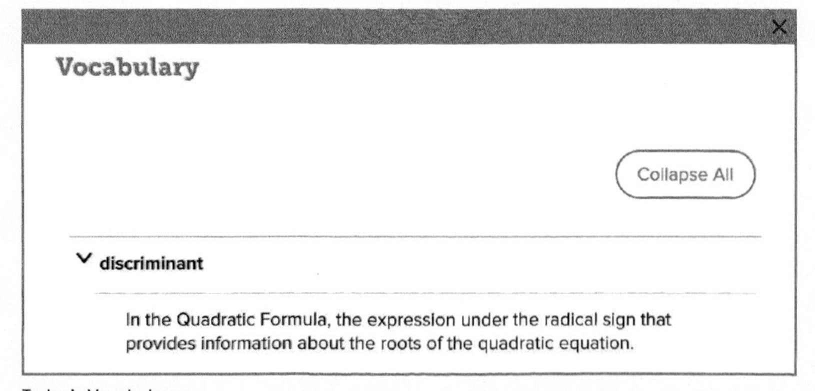

Today's Vocabulary

Warm Up

Prerequisite Skills

The Warm Up exercises address the following prerequisite skill for this lesson:

- evaluating methods for solving quadratic equations

Sample Answers:

1. D

2. A

3. B

4. C

Launch the Lesson

 Teaching the Mathematical Practices

4 Apply Mathematics In the Launch the Lesson, students will learn how the discriminant can be used to determine whether a bungee jumper is in danger.

Go Online to find additional teaching notes and questions to promote classroom discourse.

Today's Standards

Tell students that they will be addressing these content and practice standards in this lesson. You may wish to have a student volunteer read aloud *How can I meet these standards?* and *How can I use these practices?*, and connect these to the standards.

See the Interactive Presentation for I Can statement that align with the standards covered in this lesson.

Today's Vocabulary

Tell students that they will be using this vocabulary term in this lesson. You can expand the row if you wish to share the definition. Then discuss the questions below with the class.

Explore The Discriminant

Objective

Students use a graphing calculator to explore the discriminant.

 Teaching the Mathematical Practices

3 Construct Arguments In the Explore, students will use stated assumptions, definitions, and previously established results to construct an argument about the discriminant.

5 Analyze Graphs Help students analyze the graphs they have generated using graphing calculators to draw conclusions about the number of real zeros of each function.

Ideas for Use

Recommended Use Present the Inquiry Question, or have a student volunteer read it aloud. Have students work in pairs to complete the Explore activity on their devices. Pairs should discuss each of the questions. Monitor student progress during the activity. Upon completion of the Explore activity, have student volunteers share their responses to the Inquiry Question.

What if my students don't have devices? You may choose to project the activity on a whiteboard. A printable worksheet for each Explore is available online. You may choose to print the worksheet so that individuals or pairs of students can use it to record their observations.

Summary of the Activity

Students will complete guiding exercises throughout the Explore activity. The students calculate the discriminants of quadratic equations and observe the relationship between the values of the discriminant and roots of the equation. Then, students will answer the Inquiry Question.

(continued on the next page)

Interactive Presentation

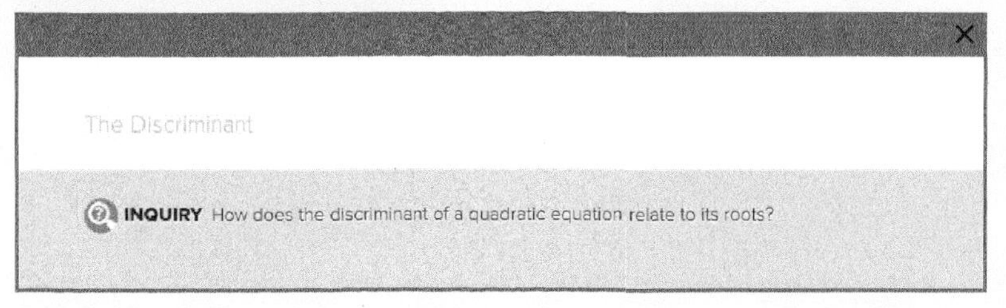

Explore

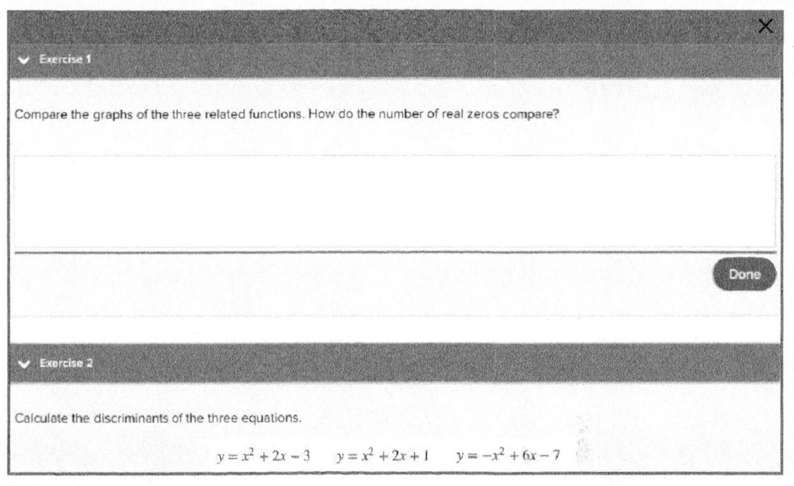

Explore

TAP

Students tap to select a calculator to help them calculate the discriminant.

TYPE

Students complete the calculations to compare the discriminants of the three equations.

Interactive Presentation

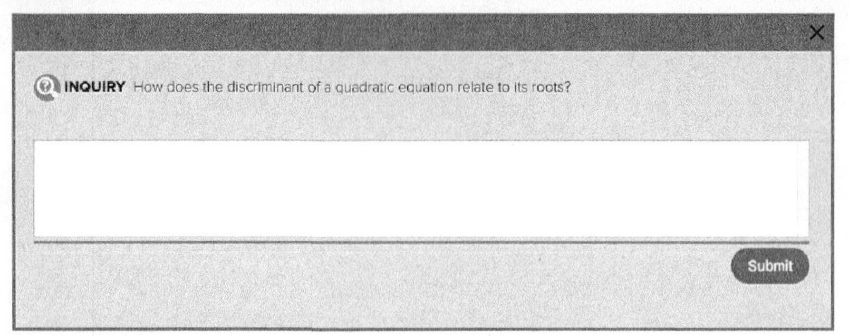

Explore

TYPE

Students respond to the Inquiry Question and can view a sample answer.

Explore The Discriminant (*continued*)

Questions

Have students complete the Explore activity.

Ask:

- What part of the quadratic formula is the discriminant? Sample answer: The discriminant is the expression inside the radical.
- How could you use the discriminant to help you sketch a graph? Sample answer: You can use the discriminant to tell you how many rational zeros the graph will have. Then you know how many times the graph should cross the x-axis, including having a graph that does not cross the x-axis.

Inquiry

How does the discriminant of a quadratic equation relate to its roots? Sample answer: If the discriminant is positive, then the roots are real. If the discriminant is a perfect square, then the roots are rational, otherwise they are irrational. If the discriminant is 0 then the equation has one real root. If the discriminant is negative, then the roots are complex.

Go Online to find additional teaching notes and sample answers for the guiding exercises.

Learn Using the Quadratic Formula

Objective

Students solve quadratic equations by using the Quadratic Formula.

Teaching the Mathematical Practices

6 Use Precision In this lesson, students learn how to calculate solutions of quadratic equations accurately and efficiently and to express solutions with a degree of precision appropriate to the problem.

DIFFERENTIATE

Reteaching Activity AL ELL

IF students substitute values into the Quadratic Formula incorrectly, **THEN** encourage students to write down the values of a, b, and c from the standard form of the quadratic equation before they begin substituting them into the formula.

Example 1 Real Roots, c Is Positive

Teaching the Mathematical Practices

4 Interpret Mathematical Results In Example 1, point out that to solve the problem, students should interpret their mathematical results in the context of the problem.

Questions for Mathematical Discourse

AL How do you identify a, b, and c? a is the coefficient of t^2, b is the coefficient of t, and c is the constant term.

OL Why is -2 seconds not a viable solution? Sample answer: Time cannot be negative.

BL If you complete the square to solve $ax^2 + bx + c$, what is the result? The Quadratic Formula

Go Online

- Find additional teaching notes.
- View performance reports of the Checks.
- Assign or present an Extra Example.

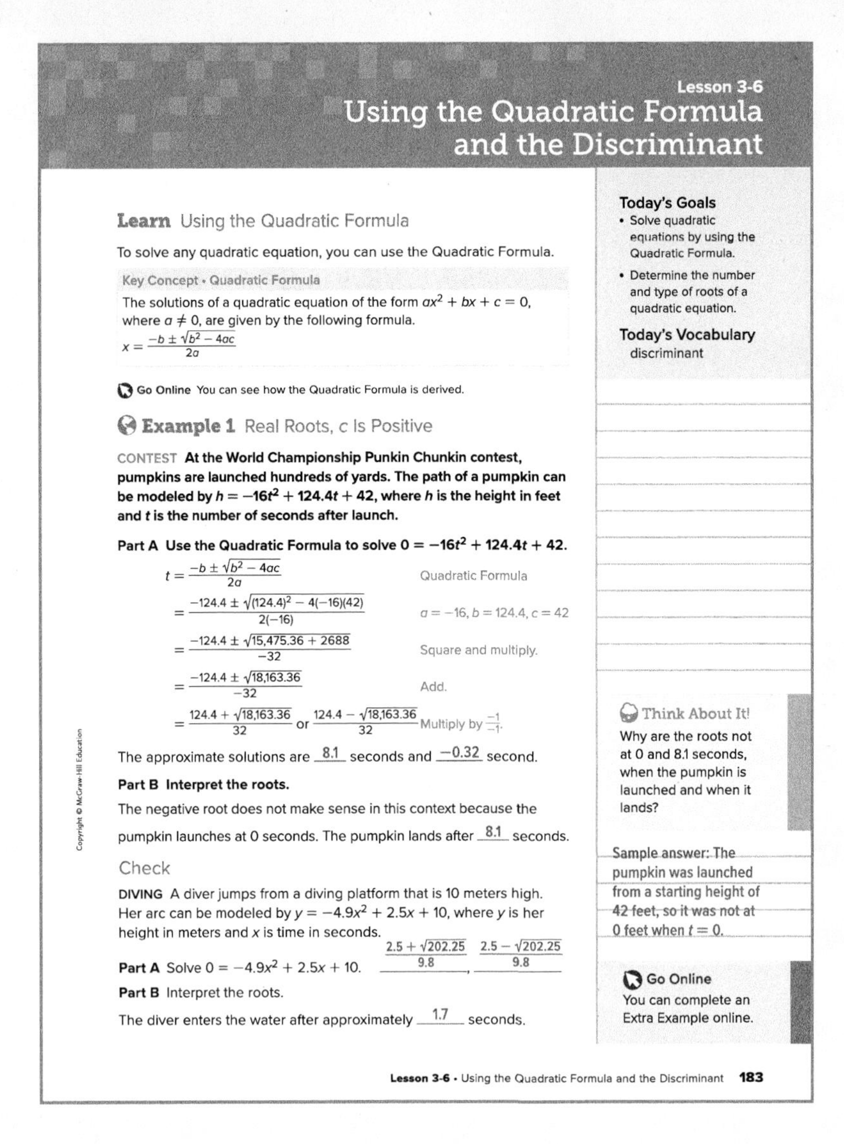

Lesson 3-6

Using the Quadratic Formula and the Discriminant

Learn Using the Quadratic Formula

To solve any quadratic equation, you can use the Quadratic Formula.

Key Concept • Quadratic Formula

The solutions of a quadratic equation of the form $ax^2 + bx + c = 0$, where $a \neq 0$, are given by the following formula.

$x = \frac{-b \pm \sqrt{b^2 - 4ac}}{2a}$

Go Online You can see how the Quadratic Formula is derived.

Example 1 Real Roots, c Is Positive

CONTEST **At the World Championship Punkin Chunkin contest, pumpkins are launched hundreds of yards. The path of a pumpkin can be modeled by $h = -16t^2 + 124.4t + 42$, where h is the height in feet and t is the number of seconds after launch.**

Part A Use the Quadratic Formula to solve $0 = -16t^2 + 124.4t + 42$.

$t = \frac{-b \pm \sqrt{b^2 - 4ac}}{2a}$ Quadratic Formula

$= \frac{-124.4 \pm \sqrt{(124.4)^2 - 4(-16)(42)}}{2(-16)}$ $a = -16$, $b = 124.4$, $c = 42$

$= \frac{-124.4 \pm \sqrt{15{,}475.36 + 2688}}{-32}$ Square and multiply.

$= \frac{-124.4 \pm \sqrt{18{,}163.36}}{-32}$ Add.

$= \frac{124.4 + \sqrt{18{,}163.36}}{32}$ or $\frac{124.4 - \sqrt{18{,}163.36}}{32}$ Multiply by $\frac{-1}{-1}$.

The approximate solutions are 8.1 seconds and −0.32 second.

Part B Interpret the roots.

The negative root does not make sense in this context because the pumpkin launches at 0 seconds. The pumpkin lands after 8.1 seconds.

Check

DIVING A diver jumps from a diving platform that is 10 meters high. Her arc can be modeled by $y = -4.9x^2 + 2.5x + 10$, where y is her height in meters and x is time in seconds.

Part A Solve $0 = -4.9x^2 + 2.5x + 10$. $\frac{2.5 + \sqrt{202.25}}{9.8}$, $\frac{2.5 - \sqrt{202.25}}{9.8}$

Part B Interpret the roots.

The diver enters the water after approximately 1.7 seconds.

Today's Goals
- Solve quadratic equations by using the Quadratic Formula.
- Determine the number and type of roots of a quadratic equation.

Today's Vocabulary
discriminant

Think About It!
Why are the roots not at 0 and 8.1 seconds, when the pumpkin is launched and when it lands?

Sample answer: The pumpkin was launched from a starting height of 42 feet, so it was not at 0 feet when $t = 0$.

Go Online
You can complete an Extra Example online.

Copyright © McGraw-Hill Education

Lesson 3-6 • Using the Quadratic Formula and the Discriminant 183

Interactive Presentation

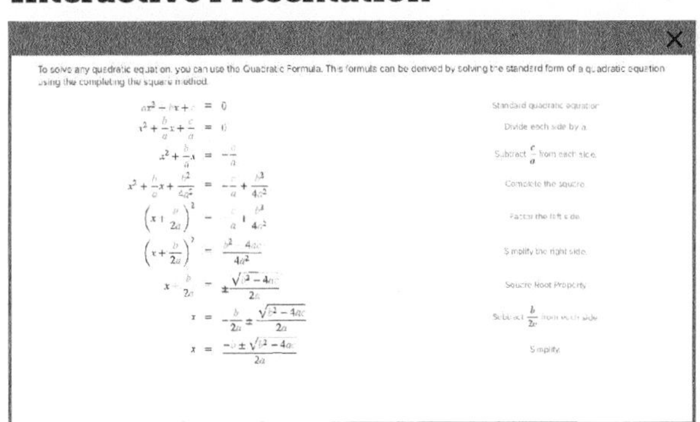

Learn

TYPE

Students answer questions to determine whether they understand using the Quadratic Formula.

N.CN.7, N.CN.8, A.SSE.1b

1 CONCEPTUAL UNDERSTANDING | **2 FLUENCY** | 3 APPLICATION

Your Notes

Example 2 Real Roots, c Is Negative

Solve $x^2 + 4x - 17 = 0$ by using the Quadratic Formula.

$x = \frac{-b \pm \sqrt{b^2 - 4ac}}{2a}$ Quadratic Formula

$= \frac{-4 \pm \sqrt{(4)^2 - 4(1)(-17)}}{2(1)}$ $a = 1, b = 4, c = -17$

$= \frac{-4 \pm \sqrt{84}}{2}$ Simplify.

$= \frac{-4 \pm \sqrt{4} \cdot \sqrt{21}}{2}$ Product Property of Square Roots

$= \frac{-4 \pm 2\sqrt{21}}{2}$ $\sqrt{4} = 2$

$= -2 \pm \sqrt{21}$ Divide the numerator and denominator by 2.

Check

Solve $3x^2 - 5x - 1 = 0$ by using the Quadratic Formula. $\frac{5 \pm \sqrt{37}}{6}$

Example 3 Complex Roots

Solve $5x^2 + 8x + 11 = 0$ by using the Quadratic Formula.

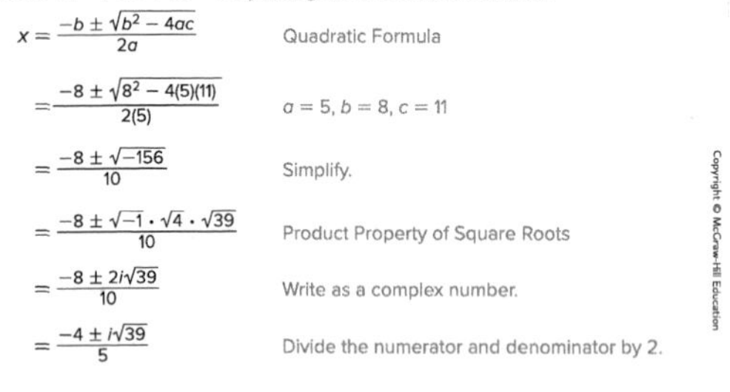

$x = \frac{-b \pm \sqrt{b^2 - 4ac}}{2a}$ Quadratic Formula

$= \frac{-8 \pm \sqrt{8^2 - 4(5)(11)}}{2(5)}$ $a = 5, b = 8, c = 11$

$= \frac{-8 \pm \sqrt{-156}}{10}$ Simplify.

$= \frac{-8 \pm \sqrt{-1} \cdot \sqrt{4} \cdot \sqrt{39}}{10}$ Product Property of Square Roots

$= \frac{-8 \pm 2i\sqrt{39}}{10}$ Write as a complex number.

$= \frac{-4 \pm i\sqrt{39}}{5}$ Divide the numerator and denominator by 2.

Check

Solve $9x^2 - 3x + 18 = 0$ by using the Quadratic Formula. $\frac{1 \pm i\sqrt{71}}{6}$

Go Online You can complete an Extra Example online.

184 Module 3 • Quadratic Functions

Interactive Presentation

Real Roots, c Is Negative

Solve $x^2 + 4x - 17 = 0$ by using the Quadratic Formula.

Example 2

CHECK

Students complete the Check online to determine whether they are ready to move on.

Example 2 Real Roots, c Is Negative

Teaching the Mathematical Practices

1 Check Answers Mathematically proficient students continually ask themselves, "Does this makes sense?" Encourage students to check their answers in Example 2 by using a different method.

Questions for Mathematical Discourse

AL Why can you divide the numerator and denominator by 2 in the last step? 2 is the common factor of the numerator and the denominator.

OL How does a negative value of c and a positive value of a affect the roots? Sample answer: The radicand will be positive; thus, the roots will be real.

BL How would a negative value of b affect the substitution in step 2? Sample answer: $-b$ would result in a positive value. The value of b^2 would remain the same.

Important to Know

Students may forget to include the sign when c is negative. As students work through Example 2, stress the importance maintaining signs.

Example 3 Complex Roots

Teaching the Mathematical Practices

6 Use Precision In Example 3, students will calculate accurately and efficiently using the quadratic formula. They will express their numerical answers exactly using radicals and complex numbers.

Questions for Mathematical Discourse

AL Why are the roots of the equation complex? The radicand in the Quadratic Formula is negative.

OL How can you write the solution as a complex number in $a + bi$ form? Write each fraction separately.

BL Why is it not possible to have one complex root and one real root? Sample answer: If one root has real and imaginary terms, the other will also have real and imaginary terms because they occur in complex conjugate pairs.

Important to Know

Remind students that they can also write their final answer in the standard form of a complex number, $a + bi$, where both a and b are real numbers.

Essential Question Follow-Up

Students have explored solving quadratic equations.

Ask:

How do you know what method to use when solving a quadratic equation? Sample answer: If the equation has terms that you know are easily factorable, you could solve by factoring. If the equation has more complex terms, you could solve by using the Quadratic Formula, completing the square, or by graphing. You could also use one method to solve and a second method to check your answer.

Learn Using the Discriminant

Objective

Students determine the number and type of roots of a quadratic equation by using the discriminant.

Teaching the Mathematical Practices

3 Analyze Cases Work with students to analyze at the cases of the discriminant. Encourage students to familiarize themselves with all of the cases.

Common Error

Because there both $-b$ and b^2 are used in the Quadratic Formula, it is easy for beginning students to confuse the locations of these two uses of the b coefficient.

DIFFERENTIATE

Language Development Activity AL ELL

Have students research the root words that form the words quadratic and discriminant. Discuss how the root words relate to the mathematical meanings of quadratic and discriminant.

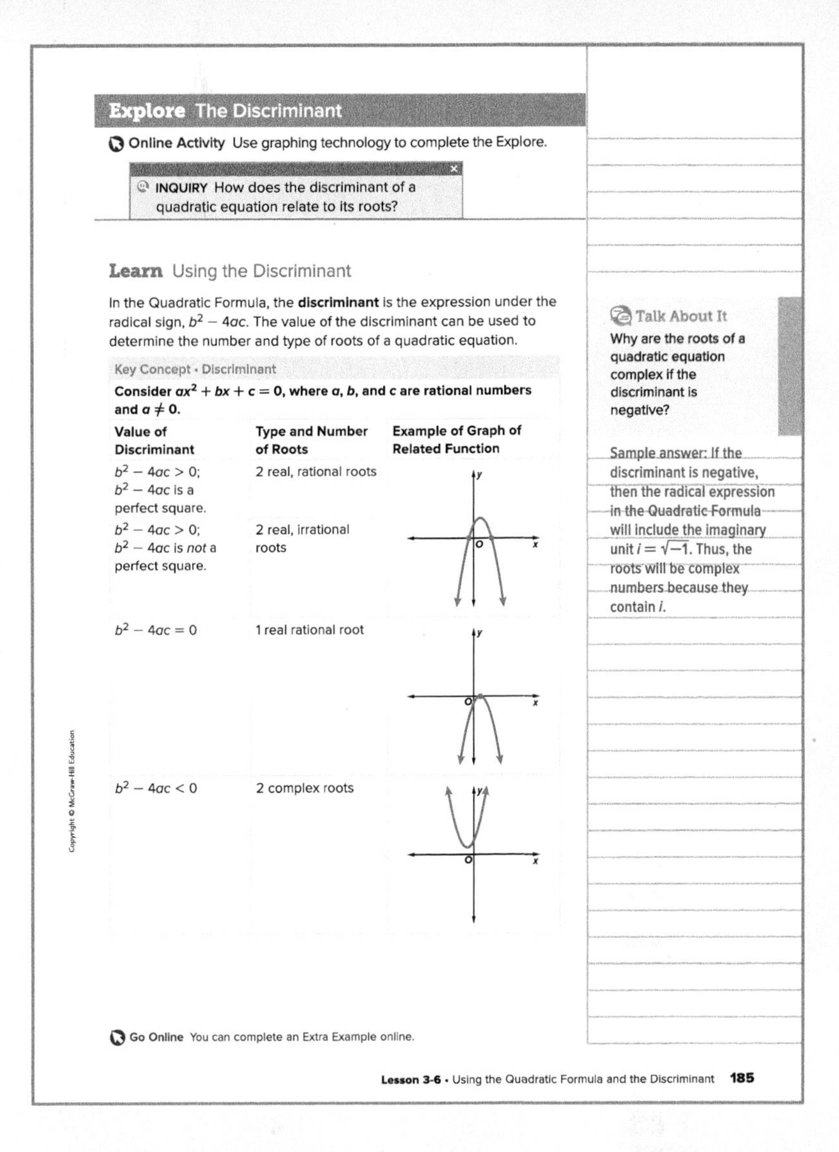

Explore The Discriminant

Online Activity Use graphing technology to complete the Explore.

INQUIRY How does the discriminant of a quadratic equation relate to its roots?

Learn Using the Discriminant

In the Quadratic Formula, the **discriminant** is the expression under the radical sign, $b^2 - 4ac$. The value of the discriminant can be used to determine the number and type of roots of a quadratic equation.

Key Concept • Discriminant

Consider $ax^2 + bx + c = 0$, where a, b, and c are rational numbers and $a \neq 0$.

Value of Discriminant	Type and Number of Roots	Example of Graph of Related Function
$b^2 - 4ac > 0$; $b^2 - 4ac$ is a perfect square.	2 real, rational roots	
$b^2 - 4ac > 0$; $b^2 - 4ac$ is *not* a perfect square.	2 real, irrational roots	
$b^2 - 4ac = 0$	1 real rational root	
$b^2 - 4ac < 0$	2 complex roots	

Talk About It

Why are the roots of a quadratic equation complex if the discriminant is negative?

Sample answer: If the discriminant is negative, then the radical expression in the Quadratic Formula will include the imaginary unit $i = \sqrt{-1}$. Thus, the roots will be complex numbers because they contain i.

Go Online You can complete an Extra Example online.

Copyright © McGraw-Hill Education

Lesson 3-6 • Using the Quadratic Formula and the Discriminant 185

Interactive Presentation

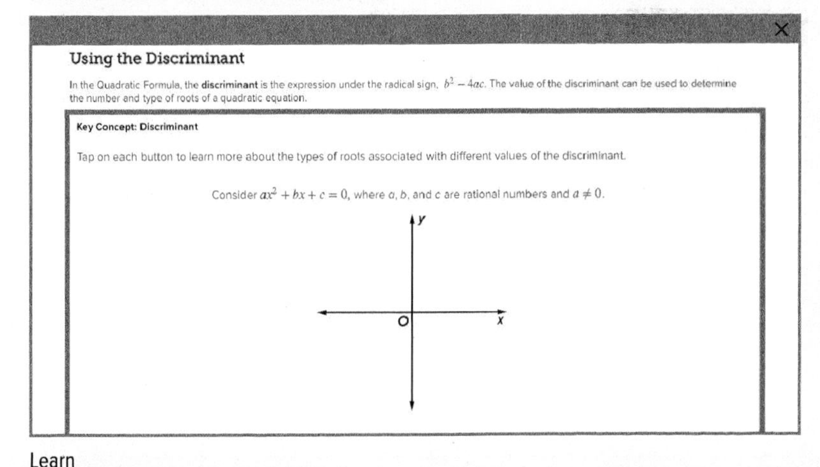

Learn

TAP

Students tap on each button to see different cases of the discriminant.

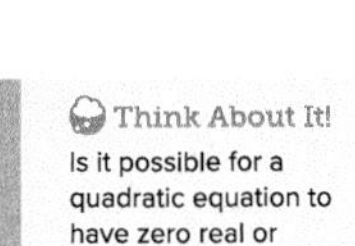

Think About It!
Is it possible for a quadratic equation to have zero real or complex roots?

No; sample answer: If the discriminant is 0, then there is exactly one solution. If it is not zero, there are two solutions.

Example 4 The Discriminant, Real Roots

Examine $2x^2 - 10x + 7 = 0$.

Part A Find the value of the discriminant for $2x^2 - 10x + 7 = 0$.

$a = 2 \quad b = -10 \quad c = 7$

$b^2 - 4ac = (-10)^2 - 4(2)(7)$
$= 100 - 56$
$= 44$

Part B Describe the number and type of roots for the equation.

The discriminant is nonzero, so there are two roots. The discriminant is positive and not a perfect square, so the roots are irrational.

Check

Examine $2x^2 + 8x + 8 = 0$.

Part A Find the value of the discriminant for $2x^2 + 8x + 8 = 0$.

$b^2 - 4ac = 0$

Part B Describe the number and type of roots for the equation.

There is/are 1 rational root(s).

Example 5 The Discriminant, Complex Roots

Examine $-5x^2 + 10x - 15 = 0$.

Part A Find the value of the discriminant for $-5x^2 + 10x - 15 = 0$.

$a = -5 \quad b = 10 \quad c = -15$

$b^2 - 4ac = (10)^2 - 4(-5)(-15)$
$= 100 - 300$
$= -200$

Part B Describe the number and type of roots for the equation.

The discriminant is nonzero, so there are two roots. The discriminant is negative, so the roots are complex.

Check

Examine $10x^2 - 4x + 7 = 0$.

Part A Find the value of the discriminant for $10x^2 - 4x + 7 = 0$.

$b^2 - 4ac = -264$

Part B Describe the number and type of roots for the equation.

There is/are 2 complex root(s).

Go Online to practice what you've learned in Lessons 3-2 and 3-4 through 3-6.

Go Online You can complete an Extra Example online.

186 Module 3 • Quadratic Functions

Interactive Presentation

The Discriminant, Real Roots

Examine $2x^2 - 10x + 7 = 0$.

Part A
Find the value of the discriminant for $2x^2 - 10x + 7 = 0$.

$a = \square \quad b = \square \quad c = \square$

$b^2 - 4ac = (\square)^2 - 4(\square)(\square)$

$= 100 - 56$

$= 44$

Example 4

CHECK

Students complete the Check online to determine whether they are ready to move on.

Example 4 The Discriminant, Real Roots

Teaching the Mathematical Practices

2 Attend to Quantities In Example 4, students must note the meaning of the discriminant. They use the meaning of the discriminant to determine the number and type of roots of the equation.

Questions for Mathematical Discourse

AL What is the value of the discriminant if the quadratic equation has exactly one real rational root? 0

OL What type(s) of roots exist if the discriminant is a nonzero perfect square? rational roots

BL What does the discriminant tell you about the solutions of a quadratic equation that looking at the graph cannot always tell you? Sample answer: It can distinguish between having rational and irrational roots.

Example 5 The Discriminant, Complex Roots

Teaching the Mathematical Practices

7 Interpret Complicated Expressions Mathematically proficient students can see complicated expressions as single objects or as being composed of several objects. In Example 5, students use the discriminant from the Quadratic Formula to determine the number and type of roots of the equation.

Questions for Mathematical Discourse

AL What about the value of the discriminant tells you the solution must be complex? It has a negative value.

OL When will the discriminant have a negative value? when $4ac > b^2$

BL If the discriminant is the negative of a perfect square, what do you know about the solution? The coefficient of the complex term will be a rational number.

Exit Ticket

Recommended Use

At the end of class, go online to display the Exit Ticket prompt and ask students to respond using a separate piece of paper. Have students hand you their responses as they leave the room.

Alternate Use

At the end of class, go online to display the Exit Ticket prompt and ask students to respond verbally or by using a mini-whiteboard. Have students hold up their whiteboards so that you can see all student responses. Tap to reveal the answer when most or all students have completed the Exit Ticket.

Practice and Homework

Suggested Assignments

Use the table below to select appropriate exercises.

DOK	Topic	Exercises
1, 2	exercises that mirror the examples	1–35
2	exercises that use a variety of skills from this lesson	36–52
3	exercises that emphasize higher-order and critical thinking skills	53–58

ASSESS AND DIFFERENTIATE

Use the data from the **Checks** to determine whether to provide resources for extension, remediation, or intervention.

IF students score 90% or more on the Checks, **THEN** assign: BL

- Practice Exercises 1–59 odd, 60–66
- Extension: Sum and Product of Roots
- **ALEKS**® Completing the Square and the Quadratic Formula

IF students score 66%–89% on the Checks, **THEN** assign: OL

- Practice Exercises 1–65 odd
- Remediation, Review Resources: Solving Quadratic Equations by Using the Quadratic Formula
- Personal Tutors
- Extra Examples 1–5
- **ALEKS**® Quadratic Equations

IF students score 65% or less on the Checks, **THEN** assign: AL

- Practice, Exercises 1–35 odd
- Remediation, Review Resources: Solving Quadratic Equations by Using the Quadratic Formula
- *Quick Review Math Handbook*: The Quadratic Formula and the Discriminant
- **ALEKS**® Quadratic Equations

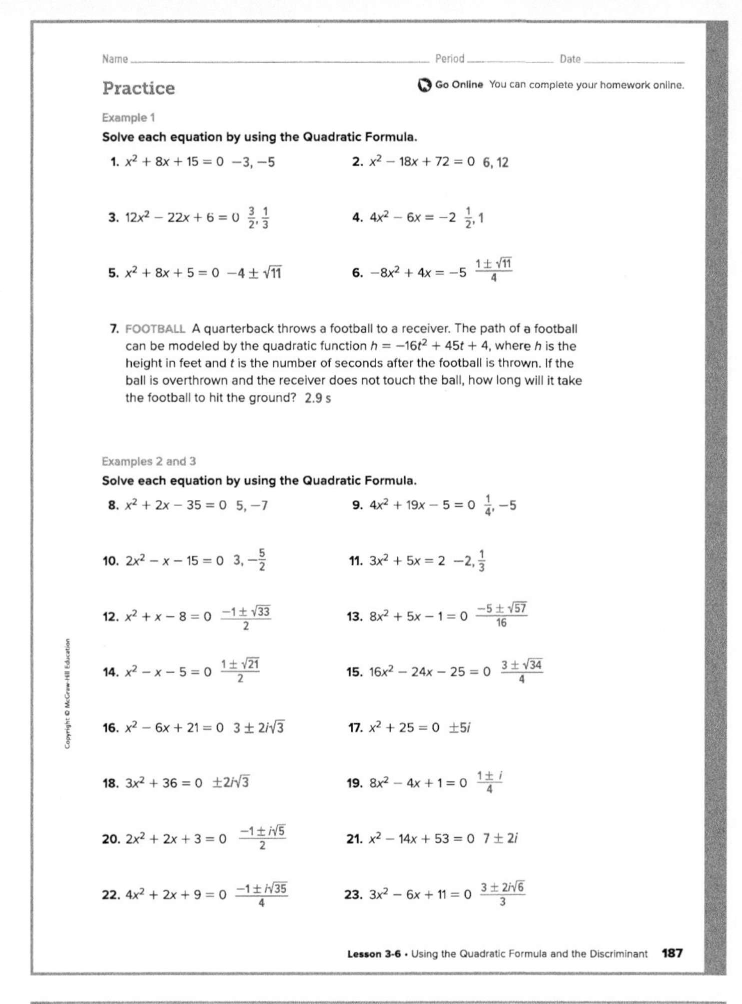

Name ______ Period ______ Date ______

Practice

Go Online You can complete your homework online.

Example 1

Solve each equation by using the Quadratic Formula.

1. $x^2 + 8x + 15 = 0$ $-3, -5$
2. $x^2 - 18x + 72 = 0$ $6, 12$
3. $12x^2 - 22x + 6 = 0$ $\frac{3}{2}, \frac{1}{3}$
4. $4x^2 - 6x = -2$ $\frac{1}{2}, 1$
5. $x^2 + 8x + 5 = 0$ $-4 \pm \sqrt{11}$
6. $-8x^2 + 4x = -5$ $\frac{1 \pm \sqrt{11}}{4}$
7. FOOTBALL A quarterback throws a football to a receiver. The path of a football can be modeled by the quadratic function $h = -16t^2 + 45t + 4$, where h is the height in feet and t is the number of seconds after the football is thrown. If the ball is overthrown and the receiver does not touch the ball, how long will it take the football to hit the ground? 2.9 s

Examples 2 and 3

Solve each equation by using the Quadratic Formula.

8. $x^2 + 2x - 35 = 0$ $5, -7$
9. $4x^2 + 19x - 5 = 0$ $\frac{1}{4}, -5$
10. $2x^2 - x - 15 = 0$ $3, -\frac{5}{2}$
11. $3x^2 + 5x = 2$ $-2, \frac{1}{3}$
12. $x^2 + x - 8 = 0$ $\frac{-1 \pm \sqrt{33}}{2}$
13. $8x^2 + 5x - 1 = 0$ $\frac{-5 \pm \sqrt{57}}{16}$
14. $x^2 - x - 5 = 0$ $\frac{1 \pm \sqrt{21}}{2}$
15. $16x^2 - 24x - 25 = 0$ $\frac{3 \pm \sqrt{34}}{4}$
16. $x^2 - 6x + 21 = 0$ $3 \pm 2i\sqrt{3}$
17. $x^2 + 25 = 0$ $\pm 5i$
18. $3x^2 + 36 = 0$ $\pm 2i\sqrt{3}$
19. $8x^2 - 4x + 1 = 0$ $\frac{1 \pm i}{4}$
20. $2x^2 + 2x + 3 = 0$ $\frac{-1 \pm i\sqrt{5}}{2}$
21. $x^2 - 14x + 53 = 0$ $7 \pm 2i$
22. $4x^2 + 2x + 9 = 0$ $\frac{-1 \pm i\sqrt{35}}{4}$
23. $3x^2 - 6x + 11 = 0$ $\frac{3 \pm 2i\sqrt{6}}{3}$

Lesson 3-6 • Using the Quadratic Formula and the Discriminant 187

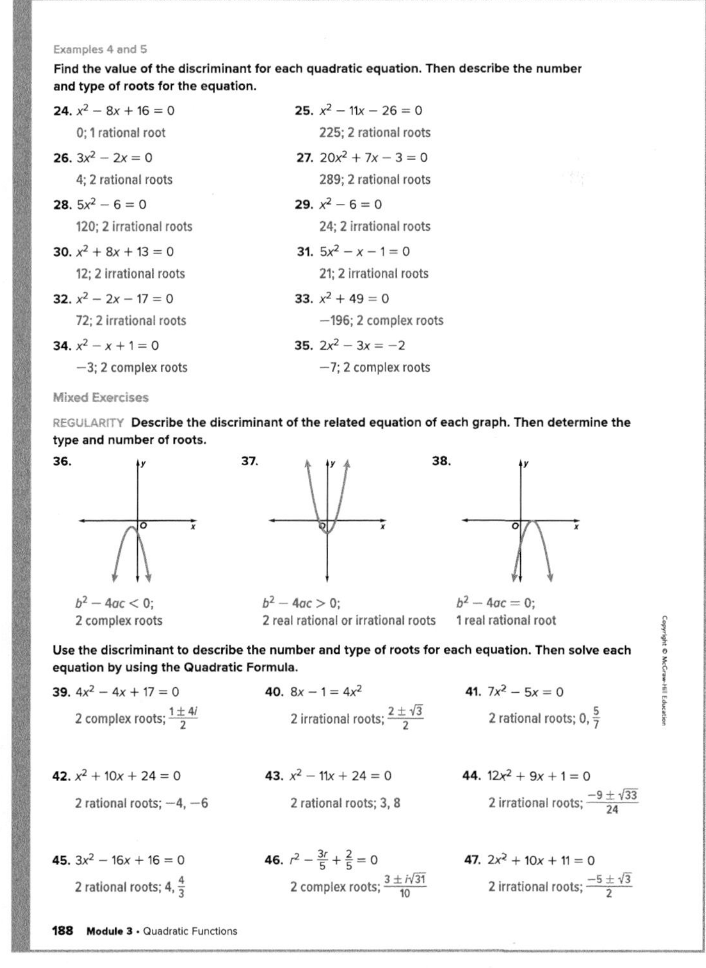

Examples 4 and 5

Find the value of the discriminant for each quadratic equation. Then describe the number and type of roots for the equation.

24. $x^2 - 8x + 16 = 0$
    0; 1 rational root
25. $x^2 - 11x - 26 = 0$
    225; 2 rational roots
26. $3x^2 - 2x = 0$
    4; 2 rational roots
27. $20x^2 + 7x - 3 = 0$
    289; 2 rational roots
28. $5x^2 - 6 = 0$
    120; 2 irrational roots
29. $x^2 - 6 = 0$
    24; 2 irrational roots
30. $x^2 + 8x + 13 = 0$
    12; 2 irrational roots
31. $5x^2 - x - 1 = 0$
    21; 2 irrational roots
32. $x^2 - 2x - 17 = 0$
    72; 2 irrational roots
33. $x^2 + 49 = 0$
    −196; 2 complex roots
34. $x^2 - x + 1 = 0$
    −3; 2 complex roots
35. $2x^2 - 3x = -2$
    −7; 2 complex roots

Mixed Exercises

REGULARITY **Describe the discriminant of the related equation of each graph. Then determine the type and number of roots.**

36. $b^2 - 4ac < 0$; 2 complex roots
37. $b^2 - 4ac > 0$; 2 real rational or irrational roots
38. $b^2 - 4ac = 0$; 1 real rational root

Use the discriminant to describe the number and type of roots for each equation. Then solve each equation by using the Quadratic Formula.

39. $4x^2 - 4x + 17 = 0$
    2 complex roots; $\frac{1 \pm 4i}{2}$
40. $8x - 1 = 4x^2$
    2 irrational roots; $\frac{2 \pm \sqrt{3}}{2}$
41. $7x^2 - 5x = 0$
    2 rational roots; $0, \frac{5}{7}$
42. $x^2 + 10x + 24 = 0$
    2 rational roots; $-4, -6$
43. $x^2 - 11x + 24 = 0$
    2 rational roots; $3, 8$
44. $12x^2 + 9x + 1 = 0$
    2 irrational roots; $\frac{-9 \pm \sqrt{33}}{24}$
45. $3x^2 - 16x + 16 = 0$
    2 rational roots; $4, \frac{4}{3}$
46. $r^2 - \frac{3r}{5} + \frac{2}{5} = 0$
    2 complex roots; $\frac{3 \pm i\sqrt{31}}{10}$
47. $2x^2 + 10x + 11 = 0$
    2 irrational roots; $\frac{-5 \pm \sqrt{3}}{2}$

188 Module 3 • Quadratic Functions

Name ______________ Period ______ Date ______

48. USE A MODEL The height $h(t)$ in feet of an object t seconds after it is propelled up from the ground with an initial velocity of 60 feet per second is modeled by the equation $h(t) = -16t^2 + 60t$. When will the object be at a height of 56 feet?
1.75 s, 2 s

49. SPORTS Natalya Lisovskaya set the women's shot put world record of 22.63 meters. Her throw can be modeled by $h = -4.9t^2 + 13.7t + 1.6$, where t is time in seconds and h is the height in meters. About how long was the shot in the air?
2.9 s

50. STOPPING DISTANCE A car's stopping distance d is the sum of the distance traveled during the time it takes the driver to react and the distance traveled while braking. This is represented as $d = vt + \frac{v^2}{2\mu g}$, where v is the initial velocity in feet per second, t is the driver's reaction time in seconds, μ is the coefficient of friction, and g is acceleration due to gravity. Use $g = 32$ ft/s^2.

a. Assume $\mu = 0.8$ for rubber tires on dry pavement and the average reaction time of 1.5 seconds to complete the table. Round to the nearest tenth.

Velocity, v (ft/s)	15	40	55	70	80
Stopping Distance, d (ft)	26.9	91.25	141.6	200.7	245

b. Make different assumptions to complete the table. Round to the nearest tenth.
coefficient of friction μ = Sample answer: 0.5 for rubber tires on wet pavement
reaction time t = Sample answer: 1.5 seconds

Velocity, v (ft/s)	15	35.1	55	59.7	67.7
Stopping Distance, d (ft)	29.5	91.25	177.0	200.7	245

c. How did your different assumptions affect the data you found? Interpret the information in the context of the situation.
Sample answer: Using the coefficient of friction for wet pavement made the stopping distance for a given velocity longer. At a given velocity, a car will take a longer distance to stop on wet pavement than it does on dry pavement.

51. GEOMETRY A rectangular box has a square base and a height that is one more than 3 times the length of a side of the base. If the sides of the base are each increased by 2 inches and the height is increased by 3 inches, the volume of the box increases by 531 cubic inches. Define a variable and write an equation to represent the situation. Then find the dimensions of the original box.
x = the length of a side of the base, $(x + 2)^2(3x + 4) = x^2(3x + 1) + 531$; 5 in. by 5 in. by 16 in.

$V_{new} = V_{original} + 531$ in^3

52. GAMES A carnival game has players hit a pad with a large rubber mallet. This fires a ball up a 20-foot vertical chute toward a target at the top. A prize is awarded if the ball hits the target. Explain how to find the initial velocity in feet per second for which the ball will fail to hit the target. Assume the height of the ball can be modeled by the function $h(t) = -16t^2 + vt$, where v is the initial velocity.
See margin.

Higher-Order Thinking Skills

53. WHICH ONE DOESN'T BELONG? Use the discriminant to determine which of these equations is different from the others. Explain your reasoning.

$x^2 - 3x - 40 = 0$ | $12x^2 - x - 6 = 0$ | $12x^2 + 2x - 4 = 0$ | $7x^2 + 6x + 2 = 0$

$7x^2 + 6x + 2 = 0$ is different from the other 3 equations because it has 2 complex roots, where the other 3 equations each have 2 rational roots.

54. FIND THE ERROR Tama and Jonathan are determining the number of solutions of $3x^2 - 5x = 7$. Is either of them correct? Explain your reasoning.

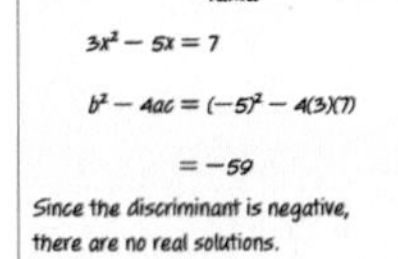

Tama
3x² − 5x = 7
b² − 4ac = (−5)² − 4(3)(7)
= −59
Since the discriminant is negative,
there are no real solutions.

Jonathan
$3x^2 - 5x = 7$
$3x^2 - 5x - 7 = 0$
$b^2 - 4ac = (-5)^2 - 4(3)(-7)$
$= 109$
Since the discriminant is positive, there are two real roots.

Jonathan; you must first write the equation in the form $ax^2 + bx + c = 0$ to determine the values of a, b, and c. Therefore, the value of c is -7, not 7.

55. ANALYZE Determine whether each statement is *sometimes*, *always*, or *never* true. Explain your reasoning. a, b. See margin.

a. In a quadratic equation written in standard form, if a and c have different signs, then the solutions will be real.

b. If the discriminant of a quadratic equation is greater than 1, the two roots are real irrational numbers.

56. CREATE Sketch the corresponding graph and state the number and type of roots for each of the following. a–e. See margin.

a. $b^2 - 4ac = 0$

b. A quadratic function in which $f(x)$ never equals zero.

c. A quadratic function in which $f(a) = 0$ and $f(b) = 0$; $a \neq b$.

d. The discriminant is less than zero.

e. a and b are both solutions and can be represented as fractions.

57. PERSEVERE Find the value(s) of m in the quadratic equation $x^2 + x + m + 1 = 0$ such that it has one solution. −0.75

58. WRITE Describe three different ways to solve $x^2 - 2x - 15 = 0$. Which method do you prefer, and why? See margin.

Answers

52. Set $h(t) = 20$; use the Quadratic Formula: $t = \frac{v \pm \sqrt{v^2 - 4(16)(20)}}{3^2}$. The ball fails to hit the target if the solutions are complex, when $v^2 - 1280 < 0$. If the initial velocity is less than approximately 35.8 feet per second, the ball will not hit the target.

55a. Always; sample answer: When a and c are opposite signs, then ac will always be negative and $-4ac$ will always be positive. Because b^2 will also always be positive, then $b^2 - 4ac$ represents the addition of two positive values, which will never be negative. Hence, the discriminant can never be negative and the solutions can never be imaginary.

55b. Sometimes; sample answers: The roots will only be irrational if $b^2 - 4ac$ is not a perfect square.

56. Sample graphs shown.

56a. 1 rational root

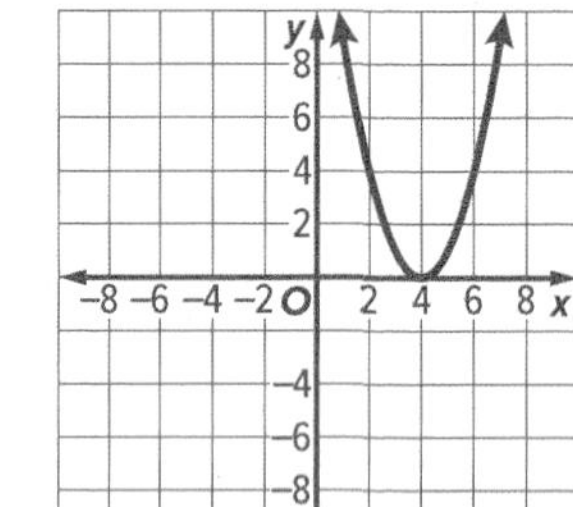
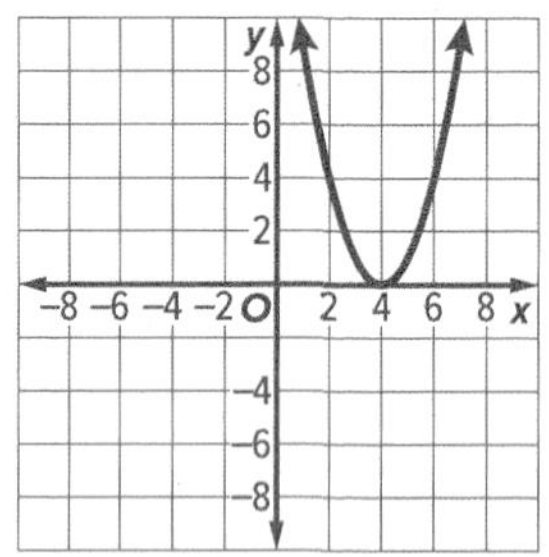

56b. 2 complex roots

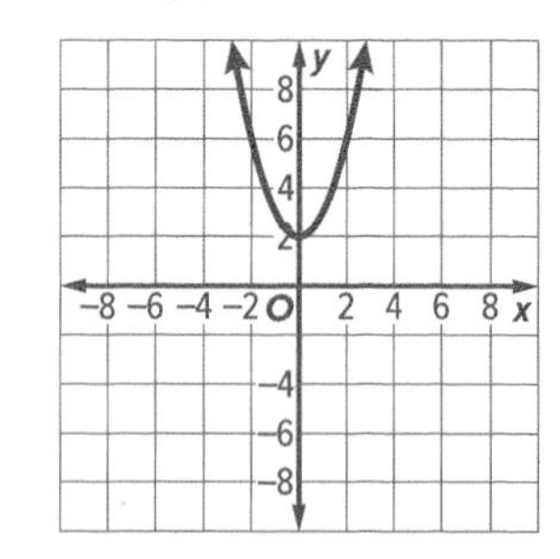

56c. 2 real roots

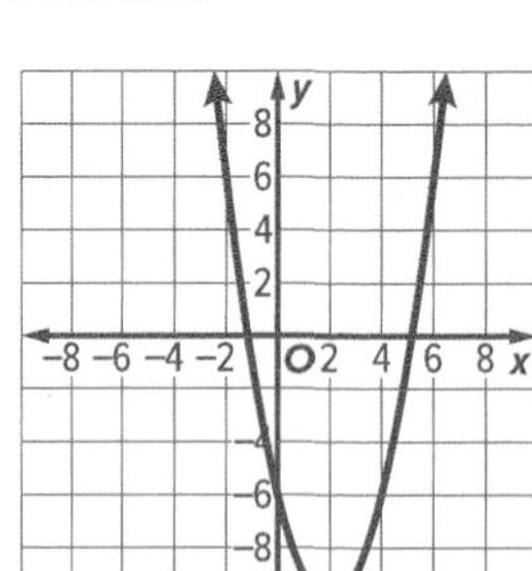
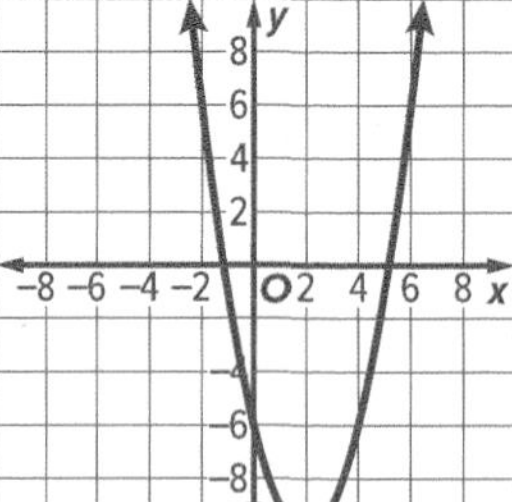

56d. 2 complex roots

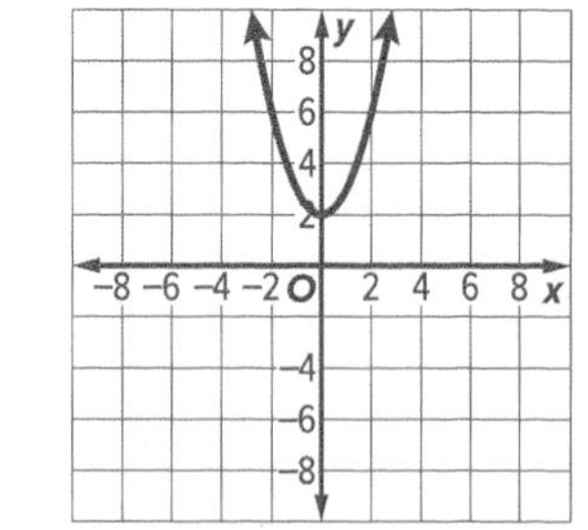

56e. 2 rational roots

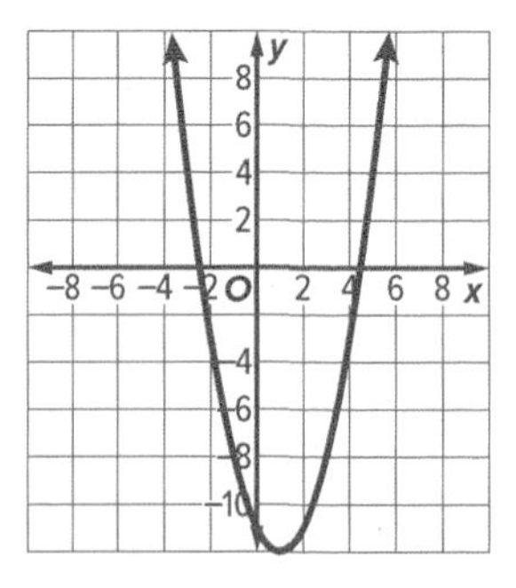

58. Sample answer: (1) Factor $x^2 - 2x - 15$ as $(x + 3)(x - 5)$. Then according to the Zero Product Property, either $x + 3 = 0$ or $x - 5 = 0$. Solving these equations, $x = -3$ or $x = 5$. (2) Rewrite the equation as $x^2 - 2x = 15$. Then add 1 to each side of the equation to complete the square on the left side. Then $(x - 1)^2 = 16$. Taking the square root of each side, $x - 1 = \pm 4$. Therefore, $x = -3$ or $x = 5$. (3) Use the Quadratic Formula. Thus, $x = \frac{2 \pm \sqrt{2^2 - 4(1)(-15)}}{2(1)}$ or $x = \frac{2 \pm \sqrt{64}}{2}$. Simplifying the expression, $x = -3$ or $x = 5$. See students' preferences.

Quadratic Inequalities

LESSON GOAL

Students graph and solve quadratic inequalities.

1 LAUNCH

Launch the lesson with a **Warm Up** and an introduction.

2 EXPLORE AND DEVELOP

Explore: Graphing Quadratic Inequalities

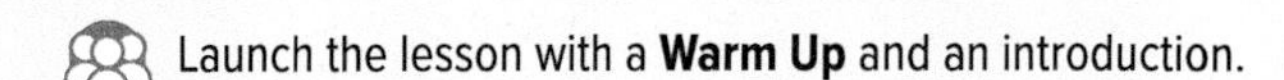

Develop:

Graphing Quadratic Inequalities
- Graph a Quadratic Inequality ($<$ or $\leq$)
- Graph a Quadratic Inequality ($>$ or $\geq$)

Solving Quadratic Inequalities
- Solve a Quadratic Inequality ($<$ or $\leq$) by Graphing
- Solve a Quadratic Inequality ($>$ or $\geq$) by Graphing
- Solve a Quadratic Inequality Algebraically

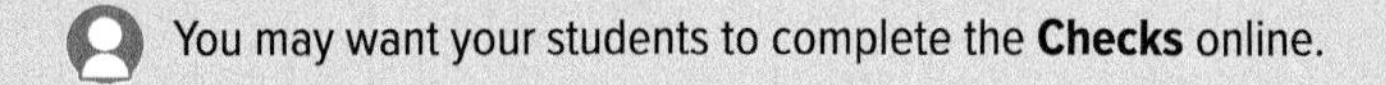

You may want your students to complete the **Checks** online.

3 REFLECT AND PRACTICE

Exit Ticket

Practice

DIFFERENTIATE

View reports of student progress on the **Checks** after each example.

Resources	AL	OL	BL	ELL
Remediation: Graphing Inequalities in Two Variables	●	●		●
Extension: Graphing Intersections of Quadratic Inequalities		●	●	●

Language Development Handbook

Assign page 23 of the *Language Development Handbook* to help your students build mathematical language related to graphing and solving quadratic inequalities.

ELL You can use the tips and suggestions on page T23 of the handbook to support students who are building English proficiency.

Suggested Pacing

90 min	0.5 day
45 min	1 day

Focus

Domain: Algebra

Standards for Mathematical Content:

A.CED.1 Create equations and inequalities in one variable and use them to solve problems.

A.CED.3 Represent constraints by equations or inequalities, and by systems of equations and/or inequalities, and interpret solutions as viable or nonviable options in a modeling context.

Standards for Mathematical Practice:

1 Make sense of problems and persevere in solving them.

4 Model with mathematics.

Coherence

Vertical Alignment

Previous
Students solved quadratic equations by graphing.
A.CED.2, F.IF.4

Now
Students graph and solve quadratic inequalities.
A.CED.1, A.CED.3

Next
Students will solve systems of linear and quadratic equations.
A.REI.11

Rigor

The Three Pillars of Rigor

1 CONCEPTUAL UNDERSTANDING	2 FLUENCY	3 APPLICATION

Conceptual Bridge In this lesson, students extend their understanding of quadratic equations to quadratic inequalities. They build fluency by solving quadratic inequalities, and they apply their understanding by solving real-world problems.

Mathematical Background

Graph quadratic inequalities in two variables using the same techniques as when graphing linear inequalities in two variables. Graph the related function. Use a dashed line if the symbol is $<$ or $>$. Test a point inside the parabola. If it is a solution, shade inside the parabola. If it is not a solution, shade outside the parabola.

Interactive Presentation

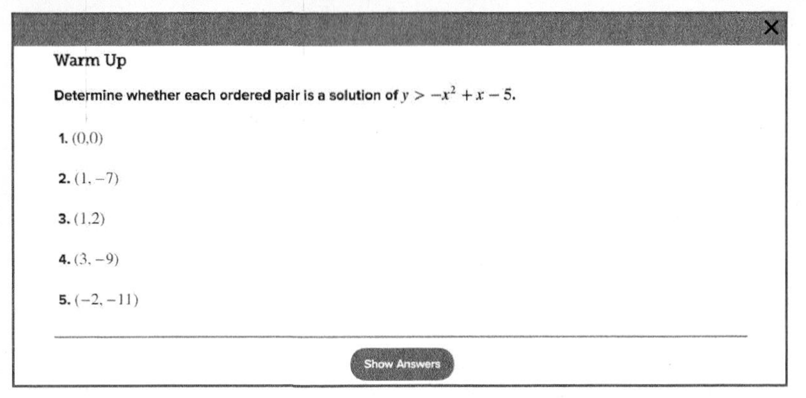

Warm Up

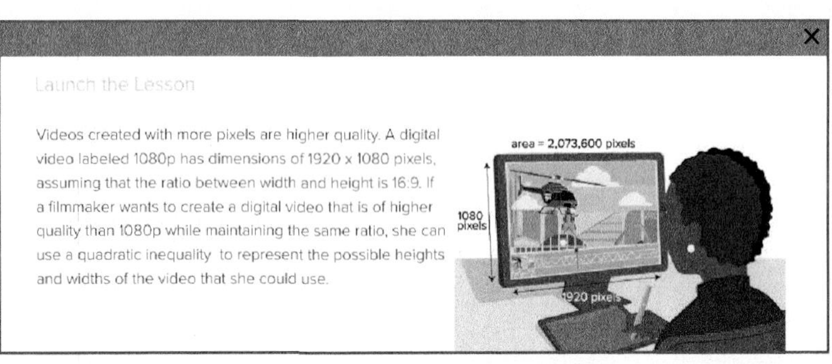

Launch the Lesson

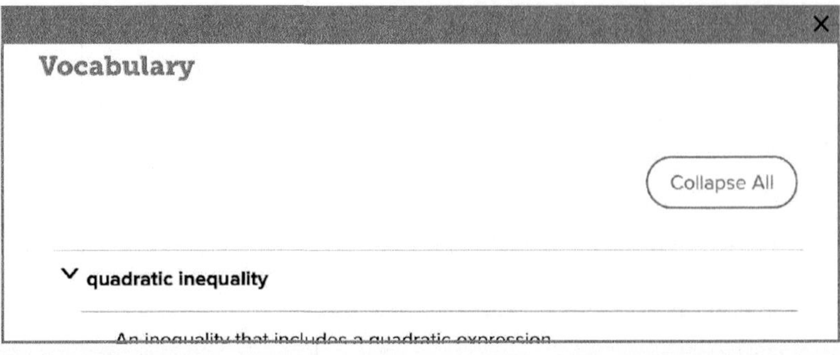

Today's Vocabulary

Warm Up

Prerequisite Skills

The Warm Up exercises address the following prerequisite skill for this lesson:

- determining whether an ordered pair is a solution

Answers:

1. yes
2. no
3. yes
4. yes
5. no

Launch the Lesson

Teaching the Mathematical Practices

4 Apply Mathematics In the Launch the Lesson, students will learn how a quadratic inequality can be used to determine the dimensions of a high quality video that maintains a given aspect ratio.

Go Online to find additional teaching notes and questions to promote classroom discourse.

Today's Standards

Tell students that they will be addressing these content and practice standards in this lesson. You may wish to have a student volunteer read aloud *How can I meet these standards?* and *How can I use these practices?*, and connect these to the standards.

See the Interactive Presentation for I Can statements that align with the standards covered in this lesson.

Today's Vocabulary

Tell students that they will be using this vocabulary term in this lesson. You can expand the row if you wish to share the definition. Then discuss the questions below with the class.

Explore Graphing Quadratic Inequalities

Objective

Students explore the graphs of quadratic inequalities.

 Teaching the Mathematical Practices

5 Use Mathematical Tools Point out that to complete the exercises in the Explore, students will need to use the sketch. Work with students to explore and deepen their understanding of quadratic inequalities.

Ideas for Use

Recommended Use Present the Inquiry Question, or have a student volunteer read it aloud. Have students work in pairs to complete the Explore activity on their devices. Pairs should discuss each of the questions. Monitor student progress during the activity. Upon completion of the Explore activity, have student volunteers share their responses to the Inquiry Question.

What if my students don't have devices? You may choose to project the activity on a whiteboard. A printable worksheet for each Explore is available online. You may choose to print the worksheet so that individuals or pairs of students can use it to record their observations.

Summary of the Activity

Students will complete guiding exercises throughout the Explore activity. Students will use a sketch to examine the value of y at various points on the curve and coordinate place to draw conclusions about graphing quadratic inequalities. Then, students will answer the Inquiry Question.

(continued on the next page)

Interactive Presentation

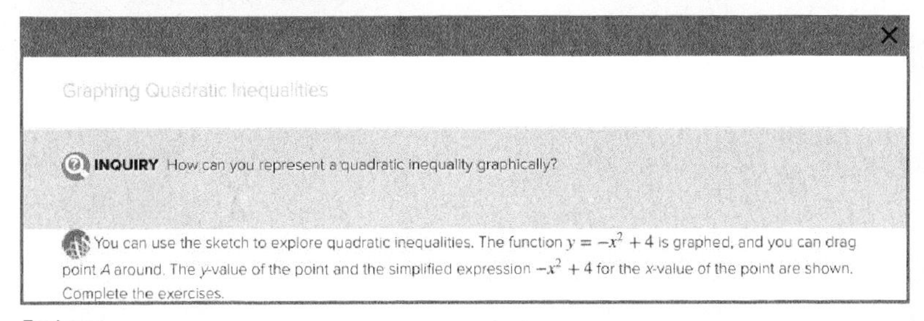

Explore

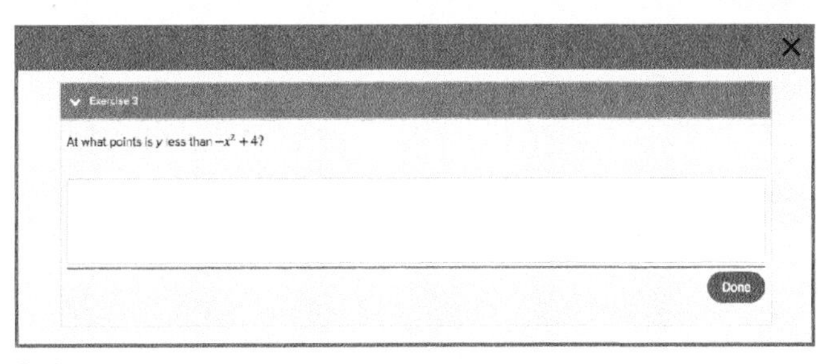

Explore

WEB SKETCHPAD

Students use a sketch to explore the graph of a quadratic inequality.

Interactive Presentation

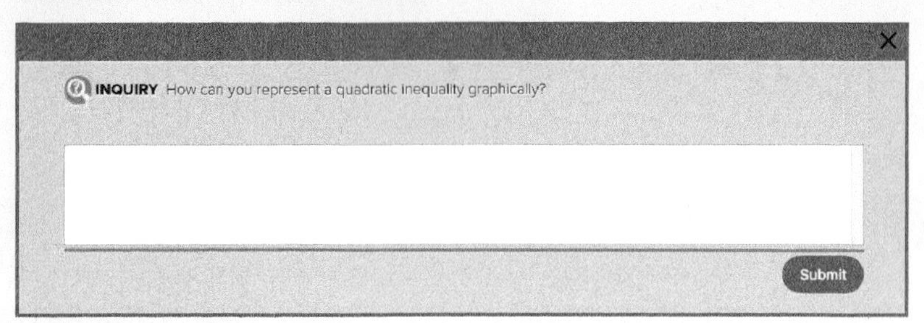

Explore

TYPE

Students respond to the Inquiry Question and can view a sample answer.

Explore Graphing Quadratic Inequalities (*continued*)

Questions

Have students complete the Explore activity.

Ask:

- How do you represent the graph of a linear inequality? Sample answer: You use a solid or dashed line based on the inequality symbol, then shade one region, usually above or below the line.
- How would the graph be different for $y > -x^2 + 4$? Sample answer: The curve would be dashed instead of solid, because the inequality symbol does not include "or equal to". Also, the shaded region would be above (outside) the curve instead of below (inside) it.

Inquiry

How can you represent a quadratic inequality graphically? Sample answer: I could graph the related function and shade to represent all the points for which that inequality is true.

Go Online to find additional teaching notes and sample answers for the guiding exercises.

1 CONCEPTUAL UNDERSTANDING | 2 FLUENCY | 3 APPLICATION

Learn Graphing Quadratic Inequalities

Objective

Students graph quadratic inequalities in two variables by using the related equations.

Teaching the Mathematical Practices

1 Explain Correspondences Encourage students to explain the relationships between graphing linear inequalities and graphing quadratic inequalities.

DIFFERENTIATE

Reteaching Activity AL

IF students are having trouble making connections between the graph of a quadratic inequality and the inequality itself,

THEN have students think about how the graph of a quadratic inequality helps them understand the meaning of the inequality. Ask them to explore whether the quadratic inequality itself or the graph of the inequality is more meaningful to them. Ask them to give explanations for their choices.

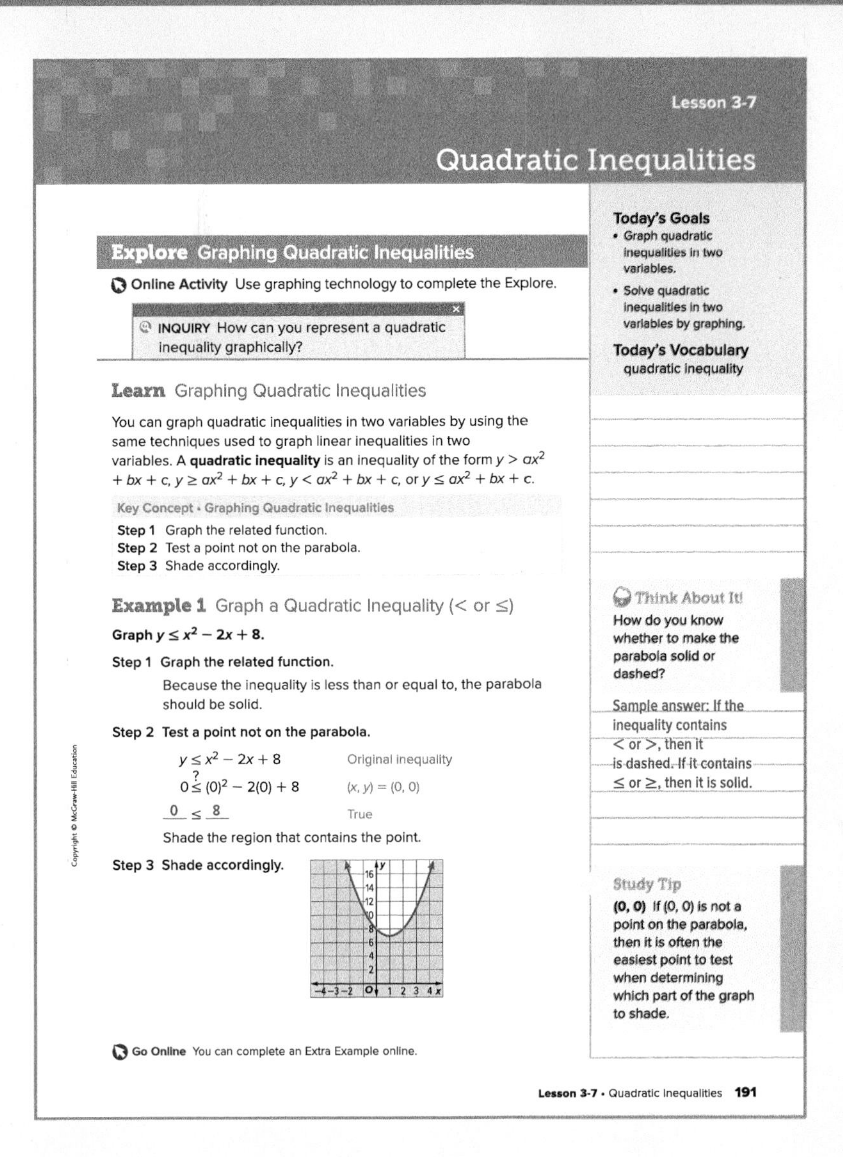

Lesson 3-7

Quadratic Inequalities

Explore Graphing Quadratic Inequalities

Online Activity Use graphing technology to complete the Explore.

INQUIRY How can you represent a quadratic inequality graphically?

Learn Graphing Quadratic Inequalities

You can graph quadratic inequalities in two variables by using the same techniques used to graph linear inequalities in two variables. A **quadratic inequality** is an inequality of the form $y > ax^2 + bx + c$, $y \geq ax^2 + bx + c$, $y < ax^2 + bx + c$, or $y \leq ax^2 + bx + c$.

Key Concept • Graphing Quadratic Inequalities

Step 1 Graph the related function.

Step 2 Test a point not on the parabola.

Step 3 Shade accordingly.

Example 1 Graph a Quadratic Inequality (< or ≤)

Graph $y \leq x^2 - 2x + 8$.

Step 1 Graph the related function.

Because the inequality is less than or equal to, the parabola should be solid.

Step 2 Test a point not on the parabola.

$y \leq x^2 - 2x + 8$ Original inequality

$0 \overset{?}{\leq} (0)^2 - 2(0) + 8$ $(x, y) = (0, 0)$

$0 \leq 8$ True

Shade the region that contains the point.

Step 3 Shade accordingly.

Go Online You can complete an Extra Example online.

Today's Goals
- Graph quadratic inequalities in two variables.
- Solve quadratic inequalities in two variables by graphing.

Today's Vocabulary
quadratic inequality

Think About It!
How do you know whether to make the parabola solid or dashed?

Sample answer: If the inequality contains < or >, then it is dashed. If it contains ≤ or ≥, then it is solid.

Study Tip
(0, 0) If (0, 0) is not a point on the parabola, then it is often the easiest point to test when determining which part of the graph to shade.

Lesson 3-7 • Quadratic Inequalities 191

Example 1 Graph a Quadratic Inequality (< or ≤)

Teaching the Mathematical Practices

1 Explain Correspondences Encourage students to explain the relationships between the inequalities and graphs used in Example 1.

Interactive Presentation

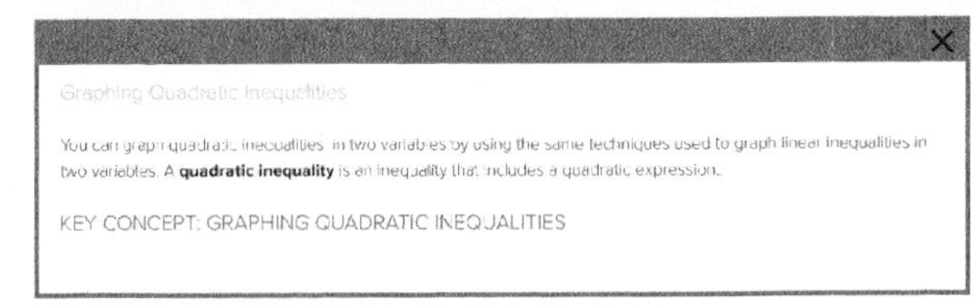

Learn

SWYPE

Students move through the step to see how to graph a quadratic inequality.

TYPE

Students answer a question to determine whether they understand how to graph a quadratic inequality.

Questions for Mathematical Discourse

AL Why are the points on the parabola valid solutions to the inequality? The inclusive symbol ≤ is used.

OL How many points are solutions to this inequality? infinitely many

BL How does changing the shape of the parabola change the number of solutions? Sample answer: The number of solutions does not change, it will still be infinite.

Important to Know

When testing a point, ask the students if there is any point on the graph that would be easier to test than others. Guide them to realizing that testing the point (0, 0) may be the easiest point to plot, unless that point is on the related function.

Go Online

- Find additional teaching notes.
- View performance reports of the Checks.
- Assign or present an Extra Example.

Your Notes

Example 2 Graph a Quadratic Inequality (> or ≥)

Graph $y > -5x^2 + 10x$.

Step 1 Graph the related function.

Because the inequality is greater than, the parabola should be dashed.

Step 2 Test a point not on the parabola.

Because (0, 0) is on the parabola, use (1, 0) as a test point.

$y > -5x^2 + 10x$ Original inequality

$0 \overset{?}{>} -5(1)^2 + 10(1)$ $(x, y) = (1, 0)$

$0 > 5$ False

So, (1, 0) is not a solution of the inequality.

Step 3 Shade accordingly.

Because (1, 0) is not a solution of the inequality, shade the region that does not contain the point.

Learn Solving Quadratic Inequalities

Key Concept • Solving Quadratic Inequalities

$ax^2 + bx + c < 0$ — $a > 0$ — $a < 0$

Graph $y = ax^2 + bx + c$ and identify the x-values for which the graph lies *below* the x-axis.

For ≤, include the x-intercepts in the solution.

$\{x \mid x_1 < x < x_2\}$ — $\{x \mid x < x_1 \text{ or } x > x_2\}$

$ax^2 + bx + c > 0$ — $a > 0$ — $a < 0$

Graph $y = ax^2 + bx + c$ and identify the x-values for which the graph lies *above* the x-axis.

For ≥, include the x-intercepts in the solution.

$\{x \mid x < x_1 \text{ or } x > x_2\}$ — $\{x \mid x_1 < x < x_2\}$

Go Online You can complete an Extra Example online.

Interactive Presentation

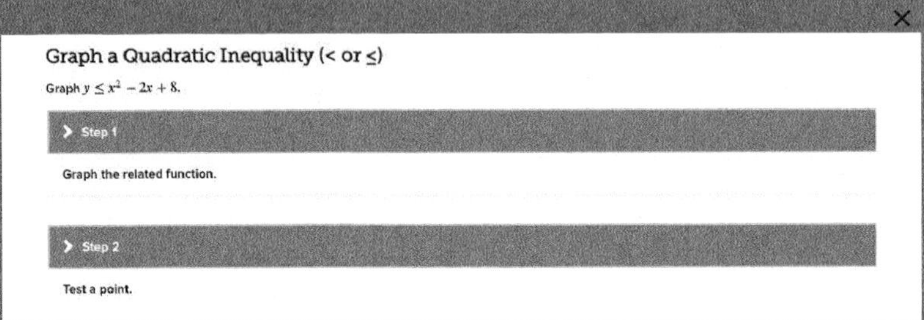

Example 1

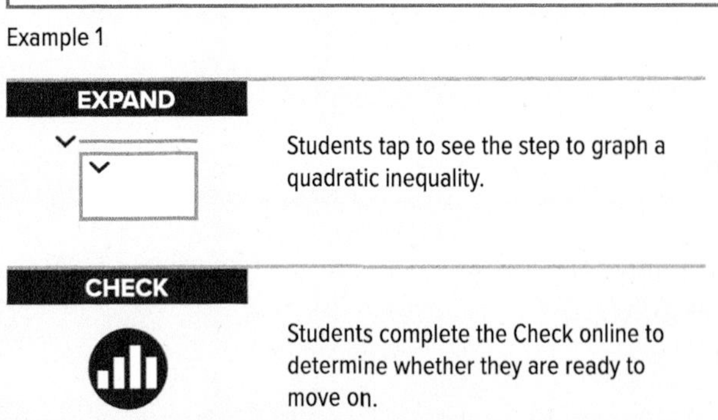

Students tap to see the step to graph a quadratic inequality.

Students complete the Check online to determine whether they are ready to move on.

Example 2 Graph a Quadratic Inequality (> or ≥)

 Teaching the Mathematical Practices

1 Explain Correspondences Encourage students to explain the relationships between the inequality and graph in Example 2. Help students to see the relationship between the inequality symbol and the shaded region.

Questions for Mathematical Discourse

AL Will the parabola of the related function open up or down? How do you know? down; $a < 0$.

OL Can point (0, 0) be used to test this inequality? Why or why not? No; the parabola passes through the point (0, 0).

BL Is it necessary to test a point that you suspect will be in the shaded region of the inequality? No, you can test any point not on the boundary line. Why might you intentionally test a point not likely to be part of the solution set? Sample answer: The calculations may be simpler when using a point outside the solution set. If the result is a false statement, you know the other region is the shaded region.

Learn Solving Quadratic Inequalities

Objective

Students solve quadratic inequalities in two variables by graphing.

 Teaching the Mathematical Practices

3 Analyze Cases Work with students to examine the cases of quadratic inequalities. Encourage students to familiarize themselves with all of the cases and how to solve each case by graphing.

DIFFERENTIATE

Reteaching Activity AL BL

IF students are having trouble figuring out how the graphing of quadratic functions in two-dimensions relates to the solution of an inequality in one variable,

THEN review some examples of linear inequalities by plotting them on a coordinate plane in a similar way. Have students relate the two-dimensional graph to a linear inequality on a number line, then compare this to the solution set from graphing a quadratic inequality.

Example 3 Solve a Quadratic Inequality ($\leq$ or $<$) by Graphing

Teaching the Mathematical Practices

1 Explain Correspondences Encourage students to explain the relationships between the inequality and graph used in Example 3.

Questions for Mathematical Discourse

AL How is this example different from Examples 1 and 2? Sample answer: the inequality has only the variable x.

OL Why do you shade a line on the x-axis rather than the whole area above the function? Sample answer: The original inequality only has x as a variable. The related function is used as a means of finding the x-values that are solutions to the original inequality.

BL How would the solution change if the inequality symbol was changed to $\leq$? -3 and 2 would be included in the solution set.

Common Error

The related equations show only a single variable, so the student may be confused about how to graph it on a 2-dimensional coordinate plane. Discuss the relationship between the related equation and the variable y.

Example 4 Solve a Quadratic Inequality ($>$ or $\geq$) by Graphing

Teaching the Mathematical Practices

3 Justify Conclusions Mathematically proficient students can explain conclusions drawn when solving a problem. The Talk About It! feature asks students to justify their conclusion about the solution set of a quadratic inequality.

Questions for Mathematical Discourse

AL Why are the roots of the quadratic included in the solution? The inequality symbol for greater than or equal is used.

OL Do you need to determine the vertex to find the solution set? Explain. No; sample answer: Because the solution set is the x-values where the function is above the x-axis, so we only need to know the intercepts and which way the parabola opens.

BL Compare and contrast the solution sets for $x^2 - 3x - 4 \geq 0$ and $x^2 - 3x - 4 \leq 0$. Sample $x^2 - 3x - 4 \geq 0$; sample answer: The solution set for $x^2 - 3x - 4 \geq 0$ is $(-\infty, -1] \cup [4, \infty)$ and the solution set for $x^2 - 3x - 4 \leq 0$ is $[-1, 4]$. The solution sets both contain -1 and 4. The solution set for $x^2 - 3x - 4 \geq 0$ is two intervals. The solution set for $x^2 - 3x - 4 \leq 0$ is one interval.

Example 3 Solve a Quadratic Inequality ($<$ or $\leq$) by Graphing

Solve $x^2 + x - 6 < 0$ by graphing.

Because the quadratic expression is less than 0, the solution consists of x-values for which the graph of the related function lies *below* the x-axis. Begin by finding the zeros of the related function.

$x^2 + x - 6 = 0$ Related equation

$(x - \underline{2})(x + \underline{3}) = 0$ Factor.

$x = \underline{2}$ or $x = \underline{-3}$ Zero Product Property

Sketch the graph of a parabola that has x-intercepts at 2 and -3. The graph should open up because $a > 0$.

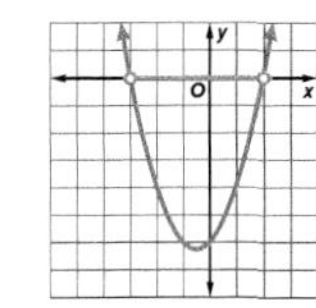

The graph lies below the x-axis between $\underline{-3}$ and $\underline{2}$. Thus, the solution set is $\{x \mid \underline{-3} < x < \underline{2}\}$ or in interval notation $(\underline{-3, 2})$.

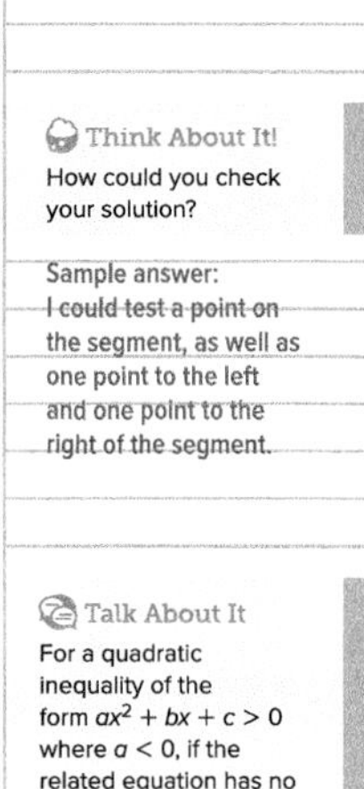

Think About It! How could you check your solution?

Sample answer: I could test a point on the segment, as well as one point to the left and one point to the right of the segment.

Example 4 Solve a Quadratic Inequality ($>$ or $\geq$) by Graphing

Solve $x^2 - 3x - 4 \geq 0$ by graphing.

Because the quadratic expression is greater than or equal to 0, the solution consists of x-values for which the graph of the related function lies *on* and *above* the x-axis. Begin by finding the zeros of the related function.

$x^2 - 3x - 4 = 0$ Related equation

$(x - \underline{4})(x + \underline{1}) = 0$ Factor.

$x = \underline{4}$ or $x = \underline{-1}$ Zero Product Property

Sketch the graph of a parabola that has x-intercepts at -1 and 4. The graph should open up because $a > 0$.

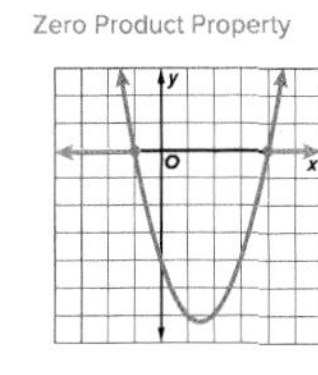

The graph lies above and on the x-axis when $x \leq \underline{-1}$ or $x \geq \underline{4}$. Thus, the solution set is $\{x \mid x \leq \underline{-1}$ or $x \geq \underline{4}\}$ or $(\underline{-\infty, -1}] \cup [\underline{4, \infty})$.

Talk About It! For a quadratic inequality of the form $ax^2 + bx + c > 0$ where $a < 0$, if the related equation has no real roots, what is the solution set? Explain your reasoning.

Sample answer: The solution set is the empty set, because there are no values of x for which $ax^2 + bx + c > 0$.

Check

Solve $-\frac{1}{4}x^2 + x + 1 > 0$ by graphing and write the solution set.

$\{x \mid \underline{-0.83} < x < \underline{4.83}\}$

Go Online You can complete an Extra Example online.

Lesson 3-7 • Quadratic Inequalities 193

Interactive Presentation

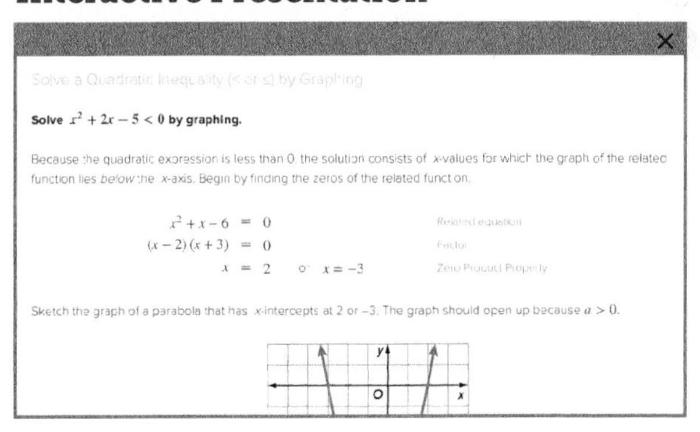

Example 3

TYPE

Students complete the calculations to solve a quadratic inequality by graphing.

TYPE

Students answer a question to determine whether they understand graphing quadratic inequalities.

Example 5 Solve a Quadratic Inequality Algebraically

GARDENING **Marcus is planning a garden. He has enough soil to cover 104 square feet, and wants the dimensions of the garden to be at least 5 feet by 10 feet. If he wants to increase the length and width by the same number of feet, by what value can he increase the dimensions of the garden without needing to buy more soil? Create a quadratic inequality and solve it algebraically.**

Step 1 **Determine the quadratic inequality.**

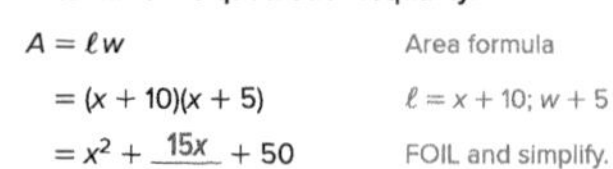

$A = \ell w$ — Area formula

$= (x + 10)(x + 5)$ — $\ell = x + 10$; $w + 5$

$= x^2 + \underline{15x} + 50$ — FOIL and simplify.

The area must be less than or equal to 104 square feet, so $x^2 + 15x + 50 \leq \underline{104}$.

Step 2 **Solve the related equation.**

$x^2 + 15x + 50 = 104$ — Related equation

$x^2 + 15x \underline{- 54} = 0$ — Subtract 104 from each side.

$(x \underline{+ 18})(x \underline{- 3}) = 0$ — Factor.

$x = -18$ or $x = 3$ — Zero Product Property

Steps 3 and 4 **Plot the solutions on a number line and test a value from each interval.**

Use dots because −18 and 3 are solutions of the original inequality.

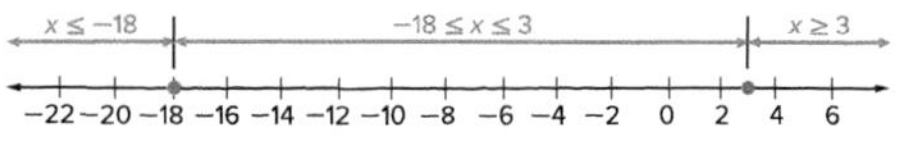

Test a value from each interval to see if it satisfies the original inequality.

Test $x = -20$, $x = 0$, and $x = 5$. The only value that satisfies the original inequality is $x = \underline{0}$, so the solution set is $[-18, \underline{3}]$. So, Marcus can increase the length and width up to $\underline{3}$ feet without needing to buy more soil. The interval $-18 \leq x \leq 0$ is not relevant because Marcus does not want to decrease the length and width or leave it as is.

Check

MANUFACTURING An electronics manufacturer can model their profits in dollars P when they sell x video players by using the function $P(x) = -0.1x^2 + 75x - 1000$. How many video players can they sell so they make \$7500 or less?

The company will make \$7500 or less if they make $\underline{139}$ video players or fewer and/or $\underline{611}$ video players or more.

Go Online You can complete an Extra Example online.

Interactive Presentation

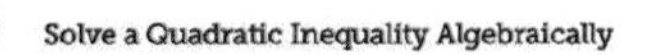

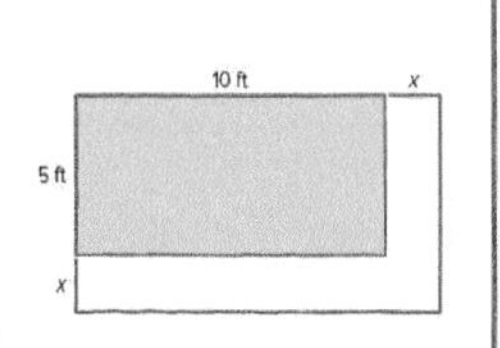

Example 5

TAP

Students tap to see the steps to solve a quadratic inequality algebraically.

CHECK

Students complete the Check online to determine whether they are ready to move on.

Example 5 Solve a Quadratic Inequality Algebraically

Teaching the Mathematical Practices

4 Analyze Relationships Mathematically Point out that to solve the problem in Exercise 5, students will need to analyze the mathematical relationships in the problem to draw a conclusion about the dimensions of the garden.

Questions for Mathematical Discourse

AL What does $A \leq 104$ mean in this context? The area of the garden that Marcus can make, without buying more soil, is 104 square feet or less.

OL How do you know from the context of the problem that the valid solutions are between the roots? Sample answer: We are looking for what length Marcus can add, which has an upper limit based on the amount of soil he has.

BL Consider the solution set if the direction of the inequality is flipped. How does the solution set compare to that of the original inequality? Sample answer: The original inequality has one solution interval that is bounded, and the flipped inequality has two solution intervals that are not bounded. Both solution sets contain an infinite number of points.

Exit Ticket

Recommended Use

At the end of class, go online to display the Exit Ticket prompt and ask students to respond using a separate piece of paper. Have students hand you their responses as they leave the room.

Alternate Use

At the end of class, go online to display the Exit Ticket prompt and ask students to respond verbally or by using a mini-whiteboard. Have students hold up their whiteboards so that you can see all student responses. Tap to reveal the answer when most or all students have completed the Exit Ticket.

Practice and Homework

Suggested Assignments

Use the table below to select appropriate exercises.

DOK	Topic	Exercises
1, 2	exercises that mirror the examples	1–20
2	exercises that use a variety of skills from this lesson	21–39
2	exercises that extend concepts learned in this lesson to new contexts	40–41
3	exercises that emphasize higher-order and critical thinking skills	42–46

ASSESS AND DIFFERENTIATE

Use the data from the **Checks** to determine whether to provide resources for extension, remediation, or intervention.

IF students score 90% or more on the Checks, **THEN** assign:

- Practice Exercises 1–41 odd, 42–46
- Extension: Graphing Intersections of Quadratic Inequalities
- ALEKS® Quadratic Inequalities

IF students score 66%–89% on the Checks, **THEN** assign: AL OL

- Practice Exercises 1–45 odd
- Remediation, Review Resources: Graphing Inequalities in Two Variables
- Personal Tutors
- Extra Examples 1–5
- ALEKS® Graphing Linear Inequalities

IF students score 65% or less on the Checks, **THEN** assign: AL

- Practice Exercises 1–19 odd
- Remediation, Review Resources: Graphing Inequalities in Two Variables
- ALEKS® Graphing Linear Inequalities

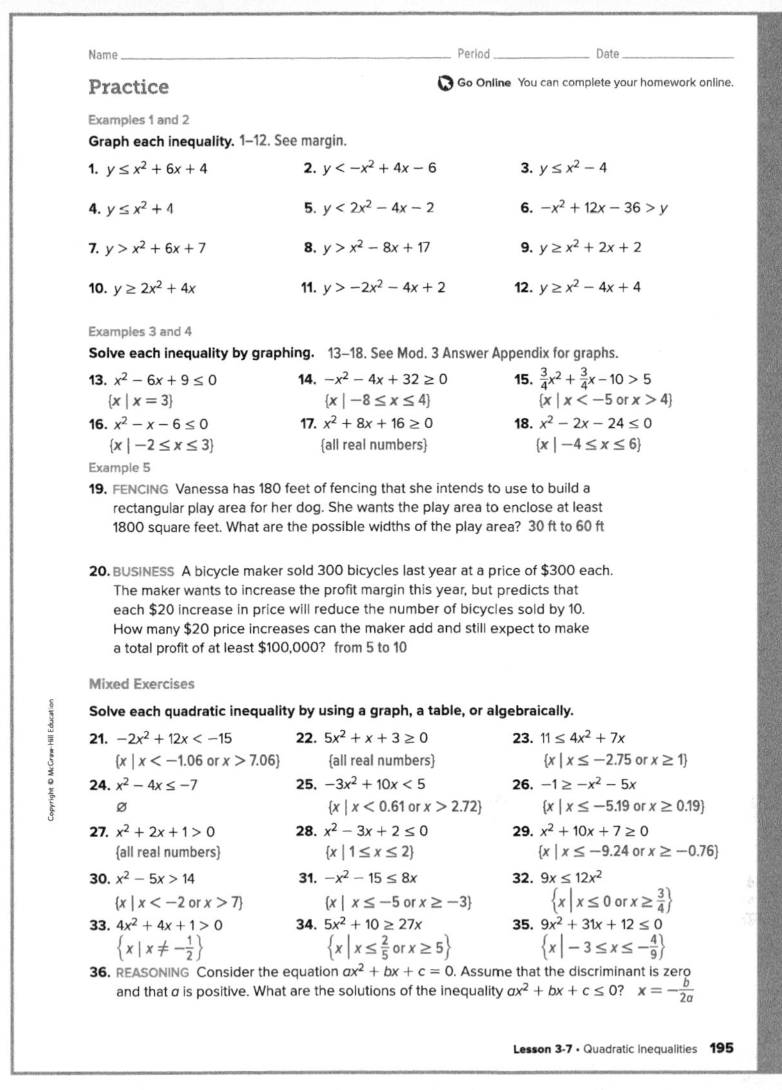

Name ______ Period ______ Date ______

Practice

Go Online You can complete your homework online.

Examples 1 and 2

Graph each inequality. 1–12. See margin.

1. $y \leq x^2 + 6x + 4$
2. $y < -x^2 + 4x - 6$
3. $y \leq x^2 - 4$
4. $y \leq x^2 + 4$
5. $y < 2x^2 - 4x - 2$
6. $-x^2 + 12x - 36 > y$
7. $y > x^2 + 6x + 7$
8. $y > x^2 - 8x + 17$
9. $y \geq x^2 + 2x + 2$
10. $y \geq 2x^2 + 4x$
11. $y > -2x^2 - 4x + 2$
12. $y \geq x^2 - 4x + 4$

Examples 3 and 4

Solve each inequality by graphing. 13–18. See Mod. 3 Answer Appendix for graphs.

13. $x^2 - 6x + 9 \leq 0$ $\{x \mid x = 3\}$
14. $-x^2 - 4x + 32 \geq 0$ $\{x \mid -8 \leq x \leq 4\}$
15. $\frac{3}{4}x^2 + \frac{3}{4}x - 10 > 5$ $\{x \mid x < -5 \text{ or } x > 4\}$
16. $x^2 - x - 6 \leq 0$ $\{x \mid -2 \leq x \leq 3\}$
17. $x^2 + 8x + 16 \geq 0$ {all real numbers}
18. $x^2 - 2x - 24 \leq 0$ $\{x \mid -4 \leq x \leq 6\}$

Example 5

19. FENCING Vanessa has 180 feet of fencing that she intends to use to build a rectangular play area for her dog. She wants the play area to enclose at least 1800 square feet. What are the possible widths of the play area? 30 ft to 60 ft

20. BUSINESS A bicycle maker sold 300 bicycles last year at a price of $300 each. The maker wants to increase the profit margin this year, but predicts that each $20 increase in price will reduce the number of bicycles sold by 10. How many $20 price increases can the maker add and still expect to make a total profit of at least $100,000? from 5 to 10

Mixed Exercises

Solve each quadratic inequality by using a graph, a table, or algebraically.

21. $-2x^2 + 12x < -15$ $\{x \mid x < -1.06 \text{ or } x > 7.06\}$
22. $5x^2 + x + 3 \geq 0$ {all real numbers}
23. $11 \leq 4x^2 + 7x$ $\{x \mid x \leq -2.75 \text{ or } x \geq 1\}$
24. $x^2 - 4x \leq -7$ $\varnothing$
25. $-3x^2 + 10x < 5$ $\{x \mid x < 0.61 \text{ or } x > 2.72\}$
26. $-1 \geq -x^2 - 5x$ $\{x \mid x \leq -5.19 \text{ or } x \geq 0.19\}$
27. $x^2 + 2x + 1 > 0$ {all real numbers}
28. $x^2 - 3x + 2 \leq 0$ $\{x \mid 1 \leq x \leq 2\}$
29. $x^2 + 10x + 7 \geq 0$ $\{x \mid x \leq -9.24 \text{ or } x \geq -0.76\}$
30. $x^2 - 5x > 14$ $\{x \mid x < -2 \text{ or } x > 7\}$
31. $-x^2 - 15 \leq 8x$ $\{x \mid x \leq -5 \text{ or } x \geq -3\}$
32. $9x \leq 12x^2$ $\left\{x \mid x \leq 0 \text{ or } x \geq \frac{3}{4}\right\}$
33. $4x^2 + 4x + 1 > 0$ $\left\{x \mid x \neq -\frac{1}{2}\right\}$
34. $5x^2 + 10 \geq 27x$ $\left\{x \mid x \leq \frac{2}{5} \text{ or } x \geq 5\right\}$
35. $9x^2 + 31x + 12 \leq 0$ $\left\{x \mid -3 \leq x \leq -\frac{4}{9}\right\}$
36. REASONING Consider the equation $ax^2 + bx + c = 0$. Assume that the discriminant is zero and that a is positive. What are the solutions of the inequality $ax^2 + bx + c \leq 0$? $x = -\frac{b}{2a}$

Lesson 3-7 • Quadratic Inequalities 195

Write a quadratic inequality for each graph.

37.

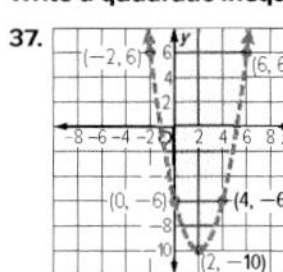

$y > x^2 - 4x - 6$

38.

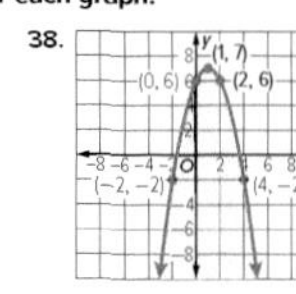

$y \leq -x^2 + 2x + 6$

39.

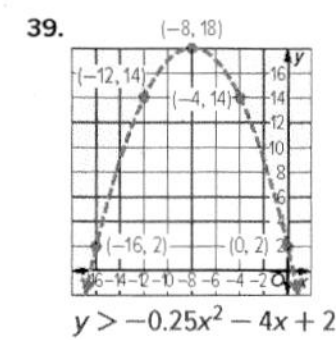

$y > -0.25x^2 - 4x + 2$

40. BASEBALL A baseball player hits a high pop-up with an initial velocity of 32 meters per second, 1.3 meters above the ground. The height $h(t)$ of the ball in meters t seconds after being hit is modeled by $h(t) = -4.9t^2 + 32t + 1.3$.

a. During what time interval is the ball higher than the camera located in the press box 43.4 meters above the ground? 1.83 to 4.70 s

b. When is the ball within 2.4 meters of the ground where the catcher can attempt to catch it? 0 to 0.035 s and 6.50 to 6.57 s

41. CONSTRUCT ARGUMENTS Are the boundaries of the solution set of $x^2 + 4x - 12 \leq 0$ twice the value of the boundaries of $\frac{1}{2}x^2 + 2x - 6 \leq 0$? Explain.
See Mod. 3 Answer Appendix.

Higher-Order Thinking Skills

42. FIND THE ERROR Don and Diego used a graph to solve the quadratic inequality $x^2 - 2x - 8 > 0$. Is either of them correct? Explain your reasoning. See Mod. 3 Answer Appendix.

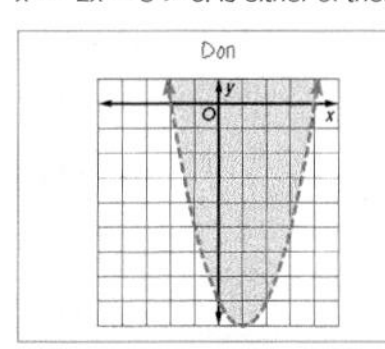

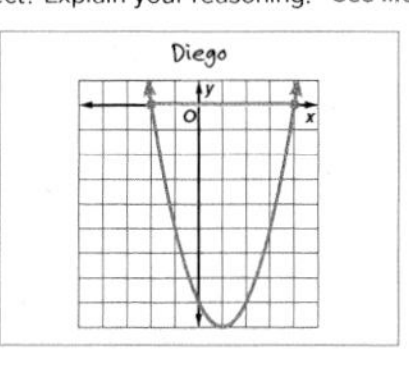

43. CREATE Write a quadratic inequality for each condition.

a. The solution set is all real numbers. Sample answer: $x^2 + 2x + 1 \geq 0$

b. The solution set is the empty set. Sample answer: $x^2 - 4x + 6 < 0$

44. ANALYZE Determine if the following statement is *sometimes*, *always*, or *never* true. Justify your reasoning. See Mod. 3 Answer Appendix.

The intersection of $y \leq -ax^2 + c$ and $y \geq ax^2 - c$ is the empty set.

45. PERSEVERE Graph the intersection of the graphs of $y \leq -x^2 + 4$ and $y \geq x^2 - 4$.
See Mod. 3 Answer Appendix.

46. WRITE How are the techniques used when solving quadratic inequalities and linear inequalities similar? How are they different? See Mod. 3 Answer Appendix.

196 Module 3 • Quadratic Functions

Answers

1.

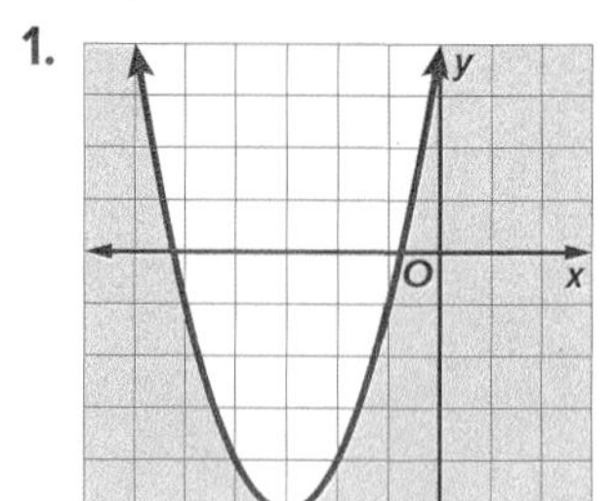

2.

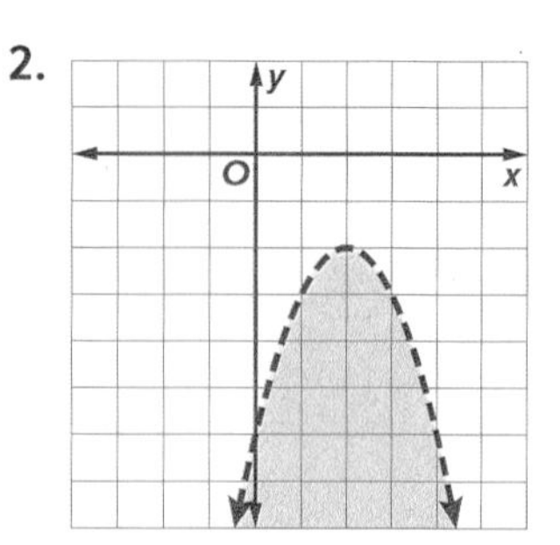

3.

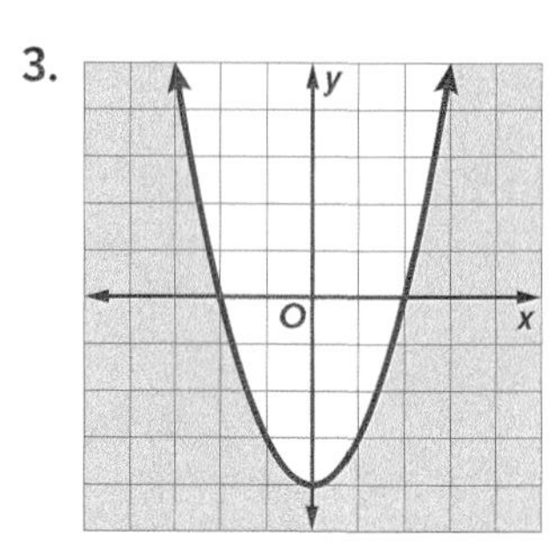

4.

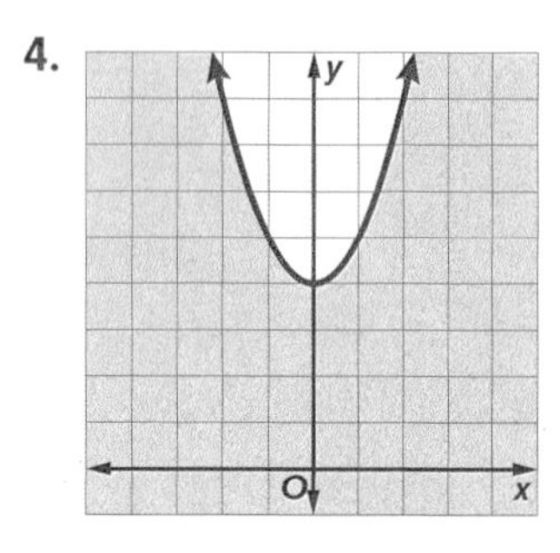

5.

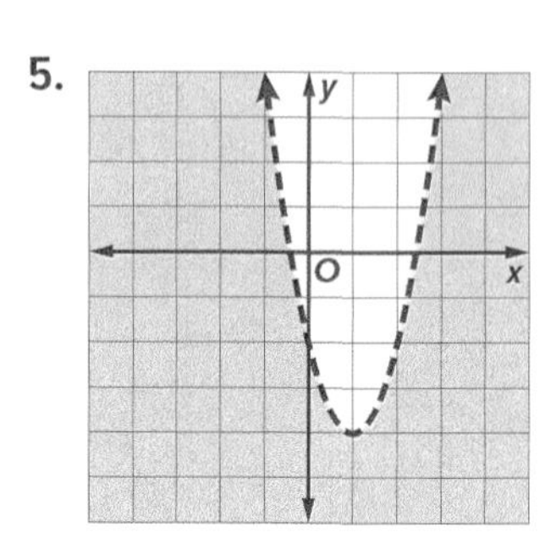

6.

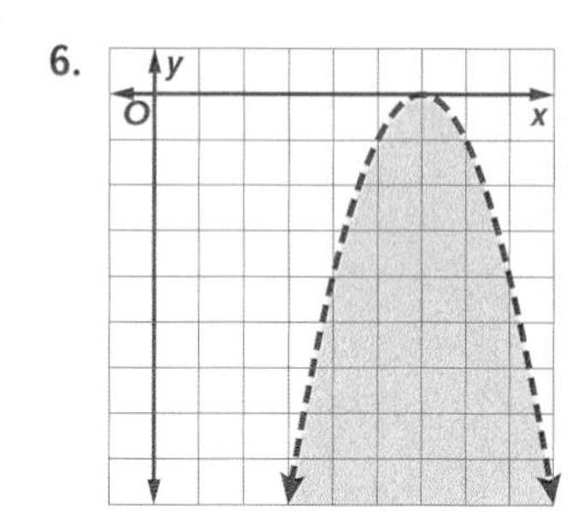

7.

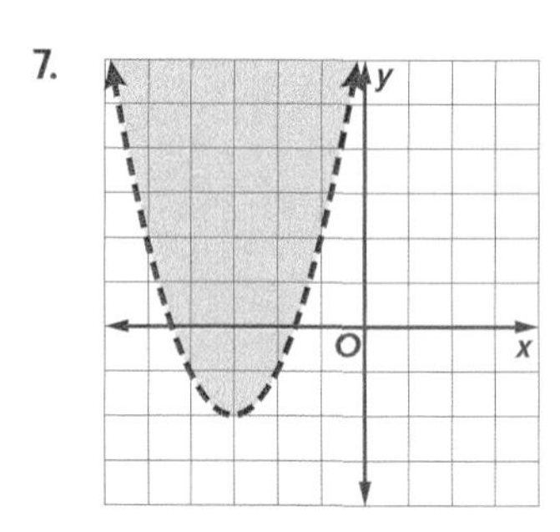

8.

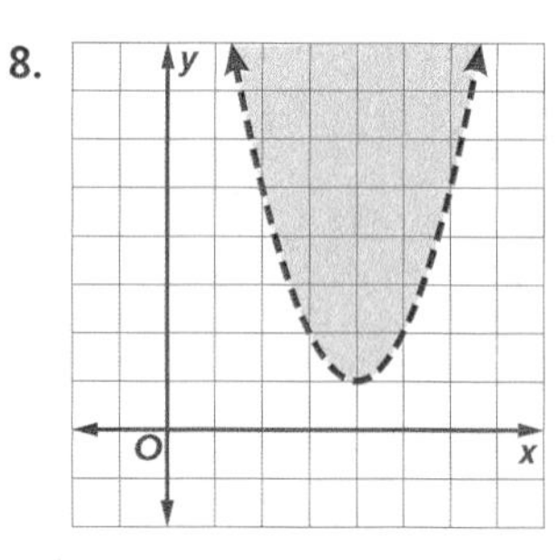

9.

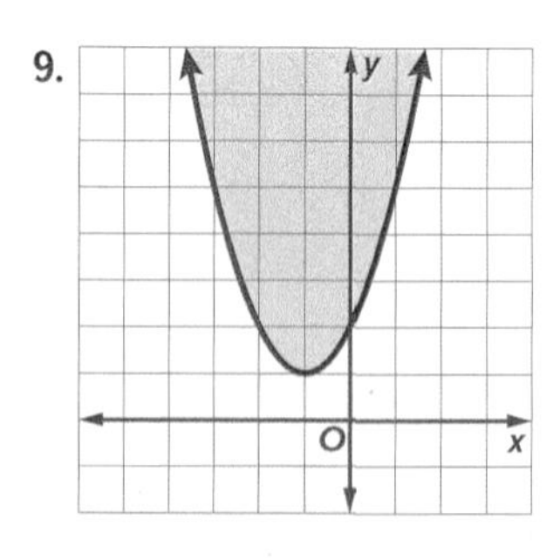

10.

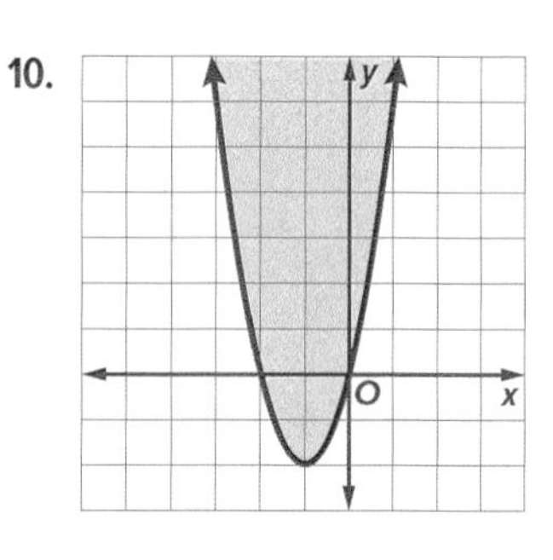

11.

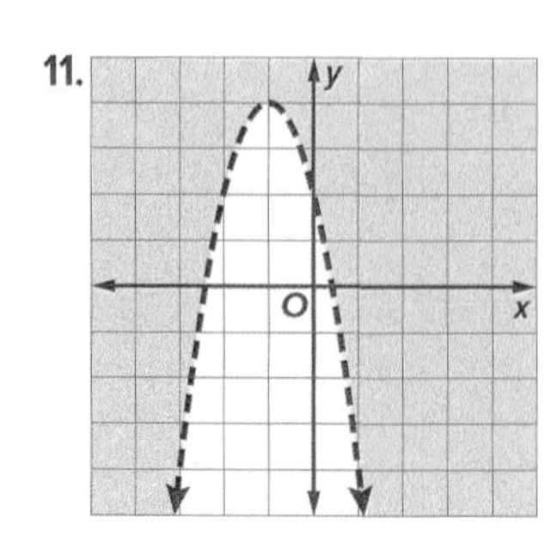

12.

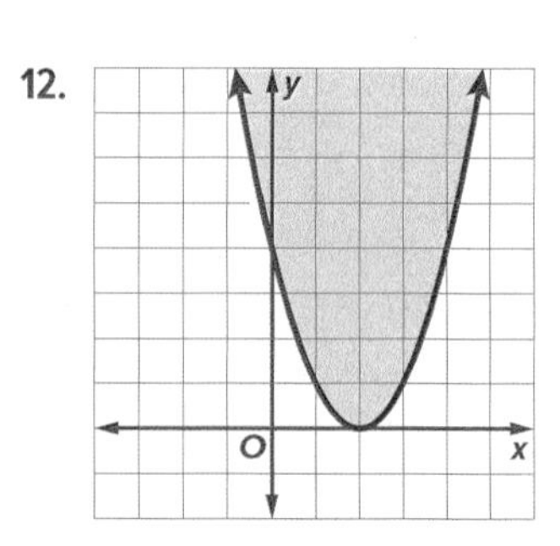

Solving Linear-Nonlinear Systems

A.REI.11

LESSON GOAL

Students solve systems of linear and quadratic equations.

1 LAUNCH

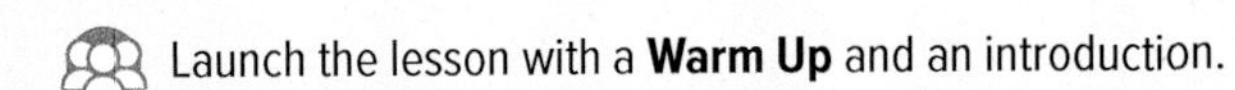
Launch the lesson with a **Warm Up** and an introduction.

2 EXPLORE AND DEVELOP

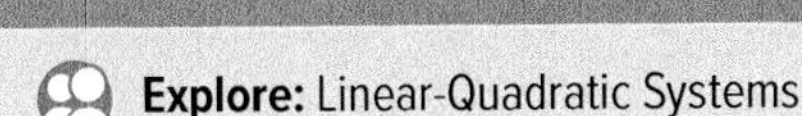
Explore: Linear-Quadratic Systems

Develop:

Solving Linear-Quadratic Systems
- Solve a Linear-Quadratic System by Using Substitution
- Solve a Linear-Quadratic System by Using Elimination
- Use a System to Solve a Quadratic Equation
- Solve a Linear-Quadratic Systems by Graphing

Solving Quadratic-Quadratic Systems
- Solve a Quadratic-Quadratic System Graphically
- Solve a Quadratic-Quadratic System Algebraically
- Use a System to Solve a Quadratic Equation

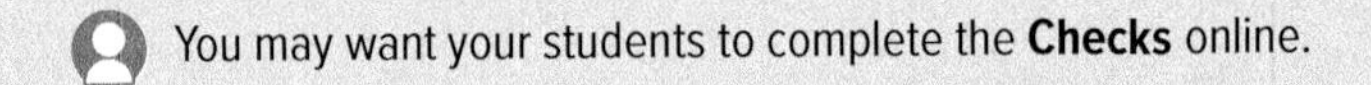
You may want your students to complete the **Checks** online.

3 REFLECT AND PRACTICE

Exit Ticket

Practice

DIFFERENTIATE

View reports of student progress on the **Checks** after each example.

Resources	AL	OL	BL	ELL
Remediation: Elimination Using Multiplication	●	●		●
Extension: Systems Involving Absolute Value Functions		●	●	●

Language Development Handbook

Assign page 24 of the *Language Development Handbook* to help your students build mathematical language related to solving systems of linear and quadratic equations.

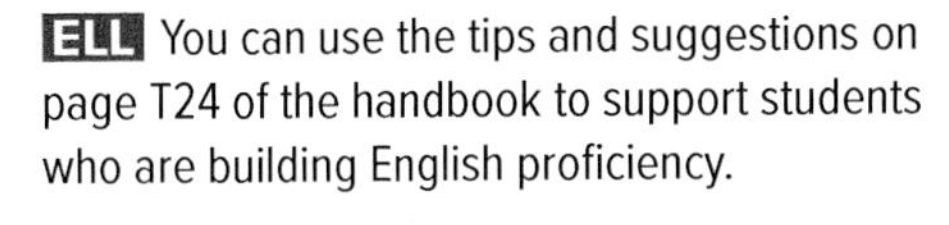
ELL You can use the tips and suggestions on page T24 of the handbook to support students who are building English proficiency.

Suggested Pacing

90 min	1 day
45 min	2 days

Focus

Domain: Algebra

Standards for Mathematical Content:

A.REI.11 Explain why the x-coordinates of the points where the graphs of the equations $y = f(x)$ and $y = g(x)$ intersect are the solutions of the equation $f(x) = g(x)$; find the solutions approximately, e.g., using technology to graph the functions, make tables of values, or find successive approximations. Include cases where $f(x)$ and/or $g(x)$ are linear, polynomial, rational, absolute value, exponential, and logarithmic functions.

Standards for Mathematical Practice:

1 Make sense of problems and persevere in solving them.
4 Model with mathematics.

Coherence

Vertical Alignment

Previous
Students solved systems of linear equations.
8.EE.8, A.CED.3 (Algebra 1, Algebra 2)

Now
Students solve systems of linear and quadratic equations.
A.REI.11

Next
Students will solve polynomial equations by graphing.
A.CED.1, A.REI.11

Rigor

The Three Pillars of Rigor

1 CONCEPTUAL UNDERSTANDING	2 FLUENCY	3 APPLICATION

Conceptual Bridge In this lesson, students extend their understanding of solving systems of linear equations to graphing systems of linear and quadratic equations. They build fluency by solving systems algebraically and graphically, and they apply their understanding by solving real-world problems.

Mathematical Background

A system of a linear and a nonlinear equation can be solved algebraically or graphically. For a linear-quadratic system, a graph indicates whether the conic section and the line intersect in 0, 1, or 2 points. For a quadratic-quadratic system, a graph indicates the number of solutions (0, 1, 2, 3, or 4).

Interactive Presentation

Warm Up

Solve each system by using substitution.

1. $3x + 2y = 12$
   $y = 2x - 1$

2. $2x + y = 6$
   $x - 2y = 8$

Solve each system by using elimination.

3. $5x + 2y = -13$
   $9x - 4y = -31$

4. $7x + 3y = 2$
   $3x - y = -6$

Warm Up

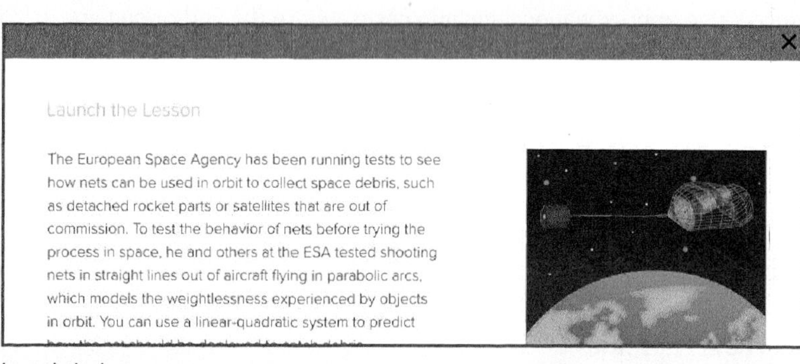

Launch the Lesson

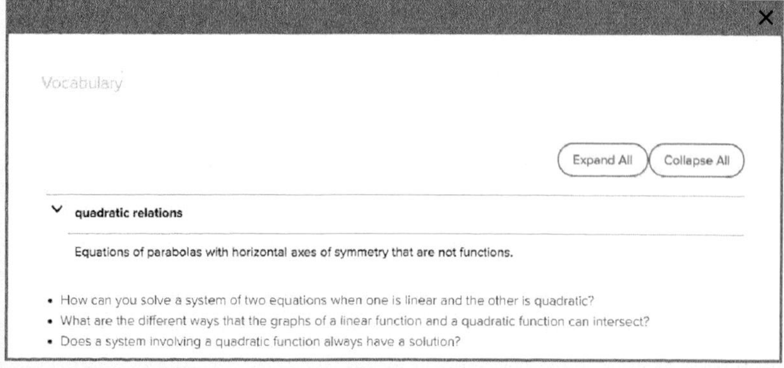

Today's Vocabulary

Warm Up

Prerequisite Skills

The Warm Up exercises address the following prerequisite skill for this lesson:

- solving linear systems of equations

Answers:

1. (2, 3)
2. (4, −2)
3. (−3, 1)
4. (−1, 3)

Launch the Lesson

 Teaching the Mathematical Practices

4 Apply Mathematics In the Launch the Lesson, students will learn how a linear-quadratic system of equations can be used predict how nets should be deployed to collect debris in space.

Go Online to find additional teaching notes and questions to promote classroom discourse.

Today's Standards

Tell students that they will be addressing these content and practice standards in this lesson. You may wish to have a student volunteer read aloud *How can I meet these standards?* and *How can I use these practices?*, and connect these to the standards.

See the Interactive Presentation for I Can statements that align with the standards covered in this lesson.

Today's Vocabulary

Tell students that they will be using this vocabulary term in this lesson. You can expand the row if you wish to share the definition. Then discuss the questions below with the class.

Explore Linear-Quadratic Systems

Objective

Students use a graphing calculator to explore linear-quadratic systems of equations.

Teaching the Mathematical Practices

5 Use Mathematical Tools Point out that to complete the exercises in the Explore, students will need to use a graphing calculator. Work with students to explore and deepen their understanding of linear-quadratic systems of equations.

Ideas for Use

Recommended Use Present the Inquiry Question, or have a student volunteer read it aloud. Have students work in pairs to complete the Explore activity on their devices. Pairs should discuss each of the questions. Monitor student progress during the activity. Upon completion of the Explore activity, have student volunteers share their responses to the Inquiry Question.

What if my students don't have devices? You may choose to project the activity on a whiteboard. A printable worksheet for each Explore is available online. You may choose to print the worksheet so that individuals or pairs of students can use it to record their observations.

Summary of the Activity

Students will complete guiding exercises throughout the Explore activity. Students will use a graphing calculator to graph the equations in a linear-quadratic system and identify the solutions of the system. Then, students will answer the Inquiry Question.

(continued on the next page)

Interactive Presentation

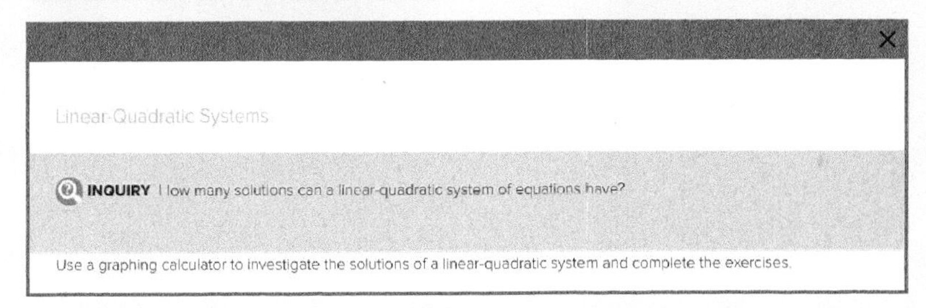

Explore

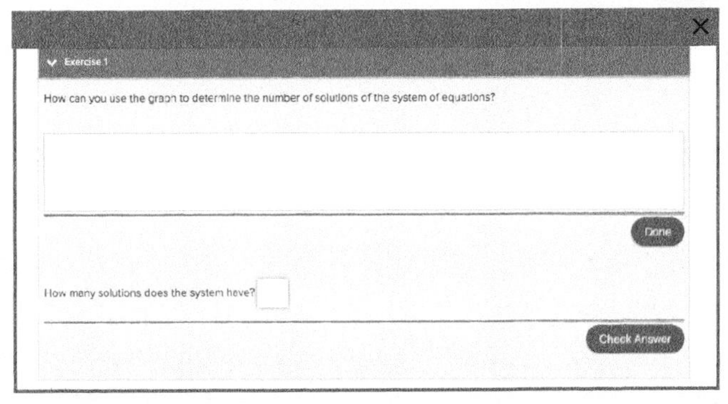

Explore

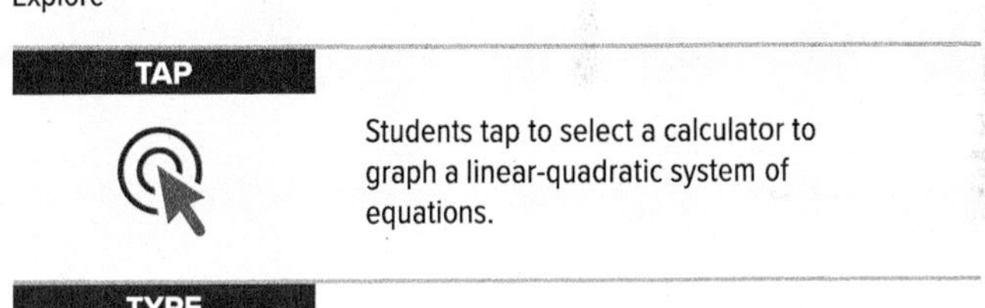

TAP

Students tap to select a calculator to graph a linear-quadratic system of equations.

TYPE

Students answer questions about the solutions of various linear-quadratic systems.

Interactive Presentation

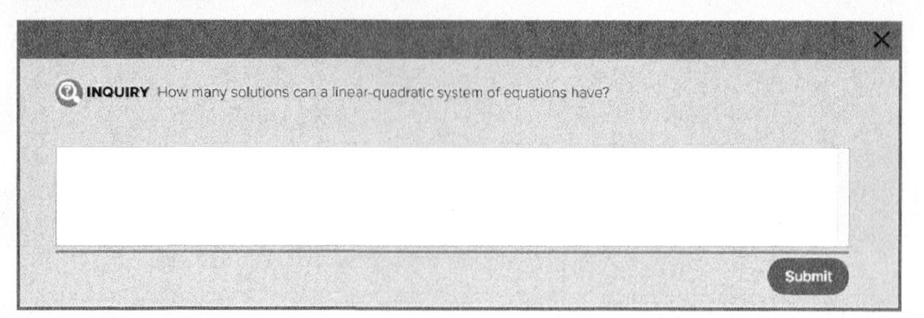

Explore

TYPE

Students respond to the Inquiry Question and can view a sample answer.

Explore Linear-Quadratic Systems (*continued*)

Questions

Have students complete the Explore activity.

Ask:

- What does it mean if a linear-quadratic system does not have a solution? Does this only happen with horizontal lines? Sample answer: If the line and parabola do not intersect, then there is no solution. This could happen with a line that has a slope not equal to zero, like moving the given line up until it no longer intersects.
- Can a linear-quadratic system containing a vertical line and a parabola that opens up or down have no solution? Why or why not? No; sample answer: If a parabola opens up or down, then the domain is all real numbers. Because a vertical line has the form $x = k$, and k is in the domain of the quadratic function, the vertical line will intersect the parabola when $x = k$.

Inquiry

How many solutions can a linear-quadratic system of equations have? A linear-quadratic system can have 0, 1, or 2 solutions.

Go Online to find additional teaching notes and sample answers for the guiding exercises.

1 CONCEPTUAL UNDERSTANDING | **2 FLUENCY** | 3 APPLICATION

Learn Solving Linear-Quadratic Systems

Objective

Students solve systems of linear and quadratic equations.

Teaching the Mathematical Practices

3 Analyze Cases To answer the question in the Talk About It! feature, students must consider the cases of linear-quadratic systems to determine whether a system can have infinitely many solutions.

Example 1 Solve a Linear-Quadratic System by Using Substitution

Teaching the Mathematical Practices

7 Use Structure Help students to use the structure of the linear and quadratic equations in the system to apply the substitution method.

Questions for Mathematical Discourse

AL How can you check your solutions? Sample answer: Substitute them into each equation and make sure they satisfy both equations.

OL What types of equations are in the system you are solving here? a quadratic equation and a linear equation

BL Describe another way to solve this system algebraically. Sample answer: Solve for y in the linear equation and substitute that expression into the quadratic equation.

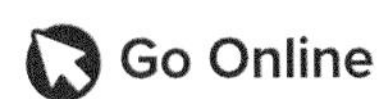

Go Online

- Find additional teaching notes.
- View performance reports of the Checks.
- Assign or present an Extra Example.

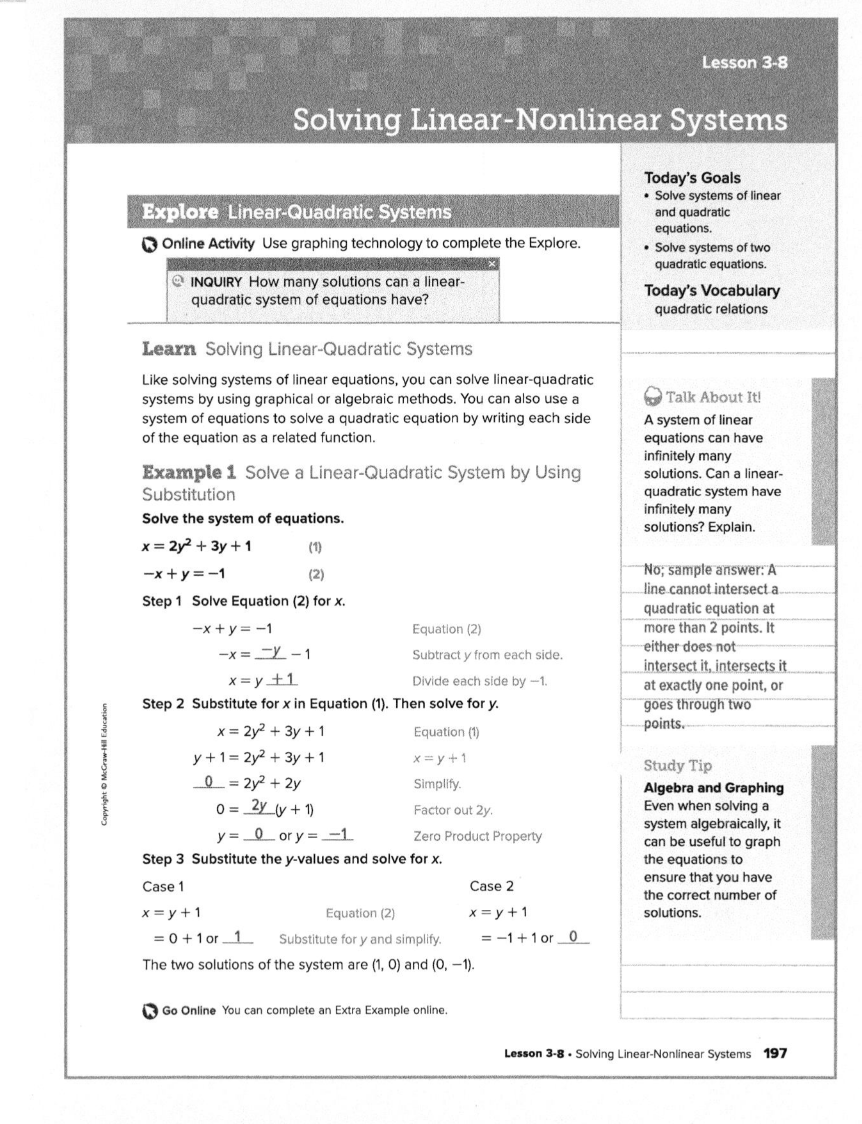

Lesson 3-8

Solving Linear-Nonlinear Systems

Today's Goals
- Solve systems of linear and quadratic equations.
- Solve systems of two quadratic equations.

Today's Vocabulary
quadratic relations

Explore Linear-Quadratic Systems

Online Activity Use graphing technology to complete the Explore.

INQUIRY How many solutions can a linear-quadratic system of equations have?

Learn Solving Linear-Quadratic Systems

Like solving systems of linear equations, you can solve linear-quadratic systems by using graphical or algebraic methods. You can also use a system of equations to solve a quadratic equation by writing each side of the equation as a related function.

Example 1 Solve a Linear-Quadratic System by Using Substitution

Solve the system of equations.

$x = 2y^2 + 3y + 1$ (1)

$-x + y = -1$ (2)

Step 1 Solve Equation (2) for x.

$-x + y = -1$ — Equation (2)

$-x = \underline{-y} - 1$ — Subtract y from each side.

$x = y \underline{+1}$ — Divide each side by -1.

Step 2 Substitute for x in Equation (1). Then solve for y.

$x = 2y^2 + 3y + 1$ — Equation (1)

$y + 1 = 2y^2 + 3y + 1$ — $x = y + 1$

$\underline{0} = 2y^2 + 2y$ — Simplify.

$0 = \underline{2y}(y + 1)$ — Factor out $2y$.

$y = \underline{0}$ or $y = \underline{-1}$ — Zero Product Property

Step 3 Substitute the y-values and solve for x.

Case 1

$x = y + 1$ — Equation (2)

$= 0 + 1$ or $\underline{1}$ — Substitute for y and simplify.

Case 2

$x = y + 1$

$= -1 + 1$ or $\underline{0}$

The two solutions of the system are (1, 0) and (0, −1).

Go Online You can complete an Extra Example online.

Talk About It! A system of linear equations can have infinitely many solutions. Can a linear-quadratic system have infinitely many solutions? Explain.

No; sample answer: A line cannot intersect a quadratic equation at more than 2 points. It either does not intersect it, intersects it at exactly one point, or goes through two points.

Study Tip

Algebra and Graphing Even when solving a system algebraically, it can be useful to graph the equations to ensure that you have the correct number of solutions.

Copyright © McGraw-Hill Education

Lesson 3-8 • Solving Linear-Nonlinear Systems 197

Interactive Presentation

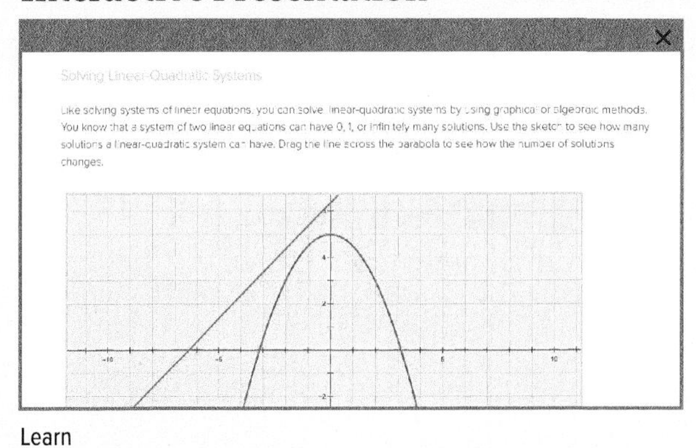

Learn

WEB SKETCHPAD

Students use a sketch to see the solutions of a linear-quadratic system.

TYPE

Students answer a question to determine whether they understand the solutions of a linear-quadratic system.

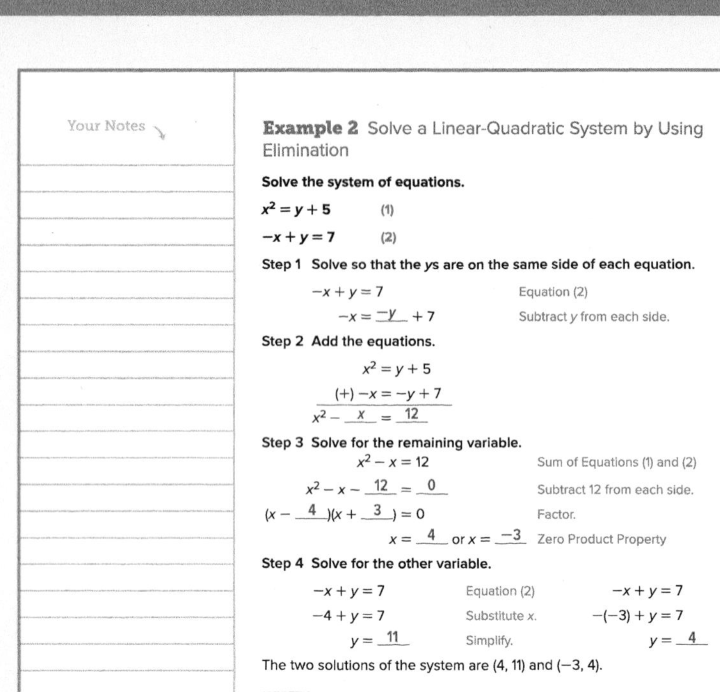

Your Notes

Example 2 Solve a Linear-Quadratic System by Using Elimination

Solve the system of equations.

$x^2 = y + 5$ (1)

$-x + y = 7$ (2)

Step 1 Solve so that the *y*s are on the same side of each equation.

$-x + y = 7$ Equation (2)

$-x = \underline{-y} + 7$ Subtract *y* from each side.

Step 2 Add the equations.

$x^2 = y + 5$

$(+)\ -x = -y + 7$

$x^2 - \underline{x} = \underline{12}$

Step 3 Solve for the remaining variable.

$x^2 - x = 12$ Sum of Equations (1) and (2)

$x^2 - x - \underline{12} = \underline{0}$ Subtract 12 from each side.

$(x - \underline{4})(x + \underline{3}) = 0$ Factor.

$x = \underline{4}$ or $x = \underline{-3}$ Zero Product Property

Step 4 Solve for the other variable.

$-x + y = 7$ Equation (2) $-x + y = 7$

$-4 + y = 7$ Substitute *x*. $-(-3) + y = 7$

$y = \underline{11}$ Simplify. $y = \underline{4}$

The two solutions of the system are (4, 11) and (−3, 4).

Example 3 Use a System to Solve a Quadratic Equation

Use a system of equations to solve $x^2 - 2x + 6 = 4x + 1$.

Step 1 Create a system of equations.

$y = x^2 - 2x + 6$ (1)

$y = 4x + 1$ (2)

Step 2 Graph the system.

The functions appear to intersect at (1, 5) and (5, 21), so the solutions of $x^2 - 2x + 6 = 4x + 1$ are $x = 1$ and $x = 5$.

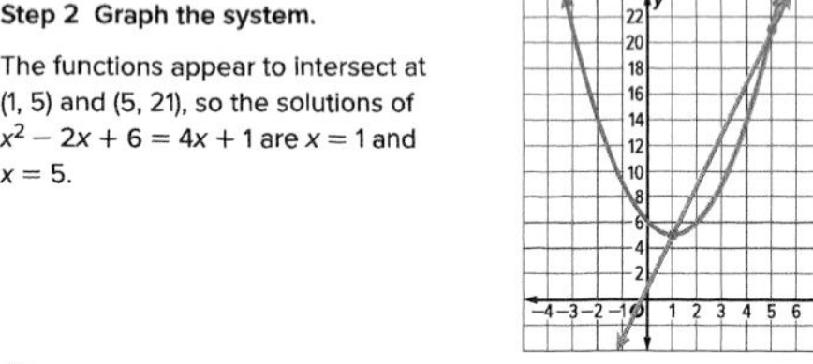

Go Online You can complete an Extra Example online.

198 Module 3 • Quadratic Functions

Interactive Presentation

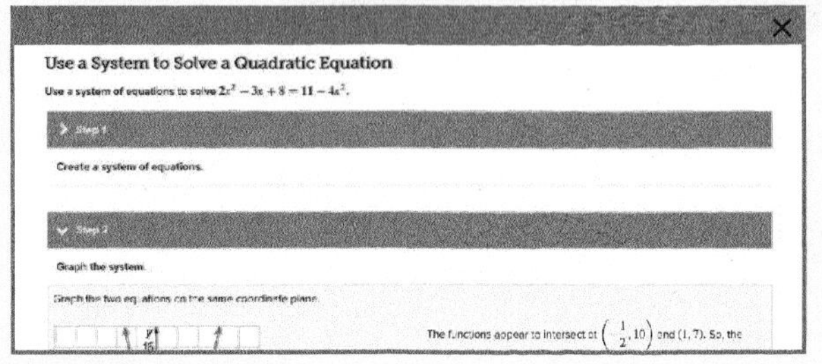

Example 3

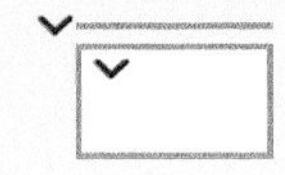

Students tap to see how to use a system to solve a quadratic equation.

Example 2 Solve a Linear-Quadratic System by Using Elimination

Teaching the Mathematical Practices

1 Seek Information Mathematically proficient students must be able to transform algebraic expressions to reach solutions. In Example 2, emphasize transforming the equations so you can eliminate the desired variable from the two equations.

Questions for Mathematical Discourse

AL What equation would you need to use to solve by substituting for *y* in the quadratic equation? $y = x + 7$

OL When using elimination, what has to be true about the variable you are eliminating? Sample answer: It must be on the same side of both equations and the absolute value of the coefficients must be equal.

BL If you rewrote the linear equation as $x + 7 = y$, what would be different about the elimination step? You would have to subtract the linear equation instead of adding it.

Example 3 Use a System to Solve a Quadratic Equation

Teaching the Mathematical Practices

1 Explain Correspondences In Example 3, encourage students to explain the relationships between the system of equations, the graph, and the solutions.

Questions for Mathematical Discourse

AL What are you looking for in the graph to identify the solutions? The points where the line and parabola intersect

OL How would the graph and/or solution set change if you rearranged the equation before writing and graphing the system? The graphs would change, but the *x*-values of the intersections would remain the same.

BL How is solving a quadratic equation using a system related to solving by graphing the related function and finding the zeros of the function? These methods are effectively the same. When you solve by graphing the related function, you are using a system to solve where one of the equations is $y = 0$.

1 CONCEPTUAL UNDERSTANDING | **2 FLUENCY** | **3 APPLICATION**

Example 4 Solve a Linear-Quadratic System by Graphing

Teaching the Mathematical Practices

4 Use Tools Point out that to solve the problem in Example 4, students will need to use a graph.

Questions for Mathematical Discourse

AL What is the maximum profit that can be obtained and for what price? $4 million; $20

OL Use the graph to determine the prices that would result in a profit of at least $3 million. $10 to $30

BL What does the model suggest as a result if the price is above $40? Sample answer: The company would lose money at that price.

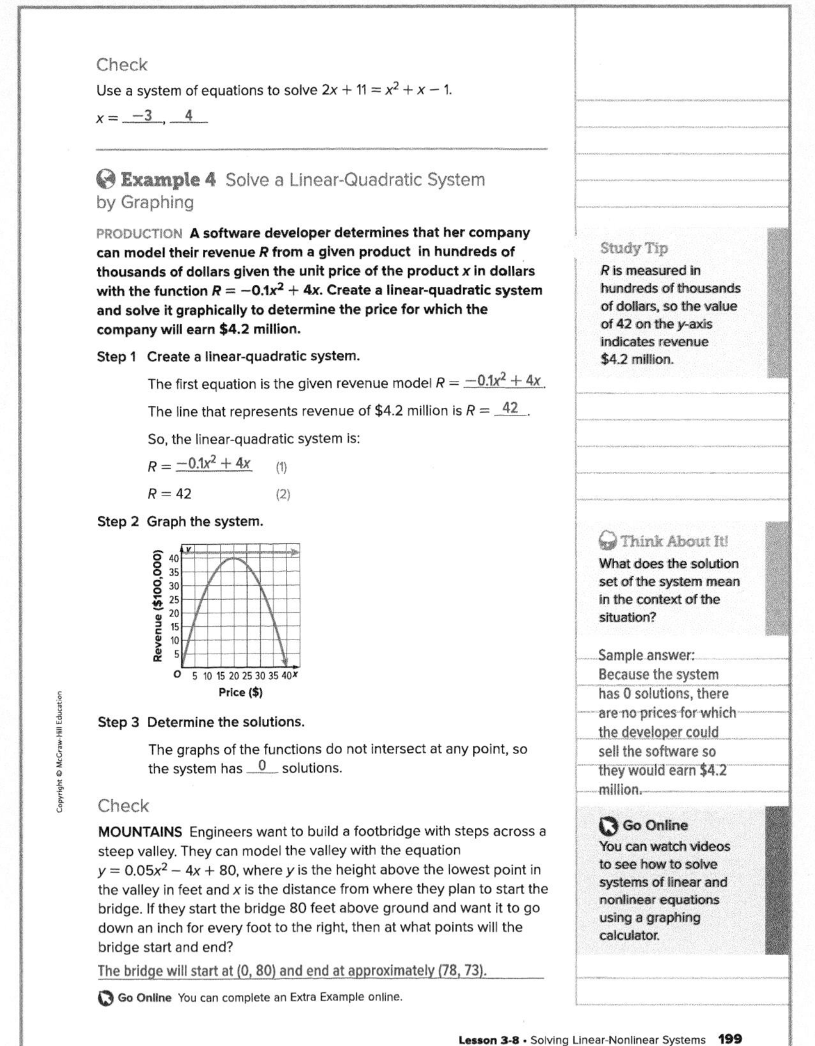

Check

Use a system of equations to solve $2x + 11 = x^2 + x - 1$.

$x =$ −3, 4

Example 4 Solve a Linear-Quadratic System by Graphing

PRODUCTION **A software developer determines that her company can model their revenue R from a given product in hundreds of thousands of dollars given the unit price of the product x in dollars with the function $R = -0.1x^2 + 4x$. Create a linear-quadratic system and solve it graphically to determine the price for which the company will earn $4.2 million.**

Step 1 Create a linear-quadratic system.

The first equation is the given revenue model $R =$ $-0.1x^2 + 4x$.

The line that represents revenue of $4.2 million is $R =$ 42.

So, the linear-quadratic system is:

$R = -0.1x^2 + 4x$ (1)

$R = 42$ (2)

Step 2 Graph the system.

Step 3 Determine the solutions.

The graphs of the functions do not intersect at any point, so the system has 0 solutions.

Check

MOUNTAINS Engineers want to build a footbridge with steps across a steep valley. They can model the valley with the equation $y = 0.05x^2 - 4x + 80$, where y is the height above the lowest point in the valley in feet and x is the distance from where they plan to start the bridge. If they start the bridge 80 feet above ground and want it to go down an inch for every foot to the right, then at what points will the bridge start and end?

The bridge will start at (0, 80) and end at approximately (78, 73).

Go Online You can complete an Extra Example online.

Study Tip

R is measured in hundreds of thousands of dollars, so the value of 42 on the y-axis indicates revenue $4.2 million.

Think About It!

What does the solution set of the system mean in the context of the situation?

Sample answer: Because the system has 0 solutions, there are no prices for which the developer could sell the software so they would earn $4.2 million.

Go Online

You can watch videos to see how to solve systems of linear and nonlinear equations using a graphing calculator.

Lesson 3-8 • Solving Linear-Nonlinear Systems 199

Interactive Presentation

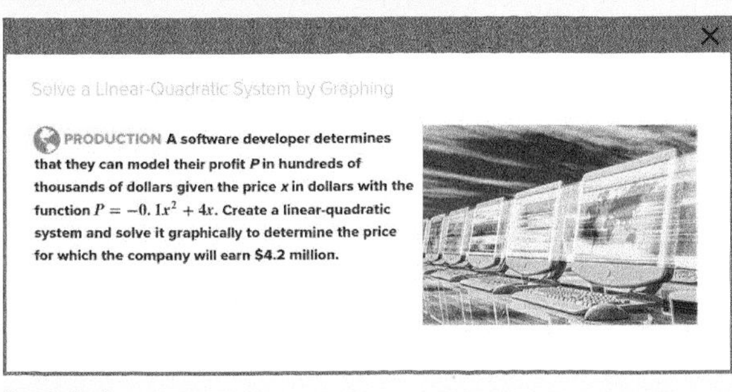

Example 4

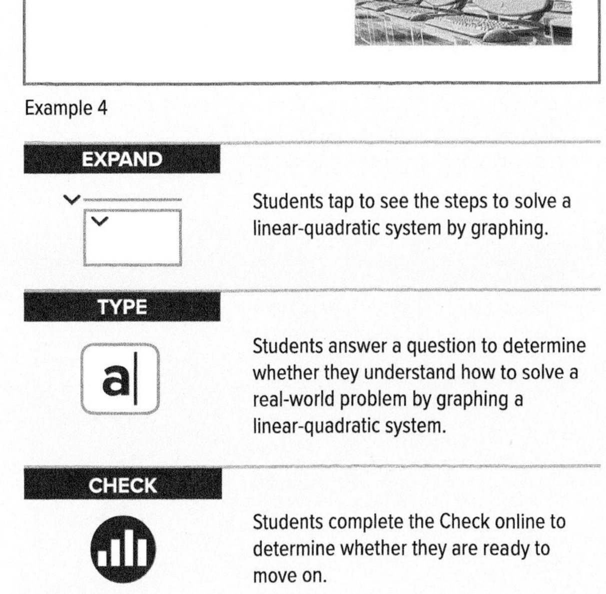

EXPAND

Students tap to see the steps to solve a linear-quadratic system by graphing.

TYPE

Students answer a question to determine whether they understand how to solve a real-world problem by graphing a linear-quadratic system.

CHECK

Students complete the Check online to determine whether they are ready to move on.

Learn Solving Quadratic-Quadratic Systems

Equations of parabolas with vertical axes of symmetry have the parent function $y = x^2$ and can be written in the form $y = a(x - h)^2 + k$. Equations of parabolas with horizontal axes of symmetry are of the form $x = a(y - k)^2 + h$ and are not functions. These are often referred to as **quadratic relations**. The graph of $x = y^2$ is the parent graph for the quadratic relations that have a horizontal axis of symmetry.

If a system contains two quadratic relations, it may have zero to four solutions. Just as with linear-quadratic systems, you can solve quadratic-quadratic systems by using graphical or algebraic methods.

Example 5 Solve a Quadratic-Quadratic System Graphically

Solve the system of equations by graphing.

$y = x^2$ (1)

$x = \frac{1}{5}(y - 6)^2 + \frac{6}{5}$ (2)

Step 1 Graph Equation (1).

Equation (1) has a vertex at (0, 0) and goes through the points (−2, 4), (−1, 1), (1, 1) and (2, 4).

Step 2 Graph Equation (2).

You can use a table of values to graph Equation (2). Because the expression on the right is set equal to x, find the value of x for several values of y.

y	$x = \frac{1}{5}(y - 6)^2 + \frac{6}{5}$	x
3	$x = \frac{1}{5}(3 - 6)^2 + \frac{6}{5}$	3
4	$x = \frac{1}{5}(4 - 6)^2 + \frac{6}{5}$	2
5	$x = \frac{1}{5}(5 - 6)^2 + \frac{6}{5}$	$\frac{7}{5}$
6	$x = \frac{1}{5}(6 - 6)^2 + \frac{6}{5}$	$\frac{6}{5}$
7	$x = \frac{1}{5}(7 - 6)^2 + \frac{6}{5}$	$\frac{7}{5}$
8	$x = \frac{1}{5}(8 - 6)^2 + \frac{6}{5}$	2
9	$x = \frac{1}{5}(9 - 6)^2 + \frac{6}{5}$	3

Step 3 Graph and solve the system.

To solve the system, graph both relations on the same coordinate plane and see where they intersect.

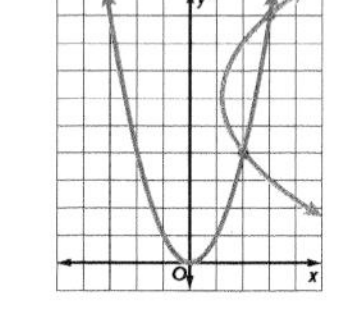

The relations intersect at (2, 4) and (3, 9), so those are the solutions of the system.

Think About It!
Is $x = \frac{1}{5}(y - 6)^2 + \frac{6}{5}$ a function? Explain your reasoning.

No; sample answer: For several values of x, there is more than one value of y.

Think About It!
How could you check your solutions?

Sample answer: I could enter the coordinates for each point into both equations to ensure that the solutions are correct.

Interactive Presentation

Solve a Quadratic-Quadratic System Graphically

Solve the system of equations by graphing.

$y = x^2$ (1)

$x = \frac{1}{5}(y - 6)^2 + \frac{6}{5}$ (2)

Example 5

TAP

Students tap to see the steps to solve a quadratic-quadratic system by graphing.

TYPE

Students answer a question to determine whether they understand how to solve a quadratic-quadratic system by graphing.

Learn Solving Quadratic-Quadratic Systems

Objective

Students solve systems of two quadratic equations.

Teaching the Mathematical Practices

1 Explain Correspondences Encourage students to explain the relationships between quadratic functions and quadratic relations. Guide students to look at how quadratic functions and relations are represented algebraically and graphically.

Common Misconception

Students may not have seen parabolas that open to the left or right before. They may confuse the x and y variable in these equations and try to graph the parabolas with a vertical axis of symmetry. Discuss the differences between a quadratic relation and a quadratic function.

Example 5 Solve a Quadratic-Quadratic System Graphically

Teaching the Mathematical Practices

1 Check Answers Mathematically proficient students continually ask themselves, "Does this make sense?" The Think About It! feature asks students to explain how they can check their solutions when solving a quadratic-quadratic system by graphing.

Questions for Mathematical Discourse

AL What can you tell about the second parabola from the fact that the squared expression contains a y? Sample answer: It will have a horizontal axis of symmetry instead of a vertical one.

OL What does the positive value of a indicate in the second equation? Sample answer: The parabola opens in the positive direction, which is to the right.

BL Could a quadratic-quadratic system ever have infinitely many solutions? What would that mean? Yes; sample answer: It would mean that both equations in the system are equivalent.

Example 6 Solve a Quadratic-Quadratic System Algebraically

Teaching the Mathematical Practices

7 Use Structure Help students to use the structure of the equations in the system to apply the substitution method.

Questions for Mathematical Discourse

AL How can you tell which quadratic equation, 1 or 2, opens down? Equation 2 has a coefficient a that is negative, so it will open down.

OL What would the graph of the system look like? Sample answer: The graph would have a parabola opening up and a parabola opening down, with two intersection points symmetric about the y-axis.

BL What inequality symbols would you use for each equation to get a bounded solution set with included boundaries? Both of the original equations would use $\leq$.

Common Error

Remind students to distribute the negative to all of the terms when performing the substitution.

Check

Solve the system of equations.

$y = -x^2 + 5x - 6$

$3y = x^2 - x - 6$

(1, −2) (3, 0)

Example 6 Solve a Quadratic-Quadratic System Algebraically

Solve the system of equations.

$2x^2 - y = 4$ (1)

$y = -\frac{1}{2}x^2 + 6$ (2)

$2x^2 - y = 4$	Equation (1)
$2x^2 - (-\frac{1}{2}x^2 + 6) = 4$	Substitution
$2x^2 + \frac{1}{2}x^2 - 6 = 4$	Distributive Property
$\frac{5}{2}x^2 - 10 = 0$	Simplify.
$5x^2 - 20 = 0$	Multiply each side by 2.
$x^2 - 4 = 0$	Divide each side by 5.
$(x + 2)(x - 2) = 0$	Difference of Two Squares
$x = \pm 2$	Zero Product Property

Substitute −2 and 2 into one of the original equations and solve for y.

Case 1

$y = -\frac{1}{2}x^2 + 6$ Equation (2)

$y = -\frac{1}{2}(2)^2 + 6$ Substitute for x.

$= 4$ Simplify.

Case 2

$y = -\frac{1}{2}x^2 + 6$

$y = -\frac{1}{2}(-2)^2 + 6$

$= 4$

The solutions are (2, 4) and (−2,4).

Check

Solve the system of equations.

$x = \frac{1}{18}y^2$

$y^2 = -18x + 72$

(2, 6) (2, −6)

Go Online You can complete an Extra Example online.

Lesson 3-8 • Solving Linear-Nonlinear Systems 201

Interactive Presentation

Solve a Quadratic-Quadratic System Algebraically

Solve the system of equations.

$2x^2 - y = 4$ (1)

$y = -\frac{1}{2}x^2 + 6$ (2)

Example 6

SWIPE

Students move through the steps to see how to solve a quadratic-quadratic system algebraically.

Example 7 Use a System to Solve a Quadratic Equation

Use a system of equations to solve $2x^2 - 3x + 8 = 11 - 4x^2$.

Step 1 Create a system of equations.

The related equations of each side of $2x^2 - 3x + 8 = 11 - 4x^2$ are:

$y = 2x^2 - 3x + 8$ (1)

$y = 11 - 4x^2$ (2)

Step 2 Graph and solve the system.

Graph the first equation.

Then graph the second equation on the same coordinate plane.

The functions appear to intersect at $\left(-\frac{1}{2}, \underline{10}\right)$ and $(1, \underline{7})$, so the solutions of $2x^2 - 3x + 8 = 11 - 4x^2$ are $\underline{-\frac{1}{2}}$ and $\underline{1}$.

Check

Use a system of equations to solve $-x^2 + 3x + 14 = -\frac{1}{4}x^2 + 5$.

System Solutions: $(\underline{-2}, \underline{4})$ $(\underline{6}, \underline{-4})$

Equation Solutions: $\underline{-2}$, $\underline{6}$

Go Online You can complete an Extra Example online.

Interactive Presentation

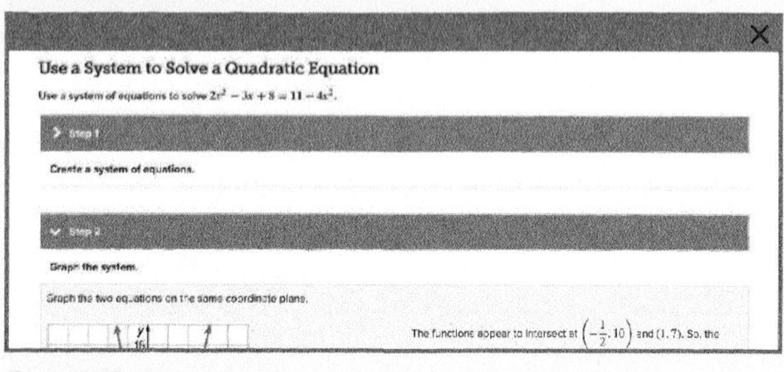

Example 7

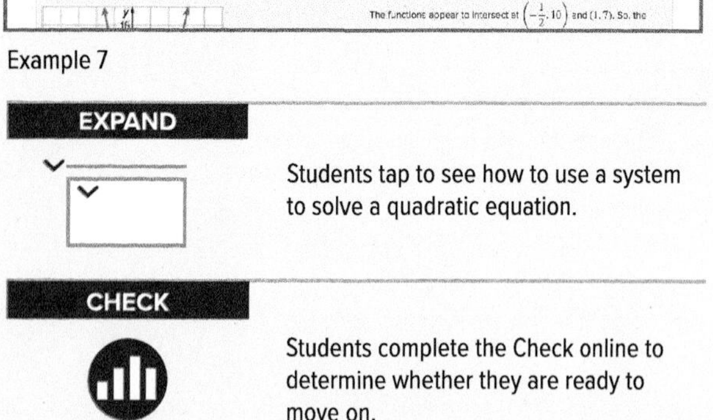

Students tap to see how to use a system to solve a quadratic equation.

Students complete the Check online to determine whether they are ready to move on.

Example 7 Use a System to Solve a Quadratic Equation

Teaching the Mathematical Practices

1 Explain Correspondences Encourage students to explain the relationships between the quadratic equation, system of equations, and graph in Example 7.

Questions for Mathematical Discourse

AL How do you know both equations will be graphed as parabolas? Both can be written in quadratic form.

OL How could you rewrite this equation to solve it as a linear-quadratic system? Sample answer: Adding $4x^2$ to each side would remove the x^2 term from the right side, turning it into a linear equation.

BL Write an equation that would only intersect with the second related equation at one point. Sample answer: $y = 11 + 4x^2$

Exit Ticket

Recommended Use

At the end of class, go online to display the Exit Ticket prompt and ask students to respond using a separate piece of paper. Have students hand you their responses as they leave the room.

Alternate Use

At the end of class, go online to display the Exit Ticket prompt and ask students to respond verbally or by using a mini-whiteboard. Have students hold up their whiteboards so that you can see all student responses. Tap to reveal the answer when most or all students have completed the Exit Ticket.

Practice and Homework

Suggested Assignments

Use the table below to select appropriate exercises.

DOK	Topic	Exercises
1, 2	exercises that mirror the examples	1–35
2	exercises that use a variety of skills from this lesson	36–68
2	exercises that extend concepts learned in this lesson to new contexts	69–71
3	exercises that emphasize higher-order and critical thinking skills	72–75

ASSESS AND DIFFERENTIATE

Use the data from the **Checks** to determine whether to provide resources for extension, remediation, or intervention.

IF students score 90% or more on the Checks, **THEN** assign: 

- Practice Exercises 1–73 odd, 74–78
- Extension: Systems Involving Absolute Value Functions
- ALEKS® Nonlinear Systems

IF students score 66%–89% on the Checks, **THEN** assign: OL

- Practice Exercises 1–77 odd
- Remediation, Review Resources: Elimination Using Multiplication
- Personal Tutors
- Extra Examples 1–7
- ALEKS® Systems of Linear Equations

IF students score 65% or less on the Checks, **THEN** assign: AL

- Practice Exercises 1–35 odd
- Remediation, Review Resources: Elimination Using Multiplication
- ALEKS® Systems of Linear Equations

Name ________ Period ________ Date ________

Practice

Go Online You can complete your homework online.

Examples 1 and 2

Solve each system of equations by using substitution or elimination.

1. $y = x^2 - 5$, $y = x - 3$ $(2, -1), (-1, -4)$
2. $y = x - 2$, $y = x^2 - 2$ $(0, -2), (1, -1)$
3. $y = x + 3$, $y = 2x^2$ $(-1, 2), (1.5, 4.5)$
4. $y = 3x$, $x = y^2$ $(0, 0), \left(\frac{1}{9}, \frac{1}{3}\right)$
5. $y = -2x + 2$, $y^2 = 2x$ $(2, -2), \left(\frac{1}{2}, 1\right)$
6. $y = 2 - x$, $y = x^2 - 4x + 2$ $(0, 2), (3, -1)$
7. $x - y + 1 = 0$, $y^2 = 4x$ $(1, 2)$
8. $y = x - 1$, $y = x^2$ no solution
9. $y = x$, $y = -2x^2 + 4$ $\left(\frac{-1+\sqrt{33}}{4}, \frac{-1+\sqrt{33}}{4}\right), \left(\frac{-1-\sqrt{33}}{4}, \frac{-1-\sqrt{33}}{4}\right)$

Example 3 10–15. See margin for graphs.

Use a system of equations to solve each quadratic equation.

10. $x^2 + 3x + 3 = -2x - 3$ $-3, -2$
11. $x^2 + 3x + 5 = 2x + 7$ $-2, 1$
12. $2x^2 + 4x - 3 = 9x$ $-\frac{1}{2}, 3$
13. $4x^2 - 6x - 3 = -3x - 2$ $-\frac{1}{4}, 1$
14. $3x^2 - 4x + 2 = 2 - 2x$ $\frac{2}{3}, 0$
15. $x^2 - 4x + 5 = 2x + 12$ $-1, 7$

Example 4

16. CIVIL ENGINEERING For safety, roads are designed to ensure that rainwater runs off to the sides. Many roads are highest in the center so that a cross section of the road is a parabola. Suppose the surface of a road can be modeled by $y = -\frac{1}{300}x^2 + \frac{3}{4}$ where x is the distance in feet from the center of the road and y is the height in feet of the road above the shoulder.

shoulder

a. How wide is the road? 30 ft

b. If the water level during a flash flood is 6 inches above the shoulder, about how wide is the portion of the road that is not under water? 17.3 ft

17. SMALL BUSINESS A small business owner determines that the profit P in dollars from sales of a specific item can be modeled by $P = 2x^2 + 30x$, where x is the selling price of the item in dollars.

a. Create a linear-quadratic system to determine the price for which the business will earn $50,000. $P = 2x^2 + 30x$ and $P = 50,000$

b. Solve the system in **part a** to determine the price for which the business will earn $50,000. Round to the nearest hundredth, if necessary. $x \approx -165.79, x \approx \$150.79$

c. What does the solution set of the system mean in the context of the situation? See margin.

Lesson 3-8 • Solving Linear-Nonlinear Systems 203

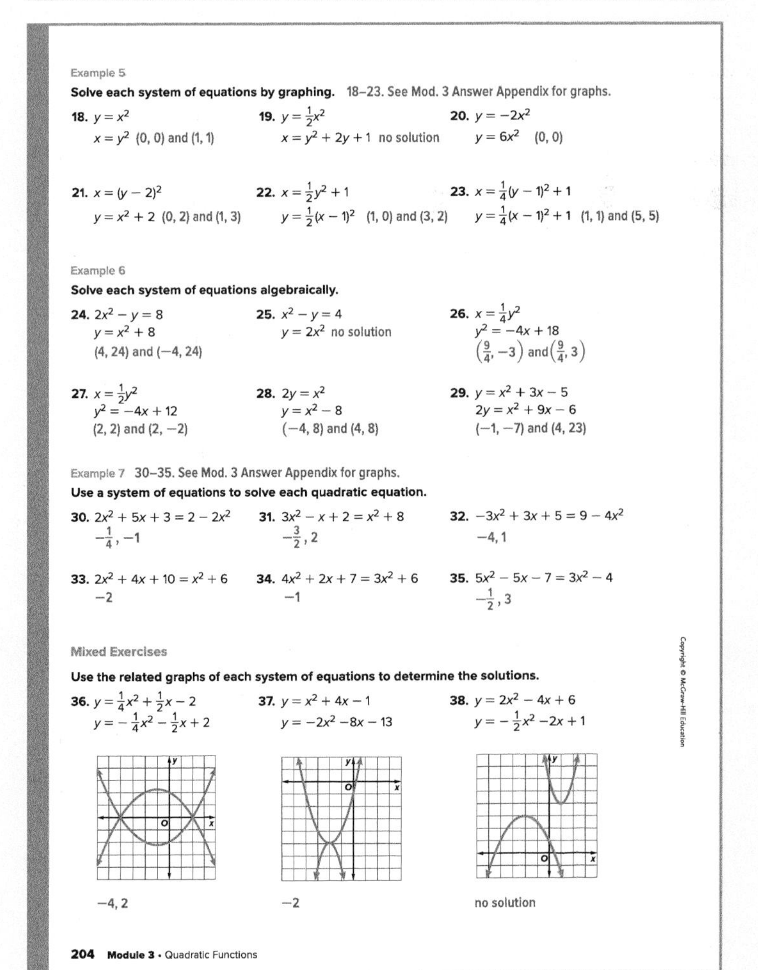

Example 5

Solve each system of equations by graphing. 18–23. See Mod. 3 Answer Appendix for graphs.

18. $y = x^2$, $x = y^2$ $(0, 0)$ and $(1, 1)$
19. $y = \frac{1}{2}x^2$, $x = y^2 + 2y + 1$ no solution
20. $y = -2x^2$, $y = 6x^2$ $(0, 0)$
21. $x = (y - 2)^2$, $y = x^2 + 2$ $(0, 2)$ and $(1, 3)$
22. $x = \frac{1}{2}y^2 + 1$, $y = \frac{1}{2}(x - 1)^2$ $(1, 0)$ and $(3, 2)$
23. $x = \frac{1}{4}(y - 1)^2 + 1$, $y = \frac{1}{4}(x - 1)^2 + 1$ $(1, 1)$ and $(5, 5)$

Example 6

Solve each system of equations algebraically.

24. $2x^2 - y = 8$, $y = x^2 + 8$ $(4, 24)$ and $(-4, 24)$
25. $x^2 - y = 4$, $y = 2x^2$ no solution
26. $x = \frac{1}{4}y^2$, $y^2 = -4x + 18$ $\left(\frac{9}{4}, -3\right)$ and $\left(\frac{9}{4}, 3\right)$
27. $x = \frac{1}{2}y^2$, $y^2 = -4x + 12$ $(2, 2)$ and $(2, -2)$
28. $2y = x^2$, $y = x^2 - 8$ $(-4, 8)$ and $(4, 8)$
29. $y = x^2 + 3x - 5$, $2y = x^2 + 9x - 6$ $(-1, -7)$ and $(4, 23)$

Example 7 30–35. See Mod. 3 Answer Appendix for graphs.

Use a system of equations to solve each quadratic equation.

30. $2x^2 + 5x + 3 = 2 - 2x^2$ $-\frac{1}{4}, -1$
31. $3x^2 - x + 2 = x^2 + 8$ $-\frac{3}{2}, 2$
32. $-3x^2 + 3x + 5 = 9 - 4x^2$ $-4, 1$
33. $2x^2 + 4x + 10 = x^2 + 6$ -2
34. $4x^2 + 2x + 7 = 3x^2 + 6$ -1
35. $5x^2 - 5x - 7 = 3x^2 - 4$ $-\frac{1}{2}, 3$

Mixed Exercises

Use the related graphs of each system of equations to determine the solutions.

36. $y = \frac{1}{4}x^2 + \frac{1}{2}x - 2$, $y = -\frac{1}{4}x^2 - \frac{1}{2}x + 2$ $-4, 2$
37. $y = x^2 + 4x - 1$, $y = -2x^2 - 8x - 13$ -2
38. $y = 2x^2 - 4x + 6$, $y = -\frac{1}{2}x^2 - 2x + 1$ no solution

204 Module 3 • Quadratic Functions

Name ______________ Period ________ Date ________

Solve each system of equations by graphing, substitution, or elimination.

39. $y = x^2 + 9x + 8$
$y = 7x + 8$ (−2, −6), (0, 8)

40. $y = x^2 + 7x + 12$
$y = x + 7$ (−5, 2), (−1, 6)

41. $y = x + 3$
$y = 2x^2 - x - 1$ (−1, 2), (2,5)

42. $y = \left(\frac{1}{2}x\right)^2$
$4x = y^2$ (0, 0) and (4, 4)

43. $y = 2x^2$
$x = y^2 - 2y + 1$
(0.42, 0.352) and (1, 2)

44. $y = -8x^2$
$y = 4x^2 + 1$ no solution

45. $2x^2 - y = 2$
$y = -x^2 + 4$
(−1.4, 2) and (1.4, 2)

46. $x^2 - y = -4$
$y = 4x^2 - 2$
(−1.4, 6) and (1.4, 6)

47. $y = x^2$
$y^2 = -3x + 5$
(−1.8, 3.2) and (1.1, 1.3)

48. $y = -x - 3$
$y = x^2 - 5$ (−2, −1), (1, −4)

49. $y = -x$
$y^2 = 2x$ (2, −2), (0, 0)

50. $y = 3x - 4$ no solution
$y = x^2 + 9x + 20$

51. $x - 2 = (y - 2)^2$
$y = x^2 + 5$ no solution

52. $3x - 1 = \frac{1}{8}y^2$ (1.4, 5.1) and
$y = x^2 + 5x - 4$ (0.4, −1.6)

53. $x = (y + 2)^2$ no solution
$y = (x + 2)^2$

54. $x = \frac{1}{4}y^2$
$y^2 = -8x + 6$
$\left(\frac{1}{2}, \sqrt{2}\right)$ and $\left(\frac{1}{2}, -\sqrt{2}\right)$

55. $4y + 3 = x^2$ $(-\sqrt{31}, 7)$
$y = x^2 - 24$ and $(\sqrt{31}, 7)$

56. $y = x^2 + x + 5$ (−3, 11) and
$y = 2x^2 + 3x + 2$ (1, 7)

Use a system of equations to solve each quadratic equation.

57. $2x^2 + x - 1 = -10x + 12$
$-6\frac{1}{2}, 1$

58. $3x^2 - 5x - 2 = -16 - 8x$
no solution

59. $x^2 + 3x + 2 = x + 5$
$-3, 1$

60. $x^2 - 2x - 3 = -\frac{1}{3}x^2 + \frac{2}{3}x + \frac{23}{3}$
$-2, 4$

61. $3x^2 + 4x + 1 = -x^2 - 2x - 1$
$-\frac{1}{2}, -1$

62. $2x^2 - 6x + 4 = -x^2 + 3x - 2$
$1, 2$

63. $2x^2 + 6x + 5 = x^2 - 4$
-3

64. $4x^2 + 2x + 7 = 3x^2 + 6$
-1

65. $5x^2 - 5x - 7 = 3x^2 - 4$
$-\frac{1}{2}, 3$

66. $x^2 + 4x + 4 = x + 4$
$-3, 0$

67. $x^2 + 5x + 12 = -4x - 6$
$-6, -3$

68. $2x^2 - x - 6 = -3x - 2$
$-2, 1$

69. ROCKETS Two model rockets are launched at the same time from different heights. The height in meters for one rocket after t seconds is modeled by $y = -4.9t^2 + 48.8t$. The height for the other rocket is modeled by $y = -4.9t^2 + 46.7t + 1.5$.

a. After how many seconds are the rockets at the same height? Round to the nearest tenth. 0.7 s

b. From what height would the slower rocket have to be launched so that the rockets land at the same time? Justify your reasoning.
about 20.9 m; The faster rocket lands after about 9.96 seconds. Solving $y = -4.9t^2 + 46.7t + c$ when $y = 0$ and $t = 9.96$, shows that $c \approx 20.9$.

70. PACKAGING A manufacturer is making two different packages, measured in inches, as shown in the figure. The manufacturer wants the surface area of the packages to be the same.

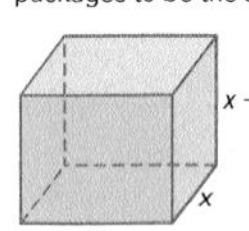

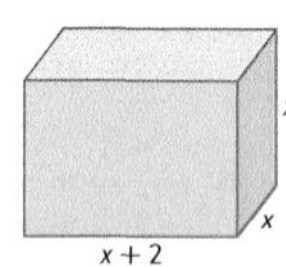

a. Create a quadratic equation that can be used to find the value of x, when the surface area of the packages is the same. $6x^2 - 8x = 6x^2 + 8x$

b. Solve the quadratic equation using any method. $x = 0$

c. Find and interpret the solution in the context of the situation. See margin.

71. VOLLEYBALL The function $h(t) = -16t^2 + vt + h_0$ models the height in feet of a volleyball, where v represents the initial velocity, h_0 represents the initial height, and t represents the time in seconds since the ball was hit. Suppose a player bumps a volleyball when it is 3 feet from the ground with an initial velocity of 25 feet per second.

a. If the net is approximately 7 feet 4 inches, could the volleyball clear the net? Justify your reasoning. Yes; the maximum height of the ball is about 12.77 feet, which is higher than the net.

b. The player on the other side of the net jumps 1 second after the ball is hit so that the path of her hands can be described by $h(t) = -16(t - 1)^2 + 12(t - 1) + 7$. If the ball clears the net, is it possible that she blocks the ball? Justify your reasoning. Yes; the player's hands could strike the ball at a height of about 9 feet. Because that is above the height of the net, the ball may be blocked.

c. What assumptions did you make?
See margin.

Higher-Order Thinking Skills

72. PERSEVERE Describe three linear-quadratic systems of equations—one with no solution, one with 1 solution, and one with 2 solutions. See Mod. 3 Answer Appendix.

73. CREATE Write a system of two quadratic equations in which there is one solution.
Sample answer: $y = x^2$ and $y = -x^2$

74. WRITE Describe a real-life situation that can be modeled by a system with a quadratic function and a linear function. See Mod. 3 Answer Appendix.

75. FIND THE ERROR Danny and Carol are solving the system $y = x^2 + 3x - 9$ and $-4x + y = 3$. Is either of them correct? Explain your reasoning.

Danny	Carol
$-4x + (x^2 + 3x - 9) = 3$	$x^2 + 3x - 9 = -4x + 3$
$x^2 - x - 9 = 3$	$x^2 + 7x - 12 = 0$
$x^2 - x - 12 = 0$	$x \approx -8.42$ or $x \approx 1.42$
$(x + 3)(x - 4) = 0$	
$x = -3$ or $x = 4$	

Danny is correct. Carol incorrectly solved for y in the second equation of the system before using the substitution method.

Answers

10. 

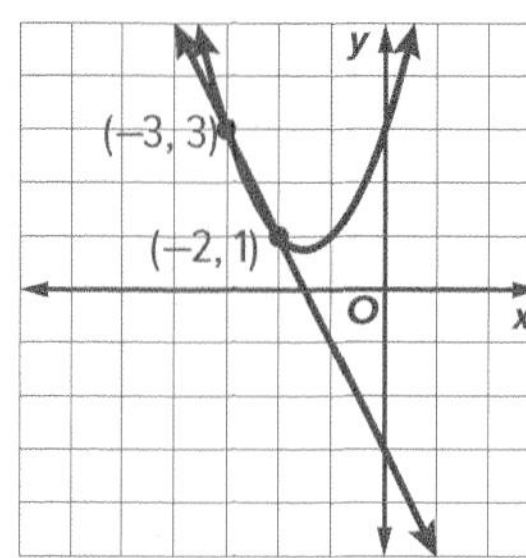

11.

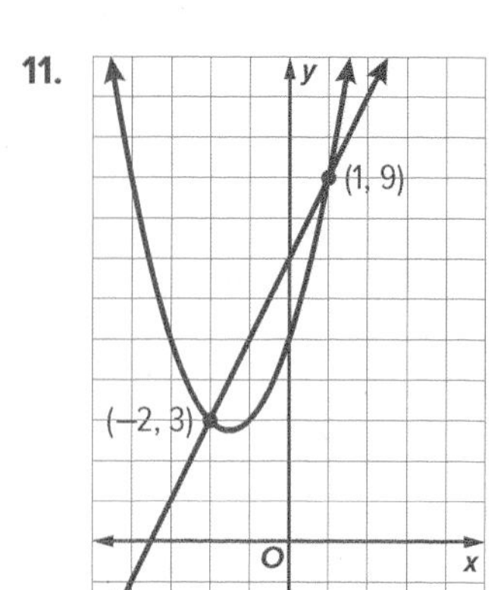

12.

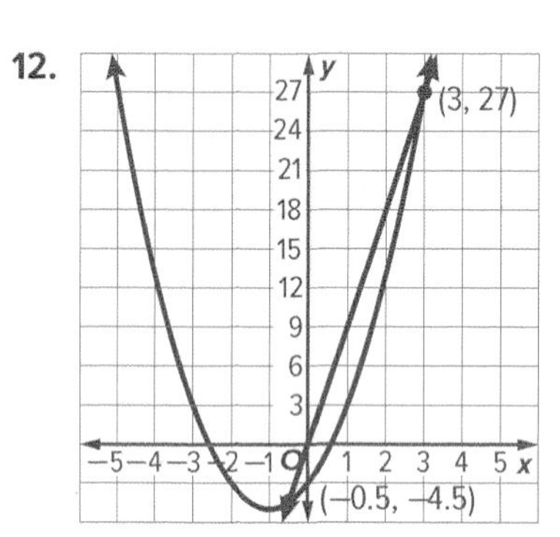

13.

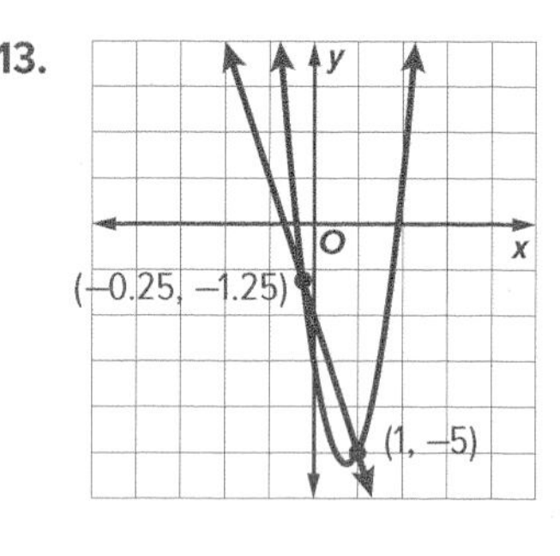

14.

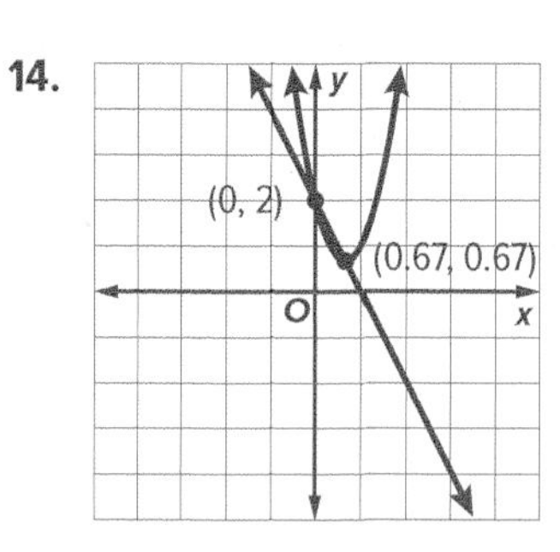

15.

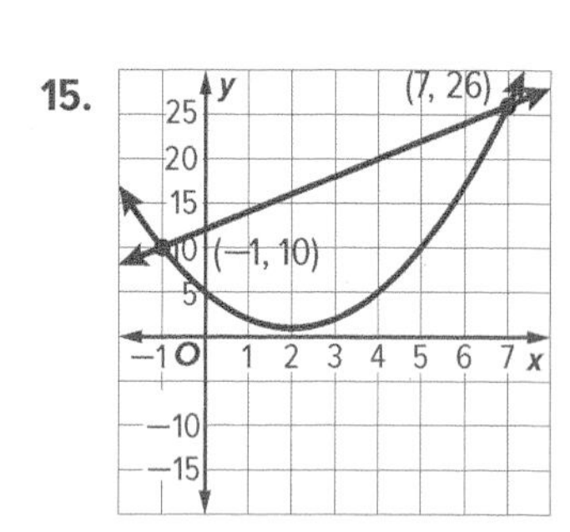

17c. Sample answer: Because the selling price cannot be negative, (150.79, 50,000) is the only viable solution in the context of the situation. This means that the business can earn a $50,000 profit when the selling price is about $150.79 per item.

70c. Sample answer: Because the side length cannot be zero, there is no viable solution in the context of the situation. This means that it is impossible to make two boxes that meet these requirements.

71c. Sample answer: I assumed that the student bumping the ball was far enough away from the net that she wouldn't hit it.

I also assumed that the player attempting to block the ball is in the correct position for the path of the ball to intersect the path of her hands.

Review

Rate Yourself! 

Have students return to the Module Opener to rate their understanding of the concepts presented in this module. They should see that their knowledge and skills have increased. After completing the chart, have them respond to the prompts in their *Interactive Student Edition* and share their responses with a partner.

Answering the Essential Question

Before answering the Essential Question, have students review their answers to the Essential Question Follow-Up questions found throughout the module.

- How can the graph of a quadratic function help you solve the corresponding quadratic equation?
- How do complex numbers relate to solving quadratic equations?
- How do you know what method to use when solving a quadratic equation?

Then have them write their answer to the Essential Question in the space provided.

DINAH ZIKE FOLDABLES

ELL A completed Foldable for this module should include the key concepts related to analyzing and graphing quadratic functions.

LS **LearnSmart** Use LearnSmart as part of your test preparation plan to measure student topic retention. You can create a student assignment in LearnSmart for additional practice on these topics for **Polynomial, Rational, and Radical Relationships.**

- Complex Numbers
- Solving Equations Graphically

Module 3 • Quadratic Functions

Review

 Essential Question

What characteristics of quadratic functions are important when analyzing real-world situations that are modeled by quadratic functions?

Sample answer: Quadratic functions have either a minimum or a maximum value. These values are often solutions when the functions model the path of a projectile or the profit made by a company.

Module Summary

Lessons 3-1 and 3-2

Graphs of Quadratic Functions

- When the coefficient on the x^2-term is positive, the parabola opens up. When it is negative, the parabola opens down.
- The average rate of change for a parabola over the interval $[a, b]$ is $\frac{f(b) - f(a)}{b - a}$.
- The solutions of an equation in one variable are the x-intercepts of the graph of a related function.

Lesson 3-3

Complex Numbers

- The imaginary unit i is the principal square root of -1. Thus, $i = \sqrt{-1}$ and $i^2 = -1$.
- Two complex numbers of the form $a + bi$ and $a - bi$ are called complex conjugates.

Lesson 3-4 through 3-6

Solving Quadratic Equations

- For any real numbers a and b, if $ab = 0$, then either $a = 0$, $b = 0$, or both a and $b = 0$.
- You can solve a quadratic equation by graphing, by factoring, by completing the square, or by using the Quadratic Formula.
- To solve a quadratic equation of the form $x^2 = n$, take the square root of each side.
- The solutions of a quadratic equation of the form $ax^2 + bx + c = 0$, where $a \neq 0$, are given by the formula $x = \frac{-b \pm \sqrt{b^2 - 4ac}}{2a}$.

Lesson 3-7

Quadratic Inequalities

- To graph a quadratic inequality, graph the related function, test a point not on the parabola and shade accordingly.
- For $ax^2 + bx + c < 0$, graph $y = ax^2 + bx + c$ and identify the x-values for which the graph lies below the x-axis. For $\leq$, include the x-intercepts in the solution.
- For $ax^2 + bx + c > 0$, graph $y = ax^2 + bx + c$ and identify the x-values for which the graph lies *above* the x-axis. For $\geq$, include the x-intercepts in the solution.

Lesson 3-8

Systems Involving Quadratic Equations

- You can use the substitution method or the elimination method to solve a system that includes a quadratic equation.
- If a system contains two quadratic relations, it may have anywhere from zero to four solutions.

Study Organizer

Foldables

Use your Foldable to review this module. Working with a partner can be helpful. Ask for clarification of concepts as needed.

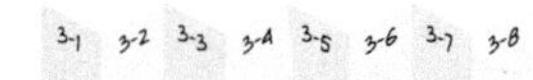

Name ______ Period ______ Date ______

Test Practice

1. **MULTIPLE CHOICE** Which is the graph of $f(x) = -x^2 - 2x + 3$? (Lesson 3-1)

Ⓐ (selected) Ⓑ Ⓒ Ⓓ

2. **OPEN RESPONSE** At a concert, a T-shirt cannon launches a T-shirt upward. The height of the T-shirt in feet a given number of seconds after the launch is shown in the table.

Time (s)	Height (ft)
1	74
2	116
3	126
4	104
5	50

Find and interpret the average rate of change in the height between 1 and 3 seconds after launch. (Lesson 3-1)

The height of the T-shirt is increasing at an average rate of 26 feet per second between 1 and 3 seconds after the launch.

3. **OPEN RESPONSE** Use a quadratic equation to find two real numbers with a sum of 31 and a product of 210. (Lesson 3-2)

10 and 21

4. **MULTI-SELECT** The graph of $f(x) = x^2 - x - 6$ is shown.

Find the solutions of $x^2 - x - 6 = 0$. Select all that apply. (Lesson 3-2)

Ⓐ -6
Ⓑ -3
Ⓒ -2 (selected)
Ⓓ 2
Ⓔ 3 (selected)
Ⓕ 6

5. **MULTIPLE CHOICE** Simplify $\sqrt{-9} \cdot \sqrt{-49}$. (Lesson 3-3)

Ⓐ $-21i$
Ⓑ $21i$
Ⓒ -21 (selected)
Ⓓ 21

6. **OPEN RESPONSE** Every complex number can be written in the form $a + bi$. For the complex number $8 - 4i$, identify the values of a and b. (Lesson 3-3)

$a = 8$; $b = -4$

Review and Assessment Options

The following online review and assessment resources are available for you to assign to your students. These resources include technology-enhanced questions that are auto-scored, as well as essay questions.

Review Resources

Put It All Together: Lessons 3-1 through 3-6
Vocabulary Activity
Module Review

Assessment Resources

Vocabulary Test
AL Module Test Form B
OL Module Test Form A
BL Module Test Form C
Performance Task*

*The module-level performance task is available online as a printable document. A scoring rubric is included.

Test Practice

You can use these pages to help your students review module content and prepare for online assessments. Exercises 1–17 mirror the types of questions your students will see on online assessments.

Question Type	Description	Exercise(s)
Multiple Choice	Students select one correct answer.	1, 5, 8, 9, 13, 14
Multi-Select	Multiple answers may be correct. Students must select all correct answers.	4, 11, 16
Table Item	Students complete a table by entering in the correct values.	18
Open Response	Students construct their own response in the area provided.	2, 3, 6, 7, 10, 13, 15, 17

To ensure that students understand the standards, check students' success on individual exercises.

Standard(s)	Lesson(s)	Exercise(s)
A.SSE.1b	3-1, 3-6	3, 13
A.SSE.2	3-4, 3-5	7, 10, 11
A.REI.11	3-8	16, 17, 18
A.CED.1	3-4, 3-6, 3-7	8, 12, 15
A.CED.2	3-1	1
A.CED.3	3-7	14, 15
A.CED.4	3-6	12, 13
F.IF.4	3-1, 3-2	2, 4
F.IF.6	3-1	2
F.IF.8a	3-4, 3-5	7, 8, 9
N.CN.1	3-3	5, 6
N.CN.2	3-3	5
N.CN.7	3-5	9
N.CN.8	3-5	9

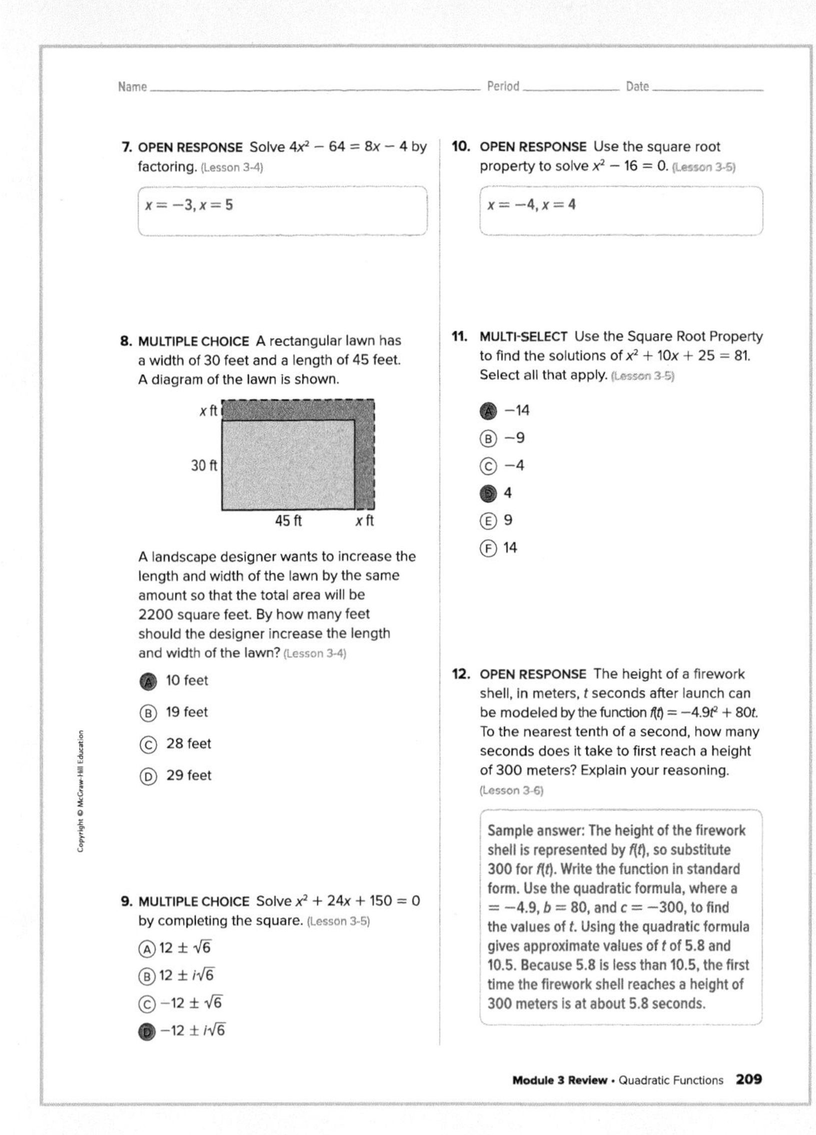
Name ______ Period ______ Date ______

7. OPEN RESPONSE Solve $4x^2 - 64 = 8x - 4$ by factoring. (Lesson 3-4)

$x = -3, x = 5$

8. MULTIPLE CHOICE A rectangular lawn has a width of 30 feet and a length of 45 feet. A diagram of the lawn is shown.

A landscape designer wants to increase the length and width of the lawn by the same amount so that the total area will be 2200 square feet. By how many feet should the designer increase the length and width of the lawn? (Lesson 3-4)

Ⓐ 10 feet
Ⓑ 19 feet
Ⓒ 28 feet
Ⓓ 29 feet

9. MULTIPLE CHOICE Solve $x^2 + 24x + 150 = 0$ by completing the square. (Lesson 3-5)

Ⓐ $12 \pm \sqrt{6}$
Ⓑ $12 \pm i\sqrt{6}$
Ⓒ $-12 \pm \sqrt{6}$
Ⓓ $-12 \pm i\sqrt{6}$

10. OPEN RESPONSE Use the square root property to solve $x^2 - 16 = 0$. (Lesson 3-5)

$x = -4, x = 4$

11. MULTI-SELECT Use the Square Root Property to find the solutions of $x^2 + 10x + 25 = 81$. Select all that apply. (Lesson 3-5)

Ⓐ −14
Ⓑ −9
Ⓒ −4
Ⓓ 4
Ⓔ 9
Ⓕ 14

12. OPEN RESPONSE The height of a firework shell, in meters, t seconds after launch can be modeled by the function $f(t) = -4.9t^2 + 80t$. To the nearest tenth of a second, how many seconds does it take to first reach a height of 300 meters? Explain your reasoning. (Lesson 3-6)

Sample answer: The height of the firework shell is represented by $f(t)$, so substitute 300 for $f(t)$. Write the function in standard form. Use the quadratic formula, where $a = -4.9$, $b = 80$, and $c = -300$, to find the values of t. Using the quadratic formula gives approximate values of t of 5.8 and 10.5. Because 5.8 is less than 10.5, the first time the firework shell reaches a height of 300 meters is at about 5.8 seconds.

Module 3 Review • Quadratic Functions 209

Name ______________ Period ________ Date ________

13. MULTIPLE CHOICE Solve $4x^2 - 6x - 5 = 0$ by using the Quadratic Formula. (Lesson 3-6)

Ⓐ $\frac{3 \pm \sqrt{11}}{4}$

Ⓑ $\frac{-3 \pm \sqrt{29}}{4}$

● $\frac{3 \pm \sqrt{29}}{4}$

Ⓓ $\frac{-3 \pm \sqrt{11}}{4}$

14. MULTIPLE CHOICE The graph of $y = x^2 - 4x + 3$ is shown. Select the values for which $x^2 - 4x + 3 < 0$. (Lesson 3-7)

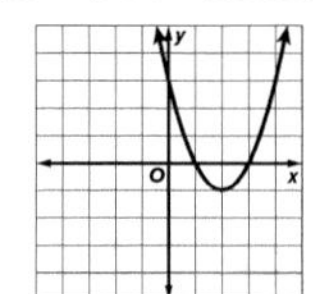

Ⓐ $x < 1, x < 3$

Ⓑ $x < 1, x > 3$

● $x > 1, x < 3$

Ⓓ $x > 1, x > 3$

15. OPEN ENDED Caleb wants to add an L-shaped deck along two sides of his garden. The garden is a rectangle 15 feet wide and 20 feet long. The deck width will be the same on both sides and the total area of the garden and deck cannot exceed 500 square feet. How wide can the deck be? (Lesson 3-7)

between 0 and 5 feet

16. MULTI-SELECT Solve the system of equations. Which ordered pair is part of the solution set? Select all that apply. (Lesson 3-8)

$$\begin{cases} y + 3 = x \\ y = x^2 - 5 \end{cases}$$

Ⓐ (−2, −1)

Ⓑ (−2, 1)

● (−1, −4)

Ⓓ (−1, 1)

Ⓔ (1, −4)

Ⓕ (1, 4)

● (2, −1)

Ⓗ (2, −4)

17. OPEN RESPONSE What are the solutions of this system of equations? (Lesson 3-8)

$$\begin{cases} y = (x - 4)^2 - 3 \\ y = x - 1 \end{cases}$$

(2, 1), (7, 6)

18. TABLE ITEM Identify if each point is a solution to the first equation only, second equation only, both or neither. (Lesson 3-8)

(1) $y = x^2 - 2x - 1$

(2) $y = x + 3$

Point	First	Second	Both	Neither
(−4, 1)				X
(−3, 0)		X		
(−1, 2)			X	
(1, 0)				X
(1, 4)		X		
(4, 7)			X	

Lesson 3-1

27a. *y*-int $= -9$; axis of symmetry: $x = 1.5$; *x*-coordinate of vertex $= 1.5$

27b.

x	$f(x)$
0	−9
1	−13
1.5	−13.5
2	−13
3	−9

27c.

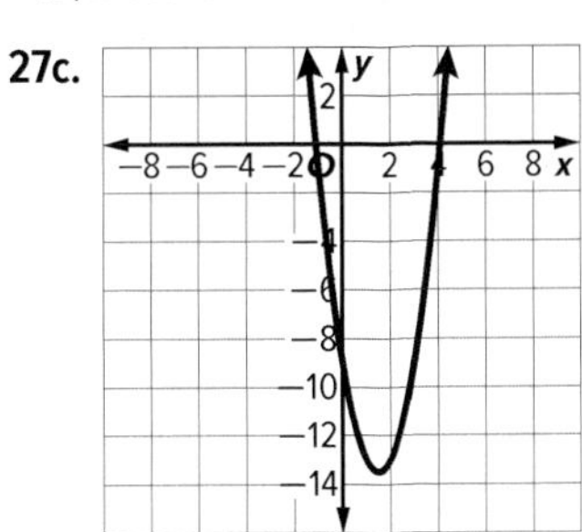

28a. *y*-inter $= 2$; axis of symmetry: $x = -1.5$; *x*-coordinate of vertex $= -1.5$

28b.

x	$f(x)$
−3	2
−2	8
−1.5	8.75
−1	8
0	2

28c.

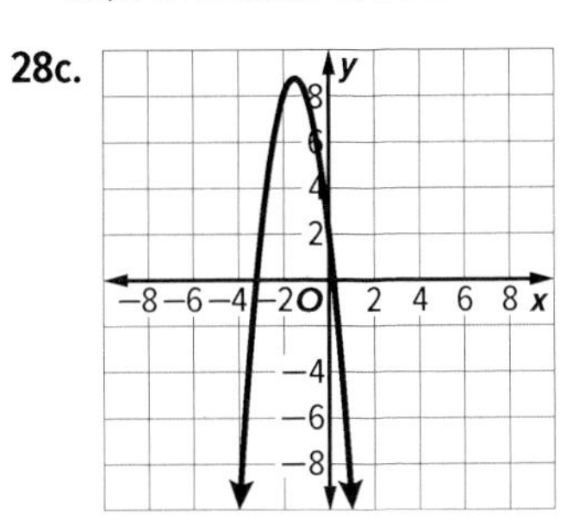

29a. *y*-int $= 0$; axis of symmetry: $x = \frac{5}{8}$; *x*-coordinate of vertex $= \frac{5}{8}$

29b.

x	$f(x)$
$-\frac{3}{4}$	−6
$\frac{1}{4}$	1
$\frac{5}{8}$	1.5625
1	1
2	−6

29c.

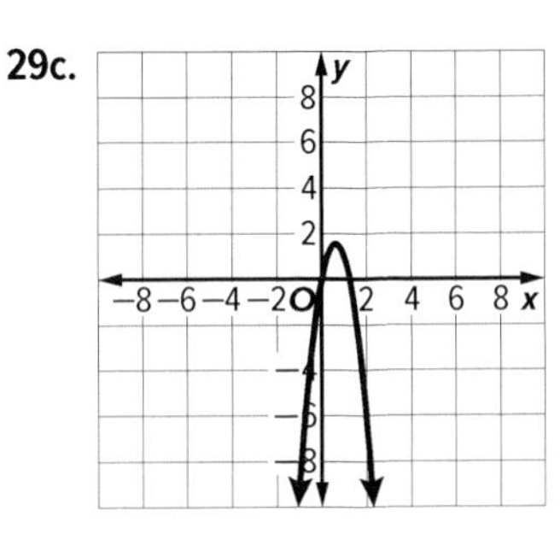

30a. *y*-int $= 0$; axis of symmetry: $x = -2.75$; *x*-coordinate of vertex $= -2.75$

30b.

x	$f(x)$
−4	−12
−3	−15
−2.75	−15.125
−2.5	−15
−1.5	−12

30c.

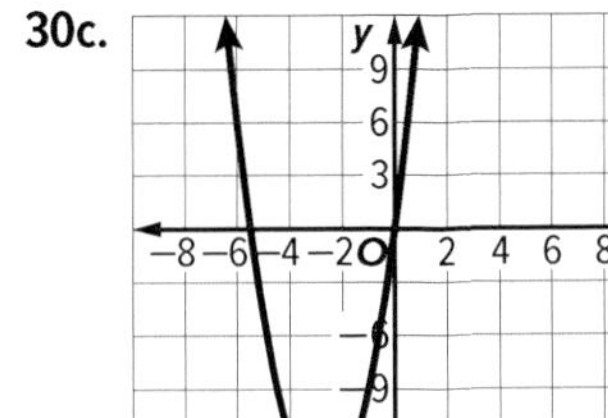

31a. *y*-int $= 4$; axis of symmetry: $x = -6$; *x*-coordinate of vertex $= -6$

31b.

x	$f(x)$
−10	−1
−8	−4
−6	−5
−4	−4
−2	−1

31c.

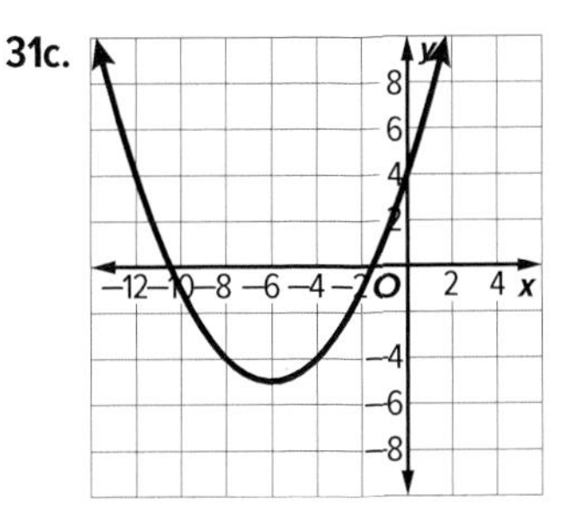

32a. *y*-int $= 6$; axis of symmetry: $x = \frac{8}{3}$; *x*-coordinate of vertex $= \frac{8}{3}$

32b.

x	$f(x)$
$\frac{4}{3}$	10
$\frac{7}{3}$	11.25
$\frac{8}{3}$	$11\frac{1}{3}$
3	11.25
4	10

32c.

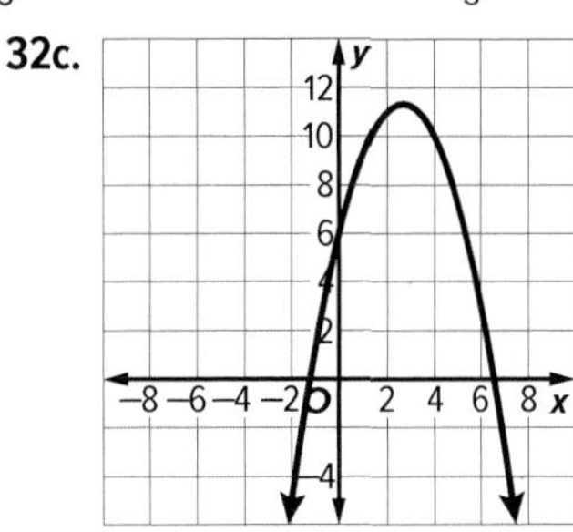

39a. Let x represent the number of \$1 price increases. Let y represent the income. Then $y = (70 - x)(20 + x)$.

x	y
0	1400
5	1625
10	1800
15	1925
20	2000
25	2025
30	2000
35	1925
40	1800
45	1625
50	1400

39b. (25, 2025); The club should increase the price by \$25 to make a maximum profit of \$2025. Sample answer: Increasing the price by \$25 is more than twice the current price, so this may be unreasonable. The club may want to revisit their assumptions.

41a. If x = number of price increases, the revenue is $R(x) = -5x^2 + 40x + 1200$. Because the price changes and the revenue will not be negative, the domain is $\{x \mid 0 \leq x \leq 20\}$.

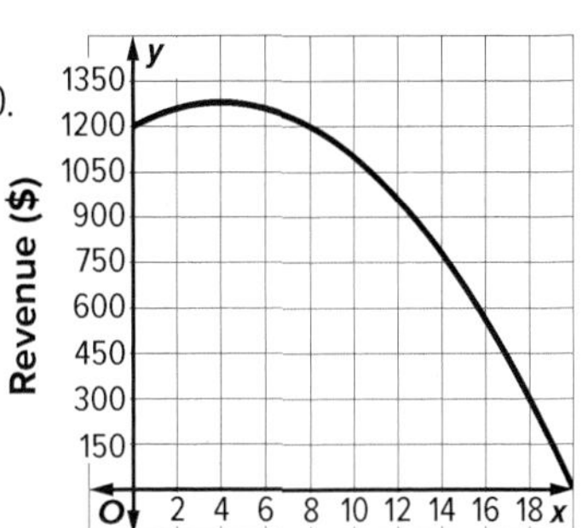

41b. The graph shows that the maximum revenue value occurs at $x = 4$, which corresponds to a ticket price of \$6.00 + 4(\$0.50) or \$8.00. The maximum revenue is $R(4) = \$1280.00$.

41c. As price increases continue, the demand for tickets will decrease. The *x*-intercept indicates a price that is too high, and no one is estimated to buy tickets for the cinema.

47. Sample answer: A function is quadratic if it has no other terms than a quadratic term, linear term, and constant term. The function has a maximum if the coefficient of the quadratic term is negative and has a minimum if the coefficient of the quadratic term is positive.

Lesson 3-2

15.

16.

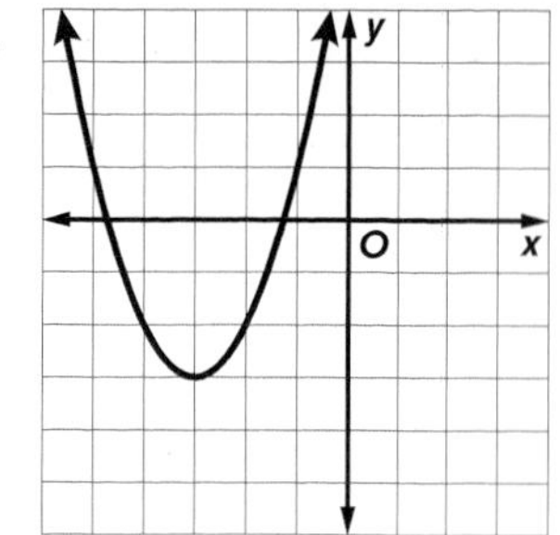

17.

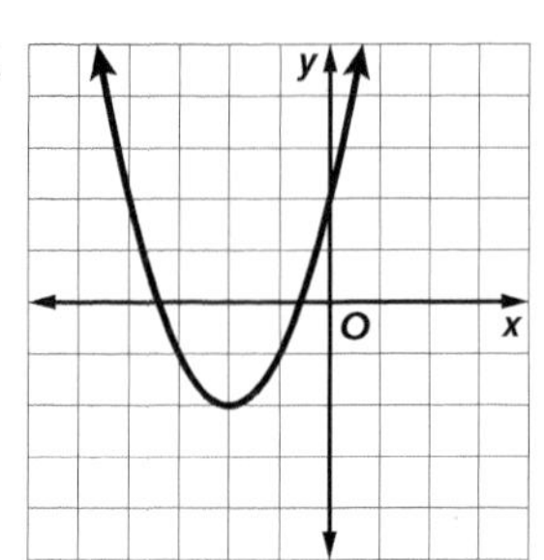

18.

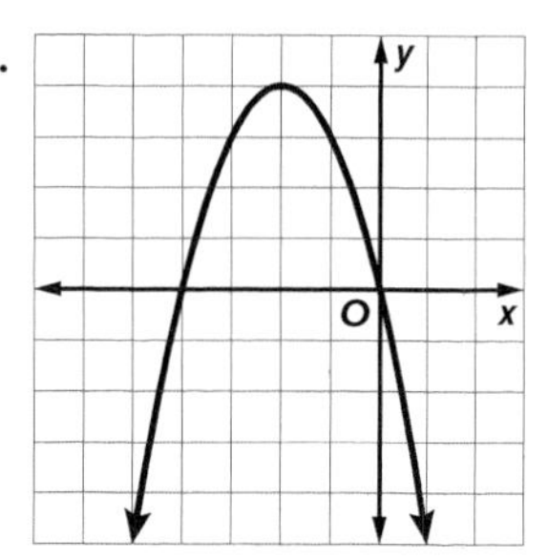

19.

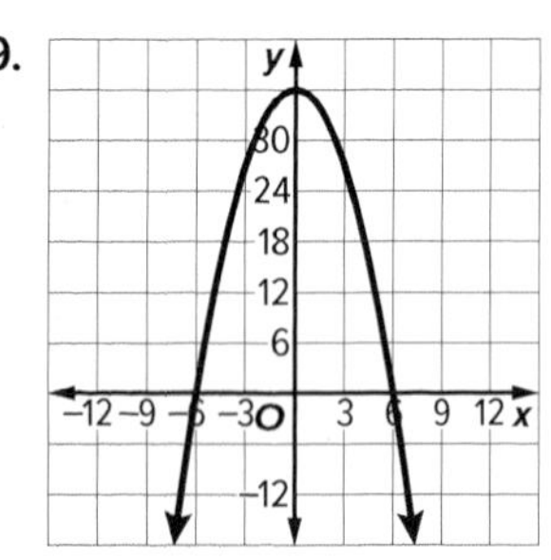

20.

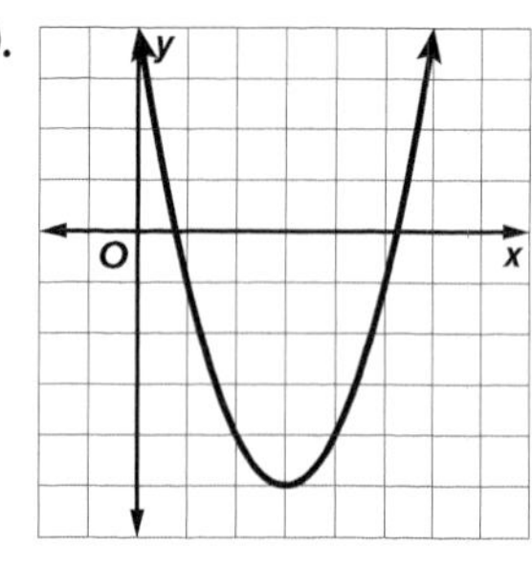

21.

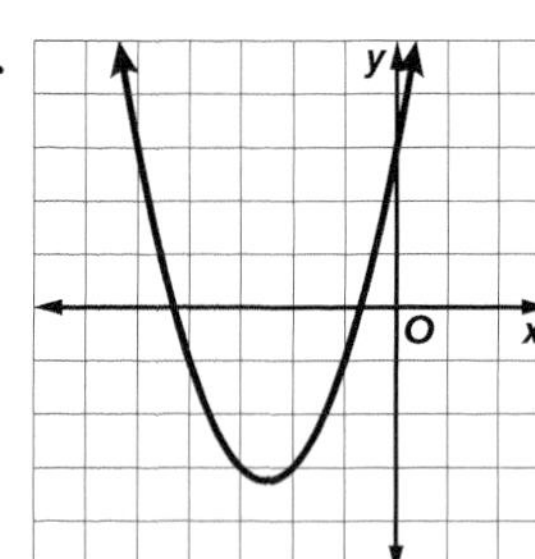

22.

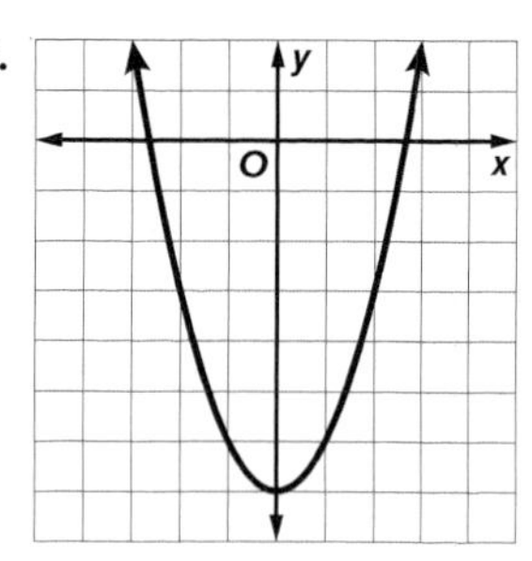

23.

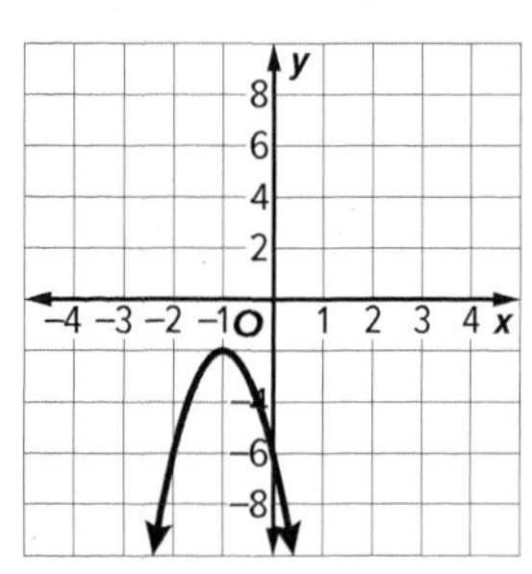

35.

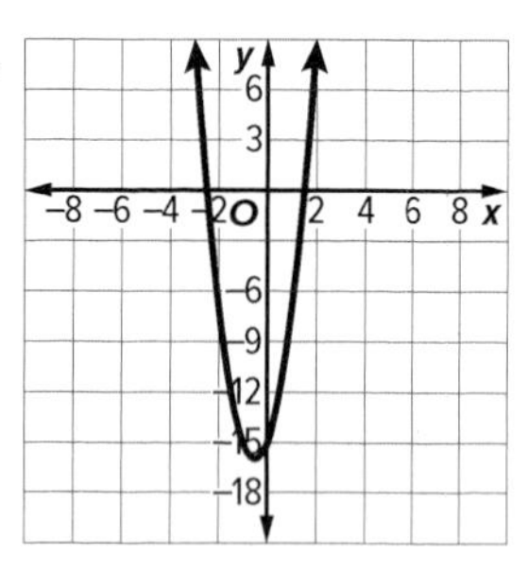

36.

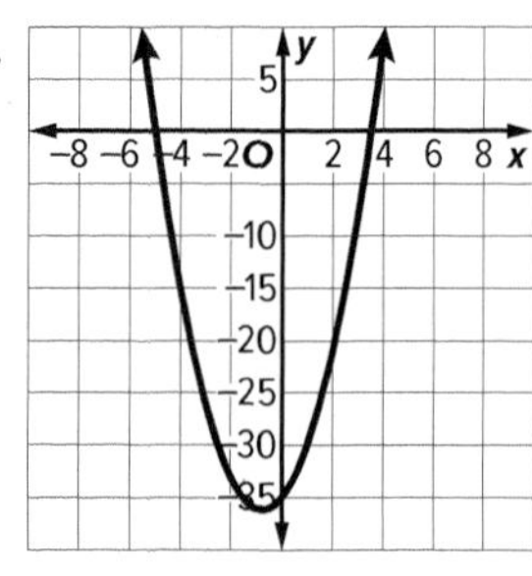

37.

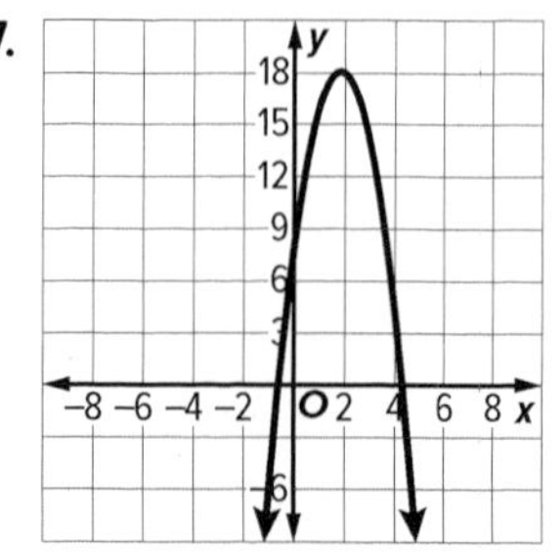

38.

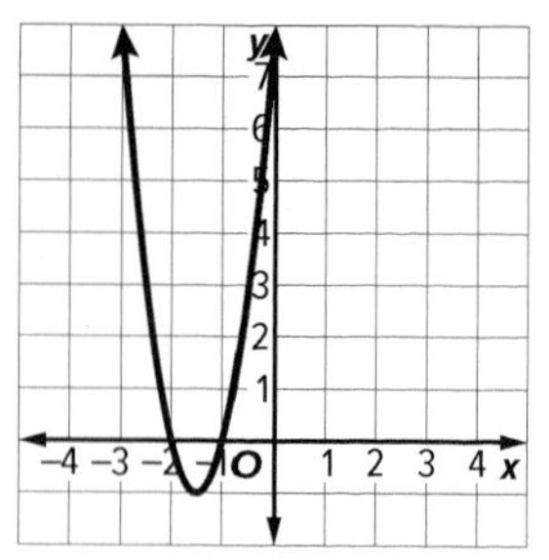

39.

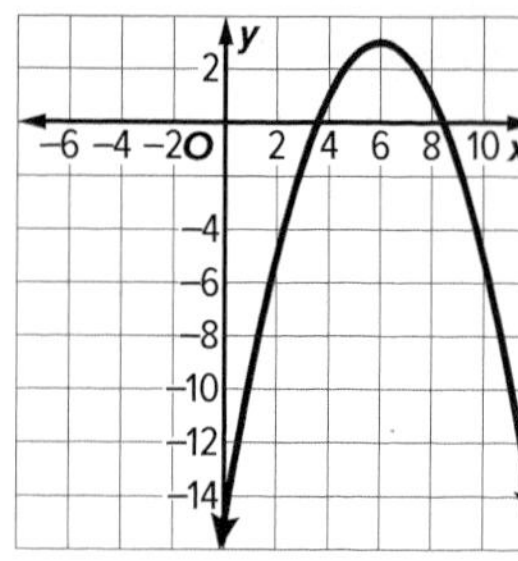

40.

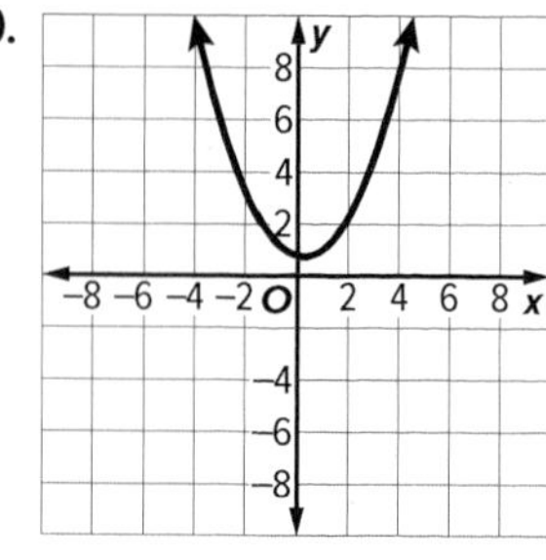

41.

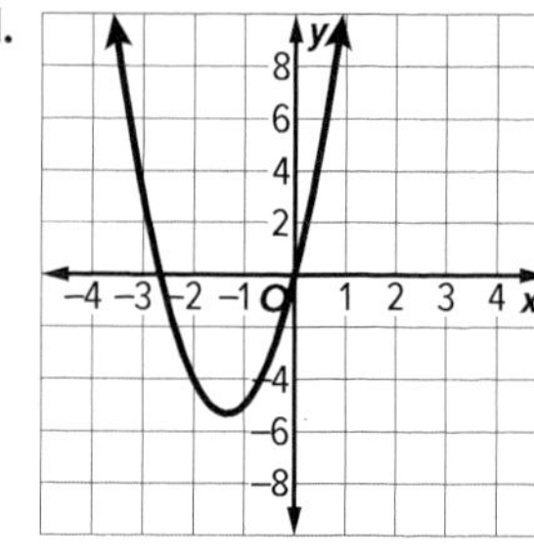

42.

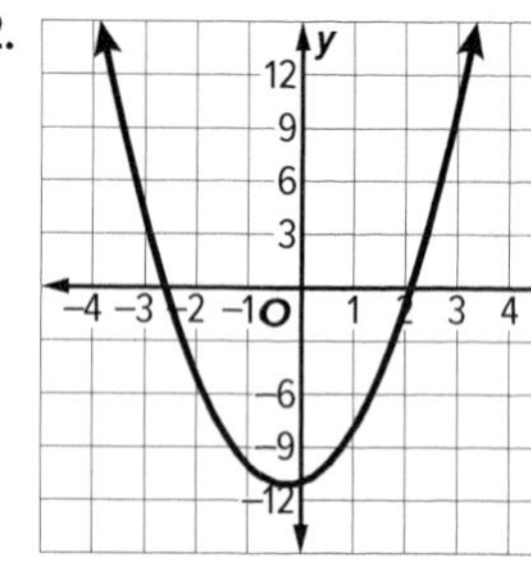

43.

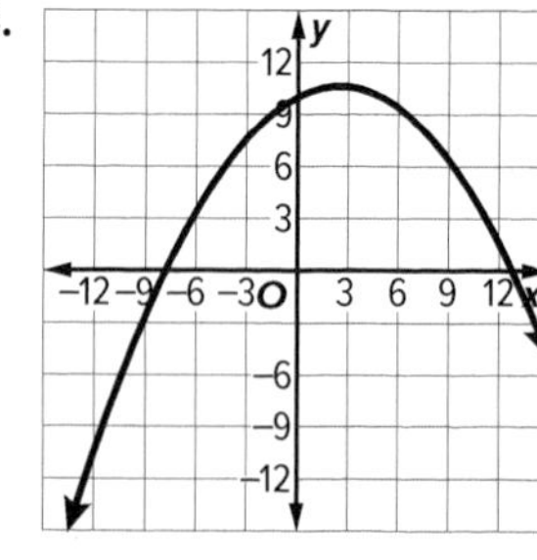

54.

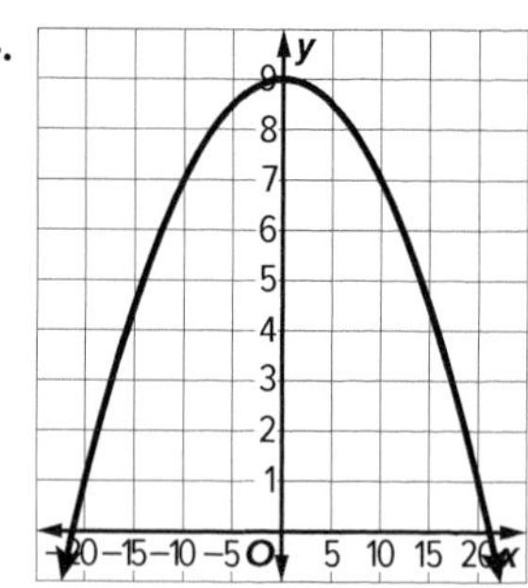

55.

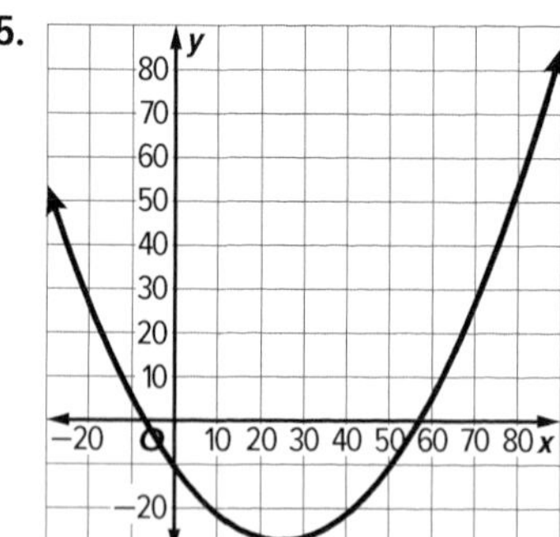

Lesson 3-7

13.

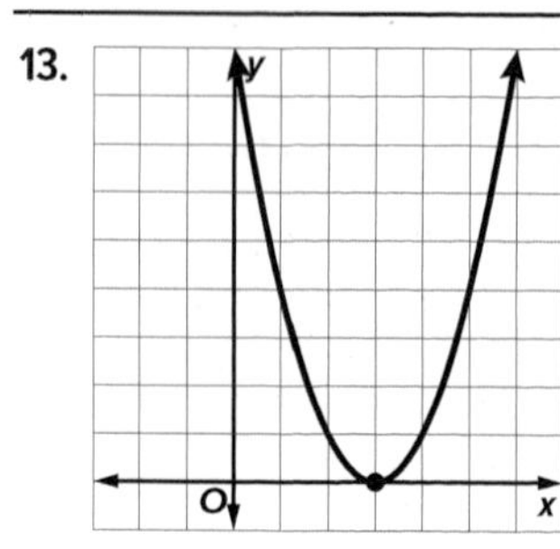

14.

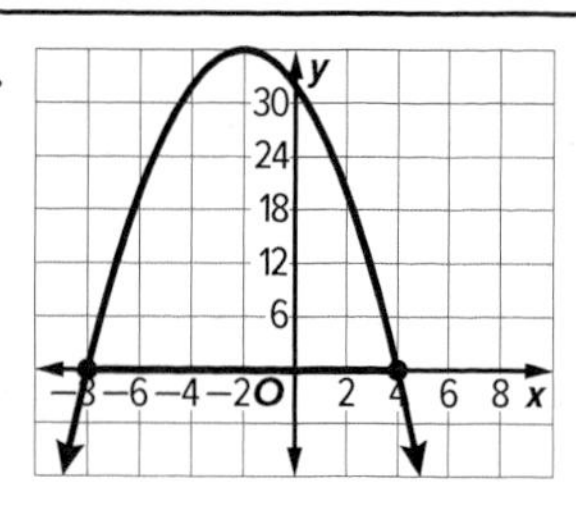

15. 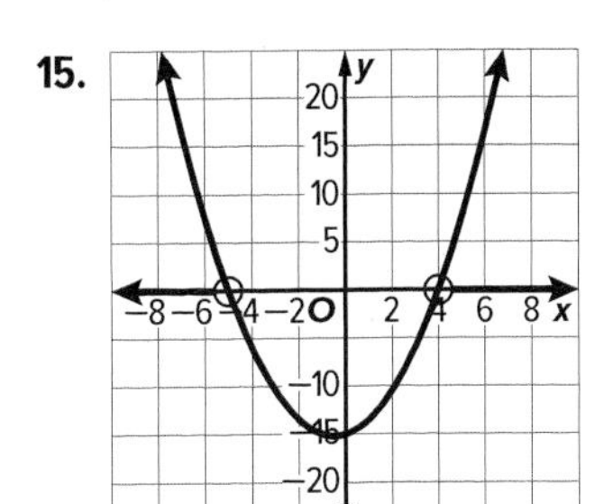

16. 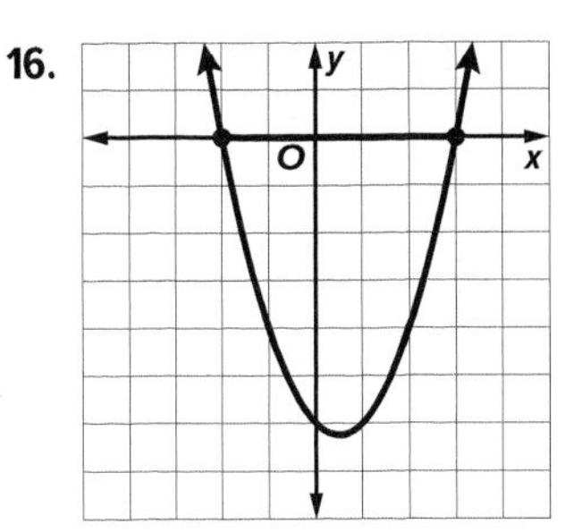

17. 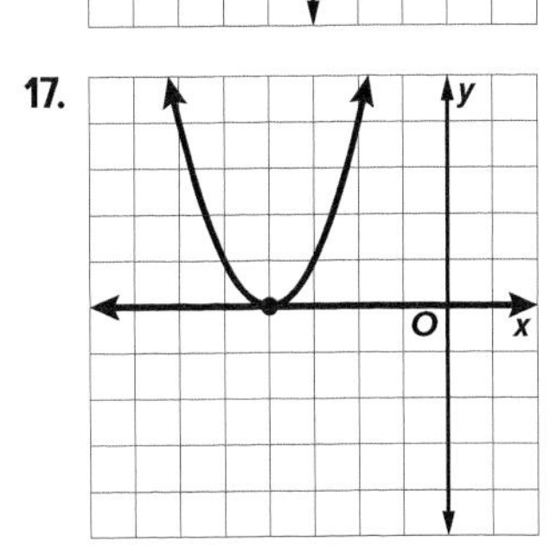

18. 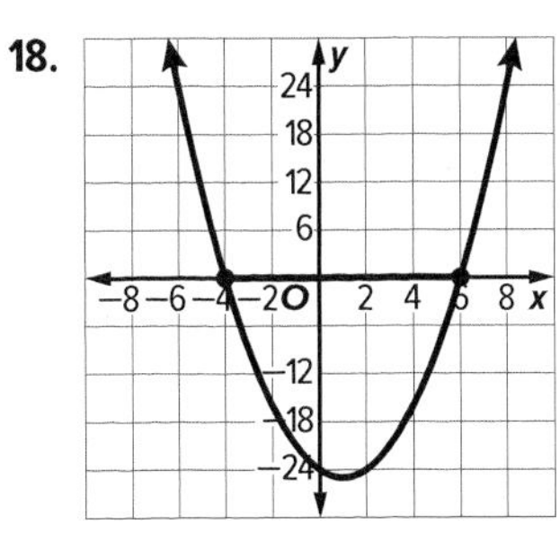

41. No; the graphs of the inequalities intersect the *x*-axis at the same points.

42. Neither; Don graphed the inequality in two variables, and Diego graphed the wrong interval.

44. Sometimes; sample answer: When *a* is positive and *c* is negative, there is no solution and when *a* is negative and *c* is positive there is a solution set.

45. 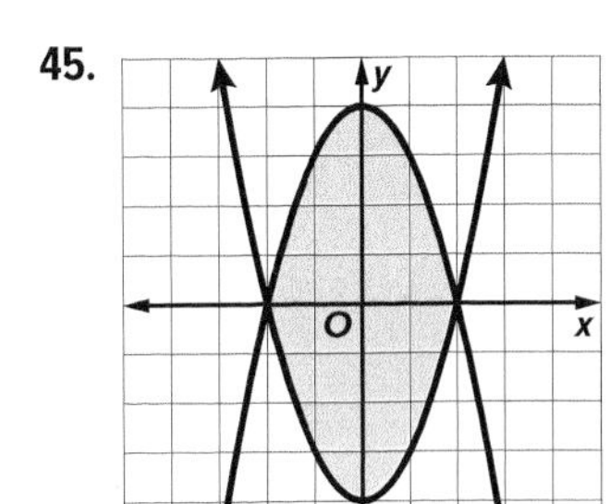

46. Sample answer: For both quadratic and linear inequalities, you must first graph the related function. You use the inequality symbol to determine if the line is dashed or solid. Then you use test points to determine where to shade. One difference is that one related function is a straight line while the other related function is a curve.

Lesson 3-8

18. 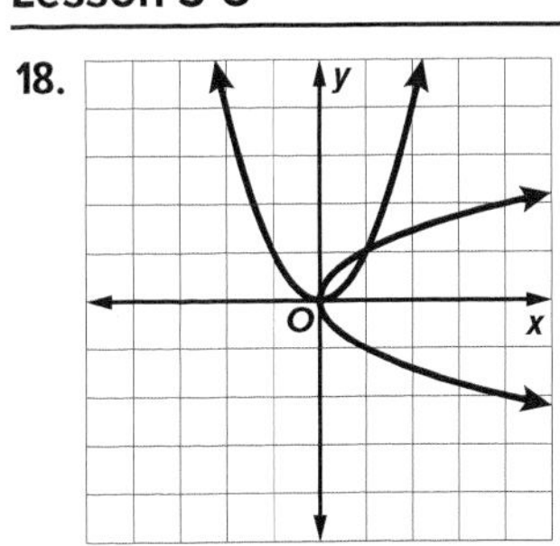

19. 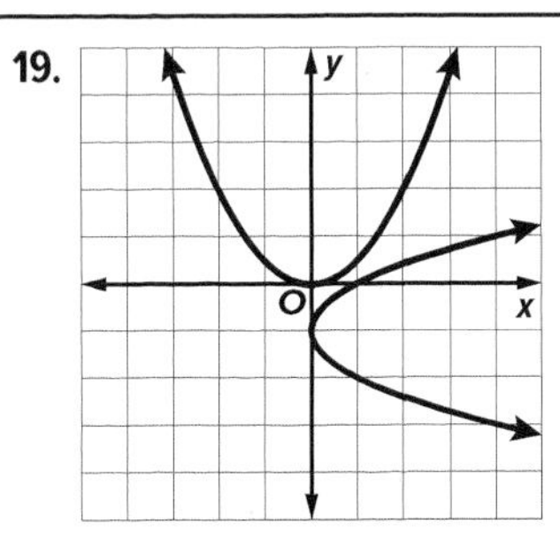

20. 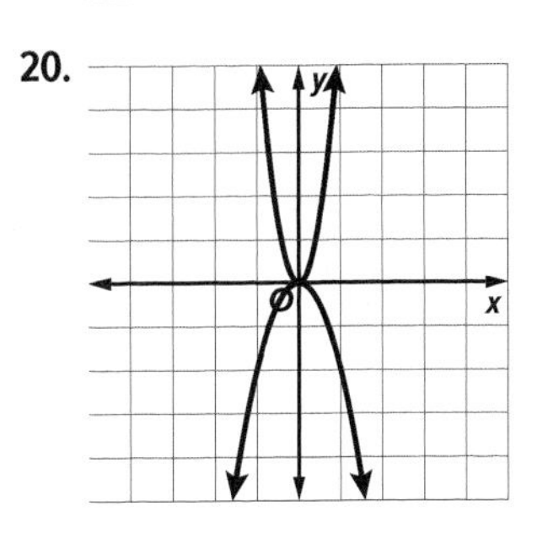

21. 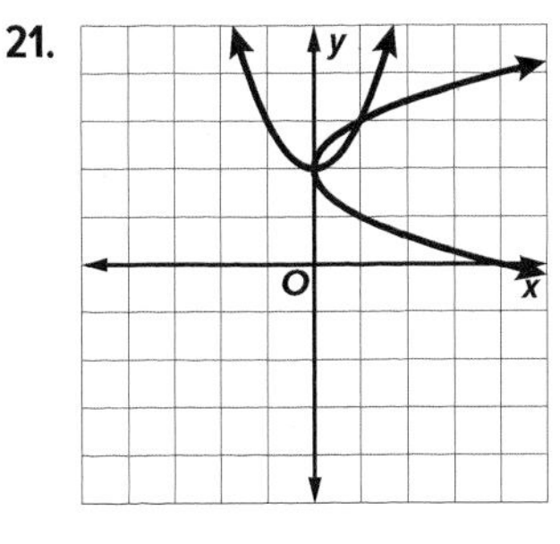

22. 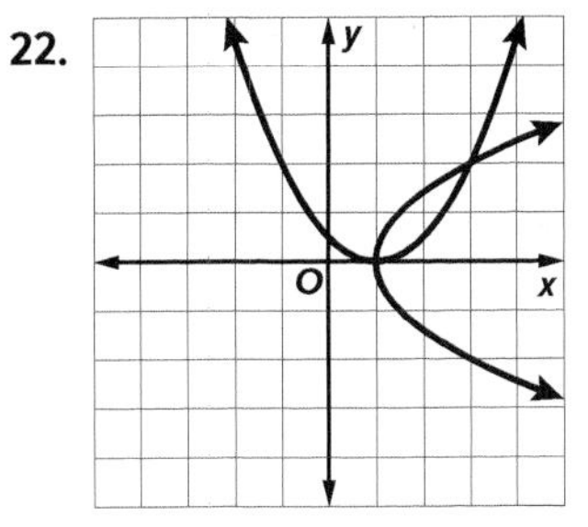

23. 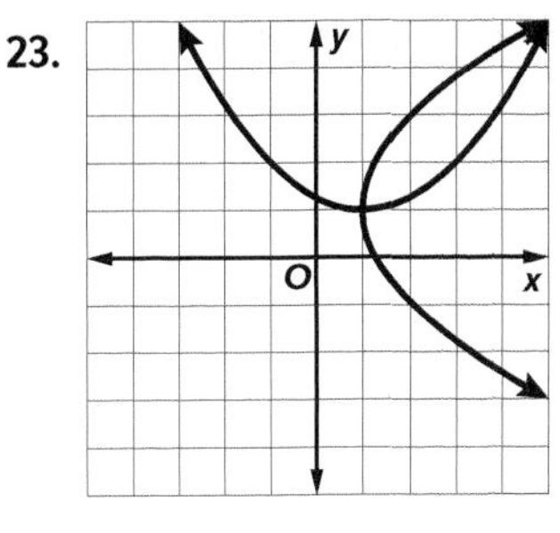

30. 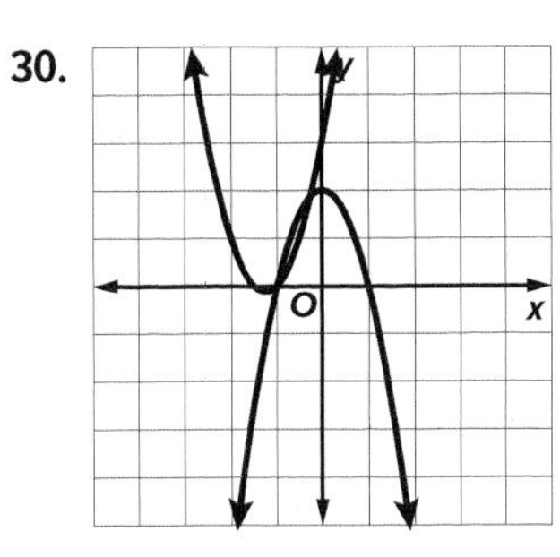

31. 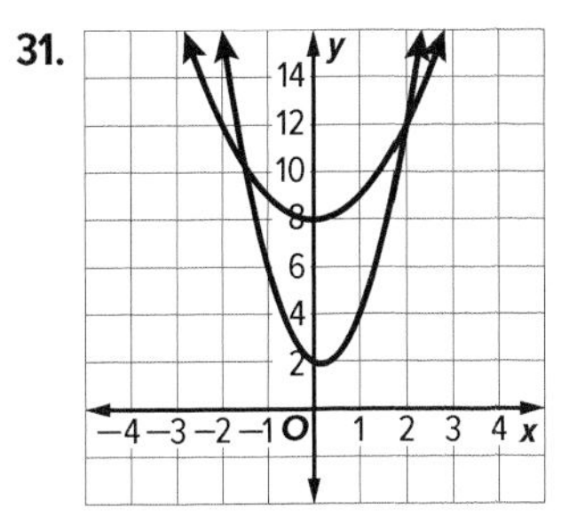

32. 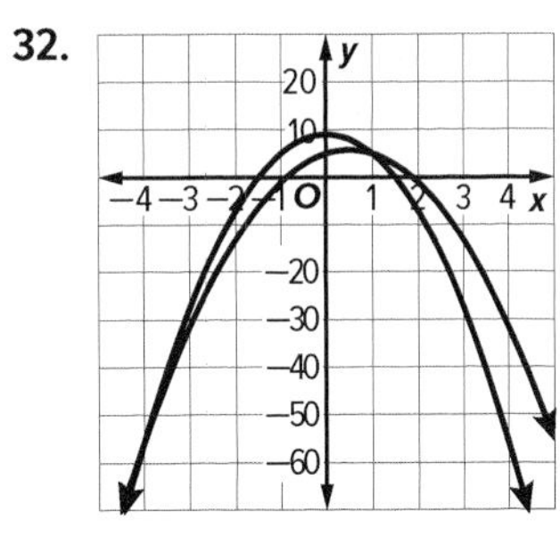

33. 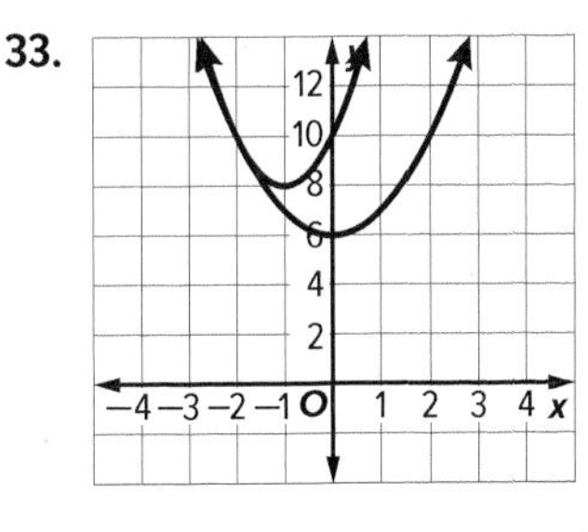

34. 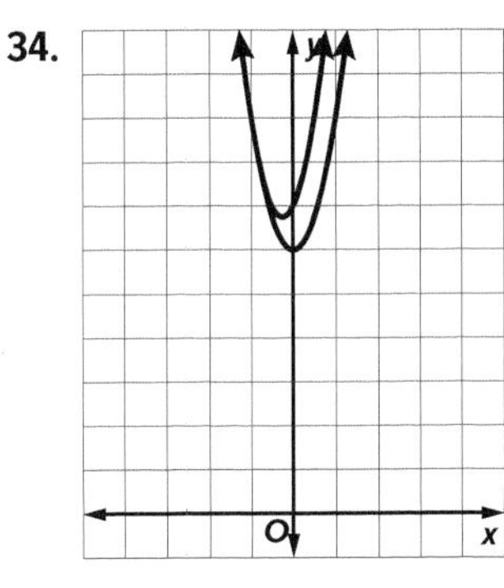

35. 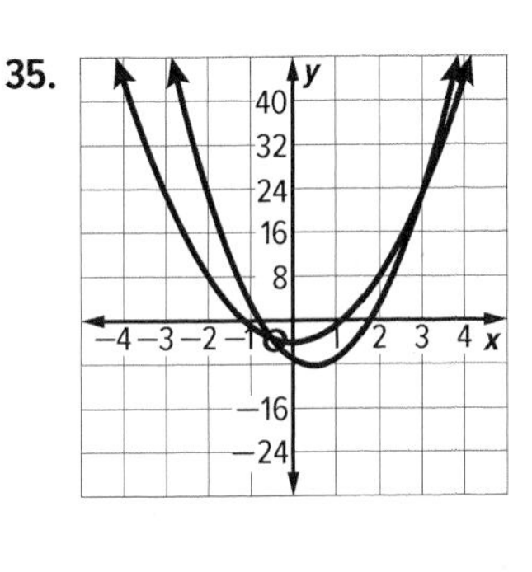

72. Sample answer: A linear-quadratic system of equations will have no points in common if the parabola opens up and the line is below the parabola with a slope that does not cause it to intersect. There will only be one solution if the line intersects the parabola at the vertex, either horizontally or vertically. There will be two solutions if the line intersects the parabola in a point other than the vertex and is not parallel to the line of symmetry of the parabola.

74. Sample answer: Company profits can be modeled by a quadratic equation and a horizontal line can be added to the model to determine specific price points or quantities of items that maximize profit for the company.

Module 4

Polynomials and Polynomial Functions

Module Goals

- Students analyze polynomial functions by examining key features and graphing.
- Students analyze the graphs of polynomial functions by identifying key features.
- Students add, subtract, and multiply polynomials.
- Students divide polynomials by using long division and synthetic division.
- Students expand powers of binomials.

Focus

Domain: Algebra, Functions

Standards for Mathematical Content:

F.IF.4 For a function that models a relationship between two quantities, interpret key features of graphs and tables in terms of the quantities, and sketch graphs showing key features given a verbal description of the relationship.

A.APR.1 Understand that polynomials form a system analogous to the integers, namely, they are closed under the operations of addition, subtraction, and multiplication; add, subtract, and multiply polynomials.

Also addresses A.APR.5, A.APR.6, and F.IF.7c.

Standards for Mathematical Practice:

All Standards for Mathematical Practice will be addressed in this module.

Coherence

Vertical Alignment

Previous

Students studied linear, exponential, and quadratic functions.
F.IF.7a, F.IF.7e (Algebra 1, Algebra 2)

Now

Students graph and analyze polynomials functions and perform operations with polynomials expressions.
A.APR.1, F.IF.7c

Next

Students will write and solve polynomial equations and identify zeros of polynomial functions.
A.CED.1, A.APR.3

Rigor

The Three Pillars of Rigor

To help students meet standards, they need to illustrate their ability to use the three pillars of rigor. Students gain conceptual understanding as they move from the Explore to Learn sections within a lesson. Once they understand the concept, they practice procedural skills and fluency and apply their mathematical knowledge as they go through the Examples and Independent Practice.

1 CONCEPTUAL UNDERSTANDING | 2 FLUENCY | 3 APPLICATION

EXPLORE → LEARN → EXAMPLE & PRACTICE

Suggested Pacing

Lessons	Standards	45-min classes	90-min classes
Module Pretest and Launch the Module Video		1	0.5
4-1 Polynomial Functions	F.IF.4; F.IF.7c	3	1.5
4-2 Analyzing Graphs of Polynomial Functions	F.IF.4; F.IF.7c	1	0.5
4-3 Operations with Polynomials	A.APR.1	2	1
4-4 Dividing Polynomials	A.APR.6	1	0.5
4-5 Powers of Binomials	A.APR.5	1	0.5
Module Review		1	0.5
Module Assessment		1	0.5
	Total Days	**11**	**5.5**

Formative Assessment Math Probe

Subtracting Polynomials

Analyze the Probe

Review the probe prior to assigning it to your students.

In this probe, students will determine which of several expressions is a simplified version of a polynomial expression involving subtraction.

Targeted Concepts Understand that polynomials can be subtracted by taking the additive inverse of terms being subtracted and then combining like terms to simplify.

Targeted Misconceptions

- Students may incorrectly distribute the subtraction across the second polynomial by:
  - completely ignoring the subtraction and treating it like an addition problem, or
  - only subtracting the first term of the second polynomial.
- Students may incorrectly combine like terms, including adding exponents.

Use the Probe after Lesson 4-3.

Module Resource

NAME ______________________ DATE ________ PERIOD ________

Cheryl Tobey Math Probe

Subtracting Polynomials

Determine which of the responses, if any, is a simplified version of $(10x^2 - 5x + 3) - (4x^2 - 6x + 7)$.

Circle *yes* or *no*.	Explain your choice.
1. $14x^2 - 11x + 10$ yes no	
2. $6x^2 - 11x + 10$ yes no	
3. $3x^6 + 10$ yes no	
4. $-5x^6 - 4$ yes no	
5. $6x^2 + x - 4$ yes no	
6. $7x^6 - 4$ yes no	
7. $7x^3 - 4$ yes no	

Cheryl Tobey Math Probe • Subtracting Polynomials

Correct Answers: 1. no 2. no 3. no 4. no 5. yes 6. no 7. no

Collect and Assess Student Answers

If the student selects these responses...	**Then** the student likely...
1. yes **2.** yes	combines like terms without correctly distributing the negative across the second polynomial by taking the additive inverse of each term. **Examples:** For Item 1, they did not take the additive inverse of any of the terms, and for Item 2, they only took the additive inverse of the first term.
3. yes **4.** yes	does not distribute the negative across the second polynomial by taking the additive inverse of each term and does not correctly combine like terms. **Example:** For Item 3, they did not take the additive inverse of any of the terms in the second polynomial and incorrectly combined all terms with a variable by adding coefficients and adding exponents.
6. yes **7.** yes	distributes the negative across the second polynomial correctly but has errors with combining like terms. **Examples:** For Item 7, they distributed the negative correctly and started to accurately combine like terms, but instead of leaving it as $6x^2 + x - 4$, they inaccurately combined terms with variables.

Take Action

After the Probe Design a plan to address any possible misconceptions. You may wish to assign the following resources.

- Lesson 4-3, Learn, Example 3

Revisit the Probe at the end of the module to be sure that your students no longer carry these misconceptions.

The Ignite! activities, created by Dr. Raj Shah, cultivate curiosity and engage and challenge students. Use these open-ended, collaborative activities, located online in the module Launch section, to encourage your students to develop a growth mindset towards mathematics and problem solving. Use the teacher notes for implementation suggestions and support for encouraging productive struggle.

Essential Question

At the end of this module, students should be able to answer the Essential Question.

How does an understanding of polynomials and polynomial functions help us understand and interpret real-world events? Sample answer: Polynomials and polynomial functions can be used to model situations where quantities increase or decrease in a nonlinear pattern.

What Will You Learn?

Prior to beginning this module, have your students rate their knowledge of each item listed. Then, at the end of the module, you will be reminded to have your students return to these pages to rate their knowledge again. They should see that their knowledge and skills have increased.

DINAH ZIKE FOLDABLES

Focus Students read about polynomial equations.

Teach Throughout the module, have students take notes under the tabs of their Foldables while working through each lesson. They should include definitions of terms and key concepts. Encourage students to record new ways of graphing, analyzing, and solving polynomial equations in their Foldable.

When to Use It Use the appropriate tabs as students cover each lesson in this module. Students should add vocabulary to the back of their Foldable during each lesson.

Launch the Module

For this module, the Launch the Module video uses energy production and usage to model polynomial functions. Exploring wind turbines, hybrid cars, and solar panels, students learn how polynomials and polynomial functions can be applied in real-world contexts.

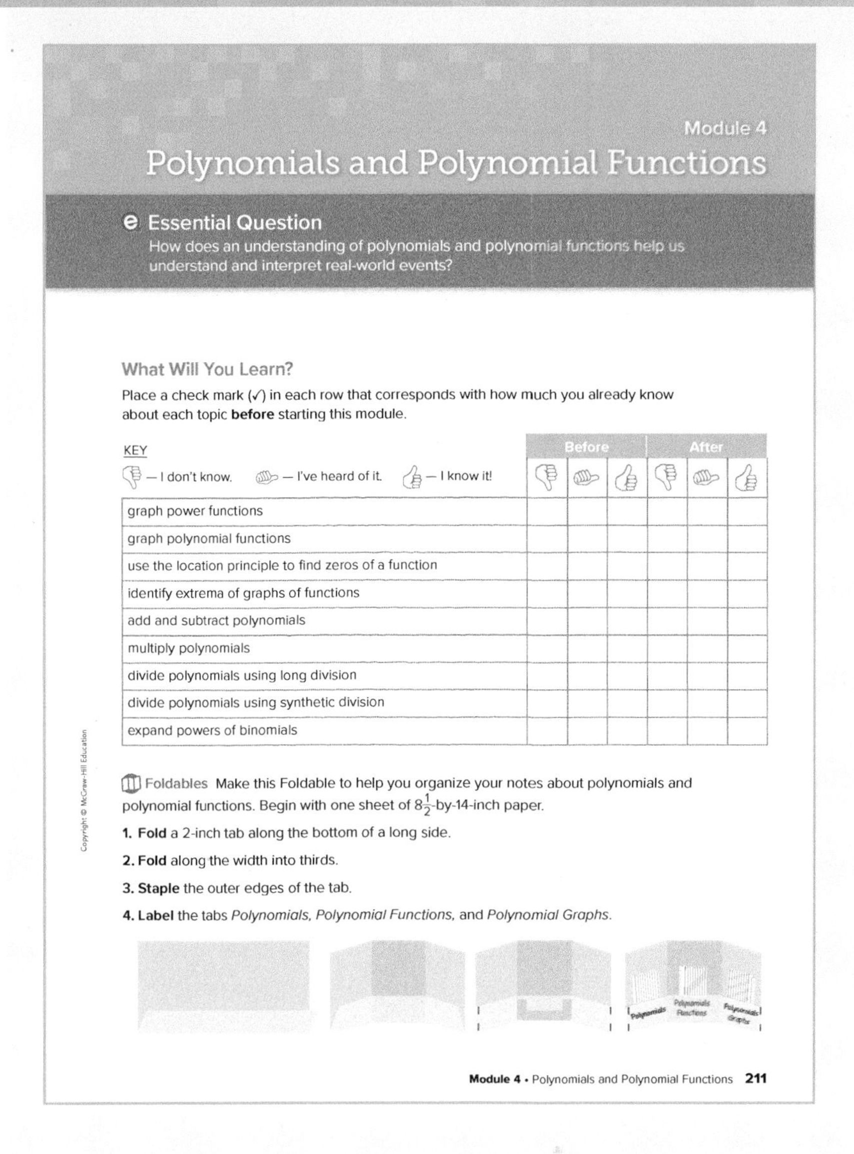

Module 4

Polynomials and Polynomial Functions

Essential Question

How does an understanding of polynomials and polynomial functions help us understand and interpret real-world events?

What Will You Learn?

Place a check mark (✓) in each row that corresponds with how much you already know about each topic **before** starting this module.

KEY: — I don't know. — I've heard of it. — I know it!

	Before			After		
	I don't know	I've heard of it	I know it	I don't know	I've heard of it	I know it
graph power functions						
graph polynomial functions						
use the location principle to find zeros of a function						
identify extrema of graphs of functions						
add and subtract polynomials						
multiply polynomials						
divide polynomials using long division						
divide polynomials using synthetic division						
expand powers of binomials						

Foldables Make this Foldable to help you organize your notes about polynomials and polynomial functions. Begin with one sheet of $8\frac{1}{2}$-by-14-inch paper.

1. **Fold** a 2-inch tab along the bottom of a long side.
2. **Fold** along the width into thirds.
3. **Staple** the outer edges of the tab.
4. **Label** the tabs *Polynomials, Polynomial Functions,* and *Polynomial Graphs.*

Interactive Presentation

What Vocabulary Will You Learn?

Check the box next to each vocabulary term that you may already know.

- ☐ binomial
- ☐ closed
- ☐ degree
- ☐ degree of a polynomial
- ☐ FOIL method
- ☐ leading coefficient
- ☐ monomial function
- ☐ Pascal's triangle
- ☐ polynomial in one variable
- ☐ polynomial function
- ☐ power function
- ☐ quartic function
- ☐ quintic function
- ☐ standard form of a polynomial
- ☐ synthetic division
- ☐ trinomial

Are You Ready?

Complete the Quick Review to see if you are ready to start this module.
Then complete the Quick Check.

Quick Review

Example 1

Rewrite $2xy - 3 - z$ as a sum.

$2xy - 3 - z$ Original expression

$= 2xy + (-3) + (-z)$ Rewrite using addition.

Example 2

Use the Distributive Property to rewrite $-3(a + b - c)$.

$-3(a + b - c)$ Original expression

$= -3(a) + (-3)(b) + (-3)(-c)$ Distributive Property

$= -3a - 3b + 3c$ Simplify.

Quick Check

Rewrite each difference as a sum.

1. $-5 - 13$ $\quad -5 + (-13)$
2. $5 - 3y$ $\quad 5 + (-3y)$
3. $5mr - 7mp$ $\quad 5mr + (-7mp)$
4. $3x^2y - 14xy^2$ $\quad 3x^2y + (-14xy^2)$

Use the Distributive Property to rewrite each expression without parentheses.

5. $-4(a + 5)$ $\quad -4a - 20$
6. $-1(3b^2 + 2b - 1)$ $\quad -3b^2 - 2b + 1$
7. $-\frac{1}{2}(2m - 5)$ $\quad -m + \frac{5}{2}$
8. $-\frac{3}{4}(3z + 5)$ $\quad -\frac{9}{4}z - \frac{15}{4}$

How Did You Do?

Which exercises did you answer correctly in the Quick Check? Shade those exercise numbers below.

① ② ③ ④ ⑤ ⑥ ⑦ ⑧

212 Module 4 • Polynomials and Polynomial Functions

What Vocabulary Will You Learn?

ELL As you proceed through the module, introduce the key vocabulary by using the following routine.

Define A polynomial function is a continuous function that can be described by a polynomial equation in one variable.

Example $f(x) = 3x^3 - 4x + 6$

Ask Is $g(x) = -4x^4 + 2$ a polynomial function? yes

Are You Ready?

Students may need to review the following prerequisite skills to succeed in this module.

- identifying values of a, b, and c
- selecting graphs by attributes
- performing simple operations with polynomials
- multiplying polynomials
- expanding perfect squares

ALEKS®

ALEKS is an adaptive, personalized learning environment that identifies precisely what each student knows and is ready to learn, ensuring student success at all levels.

You may want to use the **Quadratic and Polynomial Functions** section to ensure student success in this module.

Mindset Matters

Promote Process Over Results

The process that a student takes as they encounter a new problem is just as important—if not more important—than the result.

How Can I Apply It?

Encourage students to consider the **Think About It!** prompts in their Interactive Student Edition. Have students discuss their problem-solving strategies with a partner. Be sure to support the process and reward effort as students explore and work through problems.

Polynomial Functions

F.IF.4, F.IF.7c

LESSON GOAL

Students analyze polynomial functions by examining key features and graphing.

1 LAUNCH

Launch the lesson with a **Warm Up** and an introduction.

2 INQUIRY AND DEVELOP

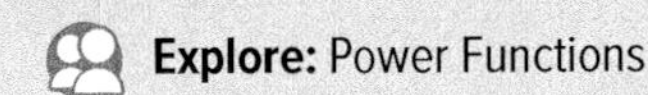

Explore: Power Functions

Develop:

Graphing Power Functions

- End Behavior and Degree of a Monomial Function
- Graph a Power Function by Using a Table

Explore: Polynomial Functions

Develop:

Graphing Polynomial Functions

- Degrees and Leading Coefficients
- Evaluate and Graph a Polynomial Function
- Zeros of a Polynomial Function
- Compare Polynomial Functions

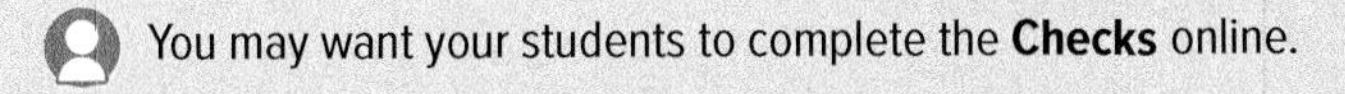

You may want your students to complete the **Checks** online.

3 REFLECT AND PRACTICE

Exit Ticket

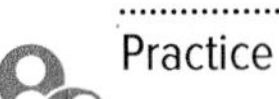

Practice

DIFFERENTIATE

View reports of student progress on the **Checks** after each example.

Resources	AL	OL	BL	ELL
Remediation: Adding and Subtracting Polynomials	●	●		●
Extension: Approximation by Means of Polynomials		●	●	●

Language Development Handbook

Assign page 25 of the *Language Development Handbook* to help your students build mathematical language related to analyzing polynomial functions.

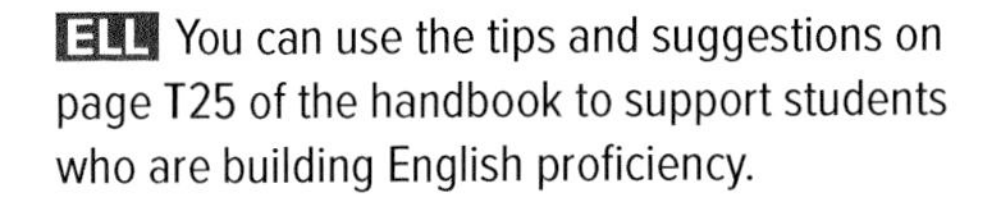

ELL You can use the tips and suggestions on page T25 of the handbook to support students who are building English proficiency.

Suggested Pacing

90 min	1.5 days
45 min	3 days

Focus

Domain: Functions

Standards for Mathematical Content:

F.IF.4 For a function that models a relationship between two quantities, interpret key features of graphs and tables in terms of the quantities, and sketch graphs showing key features given a verbal description of the relationship.

F.IF.7c Graph polynomial functions, identifying zeros when suitable factorizations are available, and showing end behavior.

Standards for Mathematical Practice:

1 Make sense of problems and persevere in solving them.

6 Attend to precision.

Coherence

Vertical Alignment

Previous
Students studied linear, exponential, and quadratic functions.
F.IF.7a, F.IF.7e (Algebra 1, Algebra 2)

Now
Students analyze polynomial functions by examining key features and graphing. **F.IF.4, F.IF.7c**

Next
Students will analyze the graphs of polynomial functions by identifying key features. **F.IF.4, F.IF.7c**

Rigor

The Three Pillars of Rigor

1 CONCEPTUAL UNDERSTANDING	2 FLUENCY	3 APPLICATION

Conceptual Bridge In this lesson, students extend their understanding of polynomials to include polynomial functions and build fluency by graphing them. They apply their understanding of graphing polynomial functions by solving real-world problems.

Mathematical Background

An expression made up of a sum of monomials that contains one variable is called a polynomial in one variable. The degree of a polynomial in one variable is the greatest exponent of its variable. The leading coefficient is the coefficient of the term with the highest degree. A polynomial function can be described by a polynomial equation in one variable.

Interactive Presentation

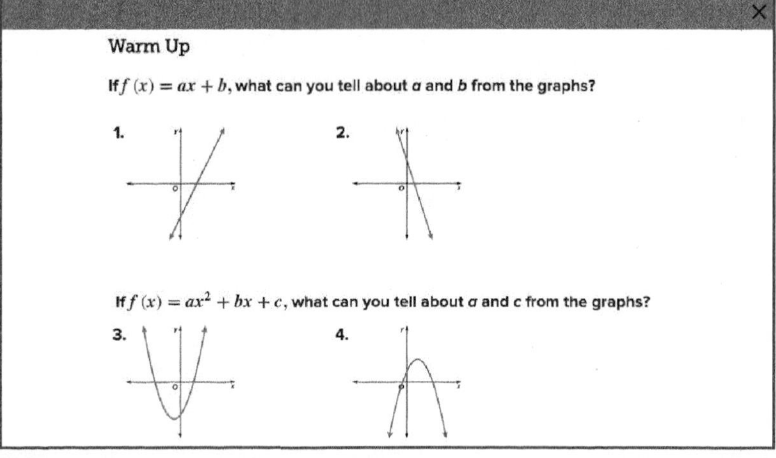

Warm Up

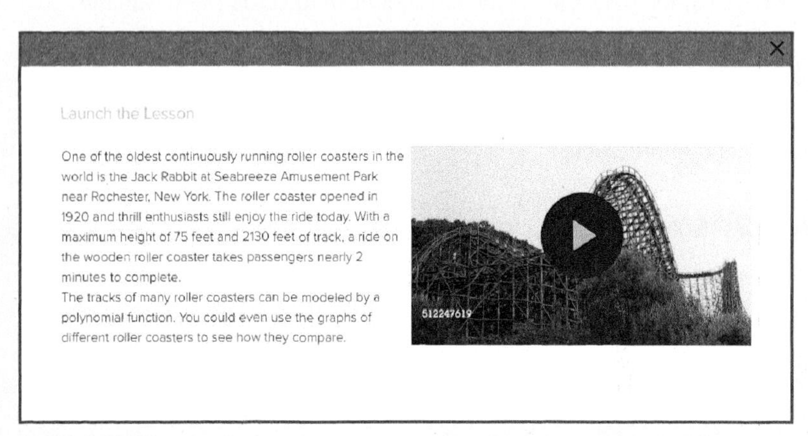

Launch the Lesson

Vocabulary

Expand All | Collapse All

- power function
- monomial function
- polynomial
- standard form of a polynomial
- degree of a polynomial
- polynomial function

- How do the operations on *polynomial expressions* follow from the operations on real numbers?
- What characteristics of *polynomial functions* can be seen on their graphs?

Today's Vocabulary

Warm Up

Prerequisite Skills

The Warm Up exercises address the following prerequisite skill for this lesson:

- evaluating values of a, b, and c

Answers:

1. $a > 0$ and $b < 0$
2. $a < 0$ and $b > 0$
3. $a > 0$ and $c < 0$
4. $a < 0$ and $c > 0$
5. 3: Since the axis of symmetry is left of the y-axis, $-\frac{b}{2a} < 0$. Since $a > 0$, $b > 0$. 4: Since the axis of symmetry is right of the y-axis, $-\frac{b}{2a} > 0$. Since $a < 0$, $b > 0$.

Launch the Lesson

 Teaching the Mathematical Practices

4 Apply Mathematics In the Launch the Lesson, students relate polynomial functions to the tracks of roller coasters.

Go Online to find additional teaching notes and questions to promote classroom discourse.

Today's Standards

Tell students that they will be addressing these content and practice standards in this lesson. You may wish to have a student volunteer read aloud *How can I meet these standards?* and *How can I use these practices?*, and connect these to the standards.

See the Interactive Presentation for I Can statements that align with the standards covered in this lesson.

Today's Vocabulary

Tell students that they will be using these vocabulary terms in this lesson. You can expand each row if you wish to share the definitions. Then discuss the questions below with the class.

Explore Power Functions

Objective

Students use a sketch to explore key features of power functions.

 Teaching the Mathematical Practices

1 Explain Correspondences Throughout the Explore, encourage students to explain the relationships between the values of the coefficient and degree of monomial function and the end behavior of the function.

Ideas for Use

Recommended Use Present the Inquiry Question, or have a student volunteer read it aloud. Have students work in pairs to complete the Explore activity on their devices. Pairs should discuss each of the questions. Monitor student progress during the activity. Upon completion of the Explore activity, have student volunteers share their responses to the Inquiry Question.

What if my students don't have devices? You may choose to project the activity on a whiteboard. A printable worksheet for each Explore is available online. You may choose to print the worksheet so that individuals or pairs of students can use it to record their observations.

Summary of the Activity

Students will complete guiding exercises throughout the Explore activity. Students will use a sketch to explore end behaviors based on different values of a and n in monomial function $f(x) = ax^n$. Then, they will make conjectures about their findings. Then, students will answer the Inquiry Question.

(continued on next page)

Interactive Presentation

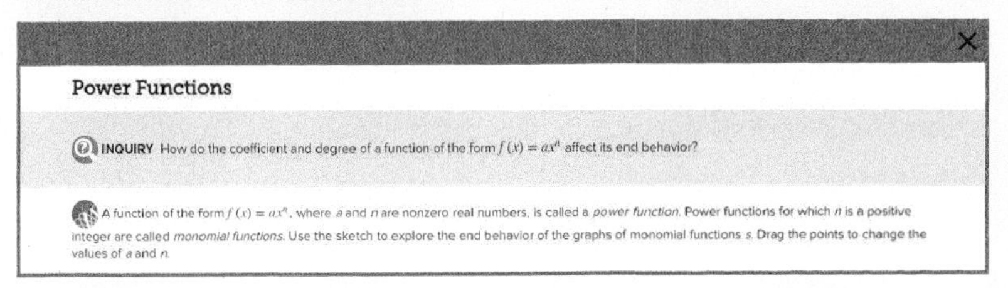

Explore

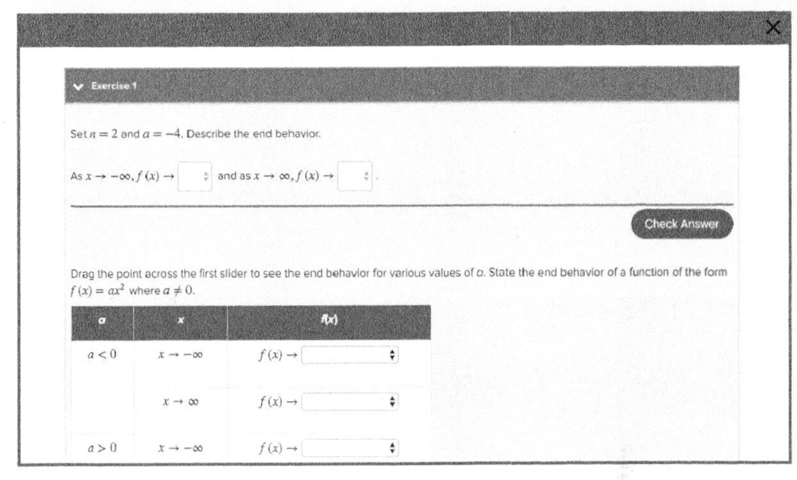

Explore

WEB SKETCHPAD

Students use a sketch to model power functions.

Interactive Presentation

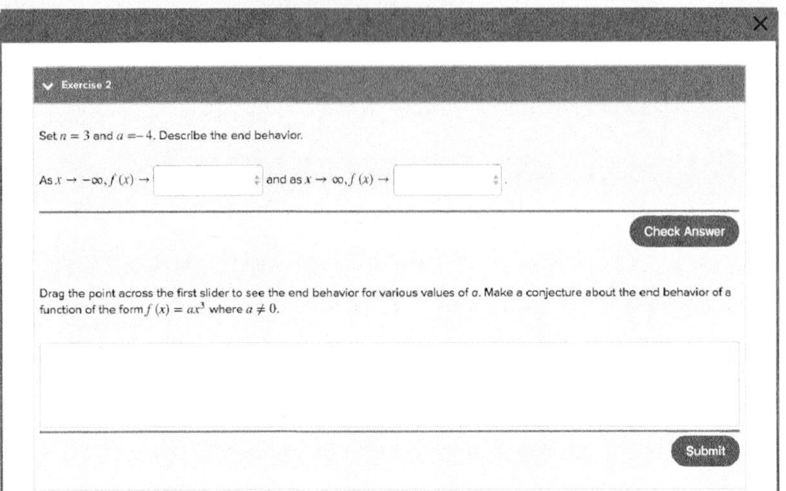

Explore

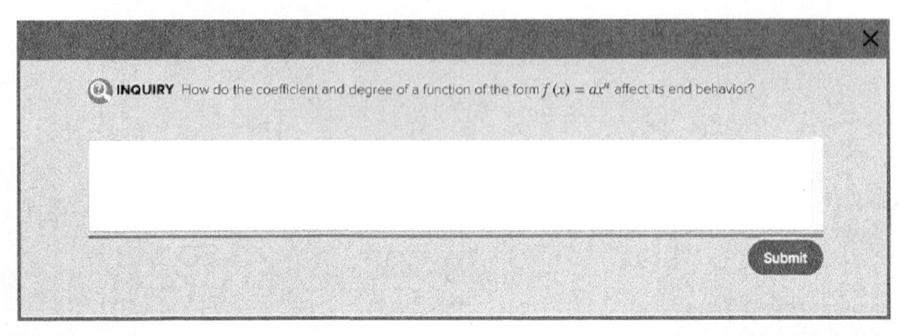

Explore

TYPE

Students respond to the Inquiry Question and can view a sample answer.

Explore Power Functions (*continued*)

Teaching the Mathematical Practices

5 Use Mathematical Tools To complete the Explore, students will need to use a sketch to analyze the end behavior of monomial functions for various coefficients and degrees. Work with students to explore and deepen their understanding of power functions.

Questions

Have students complete the Explore activity.

Ask:

- If the square of a real number is always positive, why does a negative value for a make the graph open down? Sample answer: The values for $y = x^2$ will always be positive but multiplying by a negative coefficient makes all values negative.
- What do you think is the end behavior for $y = x^6$? $y = x^9$? Sample answer: The degree is even for $y = ax^6$, so the end behavior is the same as x approaches both positive and negative infinity. The degree is odd for $y = ax^9$, so the end behavior is opposite as x approaches positive and negative infinity.

How do the coefficient and degree of a function of the form $f(x) = ax^n$ affect its end behavior? Sample answer: The end behavior of a function of the form $f(x) = ax^n$ changes based on whether the coefficient is positive or negative and whether the degree is even or odd. If n is even, then the end behavior is the same as x approaches both $-\infty$ and ∞. If n is odd, the end behavior is different as x approaches $-\infty$ and ∞.

Go Online to find additional teaching notes and sample answers for the guiding exercises.

Explore Polynomial Functions

Objective

Students use a sketch to explore key features of polynomial functions.

 Teaching the Mathematical Practices

1 Explain Correspondences Encourage students to explain the relationships between cubic, quartic, and quantic functions and key features of the graphs of the functions.

Ideas for Use

Recommended Use Present the Inquiry Question, or have a student volunteer read it aloud. Have students work in pairs to complete the Explore activity on their devices. Pairs should discuss each of the questions. Monitor student progress during the activity. Upon completion of the Explore activity, have student volunteers share their responses to the Inquiry Question.

What if my students don't have devices? You may choose to project the activity on a whiteboard. A printable worksheet for each Explore is available online. You may choose to print the worksheet so that individuals or pairs of students can use it to record their observations.

Summary of the Activity

Students will complete guiding exercises throughout the Explore activity. Students will use a sketch to explore the number of x-intercepts a polynomial function has based on coefficient values and the degree of the function. Then, students will answer the Inquiry Question.

(continued on next page)

Interactive Presentation

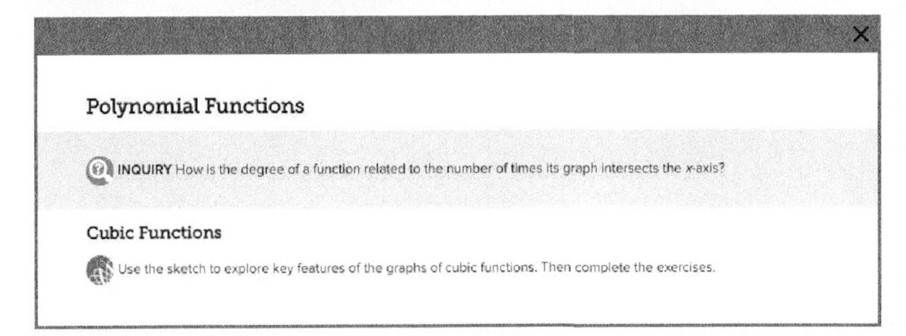

Explore

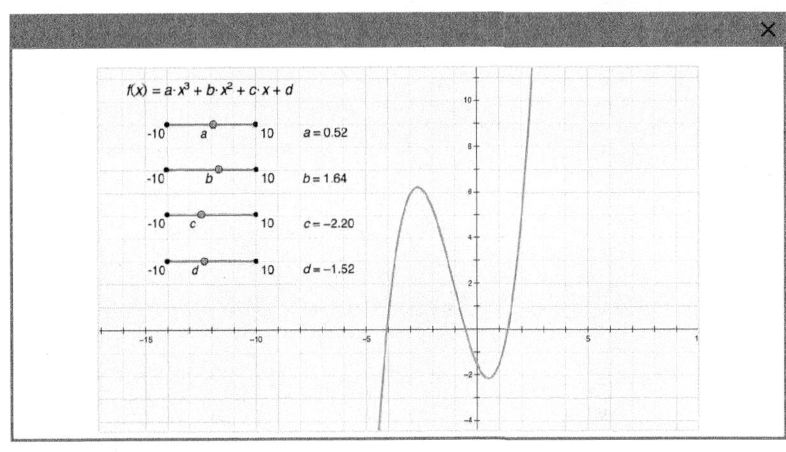

Explore

WEB SKETCHPAD

Students use a sketch to model power functions.

Interactive Presentation

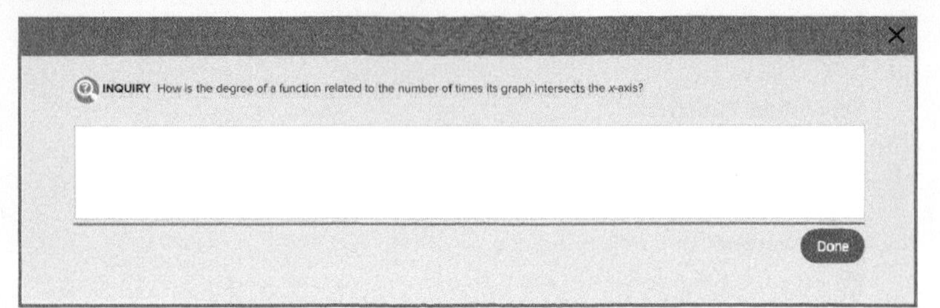

Explore

TYPE

a|

Students respond to the Inquiry Question and can view a sample answer.

Explore Polynomial Functions (*continued*)

Teaching the Mathematical Practices

3 Construct Arguments Throughout the Explore, students will use established results from the sketch to construct arguments about the key features of polynomial functions.

Questions

Have students complete the Explore activity.

Ask:

- What is the maximum number of times the graph of $y = ax^2$ can cross the x-axis? Sample answer: The graph of $y = ax^6$ can cross the x-axis 0, 1 or 2 times. The graph cannot cross more than twice.
- What is the maximum number of times the graph of $y = x^{12}$ can cross the x-axis? Sample answer: The graph of $y = x^n$ can cross a maximum of n times, so $y = x^{12}$ can cross at most 12 times.

Inquiry

How is the degree of a function related to the number of times its graph intersects the x-axis? Sample answer: The degree of a function is the maximum number of times its graph will intersect the x-axis.

Go Online to find additional teaching notes and sample answers for the guiding exercises.

1 CONCEPTUAL UNDERSTANDING | 2 FLUENCY | 3 APPLICATION

Learn Graphing Power Functions

Objective
Students graph and analyze power functions.

Teaching the Mathematical Practices
3 Construct Arguments In the Talk About It! feature, students will use definitions to determine whether $f(x) = \sqrt{x}$ is a power function and/or monomial function.

About the Key Concept
Help students understand how the graph of a monomial function changes for different values of the leading coefficient and degree of the function. Find patterns in different values, and use them to make conjectures that will help students understand the examples.

Common Misconception
Some students may not remember the vocabulary terms used to describe a function. Remind them that in $f(x) = ax^n$, a is the coefficient, and n is the degree of the function.

Essential Question Follow-Up
Students have explored power functions with different degrees.

Ask:

How does a power function's graph change as the degree of the function increases? Sample answer: As the degree increases, the ends of the graphs will alternate between going in the same and opposite directions.

DIFFERENTIATE

Language Development Activity ELL

Beginning Reinforce the use of visual context to derive meaning through examples of environmental print such as graphs of functions in news articles you find online. Pantomime or elicit one-word responses to the meaning derived from such images.

Intermediate Provide graphs from real-world articles to illustrate problems that students can work in pairs to solve. Move around the room to monitor progress.

Advanced High Provide a graph and have students write an opinion or explanation of what the graph represents. Have volunteers share their observations with their group.

Go Online
- Find additional teaching notes.
- View performance reports of the Checks.
- Assign or present an Extra Example.

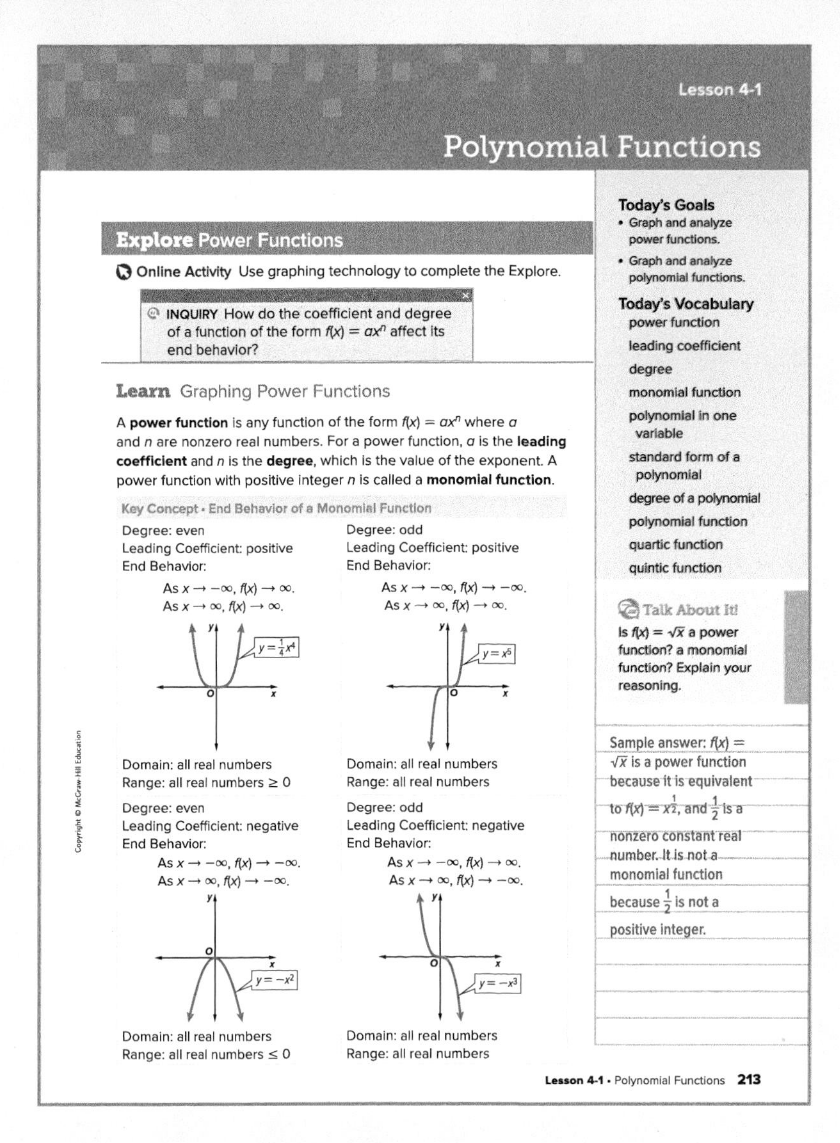

Lesson 4-1

Polynomial Functions

Explore Power Functions

Online Activity Use graphing technology to complete the Explore.

INQUIRY How do the coefficient and degree of a function of the form $f(x) = ax^n$ affect its end behavior?

Learn Graphing Power Functions

A **power function** is any function of the form $f(x) = ax^n$ where a and n are nonzero real numbers. For a power function, a is the **leading coefficient** and n is the **degree**, which is the value of the exponent. A power function with positive integer n is called a **monomial function**.

Key Concept • End Behavior of a Monomial Function

Degree: even
Leading Coefficient: positive
End Behavior:
As $x \to -\infty$, $f(x) \to \infty$.
As $x \to \infty$, $f(x) \to \infty$.
$y = \frac{1}{4}x^4$
Domain: all real numbers
Range: all real numbers ≥ 0

Degree: odd
Leading Coefficient: positive
End Behavior:
As $x \to -\infty$, $f(x) \to -\infty$.
As $x \to \infty$, $f(x) \to \infty$.
$y = x^5$
Domain: all real numbers
Range: all real numbers

Degree: even
Leading Coefficient: negative
End Behavior:
As $x \to -\infty$, $f(x) \to -\infty$.
As $x \to \infty$, $f(x) \to -\infty$.
$y = -x^2$
Domain: all real numbers
Range: all real numbers ≤ 0

Degree: odd
Leading Coefficient: negative
End Behavior:
As $x \to -\infty$, $f(x) \to \infty$.
As $x \to \infty$, $f(x) \to -\infty$.
$y = -x^3$
Domain: all real numbers
Range: all real numbers

Today's Goals
- Graph and analyze power functions.
- Graph and analyze polynomial functions.

Today's Vocabulary
power function
leading coefficient
degree
monomial function
polynomial in one variable
standard form of a polynomial
degree of a polynomial
polynomial function
quartic function
quintic function

Talk About It!
Is $f(x) = \sqrt{x}$ a power function? a monomial function? Explain your reasoning.

Sample answer: $f(x) = \sqrt{x}$ is a power function because it is equivalent to $f(x) = x^{\frac{1}{2}}$, and $\frac{1}{2}$ is a nonzero constant real number. It is not a monomial function because $\frac{1}{2}$ is not a positive integer.

Copyright © McGraw-Hill Education

Lesson 4-1 • Polynomial Functions 213

Interactive Presentation

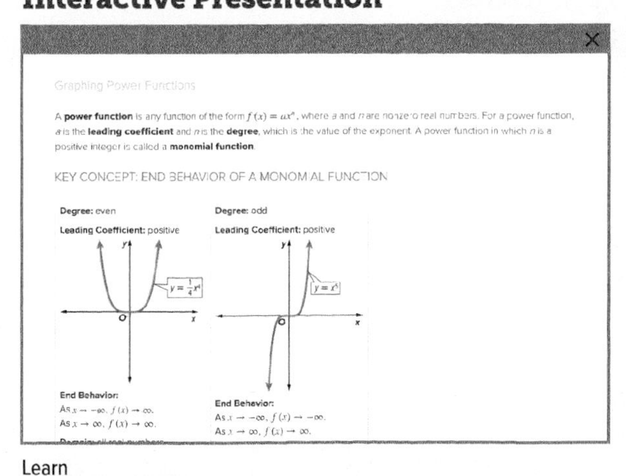

Learn

TYPE

Students answer a question to identify whether a function is a power function or a monomial function.

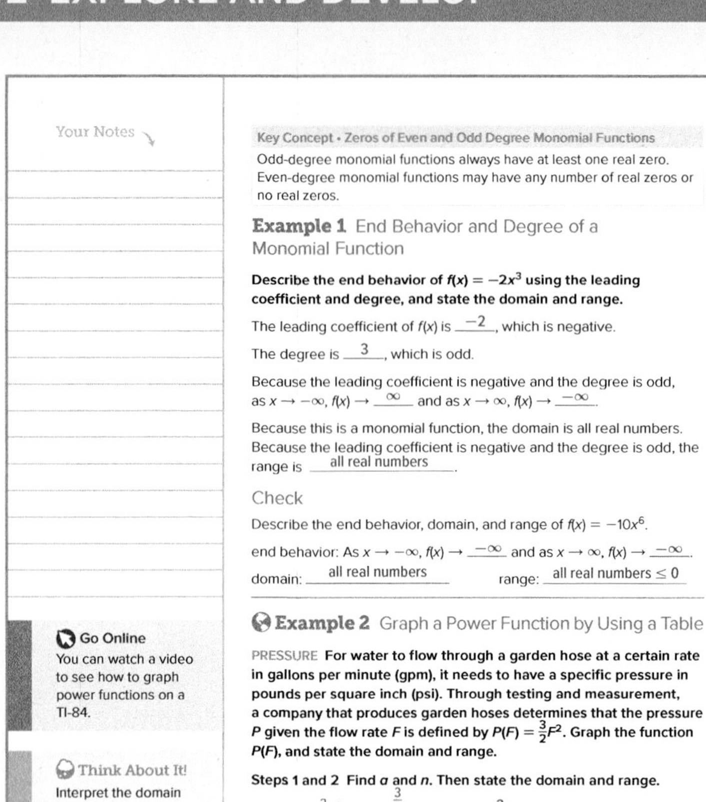

Your Notes

Key Concept • Zeros of Even and Odd Degree Monomial Functions

Odd-degree monomial functions always have at least one real zero. Even-degree monomial functions may have any number of real zeros or no real zeros.

Example 1 End Behavior and Degree of a Monomial Function

Describe the end behavior of $f(x) = -2x^3$ using the leading coefficient and degree, and state the domain and range.

The leading coefficient of $f(x)$ is $\underline{-2}$, which is negative.

The degree is $\underline{3}$, which is odd.

Because the leading coefficient is negative and the degree is odd, as $x \to -\infty$, $f(x) \to \underline{\infty}$ and as $x \to \infty$, $f(x) \to \underline{-\infty}$.

Because this is a monomial function, the domain is all real numbers. Because the leading coefficient is negative and the degree is odd, the range is $\underline{\text{all real numbers}}$.

Check

Describe the end behavior, domain, and range of $f(x) = -10x^6$.

end behavior: As $x \to -\infty$, $f(x) \to \underline{-\infty}$ and as $x \to \infty$, $f(x) \to \underline{-\infty}$.

domain: $\underline{\text{all real numbers}}$ range: $\underline{\text{all real numbers} \le 0}$

Go Online You can watch a video to see how to graph power functions on a TI-84.

Example 2 Graph a Power Function by Using a Table

PRESSURE **For water to flow through a garden hose at a certain rate in gallons per minute (gpm), it needs to have a specific pressure in pounds per square inch (psi). Through testing and measurement, a company that produces garden hoses determines that the pressure P given the flow rate F is defined by $P(F) = \frac{3}{2}F^2$. Graph the function $P(F)$, and state the domain and range.**

Think About It! Interpret the domain and range given the context of the situation.

Sample answer: Because it is unreasonable to have negative flow rate or negative pressure from a garden hose, both the domain and range are restricted to values greater than or equal to zero.

Steps 1 and 2 Find a and n. Then state the domain and range.

For $P(F) = \frac{3}{2}F^2$, $a = \underline{\frac{3}{2}}$, and $n = \underline{2}$.

The domain is $\underline{\text{all real numbers.}}$ Because a is positive and n is even, the range is all real numbers $\underline{\ge 0}$.

Steps 3–5 Create a table of values and graph the ordered pairs.

F	$\frac{3}{2}F^2$	$P(F)$
−2	$\frac{3}{2}(-2)^2$	6
−1	$\frac{3}{2}(-1)^2$	1.5
0	$\frac{3}{2}(0)^2$	0
1	$\frac{3}{2}(1)^2$	1.5
2	$\frac{3}{2}(2)^2$	6

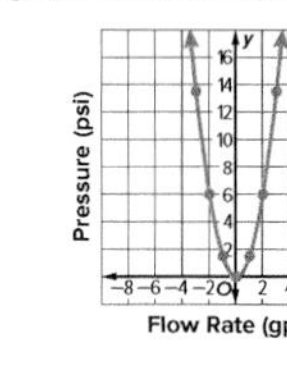

Go Online You can complete an Extra Example online.

Interactive Presentation

End Behavior and Degree of a Monomial Function

Describe the end behavior of $f(x) = -2x^3$ using the leading coefficient and degree, and state the domain and range.

$f(x) = -2x^3$

leading coefficient | degree | end behavior | domain range

Example 1

TAP

Student tap on each button to determine the end behavior of the function.

CHECK

Students complete the Check online to determine whether they are ready to move on.

Example 1 End Behavior and Degree of a Monomial Function

MP Teaching the Mathematical Practices

1 Explain Correspondences Encourage students to explain the relationships between the leading coefficient and degree of a monomial function and the key features of the function.

Questions for Mathematical Discourse

AL Why does the negative leading coefficient cause this end behavior for the odd function? The sign of x and y will always be opposite, so if $x \to \infty$, y must go to $-\infty$.

OL Through how many quadrants will this monomial function pass? two

BL What happens to the graph of ax^2 as a increases? The graph gets narrower. What happens to the graph of x^n, where n is even positive integers, as n increases? The part of the graph below $y = 1$ widens, and the part of the graph above $y = 1$ narrows.

Example 2 Graph a Power Function by Using a Table

MP Teaching the Mathematical Practices

4 Interpret Mathematical Results Use the Think About It! feature to encourage students to interpret their mathematical results in the context of the problem.

Questions for Mathematical Discourse

AL How does the graph of $P(F) = \frac{3}{2}F^2$ compare to the graph of $P(F) = F^2$? The coefficient $\frac{3}{2}$ makes the graph narrower.

OL Does the function make sense for the context of the situation? Yes; within the valid domain, an increase in flow yields an increase in pressure.

BL What could a negative flow rate represent? The flow is in the opposite direction.

Learn Graphing Polynomial Functions

Objective

Students graph and analyze polynomial functions.

Teaching the Mathematical Practices

6 Communicate Precisely Encourage students to routinely write or explain their solution methods. Students should use clear definitions when they answer the question in the Think About It! feature.

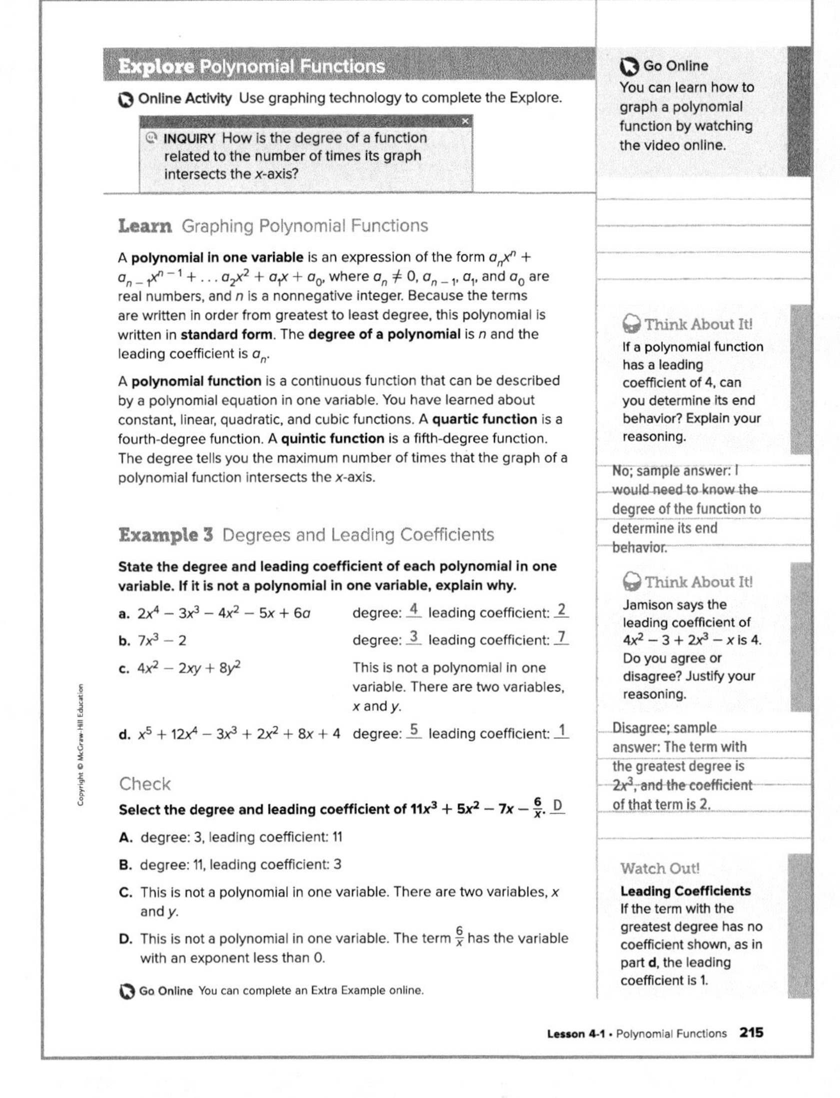

Explore Polynomial Functions

Online Activity Use graphing technology to complete the Explore.

INQUIRY How is the degree of a function related to the number of times its graph intersects the x-axis?

Learn Graphing Polynomial Functions

A **polynomial in one variable** is an expression of the form $a_nx^n + a_{n-1}x^{n-1} + \ldots a_2x^2 + a_1x + a_0$, where $a_n \neq 0$, a_{n-1}, a_1, and a_0 are real numbers, and n is a nonnegative integer. Because the terms are written in order from greatest to least degree, this polynomial is written in **standard form**. The **degree of a polynomial** is n and the leading coefficient is a_n.

A **polynomial function** is a continuous function that can be described by a polynomial equation in one variable. You have learned about constant, linear, quadratic, and cubic functions. A **quartic function** is a fourth-degree function. A **quintic function** is a fifth-degree function. The degree tells you the maximum number of times that the graph of a polynomial function intersects the x-axis.

Example 3 Degrees and Leading Coefficients

State the degree and leading coefficient of each polynomial in one variable. If it is not a polynomial in one variable, explain why.

a. $2x^4 - 3x^3 - 4x^2 - 5x + 6a$ degree: 4 leading coefficient: 2

b. $7x^3 - 2$ degree: 3 leading coefficient: 7

c. $4x^2 - 2xy + 8y^2$ This is not a polynomial in one variable. There are two variables, x and y.

d. $x^5 + 12x^4 - 3x^3 + 2x^2 + 8x + 4$ degree: 5 leading coefficient: 1

Check

Select the degree and leading coefficient of $11x^3 + 5x^2 - 7x - \frac{6}{x}$. D

A. degree: 3, leading coefficient: 11

B. degree: 11, leading coefficient: 3

C. This is not a polynomial in one variable. There are two variables, x and y.

D. This is not a polynomial in one variable. The term $\frac{6}{x}$ has the variable with an exponent less than 0.

Go Online You can complete an Extra Example online.

Go Online
You can learn how to graph a polynomial function by watching the video online.

Think About It!
If a polynomial function has a leading coefficient of 4, can you determine its end behavior? Explain your reasoning.

No; sample answer: I would need to know the degree of the function to determine its end behavior.

Think About It!
Jamison says the leading coefficient of $4x^2 - 3 + 2x^3 - x$ is 4. Do you agree or disagree? Justify your reasoning.

Disagree; sample answer: The term with the greatest degree is $2x^3$, and the coefficient of that term is 2.

Watch Out!
Leading Coefficients
If the term with the greatest degree has no coefficient shown, as in part **d**, the leading coefficient is 1.

Lesson 4-1 • Polynomial Functions 215

Example 3 Degrees and Leading Coefficients

Teaching the Mathematical Practices

7 Interpret Complicated Expressions Mathematically proficient students can see complicated expressions as single objects or as being composed of several objects. In Example 3, guide students to see what information they can gather about the expression just from looking at it.

Questions for Mathematical Discourse

AL What should you check first when determining the degree and leading coefficient of a polynomial? Check to see if the terms are in order from greatest to least degree.

OL What is the end behavior of a polynomial of degree $2n - 1$, where n is a non-negative integer? This would be an odd degree function, which results in the end behavior as $x \to \infty$, $f(x) \to -\infty$ and as $x \to -\infty$, $f(x) \to \infty$.

BL Why does the degree of the function determine the end behavior? The greatest degree term has more effect on the function for greater values of x.

Graph $y = 0.005x^4 - x^3$ using a graphing calculator or online graphing software. Does the end behavior match the degree of the polynomial? Yes, but it does not appear to until you zoom out.

Interactive Presentation

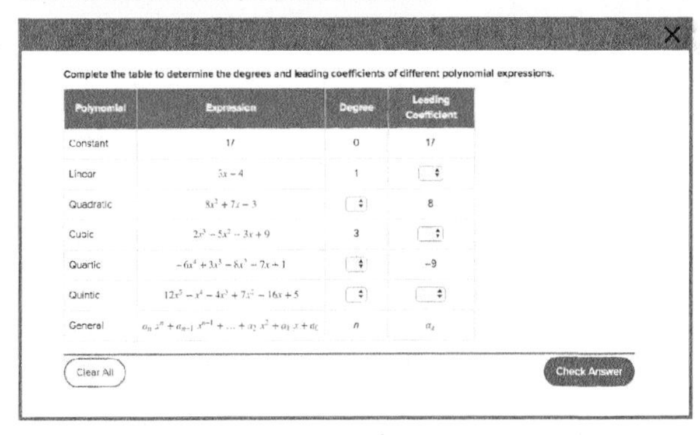

Complete the table to determine the degrees and leading coefficients of different polynomial expressions.
Polynomial
Expression
Degree
Leading Coefficient
Constant
Linear
Quadratic
Cubic
Quartic
Quintic
General
Clear All
Check Answer

Learn

SELECT

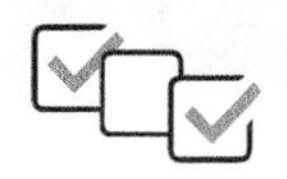

Students select the values to complete the table.

Think About It!
What values of x make sense in the context of the situation? Justify your reasoning.

Sample answer: Because x is a percent, only values between 0 and 1, inclusively, make sense in the context of the situation.

Study Tip
Axes Labels Notice that the x-axis is measuring the percent of the radius, not the actual length of the radius.

Example 4 Evaluate and Graph a Polynomial Function

SUN The density of the Sun, in grams per centimeter cubed, expressed as a percent of the distance from the core of the Sun to its surface can be modeled by the function $f(x) = 519x^4 - 1630x^3 + 1844x^2 - 889x + 155$, where x represents the percent as a decimal. At the core $x = 0$, and at the surface $x = 1$.

Part A Evaluate the function.

Find the core density of the Sun at a radius 60% of the way to the surface.

Because we need to find the core density at a radius 60% of the way to the surface, $x = 0.6$. So, replace x with 0.6 and simplify.

$f(x) = 519x^4 - 1630x^3 + 1844x^2 - 889x + 155$

$= 519(0.6)^4 - 1630(0.6)^3 + 1844(0.6)^2 - 889(0.6) + 155$

$= 67.2624 - 352.08 + 663.84 - 533.4 + 155$

$= 0.6224 \frac{g}{cm^3}$

Part B Graph the function.

Sketch a graph of the function.

Substitute values of x to create a table of values. Then plot the points, and connect them with a smooth curve.

x	$f(x)$
0.1	82.9619
0.2	38.7504
0.3	14.4539
0.4	3.4064
0.5	0.1875
0.7	1.7819
0.8	1.9824
0.9	0.7859

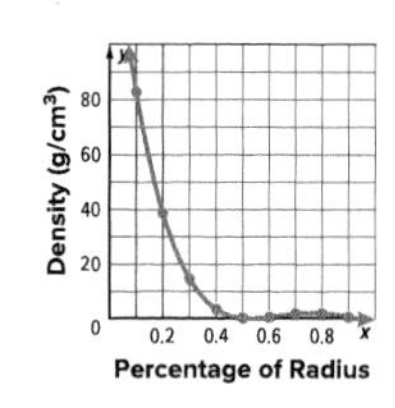

Check

CARDIOLOGY To help predict heart attacks, doctors can inject a concentration of dye in a vein near the heart to measure the cardiac output in patients. In a normal heart, the change in the concentration of dye can be modeled by $f(x) = -0.006x^4 + 0.140x^3 - 0.053x^2 + 1.79x$, where x is the time in seconds.

Part A Find the concentration of dye after 5 seconds.

$f(5) =$ 21.375

Go Online You can complete an Extra Example online.

216 Module 4 • Polynomials and Polynomial Functions

Interactive Presentation

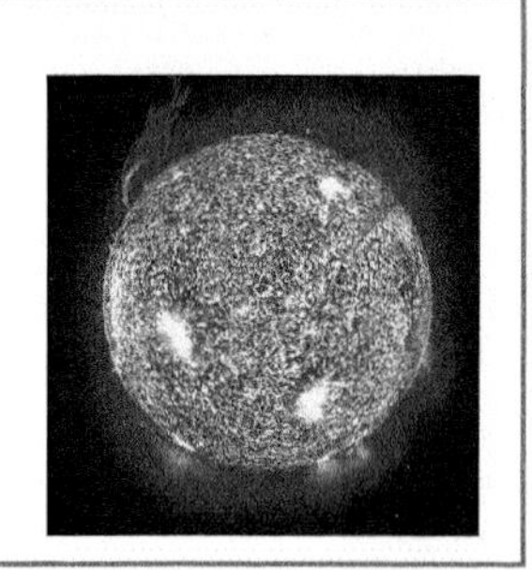

Example 4

TYPE

Students answer a question about the possible domain of the function in the context of the situation.

Example 4 Evaluate and Graph a Polynomial Function

Teaching the Mathematical Practices

6 Use Quantities In Example 4, guide students to clarify their use of quantities. Ensure that they specify the units of measure they are using in the problem and label axes appropriately.

Questions for Mathematical Discourse

AL What does the decreasing trend of the graph mean in the context of the example? As you move farther from the center of the Sun, the density decreases.

OL What do you expect the density to be at $x = 1$? Explain. The density should be zero because $x = 1$ represents the surface.
Evaluate $f(1)$. Does the result match your expectation? Explain. -1; Sample answer: No; because the density is estimated by a model, the value at $f(1)$ is close to the expected value but not exact.

BL What is the density of the Sun at its core? 155 g/cm^3 The density of air at sea level on Earth is about 1.225×10^{-3} g/cm^3. What is the difference in order of magnitude between these values? The density at the Sun's core is 5 orders of magnitude greater than the density of air on Earth at sea level.

Common Error

Students may assume that the x-axis is measuring the length of the radius, instead of a percent of the radius. Explain what the x- and y-axes represent on the coordinate plane in the context of the situation.

Example 5 Zeros of a Polynomial Function

Teaching the Mathematical Practices

1 Explain Correspondences Encourage students to explain the relationship between the graph of a polynomial function and the number of real zeros of the function.

Questions for Mathematical Discourse

AL Is the degree of the function even or odd? Explain. Odd because the end behavior in the y-direction for $x \to -\infty$ is the opposite direction of the end behavior as $x \to \infty$.

OL Can you tell the degree of the function from this graph? No, you can only tell that the degree is odd and greater than 1.

BL How would you transform this graph to only have 2 unique zeros? Sample answer: Translate the graph up until the relative minimum is on the x-axis.

Common Error

Students may confuse zeros with y-intercept(s). Remind them that zeros are the values of x when $f(x) = 0$.

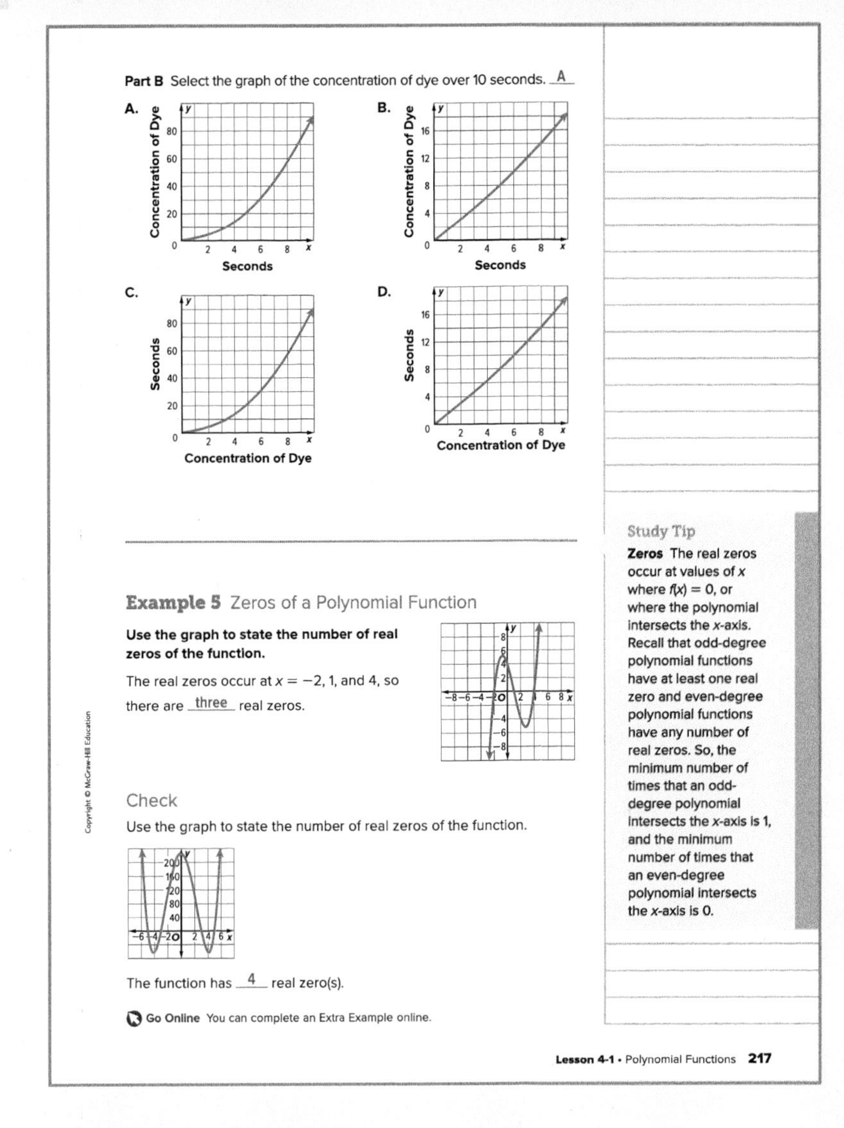

Interactive Presentation

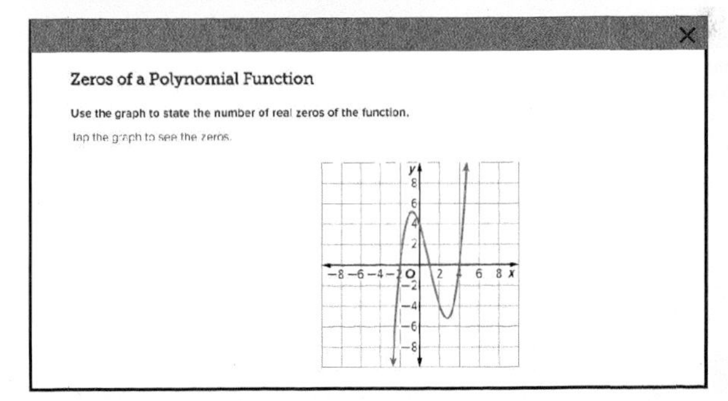

Example 5

TAP

Students tap on the graph to see the real zeros of a function.

Example 6 Compare Polynomial Functions

Examine $f(x) = x^3 + 2x^2 - 3x$ and $g(x)$ shown in the graph.

Part A Graph $f(x)$.

Substitute values for x to create a table of values. Then plot the points, and connect them with a smooth curve.

x	$f(x)$
−3	0
−2	6
−1	4
0	0
1	0
2	10
3	36

Part B Analyze the extrema.

Which function has the greater relative maximum?

$f(x)$ has a relative maximum at approximately $y = 6$, and $g(x)$ has a relative maximum between $y = 2$ and $y = 3$. So, $f(x)$ has the greater relative maximum.

Part C Analyze the key features.

Compare the zeros, x- and y-intercepts, and end behavior of $f(x)$ and $g(x)$.

zeros:

$f(x)$: −3, 0, 1

$g(x)$: The graph appears to intersect the x-axis at −1, −0.5, 1, 2

intercepts:

$f(x)$: x-intercepts: −3, 0, 1; y-intercept: 0

$g(x)$: x-intercepts: −1, −0.5, 1, 2; y-intercept: 2

end behavior:

$f(x)$: As $x \to -\infty$, $f(x) \to -\infty$, and as $x \to \infty$, $f(x) \to \infty$.

$g(x)$: As $x \to -\infty$, $g(x) \to \infty$, and as $x \to \infty$, $g(x) \to \infty$.

Think About It!

Find the domain and range of $f(x)$. Does $g(x)$ have the same domain and range? Explain.

The domain and range of $f(x)$ are both all real numbers. No; sample answer: The domain of $g(x)$ is also all real numbers, but the range is all real numbers greater than or equal to the minimum, which is between $y = -2$ and $y = -3$.

Study Tip

Zeros The zeros of a polynomial function are the x-coordinates of the points at which the graph intersects the x-axis.

Pause and Reflect

Did you struggle with anything in this lesson? If so, how did you deal with it?

Record your observations here.

See students' observations.

Go Online You can complete an Extra Example online.

218 Module 4 • Polynomials and Polynomial Functions

Interactive Presentation

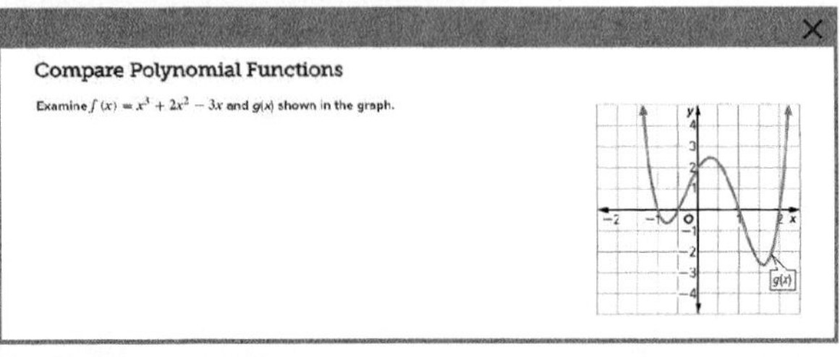

Example 6

CHECK

Students complete the Check online to determine whether they are ready to move on.

Example 6 Compare Polynomial Functions

Teaching the Mathematical Practices

1 Explain Correspondences In Example 6, students will explain the relationships between the key features of the graphs of the given functions.

Questions for Mathematical Discourse

AL How many zeros and relative extrema does each function have? $f(x)$: 3 zeros, 2 relative extrema; $g(x)$: 4 zeros; 3 relative extrema

OL Based on the end behavior of the graph of $g(x)$, what type of degree and leading coefficient would you expect the function to have? even degree, positive leading coefficient

BL Can you tell whether the degree of a function is even or odd from the number of zeros? Explain. No; sample answer: A function could intersect the x-axis at a relative extreme point, or it could continue through, and you cannot tell which is the case just by the number of unique real zeros.

Exit Ticket

Recommended Use

At the end of class, go online to display the Exit Ticket prompt and ask students to respond using a separate piece of paper. Have students hand you their responses as they leave the room.

Alternate Use

At the end of class, go online to display the Exit Ticket prompt and ask students to respond verbally or by using a mini-whiteboard. Have students hold up their whiteboards so that you can see all student responses. Tap to reveal the answer when most or all students have completed the Exit Ticket.

Practice and Homework

Suggested Assignments

Use the table below to select appropriate exercises.

DOK	Topic	Exercises
1, 2	exercises that mirror the examples	1–22
2	exercises that use a variety of skills from this lesson	23–30
2	exercises that extend concepts learned in this lesson to new contexts	31–35
3	exercises that emphasize higher-order and critical thinking skills	36–39

ASSESS AND DIFFERENTIATE

Use the data from the **Checks** to determine whether to provide resources for extension, remediation, or intervention.

IF students score 90% or more on the Checks, **THEN** assign: BL

- Practice Exercises 1–35 odd, 36–39
- Extension: Approximation by Means of Polynomials
- **ALEKS**® Real Zeros of Polynomial Functions, Graphs of Polynomial Functions

IF students score 66%–89% on the Checks, **THEN** assign: OL

- Practice, Exercises 1–39 odd
- Remediation, Review Resources: Adding and Subtracting Polynomials
- Personal Tutors
- Extra Examples 1–6
- **ALEKS**® Values of a, b, and c

IF students score less than 65% on the Checks, **THEN** assign: AL

- Practice, Exercises 1–21 odd
- Remediation, Review Resources: Adding and Subtracting Polynomials
- *Quick Review Math Handbook*: Polynomial Functions
- **ALEKS**® Values of a, b, and c

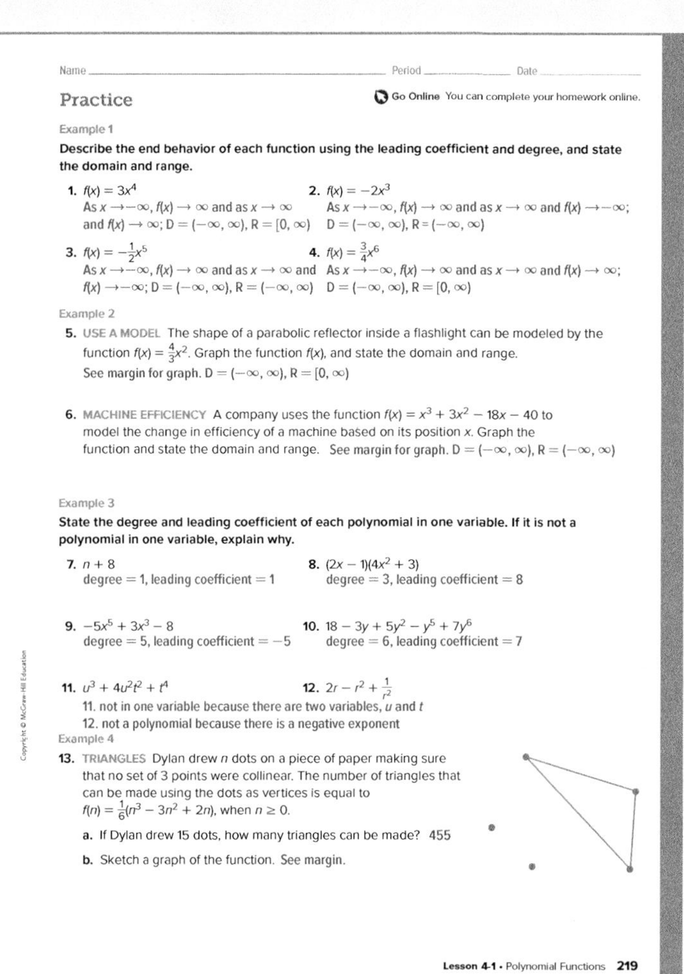

Name ______ Period ______ Date ______

Practice

Go Online You can complete your homework online.

Example 1

Describe the end behavior of each function using the leading coefficient and degree, and state the domain and range.

1. $f(x) = 3x^4$
As $x \to -\infty$, $f(x) \to \infty$ and as $x \to \infty$ and $f(x) \to \infty$; D = $(-\infty, \infty)$, R = $[0, \infty)$

2. $f(x) = -2x^3$
As $x \to -\infty$, $f(x) \to \infty$ and as $x \to \infty$ and $f(x) \to -\infty$; D = $(-\infty, \infty)$, R = $(-\infty, \infty)$

3. $f(x) = -\frac{1}{2}x^5$
As $x \to -\infty$, $f(x) \to \infty$ and as $x \to \infty$ and $f(x) \to -\infty$; D = $(-\infty, \infty)$, R = $(-\infty, \infty)$

4. $f(x) = \frac{3}{4}x^6$
As $x \to -\infty$, $f(x) \to \infty$ and as $x \to \infty$ and $f(x) \to \infty$; D = $(-\infty, \infty)$, R = $[0, \infty)$

Example 2

5. USE A MODEL The shape of a parabolic reflector inside a flashlight can be modeled by the function $f(x) = \frac{4}{3}x^2$. Graph the function $f(x)$, and state the domain and range.
See margin for graph. D = $(-\infty, \infty)$, R = $[0, \infty)$

6. MACHINE EFFICIENCY A company uses the function $f(x) = x^3 + 3x^2 - 18x - 40$ to model the change in efficiency of a machine based on its position x. Graph the function and state the domain and range. See margin for graph. D = $(-\infty, \infty)$, R = $(-\infty, \infty)$

Example 3

State the degree and leading coefficient of each polynomial in one variable. If it is not a polynomial in one variable, explain why.

7. $n + 8$
degree = 1, leading coefficient = 1

8. $(2x - 1)(4x^2 + 3)$
degree = 3, leading coefficient = 8

9. $-5x^5 + 3x^3 - 8$
degree = 5, leading coefficient = −5

10. $18 - 3y + 5y^2 - y^5 + 7y^6$
degree = 6, leading coefficient = 7

11. $u^3 + 4u^2t^2 + t^4$

12. $2r - r^2 + \frac{1}{r^2}$

11. not in one variable because there are two variables, u and t
12. not a polynomial because there is a negative exponent

Example 4

13. TRIANGLES Dylan drew n dots on a piece of paper making sure that no set of 3 points were collinear. The number of triangles that can be made using the dots as vertices is equal to $f(n) = \frac{1}{6}(n^3 - 3n^2 + 2n)$, when $n \geq 0$.
a. If Dylan drew 15 dots, how many triangles can be made? 455
b. Sketch a graph of the function. See margin.

14. DRILLING The volume of a drill bit can be estimated by the formula for a cone, $V = \frac{1}{3}\pi hr^2$, where h is the height of the bit and r is its radius. Substituting $\frac{\sqrt{3}}{3}r$ for h, the volume of the drill bit can be estimated by $V = \frac{\sqrt{3}}{9}\pi r^3$.
a. What is the volume of a drill bit with a radius of 3 centimeters? $3\pi\sqrt{3}$ cm³
b. Sketch a graph of the function in the context of the situation. See margin.

Example 5

Use the graph to state the number of real zeros of the function.

15. 1

16. 4

17. 3

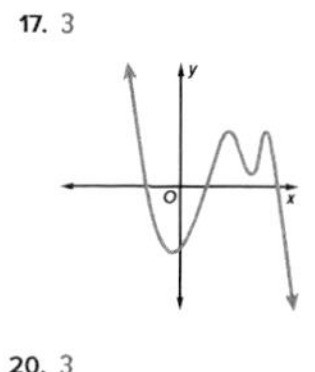

18. 2

19. 3

20. 3

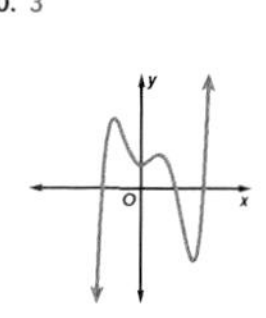

Example 6

21. Examine $f(x) = x^3 - 2x^2 - 4x + 1$ and $g(x)$ shown in the graph.
a. Which function has the greater relative maximum? $f(x)$
b. Compare the zeros, x- and y-intercepts, and end behavior of $f(x)$ and $g(x)$.
zeros: $f(x)$: −1.39, 0.23, 3.16; $g(x)$: −2.75, −0.5, 0.25
x-intercepts: $f(x)$: −1.39, 0.23, 3.16; $g(x)$: −2.75, −0.5, 0.25
y-intercept: $f(x)$: 1; $g(x)$: 1
end behavior: $f(x)$: As $x \to -\infty$, $f(x) \to -\infty$, and as $x \to \infty$, $f(x) \to \infty$; $g(x)$: As $x \to -\infty$, $g(x) \to \infty$, and as $x \to \infty$, $g(x) \to -\infty$

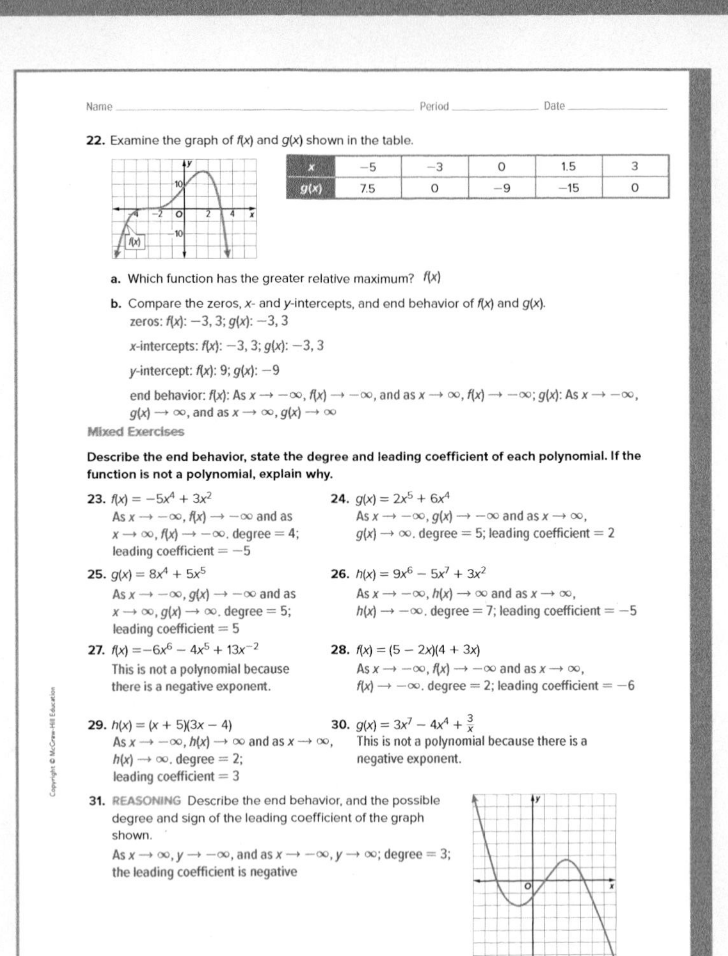

Name ______ Period ______ Date ______

22. Examine the graph of $f(x)$ and $g(x)$ shown in the table.

x	−5	−3	0	1.5	3
$g(x)$	7.5	0	−9	−15	0

a. Which function has the greater relative maximum? $f(x)$

b. Compare the zeros, x- and y-intercepts, and end behavior of $f(x)$ and $g(x)$.
zeros: $f(x)$: −3, 3; $g(x)$: −3, 3

x-intercepts: $f(x)$: −3, 3; $g(x)$: −3, 3

y-intercept: $f(x)$: 9; $g(x)$: −9

end behavior: $f(x)$: As $x \to -\infty$, $f(x) \to -\infty$, and as $x \to \infty$, $f(x) \to -\infty$; $g(x)$: As $x \to -\infty$, $g(x) \to \infty$, and as $x \to \infty$, $g(x) \to \infty$

Mixed Exercises

Describe the end behavior, state the degree and leading coefficient of each polynomial. If the function is not a polynomial, explain why.

23. $f(x) = -5x^4 + 3x^2$
As $x \to -\infty$, $f(x) \to -\infty$ and as $x \to \infty$, $f(x) \to -\infty$. degree = 4; leading coefficient = −5

24. $g(x) = 2x^5 + 6x^4$
As $x \to -\infty$, $g(x) \to -\infty$ and as $x \to \infty$, $g(x) \to \infty$. degree = 5; leading coefficient = 2

25. $g(x) = 8x^4 + 5x^5$
As $x \to -\infty$, $g(x) \to -\infty$ and as $x \to \infty$, $g(x) \to \infty$. degree = 5; leading coefficient = 5

26. $h(x) = 9x^6 - 5x^7 + 3x^2$
As $x \to -\infty$, $h(x) \to \infty$ and as $x \to \infty$, $h(x) \to -\infty$. degree = 7; leading coefficient = −5

27. $f(x) = -6x^6 - 4x^5 + 13x^{-2}$
This is not a polynomial because there is a negative exponent.

28. $f(x) = (5 - 2x)(4 + 3x)$
As $x \to -\infty$, $f(x) \to -\infty$ and as $x \to \infty$, $f(x) \to -\infty$. degree = 2; leading coefficient = −6

29. $h(x) = (x + 5)(3x - 4)$
As $x \to -\infty$, $h(x) \to \infty$ and as $x \to \infty$, $h(x) \to \infty$. degree = 2; leading coefficient = 3

30. $g(x) = 3x^7 - 4x^4 + \frac{3}{x}$
This is not a polynomial because there is a negative exponent.

31. REASONING Describe the end behavior, and the possible degree and sign of the leading coefficient of the graph shown.
As $x \to \infty$, $y \to -\infty$, and as $x \to -\infty$, $y \to \infty$; degree = 3; the leading coefficient is negative

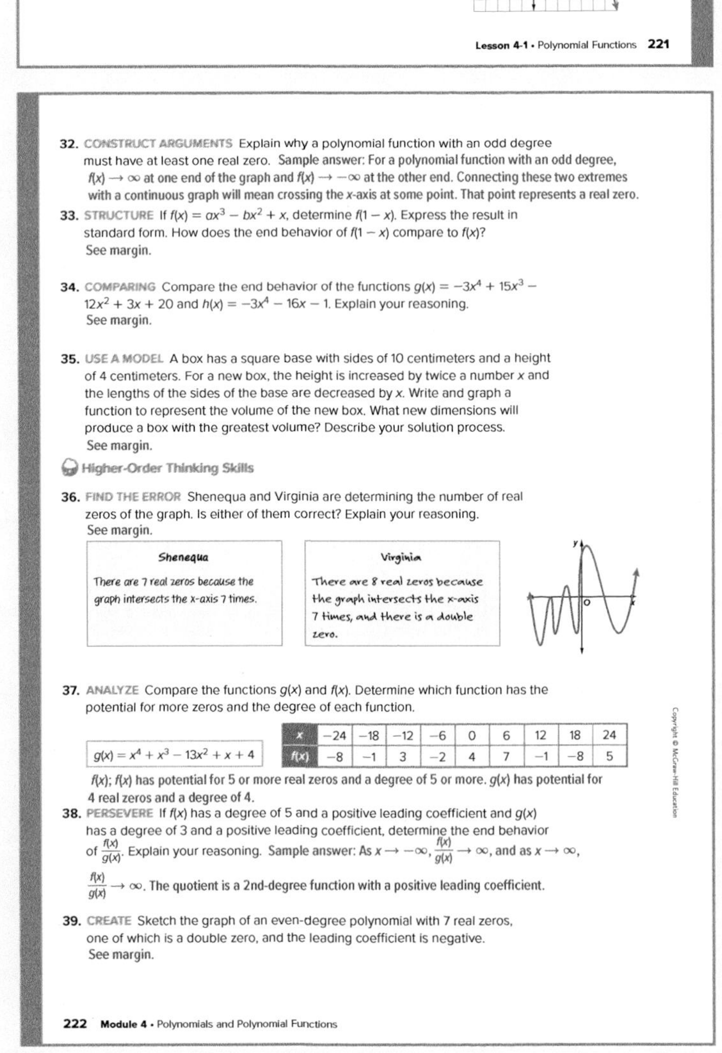

32. CONSTRUCT ARGUMENTS Explain why a polynomial function with an odd degree must have at least one real zero. Sample answer: For a polynomial function with an odd degree, $f(x) \to \infty$ at one end of the graph and $f(x) \to -\infty$ at the other end. Connecting these two extremes with a continuous graph will mean crossing the x-axis at some point. That point represents a real zero.

33. STRUCTURE If $f(x) = ax^3 - bx^2 + x$, determine $f(1 - x)$. Express the result in standard form. How does the end behavior of $f(1 - x)$ compare to $f(x)$?
See margin.

34. COMPARING Compare the end behavior of the functions $g(x) = -3x^4 + 15x^3 - 12x^2 + 3x + 20$ and $h(x) = -3x^4 - 16x - 1$. Explain your reasoning.
See margin.

35. USE A MODEL A box has a square base with sides of 10 centimeters and a height of 4 centimeters. For a new box, the height is increased by twice a number x and the lengths of the sides of the base are decreased by x. Write and graph a function to represent the volume of the new box. What new dimensions will produce a box with the greatest volume? Describe your solution process.
See margin.

Higher-Order Thinking Skills

36. FIND THE ERROR Shenequa and Virginia are determining the number of real zeros of the graph. Is either of them correct? Explain your reasoning.
See margin.

Shenequa	Virginia
There are 7 real zeros because the graph intersects the x-axis 7 times.	There are 8 real zeros because the graph intersects the x-axis 7 times, and there is a double zero.

37. ANALYZE Compare the functions $g(x)$ and $f(x)$. Determine which function has the potential for more zeros and the degree of each function.

$g(x) = x^4 + x^3 - 13x^2 + x + 4$

x	−24	−18	−12	−6	0	6	12	18	24
$f(x)$	−8	−1	3	−2	4	7	−1	−8	5

$f(x)$; $f(x)$ has potential for 5 or more real zeros and a degree of 5 or more. $g(x)$ has potential for 4 real zeros and a degree of 4.

38. PERSEVERE If $f(x)$ has a degree of 5 and a positive leading coefficient and $g(x)$ has a degree of 3 and a positive leading coefficient, determine the end behavior of $\frac{f(x)}{g(x)}$. Explain your reasoning. Sample answer: As $x \to -\infty$, $\frac{f(x)}{g(x)} \to \infty$, and as $x \to \infty$, $\frac{f(x)}{g(x)} \to \infty$. The quotient is a 2nd-degree function with a positive leading coefficient.

39. CREATE Sketch the graph of an even-degree polynomial with 7 real zeros, one of which is a double zero, and the leading coefficient is negative.
See margin.

Answers

5.

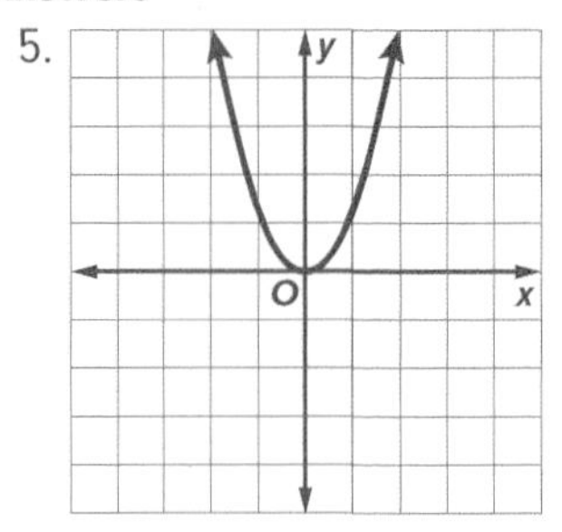

6.

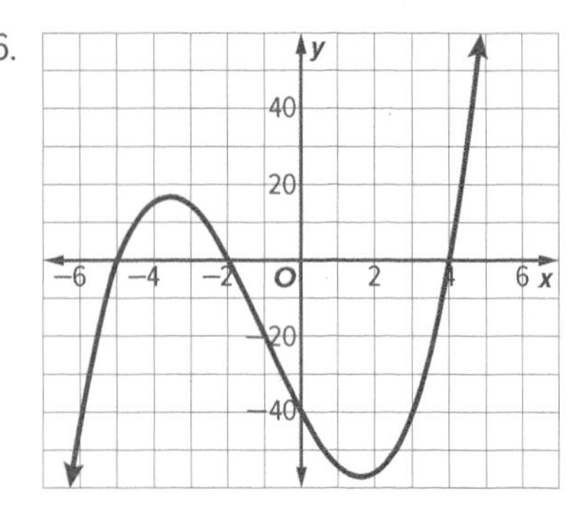

13b.

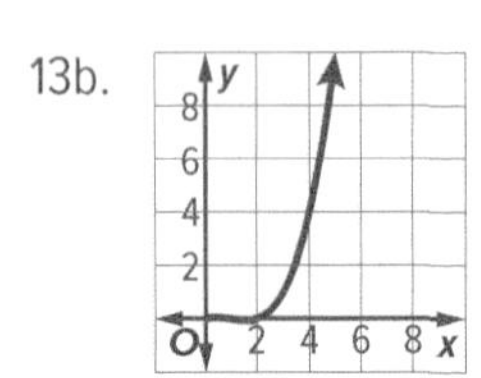

14b.

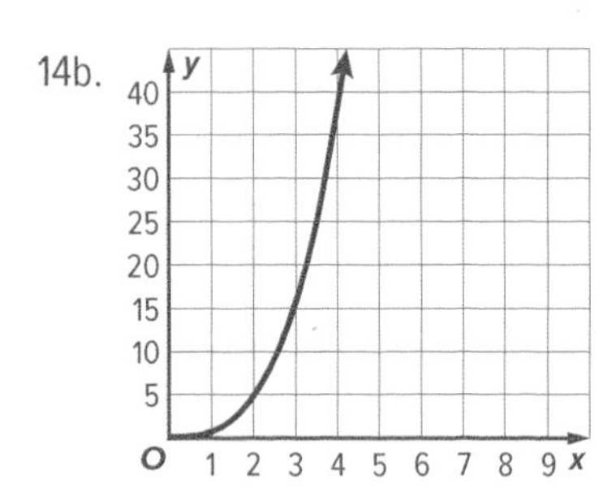

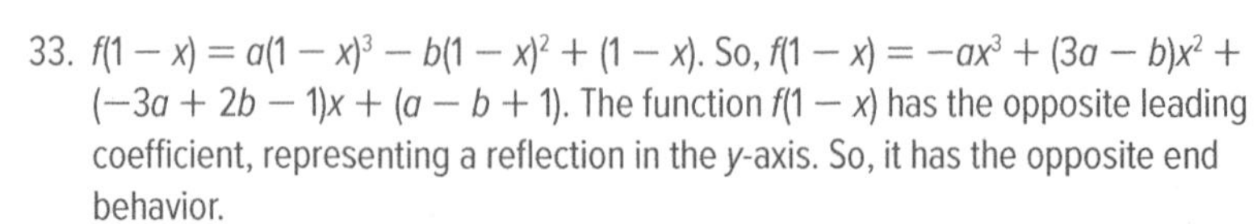

33. $f(1 - x) = a(1 - x)^3 - b(1 - x)^2 + (1 - x)$. So, $f(1 - x) = -ax^3 + (3a - b)x^2 + (-3a + 2b - 1)x + (a - b + 1)$. The function $f(1 - x)$ has the opposite leading coefficient, representing a reflection in the y-axis. So, it has the opposite end behavior.

34. The end behavior is determined by the sign of the leading coefficient and the degree of the function. These are the same for $g(x)$ and $h(x)$. As $x \to \infty$, $f(x) \to -\infty$ and as $x \to -\infty$, $f(x) \to -\infty$ for both functions.

35. Sample answer: The volume of the new box is modeled by the function $V(x) = (10 - x)^2(4 + 2x)$. The graph appears to have a relative maximum at $x = 2$ and $V(2) = 512$. So, the dimensions of the box with the greatest volume will be 8 centimeters by 8 centimeters by 8 centimeters.

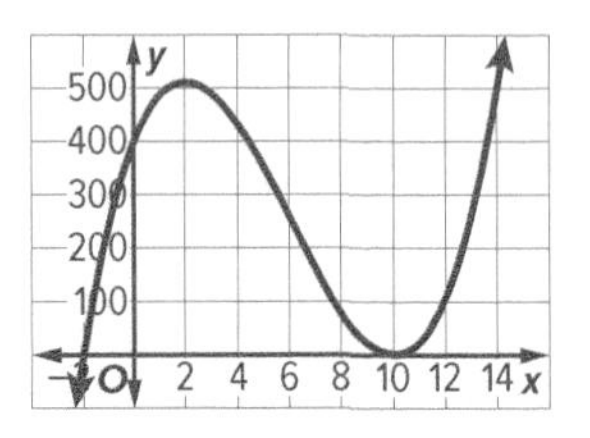

36. Shenequa is correct; sample answer: The number of real zeros is equal to exactly the number of times the graph intersects the x-axis.

39. Sample answer:

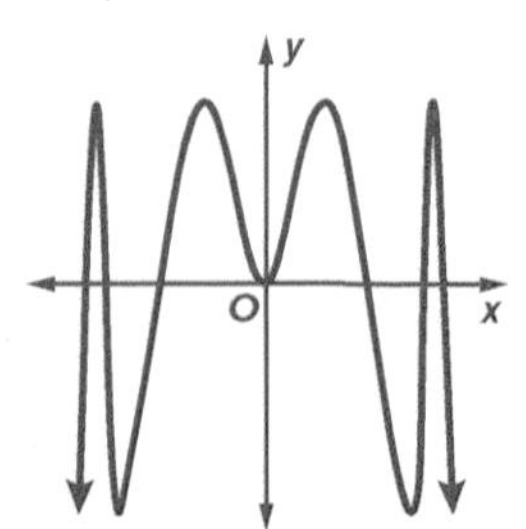

Analyzing Graphs of Polynomial Functions

LESSON GOAL

Students analyze the graphs of polynomial functions by identifying key features.

1 LAUNCH

 Launch the lesson with a **Warm Up** and an introduction.

2 EXPLORE AND DEVELOP

 **Develop:**

The Location Principle
- Locate Zeros of a Function

Extrema of Polynomials
- Identify Extrema
- Analyze a Polynomial Function

Modeling with Polynomial Functions
- Use a Polynomial Function and Technology to Model
- Find Average Rate of Change

 You may want your students to complete the **Checks** online.

3 REFLECT AND PRACTICE

 Exit Ticket

 Practice

DIFFERENTIATE

 View reports of student progress on the **Checks** after each example.

Resources	AL	OL	BL	ELL
Remediation: Polynomial Functions	●	●		●
Extension: The Bisection Method		●	●	●

Language Development Handbook

Assign page 26 of the *Language Development Handbook* to help your students build mathematical language related to analyzing the graphs of polynomial functions.

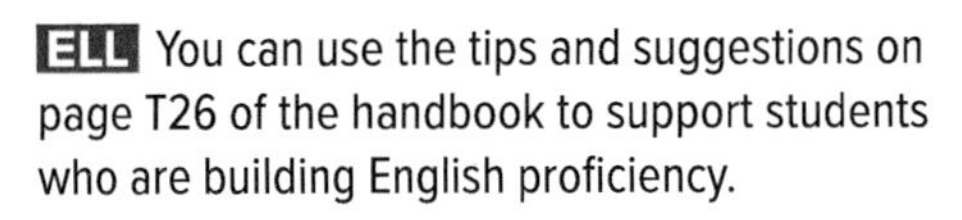
ELL You can use the tips and suggestions on page T26 of the handbook to support students who are building English proficiency.

Suggested Pacing

90 min	0.5 day
45 min	1 day

Focus

Domain: Functions

Standards for Mathematical Content:

F.IF.4 For a function that models a relationship between two quantities, interpret key features of graphs and tables in terms of the quantities, and sketch graphs showing key features given a verbal description of the relationship.

F.IF.7c Graph polynomial functions, identifying zeros when suitable factorizations are available, and showing end behavior.

Standards for Mathematical Practice:

4 Model with mathematics.

5 Use appropriate tools strategically.

Coherence

Vertical Alignment

Previous
Students analyzed polynomial functions by examining key features and graphing
F.IF.4, F.IF.7c

Now
Students analyze the graphs of polynomial functions by identifying key features, including zeros and extrema
F.IF.4, F.IF.7c

Next
Students will solve polynomial equations by graphing.
A.CED.1, A.REI.11

Rigor

The Three Pillars of Rigor

1 CONCEPTUAL UNDERSTANDING	2 FLUENCY	3 APPLICATION

Conceptual Bridge In this lesson, students build on their understanding of the graphs of polynomial functions. They build fluency by using key features of the functions to graph them, and they apply their understanding by solving real-world problems.

Mathematical Background

Tables of values can be used to explore two types of changes in the values of a polynomial function. A change of signs in the value of $f(x)$ from one value of x to the next indicates that the graph of the function crosses the x-axis between the two x-values. A change between increasing values and decreasing values indicates that the graph is turning for that interval. Extrema occur at the relative maximum or minimum points of the function.

Interactive Presentation

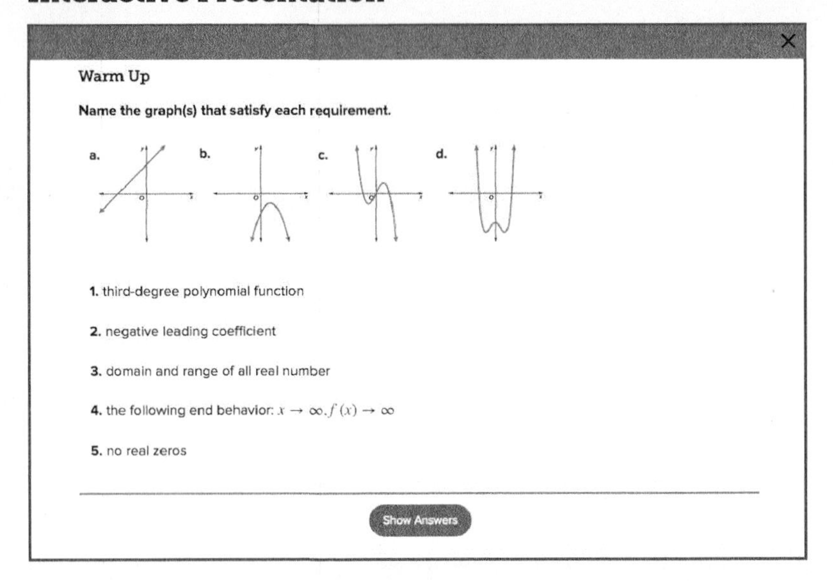

Warm Up

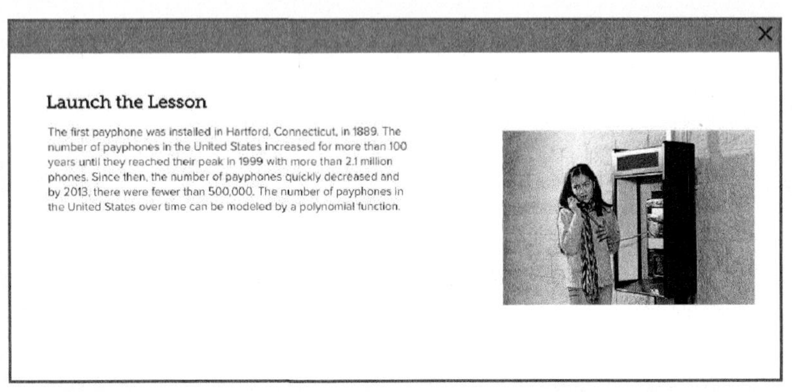

Launch the Lesson

Warm Up

Prerequisite Skills

The Warm Up exercises address the following prerequisite skill for this lesson:

- identifying polynomial functions

Answers:

1. c

2. b, c

3. a, c

4. a, d

5. b

Launch the Lesson

 Teaching the Mathematical Practices

4 Apply Mathematics In the Launch the Lesson, students apply what they have learned about polynomial functions to a real-world situation. Encourage students to identify key features of the function that models the number of payphones in the United States based on the description of the situation.

Go Online to find additional teaching notes and questions to promote classroom discourse.

Today's Standards

Tell students that they will be addressing these content and practice standards in this lesson. You may wish to have a student volunteer read aloud *How can I meet these standards?* and *How can I use these practices?*, and connect these to the standards.

See the Interactive Presentation for I Can statements that align with the standards covered in this lesson.

Learn The Location Principle

Objective

Students approximate zeros by graphing polynomial functions.

Teaching the Mathematical Practices

3 Analyze Cases The Think About It! feature guides students to examine a case where the Location Principle cannot be used to find all of the real zeros of a function. Encourage students to familiarize themselves with the cases in which the zeros of a function can and cannot be found by using the Location Principle.

Important to Know

A polynomial function must be continuous for the Location Principle to be true because if it was not continuous, then it could change signs but have discontinuity at 0.

Example 1 Locate Zeros of a Function

Teaching the Mathematical Practices

5 Use Estimation In Example 1, students must estimate the zeros of the function using the Location Principle and then find an exact answer.

Questions for Mathematical Discourse

AL Does a function of degree 4 have to have 4 zeros? Explain. No; 4 is the maximum number of possible zeros, but it can have fewer than than 4.

OL If you substitute an estimated root into the equation, will you get zero? You may, or you may get a value close to zero because the root is an estimate.

BL Suppose $g(x)$ has two real zeros; but, when using the Location Principle with integers, no zeros are found. What are two possible descriptions of a function that would have this result? Sample answer: The function could have two non-integer points that touch the axis without crossing, or the function could cross twice between two integers.

Go Online

- Find additional teaching notes.
- View performance reports of the Checks.
- Assign or present an Extra Example.

Lesson 4-2

Analyzing Graphs of Polynomial Functions

Today's Goals
- Approximate zeros by graphing polynomial functions.
- Find extrema of polynomial functions.

Learn The Location Principle

If the value of a polynomial function $f(x)$ changes signs from one value of x to the next, then there is a zero between those two x-values. This is called the Location Principle.

Key Concept • Location Principle

Suppose $y = f(x)$ represents a polynomial function, and a and b are two real numbers such that $f(a) < 0$ and $f(b) > 0$. Then the function has at least one real zero between a and b.

Think About It!
Not all real zeros can be found by using the Location Principle. Provide an example where $f(a) > 0$ and $f(b) > 0$, but there is a zero between $x = a$ and $x = b$.

Sample answer: For $f(x) = x^2$, $f(-1) = 1$ and $f(1) = 1$, but there is a zero at $x = 0$.

Example 1 Locate Zeros of a Function

Determine the consecutive integer values of x between which each real zero of $f(x) = x^4 - 2x^3 - x^2 + 1$ is located. Then draw the graph.

Step 1 Make a table.

Because $f(x)$ is a fourth-degree polynomial, it will have as many as __4__ real zeros or none at all.

x	−2	−1	0	1	2	3	4
$f(x)$	29	3	1	−1	−3	19	113

Using the Location Principle, there are zeros between $x = 0$ and $x = 1$ and between $x = 2$ and $x = 3$.

Step 2 Sketch the graph.

Use the table to sketch the graph and find the locations of the zeros.

Think About It!
How can you adjust the table on your graphing calculator to give a more precise interval for the value of each zero?

Sample answer: You can adjust the table setting to make the interval between each x-value smaller.

Check

Use technology to check the location of the zeros.

Input the function into a graphing calculator to confirm that the function crosses the x-axis between $x = 0$ and $x = 1$ and between $x = 2$ and $x = 3$.

You can find more accurate values of the zeros by using the **zero** feature in the CALC menu to find $x \approx 0.7213$ and $x \approx 2.3486$, which confirms the estimates.

Go Online You can complete an Extra Example online.

Lesson 4-2 • Analyzing Graphs of Polynomial Functions 223

Interactive Presentation

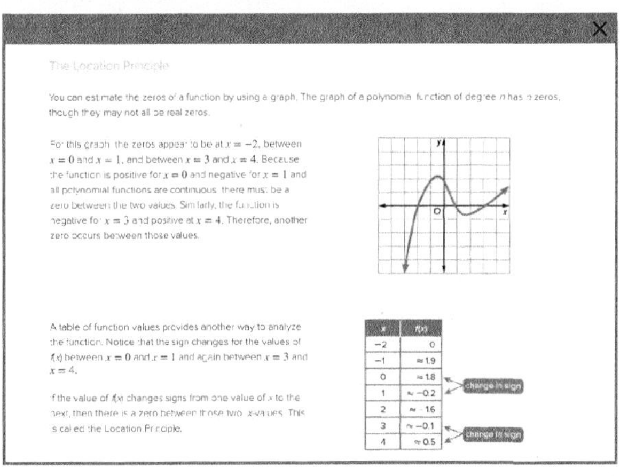

Learn

TYPE

Students answer a question to show they understand how to use the Location Principle to find zeros of a function.

CHECK

Students complete the Check online to determine whether they are ready to move on.

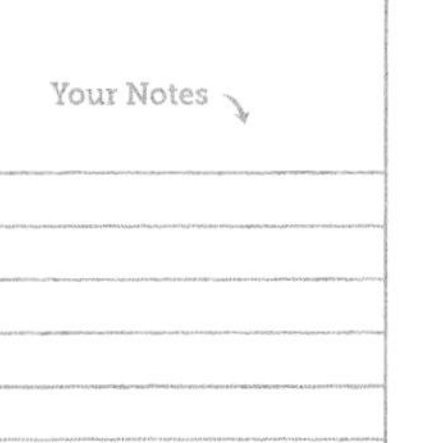

Your Notes

Check

Determine the consecutive integer values of x between which each real zero of $f(x) = 2x^4 + x^3 - 3x^2 - 2$ is located. Then draw the graph.

$x =$ <u>−2</u> and $x =$ <u>−1</u>

$x =$ <u>1</u> and $x =$ <u>2</u>

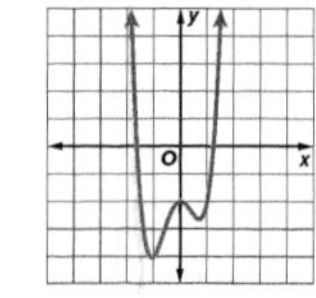

Study Tip

Turning Points Relative maxima and relative minima of a function are sometimes called turning points.

Learn Extrema of Polynomials

Extrema occur at relative maxima or minima of the function.

Point A is a relative minimum, and point B is a relative maximum. Both points A and B are extrema. The graph of a polynomial of degree n has at most $n - 1$ extrema.

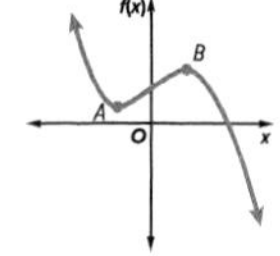

Example 2 Identify Extrema

Use a table to graph $f(x) = x^3 + x^2 - 5x - 2$. Estimate the x-coordinates at which the relative maxima and relative minima occur.

Step 1 Make a table of values and graph the function.

x	$f(x)$
−4	−30
−3	−5
−2	4
−1	3
0	−2
1	−5
2	0
3	19

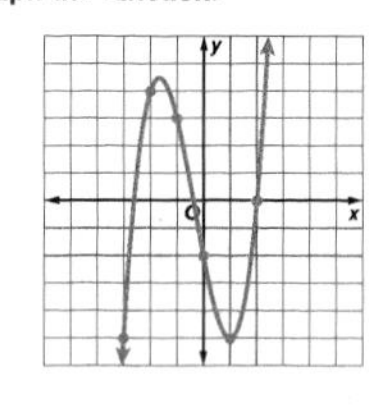

Step 2 Estimate the locations of the extrema.

The value of $f(x)$ at $x = -2$ is greater than the surrounding points indicating a maximum *near* $x =$ <u>−2</u>.

The value of $f(x)$ at $x = 1$ is less than the surrounding points indicating a minimum *near* $x =$ <u>1</u>.

You can use a graphing calculator to find the extrema of a function and confirm your estimates.

Study Tip

Extrema When graphing with a calculator, keep in mind that a polynomial of degree n has at most $n - 1$ extrema. This will help you to determine whether your viewing window is allowing you to see all of the extrema of the graph.

Go Online You can complete an Extra Example online.

224 Module 4 • Polynomials and Polynomial Functions

Interactive Presentation

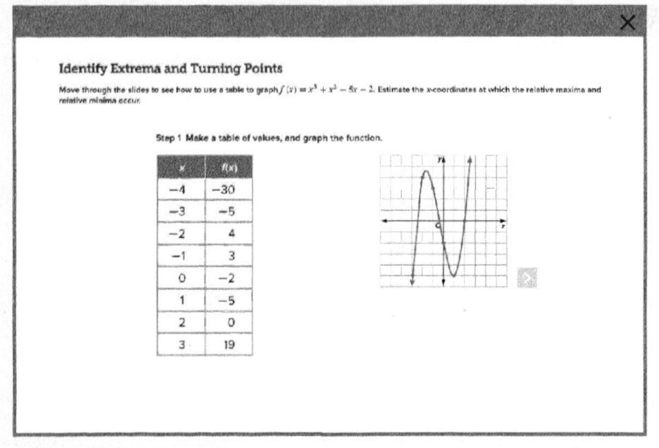

Example 2

SWIPE

Students move through the steps to estimate the locations of the extrema of a function

Learn Extrema of Polynomials

Objective

Students find the relative maxima and relative minima of polynomial functions.

Teaching the Mathematical Practices

1 Explain Correspondences Encourage students to explain the relationships between extrema and degrees of polynomial functions.

Things to Remember

- Some odd functions, like $f(x) = x^3$, have no extrema.
- Zeros and extrema will not always occur at integral values of x.

Common Misconception

Students may believe that a function has only one relative maximum or minimum. A polynomial with a degree greater than 3 may have more than one relative maximum or minimum.

Essential Question Follow-Up

Students will locate zeros of a function.

Ask:

Is it possible for a zero to be between two integers that both have positive $f(x)$ values? Sample answer: Yes, the function could become negative and come back up to positive between two integers.

Common Error

Students may assume that a function with degree 4 always has 4 zeros. Remind them that a function with degree 4 can have 0, 1, 2, 3, or 4 zeros.

Example 2 Identify Extrema

Teaching the Mathematical Practices

3 Construct Arguments In the Talk About It! feature, students will use stated assumptions and definitions to construct an argument.

1 CONCEPTUAL UNDERSTANDING | 2 FLUENCY | 3 APPLICATION

Questions for Mathematical Discourse

AL Is there a relationship between the number of extrema and the number of zeros a function will have? Explain. No; sample answer: The number of extrema does not necessarily determine the number of times a graph crosses the x-axis.

OL What is another name for the extrema of a quadratic polynomial? vertex

BL Suppose $g(x)$ has no zeros and no extrema. What is a possible type of function $g(x)$ could be? Sample answer: $g(x)$ could be a linear function with $m = 0$ and $b \neq 0$.

Things to Remember

Remind students that while the degree of the function represents the maximum number of zeros a function can have, to find the maximum number of extrema, they should calculate one less than the degree of the function.

Example 3 Analyze a Polynomial Function

Teaching the Mathematical Practices

4 Analyze Relationships Mathematically Point out that to complete Example 3, students will need to analyze the mathematical relationships in the context of the problem.

Questions for Mathematical Discourse

AL What does $f(0)$ mean in this context? the number of certified pilots, in thousands, in 1930

OL $f(0) = 7.708$. Do you need to round this number when describing the number of pilots in 1930? No, 7.708 represents 7708 pilots because x is scaled in thousands.

BL Would this model make sense for years before 1930? Explain. No; sample answer: The graph of the function crosses the x-axis at -1.501.

Check

Use a table of values of $f(x) = -x^4 - x^3 + 5x^2 + x - 3$ to estimate the x-coordinates at which the relative maxima and relative minima occur.

x	$f(x)$
−3	−15
−2	7
−1	1
0	−3
1	1
2	−5
3	−63

The relative maxima occur near $x =$ −2 and $x =$ 1.

The relative minimum occurs near $x =$ 0.

Example 3 Analyze a Polynomial Function

PILOTS **The total number of certified pilots in the United States is approximated by $f(x) = 0.0000903x^4 - 0.0166x^3 + 0.762x^2 + 6.317x + 7.708$, where x is the number of years after 1930 and $f(x)$ is the number of pilots in thousands. Graph the function and describe its key features over the relevant domain.**

Step 1 Graph the function.

Make a table of values. Plot the points and connect them with a smooth curve.

x	$f(x)$
0	7.708
10	131.381
20	320.496
30	507.961
40	648.356
50	717.933
60	714.616
70	658.001
80	589.356
90	571.621

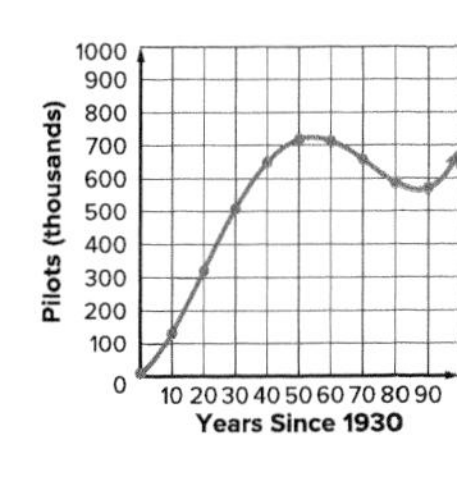

Step 2 Describe the key features.
Domain and Range

The domain and range of the function is all real numbers. Because the function models years after 1930, the relevant domain and range are $\{x \mid x \geq 0\}$ and $\{f(x) \mid f(x) \geq 7.708\}$.

(continued on the next page)

Go Online You can learn how to graph and analyze a polynomial function on a graphing calculator by watching the video online.

Think About It! What trends in the number of pilots does the graph suggest?

Sample answer: The number of pilots peaked around 1985 and declined to a relative minimum around 2015. After 2015, the number of pilots will continue to increase.

Interactive Presentation

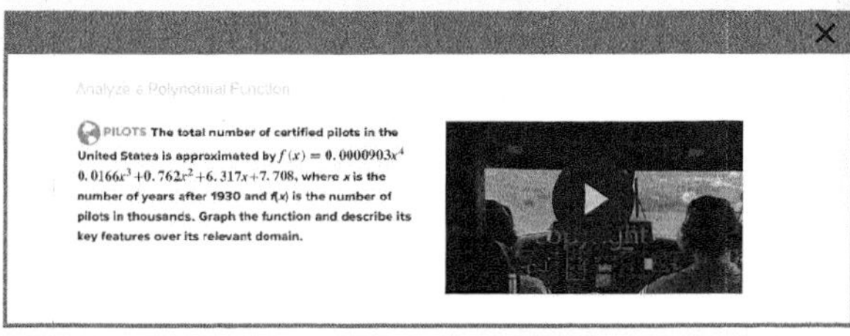

Example 3

TYPE

Students identify characteristics of a polynomial function in context.

Talk About It!
It is reasonable that the trend will continue indefinitely? Explain.

No; sample answer: It is unreasonable to assume that the number of pilots will continue to increase indefinitely. At some point, the number will stay relatively constant or begin to decline.

Study Tip
Assumptions Determining the end behavior for the graph of a polynomial that models data assumes that the trend continues and there are no other relative maxima or minima.

Extrema

There is a relative maximum between 1980 and 1990 and a relative minimum between 2010 and 2020 in the relevant domain.

End Behavior

As $x \to \infty$, $f(x) \to \infty$.

Intercepts

In the relevant domain, the y-intercept is at (0, 7.708). There is no x-intercept, or zero, because the function begins at a value greater than 0 and as $x \to \infty$, $f(x) \to \infty$.

Symmetry The graph of the function does not have symmetry.

Check

COINS The number of quarters produced by the United States Mint can be approximated by the function $f(x) = 16.4x^3 - 149.5x^2 - 148.9x + 3215.4$, where x is the number of years since 2005 and $f(x)$ is the total number of quarters produced in millions. Use a graph of the function to complete the table and describe its key features.

Part A Complete the table.

x, Years	$f(x)$, Quarters (millions)
0	3215
2	2451
4	1277
6	482
8	853
10	3176

Part B Describe the key features.

The relevant domain is $\{x \mid x \geq 0\}$.

The relevant range is $\{f(x) \mid f(x) \geq \approx 482\}$.

There is a relative minimum between 2011 and 2013.

The y-intercept is 3215.

The graph of the function does not have symmetry.

It is unreasonable to assume that the trend will continue indefinitely.

Go Online You can complete an Extra Example online.

Things to Remember

Remind students how to find the domain, range, end behavior, and intercepts from the previous lesson.

DIFFERENTIATE

Language Development Activity AL BL ELL

IF students ask how math functions can describe real-world situations, **THEN** have them discuss the appropriateness of describing real-world situations with mathematical functions. Help them to understand that a function is usually just an approximation of the real-world data, and is often only a reasonable model of a limited interval of the domain.

CHECK

Students complete the Check online to determine whether they are ready to move on.

1 CONCEPTUAL UNDERSTANDING | **2 FLUENCY** | **3 APPLICATION**

Example 4 Use a Polynomial Function and Technology to Model

Teaching the Mathematical Practices

5 Compare Predictions with Data In Example 4, students should use a graphing calculator to visualize the data, determine a function of best fit, and compare their predictions with the data.

Questions for Mathematical Discourse

AL What is the purpose of testing regressions for a set of data? The purpose is to find a polynomial function that models the data as closely as possible. The model can be used to make predictions of the dependent variable for values of the independent variable that are not part of the data.

OL Do you think testing the linear regression was necessary? Sample answer: No, the data does not appear to be well-modeled by a linear function.

BL How could you test the calculator's regression operation for accuracy? You could use points from a known function and see if the best fit is close to the original function.

Important to Know

Some students may need additional instructions to use the graphing calculator to graph polynomial functions and use it to solve problems.

Example 4 Use a Polynomial Function and Technology to Model

BACKPACKS **The table shows U.S. backpack sales in millions of dollars, according to the Travel Goods Association. Make a scatter plot and a curve of best fit to show the trend over time. Then determine the backpack sales in 2015.**

Year	Sales (million $)	Year	Sales (million $)
2000	1140	2008	1246
2001	1144	2009	1235
2002	1113	2010	1419
2003	1134	2011	1773
2004	1164	2012	1930
2005	1180	2013	2255
2006	1364	2014	2779
2007	1436		

Step 1 Enter the data.

Let the year 2000 be represented by 0. Enter the years since 2000 in List 1. Enter the backpack sales in List 2.

Step 2 Graph the scatter plot.

Choose the scatter plot feature in the **STAT PLOT** menu. Use List 1 for the **Xlist** and List 2 for the **Ylist**. Change the viewing window so that all the data are visible.

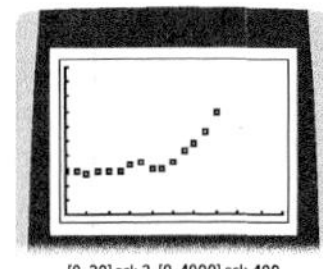
[0, 20] scl: 2, [0, 4000] scl: 400

Step 3 Determine the polynomial function of best fit.

To determine the model that best fits the data, perform linear, quadratic, cubic, and quartic regressions, and compare the coefficients of determination, r^2. The polynomial with a coefficient of determination closest to 1 will fit the data best.

A quartic function fits the data best.

The regression equation with coefficients rounded to the nearest tenths is:

$$y \approx 0.2x^4 - 3.9x^3 + 26.4x^2 - 43.6x + 1139.9.$$

Step 4 Graph and evaluate the regression function.

Assuming that the trend continues, the graph of the function can be used to predict backpack sales for a specific year. To determine the total sales in 2015, find the value of the function for $x = 15$.

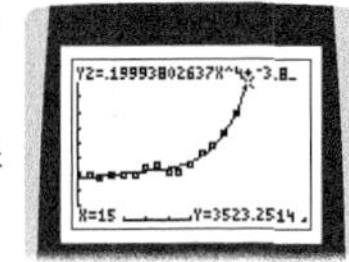
[0, 20] scl: 2, [0, 4000] scl: 400

In 2015, there were about $ 3.523 billion in backpack sales.

Math History Minute

By the age of 20, Italian mathematician **Maria Gaetana Agnesi (1718–1799)** had started working on her book *Analytical Institutions*, which was published in 1748. Early chapters included problems on maxima, minima, and turning points. Also described was a cubic curve called the "witch of Agnesi," which was translated incorrectly from the original Italian.

Think About It!

Explain the approximation that is made when using the model to determine the backpack sales in a specific year.

Sample answer: The model is an approximation of the entire set of data, but it may not be accurate at a specific year. Also, the model may not be a good representation for specific years outside of the domain of the original data because it was not considered when creating the model.

Lesson 4-2 • Analyzing Graphs of Polynomial Functions 227

Interactive Presentation

Use a Polynomial Function and Technology to Model

BACKPACKS **The table shows U.S. backpack sales in millions of dollars, according to the Travel Goods Association. Make a scatter plot and a curve of best fit to show the trend over time. Then determine the backpack sales in 2015.**

Year	Backpack Sales (million $)	Year	Backpack Sales (million $)
2000	1140	2008	1246
2001	1144	2009	1235
2002	1113	2010	1419
2003	1134	2011	1773
2004	1164	2012	1930
2005	1180	2013	2255
2006	1364	2014	2779
2007	1436		

Example 4

TAP

Students tap to move through the steps to graph and evaluate polynomial functions.

Check

TREES To estimate the amount of lumber that can be harvested from a tree, foresters measure the diameter of each tree. Determine the polynomial function of best fit, where x represents the diameter of a tree in inches and y is the estimated volume measured in board feet. Then estimate the volume of a tree with a diameter of 35 inches.

Diam (in.)	17	19	20	23	25	28	32	38	39	41
Vol (100s of board ft)	19	25	32	57	71	113	123	252	259	294

Polynomial function of best fit:
$y = -0.0006x^4 + 0.08x^3 - 3.55x^2 + 69.11x - 491.39$

The estimated volume of a 35-inch diameter tree to the nearest board foot is 188 of 100s board ft.

Study Tip

g-force One G is the acceleration due to gravity at the Earth's surface. Defined as 9.80665 meters per second squared, this is the g-force you experience when you stand still on Earth. On a roller coaster, you experience 0 Gs and feel weightless at the top of the hills, and you can experience a g-force of 6 Gs or more as you are pushed into your seat at the bottom of the hills.

Example 5 Find Average Rate of Change

ROCKETS **The Ares-V rocket was designed to carry as much as 75 tons of supplies and 4 astronauts to the Moon and possibly even to Mars. The table shows the expected g-force on the rocket over the course of its 200-second launch.**

Time (s)	Acceleration (Gs)	Time (s)	Acceleration (Gs)
0	1.34	120	1.46
20	1.26	140	1.93
40	1.12	160	2.47
60	1.01	180	2.84
80	1	200	2.2
100	1.15		

Part A Find the average rate of change.

Sketch the graph, and estimate the average rate of change of the acceleration. Then check your results algebraically.

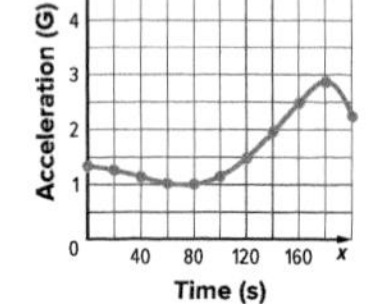

Estimate: From the graph, the change in the y-values is about 0.9, and the change in the x-values is 200. So, the rate of change is about $\frac{0.9}{200}$ or 0.0045.

Check algebraically:

The average rate of change is $\frac{f(200) - f(0)}{200 - 0} = \frac{2.2 - 1.34}{200 - 0}$ or 0.0043.

Part B Interpret the results.

From 0 to 200 seconds, the average rate of change in acceleration was an increase of 0.0043 Gs per second.

Go Online You can complete an Extra Example online.

Think About It!

Does the average rate of change from 0 to 200 seconds accurately describe the acceleration of the launch? Justify your reasoning.

No; sample answer: For the first 80 seconds, the acceleration is relatively slow. For the next 100 seconds, it increases rapidly. Then, it decreases. The average rate of change does not describe the changes in acceleration over time.

Interactive Presentation

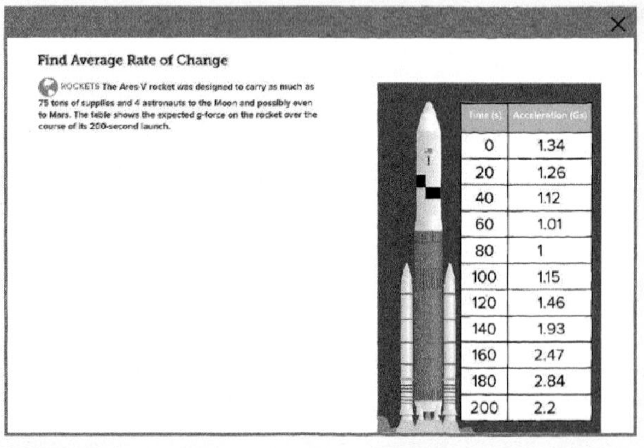

Example 5

CHECK

Students complete the Check online to determine if they are ready to move on.

Example 5 Find Average Rate of Change

Teaching the Mathematical Practices

5 Use Estimation Point out that in Example 5, students need to estimate the rate of change from the graph and then find an exact answer algebraically.

Questions for Mathematical Discourse

AL For what interval of time might the average rate of change more accurately describe the acceleration of the rocket? Explain. Sample answer: 80 seconds to 180 seconds because the acceleration appears to be increasing at a relatively consistent rate during this time.

OL If you draw a line connecting two points on the function, what is the relationship between the slope of the line and the average rate of change between the two points? The slope of the line is the same as the average rate of change between the points.

BL Consider a point P, which is a relative maximum on the curve. Now consider a point Q, which can vary. As Q approaches P, what value does the average rate of change approach? zero Using this, and that the average rate of change between two points is the slope of a line between the points, what would the instantaneous slope be at P? zero

DIFFERENTIATE

Enrichment Activity BL

The nature of polynomial functions makes them perfect for inventing funny stories about how things change over time. Challenge students to create stories about something that is increasing or decreasing (or both) and display the growth graphically. Extend the problem by having them create scales for the graphs that match the context of their stories.

Exit Ticket

Recommended Use

At the end of class, go online to display the Exit Ticket prompt and ask students to respond using a separate piece of paper. Have students hand you their responses as they leave the room.

Alternate Use

At the end of class, go online to display the Exit Ticket prompt and ask students to respond verbally or by using a mini-whiteboard. Have students hold up their whiteboards so that you can see all student responses. Tap to reveal the answer when most or all students have completed the Exit Ticket.

Practice and Homework

Suggested Assignments

Use the table below to select appropriate exercises.

DOK	Topic	Exercises
1, 2	exercises that mirror the examples	1–15
2	exercises that use a variety of skills from this lesson	16–25
2	exercises that extend concepts learned in this lesson to new contexts	26–28
3	exercises that emphasize higher-order and critical thinking skills	29–34

ASSESS AND DIFFERENTIATE

Use the data from the **Checks** to determine whether to provide resources for extension, remediation, or intervention.

IF students score 90% or more on the Checks,
THEN assign:

- Practice Exercises 1–25 odd, 26–34
- Extension: The Bisection Method
- **ALEKS**® Graphs of Polynomial Functions

IF students score 66%–89% on the Checks,
THEN assign:

- Practice Exercises 1–33 odd
- Remediation, Review Resources: Analyzing Graphs of Polynomial Functions
- Personal Tutors
- Extra Examples 1–5
- **ALEKS**® Identifying Polynomial Functions

AL

IF students score 65% or less on the Checks,
THEN assign:

- Practice Exercises 1–15 odd
- Remediation, Review Resources: Analyzing Graphs of Polynomial Functions
- *Quick Review Math Handbook:* Analyzing Graphs of Polynomial Functions
- **ALEKS**® Identifying Polynomial Functions

Name ______________________ Period ____________ Date ____________

Practice

Go Online You can complete your homework online.

Example 1

Determine the consecutive integer values of x between which each real zero of each function is located by using a table. Then sketch the graph. See margin.

1. $f(x) = x^2 + 3x - 1$

2. $f(x) = -x^3 + 2x^2 - 4$

3. $f(x) = x^3 + 4x^2 - 5x + 5$

4. $f(x) = -x^4 - x^3 + 4$

Example 2

Use a table to graph each function. Then estimate the x-coordinates at which relative maxima and relative minima occur. See Mod. 4 Answer Appendix for tables and graphs.

5. $f(x) = -2x^3 + 12x^2 - 8x$
rel. min between $x = 0$ and $x = 1$; rel. max near $x = 4$

6. $f(x) = 2x^3 - 4x^2 - 3x + 4$
rel. max: near $x = -0.3$; rel. min: near $x = 1.6$

7. $f(x) = x^4 + 2x - 1$
rel. min: near $x = -1$; no rel. max

8. $f(x) = x^4 + 8x^2 - 12$
min: near $x = 0$

Example 3

9. BUSINESS A banker models the expected value v of a company in millions of dollars by using the formula $v = n^3 - 3n^2$, where n is the number of years in business. Graph the function and describe its key features over the relevant domain. See Mod. 4 Answer Appendix.

10. HEIGHT A plant's height is modeled by the function $f(x) = 1.5x^3 - 20x^2 + 85x - 84$, where x is the number of weeks since the seed was planted and $f(x)$ is the height of the plant. Graph the function and describe its key features over its relevant domain. See Mod. 4 Answer Appendix.

Example 4

11. USE ESTIMATION The table shows U.S. car sales in millions of cars. Use a graphing calculator to make a scatter plot and a curve of best fit to show the trend over time. Then use the equation to estimate the car sales in 2017. Let 2008 be represented by year 0. Round the coefficients of the regression equation to the thousandths place.

Year	Cars (millions)	Year	Cars (millions)	Year	Cars (millions)
2008	7.659	2011	6.769	2014	6.089
2009	7.761	2012	5.400	2015	7.243
2010	7.562	2013	5.635	2016	7.780

Curve of best fit: $y = -0.012x^4 + 0.217x^3 - 1.166x^2 + 1.486x + 7.552$; car sales in 2017: 5.941 million

12. POPULATION The table shows the population in Cincinnati, Ohio, since 1960. Make a scatter plot and a curve of best fit to show the trend over the given time period. Then use the equation to estimate the population of Cincinnati in 2020.

Year	Population	Year	Population
1960	502,550	1990	364,553
1970	452,524	2000	331,258
1980	385,457	2010	296,943

Curve of best fit: $y = -0.108x^4 + 9.914x^3 - 227.160x^2 - 4070.779x + 503{,}310.647$; population in 2020: 183,032

13. VOLUNTEERS The table shows average volunteer hours per month for a local non-profit. Make a scatter plot and a curve of best fit to show the trend over the given time period. Then use the equation to estimate the volunteer hours for September. Let January be represented by month 1.

Month	Volunteer Hours	Month	Volunteer Hours
Jan.	48	May	100
Feb.	60	June	110
Mar.	72	July	105
Apr.	75	Aug.	93

Curve of best fit: $y = -0.122x^4 + 1.565x^3 - 6.325x^2 + 20.321x + 33.125$, where x is the number of months since January; average volunteer hours for September: 44

Example 5

14. FARMS The table shows the number of farms in the U.S. at various years, according to the USDA Census of Agriculture. Find the average rate of change from 1982 to 2012. Interpret the results in the context of the situation.

Year	Farms	Year	Farms
1982	2,480,000	2002	2,130,000
1987	2,340,000	2007	2,200,000
1992	2,180,000	2012	2,110,000
1997	2,220,000		

average rate of change: −12,333; From 1982 to 2012, there was an average of 12,333 fewer farms each year.

15. SALARY The table shows the annual salary of a salesperson over time. Find the average rate of change over the given time interval. Interpret the results in context of this situation.

Year	Salary	Year	Salary
2012	$45,000	2015	$55,500
2013	$49,000	2016	$73,000
2014	$47,500	2017	$67,500

average rate of change: 4500; From 2012 to 2017, the salesman's average rate of change in salary was an increase of $4500 per year.

Name ________ Period ________ Date ________

Mixed Exercises

Graph each function by using a table of values. Then, estimate the x-coordinates at which each zero and relative extrema occur, and state the domain and range.
16–19. See Mod. 4 Answer Appendix.

16. $f(x) = x^3 - 3x + 1$

17. $f(x) = 2x^3 + 9x^2 + 12x + 2$

18. $f(x) = 2x^3 - 3x^2 + 2$

19. $f(x) = x^4 - 2x^2 - 2$

20. Determine the key features for $y = \begin{cases} x^2 & \text{if } x \le -4 \\ 5 & \text{if } -4 < x \le 0. \\ x^3 & \text{if } x > 0 \end{cases}$ See Mod. 4 Answer Appendix.

USE TOOLS **Use a graphing calculator to estimate the x-coordinates at which the maxima and minima of each function occur. Round to the nearest hundredth.**

21. $f(x) = x^3 + 3x^2 - 6x - 6$
rel. max: $x \approx -2.73$; rel. min: $x \approx 0.73$

22. $f(x) = -2x^3 + 4x^2 - 5x + 8$
no relative max or min

23. $f(x) = -2x^4 + 5x^3 - 4x^2 + 3x - 7$
max: $x \approx 1.34$; no rel. min

24. $f(x) = x^5 - 4x^3 + 3x^2 - 8x - 6$
rel. max: $x \approx -1.87$; rel. min: $x \approx 1.52$

25. PRECISION Sketch the graph of a third-degree polynomial function that has a relative minimum at $x = -3$, passes through the origin, and has a relative maximum at $x = 2$. Describe the end behavior of the graph. Based on the sketch, determine whether the leading coefficient is negative or positive. See Mod. 4 Answer Appendix.

26. USE TOOLS A canister has the shape of a cylinder with spherical caps on either end. The volume of the canister in cubic millimeters is modeled by the function $V(x) = \pi(x^3 + 3x^2) + \frac{4}{3}\pi x^3$ where x represents the radius in millimeters of a canister that is $x + 3$ millimeters wide.

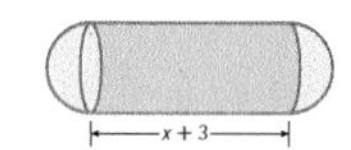

a. Use a graphing calculator to sketch the model that represents the volume of the canister. Include axes labels. See Mod. 4 Answer Appendix.

b. What is the domain of the model? Explain any restrictions that apply.
The domain is all x values greater than or equal to zero. Because x represents a radius, it cannot be negative.

27. CONSTRUCT ARGUMENTS What type of polynomial function best models the data in the graph? Explain your reasoning.
The type of polynomial function that should be used to model the graph is a function with an even degree with a negative leading coefficient. This is based on the fact that the graph is reflected in the y-axis and as $x \to -\infty$, $f(x) \to -\infty$ and as $x \to +\infty$, $f(x) \to -\infty$.

28. FORECASTING The table shows the number of deliveries a grocery store has made since they began to offer the service. Use a scatter plot and a curve of best fit to determine the number of deliveries the grocery store can expect to deliver 10 years after they begin the service. 1018 deliveries

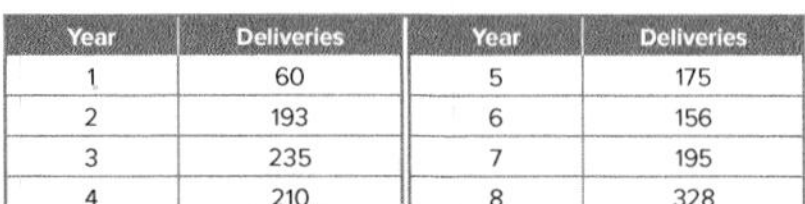

Year	Deliveries	Year	Deliveries
1	60	5	175
2	193	6	156
3	235	7	195
4	210	8	328

Higher-Order Thinking Skills

29. ANALYZE Explain why the leading coefficient and the degree are the only determining factors in the end behavior of a polynomial function.
As the x-values approach large positive or negative numbers, the term with the largest degree becomes more and more dominant in determining the value of $f(x)$.

30. ANALYZE The table below shows the values of a cubic function. Could there be a zero between $x = 2$ and $x = 3$? Justify your argument.

x	−2	−1	0	1	2	3
$g(x)$	3	−2	−1	1	−2	−2

No; sample answer: The cubic function is of degree 3 and cannot have any more than three zeros. Those zeros are located between −2 and −1, 0 and 1, and 1 and 2.

31. CREATE Sketch the graph of an odd degree polynomial function with 6 extrema and a relative extreme at $y = 0$. See Mod. 4 Answer Appendix.

32. ANALYZE Determine whether the following statement is *sometimes*, *always*, or *never* true. Justify your argument.

For any continuous polynomial function, the y-coordinate of a point where a function changes from increasing to decreasing or from decreasing to increasing is either a relative maximum or relative minimum. Always; sample answer: A point at which the graph stops increasing and begins to decrease is a maximum; a point where a graph stops decreasing and begins to increase is a minimum.

33. PERSEVERE A function is even if for every x in the domain of f, $f(x) = f(-x)$. Is every even-degree polynomial function also an even function? Explain.
No; sample answer: $f(x) = x^2 + x$ is an even degree, but $f(1) \neq f(-1)$.

34. PERSEVERE A function is odd if for every x in the domain, $-f(x) = f(-x)$. Is every odd-degree polynomial function also an odd function? Explain.
No; sample answer: $f(x) = x^3 + 2x^2$ is an odd degree, but $-f(1) \neq f(-1)$.

Answers

1. zeros between $x = -4$ and $x = -3$, and $x = 0$ and $x = 1$

x	$f(x)$
−4	3
−3	−1
−2	−3
−1	−3
0	−1
1	3

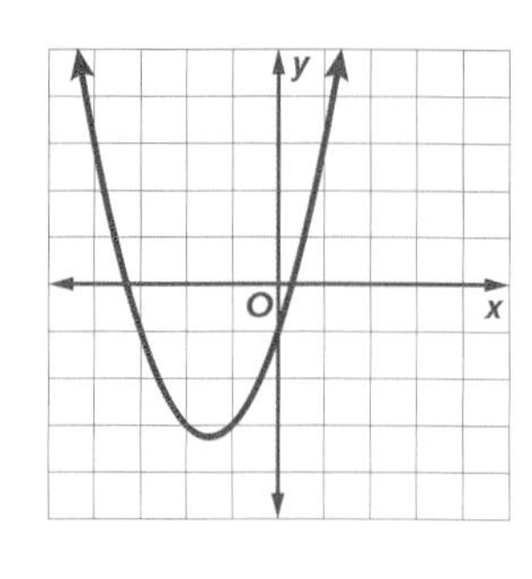

2. zero between $x = -2$ and $x = -1$

x	$f(x)$
−2	12
−1	−1
0	−4
1	−3
2	−4

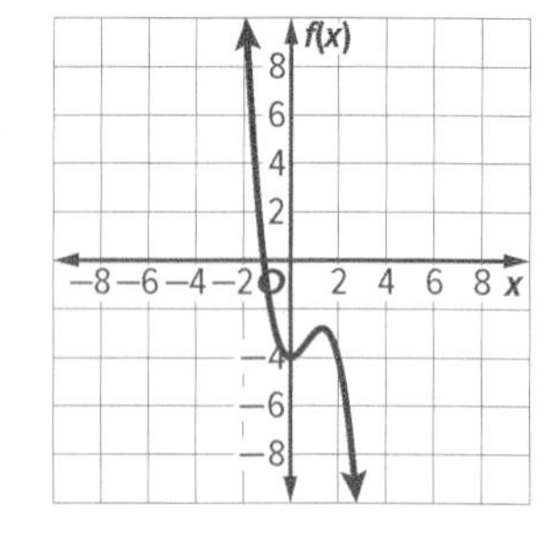

3. zero between $x = -6$ and $x = -5$

x	$f(x)$
−7	−107
−6	−37
−5	5
−4	25
−3	29

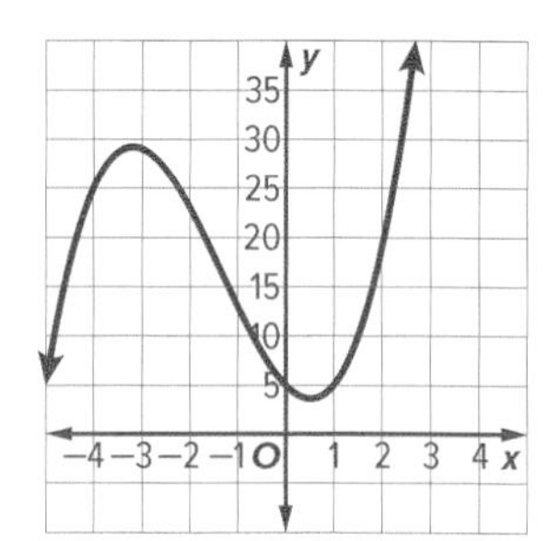

4. zeros between $x = -2$ and $x = -1$, and $x = 1$ and $x = 2$

x	$f(x)$
−3	−50
−2	−4
−1	4
0	4
1	2
2	−20
3	−104

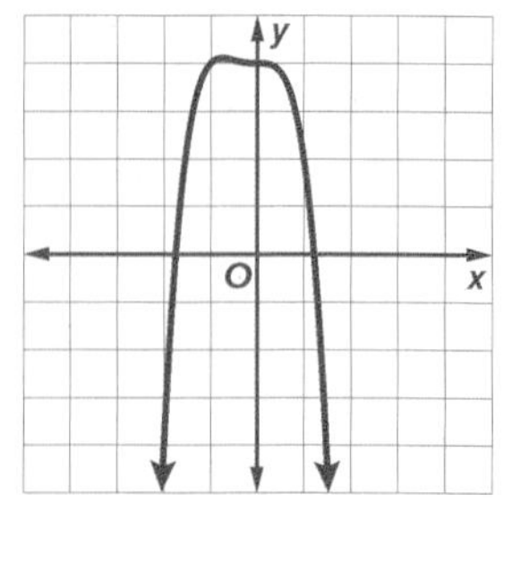

Operations with Polynomials

A.APR.1

LESSON GOAL

Students add, subtract, and multiply polynomials.

1 LAUNCH

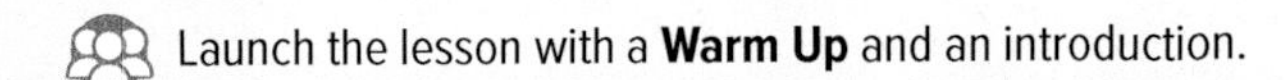
Launch the lesson with a **Warm Up** and an introduction.

2 EXPLORE AND DEVELOP

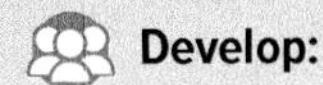
Develop:

Adding and Subtracting Polynomials
- Identify Polynomials
- Add Polynomials
- Subtract Polynomials

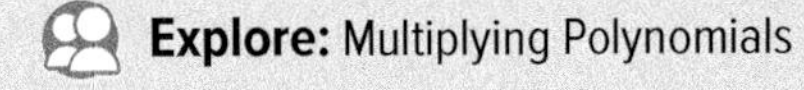
Explore: Multiplying Polynomials

Develop:

Multiplying Polynomials
- Simplify by Using the Distributive Property
- Multiply Binomials
- Multiply Polynomials (online)
- Write and Simplify a Polynomial Expression

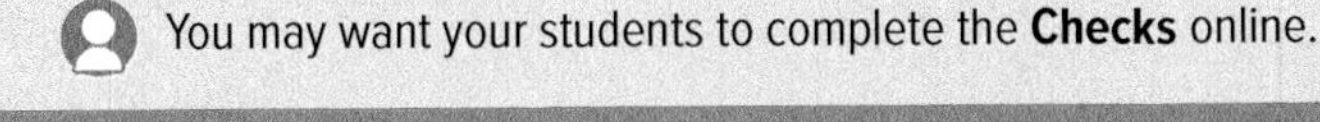
You may want your students to complete the **Checks** online.

3 REFLECT AND PRACTICE

Exit Ticket

Practice

Formative Assessment Math Probe

DIFFERENTIATE

View reports of student progress on the **Checks** after each example.

Resources	AL	OL	BL	ELL
Remediation: Adding and Subtracting Polynomials	●	●		●
Extension: Relationship Between GCF and LCM with Polynomials		●	●	●

Language Development Handbook

Assign page 27 of the Language *Development Handbook* to help your students build mathematical language related to adding, subtracting, and multiplying polynomials.

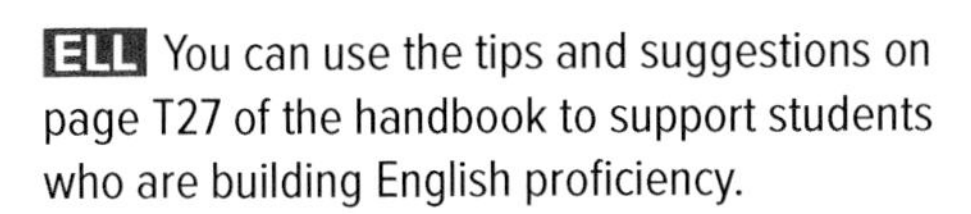
ELL You can use the tips and suggestions on page T27 of the handbook to support students who are building English proficiency.

Suggested Pacing

90 min	1 day
45 min	2 days

Focus

Domain: Algebra

Standard for Mathematical Content:

A.APR.1 Understand that polynomials form a system analogous to the integers, namely, they are closed under the operations of addition, subtraction, and multiplication; add, subtract, and multiply polynomials.

Standards for Mathematical Practice:

2 Reason abstractly and quantitatively.

8 Look for and express regularity in repeated reasoning.

Coherence

Vertical Alignment

Previous
Students performed operations on linear expressions and polynomial expressions. **7.EE.1, A.APR.1 (Algebra 1)**

Now
Students will add, subtract, and multiply polynomials.
A.APR.1

Next
Students will divide polynomials by using long division and synthetic division.
A.APR.6

Rigor

The Three Pillars of Rigor

1 CONCEPTUAL UNDERSTANDING	2 FLUENCY	3 APPLICATION

Conceptual Bridge In this lesson, students build on their understanding of polynomial operations. They build fluency by validating that polynomials are closed under arithmetic operations, and they apply their understanding by solving real-world problems.

Mathematical Background

When multiplying or dividing powers of variables, be sure that the base is the same. Add the exponents if multiplying powers of the same variable, and subtract them if dividing powers.

Interactive Presentation

Warm Up

Find the missing digits.

1. $(3x^3y^2)^3 = 3^{\square}x^{\square}y^{\square}$
   $= \square x^{\square}y^{\square}$

2. $(5 + 6x) + (-7 - 2x) = 5 + (-7) + 6x + (-2x)$
   $= \square + \square x$

3. $(x + 3)(x + 4) = x^2 + \square x + \square x + \square$
   $= x^2 + \square x + \square$

Show Answers

Warm Up

Warm Up

Prerequisite Skills

The Warm Up exercises address the following prerequisite skill for this lesson:

- performing simple operations with polynomials

Answers:

1. first row: 3; 3 • 3; 2 • 3; second row: 27; 9; 6
2. −2; 4
3. first row: 4; 3; 12; second row: 7; 12

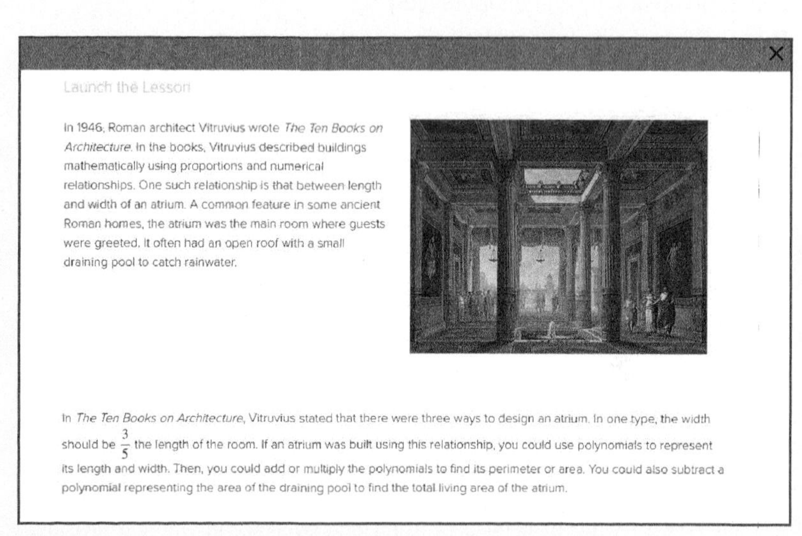

Launch the Lesson

Launch the Lesson

 **Teaching the Mathematical Practices**

2 Make Sense of Quantities Mathematically proficient students need to be able to make sense of quantities and their relationships. In the Launch the Lesson, notice the relationship between the length and width of the atrium described. Encourage students to write expressions to model the situation.

Go Online to find additional teaching notes and questions to promote classroom discourse.

Today's Standards

Tell students that they will be addressing these content and practice standards in this lesson. You may wish to have a student volunteer read aloud *How can I meet this standard?* and *How can I use these practices?*, and connect these to the standards.

See the Interactive Presentation for I Can statements that align with the standards covered in this lesson.

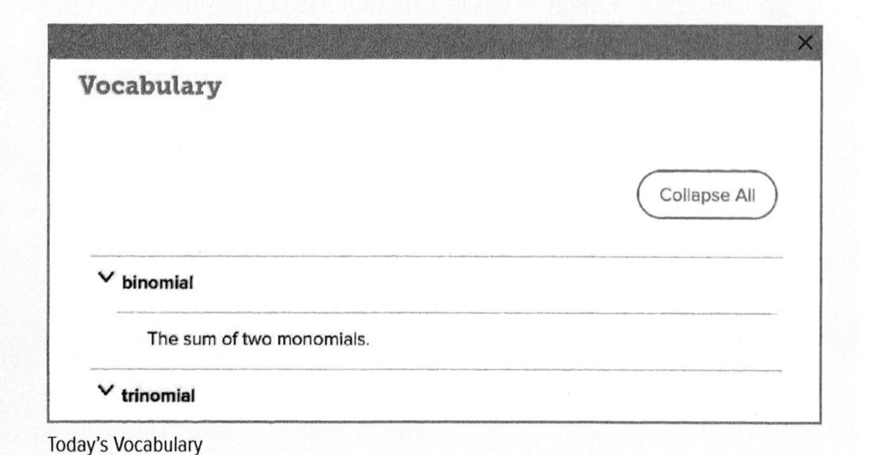

Today's Vocabulary

Today's Vocabulary

Tell students that they will be using these vocabulary terms in this lesson. You can expand each row if you wish to share the definitions. Then discuss the questions below with the class.

Explore Multiplying Polynomials

Objective

Students use a table to multiply polynomials.

Teaching the Mathematical Practices

7 Interpret Complicated Expressions Mathematically proficient students can see complicated expressions as single objects or as being composed of several objects. In the Explore, guide students to see the polynomial expressions as being composed of several terms that can be multiplied together to find the product of the two polynomials.

Ideas for Use

Recommended Use Present the Inquiry Question, or have a student volunteer read it aloud. Have students work in pairs to complete the Explore activity on their devices. Pairs should discuss each of the questions. Monitor student progress during the activity. Upon completion of the Explore activity, have student volunteers share their responses to the Inquiry Question.

What if my students don't have devices? You may choose to project the activity on a whiteboard. A printable worksheet for each Explore is available online. You may choose to print the worksheet so that individuals or pairs of students can use it to record their observations.

Summary of the Activity

Students will complete guiding exercises throughout the Explore activity. Students will be presented with two polynomials and instructed to find their product using a table. The terms of the polynomials are placed in the top row and left column of the table. Students will multiply each corresponding row and column to complete the table. Students will add all the products and combine like terms to find the product of the two polynomials. Then, students will answer the Inquiry Question.

(continued on the next page)

Interactive Presentation

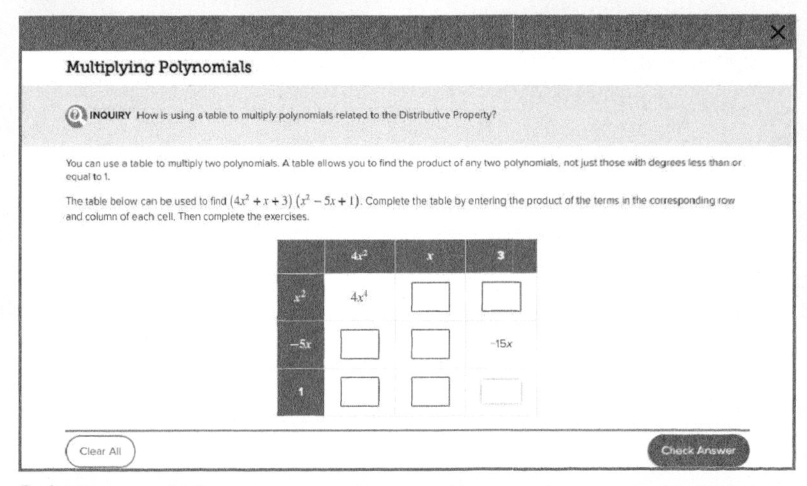

Explore

Students complete the calculations to multiply polynomials term by term.

Interactive Presentation

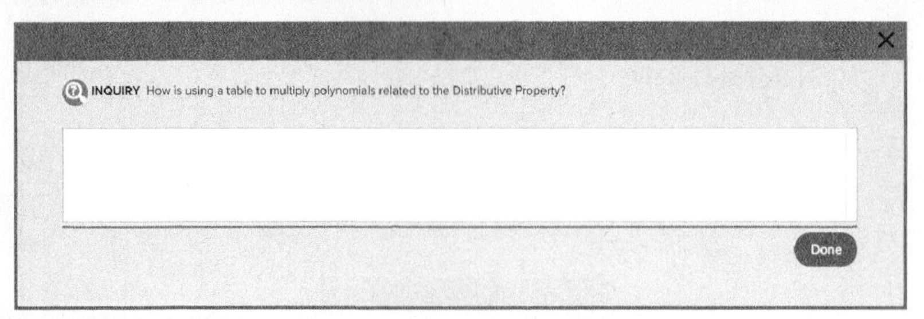

Explore

TYPE

Students respond to the Inquiry Question and can view a sample answer.

Explore Multiplying Polynomials (*continued*)

Teaching the Mathematical Practices

2 Use the Distributive Property Point out that the Distributive Property is one of the most-used properties in algebra. Guide students to see how multiplying polynomials using a table relates to the Distributive Property.

Questions

Have students complete the Explore activity.

Ask:

- How is using the table similar to multiplying polynomials with algebra tiles? Sample answer: You are creating a rectangle with the polynomials as the length and width. The area inside each cell is equal to the product of each term, just like the different sections used with algebra tiles.
- Can the table only be used to multiply polynomials with degree 2? Why or why not? Sample answer: The table could be used to multiply polynomials of any degree by making enough rows or columns for each term in the given polynomial.

Inquiry

How is using a table to multiply polynomials related to the Distributive Property? Sample answer: The table organizes the results of multiplying each term in one polynomial by each term in the other. The result will be the same as using the Distributive Property.

Go Online to find additional teaching notes and sample answers for the guiding exercises.

Learn Adding and Subtracting Polynomials

Objective

Students add and subtract polynomials by combining like terms.

Teaching the Mathematical Practices

7 Use Structure Guide students to use the structure of polynomials to see why polynomials are closed under addition and subtraction.

Important to Know

Terms and concepts students should be familiar with prior to this Learn include properties of exponents, factoring, simplifying, monomials, polynomials, degrees, and bases.

Common Misconception

Some students may believe that a common base or a common degree is enough for two terms to be like. Guide them to understand exponents and bases.

Essential Question Follow-Up

Students have begun to add and subtract polynomials.

Ask:

Are polynomials closed under addition and subtraction? Sample answer: Because adding or subtracting polynomials results in a polynomial, the set of polynomials is closed under addition and subtraction.

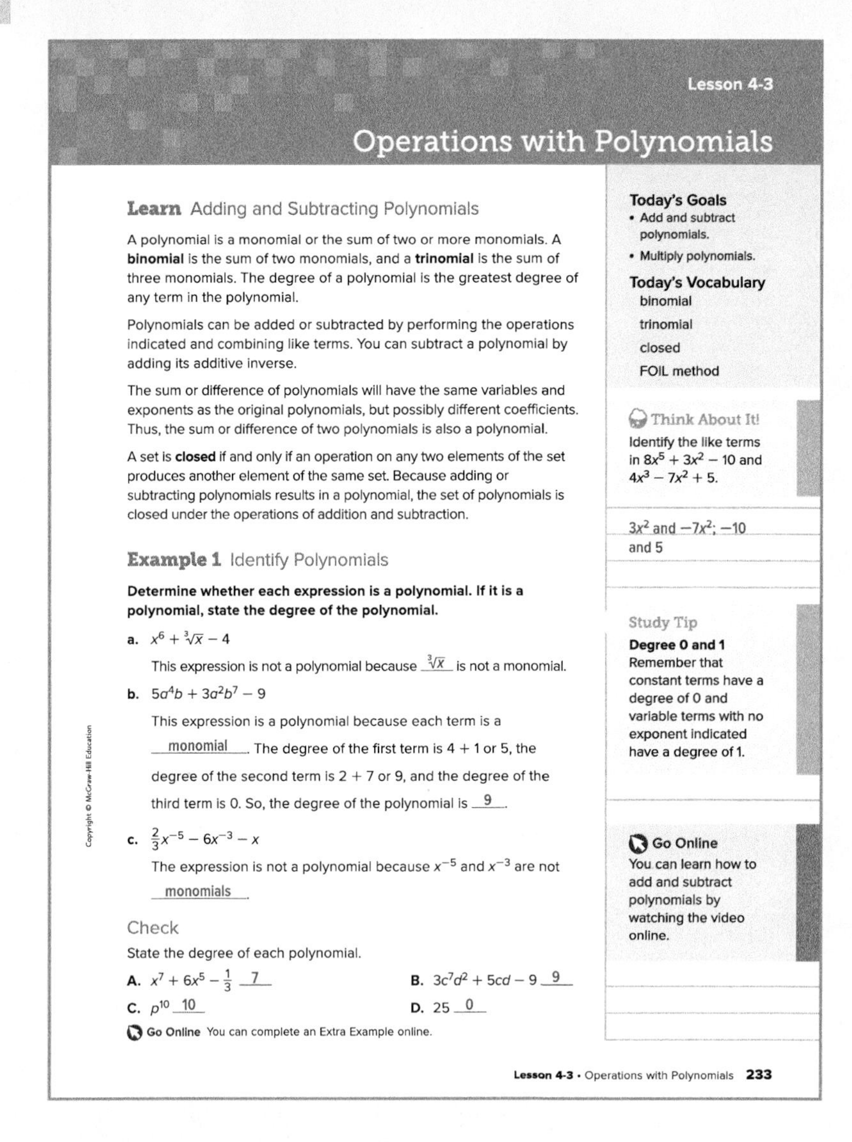

Lesson 4-3

Operations with Polynomials

Learn Adding and Subtracting Polynomials

A polynomial is a monomial or the sum of two or more monomials. A **binomial** is the sum of two monomials, and a **trinomial** is the sum of three monomials. The degree of a polynomial is the greatest degree of any term in the polynomial.

Polynomials can be added or subtracted by performing the operations indicated and combining like terms. You can subtract a polynomial by adding its additive inverse.

The sum or difference of polynomials will have the same variables and exponents as the original polynomials, but possibly different coefficients. Thus, the sum or difference of two polynomials is also a polynomial.

A set is **closed** if and only if an operation on any two elements of the set produces another element of the same set. Because adding or subtracting polynomials results in a polynomial, the set of polynomials is closed under the operations of addition and subtraction.

Example 1 Identify Polynomials

Determine whether each expression is a polynomial. If it is a polynomial, state the degree of the polynomial.

a. $x^6 + \sqrt[3]{x} - 4$

This expression is not a polynomial because $\sqrt[3]{x}$ is not a monomial.

b. $5a^4b + 3a^2b^7 - 9$

This expression is a polynomial because each term is a monomial. The degree of the first term is 4 + 1 or 5, the degree of the second term is 2 + 7 or 9, and the degree of the third term is 0. So, the degree of the polynomial is 9.

c. $\frac{2}{3}x^{-5} - 6x^{-3} - x$

The expression is not a polynomial because x^{-5} and x^{-3} are not monomials.

Check

State the degree of each polynomial.

A. $x^7 + 6x^5 - \frac{1}{3}$ 7 **B.** $3c^7d^2 + 5cd - 9$ 9

C. p^{10} 10 **D.** 25 0

Go Online You can complete an Extra Example online.

Today's Goals
- Add and subtract polynomials.
- Multiply polynomials.

Today's Vocabulary
binomial
trinomial
closed
FOIL method

Think About It!
Identify the like terms in $8x^5 + 3x^2 - 10$ and $4x^3 - 7x^2 + 5$.

$3x^2$ and $-7x^2$; -10 and 5

Study Tip
Degree 0 and 1 Remember that constant terms have a degree of 0 and variable terms with no exponent indicated have a degree of 1.

Go Online
You can learn how to add and subtract polynomials by watching the video online.

Example 1 Identify Polynomials

Teaching the Mathematical Practices

7 Interpret Complicated Expressions In Example 1, guide students to see each expression as composed of several terms to determine whether each expression is a polynomial.

Questions for Mathematical Discourse

AL Why is $\sqrt[3]{x}$ not a monomial? A monomial must have positive integer exponents.

OL What would it mean if a polynomial had a degree of 0? It is a constant polynomial.

BL Can you have a term with exponents that are not positive integers and still have a polynomial? This could only be true if the exponents simplified to positive integers.

Things to Remember

Suggest that students return to the basic definitions for exponents rather than just memorizing rules. For example, they can derive the rule for multiplying quantities such as $x^2 \cdot x^3$ by rewriting the problem as $x \cdot x \cdot x \cdot x \cdot x$.

Interactive Presentation

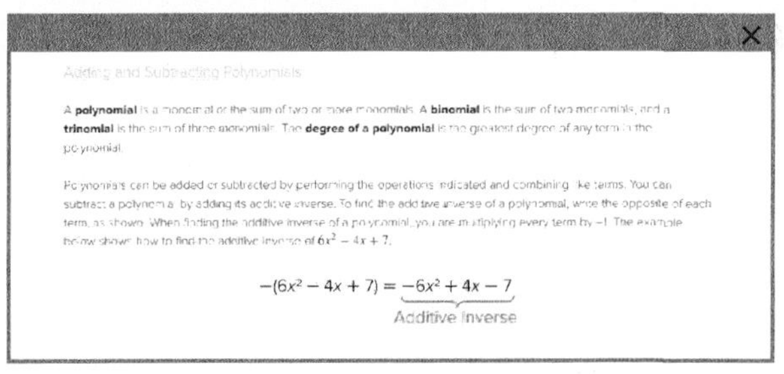

Learn

TYPE

Students identify the like terms in two polynomials.

Your Notes

Talk About It! Notice that the terms in the sum have the same variables and exponents as the terms of the original expressions: x^3, x^2, x, and a constant. Will this always be the case? Explain.

Sample answer: If the sum of any of the like terms is 0, then that degree term will not appear in the resulting polynomial.

Example 2 Add Polynomials

Find $(6x^3 + 7x^2 - 2x + 5) + (x^3 - 4x^2 - 8x + 1)$.

Method 1 Add horizontally.

Group and combine like terms.

$(6x^3 + 7x^2 - 2x + 5) + (x^3 - 4x^2 - 8x + 1)$

$= (6x^3 + x^3) + (7x^2 - 4x^2) + (-2x - 8x) + (5 + 1)$ Group like terms.

$= 7x^3 + 3x^2 - 10x + 6$ Combine like terms.

Method 2 Add vertically.

Align like terms vertically and add.

$6x^3 + 7x^2 - 2x + 5$

$(+)\ x^3 - 4x^2 - 8x + 1$ Align like terms.

$7x^3 + 3x^2 - 10x + 6$ Combine like terms.

Check

Find $(2x^3 + 9x^2 + 6x - 3) + (4x^3 - 7x^2 + 5x)$. $6x^3 + 2x^2 + 11x - 3$

Study Tip **Additive Inverse** Distributing −1 to each term in the polynomial being subtracted is the same as finding and adding its additive inverse.

Example 3 Subtract Polynomials

Find $(2x^5 + 11x^4 + 7x - 8) - (5x^4 + 9x^3 - 3x + 4)$.

Method 1 Subtract horizontally.

Group and combine like terms.

$(2x^5 + 11x^4 + 7x - 8) - (5x^4 + 9x^3 - 3x + 4)$ Original equation

$= 2x^5 + 11x^4 + 7x - 8 - 5x^4 - 9x^3 + 3x - 4$ Distribute −1.

$= 2x^5 + (11x^4 - 5x^4) + (-9x^3) + (7x + 3x) + (-8 - 4)$ Group like terms.

$= 2x^5 + 6x^4 - 9x^3 + 10x - 12$ Combine like terms.

Method 2 Subtract vertically.

Align like terms vertically and subtract.

$2x^5 + 11x^4 + 0x^3 + 7x - 8$

$(-)\ 0x^5 + 5x^4 + 9x^3 - 3x + 4$

$2x^5 + 6x^4 - 9x^3 + 10x - 12$

Think About It! Why is it helpful to insert placeholders for the $0x^3$ and $0x^5$ terms? Why were placeholders not included for the x^2-term in either polynomial?

Sample answer: Adding placeholders helps to align like terms of the expressions. Because neither polynomial has an x^2-term, the like terms can be aligned without adding a placeholder to each expression.

Check

Find $(8x^2 - 3x + 1) - (5x^3 + 2x^2 - 6x - 9)$. $-5x^3 + 6x^2 + 3x + 10$

Explore Multiplying Polynomials

Online Activity Use a table to complete the Explore.

INQUIRY How is using a table to multiply polynomials related to the Distributive Property?

234 Module 4 • Polynomials and Polynomial Functions

Interactive Presentation

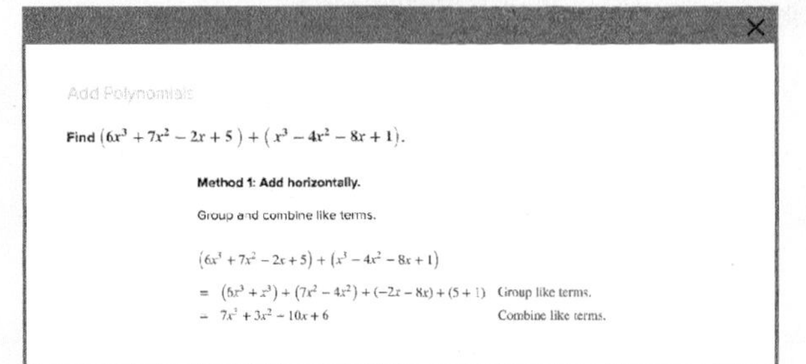

Example 2

TAP

Students tap on each button to see how to add and subtract polynomials

CHECK

Students complete the Check online to determine whether they are ready to move on.

Example 2 Add Polynomials

Teaching the Mathematical Practices

1 Special Cases Work with students to evaluate the two methods shown. Encourage students to familiarize themselves with both methods, and know the best time to use each.

Questions for Mathematical Discourse

AL Why should you group or align like terms when adding polynomials? This makes it easier to combine the like terms without error.

OL In what circumstances might each method be more useful? Sample answer: Horizontal may be more useful for combining polynomials with only a few terms, while vertical may be more useful for polynomials with several terms.

BL Can you think of another method to add two polynomials? Sample answer: You could combine from greatest to least degree while marking off each set of like terms as you record the resulting term.

Example 3 Subtract Polynomials

Teaching the Mathematical Practices

7 Use Structure Help students to use the structure of polynomial expressions to combine like terms and find the difference of polynomials.

Questions for Mathematical Discourse

AL What must you remember to do if you are subtracting polynomials? Distribute the −1.

OL What could you do in the vertical method to help you combine like terms with the correct operation? Sample answer: Change the sign of each term in the second polynomial and then add the polynomials.

BL Which method do you prefer for subtracting polynomials and why? Sample answer: I prefer the vertical method because I can use placeholders and line up the like terms without having to look back and forth between the polynomials.

Common Error

Students often only subtract the first term when subtracting one polynomial from another. Remind students to distribute −1 to *all* terms in the parentheses, not just the first term.

Go Online

- Find additional teaching notes.
- View performance reports of the Checks.
- Assign or present an Extra Example.

Learn Multiplying Polynomials

Objective

Students multiply polynomials by using the Distributive Property.

Teaching the Mathematical Practices

8 Notice Regularity Help students to see the regularity in the way that expressions are simplified when multiplying binomials using the FOIL method.

Essential Question Follow-Up

Students have begun to multiply polynomials.

Ask:

Are polynomials closed under multiplication? Sample answer: Because multiplying polynomials results in a polynomial, the set of polynomials is closed under multiplication.

DIFFERENTIATE

Reteaching Activity AL ELL

IF students have difficulty describing or using properties of exponents, **THEN** have them write their own summary of the properties of powers, such as "to multiply expressions with exponents, you add the exponents; to divide, you subtract the exponents."

Example 4 Simplify by Using the Distributive Property

Teaching the Mathematical Practices

7 Use the Distributive Property Point out that the Distributive Property is one of the most-used properties in algebra. Students should know that whenever they see a number outside of a sum or difference within parentheses, they should apply the Distributive Property.

Questions for Mathematical Discourse

AL Why is $2x(4x^3)$ equal to $8x^4$? The coefficients 2 and 4 multiply to 8. The exponents for x are added because x to a power n is the same as multiplying x by x n times. $x(x \cdot x \cdot x) = x^4$

OL Can you ever end up with negative exponents when you multiply a monomial by a polynomial? No, you are adding together two positive numbers.

BL In general, if you are multiplying a monomial by a polynomial, how many pairs of monomials will you be multiplying together? the number of terms the polynomial has

---

Learn Multiplying Polynomials

Polynomials can be multiplied by using the Distributive Property to multiply each term in one polynomial by each term in the other. When polynomials are multiplied, the product is also a polynomial. Therefore, the set of polynomials is closed under the multiplication. This is similar to the system of integers, which is also closed under multiplication. To multiply two binomials, you can use a shortcut called the **FOIL method**.

Key Concept • FOIL Method

Words: Find the sum of the products of ***F*** the *First* terms, **O** the *Outer* terms, ***I*** the *Inner* terms, and ***L*** the *Last* terms.

Symbols:

F L; Product of First Terms, Product of Outer Terms, Product of Inner Terms, Product of Last Terms; I; O

$$(2x + 4)(x - 3) = (2x)(x) + (2x)(-3) + (4)(x) + (4)(-3)$$
$$= 2x^2 - 6x + 4x - 12$$
$$= 2x^2 - 2x - 12$$

Example 4 Simplify by Using the Distributive Property

Find $2x(4x^3 + 5x^2 - x - 7)$.

$$2x(4x^3 + 5x^2 - x - 7) = 2x(4x^3) + 2x(\underline{5x^2}) + 2x(-x) + 2x(\underline{-7})$$
$$= \underline{8}\,x^4 + 10x^3 - \underline{2}\,x^2 - 14x$$

Think About It!
Why are the exponents added when you multiply the monomials?

Sample answer: Because the monomials have the same base x, their exponents are added by using the Product of Powers Property.

Example 5 Multiply Binomials

Find $(3a + 5)(a - 7)(4a + 1)$.

Step 1 Multiply any two binomials.

$(3a + 5)(a - 7) = 3a(\underline{a}) + 3a(\underline{-7}) + 5(a) + 5(-7)$ FOIL Method

$= 3a^2 - 21a + \underline{5}\,a - \underline{35}$ Multiply.

$= 3a^2 - \underline{16}\,a - 35$ Combine like terms.

Step 2 Multiply the result by the remaining binomial.

$(3a^2 - 16a - 35)(4a + 1)$

$= 3a^2(4a + 1) + (\underline{-16a})(4a + 1) + (\underline{-35})(4a + 1)$

$= \underline{12a^3} + 3a^2 - 64a^2 - 16a - 140a - 35$

$= 12a^3 - \underline{61a^2} - 156a\,\underline{-35}$

Check

Find $(-2r - 3)(5r - 1)(r + 4)$. $\underline{-10r^3 - 53r^2 - 49r + 12}$

Go Online You can complete an Extra Example online.

Go Online for an example of how to multiply two trinomials.

Interactive Presentation

Multiplying Polynomials

Polynomials can be multiplied by using the Distributive Property to multiply each term in one polynomial by each term in the other. When polynomials are multiplied, the product is also a polynomial. Therefore, the set of polynomials is closed under the multiplication. This is similar to the system of integers, which is also closed under multiplication.

The example below shows how to use the Distributive Property to multiply a monomial and a binomial.

$$5x(3x + 6) = 5x \cdot 3x + 5x \cdot 6$$
$$= 15x^2 + 30x$$

Learn

TYPE

Students answer a question to show they understand how to multiply expressions with exponents.

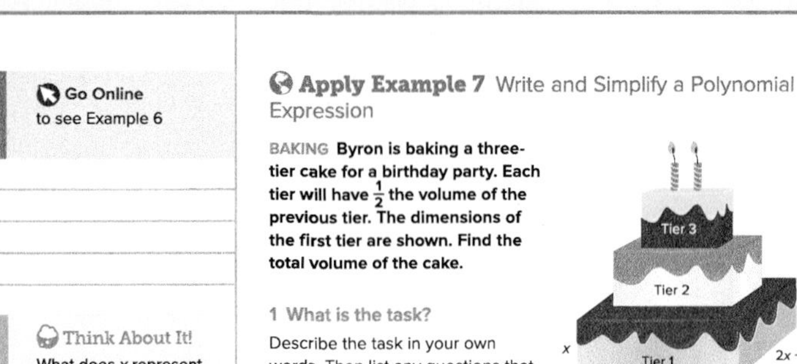

Go Online to see Example 6

Apply Example 7 Write and Simplify a Polynomial Expression

BAKING **Byron is baking a three-tier cake for a birthday party. Each tier will have $\frac{1}{2}$ the volume of the previous tier. The dimensions of the first tier are shown. Find the total volume of the cake.**

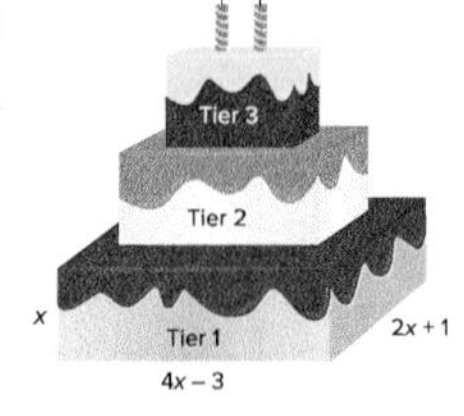

1 What is the task?
Describe the task in your own words. Then list any questions that you may have. How can you find answers to your questions?

Sample answer: I need to find the total volume of the cake, which is the sum of all 3 tiers. How can I represent the volume of each tier as a polynomial? Which properties will I need to know? I can find the answers to my questions by referencing other examples in the lesson.

Think About It! What does x represent in the polynomial expression for the volume of the cake?

the height of tier 1

2 How will you approach the task? What have you learned that you can use to help you complete the task?

Sample answer: I will find and simplify the volume of each tier and then add them together. I will use the Distributive Property and FOIL method to complete the task.

Problem-Solving Tip
Solve a Simpler Problem Some complicated problems can be more easily solved by breaking them into several simpler problems. In this case, finding the volume of each tier individually simplifies the situation and makes finding the total volume easier.

3 What is your solution?
Use your strategy to solve the problem.

What is the volume of each tier?

Tier 1: $8x^3 - 2x^2 - 3x$, Tier 2: $4x^3 - x^2 - 1.5x$, Tier 3: $2x^3 - 0.5x^2 - 0.75x$

What is the total volume of the cake?

$14x^3 - 3.5x^2 - 5.25x$

4 How can you know that your solution is reasonable?

Write About It! Write an argument that can be used to defend your solution.

Sample answer: Because all the expressions are based on the expression for the volume of Tier 1, I can check that the expression for Tier 1 is correct. I can factor the expression for volume of Tier 1 to ensure that the factors are the same as the given dimensions.

$8x^3 - 2x^2 - 3x$	Expression for Tier 1
$= (8x^2 - 2x - 3)x$	Factor x from each term.
$= (4x - 3)(2x + 1)x$ ✓	Factor $8x^2 - 2x - 3$.

236 Module 4 • Polynomials and Polynomial Functions

Interactive Presentation

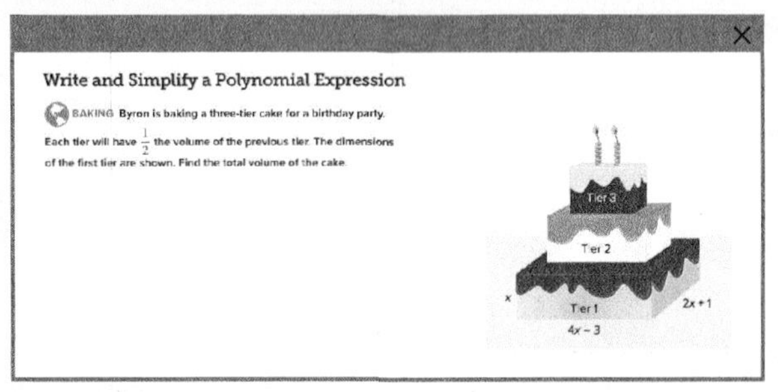

Example 6

CHECK

Students complete the Check online to determine whether they are ready to move on.

1 CONCEPTUAL UNDERSTANDING | **2 FLUENCY** | **3 APPLICATION**

Example 5 Multiply Binomials

Teaching the Mathematical Practices

7 Interpret Complicated Expressions Guide students to see the original expression as being composed of several binomials. Encourage them to break the expression into simpler problems.

Questions for Mathematical Discourse

AL When multiplying a polynomial with three terms by a polynomial with two terms, how many pairs of monomials do you multiply? 6

OL Can you tell the degree of the polynomial before multiplying? Yes; you know the leading term will be the three a-terms multiplied together, so the degree will be 3.

BL When multiplying more than two binomials, what features might you look for to decide which pair of binomials to multiply first? Sample answer: Look for a pair that results in a difference of squares because the middle terms cancel. This would make the next step of multiplication simpler.

Apply Example 7 Write and Simplify a Polynomial Expression

Teaching the Mathematical Practices

1 Make Sense of Problems and Persevere in Solving Them, 4 Model with Mathematics Students will be presented with a task. They will first seek to understand the task, and then determine possible entry points to solving it. As students come up with their own strategies, they may propose mathematical models to aid them. As they work to solve the problem, encourage them to evaluate their model and/or progress, and change direction, if necessary.

Recommended Use

Have students work in pairs or small groups. You may wish to present the task, or have a volunteer read it aloud. Then allow students the time to make sure they understand the task, think of possible strategies, and work to solve the problem.

Encourage Productive Struggle

As students work, monitor their progress. Instead of instructing them on a particular strategy, encourage them to use their own strategies to solve the problem and to evaluate their progress along the way. They may or may not find that they need to change direction or try out several strategies.

Signs of Non-Productive Struggle

If students show signs of non-productive struggle, such as feeling overwhelmed, frustrated, or disengaged, intervene to encourage them to think of alternate approaches to the problem. Some sample questions are shown.

- How can you find the the volume of Tier 1?
- Do you have to find the dimensions of Tier 2 to find its volume?

Write About It!

Have students share their responses with another pair/group of students or the entire class. Have them clearly state or describe the mathematical reasoning they can use to defend their solution.

Exit Ticket

Recommended Use

At the end of class, go online to display the Exit Ticket prompt and ask students to respond using a separate piece of paper. Have students hand you their responses as they leave the room.

Alternate Use

At the end of class, go online to display the Exit Ticket prompt and ask students to respond verbally or by using a mini-whiteboard. Have students hold up their whiteboards so that you can see all student responses. Tap to reveal the answer when most or all students have completed the Exit Ticket.

Practice and Homework

Suggested Assignments

Use the table below to select appropriate exercises.

DOK	Topic	Exercises
1, 2	exercises that mirror the examples	1–29
2	exercises that use a variety of skills from this lesson	30–45
2	exercises that extend concepts learned in this lesson to new contexts	46–47
3	exercises that emphasize higher-order and critical thinking skills	48–51

ASSESS AND DIFFERENTIATE

Use the data from the **Checks** to determine whether to provide resources for extension, remediation, or intervention.

IF students score 90% or more on the Checks, **THEN** assign: OL BL

- Practice Exercises 1–47 odd, 48–51
- Extension: Relationship Between GCF ad LCM with Polynomials
- ALEKS®

IF students score 66%–89% on the Checks, **THEN** assign: AL OL

- Practice Exercises 1–51 odd
- Remediation, Review Resources: Adding and Subtracting Polynomials
- Personal Tutors
- Extra Examples 1–6
- ALEKS® Simple Operations with Polynomials

IF students score 65% or less on the Checks, **THEN** assign: AL

- Practice Exercises 1–29 odd
- Remediation, Review Resources: Adding and Subtracting Polynomials
- *Quick Review Math Handbook*: Operations with Polynomials
- ALEKS® Simple Operations with Polynomials

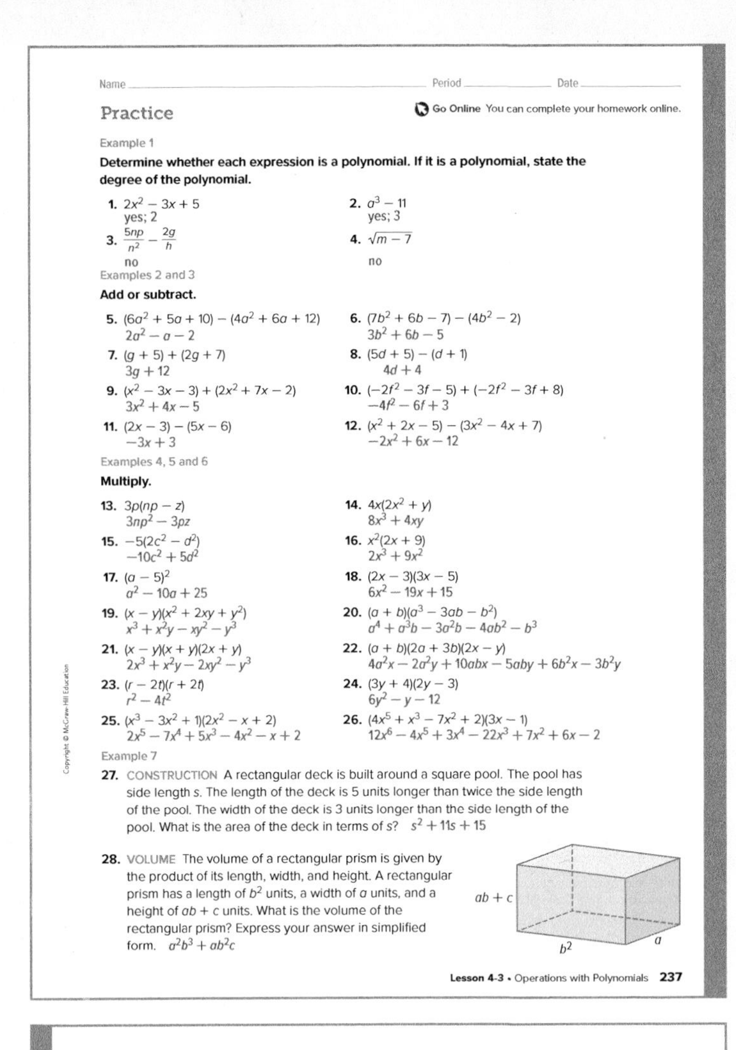

Name ______ Period ______ Date ______

Practice

Go Online You can complete your homework online.

Example 1

Determine whether each expression is a polynomial. If it is a polynomial, state the degree of the polynomial.

1. $2x^2 - 3x + 5$ yes; 2
2. $a^3 - 11$ yes; 3
3. $\frac{5np}{n^2} - \frac{2g}{h}$ no
4. $\sqrt{m - 7}$ no

Examples 2 and 3

Add or subtract.

5. $(6a^2 + 5a + 10) - (4a^2 + 6a + 12)$ $2a^2 - a - 2$
6. $(7b^2 + 6b - 7) - (4b^2 - 2)$ $3b^2 + 6b - 5$
7. $(g + 5) + (2g + 7)$ $3g + 12$
8. $(5d + 5) - (d + 1)$ $4d + 4$
9. $(x^2 - 3x - 3) + (2x^2 + 7x - 2)$ $3x^2 + 4x - 5$
10. $(-2f^2 - 3f - 5) + (-2f^2 - 3f + 8)$ $-4f^2 - 6f + 3$
11. $(2x - 3) - (5x - 6)$ $-3x + 3$
12. $(x^2 + 2x - 5) - (3x^2 - 4x + 7)$ $-2x^2 + 6x - 12$

Examples 4, 5 and 6

Multiply.

13. $3p(np - z)$ $3np^2 - 3pz$
14. $4x(2x^2 + y)$ $8x^3 + 4xy$
15. $-5(2c^2 - d^2)$ $-10c^2 + 5d^2$
16. $x^2(2x + 9)$ $2x^3 + 9x^2$
17. $(a - 5)^2$ $a^2 - 10a + 25$
18. $(2x - 3)(3x - 5)$ $6x^2 - 19x + 15$
19. $(x - y)(x^2 + 2xy + y^2)$ $x^3 + x^2y - xy^2 - y^3$
20. $(a + b)(a^3 - 3ab - b^2)$ $a^4 + a^3b - 3a^2b - 4ab^2 - b^3$
21. $(x - y)(x + y)(2x + y)$ $2x^3 + x^2y - 2xy^2 - y^3$
22. $(a + b)(2a + 3b)(2x - y)$ $4a^2x - 2a^2y + 10abx - 5aby + 6b^2x - 3b^2y$
23. $(r - 2t)(r + 2t)$ $r^2 - 4t^2$
24. $(3y + 4)(2y - 3)$ $6y^2 - y - 12$
25. $(x^3 - 3x^2 + 1)(2x^2 - x + 2)$ $2x^5 - 7x^4 + 5x^3 - 4x^2 - x + 2$
26. $(4x^5 + x^3 - 7x^2 + 2)(3x - 1)$ $12x^6 - 4x^5 + 3x^4 - 22x^3 + 7x^2 + 6x - 2$

Example 7

27. CONSTRUCTION A rectangular deck is built around a square pool. The pool has side length s. The length of the deck is 5 units longer than twice the side length of the pool. The width of the deck is 3 units longer than the side length of the pool. What is the area of the deck in terms of s? $s^2 + 11s + 15$

28. VOLUME The volume of a rectangular prism is given by the product of its length, width, and height. A rectangular prism has a length of b^2 units, a width of a units, and a height of $ab + c$ units. What is the volume of the rectangular prism? Express your answer in simplified form. $a^2b^3 + ab^2c$

$ab + c$
b^2
a

Lesson 4-3 • Operations with Polynomials 237

29. SAIL BOATS Tamara is making a sail for her sailboat.
    a. Refer to the diagram to find the area of the sail. $4x^2 + 8x + 3$
    b. If Tamara wants fabric on each side of her sail, write a polynomial to represent the total amount of fabric she will need to make the sail. $8x^2 + 16x + 6$

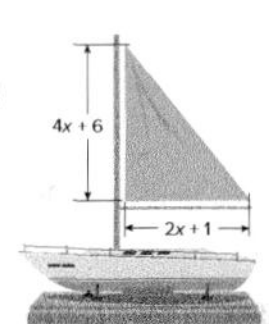

Mixed Exercises

Simplify.

30. $5xy(2x - y) + 6y^2(x^2 + 6)$ $10x^2y - 5xy^2 + 6x^2y^2 + 36y^2$
31. $3ab(4a - 5b) + 4b^2(2a^2 + 1)$ $12a^2b + 8a^2b^2 - 15ab^2 + 4b^2$
32. $\frac{1}{4}g^2(8g + 12h - 16gh^2)$ $2g^3 + 3g^2h - 4g^3h^2$
33. $\frac{1}{3}n^3(6n - 9p + 18np^4)$ $2n^4 - 3n^3p + 6n^4p^4$
34. $(g^3 - h)(g^3 + h)$ $g^6 - h^2$
35. $(n^2 - 7)(2n^3 + 4)$ $2n^5 - 14n^3 + 4n^2 - 28$
36. $(2x - 2y)^3$ $8x^3 - 24x^2y + 24xy^2 - 8y^3$
37. $(4n - 5)^3$ $64n^3 - 240n^2 + 300n - 125$
38. $(3z - 2)^3$ $27z^3 - 54z^2 + 36z - 8$
39. $\frac{1}{4}(16x - 12y) + \frac{1}{3}(9x + 3y)$ $7x - 2y$

40. STRUCTURE Use the polynomials $f(x) = -6x^3 + 2x^2 + 4$ and $g(x) = x^4 - 6x^3 - 2x$ to evaluate and simplify the given expression. Determine the degree of the resulting polynomial. Show your work.
    a. $f(x) + g(x)$ $f(x) + g(x) = (-6x^3 + 2x^2 + 4) + (x^4 - 6x^3 - 2x) = x^4 - 12x^3 + 2x^2 - 2x + 4$; 4th degree
    b. $g(x) - f(x)$ $g(x) - f(x) = (x^4 - 6x^3 - 2x) - (-6x^3 + 2x^2 + 4) = x^4 - 2x^2 - 2x - 4$; 4th degree

41. STRUCTURE Use the polynomials $f(x) = 3x^2 - 1$, $g(x) = x + 2$, and $h(x) = -x^2 - x$ to evaluate and simplify the given expressions. Determine the degree of the resulting polynomial.
    a. $f(x)g(x)$ $f(x)g(x) = (3x^2 - 1)(x + 2) = 3x^3 + 6x^2 - x - 2$; 3rd degree
    b. $h(x)f(x)$ $h(x)f(x) = (-x^2 - x)(3x^2 - 1) = -3x^4 + x^2 - 3x^3 + x$; 4th degree
    c. $[f(x)]^2$ $[f(x)]^2 = (3x^2 - 1)^2 = 9x^4 - 6x^2 + 1$; 4th degree

42. USE A MODEL Inez wants to increase the size of her rectangular garden. The original garden is 8 feet longer than it is wide. For the new garden, she will increase the length by 25% and increase the width by 5 feet.
    a. Draw and label a diagram that represents the original garden and the new garden. Define a variable and label each dimension with appropriate expressions. See margin.
    b. Write and simplify an expression for the increase in area of the garden. If the original width of the garden was 10 feet, find how many square feet the garden's area increased. See margin.

238 Module 4 • Polynomials and Polynomial Functions

Name ______________________ Period ________ Date ________

43. CONSTRUCT ARGUMENTS Complete the table to show which sets are closed under the operations. Write *yes* if the set is closed under the operation. Write *no* and provide a counterexample if the set is not closed under the operation. Assume that since division by zero is undefined, it does not affect closure.

	Addition and Subtraction	Multiplication	Division
Integers	yes	yes	No; $3 \div 2 = 1.5$
Rational Numbers	yes	yes	yes
Polynomials	yes	yes	No; $\frac{3}{x^2} = 3x^{-2}$

44. STRUCTURE The polynomial $2x^2 + 3x + 1$ can be represented by the tiles shown in the figure at the right. These tiles can be arranged to form the rectangle shown. Notice that the area of the rectangle is $2x^2 + 3x + 1$ units^2.

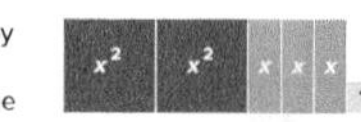

a. Find the length and width of the rectangle. See margin.

b. Find the perimeter of the rectangle. Then explain your process for finding the perimeter. See margin.

c. Select a value for x and substitute that value into each of the expressions above. For your value of x, state the length, width, perimeter, and area of the rectangle. Discuss any restrictions on the value of x. See margin.

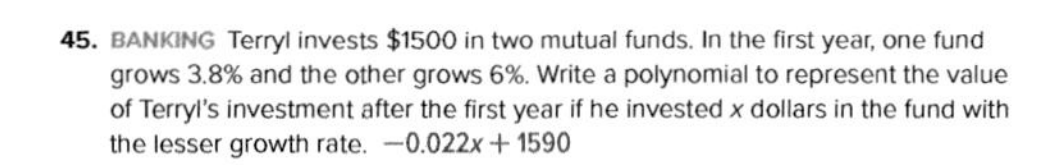

45. BANKING Terryl invests \$1500 in two mutual funds. In the first year, one fund grows 3.8% and the other grows 6%. Write a polynomial to represent the value of Terryl's investment after the first year if he invested x dollars in the fund with the lesser growth rate. $-0.022x + 1590$

46. GEOMETRY Consider a trapezoid that has one base that measures five feet greater than its height. The other base is one foot less than twice its height. Let x represent the height.

a. Write an expression for the area of the trapezoid. $A = 0.5x[(x + 5) + (2x - 1)]$ or $(1.5x^2 + 2x)$ ft^2

b. Write an expression for the area of the trapezoid if its height is increased by 4 feet. $A = 1.5x^2 + 14x + 32$ ft^2

47. URBAN DEVELOPMENT The diagram represents an aerial view of a memorial in a town center. A sidewalk that is 12 feet wide with an area of 384π square feet surrounds a statue with a circular base of radius r.

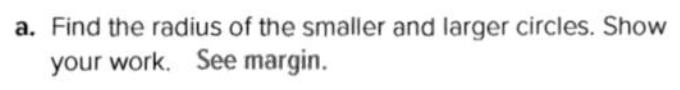

a. Find the radius of the smaller and larger circles. Show your work. See margin.

b. A nearby town wants to use the same design concept, but use two squares rather than two circles. Draw and label a diagram with two squares to represent a sidewalk with the same uniform width and area as the circular sidewalk. See margin.

c. If s represents the side length of the smaller square, write a polynomial expression for the area of the sidewalk that surrounds the smaller square. $48s + 576$ ft^2

Higher-Order Thinking Skills

48. ANALYZE Given $f(x)$ and $g(x)$ are polynomials, is the product always a polynomial? Justify your argument. See margin.

49. PERSEVERE Use your result from Exercise 48 to make conjectures about the product of a polynomial with m terms and a polynomial with n terms. Justify your conjecture.

a. How many times are two terms multiplied?
$m \cdot n$; each term of one polynomial multiplies each term of the other one time.

b. What is the least number of terms in the simplified product?
2; adding like terms may result in a sum of 0; but the first and last terms are unique.

50. FIND THE ERROR Isabella found the product of $3x^2 - 4x + 1$ and $x^2 + 5x + 6$ using vertical alignment. Is her answer correct? Explain your reasoning.

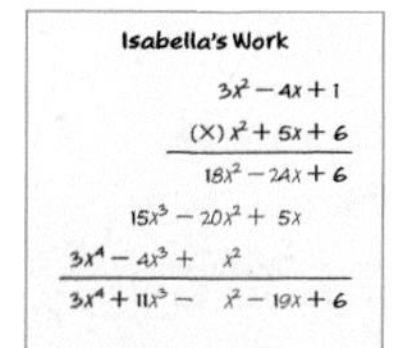

Yes; Isabella correctly aligned like terms after multiplying each term in the first polynomial by each term in the second polynomial. She then correctly added like terms.

51. CREATE Write an expression where two binomials are multiplied and have a product of $9 - 4b^2$. $(3 - 2b)(3 + 2b)$

Answers

42a. Let w represent the width of the original garden.

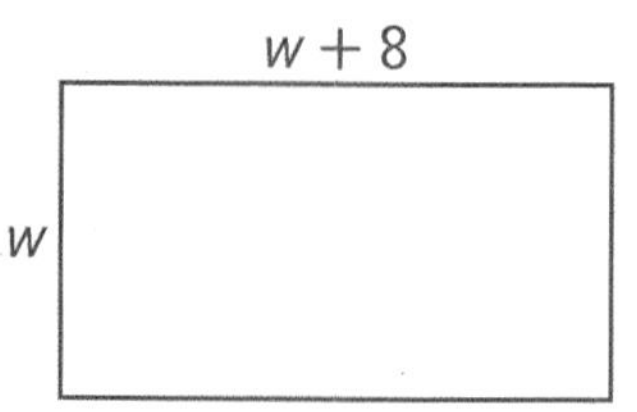

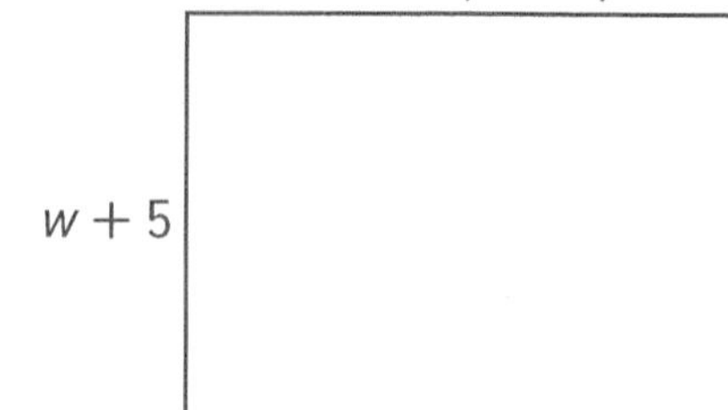

42b. $1.25(w + 8)(w + 5) - w(w + 8) = 1.25w^2 + 16.25w + 50 - w^2 - 8w = 0.25w^2 + 8.25w + 50$; if $w = 10$ feet, then the garden's area increased by $0.25(10)^2 + 8.25(10) + 50$ or 157.5 square feet.

44a. The length and width are $2x + 1$ and $x + 1$ units.

44b. The perimeter can be found using the formula $P = 2(\ell + w)$. Substituting $2x + 1$ for length and $x + 1$ for width, $P = 2(2x + 1 + x + 1) = 2(3x + 2) = 6x + 4$.

44c. Sample answer: For $x = 3$, the length is 7 units, the width is 4 units, the perimeter is 22 units, and the area is 28 units^2. The value of x must be chosen so that the length, width, perimeter, and area are all positive. The expressions $2x + 1$, $x + 1$, $6x + 4$, and $2x^2 + 3x + 1$ will all be positive only if $x > -\frac{1}{2}$. However, because x represents the side length of a tile, $x > 0$.

47a. Area of sidewalk = area of larger circle − area of smaller circle, so $384\pi = \pi(r + 12)^2 - \pi r^2 = \pi(r^2 + 24r + 144) - \pi r^2 = 24\pi r + 144\pi$. $24\pi r = 240\pi$; $r = 10$; the radius of the smaller circle is 10 feet, and the radius of the larger circle is $r + 12 = 10 + 12$ or 22 feet.

47b.

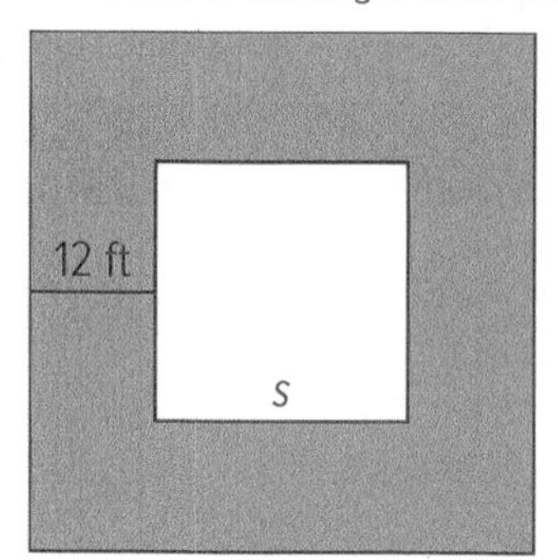

48. Yes; sample answer: For $f(x)$ and $g(x)$ to be polynomials, they must have real coefficients and whole-number exponents. Real numbers and whole numbers are closed under addition and multiplication. So, when $f(x)$ and $g(x)$ are multiplied, the coefficients of each term are multiplied, yielding new real coefficients, and the exponents of each term are added, producing new whole number exponents. By the definition of polynomials, the product of polynomials is a polynomial because the coefficients are real and the exponents are whole numbers.

Lesson 4-4

Dividing Polynomials

A.APR.6

LESSON GOAL

Students divide polynomials by using long division and synthetic division.

1 LAUNCH

Launch the lesson with a Warm Up and an introduction.

2 EXPLORE AND DEVELOP

Explore: Using Algebra Tiles to Divide Polynomials

Develop:

Dividing Polynomials by Using Long Division

- Divide a Polynomial by a Monomial
- Divide a Polynomial by a Binomial
- Find a Quotient with a Remainder

Dividing Polynomials by Using Synthetic Division

- Use Synthetic Division
- Divisor with a Coefficient Other Than 1

You may want your students to complete the **Checks** online.

3 REFLECT AND PRACTICE

Exit Ticket

Practice

DIFFERENTIATE

View reports of student progress on the **Checks** after each example.

Resources	AL	OL	BL	ELL
Remediation: Multiplying a Polynomial by a Monomial	●	●		●
Extension: Oblique Asymptotes		●	●	●

Language Development Handbook

Assign page 28 of the *Language Development Handbook* to help your students build mathematical language related to dividing polynomials by using long division and synthetic division.

ELL You can use the tips and suggestions on page T28 of the handbook to support students who are building English proficiency.

Suggested Pacing

90 min	0.5 day
45 min	1 day

Focus

Domain: Algebra

Standards for Mathematical Content:

A.APR.6 Rewrite simple rational expressions in different forms; write $a(x)/b(x)$ in the form $q(x) + r(x)/b(x)$, where $a(x)$, $b(x)$, $q(x)$, and $r(x)$ are polynomials with the degree of $r(x)$ less than the degree of $b(x)$, using inspection, long division, or, for the more complicated examples, a computer algebra system.

Standards for Mathematical Practice:

3 Construct viable arguments and critique the reasoning of others.

8 Look for and expression regularity in repeated reasoning.

Coherence

Vertical Alignment

Previous
Students performed operations on polynomial expressions.
A.APR.1 (Algebra 1, Algebra 2)

Now
Students divide polynomials by using long division and synthetic division.
A.APR.6

Next
Students will expand powers of binomials.
A.APR.5

Rigor

The Three Pillars of Rigor

1 CONCEPTUAL UNDERSTANDING	2 FLUENCY	3 APPLICATION

Conceptual Bridge In this lesson, students extend their understanding of multiplying polynomials to build fluency with dividing polynomials. They apply their understanding of dividing polynomials by solving real-world problems.

Mathematical Background

Synthetic division can be used to divide a polynomial by a binomial. The terms in both the divisor and dividend must be in descending order of power, and a coefficient of 0 must be included for any term that is missing.

Interactive Presentation

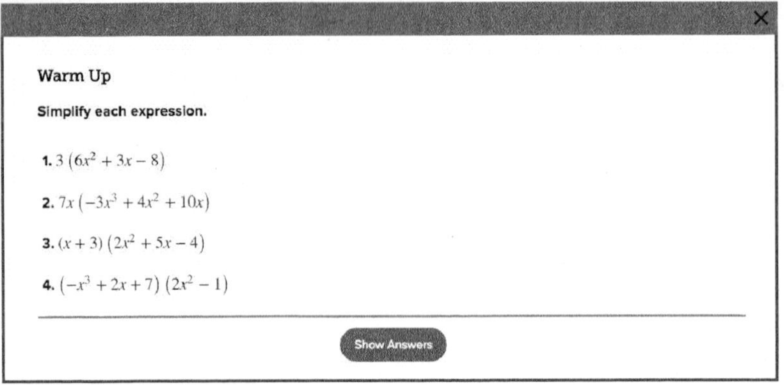

Warm Up

Warm Up

Prerequisite Skills

The Warm Up exercises address the following prerequisite skill for this lesson:

- Multiplying polynomials

Answers:

1. $18x^2 + 9x - 24$
2. $-21x^4 + 28x^3 + 70x^2$
3. $2x^3 + 11x^2 + 11x - 12$
4. $-2x^5 + 5x^3 + 14x^2 - 2x - 7$

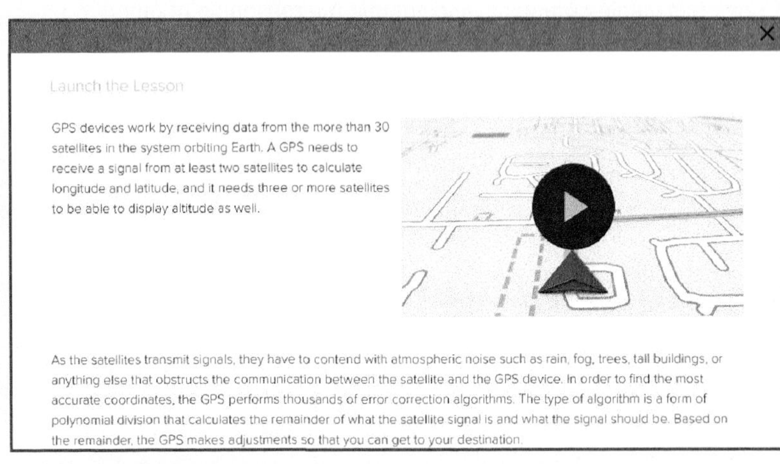

Launch the Lesson

Launch the Lesson

 **Teaching the Mathematical Practices**

4 Apply Mathematics Students see how division of polynomials is applicable to solving a real-world problem.

Go Online to find additional teaching notes and questions to promote classroom discourse.

Today's Standards

Tell students that they will be addressing these content and practice standards in this lesson. You may wish to have a student volunteer read aloud *How can I meet this standard?* and *How can I use these practices?*, and connect these to the standards.

See the Interactive Presentation for I Can statements that align with the standards covered in this lesson.

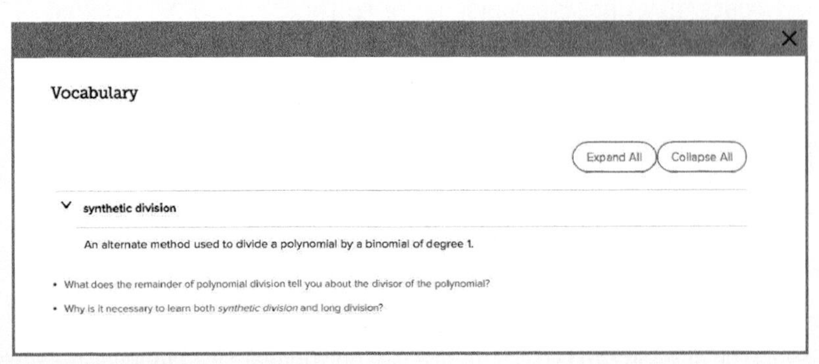

Today's Vocabulary

Today's Vocabulary

Tell students that they will be using this vocabulary term in this lesson. You can expand the row if you wish to share the definition. Then, discuss the questions below with the class.

Explore Using Algebra Tiles to Divide Polynomials

Objective

Students use algebra tiles to explore dividing a trinomial by a binomial.

Teaching the Mathematical Practices

5 Use Mathematical Tools Point out that to complete the Explore, students will need to use the Algebra Tiles eTool. Work with students to explore and deepen their understanding of polynomial division.

Ideas for Use

Recommended Use Present the Inquiry Question, or have a student volunteer read it aloud. Have students work in pairs to complete the Explore activity on their devices. Pairs should discuss each of the questions. Monitor student progress during the activity. Upon completion of the Explore activity, have student volunteers share their responses to the Inquiry Question.

What if my students don't have devices? You may choose to project the activity on a whiteboard. A printable worksheet for each Explore is available online. You may choose to print the worksheet so that individuals or pairs of students can use it to record their observations.

Summary of the Activity

Students will complete guiding exercises throughout the Explore activity. Students will use algebra tiles to model polynomials of degree 0, 1, or 2. They will use these tiles to find quotients and remainders from division of polynomials. Then, students will answer the Inquiry Question.

(continued on the next page)

Interactive Presentation

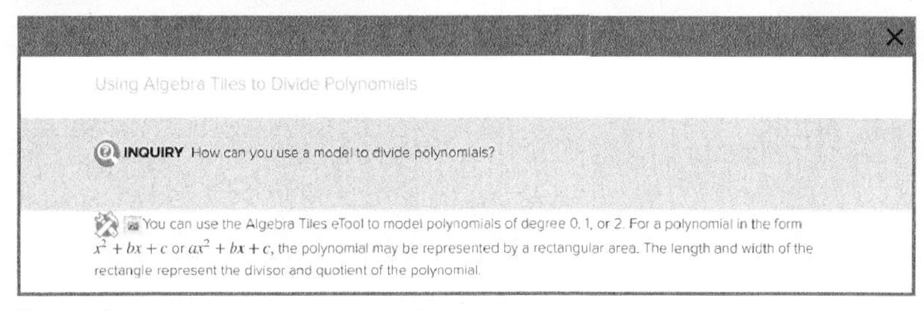

Explore

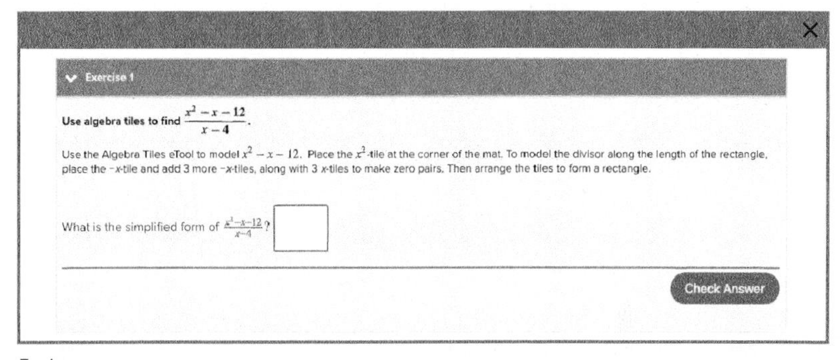

Explore

TYPE

Students simpify an expression to show they understand how to use algebra tiles to divide polynomials.

Interactive Presentation

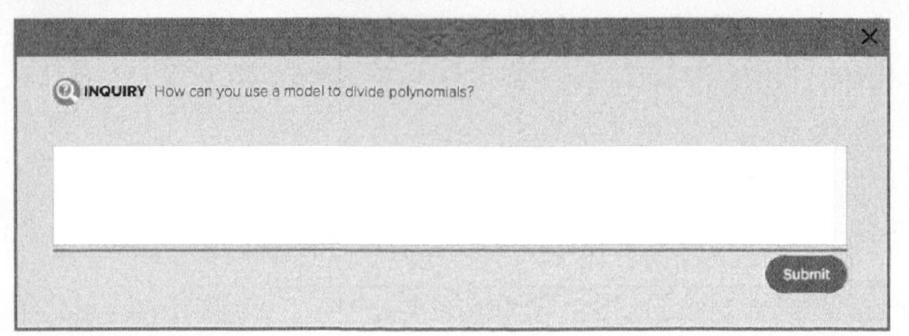

Explore

Students respond to the Inquiry Question and can view a sample answer.

Explore Using Algebra Tiles to Divide Polynomials (*continued*)

Teaching the Mathematical Practices

6 Communicate Precisely Encourage students to routinely write or explain their solution methods. Point out that they should use clear definitions and mathematical language when they discuss their solutions to the Explore activity with others.

Questions

Have students complete the Explore activity.

Ask:

- Why do you need to add x-tiles to model $\frac{x^2 - x - 12}{x - 4}$? Sample answer: The length of the rectangle is $x - 4$, so the rectangle on the mat needs to have one x^2-tile and 4 negative x-tiles. You have to add 3 positive and 3 negative x-tiles to make zero pairs and not change the value of the polynomial.
- Why did $\frac{x^2 + 2x - 9}{x - 3}$ have a remainder? What does this mean? Sample answer: The tiles couldn't be arranged into a single rectangle with length of $x - 3$. The pieces left created the remainder and shows that the polynomial does not have $x - 3$ as a factor.

Inquiry

How can you use a model to divide polynomials? Sample answer: Some polynomials can be modeled by the area of a rectangle. The length and width represent the divisor and quotient.

Go Online to find additional teaching notes and sample answers for the guiding exercises.

1 CONCEPTUAL UNDERSTANDING | 2 FLUENCY | 3 APPLICATION

Learn Dividing Polynomials by Using Long Division

Objective

Students divide polynomials by using long division.

 Teaching the Mathematical Practices

3 Make Conjectures In the Talk About It! feature, students will make a conjecture about a quotient with a remainder of zero and then build a logical progression of statements to validate the conjecture. Once students have made their conjectures, guide students to validate them.

Important to Know

The degree of the quotient when a polynomial of degree n is divided by a polynomial of degree m is $n - m$.

Common Misconception

Students may think that they can treat each term in a polynomial as an entity that can 'cancel out' with other terms incorrectly. Review how to simplify polynomials using Example 1.

Essential Question Follow-Up

Students have begun to divide polynomials by using long division.

Ask:

How do you know you have your remainder and that you are done dividing? Sample answer: When the degree of the term you are dividing into is less than the degree of the divisor.

DIFFERENTIATE

Reteaching Activity 

IF some students have trouble keeping their concentration throughout the sequence of steps required in long division,

THEN encourage them to compare intermediate results with a partner so they can ask questions and catch errors before completing the entire problem.

Go Online

- Find additional teaching notes.
- View performance reports of the Checks.
- Assign or present an Extra Example.

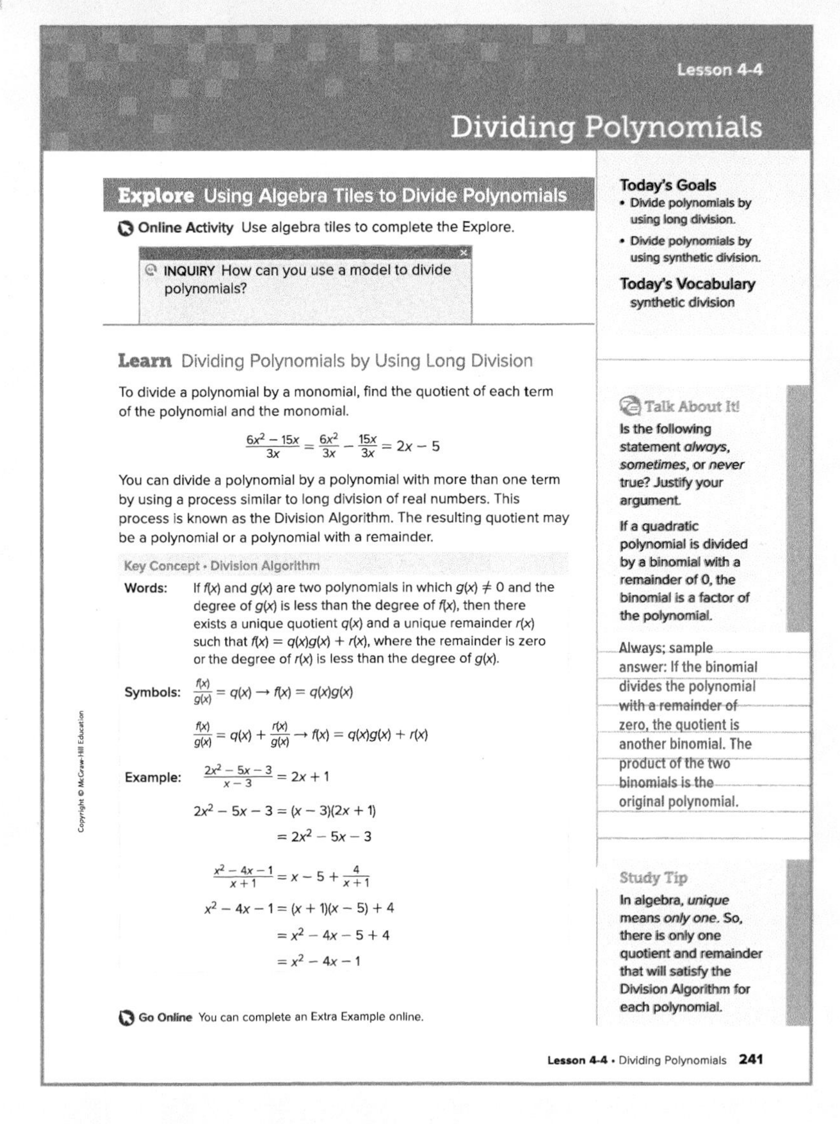

Lesson 4-4

Dividing Polynomials

Explore Using Algebra Tiles to Divide Polynomials

Online Activity Use algebra tiles to complete the Explore.

INQUIRY How can you use a model to divide polynomials?

Today's Goals
- Divide polynomials by using long division.
- Divide polynomials by using synthetic division.

Today's Vocabulary
synthetic division

Learn Dividing Polynomials by Using Long Division

To divide a polynomial by a monomial, find the quotient of each term of the polynomial and the monomial.

$$\frac{6x^2 - 15x}{3x} = \frac{6x^2}{3x} - \frac{15x}{3x} = 2x - 5$$

You can divide a polynomial by a polynomial with more than one term by using a process similar to long division of real numbers. This process is known as the Division Algorithm. The resulting quotient may be a polynomial or a polynomial with a remainder.

Key Concept • Division Algorithm

Words: If $f(x)$ and $g(x)$ are two polynomials in which $g(x) \neq 0$ and the degree of $g(x)$ is less than the degree of $f(x)$, then there exists a unique quotient $q(x)$ and a unique remainder $r(x)$ such that $f(x) = q(x)g(x) + r(x)$, where the remainder is zero or the degree of $r(x)$ is less than the degree of $g(x)$.

Symbols: $\frac{f(x)}{g(x)} = q(x) \rightarrow f(x) = q(x)g(x)$

$\frac{f(x)}{g(x)} = q(x) + \frac{r(x)}{g(x)} \rightarrow f(x) = q(x)g(x) + r(x)$

Example: $\frac{2x^2 - 5x - 3}{x - 3} = 2x + 1$

$2x^2 - 5x - 3 = (x - 3)(2x + 1)$
$= 2x^2 - 5x - 3$

$\frac{x^2 - 4x - 1}{x + 1} = x - 5 + \frac{4}{x + 1}$

$x^2 - 4x - 1 = (x + 1)(x - 5) + 4$
$= x^2 - 4x - 5 + 4$
$= x^2 - 4x - 1$

Go Online You can complete an Extra Example online.

Talk About It!
Is the following statement *always*, *sometimes*, or *never* true? Justify your argument.
If a quadratic polynomial is divided by a binomial with a remainder of 0, the binomial is a factor of the polynomial.

Always; sample answer: If the binomial divides the polynomial with a remainder of zero, the quotient is another binomial. The product of the two binomials is the original polynomial.

Study Tip
In algebra, *unique* means *only one*. So, there is only one quotient and remainder that will satisfy the Division Algorithm for each polynomial.

Lesson 4-4 • Dividing Polynomials 241

Interactive Presentation

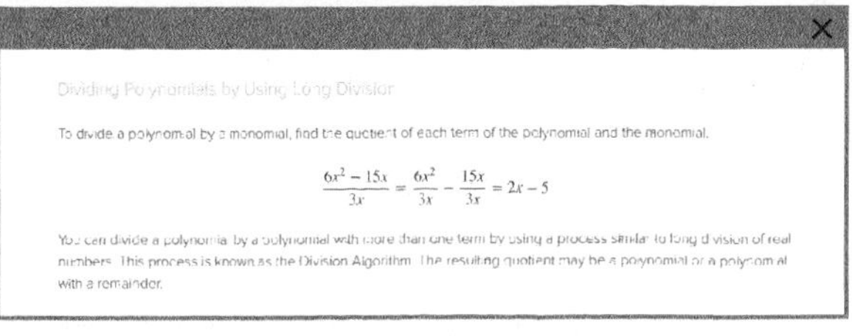

Learn

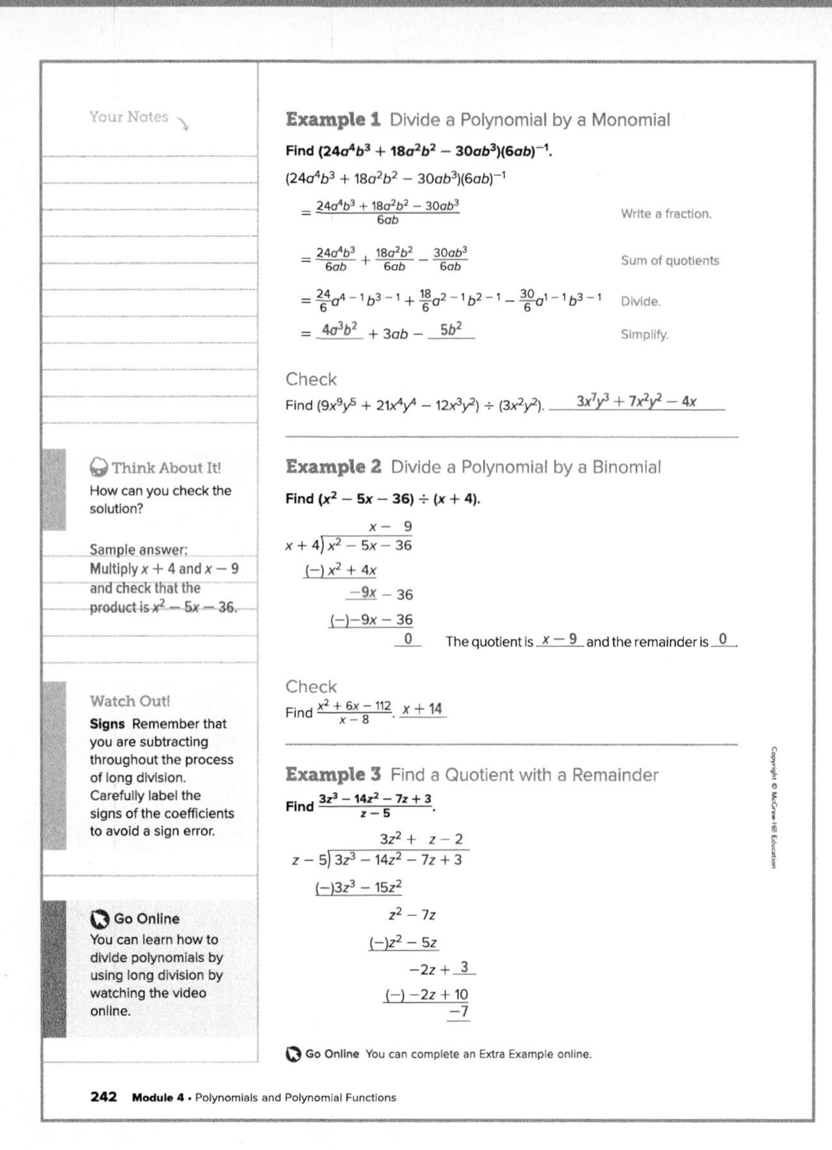

Your Notes

Example 1 Divide a Polynomial by a Monomial

Find $(24a^4b^3 + 18a^2b^2 - 30ab^3)(6ab)^{-1}$.

$(24a^4b^3 + 18a^2b^2 - 30ab^3)(6ab)^{-1}$

$= \frac{24a^4b^3 + 18a^2b^2 - 30ab^3}{6ab}$ — Write a fraction.

$= \frac{24a^4b^3}{6ab} + \frac{18a^2b^2}{6ab} - \frac{30ab^3}{6ab}$ — Sum of quotients

$= \frac{24}{6}a^{4-1}b^{3-1} + \frac{18}{6}a^{2-1}b^{2-1} - \frac{30}{6}a^{1-1}b^{3-1}$ — Divide.

$= 4a^3b^2 + 3ab - 5b^2$ — Simplify.

Check

Find $(9x^9y^5 + 21x^4y^4 - 12x^3y^2) \div (3x^2y^2)$. $3x^7y^3 + 7x^2y^2 - 4x$

Think About It!
How can you check the solution?

Sample answer: Multiply $x + 4$ and $x - 9$ and check that the product is $x^2 - 5x - 36$.

Example 2 Divide a Polynomial by a Binomial

Find $(x^2 - 5x - 36) \div (x + 4)$.

$x - 9$
$x + 4 \overline{) x^2 - 5x - 36}$
$(-) x^2 + 4x$
$-9x - 36$
$(-) -9x - 36$
0

The quotient is $x - 9$ and the remainder is 0.

Check

Find $\frac{x^2 + 6x - 112}{x - 8}$. $x + 14$

Watch Out!
Signs Remember that you are subtracting throughout the process of long division. Carefully label the signs of the coefficients to avoid a sign error.

Example 3 Find a Quotient with a Remainder

Find $\frac{3z^3 - 14z^2 - 7z + 3}{z - 5}$.

$3z^2 + z - 2$
$z - 5 \overline{) 3z^3 - 14z^2 - 7z + 3}$
$(-) 3z^3 - 15z^2$
$z^2 - 7z$
$(-) z^2 - 5z$
$-2z + 3$
$(-) -2z + 10$
-7

Go Online
You can learn how to divide polynomials by using long division by watching the video online.

Go Online You can complete an Extra Example online.

242 **Module 4** • Polynomials and Polynomial Functions

Interactive Presentation

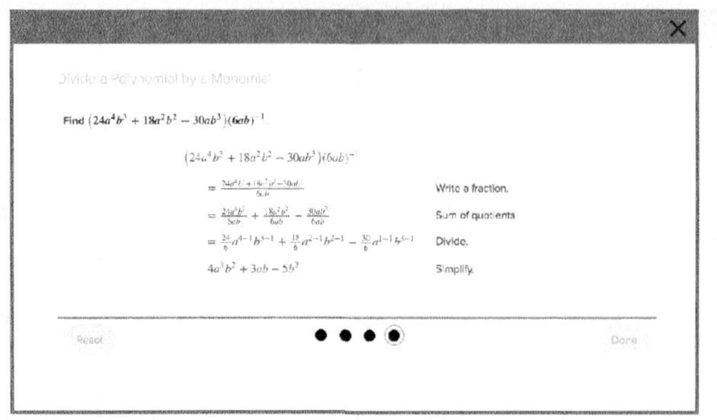

Example 1

TYPE

Students answer questions to show they understand how to divide polynomials.

CHECK

Students complete the Check online to determine whether they are ready to move on.

Example 1 Divide a Polynomial by a Monomial

Teaching the Mathematical Practices

7 Interpret Complicated Expressions Mathematically proficient students can see complicated expressions as single objects or as being composed of several objects. In Example 1, help students to see the expression as being composed of several simpler expressions.

Questions for Mathematical Discourse

AL Why is the fraction rewritten as three fractions? We want to divide each term by $6ab$.

OL What is an equivalent expression for $\frac{1}{x}$? x^{-1}

BL What is another method you could use to solve this example? Sample answer: Distribute the $(6ab)^{-1}$ to each term as multiplication and then add the exponents. Adding the negative exponent is equivalent to subtracting the positive exponent.

Example 2 Divide a Polynomial by a Binomial

Teaching the Mathematical Practices

1 Check Answers Mathematically proficient students continually ask themselves, "Does this make sense?" In the Think About It! feature, students need to explain how to check their answer.

Questions for Mathematical Discourse

AL How do you determine each term of the quotient? Look for the term that, when multiplied with the divisor, will result in canceling the first term of the remaining dividend.

OL What does it mean for the remainder to be 0? $x + 4$ divides evenly into $x^2 - 5x - 36$ and $x + 4$ is a factor of $x^2 - 5x - 36$.

BL How could you tell the remainder would be zero before doing the division? $-\frac{36}{4} = -9$, and $-9 + 4 = -5$, so $x + 4$ is a factor of the dividend, along with $x - 9$.

Example 3 Find a Quotient with a Remainder

Teaching the Mathematical Practices

7 Use Structure Help students to use the structure of the polynomials to find the quotient by using long division.

Questions for Mathematical Discourse

AL Why are there parentheses around the subtraction sign in the example? The parentheses remind us that we are subtracting the entire expression, not just the first term.

OL What statement can you make about the relationship between the divisor and the dividend for this example? The divisor is not a factor of the dividend.

BL Can you write $\frac{7}{(x-5)}$ as $\frac{7}{x} - \frac{7}{5}$? Explain. No; sample answer: Consider $\frac{7}{(3+5)} = \frac{7}{8}$, and $\frac{7}{3} + \frac{7}{5} = \frac{56}{15}$. $\frac{7}{8} \neq \frac{56}{15}$.

Learn Dividing Polynomials by Using Synthetic Division

Objective

Students divide polynomials by using synthetic division.

Teaching the Mathematical Practices

1 Understand Different Approaches Work with students to look at the alternate method for dividing polynomials. Ask students to compare and contrast long division and synthetic division.

Important to Know

You can use synthetic division to divide a polynomial by a binomial. The terms in both the divisor and dividend must be in descending order of power, and a coefficient of 0 must be included for any term that is missing.

Things to Remember

Remind students to *add* terms when performing synthetic division.

Essential Question Follow-Up

Students have begun to divide polynomials by using synthetic division.

Ask:

In what form must the divisor be to use synthetic division? What if it is not? Sample answer: The divisor needs to be in $x - a$ form. If the divisor is not in the form $x - a$, divide every term in both the divisor and dividend by the coefficient of the first term of the divisor and then use synthetic division.

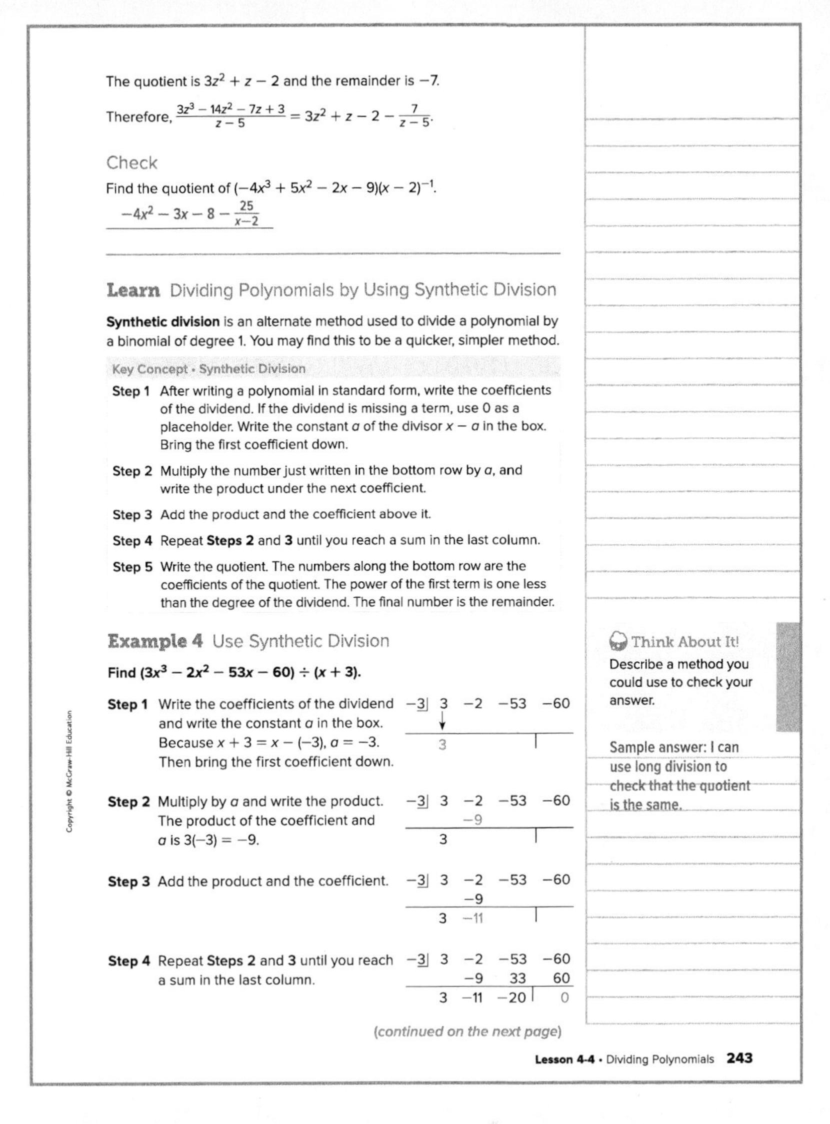

The quotient is $3z^2 + z - 2$ and the remainder is -7.

Therefore, $\frac{3z^3 - 14z^2 - 7z + 3}{z - 5} = 3z^2 + z - 2 - \frac{7}{z-5}$.

Check

Find the quotient of $(-4x^3 + 5x^2 - 2x - 9)(x - 2)^{-1}$.

$-4x^2 - 3x - 8 - \frac{25}{x-2}$

Learn Dividing Polynomials by Using Synthetic Division

Synthetic division is an alternate method used to divide a polynomial by a binomial of degree 1. You may find this to be a quicker, simpler method.

Key Concept • Synthetic Division

Step 1 After writing a polynomial in standard form, write the coefficients of the dividend. If the dividend is missing a term, use 0 as a placeholder. Write the constant a of the divisor $x - a$ in the box. Bring the first coefficient down.

Step 2 Multiply the number just written in the bottom row by a, and write the product under the next coefficient.

Step 3 Add the product and the coefficient above it.

Step 4 Repeat **Steps 2** and **3** until you reach a sum in the last column.

Step 5 Write the quotient. The numbers along the bottom row are the coefficients of the quotient. The power of the first term is one less than the degree of the dividend. The final number is the remainder.

Example 4 Use Synthetic Division

Find $(3x^3 - 2x^2 - 53x - 60) \div (x + 3)$.

Step 1 Write the coefficients of the dividend and write the constant a in the box. Because $x + 3 = x - (-3)$, $a = -3$. Then bring the first coefficient down.

-3	3	-2	-53	-60
	3			

Step 2 Multiply by a and write the product. The product of the coefficient and a is $3(-3) = -9$.

-3	3	-2	-53	-60
		-9		
	3			

Step 3 Add the product and the coefficient.

-3	3	-2	-53	-60
		-9		
	3	-11		

Step 4 Repeat **Steps 2** and **3** until you reach a sum in the last column.

-3	3	-2	-53	-60
		-9	33	60
	3	-11	-20	0

(continued on the next page)

Think About It!

Describe a method you could use to check your answer.

Sample answer: I can use long division to check that the quotient is the same.

Example 4 Use Synthetic Division

Teaching the Mathematical Practices

8 Look for a Pattern Help students to see how the calculations used in synthetic division are repeated in Example 4.

Questions for Mathematical Discourse

AL Why is $a = -3$? Synthetic division is defined for $(x - a)$, and $(x + 3) = (x - (-3))$.

OL How do you determine the degree of the terms of the quotient? The degree of the dividend minus the degree of the divisor yields the degree of the quotient, and subsequent terms go in descending order.

BL Why is this called synthetic division? Sample answer: It is a simplified process to achieve the same results as long division, derived from the pattern of operations performed in long division.

Interactive Presentation

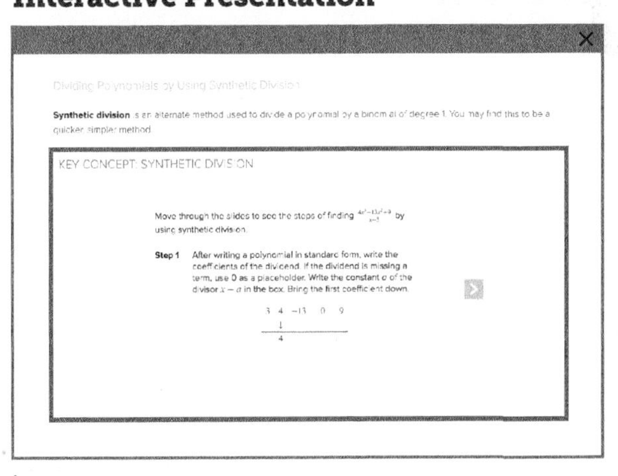

Learn

SWIPE

Students move through the steps to see how to divide polynomials using synthetic division.

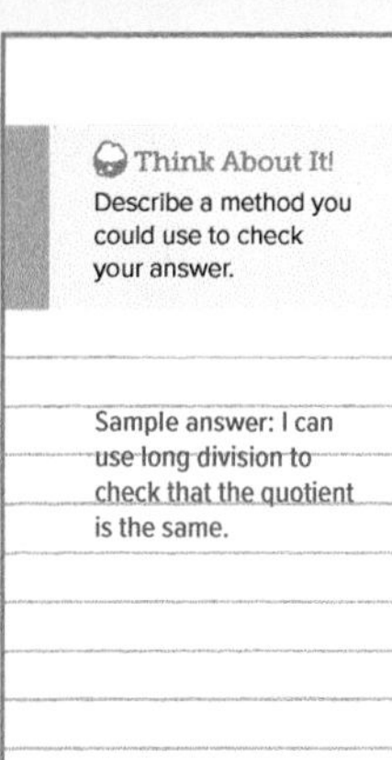

Think About It!
Describe a method you could use to check your answer.

Sample answer: I can use long division to check that the quotient is the same.

Step 5 Write the quotient. Because the degree of the dividend is 3 and the degree of the divisor is 1, the degree of the quotient is 2. The final sum in the synthetic division is 0, so the remainder is 0.

The quotient is $3x^2 - 11x - 20$.

Example 5 Divisor with a Coefficient Other Than 1

Find $\frac{4x^4 - 37x^2 + 4x + 9}{2x - 1}$.

To use synthetic division, the lead coefficient of the divisor must be 1.

$\frac{(4x^4 - 37x^2 + 4x + 9) \div 2}{(2x - 1) \div 2}$ Divide the numerator and denominator by 2.

$= \frac{2x^4 - \frac{37}{2}x^2 + 2x + \frac{9}{2}}{x - \frac{1}{2}}$ Simplify the numerator and denominator.

$x - a = x - \frac{1}{2}$, so $a = \frac{1}{2}$.

Complete the synthetic division.

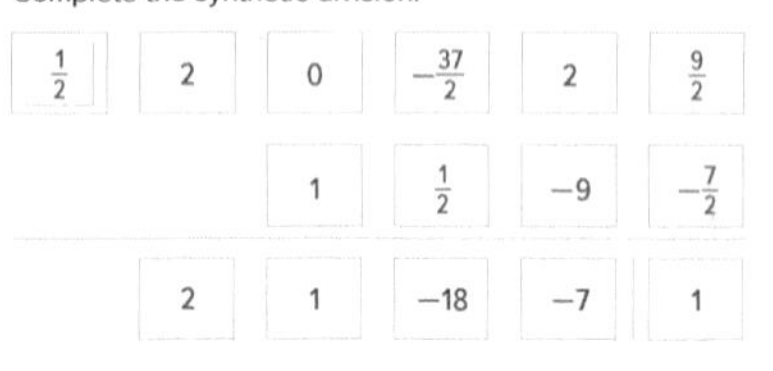

$\frac{1}{2}$	2	0	$-\frac{37}{2}$	2	$\frac{9}{2}$
		1	$\frac{1}{2}$	-9	$-\frac{7}{2}$
	2	1	-18	-7	1

Watch Out!
Missing terms Add placeholders for terms that are missing from the polynomial. In this case, there are 0 x^3-terms.

The resulting expression is $2x^3 + x^2 - 18x - 7 + \frac{1}{x - \frac{1}{2}}$. Now simplify the fraction.

$\frac{1}{x - \frac{1}{2}} = \frac{(1)2}{\left(x - \frac{1}{2}\right) \cdot 2}$ Multiply the numerator and denominator by 2.

$= \frac{2}{2x - 1}$ Simplify.

The solution is $2x^3 + x^2 - 18x - 7 + \frac{2}{2x - 1}$.

You can check your answer by using long division.

Check

Find $(4x^4 + 3x^3 - 12x^2 - x + 6)(4x + 3)^{-1}$.

$x^3 - 3x + 2$

Go Online You can complete an Extra Example online.

244 Module 4 • Polynomials and Polynomial Functions

Interactive Presentation

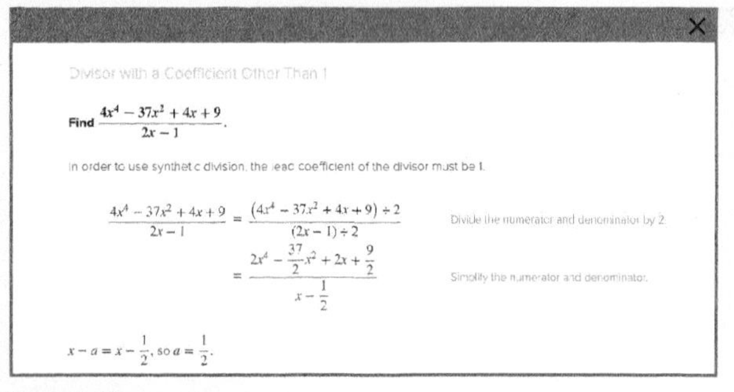

Example 5

TYPE

Students enter numbers to complete a synthetic division.

CHECK

Students complete the Check online to determine whether they are ready to move on.

Example 5 Divisor with a Coefficient Other Than 1

Teaching the Mathematical Practices

1 Seek Information Mathematically proficient students must be able to transform algebraic expression to reach solutions. In Example 5, students must transform the numerator and denominator to apply synthetic division.

Questions for Mathematical Discourse

AL What number could you multiply each term by that would be the same as dividing each term by 2? $\frac{1}{2}$

OL What type of polynomial must the divisor be to use synthetic division? linear

BL Why do you divide both the numerator and the denominator by 2? Sample answer: You divide both so that the value of the expression does not change. Dividing the numerator and denominator by the same value is the same as dividing by 1.

Things to Remember

Remind students to divide all terms in the numerator and denominator when turning the divisor to $x - a$ form.

Exit Ticket

Recommended Use

At the end of class, go online to display the Exit Ticket prompt and ask students to respond using a separate piece of paper. Have students hand you their responses as they leave the room.

Alternate Use

At the end of class, go online to display the Exit Ticket prompt and ask students to respond verbally or by using a mini-whiteboard. Have students hold up their whiteboards so that you can see all student responses. Tap to reveal the answer when most or all students have completed the Exit Ticket.

Practice and Homework

Suggested Assignments

Use the table below to select appropriate exercises.

DOK	Topic	Exercises
1, 2	exercises that mirror the examples	1–16
2	exercises that use a variety of skills from this lesson	17–34
2	exercises that extend concepts learned in this lesson to new contexts	35–37
3	exercises that emphasize higher-order and critical thinking skills	38–43

ASSESS AND DIFFERENTIATE

Use the data from the **Checks** to determine whether to provide resources for extension, remediation, or intervention.

IF students score 90% or more on the Checks, **THEN** assign: OL BL

- Practice Exercises 1–37 odd, 38–43
- Extension: Oblique Asymptotes
- ALEKS®

IF students score 66%–89% on the Checks, **THEN** assign: AL OL

- Practice Exercises 1–37 odd
- Remediation, Review Resources: Multipying a Polynomial by a Monomial
- Personal Tutors
- Extra Examples 1–5
- ALEKS® Multiplying Polynomials

IF students score 65% or less on the Checks, **THEN** assign: AL

- Practice Exercises 1–15 odd
- Remediation, Review Resources: Multipying a Polynomial by a Monomial
- *Quick Review Math Handbook*: Dividing Polynomials
- ALEKS® Multiplying Polynomials

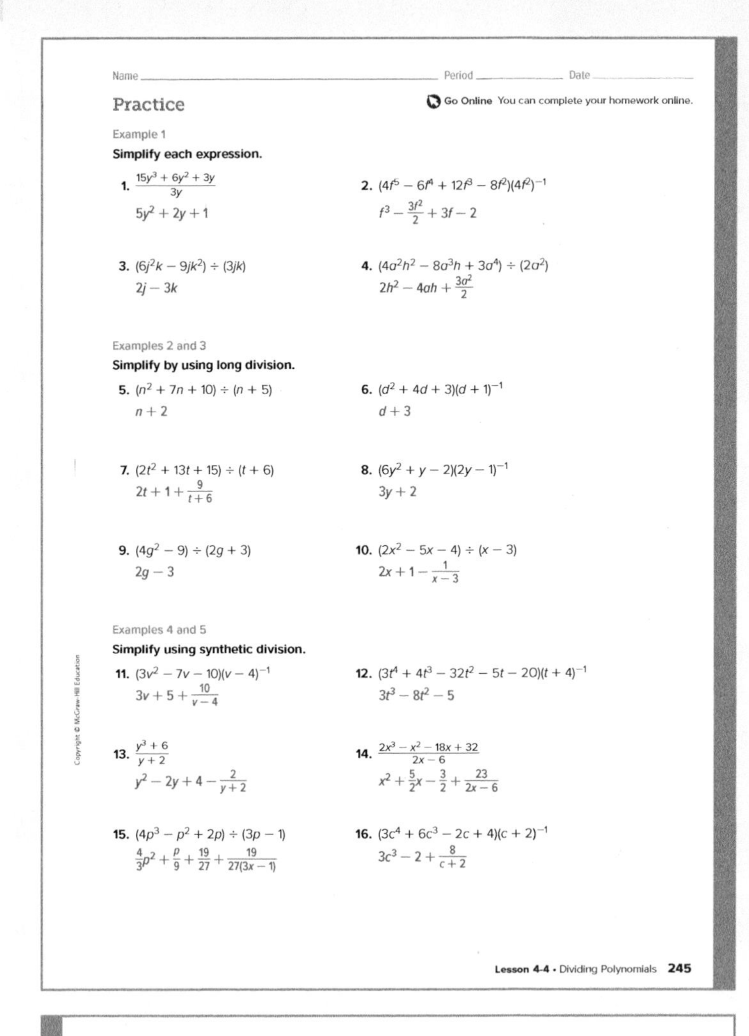

Name ______ Period ______ Date ______

Practice

Go Online You can complete your homework online.

Example 1

Simplify each expression.

1. $\frac{15y^3 + 6y^2 + 3y}{3y}$
   $5y^2 + 2y + 1$
2. $(4f^5 - 6f^4 + 12f^3 - 8f^2)(4f^2)^{-1}$
   $f^3 - \frac{3f^2}{2} + 3f - 2$
3. $(6j^2k - 9jk^2) \div (3jk)$
   $2j - 3k$
4. $(4a^2h^2 - 8a^3h + 3a^4) \div (2a^2)$
   $2h^2 - 4ah + \frac{3a^2}{2}$

Examples 2 and 3

Simplify by using long division.

5. $(n^2 + 7n + 10) \div (n + 5)$
   $n + 2$
6. $(d^2 + 4d + 3)(d + 1)^{-1}$
   $d + 3$
7. $(2t^2 + 13t + 15) \div (t + 6)$
   $2t + 1 + \frac{9}{t+6}$
8. $(6y^2 + y - 2)(2y - 1)^{-1}$
   $3y + 2$
9. $(4g^2 - 9) \div (2g + 3)$
   $2g - 3$
10. $(2x^2 - 5x - 4) \div (x - 3)$
    $2x + 1 - \frac{1}{x-3}$

Examples 4 and 5

Simplify using synthetic division.

11. $(3v^2 - 7v - 10)(v - 4)^{-1}$
    $3v + 5 + \frac{10}{v-4}$
12. $(3t^4 + 4t^3 - 32t^2 - 5t - 20)(t + 4)^{-1}$
    $3t^3 - 8t^2 - 5$
13. $\frac{y^3 + 6}{y + 2}$
    $y^2 - 2y + 4 - \frac{2}{y+2}$
14. $\frac{2x^3 - x^2 - 18x + 32}{2x - 6}$
    $x^2 + \frac{5}{2}x - \frac{3}{2} + \frac{23}{2x-6}$
15. $(4p^3 - p^2 + 2p) \div (3p - 1)$
    $\frac{4}{3}p^2 + \frac{p}{9} + \frac{19}{27} + \frac{19}{27(3x-1)}$
16. $(3c^4 + 6c^3 - 2c + 4)(c + 2)^{-1}$
    $3c^3 - 2 + \frac{8}{c+2}$

Copyright © McGraw-Hill Education

Lesson 4-4 • Dividing Polynomials 245

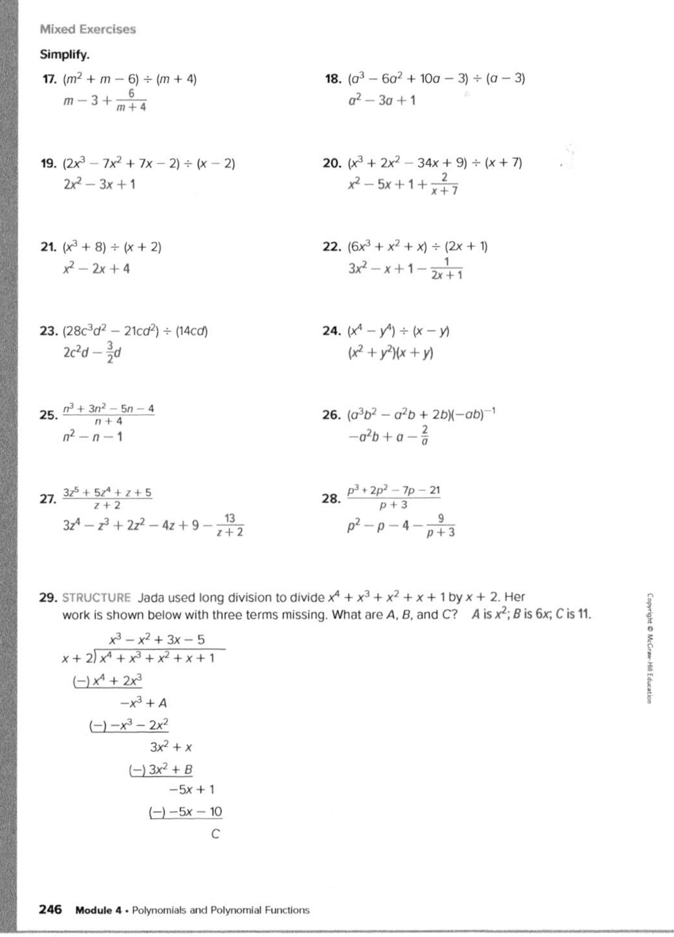

Mixed Exercises

Simplify.

17. $(m^2 + m - 6) \div (m + 4)$
    $m - 3 + \frac{6}{m+4}$
18. $(a^3 - 6a^2 + 10a - 3) \div (a - 3)$
    $a^2 - 3a + 1$
19. $(2x^3 - 7x^2 + 7x - 2) \div (x - 2)$
    $2x^2 - 3x + 1$
20. $(x^3 + 2x^2 - 34x + 9) \div (x + 7)$
    $x^2 - 5x + 1 + \frac{2}{x+7}$
21. $(x^3 + 8) \div (x + 2)$
    $x^2 - 2x + 4$
22. $(6x^3 + x^2 + x) \div (2x + 1)$
    $3x^2 - x + 1 - \frac{1}{2x+1}$
23. $(28c^3d^2 - 21cd^2) \div (14cd)$
    $2c^2d - \frac{3}{2}d$
24. $(x^4 - y^4) \div (x - y)$
    $(x^2 + y^2)(x + y)$
25. $\frac{n^3 + 3n^2 - 5n - 4}{n + 4}$
    $n^2 - n - 1$
26. $(a^3b^2 - a^2b + 2b)(-ab)^{-1}$
    $-a^2b + a - \frac{2}{a}$
27. $\frac{3z^5 + 5z^4 + z + 5}{z + 2}$
    $3z^4 - z^3 + 2z^2 - 4z + 9 - \frac{13}{z+2}$
28. $\frac{p^3 + 2p^2 - 7p - 21}{p + 3}$
    $p^2 - p - 4 - \frac{9}{p+3}$
29. STRUCTURE Jada used long division to divide $x^4 + x^3 + x^2 + x + 1$ by $x + 2$. Her work is shown below with three terms missing. What are A, B, and C? A is x^2; B is $6x$; C is 11.

$$
\begin{array}{r}
x^3 - x^2 + 3x - 5 \\
x + 2 \overline{) x^4 + x^3 + x^2 + x + 1} \\
\underline{(-)\ x^4 + 2x^3} \\
-x^3 + A \\
\underline{(-)\ -x^3 - 2x^2} \\
3x^2 + x \\
\underline{(-)\ 3x^2 + B} \\
-5x + 1 \\
\underline{(-)\ -5x - 10} \\
C
\end{array}
$$

Copyright © McGraw-Hill Education

246 Module 4 • Polynomials and Polynomial Functions

Name ______________ Period ________ Date ________

30. AVERAGES Bena has a list of $n + 1$ numbers and she needs to find their average. Two of the numbers are n^3 and 2. Each of the other $n - 1$ numbers are all equal to 1. Find the average of these numbers.
$n^2 - n + 2 - \frac{1}{n+1}$

31. VOLUME The volume of a cylinder is $\pi(x^3 + 32x^2 - 304x + 640)$. If the height of the cylinder is $x + 40$ feet, find the area of its base in terms of x and π.
$\pi(x^2 - 8x + 16)$

32. REASONING Rewrite $\frac{6x^4 + 2x^3 - 16x^2 + 24x + 32}{2x + 4}$ as $q(x) + \frac{r(x)}{d(x)}$ using long division. What does the remainder indicate in this problem?
$\frac{6x^4 + 2x^3 - 16x^2 + 24x + 32}{2x + 4} = 3x^3 - 5x^2 + 2x + 8$, remainder 0. Because the remainder is 0, $2x + 4$ is a factor of $6x^4 + 2x^3 - 16x^2 + 24x + 32$.

33. CONSTRUCT ARGUMENTS Determine whether you have enough information to fill in the missing pieces of the long division exercise shown. Justify your response. See margin.

$$\begin{array}{r} 3x - \boxed{1} \\ \boxed{3x+1}\overline{)\,9x^2 + \boxed{0x + 5}} \\ \underline{9x^2 + 3x} \\ -3x + 5 \end{array}$$

34. REGULARITY Rewrite $\frac{2x^5 - 7x^4 - 15x^3 + 2x^2 + 3x + 6}{2x + 3}$ as $q(x) + \frac{r(x)}{g(x)}$ using long division.

a. Identify $q(x)$, $r(x)$, and $g(x)$. $q(x) = x^4 - 5x^3 + x$; $r(x) = 6$; $g(x) = 2x + 3$

b. How can you check your work using the expressions of $q(x)$, $g(x)$, and $r(x)$?
Multiply $q(x)$ and $g(x)$ and then add $r(x)$. The result will equal the original dividend.

35. STRUCTURE When a polynomial is divided by $4x - 6$, the quotient is $2x^2 + x + 1$ and the remainder is -4. What is the dividend, $f(x)$? Explain.
$8x^3 - 8x^2 - 2x - 10$; I multiplied the divisor and the quotient and added the remainder: $(2x^2 + x + 1)(4x - 6) + (-4) = 8x^3 - 8x^2 - 2x - 10$.

36. USE A MODEL Luciano has a square garden. A new garden will have the same width and a length that is 3 feet more than twice the width of the original garden.

x
x
x
2x + 3

a. Define a variable and label each side of the diagrams with an expression for its length.
Let x = a side of the original garden.

b. Write a ratio to represent the percent increase in the area of the garden. Use polynomial division to simplify the expression. See margin.

c. Use your expression from **part b** to determine the percent of increase in area if the original garden was a 12-foot square. Check your answer. See margin.

37. REGULARITY Mariella makes the following claims about the degrees of the polynomials in $\frac{f(x)}{d(x)} = q(x) + \frac{r(x)}{d(x)}$. Do you agree with each claim? Justify your answers and provide examples.

a. The degree of $d(x)$ must be less than the degree of $f(x)$.
No; the degrees of $f(x)$ and $d(x)$ may be equal. For example, $\frac{3x^2 + 6}{x^2} = 3 + \frac{6}{x^2}$.

b. The degree of $r(x)$ must be at least 1 less than the degree of $d(x)$. See margin.

c. The degree of $q(x)$ must be the degree of $f(x)$ minus the degree of $d(x)$. See margin.

Higher-Order Thinking Skills

38. FIND THE ERROR Tomo and Jamal are dividing $2x^3 - 4x^2 + 3x - 1$ by $x - 3$. Tomo claims that the remainder is -100. Jamal claims that the remainder is 26. Is either of them correct? Explain your reasoning.
Sample answer: Jamal; Tomo divided by $x + 3$.

39. PERSEVERE If a polynomial is divided by a binomial and the remainder is 0, what does this tell you about the relationship between the binomial and the polynomial?
The binomial is a factor of the polynomial.

40. ANALYZE What is the relationship between the degrees of the dividend, the divisor, and the quotient in any polynomial division exercise?
Sample answer: The degree of the quotient plus the degree of the divisor equals the degree of the dividend.

41. CREATE Write a quotient of two polynomials for which the remainder is 3.
Sample answer: $\frac{x^2 + 5x + 9}{x + 2}$

42. WRITE Compare and contrast dividing polynomials using long division and using synthetic division. See margin.

43. PERSEVERE Mr. Collins has his class working with bases and polynomials. He wrote on the board that the number 1111 in base B has the value $B^3 + B^2 + B + 1$. The class was then given the following questions to answer.

a. The number 11 in base B has the value $B + 1$. What is 1111 (in base B) divided by 11 (in base B)? $B^2 + 1$

b. The number 111 in base B has the value $B^2 + B + 1$. What is 1111 (in base B) divided by 111 (in base B)? $B + \frac{1}{B^2 + B + 1}$

Answers

33. Yes; sample answer: Because $3x$ times the divisor is $9x^2 + 3x$, the divisor must be $3x + 1$. The second and third terms of the dividend must be $0x + 5$ because the first difference is $-3x + 5$.

36b. The area of the original garden is x^2. The area of the new garden is $2x^2 + 3x$. Percent of increase in the area of the new garden is $\left(\frac{2x^2 + 3x}{x^2}\right) \times 100$. Using long division and simplifying the ratio results in $200 + \frac{300}{x}$.

36c. Substitute 12 for x in the expression $200 + \frac{300}{x}$ and evaluate to determine that the percent of increase is $200 + \frac{300}{12} = 225$, or 225%. Using 12 ft for a side of the original square, the area is 144 sq ft. The area of the new design would be $(2(12) + 3)(12) = 324$ sq ft. The percent of increase is $\frac{324 - 144}{144} = \frac{180}{144} = 1.25$ or 125%.

37b. Yes; if the degree of $r(x)$ is greater than or equal to the degree of $d(x)$, then the expression $\frac{r(x)}{d(x)}$ may be simplified by division. For example, if $\frac{r(x)}{d(x)} = \frac{8x + 1}{x}$, then $8x + 1$ may be divided by x to get $8 + \frac{1}{x}$.

37c. Yes; because $\frac{f(x)}{d(x)} = q(x) + \frac{r(x)}{d(x)}$, the degree of $\frac{f(x)}{d(x)}$ must equal the degree of $q(x) + \frac{r(x)}{d(x)}$. The degree of $r(x)$ is less than the degree of $d(x)$, so the degree of $q(x) + \frac{r(x)}{d(x)}$ equals the degree of $q(x)$. This means the degree of $\frac{f(x)}{d(x)}$ equals the degree of $q(x)$, and the degree of $\frac{f(x)}{d(x)}$ is the degree of $f(x)$ minus the degree of $d(x)$. For example, in $\frac{2x^3 - 1}{x^2 + 3} = 2x + \frac{-6x - 1}{x^2 + 3}$, the degree of $q(x)$ is the degree of $f(x)$ minus the degree of $d(x)$.

42. Sample answer: A polynomial can be divided by a polynomial with more than one term using long division, which is a similar process to dividing real numbers. Synthetic division is an alternate method used to divide a polynomial by a binomial of degree 1. Synthetic division is typically a quicker, simpler method than long division.

Lesson 4-5

Powers of Binomials

A.APR.5

LESSON GOAL

Students expand powers of binomials.

1 LAUNCH

 Launch the lesson with a **Warm Up** and an introduction.

2 EXPLORE AND DEVELOP

 **Explore:** Expanding Binomials

 **Develop:**

Powers of Binomials

- Use Pascal's Triangle
- Use the Binomial Theorem
- Coefficients Other Than 1

 You may want your students to complete the **Checks** online.

3 REFLECT AND PRACTICE

 Exit Ticket

 Practice

DIFFERENTIATE

 View reports of student progress on the **Checks** after each example.

Resources	AL	OL	BL	ELL
Remediation: Special Products	●	●		●
Extension: Patterns in Pascal's Triangle		●	●	●

Language Development Handbook

Assign page 28 of the *Language Development Handbook* to help your students build mathematical language related to powers of binomials.

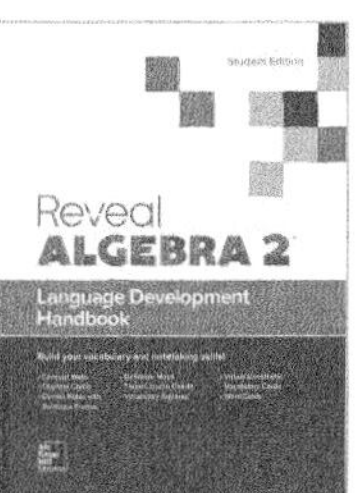

ELL You can use the tips and suggestions on page T28 of the handbook to support students who are building English proficiency.

Suggested Pacing

90 min	0.5 day
45 min	1 day

Focus

Domain: Algebra

Standards for Mathematical Content:

A.APR.5 Know and apply the Binomial Theorem for the expansion of $(x + y)^n$ in powers of x and y for a positive integer n, where x and y are any numbers, with coefficients determined for example by Pascal's Triangle.

Standards for Mathematical Practice:

4 Model with mathematics.

7 Look for and make use of structure.

Coherence

Vertical Alignment

Previous
Students performed operations on polynomial expressions.
A.APR.1 (Algebra 1, Algebra 2)

Now
Students expand powers of binomials.
A.APR.5

Next
Students will solve polynomial equations by factoring and by writing them in quadratic form. **A.CED.1**

Rigor

The Three Pillars of Rigor

1 CONCEPTUAL UNDERSTANDING	2 FLUENCY	3 APPLICATION

Conceptual Bridge In this lesson, students develop an understanding of the Binomial Theorem, and they build fluency by using the Binomial Theorem to expand binomials. They apply their understanding by solving real-world problems.

Mathematical Background

For the expansion of $(a + b)^n$, the signs of the terms depend on the signs of a and b. The sign of each term is $+$ if both a and b are positive; the sign of each even-numbered term is $-$ if b alone is negative.

Interactive Presentation

Warm Up

Expand.

1. $(x - 1)^2$
2. $(x - 3y)^2$
3. $(2x + 5y)^2$

Show Answers

Warm Up

Warm Up

Prerequisite Skills

The Warm Up exercises address the following prerequisite skill for this lesson:

- expanding perfect squares

Answers:

1. $x^2 - 2x + 1$
2. $x^2 - 6xy + 9y^2$
3. $4x^2 + 20xy + 25y^2$

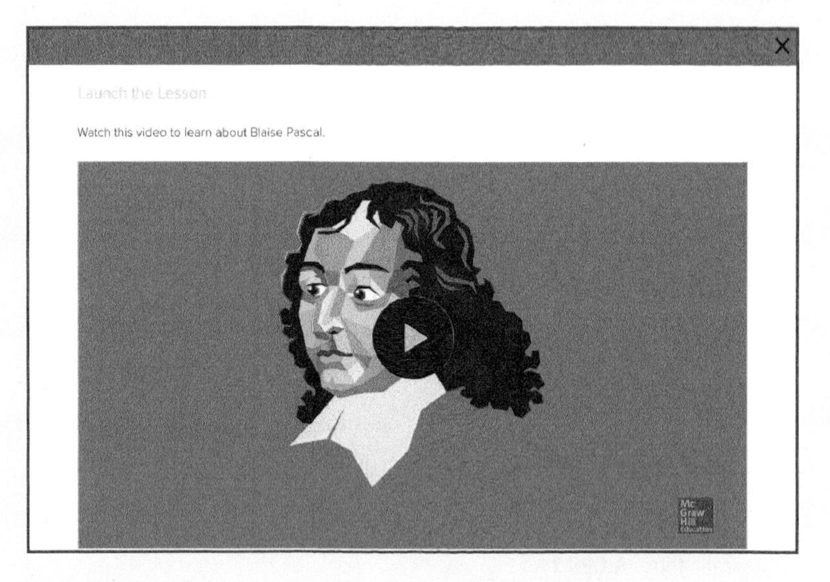

Launch the Lesson

Launch the Lesson

 **Teaching the Mathematical Practices**

7 Look for a Pattern Help students to see and use the patterns in Pascal's Triangle.

Go Online to find additional teaching notes and questions to promote classroom discourse.

Today's Standards

Tell students that they will be addressing these content and practice standards in this lesson. You may wish to have a student volunteer read aloud *How can I meet this standard?* and *How can I use these practices?*, and connect these to the standards.

See the Interactive Presentation for I Can statements that align with the standards covered in this lesson.

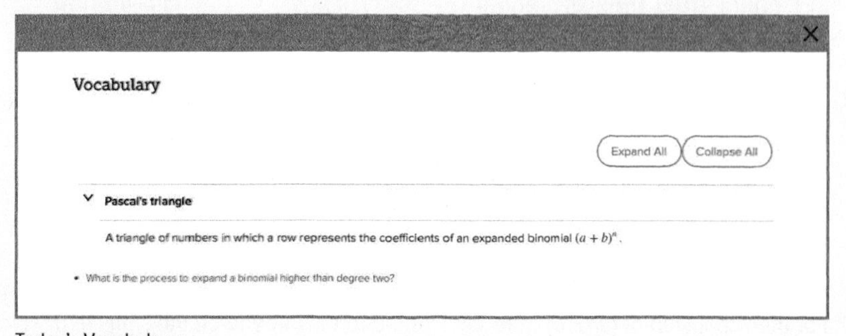

Today's Vocabulary

Today's Vocabulary

Tell students that they will be using this vocabulary term in this lesson. You can expand the row if you wish to share the definition. Then, discuss the question below with the class.

Explore Expanding Binomials

Objective

Students use a sketch to explore using Pascal's Triangle to expand binomials.

 Teaching the Mathematical Practices

3 Construct Arguments Students will use assumptions and previously established results to make a conjecture about how to use Pascal's Triangle to write expansions of binomials.

Ideas for Use

Recommended Use Present the Inquiry Question, or have a student volunteer read it aloud. Have students work in pairs to complete the Explore activity on their devices. Pairs should discuss each of the questions. Monitor student progress during the activity. Upon completion of the Explore activity, have student volunteers share their responses to the Inquiry Question.

What if my students don't have devices? You may choose to project the activity on a whiteboard. A printable worksheet for each Explore is available online. You may choose to print the worksheet so that individuals or pairs of students can use it to record their observations.

Summary of the Activity

Students will complete guiding exercises throughout the Explore activity. Students use a sketch to construct Pascal's Triangle and study its relationship with the expansion of binomial $(a + b)^n$. Then, students will answer the Inquiry Question.

(continued on the next page)

Interactive Presentation

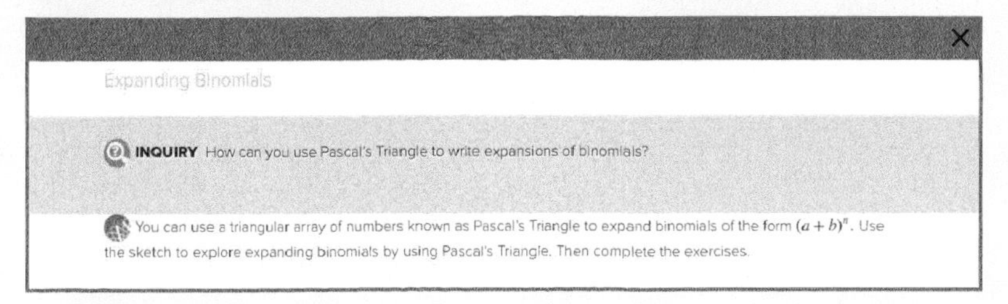

Explore

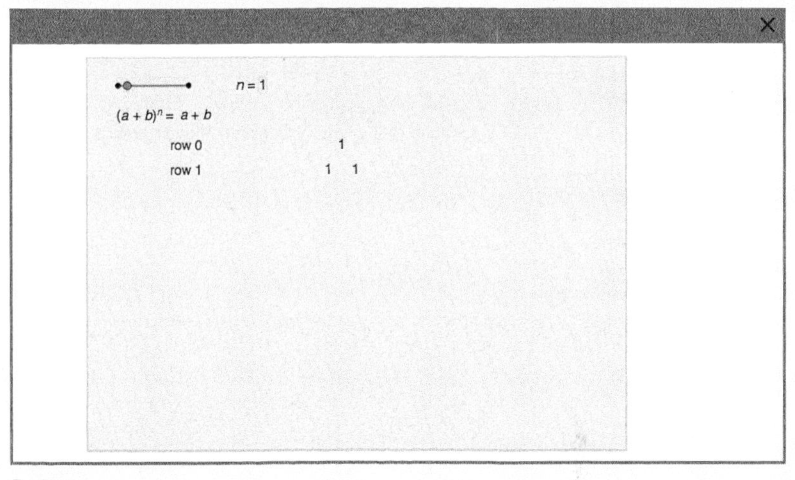

Explore

WEB SKETCHPAD

Students use a sketch to explore expanding binomials.

Interactive Presentation

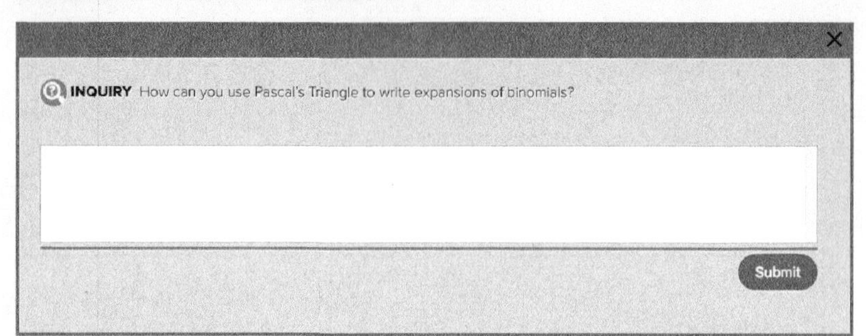

Explore

TYPE

Students respond to the Inquiry Question and can view a sample answer.

Explore Expanding Binomials (*continued*)

Questions

Have students complete the Explore activity.

Ask:

- Does it make sense to use Pascal's Triangle for rows 0 and 1? Sample answer: No, those are simple enough to find without using Pascal's Triangle.
- Which row of Pascal's Triangle do you need to use to expand $(a + b)^8$? Sample answer: The degree is 8, so you need to find and use the 8th row.

Inquiry

How can you use Pascal's Triangle to write expansions of binomials? Sample answer: Find the nth row of Pascal's triangle by following the pattern of adding the numbers in the row above. Then, use the numbers in the nth row as the coefficients of the expanded binomial. Use descending powers of a and ascending powers of b in the terms.

Go Online to find additional teaching notes and sample answers for the guiding exercises.

1 CONCEPTUAL UNDERSTANDING | 2 FLUENCY | 3 APPLICATION

Learn Powers of Binomials

Objective

Students expand powers of binomials by using Pascal's triangle and the Binomial Theorem.

Teaching the Mathematical Practices

1 Explain Correspondences Use the Think About It! feature to encourage students to explain how using Pascal's Triangle and the Binomial Theorem are related.

Essential Question Follow-Up

Students have begun to expand binomials using Pascal's triangle.

Ask:

How can you use Pascal's triangle to show that ${}_nC_k = {}_nC_{n-k}$? Sample answer: When Pascal's triangle is folded across a vertical center line, the numbers match. The corresponding numbers match k with $n - k$.

Example 1 Use Pascal's Triangle

Teaching the Mathematical Practices

7 Look for a Pattern Help students to extend the pattern in Pascal's Triangle and use it to expand the binomial.

Questions for Mathematical Discourse

AL How can you describe the pattern of Pascal's triangle in words? Sample answer: You add two terms to find the term below it in the triangle.

OL How do you determine the powers of the variables in the expansion? Sample answer: The first term has a^nb^0, and for subsequent terms the power of a goes in descending order while the power of b goes in ascending order.

BL Does the order of a and b matter? Explain. No, the expansion is symmetric.

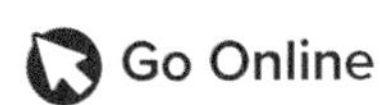

- Find additional teaching notes.
- View performance reports of the Checks.
- Assign or present an Extra Example.

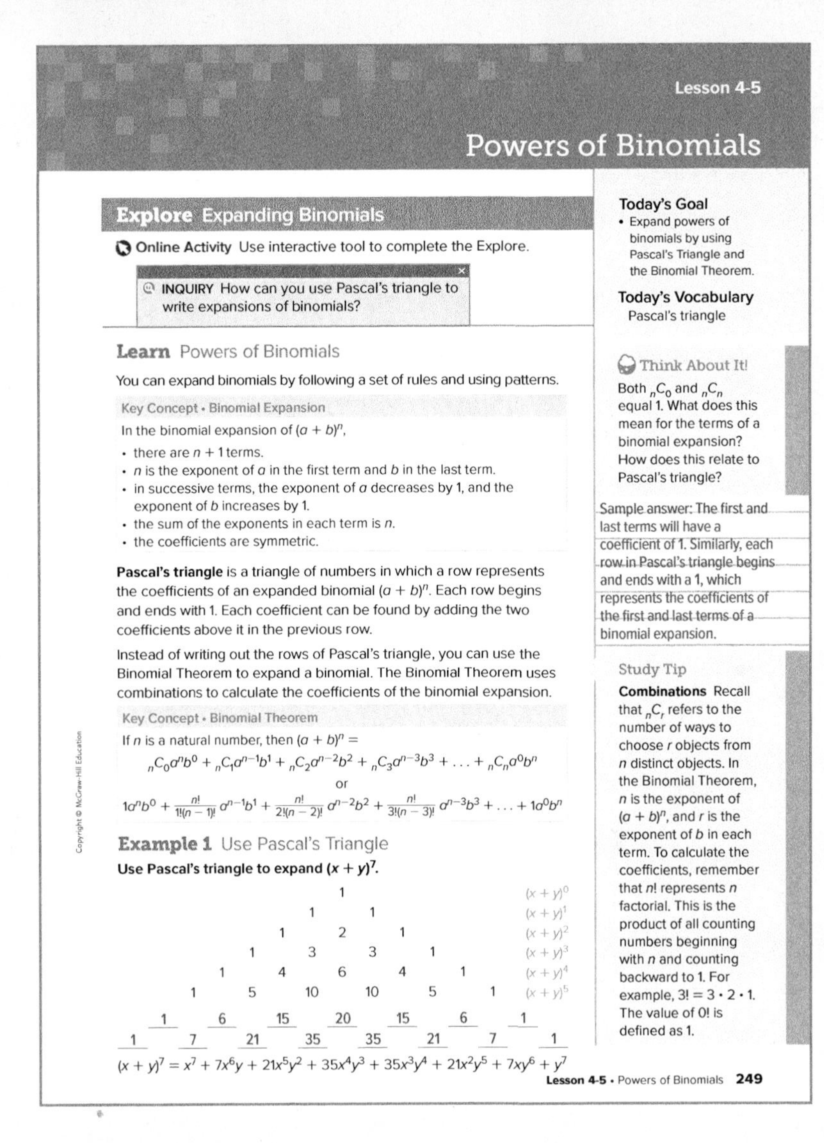

Lesson 4-5

Powers of Binomials

Explore Expanding Binomials

Online Activity Use interactive tool to complete the Explore.

INQUIRY How can you use Pascal's triangle to write expansions of binomials?

Learn Powers of Binomials

You can expand binomials by following a set of rules and using patterns.

Key Concept • Binomial Expansion

In the binomial expansion of $(a + b)^n$,

- there are $n + 1$ terms.
- n is the exponent of a in the first term and b in the last term.
- in successive terms, the exponent of a decreases by 1, and the exponent of b increases by 1.
- the sum of the exponents in each term is n.
- the coefficients are symmetric.

Pascal's triangle is a triangle of numbers in which a row represents the coefficients of an expanded binomial $(a + b)^n$. Each row begins and ends with 1. Each coefficient can be found by adding the two coefficients above it in the previous row.

Instead of writing out the rows of Pascal's triangle, you can use the Binomial Theorem to expand a binomial. The Binomial Theorem uses combinations to calculate the coefficients of the binomial expansion.

Key Concept • Binomial Theorem

If n is a natural number, then $(a + b)^n =$

$${}_nC_0a^nb^0 + {}_nC_1a^{n-1}b^1 + {}_nC_2a^{n-2}b^2 + {}_nC_3a^{n-3}b^3 + \ldots + {}_nC_na^0b^n$$

or

$$1a^nb^0 + \frac{n!}{1!(n-1)!}a^{n-1}b^1 + \frac{n!}{2!(n-2)!}a^{n-2}b^2 + \frac{n!}{3!(n-3)!}a^{n-3}b^3 + \ldots + 1a^0b^n$$

Example 1 Use Pascal's Triangle

Use Pascal's triangle to expand $(x + y)^7$.

1	$(x + y)^0$
1 1	$(x + y)^1$
1 2 1	$(x + y)^2$
1 3 3 1	$(x + y)^3$
1 4 6 4 1	$(x + y)^4$
1 5 10 10 5 1	$(x + y)^5$
1 6 15 20 15 6 1	
1 7 21 35 35 21 7 1	

$(x + y)^7 = x^7 + 7x^6y + 21x^5y^2 + 35x^4y^3 + 35x^3y^4 + 21x^2y^5 + 7xy^6 + y^7$

Today's Goal
- Expand powers of binomials by using Pascal's Triangle and the Binomial Theorem.

Today's Vocabulary
Pascal's triangle

Think About It!
Both ${}_nC_0$ and ${}_nC_n$ equal 1. What does this mean for the terms of a binomial expansion? How does this relate to Pascal's triangle?

Sample answer: The first and last terms will have a coefficient of 1. Similarly, each row in Pascal's triangle begins and ends with a 1, which represents the coefficients of the first and last terms of a binomial expansion.

Study Tip
Combinations Recall that ${}_nC_r$ refers to the number of ways to choose r objects from n distinct objects. In the Binomial Theorem, n is the exponent of $(a + b)^n$, and r is the exponent of b in each term. To calculate the coefficients, remember that $n!$ represents n factorial. This is the product of all counting numbers beginning with n and counting backward to 1. For example, $3! = 3 \cdot 2 \cdot 1$. The value of $0!$ is defined as 1.

Interactive Presentation

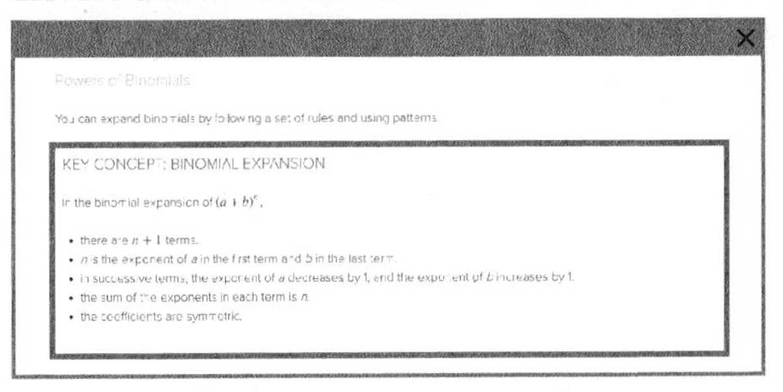

Learn

TYPE

Students explain how to write rows of Pascal's triangle.

Your Notes

Talk About It!

Describe a shortcut you could use to write out rows of Pascal's triangle instead of adding to find every number in a row. Explain your reasoning.

Sample answer: Since the coefficients in an expanded binomial are symmetric, you can just find the first or last half of the numbers in a row. Then, complete the row by using the same numbers.

Study Tip

Assumptions To use the Binomial Theorem, we assumed that the teams had an equal chance of winning and losing. Assuming that allows us to reasonably estimate the probability of an outcome with only the coefficient. To find probabilities of events that are not equally likely, substitute the probability of each event for a and b in the expansion of $(a + b)_n$.

Study Tip

Coefficients When the binomial to be expanded has coefficients other than 1, the coefficients will no longer be symmetric. In these cases, it may be easier to use the Binomial Theorem.

Check

Write the expansion of $(c + d)^4$. $c^4 + 4c^3d + 6c^2d^2 + 4cd^3 + d^4$

Example 2 Use the Binomial Theorem

BASEBALL **In 2016, the Chicago Cubs won the World Series for the first time in 108 years. During the regular season, the Cubs played the Atlanta Braves 6 times, winning 3 games and losing 3 games. If the Cubs were as likely to win as to lose, find the probability of this outcome by expanding $(w + \ell)^6$.**

$(w + \ell)^6$

$= {}_6C_0w^6 + {}_6C_1w^5\ell + {}_6C_2w^4\ell^2 + {}_6C_3w^3\ell^3 + {}_6C_4w^2\ell^4 + {}_6C_5w\ell^5 + {}_6C_6\ell^6$

$= w^6 + \frac{6!}{5!}w^5\ell + \frac{6!}{2!4!}w^4\ell^2 + \frac{6!}{3!3!}w^3\ell^3 + \frac{6!}{4!2!}w^2\ell^4 + \frac{6!}{5!}w\ell^5 + \ell^6$

$= w^6 + 6w^5\ell + 15w^4\ell^2 + 20w^3\ell^3 + 15w^2\ell^4 + 6w\ell^5 + \ell^6$

By adding the coefficients, you can determine that there were 64 combinations of wins and losses that could have occurred.

$20w^3\ell^3$ represents the number of combinations of 3 wins and 3 losses. Therefore, there was a $\frac{20}{64}$ or about a 31 % chance of the Cubs winning 3 games and losing 3 games against the Braves.

Check

GAME SHOW A group of 8 contestants are selected from the audience of a television game show. If there are an equal number of men and women in the audience, find the probability of the contestants being 5 women and 3 men by expanding $(w + m)^8$. Round to the nearest percent if necessary. 22 %

Example 3 Coefficients Other Than 1

Expand $(2c - 6d)^4$.

$(2c - 6d)^4$

$= {}_4C_0(2c)^4 + {}_4C_1(2c)^3(-6d) + {}_4C_2(2c)^2(-6d)^2 + {}_4C_3(2c)(-6d)^3 + {}_4C_4(-6d)^4$

$= 16c^4 + \frac{4!}{3!}(8c^3)(-6d) + \frac{4!}{2!2!}(4c^2)(36d^2) + \frac{4!}{3!}(2c)(-216d^3) + 1296d^4$

$= 16c^4 - 192c^3d + 864c^2d^2 - 1728cd^3 + 1296d^4$

Go Online You can complete an Extra Example online.

Example 2 Use the Binomial Theorem

Teaching the Mathematical Practices

4 Interpret Mathematical Results In Example 2, point out that to solve the problem, students should interpret their mathematical results in the context of the problem to determine the probability of the given outcome.

Questions for Mathematical Discourse

AL What do the coefficients of the expansion mean in the context of the example? Each coefficient represents the number of combinations resulting in k number of losses.

OL Why is an outcome of 3 wins and 3 losses more probable than an outcome of all wins? There are more combinations of wins and losses that have the same result of 3 wins and 3 losses, whereas all wins can only occur as one combination.

BL Was it necessary to use the binomial expansion rather than just solving the problem using combinations? Explain. No; sample answer: The solution comes only from the coefficients, but using the binomial makes it easier to keep track of which values of ${}_nC_k$ go with which type of outcome.

Example 3 Coefficients Other Than 1

Teaching the Mathematical Practices

7 Use Structure Help students to use the structure of the Binomial Theorem to expand the binomial in Example 3.

Questions for Mathematical Discourse

AL Why is the first coefficient 16? because ${}_4C_0 \times 2^4 = 16$

OL How could you still use Pascal's triangle to solve this problem? You can use Pascal's triangle to fill in the values for the combinations.

BL How could you determine a formula to find the coefficients of $(Ca + Db)^n$, where C and D are constants? Sample answer: We know from the Binomial Theorem that the coefficients for $(a + b)^n$ are $\frac{n!}{k!(n-k)!}$. We only need to determine the pattern for the coefficients C and D. The variable terms expand as $a^{n-k}b^k$, so the coefficients would also be to the same power. Thus $\frac{n!}{k!(n-k)!}C^{n-k}D^k$.

Interactive Presentation

Use the Binomial Theorem

You can use the Binomial Theorem to find the probability of a certain combination of outcomes occurring.

BASEBALL In 2016, the Chicago Cubs won the World Series for the first time in 108 years. During the regular season, they played the Atlanta Braves 6 times, winning 3 games and losing 3 games. If the Cubs were as likely to win as to lose, find the probability of this outcome by expanding $(w + \ell)^6$.

Example 2

TYPE

Students learn to use the Binomial Theorem to find the probability of a combination of outcomes occurring.

CHECK

Students complete the Check online to determine whether they are ready to move on.

Exit Ticket

Recommended Use

At the end of class, go online to display the Exit Ticket prompt and ask students to respond using a separate piece of paper. Have students hand you their responses as they leave the room.

Alternate Use

At the end of class, go online to display the Exit Ticket prompt and ask students to respond verbally or by using a mini-whiteboard. Have students hold up their whiteboards so that you can see all student responses. Tap to reveal the answer when most or all students have completed the Exit Ticket.

Practice and Homework

Suggested Assignments

Use the table below to select appropriate exercises.

DOK	Topic	Exercises
1, 2	exercises that mirror the examples	1–12
2	exercises that use a variety of skills from this lesson	13–25
3	exercises that emphasize higher-order and critical thinking skills	26–30

ASSESS AND DIFFERENTIATE

Use the data from the **Checks** to determine whether to provide resources for extension, remediation, or intervention.

IF students score 90% or more on the Checks, **THEN** assign: OL BL

- Practice Exercises 1–25 odd, 26–30
- Extension: Patterns in Pascale's Triangle
- **ALEKS**®

IF students score 66%–89% on the Checks, **THEN** assign: AL OL

- Practice Exercises 1–29 odd
- Remediation, Review Resources: Special Products
- Personal Tutors
- Extra Examples 1–3
- **ALEKS**® Expanding Perfect Squares

IF students score 65% or less on the Checks, **THEN** assign: AL

- Practice Exercises 1–11 odd
- Remediation, Review Resources: Special Products
- **ALEKS**® Expanding Perfect Squares

Name ______________ Period ________ Date ________

Practice

Go Online You can complete your homework online.

Example 1

Use Pascal's triangle to expand each binomial.

1. $(x - y)^3$
$x^3 - 3x^2y + 3xy^2 - y^3$

2. $(a + b)^4$
$a^4 + 4a^3b + 6a^2b^2 + 4ab^3 + b^4$

3. $(g - h)^4$
$g^4 - 4g^3h + 6g^2h^2 - 4gh^3 + h^4$

4. $(m + 1)^4$
$m^4 + 4m^3 + 6m^2 + 4m + 1$

5. $(y - z)^6$
$y^6 - 6y^5z + 15y^4z^2 - 20y^3z^3 + 15y^2z^4 - 6yz^5 + z^6$

6. $(d + 2)^8$
$d^8 + 16d^7 + 112d^6 + 448d^5 + 1120d^4 + 1792d^3 + 1792d^2 + 1024d + 256$

Example 2

7. BAND A school band went to 4 competitions during the year and received a superior rating 2 times. If the band is as likely to receive a superior rating as to not receive a superior rating, find the probability of this outcome by expanding $(s + n)^4$. Round to the nearest percent if necessary. 38%

8. BASKETBALL Oliver shot 8 free throws at practice, making 6 free throws and missing 2 free throws. If Oliver is equally likely to make a free throw as he is to miss a free throw, find the probability of this outcome by expanding $(m + n)^8$. Round to the nearest percent if necessary. 11%

Example 3

Expand each binomial.

9. $(3x + 4y)^5$

10. $(2c - 2d)^7$

9. $243x^5 + 1620x^4y + 4320x^3y^2 + 5760x^2y^3 + 3840xy^4 + 1024y^5$
10. $128c^7 - 896c^6d + 2688c^5d^2 - 4480c^4d^3 + 4480c^3d^4 - 2688c^2d^5 + 896cd^6 - 128d^7$

11. $(8h - 3j)^4$

12. $(4a + 3b)^6$

11. $4096h^4 - 6144h^3j + 3456h^2j^2 - 864hj^3 + 81j^4$
12. $4096a^6 + 18{,}432a^5b + 34{,}560a^4b^2 + 34{,}560a^3b^3 + 19{,}440a^2b^4 + 5832ab^5 + 729b^6$

Mixed Exercises

Expand each binomial.

13. $\left(x + \frac{1}{2}\right)^5$
$x^5 + \frac{5}{2}x^4 + \frac{5}{2}x^3 + \frac{5}{4}x^2 + \frac{5}{16}x + \frac{1}{32}$

14. $\left(x - \frac{1}{3}\right)^4$
$x^4 - \frac{4}{3}x^3 + \frac{2}{3}x^2 - \frac{4}{27}x + \frac{1}{81}$

15. $\left(2b + \frac{1}{4}\right)^5$
$32b^5 + 20b^4 + 5b^3 + \frac{5}{8}b^2 + \frac{5}{128}b + \frac{1}{1024}$

16. $\left(3c + \frac{1}{3}d\right)^3$
$27c^3 + 9c^2d + cd^2 + \frac{d^3}{27}$

17. STRUCTURE Out of 12 frames, Vince bowled 6 strikes. If Vince is as likely bowl a strike as to not bowl a strike in one frame, find the probability of this outcome. Round to the nearest percent if necessary. 23%

18. REGULARITY A group of 10 choir members are selected at random to perform solos. If there are an equal number of boys and girls in the choir, find the probability of the choir members selected being 7 boys and 3 girls. Round to the nearest percent if necessary. 12%

19. USE A MODEL A company is developing a robotic welder that produces circuit boards. At this stage in its development, the robotic welder only produces 50% of the circuit boards correctly. Use the Binomial Theorem to find the probability that 5 of 7 circuit boards chosen at random are correct. See margin.

20. USE A MODEL Diego flips a fair coin 12 times. What is the probability that the coin lands on tails 3 times? 5 times? 9 times? 5.4%; 19.3%; 5.4%

21. REASONING A test consists of 10 true-false questions. Matthew forgets to study and must guess on every question. What is the probability that he gets 8 or more correct answers on the test? Show your work using Pascal's Triangle. See margin.

22. REGULARITY Use Pascal's Triangle to find the fourth term in the expansion of $(2x + 7)^6$. Why is it the same as the fourth term in the expansion of $(7 + 2x)^6$? See margin.

23. USE A SOURCE Research the number of judges on the Supreme Court. For most rulings, a majority is needed. How many combinations of votes are possible for a majority to be reached? See margin.

24. STRUCTURE Find the term in $(a + b)^{12}$ where the exponent of a is 5. $792a^5b^7$

25. PRECISION Use the first four terms of the binomial expansion of $(1 + 0.02)^{10}$ to approximate $(1.02)^{10}$. Evaluate $(1.02)^{10}$ using a calculator and compare the value to your approximation. 1.21896; $(1.02)^{10} = 1.21899442$; the approximation differs from the value given by a calculator by 0.00003442.

Higher-Order Thinking Skills

26. PERSEVERE Find the sixth term of the expansion of $(\sqrt{a} + \sqrt{b})^{12}$. $792a^3b^2\sqrt{ab}$

27. ANALYZE Explain how the terms of $(x + y)^n$ and $(x - y)^n$ are the same and how they are different. See margin.

28. REGULARITY Each row of Pascal's triangle is like a palindrome. That is, the numbers read the same left to right as they do right to left. Explain why this is the case. See margin.

29. CREATE Write a power of a binomial for which the second term of the expansion is $6x^4y$. See margin.

30. WRITE Explain how to write out the terms of Pascal's triangle. See margin.

252 Module 4 • Polynomials and Polynomial Functions

Answers

19. If C represents a correct circuit board and N represents an incorrect circuit board, then the number of ways for the robot to produce 5 of 7 circuit boards accurately is given by the coefficient of C^5N^2 in the expansion of $(C + N)^7$. Using the Binomial Theorem, there are 21 ways for the robot to produce a correct circuit board out of 128 possibilities. So, the probability that 5 of 7 are correct is $\frac{21}{128}$ or about 16%.

21. If c represents a correct answer and w represents a wrong answer, then the coefficients expansion of $(c + w)^{10}$ can be used to represent situation. Using Pascal's triangle, there are 45 ways to get 8 questions correct, 10 ways to get 9 questions correct, and 1 way to get all correct. So there are 56 ways to get 8 or more correct. By adding all of the values in this row of Pascal's triangle, I found that there are 1024 different ways he could answer the questions in the quiz. Matthew has a $\frac{56}{1024}$ or about a 5.5% chance of getting 8 or more correct.

22. Sample answer: Using Pascal's triangle, the fourth term of the expansion $(a + b)^6$ is 20. If $a = 2x$ and $b = 7$, then the fourth term is $20a^3b^3$, or $54{,}880x^3$. If $a = 7$ and $b = 2x$, then the fourth term is also $20a^3b^3$, or $54{,}880x^3$.

23. There are 9 judges on the Supreme Court. The majority could be 5, 6, 7, 8, or 9 votes. So, there are ${}_9C_5 + {}_9C_6 + {}_9C_7 + {}_9C_8 + {}_9C_9 = 256$ combinations.

27. Sample answer: While they have the same terms, the signs for $(x + y)^n$ will all be positive, while the signs for $(x - y)^n$ will alternate.

28. Sample answer: If you switch x and y in the Binomial Theorem, the coefficients reverse their order, yet $(x + y)^n = (y + x)^n$ because $x + y = y + x$. Therefore, the coefficients must read the same forward and back.

29. Sample answer: $\left(x + \frac{6}{5}y\right)^5$

30. Sample answer: The first row is a 1. The second row is two 1s. Each new row begins and ends with 1. Each coefficient is the sum of the two coefficients above it in the previous row.

Review

Rate Yourself! 

Have students return to the Module Opener to rate their understanding of the concepts presented in this module. They should see that their knowledge and skills have increased. After completing the chart, have them respond to the prompts in their *Interactive Student Edition* and share their responses with a partner.

Answering the Essential Question

Before answering the Essential Question, have students review their answers to the Essential Question Follow-Up questions found throughout the module.

- How does a power function's graph change as the degree of the function increases?
- Are polynomials closed under addition, subtraction, and multiplication?
- How do you know you have your remainder and that you are done dividing?

Then have them write their answer to the Essential Question in the space provided.

DINAH ZIKE FOLDABLES

ELL A completed Foldable for this module should include the key concepts related to polynomials and polynomial functions.

LS **LearnSmart** Use LearnSmart as part of your test preparation plan to measure student topic retention. You can create a student assignment in LearnSmart for additional practice on these topics for **Polynomial, Rational, and Radical Relationships** and **Modeling with Functions.**

- Polynomial Expressions
- Graphing Rational, Radical, and Polynomial Functions
- Creating Function Models
- Interpreting Function Models

Module 4 • Polynomials and Polynomial Functions

Review

Essential Question

How does an understanding of polynomials and polynomial functions help us understand and interpret real-world events?

Sample answer: Polynomials and polynomial functions can be used to model situations where quantities increase or decrease in a nonlinear pattern.

Module Summary

Lessons 4-1 and 4-2

Polynomial Functions and Graphs

- A power function is any function of the form $f(x) = ax^n$, where a and n are nonzero real numbers. The leading coefficient is a and the degree is n.
- Odd-degree functions will always have at least one real zero.
- Even-degree functions may have any number of real zeros or no real zeros at all.
- A polynomial function is a continuous function that can be described by a polynomial equation in one variable.
- The degree of a polynomial function tells the maximum number of times that the graph of a polynomial function intersects the x-axis.
- If the value of $f(x)$ changes signs from one value of x to the next, then there is a zero between those two x-values.
- Extrema occur at relative maxima or minima of the function.

Lessons 4-3 and 4-4

Operations with Polynomials

- Polynomials can be added or subtracted by performing the operations indicated and combining like terms.
- To subtract a polynomial, add its additive inverse.
- Polynomials can be multiplied by using the Distributive Property to multiply each term in one polynomial by each term in the other.
- The set of polynomials is closed under the operations of addition, subtraction, and multiplication.
- To multiply two binomials, you can use a shortcut called the FOIL method.
- You can divide a polynomial by a polynomial with more than one term by using a process similar to long division of real numbers.
- Synthetic division is an alternate method used to divide a polynomial by a binomial of degree 1.

Lesson 4-5

Powers of Binomials

- Pascal's triangle is a triangle of numbers in which a row represents the coefficients of an expanded binomial $(a + b)^n$. Each row begins and ends with 1. Each coefficient can be found by adding the two coefficients above it in the previous row.
- You can also use the Binomial Theorem to expand a binomial. If n is a natural number, then $(a + b)^n = {}_nC_0a^nb^0 + {}_nC_1a^{n-1}b^1 + {}_nC_2a^{n-2}b^2 + {}_nC_3a^{n-3}b^3 + \ldots + {}_nC_na^0b^n$.

Study Organizer

Foldables

Use your Foldable to review this module. Working with a partner can be helpful. Ask for clarification of concepts as needed.

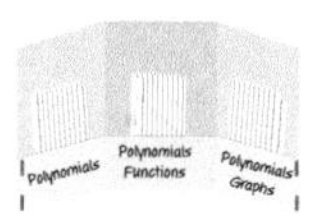

Module 4 Review • Polynomials and Polynomial Functions 253

Name ______ Period ______ Date ______

Test Practice

1. **MULTIPLE CHOICE** The weight of an ideal cut round diamond can be modeled by $f(d) = 0.0071d^3 - 0.090d^2 + 0.48d$, where d is the diameter of the diamond. Find the domain of the function in the context of the situation. (Lesson 4-1)

   (A) The domain is all real numbers.

   (B) The domain is $\{d \mid d > 0\}$.

   (C) The domain is $\{d \mid d < 0\}$.

   (D) The domain is $\{d \mid d > 0.48\}$.

2. **OPEN RESPONSE** Use the function $f(x) = 13 - 2x^2 + 6x - 9x^3$ to answer the following questions. (Lesson 4-1)

   a) What is the degree?

   b) What is the leading coefficient?

   c) Describe the end behavior using the leading coefficient and degree.

   a) 3

   b) -9

   c) Because the leading coefficient is negative and the degree is odd, the end behavior is that as $x \to -\infty$, $f(x) \to \infty$ and as $x \to \infty$, $f(x) \to -\infty$.

3. **MULTIPLE CHOICE** The revenue of a certain business can be modeled using $f(x) = -0.01(x^4 - 11x^3 + 4x^2 - 5x + 7)$, where x is the number of years since the business was started and $f(x)$ is the revenue in hundred-thousands of dollars. Which graph represents the function? (Lesson 4-1)

   (A) (B) (C) (D)

4. **MULTI-SELECT** Select all intervals in which a real zero is located for the function $f(x) = x^4 - 2x^3 + 3x^2 - 5$. (Lesson 4-2)

   (A) $x = -2$ and $x = -1$

   (B) $x = -1$ and $x = 0$

   (C) $x = 0$ and $x = 1$

   (D) $x = 1$ and $x = 2$

   (E) $x = 2$ and $x = 3$

   (F) $x = 3$ and $x = 4$

5. **OPEN RESPONSE** Describe the end behavior for $g(x) = -2x^4 - 6x^3 + 11x - 18$ as $x \to \infty$. (Lesson 4-2)

   $g(x) \to -\infty$

Review and Assessment Options

The following online review and assessment resources are available for you to assign to your students. These resources include technology-enhanced questions that are auto-scored, as well as essay questions.

Review Resources

Put It All Together: Lessons 4-1 through 4-5
Vocabulary Activity
Module Review

Assessment Resources

Vocabulary Test
AL Module Test Form B
OL Module Test Form A
BL Module Test Form C
Performance Task*

*The module-level performance task is available online as a printable document. A scoring rubric is included.

Test Practice

You can use these pages to help your students review module content and prepare for online assessments. Exercises 1-16 mirror the types of questions your students will see on online assessments.

Question Type	Description	Exercise(s)
Multiple Choice	Students select one correct answer.	1, 3, 7, 10, 11, 12, 14, 15
Multi-Select	Multiple answers may be correct. Students must select all correct answers.	4, 16
Open Response	Students construct their own response in the area provided.	2, 5, 6, 8, 9, 13

To ensure that students understand the standards, check students' success on individual exercises.

Standard(s)	Lesson(s)	Exercise(s)
A.CED.2	4-2	7
A.APR.1	4-3, 4-4	8–12
A.APR.6	4-4, 4-5	13–16
F.IF.4	4-1, 4-2	2, 4–6
F.IF.5	4-1	1
F.IF.7c	4-1	3

Name ________________ Period ________ Date ________

6. OPEN RESPONSE Marshall claims that there is only one real zero in the function $f(x) = 4x^3 + 7x^2 - 5x + 3$. Based on the table provided, determine whether you agree with Marshall. Then name the interval(s) in which the zero(s) is/are located. (Lesson 4-2)

x	$f(x)$
−3	−27
−2	9
−1	11
0	3
1	9
2	53

Sample answer: I agree with Marshall. There is one zero located between $x = -3$ and $x = -2$. I know this because the y-values change from a negative to a positive.

7. MULTIPLE CHOICE Helen started a business several years ago. The table shows her profits, in millions of dollars, for the first 7 years. Select the polynomial function of best fit that could be used to model Helen's profits. (Lesson 4-2)

x	$f(x)$
1	1.425
2	1.46
3	1.5
4	1.53
5	1.56
6	1.58
7	1.58

Ⓐ $f(x) = 0.001(-3.27x^2 + 53.51x + 1371)$

Ⓑ $f(x) = 0.0001(-6.944x^3 + 50.6x^2 + 250.4x + 13{,}957)$

Ⓒ $f(x) = 0.00001(9.47x^4 + 82.07x^3 - 312.5x^2 + 4203x + 138{,}500)$

Ⓓ $f(x) = 0.0001(-x^4 + 12x^3 - 77x^2 + 600x + 13{,}650)$

8. OPEN RESPONSE What is the difference? $(7x^4 - 3x^3 + 5x^2 + 8x - 11) - (3x^4 - 9x^3 - 4x^2 + 12x + 4)$ (Lesson 4-3)

$4x^4 + 6x^3 + 9x^2 - 4x - 15$

9. OPEN RESPONSE What is the sum? $(3x^6 - 5x^3 + 8x^2 - 4x) + (-2x^6 + 7x^5 - x^3 + 6x)$ (Lesson 4-3)

$x^6 + 7x^5 - 6x^3 + 8x^2 + 2x$

10. MULTIPLE CHOICE Enrique is designing a flag for a new school club. A smaller striped square is placed as part of the design and the rest of the flag will have chevrons. (Lesson 4-3)

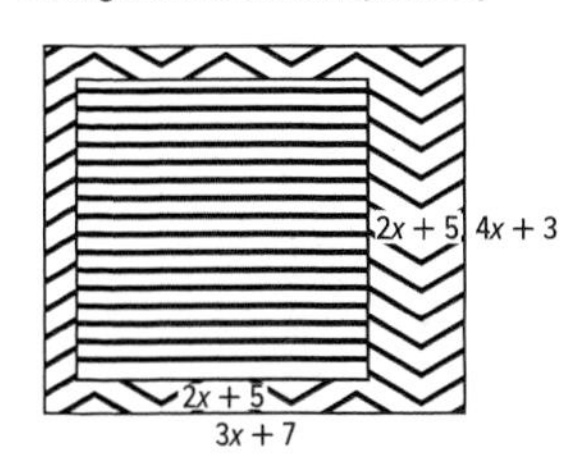

Which expression can be used to represent the area of the flag that is not striped?

Ⓐ $16x^2 + 57x + 46$

Ⓑ $8x^2 + 35x + 16$

Ⓒ $8x^2 + 17x + 4$

Ⓓ $8x^2 + 17x - 4$

11. MULTIPLE CHOICE Determine the quotient.
$(5x^4 + 12x^3 - 64x^2 - 95x + 132) \div (x - 3)$
(Lesson 4-4)

(A) $3x^3 + 21x^2 + x - 31$

(B) $3x^3 - 21x^2 - x + 31$

(C) $5x^3 + 27x^2 + 17x - 44$

(D) $5x^3 - 27x^2 - 17x + 44$

12. MULTIPLE CHOICE Determine the quotient.
(Lesson 4-4)

$$\frac{6x^3 - 71x^2 + 139x + 130}{3x + 2}$$

(A) $2x^2 - 25x + 63 + \frac{8}{3}$

(B) $2x^2 - 25x + 63 + \frac{8}{3x + 2}$

(C) $2x^2 - 25x + 63 + \frac{4}{3}$

(D) $2x^2 - 25x + 63 + \frac{4}{3x + 2}$

13. OPEN RESPONSE The volume of the rectangular prism shown is $45x^3 + 83x^2 + x - 12$.

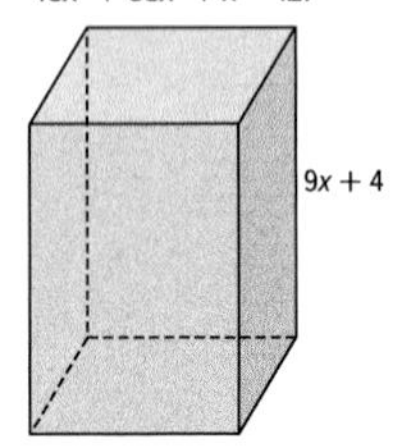

What is the area of the base? (Lesson 4-4)

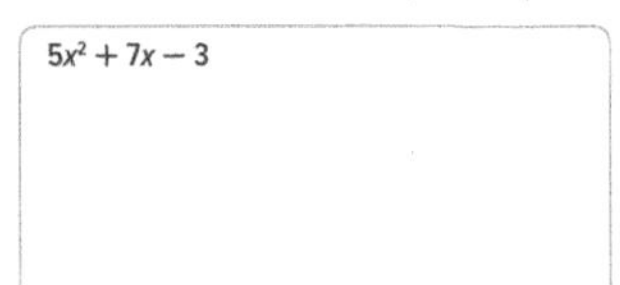

$5x^2 + 7x - 3$

14. MULTIPLE CHOICE Which of the following is the expansion of $(2h + f)^4$? (Lesson 4-5)

(A) $2h^4 + 4h^3f + 6h^2f^2 + 4hf^3 + f^4$

(B) $16h^4 + 32h^3f + 24h^2f^2 + 32hf^3 + 16f^4$

(C) $16h^4 + 32h^3f + 32h^2f^2 + 8hf^3 + f^4$

(D) $16h^4 + 32h^3f + 24h^2f^2 + 8hf^3 + f^4$

15. MULTIPLE CHOICE The first shelf on Hannah's bookshelf holds an equal number of fiction and nonfiction books. If Hannah selects 5 books randomly, what is the probability that 4 of the books will be fiction and 1 will be nonfiction?

Round your answer to the nearest tenth of a percent. (Lesson 4-5)

(A) 31.3%

(B) 15.6%

(C) 12.5%

(D) 3.1%

16. MULTI-SELECT Select all of the following that would be a coefficient of a term in the binomial expansion of $(x + y)^7$. (Lesson 4-5)

(A) 1

(B) 3

(C) 7

(D) 14

(E) 21

(F) 28

(G) 30

(H) 35

Lesson 4-2

5. relative minimum between $x = 0$ and $x = 1$; relative maximum near $x = 4$

x	f(x)
−1	22
0	0
2	16
3	30
4	32
5	10
6	−48
7	−154

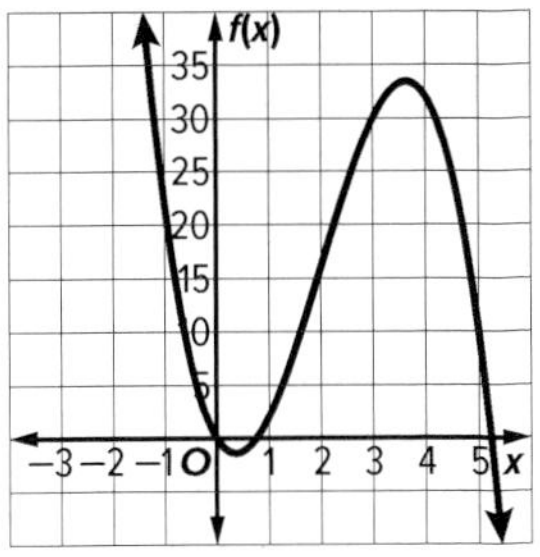

6. relative minimum near $x = 2$; relative maximum near $x = 0$

x	f(x)
−4	−176
−3	−77
−2	−22
−1	1
0	4
1	−1
2	−2
3	13
4	56

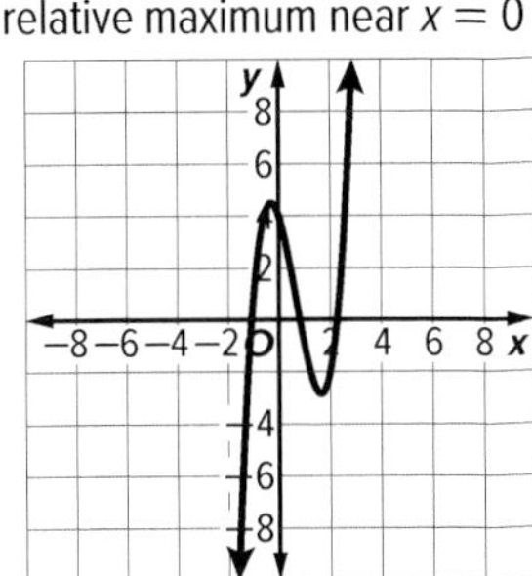

7. relative minimum near $x = -1$; no relative maximum

x	f(x)
−4	247
−3	74
−2	11
−1	−2
0	−1
1	2
2	19
3	86
4	263

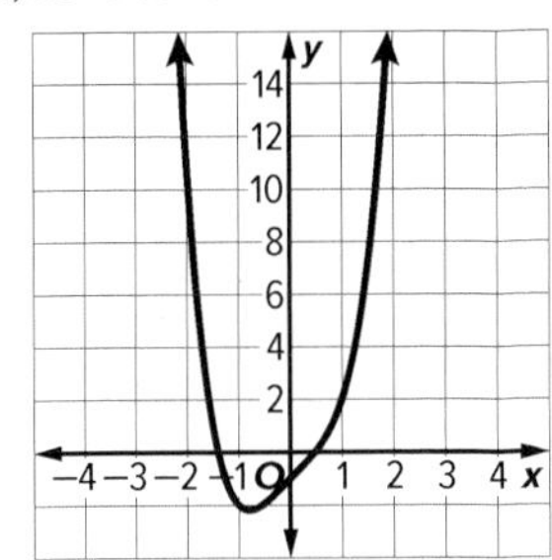

8. relative minimum near $x = 0$; no relative maximum

x	f(x)
−4	372
−3	141
−2	36
−1	−3
0	−12
1	−3
2	36
3	141
4	372

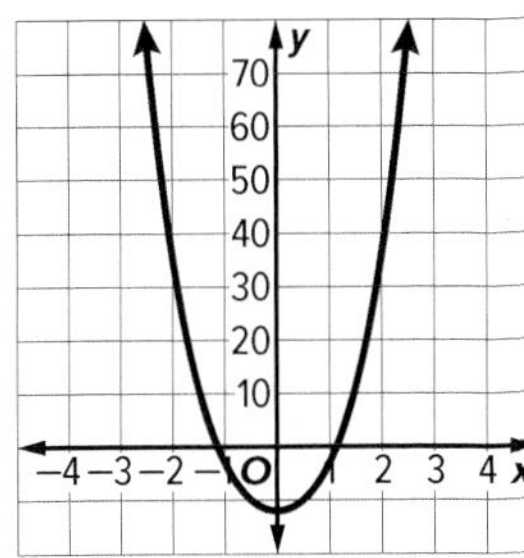

9. Domain and Range: The domain and range of the function are all real numbers. Because the function models years, the relevant domain and range are $\{n \mid n \geq 0\}$ and $\{v \mid v \geq -4\}$.

Extrema: There is a relative minimum at $n = 2$ in the relevant domain.
End Behavior: As $x \to \infty$, $f(x) \to \infty$.
Intercepts: In the relevant domain, the v-intercept is at (0, 0). The n-intercept, or zero, is at about (3, 0).
Symmetry: In the relevant domain, the graph does not have symmetry.

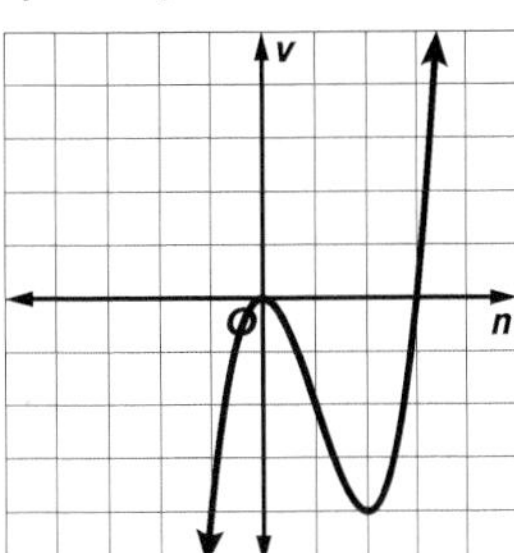

10.

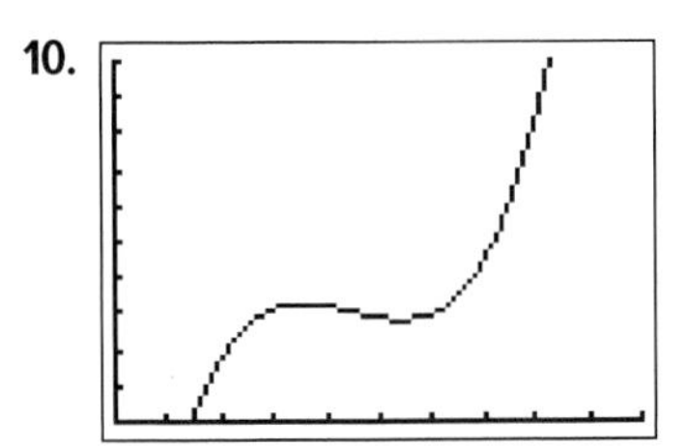

[0, 10] scl: 1 [0, 100] scl: 10

Domain and Range: The domain and range of the function are all real numbers. Because the function models weeks of growth, the relevant domain and range are $\{x \mid x \geq 1.4\}$ and $\{y \mid y \geq 0\}$.
Extrema: There is a relative maximum between $x = 3$ and $x = 4$ and a relative minimum between $x = 5$ and $x = 6$ in the relevant domain.
End Behavior: As $x \to \infty$, $f(x) \to \infty$.
Intercepts: In the relevant domain, there is no y-intercept. There is an x-intercept, or zero, at about (1.4, 0).
Symmetry: The graph does not have symmetry in the relevant domain.

16.

x	f(x)
−3	−17
−2	−1
−1	3
0	1
1	−1
2	3
3	19

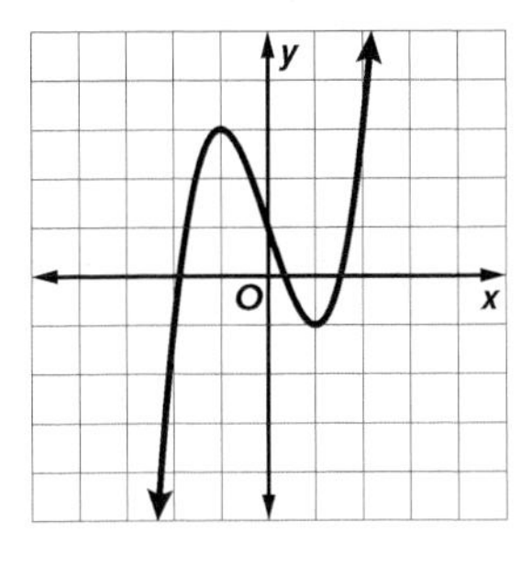

zeros between $x = -2$ and $x = -1$, between $x = 0$ and $x = 1$, and between $x = 1$ and $x = 2$; rel. max. at $x = -1$, rel. min. at $x = 1$; D = all real numbers; R = all real numbers

17.

x	f(x)
−3	−7
−2	−2
−1	−3
0	2
1	25

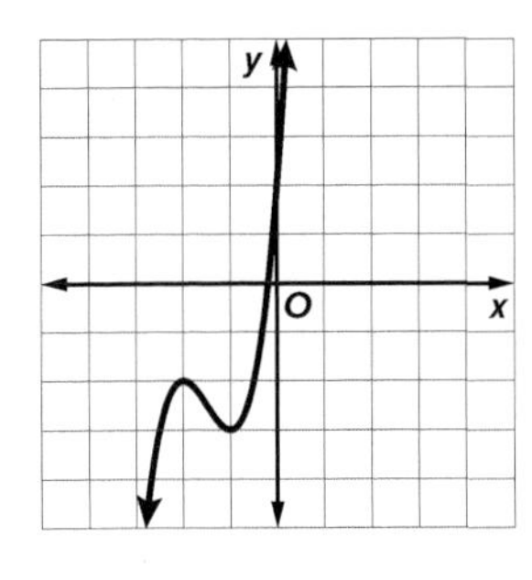

zero between $x = -1$ and $x = 0$; rel. max. at $x = -2$, rel. min. at $x = -1$; D = all real numbers; R = all real numbers

MODULE 4 ANSWER APPENDIX

18.

x	$f(x)$
−1	−3
0	2
1	1
2	6
3	29

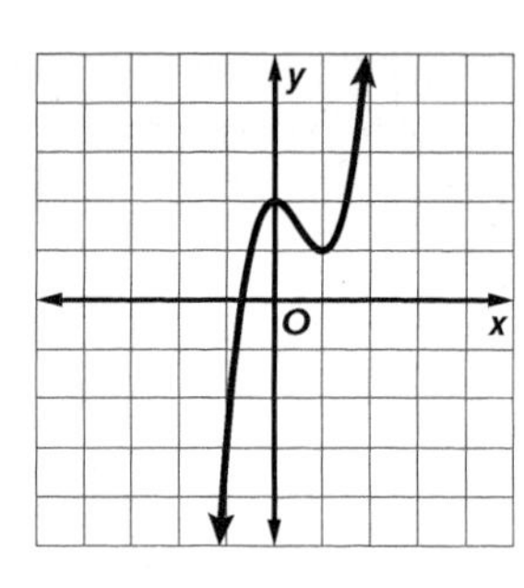

zero between $x = -1$ and $x = 0$; rel. max. at $x = 0$, rel. min. at $x = 1$; D = all real numbers; R = all real numbers

19.

x	$f(x)$
−3	61
−2	6
−1	−3
0	−2
1	−3
2	6
3	61

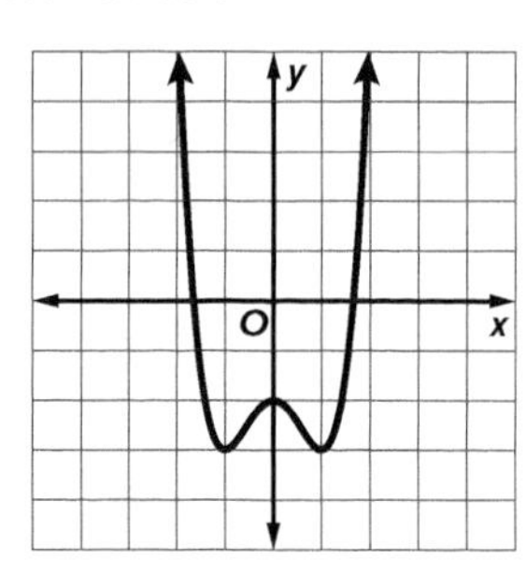

zeros between $x = -2$ and $x = -1$, and between $x = 1$ and $x = 2$; rel. max. at $x = 0$, min. at $x = -1$, and $x = 1$; D = all real numbers; R = $\{f(x) \mid f(x) \geq -3\}$

20. Domain and Range: D = all real numbers; R = $\{y|y > 0\}$

Extrema: rel. min at $(-4, 16)$; rel. min near $x = 0$

End Behavior: as $x \to \infty, y \to \infty$ and as $x \to -\infty, y \to \infty$

Intercepts: no x-intercept; y-intercept: 5

Symmetry: none

25. Sample answer: $f(x) \to -\infty$ as $x \to \infty$ and $f(x) \to \infty$ as $x \to -\infty$; the leading coefficient is negative.

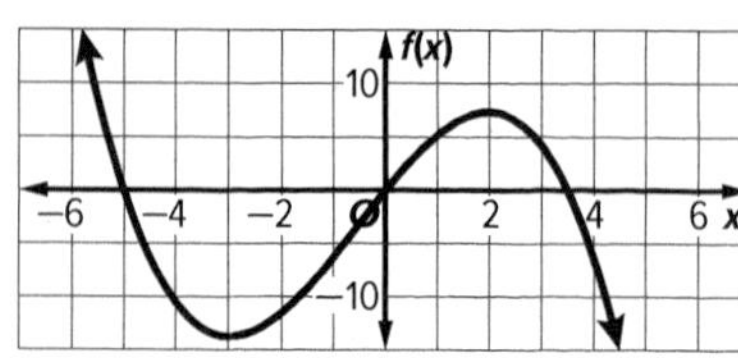

26a.

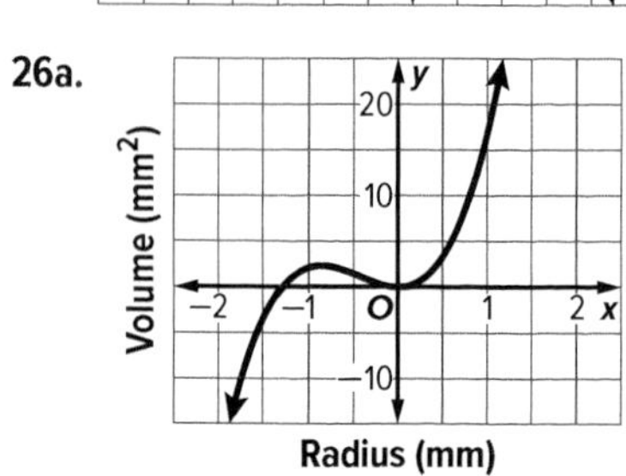

31. Sample answer:

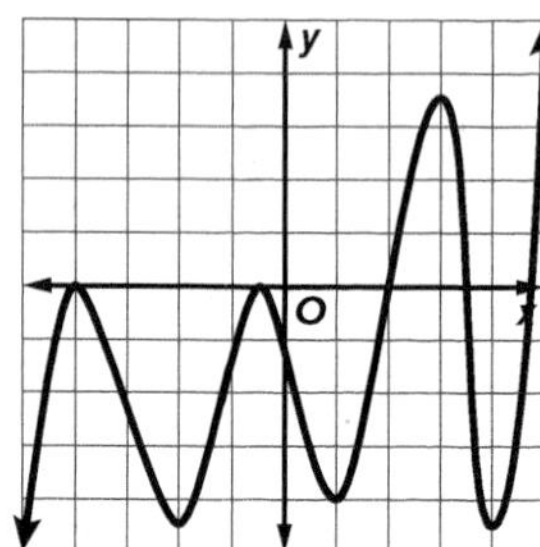

Module 5

Polynomial Equations

Module Goals

- Students solve polynomial equations by graphing, factoring, and the Fundamental Theorem of Algebra.
- Students prove polynomial identities.
- Students evaluate and factor equations using the Factor and Remainder Theorem.

Focus

Domain: Algebra

Standards for Mathematical Content:

A.CED.1 Create equations and inequalities in one variable and use them to solve problems.

A.APR.3 Identify zeros of polynomials when suitable factorizations are available, and use the zeros to construct a rough graph of the function defined by the polynomial.

Also addresses N.CN.9, A.APR.2, A.APR.4, A.REI.11, and F.IF.7c.

Standards for Mathematical Practice:

All Standards for Mathematical Practice will be addressed in this module.

Coherence

Vertical Alignment

Previous

Students graphed polynomial functions and performed operations with polynomial expressions. **A.APR.1, F.IF.7c**

Now

Students solve polynomial equations, factor polynomial expressions, and identify the zeros of polynomial functions.
A.CED.1, A.APR.3, F.IF.7c

Next

Students will solve rational equations and identify key features of rational functions.
F.IF.4, A.REI.2

Rigor

The Three Pillars of Rigor

To help students meet standards, they need to illustrate their ability to use the three pillars of rigor. Students gain conceptual understanding as they move from the Explore to Learn sections within a lesson. Once they understand the concept, they practice procedural skills and fluency and apply their mathematical knowledge as they go through the Examples and Independent Practice.

1 CONCEPTUAL UNDERSTANDING | 2 FLUENCY | 3 APPLICATION

EXPLORE → LEARN → EXAMPLE & PRACTICE

Suggested Pacing

Lessons	Standards	45-min classes	90-min classes
Module Pretest and Launch the Module Video		1	0.5
5-1 Solving Polynomial Equations by Graphing	A.CED.1, A.REI.11	1	0.5
5-2 Solving Polynomial Equations Algebraically	A.CED.1	2	1
5-3 Proving Polynomial Identities	A.APR.4	1	0.5
5-4 The Remainder and Factor Theorems	A.APR.2	1	0.5
5-5 Roots and Zeros	N.CN.9, A.APR.3, F.IF.7c	2	1
Module Review		1	0.5
Module Assessment		1	0.5
	Total Days	10	5

Formative Assessment Math Probe
Polynomial Factors

Module Resource

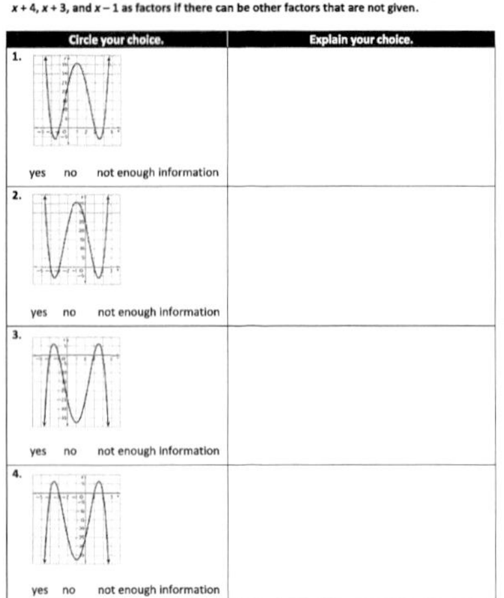
NAME ___ DATE ___ PERIOD ___

Cheryl Tobey Math Probe
Polynomial Factors

Determine whether the following graphs could represent a polynomial with $x - 2$, $x + 4$, $x + 3$, and $x - 1$ as factors if there can be other factors that are not given.

Circle your choice.	Explain your choice.
1. yes no not enough information	
2. yes no not enough information	
3. yes no not enough information	
4. yes no not enough information	

Cheryl Tobey Math Probe • Polynomial Factors © McGraw-Hill Education

Correct Answers: 1. no; 2. yes; 3. no; 4. yes

Analyze the Probe

Review the probe prior to assigning it to your students.

In this probe, students will connect polynomial factors with graphs of polynomial functions.

Targeted Concepts Understand how different representations of polynomial functions can give you information about the solutions of their related equations.

Targeted Misconceptions

- Students may not set factors equal to zero to find the roots of polynomial equations.
- Students may not consider that various polynomial equations can have the same roots.

Use the Probe after Lesson 5-1.

Collect and Assess Student Answers

If the student selects these responses...	**Then** the student likely...
1. yes **2.** no **3.** yes	does not take the additive inverse of the factors to find the roots. **Example:** The zero for the factor $x - 2$ is found by setting $x - 2 = 0$ and then solving to get $x = 2$.
4. no	does not consider the polynomial with the additive inverse of each term. **Example:** The polynomial $x^4 + 4x^3 - 7x^2 - 22x + 24$ has the same roots as $-(x^4 + 4x^3 - 7x^2 - 22x + 24)$.

Take Action

After the Probe Design a plan to address any possible misconceptions. You may wish to assign the following resources.

- ALEKS® Real Zeros of Polynomial Functions
- Lesson 5-1, Learn, all Examples

Revisit the Probe at the end of the module to be sure that your students no longer carry these misconceptions.

The Ignite! activities, created by Dr. Raj Shah, cultivate curiosity and engage and challenge students. Use these open-ended, collaborative activities, located online in the module Launch section, to encourage your students to develop a growth mindset towards mathematics and problem solving. Use the teacher notes for implementation suggestions and support for encouraging productive struggle.

Essential Question

At the end of this module, students should be able to answer the Essential Question.

What methods are useful for solving polynomial equations and finding zeros of polynomial functions? Sample answer: You can use graphing or factoring to solve polynomial equations or find roots of polynomial functions. The Remainder and Factor Theorems can also be useful.

What Will You Learn?

Prior to beginning this module, have your students rate their knowledge of each item listed. Then, at the end of the module, you will be reminded to have your students return to these pages to rate their knowledge again. They should see that their knowledge and skills have increased.

DINAH ZIKE FOLDABLES

Focus Students write notes about new terms and concepts as they are presented in each lesson of this module.

Teach Have students construct their Foldable as illustrated. Have students write an explanation of each term or concept on the appropriate section of their Foldable while working through each lesson. Encourage students to record examples of each term or concept on the back of each flap.

When to Use It Encourage students to add to their Foldable as they work through the module, and to use it to review for the module test.

Launch the Module

For this module, the Launch the Module video uses exploring Yellowstone National Park to describe modeling real-world situations with polynomial functions. Students learn about polynomial models in population growth and decay.

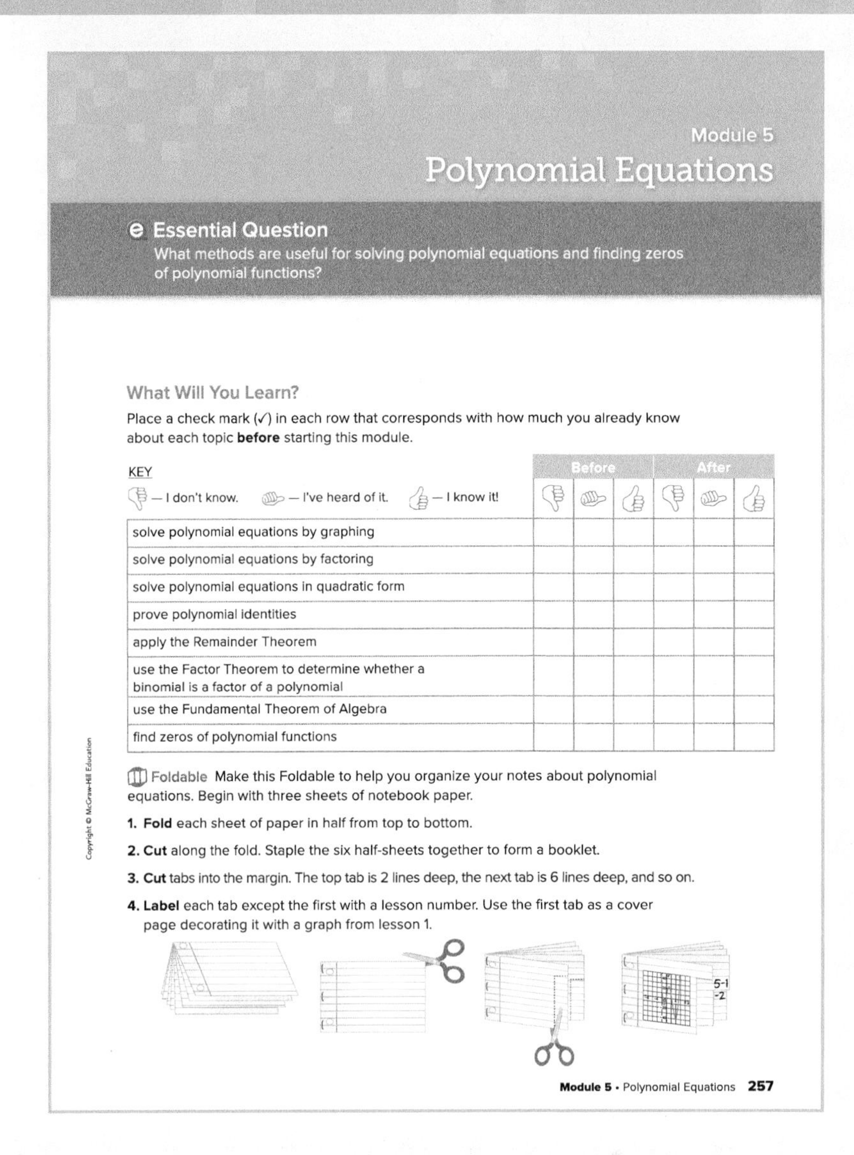

Module 5

Polynomial Equations

Essential Question
What methods are useful for solving polynomial equations and finding zeros of polynomial functions?

What Will You Learn?

Place a check mark (✓) in each row that corresponds with how much you already know about each topic **before** starting this module.

KEY
— I don't know. — I've heard of it. — I know it!

	Before			After		
	I don't know	I've heard of it	I know it!	I don't know	I've heard of it	I know it!
solve polynomial equations by graphing						
solve polynomial equations by factoring						
solve polynomial equations in quadratic form						
prove polynomial identities						
apply the Remainder Theorem						
use the Factor Theorem to determine whether a binomial is a factor of a polynomial						
use the Fundamental Theorem of Algebra						
find zeros of polynomial functions						

Foldable Make this Foldable to help you organize your notes about polynomial equations. Begin with three sheets of notebook paper.

1. **Fold** each sheet of paper in half from top to bottom.
2. **Cut** along the fold. Staple the six half-sheets together to form a booklet.
3. **Cut** tabs into the margin. The top tab is 2 lines deep, the next tab is 6 lines deep, and so on.
4. **Label** each tab except the first with a lesson number. Use the first tab as a cover page decorating it with a graph from lesson 1.

Interactive Presentation

What Vocabulary Will You Learn?

Check the box next to each vocabulary term that you may already know.

- ☐ depressed polynomial
- ☐ identity
- ☐ multiplicity
- ☐ polynomial identity
- ☐ prime polynomial
- ☐ quadratic form
- ☐ synthetic substitution

Are You Ready?

Complete the Quick Review to see if you are ready to start this module. Then complete the Quick Check.

Quick Review

Example 1

Use the Distributive Property to multiply $(x^2 - 2x - 4)(x + 5)$.

$(x^2 - 2x - 4)(x + 5)$

$= x^2(x + 5) - 2x(x + 5) - 4(x + 5)$ Distributive Property

$= x^2(x) + x^2(5) + (-2x)(x) + (-2x)(5) + (-4)(x) + (-4)(5)$ Distributive Property

$= x^3 + 5x^2 - 2x^2 - 10x - 4x - 20$ Multiply.

$= x^3 + 3x^2 - 14x - 20$ Combine like terms.

Example 2

Solve $2x^2 + 8x + 1 = 0$.

$x = \frac{-b \pm \sqrt{b^2 - 4ac}}{2a}$ Quadratic Formula

$= \frac{-8 \pm \sqrt{8^2 - 4(2)(1)}}{2(2)}$ $a = 2, b = 8, c = 1$

$= \frac{-8 \pm \sqrt{56}}{4}$ Simplify.

$= -2 \pm \frac{\sqrt{14}}{2}$ $\sqrt{56} = \sqrt{4 \cdot 14}$ or $2\sqrt{14}$

The exact solutions are $-2 + \frac{\sqrt{14}}{2}$ and $-2 - \frac{\sqrt{14}}{2}$.

The approximate solutions are −0.13 and −3.87.

Quick Check

Use the Distributive Property to multiply each set of polynomials.

1. $(6x^2 - x + 2)(4x + 2)$ $24x^3 + 8x^2 + 6x + 4$
2. $(x^2 - 2x + 7)(7x - 3)$ $7x^3 - 17x^2 + 55x - 21$
3. $(7x^2 - 6x - 6)(2x - 4)$ $14x^3 - 40x^2 + 12x + 24$
4. $(x^2 + 6x - 4)(2x - 4)$ $2x^3 + 8x^2 - 32x + 16$

Solve each equation.

5. $x^2 + 2x - 8 = 0$ $-4, 2$
6. $2x^2 + 7x + 3 = 0$ $-3, -\frac{1}{2}$
7. $6x^2 + 5x - 4 = 0$ $-\frac{4}{3}, \frac{1}{2}$
8. $4x^2 - 2x - 1 = 0$ $\frac{1 \pm \sqrt{5}}{4}$

How Did You Do?

Which exercises did you answer correctly in the Quick Check? Shade those exercise numbers below.

① ② ③ ④ ⑤ ⑥ ⑦ ⑧

258 Module 5 • Polynomial Equations

What Vocabulary Will You Learn?

ELL As you proceed through the module, introduce the key vocabulary by using the following routine.

Define The quadratic form is a form of polynomial equation, $au^2 + bu + c$, where u is an algebraic expression in x.

Example $-(2x - 1)^2 + 3(2x - 1) + 5$

Ask Is the expression in quadratic form? Explain. Yes, the expression is in the form $au^2 + bu + c$ with $a = -1$, $b = 3$, and $c = 5$ while $u = 2x - 1$.

Are You Ready?

Students may need to review the following prerequisite skills to succeed in this module.

- finding x-intercepts
- solving simple quadratic equations
- identifying equations with no solution vs. identities
- performing synthetic division
- finding zeros of functions

ALEKS®

ALEKS is an adaptive, personalized learning environment that identifies precisely what each student knows and is ready to learn, ensuring student success at all levels.

You can use the ALEKS pie report to see which students know the topics in the **Quadratic and Polynomial Functions** module—who is ready to learn these topics and who isn't quite ready to learn them yet—in order to adjust your instruction as appropriate.

Mindset Matters

Promote Growth Over Speed

Learning requires time and effort—time to think, reason, make mistakes, and learn from your mistakes and the mistakes of others. Ultimately, it is about the deep connections students make in their thinking and reasoning that matter more than the speed at which a problem is solved.

How Can I Apply It?

Have students complete the **Rate Yourself** chart before starting the module, discuss their mistakes and progress as you work through each lesson, and then reflect on their growth at the end of the module.

A.CED.1, A.REI.11

Solving Polynomial Equations by Graphing

LESSON GOAL

Students solve polynomial equations by graphing.

1 LAUNCH

Launch the lesson with a **Warm Up** and an introduction.

2 EXPLORE AND DEVELOP

Explore: Solutions of Polynomial Equations

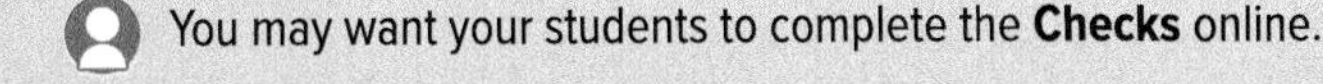

Develop:

Solving Polynomial Functions by Graphing

- Solve a Polynomial Equation by Graphing
- Solve a Polynomial Equation by Using a System

You may want your students to complete the **Checks** online.

3 REFLECT AND PRACTICE

Exit Ticket

Practice

Formative Assessment Math Probe

DIFFERENTIATE

View reports of student progress on the **Checks** after each example.

Resources	AL	OL	BL	ELL
Remediation: Multiplying Polynomials	●	●		●
Extension: Complex Roots of Polynomial Equations		●	●	●

Language Development Handbook

Assign page 30 of the *Language Development Handbook* to help your students build mathematical language related to solving polynomial equations by graphing.

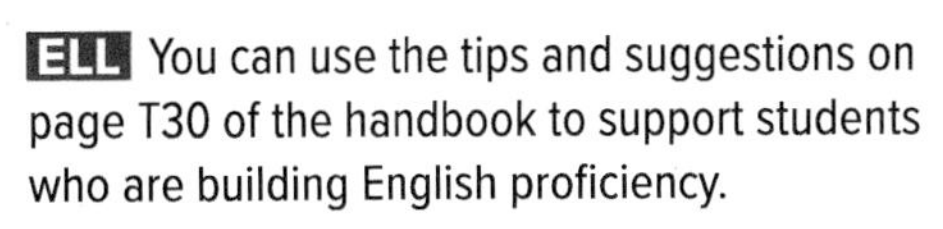

ELL You can use the tips and suggestions on page T30 of the handbook to support students who are building English proficiency.

Suggested Pacing

90 min	0.5 day
45 min	1 day

Focus

Domain: Algebra

Standards for Mathematical Content:

A.CED.1 Create equations and inequalities in one variable and use them to solve problems.

A.REI.11 Explain why the x-coordinates of the points where the graphs of the equations $y = f(x)$ and $y = g(x)$ intersect are the solutions of the equation $f(x) = g(x)$; find the solutions approximately, e.g. using technology to graph the functions, make tables of values, or find successive approximations. Include cases where $f(x)$ and/or $g(x)$ are linear, polynomial, rational, absolute value, exponential, and logarithmic functions.

Standards for Mathematical Practice:

2 Reason abstractly and quantitatively.

5 Use appropriate tools strategically.

Coherence

Vertical Alignment

Previous
Students analyzed the graphs of polynomial functions by identifying key features. **F.IF.4, F.IF.7c**

Now
Students solve polynomial equations by graphing. **A.CED.1, A.REI.11**

Next
Students will determine the numbers and types of roots of polynomial equations, find zeros, and use zeros to graph polynomial functions. **A.APR.3, F.IF.7c, N.CN.9**

Rigor

The Three Pillars of Rigor

1 CONCEPTUAL UNDERSTANDING	2 FLUENCY	3 APPLICATION

Conceptual Bridge In this lesson, students expand on their understanding of graphing polynomial functions by using the graphs to solve related polynomial equations. They build fluency by finding solutions by graphing, and they apply their understanding by using graphs to solve real-world problems.

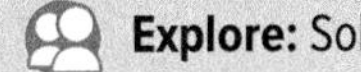

Interactive Presentation

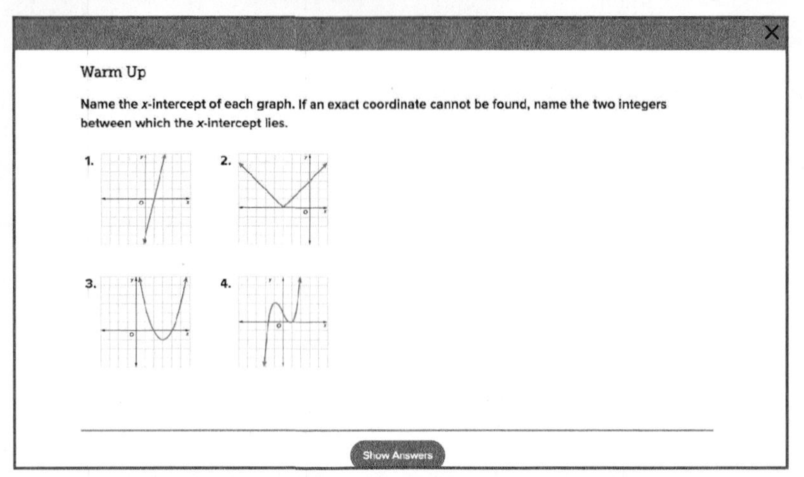

Warm Up

Launch

Warm Up

Prerequisite Skills

The Warm Up exercises address the following prerequisite skill for this lesson:

- finding x-intercepts

Answers:

1. 1
2. −3
3. 2, 4
4. between −2 and −1, between 0 and 1, 1

Launch the Lesson

Teaching the Mathematical Practices

1 Explain Correspondences Encourage students to explain the relationships between the dimensions, volume, and graph in the Launch the Lesson.

Go Online to find additional teaching notes and questions to promote classroom discourse.

Today's Standards

Tell students that they will be addressing these content and practice standards in this lesson. You may wish to have a student volunteer read aloud *How can I meet these standards*? and *How can I use these practices*?, and connect these to the standards.

See the Interactive Presentation for I Can statements that align with the standards covered in this lesson.

Mathematical Background

Tables of values can be used to explore two types of changes in the values of a polynomial function. A change of signs in the value of $f(x)$ from one value of x to the next indicates that the graph of the function crosses the x-axis between the two x-values. A change between increasing values and decreasing values indicates that the graph is turning for that interval. A turning point on a graph is a relative maximum or minimum.

Explore Solutions of Polynomial Equations

Objective

Students use a sketch to explore solving polynomial equations by graphing a related function.

 Teaching the Mathematical Practices

7 Use Structure Help students to explore how the structure of the factored form of a polynomial equation relates to the zeros of the related function.

Ideas for Use

Recommended Use Present the Inquiry Question, or have a student volunteer read it aloud. Have students work in pairs to complete the Explore activity on their devices. Pairs should discuss each of the questions. Monitor student progress during the activity. Upon completion of the Explore activity, have student volunteers share their responses to the Inquiry Question.

What if my students don't have devices? You may choose to project the activity on a whiteboard. A printable worksheet for each Explore is available online. You may choose to print the worksheet so that individuals or pairs of students can use it to record their observations.

Summary of the Activity

Students will complete guiding exercises throughout the Explore activity. They will learn that they can use related functions to solve polynomial equations. They will be presented with different polynomial equations, and will use the sketch to graph the related function. Then, students will answer the Inquiry Question.

(continued on the next page)

Interactive Presentation

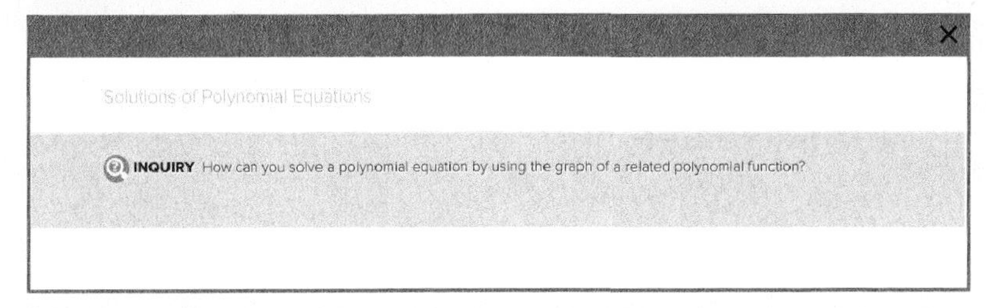

Explore

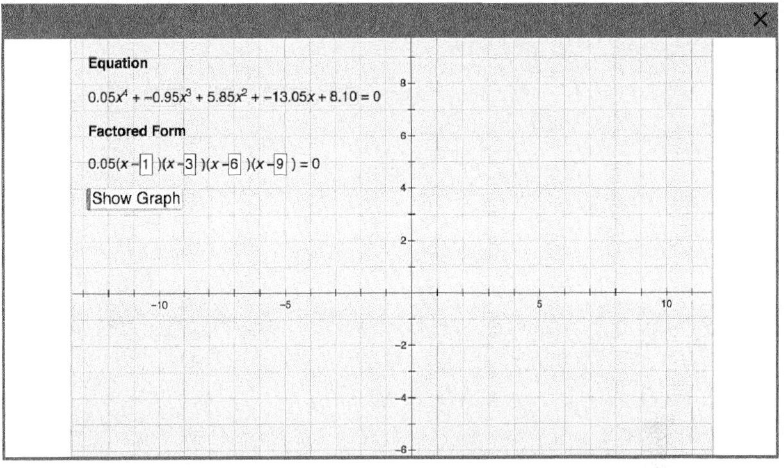

Explore

WEB SKETCHPAD

Students will use the sketch to graph the related function of a polynomial equation.

TYPE

Students answer questions pertaining to solving polynomial equations by graphing related functions.

Interactive Presentation

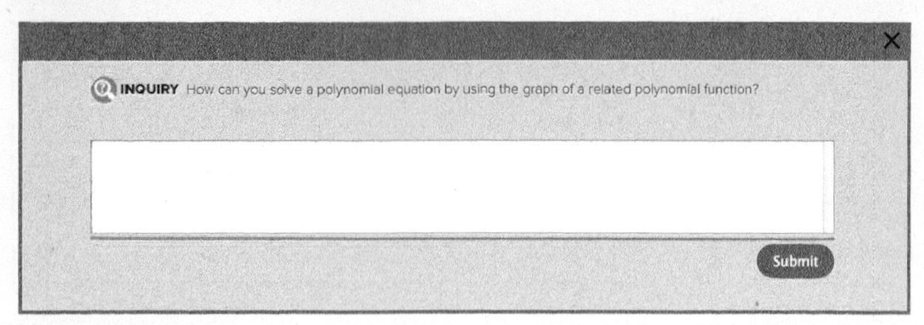

Explore

TYPE

Students respond to the Inquiry Question and can view a sample answer.

Explore Solutions of Polynomial Equations (*continued*)

Questions

Have students complete the Explore activity.

Ask:

- What would be the solution(s) of the polynomial equation $(2x + 1)(x - 3)(4x - 3) = 0$? $x = -\frac{1}{2}, 3, \frac{3}{4}$
- When graphing the related function of a polynomial equation, where are the solutions to the equation located on the graph? the x-intercepts

Inquiry

How can you solve a polynomial equation by using the graph of a related polynomial function? Sample answer: I can find the zeros of the graph of the related polynomial function. The zeros will be the same as the solutions of the equation.

Go Online to find additional teaching notes and sample answers for the guiding exercises.

Learn Solving Polynomial Equations by Graphing

Objective

Students solve polynomial equations by examining graphs of the related functions and by graphing systems of equations.

Teaching the Mathematical Practices

1 Explain Correspondences Help students to see the relationship between the solutions of a polynomial equation and the zeros of the related function.

Example 1 Solve a Polynomial Equation by Graphing

Teaching the Mathematical Practices

5 Decide When to Use Tools Mathematically proficient students make sound decisions about when to use mathematical tools such as graphing calculators. Help them see why using a graphing calculator will help solve the polynomial equation and what the limitations are of using the tool.

Questions for Mathematical Discourse

AL Why do you rewrite the equation with 0 on one side before you graph the related function? Sample answer: This makes the x-intercepts the solutions to the original equation because they are where the related function is equal to zero.

OL When finding the first zero using the TI-84 family calculator, the y-value was 1E–12. What does this value represent? Sample answer: 1E–12 is scientific notation for 0.000000000001, which is effectively zero.

BL Suppose only three x-intercepts are visible when graphing the fourth-degree related function. What might you need to do on the calculator? Sample answer: I may need to change the viewing window because the other x-intercept might be greater than 10 or less than −10.

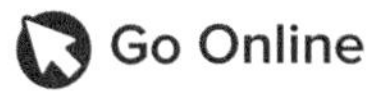

Go Online

- Find additional teaching notes.
- View performance reports of the Checks.
- Assign or present an Extra Example.

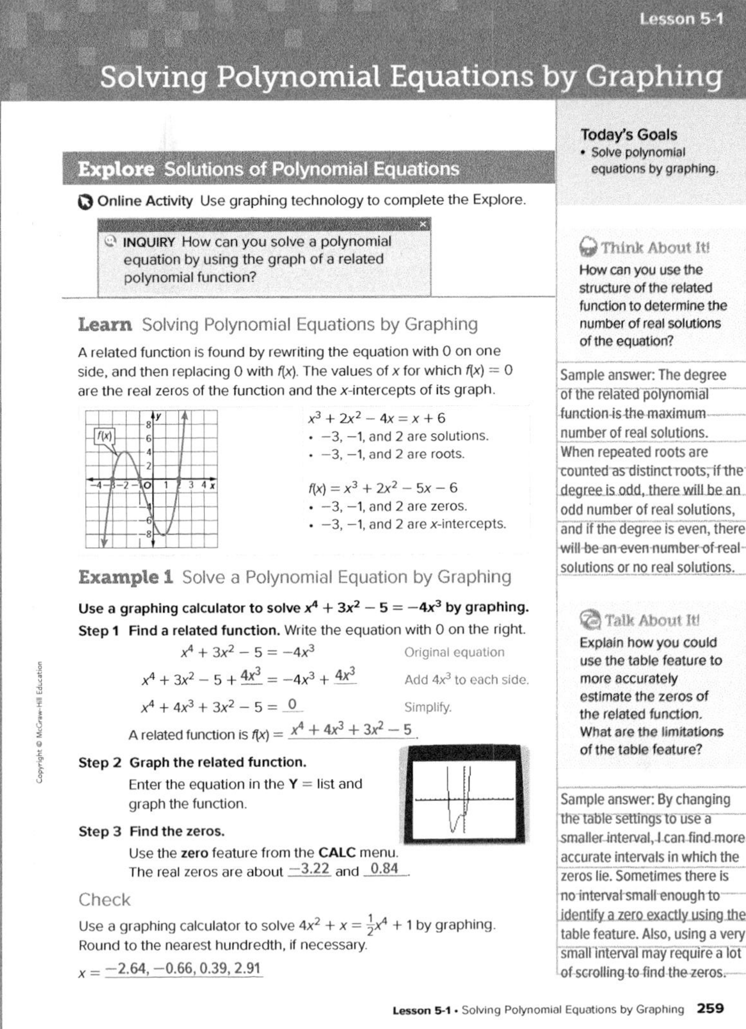

Lesson 5-1

Solving Polynomial Equations by Graphing

Today's Goals
- Solve polynomial equations by graphing.

Explore Solutions of Polynomial Equations

Online Activity Use graphing technology to complete the Explore.

INQUIRY How can you solve a polynomial equation by using the graph of a related polynomial function?

Learn Solving Polynomial Equations by Graphing

A related function is found by rewriting the equation with 0 on one side, and then replacing 0 with $f(x)$. The values of x for which $f(x) = 0$ are the real zeros of the function and the x-intercepts of its graph.

$x^3 + 2x^2 - 4x = x + 6$
- −3, −1, and 2 are solutions.
- −3, −1, and 2 are roots.

$f(x) = x^3 + 2x^2 - 5x - 6$
- −3, −1, and 2 are zeros.
- −3, −1, and 2 are x-intercepts.

Think About It! How can you use the structure of the related function to determine the number of real solutions of the equation?

Sample answer: The degree of the related polynomial function is the maximum number of real solutions. When repeated roots are counted as distinct roots, if the degree is odd, there will be an odd number of real solutions, and if the degree is even, there will be an even number of real solutions or no real solutions.

Example 1 Solve a Polynomial Equation by Graphing

Use a graphing calculator to solve $x^4 + 3x^2 - 5 = -4x^3$ by graphing.

Step 1 Find a related function. Write the equation with 0 on the right.

$x^4 + 3x^2 - 5 = -4x^3$ Original equation

$x^4 + 3x^2 - 5 + 4x^3 = -4x^3 + 4x^3$ Add $4x^3$ to each side.

$x^4 + 4x^3 + 3x^2 - 5 = 0$ Simplify.

A related function is $f(x) = x^4 + 4x^3 + 3x^2 - 5$.

Step 2 Graph the related function.

Enter the equation in the **Y =** list and graph the function.

Step 3 Find the zeros.

Use the **zero** feature from the **CALC** menu. The real zeros are about −3.22 and 0.84.

Talk About It! Explain how you could use the table feature to more accurately estimate the zeros of the related function. What are the limitations of the table feature?

Sample answer: By changing the table settings to use a smaller interval, I can find more accurate intervals in which the zeros lie. Sometimes there is no interval small enough to identify a zero exactly using the table feature. Also, using a very small interval may require a lot of scrolling to find the zeros.

Check

Use a graphing calculator to solve $4x^2 + x = \frac{1}{2}x^4 + 1$ by graphing. Round to the nearest hundredth, if necessary.

$x =$ −2.64, −0.66, 0.39, 2.91

Copyright © McGraw-Hill Education

Lesson 5-1 • Solving Polynomial Equations by Graphing 259

Interactive Presentation

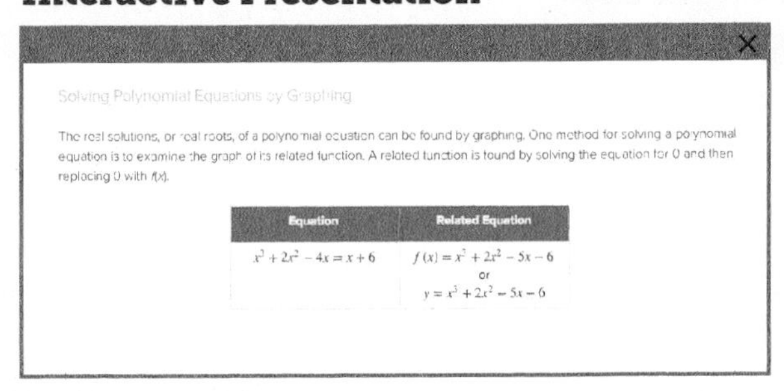

Learn

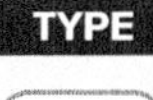

Students answer a question to determine whether they understand solving polynomial functions by graphing.

Your Notes

Example 2 Solve a Polynomial Equation by Using a System

ANIMALS **For an exhibit with six or fewer Emperor penguins, the pool must have a depth of at least 4 feet and a volume of at least 1620 gallons, or about 217 ft³, per bird. If a zoo has five Emperor penguins, what should the dimensions of the pool shown at the right be to meet the minimum requirements?**

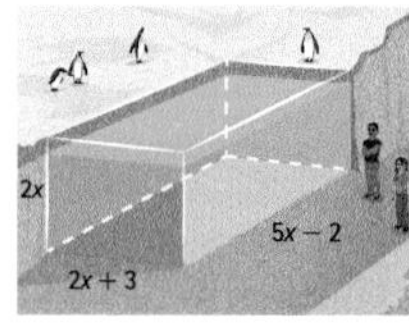

Part A Write a polynomial equation.

Use the formula for the volume of a rectangular prism, $V = \ell wh$, to write a polynomial equation that represents the volume of the pool. Let h represent the depth of the pool.

Since the minimum required volume for the pool is 217 ft³ per penguin, or 217 • 5 = 1085 ft³, the equation that represents the volume of the pool is $(2x + 3)(5x - 2)2x$ = 1085. Simplify the equation.

$(2x + 3)(5x - 2)2x = 1085$	Volume of pool
$[2x(5x) + 2x(-2) + 3(5x) + 3(-2)]2x = 1085$	FOIL
$(10x^2 - 4x + 15x - 6)2x = 1085$	Simplify.
$(10x^2 + 11x - 6)2x = 1085$	Combine like terms.
$20x^3 + 22x^2 - 12x = 1085$	Distributive Property

So, the volume of the pool is $20x^3 + 22x^2 - 12x = 1085$.

Part B Write and solve a system of equations.

Set each side equal to y to create a system of equations.

$y = 20x^3 + 22x^2 - 12x$	First equation
$y = 1085$	Second equation

Enter the equations in the **Y =** list and graph.

Use the **intersect** feature on the **CALC** menu to find the coordinates of the point of intersection.

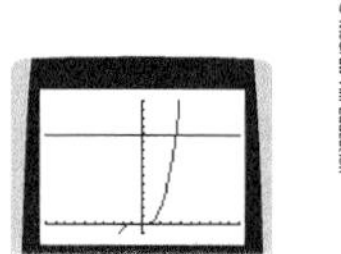

The real solution is the x-coordinate of the intersection, which is 3.5.

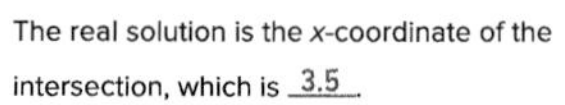

Part C Find the dimensions.

Substitute 3.5 feet for x in the length, width, and depth of the pool.

Length: $2x + 3$ = 10 ft Width: $5x - 2$ = 15.5 ft

Depth: $2x$ = 7 ft

Go Online You can complete an Extra Example online.

Think About It! Is your solution reasonable? Justify your conclusion.

Yes; sample answer: When $x = 3.5$, all of the dimensions are positive and the depth of 7 feet meets the minimum required depth of 4 feet.

260 Module 5 • Polynomial Equations

Interactive Presentation

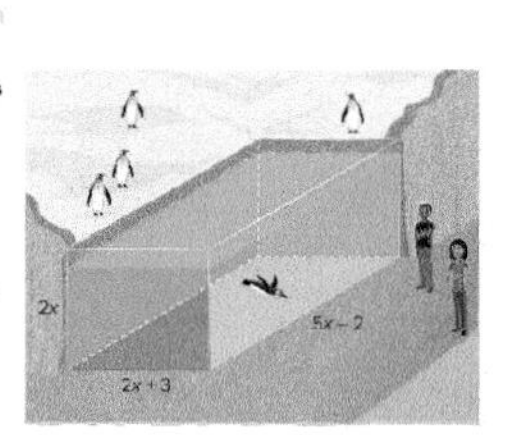

Example 2

TAP

Students tap to see the steps to write and solve a polynomial equation.

TYPE

Students complete the calculations to determine the dimensions of the pool.

CHECK

Students complete the Check online to determine whether they are ready to move on.

Example 2 Solve a Polynomial Equation by Using a System

Teaching the Mathematical Practices

2 Represent a Situation Symbolically Guide students to define variables to solve the problem in Example 2. Then work with them to find the other relationships in the problem.

Questions for Mathematical Discourse

AL Why is the total volume of the pool 1085 ft³ rather than 217 ft³? Sample answer: 217 ft³ is the volume per penguin and there are 5 penguins in the pool.

OL Could you have found the solutions by finding the zeros of a related function? Explain. Yes; sample answer: You could subtract 1085 from each side and then solve for x by finding the zeros of the related function.

BL To graph a system of equations, must the polynomial equation be simplified first? Explain. No; sample answer: The polynomial equation does not have to be simplified because the two forms are equivalent and produce the same graph.

Common Error

Many students will distribute the monomial term to both binomials before multiplying the binomials. This is incorrect as the monomial term can only be multiplied once. Encourage students to multiply binomial terms together first, then distribute any monomial terms.

Exit Ticket

Recommended Use

At the end of class, go online to display the Exit Ticket prompt and ask students to respond using a separate piece of paper. Have students hand you their responses as they leave the room.

Alternate Use

At the end of class, go online to display the Exit Ticket prompt and ask students to respond verbally or by using a mini-whiteboard. Have students hold up their whiteboards so that you can see all student responses. Tap to reveal the answer when most or all students have completed the Exit Ticket.

Practice and Homework

Suggested Assignments

Use the table below to select appropriate exercises.

DOK	Topic	Exercises
1, 2	exercises that mirror the examples	1–9
2	exercises that use a variety of skills from this lesson	10–15
2	exercises that extend concepts learned in this lesson to new contexts	16
3	exercises that emphasize higher-order and critical thinking skills	17–21

ASSESS AND DIFFERENTIATE

Use the data from the **Checks** to determine whether to provide resources for extension, remediation, or intervention.

IF students score 90% or more on the Checks,
THEN assign:

- Practice Exercises 1–17 odd, 19–23
- Extension: Complex Roots of Polynomial Equations
- ALEKS® Real Zeros of Polynomial Functions

IF students score 66%–89% on the Checks,
THEN assign:

- Practice Exercises 1–23 odd
- Remediation, Review Resources: Multiplying Polynomials
- Personal Tutors
- Extra Examples 1–2
- ALEKS® Finding x-intercepts

IF students score 65% or less on the Checks, **AL**
THEN assign:

- Practice, Exercises 1–9 odd
- Remediation, Review Resources: Multiplying Polynomials
- *Quick Review Math Handbook*: Solving Polynomial Equations
- ALEKS® Finding x-intercepts

Name ______ Period ______ Date ______

Practice

Go Online You can complete your homework online.

Example 1

Use a graphing calculator to solve each equation by graphing. If necessary, round to the nearest hundredth.

1. $\frac{2}{3}x^3 + x^2 - 5x = -9$ -4.12
2. $x^3 - 9x^2 + 27x = 20$ 1.09
3. $x^3 + 1 = 4x^2$ $-0.47, 0.54, 3.94$
4. $x^6 - 15 = 5x^4 - x^2$ $-2.31, 2.31$
5. $\frac{1}{2}x^5 = \frac{1}{5}x^2 - 2$ -1.27
6. $x^8 = -x^7 + 3$ $-1.36, 1.06$

Example 2

7. SHIPPING A shipping company will ship a package for $7.50 when the volume is no more than 15,000 cm³. Grace needs to ship a package that is $3x - 5$ cm long, $2x$ cm wide, and $x + 20$ cm tall.
   a. Write a polynomial equation to represent the situation if Grace plans to spend a maximum of $7.50. $6x^3 + 110x^2 - 200x = 15{,}000$
   b. Write and solve a system of equations. $y = 6x^3 + 110x^2 - 200x, y = 15{,}000; x = 10$ cm
   c. What should the dimensions of the package be to have the maximum volume? 25 cm by 20 cm by 30 cm
8. GARDEN A rectangular garden is 12 feet across and 16 feet long. It is surrounded by a border of mulch that is a uniform width, x. The maximum area for the garden, plus border, is 285 ft².
   a. Write a polynomial equation to represent the situation. $4x^2 + 56x + 192 = 285$
   b. Write and solve a system of equations. $y = 4x^2 + 56x + 192, y = 285; x = 1.5$ ft
   c. What are the dimensions of the garden plus border? 15 ft by 19 ft

x ft
x ft
16 ft
12 ft

9. PACKAGING A juice manufacturer is creating new cylindrical packaging. The height of the cylinder is to be 3 inches longer than the radius of the can. The cylinder is to have a volume of 628 cubic inches. Use 3.14 for π.
   a. Write a polynomial equation to support the model. $\pi x^3 + 3\pi x^2 = 628$
   b. Write and solve a system of equations. $y = \pi x^3 + 3\pi x^2, y = 628; x = 5$
   c. What are the radius and height of the new packaging? radius = 5 in., height = 8 in.

Mixed Exercises

Solve each equation. If necessary, round to the nearest hundredth.

10. $x^4 + 2x^3 = 7$ $-2.47, 1.29$
11. $x^4 - 15x^2 = -24$ $-3.63, -1.35, 1.35, 3.63$
12. $x^3 - 6x^2 + 4x = -6$ $-0.69, 1.75, 4.95$
13. $x^4 - 15x^2 + x + 65 = 0$ no solution
14. BALLOON Treyvon is standing 9 yards from the base of a hill that has a slope of $\frac{3}{4}$. He throws a water balloon from a height of 2 yards. Its path is modeled by $h(x) = -0.1x^2 + 0.8x + 2$, where h is the height of the balloon in yards and x is the distance the balloon travels in yards.
   a. Write a polynomial equation to represent the situation. $-0.1x^2 + 0.8x + 2 = \frac{3}{4}(x - 9)$
   b. How far from Treyvon will the balloon hit the hill? about 9.6 yd

15. USE TOOLS A company models its revenue in dollars using the function $P(x) = 70{,}000(x - x^4)$ on the domain (0, 1) where x is the price at which they sell their product in dollars. Use a graphing calculator to sketch a graph and find the price at which their product should be sold to make revenue of $20,000. Describe your solution process. See margin.

16. ROLLER COASTERS On a racing roller coaster, two trains start at the same time and race to see which returns to the station first. On one coaster, the height of a train on the blue track can be modeled by $f(x) = \frac{1}{20}(x^3 - 60x^2 + 900x)$ and the height of a train on the green track can be modeled by $g(x) = \frac{1}{12{,}000}(x^5 - 144x^4 + 7384x^3 - 158{,}400x^2 + 1{,}210{,}000x)$ where x is time in seconds for the first 35 seconds of the ride.

a. What equation would determine the times when the blue and green trains are at the same height? $\frac{1}{20}(x^3 - 60x^2 + 900x) = \frac{1}{12{,}000}(x^5 - 144x^4 + 7384x^3 - 158{,}400x^2 + 1{,}210{,}000x)$

b. Use a graphing calculator to sketch a graph of $f(x)$ and $g(x)$ and solve the equation from **part a**. Interpret the solution in the context of the situation. See margin.

c. Write an equation to determine the times for which the blue train modeled by $f(x)$ is at a height of 150 feet. Use a graphing calculator to solve the equation. Interpret the solution in the context of the situation. See margin.

Higher-Order Thinking Skills

17. WRITE Use a graph to explain why a function with an even degree can have zero real solutions, but a function with an odd degree must have at least one real solution. See margin.

18. CREATE Write a polynomial equation and solve it by graphing a related function and finding its zeros. Sample answer: $x^3 - 4x^2 + 12 = x - 22$; $x \approx -2.39$

19. ANALYZE Determine whether the following statement is *sometimes*, *always*, or *never* true. Justify your argument.

If a system of equations has more than one solution, then the positive solution is the only viable solution. See margin.

20. PERSEVERE During practice, a player kicks a ball from the ground with an initial velocity of 32 feet per second. The polynomial $f(x) = -16x^2 + 32x$ models the height of the ball, where x represents time in seconds. At the same time, another player heads a ball at some distance c feet off the ground with an initial velocity of 27 feet per second. The polynomial $g(x) = -16x^2 + 27x + c$ models the height of the ball.

a. If the balls are at the same height after 1.2 seconds, from what height did the second player head the ball? 6 ft

b. If $c > 0$, is it possible that the soccer balls are never at the same height? Is it reasonable in the context of the situation? Explain your reasoning. Sample answer: Only if $c > 10$ will the balls never be at the same height before they strike the ground. Because it is unreasonable for a player to head the ball more than 10 feet above the ground, it is not reasonable that the soccer balls will never be at the same height.

21. WHICH ONE DOESN'T BELONG? Which polynomial doesn't belong? Justify your conclusion. $5x^2 = -2x - 11$; It is the only one that has no real solutions.

$x - 17 = 18x^3 + 3x^2$	$x^2 = 4x^4 + 3x^2 - 8$	$5x^2 = -2x - 11$	$-4 = 2x^5 - x^2$

Answers

15. $0.29, 0.88; sample answer: graph $f(x) = 70{,}000(x - x^4)$ and $f(x) = 20{,}000$ and find the x-values of the points of intersections.

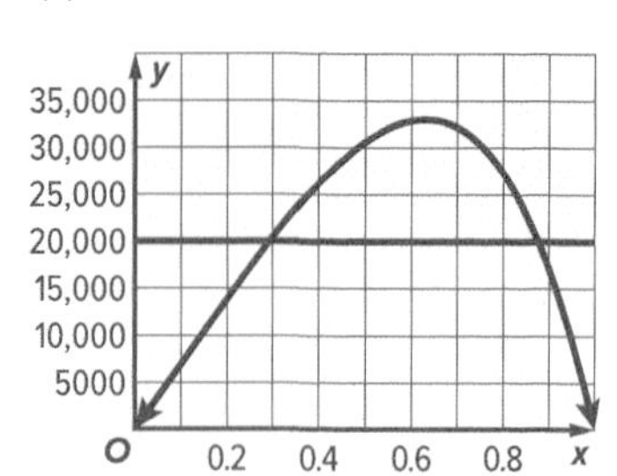

16b. The graphs of the functions intersect at $x = 0$ and $x \approx 9.6$ and $x \approx 26.0$. The passenger cars are at the same height at 0 seconds and approximately 9.6 and 26.0 seconds.

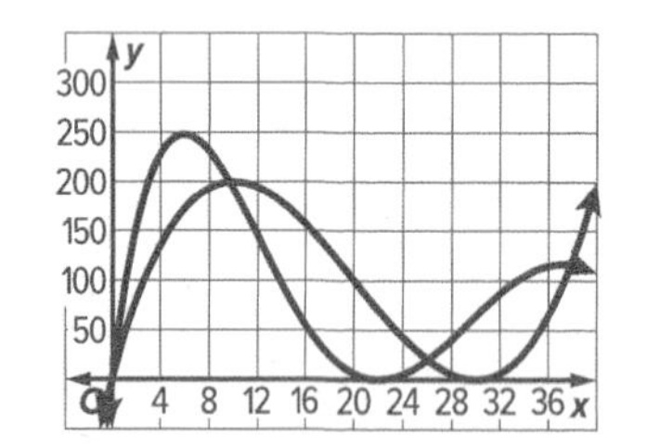

16c. $\frac{1}{20}(x^3 - 60x^2 + 900x) = 150$, which has solutions at $x \approx 16.5$ and $x \approx 4.7$. The passenger car is at 150 feet at approximately 4.7 seconds and 16.5 seconds.

17. Sample answer: A function with an even degree reverses direction, so both ends extend in the same direction. That means that if the function opens up and the vertex is above the x-axis or if the function opens down and the vertex is below the x-axis, there are no real solutions. A function with an odd degree does not reverse direction, so the ends extend in opposite directions. Therefore, it must cross the x-axis at least once.

19. Sometimes; sample answer: The positive solution is often correct when a negative value doesn't make sense, such as for distance or time; however, sometimes a negative solution is reasonable, such as for temperature or position problems. Also, sometimes there are two positive solutions and one may be unreasonable.

Lesson 5-2

A.CED.1

Solving Polynomial Equations Algebraically

LESSON GOAL

Students solve polynomial equations by factoring and by writing them in quadratic form.

1 LAUNCH

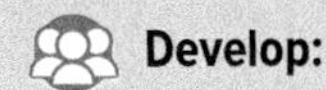

Launch the lesson with a **Warm Up** and an introduction.

2 EXPLORE AND DEVELOP

Develop:

Solving Polynomial Equations by Factoring

- Factor Sums and Differences of Cubes
- Factor by Grouping
- Combine Cubes and Squares
- Solve a Polynomial Equation by Factoring
- Write and Solve a Polynomial Equation by Factoring

Solving Polynomial Equations in Quadratic Form

- Write Expressions in Quadratic Form
- Solve Equations in Quadratic Form

You may want your students to complete the **Checks** online.

3 REFLECT AND PRACTICE

Exit Ticket

Practice

DIFFERENTIATE

View reports of student progress on the **Checks** after each example.

Resources	AL	OL	BL	ELL
Remediation: Solving Quadratic Equations by Factoring	●	●		●
Extension: History of Quadratic Equations		●	●	●

Language Development Handbook

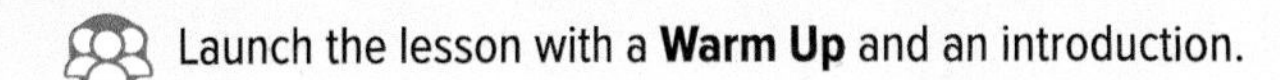

Assign page 31 of the *Language Development Handbook* to help your students build mathematical language related to solving polynomial equations by factoring.

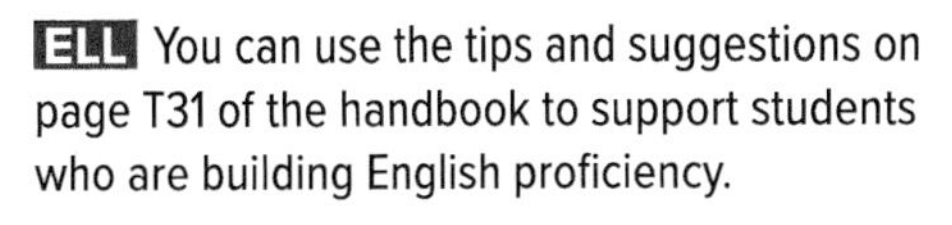

ELL You can use the tips and suggestions on page T31 of the handbook to support students who are building English proficiency.

Suggested Pacing

90 min	1 day
45 min	2 days

Focus

Domain: Algebra

Standards for Mathematical Content:

A.CED.1 Create equations and inequalities in one variable and use them to solve problems.

Standards for Mathematical Practice:

1 Make sense of problems and persevere in solving them.

7 Look for and make use of structure.

Coherence

Vertical Alignment

Previous
Students solved quadratic equations by factoring.
N.CN.7, N.CN.8, F.IF.8a

Now
Students solve polynomial equations by factoring and by writing them in quadratic form.
A.CED.1

Next
Students will evaluate and factor functions by using the Remainder and Factor Theorems.
A.APR.2

Rigor

The Three Pillars of Rigor

1 CONCEPTUAL UNDERSTANDING	2 FLUENCY	3 APPLICATION

Conceptual Bridge In this lesson, students draw on their understanding of factoring to solve polynomial equations. They build fluency by factoring to find solutions, and they apply their understanding of factoring of solving real-world problems.

Mathematical Background

If a polynomial can be written as a product of linear and quadratic factors, its zeros can be found. Set each of the factors equal to 0 and solve the resulting equations.

Interactive Presentation

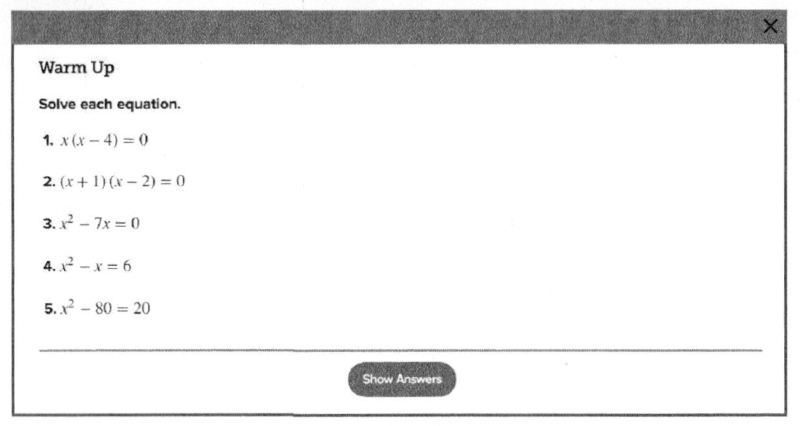

Warm Up

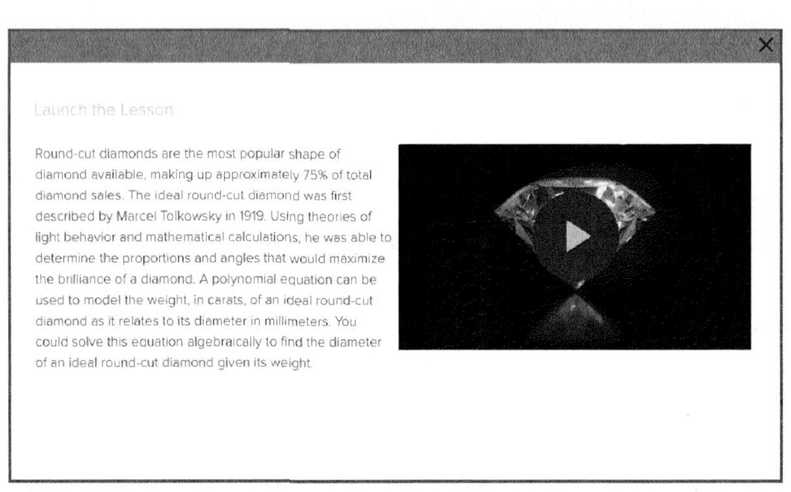

Launch the Lesson

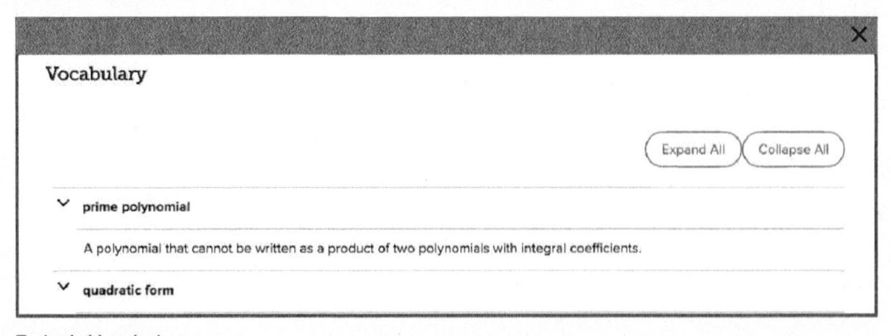

Today's Vocabulary

Warm Up

Prerequisite Skills

The Warm Up exercises address the following prerequisite skill for this lesson:

- solving simple quadratic equations

Answers:

1. 0, 4
2. −1, 2
3. 0, 7
4. −2, 3
5. −10, 10

Launch the Lesson

Teaching the Mathematical Practices

4 Apply Mathematics In the Launch the Lesson, students apply polynomial equations to a real-world problem.

Go Online to find additional teaching notes and questions to promote classroom discourse.

Today's Standards

Tell students that they will be addressing these content and practice standards in this lesson. You may wish to have a student volunteer read aloud *How can I meet this standard?* and *How can I use these practices?*, and connect these to the standards.

See the Interactive Presentation for I Can statements that align with the standards covered in this lesson.

Today's Vocabulary

Tell students that they will be using these vocabulary terms in this lesson. You can expand each row if you wish to share the definitions. Then discuss the questions below with the class.

Learn Solving Polynomial Equations by Factoring

Objective

Students solve polynomial equations by factoring.

 Teaching the Mathematical Practices

1 Understand the Approaches of Others Mathematically proficient students can explain the methods to solve a problem. The Think About It! feature asks students to justify the reasoning of Mateo.

Common Misconception

A common misconception some students may have is that a sum or difference of cubes can be factored into the prime factor of $(a \pm b)^3$. Show students that when $(a \pm b)^3$ is multiplied out, the resulting polynomial is not a sum or difference of cubes.

Example 1 Factor Sums and Differences of Cubes

 Teaching the Mathematical Practices

7 Use Structure Guide students to use the structure of each polynomial expression in Example 1 to determine which method of factoring to apply.

Questions for Mathematical Discourse

AL What is a greatest common factor? It is the greatest term that divides evenly from all terms of the algebraic expression.

OL When raising a power to another power, what operation do you perform on the exponents? multiply

BL Is the resulting trinomial in the sum or difference of cubes factorable? Explain. No; sample answer: The resulting trinomial will never be factorable since there are no two factors that can multiply to $a^2 \pm ab + b^2$.

- Find additional teaching notes.
- View performance reports of the Checks.
- Assign or present an Extra Example.

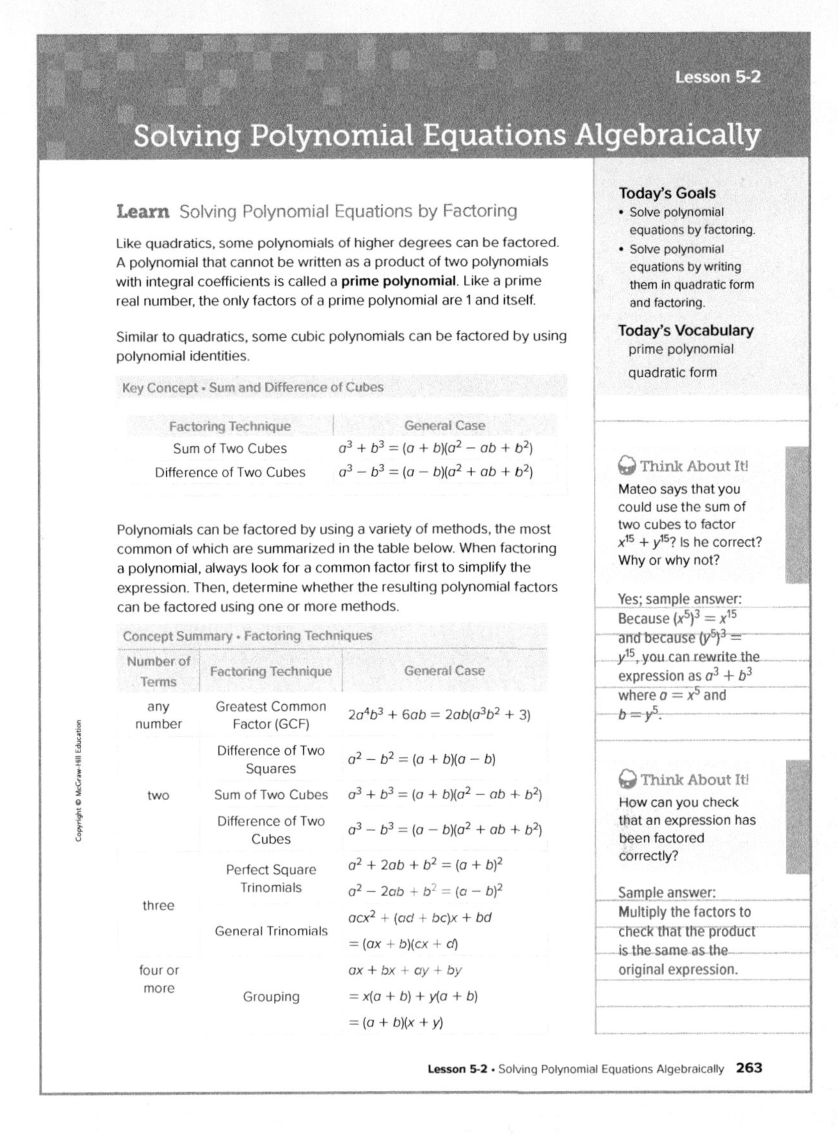

Lesson 5-2

Solving Polynomial Equations Algebraically

Learn Solving Polynomial Equations by Factoring

Like quadratics, some polynomials of higher degrees can be factored. A polynomial that cannot be written as a product of two polynomials with integral coefficients is called a **prime polynomial**. Like a prime real number, the only factors of a prime polynomial are 1 and itself.

Similar to quadratics, some cubic polynomials can be factored by using polynomial identities.

Key Concept • Sum and Difference of Cubes

Factoring Technique	General Case
Sum of Two Cubes	$a^3 + b^3 = (a + b)(a^2 - ab + b^2)$
Difference of Two Cubes	$a^3 - b^3 = (a - b)(a^2 + ab + b^2)$

Polynomials can be factored by using a variety of methods, the most common of which are summarized in the table below. When factoring a polynomial, always look for a common factor first to simplify the expression. Then, determine whether the resulting polynomial factors can be factored using one or more methods.

Concept Summary • Factoring Techniques

Number of Terms	Factoring Technique	General Case
any number	Greatest Common Factor (GCF)	$2a^4b^3 + 6ab = 2ab(a^3b^2 + 3)$
two	Difference of Two Squares	$a^2 - b^2 = (a + b)(a - b)$
	Sum of Two Cubes	$a^3 + b^3 = (a + b)(a^2 - ab + b^2)$
	Difference of Two Cubes	$a^3 - b^3 = (a - b)(a^2 + ab + b^2)$
three	Perfect Square Trinomials	$a^2 + 2ab + b^2 = (a + b)^2$ $a^2 - 2ab + b^2 = (a - b)^2$
	General Trinomials	$acx^2 + (ad + bc)x + bd = (ax + b)(cx + d)$
four or more	Grouping	$ax + bx + ay + by = x(a + b) + y(a + b) = (a + b)(x + y)$

Today's Goals
- Solve polynomial equations by factoring.
- Solve polynomial equations by writing them in quadratic form and factoring.

Today's Vocabulary
prime polynomial
quadratic form

Think About It!
Mateo says that you could use the sum of two cubes to factor $x^{15} + y^{15}$? Is he correct? Why or why not?

Yes; sample answer: Because $(x^5)^3 = x^{15}$ and because $(y^5)^3 = y^{15}$, you can rewrite the expression as $a^3 + b^3$ where $a = x^5$ and $b = y^5$.

Think About It!
How can you check that an expression has been factored correctly?

Sample answer: Multiply the factors to check that the product is the same as the original expression.

Copyright © McGraw-Hill Education

Lesson 5-2 • Solving Polynomial Equations Algebraically 263

Interactive Presentation

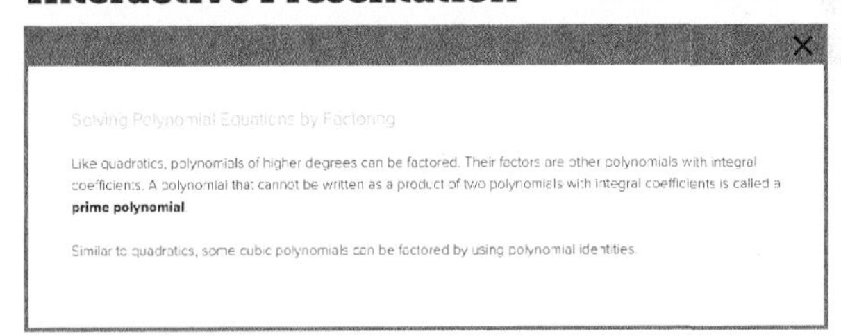

Learn

TYPE

Students explain whether a polynomial expression can be factored using the sum of two cubes.

Your Notes

Example 1 Factor Sums and Differences of Cubes

Factor each polynomial. If the polynomial cannot be factored, write *prime*.

a. $8x^3 + 125y^{12}$

The GCF of the terms is 1, but $8x^3$ and $125y^{12}$ are both perfect cubes. Factor the sum of two cubes.

$8x^3 + 125y^{12}$	Original expression
$= (2x)^3 + (\underline{5y^4})^3$	$(2x)^3 = 8x^3$; $(5y^4)^3 = 125y^{12}$
$= (2x + 5y^4)[(2x)^2 - (2x)(5y^4) + (5y^4)^2]$	Sum of two cubes
$= (2x + 5y^4)(\underline{4x^2} - 10xy^4 + \underline{25y^8})$	Simplify.

b. $54x^5 - 128x^2y^3$

$54x^5 - 128x^2y^3$	Original expression
$= 2x^2(27x^3 - 64y^3)$	Factor out the GCF.
$= 2x^2[(\underline{3x})^3 - (\underline{4y})^3]$	$(3x)^3 = 27x^3$; $(4y)^3 = 64y^3$
$= 2x^2(3x - 4y)[(3x)^2 + 3x(4y) + (4y)^2]$	Difference of two cubes
$= 2x^2(3x - 4y)(9x^2 + \underline{12xy} + \underline{16y^2})$	Simplify.

Study Tip

Grouping When grouping 6 or more terms, group the terms that have the *most* common values.

Example 2 Factor by Grouping

Factor $14ax^2 - 16by + 20cy + 28bx^2 - 35cx^2 - 8ay$. If the polynomial cannot be factored, write *prime*.

$14ax^2 - 16by + 20cy + 28bx^2 - 35cx^2 - 8ay$	Original expression
$= (14ax^2 + 28bx^2 - 35cx^2) + (-8ay - 16by + 20cy)$	Group to find a GCF.
$= \underline{7x^2}(2a + 4b - 5c) - \underline{4y}(2a + 4b - 5c)$	Factor out the GCF.
$= (7x^2 - 4y)(\underline{2a} + \underline{4b} - \underline{5c})$	Distributive Property

Think About It!

When factoring by grouping, what must be true about the expressions inside parentheses after factoring out a GCF from each group?

Sample answer: They must be the same.

Example 3 Combine Cubes and Squares

Factor $64x^6 - y^6$. If the polynomial cannot be factored, write *prime*.

This polynomial could be considered the difference of two squares or the difference of two cubes. The difference of two squares should always be done before the difference of two cubes for easy factoring.

$64x^6 - y^6$	Original expression
$= (\underline{8x^3})^2 - (\underline{y^3})^2$	$(8x^3)^2 = 64x^6$; $(y^3)^2 = y^6$
$= (8x^3 + y^3)(8x^3 - y^3)$	Difference of squares
$= [(\underline{2x})^3 + y^3][(\underline{2x})^3 - y^3]$	$(2x)^3 = 8x^3$
$= (2x + \underline{y})(4x^2 - \underline{2xy} + y^2)(2x - \underline{y})(4x^2 + \underline{2xy} + y^2)$	Sum and difference of cubes

Go Online You can complete an Extra Example online.

264 Module 5 • Polynomial Equations

Interactive Presentation

Factor by Grouping

To factor polynomials with four or more terms, the only method that can be used is factoring by grouping.

Factor $14ax^2 - 16by + 20cy + 28bx^2 - 35cx^2 - 8ay$. If the polynomial cannot be factored, write *prime*.

$14ax^2 - 16by + 20cy + 28bx^2 - 35cx^2 - 8ay$	Original expression
$= (14ax^2 + 28bx^2 - 35cx^2) + (-8ay - 16by + 20cy)$	Group to find a GCF.
$= 7x^2(2a + 4b - 5c) - 4y(2a + 4b - 5c)$	Factor out the GCF.
$= (7x^2 - 4y)(2a + 4b - 5c)$	Distributive Property

Example 2

TYPE

Students answer a question to determine whether they understand factoring by grouping.

Example 2 Factor by Grouping

Teaching the Mathematical Practices

1 Seek Information Students must transform the polynomial expression in order to apply the method of factoring by grouping.

Questions for Mathematical Discourse

AL How should an original polynomial be grouped so that it can be factored? Sample answer: so that each group has a GCF which results in two expressions that are the same after factoring out the GCF.

OL Why is the GCF for the second grouping negative? Sample answer: to make the signs match the first grouping.

BL Is there another way to group the polynomial expression into two groups and factor? Explain. No; sample answer: Even though there may be a GCF from a new grouping, the resulting polynomials will not be the same. Thus it cannot be factored by grouping.

Example 3 Combine Cubes and Squares

Teaching the Mathematical Practices

1 Analyze Givens and Constraints In Example 3, guide students through planning the solution pathway. Help students to see why the difference of squares should be factored first.

Questions for Mathematical Discourse

AL Why is the expression a difference of squares and a difference of cubes? The expression can be written as $(8x^3)^2 - (y^3)^2$ or $(4x^2)^3 - (y^2)^3$.

OL What must be true about the coefficients and exponents of an expression that is both a difference of cubes and a difference of squares? Sample answer: The coefficients must be both perfect squares and perfect cubes, and the exponents must be a multiple of both 2 and 3.

BL Why is it easier to factor by difference of squares before using the sum or difference of cubes? Sample answer: If the polynomial expression is factored using the sum or difference of cubes first, the resulting trinomial must be factored. Applying the difference of squares first, results in a simpler expression to factor.

Example 4 Solve a Polynomial Equation by Factoring

Teaching the Mathematical Practices

7 Use the Distributive Property The Distributive Property is one of the most-used properties in algebra. Students should be able use the Distributive Property to factor by grouping.

Questions for Mathematical Discourse

AL How did you know to start the factoring process by using grouping? There were 4 terms.

OL Describe another way to factor the original polynomial equation by grouping. Sample answer: I could group $(4x^3 - 9x) + (12x^2 - 27)$. This would factor to $x(4x^2 - 9) + 3(4x^2 - 9)$, which yields the same factors of $(x + 3)(4x^2 - 9)$.

BL What pattern do you see in the original equation that indicates a factor of $(x + 3)$? Sample answer: The coefficient of the second term is 3 times the coefficient of the first term, and the coefficient of the fourth term is 3 times the coefficient of the third term.

Example 5 Write and Solve a Polynomial Equation by Factoring

Teaching the Mathematical Practices

4 Use Tools Point out that to solve the problem in Example 5, students will need to use the diagram to write and solve a polynomial equation.

Questions for Mathematical Discourse

AL Why do you subtract the volume of the small cube from the volume of the large cube? You are given the volume of the figure, which is the large cube with the small cube removed.

OL Why are the solutions you found when solving $x^2 + 3x + 9 = 0$ not viable solutions? Sample answer: The solutions are imaginary and because x represents a length, it must be a real number.

BL How many and what type of cube roots does a real number have? one real cube root and a complex conjugate pair of cube roots

Learn Solving Polynomial Equations in Quadratic Form

Objective

Students solve polynomial equations by writing them in quadratic form and factoring.

Teaching the Mathematical Practices

1 Seek Information Mathematically proficient students must be able to transform algebraic expressions to reach solutions. Point out that to factor some polynomial expressions, it may be helpful to rewrite the expression in quadratic form.

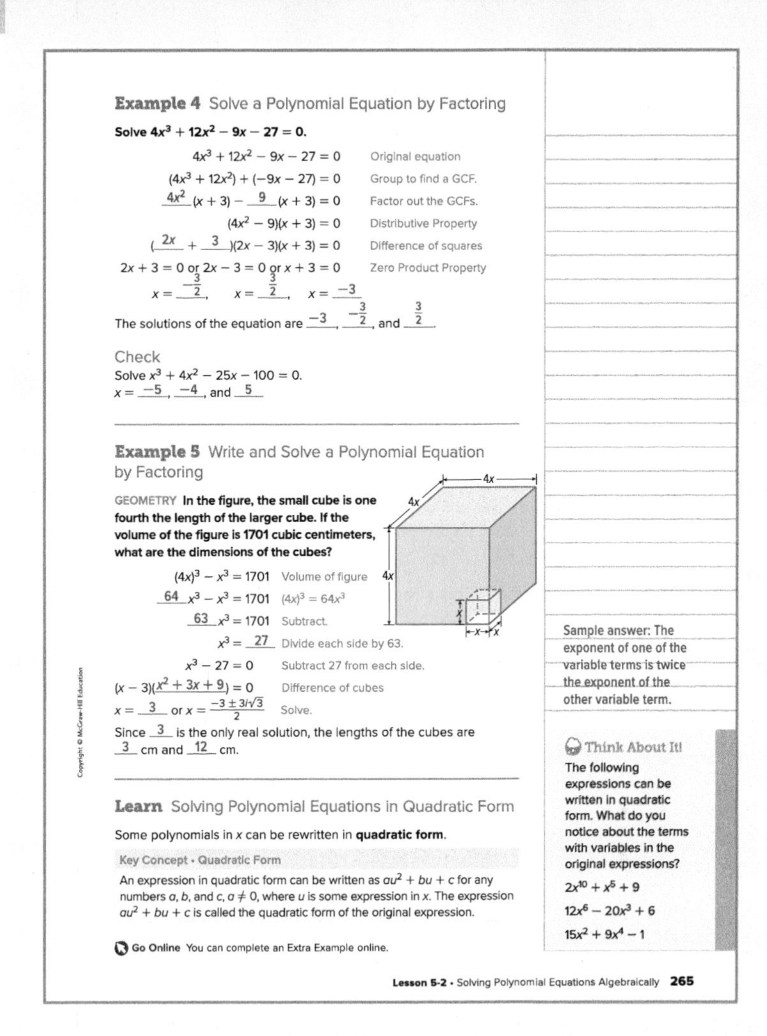

Example 4 Solve a Polynomial Equation by Factoring

Solve $4x^3 + 12x^2 - 9x - 27 = 0$.

$4x^3 + 12x^2 - 9x - 27 = 0$	Original equation
$(4x^3 + 12x^2) + (-9x - 27) = 0$	Group to find a GCF.
$\underline{4x^2}(x + 3) - \underline{9}(x + 3) = 0$	Factor out the GCFs.
$(4x^2 - 9)(x + 3) = 0$	Distributive Property
$(\underline{2x} + \underline{3})(2x - 3)(x + 3) = 0$	Difference of squares
$2x + 3 = 0$ or $2x - 3 = 0$ or $x + 3 = 0$	Zero Product Property
$x = \underline{-\frac{3}{2}}$, $x = \underline{\frac{3}{2}}$, $x = \underline{-3}$	

The solutions of the equation are $\underline{-3}$, $\underline{-\frac{3}{2}}$, and $\underline{\frac{3}{2}}$.

Check

Solve $x^3 + 4x^2 - 25x - 100 = 0$.
$x = \underline{-5}$, $\underline{-4}$, and $\underline{5}$

Example 5 Write and Solve a Polynomial Equation by Factoring

GEOMETRY **In the figure, the small cube is one fourth the length of the larger cube. If the volume of the figure is 1701 cubic centimeters, what are the dimensions of the cubes?**

$(4x)^3 - x^3 = 1701$	Volume of figure
$\underline{64}x^3 - x^3 = 1701$	$(4x)^3 = 64x^3$
$\underline{63}x^3 = 1701$	Subtract.
$x^3 = \underline{27}$	Divide each side by 63.
$x^3 - 27 = 0$	Subtract 27 from each side.
$(x - 3)(\underline{x^2 + 3x + 9}) = 0$	Difference of cubes
$x = \underline{3}$ or $x = \frac{-3 \pm 3i\sqrt{3}}{2}$	Solve.

Since $\underline{3}$ is the only real solution, the lengths of the cubes are $\underline{3}$ cm and $\underline{12}$ cm.

Learn Solving Polynomial Equations in Quadratic Form

Some polynomials in x can be rewritten in **quadratic form**.

Key Concept • Quadratic Form

An expression in quadratic form can be written as $au^2 + bu + c$ for any numbers a, b, and c, $a \neq 0$, where u is some expression in x. The expression $au^2 + bu + c$ is called the quadratic form of the original expression.

Go Online You can complete an Extra Example online.

Sample answer: The exponent of one of the variable terms is twice the exponent of the other variable term.

Think About It! The following expressions can be written in quadratic form. What do you notice about the terms with variables in the original expressions?

$2x^{10} + x^5 + 9$

$12x^6 - 20x^3 + 6$

$15x^2 + 9x^4 - 1$

Interactive Presentation

Solve a Polynomial Equation by Factoring

You can solve a polynomial equation by factoring and then using the Zero Product Property to set each factor equal to zero.

Solve $4x^3 + 12x^2 - 9x - 27 = 0$.

$4x^3 + 12x^2 - 9x - 27 = 0$	Original Equation
$(4x^3 + 12x^2) + (-9x - 27) = 0$	Group to find a GCF.
$4x^2(x + 3) - 9(x + 3) = 0$	Factor out the GCFs.
$(4x^2 - 9)(x + 3) = 0$	Distributive Property

Example 4

CHECK

Students complete the Check online to determine whether they are ready to move on.

A.CED.1

1 CONCEPTUAL UNDERSTANDING | **2 FLUENCY** | 3 APPLICATION

Talk About It!

Describe how the exponent of the expression equal to u relates to the exponents of the terms with variables.

Sample answer: The exponent of the expression equal to u is the same as the lesser exponent of the two terms with variables.

Example 6 Write Expressions in Quadratic Form

Write each expression in quadratic form, if possible.

a. $4x^{20} + 6x^{10} + 15$

Examine the terms with variables to choose the expression equal to u.

$4x^{20} + 6x^{10} + 15 = (\underline{2x^{10}})^2 + \underline{3}(2x^{10}) + \underline{15}$ — $(2x^{10})^2 = 4x^{20}$

b. $18x^4 + 180x^8 - 28$

If the polynomial is not already in standard form, rewrite it. Then examine the terms with variables to choose the expression equal to u.

$18x^4 + 180x^8 - 28 = 180x^8 + 18x^4 - 28$ — Standard form of a polynomial

$= \underline{5}(6x^4)^2 + \underline{3}(6x^4) - \underline{28}$ — $(6x^4)^2 = 36x^8$

c. $9x^6 - 4x^2 - 12$

Because $x^6 \neq (x^2)^2$, the expression _cannot_ be written in quadratic form.

Check

What is the quadratic form of $10x^4 + 100x^8 - 9$?

$4(5x^4)^2 + 2(5x^4) - 9$ or $(10x^4)^2 + 10x^4 - 9$

Example 7 Solve Equations in Quadratic Form

Solve $8x^4 + 10x^2 - 12 = 0$.

$8x^4 + 10x^2 - 12 = 0$	Original equation
$\underline{2}(2x^2)^2 + \underline{5}(2x^2) - 12 = 0$	$2(2x^2)^2 = 8x^4$
$2u^2 + 5u - 12 = 0$	Let $u = 2x^2$.
$(\underline{2u} - \underline{3})(u + 4) = 0$	Factor.
$u = \underline{\frac{3}{2}}$ or $u = \underline{-4}$	Zero Product Property
$2x^2 = \frac{3}{2}$ $\quad 2x^2 = -4$	Replace u with $2x^2$.
$x^2 = \frac{3}{4}$ $\quad x^2 = -2$	Divide each side by 2.
$x = \pm\frac{\sqrt{3}}{2}$ $\quad x = \pm i\sqrt{2}$	Take the square root of each side.

The solutions are $\frac{\sqrt{3}}{2}$, $-\frac{\sqrt{3}}{2}$, $i\sqrt{2}$, and $-i\sqrt{2}$.

Check

What are the solutions of $16x^4 + 24x^2 - 40 = 0$?

$x = \underline{-1, 1, i\frac{\sqrt{10}}{2}, -i\frac{\sqrt{10}}{2}}$

Go Online You can complete an Extra Example online.

266 Module 5 • Polynomial Equations

Interactive Presentation

Write Expressions in Quadratic Form

Write each expression in quadratic form, if possible.

a. $4x^{20} + 6x^{10} + 15$

Examine the terms with variables to choose the expression equal to u.

$4x^{20} + 6x^{10} + 15 = (2x^{10})^2 + 3(2x^{10}) + 15$ — $(2x^{10})^2 = 4x^{20}$

Example 6

TYPE

Students answer a question to determine whether they understand how to write polynomials in quadratic form.

CHECK

Students complete the Check online to determine whether they are ready to move on.

Example 6 Write Expressions in Quadratic Form

Teaching the Mathematical Practices

7 Use Structure Help students to use the structure of the expressions to rewrite them in quadratic form.

Questions for Mathematical Discourse

AL How do you square a variable and its coefficient? Multiply the exponent of the variable by two and square the coefficient.

OL How do you determine the expression equal to u? Sample answer: The coefficient of u is the principal square root of the original leading coefficient, and the exponent is half the greatest exponent.

BL What must be true about a polynomial expression to write it in quadratic form? Sample answer: The exponent of the leading term must be twice the exponent of the middle term, and the third term must be a constant.

Example 7 Solve Equations in Quadratic Form

Teaching the Mathematical Practices

6 State the Meaning of Symbols Guide students to define u in the problem. Point out that it is important to clarify the meaning of the variable.

Questions for Mathematical Discourse

AL Why are $\frac{3}{2}$ and -4 not the solutions of the equation? Sample answer: $\frac{3}{2}$ and -4 are the values of u, not x. So, you must set each value of u equal to $2x^2$ to find the solutions of the equation.

OL Why are there four solutions to the polynomial equation when there were only two factors? Sample answer: Each factor generated an equation involving x^2. When the square root of each side is taken, the plus or minus sign must be used. This creates two solutions per factor.

BL What is another first step you could take when solving this equation? Sample answer: You could factor 2 out of the equation. The solutions will remain the same.

Exit Ticket

Recommended Use

At the end of class, go online to display the Exit Ticket prompt and ask students to respond using a separate piece of paper. Have students hand you their responses as they leave the room.

Alternate Use

At the end of class, go online to display the Exit Ticket prompt and ask students to respond verbally or by using a mini-whiteboard. Have students hold up their whiteboards so that you can see all student responses. Tap to reveal the answer when most or all students have completed the Exit Ticket.

Practice and Homework

Suggested Assignments

Use the table below to select appropriate exercises.

DOK	Topic	Exercises
1, 2	exercises that mirror the examples	1–27
2	exercises that use a variety of skills from this lesson	28–67
2	exercises that extend concepts learned in this lesson to new contexts	68–71
3	exercises that emphasize higher-order and critical thinking skills	72–76

ASSESS AND DIFFERENTIATE

Use the data from the **Checks** to determine whether to provide resources for extension, remediation, or intervention.

IF students score 90% or more on the Checks,
THEN assign: BL

- Practice Exercises 1–85 odd, 87–91
- Extension: Complex Roots of Polynomial Equations
- **ALEKS**® Real Zeros of Polynomial Functions

IF students score 66%–89% on the Checks,
THEN assign: OL

- Practice Exercises 1–91 odd
- Remediation, Review Resources: Solving Quadratic Equations by Factoring
- Personal Tutors
- Extra Examples 1–7
- **ALEKS**® Solving Quadratic Equations by Factoring

IF students score 65% or less on the Checks,
THEN assign: AL

- Practice Exercises 1–27 odd
- Remediation, Review Resources: Solving Quadratic Equations by Factoring
- *Quick Review Math Handbook*: Solving Polynomial Equations
- **ALEKS**® Solving Quadratic Equations by Factoring

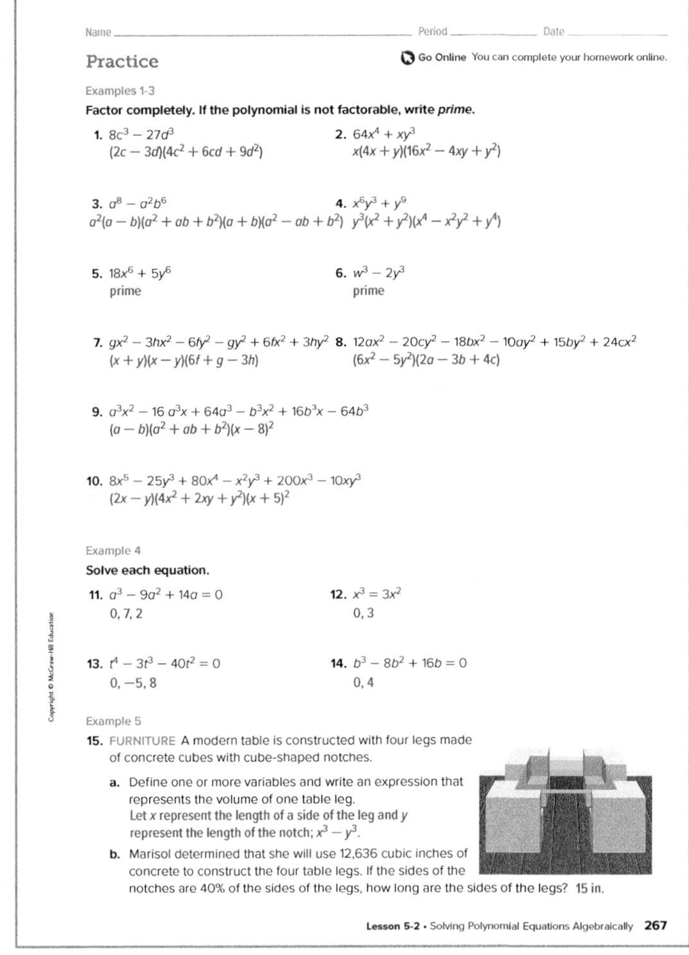

Name ______ Period ______ Date ______

Practice

Go Online You can complete your homework online.

Examples 1-3

Factor completely. If the polynomial is not factorable, write *prime*.

1. $8c^3 - 27d^3$
   $(2c - 3d)(4c^2 + 6cd + 9d^2)$
2. $64x^4 + xy^3$
   $x(4x + y)(16x^2 - 4xy + y^2)$
3. $a^8 - a^2b^6$
   $a^2(a - b)(a^2 + ab + b^2)(a + b)(a^2 - ab + b^2)$
4. $x^6y^3 + y^9$
   $y^3(x^2 + y^2)(x^4 - x^2y^2 + y^4)$
5. $18x^6 + 5y^6$
   prime
6. $w^3 - 2y^3$
   prime
7. $gx^2 - 3hx^2 - 6fy^2 - gy^2 + 6fx^2 + 3hy^2$
   $(x + y)(x - y)(6f + g - 3h)$
8. $12ax^2 - 20cy^2 - 18bx^2 - 10ay^2 + 15by^2 + 24cx^2$
   $(6x^2 - 5y^2)(2a - 3b + 4c)$
9. $a^3x^2 - 16a^3x + 64a^3 - b^3x^2 + 16b^3x - 64b^3$
   $(a - b)(a^2 + ab + b^2)(x - 8)^2$
10. $8x^5 - 25y^3 + 80x^4 - x^2y^3 + 200x^3 - 10xy^3$
    $(2x - y)(4x^2 + 2xy + y^2)(x + 5)^2$

Example 4

Solve each equation.

11. $a^3 - 9a^2 + 14a = 0$
    0, 7, 2
12. $x^3 = 3x^2$
    0, 3
13. $t^4 - 3t^3 - 40t^2 = 0$
    0, −5, 8
14. $b^3 - 8b^2 + 16b = 0$
    0, 4

Example 5

15. FURNITURE A modern table is constructed with four legs made of concrete cubes with cube-shaped notches.
    a. Define one or more variables and write an expression that represents the volume of one table leg.
       Let x represent the length of a side of the leg and y represent the length of the notch; $x^3 - y^3$.
    b. Marisol determined that she will use 12,636 cubic inches of concrete to construct the four table legs. If the sides of the notches are 40% of the sides of the legs, how long are the sides of the legs? 15 in.

Lesson 5-2 • Solving Polynomial Equations Algebraically 267

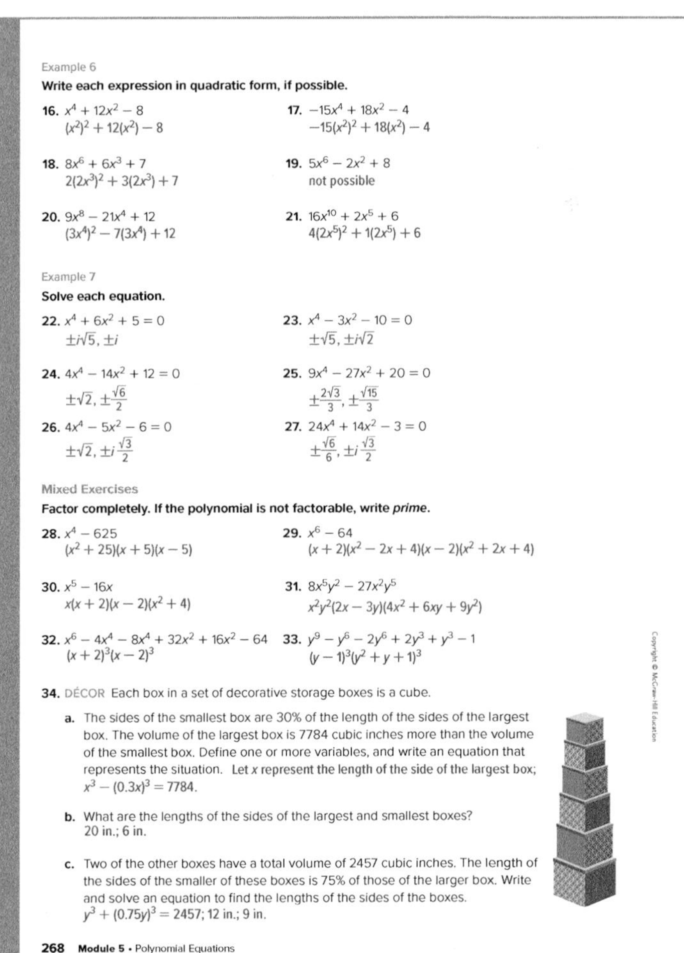

Example 6

Write each expression in quadratic form, if possible.

16. $x^4 + 12x^2 - 8$
    $(x^2)^2 + 12(x^2) - 8$
17. $-15x^4 + 18x^2 - 4$
    $-15(x^2)^2 + 18(x^2) - 4$
18. $8x^6 + 6x^3 + 7$
    $2(2x^3)^2 + 3(2x^3) + 7$
19. $5x^6 - 2x^2 + 8$
    not possible
20. $9x^8 - 21x^4 + 12$
    $(3x^4)^2 - 7(3x^4) + 12$
21. $16x^{10} + 2x^5 + 6$
    $4(2x^5)^2 + 1(2x^5) + 6$

Example 7

Solve each equation.

22. $x^4 + 6x^2 + 5 = 0$
    $\pm i\sqrt{5}, \pm i$
23. $x^4 - 3x^2 - 10 = 0$
    $\pm\sqrt{5}, \pm i\sqrt{2}$
24. $4x^4 - 14x^2 + 12 = 0$
    $\pm\sqrt{2}, \pm\frac{\sqrt{6}}{2}$
25. $9x^4 - 27x^2 + 20 = 0$
    $\pm\frac{2\sqrt{3}}{3}, \pm\frac{\sqrt{15}}{3}$
26. $4x^4 - 5x^2 - 6 = 0$
    $\pm\sqrt{2}, \pm i\frac{\sqrt{3}}{2}$
27. $24x^4 + 14x^2 - 3 = 0$
    $\pm\frac{\sqrt{6}}{6}, \pm i\frac{\sqrt{3}}{2}$

Mixed Exercises

Factor completely. If the polynomial is not factorable, write *prime*.

28. $x^4 - 625$
    $(x^2 + 25)(x + 5)(x - 5)$
29. $x^6 - 64$
    $(x + 2)(x^2 - 2x + 4)(x - 2)(x^2 + 2x + 4)$
30. $x^5 - 16x$
    $x(x + 2)(x - 2)(x^2 + 4)$
31. $8x^5y^2 - 27x^2y^5$
    $x^2y^2(2x - 3y)(4x^2 + 6xy + 9y^2)$
32. $x^6 - 4x^4 - 8x^4 + 32x^2 + 16x^2 - 64$
    $(x + 2)^3(x - 2)^3$
33. $y^9 - y^6 - 2y^6 + 2y^3 + y^3 - 1$
    $(y - 1)^3(y^2 + y + 1)^3$
34. DÉCOR Each box in a set of decorative storage boxes is a cube.
    a. The sides of the smallest box are 30% of the length of the sides of the largest box. The volume of the largest box is 7784 cubic inches more than the volume of the smallest box. Define one or more variables, and write an equation that represents the situation. Let x represent the length of the side of the largest box; $x^3 - (0.3x)^3 = 7784$.
    b. What are the lengths of the sides of the largest and smallest boxes?
       20 in.; 6 in.
    c. Two of the other boxes have a total volume of 2457 cubic inches. The length of the sides of the smaller of these boxes is 75% of those of the larger box. Write and solve an equation to find the lengths of the sides of the boxes.
       $y^3 + (0.75y)^3 = 2457$; 12 in.; 9 in.

268 Module 5 • Polynomial Equations

Name ________________ Period ________ Date ________

Solve each equation.

35. $x^4 + x^2 - 90 = 0$
$3, -3, \pm i\sqrt{10}$

36. $x^4 - 16x^2 - 720 = 0$
$6, -6, \pm 2i\sqrt{5}$

37. $x^4 - 7x^2 - 44 = 0$
$\pm\sqrt{11}, \pm 2i$

38. $x^4 + 6x^2 - 91 = 0$
$\pm\sqrt{7}, \pm i\sqrt{13}$

39. $x^3 + 216 = 0$
$-6, 3 \pm 3i\sqrt{3}$

40. $64x^3 + 1 = 0$
$-\frac{1}{4}, \frac{1 \pm i\sqrt{3}}{8}$

41. $8x^4 + 10x^2 - 3 = 0$
$\pm\frac{1}{2}, \pm i\frac{\sqrt{6}}{2}$

42. $6x^4 - 5x^2 - 4 = 0$
$\pm\frac{2\sqrt{3}}{3}, \pm i\frac{\sqrt{2}}{2}$

43. $20x^4 - 53x^2 + 18 = 0$
$\pm\frac{3}{2}, \pm\frac{\sqrt{10}}{5}$

44. $18x^4 + 43x^2 - 5 = 0$
$\pm\frac{1}{3}, \pm i\frac{\sqrt{10}}{2}$

45. $8x^4 - 18x^2 + 4 = 0$
$\pm\frac{1}{2}, \pm\sqrt{2}$

46. $3x^4 - 22x^2 - 45 = 0$
$3, -3, \pm i\frac{\sqrt{15}}{3}$

47. $x^6 + 7x^3 - 8 = 0$
$1, -2, \frac{-1 \pm i\sqrt{3}}{2}, 1 \pm i\sqrt{3}$

48. $x^6 - 26x^3 - 27 = 0$
$-1, 3, \frac{-3 \pm 3i\sqrt{3}}{2}, \frac{1 \pm i\sqrt{3}}{2}$

49. $8x^6 + 999x^3 = 125$
$-5, \frac{1}{2}, \frac{-1 \pm i\sqrt{3}}{4}, \frac{5 \pm 5i\sqrt{3}}{2}$

50. $4x^4 - 4x^2 - x^2 + 1 = 0$
$-1, 1, \pm\frac{1}{2}$

51. $x^6 - 9x^4 - x^2 + 9 = 0$
$\pm 3, \pm 1, \pm i$

52. $x^4 + 8x^2 + 15 = 0$
$\pm i\sqrt{5}, \pm i\sqrt{3}$

53. STRUCTURE Consider the equation $x^{\frac{1}{2}} - 8x^{\frac{1}{4}} + 15 = 0$.

a. How are the exponents in the equation related?
Sample answer: $\frac{1}{2} = 2\left(\frac{1}{4}\right)$

b. How could you define u so that you could rewrite the equation as a quadratic equation in terms of u? Write the quadratic equation.
$u = x^{\frac{1}{4}}; u^2 - 8u + 15 = 0$

c. Solve the original equation.
81, 625

Factor completely. If the polynomial is not factorable, write *prime*.

54. $21x^3 - 18x^2y + 24xy^2$
$3x(7x^2 - 6xy + 8y^2)$

55. $8j^3k - 4jk^3 - 7$
prime

56. $a^2 + 7a - 18$
$(a + 9)(a - 2)$

57. $2ak - 6a + k - 3$
$(2a + 1)(k - 3)$

58. $b^2 + 8b + 7$
$(b + 7)(b + 1)$

59. $z^2 - 8z - 10$
prime

60. $4f^2 - 64$
$4(f + 4)(f - 4)$

61. $d^2 - 12d + 36$
$(d - 6)^2$

62. $9x^2 + 25$
prime

63. $y^2 + 18y + 81$
$(y + 9)^2$

64. $7x^2 - 14x$
$7x(x - 2)$

65. $19x^3 - 38x^2$
$19x^2(x - 2)$

66. $n^3 - 125$
$(n - 5)(n^2 + 5n + 25)$

67. $m^4 - 1$
$(m^2 + 1)(m - 1)(m + 1)$

68. REASONING A rectangular box has dimensions of x inches, $(x + 5)$ inches, and $(x - 2)$ inches. The volume of the box is $30x$ cubic inches. Find the dimensions of the box. Explain your reasoning. See margin.

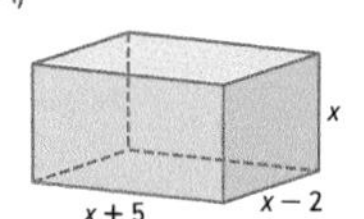

69. GEOMETRY The combined volume of a cube and a cylinder is 1000 cubic inches. If the height of the cylinder is twice the radius and the side of the cube is four times the radius, find the radius of the cylinder to the nearest tenth of an inch. 2.4 in.

Higher-Order Thinking Skills

70. ANALYZE Find the solutions of $(a + 3)^4 - 2(a + 3)^2 - 8 = 0$. Show your work. See margin.

71. WRITE If the equation $ax^2 + bx + c = 0$ has solutions $x = m$ and $x = n$, what are the solutions to $ax^4 + bx^2 + c = 0$? Explain your reasoning. See margin.

72. PERSEVERE Factor $36x^{2n} + 12x^n + 1$. $(6x^n + 1)^2$

73. PERSEVERE Solve $6x - 11\sqrt{3x} + 12 = 0$. $\frac{16}{3}, \frac{3}{4}$

74. ANALYZE Find a counterexample to the statement $a^2 + b^2 = (a + b)^2$.
Sample answer: $a = 1, b = -1$

75. CREATE The cubic form of an equation is $ax^3 + bx^2 + cx + d = 0$. Write an equation with degree 6 that can be written in cubic form. Sample answer: $12x^6 + 6x^4 + 8x^2 + 4 = 12(x^2)^3 + 6(x^2)^2 + 8(x^2) + 4$

76. WRITE Explain how the graph of a polynomial function can help you factor the polynomial. See margin.

Answers

68. 5 inches by 10 inches by 3 inches; sample answer: Volume equals the product of length, width, and height. An equation to represent the volume is $x^3 + 3x^2 - 10x = 30x$. The solutions of the equation are $x = -8, 0, 5$. Therefore, $x = 5$ and the dimensions of the box are 5 inches, 10 inches, and 3 inches.

70. $a = -1, -5, -3 + i\sqrt{2}, -3 - i\sqrt{2}$; sample answer: Use the substitution $u = (a + 3)^2$ and factor to $(u - 4)(u + 2) = 0$. Substitute for u to get $(a + 3)^2 = 4$ and $(a + 3)^2 = -2$.

71. $x = \pm\sqrt{m}$ and $x = \pm\sqrt{n}$; sample answer: Because both equations have the same coefficients, take the solutions for the quadratic equation and substitute x^2 for x.

76. Sample answer: The factors can be determined by the x-intercepts of the graph. An x-intercept of 5 represents a factor of $(x - 5)$.

Lesson 5-3

A.APR.4

Proving Polynomial Identities

LESSON GOAL

Students prove polynomial identities and use them to describe numerical relationships.

1 LAUNCH

 Launch the lesson with a **Warm Up** and an introduction.

2 EXPLORE AND DEVELOP

 **Explore:** Polynomial Identities

 **Develop:**

Proving Polynomial Identities
- Transform One Side
- Use Polynomial Identities

 You may want your students to complete the **Checks** online.

3 REFLECT AND PRACTICE

 Exit Ticket

 Practice

DIFFERENTIATE

 View reports of student progress on the **Checks** after each example.

Resources	AL	OL	BL	ELL
Remediation: Using the Distributive Property	●	●		●
Extension: Generating Pythagorean Triples		●	●	●

Language Development Handbook

Assign page 32 of the *Language Development Handbook* to help your students build mathematical language related to proving polynomial identities.

ELL You can use the tips and suggestions on page T32 of the handbook to support students who are building English proficiency.

Suggested Pacing

90 min	0.5 day
45 min	1 day

Focus

Domain: Algebra

Standards for Mathematical Content:

A.APR.4 Prove polynomial identities and use them to describe numerical relationships.

Standards for Mathematical Practice:

6 Attend to precision.

8 Look for and express regularity in repeated reasoning.

Coherence

Vertical Alignment

Previous
Students performed operations on polynomial expressions.
A.APR.1 (Algebra 1, Algebra 2)

Now
Students prove polynomial identities and use them to describe numerical relationships. **A.APR.4**

Next
Students will evaluate and factor functions by using the Remainder and Factor Theorems. **A.APR.2**

Rigor

The Three Pillars of Rigor

1 CONCEPTUAL UNDERSTANDING	2 FLUENCY	3 APPLICATION

Conceptual Bridge In this lesson, students develop an understanding of polynomial identities, and they build fluency by proving the identies and using them to describe numerical relationships. They apply their understanding by solving real-world problems.

Mathematical Background

An identity is an equation that is satisfied by any numbers that replace the variables. Thus, a polynomial identity is a polynomial equation that is true for any values that are substituted for the variables.

Interactive Presentation

Warm Up

Answer each question for both identities and equations of the form $f(x) = g(x)$.

	Identities	Equations with No Solution
1.	Is it an equation?	Is it an equation?
2.	Can you substitute $f(x)$ for $g(x)$ in another equation?	Can you substitute $f(x)$ for $g(x)$ in another equation?
3.	When is $f(x) = g(x)$ true?	When is $f(x) = g(x)$ true?
4.	What is the solution?	What is the solution?

Warm Up

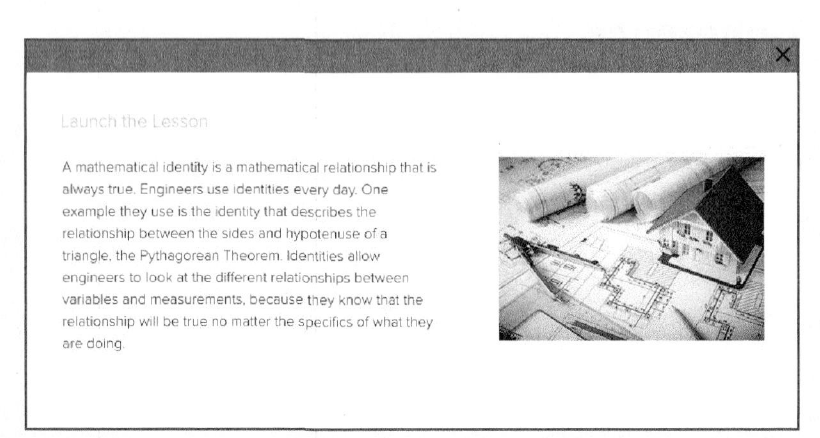

Launch the Lesson

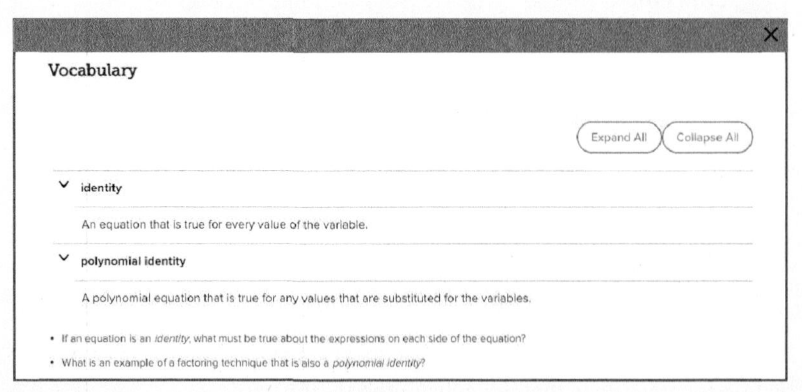

Today's Vocabulary

Warm Up

Prerequisite Skills

The Warm Up exercises address the following prerequisite skill for this lesson:

- comparing equations with no solution and identities

Answers:

Identities:	Equations with No Solution:
1. yes	1. yes
2. yes	2. no
3. always	3. never
4. all values of x	4. $\varnothing$

5. An identity is true for every value of the variable; an equation with no solution is never true.

Launch the Lesson

Teaching the Mathematical Practices

4 Apply Mathematics Students will apply polynomial identities to a real-world situation. Ask students to identify instances in which engineers may use the Pythagorean Theorem.

Go Online to find additional teaching notes and questions to promote classroom discourse.

Today's Standards

Tell students that they will be addressing these content and practice standards in this lesson. You may wish to have a student volunteer read aloud *How can I meet this standard?* and *How can I use these practices?*, and connect these to the standards.

See the Interactive Presentation for I Can statements that align with the standards covered in this lesson.

Today's Vocabulary

Tell students that they will be using these vocabulary terms in this lesson. You can expand each row if you wish to share the definitions. Then discuss the questions below with the class.

Explore Polynomial Identities

Objective

Students use a sketch to explore polynomial identities.

 Teaching the Mathematical Practices

8 Look for a Pattern Help students to see the pattern in which expressions result in the same value for any value of x and y.

Ideas for Use

Recommended Use Present the Inquiry Question, or have a student volunteer read it aloud. Have students work in pairs to complete the Explore activity on their devices. Pairs should discuss each of the questions. Monitor student progress during the activity. Upon completion of the Explore activity, have student volunteers share their responses to the Inquiry Question.

What if my students don't have devices? You may choose to project the activity on a whiteboard. A printable worksheet for each Explore is available online. You may choose to print the worksheet so that individuals or pairs of students can use it to record their observations.

Summary of the Activity

Students will complete guiding exercises throughout the Explore activity. They will be presented with different polynomial equations, and will use the sketch to determine which ones generate polynomial identities. Then, students will answer the Inquiry question.

(continued on the next page)

Interactive Presentation

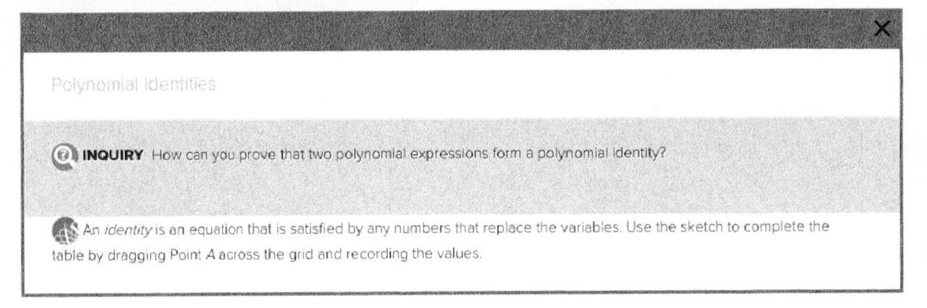

Polynomial Identities
INQUIRY How can you prove that two polynomial expressions form a polynomial identity?
An identity is an equation that is satisfied by any numbers that replace the variables. Use the sketch to complete the table by dragging Point A across the grid and recording the values.

Explore

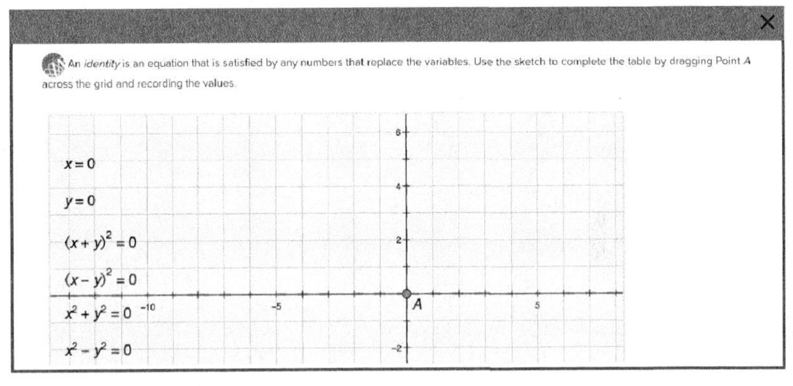

Explore

WEB SKETCHPAD

Students will use the sketch to explore polynomial identities.

TYPE

Students complete the table by finding the value of each polynomial expression for various values of x and y.

DRAG & DROP

Students drag and drop the corresponding expression to complete the polynomial identities.

Interactive Presentation

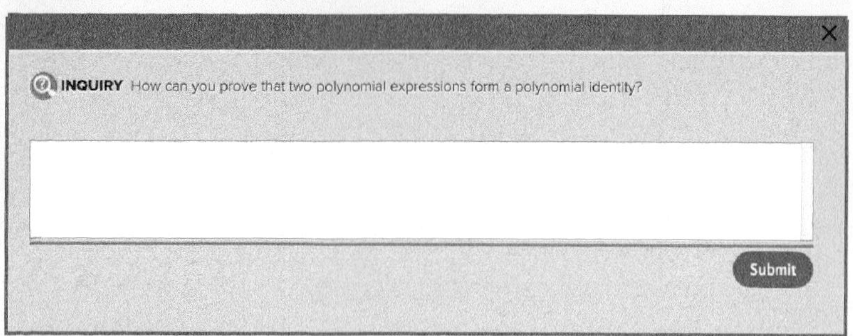

Explore

TYPE

Students respond to the Inquiry Question and can view a sample answer.

Explore Polynomial Identities (*continued*)

Questions

Have students complete the Explore activity.

Ask:

- To be considered an identity, can two expressions be equal for only certain values of x and y? Explain. No; sample answer: They must be equal for all values of x and y, not just certain pairs.
- Test your coordinates in the expression $x^2 - 2xy + y^2$. Does this appear to make an identity with another expression? If so, which one? yes; $(x - y)^2$

Inquiry

How can you prove that two polynomial expressions form a polynomial identity? Sample answer: I can take one side of the equation and simplify it until it is identical to the other side of the equation.

Go Online to find additional teaching notes and sample answers for the guiding exercises.

1 CONCEPTUAL UNDERSTANDING | 2 FLUENCY | 3 APPLICATION

Learn Proving Polynomial Identities

Objective

Students prove polynomial identities and use them to describe numerical relationships.

Teaching the Mathematical Practices

1 Seek Information Mathematically proficient students must be able to transform algebraic expressions to reach solutions. To verify polynomial identities, students must transform one side of the equation until it is identical to the other side.

About the Key Concept

A polynomial identity is a polynomial equation that is true for any values that are substituted for the variables. To verify identities by transforming one side, first simplify one side of the equation until the two sides are the same. Transform that expression into the form of the simpler side.

Example 1 Transform One Side

Teaching the Mathematical Practices

6 Communicate Precisely Encourage students to routinely write or explain their solution methods. Point out that they should use clear mathematical language when they discuss their answer to the Talk About It! feature.

Questions for Mathematical Discourse

AL How do you multiply terms with the same variables? Sample answer: Multiply the coefficients and then add the exponents of each like variable.

OL How would the polynomial identity have changed if we were given a sum of cubes? Sample answer: The identity to prove would now be $(x^3 + y^3)(x^2 - xy + y^2)$.

BL Could you have factored the difference of cubes to show the two sides are equal? Explain. No; sample answer: Factoring relies upon the identity and does not prove anything. You must multiply the factors together to prove that it equals the difference of cubes.

Common Error

While most students understand the process of simplifying one side of the polynomial equation, many may make an algebraic mistake. Encourage students to work slowly and check back over their work periodically. If the two sides are not equal then they know they have made a mistake and can try again.

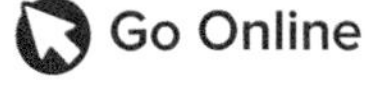

Go Online

- Find additional teaching notes.
- View performance reports of the Checks.
- Assign or present an Extra Example.

Lesson 5-3

Proving Polynomial Identities

Today's Goal
- Prove polynomial identities and use them to describe numerical relationships.

Today's Vocabulary
identity
polynomial identity

Explore Polynomial Identities

Online Activity Use graphing technology to complete the Explore.

INQUIRY How can you prove that two polynomial expressions form a polynomial identity?

Learn Proving Polynomial Identities

An **identity** is an equation that is satisfied by any numbers that replace the variables. Thus, a **polynomial identity** is a polynomial equation that is true for any values that are substituted for the variables.

Unlike solving an equation, do not begin by assuming that an identity is true. You cannot perform the same operation to both sides and assume that equality is maintained.

Key Concept • Verifying Identities by Transforming One Side
- Simplify one side of an equation until the two sides of the equation are the same. It is often easier to transform the more complicated expression into the form of the simpler side.
- Factor or multiply expressions as necessary. Simplify by combining like terms.

Study Tip
Transforming One Side It is often easier to work with the more complicated side of an equation. Look at each side and determine which requires more steps to be simplified. For example, it is often easier to work on the side that involves the square or cube of an algebraic expression.

Example 1 Transform One Side

Prove that $x^3 - y^3 = (x - y)(x^2 + xy + y^2)$.

$x^3 - y^3 \stackrel{?}{=} (x - y)(x^2 + xy + y^2)$ Original equation
$\stackrel{?}{=} x(x^2) + x(xy) + x(y^2) - y(x^2) - y(xy) - y(y^2)$ Distributive Property
$\stackrel{?}{=} \underline{x^3} + \underline{x^2y} + xy^2 - \underline{x^2y} - \underline{xy^2} - y^3$ Simplify.
$\stackrel{?}{=} x^3 + x^2y - x^2y + xy^2 - xy^2 - y^3$ Commutative Property
$= \underline{x^3} - \underline{y^3}$ True

Because the expression on the right can be simplified to be the same as the expression on the left, this proves the polynomial identity.

Go Online You can complete an Extra Example online.

Talk About It!
If you multiplied each side of the equation by a variable z, would the result still be a polynomial identity? Explain your reasoning.

Yes; sample answer: It would still be an identity because both sides of the equation would be multiplied by the same quantity z. It would be a polynomial identity because it would still consist of multiple algebraic terms.

Interactive Presentation

Proving Polynomial Identities

An **identity** is an equation that is satisfied by any numbers that replace the variables. Thus, a **polynomial identity** is a polynomial equation that is true for any values that are substituted for the variables.

Unlike solving an equation, do not begin by assuming that an identity is true. You cannot perform the same operation to both sides and assume that equality is maintained.

Key Concept: Verifying Identities by Transforming One Side

Simplify one side of an equation until the two sides of the equation are the same. It is often easier to transform the more complicated expression into the form of the simpler side.

Learn

TAP

Students tap to see a Study Tip.

Your Notes

Math History Minute:
Former quarterback **Frank Ryan (1936–)** earned his Ph.D. in mathematics about six months after he led the Cleveland Browns to the NFL championship game of 1964, where they won 27-0. During part of his academic career, Ryan studied prime numbers, including Opperman's Conjecture that there is a prime number between n^2 and $n^2 + n$, where n is an integer. This work could eventually lead to a polynomial identity that could be used to identify prime numbers.

Example 2 Use Polynomial Identities

TRIANGLES **Pedro claims that you can always create three lengths that form a right triangle by using the following method: take two positive integers x and y where $x > y$. Two legs of a right triangle are defined as $x^2 - y^2$ and $2xy$. The hypotenuse is defined as $x^2 + y^2$. Is Pedro correct? Explain your reasoning in the context of polynomial identities.**

To determine whether Pedro is correct, we can use information about right triangles and the expressions involving x and y to try to construct a polynomial identity. If $x^2 - y^2$ and $2xy$ are the legs of the triangle, and $x^2 + y^2$ is the hypotenuse, then it should be true that $(x^2 - y^2)^2 + (2xy)^2 = (x^2 + y^2)^2$.

$2xy$, $x^2 + y^2$, $x^2 - y^2$

If this is an identity, you can simplify the expressions for the sides to be the same expression.

$(x^2 - y^2)^2 + (2xy)^2 \stackrel{?}{=} (x^2 + y^2)^2$ Original equation

$x^4 - 2x^2y^2 + y^4 + 4x^2y^2 \stackrel{?}{=} x^4 + 2x^2y^2 + y^4$ Square each term.

$x^4 + 2x^2y^2 + y^4 = x^4 + 2x^2y^2 + y^4$ True

Because the identity is true, this proves that Pedro is correct. His process for creating the sides of a right triangle will always work.

Check

Write in the missing explanations to prove that $x^4 - y^4 = (x - y)(x + y)(x^2 + y^2)$.

$x^4 - y^4 \stackrel{?}{=} (x - y)(x + y)(x^2 + y^2)$ Original equation

$x^4 - y^4 \stackrel{?}{=} (x^2 - y^2)(x^2 + y^2)$ FOIL

$x^4 - y^4 \stackrel{?}{=} x^4 + x^2y^2 - x^2y^2 - y^4$ Product of a sum and a difference

$x^4 - y^4 = x^4 - y^4$ Subtract.

Use a Source
Research an application of prime numbers. How could a polynomial identity for identifying prime numbers impact the application?

Sample answer: Prime numbers are often used in cryptography. A polynomial identity for identifying prime numbers could make encoding using prime numbers obsolete.

Go Online You can complete an Extra Example online.

272 Module 5 • Polynomial Equations

Interactive Presentation

Use Polynomial Identities

TRIANGLES Pedro claims that you can always create three lengths that form a right triangle by using the following method: take two positive integers x and y where $x > y$. Two legs of a right triangle are defined as $x^2 - y^2$ and $2xy$. The hypotenuse is defined as $x^2 + y^2$. Is Pedro correct? Explain your reasoning in the context of polynomial identities.

Example 2

CHECK

Students complete the Check online to determine whether they are ready to move on.

Example 2 Use Polynomial Identities

Teaching the Mathematical Practices

1 Understand the Approaches of Others Mathematically proficient students can explain the solution methods of others. Example 2 asks students to justify the reasoning of Pedro by verifying a polynomial identity.

Questions for Mathematical Discourse

AL What does the Pythagorean Theorem state? Sample answer: $a^2 + b^2 = c^2$

OL For Pedro's method to work, why must $x > y$? Sample answer: If x is not greater than y, then $x^2 - y^2$ could give a side length of a non-positive number.

BL If x and y are not restricted to positive numbers, what would need to be true for this method to work? x and y would need to have the same sign, and $|x| > |y|$.

Common Error

Students may try to "distribute" the square over the plus or minus sign of a binomial, meaning they will only square the two terms. Remind students what squaring means and then have students write out the binomial twice to FOIL.

DIFFERENTIATE

Reteaching Activity AL ELL

IF students are struggling to prove polynomial identities,
THEN have students write each side of the equation on two different note cards. This breaks the equation into two expressions. Students simplify the more complicated expression to see if it matches the expression on the other notecard.

Exit Ticket

Recommended Use

At the end of class, go online to display the Exit Ticket prompt and ask students to respond using a separate piece of paper. Have students hand you their responses as they leave the room.

Alternate Use

At the end of class, go online to display the Exit Ticket prompt and ask students to respond verbally or by using a mini-whiteboard. Have students hold up their whiteboards so that you can see all student responses. Tap to reveal the answer when most or all students have completed the Exit Ticket.

Practice and Homework

Suggested Assignments

Use the table below to select appropriate exercises.

DOK	Topic	Exercises
1, 2	exercises that mirror the examples	1–10
2	exercises that use a variety of skills from this lesson	11–15
2	exercises that extend concepts learned in this lesson to new contexts	16–20
3	exercises that emphasize higher-order and critical thinking skills	21–25

ASSESS AND DIFFERENTIATE

Use the data from the **Checks** to determine whether to provide resources for extension, remediation, or intervention.

IF students score 90% or more on the Checks, 
THEN assign:

- Practice Exercises 1–21 odd, 22–26
- Extension: Generating Pythagorean Triples

IF students score 66%–89% on the Checks, 
THEN assign:

- Practice Exercises 1–26 odd
- Remediation, Review Resources: Using the Distributive Property
- Personal Tutors
- Extra Examples 1–2
- ALEKS® Equations With No Solutions vs. Identities

IF students score 65% or less on the Checks, AL
THEN assign:

- Practice Exercises 1–6 odd
- Remediation, Review Resources: Using the Distributive Property
- *Quick Review Math Handbook*: Polynomial Identities
- ALEKS® Equations With No Solutions vs. Identities

Name ______ Period ______ Date ______

Practice

Go Online You can complete your homework online.

Example 1

Prove each polynomial identity. 1–8. See margin.

1. $(x - y)^2 = x^2 - 2xy + y^2$
2. $(x + 5)^2 = x^2 + 10x + 25$
3. $4(x - 7)^2 = 4x^2 - 56x + 196$
4. $(2x^2 + y^2)^2 = (2x^2 - y^2)^2 + (2xy\sqrt{2})^2$
5. $a^2 - b^2 = (a + b)(a - b)$
6. $x^3 + y^3 = (x + y)(x^2 - xy + y^2)$
7. $p^4 - q^4 = (p - q)(p + q)(p^2 + q^2)$
8. $a^5 - b^5 = (a - b)(a^4 + a^3b + a^2b^2 + ab^3 + b^4)$

Example 2

9. SQUARES Aponi claims that you can find the area of a square using the following method: take two positive integers x and y. The side length of the square is defined by the expression $3x + y$. The area of the square is defined by the expression $9x^2 + 6xy + y^2$. Is Aponi correct? Explain your reasoning in the context of polynomial identities.
See margin.

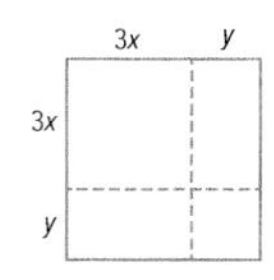

10. USE A MODEL Julio claims that you can find the area of a rectangle using the following method: take two positive integers x and y, where $x > y$. The side lengths of the rectangle are defined by the expressions $2x + y$ and $2x - y$. The area of the rectangle is defined by the expression $4x^2 - y^2$. Is Julio correct? Explain your reasoning in the context of polynomial identities.
See margin.

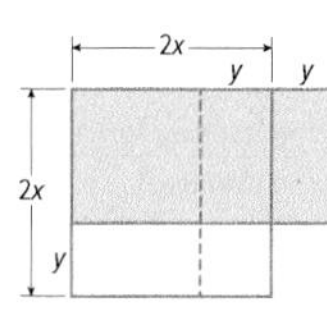

Mixed Exercises

Determine whether each equation is an identity.

11. $(x + 3)^2(x^3 + 3x^2 + 3x + 1) = (x^2 + 6x + 9)(x + 1)^3$ identity
12. $(x + 2)(x + 1)^2 = (x^2 + 3x + 2)(x + 1)$ identity
13. $(x + 3)(x - 1)^2 = (x^2 - 2x - 3)(x - 1)$ not an identity
14. $(x + 2)^2(x^3 - 3x^2 + 3x - 1) = (x^2 + 4x + 4)(x - 1)^3$ identity
15. $(a + b)^2 = a^2 - 2ab + b^2$ not an identity

16. USE TOOLS Consider the following equation.

$(x-2)^2(x^3+9x^2+27x+27)=(x^2-4x+4)(x+3)^3$

a. Evaluate the expressions for each value to complete the table.

x	$(x-2)^2(x^3+9x^2+27x+27)$	$(x^2-4x+4)(x+3)^3$
0	108	108
1	64	64
2	0	0
3	216	216
4	1372	1372

b. What conclusion can you make about the equation, based on the results in your table? Explain. The equation may be a polynomial identity because the left side of the equation equals the right side of the equation for each x-value in the table.

c. How can you prove your conclusion from **part b**? Sample answer: Transform one side of the equation to determine if it is a polynomial identity.

USE TOOLS Use a computer algebra system (CAS) to prove each identity. See Mod. 5 Answer Appendix.

17. $g^6+h^6=(g^2+h^2)(g^4-g^2h^2+h^4)$

18. $a^5+b^5=(a+b)(a^4-a^3b+a^2b^2-ab^3+b^4)$

19. $u^6-w^6=(u+w)(u-w)(u^2+uw+w^2)(u^2-uw+w^2)$

20. $(x+1)^2(x-4)^3=(x^2-3x-4)(x^3-7x^2+8x+16)$

Higher-Order Thinking Skills

21. WRITE Explain the meaning of polynomial identity and summarize the method for proving an equation is a polynomial identity. See Mod. 5 Answer Appendix.

22. CREATE Write and solve a system of equations using the identity $(x^2-y^2)^2+(2xy)^2=(x^2+y^2)^2$ to find the values of x and y that make a 3, 4, 5 Pythagorean triple. See Mod. 5 Answer Appendix.

23. ANALYZE Refer to Example 2. Notice that Pedro says x and y must be positive integers and x must be greater than y. Explain why these restrictions are necessary. See Mod. 5 Answer Appendix.

24. PERSEVERE Rebecca has a square garden with side length a that she wants to transform into a rectangle. Rebecca speculates that if she subtracts the same length b from one dimension of the garden and adds it to the other dimension the new rectangle's area will be smaller than the original garden in the amount of b^2. Draw a diagram and show algebraically that Rebecca is correct. See Mod. 5 Answer Appendix.

25. FIND THE ERROR George is proving the identity $a^3+b^3=(a+b)(a^2-ab+b^2)$ by simplifying the right side. His work is shown. Is George correct? If not, identify and correct his error. See Mod. 5 Answer Appendix.

$(a+b)(a^2-ab+b^2)$
$=a^3-a^2b+ab^2-a^2b-ab^2+b^3$
$=a^3-2a^2b+b^3$

Answers

1. $(x-y)^2 \stackrel{?}{=} x^2-2xy+y^2$ — Original equation
   $x^2-2xy+y^2=x^2-2xy+y^2$ — Distributive Property
2. $(x+5)^2 \stackrel{?}{=} x^2+10x+25$ — Original equation
   $x^2+10x+25=x^2+10x+25$ — Distributive Property
3. $4(x-7)^2 \stackrel{?}{=} 4x^2-56x+196$ — Original equation
   $4(x^2-14x+49) \stackrel{?}{=} 4x^2-56x+196$ — Distributive Property
   $4x^2-56x+196=4x^2-56x+196$ — Distributive Property
4. $(2x^2+y^2)^2 \stackrel{?}{=} (2x^2-y^2)^2+(2xy\sqrt{2})^2$ — Original equation
   $(2x^2+y^2)^2 \stackrel{?}{=} (2x^2-y^2)^2+8x^2y^2$ — Evaluate the exponent.
   $(2x^2+y^2)^2 \stackrel{?}{=} 4x^4-4x^2y^2+y^4+8x^2y^2$ — Distributive Property
   $(2x^2+y^2)^2 \stackrel{?}{=} 4x^4+4x^2y^2+y^4$ — Combine like terms.
   $(2x^2+y^2)^2=(2x^2+y^2)^2$ — Factor.
5. $a^2-b^2 \stackrel{?}{=} (a+b)(a-b)$ — Original equation
   $a^2-b^2 \stackrel{?}{=} a^2-ab+ab-b^2$ — FOIL
   $a^2-b^2=a^2-b^2$ — Simplify.
6. $x^3+y^3 \stackrel{?}{=} (x+y)(x^2-xy+y^2)$ — Original equation
   $x^3+y^3 \stackrel{?}{=} x^3-x^2y+xy^2+x^2y-xy^2+y^3$ — Distributive Property
   $x^3+y^3=x^3+y^3$ — Simplify.
7. $p^4-q^4 \stackrel{?}{=} (p-q)(p+q)(p^2+q^2)$ — Original equation
   $p^4-q^4 \stackrel{?}{=} (p^2-q^2)(p^2+q^2)$ — Difference of squares
   $p^4-q^4=p^4-q^4$ — Difference of squares
8. $a^5-b^5 \stackrel{?}{=} (a-b)(a^4+a^3b+a^2b^2+ab^3+b^4)$ — Original equation
   $a^5-b^5 \stackrel{?}{=} a^5+a^4b+a^3b^2+a^2b^3+ab^4$
   $-a^4b-a^3b^2-a^2b^3-ab^4-b^5$ — Distributive Property
   $a^5-b^5=a^5-b^5$ — Simplify.
9. $(3x+y)^2 \stackrel{?}{=} 9x^2+6xy+y^2$ — Original equation
   $9x^2+6xy+y^2 \stackrel{?}{=} 9x^2+6xy+y^2$ — Square the left side.
   $9x^2+6xy+y^2=9x^2+6xy+y^2$ — True
   Sample answer: Because the identity is true, this proves that Aponi is correct. Her process for finding the area of a square will always work.
10. $(2x+y)(2x-y) \stackrel{?}{=} 4x^2-y^2$ — Original equation
    $4x^2-2xy+2xy-y^2 \stackrel{?}{=} 4x^2-y^2$ — Use the FOIL Method.
    $4x^2-y^2=4x^2-y^2$ — True
    Sample answer: Because the identity is true, this proves that Julio is correct. His process for finding the area of a rectangle figure will always work.

The Remainder and Factor Theorems

LESSON GOAL

Students evaluate and factor functions by using the Remainder and Factor Theorems.

1 LAUNCH

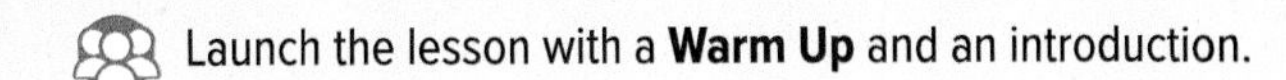

Launch the lesson with a **Warm Up** and an introduction.

2 EXPLORE AND DEVELOP

Explore: Remainders

Develop:

The Remainder Theorem

- Synthetic Substitution
- Apply the Remainder Theorem

The Factor Theorem

- Use the Factor Theorem

You may want your students to complete the **Checks** online.

3 REFLECT AND PRACTICE

Exit Ticket

Practice

DIFFERENTIATE

View reports of student progress on the **Checks** after each example.

Resources	AL	OL	BL	ELL
Remediation: Dividing Polynomials	●	●		●
Extension: Radical Notation		●	●	●

Language Development Handbook

Assign page 33 of the *Language Development Handbook* to help your students build mathematical language related to evaluating and factoring functions by using the Remainder and Faactor Theorems.

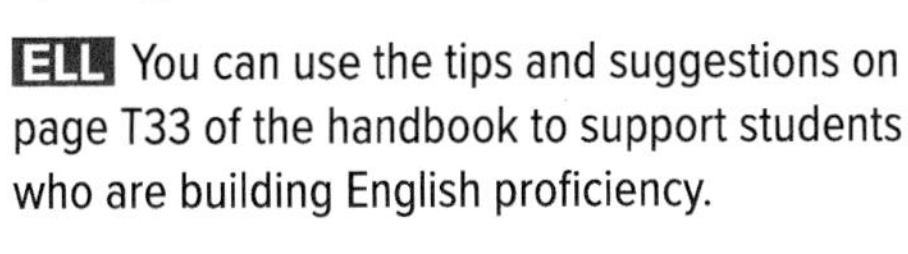

ELL You can use the tips and suggestions on page T33 of the handbook to support students who are building English proficiency.

Suggested Pacing

90 min	0.5 day
45 min	1 day

Focus

Domain: Algebra

Standards for Mathematical Content:

A.APR.2 Know and apply the Remainder Theorem: For a polynomial $p(x)$ and a number a, the remainder on division by $x - a$ is $p(a)$, so $p(a) = 0$ if and only if $(x - a)$ is a factor of $p(x)$.

Standards for Mathematical Practice:

3 Construct viable arguments and critique the reasoning of others.

4 Model with mathematics.

Coherence

Vertical Alignment

Previous
Students divided polynomials by using long division and synthetic division.
A.APR.6

Now
Students evaluate and factor functions by using the Remainder and Factor Theorems.
A.APR.2

Next
Students will determine the number and types of roots, find zeros, and use zeros to graph polynomial equations.
N.CN.9, A.APR.3, F.IF.7c

Rigor

The Three Pillars of Rigor

1 CONCEPTUAL UNDERSTANDING	2 FLUENCY	3 APPLICATION

Conceptual Bridge In this lesson, students build on their understanding of polynomial division. They build fluency by using the Remainder and Factor Theorems, and they apply their understanding by solving real-world problems.

Mathematical Background

The Remainder Theorem says that the value of $f(a)$ is the same as the remainder when the polynomial is divided by $x - a$. The Factor Theorem is a special case of the Remainder Theorem. It says: If $f(a)$ has a value of 0, then $x - a$ is a factor of the polynomial.

Interactive Presentation

Warm Up

Use synthetic division to find each quotient.

1. $2x^5 + 7x^4 + 4x^3 - x^2 + 6x + 6; x + 1$
2. $x^4 - 4x^3 - 7x^2 + 3x + 6; x - 2$

Show Answers

Warm Up

Warm Up

Prerequisite Skills

The Warm Up exercises address the following prerequisite skill for this lesson:

- using synthetic division

Answers:

1. $2x^4 + 5x^3 - x^2 + 6$
2. $x^3 - 2x^2 - 11^x + 19 - \frac{32}{x-2}$

Launch the lesson

Launch the Lesson

 Teaching the Mathematical Practices

4 Apply Mathematics Encourage students to consider how smartphone and mobile device growth can be modeled by a cubic function, and how the function can be used to estimate values.

Go Online to find additional teaching notes and questions to promote classroom discourse.

Today's Standards

Tell students that they will be addressing these content and practice standards in this lesson. You may wish to have a student volunteer read aloud *How can I meet this standard? and How can I use these practices?,* and connect these to the standards.

See the Interactive Presentation for I Can statements that align with the standards covered in this lesson.

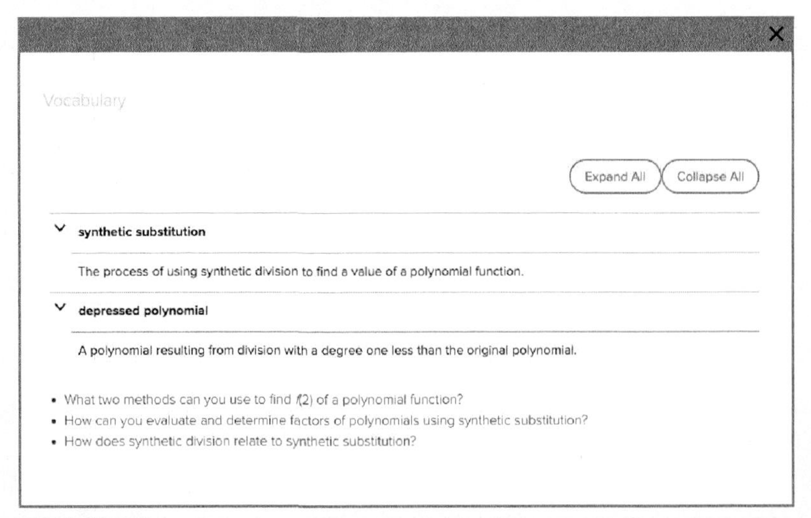

Today's Vocabulary

Today's Vocabulary

Tell students that they will be using these vocabulary terms in this lesson. You can expand each row if you wish to share the definitions. Then discuss the questions below with the class.

Explore Remainders

Objective

Students use a sketch to explore the divisor and quotient of a polynomial when the remainder is zero.

Teaching the Mathematical Practices

3 Construct Arguments In the Explore, students will use previously established results to make a conjecture about the divisor and quotient of a polynomial when the remainder is 0.

Ideas for Use

Recommended Use Present the Inquiry Question, or have a student volunteer read it aloud. Have students work in pairs to complete the Explore activity on their devices. Pairs should discuss each of the questions. Monitor student progress during the activity. Upon completion of the Explore activity, have student volunteers share their responses to the Inquiry Question.

What if my students don't have devices? You may choose to project the activity on a whiteboard. A printable worksheet for each Explore is available online. You may choose to print the worksheet so that individuals or pairs of students can use it to record their observations.

Summary of the Activity

Students will complete guiding exercises throughout the Explore activity. Students will move through a series of exercises using the sketch to determine how the quotient, divisor and remainder are related. Then, students will answer the Inquiry Question.

(continued on the next page)

Interactive Presentation

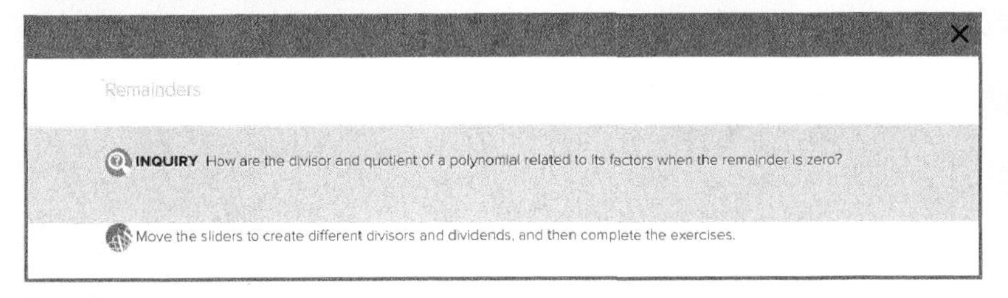

Explore

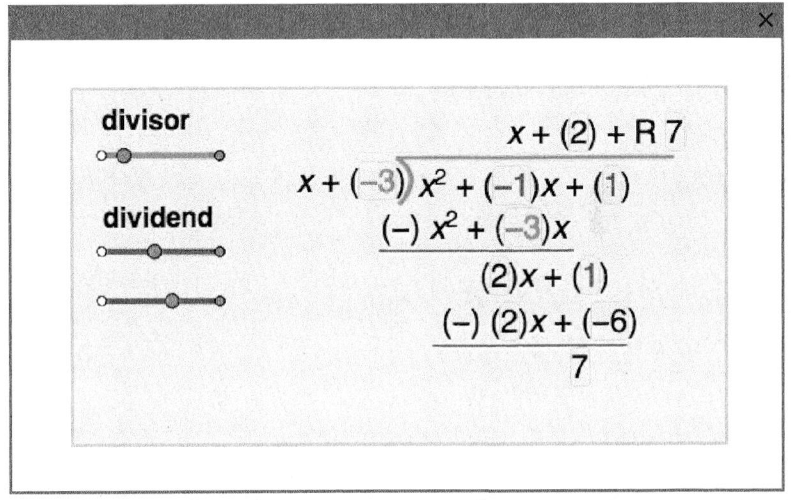

Explore

WEB SKETCHPAD

Students use the sketch to explore polynomial division.

TYPE

Students answer questions about polynomial division.

Interactive Presentation

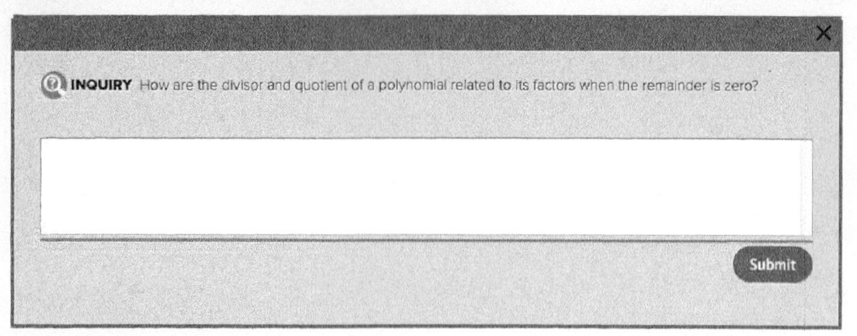

Explore

TYPE

a|

Students respond to the Inquiry Question and can view a sample answer.

Explore Remainders (*continued*)

Questions

Have students complete the Explore activity.

Ask:

- If $x^2 - 9$ were divided by $x - 3$, would the remainder be zero? Explain. Yes; sample answer: $x - 3$ is a factor of $x^2 - 9$.
- Robert says he divided $x^2 - 7x + 12$ by a binomial and got a remainder of zero. By what binomial could Robert have divided? $x - 4$ or $x - 3$

Inquiry

How are the divisor and quotient of a polynomial related to its factors when the remainder is zero? Sample answer: When the remainder is zero, the divisor and quotient are factors of the polynomial.

Go Online to find additional teaching notes and sample answers for the guiding exercises.

Learn The Remainder Theorem

Objective

Students evaluate functions by using synthetic substitution.

Teaching the Mathematical Practices

1 Explain Correspondences Guide students to analyze the relationship between the polynomial division and evaluating a function for a specific value.

Important to Know

The Remainder Theorem says that polynomial division can be used to find a function value. For any polynomial, the function value of the polynomial at $x = a$ is the value of the remainder when the polynomial is divided by $x - a$. This can be found using synthetic substitution.

Common Misconception

Some students may think that the remainder must be zero. Not all polynomials divide evenly, just like all numbers do not divide evenly. Remind students that the remainder can be any real number.

DIFFERENTIATE

Enrichment Activity BL

The remainder when $x^4 - 2x^2 + kx + 3$ is divided by $x + 3$ is 36. What is the value of k?

$k = 10$

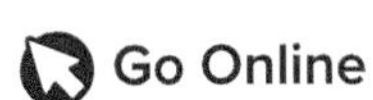

Go Online

- Find additional teaching notes.
- View performance reports of the Checks.
- Assign or present an Extra Example.

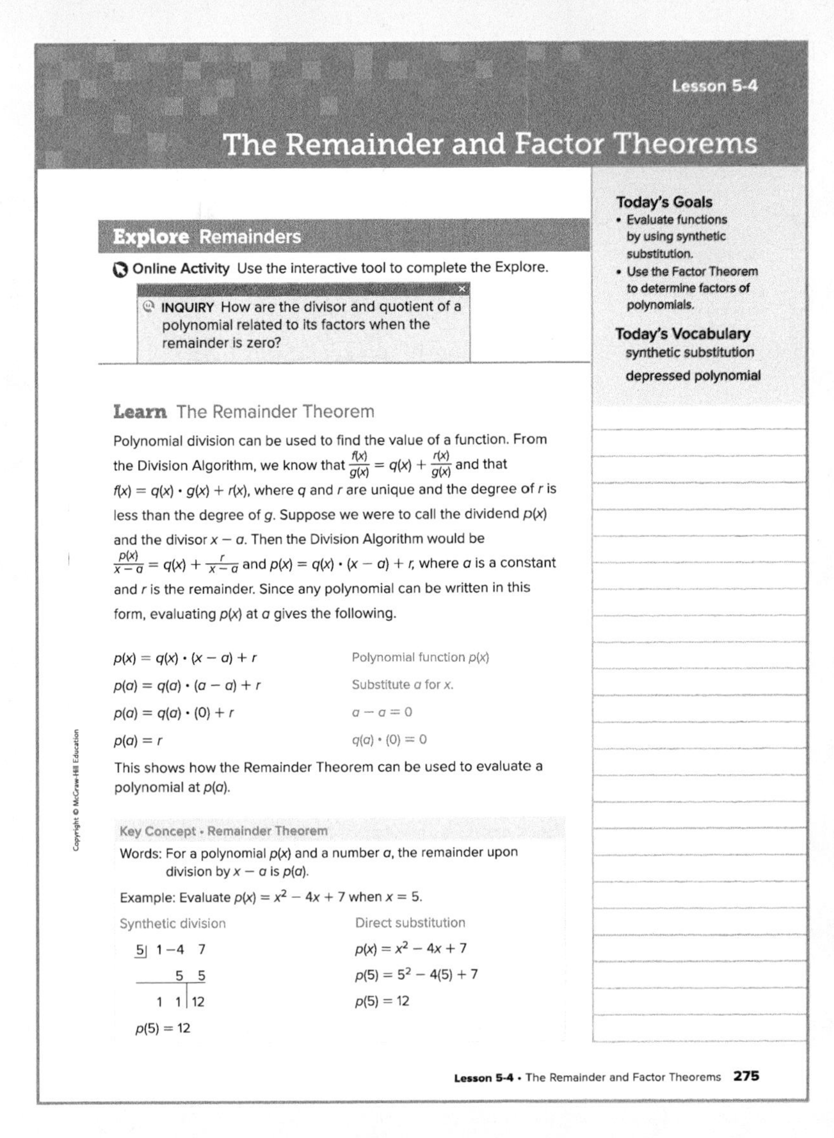

Lesson 5-4

The Remainder and Factor Theorems

Explore Remainders

Online Activity Use the interactive tool to complete the Explore.

INQUIRY How are the divisor and quotient of a polynomial related to its factors when the remainder is zero?

Today's Goals
- Evaluate functions by using synthetic substitution.
- Use the Factor Theorem to determine factors of polynomials.

Today's Vocabulary
synthetic substitution
depressed polynomial

Learn The Remainder Theorem

Polynomial division can be used to find the value of a function. From the Division Algorithm, we know that $\frac{f(x)}{g(x)} = q(x) + \frac{r(x)}{g(x)}$ and that $f(x) = q(x) \cdot g(x) + r(x)$, where q and r are unique and the degree of r is less than the degree of g. Suppose we were to call the dividend $p(x)$ and the divisor $x - a$. Then the Division Algorithm would be $\frac{p(x)}{x-a} = q(x) + \frac{r}{x-a}$ and $p(x) = q(x) \cdot (x - a) + r$, where a is a constant and r is the remainder. Since any polynomial can be written in this form, evaluating $p(x)$ at a gives the following.

$p(x) = q(x) \cdot (x - a) + r$	Polynomial function $p(x)$
$p(a) = q(a) \cdot (a - a) + r$	Substitute a for x.
$p(a) = q(a) \cdot (0) + r$	$a - a = 0$
$p(a) = r$	$q(a) \cdot (0) = 0$

This shows how the Remainder Theorem can be used to evaluate a polynomial at $p(a)$.

Key Concept • Remainder Theorem

Words: For a polynomial $p(x)$ and a number a, the remainder upon division by $x - a$ is $p(a)$.

Example: Evaluate $p(x) = x^2 - 4x + 7$ when $x = 5$.

Synthetic division	Direct substitution
5⌋ 1 −4 7	$p(x) = x^2 - 4x + 7$
5 5	$p(5) = 5^2 - 4(5) + 7$
1 1 | 12	$p(5) = 12$
$p(5) = 12$	

Copyright © McGraw-Hill Education

Lesson 5-4 • The Remainder and Factor Theorems 275

Interactive Presentation

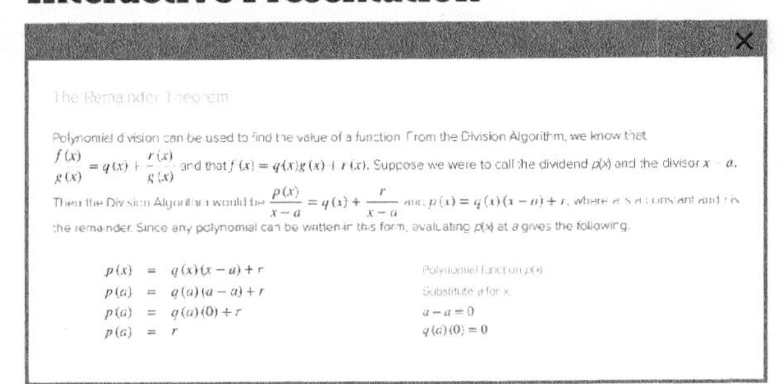

Learn

Your Notes

Applying the Remainder Theorem to evaluate a function is called **synthetic substitution**. You may find that synthetic substitution is a more convenient way to evaluate a polynomial function, especially when the degree of the function is greater than 2.

Study Tip

Missing terms Remember to include zeros as placeholders for any missing terms in the polynomial.

Example 1 Synthetic Substitution

Use synthetic substitution to find $f(-3)$ if $f(x) = -2x^4 + 3x^2 - 15x + 9$.

By the Remainder Theorem, $f(-3)$ is the remainder of $\frac{f(x)}{x-(-3)}$.

-3	-2	0	3	-15	9
		6	-18	45	-90
	-2	6	-15	30	-81

The remainder is −81. Therefore, $f(-3) =$ −81.

Use direct substitution to check.

$f(x) = -2x^4 + 3x^2 - 15x + 9$ — Original function

$f(-3) = -2(-3)^4 + 3(-3)^2 - 15(-3) + 9$ — Substitute −3 for x.

$= -162 + 27 + 45 + 9$ or -81 — True

Check

Use synthetic substitution to evaluate $f(x) = -6x^3 + 52x^2 - 27x - 31$.

$f(8) =$ 9

Think About It!

How could you use the function and synthetic substitution to estimate the number of eggs produced in 1990? What assumption would you have to make to solve this problem?

Sample answer: Because 1990 is 10 years before 2000, use synthetic substitution to find the value of $f(-10)$. I would have to assume that the production of eggs followed the same model before 2000.

Example 2 Apply the Remainder Theorem

EGG PRODUCTION **The total production of eggs in billions in the United States can be modeled by the function $f(x) = 0.007x^3 - 0.149x^2 + 1.534x + 84.755$, where x is the number of years since 2000. Predict the total production of eggs in 2025.**

Since 2025 − 2000 = 25, use synthetic substitution to determine $f(25)$.

25	0.007	-0.149	1.534	84.755
		0.175	0.65	54.6
	0.007	0.026	2.184	139.355

In 2025, approximately 139.355 billion eggs will be produced in the United States.

Go Online You can complete an Extra Example online.

276 Module 5 • Polynomial Equations

Interactive Presentation

Synthetic Substitution

Use synthetic substitution to find $f(x) = -2x^4 + 3x^2 - 15x + 9$.

By the Remainder Theorem, $f(-3)$ is the remainder of $\frac{f(x)}{x-(-3)}$.

-3	2	0	3	15	9
		6	-18	45	-90
	-2	6	-15	30	-81

The remainder is −81. Therefore, $f(-3) = -81$.

Example 1

CHECK

Students complete the Check online to determine whether they are ready to move on.

Example 1 Synthetic Substitution

Teaching the Mathematical Practices

1 Check Answers In Example 1, students should check their answer by using a different method.

Questions for Mathematical Discourse

AL If $f(a) = b$, what is the remainder of $\frac{f(x)}{x-a}$? *b*

OL When using synthetic substitution to evaluate a function, why do you find the remainder using synthetic division rather than long division? Sample answer: Synthetic substitution is usually more efficient than long division.

BL Why do you need to use 0 as a placeholder for the x^3-term? Sample answer: Synthetic division is based on the operations performed in long division. Consider solving this problem through long division without using a placeholder. When you determine your first quotient term to be $-2x^3$ and multiply it by $x + 3$ to subtract from the first two terms of the equation, you would be left with $3x^2 + 6x^3$ and would not be able to continue the process.

Example 2 Apply the Remainder Theorem

Teaching the Mathematical Practices

4 Make Assumptions Use the Think About It! feature to have students explain an assumption that is being made to solve the problem.

Questions for Mathematical Discourse

AL Why is $f(25)$ used to predict the production in 2025? Sample answer: Because x represents the number of years since 2000.

OL How many eggs are produced in 2000? Explain. 84.755 billion; sample answer: In year 2000, $x = 0$, so the y-intercept is the number of eggs produced in 2000.

BL Describe two ways that you could check your solution. Sample answer: Directly substitute 25 for x in the function or graph the function on a graphing calculator and find the value of y when $x = 25$.

Common Error

When students are setting up synthetic division, they may inadvertently leave off negative signs when the polynomial function contains subtraction. Remind students that subtraction equates to a negative coefficient.

Learn The Factor Theorem

Objective

Students determine factors of polynomials by using the Factor Theorem.

Teaching the Mathematical Practices

3 Analyze Cases Guide students to examine the how the Factor Theorem is related to the Remainder Theorem for the case where $p(a) = 0$.

Important to Know

When a binomial divides evenly into a polynomial, the binomial is a factor. A depressed polynomial is the quotient of the division and has a degree one less than the original polynomial.

Essential Question Follow-Up

Students have begun learning about the Remainder and Factor Theorems.

Ask:

Why are the Remainder and Factor Theorems important when solving polynomial equations? Sample answer: They tell you whether a binomial is a factor of a polynomial, which helps you find the roots.

Example 3 Use the Factor Theorem

Teaching the Mathematical Practices

6 Communicate Precisely Encourage students to routinely write and explain their solution methods. The Talk About It! feature asks students to explain the steps they would take to solve a problem.

Questions for Mathematical Discourse

AL What is a depressed polynomial? the quotient when a binomial evenly divides a polynomial

OL Without using synthetic division, how could you determine that $x + 8$ is a factor of $2x^3 + 15x^2 - 9x - 24$? Sample answer: Substitute -8 for x and see if the result is zero.

BL When might you want to use the Factor Theorem? Sample answer: When you are given a potential factor and need to quickly test several possible factors, or are only interested in whether a specific binomial is a factor. You can use synthetic division to verify the factor and simultaneously determine the depressed polynomial. You can then potentially factor the depressed polynomial to find the other factors of the original expression.

Check

KITTENS The ideal weight of a kitten in pounds is modeled by the function $f(x) = 0.009x^2 + 0.127x + 0.377$, where x is the age of the kitten in weeks. Determine the ideal weight of a 9-week-old kitten. Round to the nearest tenth.

2.3 pounds

Learn The Factor Theorem

When a binomial evenly divides a polynomial, the binomial is a factor of the polynomial. The quotient of this division is called a depressed polynomial. The **depressed polynomial** has a degree that is one less than the original polynomial.

A special case of the Remainder Theorem is called the Factor Theorem.

Key Concept • Factor Theorem

Words: The binomial $x - a$ is a factor of the polynomial $p(x)$ if and only if $p(a) = 0$.

Example:

$$\underbrace{x^3 - x^2 - 30x + 72}_{\text{dividend}} = \underbrace{(x^2 - 7x + 12)}_{\text{quotient}} \cdot \underbrace{(x + 6)}_{\text{divisor}} + \underbrace{0}_{\text{remainder}}$$

$x + 6$ is a factor of $x^3 - x^2 - 30x + 72$.

Go Online You may want to complete the Concept Check to check your understanding.

Example 3 Use the Factor Theorem

Show that $x + 8$ is a factor of $2x^3 + 15x^2 - 11x - 24$. Then find the remaining factors of the polynomial.

-8	2	15	-11	-24
		-16	8	24
	2	-1	-3	0

Because the remainder is 0, $x + 8$ is a factor of the polynomial by the Factor Theorem. So $2x^3 + 15x^2 - 11x - 24$ can be factored as $(x + 8)(2x^2 - x - 3)$. The depressed polynomial is $2x^2 - x - 3$.

Check to see if this polynomial can be factored.

$2x^2 - x - 3 = (2x - 3)(x + 1)$ Factor the trinomial.

Therefore, $2x^3 + 15x^2 - 11x - 24 = (x + 8)(2x - 3)(x + 1)$.

Study Tip

Factoring Some depressed polynomials may not be factorable. In that case, the only factors are the divisor and the depressed polynomial.

Interactive Presentation

The Factor Theorem

When a binomial evenly divides a polynomial, the binomial is a factor of the polynomial. The quotient of this division is called a depressed polynomial. The **depressed polynomial** has a degree that is one less than the original polynomial.

A special case of the Remainder Theorem is called the Factor Theorem.

Key Concept: Factor Theorem

Words	The binomial $x - a$ is a factor of the polynomial $p(x)$ if and only if $p(a) = 0$.
Example	$\underbrace{x^3 - x^2 - 30x + 72}_{\text{dividend}} = \underbrace{(x^2 - 7x + 12)}_{\text{quotient}} \cdot \underbrace{(x + 6)}_{\text{divisor}} + \underbrace{0}_{\text{remainder}}$
	$x + 6$ is a factor of $x^3 - x^2 - 30x + 72$.

Learn

MULTIPLE CHOICE

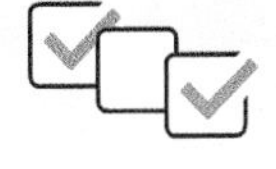

Students select features of polynomials based on specific quotients and remainders.

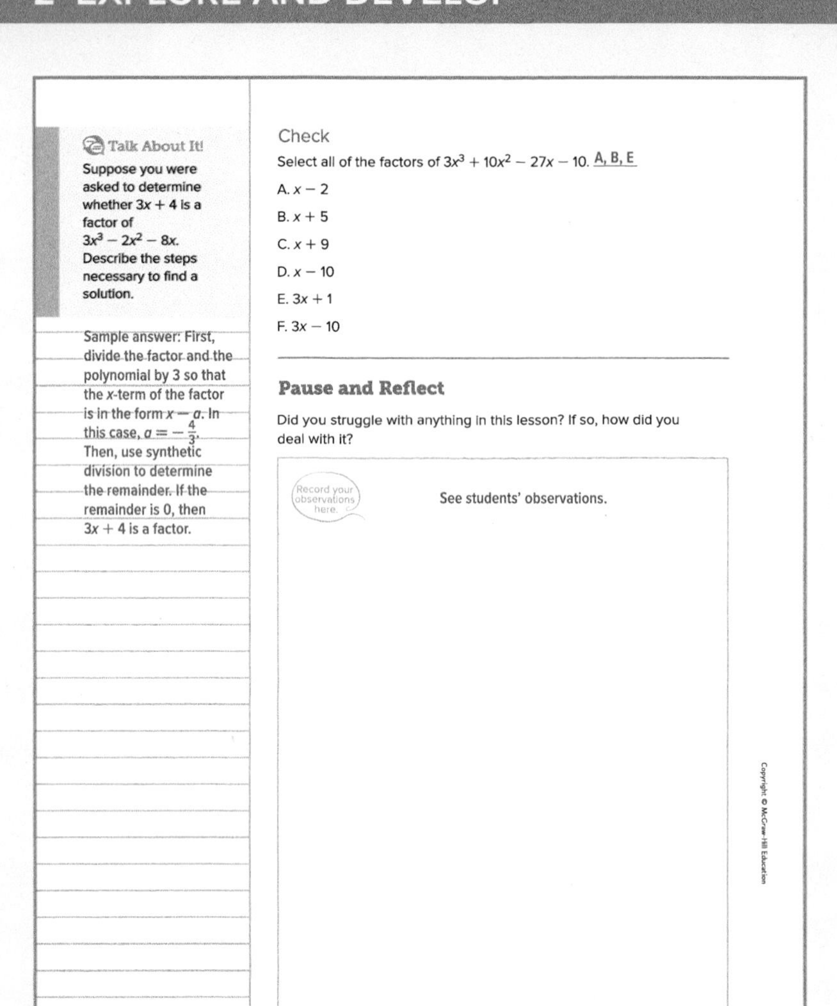

Talk About It!

Suppose you were asked to determine whether $3x + 4$ is a factor of $3x^3 - 2x^2 - 8x$. Describe the steps necessary to find a solution.

Sample answer: First, divide the factor and the polynomial by 3 so that the x-term of the factor is in the form $x - a$. In this case, $a = -\frac{4}{3}$. Then, use synthetic division to determine the remainder. If the remainder is 0, then $3x + 4$ is a factor.

Check

Select all of the factors of $3x^3 + 10x^2 - 27x - 10$. A, B, E

A. $x - 2$

B. $x + 5$

C. $x + 9$

D. $x - 10$

E. $3x + 1$

F. $3x - 10$

Pause and Reflect

Did you struggle with anything in this lesson? If so, how did you deal with it?

Record your observations here.

See students' observations.

Go Online You can complete an Extra Example online.

278 Module 5 • Polynomial Equations

Interactive Presentation

Question 1

Select all of the factors of $3x^3 + 10x^2 - 27x - 10$.

A) $x - 2$

B) $x + 5$

C) $x + 9$

D) $x - 10$

E) $3x + 1$

F) $3x - 10$

Example 3

CHECK

Students complete the Check online to determine whether they are ready to move on.

Common Error

When students are given a potential factor of a polynomial, they often forget to use the opposite sign when performing synthetic division. Remind students that factors are in the form $x - a$ so the sign of a is opposite of the factor.

DIFFERENTIATE

Language Development Activity ELL

Intermediate Use an interactive whiteboard to project an example and mark text during reading to reinforce steps and concepts. Use different colors to represent the different steps in the process.

Advanced Have students write a paragraph summarizing the sequence of steps to solving a problem using synthetic substitution. They could create a numbered flowchart to help organize their paragraphs.

Advanced High Instruct collaborative groups to create a pamphlet illustrating the sequence of steps. Have students share their pamphlets with the class.

Exit Ticket

Recommended Use

At the end of class, go online to display the Exit Ticket prompt and ask students to respond using a separate piece of paper. Have students hand you their responses as they leave the room.

Alternate Use

At the end of class, go online to display the Exit Ticket prompt and ask students to respond verbally or by using a mini-whiteboard. Have students hold up their whiteboards so that you can see all student responses. Tap to reveal the answer when most or all students have completed the Exit Ticket.

Practice and Homework

Suggested Assignments

Use the table below to select appropriate exercises.

DOK	Topic	Exercises
1, 2	exercises that mirror the examples	1–44
2	exercises that use a variety of skills from this lesson	45–60
3	exercises that emphasize higher-order and critical thinking skills	61–67

ASSESS AND DIFFERENTIATE

Use the data from the **Checks** to determine whether to provide resources for extension, remediation, or intervention.

IF students score 90% or more on the Checks, **THEN** assign: BL

- Practice Exercises 1–65 odd, 67–75
- Extension: Radical Notation
- **ALEKS**® Polynomial Division; Remainder and Factor Theorems

IF students score 66%–89% on the Checks, **THEN** assign: OL

- Practice Exercises 1–75 odd
- Remediation, Review Resources: Dividing Polynomials
- Personal Tutors
- Extra Examples 1–3
- **ALEKS**® Synthetic Division

IF students score 65% or less on the Checks, **THEN** assign: AL

- Practice Exercises 1–19 odd
- Remediation, Review Resources: Dividing Polynomials
- *Quick Review Math Handbook*: The Remainder and Factor Theorems
- **ALEKS**® Synthetic Division

Name ______________ Period ________ Date ________

Practice

Go Online You can complete your homework online.

Example 1

Use synthetic substitution to find $f(-5)$ and $f(2)$ for each function.

1. $f(x) = x^3 + 2x^2 - 3x + 1$ — -59; 11
2. $f(x) = x^2 - 8x + 6$ — 71; -6
3. $f(x) = 3x^4 + x^3 - 2x^2 + x + 12$ — 1707; 62
4. $f(x) = 2x^3 - 8x^2 - 2x + 5$ — -435; -15
5. $f(x) = x^3 - 5x + 2$ — -98; 0
6. $f(x) = x^5 + 8x^3 + 2x - 15$ — -4150; 85
7. $f(x) = x^6 - 4x^4 + 3x^2 - 10$ — $13{,}190$; 2
8. $f(x) = x^4 - 6x - 8$ — 647; -4

Use synthetic substitution to find $f(2)$ and $f(-1)$ for each function.

9. $f(x) = x^2 + 6x + 5$ — $21, 0$
10. $f(x) = x^2 - x + 1$ — $3, 3$
11. $f(x) = x^2 - 2x - 2$ — $-2, 1$
12. $f(x) = x^3 + 2x^2 + 5$ — $21, 6$
13. $f(x) = x^3 - x^2 - 2x + 3$ — $3, 3$
14. $f(x) = x^3 + 6x^2 + x - 4$ — $30, 0$
15. $f(x) = x^3 - 3x^2 + x - 2$ — $-4, -7$
16. $f(x) = x^3 - 5x^2 - x + 6$ — $-8, 1$
17. $f(x) = x^4 + 2x^2 - 9$ — $15, -6$
18. $f(x) = x^4 - 3x^3 + 2x^2 - 2x + 6$ — $2, 14$
19. $f(x) = x^5 - 7x^3 - 4x + 10$ — $-22, 20$
20. $f(x) = x^6 - 2x^5 + x^4 + x^3 - 9x^2 - 20$ — $-32, -26$

Example 2

21. BUSINESS Advertising online generates billions of dollars for global businesses each year. The revenue from online advertising in the United States from 2000 to 2015 can be modeled by $y = 0.01x^3 + 0.02x^2 + x + 6$, where x is the number of years since 2000 and y is the revenue in billions of U.S. dollars.
    a. Estimate the revenue from online advertising in 2008. **$20.4 billion**
    b. Predict the revenue from online advertising in 2022. **$144.16 billion**

22. PROFIT The profit, in thousands, of Clyde's Corporation can be modeled by $P(y) = y^4 - 4y^3 + 2y^2 + 10y - 200$, where y is the number of years after the business was started. Predict the profit of Clyde's Corporation after 10 years. **$6,100,000**

Example 3

Given a polynomial and one of its factors, find the remaining factors of the polynomial.

23. $x^3 - 3x + 2$; $x + 2$ — $(x - 1)^2$
24. $x^4 + 2x^3 - 8x - 16$; $x + 2$ — $x - 2, x^2 + 2x + 4$
25. $x^3 - x^2 - 10x - 8$; $x + 2$ — $x - 4, x + 1$
26. $x^3 - x^2 - 5x - 3$; $x - 3$ — $(x + 1)^2$
27. $2x^3 + 17x^2 + 23x - 42$; $x - 1$ — $x + 6, 2x + 7$
28. $2x^3 + 7x^2 - 53x - 28$; $x - 4$ — $x + 7, 2x + 1$
29. $x^4 + 2x^3 + 2x^2 - 2x - 3$; $x - 1$ — $x + 1, x^2 + 2x + 3$
30. $3x^3 - 19x^2 - 15x + 7$; $x - 7$ — $x + 1, 3x - 1$
31. $x^3 + 2x^2 - x - 2$; $x + 1$ — $x - 1, x + 2$
32. $3x^3 + x^2 - 5x + 3$; $x - 1$ — $x - 1, x + 3$
33. $x^3 + 3x^2 - 4x - 12$; $x + 3$ — $x - 2, x + 2$
34. $x^3 - 6x^2 + 11x - 6$; $x - 3$ — $x - 1, x - 2$
35. $x^3 + 2x^2 - 33x - 90$; $x + 5$ — $x + 3, x - 6$
36. $x^3 - 6x^2 + 32$; $x - 4$ — $x - 4, x + 2$
37. $x^3 - x^2 - 10x - 8$; $x + 2$ — $x + 1, x - 4$
38. $x^3 - 19x + 30$; $x - 2$ — $x + 5, x - 3$
39. $2x^3 + x^2 - 2x - 1$; $x + 1$ — $2x + 1, x - 1$
40. $2x^3 + x^2 - 5x + 2$; $x + 2$ — $x - 1, 2x - 1$
41. $3x^3 + 4x^2 - 5x - 2$; $3x + 1$ — $x - 1, x + 2$
42. $3x^3 + x^2 + x - 2$; $3x - 2$ — $x^2 + x + 1$
43. $6x^3 - 25x^2 + 2x + 8$; $2x + 1$ — $x - 4, 3x - 2$
44. $16x^5 - 32x^4 - 81x + 162$; $2x - 3$ — $x - 2, 2x + 3, 4x^2 + 9$

Mixed Exercises

45. REASONING Jessica evaluates the polynomial $p(x) = x^3 - 5x^2 + 3x + 5$ for a factor using synthetic substitution. Some of her work is shown below. Find the values of a and b. **11; 764**

a	1	−5	3	5
		11	66	759
	1	6	69	b

Name ______________ Period ______ Date ______

46. STATE YOUR ASSUMPTION The revenue from streaming music services in the United States from 2005 to 2016 can be modeled by $y = 0.26x^5 - 7.48x^4 + 79.20x^3 - 333.33x^2 + 481.68x + 99.13$, where x is the number of years since 2005 and y is the revenue in millions of U.S. dollars.

a. Estimate the revenue from streaming music services in 2010. **$211.78 million**

b. What might the revenue from streaming music services be in 2020? What assumption did you make to make your prediction? **$18,387.58 million; The model still represents the situation after 15 years.**

47. NATURAL EXPONENTIAL FUNCTION The natural exponential function $y = e^x$ is a special function that is applied in many fields such as physics, biology, and economics. It is not a polynomial function, however for small values of x, the value of e^x is very closely approximated by the polynomial function $f(x) = \frac{1}{6}x^3 + \frac{1}{2}x^2 + x + 1$. Use synthetic substitution to determine $f(0.1)$. **See margin.**

Find values of k so that each remainder is 3.

48. $(x^2 - x + k) \div (x - 1)$
3

49. $(x^2 + kx - 17) \div (x - 2)$
8

50. $(x^2 + 5x + 7) \div (x + k)$
1, 4

51. $(x^3 + 4x^2 + x + k) \div (x + 2)$
−3

52. If $f(-8) = 0$ and $f(x) = x^3 - x^2 - 58x + 112$, find all the zeros of $f(x)$ and use them to graph the function. **See margin.**

53. REASONING If $P(1) = 0$ and $P(x) = 10x^3 + kx^2 - 16x + 3$, find all the factors of $P(x)$ and use them to graph the function. Explain your reasoning. **See margin.**

54. GEOMETRY The volume of a box with a square base is $V(x) = 2x^3 + 15x^2 + 36x + 27$. If the height of the box is $(2x + 3)$ units, what are the measures of the sides of the base in terms of x? **See margin.**

55. SPORTS The average value of a franchise in the National Football League from 2000 to 2018 can be modeled by $y = -0.037x^5 + 1.658x^4 - 24.804x^3 + 145.100x^2 - 207.594x + 482.008$, where x is the number of years since 2000 and y is the value in millions of U.S. dollars.

a. Complete the table of estimated values. Round to the nearest million.

Year	2003	2012	2021	2025
Estimated Average Franchise Value (millions $)	621	1197	1740	−15,255

b. What assumption did you make to make your predictions? Do you think the assumption is valid? Explain. **The model still represents the situation after 25 years; no, the average value is unlikely to fall so quickly.**

56. CONSTRUCT ARGUMENTS Divide the polynomial function $f(x) = 4x^3 - 10x + 8$ by the factor $(x + 5)$. Then state and confirm the Remainder Theorem for this particular polynomial function and factor. **See margin.**

57. REGULARITY The polynomial function $P(x)$ is symmetric in the y-axis and contains the point $(2, -5)$. What is the remainder when $P(x)$ is divided by $(x + 2)$? Explain your reasoning. **See margin.**

58. STRUCTURE Verify the Remainder Theorem for the polynomial $x^2 + 3x + 5$ and the factor $(x - \sqrt{3})$ by first using synthetic division and then evaluating for $x = \sqrt{3}$. **See Mod. 5 Answer Appendix.**

59. STRUCTURE If $(x + 6)$ is a factor of $kx^3 + 15x^2 + 13x - 30$, determine the value of k, factor the polynomial and confirm the result graphically. **See Mod. 5 Answer Appendix.**

60. REGULARITY Polynomial $f(x)$ is divided by $x - c$. What can you conclude if:

a. the remainder is 0? **$x - c$ is a factor of $f(x)$.**

b. the remainder is 1? **$x - c$ is not a factor of $f(x)$.**

c. the quotient is 1, and the remainder is 0? **$f(x) = x - c$**

Higher-Order Thinking Skills

61. CREATE Write a polynomial function that has a double zero of 1 and a double zero of −5. Graph the function. **Sample answer: $f(x) = 0.1(x - 1)^2(x + 5)^2$; See Mod. 5 Answer Appendix for graph.**

62. PERSEVERE For a cubic function $P(x)$, $P(2) = -90$, $P(-8) = 0$, and $P(5) = 0$.

a. Write two possible equations for $P(x)$. Explain your answer. **See Mod. 5 Answer Appendix.**

b. Graph your equations from **part a**. What three points do these graphs have in common? **See Mod. 5 Answer Appendix.**

c. If $P(4) = 60$, write the equation for $P(x)$. **See Mod. 5 Answer Appendix.**

63. ANALYZE Review the definition for the Factor Theorem. Provide a proof of the theorem. **See Mod. 5 Answer Appendix.**

64. CREATE Write a cubic function that has a remainder of 8 for $f(2)$ and a remainder of −5 for $f(3)$. **Sample answer: $f(x) = -x^3 + x^2 + x + 10$**

65. PERSEVERE Show that the quartic function $f(x) = ax^4 + bx^3 + cx^2 + dx + e$ will always have a rational zero when the numbers 1, −2, 3, 4, and −6 are randomly assigned to replace a through e, and all of the numbers are used. **See Mod. 5 Answer Appendix.**

66. WRITE Explain how the zeros of a function can be located by using the Remainder Theorem and making a table of values for different input values and then comparing the remainders. **See Mod. 5 Answer Appendix.**

67. FIND THE ERROR The table shows x-values and their corresponding $P(x)$ values for a polynomial function. Tyrone and Nia used the Factor Theorem to find factors of $P(x)$. Is either of them correct? Explain your reasoning. **See Mod. 5 Answer Appendix.**

x	−3	−1	0	1	2	4
$P(x)$	−18	0	6	2	0	122

Tyrone	Nia
$(x + 1)$ and $(x - 2)$	$(x - 6)$

Answers

47.

0.1	$\frac{1}{6}$	$\frac{1}{2}$	1	1
		$\frac{1}{60}$	$\frac{31}{600}$	$\frac{631}{6000}$
	$\frac{1}{6}$	$\frac{31}{60}$	$\frac{631}{600}$	$\frac{6631}{6000}$

$1\frac{631}{6000}$

52. −8, 7, 2;

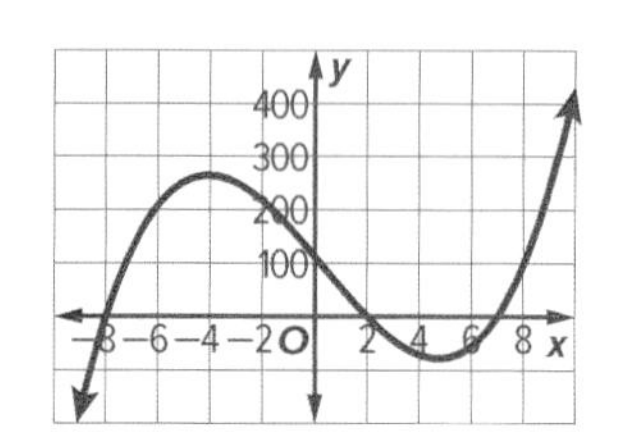

53. $(x - 1)$, $(5x - 2)$, $(2x + 3)$; Sample answer: By the Factor Theorem, $(x - 1)$ is a factor. Use synthetic division with $x = 1$. The remainder is $k - 3$. For $(x - 1)$ to be a factor, $k - 3 = 0$, so $k = 3$. The quotient is $10x^2 + 13x - 3$, which factors as $(5x - 1)(2x + 3)$. $P(x) = (x - 1)(5x - 1)(2x + 3)$. The cubic has a positive leading coefficient and zeros at −1.5, 0.2, and 1.

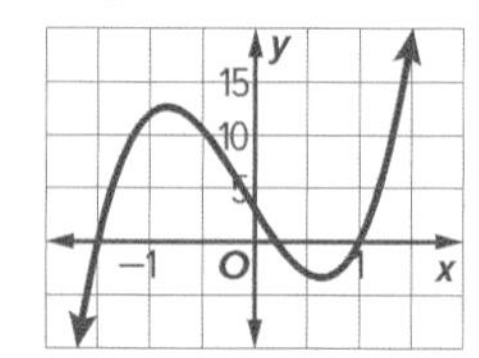

54. $x + 3$; Sample answer: Since $(2x + 3)$ is a factor of $V(x)$, use synthetic division with $x = -\frac{3}{2}$. The remainder is zero and the other factor is $(x^2 + 6x + 9)$, or $(x + 3)^2$. The sides of the base are $(x + 3)$.

56. For $f(x) = 4x^3 - 10x + 8$, the remainder on division by $(x + 5)$ is −442. This means that $f(-5) = -442$; $4(-5)^3 - 10(-5) + 8 = -500 + 50 + 8 = -442$.

57. −5; Sample answer: If $P(x)$ is symmetric to the y-axis, it must also contain the point $(-2, -5)$. According to the Remainder Theorem, the remainder when polynomial $P(x)$ is divided by $(x - r)$ is $P(r)$. Since $P(-2) = -5$, the remainder when $P(x)$ is divided by $(x + 2)$ is −5.

Lesson 5-5

Roots and Zeros

N.CN.9, A.APR.3, F.IF.7c

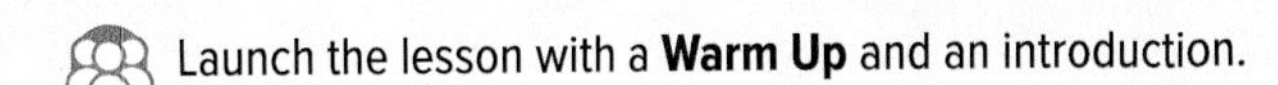

LESSON GOAL

Students determine the numbers and types of roots of polynomial equations, find zeros, and use zeros to graph polynomial functions.

1 LAUNCH

Launch the lesson with a **Warm Up** and an introduction.

2 EXPLORE AND DEVELOP

Explore: Roots of Quadratic Polynomials

Develop:

Fundamental Theorem of Algebra
- Determine Number and Type of Roots
- Find Numbers of Positive and Negative Zeros

Finding Zeros of Polynomial Functions
- Use Synthetic Substitution to Find Zeros
- Use a Graph to Write a Polynomial Function
- Use Zeros to Graph a Polynomial Function

You may want your students to complete the **Checks** online.

3 REFLECT AND PRACTICE

Exit Ticket

Practice

DIFFERENTIATE

View reports of student progress on the **Checks** after each example.

Resources	AL	OL	BL	ELL
Remediation: Intercepts of Graphs	●	●		●
Extension: The Rational Zero Theorem		●	●	●

Language Development Handbook

Assign page 34 of the *Language Development Handbook* to help your students build mathematical language related to determining the number and types of roots of polynomial equations.

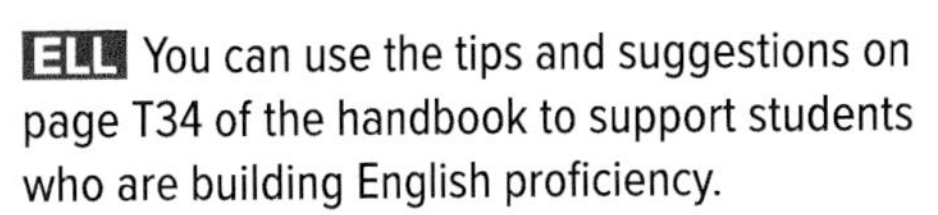

ELL You can use the tips and suggestions on page T34 of the handbook to support students who are building English proficiency.

Suggested Pacing

90 min	1 day
45 min	2 days

Focus

Domain: Number and Quantity, Algebra, Functions

Standards for Mathematical Content:

N.CN.9 Know the Fundamental Theorem of Algebra; show that it is true for quadratic polynomials.

A.APR.3 Identify zeros of polynomials when suitable factorizations are available, and use the zeros to construct a rough graph of the function defined by the polynomial.

F.IF.7c Graph polynomial functions, identifying zeros when suitable factorizations are available, and showing end behavior.

Standards for Mathematical Practice:

1 Make sense of problems and persevere in solving them.

7 Look for and make use of structure.

Coherence

Vertical Alignment

Previous
Students evaluated and factored functions by using the Remainder and Factor Theorems.
A.APR.2

Now
Students determine the numbers and types of roots, find zeros, and use zeros to graph polynomial equations.
N.CN.9, A.APR.3, F.IF.7c

Next
Students will graph and analyze reciprocal functions
F.BF.3, F.IF.5

Rigor

The Three Pillars of Rigor

1 CONCEPTUAL UNDERSTANDING	2 FLUENCY	3 APPLICATION
In this lesson, students solidify their understanding of the relationships among roots, zeros, and factors. They build fluency by identifying zeros when factorizations are available, and they apply their understanding by solving real-world problems.		

Mathematical Background

The real zeros of a polynomial function $f(x)$ are the x-intercepts of the graph $f(x)$. They are also the real solutions of the polynomial equation $f(x) = 0$. A polynomial function cannot have more zeros than its degree.

Interactive Presentation

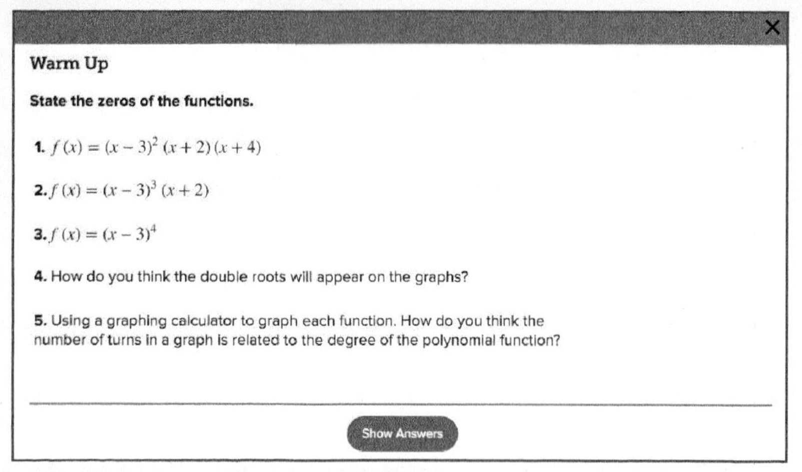

Warm Up

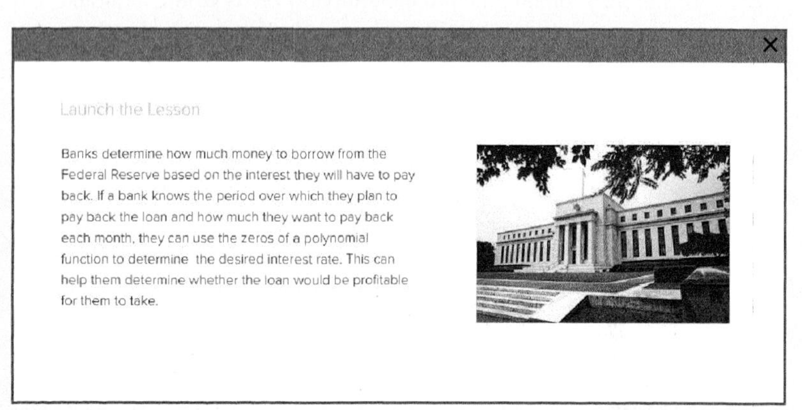

Launch the Lesson

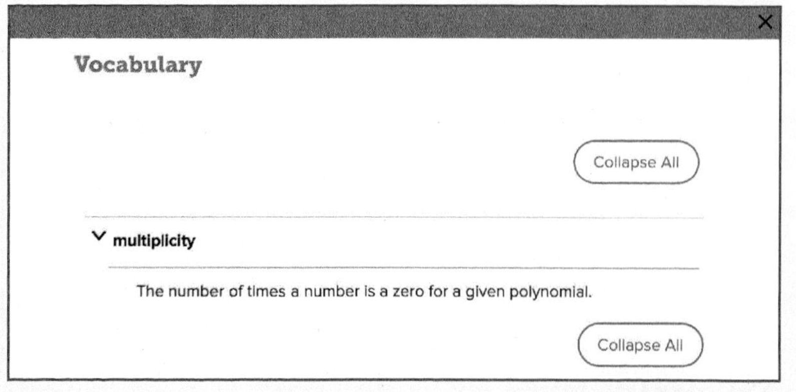

Today's Vocabulary

Warm Up

Prerequisite Skills

The Warm Up exercises address the following prerequisite skill for this lesson:

- finding zeros of functions

Answers:

1. $-4, -2, 3, 3$
2. $-2, 3, 3, 3$
3. $3, 3, 3, 3$
4. The graph will touch the x-axis at the double roots.
5. Sample answer: If the degree is n, then the number of turns is at most $n - 1$.

Launch the Lesson

Teaching the Mathematical Practices

4 Apply Mathematics In the Launch the Lesson, students will learn how the process of banks borrowing money from the Federal Reserve relates to finding the zeros of a polynomial function.

Go Online to find additional teaching notes and questions to promote classroom discourse.

Today's Standards

Tell students that they will be addressing these content and practice standards in this lesson. You may wish to have a student volunteer read aloud *How can I meet these standards?* and *How can I use these practices?*, and connect these to the standards.

See the Interactive Presentation for I Can statements that align with the standards covered in this lesson.

Today's Vocabulary

Tell students that they will be using this vocabulary term in this lesson. You can expand the row if you wish to share the definition. Then discuss the questions below with the class.

Explore Roots of Quadratic Polynomials

Objective

Students use a sketch to prove the Fundamental Theorem of Algebra for quadratic polynomials.

 Teaching the Mathematical Practices

3 Analyze Cases Throughout the Explore, work with students to examine the case where the discriminant of a quadratic equation is negative to prove that the Fundamental Theorem of Algebra holds true for quadratic equations.

Ideas for Use

Recommended Use Present the Inquiry Question, or have a student volunteer read it aloud. Have students work in pairs to complete the Explore activity on their devices. Pairs should discuss each of the questions. Monitor student progress during the activity. Upon completion of the Explore activity, have student volunteers share their responses to the Inquiry Question.

What if my students don't have devices? You may choose to project the activity on a whiteboard. A printable worksheet for each Explore is available online. You may choose to print the worksheet so that individuals or pairs of students can use it to record their observations.

Summary of the Activity

Students will complete guiding exercises throughout the Explore activity. Students use the sketch to determine how the discriminant of a quadratic equation is related to the number of zeros. Then, students will answer the Inquiry Question.

(continued on the next page)

Interactive Presentation

Roots of Quadratic Polynomials

INQUIRY Is the Fundamental Theorem of Algebra true for quadratic polynomials?

The Fundamental Theorem of Algebra states that every polynomial equation with degree greater than zero has at least one root in the set of complex numbers. Consider this theorem in terms of quadratic polynomials.

Explore

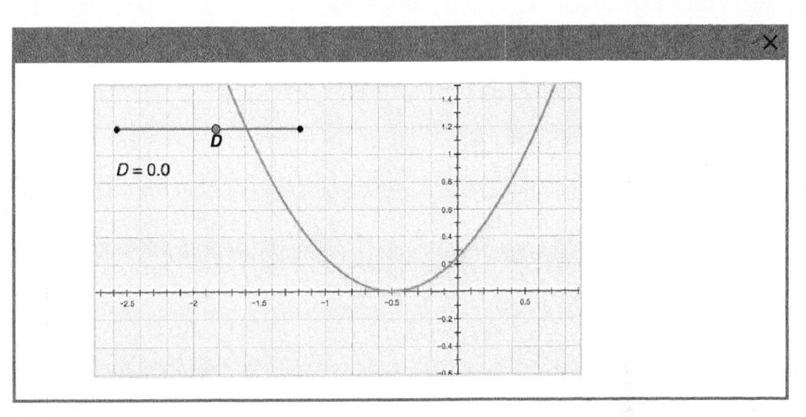

Explore

WEB SKETCHPAD

Students use the sketch to explore the relationship between the number of zeros a quadratic has and the discriminant.

TYPE

Students answer a question about determining the number of real roots a quadratic polynomial has given its graph.

Interactive Presentation

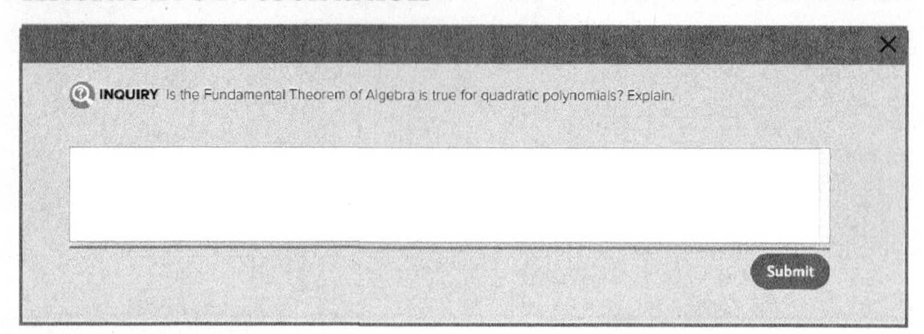

Explore

Students respond to the Inquiry Question and can view a sample answer.

Explore Roots of Quadratic Polynomials (*continued*)

Questions

Have students complete the Explore activity.

Ask:

- If a quadratic equation has a discriminant of 12, would the function have 2 real roots, 1 real root, or 2 complex roots? 2 real roots
- What type of roots would the quadratic equation $4x^2 - 12x + 9 = 0$ have? 1 real root

Inquiry

Is the Fundamental Theorem of Algebra true for quadratic polynomials? Yes; sample answer: A quadratic polynomial can have 2 real roots, 1 real root, or at least 1 complex root.

Go Online to find additional teaching notes and sample answers for the guiding exercises.

Learn Fundamental Theorem of Algebra

Objective

Students determine the numbers and types of roots of polynomial equations by using the Fundamental Theorem of Algebra.

Teaching the Mathematical Practices

1 Explain Correspondences Encourage students to explain the relationships between the zeros, factors, roots, and intercepts of polynomial functions and the relationship between the degree of a polynomial and the number of roots of the function.

About the Key Concepts

The zero of a function $f(x)$ is any value c such that $f(c) = 0$. The real zeros are the x-intercepts of the graph of the function. For a polynomial function $p(x)$, if c is a zero of $p(x)$, then c is a root or solution of $p(x) = 0$, and $x - c$ is a factor of the polynomial. Every polynomial equation with degree greater than zero has at least one root in the set of complex numbers. The degree n of a polynomial equation gives the exact number of roots in the set of complex numbers, including repeating roots. Descartes' Rule of Signs identifies the number of positive and negative zeros for a polynomial function.

Essential Question Follow-Up

Students have begun learning about the Fundamental Theorem of Algebra.

Ask:

Why is the Fundamental Theorem of Algebra important as we use polynomial equations to model and solve real-world situations? Sample answer: Many real-world situations involve multiple factors, which generate polynomial functions. Because only certain values make sense in real-world situations, the Fundamental Theorem of Algebra makes it possible to find the number and type of roots for a polynomial equation.

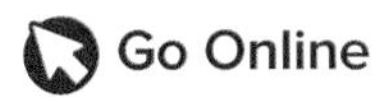

Go Online

- Find additional teaching notes.
- View performance reports of the Checks.
- Assign or present an Extra Example.

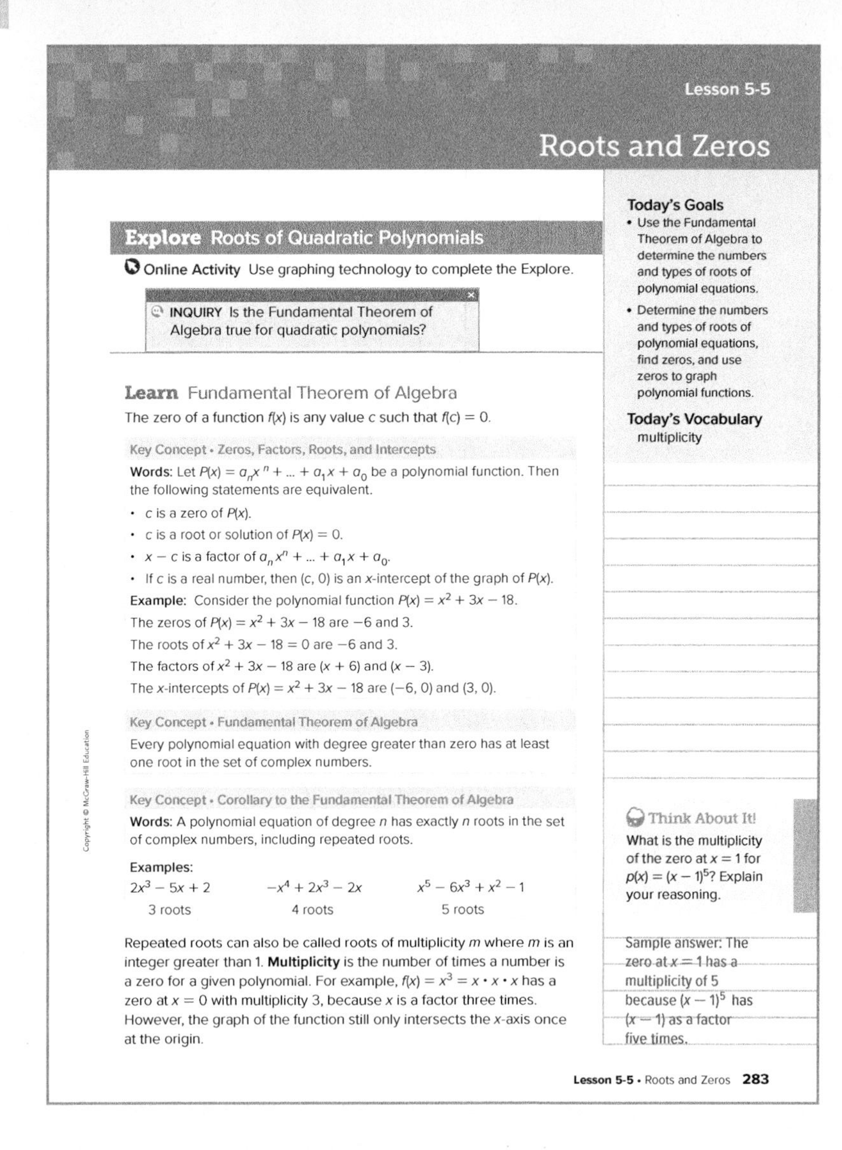

Lesson 5-5

Roots and Zeros

Explore Roots of Quadratic Polynomials

Online Activity Use graphing technology to complete the Explore.

INQUIRY Is the Fundamental Theorem of Algebra true for quadratic polynomials?

Learn Fundamental Theorem of Algebra

The zero of a function $f(x)$ is any value c such that $f(c) = 0$.

Key Concept • Zeros, Factors, Roots, and Intercepts

Words: Let $P(x) = a_nx^n + ... + a_1x + a_0$ be a polynomial function. Then the following statements are equivalent.

- c is a zero of $P(x)$.
- c is a root or solution of $P(x) = 0$.
- $x - c$ is a factor of $a_nx^n + ... + a_1x + a_0$.
- If c is a real number, then $(c, 0)$ is an x-intercept of the graph of $P(x)$.

Example: Consider the polynomial function $P(x) = x^2 + 3x - 18$.

The zeros of $P(x) = x^2 + 3x - 18$ are -6 and 3.

The roots of $x^2 + 3x - 18 = 0$ are -6 and 3.

The factors of $x^2 + 3x - 18$ are $(x + 6)$ and $(x - 3)$.

The x-intercepts of $P(x) = x^2 + 3x - 18$ are $(-6, 0)$ and $(3, 0)$.

Key Concept • Fundamental Theorem of Algebra

Every polynomial equation with degree greater than zero has at least one root in the set of complex numbers.

Key Concept • Corollary to the Fundamental Theorem of Algebra

Words: A polynomial equation of degree n has exactly n roots in the set of complex numbers, including repeated roots.

Examples:

$2x^3 - 5x + 2$	$-x^4 + 2x^3 - 2x$	$x^5 - 6x^3 + x^2 - 1$
3 roots	4 roots	5 roots

Repeated roots can also be called roots of multiplicity m where m is an integer greater than 1. **Multiplicity** is the number of times a number is a zero for a given polynomial. For example, $f(x) = x^3 = x \cdot x \cdot x$ has a zero at $x = 0$ with multiplicity 3, because x is a factor three times. However, the graph of the function still only intersects the x-axis once at the origin.

Today's Goals

- Use the Fundamental Theorem of Algebra to determine the numbers and types of roots of polynomial equations.
- Determine the numbers and types of roots of polynomial equations, find zeros, and use zeros to graph polynomial functions.

Today's Vocabulary

multiplicity

Think About It!

What is the multiplicity of the zero at $x = 1$ for $p(x) = (x - 1)^5$? Explain your reasoning.

Sample answer: The zero at $x = 1$ has a multiplicity of 5 because $(x - 1)^5$ has $(x - 1)$ as a factor five times.

Copyright © McGraw-Hill Education

Lesson 5-5 • Roots and Zeros 283

Interactive Presentation

Fundamental Theorem of Algebra

The zero of a function $f(x)$ is any value c such that $f(c) = 0$. When the function is graphed, the real zeros of the function are the x-intercepts of the graph.

KEY CONCEPT: ZEROS, FACTORS, ROOTS, AND INTERCEPTS

Words: Let $P(x) = a_nx^n + \cdots + a_1x + a_0$ be a polynomial function. Then the following statements are equivalent.
- c is a zero of $P(x)$
- c is a root or solution of $P(x) = 0$.
- $x - c$ is a factor of $a_nx^n + \cdots + a_1x + a_0$.
- If c is a real number, then $(c, 0)$ is an x-intercept of the graph of $P(x)$.

Learn

TYPE

Students answer a question about multiplicity of a zero of a polynomial function.

N.CN.9, A.APR.3, F.IF.7c

1 CONCEPTUAL UNDERSTANDING | 2 FLUENCY | 3 APPLICATION

Your Notes

Key Concept • Descartes' Rule of Signs

Let $P(x) = a_n x^n + ... + a_1 x + a_0$ be a polynomial function with real coefficients and $a_0 \neq 0$. Then the number of positive real zeros of $P(x)$ is the same as the number of changes in sign of the coefficients of the terms, or is less than this by an even number, and the number of negative real zeros of $P(x)$ is the same as the number of changes in sign of the coefficients of the terms of $P(-x)$, or is less than this by an even number.

Example 1 Determine the Number and Type of Roots

Solve $x^4 + 49x^2 = 0$. State the number and type of roots.

$x^4 + 49x^2 = 0$		Original equation
$x^2(x^2 + 49) = 0$		Factor.
$x^2 = 0$ or	$x^2 + 49 = 0$	Zero Product Property
$x = 0$	$x^2 = -49$	Subtract 49 from each side.
	$x = \pm\sqrt{-49}$	Square Root Property
	$x = \pm 7i$	Simplify.

The polynomial has degree 4, so there are four roots in the set of complex numbers. Because x^2 is a factor, $x = 0$ is a root with multiplicity 2, also called a double root. The equation has one real repeated root, 0, and two imaginary roots, $7i$ and $-7i$.

Study Tip

Repeated Roots If you factor a polynomial and a factor is raised to a power greater than 1, then there is a repeated root. The power to which the factor is raised indicates the multiplicity of the root. To be sure that you do not miss a repeated root, it can help to write out each factor. For example, you would write $x^2(x^2 + 49)$ as $x \cdot x(x^2 + 49)$ as a reminder that x^2 indicates a root of multiplicity 2.

Example 2 Find the Number of Positive and Negative Zeros

State the possible number of positive real zeros, negative real zeros, and imaginary zeros of $f(x) = x^5 - 2x^4 - x^3 + 6x^2 - 5x + 10$.

Because $f(x)$ has degree 5, it has five zeros, either real or imaginary. Use Descartes' Rule of Signs to determine the possible number and types of *real* zeros.

Part A Find the possible number of positive real zeros.

Count the number of changes in sign for the coefficients of $f(x)$.

$f(x) = x^5 - 2x^4 - x^3 + 6x^2 - 5x + 10$

yes (+ to −), no (− to −), yes (− to +), yes (+ to −), yes (− to +)

There are 4 sign changes, so there are 4, 2, or 0 positive real zeros.

Part B Find the possible number of negative real zeros.

Count the number of changes in sign for the coefficients of $f(-x)$.

$$f(-x) = (-x)^5 - 2(-x)^4 - (-x)^3 + 6(-x)^2 - 5(-x) + 10$$
$$= -x^5 - 2x^4 + x^3 + 6x^2 + 5x + 10$$

no (− to −), yes (− to +), no (+ to +), no (+ to +), no (+ to +)

There is 1 sign change, so there is 1 negative real zero.

Go Online You can complete an Extra Example online.

Interactive Presentation

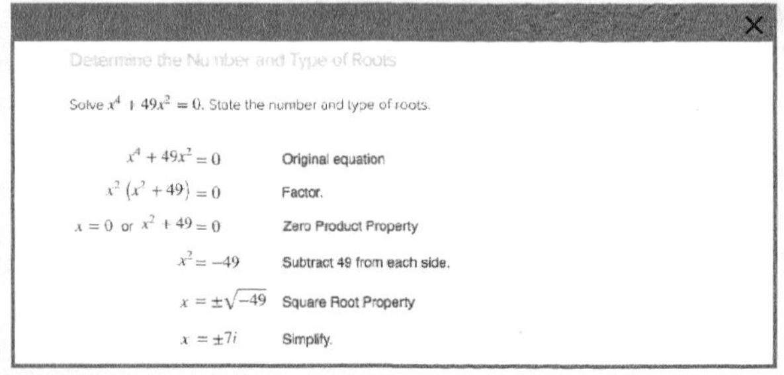

Example 1

TAP

Students tap to see a Study Tip and Common Error.

Example 1 Determine the Number and Type of Roots

Teaching the Mathematical Practices

7 Use the Distributive Property Point out that the Distributive Property is one of the most-used properties. Students will use the Distributive Property to factor and determine the roots of a polynomial equation.

Questions for Mathematical Discourse

AL Why does $x^2 = 0$ yield a double root? Sample answer: x^2 has two factors, x and x.

OL What is the value of the discriminant of the quadratic factor? What does this indicate about the solutions of the quadratic factor? −196; sample answer: It will have 2 imaginary roots.

BL Will the graph of the related function cross the x-axis? Explain. No; sample answer: Because $x = 0$ is the only real root, the graph will touch the x-axis at (0, 0), but will not cross the axis.

Example 2 Find the Number of Positive and Negative Zeros

Teaching the Mathematical Practices

7 Use Structure Help students to use the structure of the polynomial function to determine the possible number of roots of the function.

Questions for Mathematical Discourse

AL What is the smallest possible number of positive roots if a polynomial equation has 3 sign changes? 1

OL Is it possible for an odd-degree polynomial function to have no real roots? Explain in the context of Descartes' Rule of Signs. No; sample answer: An odd-degree polynomial function that does not have any sign changes and has a constant term would have sign changes for $f(-x)$. If the odd function does not have a constant term, the polynomial would have a common factor of x and result in a root of zero.

BL Why are the numbers of possible real zeros in increments of 2? Complex roots come in conjugate pairs.

Common Error

To check for possible negative zeros, students must substitute in $-x$, which may result in sign errors when simplifying. Encourage students to consider each term and the effect of a negative value to limit the number of sign errors.

Learn Finding Zeros of Polynomial Functions

Objective

Students find zeros of polynomial functions and graph polynomial functions by using zeros.

Teaching the Mathematical Practices

1 Explain Correspondences Guide students to see the relationship between the zeros of a polynomial function and its factors.

Important to Know

The Complex Conjugates Theorem states that if a polynomial function with real coefficients has an imaginary number as a root, then its conjugate is also a root.

DIFFERENTIATE

Enrichment Activity BL

A polynomial function has the roots $1 - \sqrt{5}$ and $4i$. What is the equation of the polynomial function in standard form?

$f(x) = x^4 - 2x^3 + 12x^2 - 32x - 64$

Example 3 Use Synthetic Substitution to Find Zeros

Teaching the Mathematical Practices

7 Use Structure Students will use the structure of a polynomial function to determine the number and type of zeros in order to sketch a graph of the function.

(continued on the next page)

Part C Find the possible number of imaginary zeros.

Positive Real Zeros	Negative Real Zeros	Imaginary Zeros	Total Zeros
4	1	0	$4 + 1 + 0 = 5$
2	1	2	$2 + 1 + 2 = 5$
0	1	4	$0 + 1 + 4 = 5$

Talk About It!

If a polynomial has degree n and no real zeros, then how many imaginary zeros does it have? Explain your reasoning.

n; Sample answer: A polynomial that has degree n will have n complex zeros. If there are no real zeros, then it must have $n - 0 = n$ zeros.

Check

State the possible number of positive real zeros, negative real zeros, and imaginary zeros of $f(x) = 3x^6 - x^5 + 2x^4 + x^3 - 3x^2 + 13x + 1$. Write the rows in ascending order of positive real zeros.

Number of Positive Real Zeros	Number of Negative Real Zeros	Number of Imaginary Zeros
0	0	6
0	2	4
2	0	4
2	2	2
4	0	2
4	2	0

Learn Finding Zeros of Polynomial Functions

Key Concept • Complex Conjugates Theorem

Words: Let a and b be real numbers, and $b \neq 0$. If $a + bi$ is a zero of a polynomial function with real coefficients, then $a - bi$ is also a zero of the function.

Example: If $1 + 2i$ is a zero of $f(x) = x^3 - x^2 + 3x + 5$, then $1 - 2i$ is also a zero of the function.

When you are given all of the zeros of a polynomial function and asked to determine the function, use the zeros to write the factors and multiply them together. The result will be the polynomial function.

Example 3 Use Synthetic Substitution to Find Zeros

Find all of the zeros of $f(x) = x^3 + x^2 - 7x - 15$ and use them to sketch a rough graph.

Part A Find all of the zeros.

Step 1 Determine the total number of zeros.

Since $f(x)$ has degree 3, the function has 3 zeros.

(continued on the next page)

Interactive Presentation

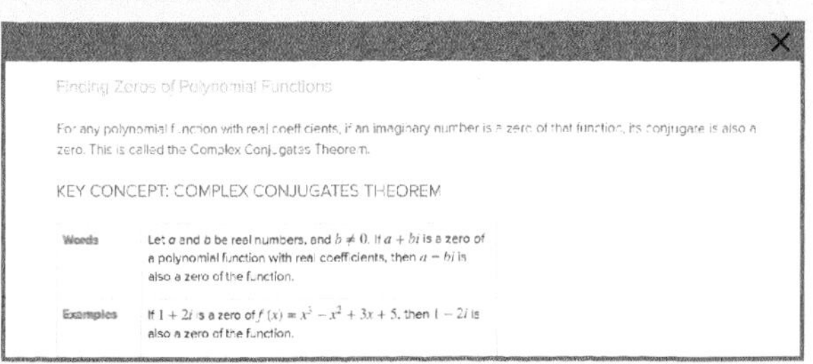

Learn

CHECK

Students complete the Check online to determine whether they are ready to move on.

Step 2 **Determine the type of zeros.**

Examine the number of sign changes for $f(x)$ and $f(-x)$.

$f(x) = x^3 + x^2 - 7x - 15$ $\qquad$ $f(-x) = -x^3 + x^2 + 7x - 15$

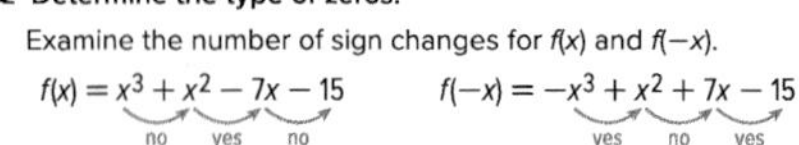

Because there is 1 sign change for the coefficients of $f(x)$, the function has 1 positive real zero. Because there are 2 sign changes for the coefficients of $f(-x)$, $f(x)$ has 2 or 0 negative real zeros. Thus, $f(x)$ has 3 real zeros, or 1 real zero and 2 imaginary zeros.

Step 3 **Determine the real zeros.**

List some possible values, and then use synthetic substitution to evaluate $f(x)$ for real values of x.

x	1	1	−7	−15
−3	1	−2	−1	−12
−2	1	−1	−5	−5
−1	1	0	−7	−8
0	1	1	−7	−15
1	1	2	−5	−20
2	1	3	−1	−17
3	1	4	5	0
4	1	5	13	37

__3__ is a zero of the function, and the depressed polynomial is $x^2 + 4x + 5$. Since it is quadratic, use the Quadratic Formula. The zeros of $f(x) = x^2 + 4x + 5$ are __$-2 - i$__ and __$-2 + i$__.

The function has zeros at 3, $-2 - i$ and $-2 + i$.

Part B Sketch a rough graph.

The function has one real zero at $x = 3$, so the function goes through (3, 0) and does not cross the x-axis at any other place.

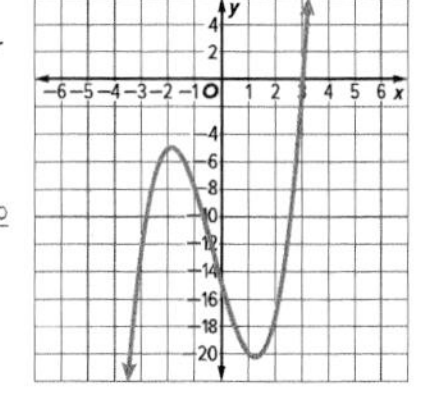

Because the degree is odd and the leading coefficient is positive, the end behavior is that as $x \to -\infty$, $f(x) \to$ __$-\infty$__ and as $x \to \infty$, $f(x) \to$ __∞__.

Use this information and points with coordinates found in the table above to sketch the graph.

Go Online You can complete an Extra Example online.

286 **Module 5** • Polynomial Equations

Interactive Presentation

Use Synthetic Substitution to Find Zeros

Find all of the zeros of $f(x) = x^3 + x^2 - 7x - 15$ and use them to sketch a rough graph.

> Part A

Find all of the zeros.

> Part B

Sketch a rough graph.

Example 3

TAP

Students tap to see the steps to find roots of a polynomial function and sketch its graph.

Questions for Mathematical Discourse

AL What is shown in the table of values in **Step 3**? Sample answer: Each value in the x-column represents a value tested with synthetic substitution. The corresponding columns show the values that result from the synthetic substitution (i.e. the bottom row when performing synthetic substitution). The values in the column under −15 are the remainders.

OL How can synthetic substitution and the Quadratic Formula be used to find all the roots of a polynomial of degree 4? Sample answer: Use synthetic substitution to find the depressed polynomial of degree 3, and then use synthetic substitution one more time to produce a quadratic polynomial. Then the Quadratic Formula can be used to find the remaining 2 roots.

BL Can you determine certain values that make sense to test as possible zeros? Explain. Yes; sample answer: You know from the Distributive Property that the constant of the binomial factor multiplied by the constant of the depressed polynomial must equal the constant of the original function. Thus, test values that are the negative of factors of the constant term.

Common Error

Many students incorrectly substitute values into the Quadratic Formula, especially the $-b$ portion. Encourage students to label a, b, and c in the quadratic function, if necessary, before using the Quadratic Formula.

DIFFERENTIATE

Reteaching Activity **AL** **ELL**

IF students are having difficulties finding the number, type, and signs of the roots of a polynomial equation,

THEN have students check each problem by graphing the polynomial. They can verify the number and sign of real roots.

Common Misconception

A common misconception some students may have is that only the given zeros are the zeros of a polynomial function. Remind students of what the Complex Conjugates Theorem states.

Example 4 Use a Graph to Write a Polynomial Function

Teaching the Mathematical Practices

1 Explain Correspondences Students must be able to explain the relationship between the graph of a polynomial and the equation of the polynomial function.

Questions for Mathematical Discourse

AL What are two features of the polynomial function that are consistent with the graph? Sample answer: When $x = 0$, $y = -8$, and the y-intercept of the graph is -8. The leading coefficient of the polynomial is positive, and the end behavior of the graph indicates a positive leading coefficient.

OL Why does the example say *a polynomial that could be represented by the graph,* rather than asking to find the exact polynomial represented by the graph? Sample answer: The graph only shows real roots, but it could represent a polynomial with pairs of imaginary roots. Thus, the polynomial of degree 3 is a possibility rather than an exact answer.

BL Does the function you wrote have any imaginary roots? Explain. No; sample answer: Because the polynomial has degree 3, it has exactly 3 roots. All 3 of those roots are real and represented on the graph.

Common Error

Many students will simply add the zero to x to create a factor instead. Remind students that if c is a root, then $x - c$ is the factor.

Check

Determine all of the zeros of $f(x) = x^4 - x^3 - 16x^2 - 4x - 80$, and use them to sketch a rough graph.

Real Zeros: -4, 5

Imaginary Zeros: $-2i$, $2i$

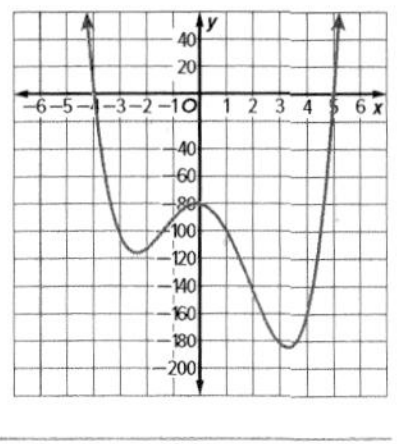

Example 4 Use a Graph to Write a Polynomial Function

Write a polynomial function that could be represented by the graph.

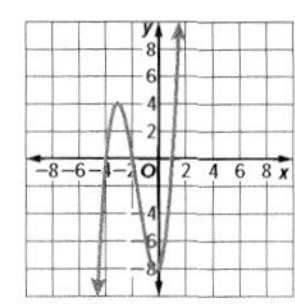

The graph crosses the x-axis 3 times, so the function is at least of degree 3. It crosses the x-axis at $x = -4$, $x = -2$, and $x = 1$, so its factors are $x + 4$, $x + 2$, and $x - 1$.

To determine a polynomial, find the product of the factors.

$y = (x + 4)(x + 2)(x - 1)$ — Set the product of the factors equal to y.

$= (x^2 + 6x + 8)(x - 1)$ — FOIL

$= x^3 + 5x^2 + 2x - 8$ — Multiply.

A polynomial function that could be represented by the graph is $y = x^3 + 5x^2 + 2x - 8$.

Check

Write a polynomial function that could be represented by the graph. C

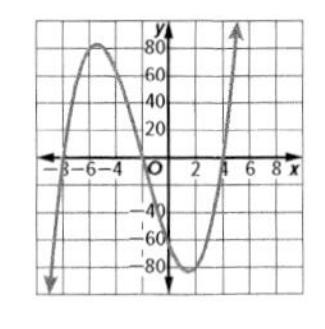

A. $y = x^3 - 6x^2 - 24x + 64$

B. $y = x^2 + 4x - 32$

C. $y = x^3 + 6x^2 - 24x - 64$

D. $y = x^3 - 64$

Go Online You can complete an Extra Example online.

Interactive Presentation

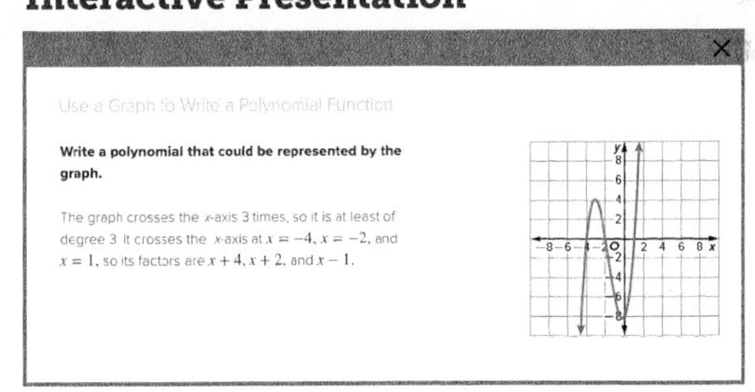

Example 4

TYPE

Students complete the statement to write a polynomial function for the graph.

Apply Example 5 Use Zeros to Graph a Polynomial Function

PROFIT MARGIN **A book publisher wants to release a special hardcover version of several Charles Dickens books. They know that if they charge \$5 or \$40, their profit will be \$0. Graph a polynomial function that could represent the company's profit in thousands of dollars given the price they charge for the book.**

1 What is the task?

Describe the task in your own words. Then list any questions that you may have. How can you find answers to your questions?

Sample answer: Let x represent the price that the publisher charges and let y represent the profit. I need to write and graph a polynomial function that relates x and y.

2 How will you approach the task? What have you learned that you can use to help you complete the task?

Sample answer: I know 5 and 40 are zeros of the function. I can use them to write factors to write an equation of the function.

3 What is your solution?

Use your strategy to solve the problem.

What is a function that represents the given information?

$y = x^2 - 45x + 200$

Graph the function.

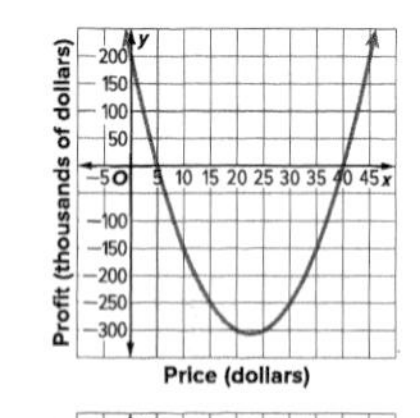

Does this function make sense in the context of the situation? If not, explain why not and write and graph a more reasonable function.

No; the graph passed through the zeros, but did not show reasonable book prices that would result in profit.

$y = -x^2 + 45x - 200$

Profit (thousands of dollars)

Price (dollars)

4 How can you know that your solution is reasonable?

Write About It! Write an argument that can be used to defend your solution.

Sample answer: With multiplying the function by −1, the new function shows that the profit is negative when charging less than \$5 per book. This makes sense in the context of the situation.

Problem-Solving Tip

Logical Reasoning When solving a problem it is important to use logical reasoning skills to analyze the problem.

Go Online You can complete an Extra Example online.

288 **Module 5** • Polynomial Equations

Interactive Presentation

PROFIT MARGIN **A book publisher wants to release a special hardcover version of several Charles Dickens books. They know that if they charge \$5 or \$40, their profit will be \$0. Graph a polynomial function that could represent the company's profit in thousands of dollars given the price they charge for the book.**

Example 5

TAP

Students tap to see the steps to graphing a polynomial using the zeros.

TYPE

Students answer a question to determine the reasonableness of their solution.

CHECK

Students complete the Check online to determine whether they are ready to move on.

Apply Example 5 Use Zeros to Graph a Polynomial Function

Teaching the Mathematical Practices

1 Make Sense of Problems and Persevere in Solving Them, 4 Model with Mathematics Students will be presented with a task. They will first seek to understand the task, and then determine possible entry points to solving it. As students come up with their own strategies, they may propose mathematical models to aid them. As they work to solve the problem, encourage them to evaluate their model and/or progress, and change direction, if necessary.

Recommended Use

Have students work in pairs or small groups. You may wish to present the task, or have a volunteer read it aloud. Then allow students the time to make sure they understand the task, think of possible strategies, and work to solve the problem.

Encourage Productive Struggle

As students work, monitor their progress. Instead of instructing them on a particular strategy, encourage them to use their own strategies to solve the problem and to evaluate their progress along the way. They may or may not find that they need to change direction or try out several strategies.

Signs of Non-Productive Struggle

If students show signs of non-productive struggle, such as feeling overwhelmed, frustrated, or disengaged, intervene to encourage them to think of alternate approaches to the problem. Some sample questions are shown.

- What factors must a polynomial have in order for it to equal zero when $x = 5$ and $x = 40$?
- Which direction must the parabola open and what must be true about the coefficients in order for that to happen?

Write About It!

Have students share their responses with another pair/group of students or the entire class. Have them clearly state or describe the mathematical reasoning they can use to defend their solution.

Exit Ticket

Recommended Use

At the end of class, go online to display the Exit Ticket prompt and ask students to respond using a separate piece of paper. Have students hand you their responses as they leave the room.

Alternate Use

At the end of class, go online to display the Exit Ticket prompt and ask students to respond verbally or by using a mini-whiteboard. Have students hold up their whiteboards so that you can see all student responses. Tap to reveal the answer when most or all students have completed the Exit Ticket.

Practice and Homework

Suggested Assignments

Use the table below to select appropriate exercises.

DOK	Topic	Exercises
1, 2	exercises that mirror the examples	1–45
2	exercises that use a variety of skills from this lesson	46–57
3	exercises that emphasize higher-order and critical thinking skills	58–66

ASSESS AND DIFFERENTIATE

Use the data from the **Checks** to determine whether to provide resources for extension, remediation, or intervention.

IF students score 90% or more on the Checks, **THEN** assign: BL

- Practice Exercises 1–65 odd, 67–72
- Extension: The Rational Root Theorem
- **ALEKS**® Real Zeros of Polynomial Functions

IF students score 66%–89% on the Checks, **THEN** assign: OL

- Practice Exercises 1–71 odd
- Remediation, Review Resources: Intercepts of Graphs
- Personal Tutors
- Extra Examples 1–5
- **ALEKS**® Zeros of Functions

IF students score 65% or less on the Checks, **THEN** assign: AL

- Practice Exercises 1–23 odd
- Remediation, Review Resources: Intercepts of Graphs
- *Quick Review Math Handbook*: Roots and Zeros
- **ALEKS**® Zeros of Functions

Name ____________ Period ________ Date ________

Practice

Go Online You can complete your homework online.

Example 1

Solve each equation. State the number and type of roots.

1. $5x + 12 = 0$
$-\frac{12}{5}$; 1 real

2. $x^2 - 4x + 40 = 0$
$2 \pm 6i$; 2 imaginary

3. $x^5 + 4x^3 = 0$
$0, 0, 0, 2i, -2i$; 3 real, 2 imaginary

4. $x^4 - 625 = 0$
$5, 5i, -5i, -5$; 2 real, 2 imaginary

5. $4x^2 - 4x - 1 = 0$
$\frac{1 \pm \sqrt{2}}{2}$; 2 real

6. $x^5 - 81x = 0$
$0, -3, 3, -3i, 3i$; 3 real, 2 imaginary

7. $2x^2 + x - 6 = 0$
$-2, \frac{3}{2}$; 2 real

8. $4x^2 + 1 = 0$
$-\frac{1}{2}i, \frac{1}{2}i$; 2 imaginary

9. $x^3 + 1 = 0$
$-1, \frac{1 \pm i\sqrt{3}}{2}$; 1 real, 2 imaginary

10. $2x^2 - 5x + 14 = 0$
$\frac{5 \pm i\sqrt{87}}{4}$; 2 imaginary

11. $-3x^2 - 5x + 8 = 0$
$-\frac{8}{3}, 1$; 2 real

12. $8x^3 - 27 = 0$
$\frac{3}{2}, \frac{-3 \pm 3i\sqrt{3}}{4}$; 1 real, 2 imaginary

13. $16x^4 - 625 = 0$
$-\frac{5}{2}, \frac{5}{2}, -\frac{5}{2}i, \frac{5}{2}i$; 2 real, 2 imaginary

14. $x^3 - 6x^2 + 7x = 0$
$0, 3 + \sqrt{2}, 3 - \sqrt{2}$; 3 real

15. $x^5 - 8x^3 + 16x = 0$
$-2, -2, 0, 2, 2$; 5 real

16. $x^5 + 2x^3 + x = 0$
$0, -i, -i, i, i$; 1 real, 4 imaginary

Example 2

State the possible number of positive real zeros, negative real zeros, and imaginary zeros of each function.

17. $g(x) = 3x^3 - 4x^2 - 17x + 6$
2 or 0; 1; 2 or 0

18. $h(x) = 4x^3 - 12x^2 - x + 3$
2 or 0; 1; 2 or 0

19. $f(x) = x^3 - 8x^2 + 2x - 4$
3 or 1; 0; 2 or 0

20. $p(x) = x^3 - x^2 + 4x - 6$
3 or 1; 0; 2 or 0

21. $q(x) = x^4 + 7x^2 + 3x - 9$
1; 1; 2

22. $f(x) = x^4 - x^3 - 5x^2 + 6x + 1$
2 or 0; 2 or 0; 4 or 2 or 0

23. $f(x) = x^4 - 5x^3 + 2x^2 + 5x + 7$
0 or 2; 0 or 2; 0, 2, or 4

24. $f(x) = 2x^3 - 7x^2 - 2x + 12$
0 or 2; 1; 0 or 2

25. $f(x) = -3x^5 + 5x^4 + 4x^2 - 8$
0 or 2; 1; 2 or 4

26. $f(x) = x^4 - 2x^2 - 5x + 19$
0 or 2; 0 or 2; 0, 2, or 4

27. $f(x) = 4x^6 - 5x^4 - x^2 + 24$
0 or 2; 0 or 2; 2, 4, or 6

28. $f(x) = -x^5 + 14x^3 + 18x - 36$
0 or 2; 1; 2 or 4

Example 3

Find all of the zeros of each function and use them to sketch a rough graph.

29–34. See margin for graphs.

29. $h(x) = x^3 - 5x^2 + 5x + 3$
$3, 1 + \sqrt{2}, 1 - \sqrt{2}$

30. $g(x) = x^3 - 6x^2 + 13x - 10$
$2, 2 + i, 2 - i$

31. $h(x) = x^3 + 4x^2 + x - 6$
$1, -2, -3$

32. $q(x) = x^3 + 3x^2 - 6x - 8$
$2, -1, -4$

33. $g(x) = x^4 - 3x^3 - 5x^2 + 3x + 4$
$-1, -1, 1, 4$

34. $f(x) = x^4 - 21x^2 + 80$
$-4, 4, -\sqrt{5}, \sqrt{5}$

35–40. See Mod. 5 Answer Appendix for graphs.

35. $f(x) = x^3 + 7x^2 + 4x - 12$
$-6, -2, 1$

36. $f(x) = x^3 + x^2 - 17x + 15$
$-5, 1, 3$

37. $f(x) = x^4 - 3x^3 - 3x^2 - 75x - 700$
$-4, 7, -5i, 5i$

38. $f(x) = x^4 + 6x^3 + 73x^2 + 384x + 576$
$-3, -3, -8i, 8i$

39. $f(x) = x^4 - 8x^3 + 20x^2 - 32x + 64$
$4, 4, -2i, 2i$

40. $f(x) = x^5 - 8x^3 - 9x$
$-3, 0, 3, -i, i$

Example 4

Write a polynomial that could be represented by each graph.

41.

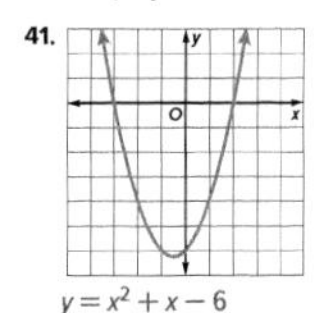

$y = x^2 + x - 6$

42.

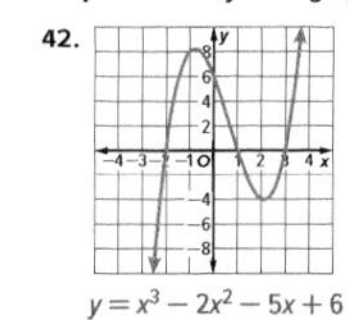

$y = x^3 - 2x^2 - 5x + 6$

43.

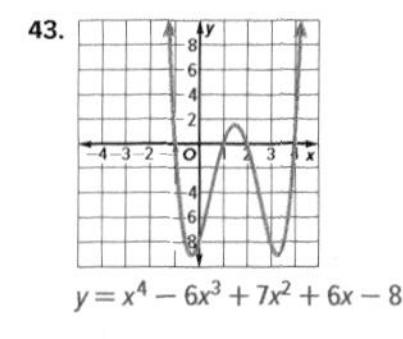

$y = x^4 - 6x^3 + 7x^2 + 6x - 8$

Example 5

44. FISH Some fish jump out of the water. When a fish is out of the water, its location is above sea level. When a fish dives back into the water, its location is below sea level. A biologist can use polynomial functions to model the location of fish compared to sea level. A biologist noticed that a fish is at sea level at −3, −2, −1, 1, 2, and 3 minutes from noon. Graph a polynomial function that could represent the location of the fish compared to sea level *y*, in centimeters, *x* seconds from noon. See Mod. 5 Answer Appendix.

45. BUSINESS After introducing a new product, a company's profit is modeled by a polynomial function. In 2012 and 2017, the company's profit on the product was \$0. Graph a polynomial function that could represent the amount of profit *p*(*x*), in thousands of dollars, *x* years since 2010. See Mod. 5 Answer Appendix.

Name ______________ Period ________ Date ________

Mixed Exercises

Write a polynomial function of least degree with integral coefficients that has the given zeros. 46–51. Sample answers given.

46. 5, −2, −1 $f(x) = x^3 - 2x^2 - 13x - 10$

47. −4, −3, 5 $f(x) = x^3 + 2x^2 - 23x - 60$

48. −1, −1, $2i$ $f(x) = x^4 + 2x^3 + 5x^2 + 8x + 4$

49. −3, 1, $-3i$ $f(x) = x^4 + 2x^3 + 6x^2 + 18x - 27$

50. 0, −5, $3 + i$ $f(x) = x^4 - x^3 - 20x^2 + 50x$

51. −2, −3, 4 − $3i$ $f(x) = x^4 - 3x^3 - 9x^2 + 77x + 150$

Sketch the graph of each function using its zeros.

52. $f(x) = x^3 - 5x^2 - 2x + 24$

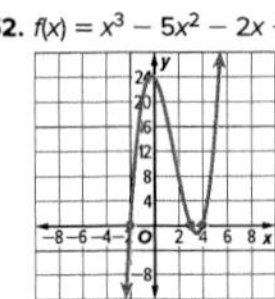

53. $f(x) = 4x^3 + 2x^2 - 4x - 2$

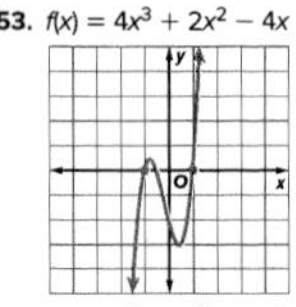

54. $f(x) = x^4 - 6x^3 + 7x^2 + 6x - 8$

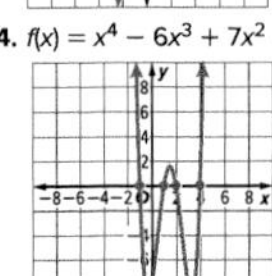

55. $f(x) = x^4 - 6x^3 + 9x^2 + 4x - 12$

56. USE A SOURCE Linear algebra is the study of linear equations. In linear algebra, the coefficients of linear equations are often organized into rectangular arrays called *matrices*. Research the eigenvalues of a matrix and how they relate to the roots of a polynomial function. What fields use linear algebra, matrices, and eigenvalues? Sample answer: The characteristic polynomial of a square matrix is a polynomial which is invariant under matrix similarity and has the eigenvalues as roots. Linear algebra matrices, and eigenvalues are used in quantum mechanics, computer engineering, geology, and other sciences.

57. SPACE The technology for a rocket that will safely return to Earth for refueling and reuse is currently being developed. The three sections of the booster that will power the flight of the payload are cylindrical with a total volume of about 234π cubic meters. If the second stage section of the booster is x meters tall, then the interstage section $x + 3$ meters tall, and the first stage section is $5x + 6.5$ meters tall. The radius of the booster is $x - 5$ meters.

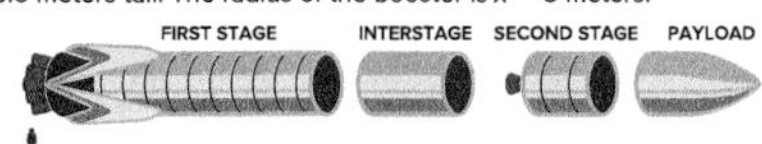

a. Write and solve an equation to represent the total volume of the booster. $(x-5)^2(7x+9.5)\pi = 234\pi$; 7, $\frac{23 \pm 3\sqrt{65}}{28}$

b. What are the dimensions of the first stage section of the booster? Explain your reasoning. height 41.5 m, radius 2 m; $x = 7$ is the only reasonable solution in the context of the situation. The other possible values of x result in negative measures.

Higher-Order Thinking Skills

58. CREATE Consider two polynomial functions, $f(x)$ and $g(x)$. a–c. See Mod. 5 Answer Appendix.

a. Write a polynomial function $f(x)$ of least degree with integral coefficients and zeros that include $-1 - 4i$ and $\frac{2}{3} + \frac{1}{3}i$. Explain how you found the function.

b. Write another polynomial function $g(x)$ with integral coefficients that has the same degree and zeros. How did you find this function?

c. Are you able to sketch the graphs of $f(x)$ and $g(x)$ based on the zeros? Explain your reasoning. Then sketch the graphs of $f(x)$ and $g(x)$.

59. ANALYZE Use the zeros to draw the graph of $P(x) = x^3 - 7x^2 + 7x + 15$ by hand. Discuss the accuracy of your graph, and what could be done to improve the accuracy. See Mod. 5 Answer Appendix.

60. PERSEVERE Let the polynomial function $f(x)$ have real coefficients, be of degree 5, and have zeros $4 + 3i$, $2 - 7i$, and $6 + bi$, where b is a real number.

a. What can be determined about b? Explain your reasoning. By the Complex Conjugates Theorem $4 - 3i$, $2 + 7i$, and $6 - bi$ are also zeros of $f(x)$. Because $f(x)$ has degree 5, it can have at most 5 zeros. This means that b must be zero.

b. Write a possible equation for $f(x)$. Sample answer: $f(x) = x^5 - 18x^4 + 182x^3 - 1184x^2 + 4469x - 7950$

61. CREATE Sketch the graph of a polynomial function with: See Mod. 5 Answer Appendix.

a. 3 real, 2 imaginary zeros b. 4 real zeros c. 2 imaginary zeros

62. PERSEVERE Write an equation in factored form of a polynomial function of degree 5 with 2 imaginary zeros, 1 real nonintegral zero, and 2 irrational zeros. Explain. Sample answer: $f(x) = (x + 2i)(x - 2i)(3x + 5)(x + \sqrt{5})(x - \sqrt{5})$; Use conjugates for the imaginary and irrational values.

63. WHICH ONE DOESN'T BELONG? Determine which equation is not like the others. Justify your conclusion. Sample answer: $r^4 + 1 = 0$; The equation has imaginary solutions and all of the others have real solutions.

$r^4 + 1 = 0$	$r^3 + 1 = 0$	$r^2 - 1 = 0$	$r^3 - 8 = 0$

64. ANALYZE Provide a counterexample for each statement.

a. All polynomial functions of degree greater than 2 have at least 1 negative real root. Sample answer: $f(x) = x^4 + 4x^2 + 4$

b. All polynomial functions of degree greater than 2 have at least 1 positive real root. Sample answer: $f(x) = x^3 + 6x^2 + 9x$

65. WRITE Explain to a friend how you would use Descartes' Rule of Signs to determine the number of possible positive real roots and the number of possible negative real roots of the polynomial function $f(x) = x^4 - 2x^3 + 6x^2 + 5x - 12$. See Mod. 5 Answer Appendix.

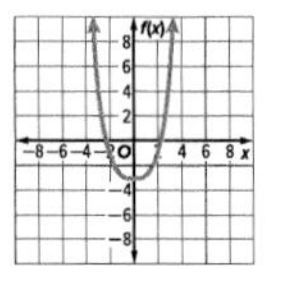

66. FIND THE ERROR The graph shows a polynomial function. Brianne says the function is a 4th degree polynomial. Amrita says the function is a 2nd degree polynomial. Is either of them correct? Explain your reasoning. See Mod. 5 Answer Appendix.

Answers

29.

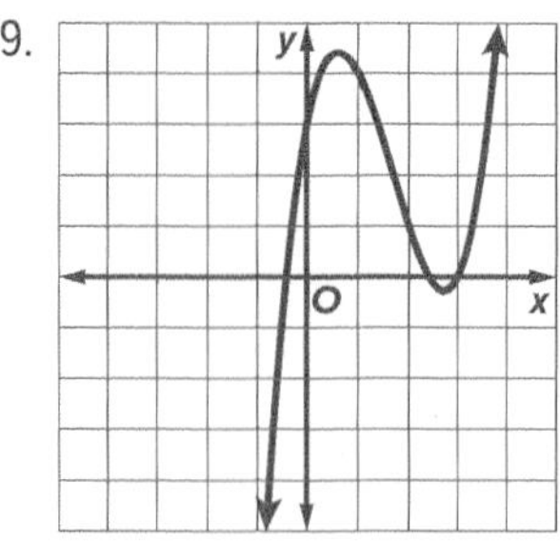

30.

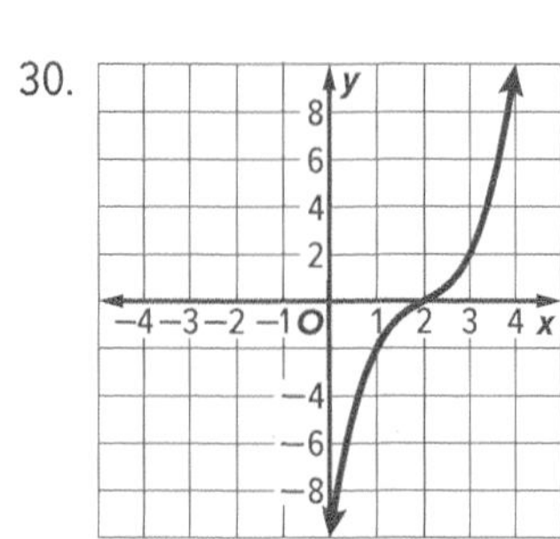

31.

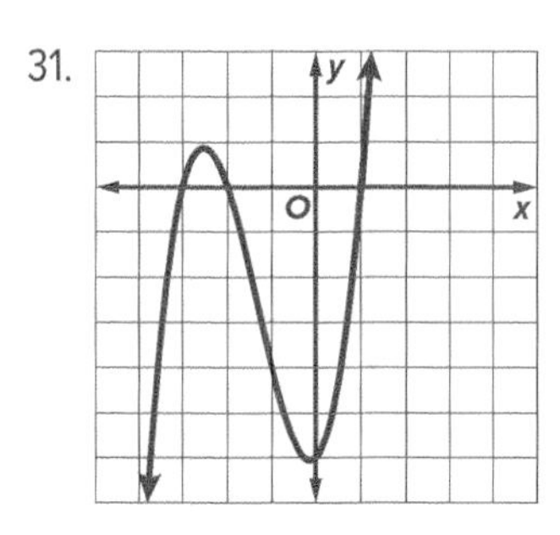

32.

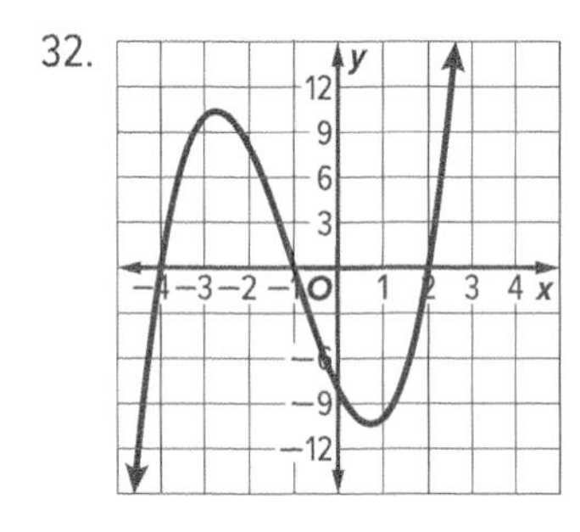

33.

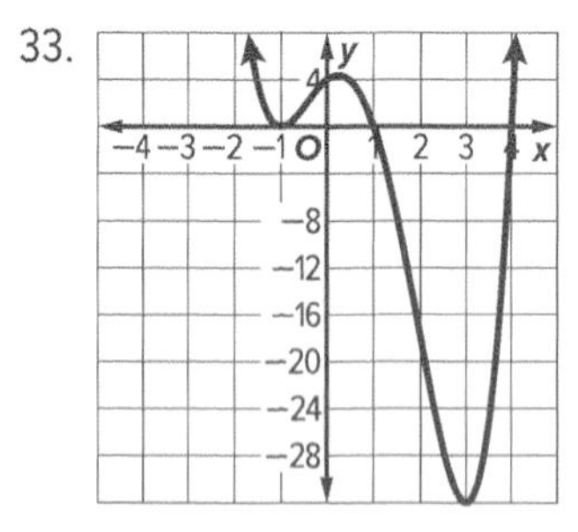

34.

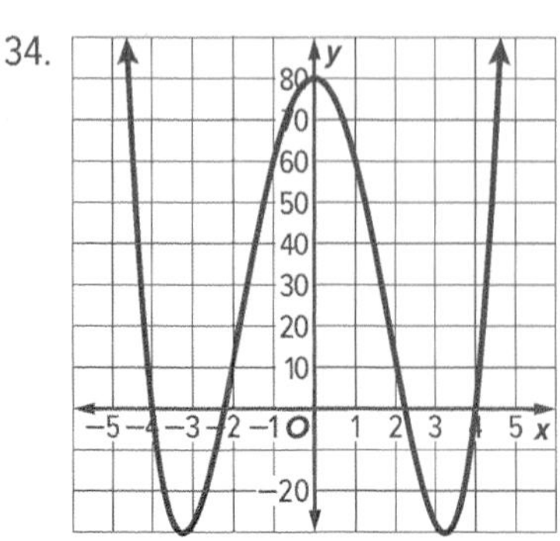

Review

Rate Yourself!

Have students return to the Module Opener to rate their understanding of the concepts presented in this module. They should see that their knowledge and skills have increased. After completing the chart, have them respond to the prompts in their *Interactive Student Edition* and share their responses with a partner.

Answering the Essential Question

Before answering the Essential Question, have students review their answers to the Essential Question Follow-Up questions found throughout the module.

- Why are the Remainder and Factor Theorems important when solving polynomial equations?
- Why is the Fundamental Theorem of Algebra important as we use polynomial equations to model and solve real-world situations?

Then have them write their answer to the Essential Question in the space provided.

DINAH ZIKE FOLDABLES

ELL A completed Foldable for this module should include the key concepts related to roots and zeros of polynomial equations.

LS **LearnSmart** Use LearnSmart as part of your test preparation plan to measure student topic retention. You can create a student assignment in LearnSmart for additional practice on these topics for **Polynomial, Rational, and Radical Relationships.**

- Solving Equations Graphically

Module 5 • Polynomial Equations

Review

Essential Question

What methods are useful for solving polynomial equations and finding zeros of polynomial functions?

Sample answer: You can use graphing or factoring to solve polynomial equations or find roots of polynomial functions. The Remainder and Factor Theorems can also be useful.

Module Summary

Lessons 5-1 and 5-2

Solving Polynomial Equations

- Polynomial equations can be solved by graphing or can be solved algebraically.
- Use patterns such as the sum or difference of two cubes to factor.
  - $a^3 + b^3 = (a + b)(a^2 - ab + b^2)$
  - $a^3 - b^3 = (a - b)(a^2 + ab + b^2)$
- When factoring a polynomial, look for a common factor to simplify the expression.
- An expression in quadratic form can be written as $au^2 + bu + c$ for any numbers a, b, and c, $a \neq 0$, where u is some expression in x.

Lesson 5-3

Polynomial Identities

- An identity is an equation that is satisfied by any numbers that replace the variables.
- A polynomial identity is a polynomial equation that is true for any values that are substituted for the variables.
- To verify an identify, simplify one side of an equation until the two sides of the equation are the same.

Lesson 5-4

The Remainder and Factor Theorems

- Remainder Theorem: For a polynomial $p(x)$ and a number a, the remainder upon division by $x - a$ is $p(a)$.
- Factor Theorem: The binomial $x - a$ is a factor of the polynomial $p(x)$ if and only if $p(a) = 0$.

Lesson 5-5

Roots and Zeros

- Let $P(x) = a_nx^n + ... + a_1x + a_0$ be a polynomial function. Then the following statements are equivalent.
  - c is a zero of $P(x)$.
  - c is a root or solution of $P(x) = 0$.
  - $x - c$ is a factor of $a_nx^n + ... + a_1x + a_0$.
- If c is a real number, then $(c, 0)$ is an x-intercept of the graph of $P(x)$.
- A polynomial equation of degree n has exactly n roots in the set of complex numbers, including repeated roots.
- The number of positive real zeros of $P(x)$ is the same as the number of changes in sign of the coefficients of the terms, or is less than this by an even number, and the number of negative real zeros of $P(x)$ is the same as the number of changes in sign of the coefficients of the terms of $P(-x)$, or is less than this by an even number.
- If $a + bi$ is a zero of a polynomial function with real coefficients, then $a - bi$ is also a zero of the function.

Study Organizer

Foldables

Use your Foldable to review this module. Working with a partner can be helpful. Ask for clarification of concepts as needed.

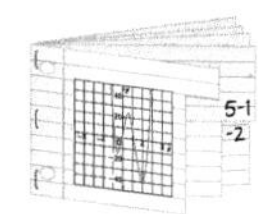

Name Period Date

Test Practice

1. MULTIPLE CHOICE Which function can be used to solve $x^3 - x = 2x^2 - 2$ by graphing? (Lesson 5-1)

Ⓐ $f(x) = x^3 - 2x^2 - x + 2$

Ⓑ $f(x) = x - 2$

Ⓒ $f(x) = x^3 + 2x^2 - x - 2$

Ⓓ $f(x) = 2x^5 - 4x^3 + 2x$

2. OPEN RESPONSE The graph of $f(x) = x^4 - 4x^2 + x + 1$ is shown.

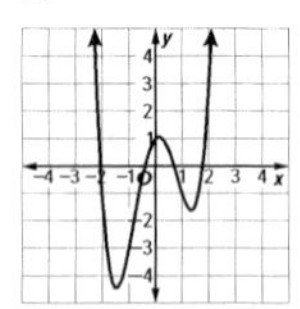

How many real solutions does the function have? (Lesson 5-1)

4

3. MULTI-SELECT Use a graphing calculator to solve $x^3 - 10x + 4 = 4 - x$. Select all of the solutions. (Lesson 5-1)

Ⓐ −3

Ⓑ 0

Ⓒ 1

Ⓓ 3

Ⓔ 4

Ⓕ 7

4. MULTIPLE CHOICE A jewelry box is 3 inches by 4 inches by 2 inches. If increasing the length of each edge by x inches doubles the volume of the jewelry box, what is the value of x? Round your answer to the nearest hundredth if necessary. (Lesson 5-1)

Ⓐ 0.40

Ⓑ 0.69

Ⓒ 0.73

Ⓓ 1.24

5. OPEN RESPONSE The volume of a figure is $x^3 - 9x$. The surface area of another figure is $8x^2$. Disregarding the units, the volume of the first figure equals the surface area of the second figure. What are the possible values of x? Explain your reasoning. (Lesson 5-2)

Set the expression for the volume of the first figure equal to the expression for the surface area of the second figure. Solve the equation $x^3 - 9x = 8x^2$. Solving the equation gives the solutions −1, 0, and 9. The value of x cannot be 0 because when substituting 0 for x in the original expressions, the volume of the first figure is $(0)^3 - 9(0) = 0 - 0 = 0$ and surface area of the second figure is $8(0)^2 = 8(0) = 0$. Volume and surface area cannot be 0, so the value of x is −1 or 9.

6. MULTI-SELECT Find the solutions of $x^4 - x^2 - 2 = 0$. Select all that apply. (Lesson 5-2)

Ⓐ $\pm\sqrt{2}$

Ⓑ $\pm i$

Ⓒ ± 1

Ⓓ ± 2

Ⓔ $\pm i\sqrt{2}$

Review and Assessment Options

The following online review and assessment resources are available for you to assign to your students. These resources include technology-enhanced questions that are auto-scored, as well as essay questions.

Review Resources

Vocabulary Activity
Module Review

Assessment Resources

Vocabulary Test
AL Module Test Form B
OL Module Test Form A
BL Module Test Form C
Performance Task*

*The module-level performance task is available online as a printable document. A scoring rubric is included.

Test Practice

You can use these pages to help your students review module content and prepare for online assessments. Exercises 1-18 mirror the types of questions your students will see on online assessments.

Question Type	Description	Exercise(s)
Multiple Choice	Students select one correct answer.	1, 4, 9, 10, 11, 13, 18
Multi-Select	Multiple answers may be correct. Students must select all correct answers.	3, 6, 8, 12
Table Item	Students complete a table by entering in the correct values.	15, 16
Open Response	Students construct their own response in the area provided.	2, 5, 7, 14

To ensure that students understand the standards, check students' success on individual exercises.

Standard(s)	Lesson(s)	Exercise(s)
A.CED.1	5-1, 5-2, 5-4	3, 4, 5, 6, 7
A.CED.2	5-1	1, 2
A.REI.11	5-1	1, 2, 3
A.APR.2	5-4	11, 12, 13, 14
A.APR.3	5-1, 5-5	15, 17, 18
A.APR.4	5-3	8, 9, 10
N.CN.9	5-5	15, 16, 18
F.IF.4	5-5	17
F.IF.7c	5-1, 5-5	17

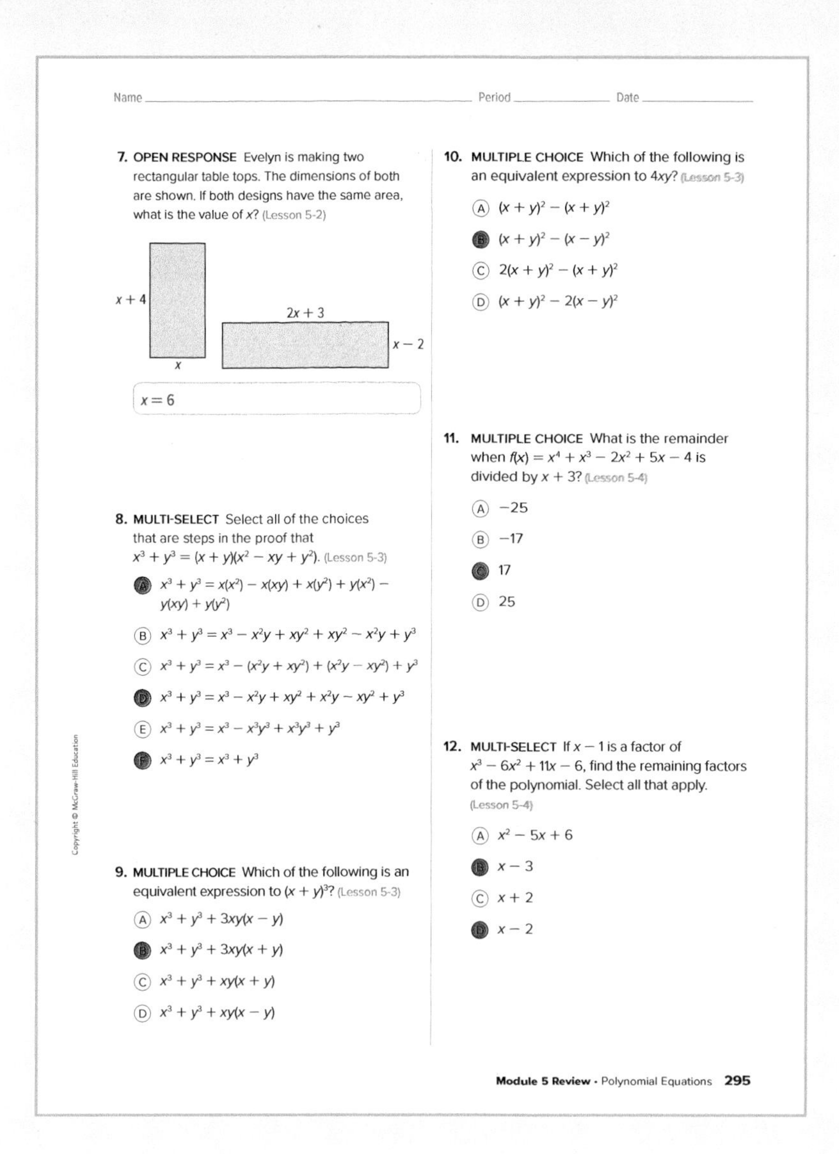

Name ______________________ Period __________ Date __________

7. OPEN RESPONSE Evelyn is making two rectangular table tops. The dimensions of both are shown. If both designs have the same area, what is the value of x? (Lesson 5-2)

$x = 6$

8. MULTI-SELECT Select all of the choices that are steps in the proof that $x^3 + y^3 = (x + y)(x^2 - xy + y^2)$. (Lesson 5-3)

Ⓐ $x^3 + y^3 = x(x^2) - x(xy) + x(y^2) + y(x^2) - y(xy) + y(y^2)$
Ⓑ $x^3 + y^3 = x^3 - x^2y + xy^2 + xy^2 - x^2y + y^3$
Ⓒ $x^3 + y^3 = x^3 - (x^2y + xy^2) + (x^2y - xy^2) + y^3$
Ⓓ $x^3 + y^3 = x^3 - x^2y + xy^2 + x^2y - xy^2 + y^3$
Ⓔ $x^3 + y^3 = x^3 - x^3y^3 + x^3y^3 + y^3$
Ⓕ $x^3 + y^3 = x^3 + y^3$

9. MULTIPLE CHOICE Which of the following is an equivalent expression to $(x + y)^3$? (Lesson 5-3)

Ⓐ $x^3 + y^3 + 3xy(x - y)$
Ⓑ $x^3 + y^3 + 3xy(x + y)$
Ⓒ $x^3 + y^3 + xy(x + y)$
Ⓓ $x^3 + y^3 + xy(x - y)$

10. MULTIPLE CHOICE Which of the following is an equivalent expression to $4xy$? (Lesson 5-3)

Ⓐ $(x + y)^2 - (x + y)^2$
Ⓑ $(x + y)^2 - (x - y)^2$
Ⓒ $2(x + y)^2 - (x + y)^2$
Ⓓ $(x + y)^2 - 2(x - y)^2$

11. MULTIPLE CHOICE What is the remainder when $f(x) = x^4 + x^3 - 2x^2 + 5x - 4$ is divided by $x + 3$? (Lesson 5-4)

Ⓐ −25
Ⓑ −17
Ⓒ 17
Ⓓ 25

12. MULTI-SELECT If $x - 1$ is a factor of $x^3 - 6x^2 + 11x - 6$, find the remaining factors of the polynomial. Select all that apply. (Lesson 5-4)

Ⓐ $x^2 - 5x + 6$
Ⓑ $x - 3$
Ⓒ $x + 2$
Ⓓ $x - 2$

Name ____________ Period ________ Date ________

13. MULTIPLE CHOICE The average price of gasoline, in dollars, from 2010 to 2016 can be modeled by the function $f(x) = 0.03x^3 - 0.4x^2 + 1.18x + 2.75$, where x represents the years since 2010. What is the estimated price of gasoline in 2011 in dollars? (Lesson 5-4)

Ⓐ \$4.70

Ⓑ \$3.56

Ⓒ \$2.42

Ⓓ \$1.14

14. OPEN RESPONSE Use synthetic substitution to find $f(-6)$ if $f(x) = -3x^4 - 20x^3 + 80x + 12$. (Lesson 5-4)

$f(-6) = -36$

15. TABLE ITEM Identify if each value of x is a real root, an imaginary root or not a root in the equation $x^4 + 3x^2 - 4 = 0$. (Lesson 5-5)

x	Real Root	Imaginary Root	Not a Root
-2			X
-1	X		
1	X		
2			X
$-2i$		X	
$-i$			X
i			X
$2i$		X	

16. TABLE ITEM What are the possible numbers of positive real roots and negative real roots of $p(x) = x^5 - 2x^4 + 3x^3 - 4x^2 + 1$? Select all that apply. (Lesson 5-5)

Number of Real Roots	Positive	Negative
0	X	
1		X
2	X	
3		
4	X	
5		

17. MULTIPLE CHOICE A template for a shipping box is made by cutting a square with side length x inches from each corner of a rectangular piece of cardboard that is 12 inches wide and 14 inches long. Which graph could represent the relationship between the volume of a shipping box y and its height? (Lesson 5-5)

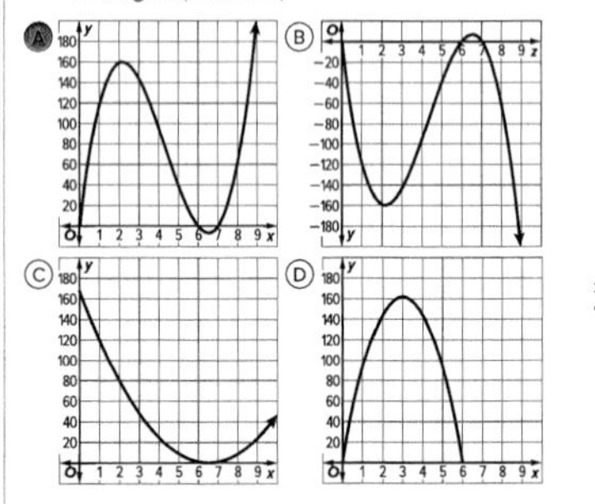

18. MULTIPLE CHOICE Consider the function $p(x) = x^5 - 3x^4 + 3x^3 - x^2$. Which of the following is a root of the function with multiplicity 3? (Lesson 5-5)

Ⓐ -1

Ⓑ 0

Ⓒ 1

Ⓓ 2

Lesson 5-3

17. Sample answer:

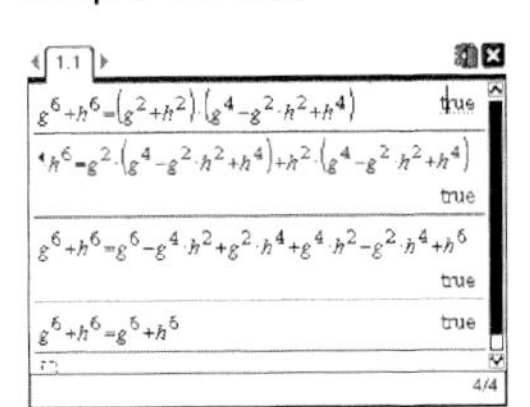

18. Sample answer:

19. Sample answer:

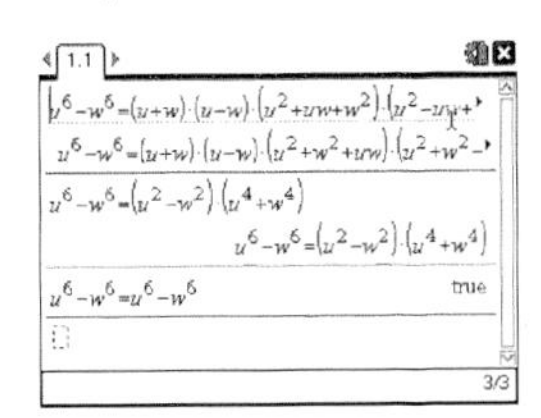

20. Sample answer:

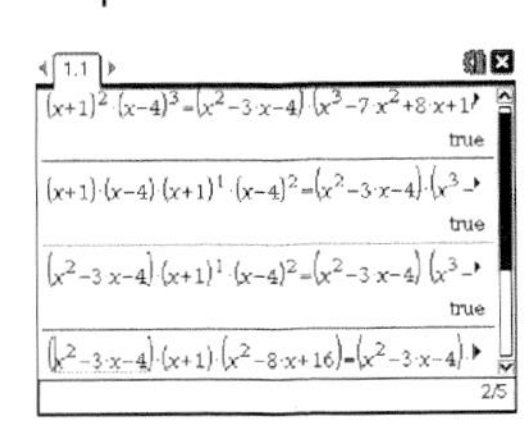

21. Sample answer: A polynomial identity is a polynomial equation that is satisfied for any values that are substituted for the variables. To prove that a polynomial equation is an identity you begin with the more complicated side of the equation and use algebra properties to transform that side of the equation until it is simplified to look like the other side.

22. $x^2 - y^2 = 3$, $2xy = 4$, $x^2 + y^2 = 5$; $x = 2$, $y = 1$

23. Sample answer: All Pythagorean triples must be made from positive integers. Therefore, x must be greater than y to prevent a negative side length and x cannot be equal to y to prevent a zero side length.

24. Sample answer:

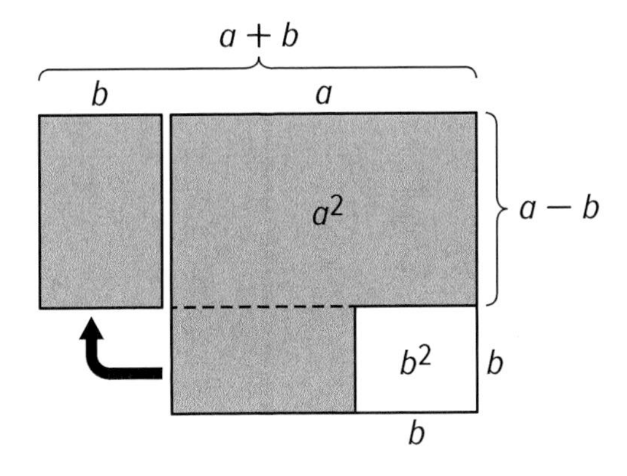

$a^2 - (a + b)(a - b) = b^2$	Original equation
$a^2 - (a^2 - ab + ab - b^2) = b^2$	Use the FOIL Method.
$a^2 - a^2 + ab - ab + b^2 = b^2$	Distribute −1.
$b^2 = b^2$	Combine like terms.
$b^2 = b^2$	True

25. Sample answer: When George multiplied b and a^2 in the second line he mistakenly made the term negative so it did not cancel when he simplified.

Lesson 5-4

58.

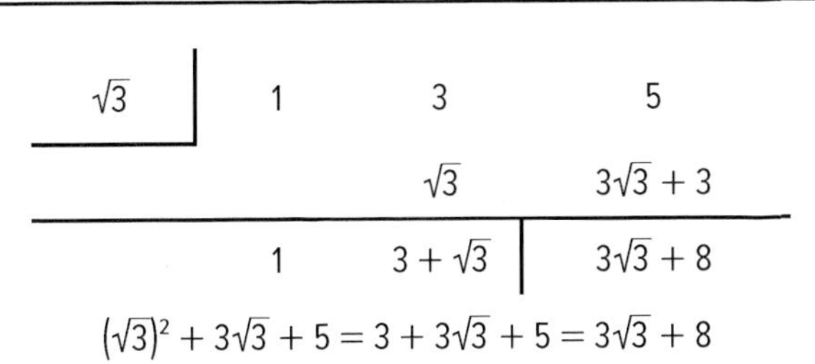

$\sqrt{3}$	1	3	5
		$\sqrt{3}$	$3\sqrt{3} + 3$
	1	$3 + \sqrt{3}$	$3\sqrt{3} + 8$

$(\sqrt{3})^2 + 3\sqrt{3} + 5 = 3 + 3\sqrt{3} + 5 = 3\sqrt{3} + 8$

59. Use synthetic substitution with $a = -6$. The remainder must be zero, so $432 - 216k = 0$, or $k = 2$. The depressed polynomial is $2x^2 + 3x - 5$, which factors to $(2x + 5)(x - 1)$. Therefore, we can write $kx^3 + 15x^2 + 13x - 30 = (x + 6)(2x + 5)(x - 1)$.

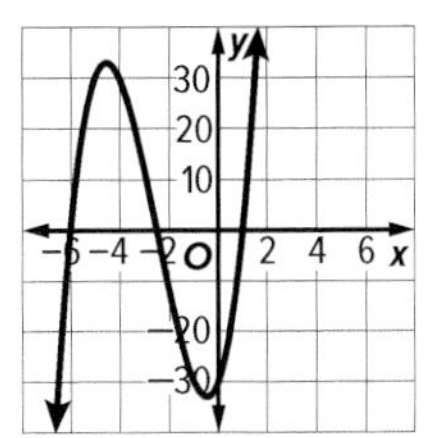

61.

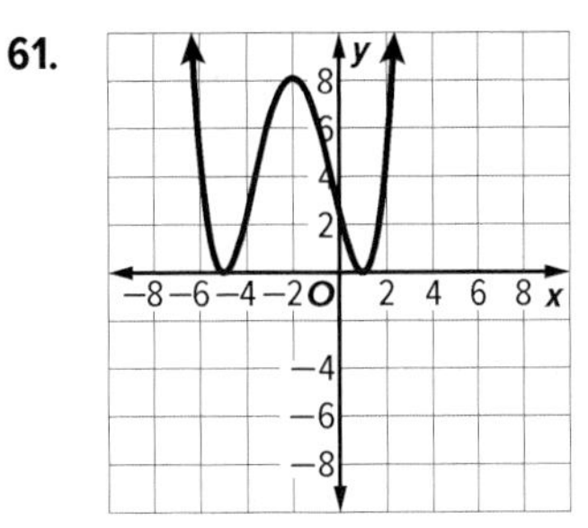

62a. Sample answer: By the Factor Theorem, $P(x) = (x + 8)(x - 5)(ax + b)$ where $a \neq 0$. Substitute for x using $P(2) = -90$ to get $-90 = (10)(-3)(2a + b)$ or $3 = 2a + b$. Two solutions are $a = -1$, $b = 5$ and $a = 1$, $b = 1$. Two possible equations for $P(x)$ are $P(x) = (x + 8)(x - 5)(-x + 5)$ and $P(x) = (x + 8)(x - 5)(x + 1)$.

62b. $(2, -90)$, $(-8, 0)$, and $(5, 0)$.

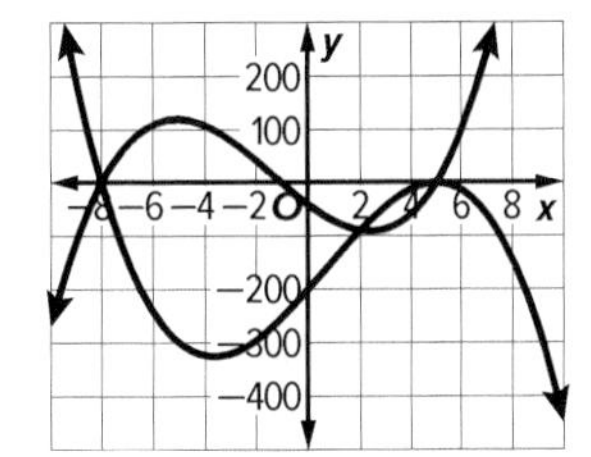

62c. $P(x) = (x + 8)(x - 5)(-4x + 11)$; sample answer: Because $3 = 2a + b$, then $b = 3 - 2a$ so $P(x) = (x + 8)(x - 5)(ax - 2a + 3)$. Substitute for x by using $P(4) = 60$ to get $60 = (12)(-1)(2a + 3)$ or $-5 = 2a + 3$ so $a = -4$ and $b = 11$.

63. Sample answer: If $x - a$ is a factor of $f(x)$, then $f(a)$ has a factor of $(a - a)$ or 0. Since a factor of $f(a)$ is 0, $f(a) = 0$. Now assume that $f(a) = 0$. If $f(a) = 0$, then the Remainder Theorem states that the remainder is 0 when $f(x)$ is divided by $x - a$. This means that $x - a$ is a factor of $f(x)$. This proves the Factor Theorem.

65. Sample answer: When $x = 1$, $f(1)$ is the sum of all of the coefficients and constants in $f(x)$, in this case, a, b, c, d, and e. The sum of a, b, c, d, and e is 0, so however the coefficients are arranged, $f(1)$ will always equal 0, and $f(x)$ will have a rational root.

66. Sample answer: A zero can be located using the Remainder Theorem and a table of values by determining when the output, or remainder, is equal to zero. For instance, if $f(6)$ leaves a remainder of 2 and $f(7)$ leaves a remainder of −1, then you know that there is a zero between $x = 6$ and $x = 7$.

67. Tyrone; sample answer: By the Factor Theorem, $(x - r)$ is a factor when $P(r) = 0$. $P(r) = 0$ when $x = -1$ and $x = 2$.

35.

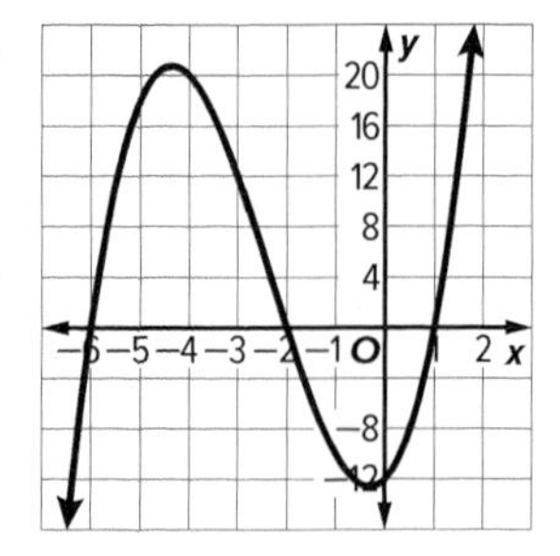

36.

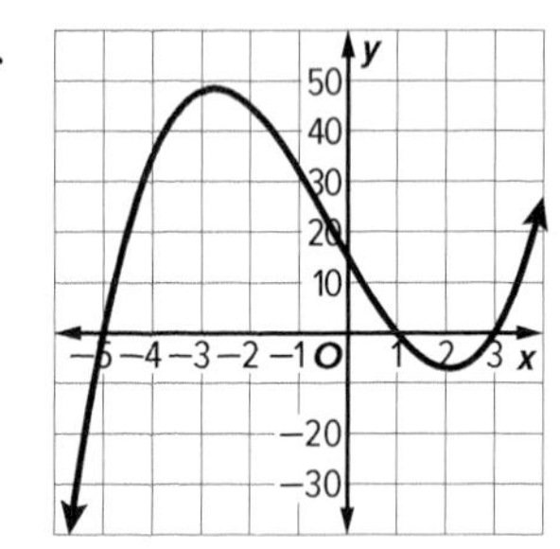

37.

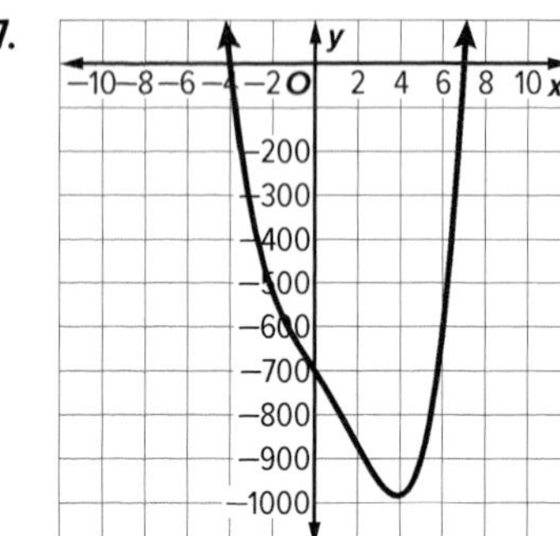

38.

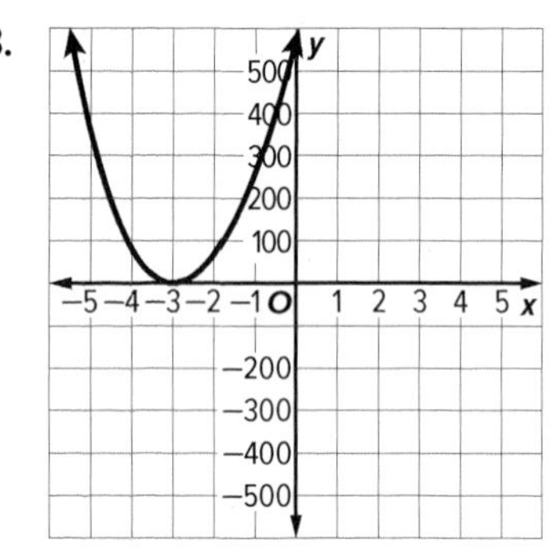

39.

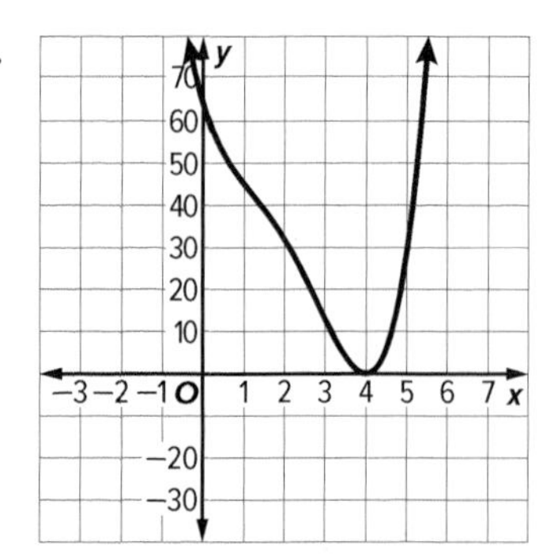

40.

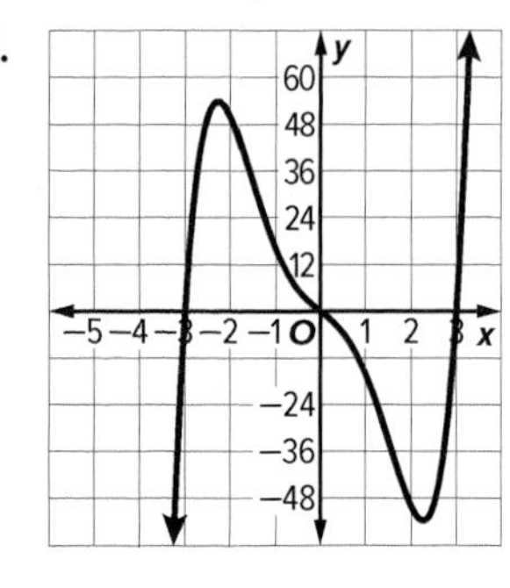

44.

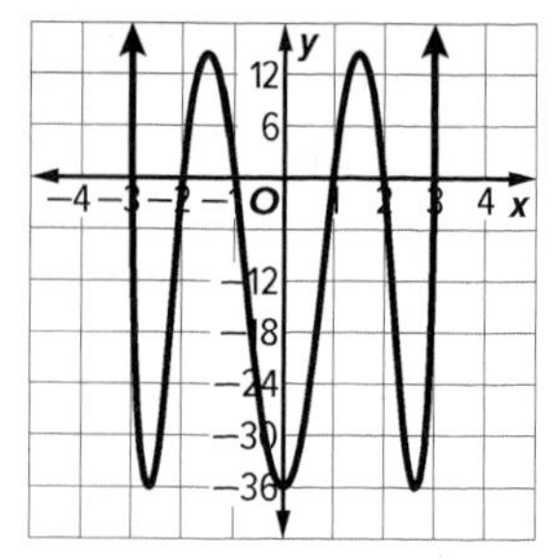

45. Sample answer:

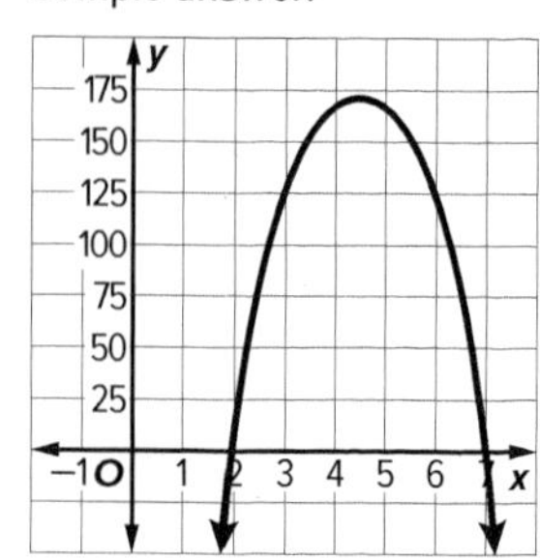

58a. $f(x) = (x^2 + 2x + 17)(9x^2 - 12x + 5)$, or $f(x) = 9x^4 + 6x^3 + 134x^2 - 194x + 85$; sample answer: If $-1 - 4i$ is a zero, $-1 + 4i$ is also a zero, and the quadratic factor is $x^2 + 2x + 17$. If $\frac{2}{3} + \frac{1}{3}i$ is a zero, then $\frac{2}{3} - \frac{1}{3}i$ is also a zero, and the quadratic factor is $x^2 - \frac{4}{3} + \frac{5}{9}x$, or $9x^2 - 12x + 5$.

58b. Sample answer: $g(x) = 2(x^2 + 2x + 17)(9x^2 - 12x + 5)$; another polynomial function with the same zeros and degree can be obtained by multiplying $f(x)$ by any nonzero whole number. For instance, $g(x) = 2(x^2 + 2x + 17)(9x^2 - 12x + 5)$.

58c. No; sample answer: The graphs cannot be sketched from only imaginary zeros because they do not show on a graph.

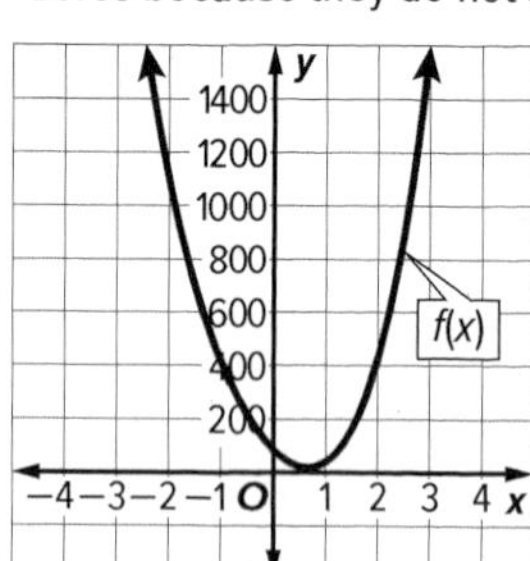

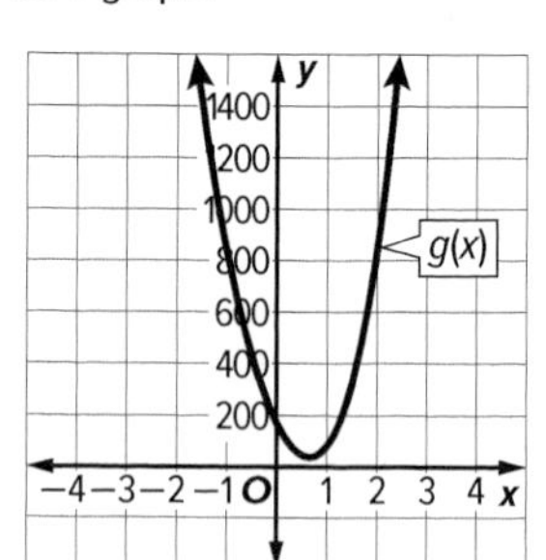

59. Sample answer: The graph is accurate as far as the location of the zeros, but does not consider any vertical dilation. It could be improved by finding more points between the roots.

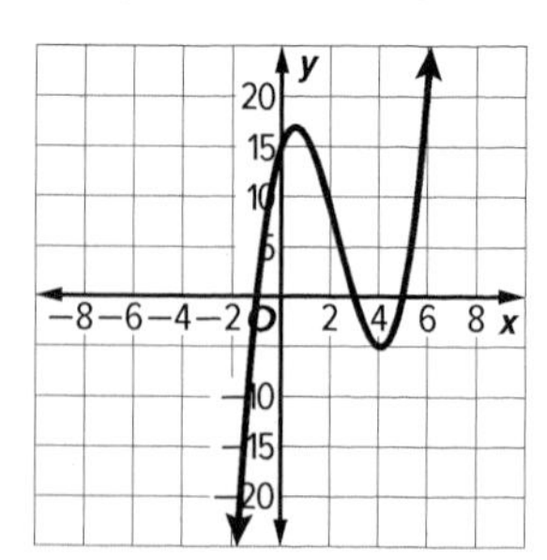

61a. Sample answer:

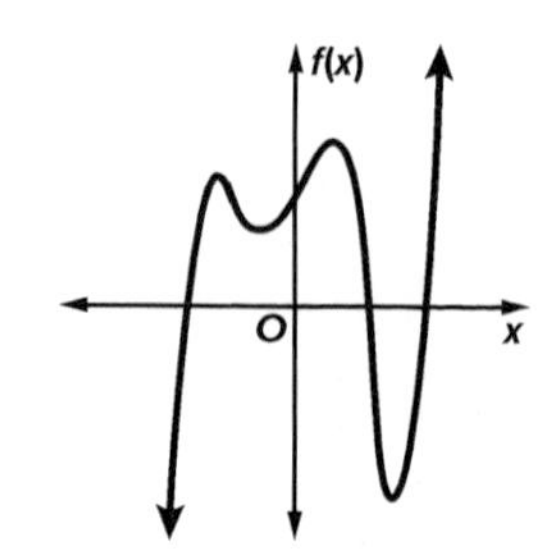

61b. Sample answer:

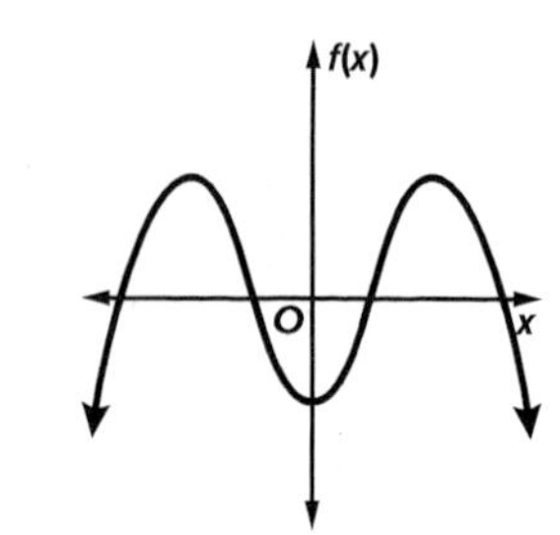

61c. Sample answer:

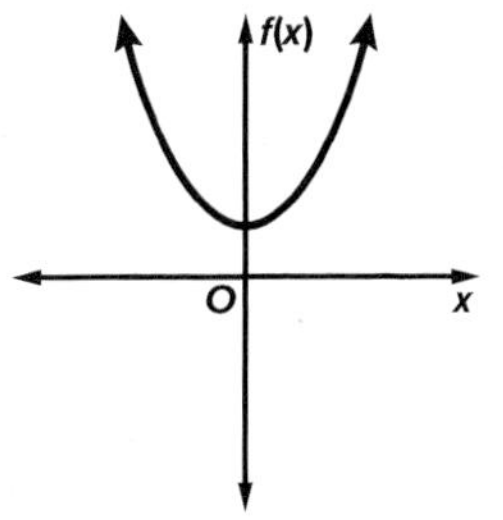

65. Sample answer: To determine the number of positive real roots, determine how many time the signs change in the polynomial as you move from left to right. In this function there are 3 changes in sign. Therefore, there may be 3 or 1 positive real roots. To determine the number of negative real roots, I would first evaluate the polynomial for $-x$. All of the terms with an odd-degree variable would change signs. Then I would again count the number of sign changes as I move from left to right. There would be only one change. Therefore there may be 1 negative root.

66. Both Brianne and Amrita could be correct; sample answer: The graph crosses the x-axis twice, so there are two real zeros, one positive and one negative. This means Amrita could be correct. There could be two other zeros of the function that are imaginary. This means Brianne could also be correct.

Module 6

Inverse and Radical Functions

Module Goals

- Students graph and verify inverses.
- Students simplify expressions with radical and rational exponents.
- Students graph, analyze, and solve radical functions and equations.

Focus

Domain: Algebra, Functions

Standards for Mathematical Content:

A.SSE.2 Use the structure of an expression to identify ways to rewrite it.

F.IF.7b Graph square root, cube root, and piecewise-defined functions, including step functions and absolute value functions.

F.BF.3 Identify the effect on the graph of replacing $f(x)$ by $f(x) + k$, $k\,f(x)$, $f(kx)$, and $f(x + k)$ for specific values of k (both positive and negative); find the value of k given the graphs. Experiment with cases and illustrate an explanation of the effects on the graph using technology. Include recognizing even and odd functions from their graphs and algebraic expressions for them.

Also addresses A.REI.2, F.IF.5, F.BF.1b, and F.BF.4a.

Standards for Mathematical Practice:

All Standards for Mathematical Practice will be addressed in this module.

Coherence

Vertical Alignment

Previous

Students understood the relationship between functions and their inverses. **F.B.4 (Algebra 1)**

Now

Students simplify radical expressions and solve radical equations. Students write and graph the inverse of a function and analyze the relationship between a function and its inverse.

A.SSE.2, F.IF.5, F.BF.4a

Next

Students will analyze the relationship between a function and its inverse (exponential and logarithmic).

A.SSE.2, F.IF.7e

Rigor

The Three Pillars of Rigor

To help students meet standards, they need to illustrate their ability to use the three pillars of rigor. Students gain conceptual understanding as they move from the Explore to Learn sections within a lesson. Once they understand the concept, they practice procedural skills and fluency and apply their mathematical knowledge as they go through the Examples and Independent Practice.

1 CONCEPTUAL UNDERSTANDING | 2 FLUENCY | 3 APPLICATION

EXPLORE → LEARN → EXAMPLE & PRACTICE

Suggested Pacing

Lessons	Standards	45-min classes	90-min classes
Module Pretest and Launch the Module Video		1	0.5
6-1 Operations on Functions	F.BF.1b	2	1
6-2 Inverse Relations and Functions	F.IF.5, F.BF.4a	2	1
6-3 *n*th Roots and Rational Exponents	A.SSE.2	2	1
6-4 Graphing Radical Functions	F.IF.7b, F.BF.3	2	1
6-5 Operations with Radical Expressions	A.SSE.2	2	1
6-6 Solving Radical Equations	A.REI.2	2	1
Module Review		1	0.5
Module Assessment		1	0.5
	Total Days	15	7.5

Formative Assessment Math Probe
Combining Functions

Analyze the Probe

Review the probe prior to assigning it to your students.

In this probe, students will combine functions by finding function values, finding inverses, finding compositions, and using operations with functions.

Targeted Concepts Understand processes to evaluate functions, find compositions of functions, and find inverse functions.

Targeted Misconceptions

- Students may treat the *f*, *g*, *h*, and so on in function notation as variables and incorrectly use properties of operations with them.
- Students may have difficulty evaluating constant functions and applying one function to the results of another (composition of functions).
- Students may have difficulty evaluating a function for a value involving a variable such as, evaluating $f(x)$ for $f(x + h)$.
- Students may misinterpret the −1 in inverse function notation as "the reciprocal of" instead of the inverse function and/or they may have difficulty recognizing that with inverse functions, they have to be able to "undo" operations in a specific order.

Use the Probe after Lesson 6-2.

Module Resource

NAME ______ DATE ______ PERIOD ______

Cheryl Tobey Math Probe
Combining Functions

Use the following functions to determine whether each statement is true.

$f(x) = 4x + 5$ $g(x) = \sqrt{x+3}$ $h(x) = 6$

Circle yes, no, or not enough information.	Explain your choice.
1. $\frac{f(2)}{f(3)} = \frac{2}{3}$ yes no not enough information	
2. $h(100) = 6$ yes no not enough information	
3. $f^{-1}(x) = \frac{1}{4x+5}$ yes no not enough information	
4. $g(x + h) = \sqrt{x+1} + h$ yes no not enough information	
5. $h \circ g(x) = \sqrt{9} = 3$ yes no not enough information	
6. $h \circ f(x) = 6$ yes no not enough information	
7. $g^{-1}(x) = x^2 - 3$ yes no not enough information	

Cheryl Tobey Math Probe • Combining Functions

Correct Answers:

1. no **2. yes** **3. no** **4. no**
5. no **6. yes** **7. yes**

Collect and Assess Student Answers

If the student selects these responses...	**Then** the student likely...
1, 3. yes or not enough information	is having difficulty with function notation. In Item 1, the student is considering the letter *f* as a variable and believes they can be "canceled." For Item 3, the –1 in the inverse function notation is being used as the reciprocal of the function rather than the inverse.
2, 6. no or not enough information **5.** yes or not enough information	does not understand how to evaluate constant functions.
4. yes or not enough information	does not understand how to evaluate functions for variable values.
7. no or not enough information	is having difficulty with the process of "undoing" operations in the correct order.

Take Action

After the Probe Design a plan to address any possible misconceptions. You may wish to assign the following resources.

- **ALEKS**® Combining Functions, Composite Functions, and/or Inverse Functions
- Lessons 6-1 and 6-2, all Learns, all Examples

The Ignite! activities, created by Dr. Raj Shah, cultivate curiosity and engage and challenge students. Use these open-ended, collaborative activities, located online in the module Launch section, to encourage your students to develop a growth mindset towards mathematics and problem solving. Use the teacher notes for implementation suggestions and support for encouraging productive struggle.

Essential Question

At the end of this module, students should be able to answer the Essential Question.

How can the inverse of a function be used to help interpret a real-world event or solve a problem? Sample answer: Similar to how it is necessary to undo operations when solving equations in algebra, an inverse of a function can be used to undo a function to solve a problem.

What Will You Learn?

Prior to beginning this module, have your students rate their knowledge of each item listed. Then, at the end of the module, you will be reminded to have your students return to these pages to rate their knowledge again. They should see that their knowledge and skills have increased.

DINAH ZIKE FOLDABLES

Focus Students write notes as they study inverses and radical functions and relations in the lessons of this module.

Teach Have students make and label their Foldables as illustrated. Have students use the appropriate tab as they cover each lesson in this module. Encourage them to apply what they have learned by writing examples as well.

When to Use It Encourage students to add to their Foldables as they work through the module and use them to review for the module test.

Launch the Module

For this module, the Launch the Module video uses world travel to introduce students to inverse and radical functions. Students learn about foreign exchange rates and Fahrenheit and Celsius temperature conversions.

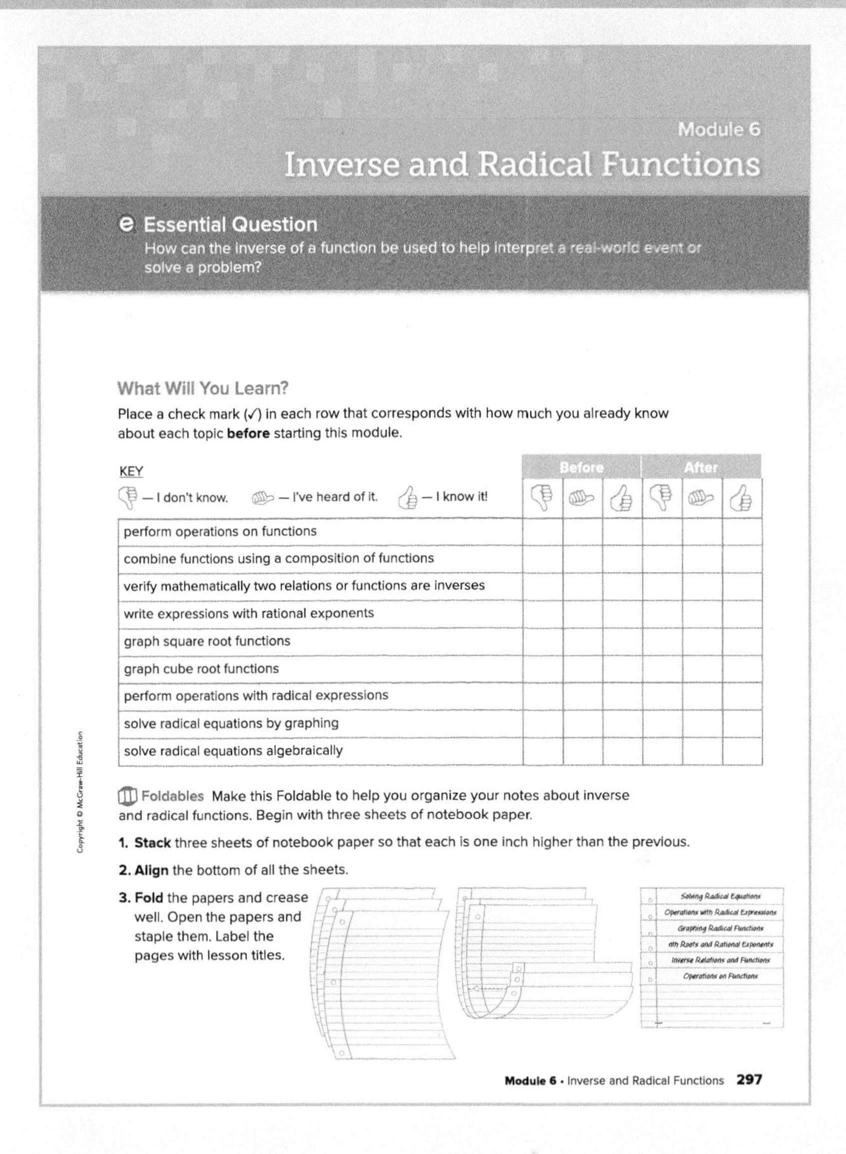

Module 6

Inverse and Radical Functions

Essential Question

How can the inverse of a function be used to help interpret a real-world event or solve a problem?

What Will You Learn?

Place a check mark (✓) in each row that corresponds with how much you already know about each topic **before** starting this module.

KEY

— I don't know. — I've heard of it. — I know it!

	Before			After		
	I don't know	I've heard of it	I know it!	I don't know	I've heard of it	I know it!
perform operations on functions						
combine functions using a composition of functions						
verify mathematically two relations or functions are inverses						
write expressions with rational exponents						
graph square root functions						
graph cube root functions						
perform operations with radical expressions						
solve radical equations by graphing						
solve radical equations algebraically						

Foldables Make this Foldable to help you organize your notes about inverse and radical functions. Begin with three sheets of notebook paper.

1. **Stack** three sheets of notebook paper so that each is one inch higher than the previous.
2. **Align** the bottom of all the sheets.
3. **Fold** the papers and crease well. Open the papers and staple them. Label the pages with lesson titles.

Solving Radical Equations
Operations with Radical Expressions
Graphing Radical Functions
nth Roots and Rational Exponents
Inverse Relations and Functions
Operations on Functions

Module 6 • Inverse and Radical Functions 297

Interactive Presentation

What Vocabulary Will You Learn?

Check the box next to each vocabulary term that you may already know.

- ☐ composition of functions
- ☐ conjugates
- ☐ cube root function
- ☐ index
- ☐ inverse functions
- ☐ inverse relations
- ☐ like radical expressions
- ☐ *n*th root
- ☐ principal root
- ☐ radical equation
- ☐ radical function
- ☐ radicand
- ☐ rational exponent
- ☐ square root function

Are You Ready?

Complete the Quick Review to see if you are ready to start this module. Then complete the Quick Check.

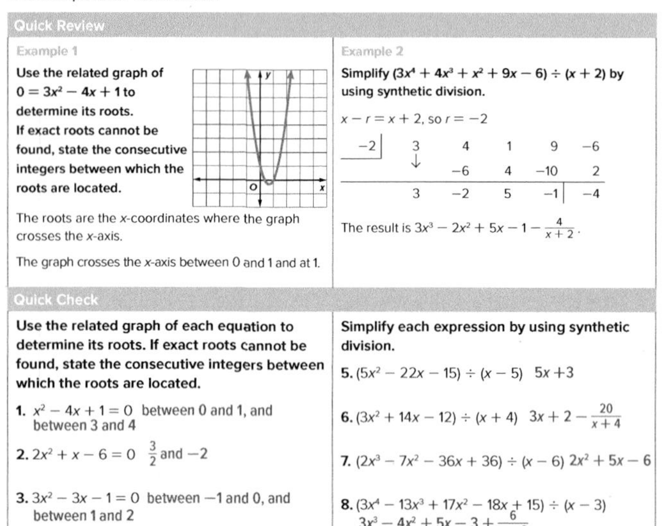

Quick Review

Example 1

Use the related graph of $0 = 3x^2 - 4x + 1$ to determine its roots. If exact roots cannot be found, state the consecutive integers between which the roots are located.

The roots are the x-coordinates where the graph crosses the x-axis.

The graph crosses the x-axis between 0 and 1 and at 1.

Example 2

Simplify $(3x^4 + 4x^3 + x^2 + 9x - 6) \div (x + 2)$ by using synthetic division.

$x - r = x + 2$, so $r = -2$

−2	3	4	1	9	−6
	↓	−6	4	−10	2
	3	−2	5	−1	−4

The result is $3x^3 - 2x^2 + 5x - 1 - \frac{4}{x+2}$.

Quick Check

Use the related graph of each equation to determine its roots. If exact roots cannot be found, state the consecutive integers between which the roots are located.

1. $x^2 - 4x + 1 = 0$ between 0 and 1, and between 3 and 4
2. $2x^2 + x - 6 = 0$ $\frac{3}{2}$ and -2
3. $3x^2 - 3x - 1 = 0$ between −1 and 0, and between 1 and 2
4. $2x^2 - 9x + 4 = 0$ $\frac{1}{2}$ and 4

Simplify each expression by using synthetic division.

5. $(5x^2 - 22x - 15) \div (x - 5)$ $5x + 3$
6. $(3x^2 + 14x - 12) \div (x + 4)$ $3x + 2 - \frac{20}{x+4}$
7. $(2x^3 - 7x^2 - 36x + 36) \div (x - 6)$ $2x^2 + 5x - 6$
8. $(3x^4 - 13x^3 + 17x^2 - 18x + 15) \div (x - 3)$ $3x^3 - 4x^2 + 5x - 3 + \frac{6}{x-3}$

How Did You Do?

Which exercises did you answer correctly in the Quick Check? Shade those exercise numbers below.

① ② ③ ④ ⑤ ⑥ ⑦ ⑧

What Vocabulary Will You Learn?

ELL As you proceed through the module, introduce the key vocabulary by using the following routine.

Define A radical equation is an equation that has a variable in the radicand.

Example $\sqrt{4x + 8} - 1 = 15$

Ask Is $7x - 11 = \sqrt{9}$ a radical equation? Explain. No; sample answer: because there is no variable in the radicand.

Are You Ready?

Students may need to review the following prerequisite skills to succeed in this module.

- finding function values
- lidentifying relations as functions
- understanding powers of rational numbers
- simplifying square roots
- simplifying radicals
- simplifying radical expressions

ALEKS®

ALEKS is an adaptive, personalized learning environment that identifies precisely what each student knows and is ready to learn, ensuring student success at all levels.

You may want to use the Radicals and Advanced Functions section to ensure student success in this module.

Mindset Matters

Model Constructive Feedback

In order for students to grow, they need to receive timely, constructive feedback that references a specific skill or area. You can also model what appropriate feedback looks and sounds like so that students can collaborate and give one another constructive feedback in a way that is positive and helpful.

How Can I Apply It?

Use the **Leveled Discussion Questions** in the Teacher Edition to ask students questions and share feedback on their thinking. This is a great opportunity to model feedback for the class so that students can give one another feedback during collaborative activities.

Operations on Functions

F.BF.1b

LESSON GOAL

Students simplify and graph operations on functions, and determine compositions of functions.

1 LAUNCH

 Launch the lesson with a **Warm Up** and an introduction.

2 EXPLORE AND DEVELOP

 **Explore:** Adding Functions

 **Develop:**

Operations on Functions
- Add and Subtract Functions
- Multiply and Divide Functions
- Graphs of Combined Functions (online)

Compositions of Functions
- Compose Functions by Using Ordered Pairs
- Compose Functions
- Use Composition of Functions

 You may want your students to complete the **Checks** online.

3 REFLECT AND PRACTICE

 Exit Ticket

 Practice

DIFFERENTIATE

 View reports of student progress on the **Checks** after each example.

Resources	AL	OL	BL	ELL
Remediation: Combining Functions	●	●		●
Extension: Iteration		●	●	●

Language Development Handbook

Assign page 35 of the *Language Development Handbook* to help your students build mathematical language related to simplifying and graphing operations on functions.

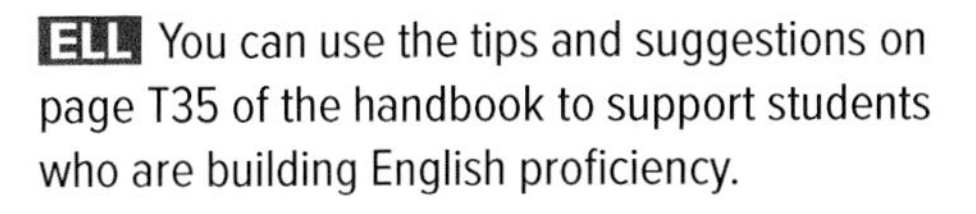
ELL You can use the tips and suggestions on page T35 of the handbook to support students who are building English proficiency.

Suggested Pacing

90 min	1 day
45 min	2 days

Focus

Domain: Functions

Standards for Mathematical Content:

F.BF.1b Combine standard function types using arithmetic operations.

Standards for Mathematical Practice:

1 Make sense of problems and persevere in solving them.

4 Model with mathematics.

7 Look for and make use of structure.

Coherence

Vertical Alignment

Previous
Students performed operations on polynomial expressions.
A.APR.1 (Algebra 1, Algebra 2)

Now
Students find the sum, difference, product, and quotient of functions.
Students compose functions.
F.BF.1b

Next
Students will describe and analyze the relationship between a function and its inverse.
F.IF.5, F.BF.4a

Rigor

The Three Pillars of Rigor

1 CONCEPTUAL UNDERSTANDING	2 FLUENCY	3 APPLICATION

Conceptual Bridge In this lesson, students develop an understanding of function operations. They build fluency by using arithmetic operations to combine functions, and they apply their understanding by solving real-world problems.

Mathematical Background

Addition, subtraction, multiplication, and division can be used to combine functions. Composition of functions is a method of combining functions, and is fundamentally different from arithmetic operations on functions. In the composition $g \circ f$, the output $f(x)$ is used as input for g.

Interactive Presentation

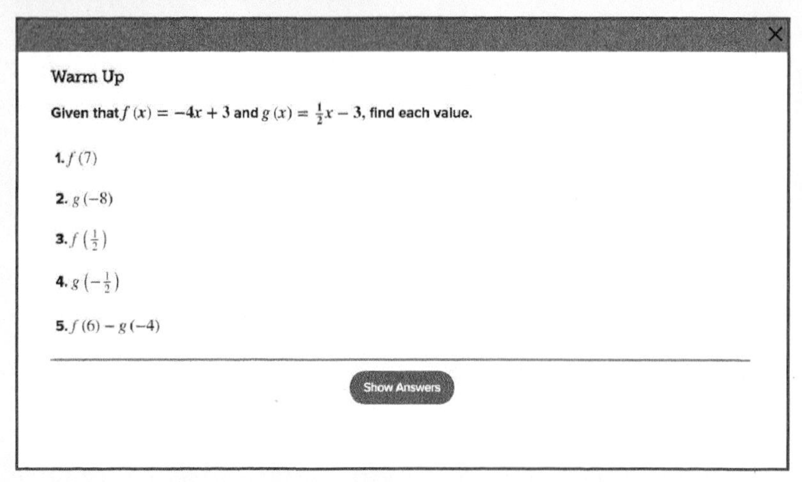

Warm Up

Launch the Lesson

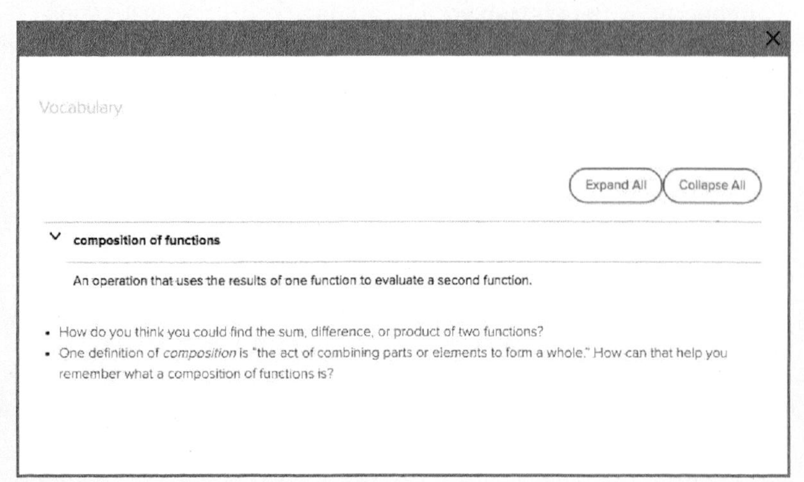

Today's Vocabulary

Warm Up

Prerequisite Skills

The Warm Up exercises address the following prerequisite skill for this lesson:

- finding function values

Answers:

1. -25
2. -7
3. 1
4. $-3\frac{1}{4}$
5. -16

Launch the Lesson

Teaching the Mathematical Practices

4 Apply Mathematics In the Launch the Lesson, students will learn how predicting the weather requires combining functions to analyze data.

Go Online to find additional teaching notes and questions to promote classroom discourse.

Today's Standards

Tell students that they will be addressing these content and practice standards in this lesson. You may wish to have a student volunteer read aloud *How can I meet this standard?* and *How can I use these practices?*, and connect these to the standards.

See the Interactive Presentation for I Can statements that align with the standards covered in this lesson.

Today's Vocabulary

Tell students that they will be using this vocabulary term in this lesson. You can expand the row if you wish to share the definition. Then discuss the questions below with the class.

Explore Adding Functions

Objective

Students analyze the slopes and y-intercepts of the sums of linear functions.

 Teaching the Mathematical Practices

3 Make Conjectures Students will make conjectures about the graph of the sum of two linear functions and then build a logical progression of statements to validate the conjectures. Once students have made their conjectures, guide the students to validate them.

5 Decide When to Use Tools Mathematically proficient students can make sound decisions about when to use mathematical tools such as a sketch. Help them see why using a sketch to analyze the graphs will help them draw conclusions.

Ideas for Use

Recommended Use Present the Inquiry Question, or have a student volunteer read it aloud. Have students work in pairs to complete the Explore activity on their devices. Pairs should discuss each of the questions. Monitor student progress during the activity. Upon completion of the Explore activity, have student volunteers share their responses to the Inquiry Question.

What if my students don't have devices? You may choose to project the activity on a whiteboard. A printable worksheet for each Explore is available online. You may choose to print the worksheet so that individuals or pairs of students can use it to record their observations.

Summary of the Activity

Students will complete guiding exercises throughout the Explore activity. Students will be presented with a sketch with three linear functions, $f(x)$, $g(x)$, and $f(x) + g(x)$, and will compare the slopes of these functions. Then, students will answer the Inquiry Question.

(continued on the next page)

Interactive Presentation

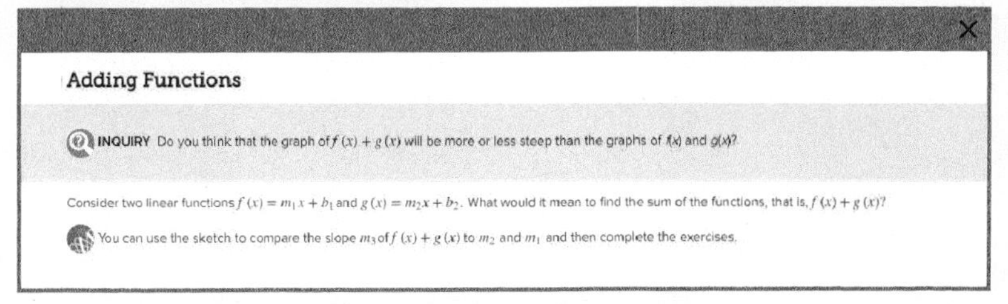

Explore

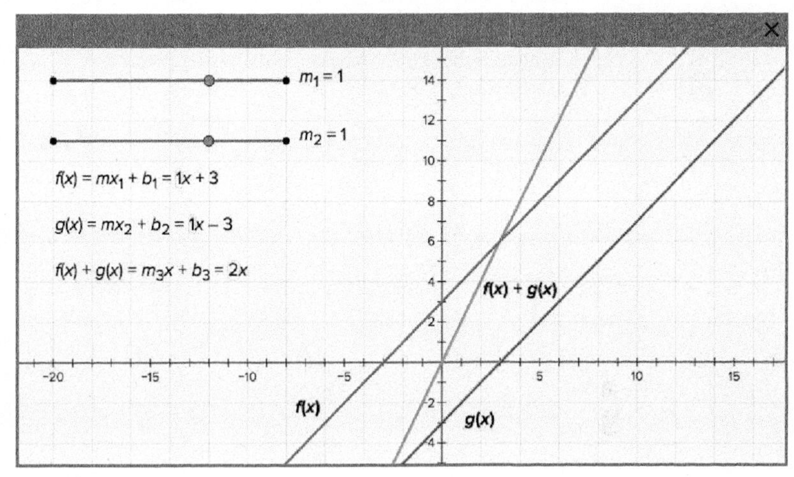

Explore

WEB SKETCHPAD

Students use the sketch to explore adding functions.

Interactive Presentation

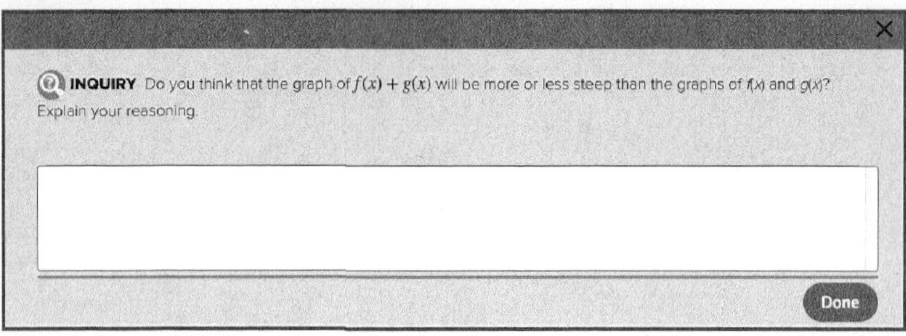

Explore

TYPE

a| Students respond to the Inquiry Question and can view a sample answer.

Explore Adding Functions (*continued*)

Questions

Have students complete the Explore activity.

Ask:

- What is the relationship between m_1, m_2, and m_3? Sample answer: m_3 is the slope of the sum equation, so $m_3 = m_1 + m_2$.
- How could you find the relationship for $f(x) - g(x)$? Sample answer: Experiment with different functions and discover how the slopes are related.

Inquiry

Do you think that the graph of $f(x) + g(x)$ will be more or less steep than the graphs of $f(x)$ and $g(x)$? Explain your reasoning. Sample answer: The slope of $f(x) + g(x)$ would be more steep than either of them if both $f(x)$ and $g(x)$ have a positive slope; it would be less steep than either of them if $m_1 < 0 < m_2$ and $m_1 + m_2 < |m_1| < m_2$. It could also be equally as steep as one of the lines if m_1 or m_2 is 0.

Go Online to find additional teaching notes and sample answers for the guiding exercises.

Learn Operations on Functions

Objective

Students find sums, differences, products, and quotients of functions.

Teaching the Mathematical Practices

1 Explain Correspondences Encourage students to explain the relationships between how sums and differences of functions are found algebraically and how they are found by graphing.

Example 1 Add and Subtract Functions

Teaching the Mathematical Practices

7 Interpret Complicated Expressions Mathematically proficient students can see complicated expressions as single objects or as being composed of several objects. Use the Study Tip to guide students to see what information they can gather about the sum or difference of two functions just by looking at the functions.

Questions for Mathematical Discourse

AL Why are parentheses used when the expressions are substituted into the function operations? Sample answer: The parentheses around each expression help you keep track of how addition or subtraction are distributed to the terms of each expression.

OL When adding two functions, is it possible for the sum to have less terms than either of the original functions? Yes; sample answer: Two like terms could result in zero when performing the operation on the functions.

BL If the domain of $f(x)$ is $(-\infty, 3) \cup (3, \infty)$ and the domain of $g(x)$ is $[-5, 7)$, what is the domain of $(f + g)(x)$? $[-5, 3) \cup (3, 7)$

Common Error

When subtracting functions, remind students to distribute the negative throughout the function.

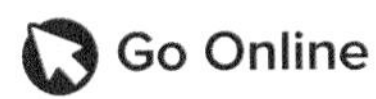

Go Online

- Find additional teaching notes.
- View performance reports of the Checks.
- Assign or present an Extra Example.

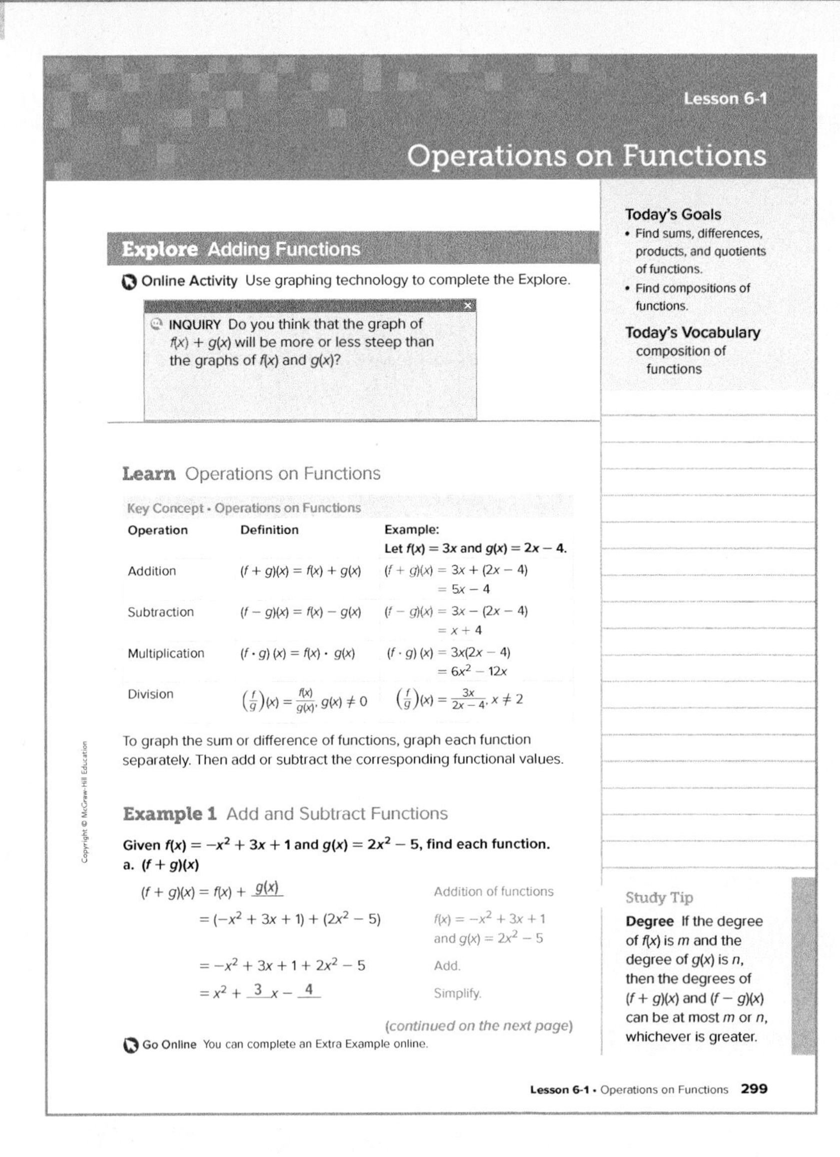

Lesson 6-1

Operations on Functions

Today's Goals
- Find sums, differences, products, and quotients of functions.
- Find compositions of functions.

Today's Vocabulary
composition of functions

Explore Adding Functions

Online Activity Use graphing technology to complete the Explore.

INQUIRY Do you think that the graph of $f(x) + g(x)$ will be more or less steep than the graphs of $f(x)$ and $g(x)$?

Learn Operations on Functions

Key Concept • Operations on Functions

Operation	Definition	Example: Let $f(x) = 3x$ and $g(x) = 2x - 4$.
Addition	$(f + g)(x) = f(x) + g(x)$	$(f + g)(x) = 3x + (2x - 4)$ $= 5x - 4$
Subtraction	$(f - g)(x) = f(x) - g(x)$	$(f - g)(x) = 3x - (2x - 4)$ $= x + 4$
Multiplication	$(f \cdot g)(x) = f(x) \cdot g(x)$	$(f \cdot g)(x) = 3x(2x - 4)$ $= 6x^2 - 12x$
Division	$\left(\frac{f}{g}\right)(x) = \frac{f(x)}{g(x)}, g(x) \neq 0$	$\left(\frac{f}{g}\right)(x) = \frac{3x}{2x - 4}, x \neq 2$

To graph the sum or difference of functions, graph each function separately. Then add or subtract the corresponding functional values.

Example 1 Add and Subtract Functions

Given $f(x) = -x^2 + 3x + 1$ and $g(x) = 2x^2 - 5$, find each function.
a. $(f + g)(x)$

$(f + g)(x) = f(x) + \underline{g(x)}$ — Addition of functions
$= (-x^2 + 3x + 1) + (2x^2 - 5)$ — $f(x) = -x^2 + 3x + 1$ and $g(x) = 2x^2 - 5$
$= -x^2 + 3x + 1 + 2x^2 - 5$ — Add.
$= x^2 + \underline{3}x - \underline{4}$ — Simplify.

(continued on the next page)

Go Online You can complete an Extra Example online.

Study Tip
Degree If the degree of $f(x)$ is m and the degree of $g(x)$ is n, then the degrees of $(f + g)(x)$ and $(f - g)(x)$ can be at most m or n, whichever is greater.

Interactive Presentation

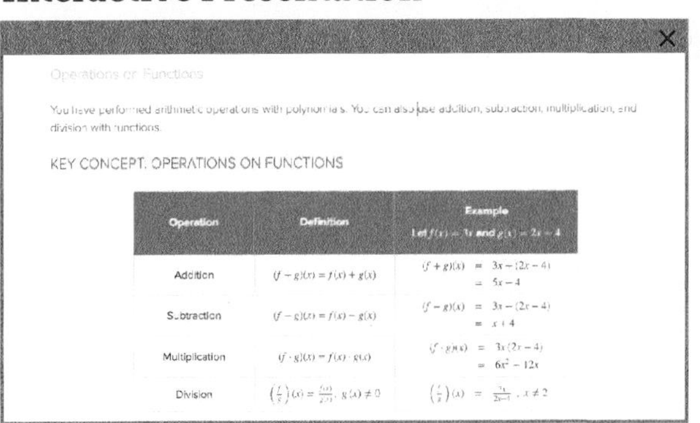

Learn

TAP

Students can tap on a Study Tip to learn more about operations on linear function.

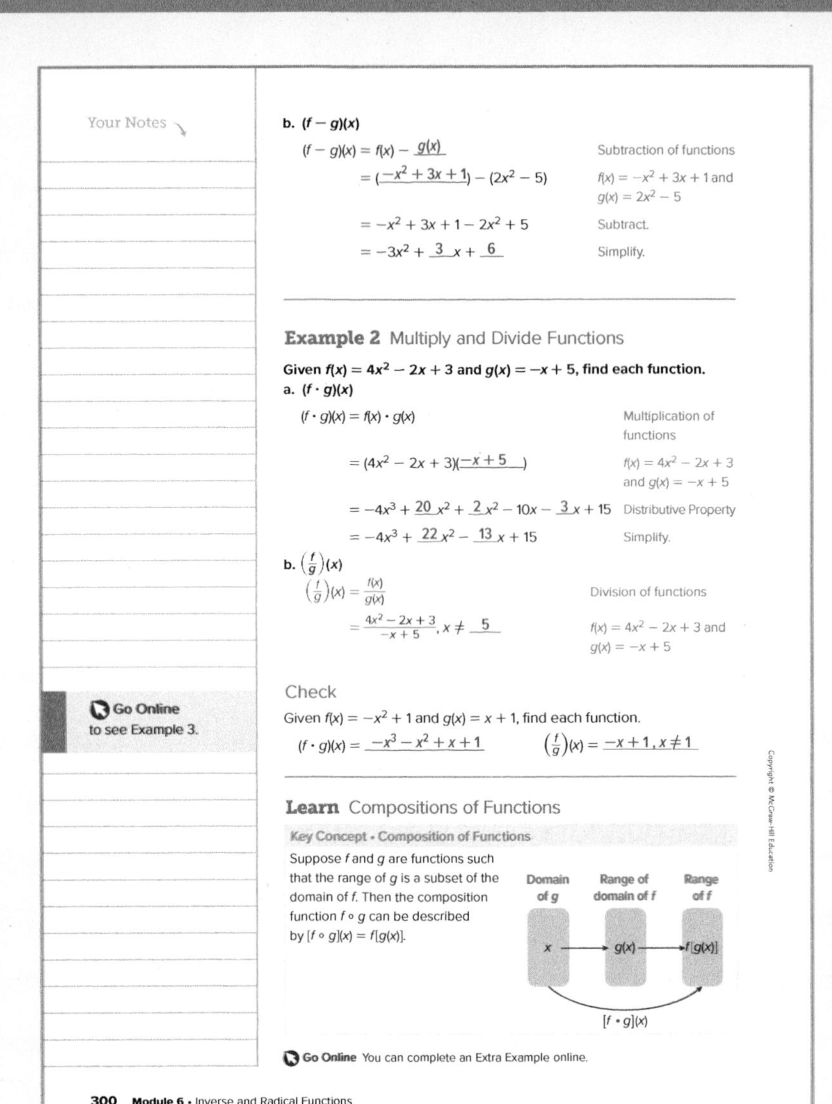

Your Notes

b. $(f - g)(x)$

$(f - g)(x) = f(x) - \underline{g(x)}$	Subtraction of functions
$= (\underline{-x^2 + 3x + 1}) - (2x^2 - 5)$	$f(x) = -x^2 + 3x + 1$ and $g(x) = 2x^2 - 5$
$= -x^2 + 3x + 1 - 2x^2 + 5$	Subtract.
$= -3x^2 + \underline{3}\,x + \underline{6}$	Simplify.

Example 2 Multiply and Divide Functions

Given $f(x) = 4x^2 - 2x + 3$ and $g(x) = -x + 5$, find each function.

a. $(f \cdot g)(x)$

$(f \cdot g)(x) = f(x) \cdot g(x)$	Multiplication of functions
$= (4x^2 - 2x + 3)(\underline{-x + 5})$	$f(x) = 4x^2 - 2x + 3$ and $g(x) = -x + 5$
$= -4x^3 + \underline{20}\,x^2 + \underline{2}\,x^2 - 10x - \underline{3}\,x + 15$	Distributive Property
$= -4x^3 + \underline{22}\,x^2 - \underline{13}\,x + 15$	Simplify.

b. $\left(\frac{f}{g}\right)(x)$

$\left(\frac{f}{g}\right)(x) = \frac{f(x)}{g(x)}$	Division of functions
$= \frac{4x^2 - 2x + 3}{-x + 5},\ x \neq \underline{5}$	$f(x) = 4x^2 - 2x + 3$ and $g(x) = -x + 5$

Go Online to see Example 3.

Check

Given $f(x) = -x^2 + 1$ and $g(x) = x + 1$, find each function.

$(f \cdot g)(x) = \underline{-x^3 - x^2 + x + 1}$ $\quad \left(\frac{f}{g}\right)(x) = \underline{-x + 1, x \neq 1}$

Learn Compositions of Functions

Key Concept • Composition of Functions

Suppose f and g are functions such that the range of g is a subset of the domain of f. Then the composition function $f \circ g$ can be described by $[f \circ g](x) = f[g(x)]$.

Domain of g | Range of g / domain of f | Range of f

$x \rightarrow g(x) \rightarrow f[g(x)]$

$[f \circ g](x)$

Go Online You can complete an Extra Example online.

300 Module 6 • Inverse and Radical Functions

Interactive Presentation

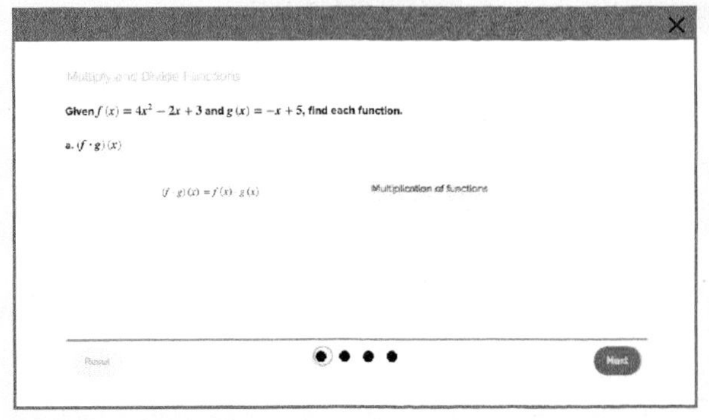

Example 2

TAP

Students move through the steps to multiply and divide functions.

CHECK

Students complete the Check online to determine whether they are ready to move on.

Example 2 Multiply and Divide Functions

Teaching the Mathematical Practices

1 Seek Information Mathematically proficient students must be able to transform algebraic expressions to reach solutions. In Example 2, students must transform the expressions to find the product or quotient of two functions.

Questions for Mathematical Discourse

AL In **part b,** why must you exclude any domain values that cause the denominator to be 0? You cannot divide by 0.

OL If the denominator was a factor of the numerator, would the domain restriction remain? Explain. Yes; the domain must exclude any values that are undefined for the quotient.

BL What would the graph of $\left(\frac{f}{g}\right)(x)$ look like near $x = 5$? As the denominator approaches zero, the function value increases more rapidly, so as x approaches 5, $\left(\frac{f}{g}\right)(x)$ would approach ∞ or $-\infty$.

Learn Compositions of Functions

Objective

Students find compositions of functions.

Teaching the Mathematical Practices

7 Use Structure Help students to explore how to use the structure of functions to compose them.

About the Key Concept

Addition, subtraction, multiplication, and division can be used to combine functions. Composition of functions is a method of combining functions, and is fundamentally different from arithmetic operations on functions. In the composition $g \circ f$, the output $f(x)$ is used as input for g.

Example 4 Compose Functions by Using Ordered Pairs

Teaching the Mathematical Practices

1 Monitor and Evaluate Point out that in Example 4, students should stop and evaluate their progress and change course if necessary. Students should evaluate whether the values they have calculated for $g(x)$ make sense before evaluating $[f \circ g](x)$.

Questions for Mathematical Discourse

AL For $f[g(x)]$, which function do you evaluate first? $g(x)$ Is $[f \circ g](x)$ the same as $f(x) \cdot g(x)$? No Does $f[g(x)] = g[f(x)]$? No

OL Why does the domain of $[g \circ f](x)$ have fewer values than $[f \circ g](x)$? Not all values of $f(x)$ are defined for $g(x)$.

BL What needs to be true for the composition of functions to have the same number of ordered pairs as the original functions? The range of the inner function must equal the domain of the outer function.

Common Error

Remind students to not confuse composition of functions, $[f \circ g](x)$, with $(f \cdot g)(x)$.

Example 5 Compose Functions

Teaching the Mathematical Practices

1 Check Answers Mathematically proficient students continually ask themselves, "Does this make sense?" Use the Study Tip to encourage students to graph the composition functions to confirm the domain and range.

Questions for Mathematical Discourse

AL In what order do you perform substitution when composing functions? Substitute into the inside function first, then substitute into the outside function.

OL What kind of function will be the result of the composition of two linear functions? a linear function

BL What would the degree of $f(g(x))$ be if $f(x)$ is a degree 5 polynomial and $g(x)$ is a degree 3 polynomial? 15

DIFFERENTIATE

Enrichment Activity BL

Ask students to prove that if $f(x)$ and $g(x)$ are linear functions with nonzero slopes, then $[f \circ g](x)$ and $[g \circ f](x)$ have domain and range of all real numbers. Let $f(x) = ax + b, a \neq 0$ and $g(x) = cx + d, c \neq 0$. Then $[f \circ g](x) = a(cx + d) + b$ or $acx + ad + b$ and $[g \circ f](x) = c(ax + b) + d$ or $acx + bc + d$. The slope of both $[f \circ g](x)$ and $[g \circ f](x)$ is ac. Because a and c are both nonzero real numbers, ac is a nonzero real number. The domain and range of a linear function with nonzero slope is all numbers. Thus the domain and range of $[f \circ g](x)$ and $[g \circ f](x)$ is all real numbers.

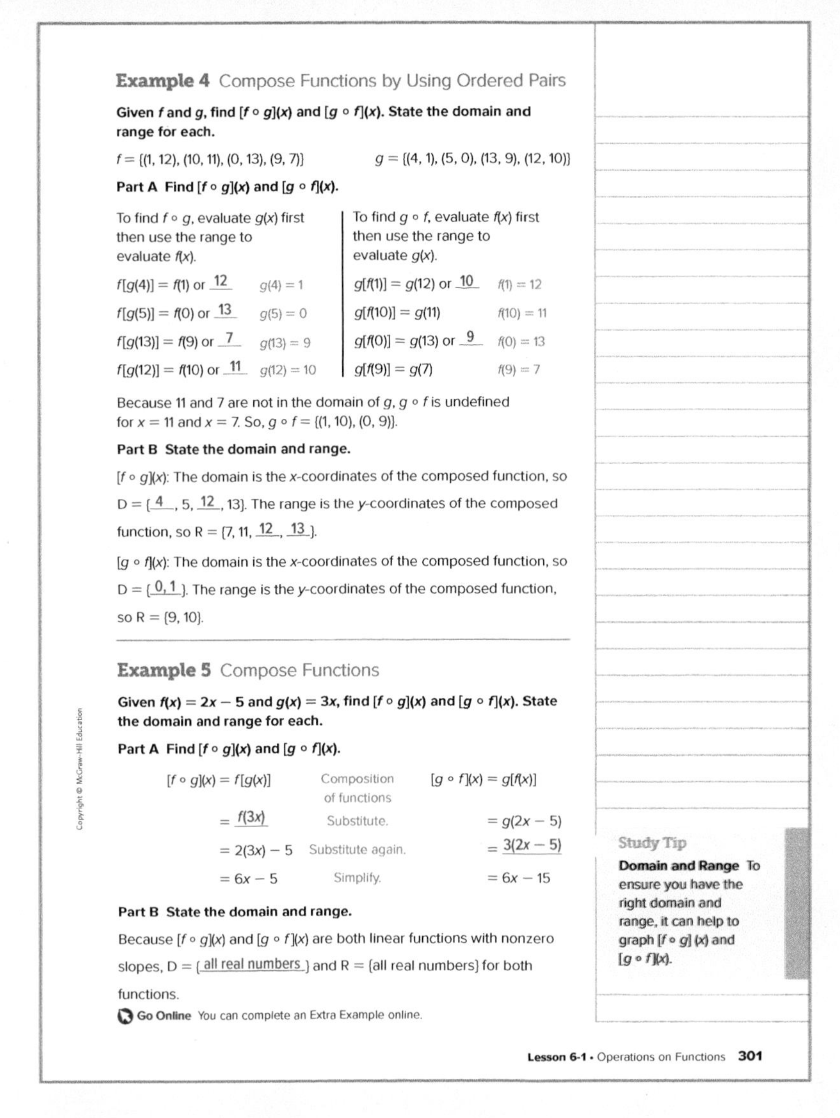

Example 4 Compose Functions by Using Ordered Pairs

Given f and g, find $[f \circ g](x)$ and $[g \circ f](x)$. State the domain and range for each.

$f = \{(1, 12), (10, 11), (0, 13), (9, 7)\}$ $g = \{(4, 1), (5, 0), (13, 9), (12, 10)\}$

Part A Find $[f \circ g](x)$ and $[g \circ f](x)$.

To find $f \circ g$, evaluate $g(x)$ first then use the range to evaluate $f(x)$.		To find $g \circ f$, evaluate $f(x)$ first then use the range to evaluate $g(x)$.	
$f[g(4)] = f(1)$ or 12	$g(4) = 1$	$g[f(1)] = g(12)$ or 10	$f(1) = 12$
$f[g(5)] = f(0)$ or 13	$g(5) = 0$	$g[f(10)] = g(11)$	$f(10) = 11$
$f[g(13)] = f(9)$ or 7	$g(13) = 9$	$g[f(0)] = g(13)$ or 9	$f(0) = 13$
$f[g(12)] = f(10)$ or 11	$g(12) = 10$	$g[f(9)] = g(7)$	$f(9) = 7$

Because 11 and 7 are not in the domain of g, $g \circ f$ is undefined for $x = 11$ and $x = 7$. So, $g \circ f = \{(1, 10), (0, 9)\}$.

Part B State the domain and range.

$[f \circ g](x)$: The domain is the x-coordinates of the composed function, so D = {4, 5, 12, 13}. The range is the y-coordinates of the composed function, so R = {7, 11, 12, 13}.

$[g \circ f](x)$: The domain is the x-coordinates of the composed function, so D = {0, 1}. The range is the y-coordinates of the composed function, so R = {9, 10}.

Example 5 Compose Functions

Given $f(x) = 2x - 5$ and $g(x) = 3x$, find $[f \circ g](x)$ and $[g \circ f](x)$. State the domain and range for each.

Part A Find $[f \circ g](x)$ and $[g \circ f](x)$.

$[f \circ g](x) = f[g(x)]$	Composition of functions	$[g \circ f](x) = g[f(x)]$
$= f(3x)$	Substitute.	$= g(2x - 5)$
$= 2(3x) - 5$	Substitute again.	$= 3(2x - 5)$
$= 6x - 5$	Simplify.	$= 6x - 15$

Part B State the domain and range.

Because $[f \circ g](x)$ and $[g \circ f](x)$ are both linear functions with nonzero slopes, D = {all real numbers} and R = {all real numbers} for both functions.

Go Online You can complete an Extra Example online.

Study Tip

Domain and Range To ensure you have the right domain and range, it can help to graph $[f \circ g](x)$ and $[g \circ f](x)$.

Copyright © McGraw-Hill Education

Lesson 6-1 • Operations on Functions 301

Interactive Presentation

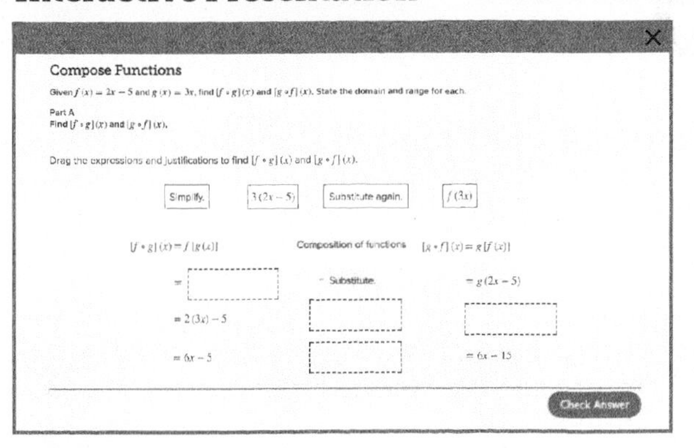

Example 5

DRAG & DROP

Students drag the expressions and justifications to compose functions.

Check

Given $f(x) = -x + 1$ and $g(x) = 2x^3 - x$, find $[f \circ g](x)$ and $[g \circ f](x)$. State the domain and range for each.

$[f \circ g](x) = -2x^3 + x + 1$ $\quad$ $[g \circ f](x) = -2x^3 + 6x^2 - 5x + 1$

Domain of $[f \circ g](x)$: all real numbers $\quad$ Domain of $[g \circ f](x)$: all real numbers

Range of $[f \circ g](x)$: all real numbers $\quad$ Range of $[g \circ f](x)$: all real numbers

Apply Example 6 Use Composition of Functions

BOX OFFICE **A movie theater charges $8.50 for each of the x tickets sold. The manager wants to determine how much the movie theater gets to keep of the ticket sales if they have to give the studios 75% of the money earned on ticket sales $t(x)$. If the amount they keep of each ticket sale is $k(x)$, which composition represents the total amount of money the theater gets to keep?**

1 What is the task?

Describe the task in your own words. Then list any questions that you may have. How can you find answers to your questions?
Sample answer: I need to write functions for $t(x)$ and $k(x)$ and use them to create a composition that represents the money that the theater keeps. If the studios get 75%, what does the theater get to keep? Should the composition be $[k \circ t](x)$ or $[t \circ k](x)$?

2 How will you approach the task? What have you learned that you can use to help you complete the task?
Sample answer: First, I will determine functions for $t(x)$ and $k(x)$. Then, I will determine the order of the composition and simplify it. I will apply what I have learned in previous examples to complete the task.

3 What is your solution?

Use your strategy to solve the problem.

What function represents the money earned on ticket sales, $t(x)$? $t(x) = 8.50x$

What function represents the amount of money the theater keeps from each ticket sale, $k(x)$? $k(x) = 0.25x$

Because the theater uses the total earnings to determine the amount they keep from the ticket sales, what composition should be used to represent the situation? $[k \circ t] = 2.125x$

4 How can you know that your solution is reasonable?

Write About It! Write an argument that can be used to defend your solution.

Sample answer: If the theater sells 1000 tickets in a weekend, they will earn $t(1000)$, or $8500. The theater will keep 25% of $8500, which is $k(8500) = \$2125$. This is the same value as $[k \circ t](1000)$.

Watch Out!

Order Remember that, for two functions $f(x)$ and $g(x)$, $[f \circ g](x)$ is not always equal to $[g \circ f](x)$. Given that the studios take their cut after the tickets have been sold, consider how that affects the order of $t(x)$ and $k(x)$.

Go Online

You can complete an Extra Example online.

302 **Module 6** • Inverse and Radical Functions

Interactive Presentation

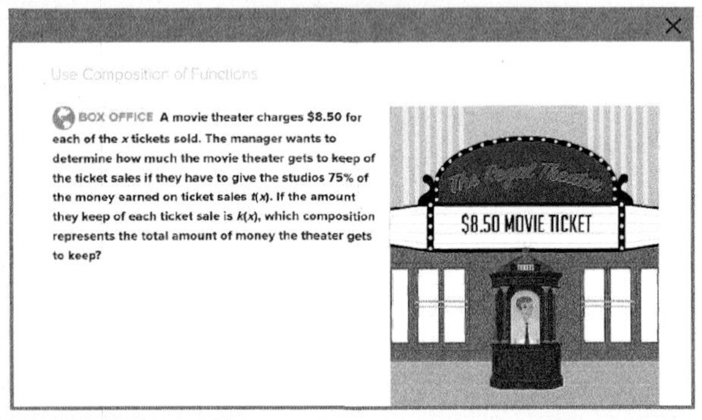

Apply Example 6

CHECK

Students complete the Check online to determine if students are ready to move on.

Apply Example 6 Use Composition of Functions

Teaching the Mathematical Practices

1 Make Sense of Problems and Persevere in Solving Them, 4 Model with Mathematics Students will be presented with a task. They will first seek to understand the task, and then determine possible entry points to solving it. As students come up with their own strategies, they may propose mathematical models to aid them. As they work to solve the problem, encourage them to evaluate their model and/or progress, and change direction, if necessary.

Recommended Use

Have students work in pairs or small groups. You may wish to present the task, or have a volunteer read it aloud. Then allow students the time to make sure they understand the task, think of possible strategies, and work to solve the problem.

Encourage Productive Struggle

As students work, monitor their progress. Instead of instructing them on a particular strategy, encourage them to use their own strategies to solve the problem and to evaluate their progress along the way. They may or may not find that they need to change direction or try out several strategies.

Signs of Non-Productive Struggle

If students show signs of non-productive struggle, such as feeling overwhelmed, frustrated, or disengaged, intervene to encourage them to think of alternate approaches to the problem. Some sample questions are shown.

- How much money would the theater keep if they sold 1 ticket?
- If the theater has to give the studio 75% of the money earned on a ticket sale, what percentage of a ticket sale do they keep?

Write About It!

Have students share their responses with another pair/group of students or the entire class. Have them clearly state or describe the mathematical reasoning they can use to defend their solution.

Exit Ticket

Recommended Use

At the end of class, go online to display the Exit Ticket prompt and ask students to respond using a separate piece of paper. Have students hand you their responses as they leave the room.

Alternate Use

At the end of class, go online to display the Exit Ticket prompt and ask students to respond verbally or by using a mini-whiteboard. Have students hold up their whiteboards so that you can see all student responses. Tap to reveal the answer when most or all students have completed the Exit Ticket.

Practice and Homework

Suggested Assignments

Use the table below to select appropriate exercises.

DOK	Topic	Exercises
1, 2	exercises that mirror the examples	1–18
2	exercises that use a variety of skills from this lesson	19–35
2	exercises that extend concepts learned in this lesson to new contexts	36–39
3	exercises that emphasize higher-order and critical thinking skills	40–44

ASSESS AND DIFFERENTIATE

Use the data from the **Checks** to determine whether to provide resources for extension, remediation, or intervention.

BL

IF students score 90% or more on the Checks,
THEN assign:

- Practice Exercises 1–39 odd, 40–44
- Extension: Iteration
- ALEKS® Function Operations Inverse Functions

OL

IF students score 66%–89% on the Checks,
THEN assign:

- Practice Exercises 1–43 odd
- Remediation, Review Resources: Combining Functions
- Personal Tutors
- Extra Examples 1–6
- ALEKS® Applications of Linear Equations with Two Variables

AL

IF students score 65% or less on the Checks,
THEN assign:

- Practice Exercises 1–17 odd
- Remediation, Review Resources: Combining Functions
- *Quick Review Math Handbook:* Operations on Functions
- ALEKS® Applications of Linear Equations with Two Variables

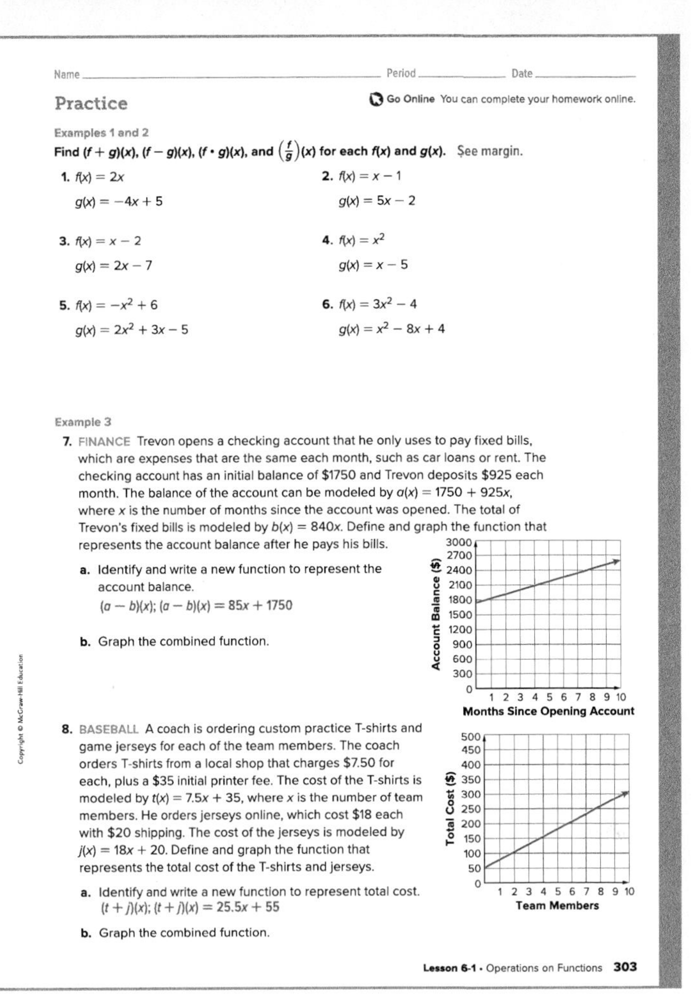

Name ____________ Period ______ Date ______

Practice — Go Online You can complete your homework online.

Examples 1 and 2

Find $(f+g)(x)$, $(f-g)(x)$, $(f \cdot g)(x)$, and $\left(\frac{f}{g}\right)(x)$ for each $f(x)$ and $g(x)$. See margin.

1. $f(x) = 2x$; $g(x) = -4x + 5$
2. $f(x) = x - 1$; $g(x) = 5x - 2$
3. $f(x) = x - 2$; $g(x) = 2x - 7$
4. $f(x) = x^2$; $g(x) = x - 5$
5. $f(x) = -x^2 + 6$; $g(x) = 2x^2 + 3x - 5$
6. $f(x) = 3x^2 - 4$; $g(x) = x^2 - 8x + 4$

Example 3

7. FINANCE Trevon opens a checking account that he only uses to pay fixed bills, which are expenses that are the same each month, such as car loans or rent. The checking account has an initial balance of \$1750 and Trevon deposits \$925 each month. The balance of the account can be modeled by $a(x) = 1750 + 925x$, where x is the number of months since the account was opened. The total of Trevon's fixed bills is modeled by $b(x) = 840x$. Define and graph the function that represents the account balance after he pays his bills.
   a. Identify and write a new function to represent the account balance. $(a - b)(x)$; $(a - b)(x) = 85x + 1750$
   b. Graph the combined function.

8. BASEBALL A coach is ordering custom practice T-shirts and game jerseys for each of the team members. The coach orders T-shirts from a local shop that charges \$7.50 for each, plus a \$35 initial printer fee. The cost of the T-shirts is modeled by $t(x) = 7.5x + 35$, where x is the number of team members. He orders jerseys online, which cost \$18 each with \$20 shipping. The cost of the jerseys is modeled by $j(x) = 18x + 20$. Define and graph the function that represents the total cost of the T-shirts and jerseys.
   a. Identify and write a new function to represent total cost. $(t + j)(x)$; $(t + j)(x) = 25.5x + 55$
   b. Graph the combined function.

Example 4

For each pair of functions, find $f \circ g$ and $g \circ f$, if they exist. State the domain and range for each. 9–12. See margin.

9. $f = \{(-8, -4), (0, 4), (2, 6), (-6, -2)\}$; $g = \{(4, -4), (-2, -1), (-4, 0), (6, -5)\}$
10. $f = \{(-7, 0), (4, 5), (8, 12), (-3, 6)\}$; $g = \{(6, 8), (-12, -5), (0, 5), (5, 1)\}$
11. $f = \{(5, 13), (-4, -2), (-8, -11), (3, 1)\}$; $g = \{(-8, 2), (-4, 1), (3, -3), (5, 7)\}$
12. $f = \{(-4, -14), (0, -6), (-6, -18), (2, -2)\}$; $g = \{(-6, 1), (-18, 13), (-14, 9), (-2, -3)\}$

Example 5

Find $[f \circ g](x)$ and $[g \circ f](x)$. State the domain and range for each. 13–16. See margin.

13. $f(x) = 2x$; $g(x) = x + 5$
14. $f(x) = -3x$; $g(x) = -x + 8$
15. $f(x) = x^2 + 6x - 2$; $g(x) = x - 6$
16. $f(x) = 2x^2 - x + 1$; $g(x) = 4x + 3$

Example 6

17. USE A MODEL Mr. Rivera wants to purchase a riding lawn mower, which is on sale for 15% off the original price. The sales tax in his area is 6.5%. Let x represent the original cost of the lawn mower. Write two functions representing the price of the lawn mower $p(x)$ after the discount and the price of the lawn mower $t(x)$ after sales tax. Write a composition of functions that represents the price of the riding lawn mower. How much will Mr. Rivera pay for a riding lawn mower that originally cost \$1350?
   $p(x) = 0.85x$; $t(x) = 1.065x$; $t[p(x)] = 1.065(0.85x)$; \$1222.09

18. REASONING A sporting goods store is offering a 20% discount on shoes. Mariana also has a \$5 off coupon that can be applied to her purchase. She is planning to buy a pair of shoes that originally costs \$89. Will the final price be lower if the discount is applied before the coupon or if the coupon is applied before the discount? Justify your response. The final price of the shoes will be less if the discount is applied before the coupon. Sample answer: If the discount is applied first, the price of the shoes is \$66.20. If the coupon is applied first, the price of the shoes is \$67.20.

Mixed Exercises

19. REASONING A bookstore that offers a 12% membership discount is currently offering 20% off each customer's total purchase when they spend more than \$50. If Keshawn has \$78 of books, should he request that the membership discount or the 20% off discount be applied first? Justify your response.
   Sample answer: The order of the discounts does not matter. Either composition results in a final cost of \$54.91.

20. CONSTRUCT ARGUMENTS Is $[f \circ g](x)$ always equal to $[g \circ f](x)$ for two functions $f(x)$ and $g(x)$? Justify your conclusions. Provide a counterexample if needed.
   No; sample answer: $[f \circ g](x)$ often does not equal $[g \circ f](x)$. For a counterexample, let $f(x) = 2x$ and $g(x) = -x + 4$. $[f \circ g](x) = -2x + 8$ and $[g \circ f](x) = -2x + 4$. Therefore, $[f \circ g](x) \neq [g \circ f](x)$.

Name ______ Period ______ Date ______

If $f(x) = 3x$, $g(x) = x + 4$, and $h(x) = x^2 - 1$, find each value.

21. $f[g(1)]$ 15

22. $g[h(0)]$ 3

23. $g[f(-1)]$ 1

24. $h[f(5)]$ 224

25. $g[h(-3)]$ 12

26. $h[f(10)]$ 899

27. $f[h(8)]$ 189

28. $[f \circ (h \circ g)](1)$ 72

29. $[f \circ (g \circ h)](-2)$ 21

30. $h[f(-6)]$ 323

31. $f[h(0)]$ −3

32. $f[g(7)]$ 33

33. $f[h(-2)]$ 9

34. $[g \circ (f \circ h)](-1)$ 4

35. $[h \circ (f \circ g)](3)$ 440

36. AREA Valeria wants to know the area of a figure made by joining an equilateral triangle and square along an edge. The function $f(s) = \frac{\sqrt{3}}{4}s^2$ gives the area of an equilateral triangle with side s. The function $g(s) = s^2$ gives the area of a square with side s. Write a function $h(s)$ that gives the area of the figure as a function of its side length s.

$h(s) = (f + g)(s) = \left(\frac{\sqrt{3}}{4} + 1\right)s^2$

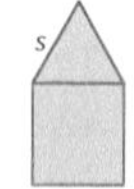

37. USE A MODEL The volume V of a weather balloon with radius r is given by $V(r) = \frac{4}{3}\pi r^3$. The balloon is being inflated so that the radius increases at a constant rate $r(t) = \frac{1}{2}t + 2$, where r is in meters and t is the number of seconds since inflation began.

a. Determine the function that represents the volume of the weather balloon in terms of time.
$V(r(t)) = \frac{\pi t^3}{6} + 2\pi t^2 + 8\pi t + \frac{32}{3}\pi$

b. Find the volume of the balloon 12 seconds after inflation begins. Round your answer to the nearest cubic meter. 2145 m^3

38. REASONING The National Center for Education Statistics reports data showing that since 2006, college enrollment for men in thousands can be modeled by $f(x) = 389x + 7500$, where x represents the number of years since 2006. Similarly, enrollment for women can be modeled by $g(x) = 480x + 10{,}075$. Write a function for $(f + g)(x)$ and interpret what it represents. See margin.

39. STRUCTURE The table shows various values of functions $f(x)$, $g(x)$, and $h(x)$.

x	−1	0	1	2	3	4
$f(x)$	7	−2	0	2	4	1
$g(x)$	−3	−4	−5	0	1	1
$h(x)$	0	4	1	1	5	5

Use the table to find the following values:

a. $(f + g)(-1)$ 4

b. $(h - g)(0)$ 8

c. $(f \cdot h)(4)$ 5

d. $\left(\frac{f}{g}\right)(3)$ 4

e. $\left(\frac{g}{h}\right)(2)$ 0

f. $\left(\frac{g}{f}\right)(1)$ undefined

Higher-Order Thinking Skills

40. PERSEVERE If $(f + g)(3) = 5$ and $(f \cdot g)(3) = 6$, find $f(3)$ and $g(3)$. Explain. See margin.

41. CREATE Write two functions $f(x)$ and $g(x)$ such that $(f \circ g)(4) = 0$.
Sample answer: $f(x) = x - 9$, $g(x) = x + 5$

42. FIND THE ERROR Chris and Tobias are finding $(f \circ g)(x)$, where $f(x) = x^2 + 2x - 8$ and $g(x) = x^2 + 8$. Is either of them correct? Explain your reasoning.
Tobias; sample answer: Chris did not substitute $g(x)$ for every x in $f(x)$.

Chris	Tobias
$(f \circ g)(x) = f[g(x)]$	$(f \circ g)(x) = f[g(x)]$
$= (x^2 + 8)^2 + 2x - 8$	$= (x^2 + 8)^2 + 2(x^2 + 8) - 8$
$= x^4 + 16x^2 + 64 + 2x - 8$	$= x^4 + 16x^2 + 64 + 2x^2 + 16 - 8$
$= x^4 + 16x^2 + 2x + 58$	$= x^4 + 18x^2 + 72$

43. PERSEVERE Given $f(x) = \sqrt{x^3}$ and $g(x) = \sqrt{x^6}$, determine each domain.

a. $g(x) \cdot g(x)$ D = {all real numbers}

b. $f(x) \cdot f(x)$ D = $\{x \mid x \geq 0\}$

44. ANALYZE State whether the following statement is *sometimes*, *always*, or *never* true. Justify your argument. See margin.

The domain of two functions $f(x)$ and $g(x)$ that are composed $g[f(x)]$ is restricted by the domain of $g(x)$.

Answers

1. $(f + g)(x) = -2x + 5$; $(f - g)(x) = 6x - 5$; $(f \cdot g)(x) = -8x^2 + 10x$; $\left(\frac{f}{g}\right)(x) = \frac{2x}{-4x + 5}$, $x \neq \frac{5}{4}$
2. $(f + g)(x) = 6x - 3$; $(f - g)(x) = -4x + 1$; $(f \cdot g)(x) = 5x^2 - 7x + 2$; $\left(\frac{f}{g}\right)(x) = \frac{x - 1}{5x - 2}$, $x \neq \frac{2}{5}$
3. $(f + g)(x) = 3x - 9$; $(f - g)(x) = -x + 5$; $(f \cdot g)(x) = 2x^2 - 11x + 14$; $\left(\frac{f}{g}\right)(x) = \frac{x - 2}{2x - 7}$, $x \neq \frac{7}{2}$
4. $(f + g)(x) = x^2 + x - 5$; $(f - g)(x) = x^2 - x + 5$; $(f \cdot g)(x) = x^3 - 5x^2$; $\left(\frac{f}{g}\right)(x) = \frac{x^2}{x - 5}$, $x \neq 5$
5. $(f + g)(x) = x^2 + 3x + 1$; $(f - g)(x) = -3x^2 - 3x + 11$; $(f \cdot g)(x) = -2x^4 - 3x^3 + 17x^2 + 18x - 30$; $\left(\frac{f}{g}\right)(x) = \frac{-x^2 + 6}{2x^2 + 3x - 5}$, $x \neq 1$ or $-\frac{5}{2}$
6. $(f + g)(x) = 4x^2 - 8x$; $(f - g)(x) = 2x^2 + 8x - 8$; $(f \cdot g)(x) = 3x^4 - 24x^3 + 8x^2 + 32x - 16$; $\left(\frac{f}{g}\right)(x) = \frac{3x^2 - 4}{x^2 - 8x + 4}$, $x \neq 4 \pm 2\sqrt{3}$

9. $f \circ g = \{(-4, 4)\}$, D = {−4}, R = {4}; $g \circ f = \{(-8, 0), (0, -4), (2, -5), (-6, -1)\}$, D = {−8, −6, 0, 2}, R = {−5, −4, −1, 0}
10. $f \circ g = \{(6, 12)\}$, D = {6}, R = {12}; $g \circ f = \{(-7, 5), (4, 1), (-3, 8)\}$, D = {−7, −3, 4}, R = {1, 5, 8}
11. $f \circ g$ is undefined, D = ∅, R = ∅; $g \circ f$ is undefined, D = ∅, R = ∅
12. $f \circ g$ is undefined, D = ∅, R = ∅; $g \circ f = \{(-4, 9), (0, 1), (-6, 13), (2, -3)\}$, D = {−6, −4, 0, 2}, R = {−3, 1, 9, 13}
13. $[f \circ g](x) = 2x + 10$, D = {all real numbers}, R = {all real numbers}; $[g \circ f](x) = 2x + 5$, D = {all real numbers}, R = {all real numbers}
14. $[f \circ g](x) = 3x - 24$, D = {all real numbers}, R = {all real numbers}; $[g \circ f](x) = 3x + 8$, D = {all real numbers}, R = {all real numbers}
15. $[f \circ g](x) = x^2 - 6x - 2$, D = {all real numbers}, R = $\{y \mid y \geq -11\}$; $[g \circ f](x) = x^2 + 6x - 8$, D = {all real numbers}, R = $\{y \mid y \geq -17\}$
16. $[f \circ g](x) = 32x^2 + 44x + 16$, D = {all real numbers}, R = $\{y \mid y \geq 0.875\}$; $[g \circ f](x) = 8x^2 - 4x + 7$, D = {all real numbers}, R = $\{y \mid y \geq 6.5\}$

38. $(f + g)(x) = 869x + 17{,}575$. This function represents the enrollment of both men and women in college since 2006.
40. Let $a = f(3)$ and $b = g(3)$. Since $(f + g)(3) = 5$, then $a + b = 5$. Solving for a, we get $a = 5 - b$. Since $(f \cdot g)(3) = 6$, $ab = 6$. Substituting a, we get $(5 - b)(b) = 6$. Distributing the left hand side, we get $5b - b^2 = 6$, or $b^2 - 5b + 6 = 0$. Factoring, we get $(b - 2)(b - 3) = 0$, so $b = 2$ or $b = 3$. If $b = 2$, then $a = 3$. If $b = 3$, then $a = 2$. So the two solutions are $f(3) = 2$, $g(3) = 3$, and $f(3) = 3$, $g(3) = 2$.
44. Sometimes; sample answer: When $f(x) = 4x$ and $g(x) = \sqrt{x}$, $g[f(x)] = \sqrt{4x}$, $x \geq 0$. The domain of $g(x)$ restricts the domain of $g[f(x)]$. When $f(x) = 4x^2$ and $g(x) = \sqrt{x}$, $g[f(x)] = \sqrt{4x^2}$. In this case, the domain of $g(x)$ does not restrict the domain of $g[f(x)]$.

Inverse Relations and Functions

LESSON GOAL

Students graph and verify inverse functions.

1 LAUNCH

Launch the lesson with a **Warm Up** and an introduction.

2 EXPLORE AND DEVELOP

Explore: Inverse Functions

Develop:

Inverse Relations and Functions
- Find an Inverse Relation
- Inverse Functions
- Inverses with Restricted Domains
- Interpret Inverse Functions

Verifying Inverses
- Use Compositions to Verify Inverses
- Verify Inverse Functions

You may want your students to complete the **Checks** online.

3 REFLECT AND PRACTICE

Exit Ticket

Practice

Formative Assessment Math Probe

DIFFERENTIATE

View reports of student progress on the **Checks** after each example.

Resources	AL	OL	BL	ELL
Remediation: Functions	●	●		●
Extension: Group Theory		●	●	●

Language Development Handbook

Assign page 36 of the *Language Development Handbook* to help your students build mathematical language related to graphing inverse functions.

ELL You can use the tips and suggestions on page T36 of the handbook to support students who are building English proficiency.

Suggested Pacing

90 min	1 day
45 min	2 days

Focus

Domain: Functions

Standards for Mathematical Content:

F.IF.5 Relate the domain of a function to its graph and, where applicable, to the quantitative relationship it describes.

F.BF.4a Solve an equation of the form $f(x) = c$ for a simple function f that has an inverse and write an expression for the inverse.

Standards for Mathematical Practice:

2 Reason abstractly and quantitatively.

6 Attend to precision.

Coherence

Vertical Alignment

Previous
Students understood the relationship between functions and their inverses.
F.B.4 (Algebra 1)

Now
Students find the inverse of a function or relation. Students determine whether two functions or relations are inverses.
F.IF.5, F.BF.4a

Next
Students will analyze the relationship between a function and its inverse (exponential and logarithmic). **A.SSE.2, F.IF.7e**

Rigor

The Three Pillars of Rigor

1 CONCEPTUAL UNDERSTANDING	2 FLUENCY	3 APPLICATION

Conceptual Bridge In this lesson, students develop an understanding of inverse relations and functions. They build fluency by finding inverses of relations and functions, and they apply their understanding by solving real-world problems.

Mathematical Background

To test whether two functions $f(x)$ and $g(x)$ are inverses, check that each of the two compositions $[f \circ g](x)$ and $[g \circ f](x)$ has the value x. When the inverse of a function is a function, then the original function is said to be one-to-one.

Interactive Presentation

Warm Up

Determine whether each relation is a function.

1. $\{(-1, -5), (3,5), (5,7), (7,11)\}$

2. $\{(0,8), (4,7), (0,9), (3,0)\}$

Write a rule for each function. State the domain and range.

3. $E = \{(-3, -15), (-1, -5), (0,0), (4,20)\}$

4. $G = \{(-4, 3), (-3,4) (1,8), (9,16)\}$

Warm Up

Warm Up

Prerequisite Skills

The Warm Up exercises address the following prerequisite skill for this lesson:

- determining whether relations are functions

Answers:

1. yes
2. no
3. $y = 5x$; D = $\{-3, -1, 0, 4\}$, R = $\{-15, -5, 0, 20\}$
4. $y = x + 7$; D = $\{-4, -3, 0, 9\}$, R = $\{3, 4, 8, 16\}$

Launch the Lesson

The sound you hear is a result of particle to particle interaction The speed of the sound varies with the temperature and mass of the medium through which it is traveling. The formula $S = 108\sqrt{\frac{T}{m}}$ can be used to find the speed of sound in some mediums, where S represents the speed of sound, T represents the temperature, and m represents the average molecular mass. You can use inverse functions to rewrite the formula for the temperature in terms of speed.

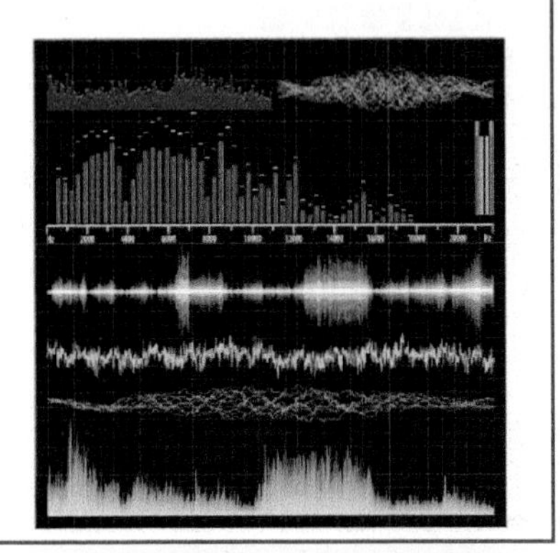

Launch the Lesson

Launch the Lesson

 Teaching the Mathematical Practices

2 Attend to Quantities Point out that it is important to note the meaning of the quantities used in the formula for the speed of sound. Once students have completed the lesson, encourage students to use inverse functions to rewrite the formula and ask students to describe the meaning of the formula.

Go Online to find additional teaching notes and questions to promote classroom discourse.

Today's Standards

Tell students that they will be addressing these content and practice standards in this lesson. You may wish to have a student volunteer read aloud *How can I meet these standards?* and *How can I use these practices?*, and connect these to the standards.

See the Interactive Presentation for I Can statements that align with the standards covered in this lesson.

Vocabulary

Collapse All

inverse relations

Two relations, one of which contains points of the form (a, b) while the other contains points of the form (b, a).

inverse functions

Two functions for which both of their compositions are the identity function.

Today's Vocabulary

Today's Vocabulary

Tell students that they will be using these vocabulary terms in this lesson. You can expand each row if you wish to share the definitions. Then discuss the questions below with the class.

Explore Inverse Functions

Objective

Students use a sketch to explore inverse functions.

Teaching the Mathematical Practices

5 Use Mathematical Tools Point out that to complete the Explore, students will need to use the sketches. Work with students to explore and deepen their understanding of inverse functions.

Ideas for Use

Recommended Use Present the Inquiry Question, or have a student volunteer read it aloud. Have students work in pairs to complete the Explore activity on their devices. Pairs should discuss each of the questions. Monitor student progress during the activity. Upon completion of the Explore activity, have student volunteers share their responses to the Inquiry Question.

What if my students don't have devices? You may choose to project the activity on a whiteboard. A printable worksheet for each Explore is available online. You may choose to print the worksheet so that individuals or pairs of students can use it to record their observations.

Summary of the Activity

Students will complete guiding exercises throughout the Explore activity. Students will be presented with a sketch of a function and its inverse. They will use the sketch to explore different points on each function and determine their relationship. Then, students will answer the Inquiry Question.

(continued on the next page)

Interactive Presentation

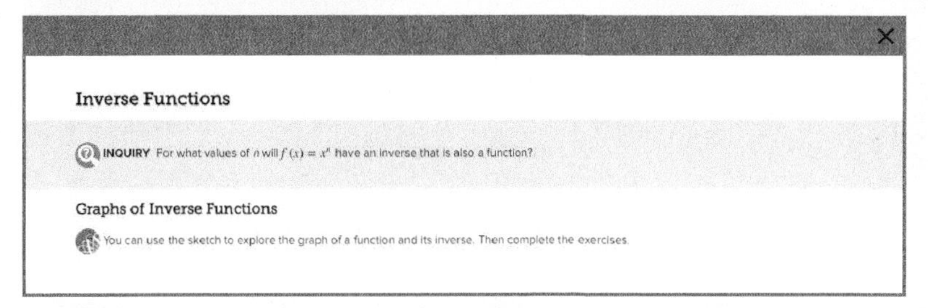

Explore

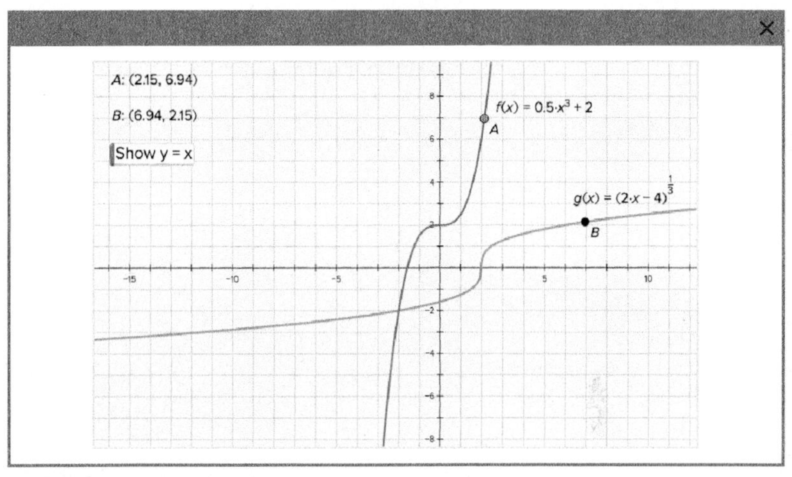

Explore

WEB SKETCHPAD

Students use the sketch to complete an activity in which they explore graphs of inverse functions.

Interactive Presentation

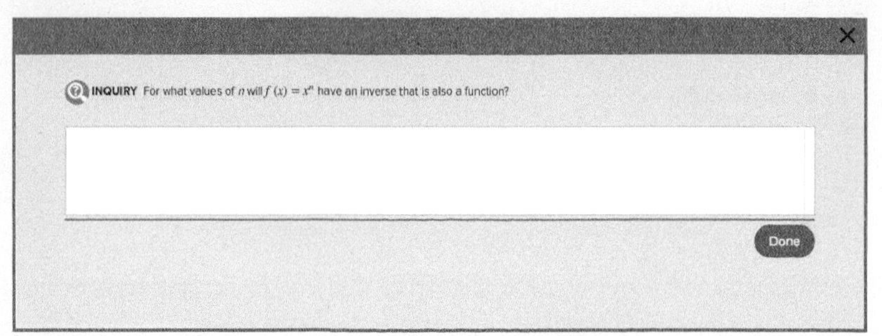

Explore

TYPE

Students respond to the Inquiry Question and can view a sample answer.

Explore Inverse Functions (*continued*)

Teaching the Mathematical Practices

3 Construct Arguments Throughout the Explore, students will use previously established results to construct arguments about graphs of functions and their inverses.

Questions

Have students complete the Explore activity.

Ask:

- What can the reflection of the graph in $y = x$ tell you about the graph of the inverse? Sample answer: You can look at the shape of the graph to make the reflection in $y = x$. Then, you can see whether the inverse would pass the vertical line test.
- Describe a function of degree 3 with an inverse that is not a function. Sample answer: If a function with degree 3 has a relative minimum and relative maximum, it will not pass the horizontal line test. For example, if it had more than one root, it would be crossing the x-axis more than once. Therefore, the inverse would not be a function.

Inquiry

For what values of n will $f(x) = x^n$ have an inverse that is also a function? Sample answer: If n is an odd whole number, the inverse of $f(x)$ will also be a function.

Go Online to find additional teaching notes and sample answers for the guiding exercises.

Learn Inverse Relations and Functions

Objective

Students find inverses of relations.

Teaching the Mathematical Practices

7 Use Structure Help students to explore the structure of inverse function. Guide them to see the pattern in the input and output values of a function and its inverse.

Important to Know

You can use a graphing calculator to compare a function and its inverse using tables and graphs.

Essential Question Follow-Up

Students have explored inverse functions.

Ask:

Why would you use the inverse of a function to model a real-world situation? Sample answer: You could use a function or the inverse of the function to model a relationship so that either quantity could be the independent variable. Then, you could use the functions to solve problems involving either quantity.

Example 1 Find an Inverse Relation

Teaching the Mathematical Practices

1 Explain Correspondences Encourage students to explain the relationships between the graph of the function and its inverse to answer the question in the Think About It! feature.

Questions for Mathematical Discourse

AL If you had a table of ordered pairs to graph a relation how could you create a table to graph the inverse? switch the x and y columns

OL For a relation of ordered pairs to be the same as its inverse, must $x = y$ for each pair? No, a relation could have points (a, b) and (b, a) and its inverse would have the same points.

BL What would be true about a relation that intersects the line $x = y$? The relation would intersect its inverse.

Go Online

- Find additional teaching notes.
- View performance reports of the Checks.
- Assign or present an Extra Example.

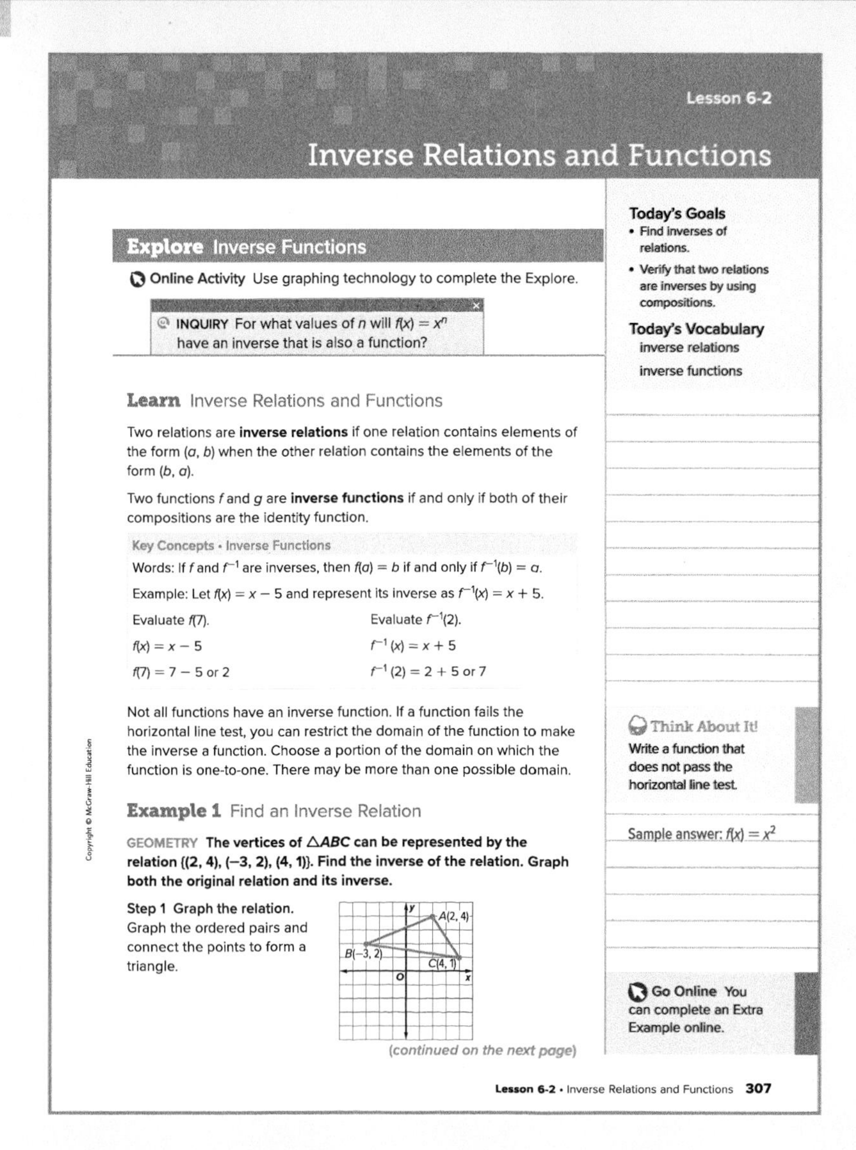

Lesson 6-2

Inverse Relations and Functions

Today's Goals
- Find inverses of relations.
- Verify that two relations are inverses by using compositions.

Today's Vocabulary
inverse relations
inverse functions

Explore Inverse Functions

Online Activity Use graphing technology to complete the Explore.

INQUIRY For what values of n will $f(x) = x^n$ have an inverse that is also a function?

Learn Inverse Relations and Functions

Two relations are **inverse relations** if one relation contains elements of the form (a, b) when the other relation contains the elements of the form (b, a).

Two functions f and g are **inverse functions** if and only if both of their compositions are the identity function.

Key Concepts • Inverse Functions

Words: If f and f^{-1} are inverses, then $f(a) = b$ if and only if $f^{-1}(b) = a$.

Example: Let $f(x) = x - 5$ and represent its inverse as $f^{-1}(x) = x + 5$.

Evaluate $f(7)$.	Evaluate $f^{-1}(2)$.
$f(x) = x - 5$	$f^{-1}(x) = x + 5$
$f(7) = 7 - 5$ or 2	$f^{-1}(2) = 2 + 5$ or 7

Not all functions have an inverse function. If a function fails the horizontal line test, you can restrict the domain of the function to make the inverse a function. Choose a portion of the domain on which the function is one-to-one. There may be more than one possible domain.

Think About It! Write a function that does not pass the horizontal line test.

Sample answer: $f(x) = x^2$

Example 1 Find an Inverse Relation

GEOMETRY **The vertices of $\triangle ABC$ can be represented by the relation $\{(2, 4), (-3, 2), (4, 1)\}$. Find the inverse of the relation. Graph both the original relation and its inverse.**

Step 1 Graph the relation.
Graph the ordered pairs and connect the points to form a triangle.

Go Online You can complete an Extra Example online.

(continued on the next page)

Interactive Presentation

Inverse Relations and Functions

Recall that a relation is a set of ordered pairs. Two relations are **inverse relations** if one relation contains elements of the form (a, b) when the other relation contains the elements of the form (b, a). The domain of a relation becomes the range of its inverse, and the range of the relation becomes the domain of its inverse. For example, if A and B are inverse relations and $A = \{(1, 9), (3, 8), (5, 7)\}$, then $B = \{(9, 1), (8, 3), (7, 5)\}$.

As with relations, the ordered pairs of inverse functions are also related. The inverse of a function can be found by exchanging the domain and range. Two functions f and g are **inverse functions** if and only if both of their compositions are the identity function.
We can write the inverse of the function $f(x)$ as $f^{-1}(x)$.

KEY CONCEPT: INVERSE FUNCTIONS

Words	If f and f^{-1} are inverses, then $f(a) = b$ if and only if $f^{-1}(b) = a$
Example	Let $f(x) = x - 5$ and represent its inverse as $f^{-1}(x) = x + 5$ Evaluate $f(7)$. Evaluate $f^{-1}(2)$. $f(x) = x - 5$ $f^{-1}(x) = x + 5$

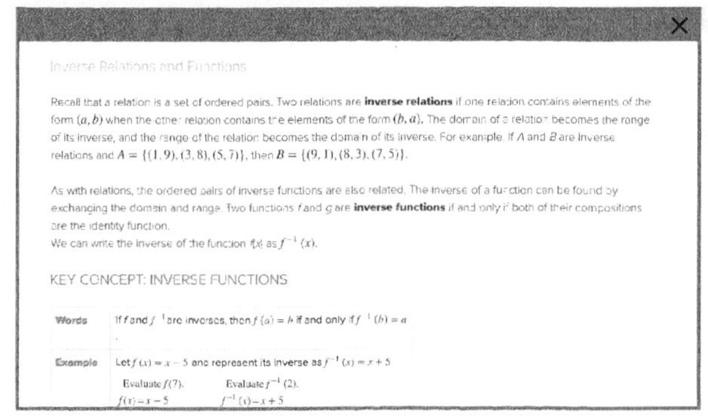

Learn

Your Notes

Think About It!
Describe the graph of the inverse relation.

Sample answer: It is a reflection in the line $y = x$.

Study Tip
Inverses If $f^{-1}(x)$ is the inverse of $f(x)$, the graph of $f^{-1}(x)$ will be a reflection of the graph of $f(x)$ in the line $y = x$.

Go Online
You can learn how to graph a relation and its inverse on a graphing calculator by watching the video online.

Step 2 Find the inverse.

To find the inverse, exchange the coordinates of the ordered pairs.
{(4, 2), (2, −3), (1, 4)}

Step 3 Graph the inverse.

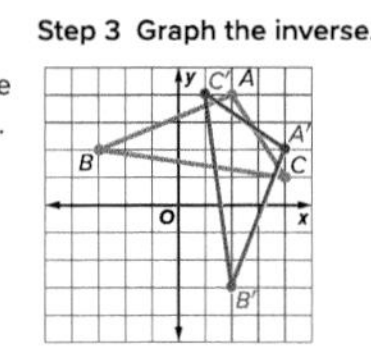

Example 2 Inverse Functions

Find the inverse of $f(x) = 3x + 2$. Then graph the function and its inverse.

Step 1 Rewrite the function.

Rewrite the function as an equation relating x and y.

$f(x) = 3x + 2 \rightarrow y = 3x + 2$

Step 2 Exchange x and y.

Exchange x and y in the equation.

$x = 3y + 2$

Step 3 Solve for y.

$$x = 3y + 2$$
$$x - 2 = 3y$$
$$\frac{x-2}{3} = y$$

Step 4 Replace y with $f^{-1}(x)$.

Replace y with $f^{-1}(x)$ in the equation.

$y = \frac{x-2}{3} \rightarrow f^{-1}(x) = \frac{x-2}{3}$

The inverse of $f(x) = 3x + 2$ is $f^{-1}(x) = \frac{x-2}{3}$.

Step 5 Graph $f(x)$ and $f^{-1}(x)$.

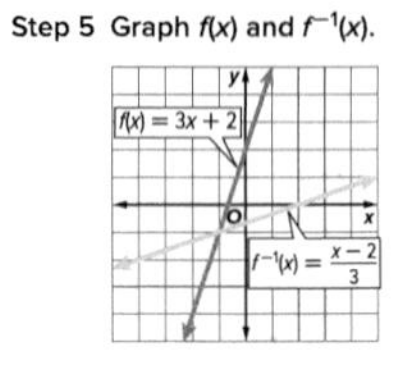

Check

Examine $f(x) = -\frac{1}{2}x + 1$.

Part A Find the inverse of $f(x) = -\frac{1}{2}x + 1$.

$f^{-1}(x) = -2x + 2$

Part B Graph $f(x) = -\frac{1}{2}x + 1$ and its inverse.

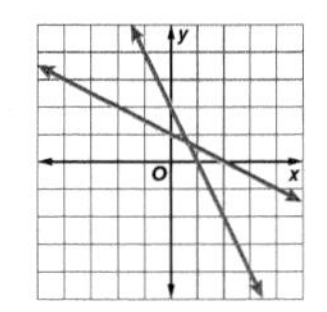

Go Online You can complete an Extra Example online.

308 Module 6 • Inverse and Radical Functions

Interactive Presentation

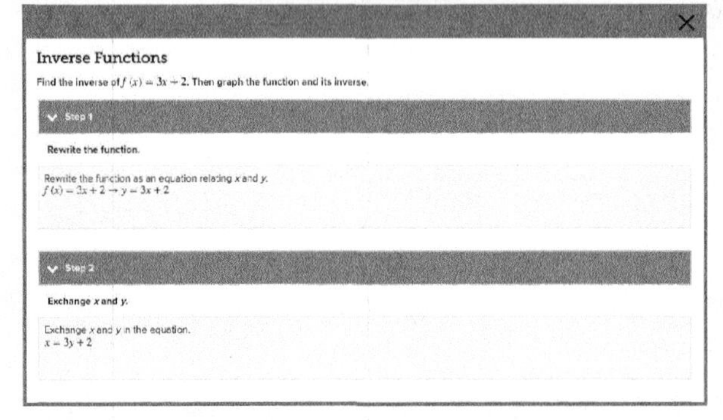

Example 2

TAP

Students move through the steps to find and graph the inverse of a function.

Example 2 Inverse Functions

Teaching the Mathematical Practices

1 Seek Information Mathematically proficient students must be able to transform algebraic expressions to reach solutions. In Example 2, students transform the function to find its inverse.

Questions for Mathematical Discourse

AL What are the slopes of $f(x)$ and $f^{-1}(x)$? The slope of $f(x)$ is 3 and the slope of $f^{-1}(x)$ is $\frac{1}{3}$.

OL Can a linear equation with a positive slope have an inverse with a negative slope? Explain. No; sample answer: the slope of the inverse of a linear equation will have the same sign because the inverse is a reflection in the line $y = x$.

BL Are any quadratic functions one-to-one? Why or why not? No; sample answer: the graph of a quadratic function is a parabola which would never pass the horizontal line test.

Common Error

Remind students that f^{-1} is read *f inverse* or *the inverse of f*. Note that −1 is not an exponent.

Example 3 Inverses with Restricted Domains

Teaching the Mathematical Practices

1 Explain Correspondences Encourage students to explain the relationships between the inverse, its graph, and the domains that make the inverse a function.

Questions for Mathematical Discourse

AL Why is the domain of $f^{-1}(x)$ restricted? $f^{-1}(x)$ isn't a function.

OL How does the axis of symmetry of $f(x)$ relate to the restricted domain of $f^{-1}(x)$? Sample answer: the axis of symmetry divides the function into two pieces that are each one-to-one. It divides the x-axis into the pieces that can be the restricted domain.

BL List the two possible restricted domains for $f(x) = ax^2 + bx + c$? $\left(-\infty, \frac{-b}{2a}\right], \left[\frac{-b}{2a}, \infty\right)$

Example 4 Interpret Inverse Functions

Teaching the Mathematical Practices

2 Attend to Quantities Point out that it is important to note the meaning of the inverse function in Example 4. Students should understand what the inverse function represents and how it can be used.

Questions for Mathematical Discourse

AL Let F equal the temperature in degrees Fahrenheit and C equal the temperature in degrees Celsius. How do F and C relate to each other through $T(x)$ and $T^{-1}(x)$? $C = T(F)$ and $F = T^{-1}(C)$

OL How do you know that $T^{-1}(x)$ will be a function? The graph of $T(x)$ passes the horizontal line test.

BL Will there be a point where $T(x)$ and $T^{-1}(x)$ intersect? Explain. Yes; sample answer: they must intersect because they are linear functions with different slopes. What does the intersection represent? It is the value for which the temperature in Celsius and in Fahrenheit are the same. What else do you need to consider when determining whether the functions intersect? Sample answer: You need to consider whether the intersection occurs in the valid domain of the functions. In this case it does, at −40. The domain is limited by absolute zero, which is 273.15 in Celsius and 459.67 in Fahrenheit.

Example 3 Inverses with Restricted Domains

Examine $f(x) = x^2 + 2x + 4$.

Part A Find the inverse of $f(x)$.

$f(x) = x^2 + 2x + 4$	Original function
$y = x^2 + 2x + 4$	Replace $f(x)$ with y.
$x = y^2 + 2y + 4$	Exchange x and y.
$x - 4 = y^2 + 2y$	Subtract 4 from each side.
$x - 4 + 1 = y^2 + 2y + 1$	Complete the square.
$x - 3 = (y + 1)^2$	Simplify.
$\pm\sqrt{x-3} = y + 1$	Take the square root of each side.
$-1 \pm \sqrt{x-3} = y$	Subtract 1 from each side.
$f^{-1}(x) = -1 \pm \sqrt{x-3}$	Replace y with $f^{-1}(x)$.

Part B If necessary, restrict the domain of the inverse so that it is a function.

Because $f(x)$ fails the horizontal line test, $f^{-1}(x)$ is not a function. Find the restricted domain of $f(x)$ so that $f^{-1}(x)$ will be a function. Look for a portion of the graph that is one-to-one. If the domain of $f(x)$ is restricted to $[-1, \infty)$ then the inverse is $f^{-1}(x) = -1 + \sqrt{x-3}$.

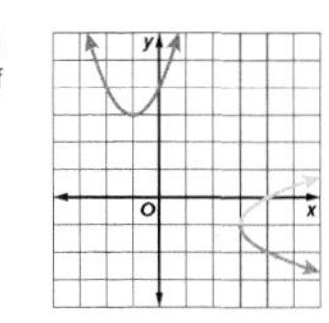

If the domain of $f(x)$ is restricted to $(-\infty, -1]$, then the inverse is $f^{-1}(x) = -1 - \sqrt{x-3}$.

Watch Out!
Inverse Functions f^{-1} is read *f inverse* or *the inverse of f*. Note that −1 is not an exponent.

Example 4 Interpret Inverse Functions

TEMPERATURE **A formula for converting a temperature in degrees Fahrenheit to degrees Celsius is $T(x) = \frac{5}{9}(x - 32)$.**

Find the inverse of $T(x)$, and describe its meaning.

$T(x) = \frac{5}{9}(x - 32)$	Original function
$y = \frac{5}{9}(x - 32)$	Replace $T(x)$ with y.
$x = \frac{5}{9}(y - 32)$	Exchange x and y.
$\frac{9x}{5} = y - 32$	Multiply each side by $\frac{9}{5}$.
$\frac{9x}{5} + 32 = y$	Add 32 to each side.
$T^{-1}(x) = \frac{9x}{5} + 32$	Replace y with $T^{-1}(x)$.

$T^{-1}(x) =$ can be used to convert a temperature in degrees Celsius to degrees Fahrenheit.

Go Online You can complete an Extra Example online.

Go Online to see Part B of the example on using the graph of $T(x)$ and $T^{-1}(x)$.

Think About It!
Find the domain of $T(x)$ and its inverse. Explain your reasoning.

Sample answer: Algebraically, the domain of both $T(x)$ and $T^{-1}(x)$ are all real numbers because temperatures can be positive and negative and do not have to be integer values.

Lesson 6-2 • Inverse Relations and Functions 309

Interactive Presentation

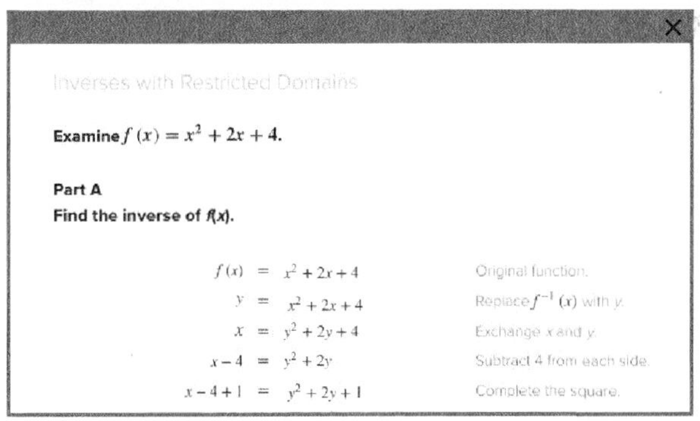

Example 3

TAP

Students move through the steps to graph the inverse of a function with a restricted domain.

CHECK

Students complete the Check online to determine whether they are ready to move on.

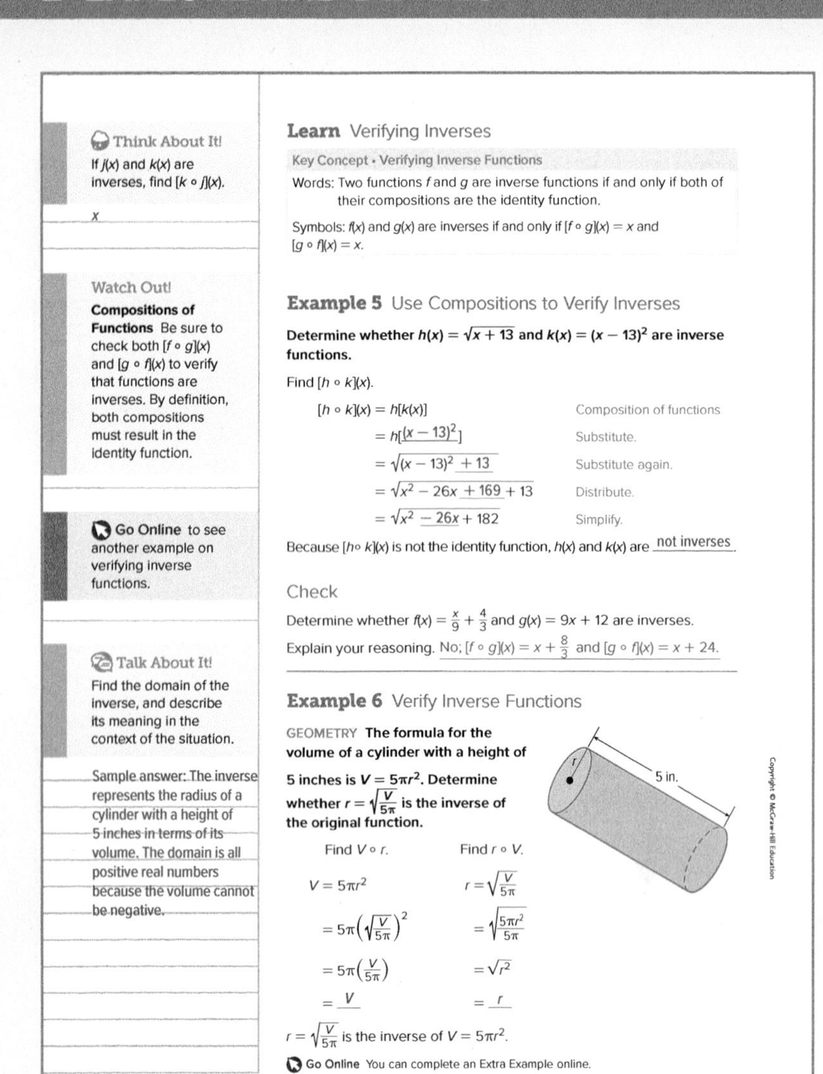

Think About It! If $f(x)$ and $k(x)$ are inverses, find $[k \circ f](x)$. x

Watch Out! **Compositions of Functions** Be sure to check both $[f \circ g](x)$ and $[g \circ f](x)$ to verify that functions are inverses. By definition, both compositions must result in the identity function.

Go Online to see another example on verifying inverse functions.

Talk About It! Find the domain of the inverse, and describe its meaning in the context of the situation.

Sample answer: The inverse represents the radius of a cylinder with a height of 5 inches in terms of its volume. The domain is all positive real numbers because the volume cannot be negative.

Learn Verifying Inverses

Key Concept • Verifying Inverse Functions

Words: Two functions f and g are inverse functions if and only if both of their compositions are the identity function.

Symbols: $f(x)$ and $g(x)$ are inverses if and only if $[f \circ g](x) = x$ and $[g \circ f](x) = x$.

Example 5 Use Compositions to Verify Inverses

Determine whether $h(x) = \sqrt{x + 13}$ and $k(x) = (x - 13)^2$ are inverse functions.

Find $[h \circ k](x)$.

$[h \circ k](x) = h[k(x)]$	Composition of functions
$= h[(x - 13)^2]$	Substitute.
$= \sqrt{(x - 13)^2 + 13}$	Substitute again.
$= \sqrt{x^2 - 26x + 169 + 13}$	Distribute.
$= \sqrt{x^2 - 26x + 182}$	Simplify.

Because $[h \circ k](x)$ is not the identity function, $h(x)$ and $k(x)$ are not inverses.

Check

Determine whether $f(x) = \frac{x}{9} + \frac{4}{3}$ and $g(x) = 9x + 12$ are inverses. Explain your reasoning. No; $[f \circ g](x) = x + \frac{8}{3}$ and $[g \circ f](x) = x + 24$.

Example 6 Verify Inverse Functions

GEOMETRY **The formula for the volume of a cylinder with a height of 5 inches is $V = 5\pi r^2$. Determine whether $r = \sqrt{\frac{V}{5\pi}}$ is the inverse of the original function.**

5 in.

Find $V \circ r$.	Find $r \circ V$.
$V = 5\pi r^2$	$r = \sqrt{\frac{V}{5\pi}}$
$= 5\pi\left(\sqrt{\frac{V}{5\pi}}\right)^2$	$= \sqrt{\frac{5\pi r^2}{5\pi}}$
$= 5\pi\left(\frac{V}{5\pi}\right)$	$= \sqrt{r^2}$
$= V$	$= r$

$r = \sqrt{\frac{V}{5\pi}}$ is the inverse of $V = 5\pi r^2$.

Go Online You can complete an Extra Example online.

310 Module 6 • Inverse and Radical Functions

Copyright © McGraw-Hill Education

Learn Verifying Inverses

Objective

Students verify that two relations are inverses by using compositions.

Teaching the Mathematical Practices

6 Use Definitions To answer the question in the Think About It! feature, students will use definition of inverse functions.

Common Misconception

Remind students to check both $[f \circ g](x)$ and $[g \circ f](x)$ to verify that functions are inverses. By definition, both compositions must be the identity function.

Example 5 Use Compositions to Verify Inverses

Teaching the Mathematical Practices

7 Use Structure Help students to use the structure of the functions to determine whether they are inverses.

Questions for Mathematical Discourse

AL Why is it unnecessary to check $k[h(x)]$? For the functions to be inverses, both $h[k(x)]$ and $k[h(x)]$ must be the identity function. Once one is determined to not be an identity function, you know they cannot be inverses.

OL What change would make $h(x)$ the inverse of $k(x)$? if + 13 was outside the radical

BL What is an example of a function that is its own inverse? Sample answer: $y = \frac{1}{x}$

Example 6 Verify Inverse Functions

Teaching the Mathematical Practices

2 Attend to Quantities Students must consider the meaning of the domain of the inverse function to answer the question in the Talk About It! feature.

Questions for Mathematical Discourse

AL Why is it useful to find inverses of real-world functions? Sample answer: You can select a variable to be the dependent variable based on the problem you are solving. In this example you have functions representing how V depends on r and for how r depends on V.

OL What is another way you can tell the functions are inverses? When you solve V for r, you get the same result as the given function r, likewise if you solve r for V.

BL Would the functions be inverses without the real-world domain restriction? Explain. No, V does not pass the horizontal line test.

Interactive Presentation

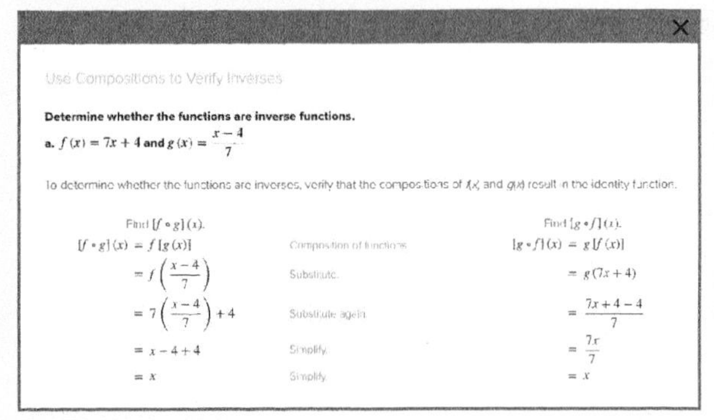

Example 5

TYPE

Students answer a question to determine whether they understand how to write the inverse of a function.

CHECK

Students complete the Check online to determine whether they are ready to move on.

Exit Ticket

Recommended Use

At the end of class, go online to display the Exit Ticket prompt and ask students to respond using a separate piece of paper. Have students hand you their responses as they leave the room.

Alternate Use

At the end of class, go online to display the Exit Ticket prompt and ask students to respond verbally or by using a mini-whiteboard. Have students hold up their whiteboards so that you can see all student responses. Tap to reveal the answer when most or all students have completed the Exit Ticket.

Practice and Homework

Suggested Assignments

Use the table below to select appropriate exercises.

DOK	Topic	Exercises
1, 2	exercises that mirror the examples	1–24
2	exercises that use a variety of skills from this lesson	25–46
2	exercises that extend concepts learned in this lesson to new contexts	47, 48
3	exercises that emphasize higher-order and critical thinking skills	49–53

ASSESS AND DIFFERENTIATE

Use the data from the **Checks** to determine whether to provide resources for extension, remediation, or intervention.

IF students score 90% or more on the Checks, **THEN** assign: BL

- Practice Exercises 1–47 odd, 49–53
- Extension: Group Theory
- ALEKS® Function Operations; Inverse Functions

IF students score 66%–89% on the Checks, **THEN** assign: OL

- Practice Exercises 1–53 odd
- Remediation, Review Resources: Functions
- Personal Tutors
- Extra Examples 1–6
- ALEKS® Introduction to Functions

IF students score 65% or less on the Checks, **THEN** assign: AL

- Practice Exercises 1–23 odd
- Remediation, Review Resources: Functions
- *Quick Review Math Handbook*: Inverse Functions and Relations
- ALEKS® Introduction to Functions

Name ______ Period ______ Date ______

Practice

Go Online You can complete your homework online.

Example 1

For each polygon, find the inverse of the relation. Then, graph both the original relation and its inverse.

1. $\triangle MNP$ with vertices at $\{(-8, 6), (6, -2), (4, -6)\}$ See margin.

2. $\triangle XYZ$ with vertices at $\{(7, 7), (4, 9), (3, -7)\}$ See margin.

3. trapezoid $QRST$ with vertices at $\{(8, -1), (-8, -1), (-2, -8), (2, -8)\}$ See margin.

4. quadrilateral $FGHJ$ with vertices at $\{(4, 3), (-4, -4), (-3, -5), (5, 2)\}$ See margin.

Examples 2 and 3

Find the inverse of each function. Then graph the function and its inverse. If necessary, restrict the domain of the inverse so that it is a function.

7–14. See Mod. 6 Answer Appendix for graphs.

5. $f(x) = x + 2$; $f^{-1}(x) = x - 2$
6. $g(x) = 5x$; $g^{-1}(x) = \frac{1}{5}x$
7. $f(x) = -2x + 1$; $f^{-1}(x) = -\frac{x}{2} + \frac{1}{2}$
8. $h(x) = \frac{x-4}{3}$; $h^{-1}(x) = 3x + 4$
9. $f(x) = -\frac{5}{3}x - 8$; $f^{-1}(x) = -\frac{3}{5}(x + 8)$
10. $g(x) = x + 4$; $g^{-1}(x) = x - 4$
11. $f(x) = 4x$; $f^{-1}(x) = \frac{1}{4}x$
12. $f(x) = -8x + 9$; $f^{-1}(x) = -\frac{x}{8} + \frac{9}{8}$
13. $f(x) = 5x^2$ If the domain of $f(x)$ is restricted to $(-\infty, 0]$, then the inverse is $f^{-1}(x) = -\frac{\sqrt{5x}}{5}$. If the domain of $f(x)$ is restricted to $[0, \infty)$, then the inverse is $f^{-1}(x) = \frac{\sqrt{5x}}{5}$.
14. $h(x) = x^2 + 4$ If the domain of $h(x)$ is restricted to $(-\infty, 0]$, then the inverse is $h^{-1}(x) = -\sqrt{x-4}$. If the domain of $h(x)$ is restricted to $[0, \infty)$, then the inverse is $h^{-1}(x) = \sqrt{x-4}$.

Example 4

15. WEIGHT The formula to convert weight in pounds to stones is $p(x) = \frac{x}{14}$, where x is the weight in pounds.
    a. Find the inverse of $p(x)$, and describe its meaning. $p^{-1}(x) = 14x$; Sample answer: The inverse converts stones to pounds, where x is the weight in stones.
    b. Graph $p(x)$ and $p^{-1}(x)$. Use the graph to find the weight in pounds of a dog that weighs about 2.5 stones. About 35 pounds; see Mod. 6 Answer Appendix for graph.

16. CRYPTOGRAPHY DeAndre is designing a code to send secret messages. He assigns each letter of the alphabet to a number, where A = 1, B = 2, C = 3, and so on. Then he uses $c(x) = 4x - 9$ to create the secret code.
    a. Find the inverse of $c(x)$, and describe its meaning. $c^{-1}(x) = \frac{x}{4} + \frac{9}{4}$; Sample answer: The inverse can be used to convert the secret code to the original message.
    b. Make tables of $c(x)$ and $c^{-1}(x)$. Use the table to decipher the message: 15, 75, 47, 3, 71, 27, 51, 47, 67. message: FUNCTIONS; See Mod. 6 Answer Appendix for table.

Example 5

Determine whether each pair of functions are inverse functions. Write *yes* or *no*.

17. $f(x) = x - 1$; $g(x) = 1 - x$ no
18. $f(x) = 2x + 3$; $g(x) = \frac{1}{2}(x - 3)$ yes
19. $f(x) = 5x - 5$; $g(x) = \frac{1}{5}x + 1$ yes
20. $f(x) = 2x$; $g(x) = \frac{1}{2}x$ yes
21. $h(x) = 6x - 2$; $g(x) = \frac{1}{6}x + 3$ no
22. $f(x) = 8x - 10$; $g(x) = \frac{1}{8}x + \frac{5}{4}$ yes

Example 6

23. GEOMETRY The formula for the volume of a right circular cone with a height of 2 feet is $V = \frac{2}{3}\pi r^2$. Determine whether $r = \sqrt{\frac{3V}{2\pi}}$ is the inverse of the original function. yes

24. GEOMETRY The formula for the area of a trapezoid is $A = \frac{h}{2}(a + b)$. Determine whether $h = 2A - (a + b)$ is the inverse of the original function. no

Mixed Exercises

Find the inverse of each function. Then graph the function and its inverse. If necessary, restrict the domain of the inverse so that it is a function.

See Mod. 6 Answer Appendix for graphs and restrictions.

25. $y = 4$; $x = 4$
26. $f(x) = 3x$; $f^{-1}(x) = \frac{1}{3}x$
27. $f(x) = x + 2$; $f^{-1}(x) = x - 2$
28. $g(x) = 2x - 1$; $g^{-1}(x) = \frac{x+1}{2}$
29. $f(x) = \frac{1}{2}x^2 - 1$; $f^{-1}(x) = \pm\sqrt{2x + 2}$ If the domain of $f(x)$ is restricted to $(-\infty, 0]$, then the inverse is $f^{-1}(x) = \sqrt{2x + 2}$. If the domain of $f(x)$ is restricted to $[0, \infty)$, then the inverse is $f^{-1}(x) = -\sqrt{2x + 2}$.
30. $f(x) = (x + 1)^2 + 3$; $f^{-1}(x) = -1 \pm \sqrt{x - 3}$ If the domain of $f(x)$ is restricted to $(-\infty, -1]$, then the inverse is $f^{-1}(x) = -1 + \sqrt{x - 3}$. If the domain of $f(x)$ is restricted to $[-1, \infty)$, then the inverse is $f^{-1}(x) = -1 - \sqrt{x - 3}$.

Name ______________ Period ______ Date ______

Determine whether each pair of functions are inverse functions. Write *yes* or *no*.

31. $f(x) = 4x^2$
$g(x) = \frac{1}{2}\sqrt{x}$ yes

32. $f(x) = \frac{1}{3}x^2 + 1$
$g(x) = \sqrt{3x - 3}$ yes

33. $f(x) = x^2 - 9$
$g(x) = x + 3$ no

34. $f(x) = \frac{2}{3}x^3$
$g(x) = \sqrt[3]{\frac{2}{3}x}$ no

35. $f(x) = (x + 6)^2$
$g(x) = \sqrt{x} - 6$ yes

36. $f(x) = 2\sqrt{x - 5}$
$g(x) = \frac{1}{4}x^2 - 5$ no

Restrict the domain of $f(x)$ so that its inverse is also a function. State the restricted domain of $f(x)$ and the domain of $f^{-1}(x)$.

37. $f(x) = x^2 + 5$
$\{x \mid x \geq 0\}$ or $\{x \mid x \leq 0\}$; $\{x \mid x \geq 5\}$

38. $f(x) = 3x^2$
$\{x \mid x \geq 0\}$ or $\{x \mid x \leq 0\}$; $\{x \mid x \geq 0\}$

39. $f(x) = \sqrt{x + 6}$
$\{x \mid x \geq -6\}$; $\{x \mid x \geq 0\}$

40. $f(x) = \sqrt{x + 3}$
$\{x \mid x \geq -3\}$; $\{x \mid x \geq 0\}$

Sketch a graph of the inverse of each function. Then state whether the inverse is a function.

41.

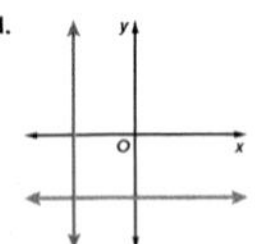

The inverse is not a function.

42.

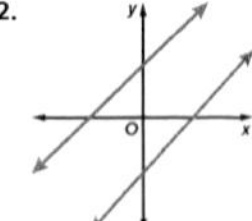

The inverse is a function.

43.

The inverse is a function.

44.

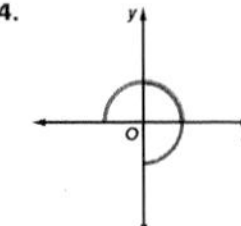

The inverse is not a function.

45. FITNESS Alejandro is a personal trainer. For his clients to gain the maximum benefit from their exercise, Alejandro calculates their maximum target heart rate using the function $f(x) = 0.85(220 - x)$, where x represents the age of the client. Find the inverse of this function and interpret its meaning in the context of the situation.
$f^{-1}(x) = 220 - \frac{x}{0.85}$; Sample answer: The inverse function gives the age of the clients based on their maximum target heart rate.

46. Graph the inverse of the piecewise function shown.

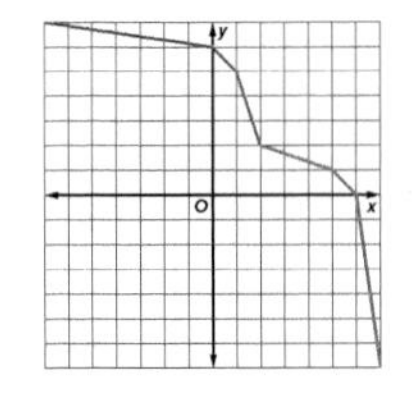

47. USE A MODEL The diagram shows a sheet of metal with squares of side length x removed from each corner which can be used to form an open box.

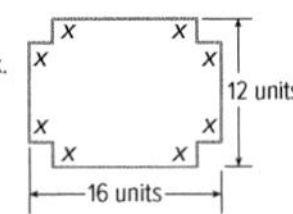

a. Write and graph the function $V(x)$ which gives the volume of the open box. Explain how the domain of this function must be restricted within the context of this scenario. See Mod. 6 Answer Appendix.

b. Does restricting the domain in **part a** allow for the inverse to also be a function? Explain your reasoning. See Mod. 6 Answer Appendix.

48. STRUCTURE Use the table to find the relationship between $(f + g)^{-1}(x)$ and $f^{-1}(x) + g^{-1}(x)$.

x	0	1	2	3
$f(x)$	0	3	1	4
$g(x)$	1	0	4	3

a. Suppose that functions $f(x)$, $g(x)$, and $(f + g)(x)$ all have inverse functions defined on the domain [0, 3]. Calculate the following values.

i. $f^{-1}(3) + g^{-1}(3)$ 4

ii. $f^{-1}(1) + g^{-1}(1)$ 2

b. Use the value of $(f + g)(1)$ to find $(f + g)^{-1}(3)$. Use the value of $(f + g)(0)$ to find $(f + g)^{-1}(1)$. See Mod. 6 Answer Appendix.

c. Joyce claims that $(f + g)^{-1}(x) = f^{-1}(x) + g^{-1}(x)$. Determine whether she is correct. Explain your reasoning. See Mod. 6 Answer Appendix.

d. Consider the functions $f(x) = 2x + 1$ and $g(x) = 2x - 1$. Compare $(f + g)^{-1}(x)$ and $f^{-1}(x) + g^{-1}(x)$. See Mod. 6 Answer Appendix.

Higher-Order Thinking Skills

49. ANALYZE If a relation is not a function, then its inverse is *sometimes*, *always*, or *never* a function. Justify your argument. Sample answer: Sometimes; $y = \pm\sqrt{x}$ is an example of a relation that is not a function, with an inverse being a function. A circle is an example of a relation that is not a function with an inverse not being a function.

50. CREATE Give an example of a function and its inverse. Verify that the two functions are inverses. Sample answer: $f(x) = 2x$, $f^{-1}(x) = 0.5x$; $f[f^{-1}(x)] = f^{-1}[f(x)] = x$

51. PERSEVERE Give an example of a function that is its own inverse.
Sample answer: $f(x) = x$ and $f^{-1}(x) = x$ or $f(x) = -x$ and $f^{-1}(x) = -x$

52. ANALYZE Can the graphs of a linear function with a slope $\neq 0$ and its inverse ever be perpendicular? Justify your answer. See Mod. 6 Answer Appendix.

53. WRITE Suppose you have a composition of two functions that are inverses. When you put in a value of 5 for x, why is the result always 5? Sample answer: One of the functions carries out an operation on 5. Then the second function that is an inverse of the first function reverses the operation on 5. Thus, the result is 5.

Answers

1. $\{(6, -8), (-2, 6), (-6, 4)\}$

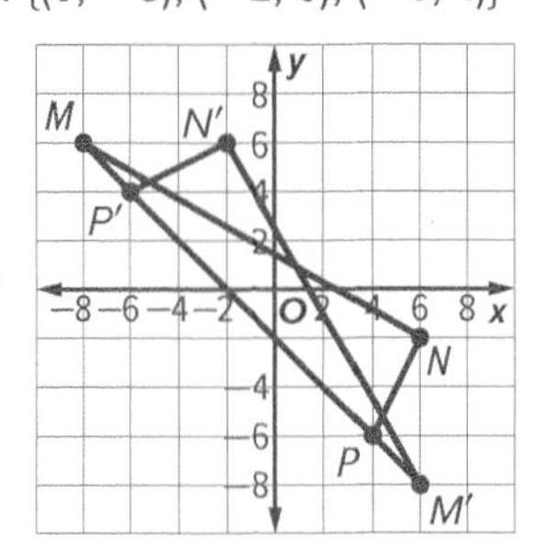

2. $\{(7, 7), (9, 4), (-7, 3)\}$

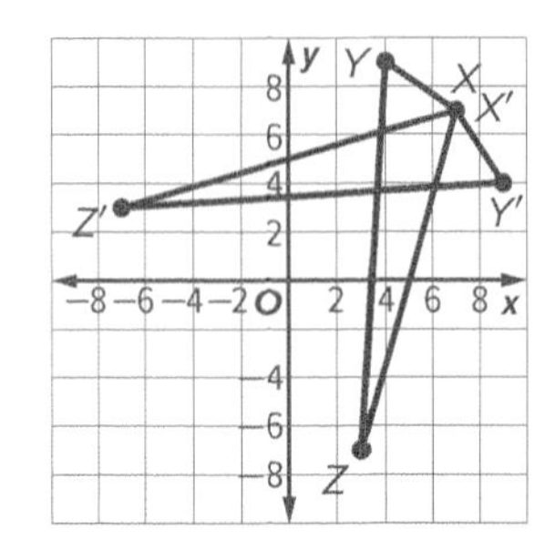

3. $\{(-1, 8), (-1, -8), (-8, -2), (-8, 2)\}$

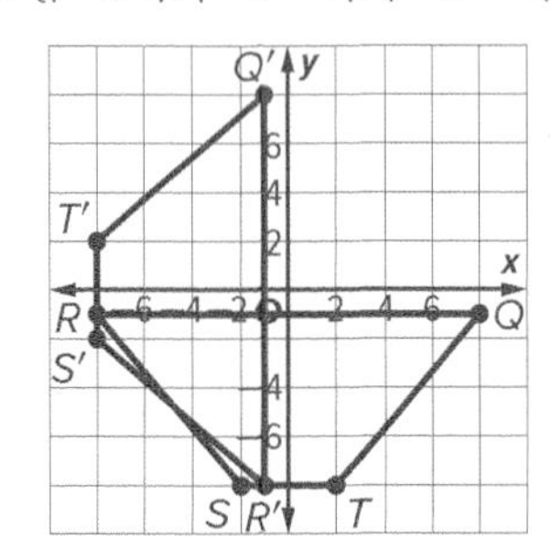

4. $\{(3, 4), (-4, -4), (-5, -3), (2, 5)\}$

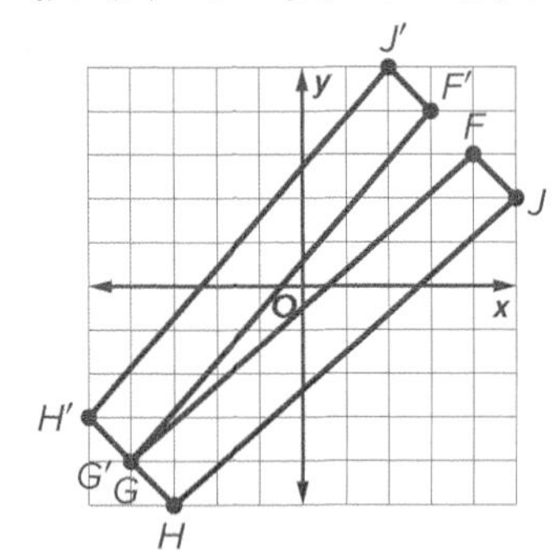

A.SSE.2

nth Roots and Rational Exponents

LESSON GOAL

Students simplify expressions involving radicals and rational exponents.

1 LAUNCH

 Launch the lesson with a **Warm Up** and an introduction.

2 EXPLORE AND DEVELOP

 **Explore:** Inverses of Rational Functions

 **Develop:**

***n*th roots**
- Find Roots
- Simplify Using Absolute Value

Rational Exponents
- Radical and Exponential Forms
- Use Rational Exponents
- Evaluate Expressions with Rational Exponents
- Simplify Expressions with Rational Exponents

 You may want your students to complete the **Checks** online.

3 REFLECT AND PRACTICE

 Exit Ticket

 Practice

DIFFERENTIATE

 View reports of student progress on the **Checks** after each example.

Resources	AL	OL	BL	ELL
Remediation: Negative Exponents	●	●		●
Extension: Approximating Square Roots		●	●	●

Language Development Handbook

Assign page 37 of the *Language Development Handbook* to help your students build mathematical language related to simplifying expressions involving radicals and rational exponents.

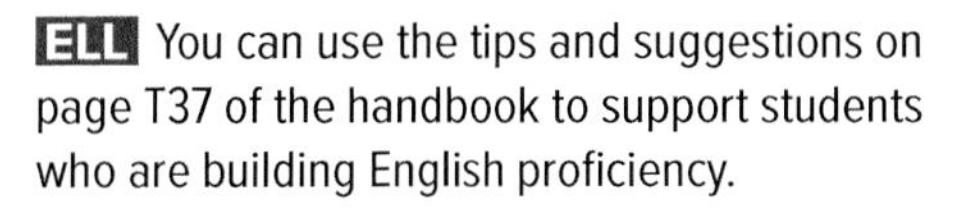

ELL You can use the tips and suggestions on page T37 of the handbook to support students who are building English proficiency.

Suggested Pacing

90 min	1 day
45 min	2 days

Focus

Domain: Algebra

Standards for Mathematical Content:

A.SSE.2 Use the structure of an expression to identify ways to rewrite it.

Standards for Mathematical Practice:

2 Reason abstractly and quantitatively.

6 Attend to precision.

7 Look for and make use of structure.

Coherence

Vertical Alignment

Previous

Students understood the relationship between functions and their inverses.

F.B.4 (Algebra 1)

Now

Students simplify expressions involving radicals and rational exponents.

A.SSE.2

Next

Students will graph and analyze square and cube root functions.

F.IF.7b, F.BF.3

Rigor

The Three Pillars of Rigor

Lessons with Explore(s)		
1 CONCEPTUAL UNDERSTANDING	2 FLUENCY	3 APPLICATION

Conceptual Bridge In this lesson, students use what they know about square roots to develop an understanding of nth roots. They build fluency by simplifying nth root expressions, and they apply their understanding by solving real-world problems related to nth roots.

Mathematical Background

In general, if n is an integer greater than or equal to 2 and a is a real number, then $\sqrt[n]{a^n} = a$ if n is odd and $\sqrt[n]{a^n} = |a|$ if n is even.

Interactive Presentation

Warm Up

1. Which are squares of a rational number?
   9 −9 0.9 0.81 0.0081
2. Which are cubes of a rational number?
   8 −8 0.8 0.08 0.008
3. Which are fourth powers of a rational number?
   16 −16 $\frac{1}{16}$ 0.0016 −625

Show Answers

Warm Up

Launch the Lesson

Astronomers are working to discover planets that could contain liquid water. The distance a planet is from its star impacts its temperature. In order for a planet to have liquid water, it must be warm enough for the water to change from a solid to a liquid, but not be so hot that the water turns to a gas. The formula

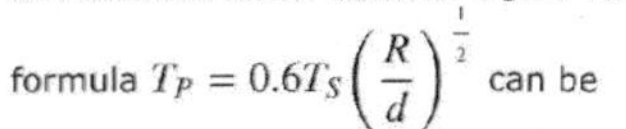

$T_P = 0.6T_S\left(\frac{R}{d}\right)^{\frac{1}{2}}$ can be used to estimate the temperature of a planet T_P, where d is the distance from the planet to its star, the temperature of the star T_S, and R is the radius of the star. You can use the properties of exponents to simplify this equation and

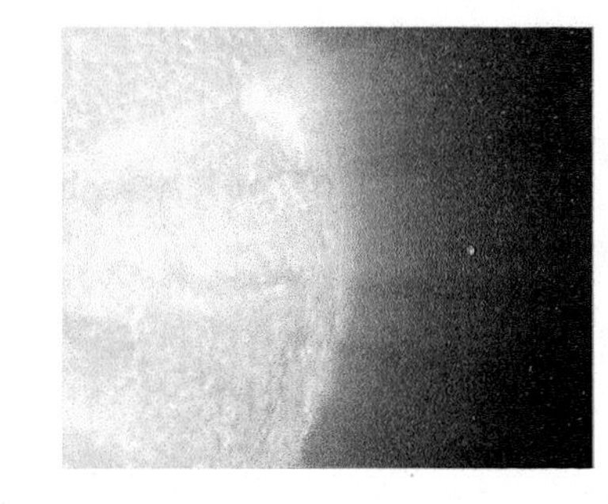

Launch the Lesson

Vocabulary

Expand All Collapse All

- nth root
- index
- radicand
- principal root
- exponential form
- radical form
- rational exponent

- What is a *rational exponent*?
- What is the connection between *rational exponents* and radicals?
- To simplify the *n*th root of an expression, what must be true about the expression?

Today's Vocabulary

Warm Up

Prerequisite Skills

The Warm Up exercises address the following prerequisite skill for this lesson:

- evaluating powers of rational numbers

Answers:

1. 9, 0.81, 0.0081

2. 8, −8, 0.008

3. 16, $\frac{1}{16}$, 0.0016

Launch the Lesson

Teaching the Mathematical Practices

4 Apply Mathematics In the Launch the Lesson, students will learn how rational exponents are used in the formula for the temperature of a planet.

Go Online to find additional teaching notes and questions to promote classroom discourse.

Today's Standards

Tell students that they will be addressing these content and practice standards in this lesson. You may wish to have a student volunteer read aloud *How can I meet these standards?* and *How can I use these practices?*, and connect these to the standards.

See the Interactive Presentation for I Can statements that align with the standards covered in this lesson.

Today's Vocabulary

Tell students that they will be using these vocabulary terms in this lesson. You can expand each row if you wish to share the definitions. Then discuss the questions below with the class.

Explore Inverses of Power Functions

Objective
Students explore *n*th roots and exponents for odd positive values of *n*.

Teaching the Mathematical Practices

3 Construct Arguments At the end of the Explore, students will use previously established results to make a conjecture about the inverses of rational functions.

7 Look for a Pattern Help students to see the pattern in the domain and range values for each pair of functions.

Ideas for Use

Recommended Use Present the Inquiry Question, or have a student volunteer read it aloud. Have students work in pairs to complete the Explore activity on their devices. Pairs should discuss each of the questions. Monitor student progress during the activity. Upon completion of the Explore activity, have student volunteers share their responses to the Inquiry Question.

What if my students don't have devices? You may choose to project the activity on a whiteboard. A printable worksheet for each Explore is available online. You may choose to print the worksheet so that individuals or pairs of students can use it to record their observations.

Summary of the Activity
Students will complete guiding exercises throughout the Explore activity. Students will work through exercises to compare two functions, *f*(*x*) and *g*(*x*), for different values of *n* to explore how the functions are related. Then, students will answer the Inquiry Question.

(continued on the next page)

Interactive Presentation

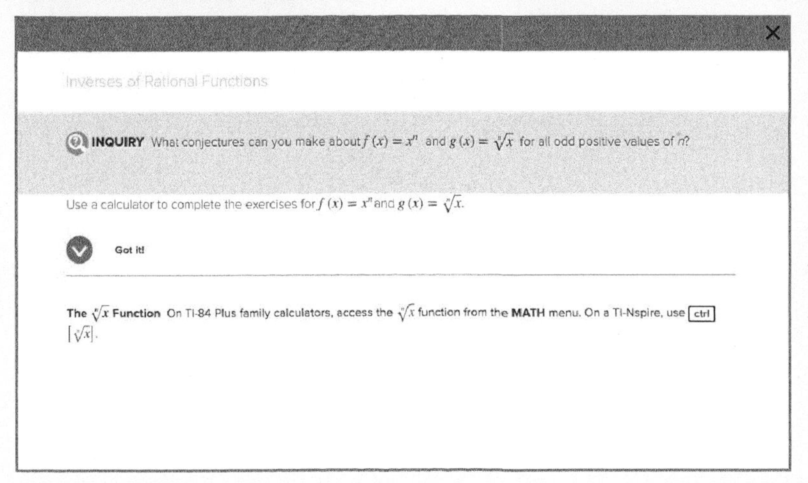

Explore

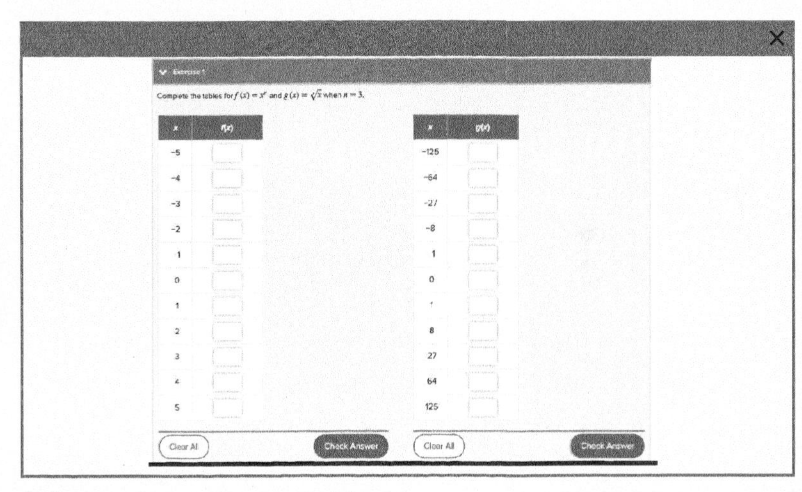

Explore

TYPE

Students complete exercises that lead them to make a power function and its inverse.

Interactive Presentation

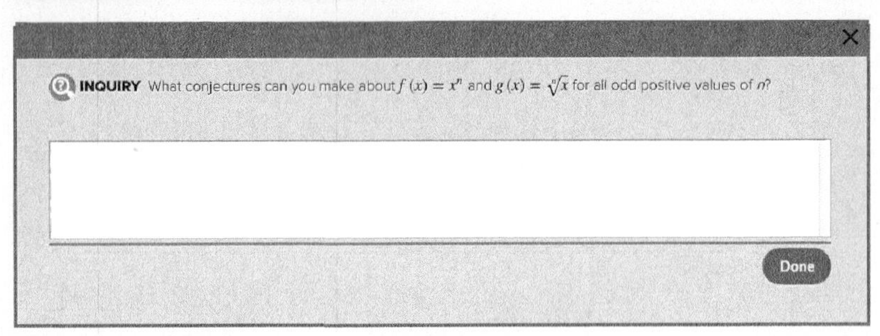

Explore

TYPE

Students respond to the Inquiry Question and can view a sample answer.

Explore Inverses of Power Functions (*continued*)

Questions

Have students complete the Explore activity.

Ask:

- Why does $\sqrt[3]{8}$ have only one answer, while $\sqrt{9}$ has two? Sample answer: For a cube root, there is only a positive root for a positive radicand and a negative root for a negative radicand. A positive number multiplied by itself three times is still positive, but a negative number cubed would be negative.
- Does the same conjecture hold true for x^n when n is even? Sample answer: The conjecture is not true when n is even because no even powered function will pass the horizontal line test. So the domain and range of the inverse cannot be the same as the domain and range of the original function.

Inquiry

What conjectures can you make about $f(x) = x^n$ and $g(x) = \sqrt[n]{x}$ for all odd positive values of n? Sample answer: For all positive odd values of n, the domain and range of $f(x)$ and $g(x)$ are all real numbers and x^n and $\sqrt[n]{x}$ are inverses of each other.

Go Online to find additional teaching notes and sample answers for the guiding exercises.

Learn nth Roots

Objective

Students simplify expressions involving radicals and rational exponents by using the properties of exponents.

Teaching the Mathematical Practices

3 Analyze Cases Work with students to analyze the number of real roots when $a > 0$, $a < 0$, and $a = 0$ for the case when n is even and when n is odd. Encourage students to familiarize themselves with all of the cases.

Things to Remember

In general, if n is an integer greater than or equal to 2 and a is a real number, then $\sqrt[n]{a^n} = a$ if n is odd and $\sqrt[n]{a^n} = |a|$ if n is even.

DIFFERENTIATE

Reteaching Activity AL

Some students tend to think that x must represent a positive number and $-x$ must represent a negative number. Reading $-x$ as "the opposite of x" should help them understand that $-x$ is 9 if $x = -9$. Also, explain that -9 has no square root that is a real number. That is, no real number can be squared to give -9. Remind students that $\sqrt{-9}$ is $3i$, an imaginary number.

Go Online

- Find additional teaching notes.
- View performance reports of the Checks.
- Assign or present an Extra Example.

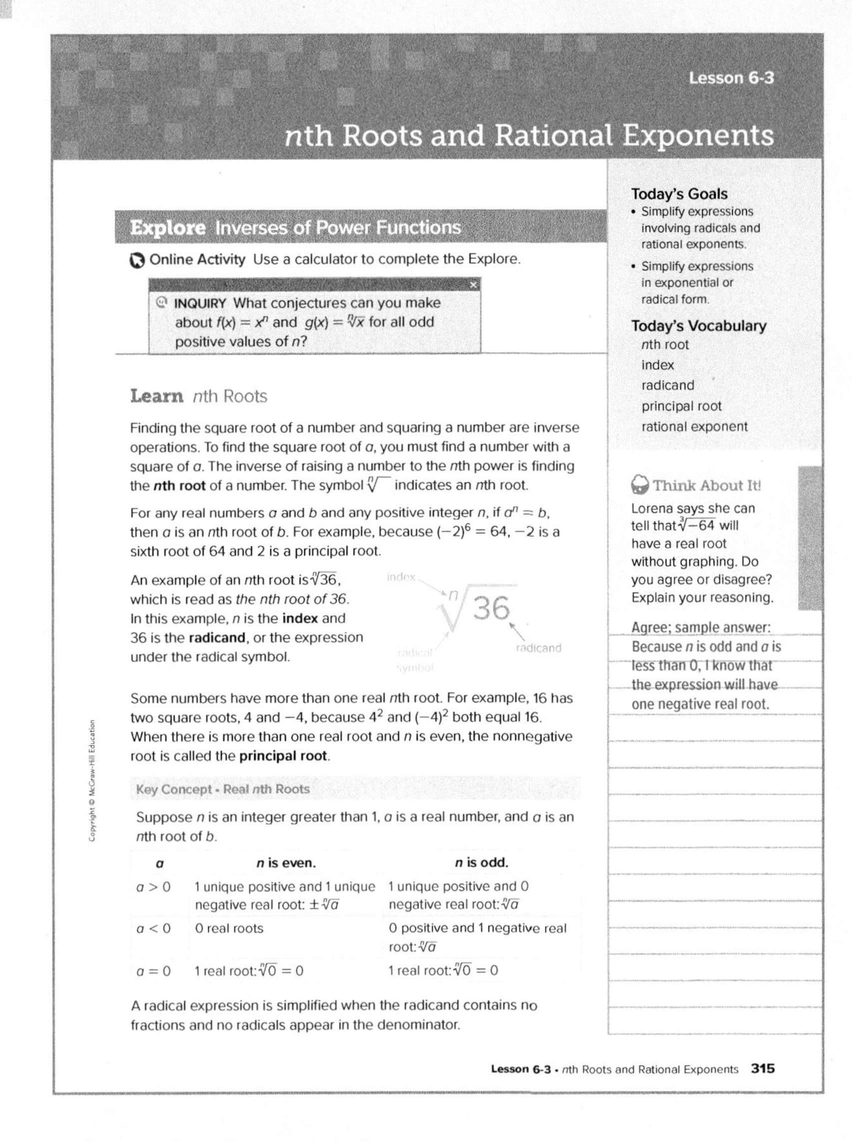

Lesson 6-3

nth Roots and Rational Exponents

Explore Inverses of Power Functions

Online Activity Use a calculator to complete the Explore.

INQUIRY What conjectures can you make about $f(x) = x^n$ and $g(x) = \sqrt[n]{x}$ for all odd positive values of n?

Today's Goals
- Simplify expressions involving radicals and rational exponents.
- Simplify expressions in exponential or radical form.

Today's Vocabulary
nth root
index
radicand
principal root
rational exponent

Learn nth Roots

Finding the square root of a number and squaring a number are inverse operations. To find the square root of a, you must find a number with a square of a. The inverse of raising a number to the nth power is finding the **nth root** of a number. The symbol $\sqrt[n]{\ }$ indicates an nth root.

For any real numbers a and b and any positive integer n, if $a^n = b$, then a is an nth root of b. For example, because $(-2)^6 = 64$, -2 is a sixth root of 64 and 2 is a principal root.

An example of an nth root is $\sqrt[n]{36}$, which is read as *the nth root of 36*. In this example, n is the **index** and 36 is the **radicand**, or the expression under the radical symbol.

index, radicand, radical symbol

Some numbers have more than one real nth root. For example, 16 has two square roots, 4 and -4, because 4^2 and $(-4)^2$ both equal 16. When there is more than one real root and n is even, the nonnegative root is called the **principal root**.

Think About It! Lorena says she can tell that $\sqrt[3]{-64}$ will have a real root without graphing. Do you agree or disagree? Explain your reasoning.

Agree; sample answer: Because n is odd and a is less than 0, I know that the expression will have one negative real root.

Key Concept • Real nth Roots

Suppose n is an integer greater than 1, a is a real number, and a is an nth root of b.

a	n is even.	n is odd.
$a > 0$	1 unique positive and 1 unique negative real root: $\pm\sqrt[n]{a}$	1 unique positive and 0 negative real root: $\sqrt[n]{a}$
$a < 0$	0 real roots	0 positive and 1 negative real root: $\sqrt[n]{a}$
$a = 0$	1 real root: $\sqrt[n]{0} = 0$	1 real root: $\sqrt[n]{0} = 0$

A radical expression is simplified when the radicand contains no fractions and no radicals appear in the denominator.

Lesson 6-3 • nth Roots and Rational Exponents 315

Interactive Presentation

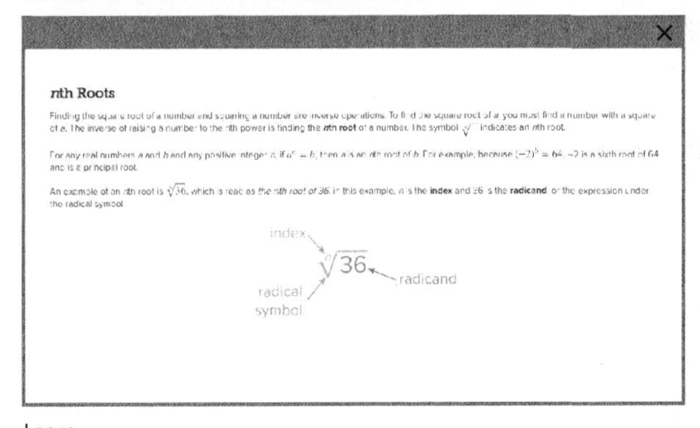

Learn

TYPE

Students answer a question to determine whether a radical expression has a real root.

Your Notes

Watch Out!

Principal Roots Because $5^2 = 25$, 25 has two square roots, 5 and -5. However, the value of $\sqrt{25}$ is 5 only. To indicate both square roots and not just the principal square root, the expression must be written as $\pm\sqrt{25}$.

Example 1 Find Roots

Simplify.

a. $\pm\sqrt{25x^4}$

$$\pm\sqrt{25x^4} = \pm\sqrt{(\underline{5x^2})^2} = \pm\ \underline{5x^2}$$

b. $-\sqrt{(y^2+7)^{12}}$

$$-\sqrt{(y^2+7)^{12}} = -\sqrt{[(y^2+7)^6]^2} = -(y^2+7)^6$$

c. $\sqrt[3]{343a^{18}b^6}$

$$\sqrt[3]{343a^{18}b^6} = \sqrt[3]{(\underline{7a^6b^2})^3} = \underline{7a^6b^2}$$

d. $\sqrt{-289c^8d^4}$

There are no real roots of $\sqrt{-289}$. However, there are two imaginary roots, $17i$ and $-17i$. Because we are only finding the principal square root, use $17i$.

$$\sqrt{-289c^8d^4} = \sqrt{\underline{-1}} \cdot \sqrt{289c^8d^4} = \underline{i} \cdot \sqrt{289c^8d^4} = \underline{17ic^4d^2}$$

Talk About It!

Compare the simplified expressions in the previous example with the ones in this example. Explain why the simplified expressions in this example require absolute value bars when the simplified expressions in the previous example did not.

Sample answer: None of the expressions in the previous example resulted in an odd power when taking an even root of an even power. However, both expressions in this example did result in an odd power when taking an even root of an even power.

Check

Write the simplified form of each expression.

a. $\pm\sqrt{196x^4}$ $\underline{\pm 14x^2}$ **b.** $-\sqrt{196x^4}$ $\underline{-14x^2}$ **c.** $\sqrt{-196x^4}$ $\underline{14ix^2}$

When you find an even root of an even power and the result is an odd power, you must use the absolute value of the result to ensure that the answer is nonnegative.

Example 2 Simplify Using Absolute Value

Simplify.

a. $\sqrt[4]{81x^4}$

$$\sqrt[4]{81x^4} = \sqrt[4]{(3x)^4} = 3|x|$$

Because x could be negative, you must use the absolute value of x to ensure that the principal square root is nonnegative.

b. $\sqrt[8]{256(y^2-2)^{24}}$

$$\sqrt[8]{256(y^2-2)^{24}} = \sqrt[8]{\underline{256}} \cdot \sqrt[8]{(y^2-2)^{24}} = \underline{2}\,|(y^2-2)^3|$$

Because $(y^2-2)^3$ could be negative, you must use the absolute value of $(y^2-2)^3$ to ensure that the principal square root is nonnegative.

316 Module 6 • Inverse and Radical Functions

Interactive Presentation

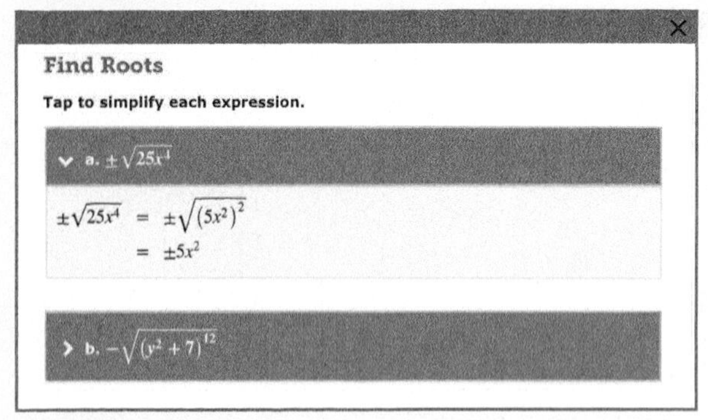

Example 1

SWIPE

Students move through the slides to simplify expressions.

TYPE

Students answer a question about principle roots.

CHECK

Students complete the Check online to determine whether they are ready to move on.

Example 1 Find Roots

Teaching the Mathematical Practices

7 Use Structure Help students to use the structure of the expressions in Example 1 to find the *n*th roots.

Questions for Mathematical Discourse

AL What should you look for when simplifying *n*th roots? Sample answer: I should check to see whether elements of the radicand can be rewritten as factors to the *n*th power.

OL Why is an odd root of a negative number a real number? When a negative number is raised to an odd power, the answer is negative.

BL What is the cube root of x^{3a}? x^a
What is the square root of x^{2a}? $|x^a|$

Example 2 Simplify Using Absolute Value

Teaching the Mathematical Practices

6 Communicate Precisely Encourage students to routinely write or explain their solutions methods. Students should use clear definitions and mathematical language when they discuss their answer to the question in the Talk About It! feature.

Questions for Mathematical Discourse

AL In part **b**, why isn't the answer $|2||(y^2-2)^3|$? because $|2| = 2$

OL Why would the absolute value be unnecessary when taking odd roots? There is only one root when taking odd roots.

BL Write a rule that explains how you find the exponent on a factor after taking the *n*th root. Sample answer: Divide the exponent in the radicand by the root.

1 CONCEPTUAL UNDERSTANDING | 2 FLUENCY | 3 APPLICATION

Learn Rational Exponents

Objective

Students simplify expressions in exponential or radical form.

1 Seek Information Mathematically proficient students must be able to transform algebraic expressions to reach solutions. Encourage students to familiarize themselves with the conditions that must be met for an expression with rational exponents to be in simplest form. Point out that when all conditions are met, they are done transforming the expression.

Example 3 Radical and Exponential Forms

 Teaching the Mathematical Practices

7 Use Structure Help students to use the structure of the expressions to write them in radical or exponential form.

Questions for Mathematical Discourse

AL What operation do you perform with the exponents when an exponent is raised to another exponent? You multiply them.

OL Why is $b^{\frac{1}{n}} = \sqrt[n]{b}$? because $\left(b^{\frac{1}{n}}\right)^n = b$

BL What is an advantage of using exponential form? Sample answer: Exponential form can make it easier to identify how to simply an expression using the rules of exponents.

Learn Rational Exponents

You can use the properties of exponents to translate expressions from exponential form to radical form or from radical form to exponential form. An expression is in **exponential form** if it is in the form x^n, where n is an exponent. An expression is in **radical form** if it contains a radical symbol.

For any real number b and a positive integer n, $b^{\frac{1}{n}} = \sqrt[n]{b}$, except where $b < 0$ and n is even. When $b < 0$ and n is even, a complex root may exist.

Examples: $125^{\frac{1}{3}} = \sqrt[3]{125}$ or 5 $\quad (-49)^{\frac{1}{2}} = \sqrt{-49}$ or $7i$

The expression $b^{\frac{1}{n}}$ has a **rational exponent**. The rules for exponents also apply to rational exponents.

Key Concept • Rational Exponents

For any nonzero number b and any integers x and y, with $y > 1$, $b^{\frac{x}{y}} = \sqrt[y]{b^x} = \left(\sqrt[y]{b}\right)^x$, except when $b < 0$ and y is even. When $b < 0$ and y is even, a complex root may exist.

Examples: $125^{\frac{2}{3}} = \left(\sqrt[3]{125}\right)^2 = 5^2$ or 25

$(-49)^{\frac{3}{2}} = (\sqrt{-49})^3 = (7i)^3$ or $-343i$

Key Concept • Simplest Form of Expressions with Rational Exponents

An expression with rational exponents is in simplest form when all of the following conditions are met.

- It has no negative exponents.
- It has no exponents that are not positive integers in the denominator.
- It is not a complex fraction.
- The index of any remaining radical is the least number possible.

Math History Minute:

Christoff Rudolff (1499–1543) wrote the first German algebra textbook. It is believed that he introduced the radical symbol $\sqrt{\ }$ in 1525 in his book *Die Coss*. Some feel that this symbol was used because it resembled a small *r*, the first letter in the Latin word *radix* or root.

Example 3 Radical and Exponential Forms

Simplify.

a. Write $x^{\frac{4}{3}}$ in radical form.

$x^{\frac{4}{3}} = \sqrt[3]{x^4}$

b. Write $\sqrt[5]{x^2}$ in exponential form.

$\sqrt[5]{x^2} = x^{\frac{2}{5}}$

Go Online You can complete an Extra Example online.

Think About It!

Write two equivalent expressions, one in radical form and one in exponential form.

Sample answer: $\sqrt[4]{16}$, $16^{\frac{1}{4}}$

Interactive Presentation

Rational Exponents

Recall that squaring a number and taking the square root of a number are inverse operations. $(\sqrt{x})^2 = x$. From the properties of exponents, you also know that $\left(x^{\frac{1}{2}}\right)^2 = x$. So, $x^{\frac{1}{2}}$ is equivalent to $\sqrt{x}$. You can use this to translate expressions from exponential form to radical form or from radical form to exponential form. An expression is in **exponential form** if it is in the form x^n, where n is an exponent. An expression is in **radical form** if it contains a radical symbol.

For any real number b and a positive integer n, $b^{\frac{1}{n}} = \sqrt[n]{b}$, except where $b < 0$ and n is even. When $b < 0$ and n is even, a complex root may exist.

$125^{\frac{1}{3}} = \sqrt[3]{125}$ or 5 $\quad (-49)^{\frac{1}{2}} = \sqrt{-49}$ or $7i$

The expression $b^{\frac{1}{n}}$ has a **rational exponent**. The rules for exponents also apply to rational exponents.

For any nonzero number b and any integers x and y with $y > 1$, $b^{\frac{x}{y}} = \sqrt[y]{b^x} = \left(\sqrt[y]{b}\right)^x$, except when $b < 0$ and y is even. When $b < 0$ and y is even, a complex root may exist.

$125^{\frac{2}{3}} = \left(\sqrt[3]{125}\right)^2 = 5^2$ or 25 $\quad (-49)^{\frac{3}{2}} = \left(\sqrt{-49}\right)^3 = (7i)^3$ or $-343i$

Learn

TYPE

Students explain how to solve an equation using the rules of exponents.

Think About It!
Why did you set t equal to $\frac{1}{4}$?

Sample answer: t is the time in years, and 3 months is equal to $\frac{1}{4}$ of a year.

Think About It!
How can you tell if $x^{\frac{y}{z}}$ will simplify to an integer?

Sample answer: If you can rewrite x as an integer to the power of z, it will simplify to an integer

Go Online to see more examples of evaluating expressions with rational exponents.

Watch Out!
Exponents Recall that when you multiply powers, the exponents are added, and when you raise a power to a power, the exponents are multiplied.

Think About It!
How would the expression in part **c** change if the exponents were $\frac{1}{3}$ and $-\frac{3}{4}$?

Sample answer: The expression would not have a positive exponent, so I would need to follow the process used in part **b**.

Example 4 Use Rational Exponents

FINANCIAL LITERACY The expression $c(1 + r)^t$ can be used to estimate the future cost of an item due to inflation, where c represents the current cost of the item, r represents the annual rate of inflation, and t represents the time in years. Write the expression in radical form for the future cost of an item 3 months from now if the annual rate of inflation is 4.7%.

$c(1 + r)^t = c(1 + 0.047)^{\frac{1}{4}}$ — $r = 0.047, t = \frac{1}{4}$

$= c(\underline{1.047})^{\frac{1}{4}}$ — Add.

$= c\sqrt[4]{1.047}$ — $b^{\frac{1}{n}} = \sqrt[n]{b}$

Example 5 Evaluate Expressions with Rational Exponents

Evaluate $32^{-\frac{2}{5}}$.

$32^{-\frac{2}{5}} = \frac{1}{32^{\frac{2}{5}}}$ — $b^{-n} = \frac{1}{b^n}$

$= \frac{1}{(2^5)^{\frac{2}{5}}}$ — $32 = 2^5$

$= \frac{1}{2^{5 \cdot \frac{2}{5}}}$ — Power of a Power

$= \frac{1}{2^2}$ or $\frac{1}{4}$ — Multiply the exponents.

Example 6 Simplify Expressions with Rational Exponents

Simplify each expression.

a. $x^{\frac{2}{3}} \cdot x^{\frac{1}{6}}$

$x^{\frac{2}{3}} \cdot x^{\frac{1}{6}} = x^{\frac{2}{3} + \frac{1}{6}}$ — Add powers.

$= x^{\frac{4}{6} + \frac{1}{6}}$ — $\frac{2}{3} = \frac{4}{6}$

$= x^{\frac{5}{6}}$ — Add the exponents.

b. $y^{-\frac{2}{3}}$

$y^{-\frac{2}{3}} = \frac{1}{y^{\frac{2}{3}}}$ — $b^{-n} = \frac{1}{b^n}$

$= \frac{1}{y^{\frac{2}{3}}} \cdot \frac{y^{\frac{1}{3}}}{y^{\frac{1}{3}}}$ — $\frac{y^{\frac{1}{3}}}{y^{\frac{1}{3}}} = 1$

$= \frac{y^{\frac{1}{3}}}{y^{\frac{3}{3}}}$ or $\underline{\frac{y^{\frac{1}{3}}}{y}}$ — $y^{\frac{2}{3}} \cdot y^{\frac{1}{3}} = y^{\frac{2}{3} + \frac{1}{3}}$

c. $z^{-\frac{1}{3}} \cdot z^{\frac{3}{4}}$

$z^{-\frac{1}{3}} \cdot z^{\frac{3}{4}} = z^{-\frac{1}{3} + \frac{3}{4}}$ — Add powers.

$= \underline{z^{-\frac{4}{12} + \frac{9}{12}}}$ or $z^{\frac{5}{12}}$ — $-\frac{1}{3} = \frac{4}{12}, \frac{3}{4} = \frac{9}{12}$

Go Online You can complete an Extra Example online.

318 **Module 6** • Inverse and Radical Functions

Interactive Presentation

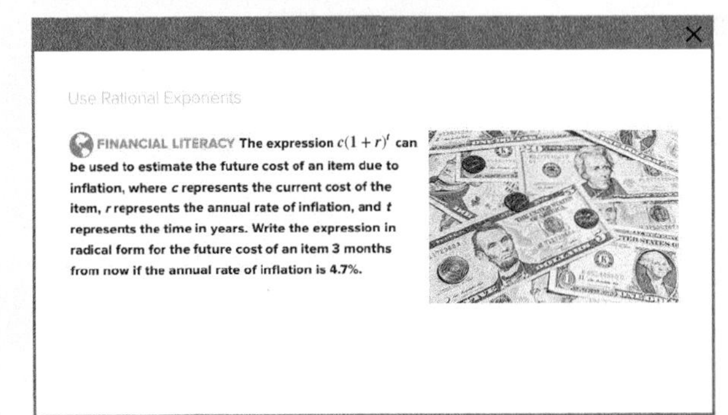

Example 4

TYPE

Students answer a question about units used in the example.

CHECK

Students complete the Check online to determine whether they are ready to move on.

Example 4 Use Rational Exponents

Teaching the Mathematical Practices

2 Consider Units Students must note the units involved in this problem to help them substitute the correct value for t in the expression. The Think About It! features asks students to explain how the units affect the value of t.

Questions for Mathematical Discourse

AL What would be the value of t if we were considering the value of the item 8 months from now? $\frac{2}{3}$

OL When does this equation have integer exponents? The exponent is an integer when the time is a whole number of years.

BL What equation could you solve to find when the cost would double? $2 = (1 + r)^t$

Example 5 Evaluate Expressions with Rational Exponents

Teaching the Mathematical Practices

8 Look for a Pattern Help student to see the pattern in how expressions with rational exponents are evaluated to generalize how you can tell if the expressions will simplify to an integer.

Questions for Mathematical Discourse

AL When evaluating a rational exponent, which part of the exponent is the index of the root? the denominator

OL Why does $32^{-\frac{2}{5}} = \frac{1}{32}^{\frac{2}{5}}$? Explain your answer. Sample answer: Multiplying either form by $32^{\frac{2}{5}}$ yields 1.

BL What is the multiplicative inverse of $a^{\frac{c}{d}}$? Explain. $a^{-\frac{c}{d}}$ because $a^{\frac{c}{d}} \cdot a^{-\frac{c}{d}} = a^0 = 1$.

Example 6 Simplify Expressions with Rational Exponents

Teaching the Mathematical Practices

7 Use Structure Guide students to use the structure of the expressions in Example 6 to write them in simplest form.

Questions for Mathematical Discourse

AL What do you need to do before adding or subtracting fractions? Ensure they have the same denominator.

OL In part **b**, why do we multiply the numerator and denominator by $y^{\frac{1}{3}}$? to rationalize the denominator so that the final form meets the rules for simplest form

BL How could you solve $x^{\frac{a}{b}} \cdot x^{\frac{c}{d}} = x$ for a? Write an equation for the exponents and solve for a. $\frac{a}{b} + \frac{c}{d} = 1$

Exit Ticket

Recommended Use

At the end of class, go online to display the Exit Ticket prompt and ask students to respond using a separate piece of paper. Have students hand you their responses as they leave the room.

Alternate Use

At the end of class, go online to display the Exit Ticket prompt and ask students to respond verbally or by using a mini-whiteboard. Have students hold up their whiteboards so that you can see all student responses. Tap to reveal the answer when most or all students have completed the Exit Ticket.

Practice and Homework

Suggested Assignments

Use the table below to select appropriate exercises.

DOK	Topic	Exercises
1, 2	exercises that mirror the examples	1–30
2	exercises that use a variety of skills from this lesson	31–56
2	exercises that extend concepts learned in this lesson to new contexts	57–60
3	exercises that emphasize higher-order and critical thinking skills	61–68

ASSESS AND DIFFERENTIATE

Use the data from the **Checks** to determine whether to provide resources for extension, remediation, or intervention.

IF students score 90% or more on the Checks, **THEN** assign: BL

- Practice Exercises 1-59 odd, 61–68
- Extension: Approximating Square Roots
- **ALEKS**® Simplifying Expressions, Rational Exponents

IF students score 66%–89% or more on the Checks, **THEN** assign: OL

- Practice Exercises 1–67 odd
- Remediation, Review Resources: Negative Exponents
- Personal Tutors
- Extra Examples 1–6
- **ALEKS**® Negative Exponents

IF students score 65% or less on the Checks, **THEN** assign: AL

- Practice Exercises 1–29 odd
- Remediation, Review Resources: Negative Exponents
- *Quick Review Math Handbook:* *n*th Roots
- **ALEKS**® Negative Exponents

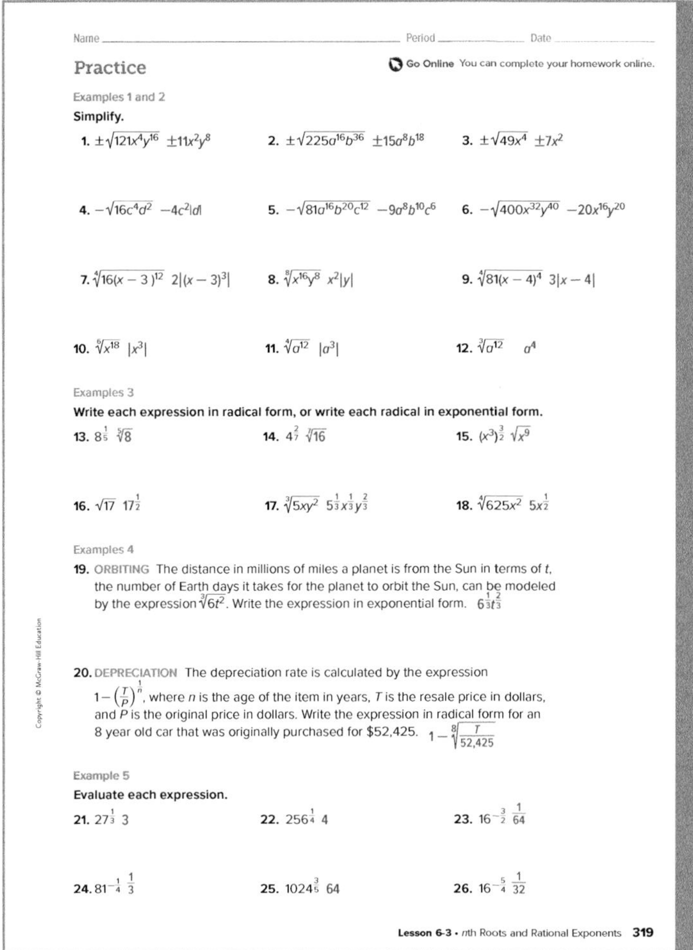

Name ______ Period ______ Date ______

Practice

Go Online You can complete your homework online.

Examples 1 and 2

Simplify.

1. $\pm\sqrt{121x^4y^{16}}$ $\pm 11x^2y^8$
2. $\pm\sqrt{225a^{16}b^{36}}$ $\pm 15a^8b^{18}$
3. $\pm\sqrt{49x^4}$ $\pm 7x^2$
4. $-\sqrt{16c^4d^2}$ $-4c^2|d|$
5. $-\sqrt{81a^{16}b^{20}c^{12}}$ $-9a^8b^{10}c^6$
6. $-\sqrt{400x^{32}y^{40}}$ $-20x^{16}y^{20}$
7. $\sqrt[4]{16(x-3)^{12}}$ $2|(x-3)^3|$
8. $\sqrt[8]{x^{16}y^8}$ $x^2|y|$
9. $\sqrt[4]{81(x-4)^4}$ $3|x-4|$
10. $\sqrt[6]{x^{18}}$ $|x^3|$
11. $\sqrt[4]{a^{12}}$ $|a^3|$
12. $\sqrt[3]{a^{12}}$ a^4

Examples 3

Write each expression in radical form, or write each radical in exponential form.

13. $8^{\frac{1}{5}}$ $\sqrt[5]{8}$
14. $4^{\frac{2}{7}}$ $\sqrt[7]{16}$
15. $(x^3)^{\frac{3}{2}}$ $\sqrt{x^9}$
16. $\sqrt{17}$ $17^{\frac{1}{2}}$
17. $\sqrt[3]{5xy^2}$ $5^{\frac{1}{3}}x^{\frac{1}{3}}y^{\frac{2}{3}}$
18. $\sqrt[4]{625x^2}$ $5x^{\frac{1}{2}}$

Examples 4

19. ORBITING The distance in millions of miles a planet is from the Sun in terms of t, the number of Earth days it takes for the planet to orbit the Sun, can be modeled by the expression $\sqrt[3]{6t^2}$. Write the expression in exponential form. $6^{\frac{1}{3}}t^{\frac{2}{3}}$

20. DEPRECIATION The depreciation rate is calculated by the expression $1-\left(\frac{T}{P}\right)^{\frac{1}{n}}$, where n is the age of the item in years, T is the resale price in dollars, and P is the original price in dollars. Write the expression in radical form for an 8 year old car that was originally purchased for \$52,425. $1-\sqrt[8]{\frac{T}{52{,}425}}$

Example 5

Evaluate each expression.

21. $27^{\frac{1}{3}}$ 3
22. $256^{\frac{1}{4}}$ 4
23. $16^{-\frac{3}{2}}$ $\frac{1}{64}$
24. $81^{-\frac{1}{4}}$ $\frac{1}{3}$
25. $1024^{\frac{3}{5}}$ 64
26. $16^{-\frac{5}{4}}$ $\frac{1}{32}$

Copyright © McGraw-Hill Education

Lesson 6-3 • *n*th Roots and Rational Exponents 319

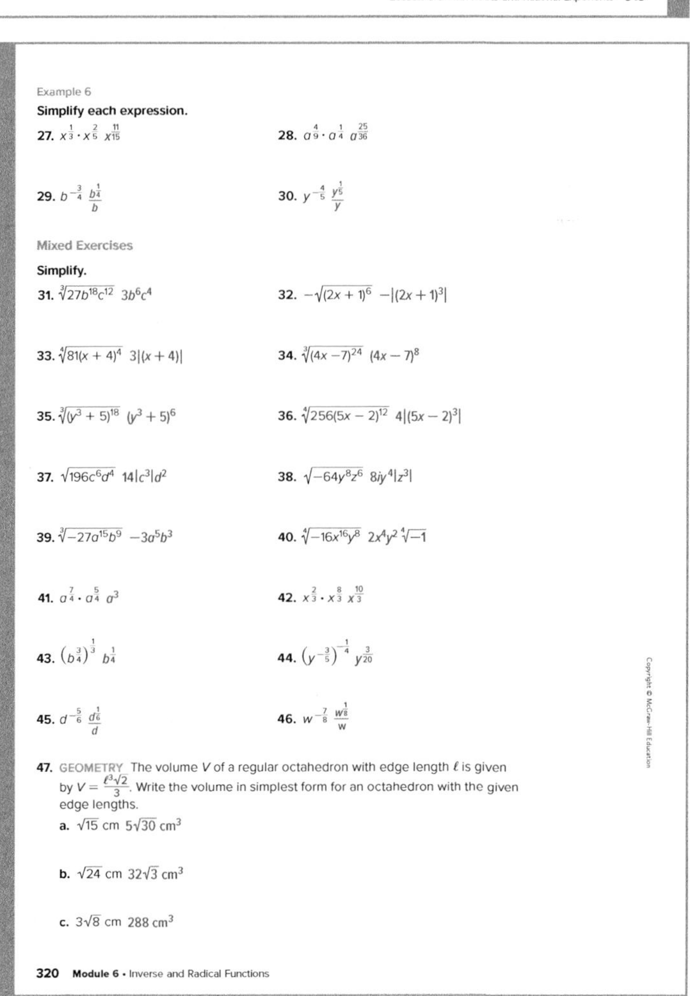

Example 6

Simplify each expression.

27. $x^{\frac{1}{3}} \cdot x^{\frac{2}{5}}$ $x^{\frac{11}{15}}$
28. $a^{\frac{4}{9}} \cdot a^{\frac{1}{4}}$ $a^{\frac{25}{36}}$
29. $b^{-\frac{3}{4}}$ $\frac{b^{\frac{1}{4}}}{b}$
30. $y^{-\frac{4}{5}}$ $\frac{y^{\frac{1}{5}}}{y}$

Mixed Exercises

Simplify.

31. $\sqrt[3]{27b^{18}c^{12}}$ $3b^6c^4$
32. $-\sqrt{(2x+1)^6}$ $-|(2x+1)^3|$
33. $\sqrt[4]{81(x+4)^4}$ $3|(x+4)|$
34. $\sqrt[3]{(4x-7)^{24}}$ $(4x-7)^8$
35. $\sqrt[3]{(y^3+5)^{18}}$ $(y^3+5)^6$
36. $\sqrt[4]{256(5x-2)^{12}}$ $4|(5x-2)^3|$
37. $\sqrt{196c^6d^4}$ $14|c^3|d^2$
38. $\sqrt{-64y^8z^6}$ $8iy^4|z^3|$
39. $\sqrt[3]{-27a^{15}b^9}$ $-3a^5b^3$
40. $\sqrt[4]{-16x^{16}y^8}$ $2x^4y^2\sqrt[4]{-1}$
41. $a^{\frac{7}{4}} \cdot a^{\frac{5}{4}}$ a^3
42. $x^{\frac{2}{3}} \cdot x^{\frac{8}{3}}$ $x^{\frac{10}{3}}$
43. $\left(b^{\frac{3}{4}}\right)^{\frac{1}{3}}$ $b^{\frac{1}{4}}$
44. $\left(y^{-\frac{3}{5}}\right)^{-\frac{1}{4}}$ $y^{\frac{3}{20}}$
45. $d^{-\frac{5}{6}}$ $\frac{d^{\frac{1}{6}}}{d}$
46. $w^{-\frac{7}{8}}$ $\frac{w^{\frac{1}{8}}}{w}$

47. GEOMETRY The volume V of a regular octahedron with edge length ℓ is given by $V = \frac{\ell^3\sqrt{2}}{3}$. Write the volume in simplest form for an octahedron with the given edge lengths.

a. $\sqrt{15}$ cm $5\sqrt{30}$ cm^3

b. $\sqrt{24}$ cm $32\sqrt{3}$ cm^3

c. $3\sqrt{8}$ cm 288 cm^3

Copyright © McGraw-Hill Education

320 Module 6 • Inverse and Radical Functions

Name ____________ Period ________ Date ________

Simplify.

48. $\dfrac{f^{-\frac{1}{4}}}{4f^{\frac{1}{2}} \cdot f^{-\frac{1}{3}}}$ $\quad \dfrac{f^{\frac{7}{12}}}{4f}$

49. $\dfrac{c^{\frac{2}{3}}}{c^{\frac{1}{6}}}$ $\quad c^{\frac{1}{2}}$

50. $\dfrac{z^{\frac{4}{5}}}{z^{\frac{1}{2}}}$ $\quad z^{\frac{3}{10}}$

51. $\sqrt[8]{36h^4j^4}$ $\quad 6^{\frac{1}{4}}h^{\frac{1}{2}}j^{\frac{1}{2}}$

52. $\dfrac{ab}{\sqrt{c}}$ $\quad \dfrac{ab\sqrt{c}}{c}$

53. $\dfrac{xy}{\sqrt[3]{z}}$ $\quad \dfrac{xy\sqrt[3]{z^2}}{z}$

54. SPORTS A volleyball has a volume of 864π cm^3. A tennis ball has a volume of 32π cm^3. By how much does the radius of the volleyball exceed that of the tennis ball? Write your answer using rational exponents.
$4\left(3^{\frac{1}{3}}\right)$cm

55. WATER TOWER One of the largest sphere water towers in the country is located in Edmond, Oklahoma. It is 218 feet tall and holds 500,000 gallons, or about 66,840 cubic feet of water. Another town is planning to build a similar water tower. However, the new water tower will hold $\frac{3}{5}$ as much water as the tower in Edmond. Determine the radius of the new water tower to the nearest foot.
21 feet

56. CELLS The number of cells in a cell culture grows exponentially. The number of cells in the culture as a function of time is given by the expression $N\left(\frac{6}{5}\right)^t$, where t is measured in hours and N is the initial size of the culture. Write the following expressions in radical form.

a. the number of cells after 20 minutes with N initial cells $N \cdot \sqrt[3]{\frac{6}{5}}$

b. the number of cells after 44 minutes with N initial cells $N \cdot \sqrt[15]{\left(\frac{6}{5}\right)^{11}}$

c. the number of cells after 1 hour and 15 minutes with 4000 initial cells $4000 \cdot \sqrt[4]{\left(\frac{6}{5}\right)^5}$

57. REASONING Simplify $\sqrt[b]{m^{3b}}$, where $b > 0$. Explain your reasoning. See margin.

58. REGULARITY There are no real *n*th roots of a number *w*. What can you conclude about the index and the number *w*? The index *n* is even and *w* is negative.

59. CONSTRUCT ARGUMENTS Determine the values of x for which $\sqrt{x^2} \neq x$. Explain your answer. $x < 0$; If $x < 0$, then $\sqrt{x^2} = -x$.

60. STRUCTURE Which of the following functions are equivalent? Justify your answer. See margin.

a. $f(x) = \sqrt[3]{x^9}$ **b.** $g(x) = \sqrt{x^6}$ **c.** $r(x) = (\sqrt[3]{x})^9$ **d.** $s(x) = (\sqrt{x})^6$

Higher-Order Thinking Skills

61. WRITE Explain how it might be easier to simplify an expression using rational exponents rather than using radicals. See margin.

62. FIND THE ERROR Destiny and Kimi are simplifying $\sqrt[4]{16x^4y^8}$. Is either of them correct? Explain your reasoning.
Kimi; sample answer: Destiny's error was keeping the y^2 inside the absolute value symbol.

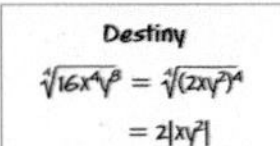

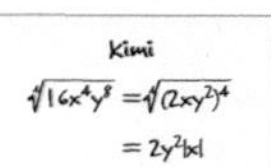

63. PERSEVERE Under what conditions is $\sqrt{x^2 + y^2} = x + y$ true?
when x or $y = 0$ and the other variable is ≥ 0

64. ANALYZE Determine whether the statement $\sqrt[4]{(-x)^4} = x$ is *sometimes, always,* or *never* true. Justify your argument.
Sometimes; sample answer: When $x = -3$, $\sqrt[4]{(-x)^4} = |(-x)|$ or 3. When $x = 3$, $\sqrt[4]{(-x)^4} = |3|$ or 3.

65. PERSEVERE For what real values of x is $\sqrt[3]{x} > x$? $0 < x < 1, x < -1$

66. WRITE Explain when and why absolute value symbols are needed when taking an *n*th root. See margin.

67. PERSEVERE Write an equivalent expression for $\sqrt[3]{2x} \cdot \sqrt[3]{8y}$. Simplify the radical. $2\sqrt[3]{2xy}$

68. CREATE Find two different expressions that equal 2 in the form $x^{\frac{1}{a}}$. Sample answer: $4^{\frac{1}{2}}$ and $16^{\frac{1}{4}}$

Answers

57. $\sqrt[b]{m^{3b}} = m^3$

60. $f(x)$ and $r(x)$ are equivalent. They simplify to x^3. $g(x)$ is different and simplifies to $|x^3|$. $s(x)$ is almost the same as x^3, but the function is not defined for negative values of x.

61. Sample answer: It may be easier to simplify an expression when it has rational exponents because all the properties of exponents apply. We do not have as many properties dealing directly with radicals. However, we can convert all radicals to rational exponents, and then use the properties of exponents to simplify.

66. Sample answer: They are needed to ensure that the answer is not a negative number. When we take any odd root of a number, we find that there is just one answer. If the number is positive, the root is positive. If the number is negative, the root is negative. Absolute value symbols are never needed when finding odd roots. Every positive real number has two *n*th roots when *n* is even; one of these roots is positive and one is negative. Negative real numbers do not have *n*th roots when *n* is even. When finding even *n*th roots, absolute value symbols are sometimes necessary to ensure the principal root is the result if it is possible the root could be negative.

Graphing Radical Functions

LESSON GOAL

Students graph and analyze square and cube root functions.

1 LAUNCH

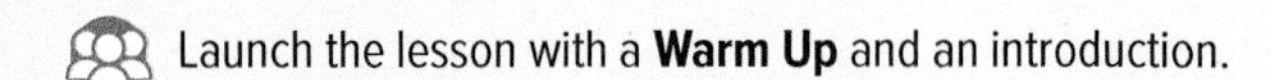

Launch the lesson with a **Warm Up** and an introduction.

2 EXPLORE AND DEVELOP

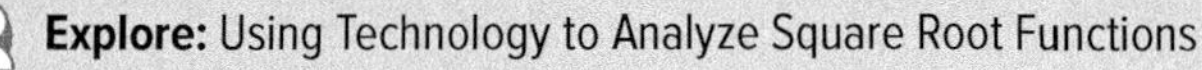

Explore: Using Technology to Analyze Square Root Functions

Develop:

Graphing Square Root Functions
- Identify Domain and Range Algebraically
- Graph a Transformed Square Root Function
- Analyze the Graph of a Square Root Function
- Graph a Square Root Inequality

Graphing Cube Root Functions
- Graph Cube Root Functions
- Compare Radical Functions
- Write a Radical Function

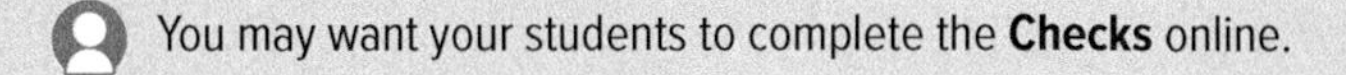

You may want your students to complete the **Checks** online.

3 REFLECT AND PRACTICE

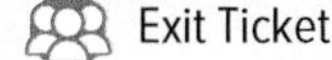

Exit Ticket

Practice

DIFFERENTIATE

View reports of student progress on the **Checks** after each example.

Resources	AL	OL	BL	ELL
Remediation: Rational Exponents	●	●		●
Extension: Relativistic Mass		●	●	●

Language Development Handbook

Assign page 38 of the *Language Development Handbook* to help your students build mathematical language related to graphing and analyzing square and cube root functions.

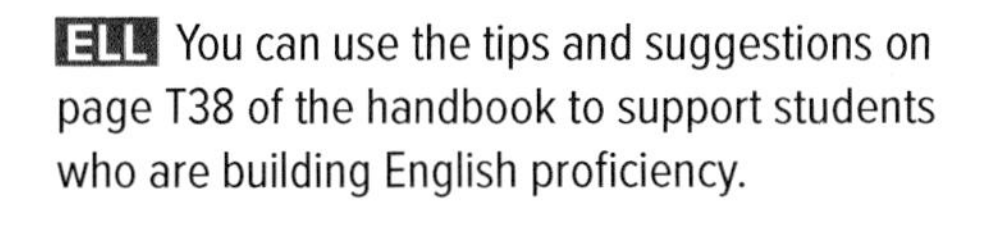

ELL You can use the tips and suggestions on page T38 of the handbook to support students who are building English proficiency.

Suggested Pacing

90 min	1 day
45 min	2 days

Focus

Domain: Functions

Standards for Mathematical Content:

F.IF.7b Graph square root, cube root, and piecewise-defined functions, including step functions and absolute value functions.

F.BF.3 Identify the effect on the graph of replacing $f(x)$ by $f(x) + k$, $k\,f(x)$, $f(kx)$, and $f(x + k)$ for specific values of k (both positive and negative); find the value of k given the graphs. Experiment with cases and illustrate an explanation of the effects on the graph using technology. Include recognizing even and odd functions from their graphs and algebraic expressions for them.

Standards for Mathematical Practice:

1 Make sense of problems and persevere in solving them.
3 Construct viable arguments and critique the reasoning of others.
5 Use appropriate tools strategically.

Coherence

Vertical Alignment

Previous
Students understood square and cube roots and simplified expressions involving radicals and rational exponents. **8.EE.2, A.SSE.2**

Now
Students graph and analyze square and cube root functions.
F.IF.7b, F.BF.3

Next
Students will simplify and perform operations with radical expressions.
A.SSE.2

Rigor

The Three Pillars of Rigor

1 CONCEPTUAL UNDERSTANDING	2 FLUENCY	3 APPLICATION

Conceptual Bridge In this lesson, students develop an understanding of rational functions and build fluency by graphing them. They apply their understanding of graphing rational functions by solving real-world problems.

Mathematical Background

A function that contains a variable inside a square root symbol is called a square root function. The domains and ranges of these functions are limited to values for which the function is defined.

Interactive Presentation

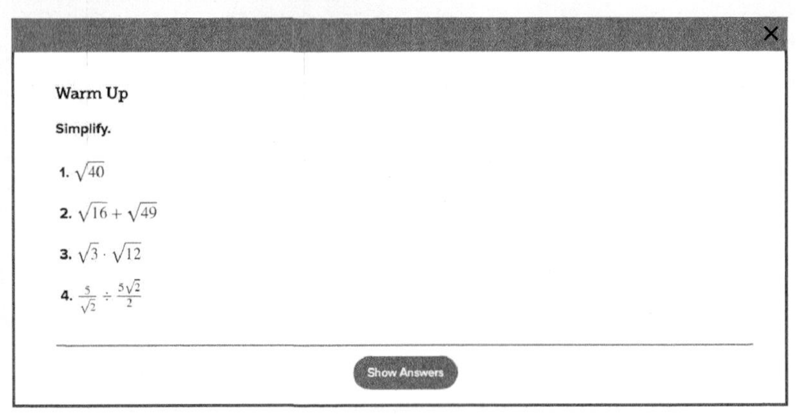

Warm Up

Warm Up

Prerequisite Skills

The Warm Up exercises address the following prerequisite skill for this lesson:

- simplifying square roots

Answers:

1. $2\sqrt{10}$

2. 11

3. 6

4. 1

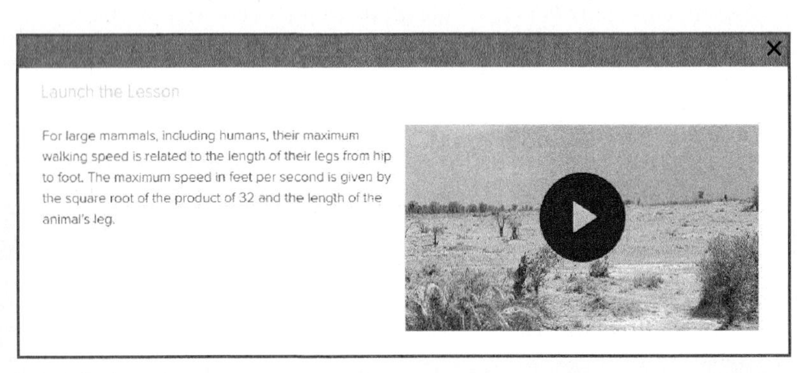

Launch the Lesson

Launch the Lesson

 Teaching the Mathematical Practices

2 Create Representations Guide students to define variables and write an equation that models the situation in the Launch the Lesson. Then ask students to use the equation to find the maximum walking speed of an animal with a given leg length.

Today's Standards

Tell students that they will be addressing these content and practice standards in this lesson. You may wish to have a student volunteer read aloud *How can I meet these standards?* and *How can I use these practices?*, and connect these to the standards.

See the Interactive Presentation for I Can statements that align with the standards covered in this lesson.

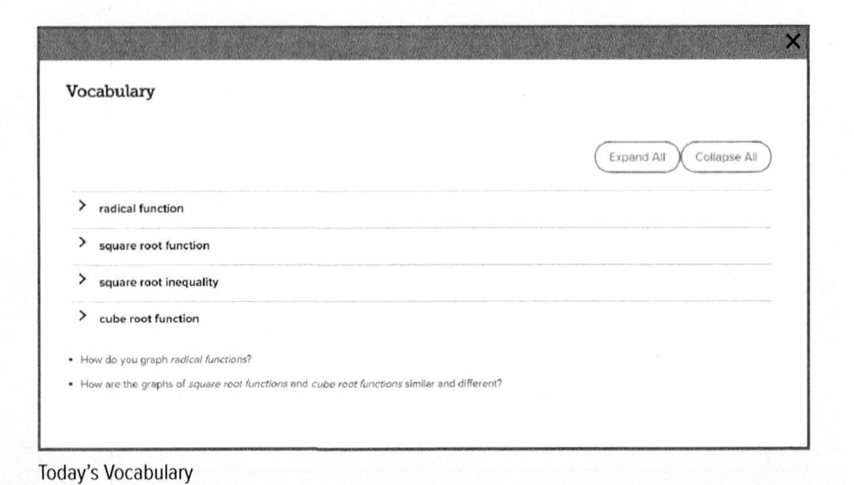

Today's Vocabulary

Today's Vocabulary

Tell students that they will be using these vocabulary terms in this lesson. You can expand each row if you wish to share the definitions. Then discuss the questions below with the class.

Explore Using Technology to Analyze Square Root Functions

Objective

Students use a graphing calculator to explore transformations and key features of square root functions.

Teaching the Mathematical Practices

5 Analyze Graphs Throughout the Explore, students will analyze the graphs they have created using graphing calculators to draw conclusions about transformations of square root functions.

Ideas for Use

Recommended Use Present the Inquiry Question, or have a student volunteer read it aloud. Have students work in pairs to complete the Explore activity on their devices. Pairs should discuss each of the questions. Monitor student progress during the activity. Upon completion of the Explore activity, have student volunteers share their responses to the Inquiry Question.

What if my students don't have devices? You may choose to project the activity on a whiteboard. A printable worksheet for each Explore is available online. You may choose to print the worksheet so that individuals or pairs of students can use it to record their observations.

Summary of the Activity

Students will complete guiding exercises throughout the Explore activity. Students will use a graphing calculator to investigate how transforming a parent function affects the key features of the graph of the function. Key features in the exercises will include *x*- and *y*-intercepts, symmetry, domain, range and end behavior. Then, students will answer the Inquiry Question.

(continued on the next page)

Interactive Presentation

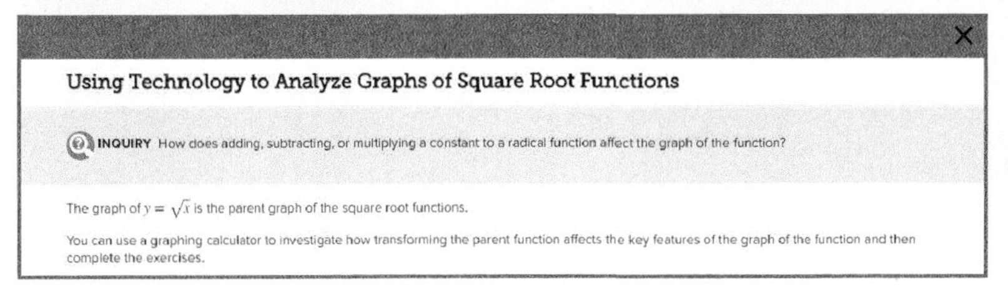

Explore

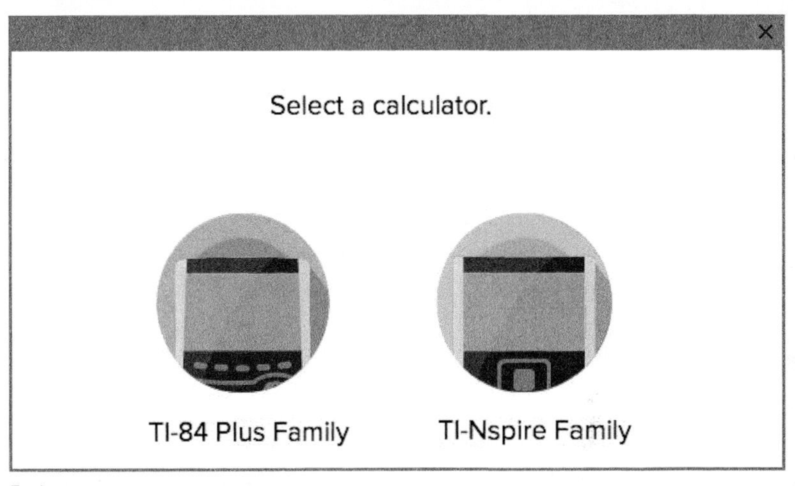

Explore

Students move through the slides to see how a graphing calculator can be used to analyze transformations.

Students analyze the key features of functions.

Interactive Presentation

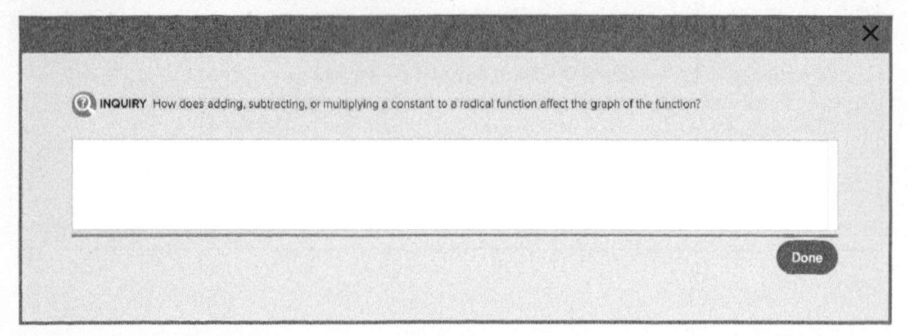

Explore

TYPE

a|

Students respond to the Inquiry Question and can view a sample answer.

Explore Using Technology to Analyze Square Root Functions (*continued*)

Questions

Have students complete the Explore activity.

Ask:

- What is an easy way to determine if a shift will be horizontal or vertical? Sample answer: A horizontal shift happens when a constant is added to or subtracted from x, so look under the radical. Otherwise, a constant added or subtracted will be a vertical shift.
- Does multiplying by a positive number ever change the end behavior of a function? Why or why not? No; sample answer: it may change the steepness of the function, but it will not change the end behavior. As the x approaches positive or negative infinity, the function is still going to have either the same or different behavior.

Inquiry

How does adding, subtracting, or multiplying a constant to a radical function affect the graph of the function? Sample answer: Adding or subtracting a constant causes the graph to be shifted horizontally or vertically. Multiplying by a constant stretches or compresses the graph. Multiplying by a negative constant causes the graph to be reflected in the x-axis.

Go Online to find additional teaching notes and sample answers for the guiding exercises.

1 CONCEPTUAL UNDERSTANDING | 2 FLUENCY | 3 APPLICATION

Learn Graphing Square Root Functions

Objective

Students graph and analyze square root functions.

Teaching the Mathematical Practices

3 Construct Arguments The Think About It! feature asks students to use stated assumptions, definitions, and previously established results to construct an argument about the domain of the parent square root function.

7 Interpret Complicated Expressions Mathematically proficient students can see complicated expressions as single objects or as being composed of several objects. Guide students to see what information they can gather about square root functions just from looking at the constants in the functions.

Essential Question Follow-Up

Students have explored square root functions.

Ask:

Why would you choose a square root function to model a set of data instead of a polynomial function? Sample answer: The end behavior of a square root function might fit the data better than a polynomial function. Also, the domain of the square root function is $x \geq 0$, which may be applicable to many real-world situations that involve quantities such as time and distance.

Go Online

- Find additional teaching notes.
- View performance reports of the Checks.
- Assign or present an Extra Example.

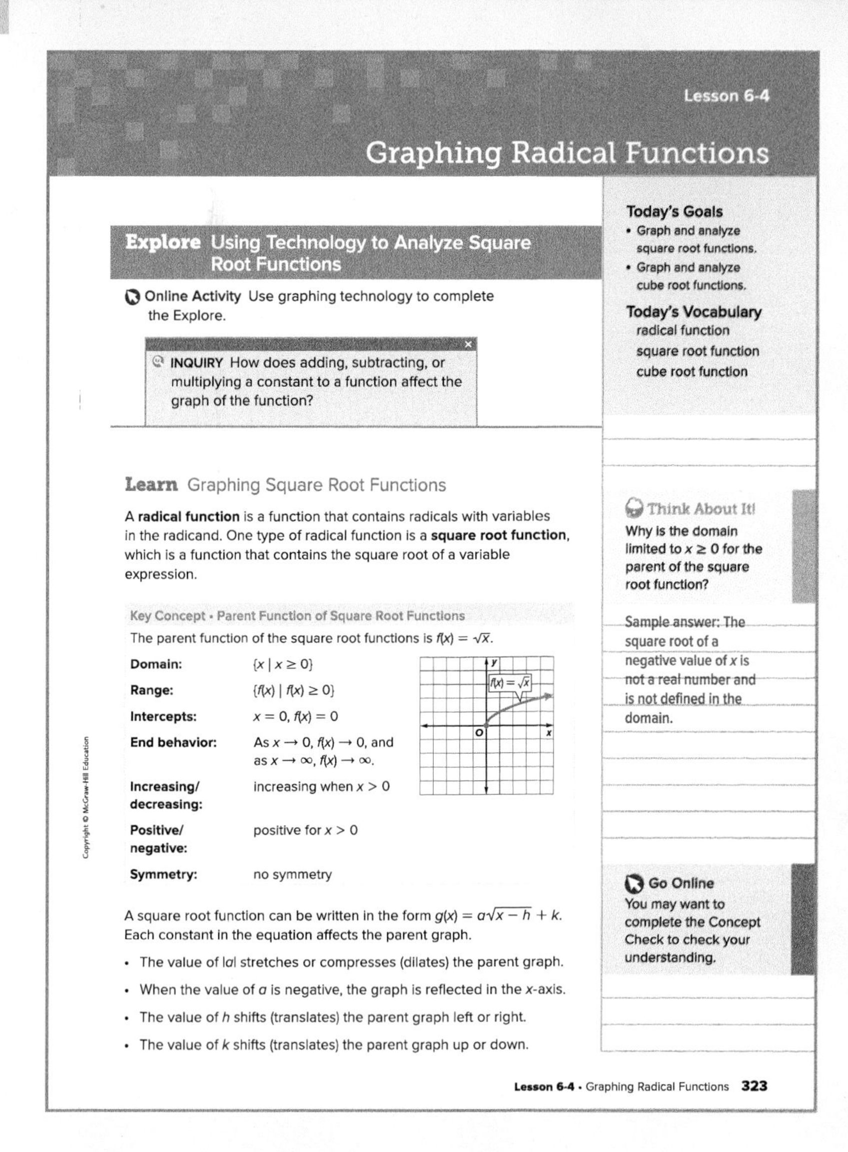

Lesson 6-4

Graphing Radical Functions

Explore Using Technology to Analyze Square Root Functions

Online Activity Use graphing technology to complete the Explore.

INQUIRY How does adding, subtracting, or multiplying a constant to a function affect the graph of the function?

Learn Graphing Square Root Functions

A **radical function** is a function that contains radicals with variables in the radicand. One type of radical function is a **square root function**, which is a function that contains the square root of a variable expression.

Key Concept • Parent Function of Square Root Functions

The parent function of the square root functions is $f(x) = \sqrt{x}$.

Domain:	$\{x \mid x \geq 0\}$
Range:	$\{f(x) \mid f(x) \geq 0\}$
Intercepts:	$x = 0, f(x) = 0$
End behavior:	As $x \to 0, f(x) \to 0$, and as $x \to \infty, f(x) \to \infty$.
Increasing/decreasing:	increasing when $x > 0$
Positive/negative:	positive for $x > 0$
Symmetry:	no symmetry

A square root function can be written in the form $g(x) = a\sqrt{x - h} + k$. Each constant in the equation affects the parent graph.

- The value of $|a|$ stretches or compresses (dilates) the parent graph.
- When the value of a is negative, the graph is reflected in the x-axis.
- The value of h shifts (translates) the parent graph left or right.
- The value of k shifts (translates) the parent graph up or down.

Today's Goals
- Graph and analyze square root functions.
- Graph and analyze cube root functions.

Today's Vocabulary
radical function
square root function
cube root function

Think About It!
Why is the domain limited to $x \geq 0$ for the parent of the square root function?

Sample answer: The square root of a negative value of x is not a real number and is not defined in the domain.

Go Online
You may want to complete the Concept Check to check your understanding.

Copyright © McGraw-Hill Education

Lesson 6-4 • Graphing Radical Functions 323

Interactive Presentation

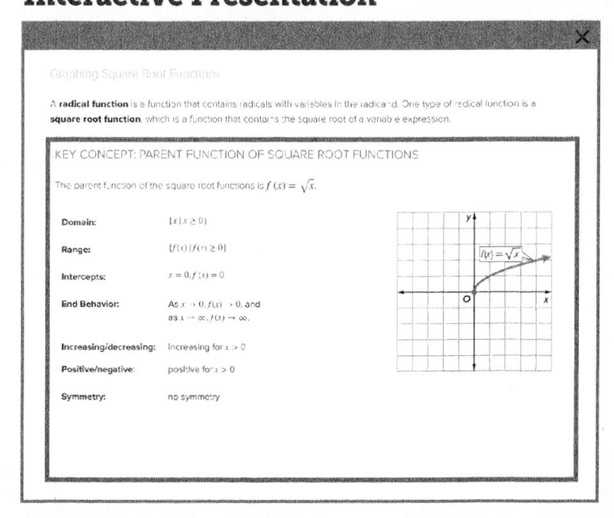

Learn

TYPE

Students answer a question to determine whether they understand the limitations on the domain for a square root function.

SELECT

Students select how adding, subtracting, or multiplying the function by a number affects the graph.

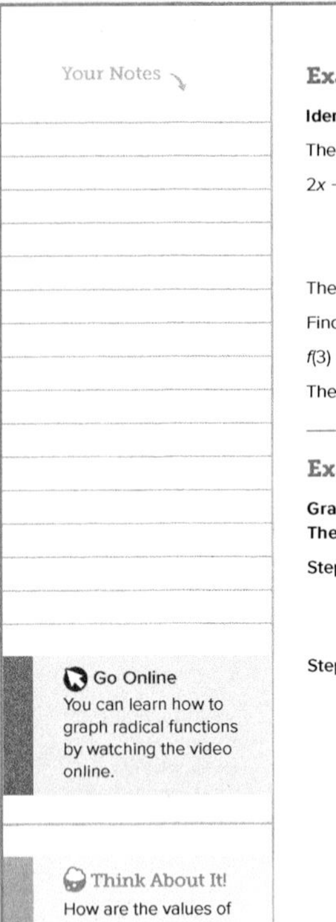

Your Notes

Example 1 Identify Domain and Range Algebraically

Identify the domain and range of $f(x) = \sqrt{2x - 6} + 1$.

The domain is restricted to values for which the radicand is nonnegative.

$2x - 6 \geq 0$ Write an inequality using the radicand.

$2x \geq 6$ Add 6 to each side.

$x \geq 3$ Divide each side by 2.

The domain is $\{x \mid x \geq 3\}$.

Find $f(3)$ to determine the lower limit of the range.

$f(3) = \sqrt{2(3) - 6} + 1$ or 1

The range is $\{f(x) \mid f(x) \geq 1\}$.

Example 2 Graph a Transformed Square Root Function

Graph $g(x) = -3\sqrt{x + 1} + 2$, and identify the domain and range. Then describe how it is related to the graph of the parent function.

Step 1 Determine the minimum domain value.

$x + 1 \geq 0$ Write an inequality using the radicand.

$x \geq -1$ Simplify.

Step 2 Make a table and graph.

Use x-values determined from **Step 1** to make a table.

x	$g(x)$
−1	2
0	−1
1	≈ −2.2
2	≈ −3.2
3	−4

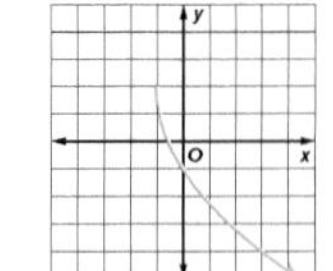

The domain is $\{x \mid x \geq -1\}$ and the range is $\{g(x) \mid g(x) \leq 2\}$.

Step 3 Compare $g(x)$ to the parent function.

The maximum is $(-1, 2)$.

Because $f(x) = \sqrt{x}$, $g(x) = a\sqrt{x - h} + k$ where $a = -3$, $h = -1$, and $k = 2$.

$a < 0$ and $|a| > 1$, so the graph of $f(x) = \sqrt{x}$ is reflected in the x-axis and stretched vertically by a factor of $|a|$, or 3.

$h < 0$, so the graph is then translated left $|h|$ units, or 1 unit.

$k > 0$, so the graph is then translated up k units, or 2 units.

Go Online You can complete an Extra Example online.

Go Online You can learn how to graph radical functions by watching the video online.

Think About It! How are the values of h and k related to the domain and range in this example?

Sample answer: In this example, the domain is $\{x \mid x \geq -1\}$ or $\{x \mid x \geq h\}$ and the range is $\{g(x) \mid g(x) \leq 2\}$ or $\{g(x) \mid g(x) \leq k\}$.

Interactive Presentation

Identify Domain and Range Algebraically

Identify the domain and range of $f(x) = \sqrt{2x - 6} + 1$.

The domain is restricted to values for which the radicand is nonnegative.

$2x - 6 \geq 0$ Write an inequality using the radicand.

$2x \geq 6$ Add 6 to each side.

$x \geq 3$ Divide each side by 2.

The domain is $\{x|x \geq 3\}$. Find $f(3)$ to determine the lower limit of the range.

$f(3) = \sqrt{2(3) - 6} + 1$ or 1

The range is $\{f(x)|f(x) \geq 1\}$.

Example 1

EXPAND

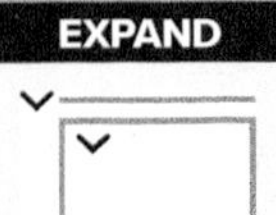

Students can tap to see how to avoid a common error when identifying domain and range of a square root function.

Example 1 Identify Domain and Range Algebraically

Teaching the Mathematical Practices

7 Interpret Complicated Expressions Mathematically proficient students can see complicated expressions as single objects or as being composed of several objects. Guide students to use the radicand of the function to determine the domain of the function.

Questions for Mathematical Discourse

AL Why do we set the radicand of a square root function greater than or equal to 0 to find the domain? The domain and range can only contain real numbers, and a negative radicand yields imaginary numbers.

OL What is another way to identify the lower limit of the range? Sample answer: The lower limit of the radicand is zero, so you can find the value of the function when the radical is zero.

BL How would the domain and range change if there were absolute value symbols around the radicand? The domain would be all real numbers, and the range would still be $\{f(x)|f(x) \geq 1\}$.

Example 2 Graph a Transformed Square Root Function

Teaching the Mathematical Practices

1 Explain Correspondences In Example 2, students must explain the relationships between the transformed function, graph, and parent function.

Questions for Mathematical Discourse

AL Why are negative values of x part of the domain? x can be negative as long as the radicand is not negative; all real numbers greater than or equal to −1 result in a nonnegative radicand.

OL What would the equation of $g(x) = -3\sqrt{x + 1} + 2$ reflected across the x-axis be? $g(x) = 3\sqrt{x + 1} - 2$

BL How would you change $g(x)$ to reflect it in the y-axis? Distribute a negative to all terms in the radicand.

Example 3 Analyze the Graph of a Square Root Function

Teaching the Mathematical Practices

4 Use Tools In Example 3, students will need to identify important quantities in the problem and map their relationships using a function, table, and graph.

Questions for Mathematical Discourse

AL What does multiplying by 104.23 do to the parent function? This stretches the function vertically.

OL Do the designations of the independent and dependent variable make sense in the context? Explain. Yes; sample answer: While the force generated depends on the RPM of the machine, in this example we are determining what RPM value is needed to achieve a given force. As the function is one-to-one, it can be considered either way.

BL Describe the end behavior in the context of the problem. As the gravitational force increases, the RPMs of the centrifuge must increase.

Example 3 Analyze the Graph of a Square Root Function

BLOOD DONATION **When blood is donated, medical professionals use a centrifuge to separate it. The centrifuge spins the blood, causing it to separate into three components, which are red cells, platelets, and plasma. In order to efficiently separate the blood, the centrifuge must spin at a specified rate, measured in rotations per minute (RPM), for the required gravitation force, or g-force, exerted on the blood. For a centrifuge with a radius of 7.8 centimeters, the RPM setting of the centrifuge is determined by the product of 104.23 and the square root of the g-force required.**

Part A Write and graph the function.

Complete the table to write the function.

Words	The RPM setting of the centrifuge	is	the product of 104.23	and the square root of the g-force required.
Variables	Let g represent the force and r represent the RPM setting.			
Function	r	$=$	$104.23 \cdot$	$\sqrt{g}$

Make a table to graph the function.

g	r
0	0
400	2085
800	2948
1200	3611
1600	4169

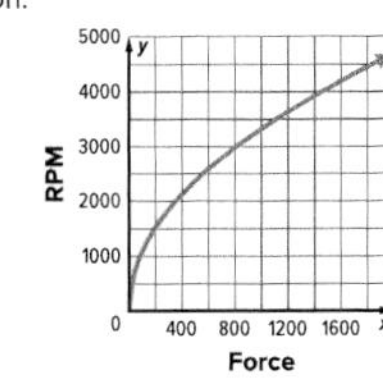

Part B Describe key features of the function.

Domain: $\{g \mid g \geq 0\}$

Range: $\{r \mid r \geq 0\}$

x-intercept: 0

y-intercept: 0

Increasing/Decreasing: increasing as $g \to \infty$

Positive/Negative: positive for $g > 0$

End Behavior: As $g \to \infty$, $r \to \infty$.

Go Online You can complete an Extra Example online.

Think About It! What do the domain and range mean in the context of the situation?

Sample answer: In this situation, the domain and range indicate that both the force and RPM must be greater than or equal to zero.

Lesson 6-4 • Graphing Radical Functions 325

Interactive Presentation

Analyze the Graph of a Square Root Function

BLOOD DONATION When blood is donated, medical professionals use a centrifuge to separate it. The centrifuge spins the blood, causing it to separate into three components, which are red cells, platelets, and plasma. In order to efficiently separate the blood, the centrifuge must spin at a specified rate, measured in rotations per minute (RPM), for the required gravitation force, or g-force, exerted on the blood. For a centrifuge with a radius of 7.8 centimeters, the RPM setting of the centrifuge is determined by the product of 104.23 and the square root of the g-force required.

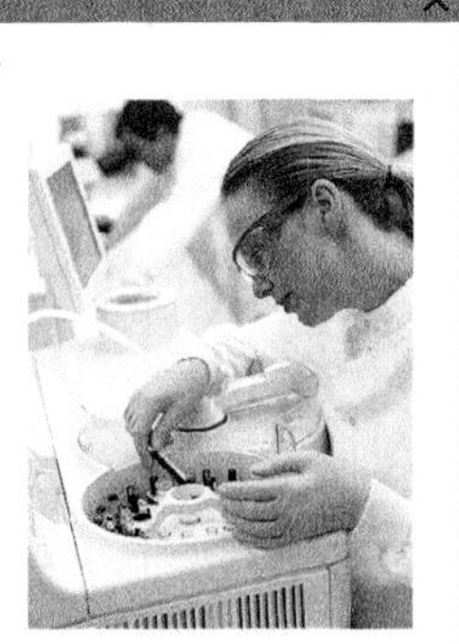

Example 3

SELECT

Students select variables and operations to write an equation to represent the situation.

TYPE

Students answer a question about interpreting the domain and range of a square root function in context.

Example 4 Graph a Square Root Inequality

Graph $y < \sqrt{2x + 5}$.

Step 1 Graph the related function.

Graph the boundary $y = \sqrt{2x + 5}$, using a dashed line because the inequality is <.

Step 2 Shade.

The domain is $\{x \mid x > -2.5\}$. Because the inequality is less than, shade the region below the boundary and within the domain.

Select a test point in the shaded region to verify the solution.

Watch Out!

Test Point When selecting a test point, make sure the point is within the domain of the related function.

Learn Graphing Cube Root Functions

A **cube root function** is a radical function that contains the cube root of a variable expression.

Key Concept • Parent Function of Cube Root Functions

The parent function of the cube root functions is $f(x) = \sqrt[3]{x}$.

Domain:	all real numbers
Range:	all real numbers
Intercepts:	$x = 0, f(x) = 0$
End behavior:	As $x \to -\infty, f(x) \to -\infty$, and as $x \to \infty, f(x) \to \infty$.
Increasing/ decreasing:	increasing as $x \to \infty$
Positive/ negative:	positive for $x > 0$ negative for $x < 0$
Symmetry:	symmetric about the origin

A cube root function can be written in the form $g(x) = a\sqrt[3]{x - h} + k$.

Talk About It!

Describe how the domain and range differ for a radical function with an odd index and a radical function with an even index.

Sample answer: The domain and range of a radical function with an odd index are unrestricted. For a radical function with an even index, the domain is restricted so that the radicand is greater than or equal to 0. The restriction of the domain causes the range to be restricted.

Example 5 Graph Cube Root Functions

Graph each function. State the domain and range.

a. $g(x) = \frac{1}{3}\sqrt[3]{x}$

In $g(x) = a\sqrt[3]{x - h} + k$, $a = \frac{1}{3}$, $h = 0$, and $k = 0$. So the function is centered at the origin and vertically compressed.

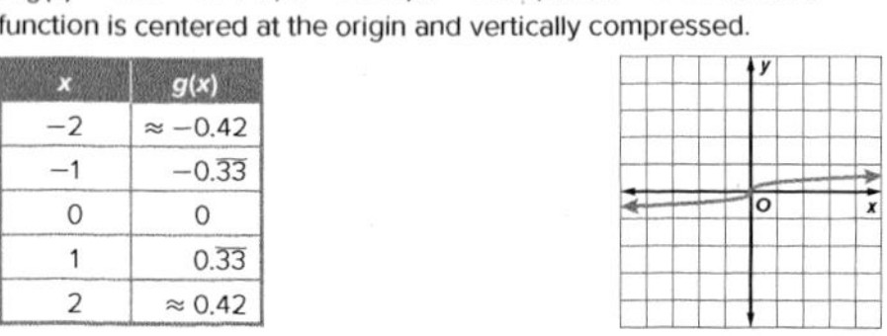

x	$g(x)$
−2	≈ −0.42
−1	−0.33
0	0
1	0.33
2	≈ 0.42

The domain is all real numbers, and the range is all real numbers.

Go Online You can learn how to graph radical functions on a graphing calculator by watching the video online.

Interactive Presentation

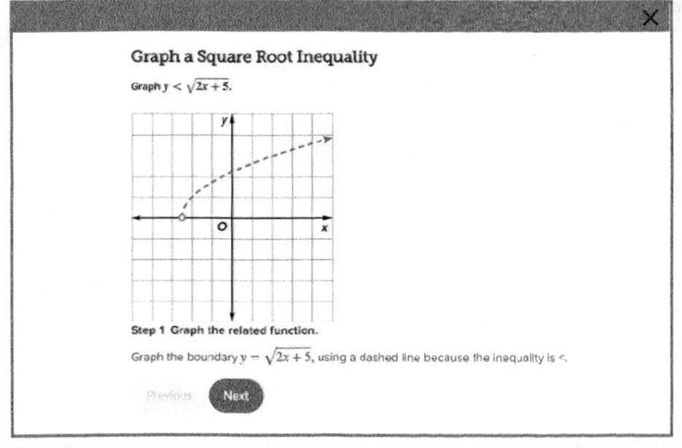

Example 4

TAP

Students move through the steps to see how to graph a square root inequality.

CHECK

Students complete the Check online to determine whether they are ready to move on.

Example 4 Graph a Square Root Inequality

Teaching the Mathematical Practices

1 Explain Correspondences Encourage students to explain the relationships between the inequality and graph in Example 4.

Questions for Mathematical Discourse

AL How do you know where to stop shading on the left side of the graph? The radical expression does not output real values less than $x = -2.5$, so the shading stops along this boundary.

OL What points $(2.5, y)$ satisfy the inequality? $y < 0$

BL What points should not be used as test points for this inequality? Points on the boundary line and points in which $x < -2.5$ should not be used.

Learn Graphing Cube Root Functions

Objective

Students graph and analyze cube root functions.

Teaching the Mathematical Practices

3 Analyze Cases The Talk About It! feature asks students to examine the domain and range of radical function for cases when the index is even and when it is odd. Encourage students to familiarize themselves with both cases.

Example 5 Graph Cube Root Functions

Teaching the Mathematical Practices

7 Use Structure Guide students to use the structure of the cube root functions to identify transformations and graph the functions.

Questions for Mathematical Discourse

AL Is a cube root an odd function or an even function? odd

OL Where would you use a negative to reflect a cube root function across the x-axis? The negative would be outside the radical. Where would you use a negative to reflect a cube root function across the y-axis? The negative would be inside the radical, distributed to all terms in the radicand.

BL Compare transformations of radical functions and transformations of polynomial functions. Sample answer: The transformations of both function types follow the same form. A difference is needing to take domain restrictions of even roots into account along with the transformations.

DIFFERENTIATE

Language Development Activity AL ELL

IF students continue to struggle with graphing square root inequalities, **THEN** have students work in pairs or small groups to discuss how to graph square root inequalities. Students should discuss how to determine the domain and range of an inequality, how to determine whether the boundary is solid or dashed, and how to determine where to shade the inequality.

Example 6 Compare Radical Functions

Teaching the Mathematical Practices

1 Explain Correspondences Encourage students to explain the relationships between the key features of the function and graph in Example 6.

Questions for Mathematical Discourse

AL What kind of function does $q(x)$ appear to be? Sample answer: square root function

OL Assuming that $q(x)$ can be written in the general form for transformed functions, what is the value of h for each function? $h = 6$ for both functions.

BL If $q(x)$ is a square root function, is $|a|$ greater than or less than 1? greater than

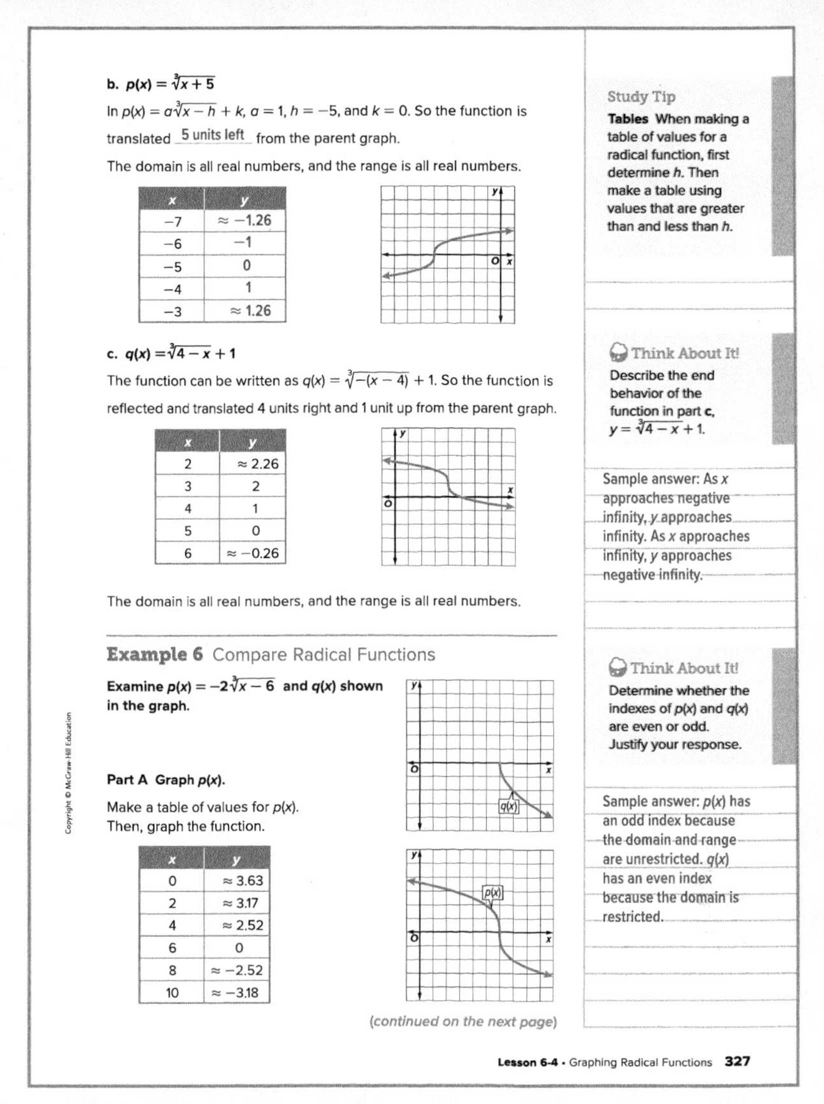

b. $p(x) = \sqrt[3]{x+5}$

In $p(x) = a\sqrt[3]{x-h} + k$, $a = 1$, $h = -5$, and $k = 0$. So the function is translated 5 units left from the parent graph.

The domain is all real numbers, and the range is all real numbers.

x	y
−7	≈ −1.26
−6	−1
−5	0
−4	1
−3	≈ 1.26

c. $q(x) = \sqrt[3]{4-x} + 1$

The function can be written as $q(x) = \sqrt[3]{-(x-4)} + 1$. So the function is reflected and translated 4 units right and 1 unit up from the parent graph.

x	y
2	≈ 2.26
3	2
4	1
5	0
6	≈ −0.26

The domain is all real numbers, and the range is all real numbers.

Example 6 Compare Radical Functions

Examine $p(x) = -2\sqrt[3]{x-6}$ and $q(x)$ shown in the graph.

Part A Graph $p(x)$.

Make a table of values for $p(x)$. Then, graph the function.

x	y
0	≈ 3.63
2	≈ 3.17
4	≈ 2.52
6	0
8	≈ −2.52
10	≈ −3.18

(continued on the next page)

Study Tip

Tables When making a table of values for a radical function, first determine h. Then make a table using values that are greater than and less than h.

Think About It!

Describe the end behavior of the function in part c, $y = \sqrt[3]{4-x} + 1$.

Sample answer: As x approaches negative infinity, y approaches infinity. As x approaches infinity, y approaches negative infinity.

Think About It!

Determine whether the indexes of $p(x)$ and $q(x)$ are even or odd. Justify your response.

Sample answer: $p(x)$ has an odd index because the domain and range are unrestricted. $q(x)$ has an even index because the domain is restricted.

Lesson 6-4 • Graphing Radical Functions 327

Interactive Presentation

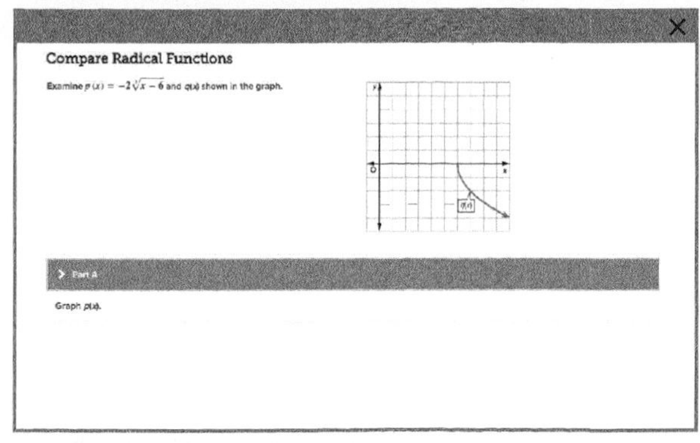

Example 6

EXPAND Students move through the steps to see how to compare the graphs of two radical functions.

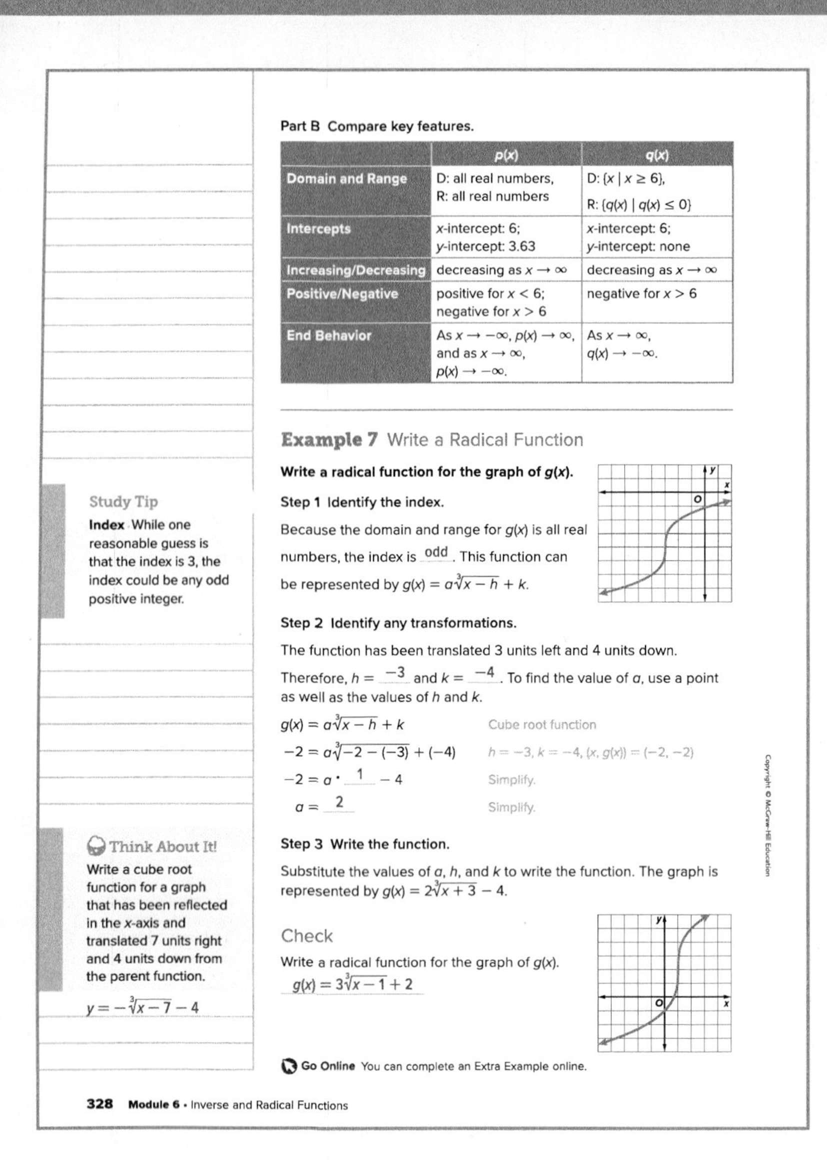

Part B Compare key features.

	$p(x)$	$q(x)$
Domain and Range	D: all real numbers, R: all real numbers	D: $\{x \mid x \geq 6\}$, R: $\{q(x) \mid q(x) \leq 0\}$
Intercepts	x-intercept: 6; y-intercept: 3.63	x-intercept: 6; y-intercept: none
Increasing/Decreasing	decreasing as $x \to \infty$	decreasing as $x \to \infty$
Positive/Negative	positive for $x < 6$; negative for $x > 6$	negative for $x > 6$
End Behavior	As $x \to -\infty$, $p(x) \to \infty$, and as $x \to \infty$, $p(x) \to -\infty$.	As $x \to \infty$, $q(x) \to -\infty$.

Example 7 Write a Radical Function

Write a radical function for the graph of $g(x)$.

Step 1 Identify the index.

Because the domain and range for $g(x)$ is all real numbers, the index is odd. This function can be represented by $g(x) = a\sqrt[3]{x-h} + k$.

Study Tip

Index While one reasonable guess is that the index is 3, the index could be any odd positive integer.

Step 2 Identify any transformations.

The function has been translated 3 units left and 4 units down. Therefore, $h = -3$ and $k = -4$. To find the value of a, use a point as well as the values of h and k.

$g(x) = a\sqrt[3]{x-h} + k$ — Cube root function

$-2 = a\sqrt[3]{-2-(-3)} + (-4)$ — $h = -3$, $k = -4$, $(x, g(x)) = (-2, -2)$

$-2 = a \cdot 1 - 4$ — Simplify.

$a = 2$ — Simplify.

Step 3 Write the function.

Substitute the values of a, h, and k to write the function. The graph is represented by $g(x) = 2\sqrt[3]{x+3} - 4$.

Think About It!

Write a cube root function for a graph that has been reflected in the x-axis and translated 7 units right and 4 units down from the parent function.

$y = -\sqrt[3]{x-7} - 4$

Check

Write a radical function for the graph of $g(x)$.

$g(x) = 3\sqrt[3]{x-1} + 2$

Go Online You can complete an Extra Example online.

Copyright © McGraw-Hill Education

328 Module 6 • Inverse and Radical Functions

Example 7 Write a Radical Function

Teaching the Mathematical Practices

2 Justify Conclusions Mathematically proficient students can explain the conclusions drawn when solving a problem. In Example 7, students should justify their conclusions as they solve the problem.

Questions for Mathematical Discourse

AL Why do we need to use a point on the graph to determine the function? We need to isolate a and determine its value. We know h and k from the translation, so if we substitute in a point, the only remaining variable is a.

OL How do you know that the index is not an even integer? Sample answer: The domain is not restricted.

BL How can you check the function to make sure you identified the correct radical n value? Sample answer: Test more points on the graph to see whether they work in the function that was written, or graph the function using a graphing calculator and compare the result to the given graph.

Interactive Presentation

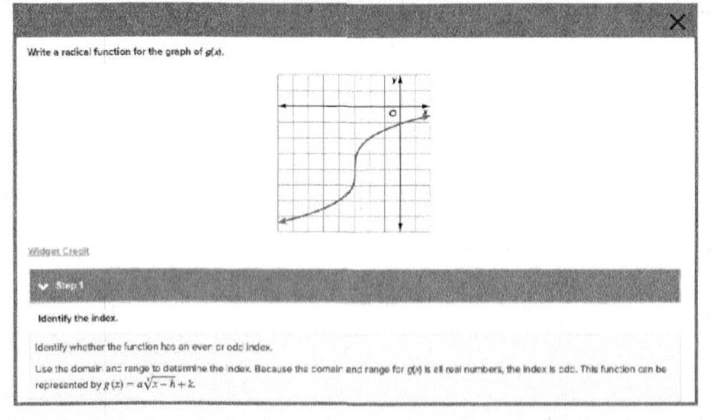

Example 7

EXPAND

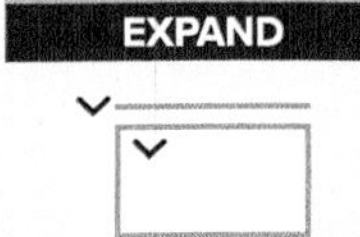

Students move through the steps to see how to write a radical function.

CHECK

Students complete the Check online to determine whether they are ready to move on.

Exit Ticket

Recommended Use

At the end of class, go online to display the Exit Ticket prompt and ask students to respond using a separate piece of paper. Have students hand you their responses as they leave the room.

Alternate Use

At the end of class, go online to display the Exit Ticket prompt and ask students to respond verbally or by using a mini-whiteboard. Have students hold up their whiteboards so that you can see all student responses. Tap to reveal the answer when most or all students have completed the Exit Ticket.

Practice and Homework

Suggested Assignments

Use the table below to select appropriate exercises.

DOK	Topic	Exercises
1, 2	exercises that mirror the examples	1–28
2	exercises that use a variety of skills from this lesson	29–37
2	exercises that extend concepts learned in this lesson to new contexts	38–42
3	exercises that emphasize higher-order and critical thinking skills	43–47

ASSESS AND DIFFERENTIATE

Use the data from the **Checks** to determine whether to provide resources for extension, remediation, or intervention.

BL

IF students score 90% or more on the Checks,
THEN assign:

- Practice Exercises 1–41 odd, 43–47
- Extension: Relativistic Mass
- ALEKS® Radical Functions

OL

IF students score 66%–89% on the Checks,
THEN assign:

- Practice Exercises 1–47 odd
- Remediation, Review Resources: Rational Exponents
- Personal Tutors
- Extra Examples 1–6
- ALEKS® Rational Exponents

AL

IF students score 65% or less on the Checks,
THEN assign:

- Practice, Exercises 1–27 odd
- Remediation, Review Resources: Rational Exponents
- *Quick Review Math Handbook*: Square Root Functions and Inequalities
- ALEKS® Rational Exponents

Answer

7. reflected in the x-axis, compressed vertically, and translated left 1 unit

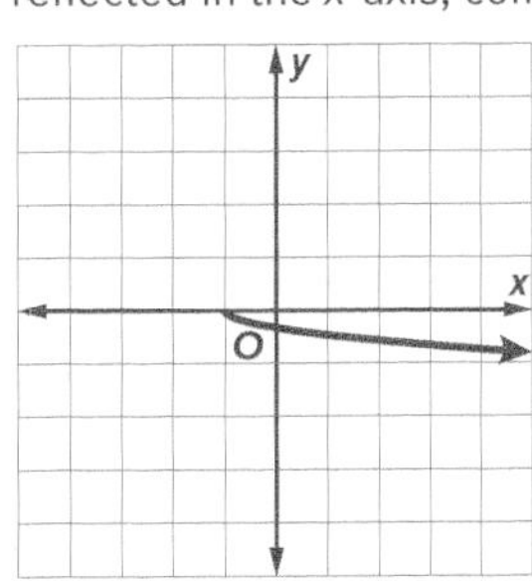

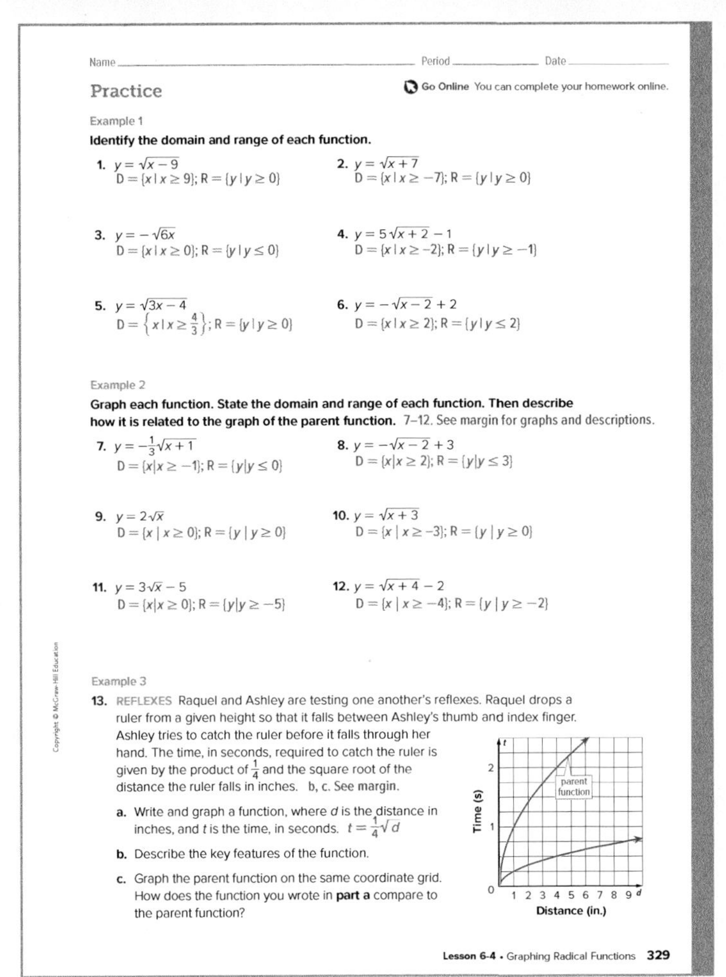

Name ______ Period ______ Date ______

Practice

Go Online You can complete your homework online.

Example 1

Identify the domain and range of each function.

1. $y = \sqrt{x - 9}$
$D = \{x \mid x \geq 9\}; R = \{y \mid y \geq 0\}$

2. $y = \sqrt{x + 7}$
$D = \{x \mid x \geq -7\}; R = \{y \mid y \geq 0\}$

3. $y = -\sqrt{6x}$
$D = \{x \mid x \geq 0\}; R = \{y \mid y \leq 0\}$

4. $y = 5\sqrt{x + 2} - 1$
$D = \{x \mid x \geq -2\}; R = \{y \mid y \geq -1\}$

5. $y = \sqrt{3x - 4}$
$D = \left\{x \mid x \geq \frac{4}{3}\right\}; R = \{y \mid y \geq 0\}$

6. $y = -\sqrt{x - 2} + 2$
$D = \{x \mid x \geq 2\}; R = \{y \mid y \leq 2\}$

Example 2

Graph each function. State the domain and range of each function. Then describe how it is related to the graph of the parent function. 7–12. See margin for graphs and descriptions.

7. $y = -\frac{1}{3}\sqrt{x + 1}$
$D = \{x \mid x \geq -1\}; R = \{y \mid y \leq 0\}$

8. $y = -\sqrt{x - 2} + 3$
$D = \{x \mid x \geq 2\}; R = \{y \mid y \leq 3\}$

9. $y = 2\sqrt{x}$
$D = \{x \mid x \geq 0\}; R = \{y \mid y \geq 0\}$

10. $y = \sqrt{x + 3}$
$D = \{x \mid x \geq -3\}; R = \{y \mid y \geq 0\}$

11. $y = 3\sqrt{x} - 5$
$D = \{x \mid x \geq 0\}; R = \{y \mid y \geq -5\}$

12. $y = \sqrt{x + 4} - 2$
$D = \{x \mid x \geq -4\}; R = \{y \mid y \geq -2\}$

Example 3

13. REFLEXES Raquel and Ashley are testing one another's reflexes. Raquel drops a ruler from a given height so that it falls between Ashley's thumb and index finger. Ashley tries to catch the ruler before it falls through her hand. The time, in seconds, required to catch the ruler is given by the product of $\frac{1}{4}$ and the square root of the distance the ruler falls in inches. b, c. See margin.

a. Write and graph a function, where d is the distance in inches, and t is the time, in seconds. $t = \frac{1}{4}\sqrt{d}$

b. Describe the key features of the function.

c. Graph the parent function on the same coordinate grid. How does the function you wrote in **part a** compare to the parent function?

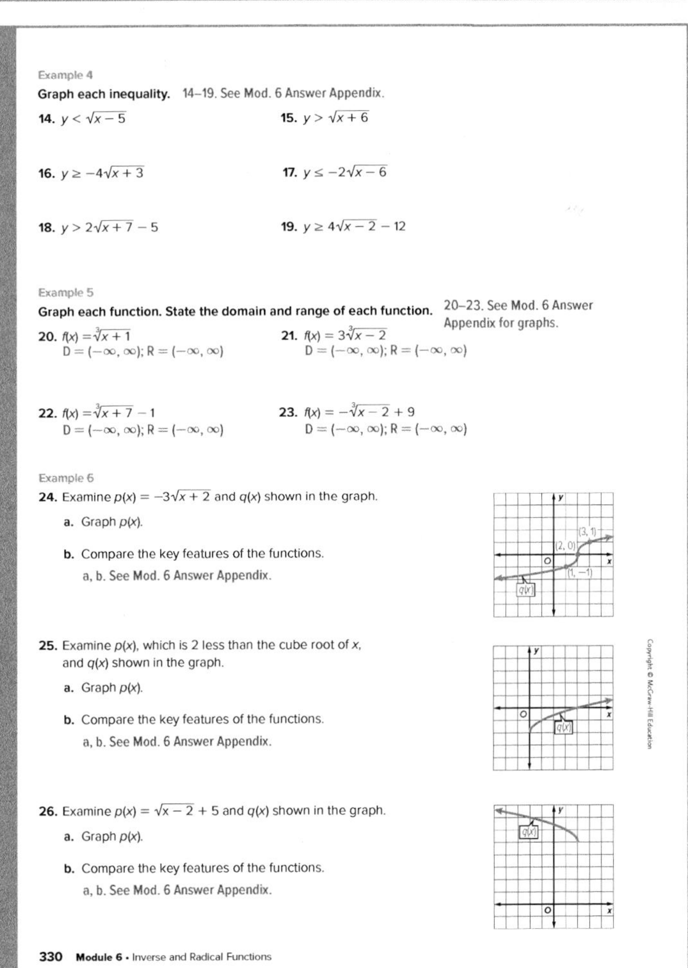

Example 4

Graph each inequality. 14–19. See Mod. 6 Answer Appendix.

14. $y < \sqrt{x - 5}$

15. $y > \sqrt{x + 6}$

16. $y \geq -4\sqrt{x + 3}$

17. $y \leq -2\sqrt{x - 6}$

18. $y > 2\sqrt{x + 7} - 5$

19. $y \geq 4\sqrt{x - 2} - 12$

Example 5

Graph each function. State the domain and range of each function. 20–23. See Mod. 6 Answer Appendix for graphs.

20. $f(x) = \sqrt[3]{x + 1}$
$D = (-\infty, \infty); R = (-\infty, \infty)$

21. $f(x) = 3\sqrt[3]{x - 2}$
$D = (-\infty, \infty); R = (-\infty, \infty)$

22. $f(x) = \sqrt[3]{x + 7} - 1$
$D = (-\infty, \infty); R = (-\infty, \infty)$

23. $f(x) = -\sqrt[3]{x - 2} + 9$
$D = (-\infty, \infty); R = (-\infty, \infty)$

Example 6

24. Examine $p(x) = -3\sqrt{x + 2}$ and $q(x)$ shown in the graph.

a. Graph $p(x)$.

b. Compare the key features of the functions.

a, b. See Mod. 6 Answer Appendix.

25. Examine $p(x)$, which is 2 less than the cube root of x, and $q(x)$ shown in the graph.

a. Graph $p(x)$.

b. Compare the key features of the functions.

a, b. See Mod. 6 Answer Appendix.

26. Examine $p(x) = \sqrt{x - 2} + 5$ and $q(x)$ shown in the graph.

a. Graph $p(x)$.

b. Compare the key features of the functions.

a, b. See Mod. 6 Answer Appendix.

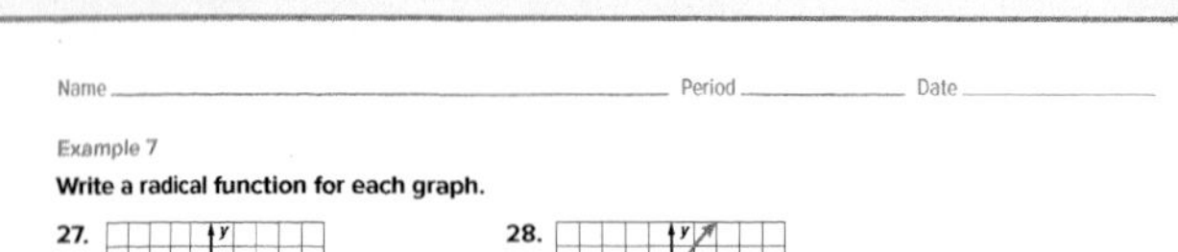

Name ______ Period ______ Date ______

Example 7

Write a radical function for each graph.

27.

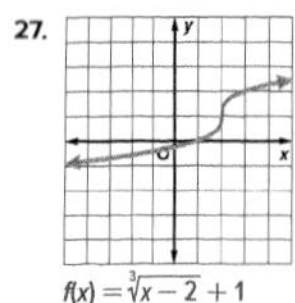

$f(x) = \sqrt[3]{x-2} + 1$

28.

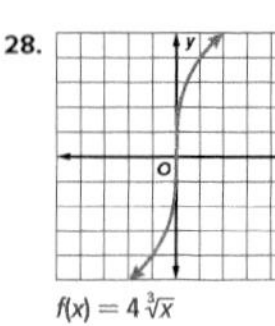

$f(x) = 4\sqrt[3]{x}$

Mixed Exercises

Graph each function and state the domain and range. Then describe how it is related to the graph of the parent function. 29–34. See Mod. 6 Answer Appendix for graphs and descriptions.

29. $f(x) = 2\sqrt{x-5} - 6$
$D = \{x \mid x \geq 5\}; R = \{f(x) \mid f(x) \geq -6\}$

30. $f(x) = \frac{3}{4}\sqrt{x+12} + 3$
$D = \{x \mid x \geq -12\}; R = \{f(x) \mid f(x) \geq 3\}$

31. $f(x) = -\frac{1}{5}\sqrt{x-1} - 4$
$D = \{x \mid x \geq 1\}; R = \{f(x) \mid f(x) \leq -4\}$

32. $f(x) = -3\sqrt{x+7} + 9$
$D = \{x \mid x \geq -7\}; R = \{f(x) \mid f(x) \leq 9\}$

33. $f(x) = -\frac{1}{3}\sqrt[3]{x+2} - 3$
$D = (-\infty, \infty); R = (-\infty, \infty)$

34. $f(x) = -\frac{1}{2}\sqrt[3]{2x-1} + 3$
$D = (-\infty, \infty); R = (-\infty, \infty)$

Graph each inequality. 35, 36. See Mod. 6 Answer Appendix.

35. $y \leq 6 - 3\sqrt{x-4}$

36. $y < \sqrt{4x-12} + 8$

Write a radical function for each graph.

37.

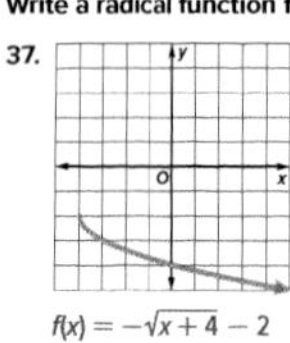

$f(x) = -\sqrt{x+4} - 2$

38.

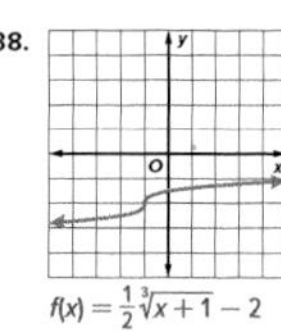

$f(x) = \frac{1}{2}\sqrt[3]{x+1} - 2$

39. STRUCTURE Consider the function $f(x) = -\sqrt{x+3} + \frac{13}{2}$ and the function $g(x)$ shown in the graph. a–c. See Mod. 6 Answer Appendix.

a. Determine which function has the greater maximum value. Explain your reasoning.

b. Compare the domains of the two functions.

c. Compare the average rates of change of the two functions over the interval [6, 13].

40. DISTANCE LaRez is standing at the side of a road watching a cyclist. The distance, in feet, between LaRez and the cyclist as a function of time is given by the square root of the sum of 9 and the product of 36 and the time squared.

a. Write and graph a function, where t is the time, in seconds, and d is the distance, in feet. $d = \sqrt{9 + 36t^2}$; See Mod. 6 Answer Appendix for graph.

b. Describe the key features of the function.
See Mod. 6 Answer Appendix for graph.

41. GEOMETRY The length of the radius of a hemisphere can be found using the formula $r = \sqrt[3]{\frac{3V}{2\pi}}$, given the volume V of the hemisphere. Graph the function. State the key features of the graph.
See Mod. 6 Answer Appendix.

42. STRUCTURE Graph $f(x) > \sqrt{-x+2} - 3$ and its inverse on the same coordinate plane as well as $y = x$. Write the inequality that defines the graph of the inverse. Determine any restrictions that must be placed on the domain of the inverse.
See Mod. 6 Answer Appendix.

Higher-Order Thinking Skills

43. PERSEVERE Write an equation for a square root function with a domain of $\{x \mid x \geq -4\}$, a range of $\{y \mid y \leq 6\}$, and that passes through (5, 3).
Sample answer: $y = -\sqrt{x+4} + 6$

44. ANALYZE For what positive values of a are the domain and range of $f(x) = \sqrt[a]{x}$ the set of real numbers? Justify your argument.
all positive odd numbers; Sample answer: For the domain to be all real numbers, a must be odd. For the domain and range to be continuous, the value of a must be positive.

45. WRITE Explain why there are limitations on the domain and range of square root functions.
See Mod. 6 Answer Appendix.

46. WRITE Explain why $y = \pm\sqrt{x}$ is not a function.
See Mod. 6 Answer Appendix.

47. CREATE Write an equation of a relation that contains a radical in its inverse such that:

a. the original relation is a function, and its inverse is not a function.

b. the original relation is not a function, and its inverse is a function.
a, b. See Mod. 6 Answer Appendix.

Answers

8. reflected in the x-axis and translated right 2 units and up 3 units

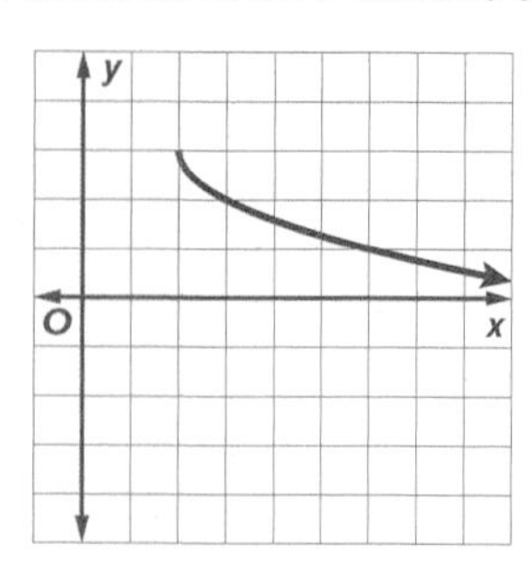

9. stretched vertically

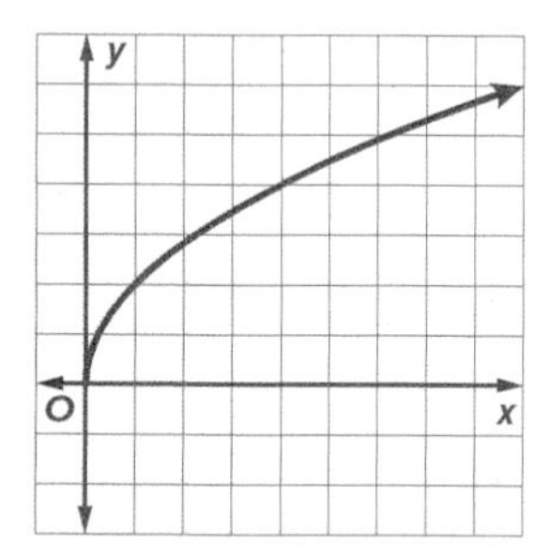

10. translated left 3 units

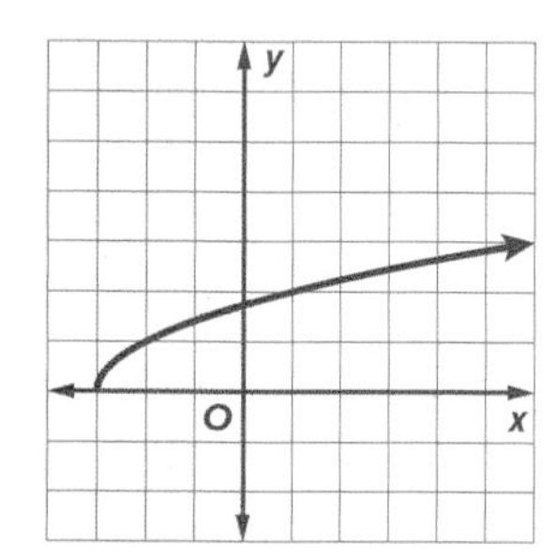

11. stretched vertically and translated down 5 units

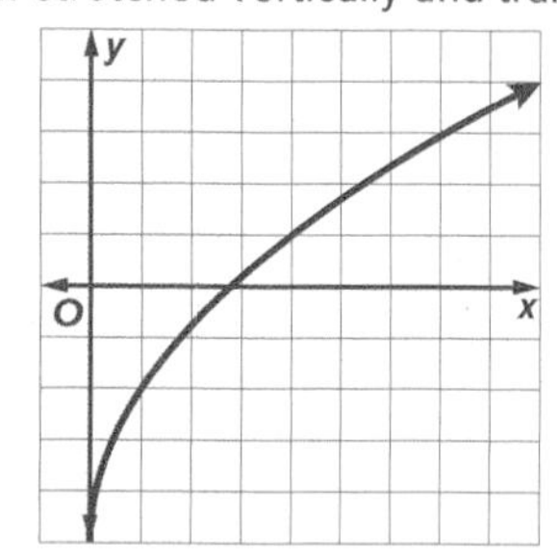

12. translated left 4 units and down 2 units

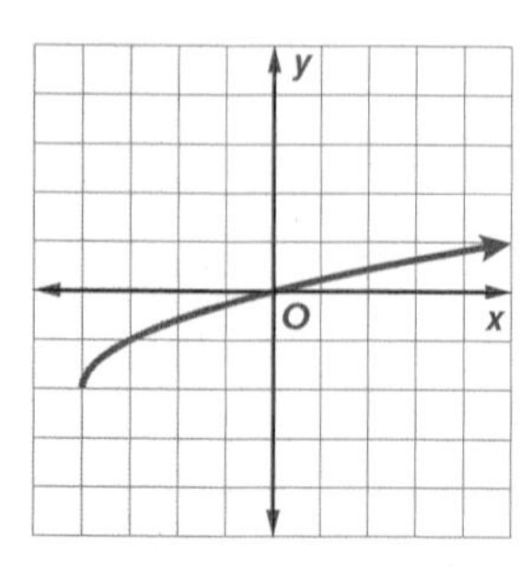

13b. $D = \{d \mid d \geq 0\}; R = \{t \mid t \geq 0\}$; increasing as $x \to \infty$; positive for $x > 0$; As $x \to \infty, y \to \infty$.

13c. compressed vertically

Operations with Radical Expressions

A.SSE.2

LESSON GOAL

Students simplify and perform operations with radical expressions.

1 LAUNCH

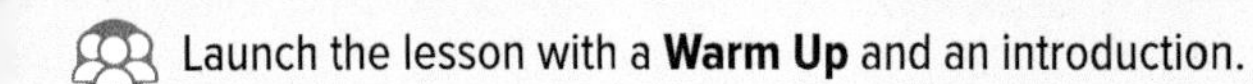

Launch the lesson with a **Warm Up** and an introduction.

2 EXPLORE AND DEVELOP

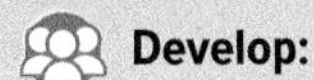

Develop:

Simplifying Radical Expressions
- Simplify Expressions with the Product Property
- Simplify Expressions with the Quotient Property

Operations with Radical Expressions
- Add and Subtract Radicals
- Multiply Radicals
- Use the Distributive Property to Multiply Radicals

Rationalizing the Denominator
- Rationalize the Denominator
- Use Conjugates to Rationalize the Denominator

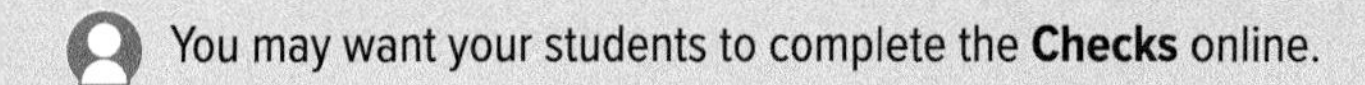

You may want your students to complete the **Checks** online.

3 REFLECT AND PRACTICE

Exit Ticket

Practice

DIFFERENTIATE

View reports of student progress on the **Checks** after each example.

Resources	AL	OL	BL	ELL
Remediation: Simplify Radical Expressions	●	●		●
Extension: Special Products with Radicals		●	●	●

Language Development Handbook

Assign page 39 of the *Language Development Handbook* to help your students build mathematical language related to simplifying and performing operations with radical functions.

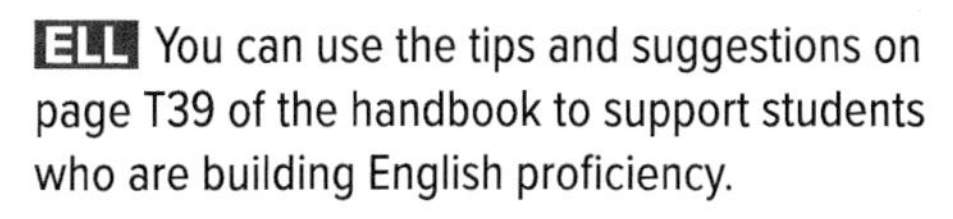

ELL You can use the tips and suggestions on page T39 of the handbook to support students who are building English proficiency.

Suggested Pacing

90 min	1 day
45 min	2 days

Focus

Domain: Algebra

Standards for Mathematical Content:

A.SSE.2 Use the structure of an expression to identify ways to rewrite it.

Standards for Mathematical Practice:

4 Model with mathematics.

7 Look for and make use of structure.

Coherence

Vertical Alignment

Previous
Students graphed and analyzed square and cube root functions.
F.IF.7b, F.BF.3

Now
Students simplify and perform operations with radical expressions.
A.SSE.2

Next
Students will solve radical equations algebraically and by graphing.
A.REI.2

Rigor

The Three Pillars of Rigor

1 CONCEPTUAL UNDERSTANDING	2 FLUENCY	3 APPLICATION

Conceptual Bridge In this lesson, students expand on their understanding of radical expressions to build fluency by adding, subtracting, and multiplying radical expressions. They apply their understanding by performing operations with radical expressions to solve real-world problems.

Mathematical Background

The Product Property of Radicals states that $\sqrt[n]{ab} = \sqrt[n]{a} \cdot \sqrt[n]{b}$. This statement is true for all real numbers a and b when n is an odd integer greater than 1. If n is even, then a and b must be nonnegative real numbers.

Interactive Presentation

Warm Up

Simplify each radical.

1. $\sqrt[4]{9}$
2. $\sqrt[5]{32}$
3. $\sqrt[3]{125x^4}$
4. $\sqrt[4]{32x^8}$

Show Answers

Warm Up

Launch the Lesson

Golden rectangles, which have side lengths with a ratio of $\frac{2}{\sqrt{5}-1}$ have long been used in art and architecture to create beautiful designs. Rectangles with sides in this ratio are considered the most pleasing to the eye and have been found in some of the world's most historic sites such as the Parthenon, the Great Pyramids of Giza, and Notre Dame Cathedral. Recently, golden rectangles were discovered in the Sun Temple at Mesa Verde National Park in Colorado.

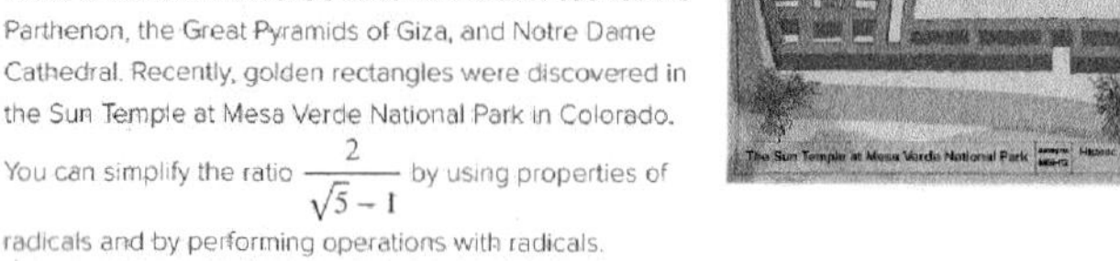

You can simplify the ratio $\frac{2}{\sqrt{5}-1}$ by using properties of radicals and by performing operations with radicals.

Launch the Lesson

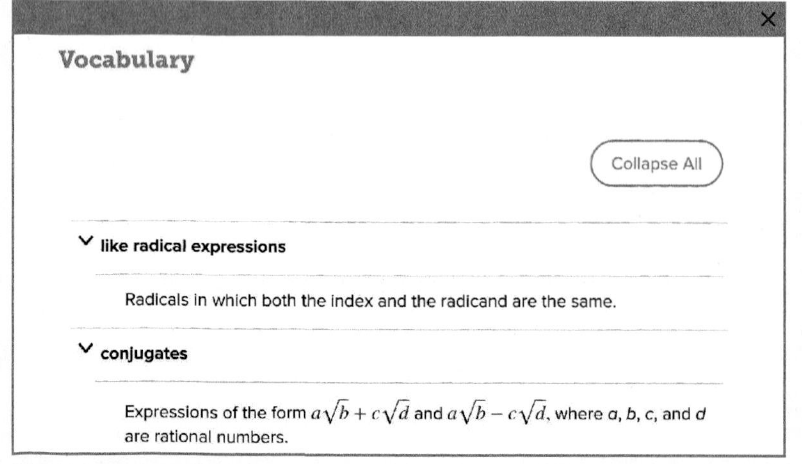

Vocabulary

Collapse All

like radical expressions

Radicals in which both the index and the radicand are the same.

conjugates

Expressions of the form $a\sqrt{b}+c\sqrt{d}$ and $a\sqrt{b}-c\sqrt{d}$, where a, b, c, and d are rational numbers.

Today's Vocabulary

Warm Up

Prerequisite Skills

The Warm Up exercises address the following prerequisite skill for this lesson:

- simplifying radicals

Answers:

1. $\sqrt{3}$
2. 2
3. $5x\sqrt[3]{x}$
4. $2x^2\sqrt[4]{2}$

Launch the Lesson

Teaching the Mathematical Practices

4 Apply Mathematics In the Launch the Lesson, students will learn how the ratio of the side lengths of a golden rectangle can be simplified by performing operations with radical expressions.

Go Online to find additional teaching notes and questions to promote classroom discourse.

Today's Standards

Tell students that they will be addressing these content and practice standards in this lesson. You may wish to have a student volunteer read aloud *How can I meet this standard?* and *How can I use these practices?*, and connect these to the standards.

See the Interactive Presentation for I Can statements that align with the standards covered in this lesson.

Today's Vocabulary

Tell students that they will be using these vocabulary terms in this lesson. You can expand each row if you wish to share the definitions. Then discuss the questions below with the class.

Learn Properties of Radicals

Objective

Students simplify radical expressions by applying the properties of radicals.

Teaching the Mathematical Practices

1 Seek Information Mathematically proficient students must be able to transform algebraic expressions to reach solutions. Encourage students to familiarize themselves with the conditions that must be met for radical expression to be in simplest form. Point out that when all conditions are met, they are done transforming the expression.

7 Use Structure Help students to understand how to use the structure of the Product Property and Quotient Property of Radicals to simplify radical expressions.

Example 1 Simplify Expressions with the Product Property

Teaching the Mathematical Practices

6 Communicate Precisely Encourage students to routinely write or explain their solution methods. The Think About It! feature asks students to explain why the solution to part **b** does not include absolute value symbols.

Questions for Mathematical Discourse

AL In part **a**, why is b^{14} written in two factors? Sample answer: We want to find perfect cubes that will simplify when applying the cube root. Though b^{14} is not a perfect cube, b^{12} is, so b^{14} is factored into the cube part and the remaining part.

OL How do you factor a coefficient in a radicand to simplify an nth root? Sample answer: First check to see if the coefficient is a perfect nth power. If not, determine the greatest factor of the coefficient that is a perfect nth power, and separate the coefficient into this factor and the remaining factor.

BL How can you use the product property of radicals to simplify $\sqrt{18a^3} \cdot \sqrt{2a}$? Multiply the terms together. Then take the square root, resulting in $\sqrt{36a^4} = 6a^2$.

Go Online

- Find additional teaching notes.
- View performance reports of the Checks.
- Assign or present an Extra Example.

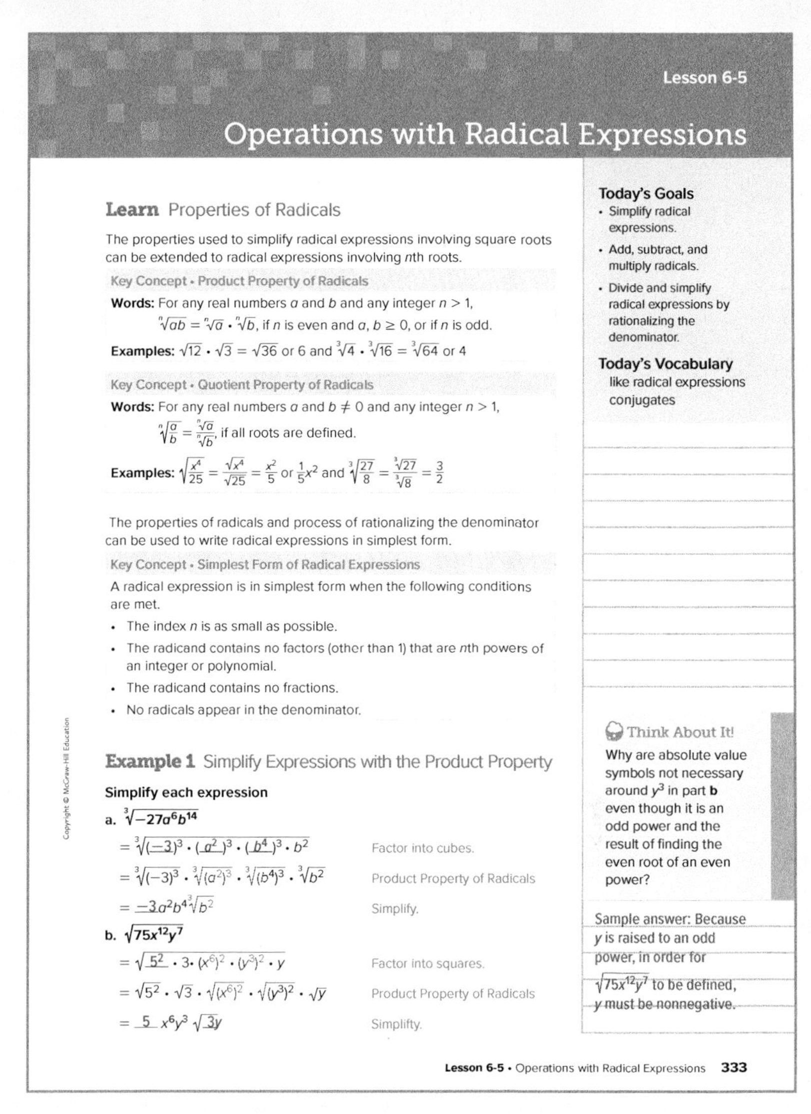

Lesson 6-5

Operations with Radical Expressions

Learn Properties of Radicals

The properties used to simplify radical expressions involving square roots can be extended to radical expressions involving nth roots.

Key Concept • Product Property of Radicals

Words: For any real numbers a and b and any integer $n > 1$, $\sqrt[n]{ab} = \sqrt[n]{a} \cdot \sqrt[n]{b}$, if n is even and $a, b \geq 0$, or if n is odd.

Examples: $\sqrt{12} \cdot \sqrt{3} = \sqrt{36}$ or 6 and $\sqrt[3]{4} \cdot \sqrt[3]{16} = \sqrt[3]{64}$ or 4

Key Concept • Quotient Property of Radicals

Words: For any real numbers a and $b \neq 0$ and any integer $n > 1$, $\sqrt[n]{\frac{a}{b}} = \frac{\sqrt[n]{a}}{\sqrt[n]{b}}$, if all roots are defined.

Examples: $\sqrt{\frac{x^4}{25}} = \frac{\sqrt{x^4}}{\sqrt{25}} = \frac{x^2}{5}$ or $\frac{1}{5}x^2$ and $\sqrt[3]{\frac{27}{8}} = \frac{\sqrt[3]{27}}{\sqrt[3]{8}} = \frac{3}{2}$

The properties of radicals and process of rationalizing the denominator can be used to write radical expressions in simplest form.

Key Concept • Simplest Form of Radical Expressions

A radical expression is in simplest form when the following conditions are met.

- The index n is as small as possible.
- The radicand contains no factors (other than 1) that are nth powers of an integer or polynomial.
- The radicand contains no fractions.
- No radicals appear in the denominator.

Example 1 Simplify Expressions with the Product Property

Simplify each expression

a. $\sqrt[3]{-27a^6b^{14}}$

$= \sqrt[3]{(-3)^3 \cdot (\underline{a^2})^3 \cdot (\underline{b^4})^3 \cdot b^2}$ Factor into cubes.

$= \sqrt[3]{(-3)^3} \cdot \sqrt[3]{(a^2)^3} \cdot \sqrt[3]{(b^4)^3} \cdot \sqrt[3]{b^2}$ Product Property of Radicals

$= \underline{-3}a^2b^4\sqrt[3]{b^2}$ Simplify.

b. $\sqrt{75x^{12}y^7}$

$= \sqrt{\underline{5^2} \cdot 3 \cdot (x^6)^2 \cdot (y^3)^2 \cdot y}$ Factor into squares.

$= \sqrt{5^2} \cdot \sqrt{3} \cdot \sqrt{(x^6)^2} \cdot \sqrt{(y^3)^2} \cdot \sqrt{y}$ Product Property of Radicals

$= \underline{5}\,x^6y^3\,\underline{\sqrt{3y}}$ Simplify.

Today's Goals
- Simplify radical expressions.
- Add, subtract, and multiply radicals.
- Divide and simplify radical expressions by rationalizing the denominator.

Today's Vocabulary
like radical expressions
conjugates

Think About It!
Why are absolute value symbols not necessary around y^3 in part **b** even though it is an odd power and the result of finding the even root of an even power?

Sample answer: Because y is raised to an odd power, in order for $\sqrt{75x^{12}y^7}$ to be defined, y must be nonnegative.

Copyright © McGraw-Hill Education

Lesson 6-5 • Operations with Radical Expressions 333

Interactive Presentation

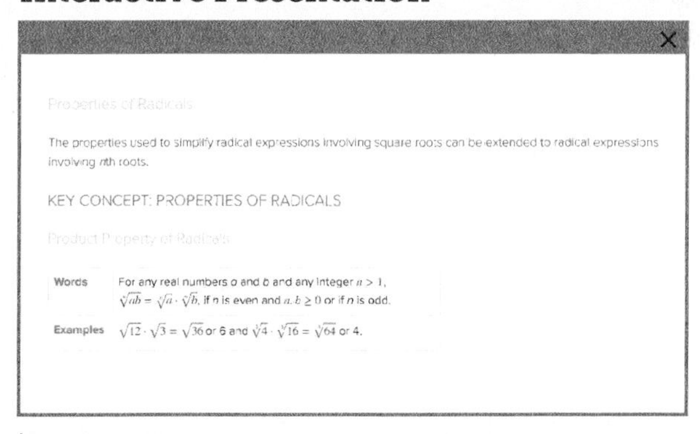

Learn

TYPE

Students answer a question applying the properties of radicals.

SELECT

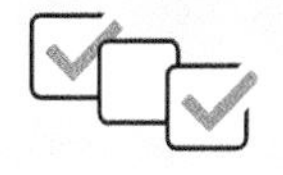

Students select whether each expression is in simplest form.

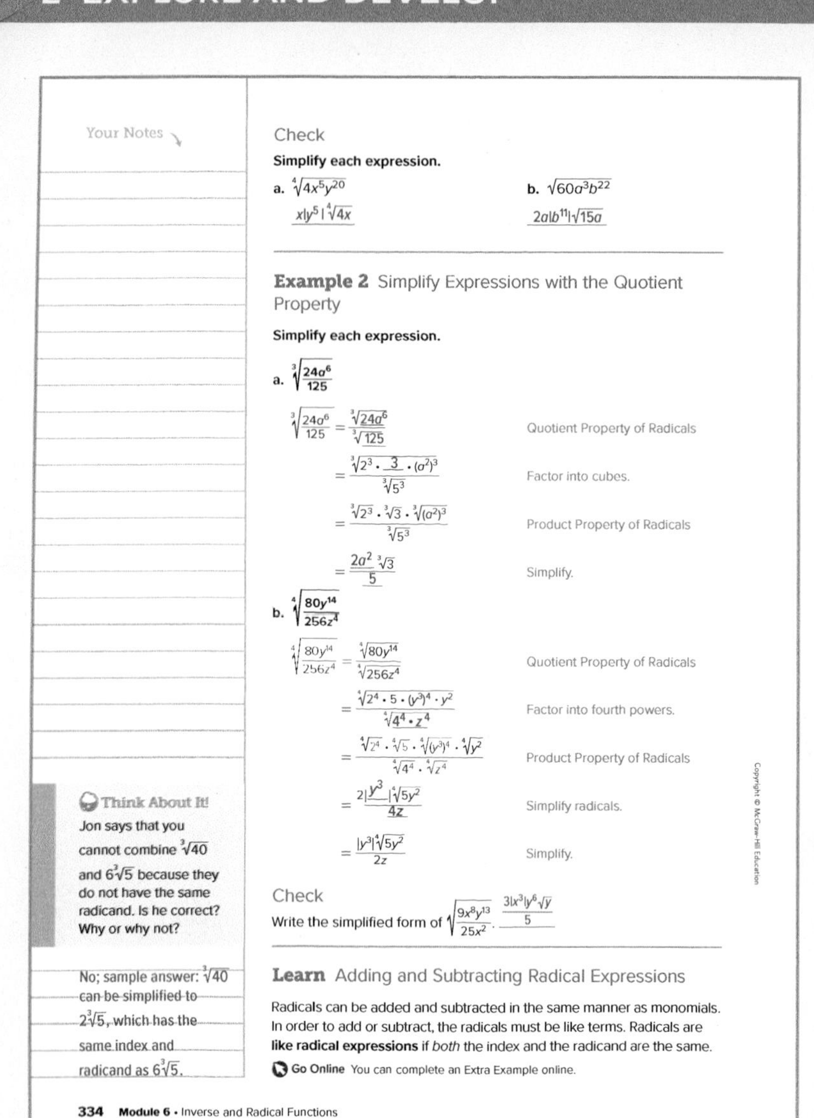

Your Notes

Check

Simplify each expression.

a. $\sqrt[4]{4x^5y^{20}}$ $|x|y^5|\sqrt[4]{4x}$

b. $\sqrt{60a^3b^{22}}$ $2a|b^{11}|\sqrt{15a}$

Example 2 Simplify Expressions with the Quotient Property

Simplify each expression.

a. $\sqrt[3]{\frac{24a^6}{125}}$

$\sqrt[3]{\frac{24a^6}{125}} = \frac{\sqrt[3]{24a^6}}{\sqrt[3]{125}}$ Quotient Property of Radicals

$= \frac{\sqrt[3]{2^3 \cdot 3 \cdot (a^2)^3}}{\sqrt[3]{5^3}}$ Factor into cubes.

$= \frac{\sqrt[3]{2^3} \cdot \sqrt[3]{3} \cdot \sqrt[3]{(a^2)^3}}{\sqrt[3]{5^3}}$ Product Property of Radicals

$= \frac{2a^2\sqrt[3]{3}}{5}$ Simplify.

b. $\sqrt[4]{\frac{80y^{14}}{256z^4}}$

$\sqrt[4]{\frac{80y^{14}}{256z^4}} = \frac{\sqrt[4]{80y^{14}}}{\sqrt[4]{256z^4}}$ Quotient Property of Radicals

$= \frac{\sqrt[4]{2^4 \cdot 5 \cdot (y^3)^4 \cdot y^2}}{\sqrt[4]{4^4 \cdot z^4}}$ Factor into fourth powers.

$= \frac{\sqrt[4]{2^4} \cdot \sqrt[4]{5} \cdot \sqrt[4]{(y^3)^4} \cdot \sqrt[4]{y^2}}{\sqrt[4]{4^4} \cdot \sqrt[4]{z^4}}$ Product Property of Radicals

$= \frac{2|y^3|\sqrt[4]{5y^2}}{4z}$ Simplify radicals.

$= \frac{|y^3|\sqrt[4]{5y^2}}{2z}$ Simplify.

Think About It!

Jon says that you cannot combine $\sqrt[3]{40}$ and $6\sqrt[3]{5}$ because they do not have the same radicand. Is he correct? Why or why not?

No; sample answer: $\sqrt[3]{40}$ can be simplified to $2\sqrt[3]{5}$, which has the same index and radicand as $6\sqrt[3]{5}$.

Check

Write the simplified form of $\sqrt[4]{\frac{9x^8y^{13}}{25x^2}}$. $\frac{3|x^3|y^6\sqrt{y}}{5}$

Learn Adding and Subtracting Radical Expressions

Radicals can be added and subtracted in the same manner as monomials. In order to add or subtract, the radicals must be like terms. Radicals are **like radical expressions** if *both* the index and the radicand are the same.

Go Online You can complete an Extra Example online.

334 Module 6 • Inverse and Radical Functions

Copyright © McGraw-Hill Education

Interactive Presentation

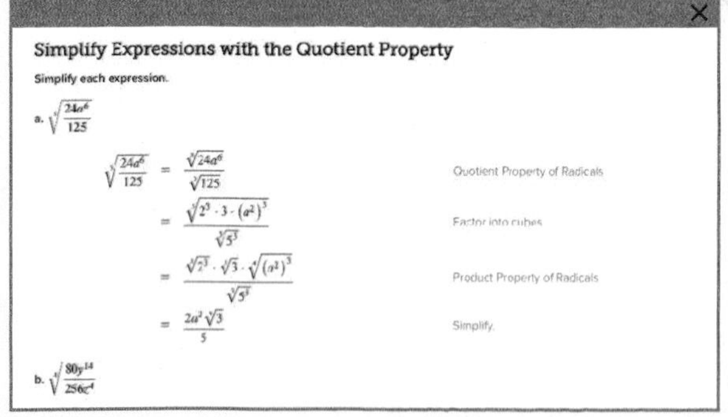

Example 2

TAP

Students move through the steps to simplify an expression.

CHECK

Students complete the Check online to determine whether they are ready to move on.

Example 2 Simplify Expressions with the Quotient Property

Teaching the Mathematical Practices

7 Interpret Complicated Expressions Mathematically proficient students can see complicated expressions as single objects or as being composed of several objects. In Example 2, guide students to simplify the numerator and denominator separately.

Questions for Mathematical Discourse

AL In **part b**, why do you need absolute value symbols around y^3? Sample answer: This ensures that the even root of an even power is a positive number and thus is the principal root.

OL In **part b**, what is another approach you could take to simplify the coefficients? Sample answer: You could simplify the fraction $\frac{80}{256}$ to $\frac{5}{16}$ first.

BL How can you check your answer? Sample answer: Raise your answer to the exponent equal to the root you took. You should get the radicand of the original expression.

Learn Adding and Subtracting Radical Expressions

Objective

Students add, subtract, and multiply radicals by using the Distributive Property and combining like terms.

Teaching the Mathematical Practices

3 Find the Error The Think About It! feature requires students to read Jon's argument, decide whether it makes sense, and explain why or why not.

8 Look for a Pattern Help students to see the pattern in the product when multiplying radical expressions with coefficients. Encourage them to write a general rule for finding the product.

About the Key Concept

Exact roots occur when the powers of the constants and variables are all identical to or multiples of the index. For example, $\sqrt[3]{2} \cdot \sqrt[3]{2^2} = \sqrt[3]{2^3}$ or 2.

1 CONCEPTUAL UNDERSTANDING | 2 FLUENCY | 3 APPLICATION

Example 3 Add and Subtract Radicals

Teaching the Mathematical Practices

1 Seek Information Mathematically proficient students must be able to transform algebraic expressions to reach solutions. In Example 3, students must transform the terms in the expression to simplify it.

Questions for Mathematical Discourse

AL Why is the first step simplifying the radicals? Simplifying the radicals results in like terms that can be combined.

OL How does the distributive property relate to being able to combine radicals with like index and radicand? Sample answer: An expression with like radicals could be factored into the radical multiplied by the expression of the coefficient terms: $a\sqrt{x} + b\sqrt{x} = (a + b)\sqrt{x}$.

BL What values of x would yield an expression with only one term? Sample answer: $\frac{3}{5}n$, where n is any perfect square, i.e. any value that makes the radicand simplify to $\sqrt{3}$.

Example 4 Multiply Radicals

Teaching the Mathematical Practices

7 Use Structure In Example 4, help students see how the radicand can be broken into several objects.

Questions for Mathematical Discourse

AL What must be true about the indices of the two roots you are multiplying together? They must be the same.

OL What is another approach you could take to simplify the expression? Sample answer: Simplify each root first and then find the product.

BL What is a potential advantage of multiplying radicands before simplifying? Multiplying first may make it possible to find factors to the nth power. If you simplify first, you would still need to multiply and simplify again.

DIFFERENTIATE

Reteaching Activity AL

Have students work in small groups. This strategy allows every student to have an opportunity to speak several times. Ask a question or give a prompt about radical expressions such as "Name one thing to keep in mind as you simplify radical expressions." Then pass a stick or other object to the student. The student speaks, everyone listens, and then passes the object to the next person. The next student speaks, everyone listens, then the student passes the object on until everyone has one or two turns.

Although two radicals, such as $\sqrt{18}$ and $\sqrt{32}$, may not appear to be like radicals, if you simplify each radical, you can see that $\sqrt{18} = 3\sqrt{2}$ and $\sqrt{32} = 4\sqrt{2}$. These simplified expressions are like radical expressions and can be combined.

Radicals with the same index can be multiplied by using the Product Property of Radicals. If the radicals have coefficients before the radical symbol, multiply the coefficients. Then, multiply the radicands of each expression. To multiply radical expressions with more than one term, you can use the Distributive Property or FOIL method.

$(\sqrt[3]{6} + 1)(\sqrt[3]{2} - \sqrt[3]{7}) = \sqrt[3]{12} - \sqrt[3]{42} + \sqrt[3]{2} - \sqrt[3]{7}$

Think About It!

Complete the statement to write a general method for multiplying radicals with coefficients.

For any real numbers a, b, c, and d and any integer $n > 1$, $c\sqrt[n]{a} \cdot d\sqrt[n]{b} =$ $cd\sqrt[n]{ab}$.

Example 3 Add and Subtract Radicals

Simplify $6\sqrt{45x} + \sqrt{12} - 3\sqrt{20x}$.

$= 6\sqrt{3^2 \cdot 5x} + \sqrt{2^2 \cdot 3} - 3\sqrt{2^2 \cdot 5x}$ — Factor using squares.

$= 6(\sqrt{3^2} \cdot \sqrt{5x}) + (\sqrt{2^2} \cdot \sqrt{3}) - 3(\sqrt{2^2} \cdot \sqrt{5x})$ — Product Property

$= 6(3\sqrt{5x}) + (2\sqrt{3}) - 3(2\sqrt{5x})$ — Simplify radicals.

$= 18\sqrt{5x} + 2\sqrt{3} - 6\sqrt{5x}$ — Multiply.

$= 12\sqrt{5x} + 2\sqrt{3}$ — Simplify.

Check

Write the simplified form of $\sqrt{18x} - 5\sqrt{28} - 3\sqrt{98x} + 3\sqrt{7x}$.

$3\sqrt{7x} - 18\sqrt{2x} - 10\sqrt{7}$

Think About It!

In Example 4, $\sqrt[5]{-1}$ simplifies to -1 because 5 is an odd root. What would happen if the example used an even root?

Sample answer: If n were even, $\sqrt[n]{-1}$ would be imaginary.

Example 4 Multiply Radicals

Simplify $4\sqrt[5]{-10x^2y^6} \cdot 3\sqrt[5]{16x^4y^4}$.

$= 4 \cdot 3 \cdot \sqrt[5]{-10x^2y^6 \cdot 16x^4y^4}$ — Product Property of Radicals

$= 12 \cdot \sqrt[5]{-1 \cdot 2 \cdot 5 \cdot x^2y^6 \cdot 2^4 \cdot x^4y^4}$ — Factor the constants.

$= 12 \cdot \sqrt[5]{-1 \cdot 2^5 \cdot 5 \cdot x^5 \cdot x \cdot y^{10}}$ — Group into powers of 5.

$= 12 \cdot \sqrt[5]{-1} \cdot \sqrt[5]{2^5} \cdot \sqrt[5]{5} \cdot \sqrt[5]{x^5} \cdot \sqrt[5]{x} \cdot \sqrt[5]{y^{10}}$ — Product Power of Radicals

$= 12 \cdot (-1) \cdot 2 \cdot \sqrt[5]{5} \cdot x \cdot \sqrt[5]{x} \cdot y^2$ — Simplify.

$= -24xy^2\sqrt[5]{5x}$ — Multiply.

Check

Write the simplified form of $5\sqrt[5]{-9x^3y^5} \cdot 3\sqrt[5]{27x^4y^5}$.

$-45xy^2\sqrt[5]{x^2}$

Study Tip

Negative Radicands If a radicand has a negative constant, it may be helpful to use -1 as a factor. Then you can simplify the nth root of -1, which will be -1 if n is odd and i if n is even.

Go Online You can complete an Extra Example online.

Interactive Presentation

Add and Subtract Radicals

Simplify $6\sqrt{45x} + \sqrt{12} - 3\sqrt{20x}$.

$6\sqrt{45x} + \sqrt{12} - 3\sqrt{20x}$

$= 6\sqrt{3^2 \cdot 5x} + \sqrt{2^2 \cdot 3} - 3\sqrt{2^2 \cdot 5x}$ — Factor using squares.

$= 6\left(\sqrt{3^2} \cdot \sqrt{5x}\right) + \left(\sqrt{2^2} \cdot \sqrt{3}\right) - 3\left(\sqrt{2^2} \cdot \sqrt{5x}\right)$ — Product Property

$= 6\left(3\sqrt{5x}\right) + \left(2\sqrt{3}\right) - 3\left(2\sqrt{5x}\right)$ — Simplify radicals.

$= 18\sqrt{5x} + 2\sqrt{3} - 6\sqrt{5x}$ — Multiply.

$= 12\sqrt{5x} + 2\sqrt{3}$ — $(18 - 6)\sqrt{5x} = 12\sqrt{5x}$

Example 3

TYPE

Students answer a question about even and odd roots.

Example 5 Use the Distributive Property to Multiply Radicals

SPORTS **A sports pennant has the dimensions shown. Find the area, in square inches.**

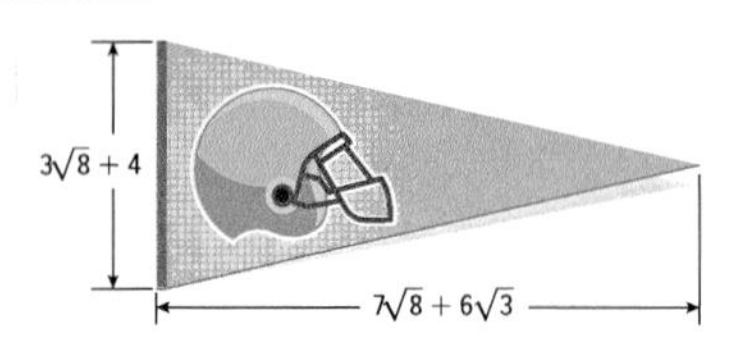

Area $= \frac{1}{2} \cdot$ base $\cdot$ height, so the area is $\frac{1}{2}(3\sqrt{8} + 4)(\underline{7\sqrt{8}} + \underline{6\sqrt{3}})$.

$= \frac{1}{2} \cdot [3\sqrt{8} \cdot 7\sqrt{8} + 3\sqrt{8} \cdot 6\sqrt{3} + 4 \cdot 7\sqrt{8} + 4 \cdot 6\sqrt{3}]$

$= \frac{1}{2} \cdot [21\sqrt{8^2} + 18\sqrt{24} + 28\sqrt{8} + 24\sqrt{3}]$

$= \frac{1}{2} \cdot [21\sqrt{8^2} + 18\sqrt{2^2 \cdot 6} + 28\sqrt{2^2 \cdot 2} + 24\sqrt{3}]$

$= \frac{1}{2} \cdot [21\sqrt{8^2} + 18 \cdot \sqrt{2^2} \cdot \sqrt{6} + 28 \cdot \sqrt{2^2} \cdot \sqrt{2} + 24\sqrt{3}]$

$= \frac{1}{2} \cdot [\underline{168} + \underline{36}\sqrt{6} + \underline{56}\sqrt{2} + \underline{24}\sqrt{3}]$

$= \underline{84} + 18\sqrt{6} + \underline{28}\sqrt{2} + 12\sqrt{3}$

The area of the pennant is $84 + 18\sqrt{6} + 28\sqrt{2} + 12\sqrt{3}$ in^2, or about $\underline{188.5}$ in^2.

Watch Out!

Simplest Form Do not forget to check that your result is in simplest form. Make sure that none of the individual radicals can be further simplified or combined.

Check

POOLS A rectangular pool safety cover has a length of $7\sqrt{10} - 4$ feet and a width of $6\sqrt{10} + 8\sqrt{5}$ feet. Which expression represents the area of the pool cover in simplest form? $\underline{\text{C}}$

A. $420 + 280\sqrt{2} + 24\sqrt{10} + 32\sqrt{5}$ ft^2

B. $42\sqrt{100} + 280\sqrt{2} - 24\sqrt{10} - 32\sqrt{5}$ ft^2

C. $420 + 280\sqrt{2} - 24\sqrt{10} - 32\sqrt{5}$ ft^2

D. $420 + 56\sqrt{50} - 24\sqrt{10} - 32\sqrt{5}$ ft^2

Learn Rationalizing the Denominator

If a radical expression contains a radical in the denominator, you can rationalize the denominator to simplify the expression. Recall that to rationalize a denominator, you should multiply the numerator and denominator by a quantity so that the radicand has an exact root.

336 Module 6 • Inverse and Radical Functions

Interactive Presentation

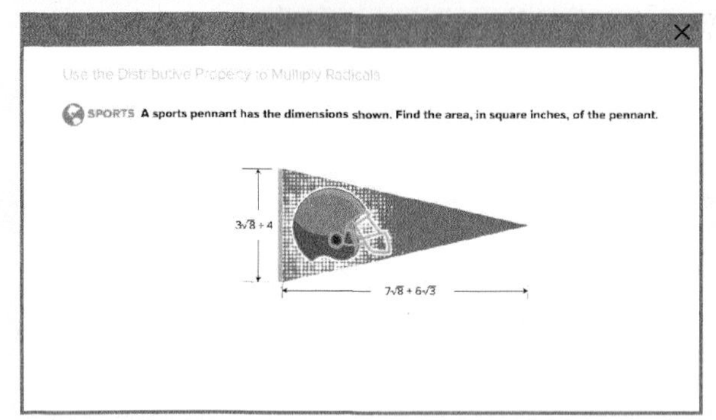

Example 5

TAP

Students can tap to see a tip about writing results in simplest form.

CHECK

Students complete the Check online to determine whether they are ready to move on.

Example 5 Use the Distributive Property to Multiply Radicals

Teaching the Mathematical Practices

4 Apply Mathematics In Example 5, students will apply what they have learned about adding, subtracting, and multiplying radical expressions to solve a real-world problem.

Questions for Mathematical Discourse

AL Why does the exact solution have four terms? None of the terms have the same radicand, so they cannot be combined. 188.5 is an approximation.

OL How would you multiply a radical expression with three terms by a radical expression with two terms? Use the Distributive Property and multiply each term of the first expression by each term in the second expression.

BL What is the maximum number of terms that will result from multiplying two radical expressions? The maximum number of terms will be the number of terms of the first expression times the number of terms of the second expression.

DIFFERENTIATE

Reteaching Activity **AL**

IF when presented with a radical expression such as $11 + 6\sqrt{3}$, some students persist in trying to add 11 and 6,

THEN to help them understand why this cannot be done, compare the radical expression $11 + 6\sqrt{3}$ to the expression $11 + 6x$. Stress that the radical $6\sqrt{3}$ is a multiplication expression just like $6x$. Remind students that the order of operations requires that multiplication be performed before addition.

Learn Rationalizing the Denominator

Objective

Students divide and simplify radical expressions by rationalizing the denominator.

Teaching the Mathematical Practices

3 Analyze Cases Guide students to examine the cases of rationalizing the denominator. Encourage students to familiarize themselves with each case and know when to apply each.

7 Use Structure Help students to understand how the structure of conjugates is used to rationalize the denominator.

Things to Remember

The product of conjugates is always a rational number.

Example 6 Rationalize the Denominator

Teaching the Mathematical Practices

1 Seek Information Mathematically proficient students must be able to transform algebraic expressions to reach solutions. In Example 6, students must use properties of radicals and the process of rationalizing the denominator to transform and simplify the expression.

Questions for Mathematical Discourse

AL When rationalizing the denominator, why are the numerator and denominator both multiplied by $\sqrt[3]{(7^2a^2)}$? Sample answer: this is done so that there will no longer be a radical in the denominator. The term must be multiplied to both the numerator and denominator so that the entire expression is only being multiplied by 1.

OL Why is the radicand of the expression used to rationalize the denominator $\sqrt[3]{7^2a^2}$? To simplify the radical from the denominator, the resulting radicand after the multiplication must be cubed factors.

BL What is another way that you could have begun to simplify the expression? cancel a from numerator and denominator

If the denominator is:	Multiply the numerator and denominator by:	Examples
$\sqrt{b}$	$\sqrt{b}$	$\frac{4}{\sqrt{7}} = \frac{4}{\sqrt{7}} \cdot \frac{\sqrt{7}}{\sqrt{7}}$ or $\frac{4\sqrt{7}}{7}$
$\sqrt[n]{b^x}$	$\sqrt[n]{b^{n-x}}$	$\frac{3}{\sqrt[5]{2}} = \frac{3}{\sqrt[5]{2}} \cdot \frac{\sqrt[5]{2^4}}{\sqrt[5]{2^4}}$ or $\frac{3\sqrt[5]{16}}{2}$

Binomials of the form $a\sqrt{b} + c\sqrt{d}$ and $a\sqrt{b} - c\sqrt{d}$, where a, b, c, and d are rational numbers, are called **conjugates** of each other. Multiplying the numerator and denominator by the conjugate of the denominator will eliminate the radical from the denominator of the expression.

Example 6 Rationalize the Denominator

Simplify $\sqrt[3]{\frac{250a^6}{7a}}$.

$\sqrt[3]{\frac{250a^6}{7a}} = \frac{\sqrt[3]{250a^6}}{\sqrt[3]{7a}}$ Quotient Property of Radicals

$= \frac{\sqrt[3]{5^3 \cdot 2 \cdot (a^2)^3}}{\sqrt[3]{7a}}$ Factor into cubes.

$= \frac{\sqrt[3]{5^3} \cdot \sqrt[3]{2} \cdot \sqrt[3]{(a^2)^3}}{\sqrt[3]{7a}}$ Product Property of Radicals

$= \frac{5a^2\sqrt[3]{2}}{\sqrt[3]{7a}}$ Simplify.

$= \frac{5a^2\sqrt[3]{2}}{\sqrt[3]{7a}} \cdot \frac{\sqrt[3]{7^2a^2}}{\sqrt[3]{7^2a^2}}$ Rationalize the denominator.

$= \frac{5a^2\sqrt[3]{2 \cdot 7^2a^2}}{\sqrt[3]{7a \cdot 7^2a^2}}$ Product Property of Radicals

$= \frac{5a^2\sqrt[3]{98a^2}}{\sqrt[3]{7^3a^3}}$ Multiply.

$= \frac{5a^2\sqrt[3]{98a^2}}{7a}$ $\sqrt[3]{7^3a^3} = 7a$

$= \frac{5a\sqrt[3]{98a^2}}{7}$ Simplify.

Check

Write the simplified form of $\sqrt{\frac{20b}{3b^5}}$.

$\frac{2\sqrt{15}}{3b^2}$

Go Online You can complete an Extra Example online.

Think About It!
Why is the product of $\sqrt[n]{b^x}$ and $\sqrt[n]{b^{n-x}}$ an exact root?

Sample answer: Using the Product Property of Radicals, $\sqrt[n]{b^x} \cdot \sqrt[n]{b^{n-x}} = \sqrt[n]{b^x \cdot b^{n-x}}$. Then, by the Product of Powers, $\sqrt[n]{b^x \cdot b^{n-x}} = \sqrt[n]{b^{x+n-x}} = \sqrt[n]{b^n} = b$ or $|b|$ if both x and n are even.

Think About It!
How does multiplying conjugates relate to the difference of squares identity that can be used when multiplying binomials?

Sample answer: The difference of squares says that $(a + b)(a - b) = a^2 - b^2$. Similarly, when two conjugates are multiplied, the products of the *outer* and *inner* terms are subtracted and you are left with $(a\sqrt{b})^2 - (c\sqrt{d})^2$, or $a^2b - c^2d$.

Watch Out!
Rationalizing the Denominator When determining the quantity to multiply by when rationalizing the denominator, make sure you raise the entire term under the radical to the power of $n - x$.

Interactive Presentation

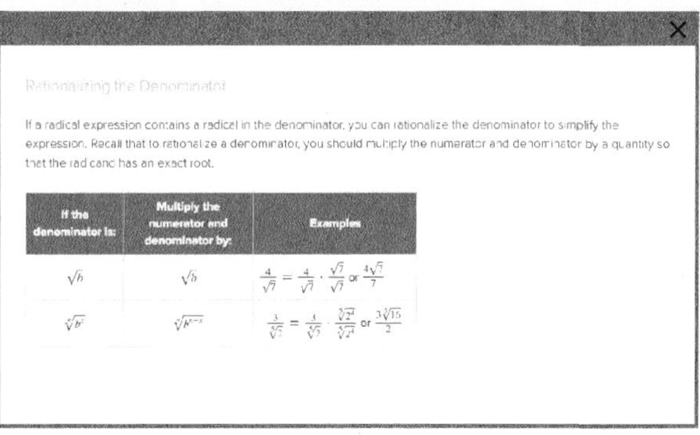

Learn

Students use the properties of radicals to simplify an expression.

Example 7 Use Conjugates to Rationalize the Denominator

Simplify $\frac{4x}{2\sqrt{7}-5}$.

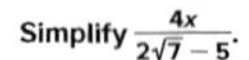

To rationalize the denominator, multiply the numerator and denominator by the conjugate of $2\sqrt{7}-5$.

$$\frac{4x}{2\sqrt{7}-5} = \frac{4x}{2\sqrt{7}-5} \cdot \frac{2\sqrt{7}+5}{2\sqrt{7}+5}$$ $2\sqrt{7}+5$ is the conjugate of $2\sqrt{7}-5$.

$$= \frac{4x(2\sqrt{7}) + 4x(5)}{(2\sqrt{7})^2 + 5(2\sqrt{7}) - 5(2\sqrt{7}) - 5(5)}$$ Multiply.

$$x = \frac{8x\sqrt{7}+20x}{28+10\sqrt{7}-10\sqrt{7}-25}$$ Simplify.

$$= \frac{8x\sqrt{7}+20x}{3}$$ Subtract.

Check

Write the simplified form of $\frac{6x}{4\sqrt{5}-4}$.

$\frac{3x\sqrt{5}+3x}{8}$

Pause and Reflect

Did you struggle with anything in this lesson? If so, how did you deal with it?

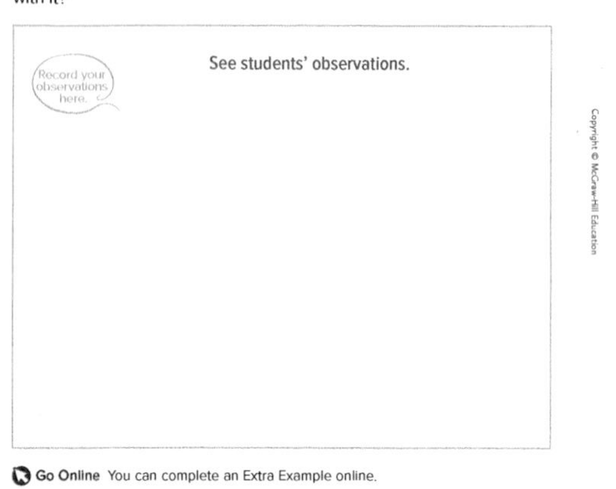

Go Online You can complete an Extra Example online.

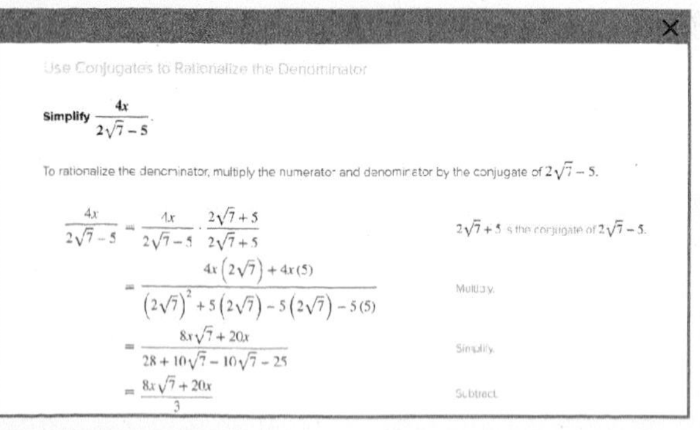

Example 7

TYPE

Students compare and contrast the method used in Example 7 with an alternative method.

CHECK

Students complete the Check online to determine whether they are ready to move on.

Example 7 Use Conjugates to Rationalize the Denominator

Teaching the Mathematical Practices

7 Use Structure Help students to use the structure of the denominator to determine what to multiply the expression by to rationalize the denominator and simplify the expression.

Questions for Mathematical Discourse

AL Why is a conjugate used to rationalize the denominator? When the denominator is multiplied with the conjugate, the radical terms cancel, and the resulting denominator is rational.

OL Why are both the numerator and denominator multiplied by the conjugate when it is the denominator we want to rationalize? Sample answer: Multiplying both the numerator and denominator by the conjugate means the expression is being multiplied by 1, so it does not change the value of the expression.

BL Does the order of the terms matter for determining the conjugate? Explain. No; sample answer: as long as one term keeps the same sign while the other term changes sign, and the numerator and denominator are multiplied by the same expression, the result will be the same.

Common Error

Some students may try to multiply the numerator and denominator by the denominator. Remind them to multiply the numerator and denominator by the conjugate of the denominator.

Exit Ticket

Recommended Use

At the end of class, go online to display the Exit Ticket prompt and ask students to respond using a separate piece of paper. Have students hand you their responses as they leave the room.

Alternate Use

At the end of class, go online to display the Exit Ticket prompt and ask students to respond verbally or by using a mini-whiteboard. Have students hold up their whiteboards so that you can see all student responses. Tap to reveal the answer when most or all students have completed the Exit Ticket.

Practice and Homework

Suggested Assignments

Use the table below to select appropriate exercises.

DOK	Topic	Exercises
1, 2	exercises that mirror the examples	1–38
2	exercises that use a variety of skills from this lesson	39–51
2	exercises that extend concepts learned in this lesson to new contexts	52–55
3	exercises that emphasize higher-order and critical thinking skills	56–60

ASSESS AND DIFFERENTIATE

Use the data from the **Checks** to determine whether to provide resources for extension, remediation, or intervention.

IF students score 90% or more on the Checks, **THEN** assign: BL

- Practice Exercises 1–55 odd, 56–60
- Extension: Special Products with Radicals
- ALEKS® Simplifying Expressions

IF students score 66%–89% or more on the Checks, **THEN** assign: OL

- Practice Exercises 1–59 odd
- Remediation, Review Resources: Simplifying Radical Expressions
- Personal Tutors
- Extra Examples 1–7
- ALEKS® Other Topics Available: Radicals

IF students score 65% or less on the Checks, **THEN** assign: AL

- Practice Exercises 1–37 odd
- Remediation, Review Resources: Simplifying Radical Expressions
- *Quick Review Math Handbook*: Operations with Radical Expressions
- ALEKS® Other Topics Available: Radicals

Name ______ Period ______ Date ______

Practice

Go Online You can complete your homework online.

Examples 1 and 2

Simplify.

1. $\sqrt{72a^8b^5}$ $6a^4b^2\sqrt{2b}$
2. $\sqrt{9a^{15}b^3}$ $3a^7b\sqrt{ab}$
3. $\sqrt{24a^{16}b^8c}$ $2a^8b^4\sqrt{6c}$
4. $\sqrt{18a^6b^3c^5}$ $3|a^3|bc^2\sqrt{2bc}$
5. $\sqrt[4]{64a^4b^4}$ $2|ab|\sqrt[4]{4}$
6. $\sqrt[3]{-8d^2f^5}$ $-2f\sqrt[3]{d^2f^2}$
7. $\sqrt{\frac{25}{36}r^2t}$ $\frac{5}{6}|r|\sqrt{t}$
8. $\sqrt{\frac{192k^4}{64}}$ $k^2\sqrt{3}$
9. $\sqrt[5]{\frac{3072h^8}{243f^5}}$ $\frac{4h \cdot \sqrt[5]{3h^3}}{3f}$
10. $\sqrt[3]{\frac{432n^{12}}{64q^6}}$ $\frac{3n^4 \cdot \sqrt[3]{2}}{2q^2}$

Example 3

Simplify.

11. $\sqrt{2} + \sqrt{8} + \sqrt{50}$ $8\sqrt{2}$
12. $\sqrt{12} - 2\sqrt{3} + \sqrt{108}$ $6\sqrt{3}$
13. $8\sqrt{5} - \sqrt{45} - \sqrt{80}$ $\sqrt{5}$
14. $2\sqrt{48} - \sqrt{75} - \sqrt{12}$ $\sqrt{3}$
15. $\sqrt{28x} - \sqrt{14} + \sqrt{63x}$ $5\sqrt{7x} - \sqrt{14}$
16. $\sqrt{135} + 5\sqrt{10d} - 3\sqrt{60}$ $5\sqrt{10d} - 3\sqrt{15}$

Example 4

Simplify.

17. $3\sqrt{5y} \cdot 8\sqrt{10yz}$ $120y\sqrt{2z}$
18. $2\sqrt{32a^3b^5} \cdot \sqrt{8a^7b^2}$ $32a^5b^3\sqrt{b}$
19. $6\sqrt{3ab} \cdot 4\sqrt{24ab^3}$ $144ab^2\sqrt{2}$
20. $5\sqrt{x^8y^3} \cdot 5\sqrt{2x^5y^4}$ $25x^6y^3\sqrt{2xy}$
21. $5\sqrt{2x} \cdot 3\sqrt{7x^2y^3}$ $15xy\sqrt{14xy}$
22. $3\sqrt{a^5b^7} \cdot 2\sqrt{5a^7b^3}$ $6a^6b^5\sqrt{5}$

Example 5

23. TRAMPOLINE There are two trampoline runways at a gymnastics practice facility. Both runways are $\sqrt{3}$ meters wide. One is $6\sqrt{3}$ meters long and the other is $5\sqrt{2}$ meters long. What is the total area of the trampoline runways? $18 + 5\sqrt{6}$ m²

24. DISTANCE Jayla walks 5 blocks north, then 8 blocks east to get to the library. Each block is $5\sqrt{10}$ yards long. If Jayla could walk in a straight line to the library instead, how far would the walk be, in yards? $5\sqrt{890}$ yd

Simplify.

25. $(7\sqrt{2} - 3\sqrt{3})(4\sqrt{6} + 3\sqrt{12})$ $56\sqrt{3} + 42\sqrt{6} - 36\sqrt{2} - 54$
26. $(8\sqrt{5} - 6\sqrt{3})(8\sqrt{5} + 6\sqrt{3})$ 212
27. $(12\sqrt{10} - 6\sqrt{5})(12\sqrt{10} + 6\sqrt{5})$ 1260
28. $(6\sqrt{3} + 5\sqrt{2})(2\sqrt{6} + 3\sqrt{8})$ $36\sqrt{2} + 36\sqrt{6} + 20\sqrt{3} + 60$

Examples 6 and 7

Simplify.

29. $\frac{\sqrt{5a^5}}{\sqrt{b^{13}}}$ $\frac{a^2\sqrt{5ab}}{b^7}$
30. $\frac{\sqrt{7x}}{\sqrt{10x^3}}$ $\frac{\sqrt{70}}{10x}$
31. $\frac{3\sqrt[3]{6x^2}}{3\sqrt[3]{5y}}$ $\frac{\sqrt[3]{150x^2y^2}}{5y}$
32. $\sqrt[4]{\frac{7x^3}{4b^2}}$ $\frac{\sqrt[4]{28b^2x^3}}{2|b|}$
33. $\frac{6}{\sqrt{3} - \sqrt{2}}$ $6\sqrt{3} + 6\sqrt{2}$
34. $\frac{\sqrt{2}}{\sqrt{5} - \sqrt{3}}$ $\frac{\sqrt{10} + \sqrt{6}}{2}$
35. $\frac{9 - 2\sqrt{3}}{\sqrt{3} + 6}$ $\frac{20 - 7\sqrt{3}}{11}$
36. $\frac{2\sqrt{2} + 2\sqrt{5}}{\sqrt{5} + \sqrt{2}}$ 2
37. $\frac{3\sqrt{7}}{\sqrt{5} - 1}$ $\frac{3\sqrt{7} + 3\sqrt{35}}{4}$
38. $\frac{7x}{3 - \sqrt{2}}$ $3x + x\sqrt{2}$

Mixed Exercises

Simplify.

39. $\sqrt[3]{16y^4z^{12}}$ $2yz^4\sqrt[3]{2y}$
40. $\sqrt[3]{-54x^6y^{11}}$ $-3x^2y^3\sqrt[3]{2y^2}$
41. $\frac{x + 1}{\sqrt{x} - 1}$ $\frac{(x + 1)(\sqrt{x} + 1)}{x - 1}$ or $\frac{x\sqrt{x} + \sqrt{x} + x + 1}{x - 1}$
42. $\frac{x - 2}{\sqrt{x^2 - 4}}$ $\frac{\sqrt{x^2 - 4}}{x + 2}$
43. $3\sqrt{24x} - 2\sqrt{54x} + \sqrt{48}$ $4\sqrt{3}$
44. $5\sqrt{18c} + 3\sqrt{72c} + 6\sqrt{76}$ $33\sqrt{2c} + 12\sqrt{19}$
45. $10\sqrt{175a} - 4\sqrt{112a} - 2\sqrt{63a}$ $28\sqrt{7a}$
46. $7\sqrt{204y} + 4\sqrt{459y} - 8\sqrt{140y}$ $26\sqrt{51y} - 16\sqrt{35y}$

47. VOLUME McKenzie has a rectangular prism with dimensions 20 inches by 35 inches by 40 inches. She would like to replace it with a cube with the same volume. What should the length of a side of the cube be? Express your answer as a radical expression in simplest form. $10\sqrt[3]{28}$ in.

48. MUSIC Traditionally, musical instruments are tuned so that the note A has a frequency of 440 Hertz. With each note higher on the instrument, the frequency of the pitch is multiplied by a factor of $\sqrt[12]{2}$. What is the ratio of the frequencies of two notes that are 6 steps apart on the instrument? What is the ratio of the frequencies of two notes that are 9 steps apart on the instrument? Express your answers in simplest form. $\sqrt{2}$; $\sqrt[4]{8}$

1 CONCEPTUAL UNDERSTANDING 2 FLUENCY 3 APPLICATION

Name ______ Period ______ Date ______

49. PHYSICS The speed of a wave traveling over a string is given by $\frac{\sqrt{t}}{\sqrt{u}}$, where t is the tension of the string and u is the density. Simplify the expression. $\frac{\sqrt{tu}}{u}$

50. LIGHT Suppose a light has a brightness intensity of I_1 when it is at a distance of d_1 and a brightness intensity of I_2 when it is at a distance of d_2. These quantities are related by the equation $\frac{d_2}{d_1} = \sqrt{\frac{I_1}{I_2}}$. If $I_1 = 50$ units and $I_2 = 24$ units, find $\frac{d_2}{d_1}$. Express your answer in simplest form. $\frac{5\sqrt{3}}{6}$

51. RACING Jay likes to race his younger brother while running. To make the race fair, John and Jay start at different locations, but finish at the same point. The diagram shows their running paths. Both of them finished the race in exactly 4 minutes.

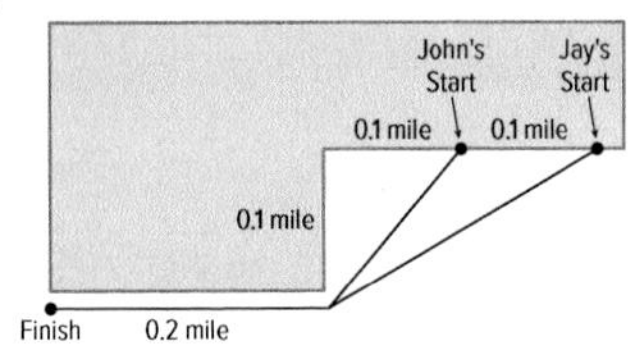

a. If John and Jay continued at their average paces during the race, exactly how many minutes would it take them each to run a mile? Express your answer as a radical expression in simplest form.
John: $\frac{0.8 - 4\sqrt{0.02}}{0.02}$ or $40 - 20\sqrt{2}$ min; Jay: $\frac{0.8 - 4\sqrt{0.05}}{-0.01}$ or $40\sqrt{5} - 80$ min

b. Exactly how many times faster is Jay compared to John? Express your answer as a radical expression in simplest form. $2 - \sqrt{2} + \sqrt{5} - \frac{\sqrt{10}}{2}$

52. STRUCTURE Write the ratio of the side lengths of the two cubes described in simplest form.

a. The volumes of the two cubes are $270x$ cubic inches and $32x^2$ cubic inches. $\frac{3\sqrt[3]{20x^2}}{4x}$

b. The surface areas of the two cubes are $6x^4$ square feet and $6(x + 1)$ square feet. $\frac{x^2\sqrt{x+1}}{x+1}$

53. REGULARITY Rewrite each of the following expressions as a single expression in the form ax^m for appropriate choices of a and m. Show your work.

a. $\sqrt{x}(\sqrt{x} + \sqrt{4x})$ $3x$; $(\sqrt{x} \cdot \sqrt{x}) + (\sqrt{x} \cdot \sqrt{4x}) = (\sqrt{x^2} + \sqrt{4x^2}) = x + 2x = 3x$

b. $\sqrt{x}(\sqrt{x} + \sqrt{4x} + \sqrt{9x})$ $6x$; $(\sqrt{x} \cdot \sqrt{x}) + (\sqrt{x} \cdot \sqrt{4x}) + (\sqrt{x} + \sqrt{9x}) = (\sqrt{x^2} + \sqrt{4x^2} + \sqrt{9x^2}) = x + 2x + 3x = 6x$

c. $\sqrt{x}(\sqrt{x} + \sqrt{4x} + \sqrt{9x} + \sqrt{16x})$ $10x$; $(\sqrt{x} \cdot \sqrt{x}) + (\sqrt{x} \cdot \sqrt{4x}) + (\sqrt{x} \cdot \sqrt{9x}) + (\sqrt{x} \cdot \sqrt{16x}) = (\sqrt{x^2} + \sqrt{4x^2} + \sqrt{9x^2} + \sqrt{16x^2}) = x + 2x + 3x + 4x = 10x$

d. More generally, simplify $\sqrt{x}(\sqrt{x} + \sqrt{4x} + \cdots + \sqrt{n^2x})$ for any positive integer n.
Use the fact that $1 + 2 + \cdots + n = \frac{n(n+1)}{2}$.
$(\sqrt{x} \cdot \sqrt{x}) + (\sqrt{x} \cdot \sqrt{4x}) + \cdots + (\sqrt{x} + \sqrt{n^2x}) = x + 2x + \cdots + nx = (1 + 2 + \cdots + n)x = \left(\frac{n(n+1)}{2}\right)x$

54. USE A MODEL If the area of the trapezoid shown is 200 square feet, what is the height h of the trapezoid? $\frac{1400\sqrt{3} - 1000\sqrt{2}}{97}$

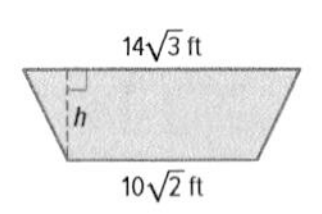

55. CONSTRUCT ARGUMENTS A spherical paperweight with a volume of 72π cubic centimeters is to be packaged in a gift box that is a cube. There must be at least 2 centimeters of packing material around the paperweight to protect it during shipping. The formula for the volume of a sphere is $V = \frac{4}{3}\pi r^3$.

a. Write an expression for the minimum length of a side of the gift box. Show your work.
$6\sqrt[3]{2} + 4$; $r = \sqrt[3]{\frac{3}{4\pi} \cdot 72\pi} = \sqrt[3]{54} = 3\sqrt[3]{2}$; $s = 2 \cdot 3\sqrt[3]{2} + 2(2)$

b. The shipper wants to use a box with a volume of 384 cubic centimeters that they already have in inventory. Is this box suitable? Justify your argument. No; sample answer: A box with a volume of 384 cm³ has a side that is $\sqrt[3]{384}$, or about 7.27 cm; $s = 6\sqrt[3]{2} + 4 \approx 11.55$ is the least possible value for a side of the gift box.

Higher-Order Thinking Skills

56. FIND THE ERROR Twyla and Brandon are simplifying $4\sqrt{32} + 6\sqrt{18}$. Is either of them correct? Explain your reasoning. See margin.

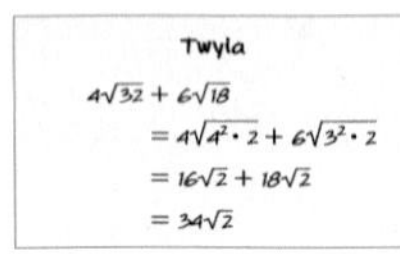

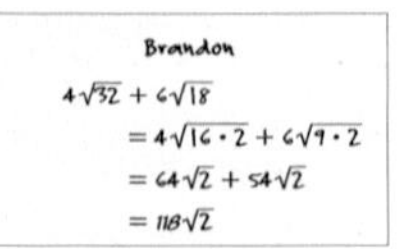

57. PERSEVERE Find four combinations of whole numbers that satisfy $\sqrt[a]{256} = b$.
$a = 1, b = 256$; $a = 2, b = 16$; $a = 4, b = 4$; $a = 8, b = 2$

58. PERSEVERE Show that $\frac{-1 - i\sqrt{3}}{2}$ is a cube root of 1. See margin.

59. WRITE Explain why absolute values may be unnecessary when an nth root of an even power results in an odd power. See margin.

60. WHICH ONE DOESN'T BELONG? Determine which of the radical expressions doesn't belong. Justify your conclusion.
$\sqrt[8]{256g^4h^{16}}$; Sample answer: The other two simplify to $2g^2\sqrt{h}$.

$\sqrt[4]{16g^8h^2}$ $\sqrt[6]{64g^{12}h^3}$

Answers

56. Twyla; Brandon's mistakes were multiplying the 4 by 16 instead of 4 and multiplying the 6 by 9 instead of 3.

58.
$$\left(\frac{-1 - i\sqrt{3}}{2}\right)^3 = \left(\frac{-1 - i\sqrt{3}}{2}\right)\left(\frac{-1 - i\sqrt{3}}{2}\right)\left(\frac{-1 - i\sqrt{3}}{2}\right)$$
$$= \frac{(-1 - i\sqrt{3})(-1 - i\sqrt{3})(-1 - i\sqrt{3})}{8}$$
$$= \frac{(1 + i\sqrt{3} + i\sqrt{3} + 3i^2)(-1 - i\sqrt{3})}{8}$$
$$= \frac{(2i\sqrt{3} - 2)(-1 - i\sqrt{3})}{8}$$
$$= \frac{-2i\sqrt{3} - 6i^2 + 2 + 2i\sqrt{3}}{8}$$
$$= \frac{-6i^2 + 2}{8} = \frac{8}{8} \text{ or } 1$$

59. Sample answer: It is only necessary to use absolute values when it is possible that n could be odd or even and still be defined. It is when the radicand must be nonnegative in order for the root to be defined that the absolute values are not necessary.

Solving Radical Equations

LESSON GOAL

Students solve radical equations algebraically and by graphing.

1 LAUNCH

 Launch the lesson with a **Warm Up** and an introduction.

2 EXPLORE AND DEVELOP

 **Explore:** Solutions of Radical Equations

Develop:

Solving Radical Equations Algebraically
- Solve a Square Root Equation
- Solve a Cube Root Equation
- Identify Extraneous Solutions
- Solve a Radical Equation

Solving Radical Equations by Graphing
- Solve a Radical Equation by Graphing
- Solve a Radical Equation by Using a System
- Confirm Solutions by Using Technology

 You may want your students to complete the **Checks** online.

3 REFLECT AND PRACTICE

 Exit Ticket

 Practice

DIFFERENTIATE

 View reports of student progress on the **Checks** after each example.

Resources	AL	OL	BL	ELL
Remediation: Operations with Radical Expressions	●	●		●
Extension: Lesser-Known Geometric Formulas		●	●	●

Language Development Handbook

Assign page 40 of the *Language Development Handbook* to help your students build mathematical language related to solving radical equations.

ELL You can use the tips and suggestions on page T40 of the handbook to support students who are building English proficiency.

Suggested Pacing

90 min	1 day
45 min	2 days

Focus

Domain: Algebra

Standards for Mathematical Content:

A.REI.2 Solve simple rational and radical equations in one variable, and give examples showing how extraneous solutions may arise.

Standards for Mathematical Practice:

3 Construct viable arguments and critique the reasoning of others.
5 Use appropriate tools strategically.
8 Look for and express regularity in repeated reasoning.

Coherence

Vertical Alignment

Previous
Students simplified and performed operations with radical expressions.
A.SSE.2

Now
Students solve radical equations algebraically and by graphing.
A.REI.2

Next
Students will solve exponential and logarithmic equations.
A.CED.1, A.REI.11

Rigor

The Three Pillars of Rigor

1 CONCEPTUAL UNDERSTANDING	2 FLUENCY	3 APPLICATION

Conceptual Bridge In this lesson, students expand on their understanding of radical functions to build fluency by solving radical equations. They apply their understanding of radical equations by solving real-world problems.

Mathematical Background

It is always important to check the solution to an equation in the original equation, but it is especially important when both sides of an equation are raised to a power. If a solution is an approximation, it is sometimes difficult to determine whether a discrepancy is due to rounding or if it is an incorrect solution. Students should check exact solutions whenever possible.

Interactive Presentation

Warm Up

Simplify each expression.

1. $\left(\sqrt{3x-1}\right)^2$
2. $\left(\sqrt[3]{2x+5}\right)^3$
3. $\left(7\sqrt{x+6}\right)^2$
4. $\left(\sqrt{x-5}+3\right)^2$

Show Answers

Warm Up

Warm Up

Prerequisite Skills

The Warm Up exercises address the following prerequisite skill for this lesson:

- simplifying radical expressions

Answers:

1. $3x - 1$
2. $2x + 5$
3. $49x + 294$
4. $x + 4 + 6\sqrt{x-5}$

Launch the Lesson

Launch the Lesson

 Teaching the Mathematical Practices

4 Analyze Relationships Mathematically In the Launch the Lesson, encourage students to analyze the mathematical relationships in the formula for the speed of a car.

Go Online to find additional teaching notes and questions to promote classroom discourse.

Today's Standards

Tell students that they will be addressing these content and practice standards in this lesson. You may wish to have a student volunteer read aloud *How can I meet these standards?* and *How can I use these practices?*, and connect these to the standards.

See the Interactive Presentation for I Can statements that align with the standards covered in this lesson.

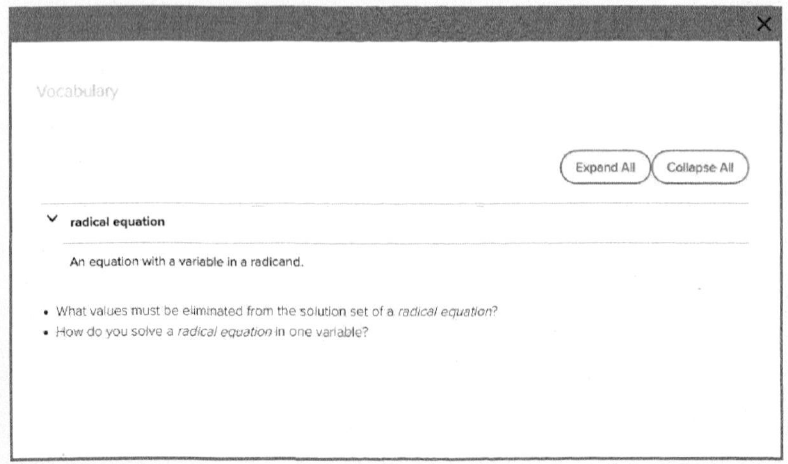

Today's Vocabulary

Today's Vocabulary

Tell students that they will be using this vocabulary term in this lesson. You can expand the row if you wish to share the definition. Then discuss the questions below with the class.

Explore Solutions of Radical Equations

Objective

Students use a sketch to explore solutions of radical equations.

Teaching the Mathematical Practices

3 Reason Inductively Throughout the Explore, students will use inductive reasoning to analyze data and make arguments about when radical equations will have a solution.

5 Use Mathematical Tools Point out that to complete the Explore, students will need to use the sketch. Work with students to deepen their understanding of solutions of radical equations.

Ideas for Use

Recommended Use Present the Inquiry Question, or have a student volunteer read it aloud. Have students work in pairs to complete the Explore activity on their devices. Pairs should discuss each of the questions. Monitor student progress during the activity. Upon completion of the Explore activity, have student volunteers share their responses to the Inquiry Question.

What if my students don't have devices? You may choose to project the activity on a whiteboard. A printable worksheet for each Explore is available online. You may choose to print the worksheet so that individuals or pairs of students can use it to record their observations.

Summary of the Activity

Students will complete guiding exercises throughout the Explore activity. Students will use a sketch to explore when radical equations will have a solution and when they will not have a solution. Then, students will answer the Inquiry Question.

(continued on the next page)

Interactive Presentation

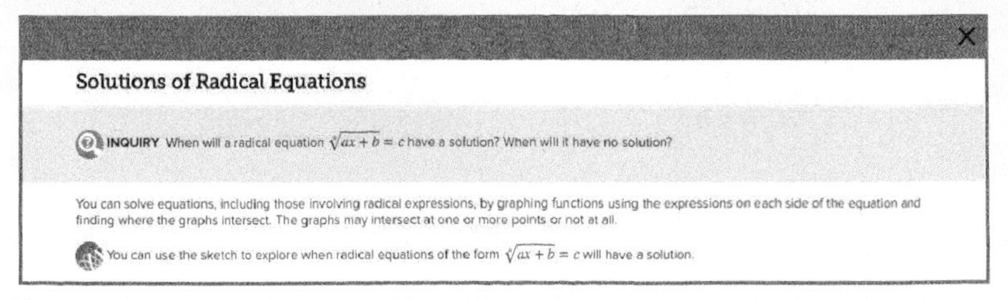

Explore

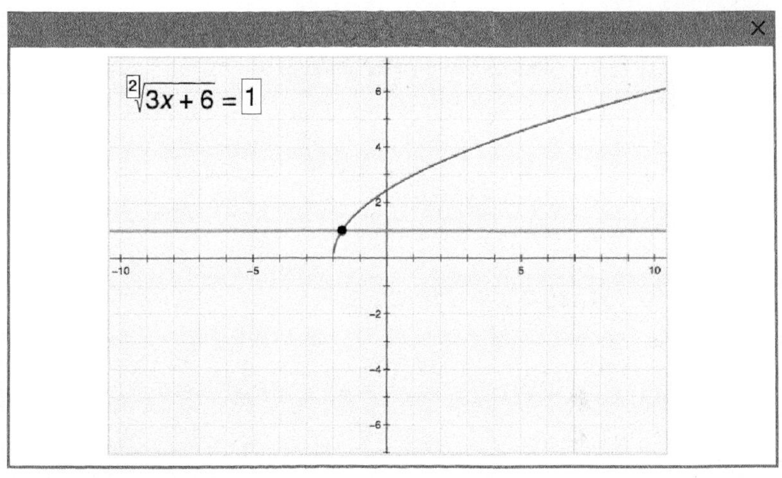

Explore

WEB SKETCHPAD

Students use the sketch to complete an activity in which they explore the solutions of radical equations.

Interactive Presentation

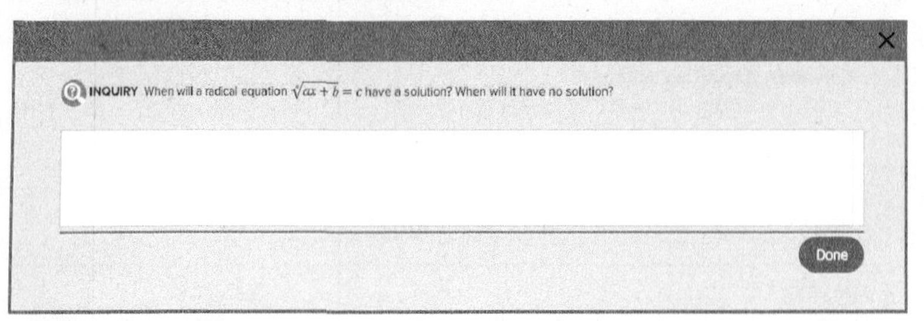

Explore

TYPE

Students respond to the Inquiry Question and can view a sample answer.

Explore Solutions of Radical Equations (*continued*)

Questions

Have students complete the Explore activity.

Ask:

- What must be true about the radicand of $\sqrt{ax + b} = c$ in order to have a real solution? Sample answer: The radicand of a square root must be greater than or equal to zero in order to have a real root.
- How many solutions will $\sqrt[4]{3x + 7} = -2$ have? Why? Sample answer: There will be no real solution because the index is even, and the value of c is negative

Inquiry

When will a radical equation $\sqrt[n]{ax + b} = c$ have a solution? When will it have no solution? Sample answer: If n is odd, there will always be a solution. If n is even, there will be a solution if $c \geq 0$, and there will be no solution if $c < 0$.

Go Online to find additional teaching notes and sample answers for the guiding exercises.

1 CONCEPTUAL UNDERSTANDING | 2 FLUENCY | 3 APPLICATION

Learn Solving Radical Equations Algebraically

Objective

Students solve radical equations in one variable and identify extraneous solutions.

Teaching the Mathematical Practices

8 Look for a Pattern Guide students to see the pattern in how the radical is eliminated when solving a radical equation. Ask students to explain the relationship between the index and the power to which the radical should be raised.

Example 1 Solve a Square Root Equation

Teaching the Mathematical Practices

7 Interpret Complicated Expressions Mathematically proficient students can see complicated expressions as single objects or as being composed of several objects. In Example 1, guide students to see the radical expression as a single object that should be isolated.

Questions for Mathematical Discourse

AL Why are both sides of the equation squared in the second step? Sample answer: the same operation must be performed on both sides of the equation to maintain equality.

OL What is the domain of the related function for this equation? Explain. $\left\{x \mid x \geq \frac{5}{3}\right\}$; the radicand must be positive.

BL What is the y-intercept of the related function for this equation? There is no y-intercept because $x = 0$ yields a negative radicand.

Example 2 Solve a Cube Root Equation

Teaching the Mathematical Practices

1 Check Answers Point out that when solving radical equation, extraneous solutions may be introduced. In Example 2, students need to check their answer.

Questions for Mathematical Discourse

AL If you wrote the equation as a radical equation, what would be the radicand, and what would be the index? The radicand would be $2x + 6$, and the index would be 3.

OL What rule of exponents is used when cubing $(2x + 6)^{\frac{1}{3}}$? Power of a Power Property; to raise an exponent to an exponent, multiply the exponents: $\frac{1}{3} \cdot 3 = 1$.

BL To what exponent would you raise both sides if the exponent of the expression was $\frac{a}{b}$? $\frac{b}{a}$

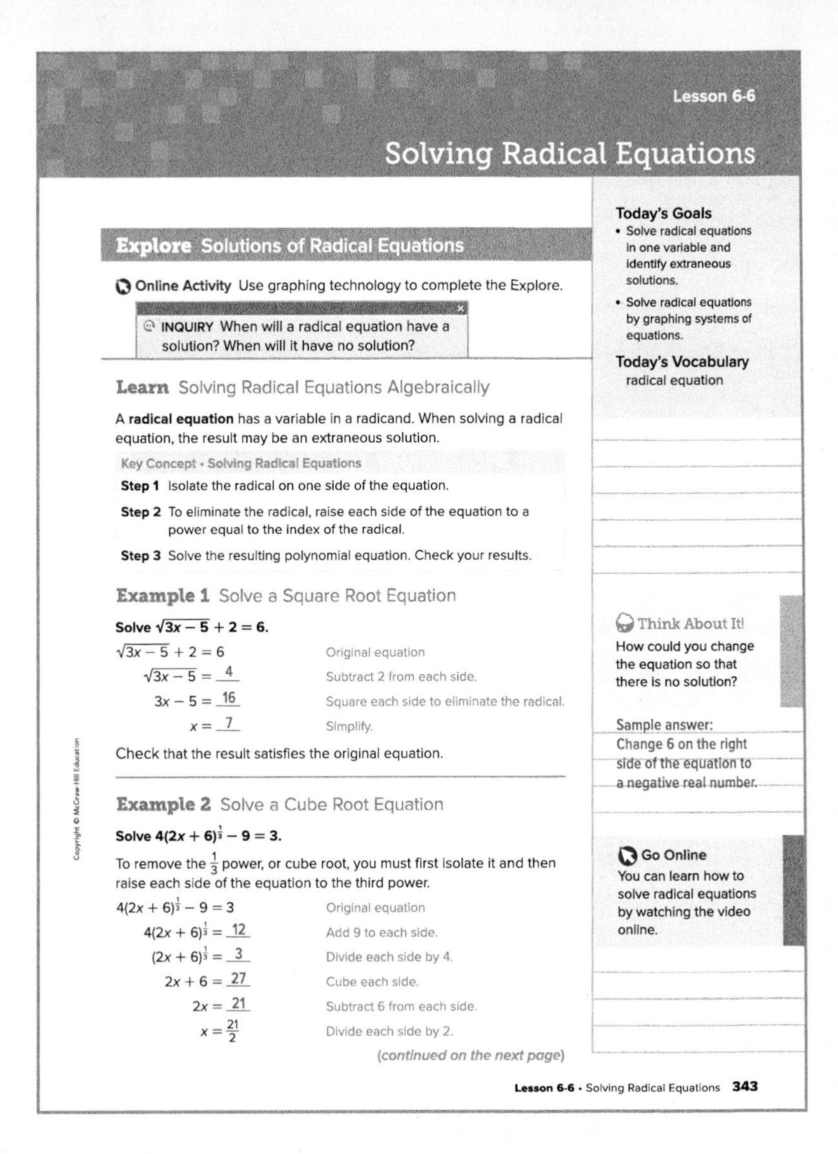

Lesson 6-6

Solving Radical Equations

Today's Goals
- Solve radical equations in one variable and identify extraneous solutions.
- Solve radical equations by graphing systems of equations.

Today's Vocabulary
radical equation

Explore Solutions of Radical Equations

Online Activity Use graphing technology to complete the Explore.

INQUIRY When will a radical equation have a solution? When will it have no solution?

Learn Solving Radical Equations Algebraically

A **radical equation** has a variable in a radicand. When solving a radical equation, the result may be an extraneous solution.

Key Concept • Solving Radical Equations

Step 1 Isolate the radical on one side of the equation.

Step 2 To eliminate the radical, raise each side of the equation to a power equal to the index of the radical.

Step 3 Solve the resulting polynomial equation. Check your results.

Example 1 Solve a Square Root Equation

Solve $\sqrt{3x - 5} + 2 = 6$.

$\sqrt{3x-5} + 2 = 6$	Original equation
$\sqrt{3x-5} = 4$	Subtract 2 from each side.
$3x - 5 = 16$	Square each side to eliminate the radical.
$x = 7$	Simplify.

Check that the result satisfies the original equation.

Think About It! How could you change the equation so that there is no solution?

Sample answer: Change 6 on the right side of the equation to a negative real number.

Example 2 Solve a Cube Root Equation

Solve $4(2x + 6)^{\frac{1}{3}} - 9 = 3$.

To remove the $\frac{1}{3}$ power, or cube root, you must first isolate it and then raise each side of the equation to the third power.

$4(2x+6)^{\frac{1}{3}} - 9 = 3$	Original equation
$4(2x+6)^{\frac{1}{3}} = 12$	Add 9 to each side.
$(2x+6)^{\frac{1}{3}} = 3$	Divide each side by 4.
$2x + 6 = 27$	Cube each side.
$2x = 21$	Subtract 6 from each side.
$x = \frac{21}{2}$	Divide each side by 2.

(continued on the next page)

Go Online You can learn how to solve radical equations by watching the video online.

Copyright © McGraw-Hill Education

Lesson 6-6 • Solving Radical Equations 343

Interactive Presentation

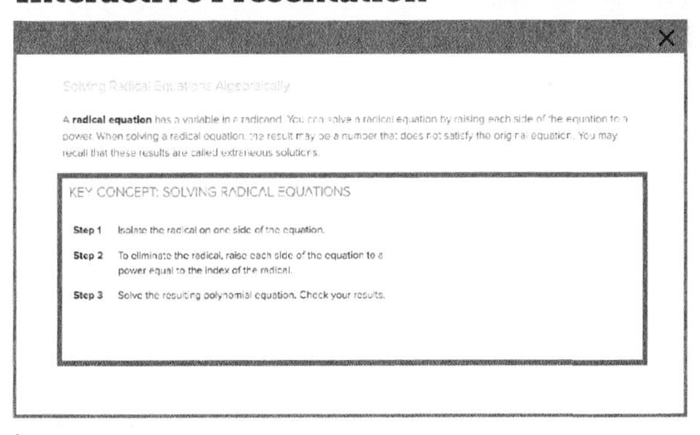

Learn

TYPE

Students determine the power to raise each expression to eliminate the radical.

Your Notes

CHECK

$4(2x+6)^{\frac{1}{3}} - 9 = 3$	Original equation
$4\left(2 \cdot \underline{\frac{21}{2}} + 6\right)^{\frac{1}{3}} - 9 \stackrel{?}{=} 3$	Replace x with $\frac{21}{2}$.
$4(\underline{27})^{\frac{1}{3}} - 9 \stackrel{?}{=} 3$	Simplify.
$4(\underline{3}) - 9 \stackrel{?}{=} 3$	The cube root of 27 is 3.
$\underline{3} = 3$	True

Example 3 Identify Extraneous Solutions

Solve $\sqrt{x+21} = 3 - \sqrt{x}$.

$\sqrt{x+21} = 3 - \sqrt{x}$	Original equation
$x + 21 = \underline{9} - \underline{6}\sqrt{x} + \underline{x}$	Square each side.
$\underline{12} = -6\sqrt{x}$	Isolate the radical.
$\underline{-2} = \sqrt{x}$	Divide each side by −6.
$\underline{4} = x$	Square each side.

CHECK

$\sqrt{x+21} = 3 - \sqrt{x}$	Original equation
$\sqrt{4+21} \stackrel{?}{=} 3 - \sqrt{4}$	Replace x with 4.
$\sqrt{25} \stackrel{?}{=} 3 - 2$	Simplify.
$\underline{5} \neq \underline{1}$	False

The result does not satisfy the original equation, so it is an extraneous solution. Therefore, there is no real solution.

Talk About It! In Example 3, could you tell that 4 was an extraneous solution before checking the result? Explain your reasoning.

Yes; sample answer: Because a square root is never negative, once the equation simplified to $\sqrt{x} = -2$, I could tell that there was no solution.

Example 4 Solve a Radical Equation

Solve $\frac{2}{3}(11x+14)^{\frac{1}{6}} + 8 = 10$.

$\frac{2}{3}(11x+14)^{\frac{1}{6}} + 8 = 10$	Original equation
$\frac{2}{3}(11x+14)^{\frac{1}{6}} = 2$	Subtract 8 from each side.
$(11x+14)^{\frac{1}{6}} = 3$	Multiply each side by $\frac{3}{2}$.
$11x + 14 = 729$	Raise each side to the sixth power.
$11x = 715$	Subtract 14 from each side.
$x = 65$	Divide each side by 11.

The value of 65 does make the equation true.

Go Online You can complete an Extra Example online.

Example 3 Identify Extraneous Solutions

Teaching the Mathematical Practices

3 Justify Conclusions Mathematically proficient students can explain the conclusions drawn when solving a problem. In Example 3 and the Talk About It! feature, students must justify their conclusions about the solution of the equation.

Questions for Mathematical Discourse

AL Why does the right side of the equation have three terms after the first step of squaring each side? Sample answer: The right side of the original equation has two terms. Squaring the expression is the same as multiplying the expression by itself, which requires using the Distributive Property.

OL What is an extraneous solution, and what makes one occur in this case? An extraneous solution is a result that does not satisfy the original equation. In this case it is the result of an even radical being equal to a negative value, when the even radical is defined to yield the nonnegative root.

BL Can an odd radical equation have an extraneous solution? Explain. No; sample answer: odd radicals yield one root that can be positive or negative.

Example 4 Solve a Radical Equation

Teaching the Mathematical Practices

1 Seek Information Mathematically proficient students must be able to transform algebraic expressions to reach solutions. In Example 4, students must transform the radical equation to reach a solution.

Questions for Mathematical Discourse

AL Why should the radical be isolated before raising both sides to the index? Sample answer: It simplifies the process to raise single terms to the power of the index.

OL What would happen to the solution if the right side of the original equation was −10? The solution would be extraneous.

BL What would happen to the related function if the right side of the original equation was changed to −10? The graph would be translated up such that it would not intersect the x-axis.

Common Error

Have a discussion with students about which operations may introduce extraneous solutions when solving an equation that contains a radical. Remind students that the square root sign in an equation means the principal root.

Go Online

- Find additional teaching notes.
- View performance reports of the Checks.
- Assign or present an Extra Example.

Interactive Presentation

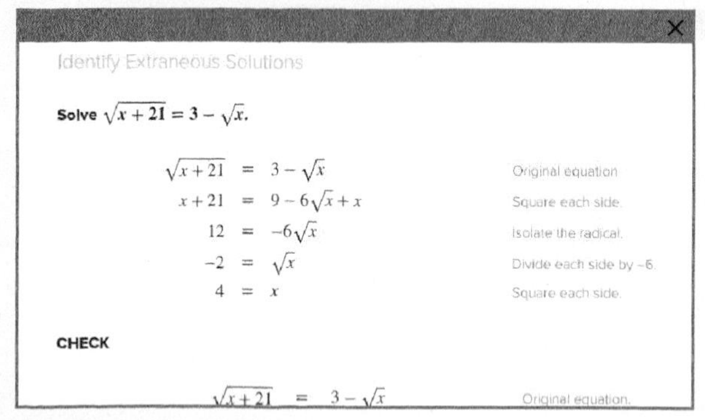

Example 3

TYPE

Students complete the steps of solving a radical equation.

CHECK

Students complete the Check online to determine whether they are ready to move on.

Learn Solving Radical Equations by Graphing

Objective

Students solve radical equations by examining graphs of related functions and graphing systems of equations.

Teaching the Mathematical Practices

1 Understand Different Approaches Work with students to look at both methods shown. Ask students to compare and contrast the methods.

Important to Know

Students should know the constraints on the values of the variables in a radical equation so that the solutions are real numbers.

Example 5 Solve a Radical Equation by Graphing

Teaching the Mathematical Practices

5 Decide When to Use Tools Mathematically proficient students can make sound decision about when to use mathematical tools such as graphing calculators. Help them see what using these tools will help to solve radical equations and what the limitations are of using the tools.

Questions for Mathematical Discourse

AL Why did we rewrite the equation with 0 on one side? Sample answer: This allows the solution(s) of the equation to correspond to the zeros of the related function.

OL Do we need to check if our solution is an extraneous solution while using a graphing method to find a solution? Explain. No; sample answer: only real solutions can be represented as a coordinate of a zero.

BL If you graphed an even root with an extraneous solution, how could you use the graph to determine the extraneous solution? Sample answer: The extraneous solution occurs at the x-value where the reflection of the function across $x = k$ intersects the x-axis, where k is the vertical translation from the parent function. Thus, find the x-value for which $y = 2k$.

Learn Solving Radical Equations by Graphing

To solve a radical equation using the graph of a related function, rewrite the equation with 0 on one side and then replace 0 with $f(x)$.

Equation: $\sqrt{2x+5} + 1 = 4$

Related Function: $f(x) = \sqrt{2x+5} - 3$ or $y = \sqrt{2x+5} - 3$

The values of x for which $f(x) = 0$ are the zeros of the function and occur at the x-intercepts of its graph. The solutions or roots of an equation are the zeros or x-intercepts of its related function.

You can also solve a radical equation by writing and solving a system of equations based on the equation. Set the expressions on each side of the equation equal to y to create the system of equations.

Equation: $\sqrt{2x+5} + 1 = 4$

System of Equations: $y = \sqrt{2x+5} + 1$ $\quad y = 4$

The x-coordinate of the intersection of the system of equations is the value of x where the two equations are equal. Thus, the x-coordinate of the point of intersection is the solution of the radical equation.

Example 5 Solve a Radical Equation by Graphing

Use a graphing calculator to solve $2\sqrt[3]{3x-4} + 10 = 9$ by graphing.

Step 1 **Find a related function.** Rewrite equation with 0 on right side.

$2\sqrt[3]{3x-4} + 10 = 9$ Original equation

$2\sqrt[3]{3x-4} + \underline{1} = \underline{0}$ Subtract 9 from each side.

Replacing 0 with $f(x)$ gives the related function $f(x) = 2\sqrt[3]{3x-4} + 1$.

Step 2 **Graph the related function.**

Use the Y= list to graph.

Step 3 **Use a table.**

You can use the **TABLE** feature to find the interval where the zero lies.

The function changes sign between $x = 1$ and $x = 2$ which indicates that there is a zero between $\underline{1}$ and $\underline{2}$.

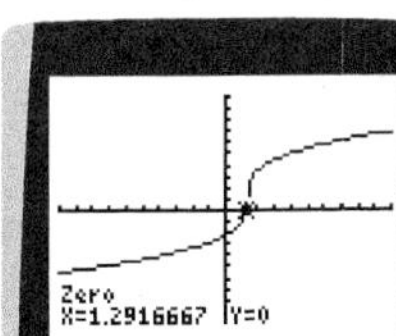

$[-10, 10]$ scl: 1 by $[-10, 10]$ scl: 1

Step 4 **Find the zero.**

Use the **zero** feature from the **CALC** menu to find the zero of the function.

The zero is about $\underline{1.29}$. This is between 1 and 2, which is consistent with the interval we found using the table.

Go Online You can complete an Extra Example online.

Think About It! What would the graph of the related function of a radical equation with no solution look like?

Sample answer: The graph would not intersect the x-axis.

Watch Out!

Misleading Graphs Although the TI-84 may show what appears to be a discontinuity in the graphs of radical functions with odd roots, these functions are in fact continuous for all real numbers.

Interactive Presentation

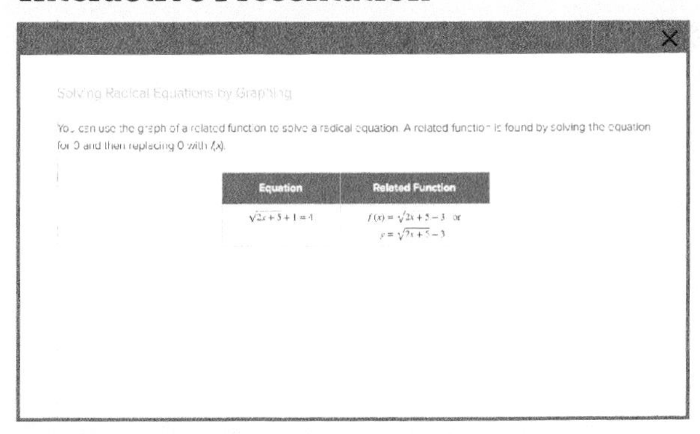

Learn

TYPE

Students consider the graph of a radical equation with no solution.

Think About It!

How can you use the table feature on your calculator to find the intersection?

Sample answer: I can scroll through the table to find the value of x when the two equations have the same y-value.

Example 6 Solve a Radical Equation by Using a System

Use a graphing calculator to solve $\sqrt{x+6} - 5 = -\sqrt{2x} + 1$ by using a system of equations.

Step 1 Write a system. Set each side of $\sqrt{x+6} - 5 = -\sqrt{2x} + 1$ equal to y to create a system of equations.

$y = \sqrt{x+6} - 5$ First equation

$y = -\sqrt{2x} + 1$ Second equation

Step 2 Graph the system. Enter the equations in the **Y=** list and graph in the standard viewing window.

Step 3 Find the intersection.

Use the **intersect** feature from the **CALC** menu to find the coordinates of the point of intersection.

The solution is the x-coordinate of the intersection, which is about 4.02.

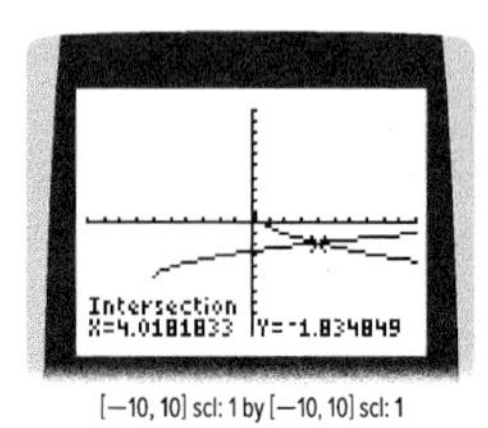

[−10, 10] scl: 1 by [−10, 10] scl: 1

Check

Use a graphing calculator to solve $-4(\sqrt[3]{x-2}) = \sqrt[4]{x-3} - 6$ by using a system of equations. Round to the nearest hundredth if necessary.

$x \approx$ 3.96

Use a Source

Research the time it takes another planet to orbit the Sun. Write and solve a radical equation to find that planet's mean distance from the Sun.

Sample answer: It takes Mercury about 0.241 years to orbit the Sun, so $0.241 = \sqrt{a^3}$. Therefore, Mercury has a mean distance of about 0.387 AU from the Sun.

Example 7 Confirm Solutions by Using Technology

SPACE **The square of the time it takes a planet to orbit the Sun T is proportional to the cube of the planet's mean distance from the Sun a. This relationship can also be written as $T = \sqrt{a^3}$, where T is measured in years and a is measured in astronomical units (AU). If it takes Mars 1.88 years to orbit the Sun, use a graphing calculator to find the mean distance from Mars to the Sun.**

$1.88 = \sqrt{a^3}$ $T = 1.88$

$3.5344 = a^3$ Square each side.

$1.5233 \approx a$ Take the cube root of each side.

So, the mean distance from Mars to the Sun is about 1.5233 AU. Use a graphing calculator to confirm this solution by graphing.

[−10, 10] scl: 1 by [−10, 10] scl: 1

Go Online You can complete an Extra Example online.

Interactive Presentation

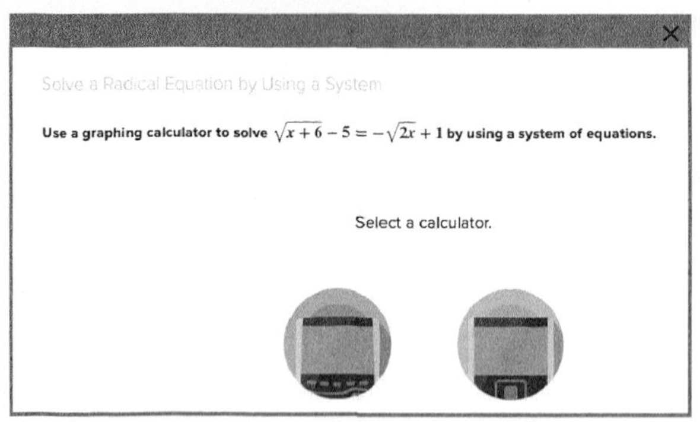

Example 6

TAP

Students move through the slides to see how to use a graphing calculator to solve a radical equation by using a system.

TYPE

Students consider how the table feature on a graphing calculator could be used to find the intersection.

CHECK

Students complete the Check online to determine whether they are ready to move on.

Example 6 Solve a Radical Equation by Using a System

Teaching the Mathematical Practices

5 Analyze Graphs Help students analyze the graph they have generated using graphing calculators. Point out that to see the entire graph, students may need to adjust the viewing window.

Questions for Mathematical Discourse

AL Why does graphing the two sides of the equation and finding the intersection yield the value for x? Sample answer: The intersection shows the value of x where the two functions are equal, and thus makes the two sides of the original equation equal.

OL How can you graph the two sides without a graphing calculator? Sample answer: The equations can be graphed using transformations from the parent function and a table of values.

BL Would a reflection of either of the equations in $x = k$ result in a system with a solution? Explain. Sample answer: Reflecting $y = \sqrt{(x+6)} - 5$ across $x = -5$ would also have a solution. Both functions would be decreasing, but $y = -\sqrt{(2x)} + 1$, which has a higher starting point, decreases faster due to the coefficient of 2.

Example 7 Confirm Solutions by Using Technology

Teaching the Mathematical Practices

5 Use a Source Guide students to find external information about the time it takes other planets to orbit the Sun to answer the questions posed in the Use a Source feature.

Questions for Mathematical Discourse

AL How do you confirm the solution by graphing? Either rewrite the equation equal to zero and find the x-intercept of the related function, or use each side to create a system and find the intersection.

OL What would be the exponent if this equation was written in exponential form? $\frac{3}{2}$

BL What would you input into a calculator to find the approximate solution in one step instead of two? $1.88^{\frac{2}{3}}$

Exit Ticket

Recommended Use

At the end of class, go online to display the Exit Ticket prompt and ask students to respond using a separate piece of paper. Have students hand you their responses as they leave the room.

Alternate Use

At the end of class, go online to display the Exit Ticket prompt and ask students to respond verbally or by using a mini-whiteboard. Have students hold up their whiteboards so that you can see all student responses. Tap to reveal the answer when most or all students have completed the Exit Ticket.

Practice and Homework

Suggested Assignments

Use the table below to select appropriate exercises.

DOK	Topic	Exercises
1, 2	exercises that mirror the examples	1–30
2	exercises that use a variety of skills from this lesson	31–47
2	exercises that extend concepts learned in this lesson to new contexts	48–50
3	exercises that emphasize higher-order and critical thinking skills	51–56

ASSESS AND DIFFERENTIATE

Use the data from the **Checks** to determine whether to provide resources for extension, remediation, or intervention.

IF students score 90% or more on the Checks, **THEN** assign: BL

- Practice Exercises 1–49 odd, 51–56
- Extension: Lesser-Known Geometric Formulas
- **ALEKS**® Radical Equations and Applications

IF students score 66%–89% or more on the Checks, **THEN** assign: OL

- Practice Exercises 1–55 odd
- Remediation, Review Resources: Operations with Radical Expressions
- Personal Tutors
- Extra Examples 1–7
- **ALEKS**® Other Topics Available: Radicals

IF students score 65% or less on the Checks, **THEN** assign: AL

- Practice, Exercises 1–29 odd
- Remediation, Review Resources: Operations with Radical Expressions
- *Quick Review Math Handbook*: Solving Radical Equations and Inequalities
- **ALEKS**® Other Topics Available: Radicals

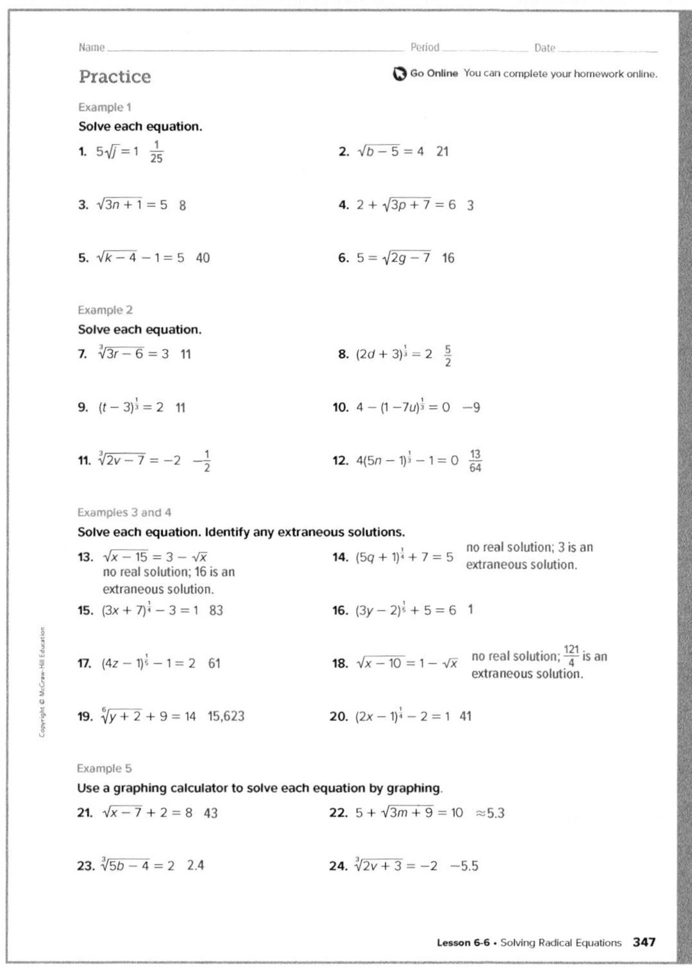

Name ______ Period ______ Date ______

Practice

Go Online You can complete your homework online.

Example 1

Solve each equation.

1. $5\sqrt{j} = 1$ $\frac{1}{25}$
2. $\sqrt{b-5} = 4$ 21
3. $\sqrt{3n+1} = 5$ 8
4. $2 + \sqrt{3p+7} = 6$ 3
5. $\sqrt{k-4} - 1 = 5$ 40
6. $5 = \sqrt{2g-7}$ 16

Example 2

Solve each equation.

7. $\sqrt[3]{3r-6} = 3$ 11
8. $(2d+3)^{\frac{1}{3}} = 2$ $\frac{5}{2}$
9. $(t-3)^{\frac{1}{3}} = 2$ 11
10. $4 - (1-7u)^{\frac{1}{3}} = 0$ −9
11. $\sqrt[3]{2v-7} = -2$ $-\frac{1}{2}$
12. $4(5n-1)^{\frac{1}{3}} - 1 = 0$ $\frac{13}{64}$

Examples 3 and 4

Solve each equation. Identify any extraneous solutions.

13. $\sqrt{x-15} = 3 - \sqrt{x}$ no real solution; 16 is an extraneous solution.
14. $(5q+1)^{\frac{1}{4}} + 7 = 5$ no real solution; 3 is an extraneous solution.
15. $(3x+7)^{\frac{1}{4}} - 3 = 1$ 83
16. $(3y-2)^{\frac{1}{5}} + 5 = 6$ 1
17. $(4z-1)^{\frac{1}{5}} - 1 = 2$ 61
18. $\sqrt{x-10} = 1 - \sqrt{x}$ no real solution; $\frac{121}{4}$ is an extraneous solution.
19. $\sqrt[6]{y+2} + 9 = 14$ 15,623
20. $(2x-1)^{\frac{1}{4}} - 2 = 1$ 41

Example 5

Use a graphing calculator to solve each equation by graphing.

21. $\sqrt{x-7} + 2 = 8$ 43
22. $5 + \sqrt{3m+9} = 10$ ≈5.3
23. $\sqrt[3]{5b-4} = 2$ 2.4
24. $\sqrt[3]{2v+3} = -2$ −5.5

Lesson 6-6 • Solving Radical Equations 347

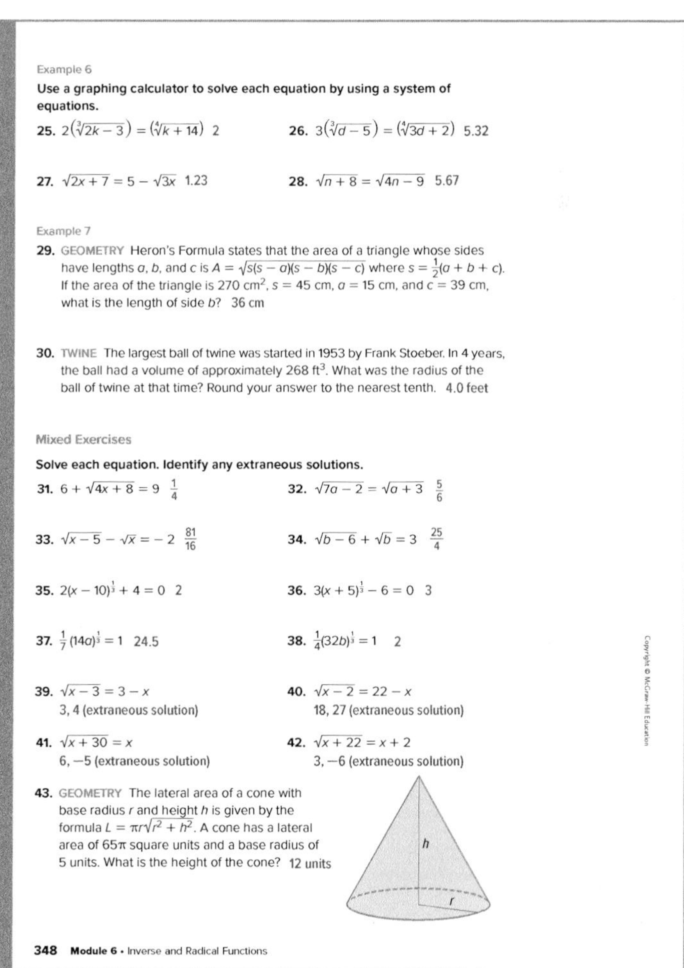

Example 6

Use a graphing calculator to solve each equation by using a system of equations.

25. $2(\sqrt[3]{2k-3}) = (\sqrt[4]{k+14})$ 2
26. $3(\sqrt[3]{d-5}) = (\sqrt[4]{3d+2})$ 5.32
27. $\sqrt{2x+7} = 5 - \sqrt{3x}$ 1.23
28. $\sqrt{n+8} = \sqrt{4n-9}$ 5.67

Example 7

29. GEOMETRY Heron's Formula states that the area of a triangle whose sides have lengths a, b, and c is $A = \sqrt{s(s-a)(s-b)(s-c)}$ where $s = \frac{1}{2}(a+b+c)$. If the area of the triangle is 270 cm², $s = 45$ cm, $a = 15$ cm, and $c = 39$ cm, what is the length of side b? 36 cm

30. TWINE The largest ball of twine was started in 1953 by Frank Stoeber. In 4 years, the ball had a volume of approximately 268 ft³. What was the radius of the ball of twine at that time? Round your answer to the nearest tenth. 4.0 feet

Mixed Exercises

Solve each equation. Identify any extraneous solutions.

31. $6 + \sqrt{4x+8} = 9$ $\frac{1}{4}$
32. $\sqrt{7a-2} = \sqrt{a+3}$ $\frac{5}{6}$
33. $\sqrt{x-5} - \sqrt{x} = -2$ $\frac{81}{16}$
34. $\sqrt{b-6} + \sqrt{b} = 3$ $\frac{25}{4}$
35. $2(x-10)^{\frac{1}{3}} + 4 = 0$ 2
36. $3(x+5)^{\frac{1}{3}} - 6 = 0$ 3
37. $\frac{1}{7}(14a)^{\frac{1}{3}} = 1$ 24.5
38. $\frac{1}{4}(32b)^{\frac{1}{3}} = 1$ 2
39. $\sqrt{x-3} = 3 - x$ 3, 4 (extraneous solution)
40. $\sqrt{x-2} = 22 - x$ 18, 27 (extraneous solution)
41. $\sqrt{x+30} = x$ 6, −5 (extraneous solution)
42. $\sqrt{x+22} = x + 2$ 3, −6 (extraneous solution)
43. GEOMETRY The lateral area of a cone with base radius r and height h is given by the formula $L = \pi r\sqrt{r^2 + h^2}$. A cone has a lateral area of 65π square units and a base radius of 5 units. What is the height of the cone? 12 units

348 Module 6 • Inverse and Radical Functions

Name ____________ Period ______ Date ______

44. TETHERS A tether of length y secures a telephone pole at 25 feet off the ground. The distance from the tether to the pole along the ground is represented by x. By the Pythagorean Theorem, the length of the tether is given by $y = \sqrt{x^2 + 25^2}$. If $x + y = 50$ feet, what is the measure of x? 18.75 ft

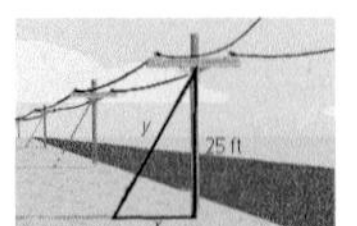

45. SPACE NASA's Near-Earth Asteroid Tracking project tracked more than 300 asteroids. An asteroid is passing near Earth. If Earth is located at the origin of a coordinate plane, the path that the asteroid will trace out is given by $y = \frac{17}{x}, x > 0$, where unit corresponds to one million miles. One asteroid will be visible by telescope when it is within $\frac{145}{12}$ million miles of Earth.

a. Write an expression that gives the distance of the asteroid from Earth as a function of x. $\sqrt{x^2 + \frac{289}{x^2}}$

b. For what values of x will the asteroid be in range of a telescope? $\frac{17}{12} \leq x \leq 12$

46. DRIVING To determine the speed of a car when it begins to skid to a stop, the formula $s = \sqrt{30fd}$ can be used, where s is the speed of the car, f is the coefficient of friction, and d is the length of the skid marks in feet. If the speed limit is 25 mph and the coefficient of friction is 0.6, what is the length of the skid marks if the driver is driving the speed limit? 34.7 feet

47. CHEMISTRY The nuclear radius of an element can be approximated by $r = (1.2 \times 10^{-15})A^{\frac{1}{3}}$ where r is the length of the radius in meters and A is the molecular mass of the element.

a. The nuclear radius of neon is about 3.267×10^{-15} meter. Find its molecular mass. 20.2

b. Which element has a molecular radius of approximately 5.713×10^{-15} meter? Justify your conclusion. Silver; $5.713 \times 10^{-15} \approx (1.2 \times 10^{-15})A^{\frac{1}{3}}$; $4.761 \approx A^{\frac{1}{3}}$; $107.917 \approx A$

Element	Molecular Mass
copper (Cu)	63.5
gold (Au)	197.0
magnesium (Mg)	24.3
neon (Ne)	20.2
silver (Ag)	107.9
titanium (Ti)	47.9

48. USE A MODEL Explain how to find the solutions to $\sqrt[4]{10x + 11} - \sqrt{x + 2} = 0$ graphically and confirm your results algebraically. What are the solutions? See margin.

49. USE TOOLS The surface area of a sphere is 20 cm² greater than the surface area of a cube. Find functions to represent the radius of the sphere and the side length of the cube, in terms of the surface area of the cube. If the radius of the sphere equals the side length of the cube, describe how to use a graphing calculator to find the surface area of the cube and sphere. Sketch the graph. Find the surface area of the cube and the sphere. See margin.

50. CONSTRUCT ARGUMENTS Explain how we know that the equation $\sqrt{x - 5} + 1 = \sqrt{(2 - x)}$ has no solutions without having to actually solve it. Confirm this by graphing the two sides of the equation. See margin.

Higher-Order Thinking Skills

51. WHICH ONE DOESN'T BELONG? Determine which of the equations doesn't belong. Justify your conclusion.

$\sqrt{x-1} + 3 = 4$	$\sqrt{x+1} + 3 = 4$	$\sqrt{x-2} + 7 = 10$	$\sqrt{x+2} - 7 = -10$

$\sqrt{x+2} - 7 = -10$; Sample answer: This equation does not have a real solution while the other three equations do.

52. PERSEVERE Haruko is working to solve $(x + 5)^{\frac{1}{4}} = -4$. He said that he could tell there was no real solution without even working the problem. Is Haruko correct? Explain your reasoning.
Yes; sample answer: Because $\sqrt[4]{x + 5} \geq 0$, the left side of the equation is nonnegative. Therefore, the left side of the equation cannot equal -4. Thus the equation has no solution.

53. ANALYZE Determine whether $\frac{\sqrt{(x^2)^2}}{-x} = x$ is *sometimes*, *always*, or *never* true when x is a real number. Justify your argument. See margin.

54. CREATE Write an equation that can be solved by raising each side of the equation to the given power.

a. $\frac{3}{2}$ power   **b.** $\frac{5}{4}$ power   **c.** $\frac{7}{8}$ power

Sample answer: $0 = 6x^{\frac{2}{3}} - 5$ Sample answer: $0 = x^{\frac{4}{5}} - 9$ Sample answer: $10x^{\frac{8}{7}} = -1$

ANALYZE **Determine whether the following statements are *sometimes*, *always*, or *never* true for $x^{\frac{1}{n}} = a$. Explain your reasoning.**

55. If n is odd, there will be extraneous solutions.
Never; sample answer: The radicand can be negative.

56. If n is even, there will be extraneous solutions.
Sometimes; sample answer: when the radicand is negative, then there will be extraneous roots.

Answers

48. Isolate the radicals and write each side of the equation as a function. Then graph the functions and find their intersection(s). The intersections appear to be at $x = -1$ and $x = 7$. To verify algebraically, raise each side of the equation to the 4th power. The result is $(x + 2)^2 = 10x + 11$, or $(x - 7)(x + 1) = 0$. The solutions are $x = -1$ and $x = 7$.

49. Graph the functions $f(x) = \sqrt{\frac{x}{6}}$ and $g(x) = \sqrt{\frac{x + 20}{4\pi}}$, where x is the surface area of the cube. The graphs intersect at $x \approx 18.27$. The surface areas of the cube and sphere are approximately 18.27 cm² and 38.27 cm².

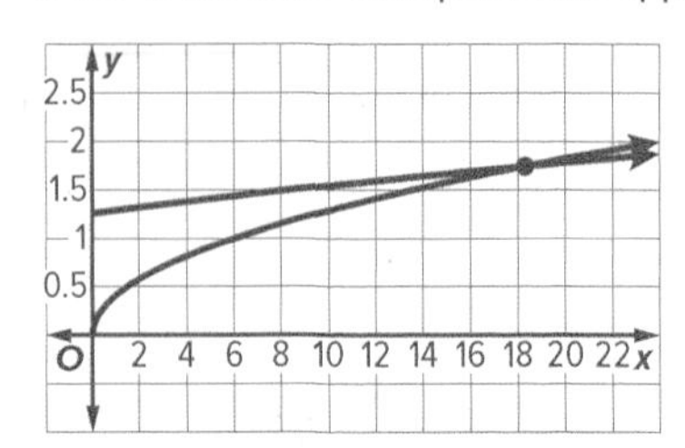

50. The domain of the left-hand side is $x \geq 5$. The domain of the right-hand side is $x \leq 2$. These two domains do not overlap, so no real x-value could simultaneously satisfy both sides of the equation. The graph confirms this.

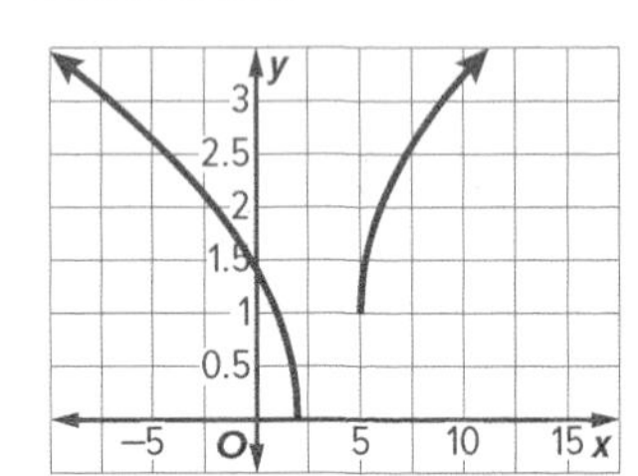

53. never;

$$\frac{\sqrt{(x^2)^2}}{-x} = x$$

$$\frac{x^2}{-x} = x$$

$$x^2 = (x)(-x)$$

$$x^2 \neq -x^2$$

Review

Rate Yourself!

Have students return to the Module Opener to rate their understanding of the concepts presented in this module. They should see that their knowledge and skills have increased. After completing the chart, have them respond to the prompts in their *Interactive Student Edition* and share their responses with a partner.

Answering the Essential Question

Before answering the Essential Question, have students review their answers to the Essential Question Follow-Up questions found throughout the module.

- Why would you use the inverse of a function to model a real-world situation?
- Why would you choose a square root function to model a set of data instead of a polynomial function?

Then have them write their answer to the Essential Question in the space provided.

DINAH ZIKE FOLDABLES

ELL A completed Foldable for this module should include the key concepts related to inverses and radical functions.

LS **LearnSmart** Use LearnSmart as part of your test preparation plan to measure student topic retention. You can create a student assignment in LearnSmart for additional practice on these topics for **Polynomial, Rational, and Radical Relationships.**

- Solving Rational and Radical Equations
- Graphing Rational, Radical, and Polynomials Functions

Module 6 • Inverses and Radical Functions

Review

Essential Question

How can the inverse of a function be used to help interpret a real-world event or solve a problem?

Sample answer: Much like it is necessary to undo operations when solving equations in algebra, an inverse of a function can be used to undo a function to solve a problem.

Module Summary

Lesson 6-1

Operations on Functions

- $(f + g)(x) = f(x) + g(x)$
- $(f - g)(x) = f(x) - g(x)$
- $(f \cdot g)(x) = f(x) \cdot g(x)$
- $\left(\frac{f}{g}\right)(x) = \frac{f(x)}{g(x)}, g(x) \neq 0$
- $[f \circ g](x) = f[g(x)]$

Lesson 6-2

Inverse Relations and Functions

- If f and f^{-1} are inverses, then $f(a) = b$ if and only if $f^{-1}(b) = a$.
- $f(x)$ and $g(x)$ are inverses if and only if $[f \circ g](x) = x$ and $[g \circ f](x) = x$.

Lesson 6-3

*n*th Roots and Rational Exponents

- For any real numbers a and b and any positive integer n, if $a^n = b$, then a is an nth root of b.
- When there is more than one real root and n is even, the nonnegative root is called the principal root.
- An expression with rational exponents is in simplest form when it has no negative exponents, it has no exponents that are not positive integers in the denominator, it is not a complex fraction, and the index of any remaining radical is the least number possible.

Lesson 6-4

Graphing Radical Functions

- The parent function of the square root functions is $f(x) = \sqrt{x}$. The parent function of the cube root functions is $f(x) = \sqrt[3]{x}$.
- A square root function can be written in the form $g(x) = a\sqrt{x - h} + k$. A cube root function can be written in the form $g(x) = a\sqrt[3]{x - h} + k$.

Lessons 6-5 and 6-6

Radical Expressions and Equations

- For any real numbers a and b and any integer $n > 1$, $\sqrt[n]{ab} = \sqrt[n]{a} \cdot \sqrt[n]{b}$, if n is even and $a, b \geq 0$ or if n is odd.
- For any real numbers a and $b \neq 0$ and any integer $n > 1$, $\sqrt[n]{\frac{a}{b}} = \frac{\sqrt[n]{a}}{\sqrt[n]{b}}$, if all roots are defined.
- $a\sqrt{b} + c\sqrt{d}$ and $a\sqrt{b} - c\sqrt{d}$ are conjugates of each other.
- To solve a radical equation, isolate the radical on one side of the equation. Raise each side of the equation to a power equal to the index of the radical. Solve the resulting polynomial equation. Check your results.

Study Organizer

Foldables

Use your Foldable to review this module. Working with a partner can be helpful. Ask for clarification of concepts as needed.

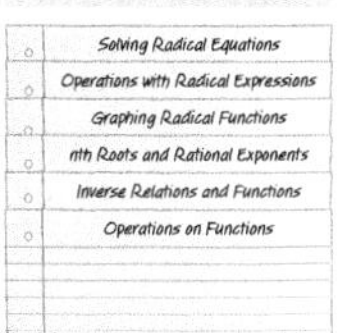

Name ______________ Period ________ Date ________

Test Practice

1. MULTIPLE CHOICE Given $f(x) = 4x^3 - 5x^2 + 8$ and $g(x) = 2x^3 - 9x^2 - 7x$, find $(f - g)(x)$. (Lesson 6-1)

Ⓐ $(f - g)(x) = 2x^3 - 14x^2 + x$

Ⓑ $(f - g)(x) = 2x^3 + 4x^2 + 15x$

Ⓒ $(f - g)(x) = 2x^3 - 14x^2 - 7x + 8$

● $(f - g)(x) = 2x^3 + 4x^2 + 7x + 8$

2. OPEN RESPONSE Given $f(x) = 2x^3 + 7x^2 - 7x - 30$ and $g(x) = x - 2$, find $\left(\frac{f}{g}\right)(x)$. (Lesson 6-1)

$2x^2 + 11x + 15$

3. MULTIPLE CHOICE Given $f(x) = 4x^3 - 2x^2$ and $g(x) = -x^2 + 3x - 2$, find $(f \cdot g)(x)$. (Lesson 6-1)

Ⓐ $(f \cdot g)(x) = 4x^3 - x^2 - 3x + 2$

Ⓑ $(f \cdot g)(x) = 4x^3 - x^2 + 3x - 2$

● $(f \cdot g)(x) = -4x^5 + 14x^4 - 14x^3 + 4x^2$

Ⓓ $(f \cdot g)(x) = 4x^5 - 14x^4 + 14x^3 - 4x^2$

4. MULTIPLE CHOICE Given $f(x) = 14x^3 - x^2 + x + 5$ and $g(x) = 7x^3 + 4x^2 - 2x - 1$, find $(f - g)(x)$. (Lesson 6-1)

Ⓐ $(f - g)(x) = 7x^3 + 3x^2 + x + 4$

● $(f - g)(x) = 7x^3 - 5x^2 + 3x + 6$

Ⓒ $(f - g)(x) = 7x^3 + 3x^2 + 3x + 6$

Ⓓ $(f - g)(x) = 7x^3 - 5x^2 + x + 4$

5. OPEN RESPONSE Given $f(x) = 3x - 7$ and $g(x) = 4x + 5$, find $(f \circ g)(x)$. (Lesson 6-1)

$(f \circ g)(x) = 12x + 8$

6. MULTIPLE CHOICE The graph shows $f(x)$. Which of the following represents $f^{-1}(x)$? (Lesson 6-2)

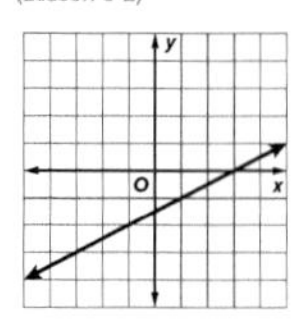

● $f^{-1}(x) = 2x + 3$

Ⓑ $f^{-1}(x) = 3x + 2$

Ⓒ $f^{-1}(x) = x + 4$

Ⓓ $f^{-1}(x) = 4x$

7. OPEN RESPONSE Find the inverse of $f(x) = x^2 + 4x + 3$. Restrict the domain, if necessary. (Lesson 6-2)

If the domain of $f(x)$ is restricted to $(-\infty, -2]$, then the inverse is $f^{-1}(x) = -2 - \sqrt{x + 1}$. If the domain of $f(x)$ is restricted to $[-2, \infty)$, then the inverse is $f^{-1}(x) = -2 + \sqrt{x + 1}$.

Review and Assessment Options

The following online review and assessment resources are available for you to assign to your students. These resources include technology-enhanced questions that are auto-scored, as well as essay questions.

Review Resources

Vocabulary Activity
Module Review

Assessment Resources

Vocabulary Test
AL Module Test Form B
OL Module Test Form A
BL Module Test Form C
Performance Task*

*The module-level performance task is available online as a printable document. A scoring rubric is included.

Test Practice

You can use these pages to help your students review module content and prepare for online assessments. Exercises 1–19 mirror the types of questions your students will see on online assessments.

Question Type	Description	Exercise(s)
Multiple Choice	Students select one correct answer.	1, 3, 4, 6, 8, 9, 10, 13, 14, 16–20
Open Response	Students construct their own response in the area provided.	2, 5, 7, 10–12, 15

To ensure that students understand the standards, check students' success on individual exercises.

Standard(s)	Lesson(s)	Exercise(s)
F.BF.1b	6-1	1–5
F.IF.7b	6-4	11, 13, 14
F. BF.4a	6-2	6, 7
A.SSE.2	6-3, 6-5	8–10, 12, 15–17
A.REI.2	6-6	18–20

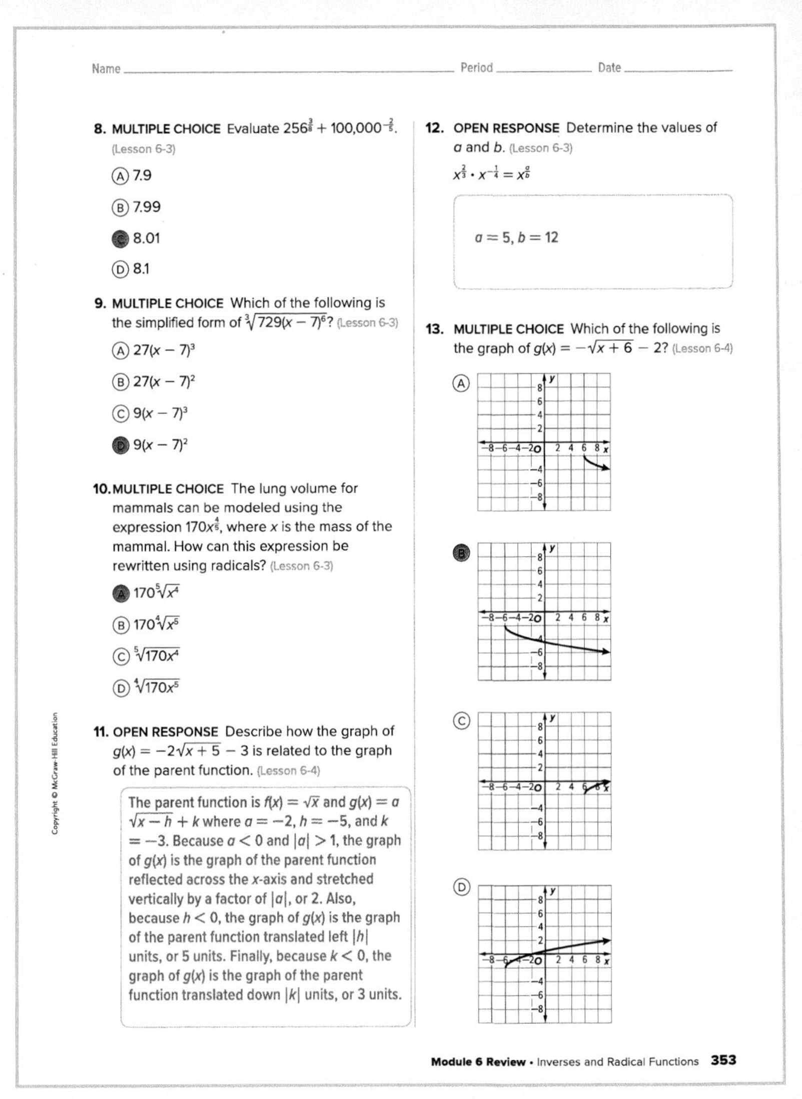

Name ____________ Period ________ Date ________

14. MULTIPLE CHOICE Which function is shown on the graph? (Lesson 6-4)

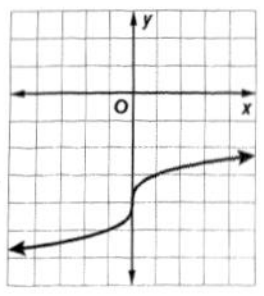

Ⓐ $g(x) = (x + 4)^{\frac{1}{3}}$

Ⓑ $g(x) = (x - 4)^{\frac{1}{3}}$

● $g(x) = x^{\frac{1}{3}} - 4$

Ⓓ $g(x) = x^{\frac{1}{3}} + 4$

15. OPEN RESPONSE Determine the sum.
$7\sqrt{12} + 2\sqrt{48} - 4\sqrt{75}$ (Lesson 6-5)

$2\sqrt{3}$

16. MULTIPLE CHOICE Heather drew a rectangle and labeled the length as $3\sqrt[4]{162x^4y^7}$ and the width as $2\sqrt[4]{40x^3y^5}$. Which is the simplified form of the area of the rectangle? (Lesson 6-5)

Ⓐ $36\sqrt[3]{5x^{12}y^{35}}$

● $36xy^3\sqrt[4]{5x^3}$

Ⓒ $36xy^3\sqrt[4]{5x^3y}$

Ⓓ $36\sqrt[4]{5y^{12}}$

17. MULTIPLE CHOICE Latasia wants to rationalize the denominator of $\frac{7x^4y^5}{\sqrt[3]{z}}$. What should she multiply the numerator and denominator by? (Lesson 6-5)

Ⓐ $\sqrt[3]{z}$

● $\sqrt[3]{z^2}$

Ⓒ z^2

Ⓓ z^3

18. MULTIPLE CHOICE Solve for x, $3(6x - 8)^{\frac{1}{4}} + 3 = 9$. (Lesson 6-6)

Ⓐ $x = 2$

● $x = 4$

Ⓒ $x = 6$

Ⓓ $x = 8$

19. MULTIPLE CHOICE Select the solution(s) of the equation $\sqrt{3x + 7} = x - 1$. (Lesson 6-6)

● 6

Ⓑ 3

Ⓒ 2 and 3

Ⓓ 6 and −1

20. MULTIPLE CHOICE The distance in miles, d, a pilot can see to the horizon is 123% of the square root of the altitude in feet above sea level, a, of the plane. How many miles to the horizon can a pilot flying a plane at an altitude of 30,000 feet see? Round your answer to the nearest mile. (Lesson 6-6)

● 213 miles

Ⓑ 369 miles

Ⓒ 21,304 miles

Ⓓ 36,900 miles

Lesson 6-2

5.

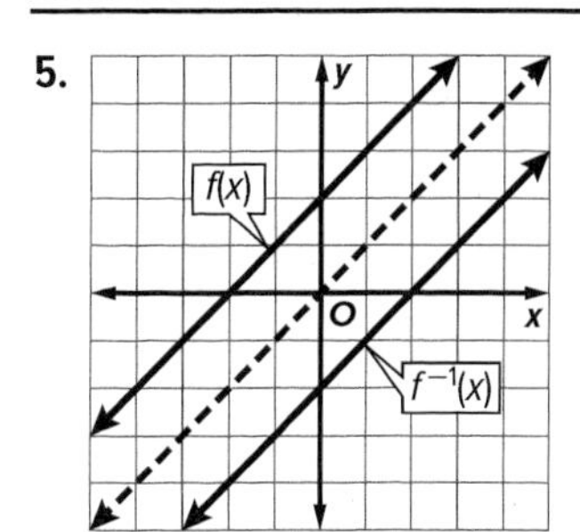

6.

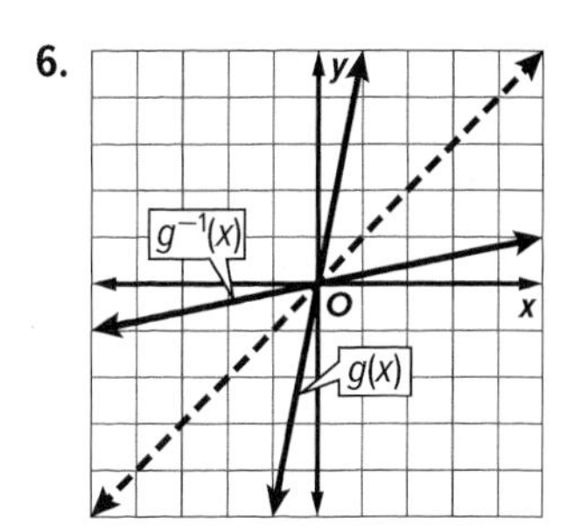

7.

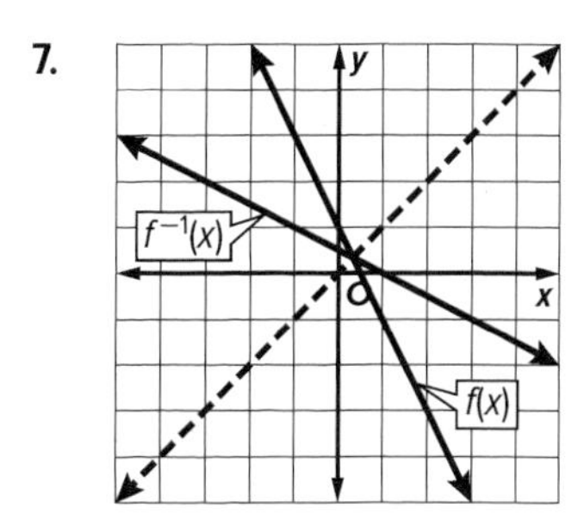

8.

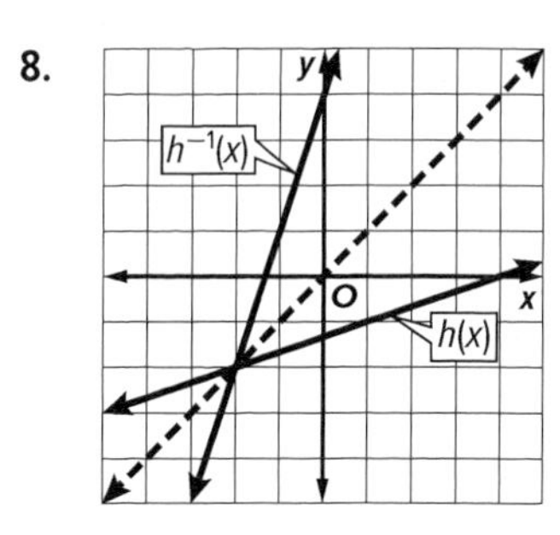

9.

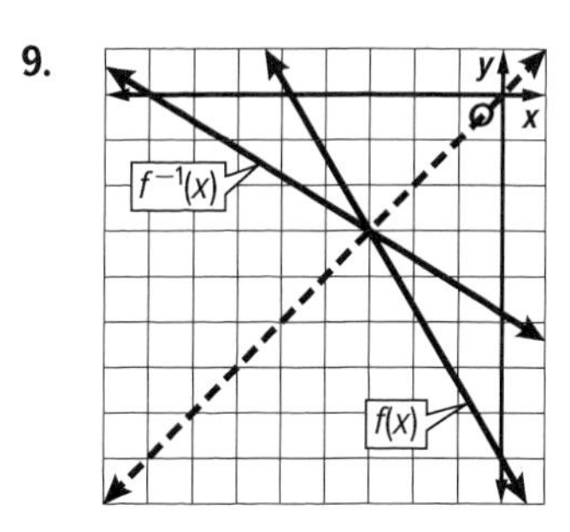

10.

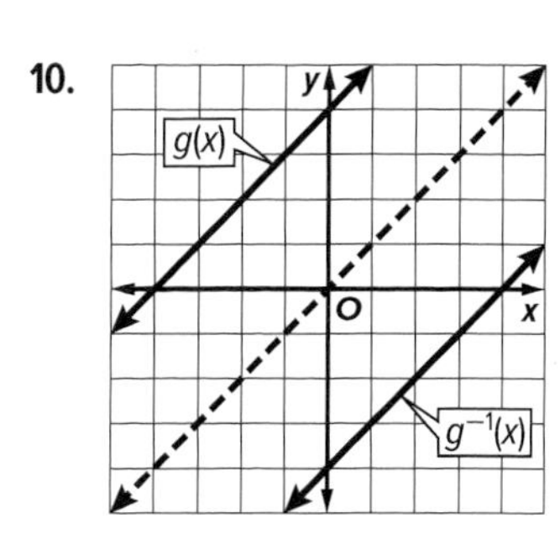

11.

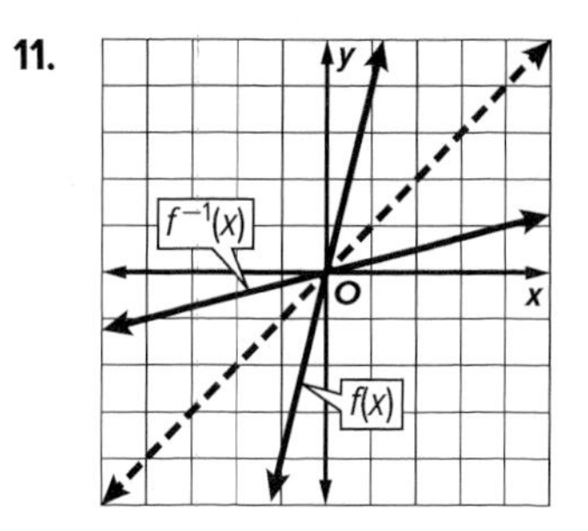

12.

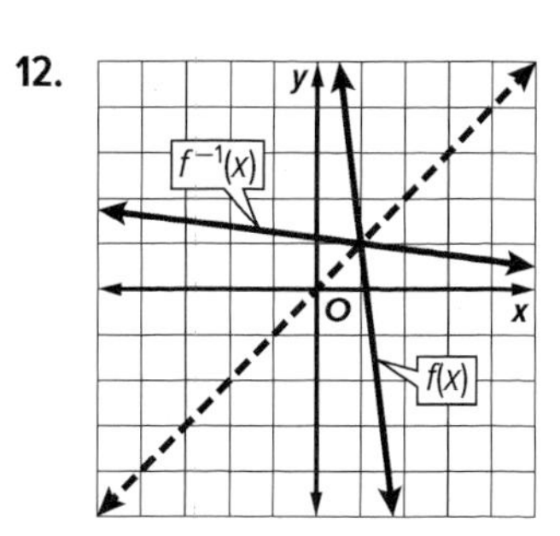

13.

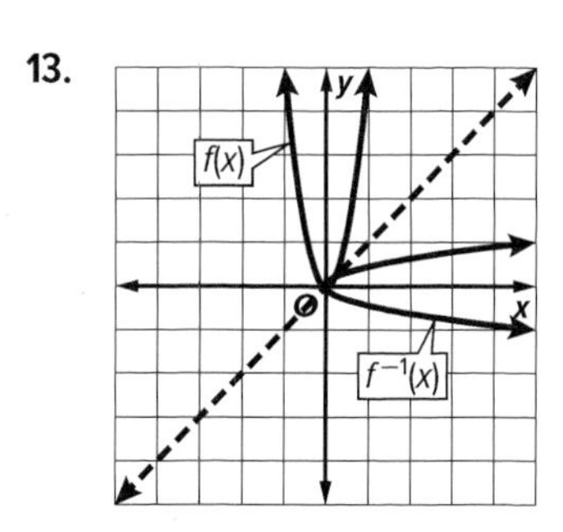

14.

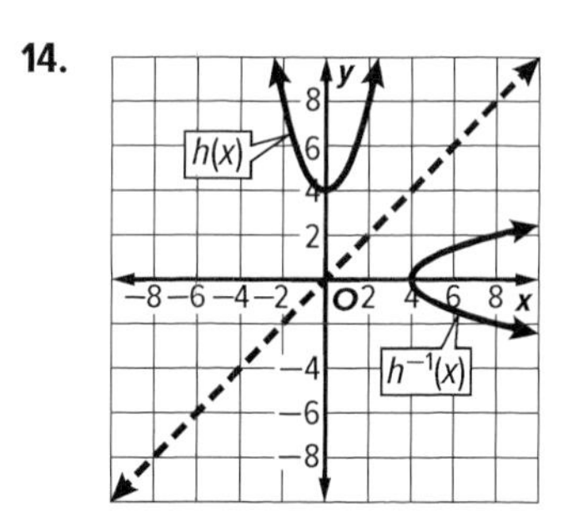

15b.

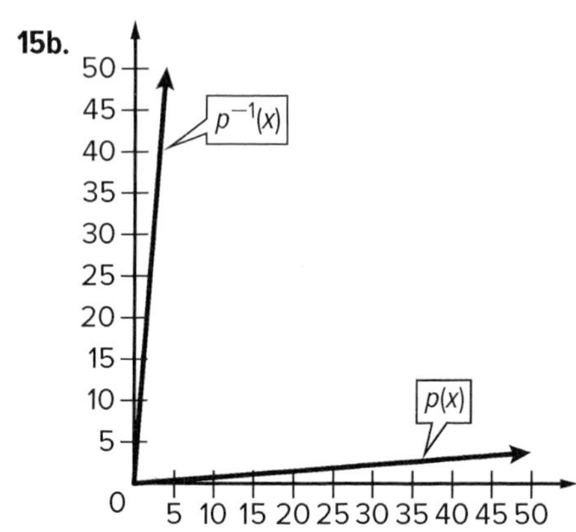

16b.

Letter	$c(x)$	Letter	$c(x)$	Letter	$c(x)$
A = 1	−5	J = 10	31	S = 19	67
B = 2	−1	K = 11	35	T = 20	71
C = 3	3	L = 12	39	U = 21	75
D = 4	7	M = 13	43	V = 22	79
E = 5	11	N = 14	47	W = 23	83
F = 6	15	O = 15	51	X = 24	87
G = 7	19	P = 16	55	Y = 25	91
H = 8	23	Q = 17	59	Z = 26	95
I = 9	27	R = 18	63		

$c^{-1}(x)$	Letter	$c^{-1}(x)$	Letter	$c^{-1}(x)$	Letter
−5	A = 1	31	J = 10	67	S = 19
−1	B = 2	35	K = 11	71	T = 20
3	C = 3	39	L = 12	75	U = 21
7	D = 4	43	M = 13	79	V = 22
11	E = 5	47	N = 14	83	W = 23
15	F = 6	51	O = 15	87	X = 24
19	G = 7	55	P = 16	91	X = 25
23	H = 8	59	Q = 17	95	Z = 26
27	I = 9	63	R = 18		

25.

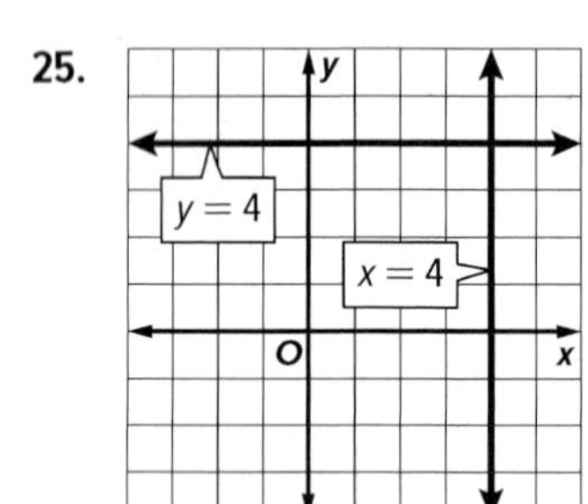

26.

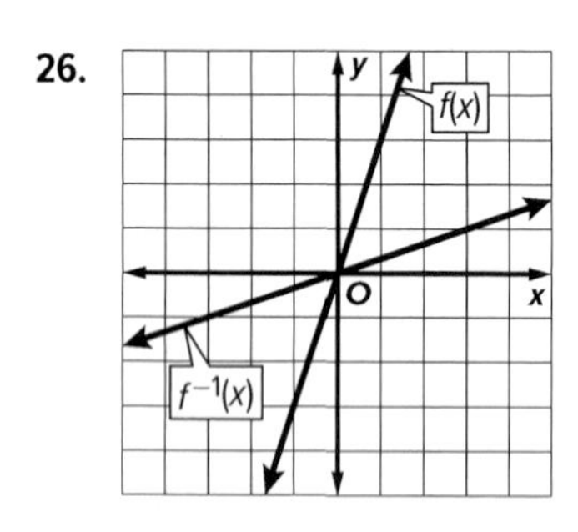

27.

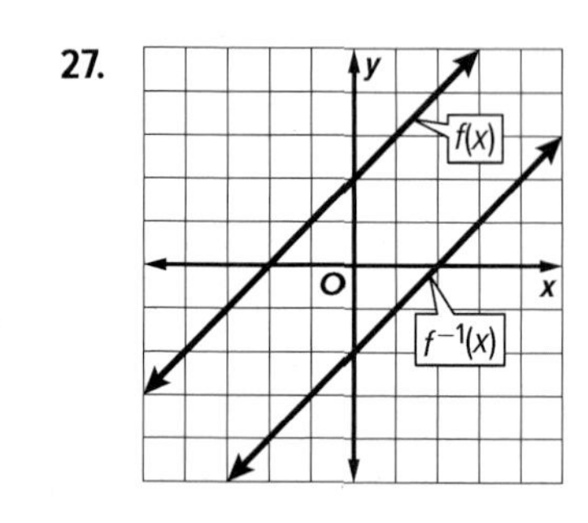

28.

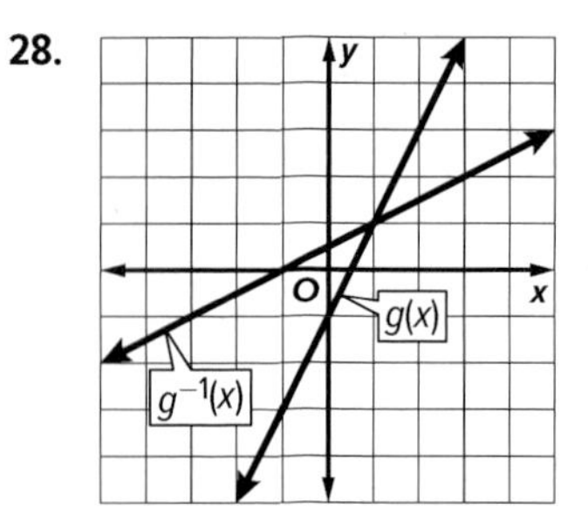

29. If the domain of $f(x)$ is restricted to $(-\infty, 0]$, then the inverse $f^{-1} = \sqrt{2x + 2}$. If the domain of $f(x)$ is restricted to $[0, \infty)$, then the inverse is $f^{-1} = -\sqrt{2x + 2}$.

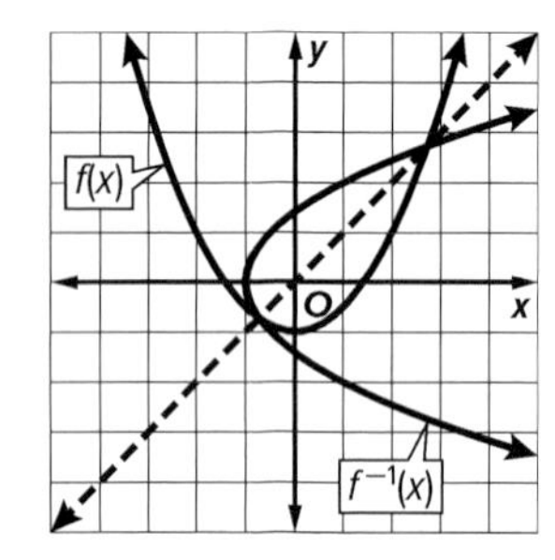

30. If the domain of $f(x)$ is restricted to $(-\infty, -1]$, then the inverse $f^{-1} = -1 + \sqrt{x - 3}$. If the domain of $f(x)$ is restricted to $[-1, \infty)$, then the inverse is $f^{-1} = -1 - \sqrt{x - 3}$.

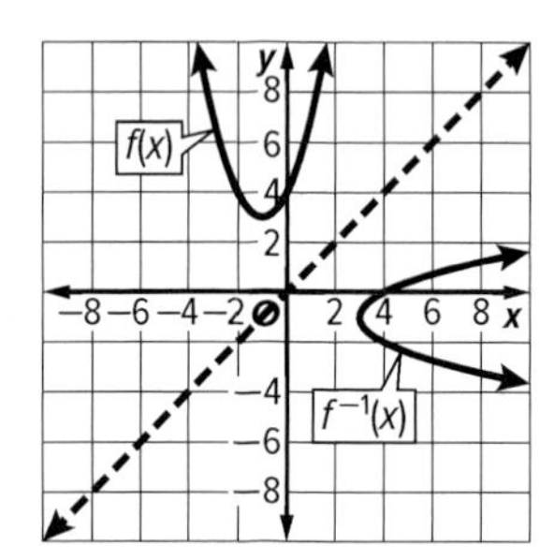

47a. $V(x) = x(12 - 2x)(16 - 2x)$; the value of x must be between 0 and 6 since the length x must be more than 0, but the length of the side of the box $(12 - 2x)$ cannot be less than 0.

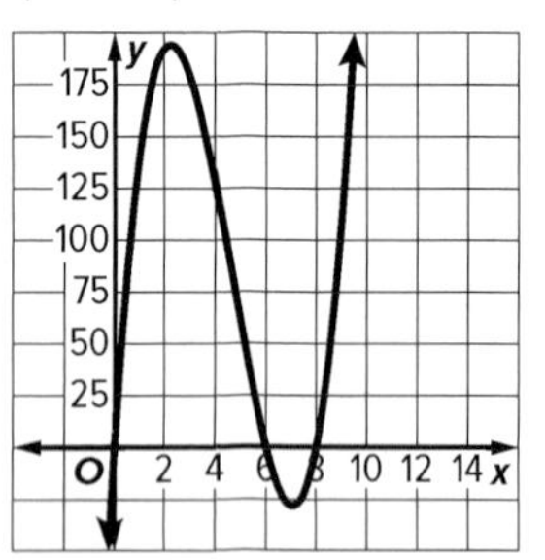

47b. No; sample answer: even with the restriction, the function does not pass the horizontal line test, so its inverse will not be a function.

48b. $(f + g)(1) = f(1) + g(1) = 3 + 0 = 3$. Since $(f + g)(1) = 3$, $(f + g)^{-1}(3) = 1$. $(f + g)(0) = f(0) + g(0) = 0 + 1 = 1$. Since $(f + g)(0) = 1$, $(f + g)^{-1}(1) = 0$.

48c. Part b provides two counterexamples to this property. $(f + g)^{-1}(3) = 1$, but $f^{-1}(3) + g^{-1}(3) = 4$. $(f + g)^{-1}(1) = 0$, but $f^{-1}(1) + g^{-1}(1) = 2$.

48d. $f^{-1}(x) = \frac{1}{2}x - \frac{1}{2}$, and $g^{-1}(x) = \frac{1}{2}x + \frac{1}{2}$. So $f^{-1}(x) + g^{-1}(x) = x$. However, since $(f + g)(x) = 4x$, $(f + g)^{-1}(x) = \frac{1}{4}x$. The two are not the same.

52. No; sample answer: If $f(x) = mx + b$ then $f^{-1} = \frac{1}{m}x - \frac{b}{m}$. Perpendicular lines have slopes that are opposite reciprocals. Because $m \cdot \frac{1}{m} \neq -1$, a function with slope $\neq 0$ is not perpendicular to its inverse.

Lesson 6-4

14.

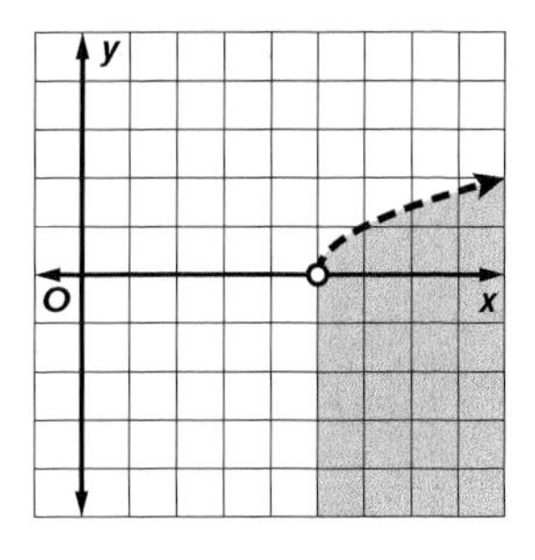

15.

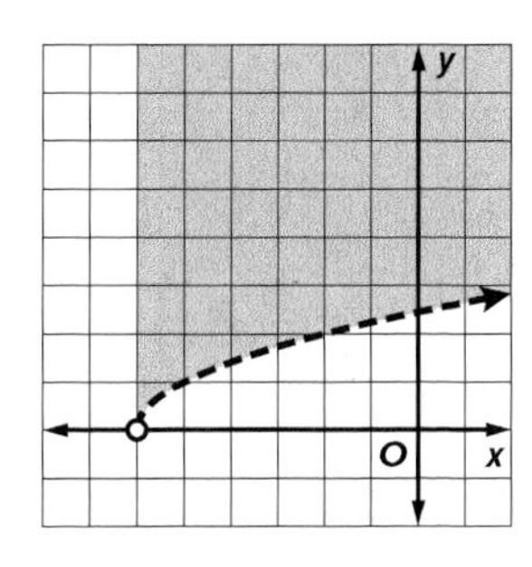

16.

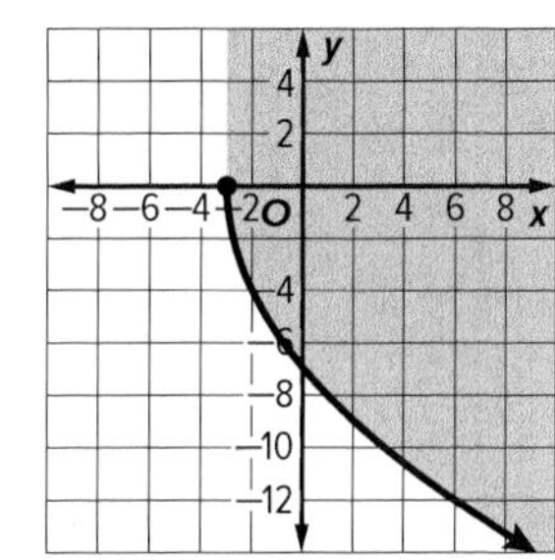

17.

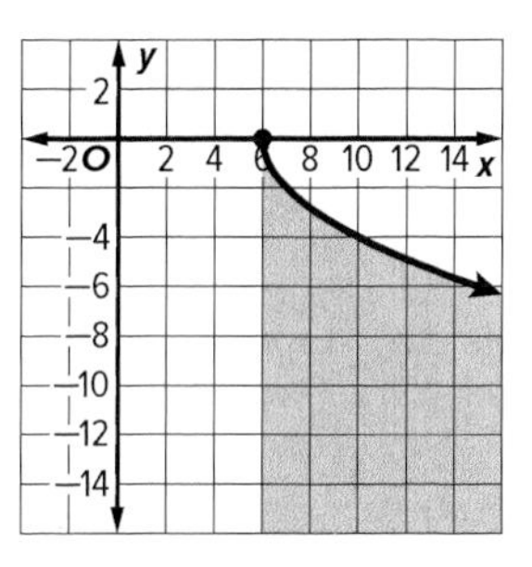

18.

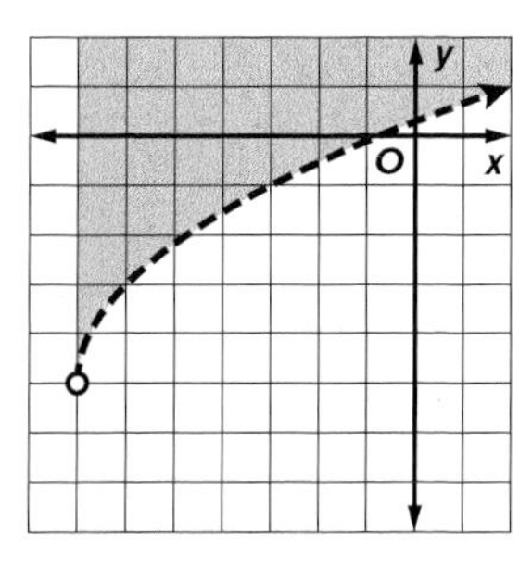

19.

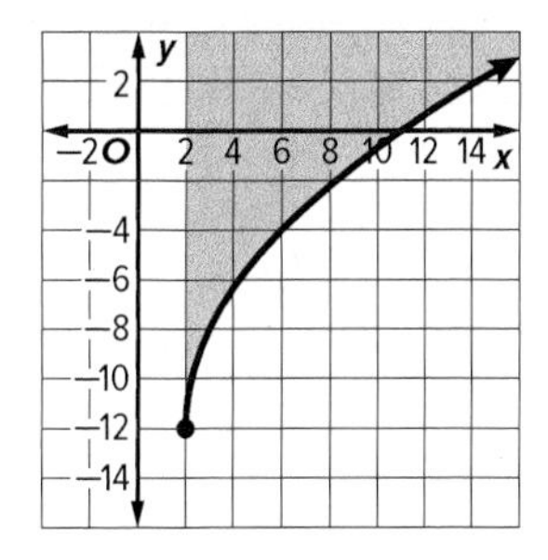

20.

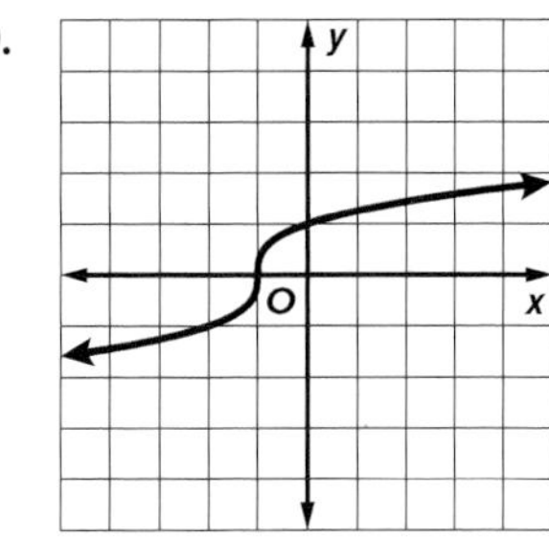

21.

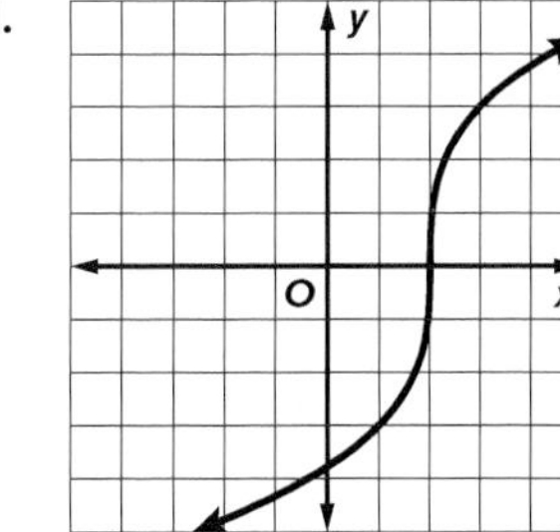

22.

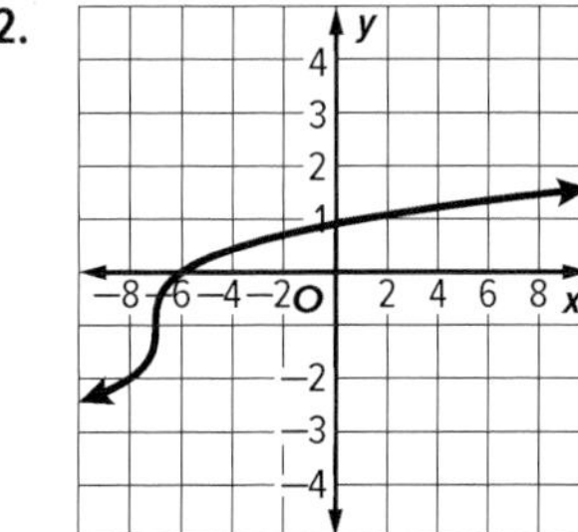

23.

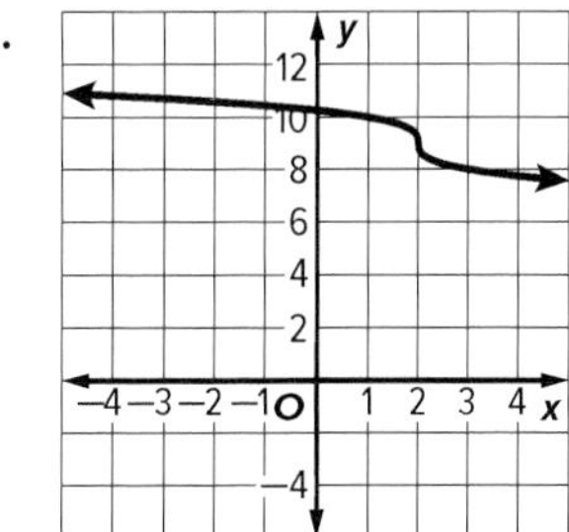

24a.

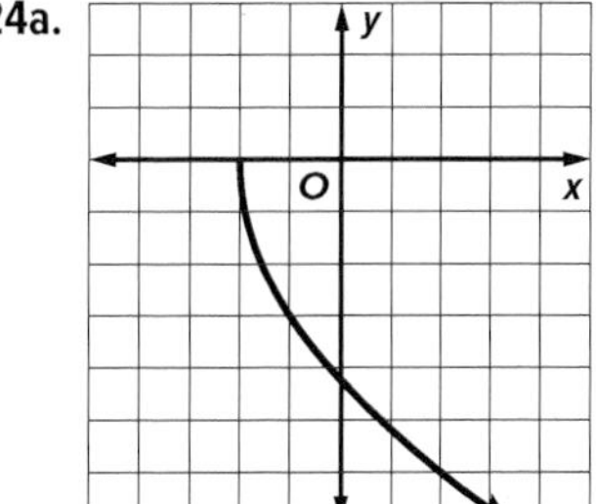

24b.

	$p(x) = -3\sqrt{x+2}$	$q(x)$
Domain	$\{x \mid x \geq -2\}$	all real numbers
Range	$\{y \mid y \leq 0\}$	all real numbers
Intercepts	x-int: -2; y-int: -4.24	x-int: 2; y-int: -1.25
Increasing/ Decreasing	decreasing as $x \to \infty$	increasing as $x \to \infty$
Positive/Negative	negative for $x > -2$	negative for $x < 2$; positive for $x > 2$
End Behavior	as $x \to \infty$, $p(x) \to -\infty$	as $x \to -\infty$, $q(x) \to -\infty$ as $x \to \infty$, $q(x) \to \infty$

25a.

25b.

	$p(x) = \sqrt[3]{x} - 2$	$q(x)$
Domain	all real numbers	$\{x \mid x \geq 0\}$
Range	all real numbers	$\{y \mid y \geq -2\}$
Intercepts	x-int: 8; y-int: -2	x-int: 4; y-int: -2
Increasing/ Decreasing	increasing as $x \to \infty$	increasing as $x \to \infty$
Positive/Negative	negative for $x < 8$; positive for $x > 8$	negative for $x < 4$; positive for $x > 4$
End Behavior	as $x \to -\infty$, $p(x) \to -\infty$ as $x \to \infty$, $p(x) \to \infty$	as $x \to \infty$, $q(x) \to \infty$

26a.

26b.

	$p(x) = \sqrt{x-2} + 5$	$q(x)$
Domain	$\{x \mid x \geq 2\}$	$\{x \mid x \leq 2\}$
Range	$\{y \mid y \geq 5\}$	$\{y \mid y \geq 5\}$
Intercepts	x-int: none; y-int: none	x-int: none; y-int: about 6.5
Increasing/ Decreasing	increasing as $x \to \infty$	decreasing as $x \to 2$
Positive/Negative	positive for $x \geq 2$	positive for $x \leq 2$
End Behavior	as $x \to \infty$, $p(x) \to \infty$	as $x \to -\infty$, $q(x) \to \infty$

29. stretched vertically, translated 5 units right and 6 units down

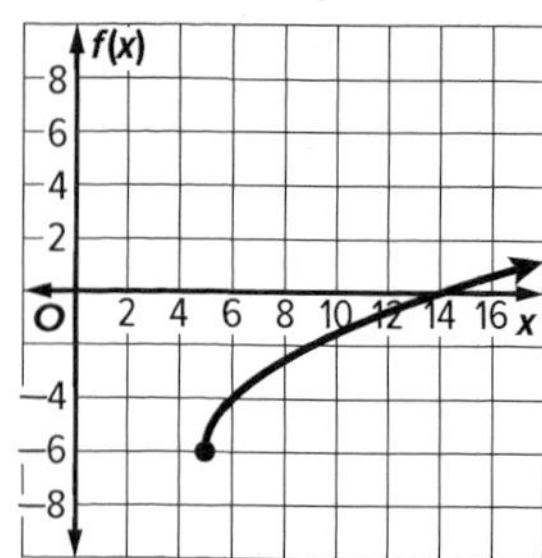

30. compressed vertically, translated 12 units left and 3 units up

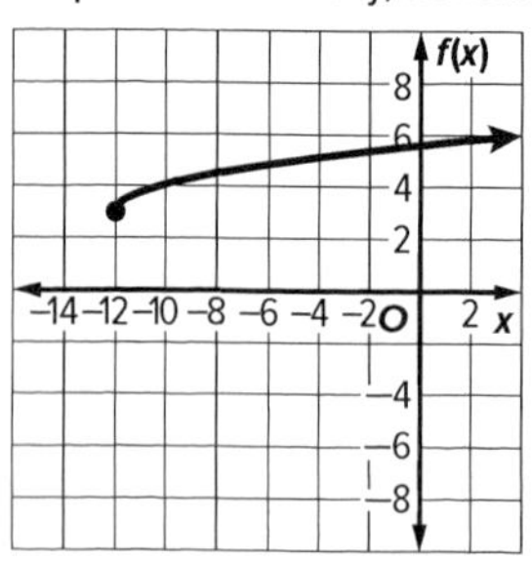

31. compressed vertically, translated 1 unit right and 4 units down, reflected in the x-axis

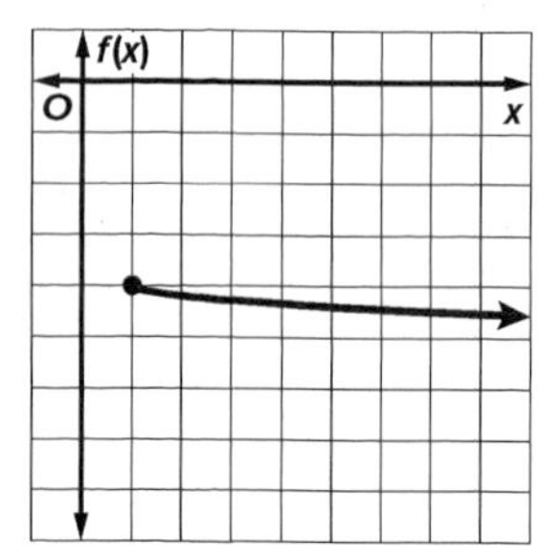

32. stretched vertically, translated 7 units left and 9 units up, reflected in the x-axis

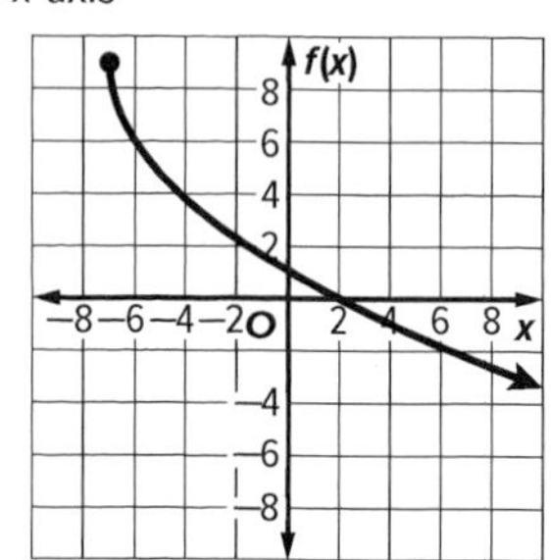

33. compressed vertically, translated 2 units left and 3 units down, reflected in the x-axis

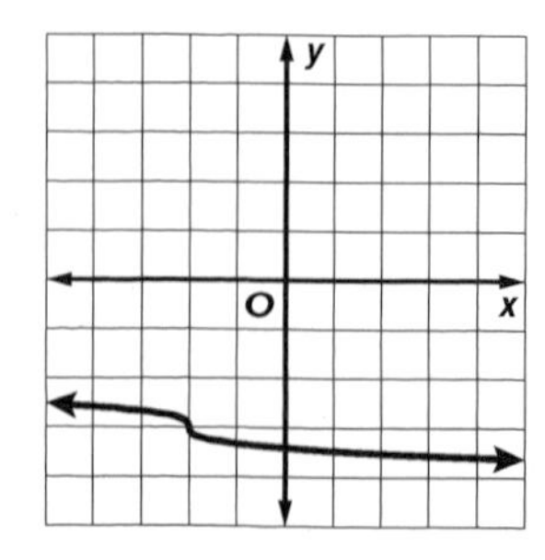

34. compressed vertically and horizontally, translated $\frac{1}{2}$ unit right and 3 units up, reflected in the x-axis

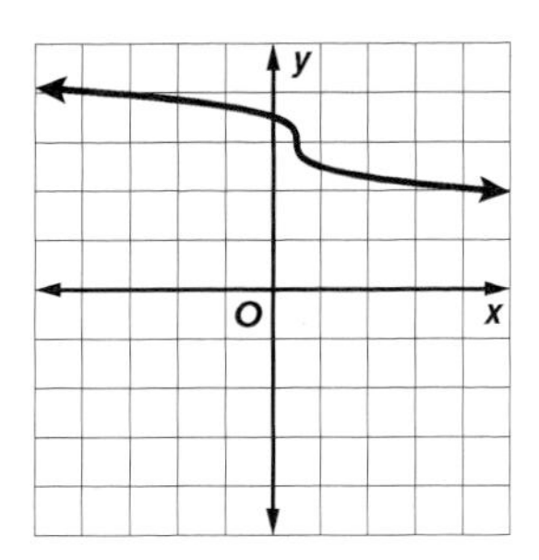

35.

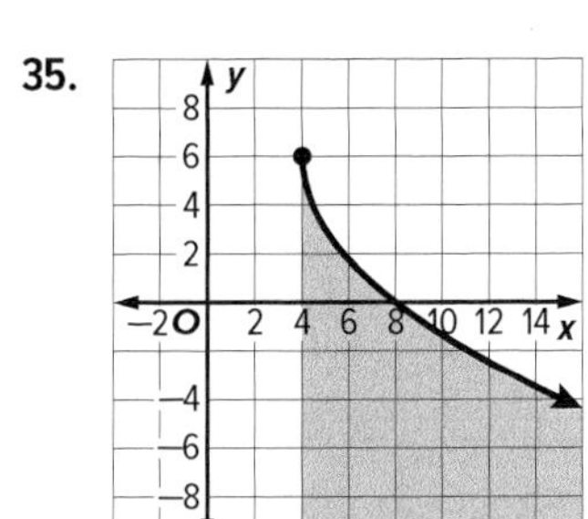

36.

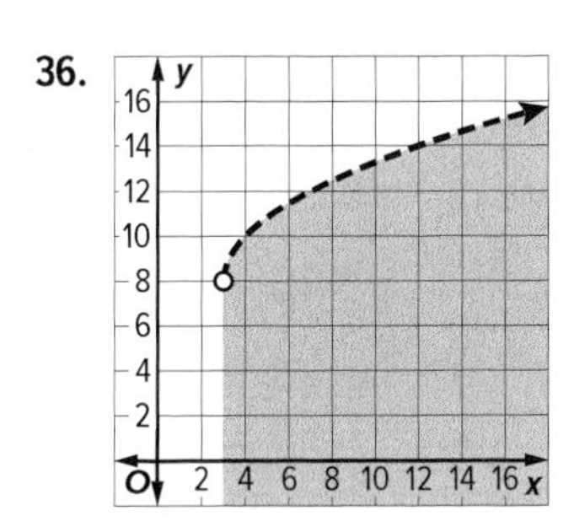

39a. $f(x)$ has the greater maximum value because its maximum, $\frac{13}{2}$, is greater than 6, the maximum value of $g(x)$.

39b. The domain of $f(x)$ is $x \geq -3$, since any values less than -3 produce a negative value under the radical. The domain of $g(x)$ is $x \geq \frac{5}{2}$.

39c. The average rate of change over the interval is $-\frac{1}{7}$ for $f(x)$. It appears that the rate of change for $g(x)$ is the same.

40a.

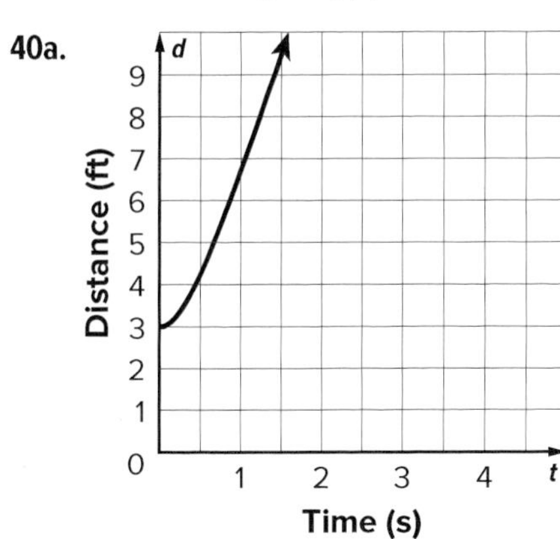

40b. D = $\{t \mid t \geq 0\}$; R = $\{d \mid d \geq 3\}$; increasing as $t \to \infty$; positive for $t > 0$; As $t \to \infty$, $d \to \infty$.

41. D = $\{V \mid -\infty < V < \infty\}$ or $(-\infty, \infty)$; R = $\{r(V) \mid -\infty < r(V) < \infty\}$ or $(-\infty, \infty)$; End behavior: The values of r increase as the values of V increase.

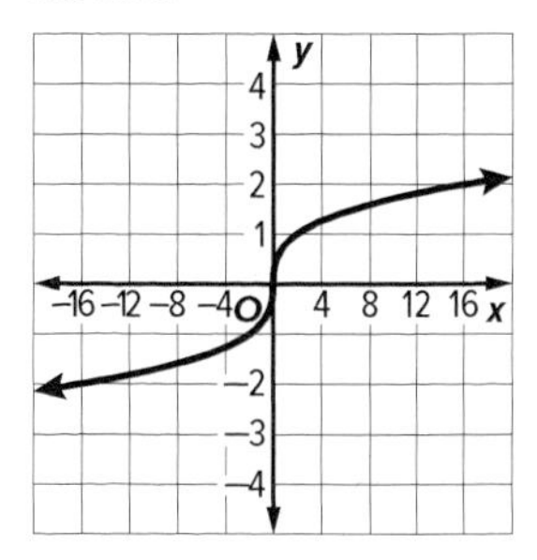

42. The inverse inequality is $f^{-1}(x) > -(x + 3)^2 + 2$. The domain of the inverse is restricted to $x > -3$.

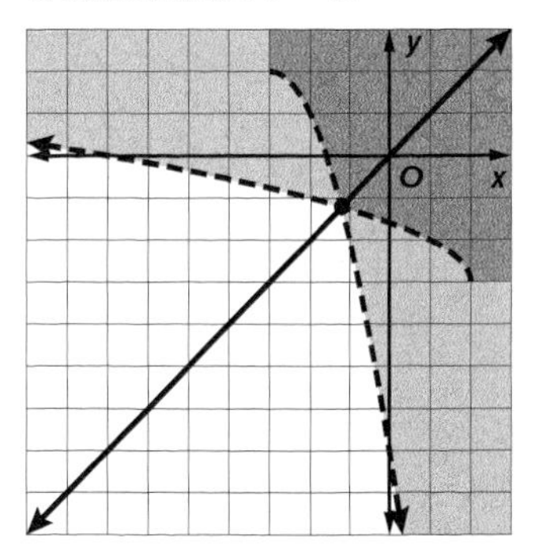

45. Sample answer: The domain is limited because square roots of negative numbers are imaginary. The range is limited due to the limitation of the domain.

46. To be a function, for every x-value there must be exactly one y-value. For every x in this equation there are two y-values, one that is negative and one that is positive. Also, the graph of $y = \pm\sqrt{x}$ does not pass the vertical line test.

47a. Sample answer: The original is $y = x^2 + 2$ and the inverse is $y = \pm\sqrt{x - 2}$.

47b. Sample answer: The original is $y = \pm\sqrt{x} + 4$ and the inverse is $y = (x - 4)^2$.

Selected Answers

Module 1

Quick Check

1. −15 **3.** 10 **5.** $b = \frac{a}{3} - 3$ **7.** $x = \frac{4}{3}y + \frac{8}{3}$

Lesson 1-1

1. D = {all real numbers}; R = {all real numbers}; Codomain = {all real numbers}; onto
3. D = {all real numbers}, R = $\{y \mid y \geq 0\}$, Codomain = {all real numbers}; not onto
5. D = {1, 2, 3, 4, 5, 6, 7}; R = {56, 52, 44, 41, 43, 46, 53}; one-to-one
7. neither **9.** both
11. continuous; D = {all real numbers}, R = {all real numbers}
13. discrete; D = {1, 2, 3, 4, 5, 6}, R = {3, 4, 5, 6}
15. continuous; D = all positive real numbers, R = $\{y \mid y \geq 0\}$
17. D = $\{x \mid x \in \mathbb{R}\}$ or $(-\infty, \infty)$; R = $\{y \mid y \leq 0\}$ or $(-\infty, 0]$
19. D = $\{x \mid x \leq -1 \text{ or } x \geq 1\}$ or $(-\infty, -1] \cup [1, \infty)$; R = $\{y \mid y \in \mathbb{R}\}$ or $[-\infty, \infty)$
21. D = $\{x \mid x \leq -2 \text{ or } x \geq 1\}$ or $(-\infty, -2] \cup [1, \infty)$; R = $\{y \mid y \geq -2\}$ or $[-2, \infty)$
23. D = $\{x \mid x \in \mathbb{R}\}$ or $(-\infty, \infty)$; R = $\{y \mid y \geq -4\}$ or $[-4, \infty)$; neither one-to-one nor onto; continuous
25. D = $\{x \mid x \in \mathbb{R}\}$ or $(-\infty, \infty)$; R = $\{y \mid y \in \mathbb{R}\}$ or $[-\infty, \infty)$; both; continuous
27. D = $\{x \mid x \in \mathbb{R}\}$ or $(-\infty, \infty)$; R = $\{y \mid y \in \mathbb{R}\}$ or $[-\infty, \infty)$; onto; continuous
29. D = $\{w \mid 0 \leq w \leq 15\}$; R = $\{L \mid 4 \leq L \leq 11.5\}$; one-to-one; continuous
31. neither one-to-one nor onto; discrete
33. D = $\{n \mid n \geq 0\}$; R = $\{T(n) \mid T(n) \geq 0\}$; both (within the restrictions of the domain and codomain); continuous
35.

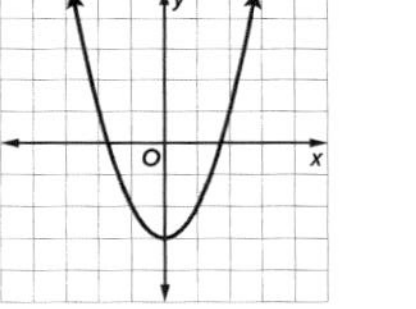

37. D = $\{x \mid x \neq 0\}$ or $(-\infty, 0) \cup (0, \infty)$; R = $\{y \mid y \neq 0\}$ or $(-\infty, 0) \cup (0, \infty)$; neither; continuous
39. Sample answer: The vertical line test is used to determine whether a relation is a function. If no vertical line intersects a graph in more than one point, the graph represents a function. The horizontal line test is used to determine whether a function is one-to-one. If no horizontal line intersects the graph more than once, then the function is one-to one. The horizontal line test can also be used to determine whether a function is onto. If every horizontal line intersects the graph at least once, then the function is onto.

Lesson 1-2

1. Yes; it can be written in $y = mx + b$ form.
3. Yes; it can be written in $y = mx + b$ form.
5. nonlinear **7.** nonlinear **9.** The number of inches and corresponding number of feet is a linear function because when graphed, a line contains all of the points shown in the table.
11. x-int: 12; y-int: −18 **13.** x-int: 6; y-int: −18
15. x-int: 0, 4; y-int: 0 **17a.** x-int: 4; y-int: 20
17b. The x-intercept represents the number of days until Aksa will run out of money. The y-intercept represents the total amount Aksa had in her lunch account at the beginning of the week. **19.** point symmetry
21. point symmetry **23.** neither; $f(-x) = (-x)^3 + (-x)^2 = -x^3 + x^2 \neq f(x)$ and $\neq -f(-x)$
25. No; x is in a denominator. The equation is neither even nor odd. **27.** Yes; it can be written in $y = mx + b$ form. The equation is even. **29.** No; there is an x^2 term. The equation is even. **31.** linear; x-int: $\frac{10}{3}$; y-int: $\frac{30}{7}$; point symmetry **33.** nonlinear; x-int: −3, −2, −1, 1, 2; y-int: 12; neither point nor line symmetry
35. line symmetry; $x = 0.4$
37a. $2x + 2y + 10 = 500$
37b. Yes; it can be written in $y = mx + b$ form.

Selected Answers

Selected Answers

Selected Answers SA1

37c. point symmetry;

Measure of *y* (ft)

Measure of *x* (ft)

39. Odd; $f(-r) = -f(r)$
41. No; sample answer: $f(x) = x^3 + 2x - 5$
43. Never; sample answer: the graph of $x = a$ is a vertical line, so it is not a function.

Lesson 1-3

1. rel. max. at $(-2, -2)$ rel. min. at $(-1, -3)$
3. rel. max. at $(0, -2)$, min. at $(-1, -3)$ and $(1, -3)$
5. The relative maxima occur at $x = -3.7$ and $x = 4.5$, and the relative minimum occurs at $x = 0$. The relative maxima at $x = -3.7$ and $x = 4.5$ represents the top of two hills. The relative minimum at $x = 0$ represents a valley between the hills.
7. As $x \to -\infty, y \to -\infty$ and as $x \to \infty, y \to -\infty$.
9. As $x \to -\infty, y \to \infty$ and as $x \to \infty, y \to -\infty$.
11a. $t = 1.5$; The fish reaches its maximum height 1.5 seconds after it is thrown.
11b. As $t \to -\infty$, $h(t) \to -\infty$ and as $t \to \infty$, $h(t) \to -\infty$; The height cannot be negative because we are considering the path of the fish above the surface of the water, $h = 0$. Time cannot be negative because we're measuring from the initial time the fish was thrown at $t = 0$.
13. rel. max. at $x = 2.7$, rel. min. at $x = -1.2$; As $x \to -\infty$, $f(x) \to \infty$ and as $x \to \infty$, $f(x) \to -\infty$.
15. rel. max. at $x = -2$, rel. min. at $x = -1$; As $x \to -\infty$, $f(x) \to \infty$ and as $x \to \infty$, $f(x) \to \infty$.
17. no rel. max, min: $x = 0$; As $x \to -\infty$, $f(x) \to \infty$ and as $x \to \infty$, $f(x) \to \infty$.
19. As temperature increases, density decreases.
21. rel. max: $(-2.8, 6)$, $(1.8, 3)$; rel. min: $(0, 2)$, $(5, -6)$; As $x \to -\infty, y \to -\infty$ and as $x \to \infty$, $y \to \infty$.
23. no relative max or min
25. rel. max: $x = -1.87$; rel. min: $x = 1.52$

27. The dynamic pressure would approach ∞.
29. as $r \to \infty$, $V \to \infty$
31. Sample answer: The end behavior of a graph describes the output values the input values approach negative and positive infinity. It can be determined by examining the graph.
33. As the concentration of the catalyst is increased, the reaction rate approaches 0.5.
35. Joshua switched the $f(x)$ values. He read the graph from right to left instead of left to right.

Lesson 1-4

1. **3.**

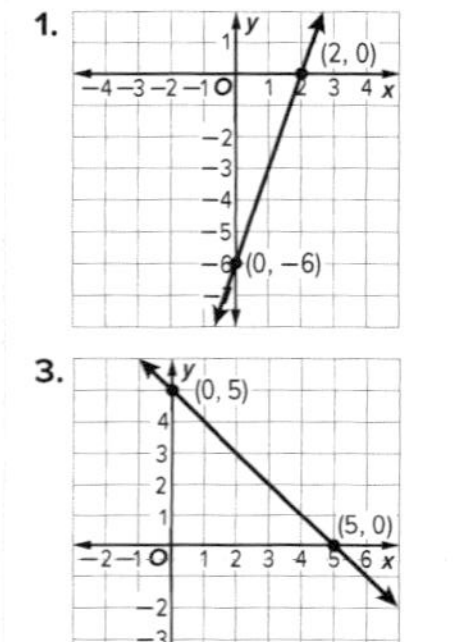

5.

7.

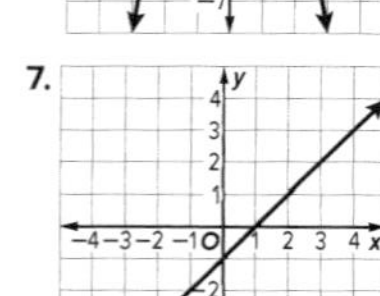

9. Pelican's Height

Height (ft)

Time (s)

11. The *x*-intercept of $f(x)$ is 2, and the *x*-intercept of $g(x)$ is $-\frac{2}{3}$. The *x*-intercept of $f(x)$ is greater than the *x*-intercept of $g(x)$. So, $f(x)$ intersects the *x*-axis at a point farther to the right than $g(x)$. The *y*-intercept of $f(x)$ is -1, and the *y*-intercept of $g(x)$ is 2. The *y*-intercept of $g(x)$ is greater than the *y*-intercept of $f(x)$. So, $g(x)$ intersects the *y*-axis at a higher point than $f(x)$. The slope of $f(x)$ is $\frac{1}{2}$ and the slope of $g(x)$ is 3. Each function is increasing, but the slope of $g(x)$ is greater than the slope of $f(x)$. So, $g(x)$ increases faster than $f(x)$. **13.** Both *x*-intercepts of $f(x)$ are less than the *x*-intercept of $g(x)$. The graph of $f(x)$ intersects the *x*-axis more times than $g(x)$. The *y*-intercept of $g(x)$ is less than the *y*-intercept of $f(x)$. So, $f(x)$ intersects the *y*-axis at a higher point than $g(x)$. Neither function has a relative maximum. $f(x)$ has a minimum at $(-2, -4)$. The two functions have the opposite end behaviors as $x \to -\infty$. The two functions have the same end behavior as $x \to \infty$.
15. The *x*-intercept of $f(x)$ is $\frac{2}{3}$, and the *x*-intercept of $g(x)$ is $\frac{3}{8}$. The *x*-intercept of $f(x)$ is greater than the *x*-intercept of $g(x)$. So, $f(x)$ intersects the *x*-axis at a point farther to the right than $g(x)$. The *y*-intercept of $f(x)$ and $g(x)$ is $\frac{1}{2}$. So, $f(x)$ and $g(x)$ intersect the *y*-axis at the same point. The slope of $f(x)$ is $\frac{3}{4}$ and the slope of $g(x)$ is $-\frac{4}{3}$. Each function is decreasing, but the slope of $g(x)$ is less than the slope of $f(x)$. So, $g(x)$ decreases faster than $f(x)$.
17.

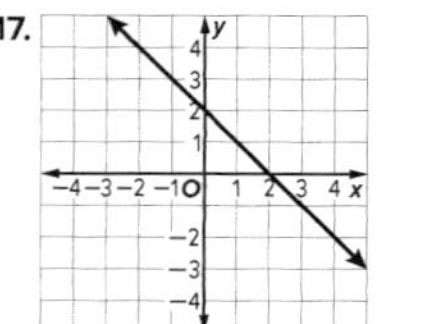

19. Monica's Walk

Distance From Monica's House

Time (min)

21a. linear; Sample answer: It is linear because it makes no stops along the way, and it descends at a steady pace, which indicates a constant rate of change, or slope.
21b. Ski Lift Height

Height (ft)

Time (min)

23. Sample answer: The function is continuous. The function has a *y*-intercept at -3. The function has a maximum at $(-3, 1)$. The function has a minimum at $(1.4, -4)$. As $x \to -\infty$, $f(x) \to -\infty$ and as $x \to \infty$, $f(x) \to \infty$.

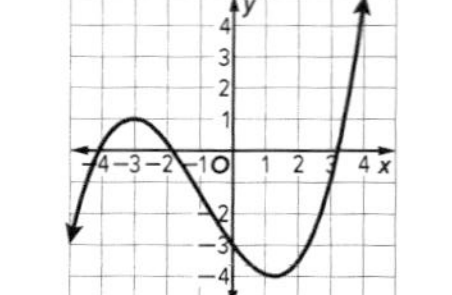

25. Always; Sample answer: A linear function cannot cross the *x*-axis more than once. So, if a function has more than one *x*-intercept, it is a nonlinear function.
27. Both Linda and Rubio sketched correct graphs. Both graphs have an *x*-intercept at 2, a *y*-intercept at -9, are positive for $x > 2$, and have an end behavior of as $x \to -\infty$, $f(x) \to -\infty$ and as $x \to \infty$, $f(x) \to \infty$.

Lesson 1-5

1.
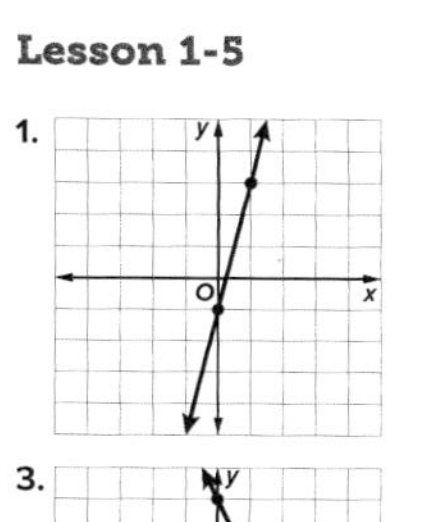

3.
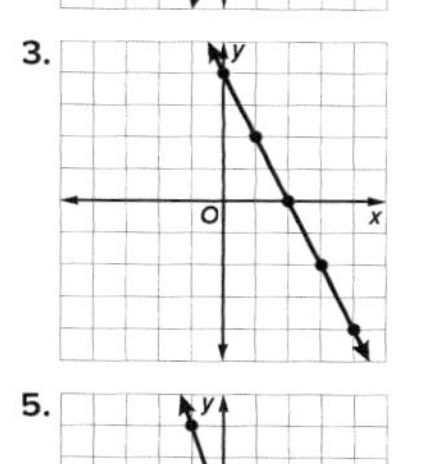

5.
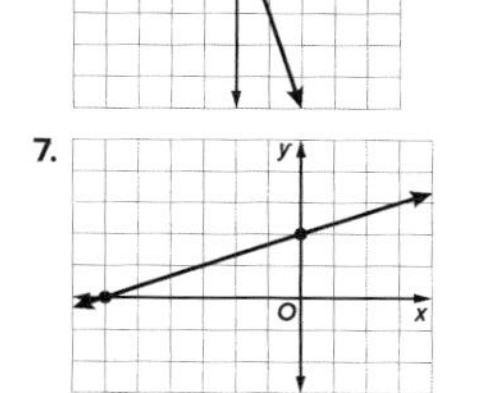

7.
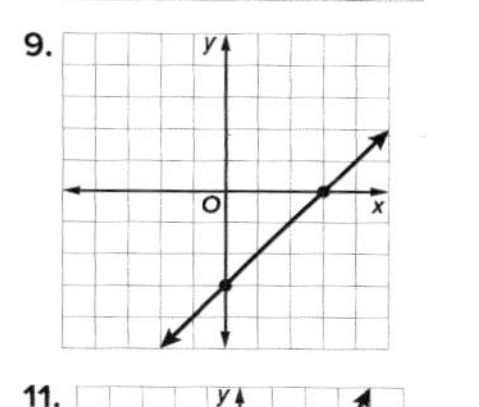

9.
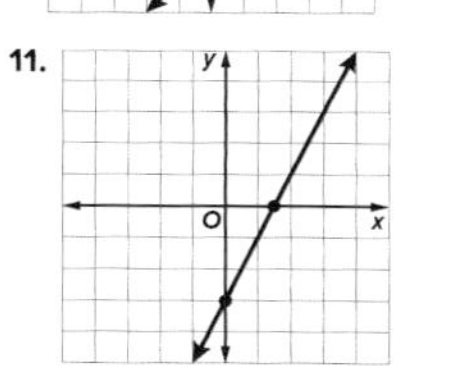

11.

13.
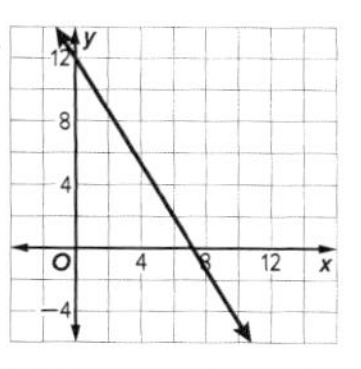

15.
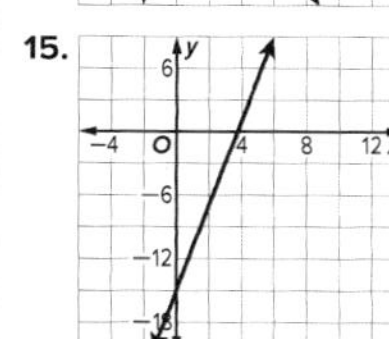

17.
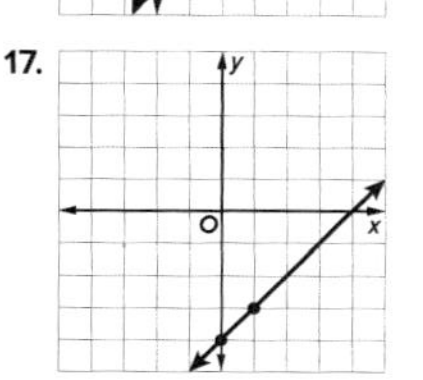

19.
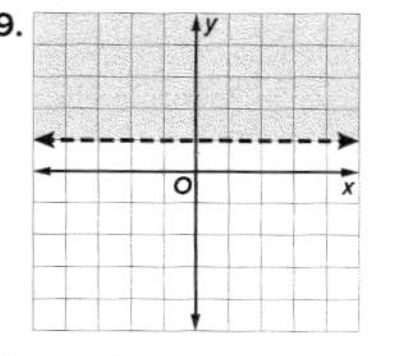

21.
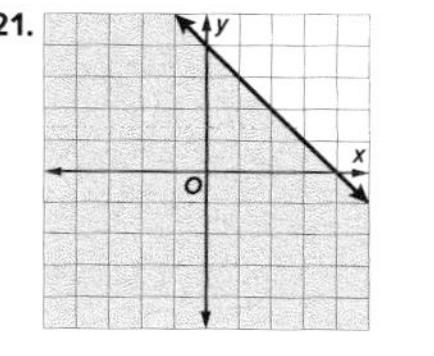

23.
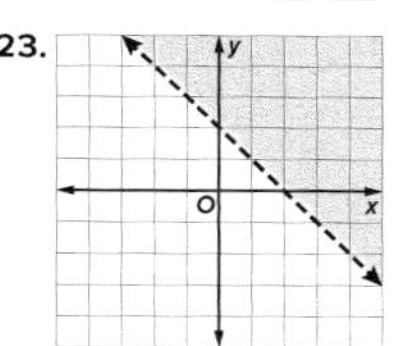

25. 27. 29.
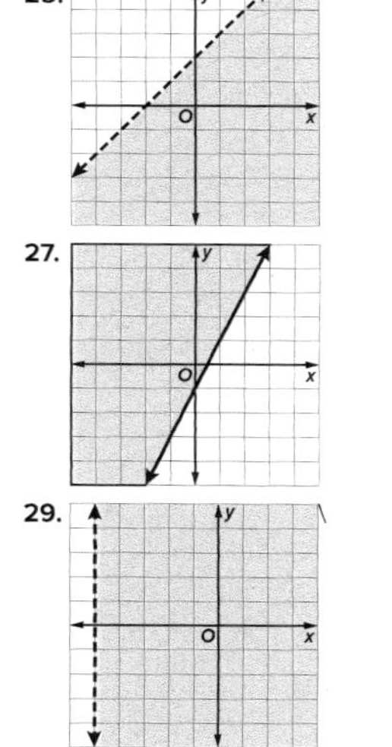

31a. $x + y \geq 400$

31b.
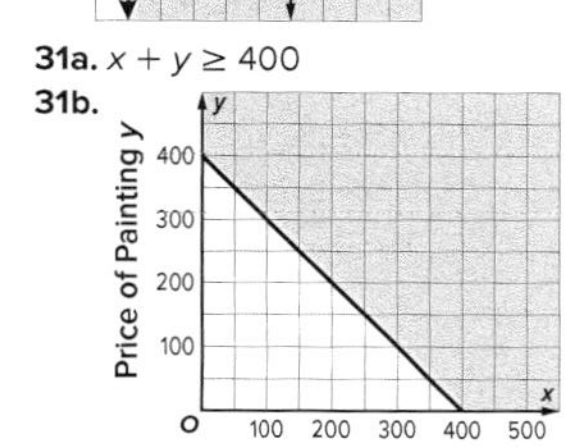

33. 35.
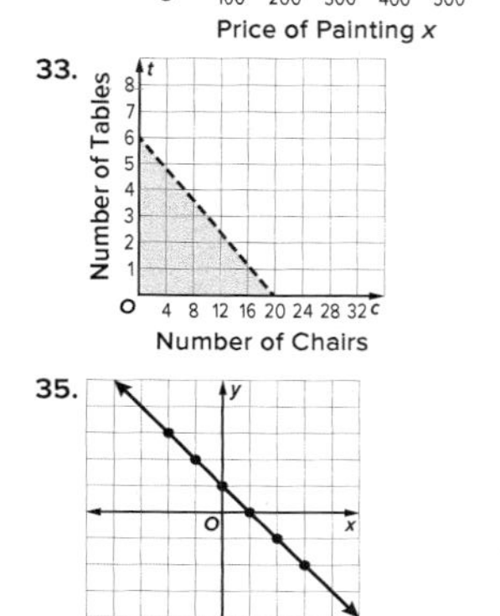

37.
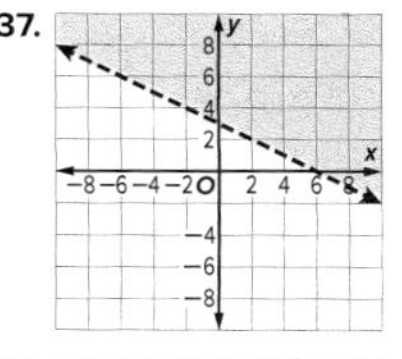

39.
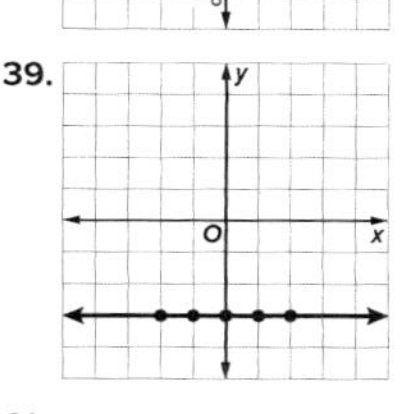

41.
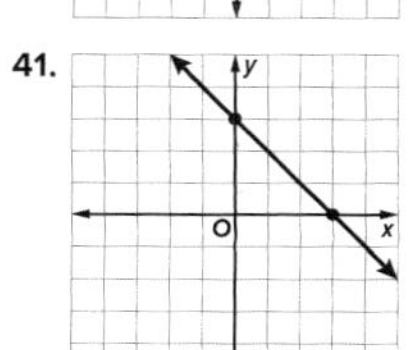

43.
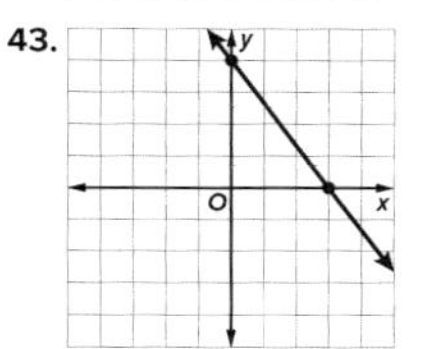

45.
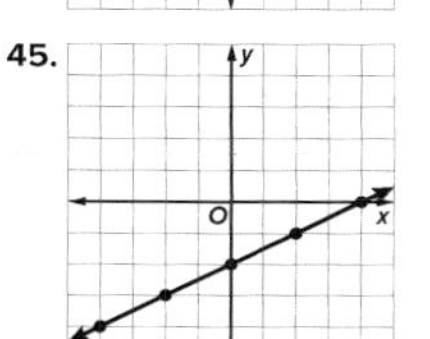

47.
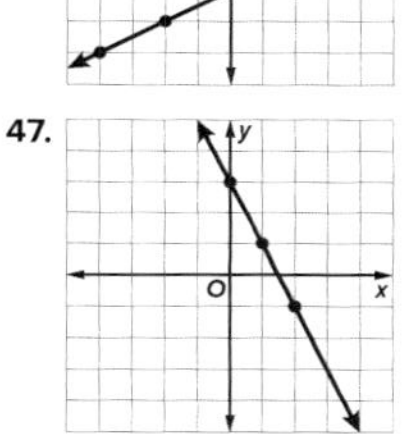

49.

51.

53. x-int = 2, y-int = −6; graph is increasing;

55a. Let x = number of desktops; let y = number of notebooks; $1000x + 1200y \leq 80{,}000$.

55b. Computers Purchased

55c. Yes; Sample answer: the point (50, 25) is on the line, which is part of the viable region.

57a. Let x = cost of a student ticket; let y = cost of an adult ticket; $300x + 150y = 1800$.

57b. Sample answer:
$x = \$2.00, y = \$8.00$;
$x = \$3.00, y = \$6.00$;
$x = \$4.60, y = \$2.80$;
$x = \$6.00, y = \$0.00$

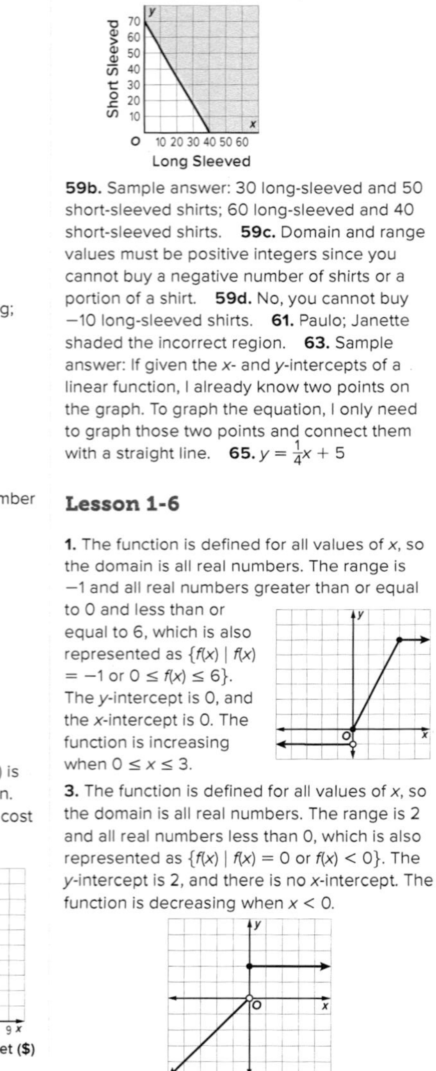

59a. Let x = long-sleeved shirts; let y = short-sleeved shirts; $7x + 4y \geq 280$;

59b. Sample answer: 30 long-sleeved and 50 short-sleeved shirts; 60 long-sleeved and 40 short-sleeved shirts. **59c.** Domain and range values must be positive integers since you cannot buy a negative number of shirts or a portion of a shirt. **59d.** No, you cannot buy −10 long-sleeved shirts. **61.** Paulo; Janette shaded the incorrect region. **63.** Sample answer: If given the x- and y-intercepts of a linear function, I already know two points on the graph. To graph the equation, I only need to graph those two points and connect them with a straight line. **65.** $y = \frac{1}{4}x + 5$

Lesson 1-6

1. The function is defined for all values of x, so the domain is all real numbers. The range is −1 and all real numbers greater than or equal to 0 and less than or equal to 6, which is also represented as $\{f(x) \mid f(x) = -1 \text{ or } 0 \leq f(x) \leq 6\}$. The y-intercept is 0, and the x-intercept is 0. The function is increasing when $0 \leq x \leq 3$.

3. The function is defined for all values of x, so the domain is all real numbers. The range is 2 and all real numbers less than 0, which is also represented as $\{f(x) \mid f(x) = 0 \text{ or } f(x) < 0\}$. The y-intercept is 2, and there is no x-intercept. The function is decreasing when $x < 0$.

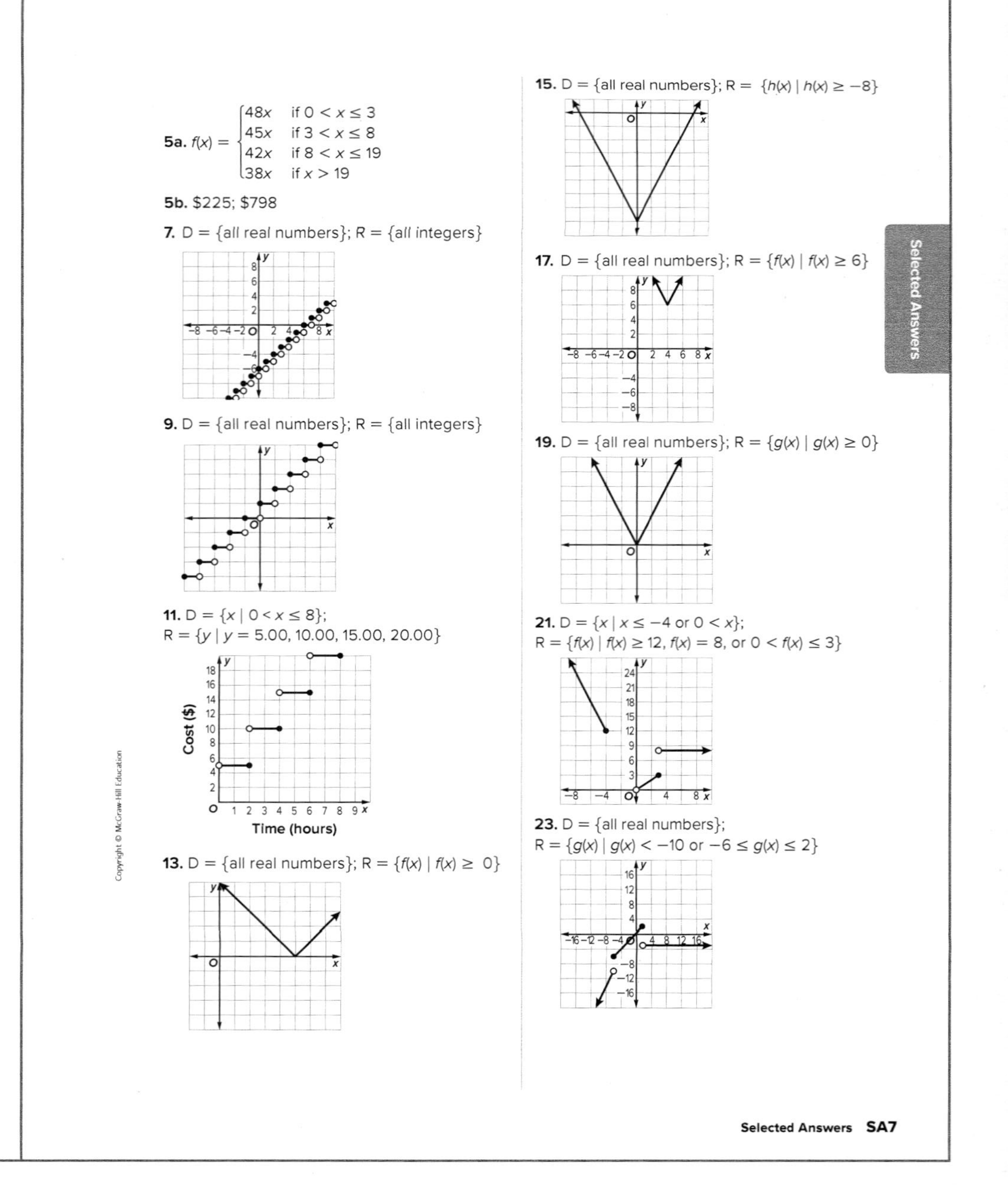

5a. $f(x) = \begin{cases} 48x & \text{if } 0 < x \leq 3 \\ 45x & \text{if } 3 < x \leq 8 \\ 42x & \text{if } 8 < x \leq 19 \\ 38x & \text{if } x > 19 \end{cases}$

5b. \$225; \$798

7. D = {all real numbers}; R = {all integers}

9. D = {all real numbers}; R = {all integers}

11. D = $\{x \mid 0 < x \leq 8\}$;
R = $\{y \mid y = 5.00, 10.00, 15.00, 20.00\}$

13. D = {all real numbers}; R = $\{f(x) \mid f(x) \geq 0\}$

15. D = {all real numbers}; R = $\{h(x) \mid h(x) \geq -8\}$

17. D = {all real numbers}; R = $\{f(x) \mid f(x) \geq 6\}$

19. D = {all real numbers}; R = $\{g(x) \mid g(x) \geq 0\}$

21. D = $\{x \mid x \leq -4 \text{ or } 0 < x\}$;
R = $\{f(x) \mid f(x) \geq 12, f(x) = 8, \text{ or } 0 < f(x) \leq 3\}$

23. D = {all real numbers};
R = $\{g(x) \mid g(x) < -10 \text{ or } -6 \leq g(x) \leq 2\}$

25. D = {all real numbers}; R = $\{f(x) \mid f(x) > -3\}$

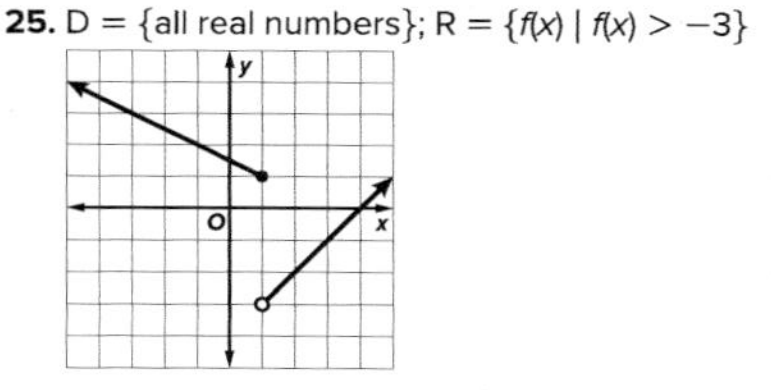

27. D = {all real numbers}; R = {all integers}

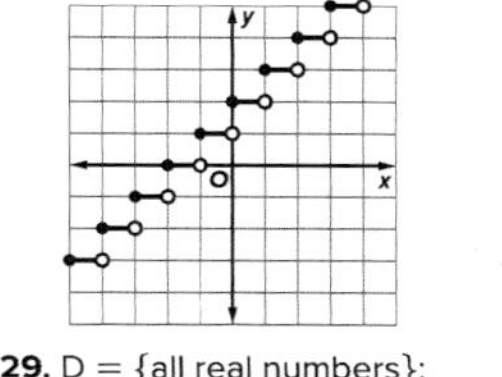

29. D = {all real numbers}; R = {all whole numbers}

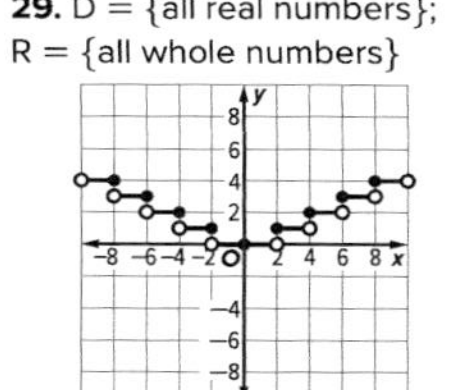

31. D = {all real numbers}; R = $\{g(x) \mid g(x) \le 4\}$

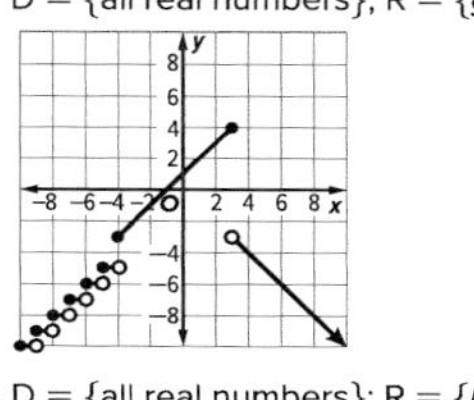

33. D = {all real numbers}; R = $\{h(x) \mid h(x) \ge 2\}$

35. $C(x) = \begin{cases} 5 & \text{if } 0 \le x \le 1 \\ 7.5 & \text{if } 1 < x \le 2 \\ 10 & \text{if } 2 < x \le 3 \\ 12.5 & \text{if } 3 < x \le 4 \\ 15 & \text{if } 4 < x \le 24 \end{cases}$

37a. $C(x) = \begin{cases} 500 + 17.50x \text{ if } 0 \le x \le 40 \\ 1200 + 14.75(x - 40) \text{ if } x \ge 41 \end{cases}$

37b.

[0, 50] scl: 5 by [500, 1500] scl: 100

37c. Because it costs $1200 for 40 guests to attend, use the first expression in the function $C(x)$. Solve the equation $500 + 17.50x = 900$ to obtain about 22.9. Because there cannot be a fraction of a guest, at most 22 guests can be invited to the reunion.

39a. $R(t) = \begin{cases} \frac{20}{3}t + 30 & \text{if } 0 \le t \le 3 \\ 60 & \text{if } 3 < t < 4 \\ 80 & \text{if } 4 \le t \le 5 \\ 60 & \text{if } 5 < t < 6 \\ -\frac{50}{3} + 160 & \text{if } 6 \le t \le 9 \end{cases}$

Range = [10, 60] ∪ {80}

39b. The graph is increasing from $t = 0$ to $t = 3$. This corresponds to the months of September, October, and November.

41. Because the absolute value takes negative f(x)-values and makes them positive, the graph retains the step-like nature of the greatest integer function, but it also has the "v" shape of the absolute value.

[−10, 10] scl: 1 by [−10, 10] scl: 1

43. $f(x) = \begin{cases} -x + 2 & \text{if } x \le 0 \\ -x - 2 & \text{if } x > 0 \end{cases}$

45. $f(x) = \begin{cases} \frac{1}{2}x + 1 & \text{if } x < 2 \\ x - 4 & \text{if } x > 2 \end{cases}$

47. D = $\{m \mid 0 < m \le 6\}$; R = $\{C \mid C = 2.00, 4.00, 6.00, 8.00, 10.00, 12.00\}$

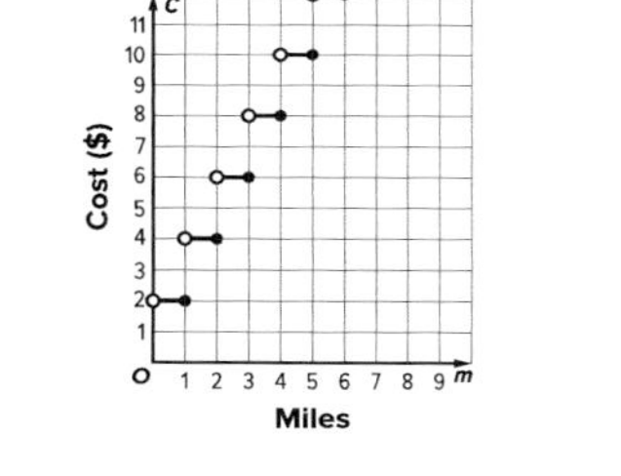

49. Sample answer: $|y| = x$ **51.** Sample answer: 8.6; The greatest integer function asks for the greatest integer less than or equal to the given value; thus 8 is the greatest integer. If we were to round this value to the nearest integer, we would round up to 9. **53.** Sample answer: Piecewise functions can be used to represent the cost of items when purchased in quantities, such as a dozen eggs.

Lesson 1-7

1. translation of the graph of $y = x^2$ up 4 units **3.** translation of the graph of $y = x$ down 1 unit **5.** translation of the graph of $y = x^2$ right 5 units **7.** $y = x^2 - 2$ **9.** $y = x + 1$ **11.** $y = |x + 3| + 1$ **13.** compressed horizontally and reflected in the y-axis **15.** stretched vertically and reflected in the x-axis **17.** compressed vertically and reflected in the x-axis **19.** stretched horizontally and reflected in the y-axis **21.** reflected in the x-axis, stretched vertically, and translated down 4 units **23.** compressed vertically and translated down 2 units **25.** reflected in the x-axis, compressed vertically, and translated left 3 units **27.** stretched horizontally by a factor of 0.5; The absolute value function shows the ball bouncing off the edge of the pool table and the stretch shows the wide angle. **29.** compressed vertically by a factor of 0.75 and translated up 25; There is a $25 fixed cost, plus $0.75 per mile, regardless of direction. **31.** $y = 2|x + 2| + 5$ **33.** $y = -|x| - 3$ **35.** $y = -(x - 4)^2$ **37.** translation of $y = |x|$ down 2 units

39. reflection of $y = x$ in the x-axis

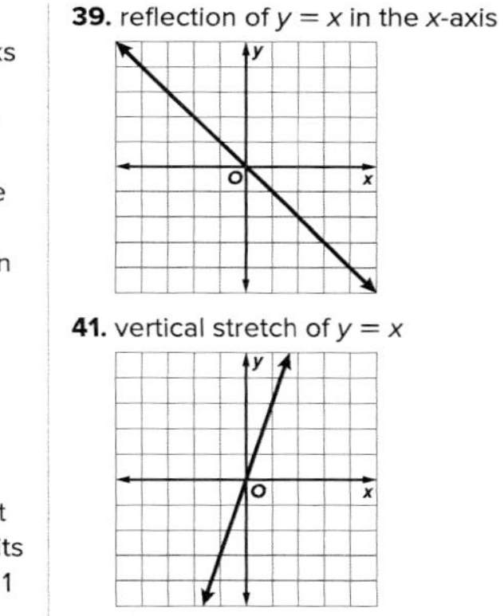

41. vertical stretch of $y = x$

43. translation of $y = x^2$ down 4 units **45.** horizontal compression of the graph of $y = x^2$

47a. translation of $y = |x|$ right 8 units

47b. Sample answer: The speedometer is stuck at 8 mph.

49. stretched vertically by a factor of 4 **51.** Maria stretched the function vertically by a factor of 10. **53a.** quadratic **53b.** x-axis **53c.** right 25 units and up 81 units **53d.** $y = -(x - 25)^2 + 81$

55a.

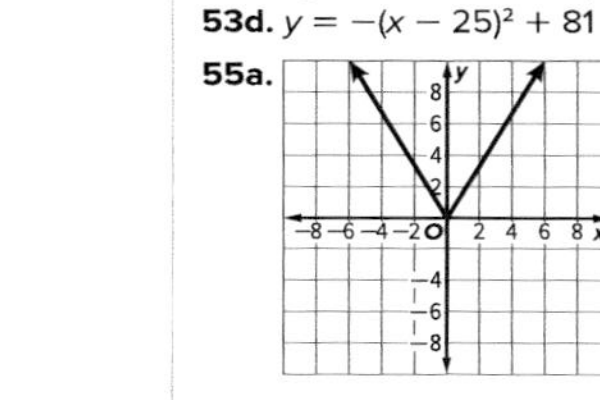

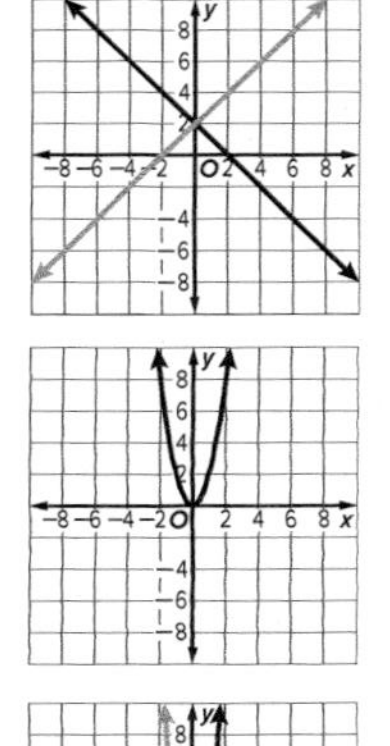

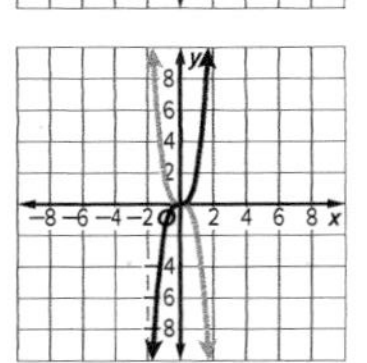

55b. $f(x)$ and $h(x)$ are even, $g(x)$ is neither, and $k(x)$ is odd. **55c.** Even functions are symmetric in the y-axis. If $f(-x) = f(x)$, then the graphs of $f(-x)$ and $f(x)$ coincide. If the graph of a function coincides with its own reflection in the y-axis, then the graph is symmetric in the y-axis. Odd functions are symmetric in the origin, which means that the graph of an odd function coincides with its rotation of 180° about the origin. A rotation of 180° is equivalent to reflection in two perpendicular lines. $f(-x)$ is a reflection in the y-axis and $-f(x)$ is a reflection in the x-axis. Thus if the graphs of $f(-x)$ and $-f(x)$ coincide, $f(x)$ is symmetric about the origin.

57. Sample answer: The graph in Quadrant II has been reflected in the x-axis and moved right 10 units.

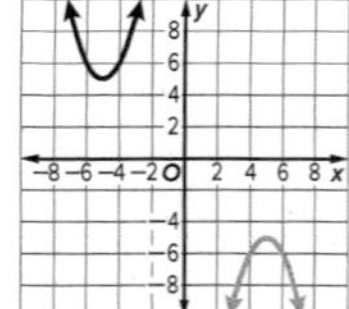

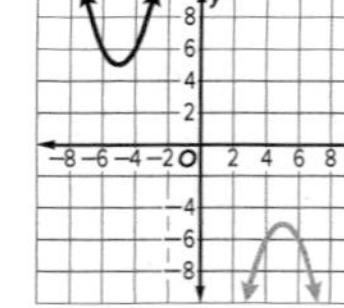

59. Sample answer: Because the graph of $g(x)$ is symmetric about the y-axis, reflecting in the y-axis results in a graph that appears the same. It is not true for all quadratic functions. When the axis of symmetry of the parabola is not along the y-axis, the graph and the graph reflected in the y-axis will be different.

Module 1 Review

1. C **3.** Sample answer: The x-intercept is (6, 0). This means that after 6 weeks, Tia owes her friend \$0. The y-intercept is (0, 80). This means that Tia initially owed her friend \$80, or that Tia borrowed \$80 from her friend to go to a theme park.

5.

x	Relative Maximum	Relative Minimum	Neither
−5	X		
−4			X
−2		X	
0			X
1	X		
5		X	

7. The x-intercept of $f(x)$ is 2, and the x-intercept of $g(x)$ is 1. The x-intercept of $f(x)$ is greater than the x-intercept of $g(x)$. So, $f(x)$ intersects the x-axis at a point farther to the right than $g(x)$. The y-intercept of $f(x)$ is 4, and the y-intercept of $g(x)$ is −5. The y-intercept of $f(x)$ is greater than the y-intercept of $g(x)$. So, $f(x)$ intersects the y-axis at a higher point than $g(x)$. The slope of $f(x)$ is −2 and the slope of $g(x)$ is 5. $f(x)$ is decreasing and $g(x)$ is increasing. The slope of $g(x)$ is greater than the slope of $f(x)$. **9.** A

11.

13. Sample answer: The graph of the parent function $f(x) = x^2$ has been stretched vertically, translated 2 units to the right, and translated up 9 units. **15.** C

Module 2

Quick Check

1.

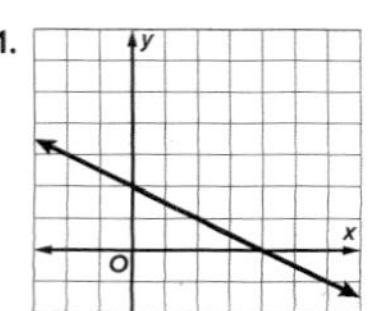

3.

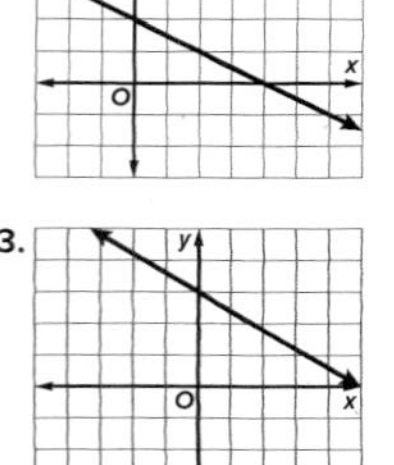

5.

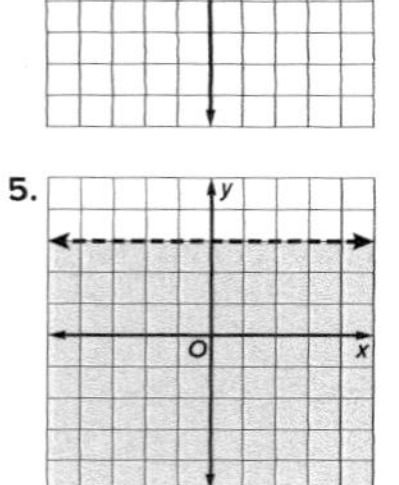

7.

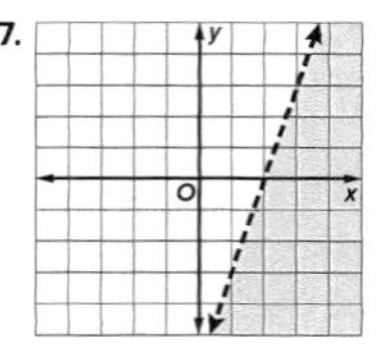

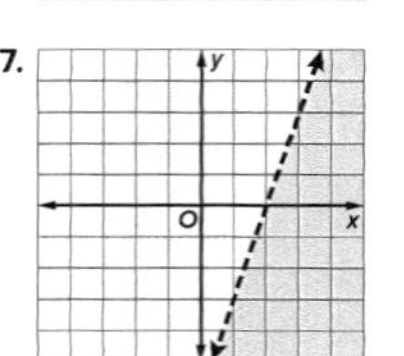

Lesson 2-1

1. $\frac{4}{5}$ **3.** $\frac{1}{4}$ **5.** 5 **7.** 0 **9.** −3

11. g = green fees per person; $6(2) + 4g = 76$; \$16

13. $y = 2A - x$

15. $h = \frac{A - 2\pi r^2}{2\pi r}$

17. $f(x) = 2x + 12$ The solution is −6.

19. $f(x) = \frac{1}{2}x - 6$ The solution is 12.

21. $f(x) = -3x - 2$ The solution is $x = -\frac{2}{3}$.

23a. Sample answer: about 5.5 weeks

23b. $w \approx 5.56$

25. $\{z \mid z \leq -8\}$

−9 −8 −7 −6 −5 −4 −3 −2 −1

27. $\{n \mid n < 2\}$

−4 −3 −2 −1 0 1 2 3 4

29. $\{m \mid m \leq 4\}$

−2 −1 0 1 2 3 4 5 6

31. $15P + 300 \geq 1500$; $P \geq 80$; Manuel must translate at least 80 pages.

33. $\frac{22}{5}$ **35.** 5

37. $\{x \mid x \le 6\}$

0 1 2 3 4 5 6 7 8

39. Sample answer: Let x represent the number of skating session. With a membership: $6x + 60 \le 90$; $x \le 5$. Without a membership: $10x \le 90$; $x \le 9$. She should not buy a membership.

41. Jade; Sample answer: in the last step, when Steven subtracted b_1 from each side, he mistakenly put b_1 in the numerator instead of subtracting it from the fraction.

43. $y_1 = y_2 - \sqrt{d^2 - (x_2 - x_1)^2}$

45. Sample answer: When one number is greater than another number, it is either more positive or less negative than that number. When these numbers are multiplied by a negative value, their roles are reversed. That is, the number that was more positive is now more negative than the other number. Thus, it is now *less than* that number and the inequality symbol needs to be reversed.

Lesson 2-2

1. $\{11\}$ **3.** $\{3, 4\}$ **5.** $\{-8\}$ **7.** $\{-2, -1\}$

9. $|x - 87.4| \le 1.5$; 85.9° F; 88.9° F

11. $\{x \mid -2 \le x \le 0\}$

−4 −3 −2 −1 0 1 2 3 4

13. ∅

−4 −3 −2 −1 0 1 2 3 4

15. $\{x \mid -1.8 < x < 3.4\}$

−3 −2 −1 0 1 2 3 4 5 6

17. $\left\{x \mid x < -1 \text{ or } x > \frac{1}{3}\right\}$

−1 $-\frac{2}{3}$ $-\frac{1}{3}$ 0 $\frac{1}{3}$ $\frac{2}{3}$ 1

19. $|r - 24| > 6.5$; $\{r \mid r < 17.5 \text{ or } r > 30.5\}$

15 20 25 30 35

21. $\left\{1, \frac{1}{5}\right\}$

23. $\left\{z \mid z > -\frac{8}{3}\right\}$

25. $|4x + 7| = 2x + 3$; $x = -2$, $x = -\frac{5}{3}$; The absolute value equation is valid when $2x + 3 \ge 0$, so the equation is valid when $x \ge -\frac{3}{2}$. Since neither value of x is greater than or equal to $-\frac{3}{2}$, both solutions are extraneous.

27a. $|x - 36| \le 0.125$; Sample answer: The inequality shows that the length of the lumber x could be as much as 0.125 inch greater than 36 inches or 0.125 inch less than 36 inches.

27b. $\{x \mid 35.875 \le x \le 36.125\}$; The length of the lumber can range from 35.875 inches to 36.125 inches.

29. Yuki; Sample answer: Yuki is correct because if $|a| = |b|$, then either $a = b$ or $a = -b$. They will get the same answers because $a = -b$ and $b = -a$ and $a = b$ and $-a = -b$ are equivalent equations.

31. $\left\{p \mid -\frac{9}{4} < p < \frac{5}{4}\right\}$

−3 −2 −1 0 1 2 3

33. $\left\{w \mid w \le -\frac{23}{2} \text{ or } w \ge \frac{7}{2}\right\}$

−16 −14 −12 −10 −8 −6 −4 −2 0 2 4

35. ∅

37. $40 < |200 - 32t| < 88$; $3.5 < t < 5$ or $7.5 < t < 9$; The speed is between 40 and 88 feet per second in the intervals from 3.5 to 5 seconds going up and from 7.5 to 9 seconds coming down.

39. Roberto is correct. Sample answer: The solution set for each inequality is all real numbers. For any value of c (positive, negative, or zero), each inequality will be true.

41. $x > 5$ and $x < 1$; Sample answer: Each of these has a non-empty solution set except for $x > 5$ and $x < 1$. There are no values of x that are simultaneously greater than 5 and less than 1.

43. The 4 potential solutions are:

1. $(2x - 1) \ge 0$ and $(5 - x) \ge 0$
2. $(2x - 1) \ge 0$ and $(5 - x) < 0$
3. $(2x - 1) < 0$ and $(5 - x) \ge 0$
4. $(2x - 1) < 0$ and $(5 - x) < 0$

The resulting equations corresponding to these cases are:

1. $2x - 1 + 3 = 5 - x : x = 1$
2. $2x - 1 + 3 = x - 5 : x = -7$
3. $1 - 2x + 3 = 5 - x : x = -1$
4. $1 - 2x + 3 = x - 5 : x = 3$

The solutions from case 1 and case 3 work. The others are extraneous. The solution set is $\{-1, 1\}$.

45. Always; if $|x| < 3$, then x is between −3 or 3. Adding 3 to the absolute value of any of the numbers in this set will produce a positive number.

47. Sample answer: $\left|x - \frac{a+b}{2}\right| \le b - \frac{a+b}{2}$

Lesson 2-3

1. $7x + 5y = -35$; $A = 7$, $B = 5$, $C = -35$

3. $3x - 10y = -5$; $A = 3$, $B = -10$, $C = -5$

5. $5x + 32y = 160$; $A = 5$, $B = 32$, $C = 160$

7. $y = -2x + 4$; $m = -2$, $b = 4$

9. $y = -4x + 12$; $m = -4$, $b = 12$

11. $y = -\frac{2}{3}x + \frac{5}{3}$; $m = -\frac{2}{3}$, $b = \frac{5}{3}$

13. $y = 20x + 83$; There were 83 shirts collected before noon. There were 20 shirts collected each hour after noon.

15a. Let x represent the number of hours the plumber spends working at a job site, and y represent the total cost for the services.

15b. slope: 42; y-intercept: 65; $y = 42x + 65$

15c. $275

17. $y + 8 = -5(x + 3)$

19. $y + 8 = -\frac{2}{3}(x - 6)$

21. $y + 3 = -8(x - 2)$ or $y - 5 = -8(x - 1)$

23. $y + 2 = -\frac{3}{2}(x + 1)$ or $y - 1 = -\frac{3}{2}(x + 3)$

25. $y - 5.919 = 0.856(x - 1)$ or $y - 11.055 = 0.856(x - 7)$

27. $2x + y = 5$; $y = -2x + 5$; $y + 7 = -2(x - 6)$

29. $x - y = 5$; $y = x - 5$; $y + 1 = 1(x - 4)$ or $y - 3 = 1(x - 8)$

31. $y = -0.25x + 648$; Sample answer: I assumed that the water level continues to drop at a constant rate.

33. $16x - 19y = -41$

35a. Sample answer: (8, −9); I used the given two points to write an equation. Then, I substituted 8 for x and solved for y.

35b. Sample answer: The equations are equivalent when simplified.

37. Never; sample answer: The graph of $x = a$ is a vertical line.

39. Sample answer: $y - 0 = 2(x - 3)$

41. No; Sample answer: You can choose points on the graph and show on a coordinate plane that they do not fall on a single line. For instance, the points (0, 2), (1, 10), (2, 24), and (3, 49) do not lie on a straight line.

43. Sample answer: Depending on what information is given and what the problem is, it might be easier to represent a linear equation in one form over another. For example, if you are given the slope and the y-intercept, you could represent the equation in slope-intercept form. If you are given a point and the slope, you could represent the equation in point-slope form. If you are trying to graph an equation using the x- and y-intercepts, you could represent the equation in standard form.

Lesson 2-4

1. 1; consistent and independent

3. 1; consistent and independent

5. infinitely many; consistent and dependent

7. (2, 1)

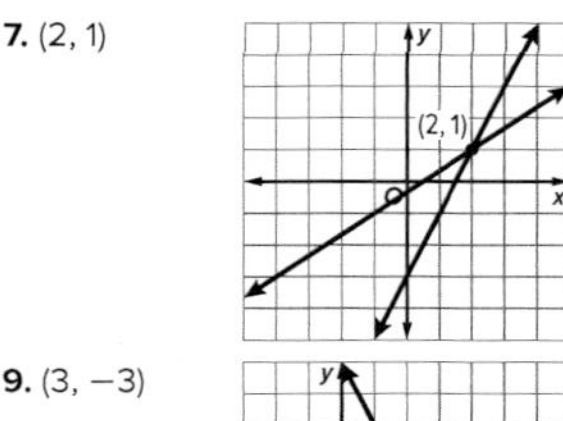

9. (3, −3)

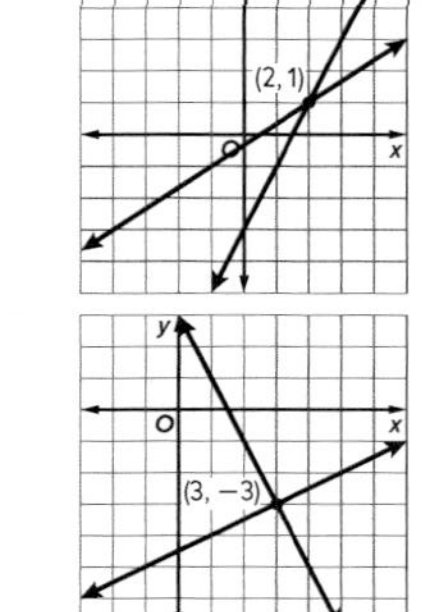
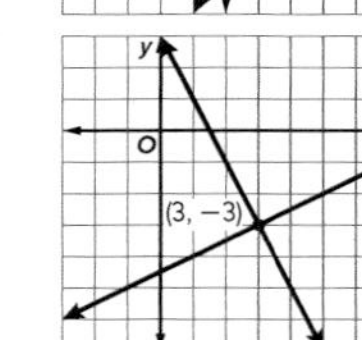

11. no solution

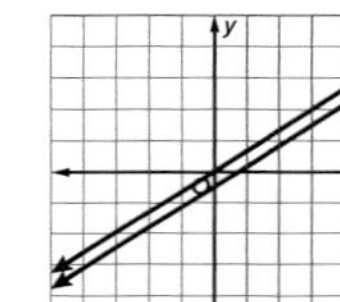
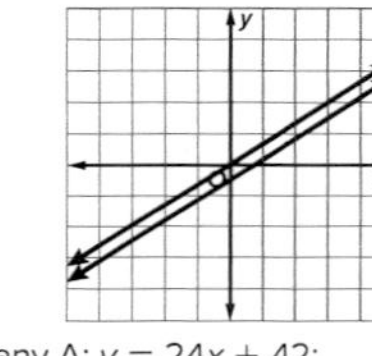

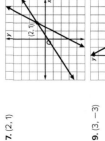

13a. Company A: $y = 24x + 42$;
Company B: $y = 28x + 25$

13b. about 4 containers

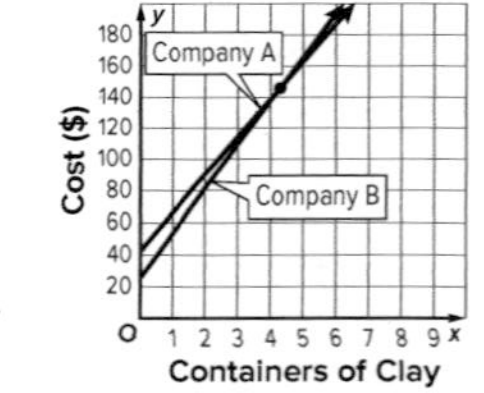

13c. Sample answer: I estimated that the cost would be the same when ordering 4 containers. By substituting $x = 4$ in the equations, the cost at Company A is \$138 and the cost at Company B is \$137. These values are approximately equal, so the estimate is reasonable.

15. (2.07, −0.39)

17. (15.03, 10.98)

19. 2.76

21. −0.99

23. (2, 1)

25. (3.78, 5.04)

27. Always; Sample answer: *a* and *b* are the same line. *b* is parallel to *c*, so *a* is also parallel to *c*. Since *c* and *d* are consistent and independent, then *c* is not parallel to *d* and, thus, intersects *d*. Since *a* and *c* are parallel, then *a* cannot be parallel to *d*, so, *a* must intersect *d* and must be consistent and independent with *d*.

29. Never; Sample answer: Lines cannot intersect at exactly two distinct points. Lines intersect once (one solution), coincide (infinite solutions), or never intersect (no solution).

Lesson 2-5

1. (3, −3) **3.** (2, 1) **5.** no solution

7a. Cassandra: $3x + 14y = 203$;
Alberto: $11x + 11y = 220$; $x = 7$, $y = 13$

7b. The cost of each small pie is \$7. The cost of each large pie is \$13.

7c. Sample answer: By substituting the solution into each equation in the system, you can verify that it is correct. $3(7) + 14(13) = 203$, and $11(7) + 11(13) = 220$.

9. (−2, −5) **11.** no solution **13.** (3, 5)
15. (−2, 3) **17.** (1, 6) **19.** (6, −5) **21.** 4, 8

23. adult ticket \$5.50; student ticket \$2.75.

25. Gloria is correct.; Sample answer: Syreeta subtracted 26 from 17 instead of 17 from 26 and got $3x = -9$ instead of $3x = 9$. She proceeded to get a value of −11 for *y*. She would have found her error if she had substituted the solution into the original equations.

27. Sample answer: It is more helpful to use substitution when one of the variables has a coefficient of 1 or if a coefficient can easily be reduced to 1.

Lesson 2-6

1.

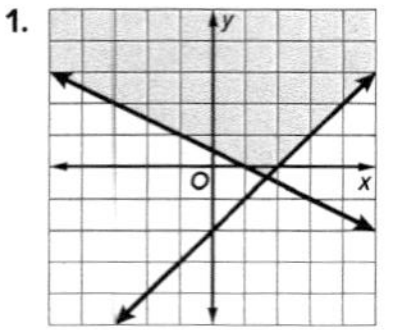
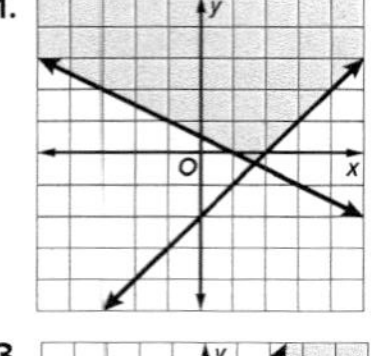

3.

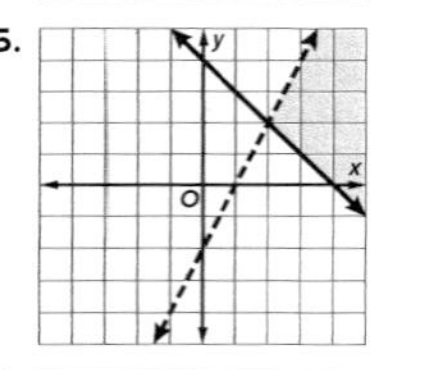

5.

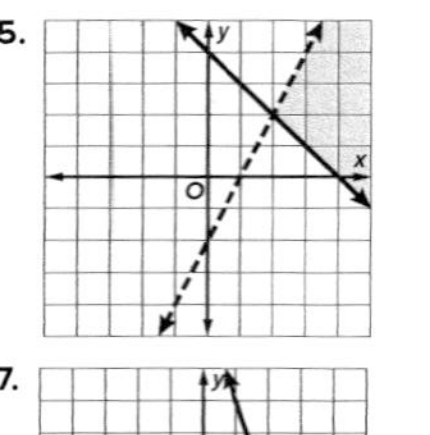

7.

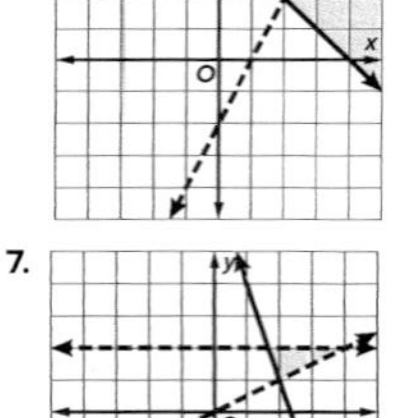

9.

11.

13a. Let *x* be student tickets and *y* be adult tickets. $x + y \leq 800$; $4x + 7y \geq 3400$

13b. Quadrant I

13c. No; they would only make \$3300.

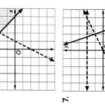

15.

17.

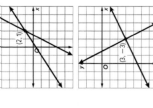

19a. Let x represent the low risk investment and y represent the high risk investment. $x + y \le 2000$; $0.03x + 0.12y \ge 150$

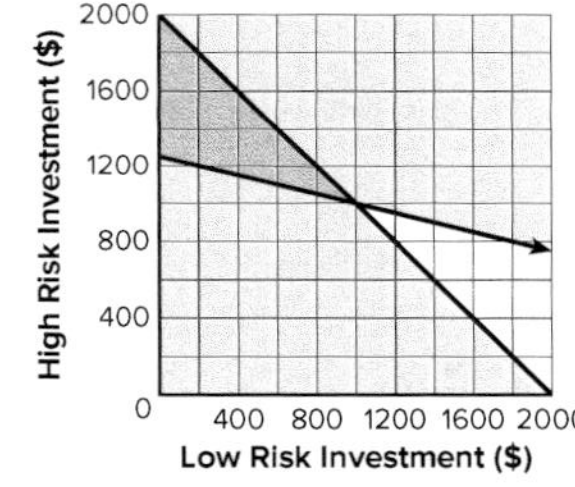

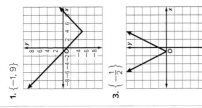

19b. Sample answer: Because Sheila cannot invest a negative amount of money, the graph is limited to positive values of x and y.

19c. Sample answer: $200 in low risk and $1700 in high risk; $400 in low risk and $1600 in high risk; $1000 in low risk and $1000 in high risk

21. 75 units^2

23. True; sample answer: The feasible region is the intersection of the graph of the inequalities. If the graphs intersect, there are infinitely many points in the feasible region. If the graphs do not intersect, they contain no common points and there is no solution.

Lesson 2-7

1. vertices: (1, 2), (1, 4), (5, 8), (5, 2); max: 11; min: −5

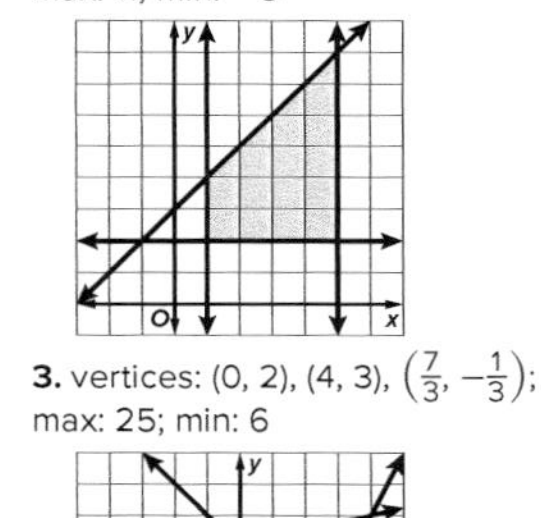

3. vertices: (0, 2), (4, 3), $\left(\frac{7}{3}, -\frac{1}{3}\right)$; max: 25; min: 6

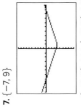

5. vertices: (1, −1), (1, 6), (8, 6); max: 2; min: −5

7. vertex: (−1, 7); max: 13; no min.

9. vertices: (−2, 0); $\left(-\frac{7}{5}, \frac{9}{5}\right)$; max: $-\frac{34}{4}$; no min.

11a. Let x represent clay beads and y represent glass beads.; $0 \le x \le 10$; $y \ge 4$; $4y \le 2x + 8$ $C = 0.20x + 0.40y$; The total cost equals 0.20 times the number of clay beads plus 0.40 times the number of glass beads.

11b. (4, 4), (10, 4), and (10, 7)

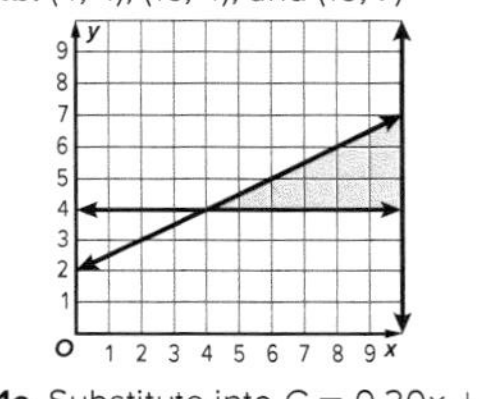

11c. Substitute into $C = 0.20x + 0.40y$: (4, 4) yields $2.40, (10, 4) yields $3.60, and (10, 7) yields $4.80. The minimum cost would be $2.40 at (4, 4), which represents 4 clay beads and 4 glass beads.

13. (5, 2); 19 feet

15. Always; Sample answer: if a point on the unbounded region forms a minimum, then a maximum cannot also be formed because of the unbounded region. There will always be a value in the solution that will produce a higher value than any projected maximum.

17. Sample answer: Even though the region is bounded, multiple maximums occur at A and B and all of the points on the boundary of the feasible region containing both A and B. This happened because that boundary of the region has the same slope as the function.

Lesson 2-8

1. (4, −3, −1) **3.** infinitely many solutions
5. no solution **7.** (5, −5, −20)
9. (−3, 2, 1) **11.** (2, −1, 3)
13. 60 grams of Mix A, 50 grams of Mix B, and 40 grams of Mix C
15. 3 oz of apples, 7 oz of raisins, 6 oz of peanuts
17. 9, 6, 3 **19.** fastest press: 1700 papers; slower press: 1000 papers; slowest press: 800 papers
21. orchestra ticket: $10; mezzanine ticket: $8; balcony ticket: $7
23. $a = -3$, $b = 4$, $c = -6$; $y = -3x^2 + 4x - 6$
25. Sample answer:

$$3x + 4y + z = -17$$
$$2x - 5y - 3z = -18$$
$$-x + 3y + 8z = 47$$

$$3x + 4y + z = -17$$
$$3(-5) + 4(-2) + 6 = -17$$
$$-15 + (-8) + 6 = -17$$
$$-17 = -17 \checkmark$$

$$2x - 5y - 3z = -18$$
$$2(-5) - 5(-2) - 3(6) = -18$$
$$-10 + 10 - 18 = -18$$
$$-18 = -18 \checkmark$$

$$-x + 3y + 8z = 47$$
$$-(-5) + 3(-2) + 8(6) = 47$$
$$5 - 6 + 48 = 47$$
$$47 = 47 \checkmark$$

Copyright © McGraw-Hill Education

Lesson 2-9

1. {−1, 9}

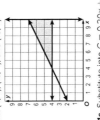

3. $\left\{-\frac{1}{2}\right\}$

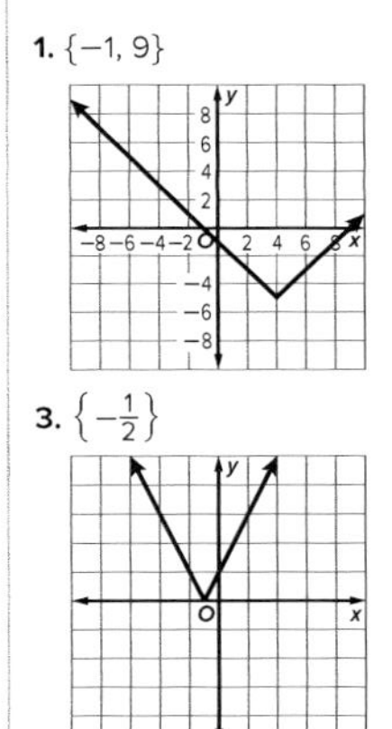

5. $\left\{\frac{1}{3}\right\}$

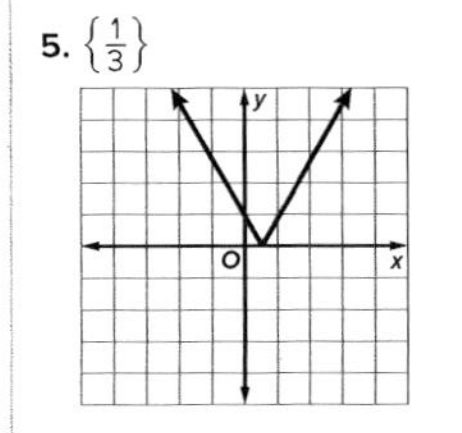

7. {−7, 9}

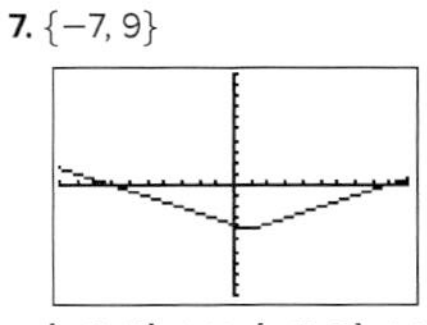

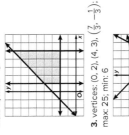

[−10, 10] scl: 1 by [−10, 10] scl: 1

9. {−10, 14}

[−15, 15] scl: 1 by [−10, 10] scl: 1

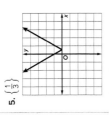

11. $\{-56, 44\}$

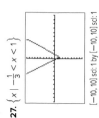

$[-60, 45]$ scl: 5 by $[-10, 10]$ scl: 1

13. $\{-12, 16\}$ 15. $\varnothing$ 17. $\{-2, -1\}$

19. $\{x \mid 1 \le x \le 5\}$

21. $\left\{x \mid x \le -\frac{3}{2} \text{ or } x \ge \frac{5}{2}\right\}$

23. $\{x \mid -6 < x < 2\}$

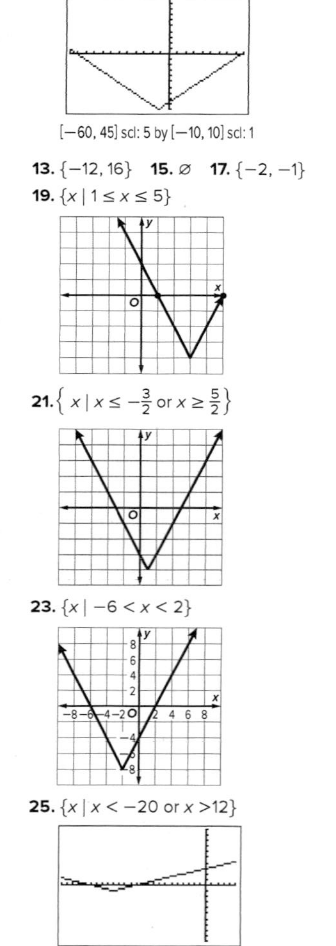

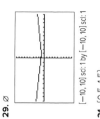

25. $\{x \mid x < -20 \text{ or } x > 12\}$

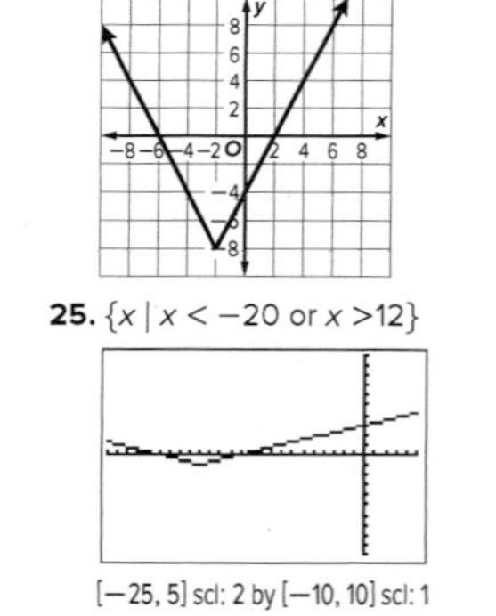

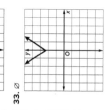

$[-25, 5]$ scl: 2 by $[-10, 10]$ scl: 1

27. $\left\{x \mid -\frac{1}{3} < x < 1\right\}$

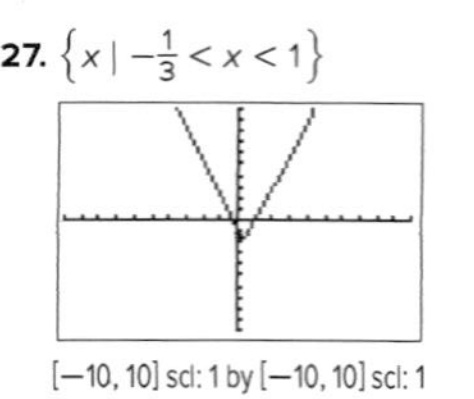

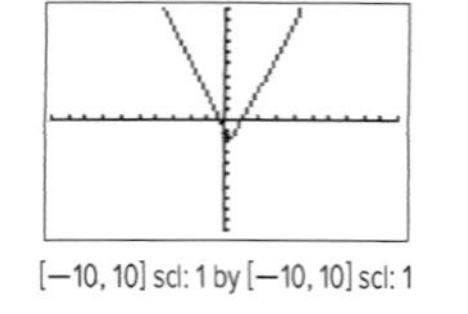

$[-10, 10]$ scl: 1 by $[-10, 10]$ scl: 1

29. $\varnothing$

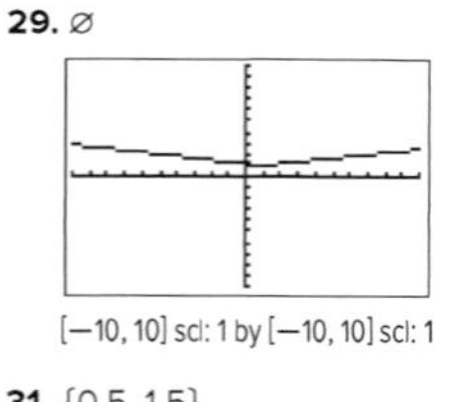

$[-10, 10]$ scl: 1 by $[-10, 10]$ scl: 1

31. $\{0.5, 1.5\}$

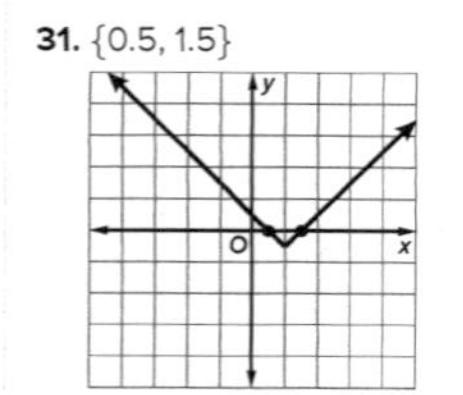

33. $\varnothing$

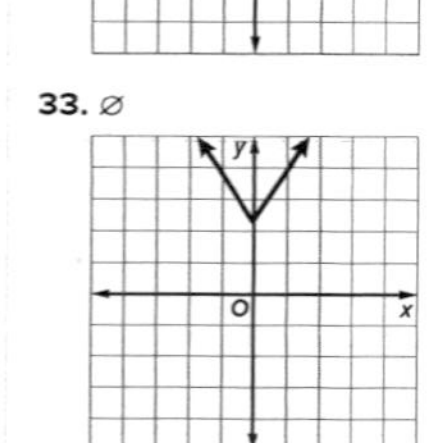

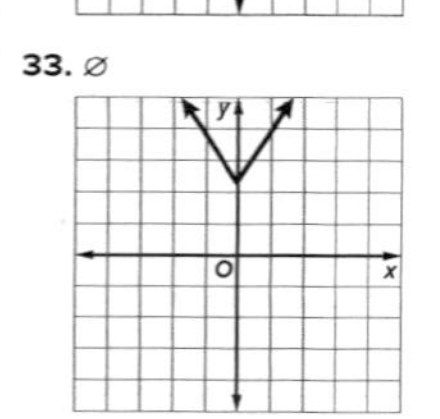

35. $\{x \mid 2 \le x \le 4\}$

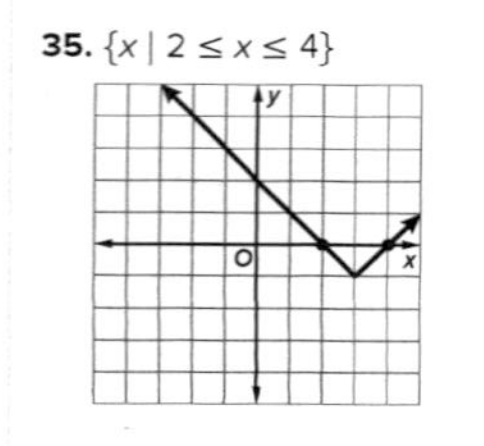

37. $\left\{x \mid x < -10\frac{4}{5} \text{ or } x > 7\frac{1}{5}\right\}$

39. $\{-85, 95\}$

41. $|x - 1.524| \le 0.147$; $\{x \mid 1.377 \le x \le 1.671\}$

$[0, 2]$ scl: 0.25 by $[-1, 1]$ scl: 0.25

43. Sample answer: The process by which the equation or inequality is set to zero to represent $f(x)$ and making a table of values is the same. However, the graph of an absolute value equation is restricted to having either 0, 1, or 2 solutions; whereas the absolute value inequalities can have infinitely many solutions.

45. Sample answer: $|x - 10| = 1$

Module 2 Review

1. Sample answer: First, use the Distributive Property: $6x + 27 + 2x - 4 = 55$. Combine like terms: $8x + 23 = 55$. Then use the Subtraction Property of Equality: $8x = 32$. Finally, use the Division Property of Equality: $x = 4$.

3. $C = \frac{5}{9}(F - 32)$; 77°F

5. $y = -2x + 10$; -2 represents the miles Allie is running each day. 10 represents that Allie's goal at the beginning of the week is to jog 10 miles.

7. B 9. $(0.5, 0.75)$

11. No; Sample answer: Jazmine would only earn $210.

13. A, D, G

15. $2x + 3y + z = 285$
$x + 5y + 2z = 400$
$3x + 2y + 4z = 440$

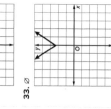

Module 3

Quick Check

1. 22 **3.** − 21 **5.** $(x - 3)(x - 7)$
7. $(2x + 3)(x - 5)$

Lesson 3-1

1. D = {all real numbers}, R = $\{y \mid y \geq -1\}$

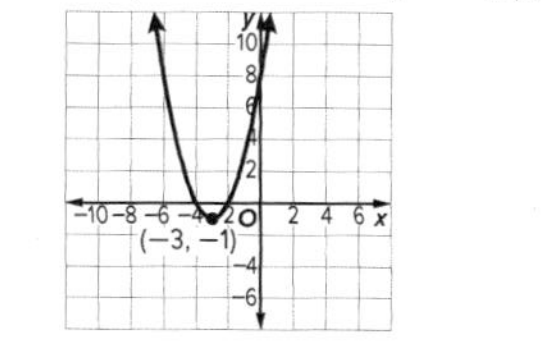

3. D = {all real numbers}, R = $\{y \mid y \geq 1\}$

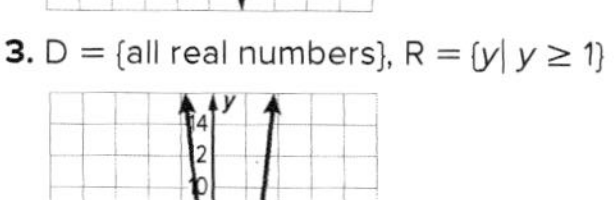

5. D = {all real numbers}, R = $\{y \mid y \geq 0\}$

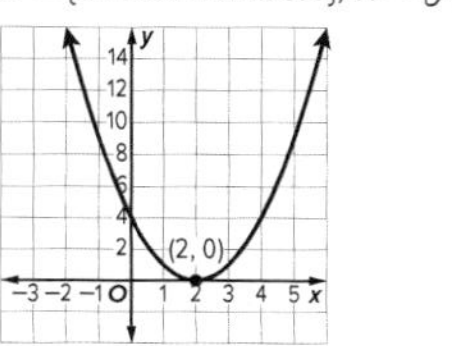

7. $g(x)$; Sample answer: Its vertex is a maximum point at (1, 3), which is 4 units above the vertex of $f(x)$ which is (1, −1).

9. $f(x)$; Sample answer: Its vertex is a minimum point at (5, −20), which is 10 units below the vertex of $g(x)$ which is (5, −10).

11a. For x = the number of price increases and y = revenue, $y = (80 + 4x)(480 - 16x)$.

11b. $100 per permit

11c. $D = \{x \mid 0 \leq x \leq 30\}$, $R = \{y \mid 0 \leq y \leq 40{,}000\}$

13. −10 **15.** −3 **17.** 0 **19.** 2 **21.** 3 **23.** 0

25a. $\approx \frac{12.5}{5}$ or 2.5; $\frac{55.35 - 42.97}{5} = 2.476$

25b. Sample answer: From 2011–2016, the number of people in the U.S. who consumed between 8 and 11 bags of potato chips annually increased by 2.476 million people per year.

27a. y-int = −9; axis of symmetry: $x = 1.5$; x-coordinate of vertex = 1.5

27b.

x	$f(x)$
0	9
1	−13
1.5	−13.5
2	−13
3	−9

27c.

29a. y-int = 0; axis of symmetry: $x = \frac{5}{8}$; x-coordinate of vertex = $\frac{5}{8}$

29b.

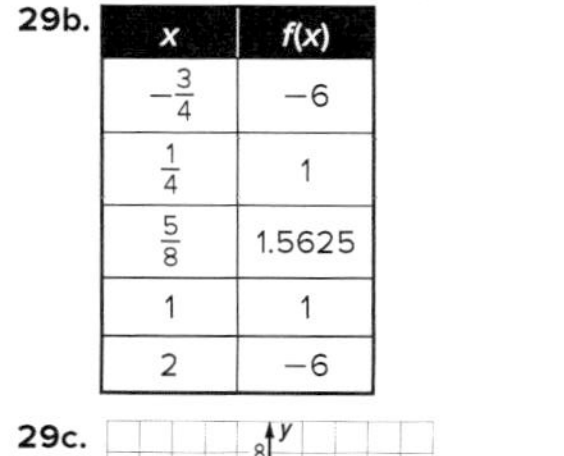

x	$f(x)$
$-\frac{3}{4}$	−6
$\frac{1}{4}$	1
$\frac{5}{8}$	1.5625
1	1
2	−6

29c.

31a. y-int = 4; axis of symmetry: $x = -6$; x-coordinate of vertex = −6

31b.

x	$f(x)$
−10	−1
−8	−4
−6	−5
−4	−4
−2	−1

31c.

33. max; $\frac{2}{3}$; 23

35. max; 1.4; 3.8 **37.** max ≈ −4.11

39a. Let x represent the number of $1 price increases. Let y represent the income. Then $y = (70 - x)(20 + x)$.

x	y
0	1400
5	1625
10	1800
15	1925
20	2000
25	2025
30	2000
35	1925
40	1800
45	1625
50	1400

39b. (25, 2025); The club should increase the price by $25 to make a maximum profit of $2025. Sample answer: Increasing the price by $25 is more than twice the current price, so this may be unreasonable. The club may want to revisit their assumptions.

41a. If x = number of price increases, the revenue is $R(x) = -5x^2 + 40x + 1200$. Because the price changes and the revenue will not be negative, the domain is $\{x \mid 0 \leq x \leq 20\}$.

41b. The graph shows that the maximum revenue value occurs at $x = 4$, which corresponds to a ticket price of $6.00 + 4($0.50) or $8.00. The maximum revenue is $R(4) = \$1280.00$.

41c. As price increases continue, the demand for tickets will decrease. The x-intercept indicates a price that is too high, and no one is estimated to buy tickets for the cinema.

43. 20 ft

45. Madison; sample answer: $f(x)$ has a maximum of −2. $g(x)$ has a maximum of 1.

47. Sample answer: A function is quadratic if it has no other terms than a quadratic term, linear term, and constant term. The function has a maximum if the coefficient of the quadratic term is negative and has a minimum if the coefficient of the quadratic term is positive.

Lesson 3-2

1. no real solution **3.** −4

5. $-\frac{1}{2}$

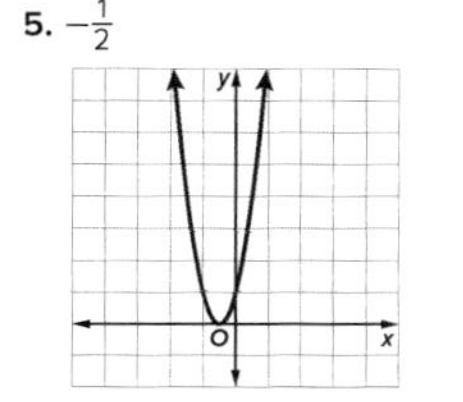

7. −3, 1

9. 1, 5

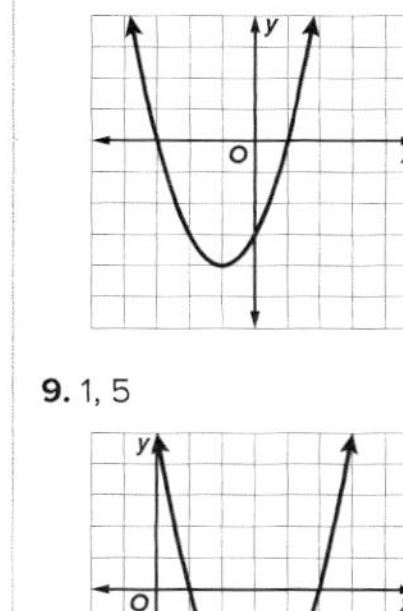

11. $-2, 5$

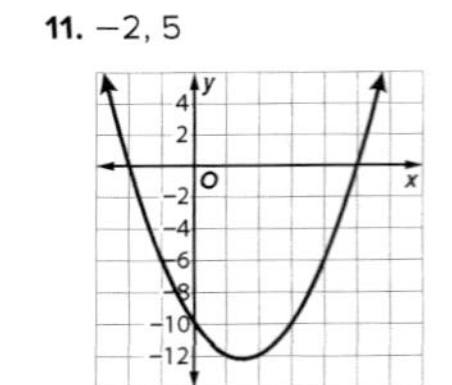

13. 6 and -4

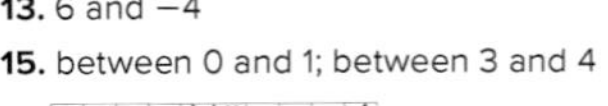

15. between 0 and 1; between 3 and 4

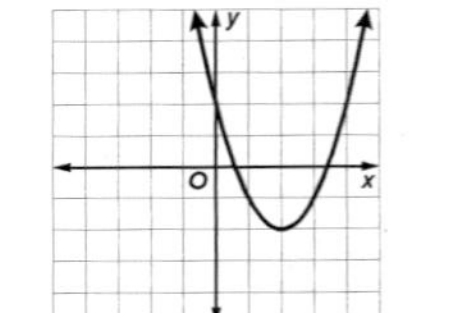

17. between -1 and 0; between -4 and -3

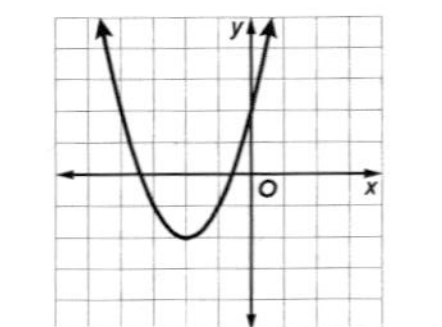

19. $-6, 6$

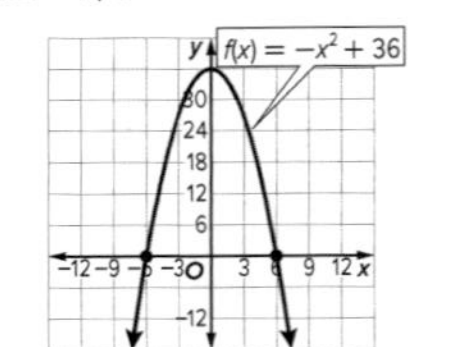

21. between -1 and 0; between -5 and -4

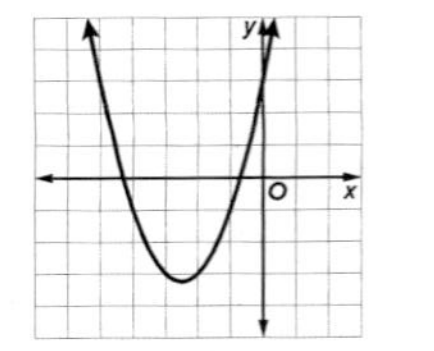

23. no real solution

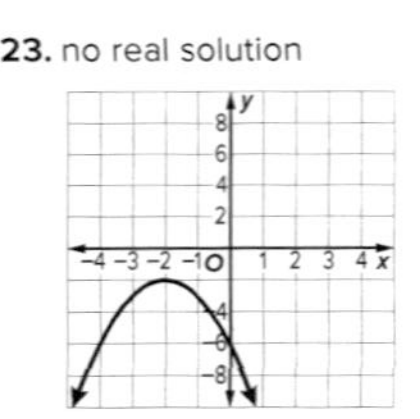

25. between 0 and 1; between 2 and 3

27. $-1, 1$ **29.** $\approx -1.45, \approx 3.45$

31. $\approx 3.24, \approx -1.24$ **33.** 2.1 seconds

35. between -3 and -2, between 1 and 2

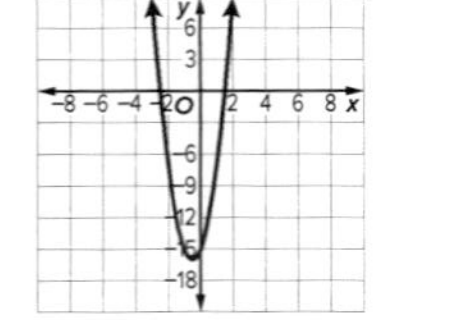

37. between -1 and 0, between 4 and 5

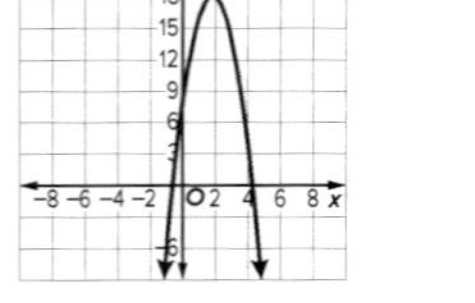

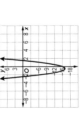

39. between 3 and 4, between 8 and 9

41. between -3 and -2; 0

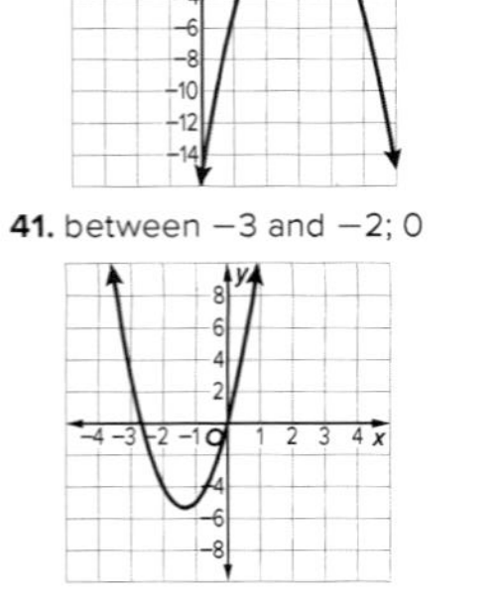

43. between -8 and -7; between 12 and 13

45. no real solution **47.** -2

49. between -3 and -2, between 4 and 5

51. about -5 and 17 **53.** 11 and -19

55. 64 m

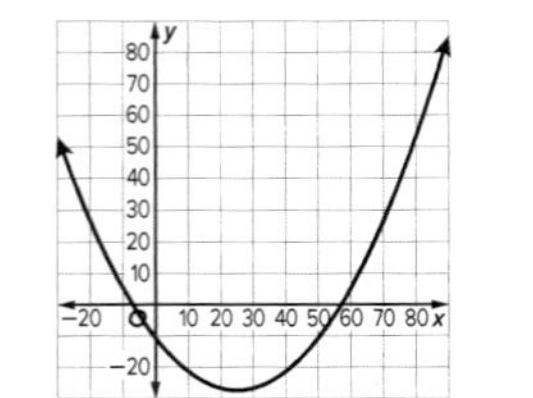

57. 5 seconds

59. No; sample answer: Hakeem is right about the location of one of the roots, but his reason is not accurate. The roots are located where $f(x)$ changes signs.

61. 5; Sample answer: The intercepts are equidistant from the axis of symmetry.

63. Sample answer: Graph the function using the axis of symmetry. Determine where the graph intersects the x-axis. The x-coordinates of those points are solutions of the quadratic equation.

Lesson 3-3

1. $4i\sqrt{3}$ **3.** $6i\sqrt{2}$ **5.** $2i\sqrt{21}$ **7.** $-23\sqrt{2}$

9. $-5\sqrt{2}$ **11.** $-i$ **13.** $\pm 3i$ **15.** $\pm i$ **17.** $\pm 5i$

19. 3, 3 **21.** $-\frac{11}{2}, -3$ **23.** 4, -3 **25.** $10 - 4i$

27. $2 + i$ **29.** $10 - 5i$ **31.** $7 + i$ **33.** $-10i$

35. $\frac{3}{2} - \frac{1}{2}i$ **37.** $-\frac{5}{3} - 2i$ **39.** $\frac{7}{9} - \frac{4i\sqrt{2}}{9}$

41. $(3 + i)x^2 + (-2 + i)x - 8i + 7$

43. $10 + 10j$ volts **45.** $8 - 2j$ ohms

47. Zoe; $i^3 = -i$, not -1.

49. Always; Sample answer: the value of 5 can be represented by $5 + 0i$, and the value of $3i$ can be represented by $0 + 3i$.

51. Some quadratic equations have complex solutions and cannot be solved using only the real numbers.

Lesson 3-4

1. $0, \frac{1}{3}$ **3.** $0, -\frac{5}{4}$ **5.** $-11, -3$ **7.** 7 ft by 9 ft

9. $\frac{3}{2}, -1$ **11.** $\frac{2}{5}, -6$ **13.** $8, -\frac{5}{2}$

15. $-8, 8$ **17.** $-17, 17$ **19.** $-13, 13$ **21.** $\frac{7}{2}$

23. $\frac{3}{4}$ **25.** $-\frac{8}{5}$ **27.** $5i, -5i$

29. $15i, -15i$ **31.** $\frac{5}{6}i, -\frac{5}{6}i$ **33.** $-3, \frac{1}{2}$

35. $9i, -9i$ **37.** $\frac{1}{5}i, -\frac{1}{5}i$

39. 13 inches by 16 inches

41. 24, 26

43. Neither; both students made a mistake in Step 3.

45. Sample answer: 3 and 6 $\rightarrow x^2 - 9x + 18 = 0$. -3 and $-6 \rightarrow x^2 + 9x + 18 = 0$. The linear term changes sign.

47. Sample answer: Standard form is $ax^2 + bx + c$. Multiply a and c. Then find a pair of integers, g and h, that multiply to equal ac and add to equal b. Then write the quadratic expression, substituting the middle term, bx, with $gx + hx$. The expression is now $ax^2 + gx + hx + c$. Then factor the GCF from the first two terms and factor the GCF from the second two terms. So, the expression becomes $GCF(x - q) + GCF_2(x - q)$. Simplify to get $(GCF + GCF_2)(x - q)$ or $(x - p)(x - q)$.

Lesson 3-5

1. 2, 16 **3.** $0, \frac{4}{3}$ **5.** $\frac{15}{2}, -\frac{1}{2}$

7. $\frac{-1 \pm 3\sqrt{2}}{6}$ **9.** $\frac{-2 \pm 5\sqrt{3}}{5}$ **11.** $\frac{3 \pm 4\sqrt{6}}{5}$

13. $8 \pm 8i$ **15.** $-7 \pm 10i$

17. $-4 \pm 6i$ **19.** 25; $(x + 5)^2$

21. 144; $(x + 12)^2$ **23.** $\frac{81}{4}; \left(x - \frac{9}{2}\right)^2$

25. 4, 9 **27.** $2 \pm \sqrt{17}$ **29.** $\frac{1 \pm \sqrt{13}}{2}$

31. 16 in. by 16 in. by 16 in.

33. $1, \frac{1}{2}$ **35.** $\frac{1}{5}, -\frac{9}{5}$ **37.** $\frac{1}{3}, -1$

$39.\ \frac{3 \pm i\sqrt{31}}{4}$ **41.** $1 \pm i\sqrt{2}$ **43.** $1 \pm i\sqrt{3}$

45. $y = (x + 3)^2 - 8$; $x = -3$; $(h, k) = (-3, -8)$; minimum

47. $y = -(x + 4)^2 + 11$; $x = -4$; $(h, k) = (-4, 11)$; maximum

49. $y = 3(x + 1)^2 - 4$; $x = -1$; $(h, k) = (-1, -4)$; minimum

51a. $h(t) = -4.9(t - 0.427)^2 + 8.931$

51b. $t = 0.427$; points equidistant from the axis of symmetry represent the times when the diver will be at the same height during his dive. Vertex = (0.427, 8.391); Malik reaches a maximum height of about 8.391 meters approximately 0.427 second after he begins his dive.

53. −2.77, −0.56 **55.** −1.44, 0.24

57. $1.6 \pm 0.9i$ **59.** $-1.3 \pm 4.09i$

61. $h(t) = -4.9t^2 + 25.8$; 2.3 s **63.** 5%

65. $y = (x - 5)^2 + 3$; (5, 3)

67. $y = (x - 10)^2 + 4$; (10, 4)

69a. $n^2 + 90n$ **69b.** 10

71a. w = width, $V(w)$ = the volume; $V(w) = 6w^2 - 32w$

71b. 20.7 by 62.1 in.

73. Alsonso; Aika did not add 16 to each side; she added it only to the left side.

75a. 2; rational; 16 is a perfect square, so $x + 2$ and x are rational.

75b. 2; rational; 16 is a perfect square, so $x - 2$ and x are rational.

75c. 2; complex; if the opposite of a square is positive, the square is negative. The square root of a negative number is complex.

75d. 2; real; the square must equal 20. Since that is positive but not a perfect square, the solutions will be real but not rational.

75e. 1; rational; the expression must be equal to 0 and only −2 makes the expression equal to 0.

75f. 1; rational; the expressions $(x + 4)$ and $(x + 6)$ must either be equal or opposites. No value makes them equal, −5 makes them opposites. The only solution is −5.

77. Sample answer: Completing the square is rewriting one side of a quadratic equation in the form of a perfect square. Once in this form, the equation can be solved by using the Square Root Property.

Lesson 3-6

1. −3, −5 **3.** $\frac{3}{2}, \frac{1}{3}$ **5.** $-4 \pm \sqrt{11}$

7. 2.9 s

9. $\frac{1}{4}, -5$ **11.** $-2, \frac{1}{3}$ **13.** $\frac{-5 \pm \sqrt{57}}{16}$

15. $\frac{3 \pm \sqrt{34}}{4}$ **17.** $\pm 5i$ **19.** $\frac{1 \pm i}{4}$

21. $7 \pm 2i$ **23.** $\frac{3 \pm 2i\sqrt{6}}{3}$

25. 225; 2 rational roots

27. 289; 2 rational roots

29. 24; 2 irrational roots

31. 21; 2 irrational roots

33. −196; 2 complex roots

35. −7; 2 complex roots

37. $b^2 - 4ac > 0$; 2 real rational or irrational roots

39. 2 complex roots; $\frac{1 \pm 4i}{2}$

41. 2 rational roots; 0, $\frac{5}{7}$

43. 2 rational roots; 3, 8

45. 2 rational roots; 4, $\frac{4}{3}$

47. 2 irrational roots; $\frac{-5 \pm \sqrt{3}}{2}$

49. about 2.9 seconds

51. x = the length of a side of the base, $(x + 2)^2(3x + 4) = x^2(3x + 1) + 531$; 5 in. by 5 in. by 16 in.

53. $7x^2 + 6x + 2 = 0$ is different from the other 3 equations because it has 2 complex roots, where the other 3 equations each have 2 rational roots.

55a. Sample answer: Always; when a and c are opposite signs, then ac will always be negative and $-4ac$ will always be positive. Because b^2 will also always be positive, then $b^2 - 4ac$ represents the addition of two positive values, which will never be negative. Hence, the discriminant can never be negative and the solutions can never be imaginary.

55b. Sample answer: Sometimes; the roots will only be irrational if $b^2 - 4ac$ is not a perfect square.

57. −0.75

Lesson 3-7

1.

3.

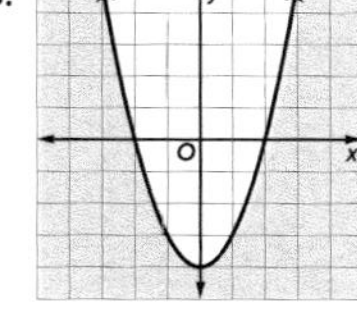
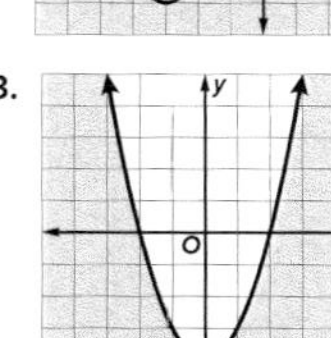

5.

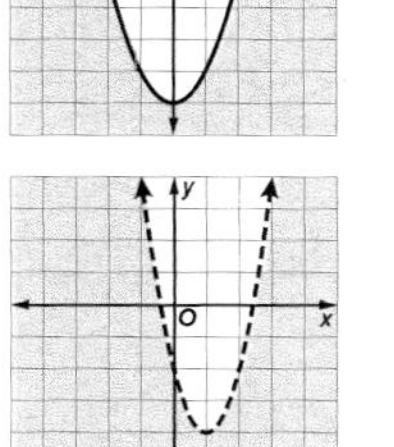
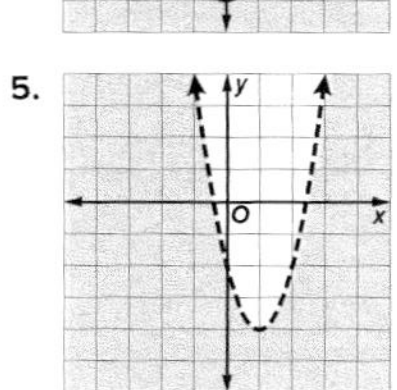
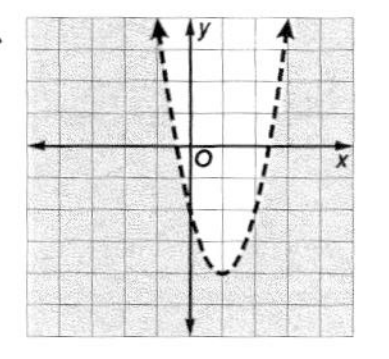
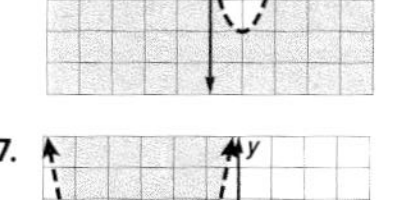

7.

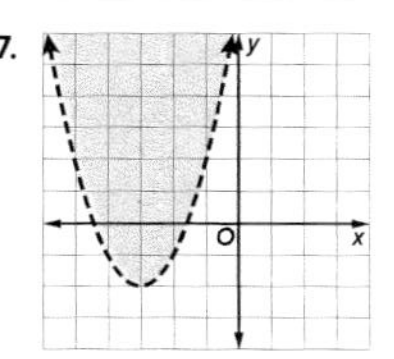
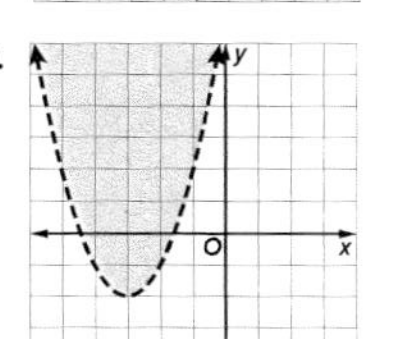

9.

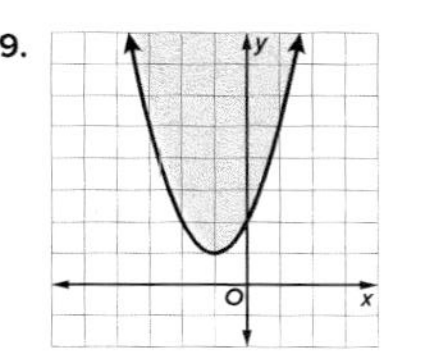
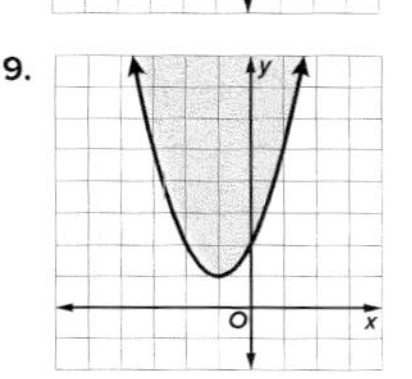
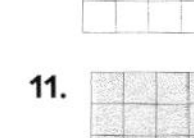

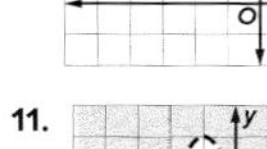
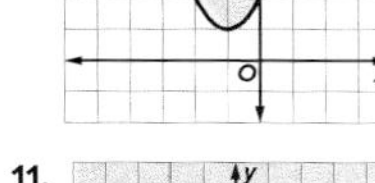

11.

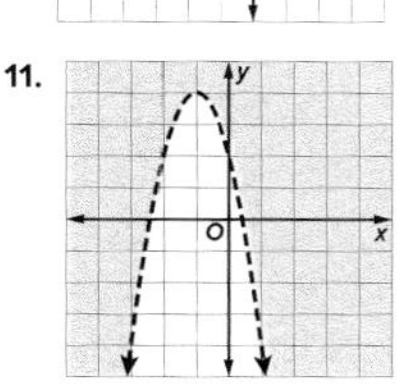
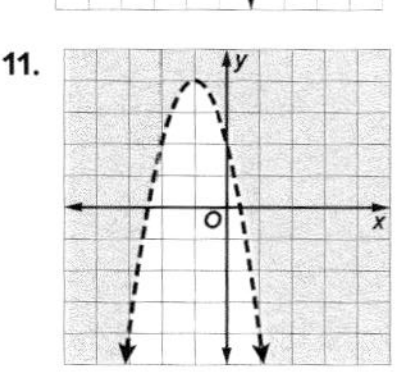

13. $\{x \mid x = 3\}$

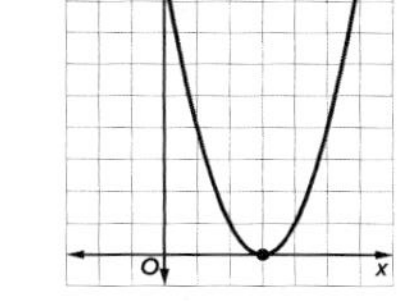
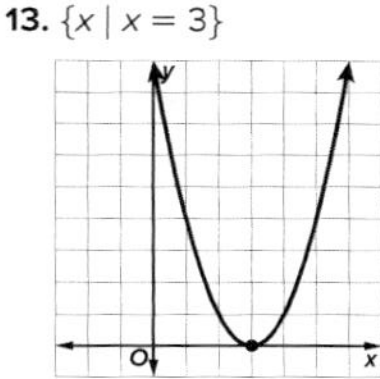
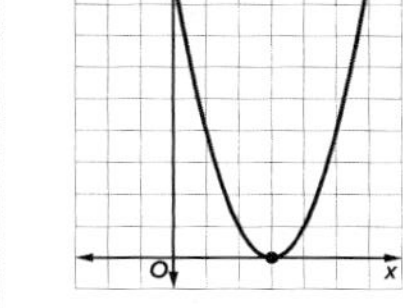
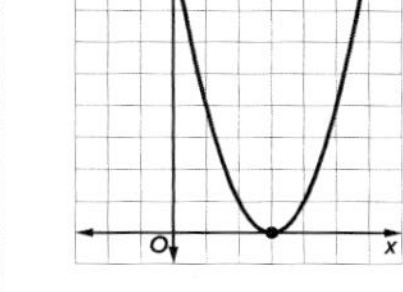

15. $\{x \mid x < -5 \text{ or } x > 4\}$

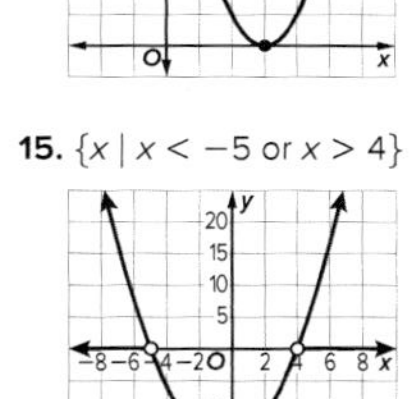

17. {all real numbers}

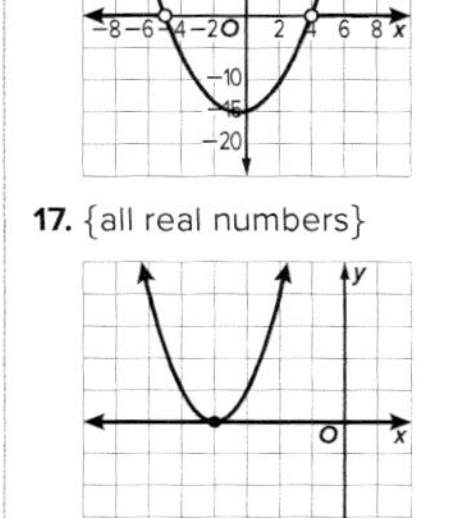
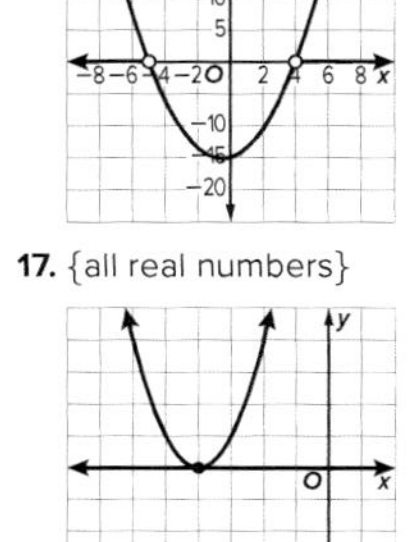
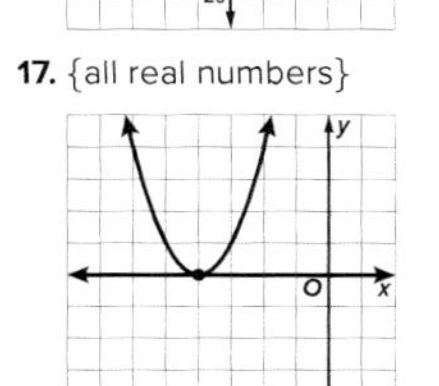
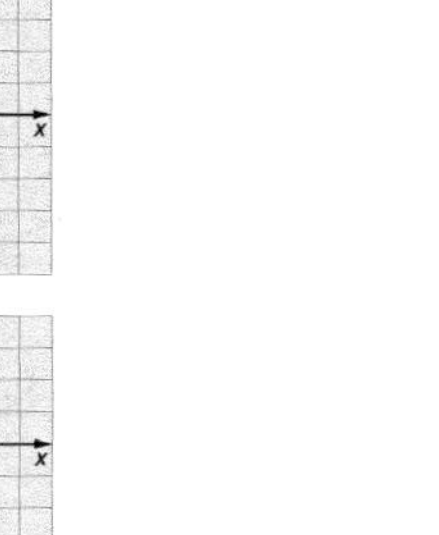

19. 30 ft to 60 ft **21.** $\{x \mid x < -1.06 \text{ or } x > 7.06\}$

23. $\{x \mid x \le -2.75 \text{ or } x \ge 1\}$

25. $\{x \mid x < 0.61 \text{ or } x > 2.72\}$

27. {all real numbers}

29. $\{x \mid x \le -9.24 \text{ or } x \ge -0.76\}$

31. $\{x \mid -5 \le x \le -3\}$

33. $\left\{x \mid x \ne -\frac{1}{2}\right\}$ **35.** $\left\{x \mid -3 \le x \le -\frac{4}{9}\right\}$

37. $y > x^2 - 4x - 6$ **39.** $y > -0.25x^2 - 4x + 2$

41. No; the graphs of the inequalities intersect the x-axis at the same points.

43a. Sample answer: $x^2 + 2x + 1 \ge 0$

43b. Sample answer: $x^2 - 4x + 6 < 0$

45.

Selected Answers

Lesson 3-8

1. (2, −1), (−1, −4) **3.** (−1, 2), (1.5, 4.5)
5. (2, −2), $\left(\frac{1}{2}, 1\right)$ **7.** (1, 2)
9. $\left(\frac{-1+\sqrt{33}}{4}, \frac{-1+\sqrt{33}}{4}\right), \left(\frac{-1-\sqrt{33}}{4}, \frac{-1-\sqrt{33}}{4}\right)$

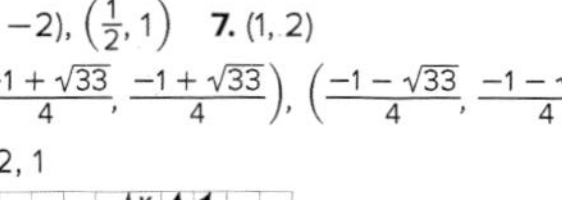

11. −2, 1

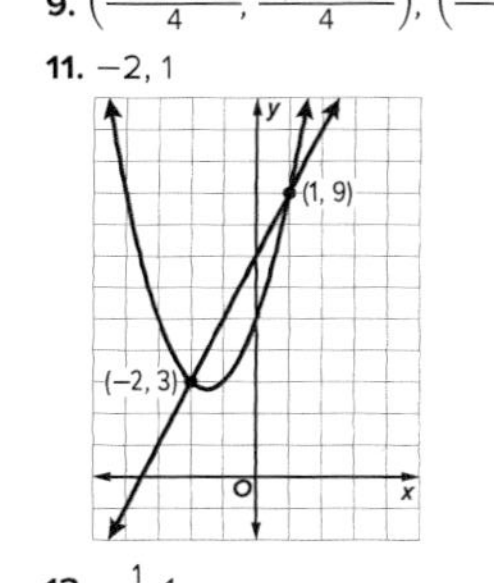

13. $-\frac{1}{4}$, 1

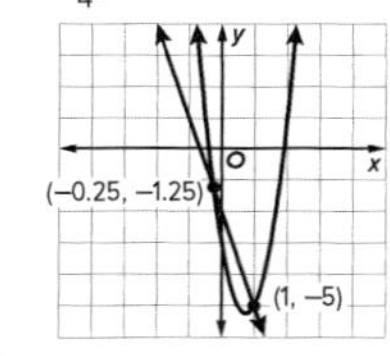

15. −1, 7

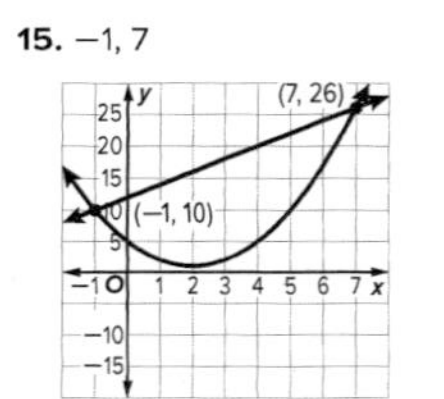

17a. $P = 2x^2 + 30x$ and $P = 50{,}000$

17b. $x \approx -165.79$, $x \approx \$150.79$

17c. Sample answer: Because the selling price cannot be negative, (150.79, 50,000) is the only viable solution in the context of the situation. This means that the business can earn a $50,000 profit when the selling price is about $150.79 per item.

19. no solution

21. (0, 2) and (1, 3)

23. (1, 1) and (5, 5)

25. no solution **27.** (2, 2) and (2, −2)

29. (−1, −7) and (4, 23)

31. $-\frac{3}{2}$, 2

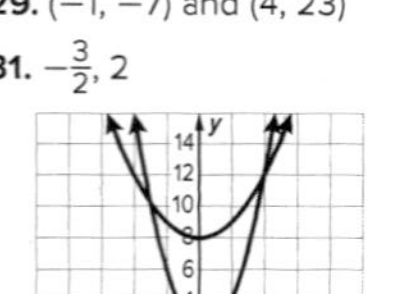

33. −2

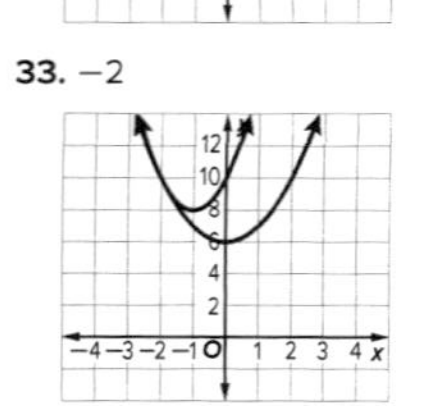

35. $-\frac{1}{2}$, 3

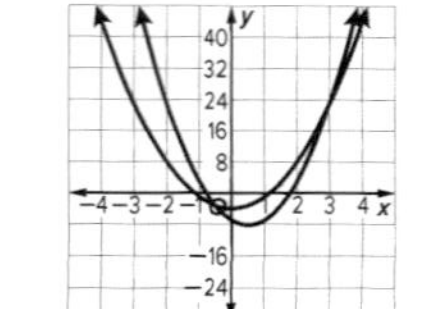

37. −2 **39.** (−2, −6), (0, 8)
41. (−1, 2), (2, 5) **43.** (0.42, 0.352) and (1, 2)
45. (−1.4, 2) and (1.4, 2)
47. (−1.8, 3.2) and (1.1, 1.3)
49. (2, −2), (0, 0) **51.** no solution
53. no solution **55.** $(-\sqrt{31}, 7)$ and $(\sqrt{31}, 7)$
57. $-6\frac{1}{2}$, 1 **59.** −3, 1 **61.** $-\frac{1}{2}$, −1
63. −3 **65.** $-\frac{1}{2}$, 3
67. −6, −3 **69a.** 0.7 s

69b. about 20.9 m; The faster rocket lands after about 9.96 seconds. Solving $y = -4.9t^2 + 46.7t + c$ when $y = 0$ and $t = 9.96$, shows that $c \approx 20.9$.

71a. Yes; the maximum height of the ball is about 12.77 feet, which is higher than the net.

71b. Yes; the player's hands could strike the ball at a height of about 9 feet. Because that is above the height of the net, the ball may be blocked.

71c. Sample answer: I assumed that the student bumping the ball was far enough away from the net that she wouldn't hit it. I also assumed that the player attempting to block the ball is in the correct position for the path of the ball to intersect the path of her hands.

73. Sample answer: $y = x^2$ and $y = -x^2$

75. Danny is correct. Carol incorrectly solved for y in the second equation of the system before using the substitution method.

Module 3 Review

1. A **3.** 10 and 21 **5.** C **7.** $x = -3$, $x = 5$
9. D **11.** A, D **13.** C
15. between 0 and 5 feet
17. (2, 1), (7, 6)

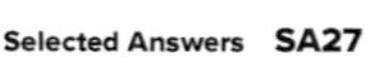

Module 4

Quick Check

1. $-5 + (-13)$ **3.** $5mr + (-7mp)$
5. $-4a - 20$ **7.** $-m + \frac{5}{2}$

Lesson 4-1

1. As $x \to -\infty$, $f(x) \to \infty$ and as $x \to \infty$ and $f(x) \to \infty$; D = $(-\infty, \infty)$, R = $[0, \infty)$

3. As $x \to -\infty$, $f(x) \to \infty$ and as $x \to \infty$ and $f(x) \to -\infty$; D = $(-\infty, \infty)$, R = $(-\infty, \infty)$

5. D = $(-\infty, \infty)$, R = $[0, \infty)$

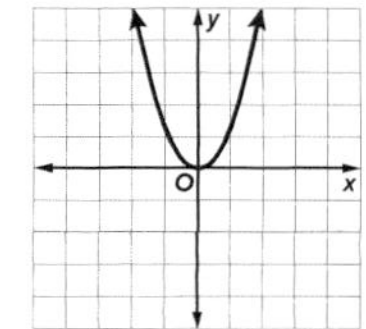

7. degree = 1, leading coefficient = 1

9. degree = 5, leading coefficient = −5

11. not in one variable because there are two variables, *u* and *t*

13a. 455

13b.

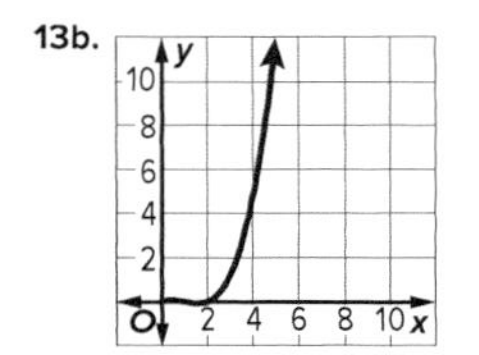

15. 1 **17.** 3 **19.** 3 **21a.** $f(x)$

21b. zeros: $f(x)$: −1.39, 0.23, 3.16; $g(x)$: −2.75, −0.5, 0.25

x-intercepts: $f(x)$: −1.39, 0.23, 3.16; $g(x)$: −2.75, −0.5, 0.25 *y*-intercept: $f(x)$: 1; $g(x)$: 1 end behavior: $f(x)$: As $x \to -\infty$, $f(x) \to -\infty$, and as $x \to \infty$, $f(x) \to \infty$; $g(x)$: As $x \to -\infty$, $g(x) \to \infty$, and as $x \to \infty$, $g(x) \to -\infty$.

23. As $x \to -\infty$, $f(x) \to -\infty$ and as $x \to \infty$, $f(x) \to -\infty$. degree = 4; leading coefficient = −5

25. As $x \to -\infty$, $g(x) \to -\infty$ and as $x \to \infty$, $g(x) \to \infty$. degree = 5; leading coefficient = 5

27. This is not a polynomial because there is a negative exponent.

29. As $x \to -\infty$, $h(x) \to \infty$ and as $x \to \infty$, $h(x) \to \infty$. degree = 2, leading coefficient = 3

31. As $x \to \infty$, $y \to -\infty$, and as $x \to -\infty$, $y \to \infty$; degree = 3; the leading coefficient is negative.

33. $f(1 - x) = a(1 - x)^3 - b(1 - x)^2 + (1 - x)$. So, $f(1 - x) = -ax^3 + (3a - b)x^2 + (-3a + 2b - 1)x + (a - b + 1)$. The function $f(1 - x)$ has the opposite leading coefficient, representing a reflection in the *y*-axis. So, it has the opposite end behavior.

35. Sample answer: The volume of the new box is modeled by the function $V(x) = (10 - x)^2(4 + 2x)$. The graph appears to have a relative maximum at $x = 2$ and $V(2) = 512$. So, the dimensions of the box with the greatest volume will be 8 centimeters by 8 centimeters by 8 centimeters.

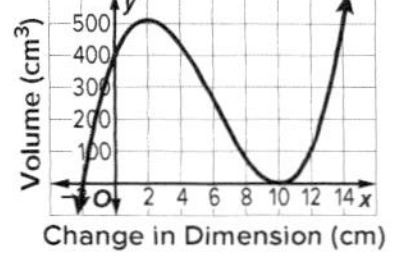

37. $f(x)$; $f(x)$ has potential for 5 or more real zeros and a degree of 5 or more. $g(x)$ has potential for 4 real zeros and a degree of 4.

39. Sample answer:

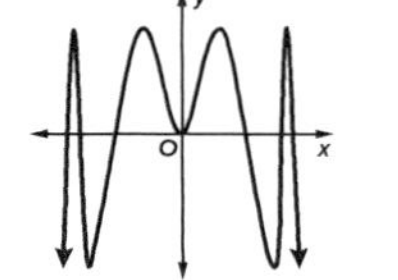

Lesson 4-2

1. zeros between $x = -4$ and $x = -3$, and between $x = 0$ and $x = 1$

x	*f(x)*
−4	3
−3	−1
−2	−3
−1	−3
0	−1
1	3
2	9

3a. zero between $x = -6$ and $x = -5$

x	*f(x)*
−6	37
−5	5
−4	25
−3	29
−2	23
−1	13
0	5
1	5
2	19
3	53

5. relative minimum between $x = 0$ and $x = 1$; relative maximum near $x = 4$

x	*f(x)*
−1	22
0	0
1	2
2	16
3	30
4	32
5	10
6	−48
7	−154

7. relative minimum near $x = -1$; no relative maximum

x	*f(x)*
−4	247
−3	74
−2	11
−1	−2
0	−1
1	2
2	19
3	86
4	263

9. Domain and Range: The domain and range of the function are all real numbers. Because the function models years, the relevant domain is $\{n \mid n \geq 0\}$ and the relevant range is $\{v \mid v \geq -4\}$.

Extrema: There is a relative minimum at $n = 2$ in the relevant domain.

End Behavior: As $x \to \infty$, $v \to \infty$.

Intercepts: In the relevant domain, the *v*-intercept is at (0, 0). The *n*-intercept, or zero, is at about (3, 0).

Symmetry: In the relevant domain, the graph does not have symmetry.

11. Curve of best fit: $y = -0.012x^4 + 0.217x^3 - 1.116x^2 + 1.486x + 7.552$; car sales in 2017: 9.991 million

[0, 10] scl: 1 by [0, 10] scl: 1

Selected Answers

13. Curve of best fit: $y = -0.122x^4 + 1.565x^3 - 6.325x^2 + 20.321x + 33.125$, where x is the number of months since January; average volunteer hours for September: 44

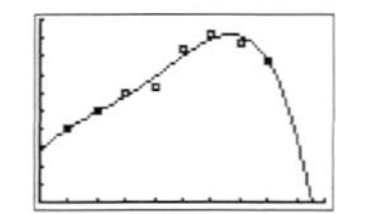

[0, 10] scl: 1 by [0, 120] scl: 12

15. average rate of change: 4500; From 2012 to 2017, the salesman's average rate of change in salary was an increase of $4500 per year.

17.

x	$f(x)$
−3	−7
−2	−2
−1	−3
0	2
1	25

zero between $x = -1$ and $x = 0$;
rel. max. at $x = -2$, rel. min. at $x = -1$;
D = all real numbers; R = all real numbers

19.

x	$f(x)$
−3	61
−2	6
−1	−3
0	−2
1	−3
2	6
3	61

zeros between $x = -2$ and $x = -1$, and between $x = 1$ and $x = 2$;
rel. max. at $x = 0$, min. at $x = -1$, and $x = 1$;
D = all real numbers; R = $\{f(x) \mid f(x) \geq -3\}$

21. rel. max: $x \approx -2.73$; rel. min: $x \approx 0.73$

23. max: $x \approx 1.34$; no rel. min

25. Sample answer: $f(x) \to -\infty$ as $x \to \infty$ and $f(x) \to \infty$ as $x \to -\infty$; the leading coefficient is negative.

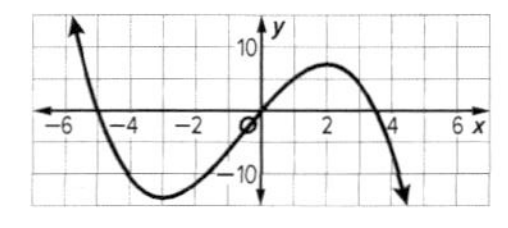

27. The type of polynomial function that should be used to model the graph is a function with an even degree with a negative leading coefficient. This is based on the fact that the graph is reflected in the x-axis and as $x \to -\infty$, $f(x) \to -\infty$ and as $x \to \infty$, $f(x) \to -\infty$.

29. As the x-values approach large positive or negative numbers, the term with the largest degree becomes more and more dominant in determining the value of $f(x)$.

31. Sample answer:

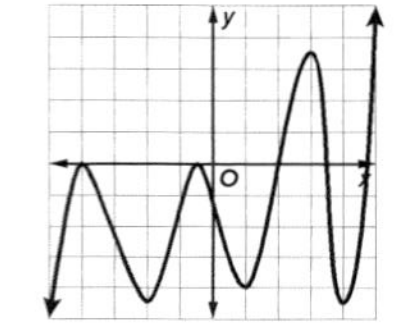

33. Sample answer: No; $f(x) = x^2 + x$ is an even degree, but $f(1) \neq f(-1)$.

Lesson 4-3

1. yes; 2 **3.** no **5.** $2a^2 - a - 2$ **7.** $3g + 12$ **9.** $3x^2 + 4x - 5$ **11.** $-3x + 3$ **13.** $3np^2 - 3pz$ **15.** $-10c^2 + 5d^2$ **17.** $a^2 - 10a + 25$ **19.** $x^3 + x^2y - xy^2 - y^3$ **21.** $2x^3 + x^2y - 2xy^2 - y^3$ **23.** $r^2 - 4t^2$ **25.** $2x^5 - 7x^4 + 5x^3 - 4x^2 - x + 2$ **27.** $s^2 + 11s + 15$ **29a.** $4x^2 + 8x + 3$ **29b.** $8x^2 + 16x + 6$ **31.** $12a^2b + 8a^2b^2 - 15ab^2 + 4b^2$

33. $2n^4 - 3n^3p + 6n^4p^4$
35. $2n^5 - 14n^3 + 4n^2 - 28$
37. $64n^3 - 240n^2 + 300n - 125$
39. $7x - 2y$
41a. $f(x)g(x) = (3x^2 - 1)(x + 2) = 3x^3 + 6x^2 - x - 2$; 3rd degree
41b. $h(x)f(x) = (-x^2 - x)(3x^2 - 1) = -3x^4 + x^2 - 3x^3 + x$; 4th degree
41c. $[f(x)]^2 = (3x^2 - 1)^2 = 9x^4 - 6x^2 + 1$; 4th degree
43.

	Addition and Subtraction	Multiplication	Division
Integers	yes	yes	No; $3 \div 2 = 1.5$
Rational Numbers	yes	yes	yes
Polynomials	yes	yes	No; $\frac{3}{x^2} = 3x^{-2}$

45. $-0.022x + 1590$
47a. Area of sidewalk = area of larger circle − area of smaller circle, so $384\pi = \pi(r + 12)^2 - \pi r^2 = \pi(r^2 + 24r + 144) - \pi r^2 = 24\pi r + 144\pi$. $24\pi r = 240\pi$; $r = 10$; the radius of the smaller circle is 10 feet, and the radius of the larger circle is $r + 12 = 10 + 12$ or 22 feet.
47b.
47c. $48s + 576$ ft²
49a. $m \cdot n$; each term of one polynomial multiplies each term of the other one time.
49b. 2; adding like terms may result in a sum of 0; but the first and last terms are unique.
51. $(3 - 2b)(3 + 2b)$

Lesson 4-4

1. $5y^2 + 2y + 1$ **3.** $2j - 3k$ **5.** $n + 2$ **7.** $\frac{2t + 1 + 9}{t + 6}$ **9.** $2g - 3$ **11.** $3v + 5 + \frac{10}{v - 4}$ **13.** $y^2 - 2y + 4 - \frac{2}{y + 2}$ **15.** $\frac{4}{3}p^2 + \frac{p}{9} + \frac{19}{27} + \frac{19}{27(3x - 1)}$ **17.** $m - 3 + \frac{6}{m + 4}$ **19.** $2x^2 - 3x + 1$ **21.** $x^2 - 2x + 4$ **23.** $2c^2d - \frac{3}{2}d$ **25.** $n^2 - n - 1$ **27.** $3z^4 - z^3 + 2z^2 - 4z + 9 - \frac{13}{z + 2}$ **29.** A is x^2; B is $6x$; C is 11. **31.** $\pi(x^2 - 8x + 16)$

33. Yes; Because $3x$ times the divisor is $9x^2 + 3x$, the divisor must be $3x + 1$. The second and third terms of the dividend must be $0x + 5$ because the first difference is $-3x + 5$.

35. $8x^3 - 8x^2 - 2x - 10$; I multiplied the divisor and the quotient and added the remainder: $(2x^2 + x + 1)(4x - 6) + (-4) = 8x^3 - 8x^2 - 2x - 10$.

37a. No; the degrees of $f(x)$ and $d(x)$ may be equal. For example, $\frac{3x^2 + 6}{x^2} = 3 + \frac{6}{x^2}$.

37b. Yes; if the degree of $r(x)$ is greater than or equal to the degree of $d(x)$, then the expression $\frac{r(x)}{d(x)}$ may be simplified by division. For example, if $\frac{r(x)}{d(x)} = \frac{8x + 1}{x}$, then $8x + 1$ may be divided by x to get $8 + \frac{1}{x}$.

37c. Yes; because $\frac{f(x)}{d(x)} = q(x) + \frac{r(x)}{d(x)}$, the degree of $\frac{f(x)}{d(x)}$ must equal the degree of $q(x) + \frac{r(x)}{d(x)}$. The degree of $r(x)$ is less than the degree of $d(x)$, so the degree of $q(x) + \frac{r(x)}{d(x)}$ equals the degree of $q(x)$. This means the degree of $\frac{f(x)}{d(x)}$ equals the degree of $q(x)$, and the degree of $\frac{f(x)}{d(x)}$ is the degree of $f(x)$ minus the degree of $d(x)$. For example, in $\frac{2x^3 - 1}{x^2 + 3} = 2x + \frac{-6x - 1}{x^2 + 3}$, the degree of $q(x)$ is the degree of $f(x)$ minus the degree of $d(x)$.

39. The binomial is a factor of the polynomial.

41. Sample answer: $\frac{x^2 + 5x + 9}{x + 2}$

43a. $B^2 + 1$

43b. $B + \frac{1}{B^2 + B + 1}$

Lesson 4-5

1. $x^3 - 3x^2y + 3xy^2 - y^3$

3. $g^4 - 4g^3h + 6g^2h^2 - 4gh^3 + h^4$

5. $y^3 - 21y^2 + 147y - 343$

7. 38%

9. $243x^5 + 1620x^4y + 4320x^3y^2 + 5760x^2y^3 + 3840xy^4 + 1024y^5$

11. $4096h^4 - 6144h^3j + 3456h^2j^2 - 864hj^3 + 81j^4$

13. $x^5 + \frac{5}{2}x^4 + \frac{5}{2}x^3 + \frac{5}{4}x^2 + \frac{5}{16}x + \frac{1}{32}$

15. $32b^5 + 20b^4 + 5b^3 + \frac{5}{8}b^2 + \frac{5}{128}b + \frac{1}{1024}$

17. 23%

19. If C represents a correct circuit board and N represents an incorrect circuit board, then the number of ways for the robot to produce 5 of 7 circuit boards accurately is given by the coefficient of C^5N^2 in the expansion of $(C + N)^7$. Using the Binomial Theorem, there are 21 ways for the robot to produce a correct circuit board out of 128 possibilities. So, the probability that 5 of 7 are correct is $\frac{21}{128}$ or about 16% probability.

21. If c represents a correct answer and w represents a wrong answer, then the coefficients expansion of $(c + w)^{10}$ can be used to represent situation. Using Pascal's triangle, there are 45 ways to get 8 questions correct, 10 ways to get 9 questions correct, and 1 way to get all correct. So there are 56 ways to get 8 or more correct. By adding all of the values in this row of Pascal's triangle, I found that there are 1024 different ways he could answer the questions in the quiz. Matthew has a $\frac{56}{1024}$ or about a 5.5% chance of getting 8 or more correct.

23. There are 9 judges on the Supreme Court. The majority could be 5, 6, 7, 8, or 9 votes. So, there are ${}_9C_5 + {}_9C_6 + {}_9C_7 + {}_9C_8 + {}_9C_9 = 256$ combinations.

25. 1.21896; $(1.02)^{10} + 1.21899442$; the approximation differs from the value given by a calculator by 0.00003442.

27. Sample answer: While they have the same terms, the signs for $(x + y)^n$ will all be positive, while the signs for $(x - y)^n$ will alternate.

29. Sample answer: $\left(x + \frac{6}{5}y\right)^5$

Module 4 Review

1. B **3.** B **5.** $g(x) \to -\infty$ **7.** C

9. $x^6 + 7x^5 - 6x^3 + 8x^2 + 2x$

11. C **13.** $5x^2 + 7x - 3$ **15.** B

Module 5

Quick Check

1. $24x^3 + 8x^2 + 6x + 4$

3. $14x^3 - 40x^2 + 12x + 24$

5. $-4, 2$ **7.** $-\frac{4}{3}, \frac{1}{2}$

Lesson 5-1

1. -4.12 **3.** $-0.47, 0.54, 3.94$ **5.** -1.27

7a. $6x^3 + 110x^2 - 200x = 15{,}000$

7b. $y = 6x^3 + 110x^2 - 200x, y = 15{,}000$; $x = 10$ cm

7c. 25 cm by 20 cm by 30 cm

9a. $\pi x^3 + 3\pi x^2 = 628$

9b. $y = \pi x^3 + 3\pi x^2, y = 628; x = 5$

9c. radius = 5 in., height = 8 in.

11. $-3.63, -1.35, 1.35, 3.63$ **13.** no solution

15. \$0.29, 0.88; sample answer: graph $f(x) = 70{,}000(x - x^4)$ and $f(x) = 20{,}000$ and find the x-values of the points of intersections.

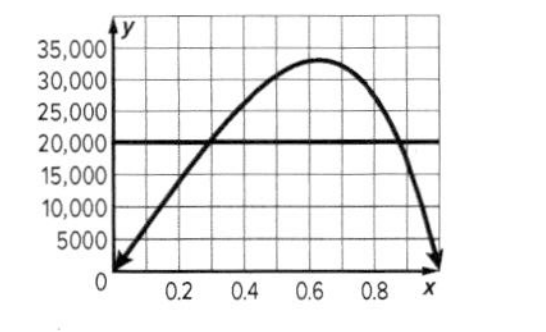

17. Sample answer: A function with an even degree reverses direction, so both ends extend in the same direction. That means that if the function opens up and the vertex is above the x-axis or if the function opens down and the vertex is below the x-axis, there are no real solutions. A function with an odd degree does not reverse direction, so the ends extend in opposite directions. Therefore, it must cross the x-axis at least once.

19. Sometimes; sample answer: The positive solution is often correct when a negative value doesn't make sense, such as for distance or time; however, sometimes a negative solution is reasonable, such as for temperature or position problems. Also, sometimes there are two positive solutions and one may be unreasonable.

21. $5x^2 = -2x - 11$; It is the only one that has no real solutions.

Lesson 5-2

1. $(2c - 3d)(4c^2 + 6cd + 9d^2)$

3. $a^2(a - b)(a^2 + ab + b^2)(a + b)(a^2 - ab + b^2)$

5. prime **7.** $(x + y)(x - y)(6f + g - 3h)$

9. $(a - b)(a^2 + ab + b^2)(x - 8)^2$ **11.** 0, 7, 2

13. 0, −5, 8

15a. Let x represent the length of a side of the leg and y represent the length of the notch; $x^3 - y^3$.

15b. 15 in.

17. $-15(x^2)^2 + 18(x^2) - 4$ **19.** not possible

21. $4(2x^5)^2 + 1(2x^5) + 6$ **23.** $\pm\sqrt{5}, \pm i\sqrt{2}$

25. $\pm\frac{2\sqrt{3}}{3}, \pm\frac{\sqrt{15}}{3}$ **27.** $\pm\frac{\sqrt{6}}{6}, \pm i\frac{\sqrt{3}}{2}$

29. $(x + 2)(x^2 - 2x + 4)(x - 2)(x^2 + 2x + 4)$

31. $x^2y^2(2x + 3y)(4x^2 + 6xy + 9y^2)$

33. $(y - 1)^3(y^2 + y + 1)^3$

35. $3, -3, \pm i\sqrt{10}$ **37.** $\pm\sqrt{11}, \pm 2i$

39. $-6, 3 \pm 3i\sqrt{3}$

41. $\pm\frac{1}{2}, \pm i\frac{\sqrt{6}}{2}$ **43.** $\pm\frac{3}{2}, \pm\frac{\sqrt{10}}{5}$

45. $\pm\frac{1}{2}, \pm\sqrt{2}$ **47.** $1, -2, \frac{-1 \pm i\sqrt{3}}{2}, 1 \pm i\sqrt{3}$

49. $-5, \frac{1}{2}, \frac{-1 \pm i\sqrt{3}}{2}, \frac{5 \pm 5i\sqrt{3}}{2}$ **51.** $\pm 3, \pm 1, \pm i$

53a. Sample answer: $\frac{1}{2} = 2\left(\frac{1}{4}\right)$

53b. $u = x^{\frac{1}{4}}$; $u^2 - 8u + 15 = 0$

53c. 81, 625

55. prime **57.** $(2a + 1)(k - 3)$ **59.** prime

61. $(d - 6)^2$ **63.** $(y + 9)^2$ **65.** $19x^2(x - 2)$

67. $(m^2 + 1)(m - 1)(m + 1)$ **69.** 2.4 in.

71. The solutions are $x = \pm\sqrt{m}$ and $\pm\sqrt{n}$. because both equations have the same coefficients, take the solutions for the quadratic equation and substitute x^2 for x.

73. $\frac{16}{3}, \frac{3}{4}$

75. Sample answer: $12x^6 + 6x^4 + 8x^2 + 4 = 12(x^2)^3 + 6(x^2)^2 + 8(x^2) + 4$

Lesson 5-3

1. $(x - y)^2 = x^2 - 2xy + y^2$ (Original equation)
$x^2 - 2xy + y^2 = x^2 - 2xy + y^2$ (Distributive Property)

3. $4(x - 7)^2 = 4x^2 - 56x + 196$ (Original equation)
$4(x^2 - 14x + 49) = 4x^2 - 56x + 196$ (Distributive Property)
$4x^2 - 56x + 196 = 4x^2 - 56x + 196$ (Distributive Property)

5. $a^2 - b^2 = (a + b)(a - b)$ (Original equation)
$= a^2 - ab + ab - b^2$ (Distributive Property)
$= a^2 - b^2$ (Simplify.)

7. $p^4 - q^4 = (p - q)(p + q)(p^2 + q^2)$ (Original equation)
$= (p^2 - q^2)(p^2 + q^2)$ (Difference of squares)
$= p^4 - q^4$ (Difference of squares)

9. $(3x + y)^2 = 9x^2 + 6xy + y^2$ (Original equation)
$9x^2 + 6xy + y^2 = 9x^2 + 6xy + y^2$ (Square the left side.)
$9x^2 + 6xy + y^2 = 9x^2 + 6xy + y^2$ (True)
Because the identity is true, this proves that Aponi is correct. Her process for finding the area of a square will always work.

11. identity **13.** not an identity **15.** not an identity

17. $g^6 + h^6$
$= (g^2 + h^2)(g^4 - g^2h^2 + h^4)$ (Original equation)
$= g^6 + g^4h^2 + g^2h^4 + g^4h^2 - g^2h^4 + h^6$ (Distributive Property)
$= g^6 + h^6$ (Simplify.)

19. $u^6 - w^6$
$= (u + w)(u - w)(u^2 + uw + w^2)(u^2 - uw + w^2)$ (Original equation)
$= (u^2 + w^2)(u^2 + uw + w^2)(u^2 - uw + w^2)$ (FOIL)
$= (u^4 + u^3w - uw^3 - w^4)(u^2 - uw + w^2)$ (Distributive Property)
$= u^6 + u^5w - u^3w^3 - u^2w^4 - u^5w - u^4w^2 + u^2w^4 + uw^5 + u^4w^2 + u^3w^3 - uw^5 - w^6$ (Distributive Property)
$= u^6 - w^6$ (Simplify.)

21. A polynomial identity is a polynomial equation that is satisfied for any values that are substituted for the variables. To prove that a polynomial equation is an identity you begin with the more complicated side of the equation and use algebra properties to transform that side of the equation until it is simplified to look like the other side.

23. $x^2 - y^2 = 3, 2xy = 4, x^2 + y^2 = 5; x = 2, y = 1$

25. When George multiplied b and a^2 in the second line he mistakenly made the term negative so it did not cancel when he simplified.

Lesson 5-4

1. −59; 11 **3.** 1707; 62 **5.** −98; 0 **7.** 13,190; 2
9. 21, 0 **11.** −2, 1 **13.** 3, 3 **15.** −4, −7
17. 15, −6 **19.** −22, 20
21a. \$20.4 billion **21b.** \$144.16 billion
23. $(x - 1)^2$ **25.** $x - 4, x + 1$
27. $x + 6, 2x + 7$ **29.** $x + 1, x^2 + 2x + 3$
31. $x - 1, x + 2$ **33.** $x - 2, x + 2$
35. $x + 3, x - 6$ **37.** $x + 1, x - 4$
39. $2x + 1, x - 1$ **41.** $x - 1, x + 2$
43. $x - 4, 3x - 2$ **45.** 11; 764

47.

0.1	$\frac{1}{6}$	$\frac{1}{2}$	1	1
		$\frac{1}{60}$	$\frac{31}{600}$	$\frac{631}{6000}$
	$\frac{1}{6}$	$\frac{31}{60}$	$\frac{631}{600}$	$\frac{6631}{6000}$

$1\frac{631}{6000}$

49. 8 **51.** −3

53. $(x - 1), (5x - 2), (2x + 3)$; Sample answer: By the Factor Theorem, $(x - 1)$ is a factor. Use synthetic division with $x = 1$. The remainder is $k - 3$. For $(x - 1)$ to be a factor, $k - 3 = 0$, so $k = 3$. The quotient is $10x^2 + 13x - 3$, which factors as $(5x - 1)(2x + 3)$. $P(x) = (x - 1)(5x - 1)(2x + 3)$. The cubic has a positive leading coefficient and zeros at −1.5, 0.2, and 1.

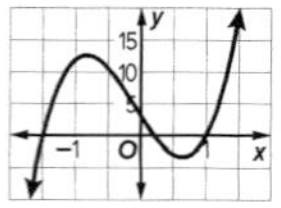

55a. (2003, 621), (2012, 1197), (2021, 1740), (2025, −15,255)

55b. The model still represents the situation after 25 years.; Sample answer: No, the average value is unlikely to fall so quickly.

57. −5; If $P(x)$ is symmetric to the y-axis, it must also contain the point (−2, −5). According to the Remainder Theorem, the remainder when polynomial $P(x)$ is divided by $(x - r)$ is $P(r)$. Since $P(-2)$, = −5, the remainder when $P(x)$ is divided by $(x + 2)$ is −5.

59. Use synthetic substitution with $a = -6$. The remainder must be zero, so $432 - 216k = 0$, or $k = 2$. The depressed polynomial is $2x^2 + 3x - 5$, which factors to $(2x + 5)(x - 1)$. Therefore, we can write $kx^3 + 15x^2 + 13x - 30 = (x + 6)(2x + 5)(x - 1)$.

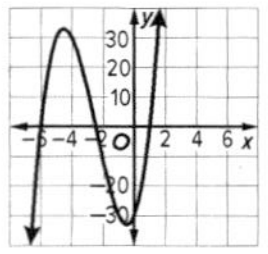

61. $f(x) = 0.1(x - 1)^2(x + 5)^2$

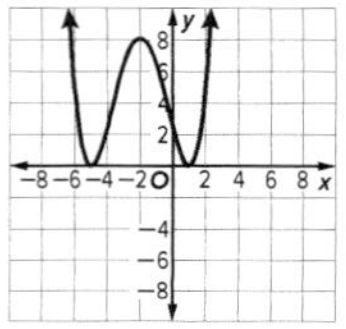

63. If $x - a$ is a factor of $f(x)$, then $f(a)$ has a factor of $(a - a)$ or 0. Since a factor of $f(a)$ is 0, $f(a) = 0$. Now assume that $f(a) = 0$. If $f(a) = 0$, then the Remainder Theorem states that the remainder is 0 when $f(x)$ is divided by $x - a$. This means that $x - a$ is a factor of $f(x)$. This proves the Factor Theorem.

65. Sample answer: When $x = 1$, $f(1)$ is the sum of all of the coefficients and constants in $f(x)$, in this case, a, b, c, d, and e. The sum of a, b, c, d, and e is 0, so however the coefficients are arranged, $f(1)$ will always equal 0, and $f(x)$ will have a rational root.

67. Tyrone; Sample answer: By the Factor Theorem, $(x - r)$ is a factor when $P(r) = 0$. $P(r) = 0$ when $x = -1$ and $x = 2$.

Lesson 5-5

1. $-\frac{12}{5}$; 1 real

3. $0, 0, 0, 2i, -2i$; 3 real, 2 imaginary

5. $\frac{1 \pm \sqrt{2}}{2}$; 2 real

7. $-2, \frac{3}{2}$; 2 real

9. $-1, \frac{1 \pm i\sqrt{3}}{2}$; 1 real, 2 imaginary

11. $-\frac{8}{3}$, 1; 2 real

13. $-\frac{5}{2}, \frac{5}{2}, -\frac{5}{2}i, \frac{5}{2}i$; 2 real, 2 imaginary

15. −2, −2, 0, 2, 2; 5 real

17. 2 or 0; 1; 2 or 0

19. 3 or 1; 0; 2 or 0

21. 1; 1; 2

23. 0 or 2; 0 or 2; 0, 2, or 4

25. 0 or 2; 1; 2 or 4

27. 0 or 2; 0 or 2; 2, 4, or 6

29. $3, 1 + \sqrt{2}, 1 - \sqrt{2}$

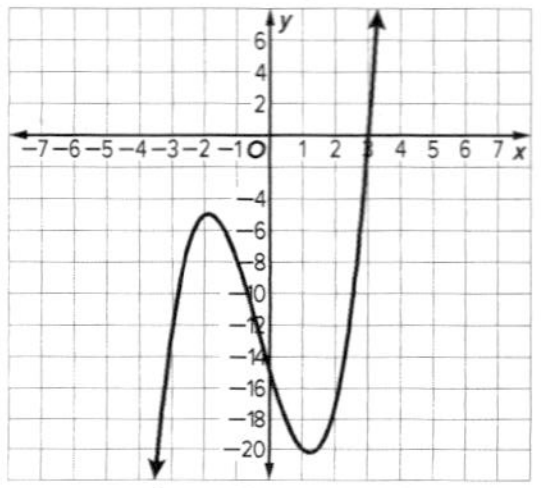

31. 1, −2, −3

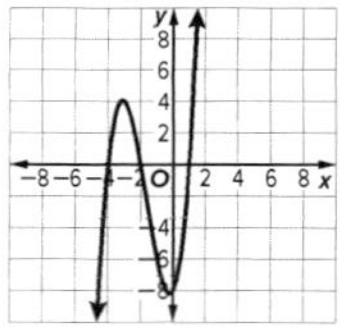

33. −1, −1, 1, 4

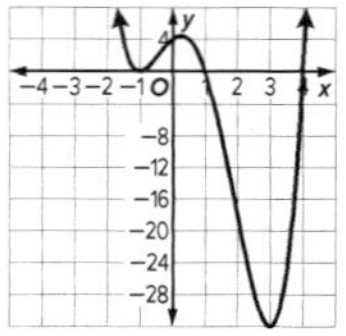

35. $-6, -2, 1$

37. $-4, 7, -5i, 5i$

39. $4, 4, -2i, 2i$

41. $y = x^2 + x - 6$

43. $y = x^4 - 6x^3 + 7x^2 + 6x - 8$

45. Sample Answer:

47. Sample answer: $f(x) = x^3 + 2x^2 - 23x - 60$

49. Sample answer: $f(x) = x^4 + 2x^3 + 6x^2 + 18x - 27$

51. Sample answer: $f(x) = x^4 - 3x^3 - 9x^2 + 77x + 150$

53.

55.

57a. $(x - 5)^2(7x + 9.5)\pi = 234\pi$; 7, $\frac{23 \pm 3\sqrt{65}}{28}$

57b. height 41.5 m, radius 2 m; $x = 7$ is the only reasonable solution in the context of the situation. The other possible values of x result in negative measures.

59. The graph is accurate as far as the location of the zeros, but does not consider any vertical dilation. It could be improved by finding more points between the roots.

61a.

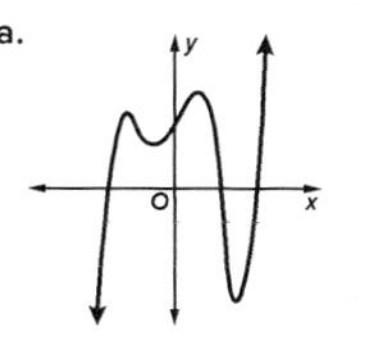

61b.

61c.

63. $r^4 + 1 = 0$; Sample answer: The equation has imaginary solutions and all of the others have real solutions.

65. Sample answer: To determine the number of positive real roots, determine how many time the signs change in the polynomial as you move from left to right. In this function there are 3 changes in sign. Therefore, there may be 3 or 1 positive real roots. To determine the number of negative real roots, I would first evaluate the polynomial for $-x$. All of the terms with an odd-degree variable would change signs. Then I would again count the number of sign changes as I move from left to right. There would be only one change. Therefore there may be 1 negative root.

Module 5 Review

1. A **3.** A, B, D

5. Set the expression for the volume of the first figure equal to the expression for the surface area of the second figure. Solve the equation $x^3 - 9x = 8x^2$. Solving the equation gives the solutions -1, 0, and 9. The value of x cannot be 0 because when substituting 0 for x in the original expressions, the volume of the first figure is $(0)3 - 9(0) = 0 - 0 = 0$ and surface area of the second figure is $8(0)^2 = 8(0) = 0$. Volume and surface area cannot be 0, so the value of x is -1 or 9.

7. $x = 6$

9. B **11.** C **13.** B

15.

x	Real Root	Imaginary Root	Not a Root
-2			X
-1	X		
1	X		
2			X
$-2i$		X	
$-i$			X
i			X
$2i$		X	

17. A

Module 6

Quick Check

1. between 0 and 1, and between 3 and 4
3. between −1 and 0, and between 1 and 2
5. $5x + 3$ **7.** $2x^2 + 5x - 6$

Lesson 6-1

1. $(f + g)(x) = -2x + 5$; $(f - g)(x) = 6x - 5$; $(f \cdot g)(x) = -8x^2 + 10x$; $\left(\frac{f}{g}\right)(x) = \frac{2x}{-4x + 5}, x \neq \frac{5}{4}$

3. $(f + g)(x) = 3x - 9$; $(f - g)(x) = -x + 5$; $(f \cdot g)(x) = 2x^2 - 11x + 14$; $\left(\frac{f}{g}\right)(x) = \frac{x - 2}{2x - 7}, x \neq \frac{7}{2}$

5. $(f + g)(x) = x^2 + 3x + 1$; $(f - g)(x) = -3x^2 - 3x + 11$; $(f \cdot g)(x) = -2x^4 - 3x^3 + 17x^2 + 18x - 30$; $\left(\frac{f}{g}\right)(x) = \frac{-x^2 + 6}{2x^2 + 3x - 5}, x \neq 1 \text{ or } -\frac{5}{2}$

7a. $(a - b)(x)$; $(a - b)(x) = 85x + 1750$

7b.

Account Balance ($): 0, 600, 1200, 1800, 2400, 3000
Months Since Opening Account: 1 2 3 4 5 6 7 8 9 10

9. $f \circ g = \{(-4, 4)\}$, D = {−4}, R = {4}; $g \circ f = \{(-8, 0), (0, -4), (2, -5), (-6, -1)\}$, D = {−6, 0, 2}, R = {−5, −4, −1, 0}

11. $f \circ g$ is undefined, D = ∅, R = ∅; $g \circ f$ is undefined, D = ∅, R = ∅

13. $[f \circ g](x) = 2x + 10$, D = {all real numbers}, R = {all even numbers}; $[g \circ f](x) = 2x + 5$, D = {all real numbers}, R = {all odd numbers}

15. $[f \circ g](x) = x^2 - 6x - 2$, D = {all real numbers}, R = $\{y \mid y \geq -11\}$; $[g \circ f](x) = x^2 + 6x - 8$, D = {all real numbers}, R = $\{y \mid y \geq -17\}$

17. $p(x) = 0.85x$; $t(x) = 1.065x$; $t[p(x)] = 1.065(0.85x)$; $1222.09

19. Sample answer: The order of the discounts does not matter. Either composition results in a final cost of $54.91.

21. 15 **23.** 1 **25.** 12 **27.** 189 **29.** 21
31. −3 **33.** 9 **35.** 440

37a. $V[r(t)] = \frac{\pi t^3}{6} + 2\pi t^2 + 8\pi t + \frac{32}{3}\pi$

37b. 2145 m³

39a. $(f + g)(-1) = 4$

39b. $(h - g)(0) = 8$

39c. $(f \cdot h)(4) = 5$

39d. $\left(\frac{f}{g}\right)(3) = 4$

39e. $\left(\frac{g}{h}\right)(2) = 0$

39f. $\left(\frac{g}{f}\right)(1) =$ undefined

41. Sample answer: $f(x) = x - 9$, $g(x) = x + 5$

43a. D = {all real numbers}

43b. D = $\{x \mid x \geq 0\}$

Lesson 6-2

1. {(6, −8), (−2, 6), (−3, 7)}

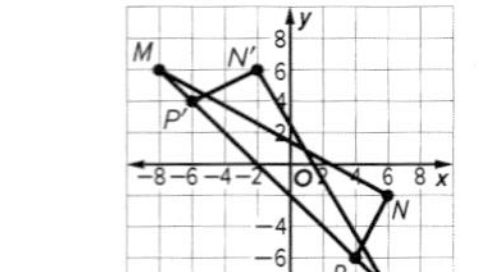
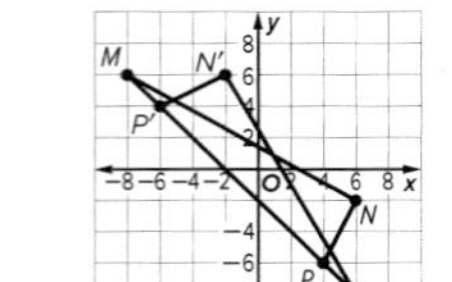

3. {(−1, 8), (−1, −8), (−8, −2), (−8, 2)}

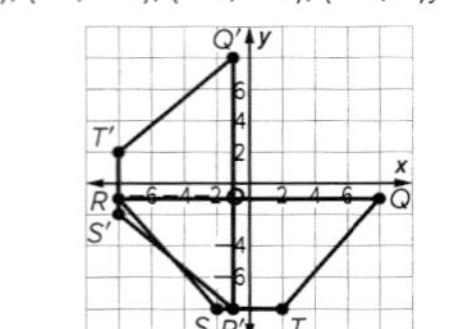
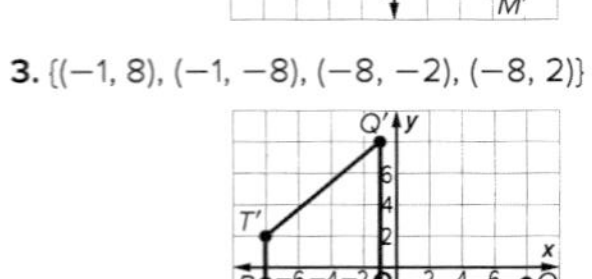
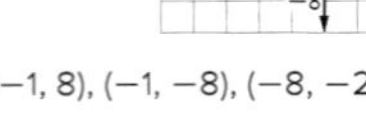
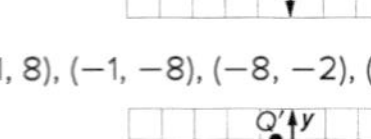

5. $f^{-1}(x) = x - 2$

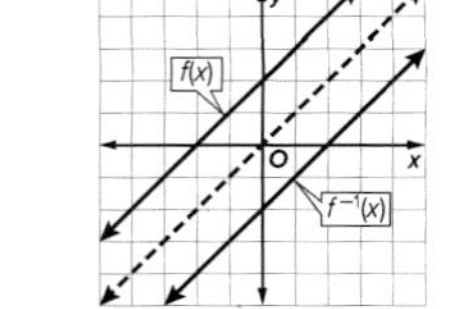
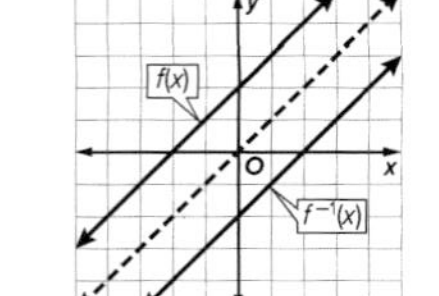
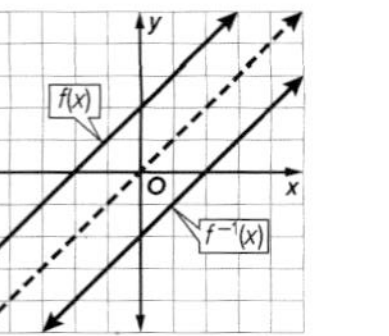

7. $f^{-1}(x) = -\frac{x}{2} + \frac{1}{2}$

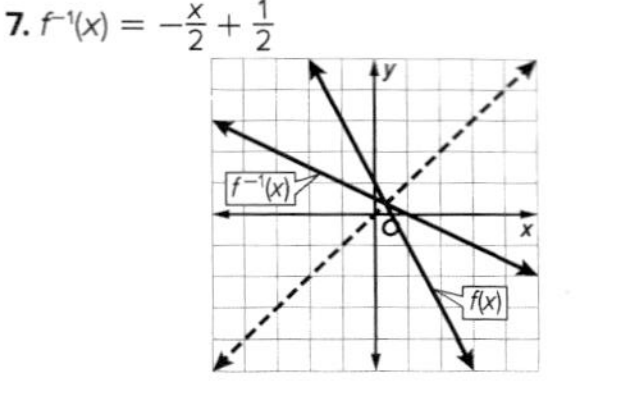

9. $f^{-1}(x) = -\frac{3}{5}(x + 8)$

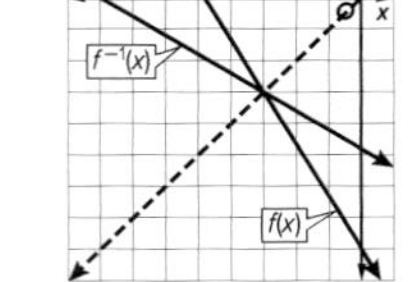

11. $f^{-1}(x) = \frac{1}{4}x$

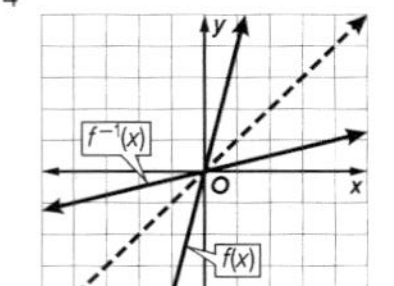

13. $f^{-1}(x) = \pm\frac{\sqrt{5x}}{5}$

If the domain of $f(x)$ is restricted to $(-\infty, 0]$, then the inverse is $f^{-1}(x) = -\frac{\sqrt{5x}}{5}$.
If the domain of $f(x)$ is restricted to $[0, \infty)$, then the inverse is $f^{-1}(x) = \frac{\sqrt{5x}}{5}$.

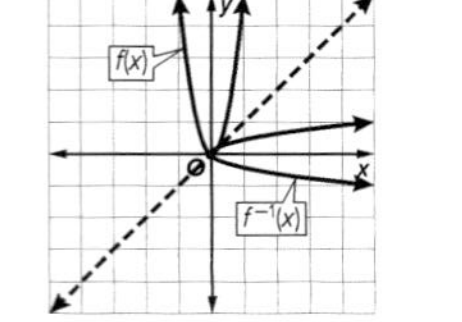

15a. $p^{-1}(x) = 14x$; Sample answer: The inverse converts stones to pounds, where x is the weight in stones.

15b. About 35 pounds

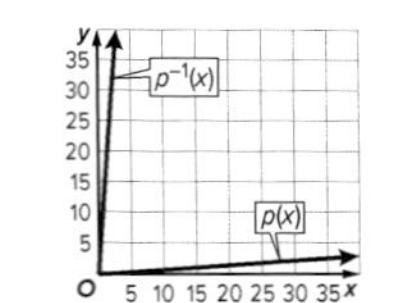

17. no **19.** yes **21.** no **23.** yes
25. $x = 4$

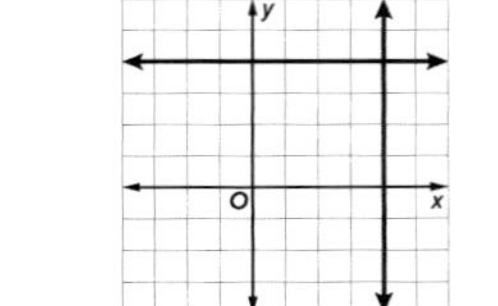

27. $f^{-1}(x) = x - 2$

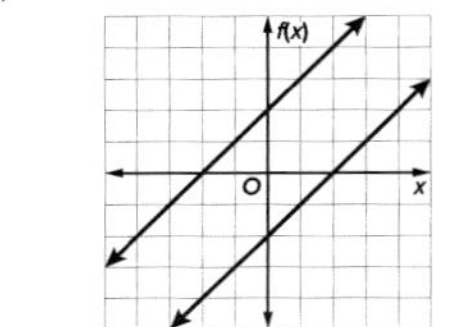

29. $f^{-1}(x) = \pm\sqrt{2x + 2}$
If the domain of $f(x)$ is restricted to $(-\infty, 0]$, then the inverse is $f^{-1}(x) = \sqrt{2x + 2}$. If the domain of $f(x)$ is restricted to $[0, \infty)$, then the inverse is $f^{-1}(x) = -\sqrt{2x + 2}$.

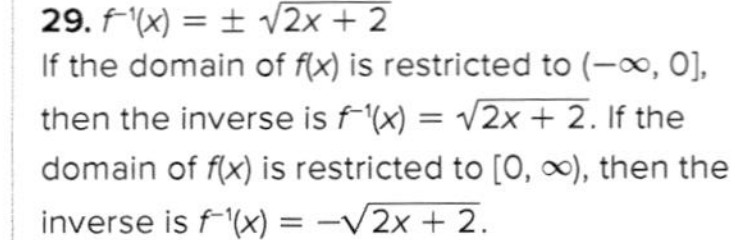
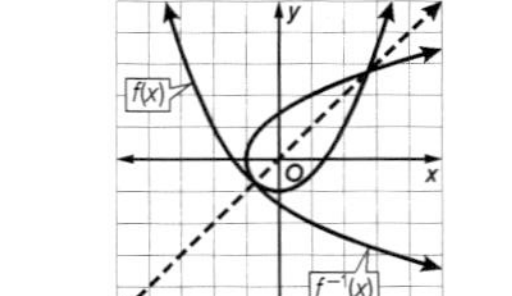

31. yes **33.** no **35.** yes
37. $\{x \mid x \geq 0\}$ or $\{x \mid x \leq 0\}$; $\{x \mid x \geq 5\}$
39. $\{x \mid x \geq -6\}$; $\{x \mid x \geq 0\}$

41.

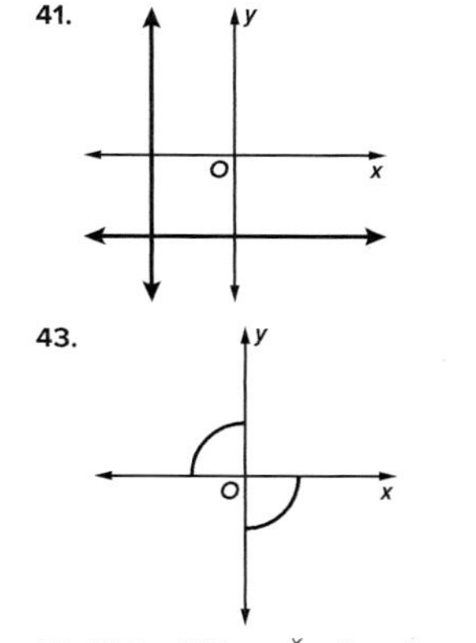

43.

45. $f^{-1}(x) = 220 - \frac{x}{0.85}$; Sample answer: The inverse function gives the age of the clients based on their maximum target heart rate.

47a. $V(x) = x(12 - 2x)(16 - 2x)$; Sample answer: The value of x must be between 0 and 6 because the length of x must be positive, but the length of the side of the box $(12 - 2x)$ cannot be less than 0.

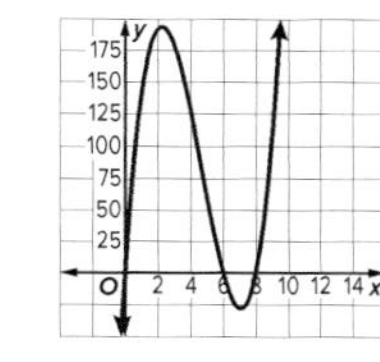

47b. No; sample answer: even with the restriction, the function does not pass the horizontal line test, so its inverse will not be a function.

49. Sample answer: Sometimes; $y = \pm\sqrt{x}$ is an example of a relation that is not a function, with an inverse being a function. A circle is an example of a relation that is not a function with an inverse not being a function.

51. Sample answer: $f(x) = x$ and $f^{-1}(x) = x$ or $f(x) = -x$ and $f^{-1}(x) = -x$

53. Sample answer: One of the functions carries out an operation on 5. Then the second function that is an inverse of the first function reverses the operation on 5. Thus, the result is 5.

Lesson 6-3

1. $\pm 11x^2y^8$ **3.** $\pm 7x^2$ **5.** $-9a^8b^{10}c^6$ **7.** $2|(x-3)^3|$ **9.** $3|x-4|$ **11.** $|a^3|$ **13.** $\sqrt[5]{8}$ **15.** $\sqrt{x^9}$ **17.** $5^{\frac{1}{3}}x^{\frac{1}{3}}y^{\frac{2}{3}}$ **19.** $6^{\frac{1}{3}}t^{\frac{2}{3}}$ **21.** 3 **23.** $\frac{1}{64}$ **25.** 64 **27.** $x^{\frac{11}{15}}$ **29.** $\frac{b^{\frac{1}{4}}}{b}$ **31.** $3b^6c^4$ **33.** $3|(x+4)|$ **35.** $(y^3+5)^6$ **37.** $14|c^3|d^2$ **39.** $-3a^5b^3$ **41.** a^3 **43.** $b^{\frac{1}{4}}$ **45.** $\frac{d^{\frac{1}{6}}}{d}$

47a. $5\sqrt{30}$ cm³

47b. $32\sqrt{3}$ cm³

47c. 288 cm³

49. $c^{\frac{1}{2}}$ **51.** $6^{\frac{1}{4}}h^{\frac{1}{2}}j^{\frac{1}{2}}$ **53.** $\frac{xy\sqrt[3]{z^2}}{z}$ **55.** 21 feet

57. $\sqrt[b]{m^{3b}} = m^3$.

59. $x < 0$; If $x < 0$, then $\sqrt{x^2} = -x$.

61. Sample answer: It may be easier to simplify an expression when it has rational exponents because all the properties of exponents apply. We do not have as many properties dealing directly with radicals. However, we can convert all radicals to rational exponents, and then use the properties of exponents to simplify.

63. when x or $y = 0$ and the other variable is ≥ 0

65. $0 < x < 1, x < -1$ **67.** $2\sqrt[3]{2xy}$

Lesson 6-4

1. $D = \{x \mid x \geq 9\}$; $R = \{y \mid y \geq 0\}$

3. $D = \{x \mid x \geq 0\}$; $R = \{y \mid y \leq 0\}$

5. $D = \left\{x \mid x \geq \frac{4}{3}\right\}$; $R = \{y \mid y \geq 0\}$

7. $D = \{x \mid x \geq -1\}$; $R = \{y \mid y \leq 0\}$; reflected in the x-axis, compressed vertically, and translated left 1 unit

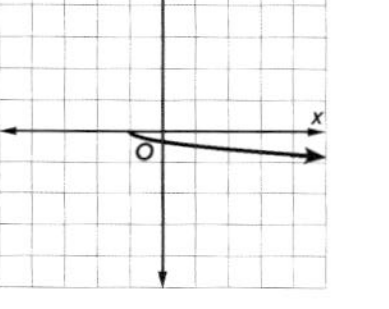

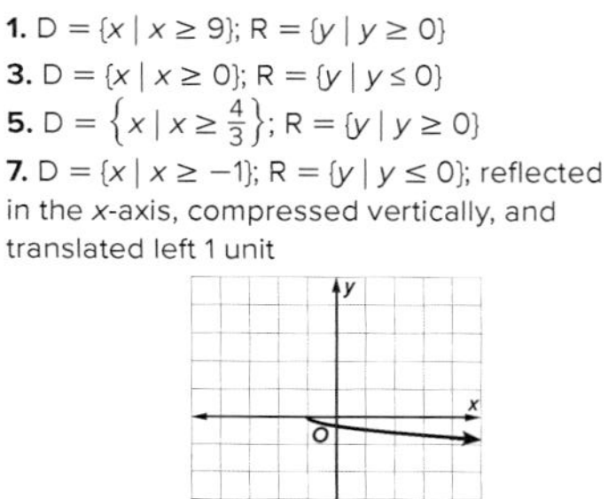

9. $D = \{x \mid x \geq 0\}$; $R = \{y \mid y \geq 0\}$ stretched vertically

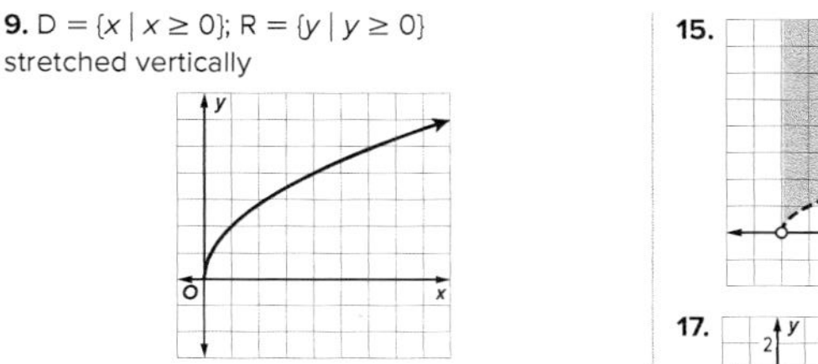

11. $D = \{x \mid x \geq 0\}$; $R = \{y \mid y \geq -5\}$; stretched vertically and translated down 5 units

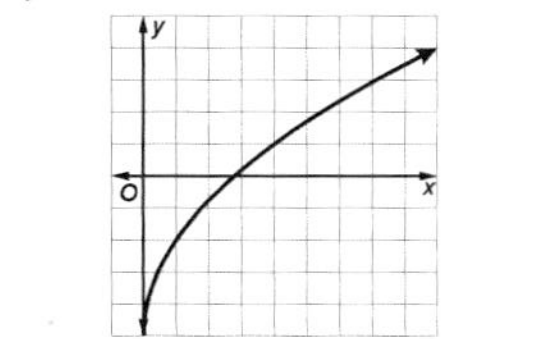

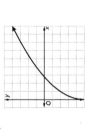

13a. $t = \frac{1}{4}\sqrt{d}$

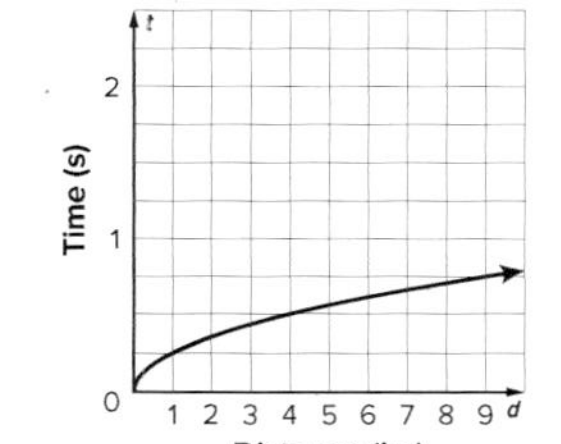

13b. $D = \{d \mid d \geq 0\}$; $R = \{t \mid t \geq 0\}$; increasing as $x \to \infty$; positive for $x > 0$; As $x \to \infty$, $y \to \infty$.

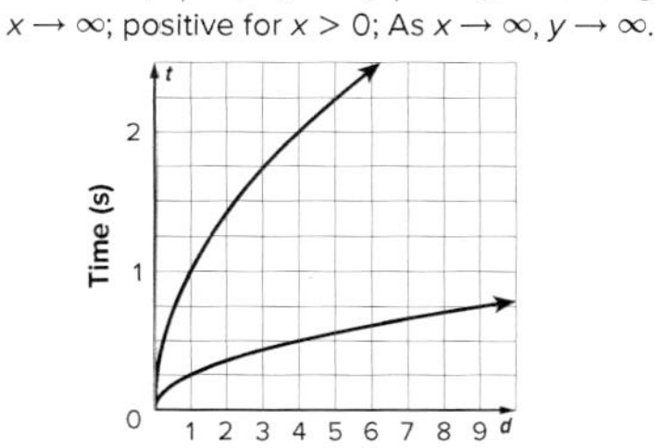

13c. compressed vertically

15.

17.

19.

21. $D = (-\infty, \infty)$; $R = (-\infty, \infty)$

23. $D = (-\infty, \infty)$; $R = (-\infty, \infty)$

25a.

25b.

	$p(x) = \sqrt[3]{x} - 2$	$q(x)$
Domain	all real numbers	$\{x \mid x \geq 0\}$
Range	all real numbers	$\{y \mid y \leq -3\}$
Intercepts	x-int: 8; y-int: −2	x-int: 9; y-int: −3
Increasing/ Decreasing	increasing as $x \to \infty$	increasing as $x \to \infty$
Positive/ Negative	negative for $x < 8$; positive for $x > 8$	negative for $x < 9$; positive for $x > 9$
End Behavior	as $x \to -\infty$, $p(x) \to -\infty$ as $x \to \infty$, $p(x) \to \infty$	as $x \to -\infty$, $q(x) \to -2$ as $x \to \infty$, $q(x) \to \infty$

27. $f(x) = \sqrt[3]{x-2} + 1$

29. D = $\{x \mid x \geq 5\}$; R = $\{f(x) \mid f(x) \geq -6\}$ stretched vertically translated 5 units right and 6 units down

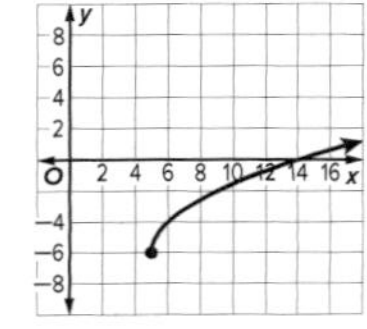

31. D = $\{x \mid x \geq 1\}$; R = $\{f(x) \mid f(x) \leq -4\}$ compressed vertically, translated 1 unit right and 4 units down, reflected across the x-axis

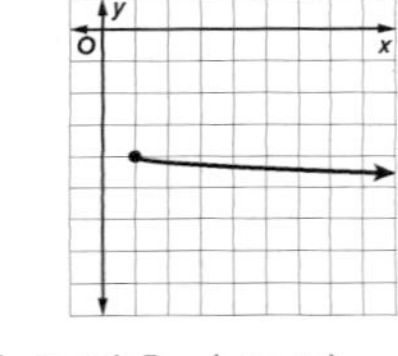

33. D = $(-\infty, \infty)$; R = $(-\infty, \infty)$

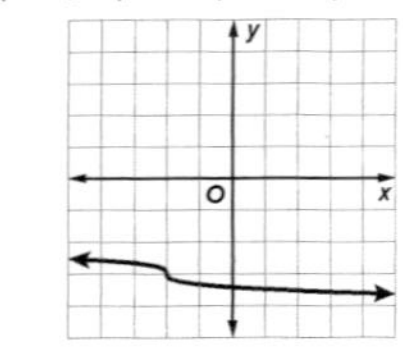
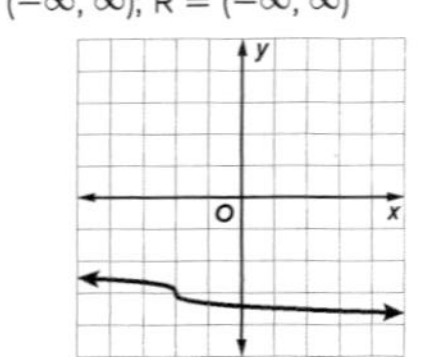
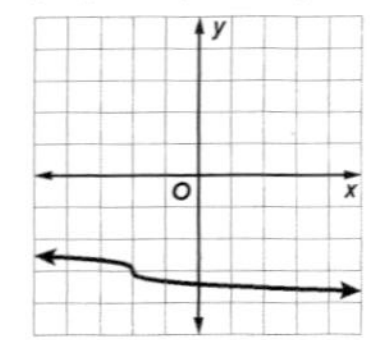

35.

37. $f(x) = -\sqrt{x+4} - 2$

39a. $f(x)$ has the greater maximum value because its maximum, $\frac{13}{2}$, is greater than 6, the maximum value of $g(x)$.

39b. The domain of $f(x)$ is $x \geq -3$, since any values less than 3 produce a negative value under the radical. The domain of $g(x)$ is $x \geq \frac{5}{2}$.

39c. The average rate of change over the interval is $-\frac{1}{7}$ for $f(x)$. It appears that the rate of change for $g(x)$ is the same.

41. D = $\{V \mid -\infty < V < \infty\}$ or $(-\infty, \infty)$; R = $\{r(V) \mid -\infty < r(V) < \infty\}$ or $(-\infty, \infty)$; End behavior: The values of r increase as the values of V increase.

43. Sample answer: $y = -\sqrt{x+4} + 6$

45. Sample answer: The domain is limited because square roots of negative numbers are imaginary. The range is limited due to the limitation of the domain.

47a. Sample answer: The original is $y = x^2 + 2$ and inverse is $y = \pm\sqrt{x-2}$.

47b. Sample answer: The original is $y = \pm\sqrt{x+4}$ and inverse is $y = (x-4)^2$.

Lesson 6-5

1. $6a^4b^2\sqrt{2b}$ **3.** $2a^8b^4\sqrt{6cb}$ **5.** $2|ab|\sqrt[4]{4}$ **7.** $\frac{5}{6}|r|\sqrt{t}$ **9.** $\frac{4h\sqrt[5]{3h^3}}{3f}$ **11.** $8\sqrt{2}$ **13.** $\sqrt{5}$ **15.** $5\sqrt{7x} - \sqrt{14}$ **17.** $120y\sqrt{2z}$ **19.** $144ab^2\sqrt{2}$ **21.** $15xy\sqrt{14xy}$

21. $15xy\sqrt{14xy}$
23. $18 + 5\sqrt{6}m^2$
25. $56\sqrt{3} + 42\sqrt{6} - 36\sqrt{2} - 54$ **27.** 1260
29. $\frac{a^2\sqrt{5ab}}{b^7}$ **31.** $\frac{\sqrt[3]{150x^2y^2}}{5y}$
33. $6\sqrt{3} + 6\sqrt{2}$ **35.** $\frac{20 - 7\sqrt{3}}{11}$ **37.** $\frac{3\sqrt{7} + 3\sqrt{35}}{4}$
39. $2yz^4\sqrt[3]{2y}$
41. $\frac{(x+1)(\sqrt{x}+1)}{x-1}$ or $\frac{x\sqrt{x} + \sqrt{x} + x + 1}{x-1}$
43. $4\sqrt{3}$ **45.** $28\sqrt[3]{7a}$ **47.** $10\sqrt[3]{28}$ in. **49.** $\frac{\sqrt{tu}}{u}$
51a. John: $\frac{0.8 - 4\sqrt{0.02}}{0.02}$ or $40 - 20\sqrt{2}$ min;
Jay: $\frac{0.8 - 4\sqrt{0.05}}{-0.01}$ or $40\sqrt{5} - 80$ min
51b. $2 - \sqrt{2} + \sqrt{5} + \frac{\sqrt{10}}{2}$
53a. $3x$; $(\sqrt{x} \cdot x) + (\sqrt{x} \cdot \sqrt{4x}) = (\sqrt{x^2} + \sqrt{4x^2}) = x + 2x = 3x$
53b. $6x$; $(\sqrt{x} \cdot \sqrt{x}) + (\sqrt{x} \cdot \sqrt{4x}) + (\sqrt{x} + \sqrt{9x}) = (\sqrt{x^2} + \sqrt{4x^2} + \sqrt{9x^2}) = x + 2x + 3x = 6x$
53c. $10x$; $(\sqrt{x} \cdot \sqrt{x}) + (\sqrt{x} \cdot \sqrt{4x}) + (\sqrt{x} \cdot \sqrt{9x}) + (\sqrt{x} \cdot \sqrt{16x}) = (\sqrt{x^2} + \sqrt{4x^2} + \sqrt{9x^2} + \sqrt{16x^2}) = x + 2x + 3x + 4x = 10x$
53d. $(\sqrt{x} \cdot \sqrt{x}) + (\sqrt{x} \cdot \sqrt{4x}) + \ldots + (\sqrt{x} + \sqrt{n^2x}) = x + 2x + \ldots + nx = (1 + 2 + \ldots + n)x = \left(\frac{n(n+1)}{2}\right)x$
55a. $6\sqrt[3]{2} + 1$; $r = \sqrt[3]{\frac{3}{4\pi} \cdot 72\pi} = \sqrt[3]{54} = 3\sqrt[3]{2}$; $S = 3\sqrt[3]{2} + 2(2)$
55b. No; sample answer: a box with a volume of 384 cm³ has a side that is $\sqrt[3]{384}$, or about 7.27 cm; $s = 6\sqrt[3]{2} + 4 \approx 11.55$ is the least possible value for a side of the gift box.
57. $a = 1, b = 256$; $a = 2, b = 16$; $a = 4, b = 4$; $a = 8, b = 2$
59. Sample answer: It is only necessary to use absolute values when it is possible that n could be odd or even and still be defined. It is when the radicand must be nonnegative in order for the root to be defined that the absolute values are not necessary.

Lesson 6-6

1. $\frac{1}{25}$ **3.** 8 **5.** 40 **7.** 11 **9.** 11 **11.** $-\frac{1}{2}$
13. no real solution; 16 is an extraneous solution.
15. 83 **17.** 61 **19.** 15,623 **21.** 43
23. 2.4 **25.** 2 **27.** 1.23 **29.** 36 cm
31. $\frac{1}{4}$ **33.** $\frac{81}{16}$
35. 2 **37.** 24.5
39. 3, 4 (extraneous solution)
41. 6, −5 (extraneous solution)
43. 12 units
45a. $\sqrt{x^2 + \frac{289}{x^2}}$
45b. $\frac{17}{12} \leq x \leq 12$
47a. 20.2
47b. Silver; $5.713 \times 10^{-15} \approx (1.2 \times 10^{-15})(A)^{\frac{1}{3}}$; $4.761 \approx (A)^{\frac{1}{3}}$; $107.917 \approx A$
49. Graph functions $f(x) = \sqrt{\frac{x}{6}}$ and $g(x) = \sqrt{\frac{x + 20}{4\pi}}$, where x is surface area of the cube. The graphs intersect at $x \approx 18.27$. The surface areas of the cube and sphere are approximately 18.27 cm² and 38.27 cm².

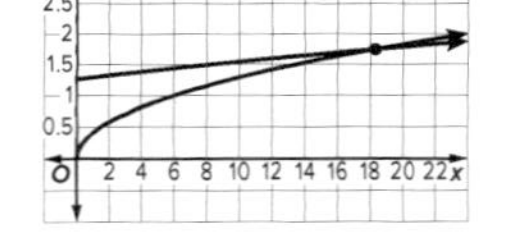

51. $\sqrt{x+2} - 7 = -10$; Sample answer: This equation does not have a real solution while the other three equations do.
53. never;
$\frac{\sqrt{(x^2)^2}}{-x} = x$
$\frac{x^2}{-x} = x$
$x^2 = (x)(-x)$
$x^2 \neq -x^2$
55. Never; sample answer: The radicand can be negative.

Module 6 Review

1. D

3. C

5. $(f \circ g)(x) = 12x + 8$

7. If the domain of $f(x)$ is restricted to $(-\infty, -2]$, then the inverse is $f^{-1}(x) = -2 - \sqrt{x+1}$. If the domain of $f(x)$ is restricted to $[-2, \infty)$, then the inverse is $f^{-1}(x) = -2 + \sqrt{x+1}$.

9. D

11. The parent function is $f(x) = \sqrt{x}$ and $g(x) = a\sqrt{x-h} + k$ where $a = -2$, $h = -5$, and $k = -3$. Because $a < 0$ and $|a| > 1$, the graph of $g(x)$ is the graph of the parent function reflected across the x-axis and stretched vertically by a factor of $|a|$, or 2. Also, because $h < 0$, the graph of $g(x)$ is the graph of the parent function translated left $|h|$ units, or 5 units. Finally, because $k < 0$, the graph of $g(x)$ is the graph of the parent function translated down $|k|$ units, or 3 units.

13. B **15.** $2\sqrt{3}$ **17.** B **19.** A

Glossary

English	Español
A	
absolute value (Lesson 2-2) The distance a number is from zero on the number *line*.	valor absoluto La distancia que un número es de cero en la *línea* numérica.
absolute value function (Lesson 1-6) A function written as $f(x) = \|x\|$, in which $f(x) \geq 0$ for all values of x.	función del valor absoluto Una función que se escribe $f(x) = \|x\|$, donde $f(x) \geq 0$, para todos los valores de x.
algebraic notation (Lesson 1-1) Mathematical notation that describes a set by using algebraic expressions.	notación algebraica Notación matemática que describe un conjunto usando expresiones algebraicas.
amplitude (Lesson 11-4) For functions of the form $y = a \sin b\theta$ or $y = a \cos b\theta$, the amplitude is $\|a\|$.	amplitud Para funciones de la forma $y = a \text{ sen } b\theta$ o $y = a \cos b\theta$, la amplitud es $\|a\|$.
asymptote (Lesson 7-1) A line that a graph approaches.	asíntota Una línea que se aproxima a un gráfico.
average rate of change (Lesson 3-1) The change in the value of the dependent variable divided by the change in the value of the independent variable.	tasa media de cambio El cambio en el valor de la variable dependiente dividido por el cambio en el valor de la variable independiente.
axis of symmetry (Lesson 3-1) The line about which a graph is symmetric.	eje de simetría Una línea sobre la cual un gráfica es simétrico.
B	
bias (Lesson 10-1) An error that results in a misrepresentation of a population.	sesgo Un error que resulta en una tergiversación de una población.
binomial (Lesson 4-3) The sum of two monomials.	binomio La suma de dos monomios.
boundary (Lesson 1-5) The edge of the graph of an inequality that separates the coordinate plane into regions.	frontera El borde de la gráfica de una desigualdad que separa el plano de coordenadas en regiones.
bounded (Lesson 2-6) When the graph of a system of constraints is a polygonal region.	acotada Cuando la gráfica de un sistema de restricciones es una región poligonal.
C	
central angle of a circle (Lesson 11-1) An angle with a vertex at the center of a circle and sides that are radii.	ángulo central de un círculo Un ángulo con un vértice en el centro de un círculo y los lados que son radios.

Glossary

circular function (Lesson 11-3) A function that describes a point on a circle as the function of an angle defined in radians.

función circular Función que describe un punto en un círculo como la función de un ángulo definido en radianes.

closed (Lesson 4-3) If for any members in a set, the result of an operation is also in the set.

cerrado Si para cualquier número en el conjunto, el resultado de la operación es también en el conjunto.

closed half-plane (Lesson 1-5) The solution of a linear inequality that includes the boundary line.

semi-plano cerrado La solución de una desigualdad linear que incluye la línea de limite.

codomain (Lesson 1-1) The set of all the y-values that could possibly result from the evaluation of the function.

codominar El conjunto de todos los valores y que podrían resultar de la evaluación de la función.

coefficient of determination (Lesson 7-5) An indicator of how well a function fits a set of data.

coeficiente de determinación Un indicador de lo bien que una función se ajusta a un conjunto de datos.

cofunction identities (Lesson 12-1) Identities that show the relationships between sine and cosine, tangent and cotangent, and secant and cosecant.

identidades de cofunción Identidades que muestran las relaciones entre seno y coseno, tangente y cotangente, y secante y cosecante.

combined variation (Lesson 9-5) When one quantity varies directly and/or inversely as two or more other quantities.

variación combinada Cuando una cantidad varía directamente y / o inversamente como dos o más cantidades.

common logarithms (Lesson 8-3) Logarithms of base 10.

logaritmos comunes Logaritmos de base 10.

common ratio (Lesson 7-4) The ratio of consecutive terms of a geometric sequence.

razón común El razón de términos consecutivos de una secuencia geométrica.

completing the square (Lesson 3-5) A process used to turn a quadratic expression into a perfect square trinomial.

completar el cuadrado Un proceso usado para hacer una expresión cuadrática en un trinomio cuadrado perfecto.

complex conjugates (Lesson 3-3) Two complex numbers of the form $a + bi$ and $a - bi$.

conjugados complejos Dos números complejos de la forma $a + bi$ y $a - bi$.

complex fraction (Lesson 9-1) A rational expression with a numerator and/or denominator that is also a rational expression.

fracción compleja Una expresión racional con un numerador y / o denominador que también es una expresión racional.

complex number (Lesson 3-3) Any number that can be written in the form $a + bi$, where a and b are real numbers and i is the imaginary unit.

número complejo Cualquier número que se puede escribir en la forma $a + bi$, donde a y b son números reales e i es la unidad imaginaria.

composition of functions (Lesson 6-1) An operation that uses the results of one function to evaluate a second function.

composición de funciones Operación que utiliza los resultados de una función para evaluar una segunda función.

compound interest (Lesson 7-2) Interest calculated on the principal and on the accumulated interest from previous periods.

interés compuesto Intereses calculados sobre el principal y sobre el interés acumulado de períodos anteriores.

confidence interval (Lesson 10-5) An estimate of the population parameter stated as a range with a specific degree of certainty.

intervalo de confianza Una estimación del parámetro de población se indica como un rango con un grado específico de certeza.

conjugates (Lesson 6-5) Two expressions, each with two terms, in which the second terms are opposites.

conjugados Dos expresiones, cada una con dos términos, en la que los segundos términos son opuestos.

consistent (Lesson 2-4) A system of equations with at least one ordered pair that satisfies both equations.

consistente Una sistema de ecuaciones para el cual existe al menos un par ordenado que satisface ambas ecuaciones.

constant function (Lesson 1-7) The function $f(x) = a$, where a is any number.

función constante La función $f(x) = a$, donde a es cualquier número.

constant of variation (Lesson 9-5) The constant in a variation function.

constante de variación La constante en una función de variación.

constraint (Lesson 1-5) A condition that a solution must satisfy.

restricción Una condición que una solución debe satisfacer.

continuous function (Lesson 1-1) A function that can be graphed with a line or unbroken curve.

función continua Una función que se puede representar gráficamente con una línea o curva ininterrumpida.

continuous random variable (Lesson 10-4) The numerical outcome of a random event that can take on any value.

variable aleatoria continua El resultado numérico de un evento aleatorio que puede tomar cualquier valor.

convenience sample (Lesson 10-1) Members that are readily available or easy to reach are selected.

muestra conveniente Se seleccionan los miembros que están fácilmente disponibles o de fácil acceso.

cosecant (Lesson 11-2) The ratio of the length of a hypotenuse to the length of the leg opposite the angle.

cosecante Relación entre la longitud de la hipotenusa y la longitud de la pierna opuesta al ángulo.

cosine (Lesson 11-2) The ratio of the length of the leg adjacent to an angle to the length of the hypotenuse.

coseno Relación entre la longitud de la pierna adyacente a un ángulo y la longitud de la hipotenusa.

cotangent (Lesson 11-2) The ratio of the length of the leg adjacent to an angle to the length of the leg opposite the angle.

cotangente La relación entre la longitud de la pata adyacente a un ángulo y la longitud de la pata opuesta al ángulo.

coterminal angles (Lesson 11-1) Angles in standard position that have the same terminal side.

ángulos coterminales Ángulos en posición estándar que tienen el mismo lado terminal.

critical values (Lesson 10-5) The z-values corresponding to the most common degrees of certainty.

valores críticos Los valores z correspondientes a los grados de certeza más comunes.

cube root function (Lesson 6-4) A radical function that contains the cube root of a variable expression.

función de la raíz del cubo Función radical que contiene la raíz cúbica de una expresión variable.

cycle (Lesson 11-3) One complete pattern of a periodic function.

ciclo Un patron completo de una función periódica.

D

decay factor (Lesson 7-1) The base of an exponential expression, or $1 - r$.

factor de decaimiento La base de una expresión exponencial, o $1 - r$.

degree (Lesson 4-1) The value of the exponent in a power function.

grado Valor del exponente en una función de potencia.

degree of a polynomial (Lesson 4-1) The greatest degree of any term in the polynomial.

grado de un polinomio El grado mayor de cualquier término del polinomio.

dependent (Lesson 2-4) A consistent system of equations with an infinite number of solutions.

dependiente Una sistema consistente de ecuaciones con un número infinito de soluciones.

depressed polynomial (Lesson 5-4) A polynomial resulting from division with a degree one less than the original polynomial.

polinomio reducido Un polinomio resultante de la división con un grado uno menos que el polinomio original.

descriptive statistics (Lesson 10-3) The branch of statistics that focuses on collecting, summarizing, and displaying data.

estadística descriptiva Rama de la estadística cuyo enfoque es la recopilación, resumen y demostración de los datos.

difference of squares (Lesson 3-4) A binomial in which the first and last terms are perfect squares.

diferencia de cuadrados Un binomio en el que los términos primero y último son cuadrados perfectos.

dilation (Lesson 1-7) A transformation that stretches or compresses the graph of a function.

homotecia Una transformación que estira o comprime el gráfico de una función.

direct variation (Lesson 9-5) When one quantity is equal to a constant times another quantity.

variación directa Cuando una cantidad es igual a una constante multiplicada por otra cantidad.

discontinuous function (Lesson 1-1) A function that is not continuous.

función discontinua Una función que no es continua.

discrete function (Lesson 1-1) A function in which the points on the graph are not connected.

función discreta Una función en la que los puntos del gráfico no están conectados.

discrete random variable (Lesson 10-4) The numerical outcome of a random event that is finite and can be counted.

variable aleatoria discreta El resultado numérico de un evento aleatorio que es finito y puede ser contado.

discriminant (Lesson 3-6) In the Quadratic Formula, the expression under the radical sign that provides information about the roots of the quadratic equation.

discriminante En la Fórmula cuadrática, la expresión bajo el signo radical que proporciona información sobre las raíces de la ecuación cuadrática.

distribution (Lesson 10-3) A graph or table that shows the theoretical frequency of each possible data value.

distribución Un gráfico o una table que muestra la frecuencia teórica de cada valor de datos posible.

domain (Lesson 1-1) The set of x-values to be evaluated by a function.

dominio El conjunto de valores x para ser evaluados por una función.

E

e (Lesson 7-3) An irrational number that approximately equals 2.7182818....

e Un número irracional que es aproximadamente igual a 2.7182818 ....

elimination (Lesson 2-5) A method that involves eliminating a variable by combining the individual equations within a system of equations.

eliminación Un método que consiste en eliminar una variable combinando las ecuaciones individuales dentro de un sistema de ecuaciones.

empty set (Lesson 2-2) The set that contains no elements, symbolized by { } or ∅.

conjunto vacio El conjunto que no contiene elementos, simbolizado por { } o ∅.

end behavior (Lesson 1-3) The behavior of a graph at the positive and negative extremes in its domain.

comportamiento del fin El comportamiento de un gráfico en los extremos positivo y negativo en su dominio.

equation (Lesson 2-1) A mathematical statement that contains two expressions and an equals sign, =.

ecuación Un enunciado matemático que contiene dos expresiones y un signo igual, =.

even functions (Lesson 1-2) Functions that are symmetric in the y-axis.

incluso funciones Funciones que son simétricas en el eje y.

excluded values (Lesson 9-2) Values for which a function is not defined.

valores excluidos Valores para los que no se ha definido una función.

experiment (Lesson 10-1) A sample is divided into two groups. The experimental group undergoes a change, while there is no change to the control group. The effects on the groups are then compared.

experimento Una muestra se divide en dos grupos. El grupo experimental experimenta un cambio, mientras que no hay cambio en el grupo de control. A continuación se comparan los efectos sobre los grupos.

experimental probability (Lesson 10-2) Probability calculated by using data from an actual experiment.

probabilidad experimental Probabilidad calculada utilizando datos de un experimento real.

explicit formula (Lesson 7-4) A formula that allows you to find any term a_n of a sequence by using a formula written in terms of n.

fórmula explícita Una fórmula que le permite encontrar cualquier término a_n de una secuencia usando una fórmula escrita en términos de n.

exponential decay (Lesson 7-1) Change that occurs when an initial amount decreases by the same percent over a given period of time.

desintegración exponencial Cambio que ocurre cuando una cantidad inicial disminuye en el mismo porcentaje durante un período de tiempo dado.

exponential equation (Lesson 7-2) An equation in which the independent variable is an exponent.

ecuación exponencial Una ecuación en la que la variable independiente es un exponente.

exponential form (Lesson 6-3) When an expression is in the form x^n.

forma exponencial Cuando una expresión está en la forma x^n.

exponential function (Lesson 7-1) A function in which the independent variable is an exponent.

función exponencial Una función en la que la variable independiente es el exponente.

exponential growth (Lesson 7-1) Change that occurs when an initial amount increases by the same percent over a given period of time.

crecimiento exponencial Cambio que ocurre cuando una cantidad inicial aumenta por el mismo porcentaje durante un período de tiempo dado.

exponential inequality (Lesson 7-2) An inequality in which the independent variable is an exponent.

desigualdad exponencial Una desigualdad en la que la variable independiente es un exponente.

extraneous solution (Lesson 2-2) A solution of a simplified form of an equation that does not satisfy the original equation.

solución extraña Una solución de una forma simplificada de una ecuación que no satisface la ecuación original.

extrema (Lesson 1-3) Points that are the locations of relatively high or low function values.

extrema Puntos que son las ubicaciones de valores de función relativamente alta o baja.

F

factored form (Lesson 3-4) A form of quadratic equation, $0 = a(x - p)(x - q)$, where $a \neq 0$, in which p and q are the x-intercepts of the graph of the related function.

forma factorizada Una forma de ecuación cuadrática, $0 = a(x - p)(x - q)$, donde $a \neq 0$, en la que p y q son las intercepciones x de la gráfica de la función relacionada.

family of graphs (Lesson 1-7) Graphs and equations of graphs that have at least one characteristic in common.

familia de gráficas Gráficas y ecuaciones de gráficas que tienen al menos una característica común.

feasible region (Lesson 2-6) The intersection of the graphs in a system of constraints.

región factible La intersección de los gráficos en un sistema de restricciones.

finite sequence (Lesson 7-4) A sequence that contains a limited number of terms.

secuencia finita Una secuencia que contiene un número limitado de términos.

frequency (Lesson 11-4) The number of cycles in a given unit of time.

frecuencia El número de ciclos en una unidad del tiempo dada.

G

geometric means (Lesson 7-4) The terms between two nonconsecutive terms of a geometric sequence.

medios geométricos Los términos entre dos términos no consecutivos de una secuencia geométrica.

geometric sequence (Lesson 7-4) A pattern of numbers that begins with a nonzero term and each term after is found by multiplying the previous term by a nonzero constant r.

secuencia geométrica Un patrón de números que comienza con un término distinto de cero y cada término después se encuentra multiplicando el término anterior por una constante no nula r.

geometric series (Lesson 7-4) The indicated sum of the terms in a geometric sequence.

series geométricas La suma indicada de los términos en una secuencia geométrica.

greatest integer function (Lesson 1-6) A step function in which $f(x)$ is the greatest integer less than or equal to x.

función más grande del número entero Una función del paso en que $f(x)$ es el número más grande menos que o igual a x.

growth factor (Lesson 7-1) The base of an exponential expression, or $1 + r$.

factor de crecimiento La base de una expresión exponencial, o $1 + r$.

H

horizontal asymptote (Lesson 9-2) A horizontal line that a graph approaches.

asíntota horizontal Una línea horizontal que se aproxima a un gráfico.

hyperbola (Lesson 9-2) The graph of a reciprocal function.

hipérbola La gráfica de una función recíproca.

I

identity (Lesson 5-3) An equation that is true for every value of the variable.

identidad Una ecuación que es verdad para cada valor de la variable.

identity function (Lesson 1-7) The function $f(x) = x$.

función identidad La función $f(x) = x$.

imaginary unit i (Lesson 3-3) The principal square root of -1.

unidad imaginaria i La raíz cuadrada principal de -1.

inconsistent (Lesson 2-4) A system of equations with no ordered pair that satisfies both equations.

inconsistente Una sistema de ecuaciones para el cual no existe par ordenado alguno que satisfaga ambas ecuaciones.

independent (Lesson 2-4) A consistent system of equations with exactly one solution.

independiente Un sistema consistente de ecuaciones con exactamente una solución.

index (Lesson 6-3) In nth roots, the value that indicates to what root the value under the radicand is being taken.

índice En enésimas raíces, el valor que indica a qué raíz está el valor bajo la radicand.

inequality (Lesson 2-1) A mathematical sentence that contains $<$, $>$, $\leq$, $\geq$, or $\neq$.

desigualdad Una oración matemática que contiene uno o más de $<$, $>$, $\leq$, $\geq$, o $\neq$.

inferential statistics (Lesson 10-5) When the data from a sample is used to make inferences about the corresponding population.

estadísticas inferencial Cuando los datos de una muestra se utilizan para hacer inferencias sobre la población correspondiente.

infinite sequence (Lesson 7-4) A sequence that continues without end.

secuencia infinita Una secuencia que continúa sin fin.

initial side (Lesson 11-1) The part of an angle that is fixed on the x-axis.

lado inicial La parte de un ángulo que se fija en el eje x.

intercept (Lesson 1-2) A point at which the graph of a function intersects an axis.

interceptar Un punto en el que la gráfica de una función corta un eje.

interval notation (Lesson 1-1) Mathematical notation that describes a set by using endpoints with parentheses or brackets.

notación de intervalo Notación matemática que describe un conjunto utilizando puntos finales con paréntesis o soportes.

Glossary · Glosario

inverse functions (Lesson 6-2) Two functions, one of which contains points of the form (a, b) while the other contains points of the form (b, a).

funciones inversas Dos funciones, una de las cuales contiene puntos de la forma (a, b) mientras que la otra contiene puntos de la forma (b, a).

inverse relations (Lesson 6-2) Two relations, one of which contains points of the form (a, b) while the other contains points of the form (b, a).

relaciones inversas Dos relaciones, una de las cuales contiene puntos de la forma (a, b) mientras que la otra contiene puntos de la forma (b, a).

inverse trigonometric functions (Lesson 11-7) Arcsine, Arccosine, and Arctangent.

funciones trigonométricas inversas Arcsine, Arccosine y Arctangent.

inverse variation (Lesson 9-5) When the product of two quantities is equal to a constant k.

variación inversa Cuando el producto de dos cantidades es igual a una constante k.

J

joint variation (Lesson 9-5) When one quantity varies directly as the product of two or more other quantities.

variación conjunta Cuando una cantidad varía directamente como el producto de dos o más cantidades.

L

leading coefficient (Lesson 4-1) The coefficient of the first term when a polynomial is in standard form.

coeficiente líder El coeficiente del primer término cuando un polinomio está en forma estándar.

like radical expressions (Lesson 6-5) Radicals in which both the index and the radicand are the same.

expresiones radicales semejantes Radicales en los que tanto el índice como el radicand son iguales.

line of reflection (Lesson 1-7) The line in which a reflection flips the graph of a function.

línea de reflexión La línea en la que una reflectión voltea la gráfica de una función.

line of symmetry (Lesson 1-2) An imaginary line that separates a figure into two congruent parts.

línea de simetría Una línea imaginaria que separa una figura en dos partes congruentes.

line symmetry (Lesson 1-2) A graph has line symmetry if it can be reflected in a vertical line so that each half of the graph maps exactly to the other half.

simetría de línea Un gráfico tiene simetría de línea si puede reflejarse en una línea vertical, de modo que cada mitad del gráfico se asigna exactamente a la otra mitad.

linear equation (Lesson 1-2) An equation that can be written in the form $Ax + By = C$ with a graph that is a straight line.

ecuación lineal Una ecuación que puede escribirse de la forma $Ax + By = C$ con un gráfico que es una línea recta.

linear function (Lesson 1-2) A function in which no independent variable is raised to a power greater than 1.

función lineal Una función en la que ninguna variable independiente se eleva a una potencia mayor que 1.

linear inequality (Lesson 1-5) A half-plane with a boundary that is a straight line.

desigualdad lineal Un medio plano con un límite que es una línea recta.

linear programming (Lesson 2-7) The process of finding the maximum or minimum values of a function for a region defined by a system of linear inequalities.

programación lineal El proceso de encontrar los valores máximos o mínimos de una función para una región definida por un sistema de desigualdades lineales.

logarithm (Lesson 8-1) In $x = b^y$, y is called the logarithm, base b, of x.

logaritmo En $x = b^y$, y se denomina logaritmo, base b, de x.

logarithmic equation (Lesson 8-2) An equation that contains one or more logarithms.

ecuación logarítmica Una ecuación que contiene uno o más logaritmos.

logarithmic function (Lesson 8-1) A function of the form $f(x) = \log$ base b of x, where $b > 0$ and $b \neq 1$.

función logarítmica Una función de la forma $f(x) =$ base $\log b$ de x, donde $b > 0$ y $b \neq 1$.

M

maximum (Lesson 1-3) The highest point on the graph of a function.

máximo El punto más alto en la gráfica de una función.

maximum error of the estimate (Lesson 10-5) The maximum difference between the estimate of the population mean and its actual value.

error máximo de la estimación La diferencia máxima entre la estimación de la media de la población y su valor real.

midline (Lesson 11-4) The line about which the graph of a function oscillates.

línea media La línea sobre la cual oscila la gráfica de una función periódica.

minimum (Lesson 1-3) The lowest point on the graph of a function.

mínimo El punto más bajo en la gráfica de una función.

mixture problems (Lesson 9-6) Problems that involve creating a mixture of two or more kinds of things and then determining some quantity of the resulting mixture.

problemas de mezcla Problemas que implican crear una mezcla de dos o más tipos de cosas y luego determinar una cierta cantidad de la mezcla resultante.

monomial function (Lesson 4-1) A function of the form $f(x) = ax^n$, for which a is a nonzero real number and n is a positive integer.

función monomial Una función de la forma $f(x) = ax^n$, para la cual a es un número real no nulo y n es un entero positivo.

multiplicity (Lesson 5-5) The number of times a number is a zero for a given polynomial.

multiplicidad El número de veces que un número es cero para un polinomio dado.

N

natural base exponential function (Lesson 8-4) An exponential function with base e, written as $y = e^x$.

función exponencial de base natural Una función exponencial con base e, escrita como $y = e^x$.

natural logarithm (Lesson 8-4) The inverse of the natural base exponential function, most often abbreviated as $\ln x$.

logaritmo natural La inversa de la función exponencial de base natural, más a menudo abreviada como $\ln x$.

nonlinear function (Lesson 1-2) A function that has a graph with points that cannot all lie on the same line.	función no lineal Una función en la que un conjunto de puntos no puede estar en la misma línea.
normal distribution (Lesson 10-4) A continuous, symmetric, bell-shaped distribution of a random variable.	distribución normal Distribución con forma de campana, simétrica y continua de una variable aleatoria.
*n*th root (Lesson 6-3) If $a^n = b$ for a positive integer n, then a is the nth root of b.	raíz enésima Si $a^n = b$ para cualquier entero positive n, entonces a se llama una raíz enésima de b.

O

oblique asymptote (Lesson 9-4) An asymptote that is neither horizontal nor vertical.	asíntota oblicua Una asíntota que no es ni horizontal ni vertical.
observational study (Lesson 10-1) Members of a sample are measured or observed without being affected by the study.	estudio de observación Los miembros de una muestra son medidos o observados sin ser afectados por el estudio.
odd functions (Lesson 1-2) Functions that are symmetric in the origin.	funciones extrañas Funciones que son simétricas en el origen.
one-to-one function (Lesson 1-1) A function for which each element of the range is paired with exactly one element of the domain.	función biunívoca Función para la cual cada elemento del rango está emparejado con exactamente un elemento del dominio.
onto function (Lesson 1-1) A function for which the codomain is the same as the range.	sobre la función Función para la cual el codomain es el mismo que el rango.
open half-plane (Lesson 1-5) The solution of a linear inequality that does not include the boundary line.	medio plano abierto La solución de una desigualdad linear que no incluye la línea de limite.
optimization (Lesson 2-7) The process of seeking the optimal value of a function subject to given constraints.	optimización El proceso de buscar el valor óptimo de una función sujeto a restricciones dadas.
ordered triple (Lesson 2-8) Three numbers given in a specific order used to locate points in space.	triple ordenado Tres números dados en un orden específico usado para localizar puntos en el espacio.
oscillation (Lesson 11-4) How much the graph of a function varies between its extreme values as it approaches positive or negative infinity.	oscilación Cuánto la gráfica de una función varía entre sus valores extremos cuando se acerca al infinito positivo o negativo.
outcome (Lesson 10-4) The result of a single event.	resultado El resultado de un solo evento.
outlier (Lesson 10-3) A value that is more than 1.5 times the interquartile range above the third quartile or below the first quartile.	parte aislada Un valor que es más de 1,5 veces el rango intercuartílico por encima del tercer cuartil o por debajo del primer cuartil.

P

parabola (Lesson 1-2) The graph of a quadratic function.	parábola La gráfica de una función cuadrática.
parameter (Lesson 10-1) A measure that describes a characteristic of a population.	parámetro Una medida que describe una característica de una población.
parent function (Lesson 1-6) The simplest of functions in a family.	función basica La función más fundamental de un familia de funciones.
Pascal's triangle (Lesson 4-5) A triangle of numbers in which a row represents the coefficients of an expanded binomial $(a + b)^n$.	triángulo de Pascal Un triángulo de números en el que una fila representa los coeficientes de un binomio expandido $(a + b)^n$.
percent rate of change (Lesson 7-1) The percent of increase per time period.	por ciento tasa de cambio El porcentaje de aumento por período de tiempo.
perfect square trinomials (Lesson 3-4) Squares of binomials.	trinomio cuadrado perfecto Cuadrados de los binomios.
period (Lesson 11-3) The horizontal length of one cycle.	periodo La longitud horizontal de un ciclo.
periodic function (Lesson 11-3) A function with y-values that repeat at regular intervals.	función periódica Una función con y-valores aquella repetición con regularidad.
phase shift (Lesson 11-6) A horizontal translation of the graph of a trigonometric function.	cambio de fase Una traducción horizontal de la gráfica de una función trigonométrica.
piecewise-defined function (Lesson 1-6) A function defined by at least two subfunctions, each of which is defined differently depending on the interval of the domain.	función definida por piezas Una función definida por al menos dos subfunciones, cada una de las cuales se define de manera diferente dependiendo del intervalo del dominio.
point discontinuity (Lesson 9-4) An area that appears to be a hole in a graph.	discontinuidad de punto Un área que parece ser un agujero en un gráfico.
point of symmetry (Lesson 1-2) The point about which a graph is rotated.	punto de simetría El punto sobre el que se gira un gráfico.
point symmetry (Lesson 1-2) A figure or graph has this when a figure is rotated 180° about a point and maps exactly onto the other part.	simetría de punto Una figura o gráfica tiene esto cuando una figura se gira 180° alrededor de un punto y se mapea exactamente sobre la otra parte.
polynomial (Lesson 4-1) A monomial or the sum of two or more monomials.	polinomio Un monomio o la suma de dos o más monomios.
polynomial function (Lesson 4-1) A continuous function that can be described by a polynomial equation in one variable.	función polinómica Función continua que puede describirse mediante una ecuación polinómica en una variable.

polynomial identity (Lesson 5-3) A polynomial equation that is true for any values that are substituted for the variables.

identidad polinomial Una ecuación polinómica que es verdadera para cualquier valor que se sustituya por las variables.

population (Lesson 10-1) All of the members of a group of interest about which data will be collected.

población Todos los miembros de un grupo de interés sobre cuáles datos serán recopilados.

population proportion (Lesson 10-5) The number of members in the population with a particular characteristic divided by the total number of members in the population.

proporción de la población El número de miembros en la población con una característica particular dividida por el número total de miembros en la población.

power function (Lesson 4-1) A function of the form $f(x) = ax^n$, where a and n are nonzero real numbers.

función de potencia Una ecuación polinomial que es verdadera para una función de la forma $f(x) = ax^n$, donde a y n son números reales no nulos.

prime polynomial (Lesson 5-2) A polynomial that cannot be written as a product of two polynomials with integer coefficients.

polinomio primo Un polinomio que no puede escribirse como producto de dos polinomios con coeficientes enteros.

principal root (Lesson 6-3) The nonnegative root of a number.

raíz principal La raíz no negativa de un número.

principal values (Lesson 11-7) The values in the restricted domains of trigonometric functions.

valores principales Valores de los dominios restringidos de las functiones trigonométricas.

probability distribution (Lesson 10-4) A function that maps the sample space to the probabilities of the outcomes in the sample space for a particular random variable.

distribución de probabilidad Una función que mapea el espacio de muestra a las probabilidades de los resultados en el espacio de muestra para una variable aleatoria particular.

probability model (Lesson 10-2) A mathematical representation of a random event that consists of the sample space and the probability of each outcome.

modelo de probabilidad Una representación matemática de un evento aleatorio que consiste en el espacio muestral y la probabilidad de cada resultado.

projectile motion problems (Lesson 3-5) Problems that involve objects being thrown or dropped.

problemas de movimiento del proyectil Problemas que involucran objetos que se lanzan o caen.

pure imaginary number (Lesson 3-3) A number of the form bi, where b is a real number and i is the imaginary unit.

número imaginario puro Un número de la forma bi, donde b es un número real e i es la unidad imaginaria.

Pythagorean identities (Lesson 12-1) Identities that express the Pythagorean Theorem in terms of the trigonometric functions.

identidades pitagóricas Identidades que expresan el Teorema de Pitágoras en términos de las funciones trigonométricas.

Q

quadrantal angle (Lesson 11-2) An angle in standard position with a terminal side that coincides with one of the axes.

ángulo de cuadrante Un ángulo en posición estándar con un lado terminal que coincide con uno de los ejes.

quadratic equation (Lesson 3-2) An equation that includes a quadratic expression.

ecuación cuadrática Una ecuación que incluye una expresión cuadrática.

quadratic form (Lesson 5-2) A form of polynomial equation, $au^2 + bu + c$, where u is an algebraic expression in x.

forma cuadrática Una forma de ecuación polinomial, $au^2 + bu + c$, donde u es una expresión algebraica en x.

quadratic function (Lesson 3-1) A function with an equation of the form $y = ax^2 + bx + c$, where $a \neq 0$.

función cuadrática Una función con una ecuación de la forma $y = ax^2 + bx + c$, donde $a \neq 0$.

quadratic inequality (Lesson 3-7) An inequality that includes a quadratic expression.

desigualdad cuadrática Una desigualdad que incluye una expresión cuadrática.

quadratic relations (Lesson 3-8) Equations of parabolas with horizontal axes of symmetry that are not functions.

relaciones cuadráticas Ecuaciones de parábolas con ejes horizontales de simetría que no son funciones.

quartic function (Lesson 4-1) A fourth-degree function.

función cuartica Una función de cuarto grado.

quintic function (Lesson 4-1) A fifth-degree function.

función quíntica Una función de quinto grado.

R

radian (Lesson 11-1) A unit of angular measurement equal to $\frac{180°}{\pi}$ or about 57.296°.

radián Una unidad de medida angular igual o $\frac{180°}{\pi}$ alrededor de 57.296°.

radical equation (Lesson 6-6) An equation with a variable in a radicand.

ecuación radical Una ecuación con una variable en un radicand.

radical form (Lesson 6-3) When an expression contains a radical symbol.

forma radical Cuando una expresión contiene un símbolo radical.

radical function (Lesson 6-4) A function that contains radicals with variables in the radicand.

función radical Función que contiene radicales con variables en el radicand.

radicand (Lesson 6-3) The expression under a radical sign.

radicando La expresión debajo del signo radical.

range (Lesson 1-1) The set of y-values that actually result from the evaluation of the function.

rango El conjunto de valores y que realmente resultan de la evaluación de la función.

rate of change (Lesson 3-1) How a quantity is changing with respect to a change in another quantity.

tasa de cambio Cómo cambia una cantidad con respecto a un cambio en otra cantidad.

rational equation (Lesson 9-6) An equation that contains at least one rational expression.

ecuación racional Una ecuación que contiene al menos una expresión racional.

rational exponent (Lesson 6-3) An exponent that is expressed as a fraction.

exponente racional Un exponente que se expresa como una fracción.

rational expression (Lesson 9-1) A ratio of two polynomial expressions.

expresión racional Una relación de dos expresiones polinomiales.

rational function (Lesson 9-4) An equation of the form $f(x) = \frac{a(x)}{b(x)}$, where $a(x)$ and $b(x)$ are polynomial expressions and $b(x) \neq 0$.

función racional Una ecuación de la forma $f(x) = \frac{a(x)}{b(x)}$, donde $a(x)$ y $b(x)$ son expresiones polinomiales y $b(x) \neq 0$.

rational inequality (Lesson 9-6) An inequality that contains at least one rational expression.

desigualdad racional Una desigualdad que contiene al menos una expresión racional.

rationalizing the denominator (Lesson 3-3) A method used to eliminate radicals from the denominator of a fraction or fractions from a radicand.

racionalizando el denominador Método utilizado para eliminar radicales del denominador de una fracción o fracciones de una radicand.

reciprocal function (Lesson 9-2) An equation of the form $f(x) = \frac{n}{b(x)}$, where n is a real number and $b(x)$ is a linear expression that cannot equal 0.

función recíproca Una ecuación de la forma $f(x) = \frac{n}{b(x)}$, donde n es un número real y $b(x)$ es una expresión lineal que no puede ser igual a 0.

reciprocal trigonometric functions (Lesson 11-5) Trigonometric functions that are reciprocals of each other.

funciones trigonométricas recíprocas Funciones trigonométricas que son reciprocales entre sí.

recursive formula (Lesson 7-4) A formula that gives the value of the first term in the sequence and then defines the next term by using the preceding term.

fórmula recursiva Una fórmula que da el valor del primer término en la secuencia y luego define el siguiente término usando el término anterior.

reference angle (Lesson 11-2) The acute angle formed by the terminal side of an angle and the x-axis.

ángulo de referencia El ángulo agudo formado por el lado terminal de un ángulo en posición estándar y el eje x.

reflection (Lesson 1-7) A transformation in which a figure, line, or curve is flipped across a line.

reflexión Una transformación en la que una figura, línea o curva se voltea a través de una línea.

regression function (Lesson 7-5) A function generated by an algorithm to find a line or curve that fits a set of data.

función de regresión Función generada por un algoritmo para encontrar una línea o curva que se ajuste a un conjunto de datos.

relative maximum (Lesson 1-3) A point on the graph of a function where no other nearby points have a greater y-coordinate.

máximo relativo Un punto en la gráfica de una función donde ningún otro punto cercano tiene una coordenada y mayor.

relative minimum (Lesson 1-3) A point on the graph of a function where no other nearby points have a lesser y-coordinate.

mínimo relativo Un punto en la gráfica de una función donde ningún otro punto cercano tiene una coordenada y menor.

Copyright © McGraw-Hill Education

root (Lesson 2-1) A solution of an equation.

raíz Una solución de una ecuación.

S

sample (Lesson 10-1) A subset of a population.

muestra Un subconjunto de una población.

sample space (Lesson 10-4) The set of all possible outcomes.

espacio muestral El conjunto de todos los resultados posibles.

sampling error (Lesson 10-5) The variation between samples taken from the same population.

error de muestreo La variación entre muestras tomadas de la misma población.

secant (Lesson 11-2) The ratio of the length of the hypotenuse to the length of the leg adjacent to the angle.

secante Relación entre la longitud de la hipotenusa y la longitud de la pierna adyacente al ángulo.

self-selected sample (Lesson 10-1) Members volunteer to be included in the sample.

muestra auto-seleccionada Los miembros se ofrecen como voluntarios para ser incluidos en la muestra.

sequence (Lesson 7-4) A list of numbers in a specific order.

secuencia Una lista de números en un orden específico.

series (Lesson 7-4) The indicated sum of the terms in a sequence.

serie La suma indicada de los términos en una secuencia.

set-builder notation (Lesson 1-1) Mathematical notation that describes a set by stating the properties that its members must satisfy.

notación de construción de conjuntos Notación matemática que describe un conjunto al declarar las propiedades que sus miembros deben satisfacer.

sigma notation (Lesson 7-4) A notation that uses the Greek uppercase letter S to indicate that a sum should be found.

notación de sigma Una notación que utiliza la letra mayúscula griega S para indicar que debe encontrarse una suma.

simple random sample (Lesson 10-1) Each member of the population has an equal chance of being selected as part of the sample.

muestra aleatoria simple Cada miembro de la población tiene la misma posibilidad de ser seleccionado como parte de la muestra.

simulation (Lesson 10-2) The use of a probability model to imitate a process or situation so it can be studied.

simulación El uso de un modelo de probabilidad para imitar un proceso o situación para que pueda ser estudiado.

sine (Lesson 10-2) The ratio of the length of the leg opposite an angle to the length of the hypotenuse.

seno La relación entre la longitud de la pierna opuesta a un ángulo y la longitud de la hipotenusa.

sinusoidal function (Lesson 11-4) A function that can be produced by translating, reflecting, or dilating the sine function.

función sinusoidal Función que puede producirse traduciendo, reflejando o dilatando la función sinusoidal.

solution (Lesson 2-1) A value that makes an equation true.

solución Un valor que hace que una ecuación sea verdadera.

square root function (Lesson 6-4) A radical function that contains the square root of a variable expression.

función raíz cuadrada Función radical que contiene la raíz cuadrada de una expresión variable.

square root inequality (Lesson 6-4) An inequality that contains the square root of a variable expression.

square root inequality Una desigualdad que contiene la raíz cuadrada de una expresión variable.

standard deviation (Lesson 10-3) A measure that shows how data deviate from the mean.

desviación estándar Una medida que muestra cómo los datos se desvían de la media.

standard error of the mean (Lesson 10-5) The standard deviation of the distribution of sample means taken from a population.

error estandar de la media La desviación estándar de la distribución de los medios de muestra se toma de una población.

standard form of a polynomial (Lesson 4-1) The terms of a polynomial written in order from greatest to least degree.

forma estándar de un polinomio Un polinomio que se escribe con los términos en orden del grado más grande a menos grado.

standard form of a quadratic equation (Lesson 3-2) A quadratic equation can be written in the form $ax^2 + bx + c = 0$, where $a \neq 0$ and a, b, and c are integers.

forma estándar de una ecuación cuadrática Una ecuación cuadrática puede escribirse en la forma $ax^2 + bx + c = 0$, donde $a \neq 0$ y a, b, y c son enteros.

standard normal distribution (Lesson 10-4) A normal distribution with a mean of 0 and a standard deviation of 1.

distribución normal estándar Distribución normal con una media de 0 y una desviación estándar de 1.

standard position (Lesson 11-1) An angle positioned so that the vertex is at the origin and the initial side is on the positive x-axis.

posición estándar Un ángulo colocado de manera que el vértice está en el origen y el lado inicial está en el eje x positivo.

statistic (Lesson 10-1) A measure that describes a characteristic of a sample.

estadística Una medida que describe una característica de una muestra.

statistics (Lesson 10-1) An area of mathematics that deals with collecting, analyzing, and interpreting data.

estadísticas El proceso de recolección, análisis e interpretación de datos.

step function (Lesson 1-6) A type of piecewise-linear function with a graph that is a series of horizontal line segments.

función escalonada Un tipo de función lineal por piezas con un gráfico que es una serie de segmentos de línea horizontal.

stratified sample (Lesson 10-1) The population is first divided into similar, nonoverlapping groups. Then members are randomly selected from each group.

muestra estratificada La población se divide primero en grupos similares, sin superposición. A continuación, los miembros se seleccionan aleatoriamente de cada grupo.

substitution (Lesson 2-5) A process of solving a system of equations in which one equation is solved for one variable in terms of the other.

sustitución Un proceso de resolución de un sistema de ecuaciones en el que una ecuación se resuelve para una variable en términos de la otra.

survey (Lesson 10-1) Data are collected from responses given by members of a group regarding their characteristics, behaviors, or opinions.

encuesta Los datos se recogen de las respuestas dadas por los miembros de un grupo con respecto a sus características, comportamientos u opiniones.

symmetric distribution (Lesson 10-3) A distribution in which the mean and median are approximately equal.

distribución simétrica Un distribución en la que la media y la mediana son aproximadamente iguales.

symmetry (Lesson 1-2) A figure has this if there exists a rigid motion—reflection, translation, rotation, or glide reflection—that maps the figure onto itself.

simetría Una figura tiene esto si existe una reflexión-reflexión, una traducción, una rotación o una reflexión de deslizamiento rígida-que mapea la figura sobre sí misma.

synthetic division (Lesson 4-4) An alternate method used to divide a polynomial by a binomial of degree 1.

división sintética Un método alternativo utilizado para dividir un polinomio por un binomio de grado 1.

synthetic substitution (Lesson 5-4) The process of using synthetic division to find a value of a polynomial function.

sustitución sintética El proceso de utilizar la división sintética para encontrar un valor de una función polynomial.

systematic sample (Lesson 10-1) Members are selected according to a specified interval from a random starting point.

muestra sistemática Los miembros se seleccionan de acuerdo con un intervalo especificado desde un punto de partida aleatorio.

system of equations (Lesson 2-4) A set of two or more equations with the same variables.

sistema de ecuaciones Un conjunto de dos o más ecuaciones con las mismas variables.

system of inequalities (Lesson 2-6) A set of two or more inequalities with the same variables.

sistema de desigualdades Un conjunto de dos o más desigualdades con las mismas variables.

T

tangent (Lesson 11-2) The ratio of the length of the leg opposite an angle to the length of the leg adjacent to the angle.

tangente La relación entre la longitud de la pata opuesta a un ángulo y la longitud de la pata adyacente al ángulo.

term of a sequence (Lesson 7-4) A number in a sequence.

término de una sucesión Un número en una secuencia.

terminal side (Lesson 11-1) The part of an angle that rotates about the center.

lado terminal La parte de un ángulo que gira alrededor de un centro.

theoretical probability (Lesson 10-2) Probability based on what is expected to happen.

probabilidad teórica Probabilidad basada en lo que se espera que suceda.

transformation (Lesson 1-7) The movement of a graph on the coordinate plane.

transformación El movimiento de un gráfico en el plano de coordenadas.

translation (Lesson 1-7) A transformation in which a figure is slid from one position to another without being turned.

translación El movimiento de un gráfico en el plano de coordenadas.

trigonometric equation (Lesson 12-5) An equation that includes at least one trigonometric function.

ecuación trigonométrica Una ecuación que incluye al menos una función trigonométrica.

trigonometric function (Lesson 11-2) A function that relates the measure of one nonright angle of a right triangle to the ratios of the lengths of any two sides of the triangle.

función trigonométrica Función que relaciona la medida de un ángulo no recto de un triángulo rectángulo con las relaciones de las longitudes de cualquiera de los dos lados del triángulo.

trigonometric identity (Lesson 12-1) An equation involving trigonometric functions that is true for all values for which every expression in the equation is defined.

identidad trigonométrica Una ecuación que implica funciones trigonométricas que es verdadera para todos los valores para los cuales se define cada expresión en la ecuación.

trigonometric ratio (Lesson 11-2) A ratio of the lengths of two sides of a right triangle.

relación trigonométrica Una relación de las longitudes de dos lados de un triángulo rectángulo.

trigonometry (Lesson 11-2) The study of the relationships between the sides and angles of triangles.

trigonometría El estudio de las relaciones entre los lados y los ángulos de los triángulos.

trinomial (Lesson 4-3) The sum of three monomials.

trinomio La suma de tres monomios.

U

unbounded (Lesson 2-6) When the graph of a system of constraints is open.

no acotado Cuando la gráfica de un sistema de restricciones está abierta.

uniform motion problems (Lesson 9-6) Problems that use the formula $d = rt$, where d is the distance, r is the rate, and t is the time.

problemas de movimiento uniforme Problemas que utilizan la fórmula $d = rt$, donde d es la distancia, r es la velocidad y t es el tiempo.

unit circle (Lesson 11-3) A circle with a radius of 1 unit centered at the origin on the coordinate plane.

círculo unitario Un círculo con un radio de 1 unidad centrado en el origen en el plano de coordenadas.

V

variance (Lesson 10-3) The square of the standard deviation.

varianza El cuadrado de la desviación estándar.

vertex (Lesson 3-1) Either the lowest point or the highest point on a parabola.

vértice El punto más bajo o el punto más alto de una parábola.

vertex form (Lesson 3-5) A quadratic function written in the form $f(x) = a(x - h)^2 + k$.

forma de vértice Una función cuadrática escribirse de la forma $f(x) = a(x - h)^2 + k$.

vertical asymptote (Lesson 9-2) A vertical line that a graph approaches.

asíntota vertical Una línea vertical que se aproxima a un gráfico.

vertical shift (Lesson 11-6) A vertical translation of the graph of a trigonometric function.

cambio vertical Una traducción vertical de la gráfica de una función trigonométrica.

W

work problems (Lesson 9-6) Problems that involve two people working at different rates who are trying to complete a single job.

problemas de trabajo Problemas que involucran a dos personas trabajando a diferentes ritmos que están tratando de completar un solo trabajo.

X

x-intercept (Lesson 1-2) The x-coordinate of a point where a graph crosses the x-axis.

intercepción x La coordenada x de un punto donde la gráfica corte al eje de x.

Y

y-intercept (Lesson 1-2) The y-coordinate of a point where a graph crosses the y-axis.

intercepción y La coordenada y de un punto donde la gráfica corte al eje de y.

Z

zero (Lesson 2-1) An x-intercept of the graph of a function; a value of x for which $f(x) = 0$.

cero Una intercepción x de la gráfica de una función; un punto x para los que $f(x) = 0$.

z-value (Lesson 10-4) The number of standard deviations that a given data value is from the mean.

valor z El número de variaciones estándar que separa un valor dado de la media.

Index

Index

E

G

G

H

L

P

Q

R

S

Index

M

Index

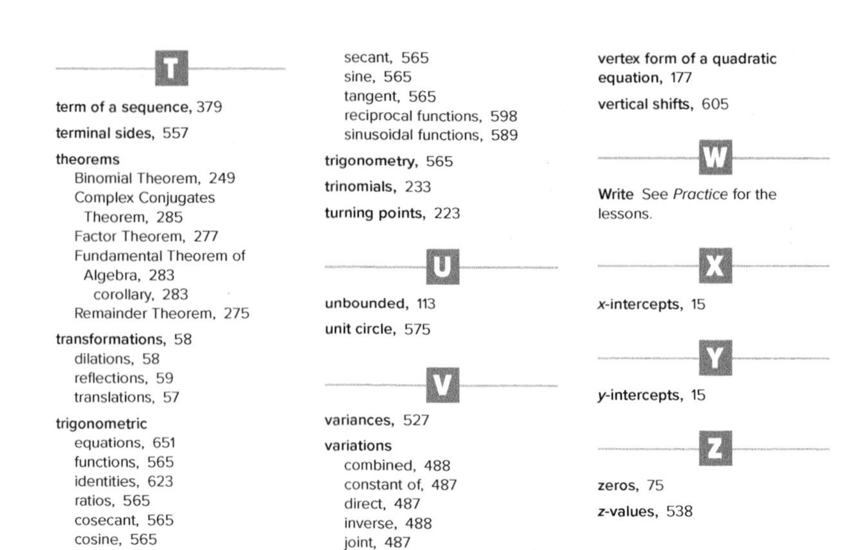

Standard Normal Cumulative Probability Table

Cumulative probabilities for negative z-values are shown in the following table.

z	0.00	0.01	0.02	0.03	0.04	0.05	0.06	0.07	0.08	0.09
−3.4	0.0003	0.0003	0.0003	0.0003	0.0003	0.0003	0.0003	0.0003	0.0003	0.0002
−3.3	0.0005	0.0005	0.0005	0.0004	0.0004	0.0004	0.0004	0.0004	0.0004	0.0003
−3.2	0.0007	0.0007	0.0006	0.0006	0.0006	0.0006	0.0006	0.0005	0.0005	0.0005
−3.1	0.0010	0.0009	0.0009	0.0009	0.0008	0.0008	0.0008	0.0008	0.0007	0.0007
−3.0	0.0013	0.0013	0.0013	0.0012	0.0012	0.0011	0.0011	0.0011	0.0010	0.0010
−2.9	0.0019	0.0018	0.0018	0.0017	0.0016	0.0016	0.0015	0.0015	0.0014	0.0014
−2.8	0.0026	0.0025	0.0024	0.0023	0.0023	0.0022	0.0021	0.0021	0.0020	0.0019
−2.7	0.0035	0.0034	0.0033	0.0032	0.0031	0.0030	0.0029	0.0028	0.0027	0.0026
−2.6	0.0047	0.0045	0.0044	0.0043	0.0041	0.0040	0.0039	0.0038	0.0037	0.0036
−2.5	0.0062	0.0060	0.0059	0.0057	0.0055	0.0054	0.0052	0.0051	0.0049	0.0048
−2.4	0.0082	0.0080	0.0078	0.0075	0.0073	0.0071	0.0069	0.0068	0.0066	0.0064
−2.3	0.0107	0.0104	0.0102	0.0099	0.0096	0.0094	0.0091	0.0089	0.0087	0.0084
−2.2	0.0139	0.0136	0.0132	0.0129	0.0125	0.0122	0.0119	0.0116	0.0113	0.0110
−2.1	0.0179	0.0174	0.0170	0.0166	0.0162	0.0158	0.0154	0.0150	0.0146	0.0143
−2.0	0.0228	0.0222	0.0217	0.0212	0.0207	0.0202	0.0197	0.0192	0.0188	0.0183
−1.9	0.0287	0.0281	0.0274	0.0268	0.0262	0.0256	0.0250	0.0244	0.0239	0.0233
−1.8	0.0359	0.0351	0.0344	0.0336	0.0329	0.0322	0.0314	0.0307	0.0301	0.0294
−1.7	0.0446	0.0436	0.0427	0.0418	0.0409	0.0401	0.0392	0.0384	0.0375	0.0367
−1.6	0.0548	0.0537	0.0526	0.0516	0.0505	0.0495	0.0485	0.0475	0.0465	0.0455
−1.5	0.0668	0.0655	0.0643	0.0630	0.0618	0.0606	0.0594	0.0582	0.0571	0.0559
−1.4	0.0808	0.0793	0.0778	0.0764	0.0749	0.0735	0.0721	0.0708	0.0694	0.0681
−1.3	0.0968	0.0951	0.0934	0.0918	0.0901	0.0885	0.0869	0.0853	0.0838	0.0823
−1.2	0.1151	0.1131	0.1112	0.1093	0.1075	0.1056	0.1038	0.1020	0.1003	0.0985
−1.1	0.1357	0.1335	0.1314	0.1292	0.1271	0.1251	0.1230	0.1210	0.1190	0.1170
−1.0	0.1587	0.1562	0.1539	0.1515	0.1492	0.1469	0.1446	0.1423	0.1401	0.1379
−0.9	0.1841	0.1814	0.1788	0.1762	0.1736	0.1711	0.1685	0.1660	0.1635	0.1611
−0.8	0.2119	0.2090	0.2061	0.2033	0.2005	0.1977	0.1949	0.1922	0.1894	0.1867
−0.7	0.2420	0.2389	0.2358	0.2327	0.2296	0.2266	0.2236	0.2206	0.2177	0.2148
−0.6	0.2743	0.2709	0.2676	0.2643	0.2611	0.2578	0.2546	0.2514	0.2483	0.2451
−0.5	0.3085	0.3050	0.3015	0.2981	0.2946	0.2912	0.2877	0.2843	0.2810	0.2776
−0.4	0.3446	0.3409	0.3372	0.3336	0.3300	0.3264	0.3228	0.3192	0.3156	0.3121
−0.3	0.3821	0.3783	0.3745	0.3707	0.3669	0.3632	0.3594	0.3557	0.3520	0.3483
−0.2	0.4207	0.4168	0.4129	0.4090	0.4052	0.4013	0.3974	0.3936	0.3897	0.3859
−0.1	0.4602	0.4562	0.4522	0.4483	0.4443	0.4404	0.4364	0.4325	0.4286	0.4247
0.0	0.5000	0.4960	0.4920	0.4880	0.4840	0.4801	0.4761	0.4721	0.4681	0.4641

Standard Normal Cumulative Probability Table

Cumulative probabilities for positive z-values are shown in the following table.

z	0.00	0.01	0.02	0.03	0.04	0.05	0.06	0.07	0.08	0.09
0.0	0.5000	0.5040	0.5080	0.5120	0.5160	0.5199	0.5239	0.5279	0.5319	0.5359
0.1	0.5398	0.5438	0.5478	0.5517	0.5557	0.5596	0.5636	0.5675	0.5714	0.5753
0.2	0.5793	0.5832	0.5871	0.5910	0.5948	0.5987	0.6026	0.6064	0.6103	0.6141
0.3	0.6179	0.6217	0.6255	0.6293	0.6331	0.6368	0.6406	0.6443	0.6480	0.6517
0.4	0.6554	0.6591	0.6628	0.6664	0.6700	0.6736	0.6772	0.6808	0.6844	0.6879
0.5	0.6915	0.6950	0.6985	0.7019	0.7054	0.7088	0.7123	0.7157	0.7190	0.7224
0.6	0.7257	0.7291	0.7324	0.7357	0.7389	0.7422	0.7454	0.7486	0.7517	0.7549
0.7	0.7580	0.7611	0.7642	0.7673	0.7704	0.7734	0.7764	0.7794	0.7823	0.7852
0.8	0.7881	0.7910	0.7939	0.7967	0.7995	0.8023	0.8051	0.8078	0.8106	0.8133
0.9	0.8159	0.8186	0.8212	0.8238	0.8264	0.8289	0.8315	0.8340	0.8365	0.8389
1.0	0.8413	0.8438	0.8461	0.8485	0.8508	0.8531	0.8554	0.8577	0.8599	0.8621
1.1	0.8643	0.8665	0.8686	0.8708	0.8729	0.8749	0.8770	0.8790	0.8810	0.8830
1.2	0.8849	0.8869	0.8888	0.8907	0.8925	0.8944	0.8962	0.8980	0.8997	0.9015
1.3	0.9032	0.9049	0.9066	0.9082	0.9099	0.9115	0.9131	0.9147	0.9162	0.9177
1.4	0.9192	0.9207	0.9222	0.9236	0.9251	0.9265	0.9279	0.9292	0.9306	0.9319
1.5	0.9332	0.9345	0.9357	0.9370	0.9382	0.9394	0.9406	0.9418	0.9429	0.9441
1.6	0.9452	0.9463	0.9474	0.9484	0.9495	0.9505	0.9515	0.9525	0.9535	0.9545
1.7	0.9554	0.9564	0.9573	0.9582	0.9591	0.9599	0.9608	0.9616	0.9625	0.9633
1.8	0.9641	0.9649	0.9656	0.9664	0.9671	0.9678	0.9686	0.9693	0.9699	0.9706
1.9	0.9713	0.9719	0.9726	0.9732	0.9738	0.9744	0.9750	0.9756	0.9761	0.9767
2.0	0.9772	0.9778	0.9783	0.9788	0.9793	0.9798	0.9803	0.9808	0.9812	0.9817
2.1	0.9821	0.9826	0.9830	0.9834	0.9838	0.9842	0.9846	0.9850	0.9854	0.9857
2.2	0.9861	0.9864	0.9868	0.9871	0.9875	0.9878	0.9881	0.9884	0.9887	0.9890
2.3	0.9893	0.9896	0.9898	0.9901	0.9904	0.9906	0.9909	0.9911	0.9913	0.9916
2.4	0.9918	0.9920	0.9922	0.9925	0.9927	0.9929	0.9931	0.9932	0.9934	0.9936
2.5	0.9938	0.9940	0.9941	0.9943	0.9945	0.9946	0.9948	0.9949	0.9951	0.9952
2.6	0.9953	0.9955	0.9956	0.9957	0.9959	0.9960	0.9961	0.9962	0.9963	0.9964
2.7	0.9965	0.9966	0.9967	0.9968	0.9969	0.9970	0.9971	0.9972	0.9973	0.9974
2.8	0.9974	0.9975	0.9976	0.9977	0.9977	0.9978	0.9979	0.9979	0.9980	0.9981
2.9	0.9981	0.9982	0.9982	0.9983	0.9984	0.9984	0.9985	0.9985	0.9986	0.9986
3.0	0.9987	0.9987	0.9987	0.9988	0.9988	0.9989	0.9989	0.9989	0.9990	0.9990
3.1	0.9990	0.9991	0.9991	0.9991	0.9992	0.9992	0.9992	0.9992	0.9993	0.9993
3.2	0.9993	0.9993	0.9994	0.9994	0.9994	0.9994	0.9994	0.9995	0.9995	0.9995
3.3	0.9995	0.9995	0.9995	0.9996	0.9996	0.9996	0.9996	0.9996	0.9996	0.9997
3.4	0.9997	0.9997	0.9997	0.9997	0.9997	0.9997	0.9997	0.9997	0.9997	0.9998